Intermediate Algebra

Ninth Edition

Intermediate Algebra

Ninth Edition

Margaret L. Lial
American River College

John Hornsby
University of New Orleans

Terry McGinnis

Addison-Wesley
Boston • New York • San Francisco
London • Toronto • Sydney • Tokyo • Singapore • Madrid
Mexico City • Munich • Paris • Cape Town • Hong Kong • Montreal

Editorial Director	Christine Hoag
Editor in Chief	Maureen O'Connor
Executive Project Manager	Kari Heen
Project Editor	Courtney Slade
Editorial Assistant	Mary Gallagher
Senior Managing Editor	Karen Wernholm
Senior Production Supervisor	Kathleen A. Manley
Senior Designer	Barbara T. Atkinson
Photo Researcher	Beth Anderson
Supplements Production	Marianne Groth and Kayla Smith-Tarbox
Media Producers	Ceci Fleming, Lin Mahoney, and Jean Choe
Software Development	Rebecca Williams, MathXL; Mary Durnwald, TestGen
Senior Marketing Manager	Michelle Renda
Marketing Assistant	Nathaniel Koven
Senior Author Support/Technology Specialist	Joe Vetere
Senior Prepress Supervisor	Caroline Fell
Senior Media Buyer	Ginny Michaud
Rights and Permissions Advisor	Dana Weightman
Senior Manufacturing Buyer	Carol Melville
Composition/Production Coordination	Nesbitt Graphics, Inc.
Cover Image	Birch Woods in Spring Copyright © Lorraine Cota Manley

Library of Congress Cataloging-in-Publication Data

Lial, Margaret L.

Intermediate algebra / Margaret L. Lial, John Hornsby, Terry McGinnis.—9th ed.

p. cm.

Includes index.

ISBN-13: 978-0-321-57497-8 (student edition)

ISBN-10: 0-321-57497-4 (student edition)

1. Algebra—Textbooks. I. Hornsby, E. John. II. McGinnis, Terry. III. Title.

QA152.3.L534 2010

512.9—dc22 2008024475

For permission to use copyrighted material, grateful acknowledgment is made to the copyright holders on page P-1, which is hereby made part of this copyright page.

2 3 4 5 6 7 8 9 10—RRDJC—12 11 10 09

Addison-Wesley is an imprint of

PEARSON

www.pearsonhighered.com

ISBN 10: 0-321-57497-4
ISBN 13: 978-0-321-57497-8

To Bill Hebert—my mentor, my colleague, my friend

E.J.H.

To my parents, Tom and Marge Ruhberg, for their love and support

T.R.M.

CONTENTS

Preface ix

Study Skills Your Brain *Can* Learn Mathematics 1

Chapter 1 Review of the Real Number System **3**

1.1 Basic Concepts 4
1.2 Operations on Real Numbers 17
1.3 Exponents, Roots, and Order of Operations 29
1.4 Properties of Real Numbers 39

Summary 47 ▲ Review Exercises 51 ▲ Test 55

Study Skills Using Your Textbook 57

Chapter 2 Linear Equations and Applications **59**

2.1 Linear Equations in One Variable 60
Study Skills Homework: How, Why, and When 71
2.2 Formulas and Percent 73
2.3 Applications of Linear Equations 85
Study Skills Taking Lecture Notes 99
2.4 Further Applications of Linear Equations 101
***Summary Exercises* on Solving Applied Problems** 109
Study Skills Reviewing a Chapter 111

Summary 113 ▲ Review Exercises 117 ▲ Test 121

Study Skills Preparing for Tests 123

Cumulative Review Exercises Chapters 1–2 125

Chapter 3 Linear Inequalities and Absolute Value **127**

3.1 Linear Inequalities in One Variable 128
Study Skills Using Study Cards 143
3.2 Set Operations and Compound Inequalities 145
Study Skills Using Study Cards Revisited 155
3.3 Absolute Value Equations and Inequalities 157
***Summary Exercises* on Solving Linear and Absolute Value Equations and Inequalities** 169
Study Skills Managing Your Time 171

Summary 173 ▲ Review Exercises 177 ▲ Test 181

Study Skills Tips for Taking Math Tests 183

Cumulative Review Exercises Chapters 1–3 185

Chapter 4 Graphs, Linear Equations, and Functions **187**

4.1 The Rectangular Coordinate System 188
4.2 Slope of a Line 199
4.3 Linear Equations in Two Variables 213
***Summary Exercises* on Slopes and Equations of Lines** 227
4.4 Linear Inequalities in Two Variables 229
4.5 Introduction to Functions 237

Summary 253 ▲ Review Exercises 257 ▲ Test 261

Study Skills Analyzing Your Test Results 263

Cumulative Review Exercises Chapters 1–4 265

Chapter 5 Systems of Linear Equations **267**

5.1 Systems of Linear Equations in Two Variables 268
5.2 Systems of Linear Equations in Three Variables 283
5.3 Applications of Systems of Linear Equations 293
5.4 Solving Systems of Linear Equations by Matrix Methods 307

Summary 315 ▲ Review Exercises 319 ▲ Test 323

Cumulative Review Exercises Chapters 1–5 325

Chapter 6 Exponents, Polynomials, and Polynomial Functions **327**

6.1 Integer Exponents and Scientific Notation 328
6.2 Adding and Subtracting Polynomials 345
6.3 Polynomial Functions 351
6.4 Multiplying Polynomials 359
6.5 Dividing Polynomials 369

Summary 375 ▲ Review Exercises 379 ▲ Test 383

Cumulative Review Exercises Chapters 1–6 385

Chapter 7 Factoring **387**

7.1 Greatest Common Factors; Factoring by Grouping 388

7.2 Factoring Trinomials 395
7.3 Special Factoring 403
7.4 A General Approach to Factoring 409
7.5 Solving Equations by Factoring 413

Summary 423 ▲ Review Exercises 425 ▲ Test 427
Cumulative Review Exercises Chapters 1–7 429

Chapter 8 Rational Expressions and Functions 431

8.1 Rational Expressions and Functions; Multiplying and Dividing 432
Study Skills Making a Mind Map 443
8.2 Adding and Subtracting Rational Expressions 445
8.3 Complex Fractions 455
8.4 Equations with Rational Expressions and Graphs 461
***Summary Exercises* on Rational Expressions and Equations 469**
8.5 Applications of Rational Expressions 471
8.6 Variation 485

Summary 495 ▲ Review Exercises 499 ▲ Test 503
Cumulative Review Exercises Chapters 1–8 505

Chapter 9 Roots, Radicals, and Root Functions 507

9.1 Radical Expressions and Graphs 508
9.2 Rational Exponents 517
9.3 Simplifying Radical Expressions 527
9.4 Adding and Subtracting Radical Expressions 541
9.5 Multiplying and Dividing Radical Expressions 545
***Summary Exercises* on Operations with Radicals and Rational Exponents 555**
9.6 Solving Equations with Radicals 557
9.7 Complex Numbers 565

Summary 575 ▲ Review Exercises 579 ▲ Test 583
Cumulative Review Exercises Chapters 1–9 585
Study Skills Preparing for Your Math Final Exam 587

Chapter 10 Quadratic Equations, Inequalities, and Functions 589

10.1 The Square Root Property and Completing the Square 590
10.2 The Quadratic Formula 601
10.3 Equations Quadratic in Form 609
***Summary Exercises* on Solving Quadratic Equations 621**
10.4 Formulas and Further Applications 623
10.5 Graphs of Quadratic Functions 631
10.6 More about Parabolas and Their Applications 643
10.7 Polynomial and Rational Inequalities 655

Summary 665 ▲ Review Exercises 669 ▲ Test 673
Cumulative Review Exercises Chapters 1–10 677

Chapter 11 Inverse, Exponential, and Logarithmic Functions 679

11.1 Inverse Functions 680
11.2 Exponential Functions 689
11.3 Logarithmic Functions 697
11.4 Properties of Logarithms 705
11.5 Common and Natural Logarithms 713
11.6 Exponential and Logarithmic Equations; Further Applications 719

Summary 729 ▲ Review Exercises 733 ▲ Test 737
Cumulative Review Exercises Chapters 1–11 739

Chapter 12 Nonlinear Functions, Conic Sections, and Nonlinear Systems 741

12.1 Additional Graphs of Functions; Composition 742
12.2 The Circle and the Ellipse 753
12.3 The Hyperbola and Other Functions Defined by Radicals 763
12.4 Nonlinear Systems of Equations 773
12.5 Second-Degree Inequalities and Systems of Inequalities 779

Summary 785 ▲ Review Exercises 789 ▲ Test 793
Cumulative Review Exercises Chapters 1–12 797

Appendix A Strategies for Problem Solving 799

Appendix B Review of Fractions 809

Appendix C Synthetic Division 821

Answers to Selected Exercises A-1

Solutions to Selected Exercises S-1

Index I-1

Photo and Text Credits P-1

PREFACE

It is with pleasure that we offer the ninth edition of *Intermediate Algebra*. With each new edition, the text has been shaped and adapted to meet the changing needs of both students and educators, and this edition faithfully continues that process. As always, we have taken special care to respond to the specific suggestions of users and reviewers through enhanced discussions, new and updated examples and exercises, helpful features, updated figures and graphs, and an extensive package of supplements and study aids. We believe the result is an easy-to-use, comprehensive text that is the best edition yet.

Students who have never studied algebra—as well as those who require further review of basic algebraic concepts before taking additional courses in mathematics, business, science, nursing, or other fields—will benefit from the text's student-oriented approach. Of particular interest to students and instructors will be the **NEW** pointers in examples, Study Skills activities, Math in the Media feature, and Solutions section.

This text is part of a series that also includes the following books:

- *Essential Mathematics*, Third Edition, by Lial and Salzman
- *Basic College Mathematics*, Eighth Edition, by Lial, Salzman, and Hestwood
- *Prealgebra*, Fourth Edition, by Lial and Hestwood
- *Introductory Algebra*, Ninth Edition, by Lial, Hornsby, and McGinnis
- *Introductory and Intermediate Algebra*, Fourth Edition, by Lial, Hornsby, and McGinnis
- *Prealgebra and Introductory Algebra*, Third Edition, by Lial, Hestwood, Hornsby, and McGinnis
- *Developmental Mathematics: Basic Mathematics and Algebra,* Second Edition, by Lial, Hornsby, McGinnis, Salzman, and Hestwood

Hallmark Features

We are pleased to offer the following features, each of which is designed to increase ease-of-use by students and actively engage them in learning mathematics.

▶ ***Chapter Openers*** New and updated chapter openers feature real-world applications of mathematics that are relevant to students and tied to specific material within the chapters. Examples of topics include television, higher education costs, and temperature. (See pp. 59, 127, and 187—Chapters 2, 3, and 4.)

▶ ***Real-Life Applications*** We are always on the lookout for interesting data to use in real-life applications. As a result, we have included new or updated examples and exercises from fields such as business, pop culture, sports, the life sciences, and technology that show the relevance of algebra to daily life. (See pp. 83, 88, and 220.)

▶ ***Figures, Photos, and* NEW *Hand-Drawn Graphs*** Today's students are more visually oriented than ever. Thus, we have made a concerted effort to include mathematical figures, diagrams, tables, and graphs, including the new "hand-drawn" style of graphs, whenever possible. (See pp. 77, 188, and 230.) Many of the graphs use a style similar to that seen by students in today's print and electronic media. Even more photos have been incorporated to enhance applications in examples and exercises. (See pp. 141 and 219.)

- ***Emphasis on Problem Solving*** Introduced in Chapter 2, our six-step problem-solving method is integrated throughout the text. The six steps, *Read, Assign a Variable, Write an Equation, Solve, State the Answer*, and *Check*, are emphasized in boldface type and repeated in examples and exercises to reinforce the problem-solving process for students. (See pp. 87 and 101.) **Problem-Solving Hint** boxes provide students with helpful problem-solving tips and strategies. (See pp. 102 and 104.)

  Appendix A: Strategies for Problem Solving provides examples of additional problem-solving techniques, such as working backward, using trial and error, and looking for patterns. A wide variety of applications are included. (See pp. 799–808.)

- ***Learning Objectives*** Each section begins with clearly stated, numbered objectives, and the included material is directly keyed to these objectives so that students know exactly what is covered in each section. (See pp. 128 and 145.)

- ***Examples*** The new edition of the text features a multitude of step-by-step, worked-out examples that include pedagogical color, helpful side comments, and **NEW** pointers. We give increased attention to checking example solutions—more checks, designated using a special **Check** tag, are included than in past editions. (See pp. 61 and 64.)

- ***Margin Problems*** Margin problems, with answers immediately available at the bottom of the page, are found in every section of the text. (See pp. 76 and 90.) This popular feature allows students to immediately practice the material covered in the examples in preparation for the exercise sets. We have added more margin problems in this edition.

- ***Cautions and Notes*** One of the most popular features of previous editions, **CAUTION** and **Note** boxes warn students about common errors and emphasize important ideas throughout the exposition. (See pp. 63 and 65.) The text design makes them easy to spot: Cautions are highlighted in bright yellow and Notes are highlighted in purple.

- ***Calculator Tips*** Optional Calculator Tips, marked with calculator icons, offer basic information and instruction for students using calculators in the course. (See p. 269.)

- ***Ample and Varied Exercise Sets*** One of the most commonly mentioned strengths of this text is its exercise sets. The text contains a wealth of exercises to provide students with opportunities to practice, apply, connect, and extend the algebraic concepts and skills they are learning. Numerous illustrations, tables, graphs, and photos have been added to the exercise sets to help students visualize the problems they are solving. Problem types include writing, estimation, and calculator exercises as well as applications and multiple-choice, matching, true/false, and fill-in-the-blank problems. In the *Annotated Instructor's Edition* of the text, writing exercises are marked with icons so that teachers may assign these problems at their discretion. Exercises suitable for calculator work are marked in both the student and teacher editions with icons. (See pp. 67, 82, and 84.) Students can watch an instructor work through the solutions for exercises marked with the DVD icon on the Videos on DVD or in MyMathLab.

- ***Relating Concepts Exercises*** These sets of exercises help students tie together topics and develop problem-solving skills as they compare and contrast ideas, identify and describe patterns, and extend concepts to new situations. (See pp. 79 and 98.) These exercises make great collaborative activities for pairs or small groups of students.

- ***Summary Exercises*** Every chapter includes at least one set of these popular in-chapter summary exercises. These special exercise sets provide students with the all-important *mixed* review problems they need to master topics. Summaries of solution methods or additional examples are often included. (See pp. 109 and 169.)

- ***Test Your Word Power*** To help students understand and master mathematical vocabulary, this feature can be found in each chapter summary. Key terms from the chapter are presented along with four possible definitions in a multiple-choice format. Answers and examples illustrating each term are provided. (See pp. 113 and 173.)

▶ ***Ample Opportunity for Review*** Each chapter concludes with a Chapter Summary that features Key Terms with definitions and helpful graphics, New Symbols, Test Your Word Power, and a Quick Review of each section's content with additional examples. A comprehensive set of Chapter Review Exercises, keyed to individual sections, is included, as are Mixed Review Exercises and a Chapter Test. Beginning with Chapter 2, each chapter concludes with a set of Cumulative Review Exercises that cover material going back to Chapters R and 1. (See pp. 113–126.) Students can watch an instructor work through the full solutions for all Chapter Test exercises on the Chapter Test Prep Video CD that accompanies each new copy of the text.

What's New in This Edition?

Throughout this edition of the text, we are pleased to offer the following new student-oriented features:

NEW ***Math in the Media*** These new one-page activities provide a relevant application of mathematics as it is found in various media forms, such as newspapers, movies, and TV. Designed to help teachers answer the often-asked question, "Why do I need to learn this?", these activities are well-suited for individual or collaborative work, as well as class discussions. We hope both students and instructors will enjoy them. They include the following:

- Learn Math, Lose Weight (p. 116)
- An Application of Mathematics Any Student Can Appreciate (p. 176)
- How Can 8.0 Be One Thousand Times 5.0? (p. 356)
- Prime Numbers in Prime Time (p. 378)
- A "Big League" Application of Rational Expressions and Equations (p. 478)
- So, Did the Scarecrow Really Get a Brain? (p. 534)

NEW ***Study Skills*** Poor study skills are a major reason why students do not succeed in mathematics. These new two-page activities provide helpful information, tips, and strategies on a variety of essential study skills, including *Using Your Textbook, Taking Lecture Notes, Tips for Taking Math Tests*, and *Managing Your Time*. While most of the activities are concentrated in the early chapters of the text, each has been designed independently and can be used at most any point in your course with individuals or small groups of students, or as a source of material for in-class discussions. (See pp. 99 and 100.)

NEW ***Solutions to Selected Exercises*** Exercise numbers enclosed in a blue square, such as **11.**, indicate that a step-by-step, worked-out solution for the problem is included at the back of the text. These solutions are given for selected exercises that extend the skills and concepts presented in the section examples—actually providing students with a pool of examples for exercises that include some kind of twist or are a bit more difficult. (See pp. S-1 through S-17.)

NEW ***Pointers*** Pointers from the authors have been added to examples and provide students with important on-the-spot reminders and warnings about common pitfalls. (See pp. 62 and 132.)

NEW ***Chapter Test Prep Video CD*** The Chapter Test Prep Video CD provides students with the opportunity to watch instructors work through step-by-step solutions to all the Chapter Test exercises from the textbook. The Chapter Test Video CD is included with each new student text.

A primary focus of this revision of the text was to polish and enhance individual presentations of topics and exercise sets, based on user and reviewer feedback, and we have worked hard to do this throughout the book. Some of the specific content changes you may notice include the following:

- The exercise sets received special attention. There are approximately 450 new and updated exercises, including problems that check conceptual understanding, focus on skill development, and provide review.
- Real-world data in over 230 applications in the examples and exercises have been updated.
- There is an increased emphasis on the difference between expressions and equations. In particular, we have added exposition and a margin exercise at the beginning of Section 2.1. Throughout the text, we have reformatted many example solutions to use a "drop down" layout in order to further emphasize for students the difference between simplifying expressions and solving equations.
- There is an increased emphasis on checking solutions and answers, as indicated by the new **Check** tag in the exposition and examples.
- Explanation, examples, and exercises involving percent increase and percent decrease are included in Section 2.2.
- The midpoint formula is now introduced and covered in the exposition, examples, and exercises in Section 4.1.
- When a new type of graph is introduced (Sections 4.1, 4.4, 6.3, 8.4, 9.1, 10.5, 10.6, 11.2, 11.3, 12.1), a new "hand-drawn" graph style is used to simulate what a student might actually sketch on graph paper.
- Chapter 4 includes a new set of *Summary Exercises on Slopes and Equations of Lines*.
- The *Summary Exercises on Factoring* (from the eighth edition) in Chapter 7 have been expanded into new Section 7.4 that now features examples and a comprehensive review of factoring strategies, plus additional randomly organized exercises.
- While variation, formerly covered in Section 4.6, is now presented in Section 8.6, coverage remains flexible and allows this topic to be included at various places in course syllabi.
- Presentations of the following topics have also been enhanced and expanded:

  Review of fractions (Section 1.2 and Appendix B)
  Solving linear equations in two variables for y (Section 2.2)
  Average rate of change (Section 4.2)
  Finding equations of lines using slope-intercept form (Section 4.3)
  Solving systems of linear equations in three variables (Section 5.2)
  Solving equations quadratic in form (Section 10.3)
  Graphing logarithmic functions (Section 11.3)
  Applying the laws of logarithms (Section 11.4)

What Supplements Are Available?

For a comprehensive list of the supplements and study aids that accompany *Intermediate Algebra*, Ninth Edition, see pages xiv and xv.

Acknowledgments

The comments, criticisms, and suggestions of users, nonusers, instructors, and students have positively shaped this textbook over the years, and we are most grateful for the many responses we have received. We especially wish to thank the following reviewers whose valuable contributions have helped to refine this edition of the text.

Mary Kay Abbey, *Montgomery College*
Randall Allbritton, *Daytona Beach College*
Theresa Allen, *University of Idaho*
Sonya Armstrong, *West Virginia State College*
Linda Beller, *Brevard Community College*
Carla J. Bissell, *University of Nebraska at Omaha*
Vernon Bridges, *Durham Technical Community College*
Dawn Cox, *Cochise College*
Julie Dewan, *Mohawk Valley Community College*
Lucy Edwards, *Las Positas College*
Rob Farinelli, *Community College of Allegheny–Boyce Campus*
Anthony Hearn, *Community College of Philadelphia*
Jeffrey Kroll, *Brazosport College*
Barbara Krueger, *Cochise College*
Sandy Lofstock, *California Lutheran University*
Janice Rech, *University of Nebraska at Omaha*
Dwight Smith, *Prestonburg Community College*
Theresa Stalder, *University of Illinois–Chicago*
Mark Tom, *College of the Sequoias*

Over the years, we have come to rely on an extensive team of experienced professionals. Our sincere thanks go to these dedicated individuals at Addison-Wesley, who worked long and hard to make this revision a success: Greg Tobin, Maureen O'Connor, Michelle Renda, Kari Heen, Courtney Slade, Kathy Manley, Barbara Atkinson, Beth Anderson, Lin Mahoney, Ceci Fleming, Nathaniel Koven, and Mary Gallagher.

Abby Tanenbaum did an outstanding job helping us with manuscript preparation. We are truly grateful for her contributions to so many of our books over the years. Janette Krauss, Bonnie Boehme, and Nesbitt Graphics, Inc. provided excellent production work on the challenging format of these books. Special thanks are due Diana Hestwood and Linda Russell for their hard work on the Study Skills; Jeff Cole, who continues to provide accurate, helpful solutions manuals; Barb Brown, who helped us update the real-data applications; and Lucie Haskins for another useful index. Janis Cimperman, Perian Herring, Paul Lorczak, and Sarah Sponholz did a wonderful and timely job accuracy checking.

As an author team, we are committed to providing the best possible text and supplements package to help students succeed and instructors teach. As we continue to work toward this goal, we would welcome any comments or suggestions you might have via e-mail to *math@pearson.com.*

Margaret L. Lial
John Hornsby
Terry McGinnis

Student Supplements

Student's Solutions Manual
- By Jeffery A. Cole, *Anoka-Ramsey Community College*
- Provides detailed solutions to the odd-numbered, section-level exercises and to all margin, Relating Concepts, Summary, Chapter Review, Chapter Test, and Cumulative Review Exercises

ISBNs: 0-321-57629-2, 978-0-321-57629-3

Videos on DVD
- Feature an engaging team of lecturers
- Include a complete set of lectures for each section of the text on DVD for student use at home or on campus
- Ideal for distance learning or supplemental instruction
- Include optional English and Spanish subtitles
- Watch an instructor work through the complete solution for all exercises marked with a DVD icon

ISBNs: 0-321-57628-4, 978-0-321-57628-6

Worksheets for Classroom or Lab Practice
- Provide extra practice exercises for every section of the text with ample space for students to show their work
- List the learning objectives and key vocabulary terms for every text section, along with vocabulary practice problems

ISBNs: 0-321-57635-7, 978-0-321-57635-4

InterAct Math Tutorial Web site *www.interactmath.com*
- Offers online practice and tutorial help
- Retry an exercise with new values each time for unlimited practice and mastery
- Every exercise accompanied by an interactive guided solution that gives helpful feedback when an incorrect answer is entered
- Allows students to view steps of a worked-out sample problem similar to those in the text

Chapter Test Prep Video CD
- Allows students to watch instructors work through step-by-step solutions to all the Chapter Test exercises from the textbook
- Included with each new student text
- Available with optional English subtitles

Instructor Supplements

Annotated Instructor's Edition
- Provides answers to all text exercises in color next to the corresponding problem
- Includes icons to identify writing and calculator exercises

ISBNs: 0-321-57622-5, 978-0-321-57622-4

Instructor's Solutions Manual
- By Jeffery A. Cole, *Anoka-Ramsey Community College*
- Provides complete answers to all the exercises in the text

ISBNs: 0-321-57623-3, 978-0-321-57623-1

Additional Teaching Resources
Includes resources to help both new and adjunct faculty with course preparation and classroom management by offering helpful teaching tips correlated to the sections of the text
Available for download at *www.pearsonhighered.com*

Instructor's Resource Manual with Tests
- By James Ball, *Indiana State University*
- Contains a test bank with two diagnostic pretests, six free-response and two multiple-choice test forms per chapter, and two final exams
- Also contains a mini-lecture for each section of the text with objectives, key examples, and teaching tips
- Includes a correlation guide from the eighth to the ninth edition and phonetic spellings for all key terms in the text

ISBNs: 0-321-57624-1, 978-0-321-57624-8

PowerPoint® Lecture Slides
- Present key concepts and definitions from the text
- Available for download at *www.pearsonhighered.com*

ISBNs: 0-321-57636-5, 978-0-321-57636-1

TestGen® *www.pearsonhighered.com/testgen*
- Enables instructors to build, edit, print, and administer tests using a computerized bank of questions developed to cover all text objectives
- Allows instructors to create multiple but equivalent versions of the same question or test with the click of a button
- Allows instructors to modify test bank questions or add new questions
- Tests can be printed or administered online

Pearson Math Adjunct Support Center
http://www.pearsontutorservices.com/math-adjunct.html
Staffed by qualified instructors with more than 50 years of combined experience at both the community college and university levels. Assistance provided for faculty in the following areas:
- Suggested syllabus consultation
- Tips on using materials packed with your book
- Book-specific content assistance
- Teaching suggestions, including advice on classroom strategies

Available for Students and Instructors

MyMathLab® MyMathLab is a series of text-specific, easily customizable online courses for Pearson Education's textbooks in mathematics and statistics. Powered by CourseCompass™ (our online teaching and learning environment) and MathXL® (our online homework, tutorial, and assessment system), MyMathLab provides the tools needed to deliver all or a portion of a course online, whether students are in a lab setting or working from home. MyMathLab provides a rich and flexible set of course materials, featuring free-response exercises that are algorithmically generated for unlimited practice and mastery. Students can also use online tools, such as video lectures, animations, and a multimedia textbook, to independently improve their understanding and performance. Instructors can use MyMathLab's homework and test managers to select and assign online exercises correlated directly to the textbook, and they can also create and assign their own online exercises and import TestGen tests for added flexibility. MyMathLab's online gradebook—designed specifically for mathematics and statistics—automatically tracks students' homework and test results and gives the instructor control over how to calculate final grades. Instructors can also add offline (paper-and-pencil) grades to the gradebook. MyMathLab also includes access to the **Pearson Tutor Center** (*www.pearsontutorservices.com*). The Tutor Center is staffed by qualified mathematics instructors who provide textbook-specific tutoring for students via toll-free phone, fax, email, and interactive Web sessions. MyMathLab is available to qualified adopters. For more information, visit our Web site at *www.mymathlab.com,* or contact your sales representative.

MathXL® MathXL is a powerful online homework, tutorial, and assessment system that accompanies Pearson Education's textbooks in mathematics or statistics. With MathXL, instructors can create, edit, and assign online homework and tests using algorithmically generated exercises correlated at the objective level to the textbook. They can also create and assign their own online exercises and import TestGen tests for added flexibility. All student work is tracked in MathXL's online gradebook. Students can take chapter tests in MathXL and receive personalized study plans based on their test results. The study plan diagnoses weaknesses and links students directly to tutorial exercises for the objectives they need to study and retest. Students can also access supplemental animations and video clips directly from selected exercises. MathXL is available to qualified adopters. For more information, visit our Web site at *www.mathxl.com,* or contact your sales representative.

MathXL® Tutorials on CD This interactive tutorial CD-ROM provides algorithmically generated practice exercises that are correlated at the objective level to the exercises in the textbook. Every practice exercise is accompanied by an example and a guided solution designed to involve students in the solution process. Selected exercises may also include a video clip to help students visualize concepts. The software provides helpful feedback for incorrect answers and can generate printed summaries of students' progress.

ISBNs:
0-321-57627-6
978-0-321-57627-9

Study Skills

YOUR BRAIN *CAN* LEARN MATHEMATICS

OBJECTIVES

1. **Describe how practice fosters dendrite growth.**
2. **Explain the effect of anxiety on the brain.**

Your brain knows how to learn, just as your lungs know how to breathe; however, there are important things you can do to maximize your brain's ability to do its work. This short introduction will help you choose effective strategies for learning mathematics. This is a simplified explanation of a complex process.

Your brain's outer layer, called the **neocortex,** is where higher level thinking, language, reasoning, and purposeful behavior occur. The neocortex has about 100 billion (100,000,000,000) brain cells called **neurons.**

Learning Something New

- As you learn something new, threadlike branches grow out of each neuron. These branches are called **dendrites.**
- When the dendrite from one neuron grows close enough to the dendrite from another neuron, a connection is made. There is a small gap at the connection point called a **synapse.** One dendrite sends an electrical signal across the gap to another dendrite.
- ***Learning = growth and connecting of dendrites.***

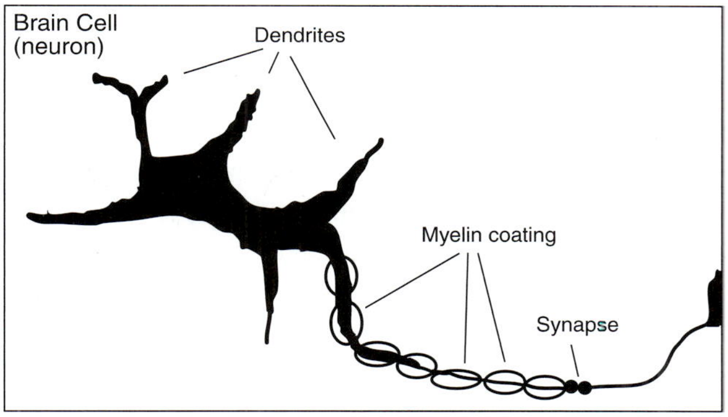

A neuron with several dendrites: one dendrite has developed a myelin coating through repeated practice

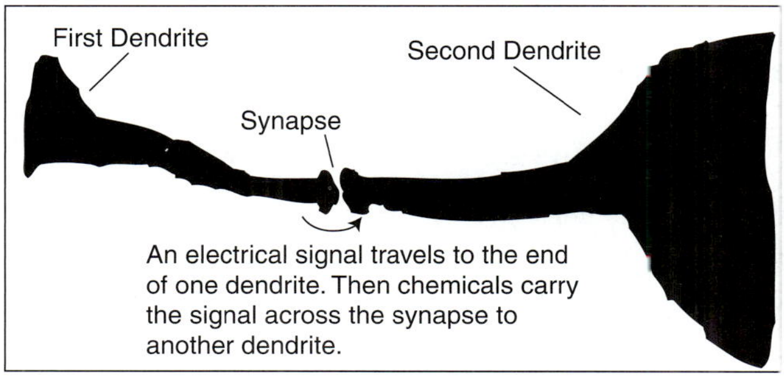

A close up view of the connection (synapse) between two dendrites

Remembering New Skills

- When you practice a skill just once or twice, the connections between neurons are very weak. If you do not practice the skill again, the dendrites at the connection points wither and die back. You have forgotten the new skill.
- If you practice a new skill many times, the dendrites for that skill become coated with a fatty protein called **myelin.** Each time one dendrite sends a signal to another dendrite, the myelin coating becomes thicker and smoother, allowing the signals to move faster and with less interference. Thinking can now occur more quickly and easily, and you will remember the skill for a long time because the dendrite connections are strong.

Become An Effective Student

- You grow dendrites specifically for the topic you are studying. ***So, be sure you are actively learning and practicing.***
- If you practice something the wrong way, you will develop connections for doing it the wrong way. ***So, as you study, check frequently that you are getting correct answers.***
- As you study a new topic that is related to things you already know, your brain sends signals throughout the network of dendrites for the related topics. In this way, you build a complex **neural network** that allows you to apply concepts, see differences and similarities between ideas, and understand relationships between concepts.

In the first few chapters of this textbook you will find activities that are designed to help you grow and develop your own reliable neural networks for mathematics. When you incorporate the activities into your regular study routine, you will discover that you understand better, remember longer, and forget less.

Also remember that it takes time to learn something new. Trying to cram in several new concepts and skills at the last minute is not possible. You can't expect to develop huge muscles by lifting weights for just one evening before a body building competition. In the same way, ***practice the study techniques throughout the course*** to facilitate strong growth of dendrites.

When Anxiety Strikes

If you are under stress or feeling anxious, such as during a test, your body secretes **adrenaline** into your system. Adrenaline in the brain blocks connections between neurons, so you can't think clearly. If you've ever experienced "blanking out" on a test, you know what adrenaline does. You'll learn several solutions to that problem in later activities.

Start Your Course Right!

- ***Attend all class sessions (especially the first one).***
- ***Gather the necessary supplies.***
- ***Carefully read the syllabus for the course, and ask questions if you don't understand.***

1

1.1 Basic Concepts
1.2 Operations on Real Numbers
1.3 Exponents, Roots, and Order of Operations
1.4 Properties of Real Numbers

Review of the Real Number System

Americans are crazy about their pets. Over 71 million U.S. households owned pets in 2007. Combined, these households spent more than $41 billion pampering their animal friends. The fastest-growing segment of the pet industry is the high-end luxury area, which includes everything from gourmet pet foods, designer toys, and specialty furniture to groomers, dog walkers, boarding in posh pet hotels, and even pet therapists. (*Source:* American Pet Products Manufacturers Association.)

In Example 9 of Section 1.3, we use an *algebraic expression,* one of the topics of this chapter, to determine how much Americans have spent annually on their pets in recent years.

1.1 Basic Concepts

OBJECTIVES

1. Write sets using set notation.
2. Use number lines.
3. Know the common sets of numbers.
4. Find additive inverses.
5. Use absolute value.
6. Use inequality symbols.

OBJECTIVE 1 Write sets using set notation. A **set** is a collection of objects called the **elements,** or **members,** of the set. In algebra, the elements of a set are usually numbers. Set braces, { }, are used to enclose the elements. For example, 2 is an element of the set $\{1, 2, 3\}$. Since we can count the number of elements in the set $\{1, 2, 3\}$ and the counting process comes to an end, it is a **finite set.**

In algebra, we refer to certain sets of numbers by name. The set

$$N = \{1, 2, 3, 4, 5, 6, \ldots\} \quad \text{Natural (counting) numbers}$$

is called the **natural numbers,** or the **counting numbers.** The three dots (called *ellipsis points*) show that the list continues in the same pattern indefinitely. We cannot list all of the elements of the set of natural numbers, so it is an **infinite set.**

When 0 is included with the set of natural numbers, we have the set of **whole numbers,** written

$$W = \{0, 1, 2, 3, 4, 5, 6, \ldots\}. \quad \text{Whole numbers}$$

A set containing no elements, such as the set of whole numbers less than 0, is called the **empty set,** or **null set,** usually written $\emptyset$.

CAUTION
Do *not* write $\{\emptyset\}$ for the empty set; $\{\emptyset\}$ is a set with one element, $\emptyset$. Use only the notation $\emptyset$ for the empty set.

◀ Work Problem 1 at the Side.

1 Consider the set
$$\left\{0, 10, \frac{3}{10}, 52, 98.6\right\}.$$
(a) Which elements of the set are natural numbers?

(b) Which elements of the set are whole numbers?

In algebra, letters called **variables** are often used to represent numbers or to define sets of numbers. For example,

$$\{x \mid x \text{ is a natural number between 3 and 15}\}$$

(read "the set of all elements x such that x is a natural number between 3 and 15") defines the set

$$\{4, 5, 6, 7, \ldots, 14\}.$$

The notation $\{x \mid x \text{ is a natural number between 3 and 15}\}$ is an example of **set-builder notation.**

$$\{x \mid x \text{ has property } P\}$$

the set of — all elements x — such that — x has a given property P

EXAMPLE 1 Listing the Elements in Sets

List the elements in each set.

(a) $\{x \mid x \text{ is a natural number less than 4}\}$
The natural numbers less than 4 are 1, 2, and 3. This set is $\{1, 2, 3\}$.

(b) $\{y \mid y \text{ is one of the first five even natural numbers}\}$ is $\{2, 4, 6, 8, 10\}$.

(c) $\{z \mid z \text{ is a natural number greater than or equal to 7}\}$
The set of natural numbers greater than or equal to 7 is an infinite set, written with ellipsis points as $\{7, 8, 9, 10, \ldots\}$.

◀ Work Problem 2 at the Side.

2 List the elements in each set.
(a) $\{x \mid x \text{ is a whole number less than 5}\}$

(b) $\{y \mid y \text{ is a whole number greater than 12}\}$

ANSWERS

1. (a) 10 and 52 (b) 0, 10, and 52
2. (a) $\{0, 1, 2, 3, 4\}$ (b) $\{13, 14, 15, \ldots\}$

EXAMPLE 2 Using Set-Builder Notation to Describe Sets

Use set-builder notation to describe each set.

(a) $\{1, 3, 5, 7, 9\}$

There are often several ways to describe a set with set-builder notation. One way is $\{y \mid y$ is one of the first five odd natural numbers$\}$.

(b) $\{5, 10, 15, \ldots\}$

This set can be described as $\{x \mid x$ is a multiple of 5 greater than $0\}$.

Work Problem **3** *at the Side.* ▶

OBJECTIVE 2 Use number lines. A good way to get a picture of a set of numbers is by using a **number line.** To construct a number line, choose any point on a horizontal line and label it 0. Next, choose a point to the right of 0 and label it 1. The distance from 0 to 1 establishes a scale that can be used to locate more points, with positive numbers to the right of 0 and negative numbers to the left of 0. The number 0 is neither positive nor negative. A number line is shown in Figure 1.

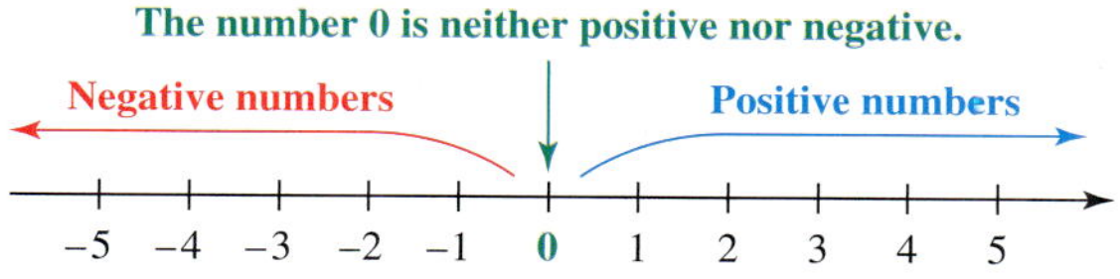

Figure 1

The set of numbers identified on the number line in Figure 1, including positive and negative numbers and 0, is part of the set of **integers,** written

$$I = \{\ldots, -3, -2, -1, 0, 1, 2, 3, \ldots\}. \quad \text{Integers}$$

Each number on a number line is called the **coordinate** of the point that it labels, while the point is the **graph** of the number. Figure 2 shows a number line with several selected points graphed on it.

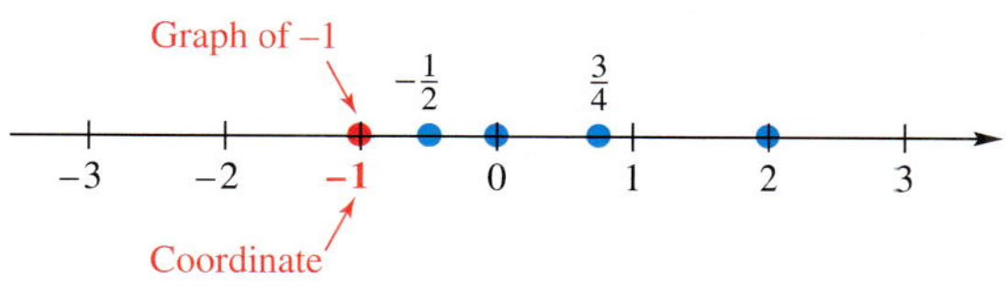

Figure 2

Work Problem **4** *at the Side.* ▶

The fractions $-\frac{1}{2}$ and $\frac{3}{4}$, graphed on the number line in Figure 2, are examples of *rational numbers*. A **rational number** can be expressed as the quotient of two integers, with denominator not 0. The set of all rational numbers is written

$$\left\{\frac{p}{q} \,\middle|\, p \text{ and } q \text{ are integers, } q \neq 0\right\}. \quad \text{Rational numbers}$$

The set of rational numbers includes the natural numbers, whole numbers, and integers, since these numbers can be written as fractions. For example, $14 = \frac{14}{1}$, $-3 = \frac{-3}{1}$, and $0 = \frac{0}{1}$. A rational number written as a fraction, such as $\frac{1}{8}$ or $\frac{2}{3}$, can also be expressed as a decimal by dividing the numerator by the denominator, as shown on the next page.

3 Use set-builder notation to describe each set.

(a) $\{0, 1, 2, 3, 4, 5\}$

(b) $\{7, 14, 21, 28, \ldots\}$

4 Graph the elements of each set.

(a) $\{-4, -2, 0, 2, 4, 6\}$

(b) $\left\{-1, 0, \frac{2}{3}, 2.5\right\}$

(c) $\left\{5, \frac{16}{3}, 6, \frac{13}{2}, 7, \frac{29}{4}\right\}$

ANSWERS

3. (a) One answer is $\{x \mid x$ is a whole number less than $6\}$. (b) One answer is $\{x \mid x$ is a multiple of 7 greater than $0\}$.

4. (a) 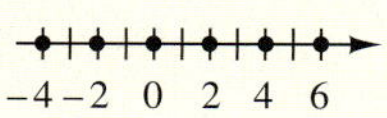

(b) 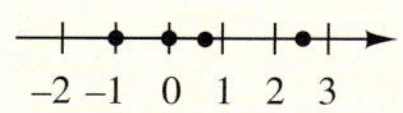

(c) 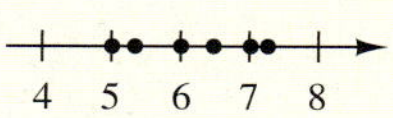

$$\begin{array}{r} 0.125 \\ 8\overline{)1.000} \\ \underline{8} \\ 20 \\ \underline{16} \\ 40 \\ \underline{40} \\ 0 \end{array}$$

0.125 ← Terminating decimal; 0 ← Remainder is 0.

$$\frac{1}{8} = 0.125$$

$$\begin{array}{r} 0.666\ldots \\ 3\overline{)2.000\ldots} \\ \underline{18} \\ 20 \\ \underline{18} \\ 20 \\ \underline{18} \\ 2 \end{array}$$

0.666... ← Repeating decimal; 2 ← Remainder is never 0.

$$\frac{2}{3} = 0.\overline{6}$$ ← A bar is written over the repeating digit(s).

Thus, terminating decimals, such as $0.125 = \frac{1}{8}$, $0.8 = \frac{4}{5}$, and $2.75 = \frac{11}{4}$, and repeating decimals, such as $0.\overline{6} = \frac{2}{3}$ and $0.\overline{27} = \frac{3}{11}$, are rational numbers.

Decimal numbers that neither terminate nor repeat are *not* rational, and thus are called **irrational numbers.** Many square roots are irrational numbers; for example, $\sqrt{2} = 1.4142135\ldots$ and $-\sqrt{7} = -2.6457513\ldots$ repeat indefinitely without pattern. (Some square roots *are* rational: $\sqrt{16} = 4$, $\sqrt{100} = 10$, and so on.) Another irrational number is π (pronounced "pie"), the ratio of the distance around, or circumference, C of a circle to its diameter d. See Figure 3.

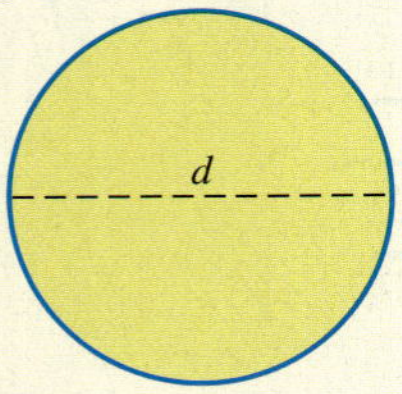

$\pi = \frac{C}{d}$

Figure 3

Some of the rational and irrational numbers discussed above are graphed on the number line in Figure 4. The rational numbers together with the irrational numbers make up the set of **real numbers.** ***Every point on a number line corresponds to a real number, and every real number corresponds to a point on the number line.***

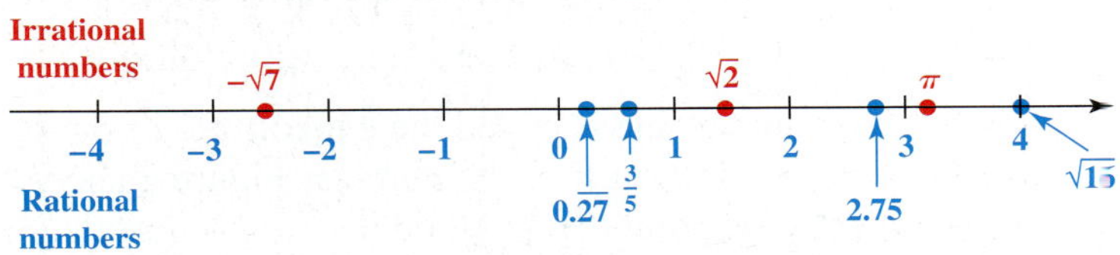

Figure 4

OBJECTIVE 3 Know the common sets of numbers. The sets of numbers listed below will be used throughout the rest of this text.

Sets of Numbers

Natural numbers	$\{1, 2, 3, 4, 5, 6, \ldots\}$
Whole numbers	$\{0, 1, 2, 3, 4, 5, 6, \ldots\}$
Integers	$\{\ldots, -3, -2, -1, 0, 1, 2, 3, \ldots\}$
Rational numbers	$\left\{\frac{p}{q} \middle\vert\, p \text{ and } q \text{ are integers}, q \neq 0\right\}$ *Examples:* $\frac{4}{1}$, 1.3, $-\frac{9}{2}$, $\frac{16}{8}$ or 2, $\sqrt{9}$ or 3, $0.\overline{6}$
Irrational numbers	$\{x \mid x$ **is a real number that is not rational**$\}$ *Examples:* $\sqrt{3}$, $-\sqrt{2}$, π
Real numbers	$\{x \mid x$ **is a rational number or an irrational number**$\}$*

*An example of a number that is not real is $\sqrt{-1}$. This number, part of the *complex number system,* is discussed in **Section 9.7.**

Figure 5 shows that the set of real numbers includes both the rational and irrational numbers. ***Every real number is either rational or irrational.*** Also, notice that the integers are elements of the set of rational numbers, and that the whole numbers and natural numbers are elements of the set of integers.

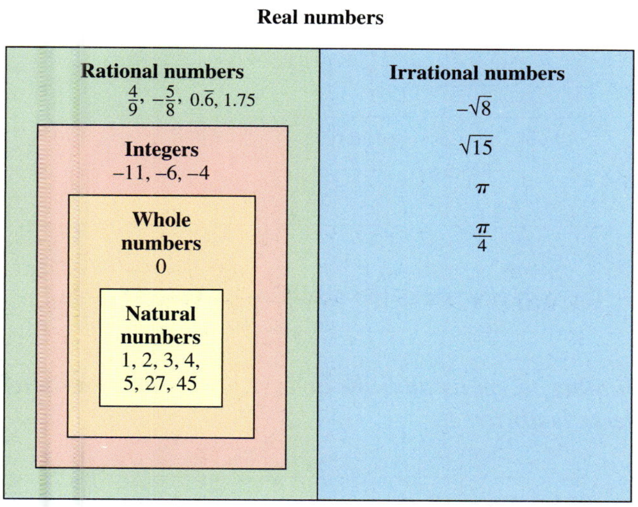

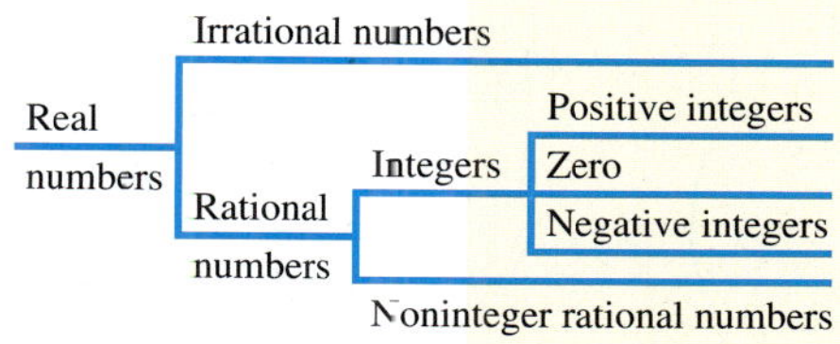

Figure 5 The Real Numbers

EXAMPLE 3 Identifying Examples of Number Sets

Which numbers in

$$\left\{-8, -\sqrt{2}, -\frac{9}{64}, 0, 0.5, \frac{2}{3}, 1.\overline{12}, \sqrt{3}, 2\right\}$$

are elements of each set?

(a) Integers
-8, 0, and 2 are integers.

(b) Rational numbers
$-8, -\frac{9}{64}, 0, 0.5, \frac{2}{3}, 1.\overline{12}$, and 2 are rational numbers.

(c) Irrational numbers
$-\sqrt{2}$ and $\sqrt{3}$ are irrational numbers.

(d) Real numbers
All the numbers in the given set are real numbers.

Work Problem **5** *at the Side.* ▶

EXAMPLE 4 Determining Relationships between Sets of Numbers

Decide whether each statement is *true* or *false*.

(a) All irrational numbers are real numbers.
This is true. As shown in Figure 5, the set of real numbers includes all irrational numbers.

(b) Every rational number is an integer.
This statement is false. Although some rational numbers are integers, other rational numbers, such as $\frac{2}{3}$ and $-\frac{1}{4}$, are not.

Work Problem **6** *at the Side.* ▶

5 Select all the sets from the following list that apply to each number.
Whole number
Rational number
Irrational number
Real number

(a) -6 **(b)** 12

(c) $0.\overline{3}$ **(d)** $-\sqrt{15}$

(e) π **(f)** $\frac{22}{7}$

(g) 3.14

6 Decide whether the statement is *true* or *false*. If *false,* tell why.

(a) All whole numbers are integers.

(b) Some integers are whole numbers.

(c) Every real number is irrational.

ANSWERS

5. **(a)** rational, real **(b)** whole, rational, real **(c)** rational, real **(d)** irrational, real **(e)** irrational, real **(f)** rational, real **(g)** rational, real

6. **(a)** true **(b)** true **(c)** false; Some real numbers are irrational, but others are rational numbers.

OBJECTIVE 4 Find additive inverses. Look at the number line in Figure 6. For each positive number, there is a negative number on the opposite side of 0 that lies the same distance from 0. These pairs of numbers are called *additive inverses, negatives,* or *opposites* of each other. For example, 3 is the additive inverse of -3, and -3 is the additive inverse of 3.

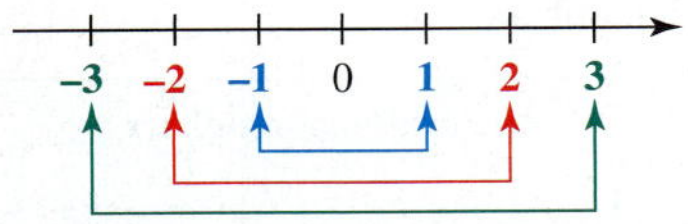

Additive inverses (opposites)

Figure 6

> **Additive Inverse**
>
> For any real number a, the number $-a$ is the **additive inverse** of a.

Change the sign of a number to get its additive inverse. The sum of a number and its additive inverse is always 0.

> **Uses of the Symbol −**
>
> The symbol "−" can be used to indicate any of the following:
>
> **1.** a negative number, such as -9 or -15;
>
> **2.** the additive inverse of a number, as in "-4 is the additive inverse of 4";
>
> **3.** subtraction, as in $12 - 3$.

In the expression $-(-5)$, the symbol "−" is being used in two ways: the first − indicates the additive inverse (or opposite) of -5, and the second indicates a negative number, -5. Since the additive inverse of -5 is 5, then

$$-(-5) = 5.$$

This example suggests the following property.

> **$-(-a)$**
>
> For any real number a, $\quad -(-a) = a.$

Numbers written with positive or negative signs, such as $+4$, $+8$, -9, and -5, are called **signed numbers.** A positive number can be called a signed number even though the positive sign is usually left off. The following table shows the additive inverses of several signed numbers. The number 0 is its own additive inverse.

Number	Additive Inverse
6	-6
-4	4
$\frac{2}{3}$	$-\frac{2}{3}$
-8.7	8.7
0	0

◀ Work Problem 7 at the Side.

7 Give the additive inverse of each number.

(a) 9

(b) -12

(c) $-\frac{6}{5}$

(d) 0

(e) 1.5

Answers

7. (a) -9 **(b)** 12 **(c)** $\frac{6}{5}$ **(d)** 0 **(e)** -1.5

OBJECTIVE 5 Use absolute value. Geometrically, the **absolute value** of a number a, written $|a|$, is the distance on the number line from 0 to a. For example, the absolute value of 5 is the same as the absolute value of -5 because each number lies five units from 0. See Figure 7. That is,

$$|5| = 5 \quad \text{and} \quad |-5| = 5.$$

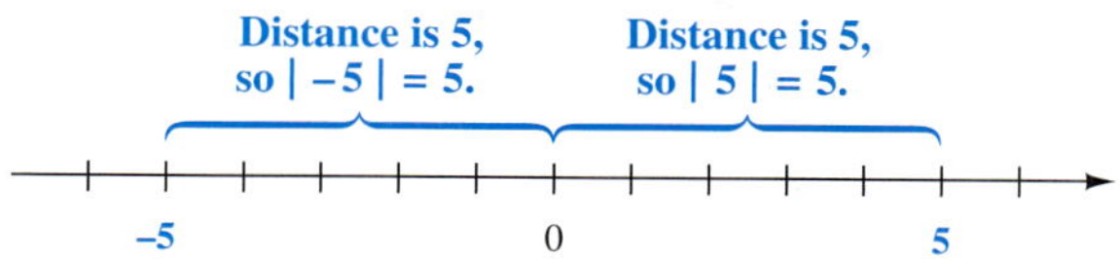

Figure 7

> **CAUTION**
> ***Because absolute value represents distance, and distance is never negative, the absolute value of a number is always positive or 0.***

The formal definition of absolute value follows.

> **Absolute Value**
>
> $$|a| = \begin{cases} a & \textbf{if } a \textbf{ is positive or 0} \\ -a & \textbf{if } a \textbf{ is negative} \end{cases}$$

The second part of this definition, $|a| = -a$ if a is negative, requires careful thought. If a is a *negative* number, then $-a$, the additive inverse or opposite of a, is a positive number. Thus, $|a|$ is positive. For example, if $a = -3$, then

$$|a| = |-3| = -(-3) = 3. \qquad |a| = -a \text{ if } a \text{ is negative.}$$

EXAMPLE 5 Finding Absolute Value

Simplify by finding each absolute value.

(a) $|13| = 13$

(b) $|-2| = -(-2) = 2$

(c) $|0| = 0$

(d) $|-0.75| = 0.75$

(e) $-|8|$

Evaluate the absolute value first. Then find the additive inverse.

$$-|8| = -(8) = -8$$

(f) $-|-8|$

Work as in part (e): $|-8| = 8$, so

$$-|-8| = -(8) = -8.$$

(g) $|-2| + |5|$

Evaluate each absolute value first, and then add.

$$|-2| + |5| = 2 + 5 = 7$$

(h) $-|5 - 2| = -|3| = -3$

Work Problem 8 at the Side. ▶

8 Find the value of each expression.

(a) $|6|$

(b) $|-3|$

(c) $-\left|\frac{1}{4}\right|$

(d) $-|-2|$

(e) $-|-7.25|$

(f) $|-6| + |-3|$

(g) $|-9| - |-4|$

(h) $-|9 - 4|$

ANSWERS

8. (a) 6 **(b)** 3 **(c)** $-\frac{1}{4}$ **(d)** -2 **(e)** -7.25 **(f)** 9 **(g)** 5 **(h)** -5

Absolute value is useful when comparing size without regard to sign.

9 Refer to the table in Example 6. Of the software publishers and fabric mills industries, which will show the greater change (without regard to sign)?

EXAMPLE 6 Comparing Rates of Change in Industries

The projected annual rates of change in employment (in percent) in some of the fastest-growing and in some of the most rapidly declining industries from 2002 through 2012 are shown in the table.

Industry (2002–2012)	Annual Rate of Change (in percent)
Software publishers	5.3
Care services for the elderly	4.5
Child day-care services	3.6
Cut-and-sew apparel manufacturing	−12.2
Fabric mills	−5.9
Metal ore mining	−4.8

Source: U.S. Bureau of Labor Statistics.

What industry in the list is expected to see the greatest change? the least change?

We want the greatest *change,* without regard to whether the change is an increase or a decrease. Look for the number in the list with the largest absolute value. That number is found in cut-and-sew apparel manufacturing, since

$$|-12.2| = 12.2.$$

Similarly, the least change is in the child day-care services industry:

$$|3.6| = 3.6.$$

◀ Work Problem 9 at the Side.

OBJECTIVE 6 Use inequality symbols. The statement

$$4 + 2 = 6$$

is an **equation**—a statement that two quantities are equal. The statement

$$4 \neq 6$$

(read "4 is not equal to 6") is an **inequality**—a statement that two quantities are *not* equal. When two numbers are not equal, one must be less than the other. The symbol $<$ means "is less than." For example,

$$8 < 9, \quad -6 < 15, \quad -6 < -1, \quad 0.5 < 0.9, \quad \text{and} \quad 0 < \frac{4}{3}.$$

The symbol $>$ means "is greater than." For example,

$$12 > 5, \quad 9 > -2, \quad -4 > -6, \quad 1.25 > 1.2, \quad \text{and} \quad \frac{6}{5} > 0.$$

In each case, the symbol "points" toward the lesser number.

ANSWER

9. Fabric mills

The number line in Figure 8 shows the graphs of the numbers 4 and 9. We know that $4 < 9$. On the graph, 4 is to the left of 9. ***The lesser of two numbers is always to the left of the other on a number line.***

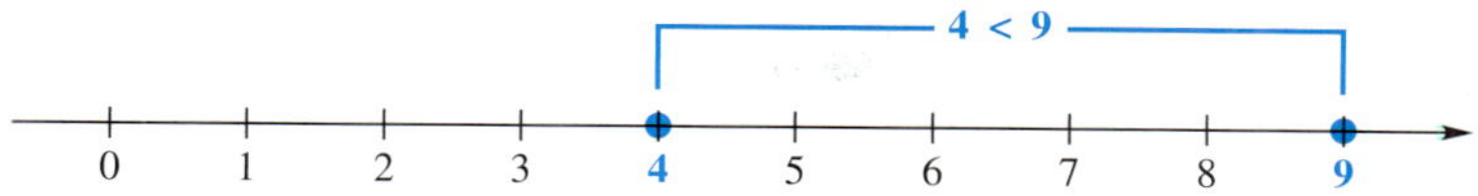

Figure 8

Inequalities on a Number Line

On a number line,

$a < b$ if a is to the left of b; $a > b$ if a is to the right of b.

We can use a number line to determine order. As shown on the number line in Figure 9, -6 is located to the left of 1. For this reason, $-6 < 1$. Also, $1 > -6$. From the same number line, $-5 < -2$, or $-2 > -5$.

Figure 9

CAUTION

Be careful when ordering negative numbers. Since -5 is to the left of -2 on the number line in Figure 9, $-5 < -2$, or $-2 > -5$. In each case, the symbol points to -5, the lesser number.

Work Problem **10** *at the Side.* ▶

The following table summarizes results about positive and negative numbers in both words and symbols.

Words	Symbols
Every negative number is less than 0.	If a is negative, then $a < 0$.
Every positive number is greater than 0.	If a is positive, then $a > 0$.
0 is neither positive nor negative.	

In addition to the symbols $\neq$, $<$, and $>$, the symbols $\leq$ and $\geq$ are often used.

INEQUALITY SYMBOLS

Symbol	Meaning	Example
$\neq$	is not equal to	$3 \neq 7$
$<$	is less than	$-4 < -1$
$>$	is greater than	$3 > -2$
$\leq$	is less than or equal to	$6 \leq 6$
$\geq$	is greater than or equal to	$-8 \geq -10$

10 Insert $<$ or $>$ in each blank to make a true statement.

(a) 3 ______ 7

(b) 9 ______ 2

(c) -4 ______ -8

(d) -2 ______ -1

(e) 0 ______ -3.5

(f) $\frac{5}{8}$ ______ $\frac{3}{4}$

(g) -0.3 ______ -0.5

ANSWERS

10. (a) $<$ **(b)** $>$ **(c)** $>$ **(d)** $<$ **(e)** $>$ **(f)** $<$ **(g)** $>$

The following table shows several inequalities and why each is true.

Inequality	Why It Is True
$6 \leq 8$	$6 < 8$
$-2 \leq -2$	$-2 = -2$
$-9 \geq -12$	$-9 > -12$
$-3 \geq -3$	$-3 = -3$
$6 \cdot 4 \leq 5(5)$	$24 < 25$

Notice the reason why $-2 \leq -2$ is true. ***With the symbol ≤, if either the < part or the = part is true, then the inequality is true. This is also the case with the ≥ symbol.***

In the last row of the table, recall that the dot in $6 \cdot 4$ indicates the product 6×4, or 24, and $5(5)$ means 5×5, or 25. Thus, the inequality $6 \cdot 4 \leq 5(5)$ becomes $24 \leq 25$, which is true.

◀ Work Problem 11 at the Side.

11 Answer *true* or *false*.

(a) $-2 \leq -3$

(b) $0.5 \leq 0.5$

(c) $-9 \geq -1$

(d) $5 \cdot 8 \leq 7 \cdot 7$

(e) $3(4) > 2(6)$

Answers

11. **(a)** false **(b)** true **(c)** false **(d)** true **(e)** false

1.1 Exercises

FOR EXTRA HELP

MyMathLab | MathXL PRACTICE | WATCH | DOWNLOAD | READ | REVIEW

Write each set by listing its elements. See Example 1.

1. $\{x \mid x$ is a natural number less than $6\}$
2. $\{m \mid m$ is a natural number less than $9\}$
3. $\{z \mid z$ is an integer greater than $4\}$
4. $\{y \mid y$ is an integer greater than $8\}$
5. $\{a \mid a$ is an even integer greater than $8\}$
6. $\{k \mid k$ is an odd integer less than $1\}$
7. $\{x \mid x$ is an irrational number that is also rational$\}$
8. $\{r \mid r$ is a number that is both positive and negative$\}$
9. $\{p \mid p$ is a number whose absolute value is $4\}$
10. $\{w \mid w$ is a number whose absolute value is $7\}$

Write each set using set-builder notation. See Example 2. (More than one description is possible.)

11. $\{2, 4, 6, 8\}$
12. $\{11, 12, 13, 14\}$
13. $\{4, 8, 12, 16, \ldots\}$
14. $\{\ldots, -6, -3, 0, 3, 6, \ldots\}$

Graph the elements of each set on a number line.

15. $\{-3, -1, 0, 4, 6\}$
16. $\{-4, -2, 0, 3, 5\}$
17. $\left\{-\frac{2}{3}, 0, \frac{4}{5}, \frac{12}{5}, \frac{9}{2}, 4.8\right\}$
18. $\left\{-\frac{6}{5}, -\frac{1}{4}, 0, \frac{5}{6}, \frac{13}{4}, 5.2, \frac{11}{2}\right\}$

*Which elements of each set are **(a)** natural numbers, **(b)** whole numbers, **(c)** integers, **(d)** rational numbers, **(e)** irrational numbers, **(f)** real numbers? See Example 3.*

19. $\left\{-8, -\sqrt{5}, -0.6, 0, \frac{3}{4}, \sqrt{3}, \pi, 5, \frac{13}{2}, 17, \frac{40}{2}\right\}$
20. $\left\{-9, -\sqrt{6}, -0.7, 0, \frac{6}{7}, \sqrt{7}, 4.\overline{6}, 8, \frac{21}{2}, 13, \frac{75}{5}\right\}$

Decide whether each statement is true *or* false. *If* false, *tell why. See Example 4.*

21. Every rational number is an integer.

22. Every natural number is an integer.

23. Every irrational number is an integer.

24. Every integer is a rational number.

25. Every natural number is a whole number.

26. Some rational numbers are irrational.

27. Some rational numbers are whole numbers.

28. Some real numbers are integers.

29. The absolute value of any number is the same as the absolute value of its additive inverse.

30. The absolute value of any nonzero number is positive.

Give **(a)** *the additive inverse and* **(b)** *the absolute value of each number. See the discussion of additive inverses and Example 5.*

31. 6

32. 8

33. -12

34. -15

35. $\frac{6}{5}$

36. 0.13

Find the value of each expression. See Example 5.

37. $|-8|$

38. $|-11|$

39. $\left|\frac{3}{2}\right|$

40. $\left|\frac{7}{4}\right|$

41. $-|5|$

42. $-|17|$

43. $-|-2|$

44. $-|-8|$

45. $-|4.5|$

46. $-|12.6|$

47. $|-2| + |3|$

48. $|-16| + |12|$

49. $|-9| - |-3|$

50. $|-10| - |-5|$

51. $|-1| + |-2| - |-3|$

52. $|-6| + |-4| - |-10|$

Solve each problem. See Example 6.

53. The table shows the percent change in population from 2000 through 2006 for selected states.

State	Percent Change
Alabama	3.4
Iowa	1.9
Louisiana	−4.1
Michigan	1.6
North Dakota	−1.0
West Virginia	0.6

Source: U.S. Census Bureau.

(a) Which state had the greatest change in population? What was this change? Was it an increase or a decrease?

(b) Which state had the least change in population? What was this change? Was it an increase or a decrease?

54. The table gives the net trade balance, in millions of dollars, for selected U.S. trade partners for January 2006.

Country	Trade Balance (in millions of dollars)
India	−1257
China	−17,911
Netherlands	756
France	−85
Turkey	−78

Source: U.S. Census Bureau.

A negative balance means that imports to the United States exceeded exports from the United States, while a positive balance means that exports exceeded imports.

(a) Which country had the greatest discrepancy between exports and imports? Explain.

(b) Which country had the least discrepancy between exports and imports? Explain.

Sea level refers to the surface of the ocean. The depth of a body of water such as an ocean or sea can be expressed as a negative number, representing average depth in feet below sea level. On the other hand, the altitude of a mountain can be expressed as a positive number, indicating its height in feet above sea level. The table gives selected depths and heights.

Body of Water	Average Depth in Feet (as a negative number)	Mountain	Altitude in Feet (as a positive number)
Pacific Ocean	−12,925	McKinley	20,320
South China Sea	−4,802	Point Success	14,158
Gulf of California	−2,375	Matlalcueyetl	14,636
Caribbean Sea	−8,448	Rainier	14,410
Indian Ocean	−12,598	Steele	16,644

Source: World Almanac and Book of Facts.

55. List the bodies of water in order, starting with the deepest and ending with the shallowest.

56. List the mountains in order, starting with the shortest and ending with the tallest.

57. *True* or *false:* The absolute value of the depth of the Pacific Ocean is greater than the absolute value of the depth of the Indian Ocean.

58. *True* or *false:* The absolute value of the depth of the Gulf of California is greater than the absolute value of the depth of the Caribbean Sea.

Use order on a number line to answer true *or* false *to each statement.*

59. $-6 < -2$

60. $-4 < -3$

61. $-4 > -3$

62. $-2 > -1$

63. $3 > -2$

64. $5 > -3$

65. $-3 \geq -3$

66. $-4 \leq -4$

Use an inequality symbol to write each statement.

67. 7 is greater than y.

68. -4 is less than 12.

69. 5 is greater than or equal to 5.

70. -3 is less than or equal to -3.

71. $3t - 4$ is less than or equal to 10.

72. $5x + 4$ is greater than or equal to 19.

73. $5x + 3$ is not equal to 0.

74. $6x + 7$ is not equal to -3.

First simplify each side of the inequality. Then tell whether the resulting statement is true *or* false.

75. $-6 < 7 + 3$

76. $-7 < 4 + 2$

77. $2 \cdot 5 \geq 4 + 6$

78. $8 + 7 \leq 3 \cdot 5$

79. $-|-3| \geq -3$

80. $-|-5| \leq -5$

81. $-8 > -|-6|$

82. $-9 > -|-4|$

The graph shows egg production in millions of eggs in selected states for 2005 and 2006. Use this graph to work Exercises 83–87.

83. In 2005, was egg production in Iowa (IA) less than or greater than egg production in California (CA)?

84. In 2006, which states had production greater than 6000 million eggs?

85. In which states was 2006 egg production less than 2005 production?

86. If x represents 2006 egg production for Texas (TX) and y represents 2006 egg production for California (CA), which is true: $x < y$ or $x > y$?

87. If x represents 2005 egg production for Ohio (OH) and y represents 2006 egg production for Ohio (OH), which is true: $x < y$ or $x > y$?

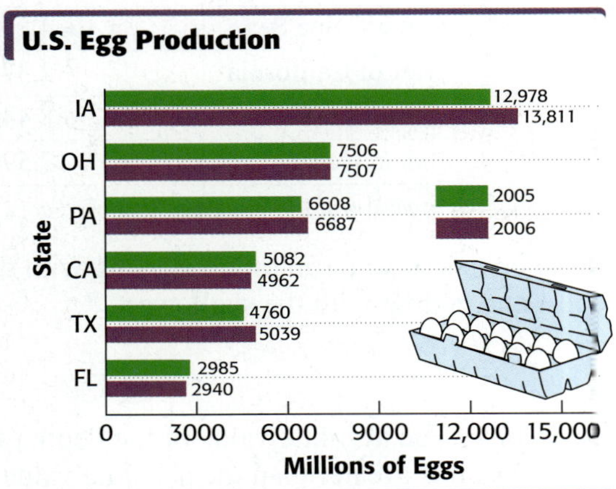

Source: U.S. Department of Agriculture.

1.2 Operations on Real Numbers

OBJECTIVES

1. **Add real numbers.**
2. **Subtract real numbers.**
3. **Multiply real numbers.**
4. **Find the reciprocal of a number.**
5. **Divide real numbers.**

OBJECTIVE 1 Add real numbers. Number lines can be used to illustrate addition and subtraction of real numbers. To add two real numbers on a number line, start at 0. Move right (the *positive* direction) to add a positive number or left (the *negative* direction) to add a negative number. See Figure 10.

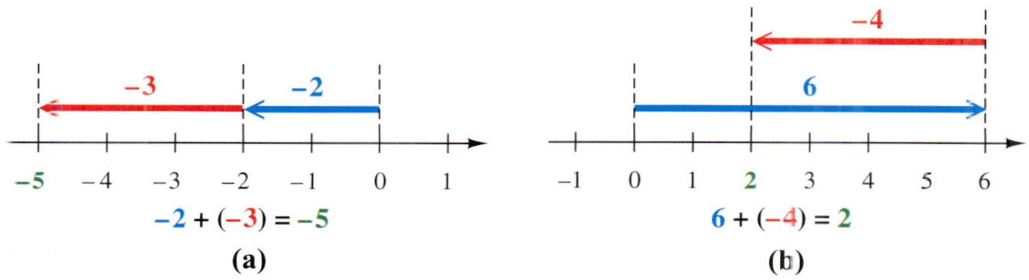

Figure 10

Recall that the answer to an addition problem is called the **sum.** The procedure for adding real numbers can be generalized in the following rules.

Adding Real Numbers

Same sign To add two numbers with the *same* sign, add their absolute values. The sum has the same sign as the given numbers.

Different signs To add two numbers with *different* signs, find the absolute values of the numbers, and subtract the lesser absolute value from the greater. The sum has the same sign as the number with the greater absolute value.

EXAMPLE 1 Adding Two Negative Real Numbers

Find each sum.

(a) $-12 + (-8)$

First find the absolute values.

$$|-12| = 12 \quad \text{and} \quad |-8| = 8$$

Because -12 and -8 have the *same* sign, add their absolute values.

Both numbers are negative, so the answer will be negative. $\quad -12 + (-8) = -(12 + 8) = -(20) = -20$

(b) $-6 + (-3) = -(|-6| + |-3|) = -(6 + 3) = -9$

(c) $-1.2 + (-0.4) = -(1.2 + 0.4) = -1.6$

(d) $-\frac{5}{6} + \left(-\frac{1}{3}\right) = -\left(\frac{5}{6} + \frac{1}{3}\right)$ Add the absolute values. Both numbers are negative, so the answer will be negative.

$= -\left(\frac{5}{6} + \frac{2}{6}\right)$ The least common denominator is 6; $\frac{1 \cdot 2}{3 \cdot 2} = \frac{2}{6}$

$= -\frac{7}{6}$ Add numerators; keep the same denominator.

Work Problem 1 at the Side. ▶

1 Find each sum.

(a) $-2 + (-7)$

(b) $-15 + (-6)$

(c) $-1.1 + (-1.2)$

(d) $-\frac{3}{4} + \left(-\frac{1}{2}\right)$

ANSWERS

1. (a) -9 **(b)** -21 **(c)** -2.3 **(d)** $-\frac{5}{4}$

2 Find each sum.

(a) $12 + (-1)$

(b) $3 + (-7)$

(c) $-17 + 5$

(d) $-1.5 + 3.2$

(e) $-\frac{3}{4} + \frac{1}{2}$

EXAMPLE 2 Adding Real Numbers with Different Signs

Find each sum.

(a) $-17 + 11$

First find the absolute values.

$$|-17| = 17 \quad \text{and} \quad |11| = 11$$

Because -17 and 11 have *different* signs, subtract their absolute values.

$$17 - 11 = 6$$

The number -17 has a greater absolute value than 11, so the answer is negative.

$$-17 + 11 = -6$$

The sum is negative because $|-17| > |11|$.

(b) $4 + (-1)$

Subtract the absolute values, 4 and 1. Because 4 has the greater absolute value, the sum must be positive.

$$4 + (-1) = 4 - 1 = 3$$

The sum is positive because $|4| > |-1|$.

(c) $-9 + 17 = 17 - 9 = 8$

(d) $-2.3 + 5.6 = 5.6 - 2.3 = 3.3$

(e) $-16 + 12$

The absolute values are 16 and 12. Subtract the absolute values. The negative number has the greater absolute value, so the answer is negative.

$$-16 + 12 = -(16 - 12) = -4$$

(f) $-\frac{4}{5} + \frac{2}{3}$

Write each fraction with the least common denominator, 15.

$$-\frac{4}{5} + \frac{2}{3} = -\frac{12}{15} + \frac{10}{15} \qquad -\frac{4 \cdot 3}{5 \cdot 3} = -\frac{12}{15}; \frac{2 \cdot 5}{3 \cdot 5} = \frac{10}{15}$$

$$= -\left(\frac{12}{15} - \frac{10}{15}\right)$$

Subtract the absolute values. $-\frac{12}{15}$ has the greater absolute value, so the answer will be negative.

$$= -\frac{2}{15}$$

Subtract numerators; keep the same denominator.

◀ Work Problem **2** at the Side.

OBJECTIVE 2 Subtract real numbers. Recall that the answer to a subtraction problem is called the **difference.** Consider the following:

$$6 - 4 = 2$$
$$6 + (-4) = 2.$$

(The second statement is pictured on the number line in Figure 10(b) at the beginning of this section). Thus,

$$6 - 4 = 6 + (-4).$$

To subtract 4 from 6, we add the additive inverse of 4 to 6. This example suggests the following definition of subtraction of real numbers.

Answers

2. **(a)** 11 **(b)** -4 **(c)** -12 **(d)** 1.7 **(e)** $-\frac{1}{4}$

Subtraction

For all real numbers a and b,

$$a - b = a + (-b).$$

In words, to subtract b from a, add the additive inverse (or opposite) of b to a.

EXAMPLE 3 Subtracting Real Numbers

Find each difference.

Change to addition.
The additive inverse of 8 is −8.

(a) $6 - 8 = 6 + (-8) = -2$

Change to addition.
The additive inverse of 4 is −4.

(b) $-12 - 4 = -12 + (-4) = -16$

(c) $-10 - (-7) = -10 + 7$ Add the additive inverse of −7, which is 7.
$= -3$

(d) $-2.4 - (-8.1) = -2.4 + 8.1 = 5.7$

(e) $\frac{5}{6} - \left(-\frac{3}{8}\right) = \frac{5}{6} + \frac{3}{8}$ To subtract, add the additive inverse (opposite).

$= \frac{20}{24} + \frac{9}{24}$ Write each fraction with the least common denominator, 24.

$= \frac{29}{24}$ Add numerators; keep the same denominator.

Work Problem **3** *at the Side.* ▶

In a problem that involves both addition and subtraction, add and subtract in order from left to right. Work inside brackets or parentheses first.

EXAMPLE 4 Adding and Subtracting Real Numbers

Perform the indicated operations.

(a) $-8 + 5 - 6$
$= (-8 + 5) - 6$ Work from left to right.
$= -3 - 6$
$= -3 + (-6)$ To subtract, add the additive inverse.
$= -9$

(b) $15 - (-3) - 5 - 12$
$= (15 + 3) - 5 - 12$
$= 18 - 5 - 12$
$= 13 - 12$
$= 1$

Continued on Next Page

3 Find each difference.

(a) $9 - 12$

(b) $-7 - 2$

(c) $-8 - (-2)$

(d) $12 - (-5)$

(e) $-6.3 - (-11.5)$

(f) $\frac{3}{4} - \left(-\frac{2}{3}\right)$

ANSWERS

3. (a) −3 (b) −9 (c) −6 (d) 17 (e) 5.2 (f) $\frac{17}{12}$

4 Perform the indicated operations.

(a) $-6 + 9 - 2$

(b) $12 - (-4) + 8$

(c) $-6 - (-2) - 8 - 1$

(d) $-3 - [(-7) + 15] + 6$

5 Find each product.

(a) $-7(-5)$

(b) $-0.9(-15)$

(c) $-\frac{4}{7}\left(-\frac{14}{3}\right)$

(d) $7(-2)$

(e) $-0.8(0.006)$

(f) $\frac{5}{8}(-16)$

(g) $-\frac{2}{3}(12)$

Answers

4. **(a)** 1 **(b)** 24 **(c)** −13 **(d)** −5

5. **(a)** 35 **(b)** 13.5 **(c)** $\frac{8}{3}$ **(d)** −14 **(e)** −0.0048 **(f)** −10 **(g)** −8

(c) $-9 - [-8 - (-4)] + 6$

$= -9 - [-8 + 4] + 6$ Work inside the brackets.

$= -9 - [-4] + 6$

$= -9 + 4 + 6$ To subtract, add the additive inverse.

$= -5 + 6$ Work from left to right.

$= 1$

◀ *Work Problem* **4** *at the Side.*

OBJECTIVE 3 **Multiply real numbers.** The answer to a multiplication problem is called the **product.** For example, 24 is the product of 8 and 3.

Multiplying Real Numbers

Same sign The product of two numbers with the *same* sign is positive.

Different signs The product of two numbers with *different* signs is negative.

EXAMPLE 5 **Multiplying Real Numbers**

Find each product.

(a) $-3(-9) = 27$ Same sign; product is positive.

(b) $-0.5(-0.4) = 0.2$

(c) $-\frac{3}{4}\left(-\frac{5}{6}\right) = \frac{15}{24}$ Multiply numerators; multiply denominators.

$= \frac{5 \cdot 3}{8 \cdot 3}$ Factor to write in lowest terms.

$= \frac{5}{8}$ Divide out the common factor, 3.

(d) $6(-9) = -54$ Different signs; product is negative.

(e) $-0.05(0.3) = -0.015$

(f) $-\frac{3}{4}\left(\frac{2}{9}\right) = -\frac{1}{6}$

(g) $\frac{2}{3}(-6) = -4$ ($-6 = -\frac{6}{1}$)

◀ *Work Problem* **5** *at the Side.*

OBJECTIVE 4 **Find the reciprocal of a number.** The definition of division depends on the idea of a **multiplicative inverse** or *reciprocal.* Two numbers are *reciprocals* if they have a product of 1.

Reciprocal

The **reciprocal** of a nonzero number a is $\frac{1}{a}$.

Calculator Tip Reciprocals (in decimal form) can be found with a calculator that has a key labeled 1/x or x^{-1}. For example, a calculator shows that the reciprocal of 25 is 0.04.

The table gives several numbers and their reciprocals.

Number	Reciprocal
$-\frac{2}{5}$	$-\frac{5}{2}$
-6, or $-\frac{6}{1}$	$-\frac{1}{6}$
$\frac{7}{11}$	$\frac{11}{7}$
0.05	20
0	None

$-\frac{2}{5}\left(-\frac{5}{2}\right) = 1$

$-6\left(-\frac{1}{6}\right) = 1$

$\frac{7}{11}\left(\frac{11}{7}\right) = 1$

$0.05(20) = 1$

Reciprocals have a product of 1.

There is no reciprocal for 0 because there is no number that can be multiplied by 0 to give a product of 1.

CAUTION

A number and its additive inverse have opposite signs; however, a number and its reciprocal always have the same sign.

Work Problem 6 at the Side. ▶

OBJECTIVE 5 Divide real numbers. The result of dividing one number by another is called the **quotient.** For example, when 45 is divided by 3, the quotient is 15. To define division of real numbers, we can write the quotient of 45 and 3 as $\frac{45}{3}$, which equals 15. The same answer will be obtained if 45 and $\frac{1}{3}$ are multiplied, as follows.

$$45 \div 3 = \frac{45}{3} = 45 \cdot \frac{1}{3} = 15$$

This suggests the following definition of division of real numbers.

Division

For all real numbers a and b (where $b \neq 0$),

$$\boldsymbol{a \div b = \frac{a}{b} = a \cdot \frac{1}{b}}.$$

In words, multiply the first number (the **dividend**) by the reciprocal of the second number (the **divisor**).

There is no reciprocal for the number 0, so ***division by 0 is undefined.*** For example, $\frac{15}{0}$ is undefined and $-\frac{1}{0}$ is undefined.

CAUTION

Division by 0 is undefined. However, dividing 0 by a nonzero number gives the quotient 0. For example,

$$\frac{6}{0} \text{ is undefined, but } \frac{0}{6} = 0 \quad (\text{since } 0 \cdot 6 = 0).$$

Be careful when 0 is involved in a division problem.

Work Problem 7 at the Side. ▶

6 Give the reciprocal of each number.

(a) 15

(b) -7

(c) $\frac{8}{9}$

(d) $-\frac{1}{3}$

(e) 0.125

7 Divide where possible.

(a) $\frac{9}{0}$

(b) $\frac{0}{9}$

(c) $\frac{-9}{0}$

(d) $\frac{0}{-9}$

Answers

6. (a) $\frac{1}{15}$ **(b)** $-\frac{1}{7}$ **(c)** $\frac{9}{8}$ **(d)** -3 **(e)** 8

7. (a) undefined **(b)** 0 **(c)** undefined **(d)** 0

Since division is defined as multiplication by the reciprocal, the rules for signs of quotients are the same as those for signs of products.

Dividing Real Numbers

Same sign The quotient of two nonzero real numbers with the *same* sign is positive.

Different signs The quotient of two nonzero real numbers with *different* signs is negative.

EXAMPLE 6 Dividing Real Numbers

Find each quotient.

(a) $\dfrac{-12}{4} = -12 \cdot \dfrac{1}{4} = -3$ $\quad \frac{a}{b} = a \cdot \frac{1}{b}$

(b) $\dfrac{6}{-3} = 6\left(-\dfrac{1}{3}\right) = -2$ $\quad$ The reciprocal of -3 is $-\frac{1}{3}$.

(c) $\dfrac{-\frac{2}{3}}{-\frac{5}{9}} = -\dfrac{2}{3}\left(-\dfrac{9}{5}\right) = \dfrac{6}{5}$ $\quad$ The reciprocal of $-\frac{5}{9}$ is $-\frac{9}{5}$.

This is a *complex fraction* **(Section 8.3)**—a fraction that has a fraction in the numerator, the denominator, or both.

(d) $-\dfrac{9}{14} \div \dfrac{3}{7} = -\dfrac{9}{14} \cdot \dfrac{7}{3}$ $\quad$ Multiply by the reciprocal.

$= -\dfrac{3 \cdot 3 \cdot 7}{2 \cdot 7 \cdot 3}$ $\quad$ Factor; multiply numerators and multiply denominators.

$= -\dfrac{3}{2}$ $\quad$ Lowest terms

◀ Work Problem 8 at the Side.

Every fraction has three signs: the sign of the numerator, the sign of the denominator, and the sign of the fraction itself. The rules for multiplication and division suggest the following results.

Equivalent Forms of a Fraction

The fractions $\frac{-x}{y}$, $\frac{x}{-y}$, and $-\frac{x}{y}$ are equivalent $(y \neq 0)$.

Example: $\frac{-4}{7} = \frac{4}{-7} = -\frac{4}{7}$

The fractions $\frac{x}{y}$ and $\frac{-x}{-y}$ are equivalent $(y \neq 0)$.

Example: $\frac{4}{7} = \frac{-4}{-7}$

◀ Work Problem 9 at the Side.

8 Find each quotient.

(a) $\dfrac{-16}{4}$

(b) $\dfrac{8}{-2}$

(c) $\dfrac{-15}{-3}$

(d) $\dfrac{\frac{3}{8}}{-\frac{11}{16}}$

(e) $-\dfrac{3}{4} \div \dfrac{7}{16}$

9 Which of the following fractions equal $\frac{-3}{5}$?

A. $\dfrac{3}{5}$ **B.** $\dfrac{3}{-5}$

C. $-\dfrac{3}{5}$ **D.** $\dfrac{-3}{-5}$

ANSWERS

8. **(a)** -4 **(b)** -4 **(c)** 5 **(d)** $-\frac{6}{11}$ **(e)** $-\frac{12}{7}$

9. B, C

1.2 Exercises

FOR EXTRA HELP

MyMathLab

MathXL PRACTICE | WATCH | DOWNLOAD | READ | REVIEW

Complete each statement and give an example.

1. The sum of a positive number and a negative number is 0 if ____________.

2. The sum of two positive numbers is a ____________ number.

3. The sum of two negative numbers is a ____________ number.

4. The sum of a positive number and a negative number is negative if ____________.

5. The sum of a positive number and a negative number is positive if ____________.

6. The difference between two positive numbers is negative if ____________.

7. The difference between two negative numbers is negative if ____________.

8. The product of two numbers with the same sign is ____________.

9. The product of two numbers with different signs is ____________.

10. The quotient formed by any nonzero number divided by 0 is ____________, and the quotient formed by 0 divided by any nonzero number is ____________.

Add or subtract as indicated. See Examples 1–3.

11. $13 + (-4)$

12. $19 + (-13)$

13. $-6 + (-13)$

14. $-8 + (-15)$

15. $-\frac{7}{3} + \frac{3}{4}$

16. $-\frac{5}{6} + \frac{3}{8}$

17. $-2.3 + 0.45$

18. $-0.238 + 4.55$

19. $-6 - 5$ **20.** $-8 - 13$ **21.** $8 - (-13)$ **22.** $13 - (-22)$

23. $-16 - (-3)$ **24.** $-21 - (-8)$ **25.** $-12.31 - (-2.13)$ **26.** $-15.88 - (-9.22)$

27. $\frac{9}{10} - \left(-\frac{4}{3}\right)$ **28.** $\frac{3}{14} - \left(-\frac{1}{4}\right)$ **29.** $|-8 - 6|$ **30.** $|-7 - 9|$

31. $-|-4 + 9|$ **32.** $-|-5 + 7|$ **33.** $-2 - |-4|$ **34.** $9 - |-13|$

Perform the indicated operations. See Example 4.

35. $-7 + 5 - 9$ **36.** $-12 + 13 - 19$ **37.** $6 - (-2) + 8$

38. $7 - (-3) + 12$ **39.** $-9 - 4 - (-3) + 6$ **40.** $-10 - 5 - (-12) + 8$

41. $-0.382 + 4 - (-0.6)$ **42.** $3 - 2.94 - (-0.63)$ **43.** $-\frac{3}{4} - \left(\frac{1}{2} - \frac{3}{8}\right)$

44. $\frac{7}{5} - \left(\frac{9}{10} - \frac{3}{2}\right)$ **45.** $-4 - [(-4 - 6) + 12] - 13$ **46.** $-10 - [(-2 + 3) - 4] - 17$

47. $|-11| - |-5| - |7| + |-2|$ **48.** $|-6| + |-3| - |4| - |-8|$

Multiply. See Example 5.

49. $5(-7)$ **50.** $6(-6)$ **51.** $-8(-5)$ **52.** $-10(-4)$

53. $-10\left(-\frac{1}{5}\right)$

54. $-\frac{1}{2}(-12)$

55. $\frac{3}{4}(-16)$

56. $\frac{4}{5}(-35)$

57. $-\frac{5}{2}\left(-\frac{12}{25}\right)$

58. $-\frac{9}{7}\left(-\frac{35}{36}\right)$

59. $-\frac{3}{8}\left(-\frac{24}{9}\right)$

60. $-\frac{2}{11}\left(-\frac{99}{4}\right)$

61. $-2.4(-2.45)$

62. $-3.45(-2.14)$

63. $3.4(-3.14)$

64. $5.66(-2.1)$

Give the reciprocal of each number. See Objective 4.

65. 5

66. 8

67. -7

68. -11

69. $-\frac{2}{3}$

70. $-\frac{7}{8}$

71. $\frac{1}{5}$

72. $\frac{1}{4}$

73. 0.02

74. 0.45

75. -0.001

76. -0.0003

Divide where possible. See Example 6.

77. $\frac{-14}{2}$

78. $\frac{-26}{13}$

79. $\frac{-24}{-4}$

80. $\frac{-36}{-9}$

81. $\frac{100}{-25}$

82. $\frac{300}{-60}$

83. $\frac{0}{-8}$

84. $\frac{0}{-10}$

85. $\frac{5}{0}$

86. $\frac{12}{0}$

87. $-\frac{10}{17} \div \left(-\frac{12}{5}\right)$

88. $-\frac{22}{23} \div \left(-\frac{33}{4}\right)$

89. $\frac{\frac{12}{13}}{-\frac{4}{3}}$

90. $\frac{\frac{5}{6}}{-\frac{1}{30}}$

91. $-\frac{27.72}{13.2}$

92. $\frac{-126.7}{36.2}$

93. $\frac{-100}{-0.01}$

94. $\frac{-50}{-0.05}$

Exercises 95–120 provide more practice on operations with fractions and decimals. Perform the indicated operations.

95. $\frac{1}{6} - \left(-\frac{7}{9}\right)$

96. $\frac{7}{10} - \left(-\frac{5}{6}\right)$

97. $-\frac{1}{9} + \frac{7}{12}$

98. $-\frac{1}{12} + \frac{11}{16}$

99. $-\frac{3}{8} - \frac{5}{12}$

100. $-\frac{11}{15} - \frac{2}{9}$

101. $-\frac{7}{30} + \frac{2}{45} - \frac{3}{10}$

102. $-\frac{8}{15} - \frac{3}{20} + \frac{5}{6}$

103. $\frac{8}{25}\left(-\frac{5}{12}\right)$

104. $\frac{9}{20}\left(-\frac{4}{15}\right)$

105. $\frac{5}{6}\left(-\frac{9}{10}\right)\left(-\frac{4}{5}\right)$

106. $\frac{2}{3}\left(-\frac{9}{20}\right)\left(-\frac{5}{12}\right)$

107. $\frac{7}{6} \div \left(-\frac{9}{10}\right)$

108. $\frac{8}{5} \div \left(-\frac{18}{25}\right)$

109. $\frac{-\frac{8}{9}}{2}$

110. $\frac{-\frac{15}{16}}{3}$

111. $-8.6 - 3.751$

112. $-27.8 - 13.582$

113. $(-4.2)(1.4)(2.7)$

114. $(1.9)(-10.3)(0.04)$

115. $-24.84 \div 6$

116. $-32.84 \div 4$

117. $-2496 \div (-0.52)$

118. $-161.7 \div (-0.75)$

119. $-14.23 + 9.81 + 74.63 - 18.715$

120. $-89.416 + 21.32 - 478.91 + 298.213$

Solve each problem.

121. The highest temperature ever recorded in Juneau, Alaska, was 90°F. The lowest temperature ever recorded there was −22°F. What is the difference between these two temperatures? (*Source: World Almanac and Book of Facts.*)

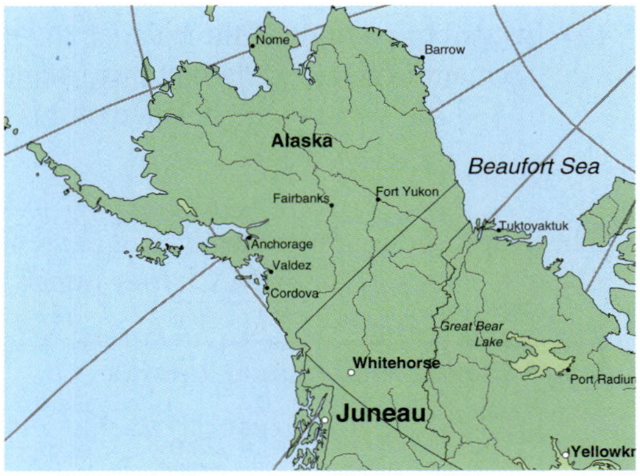

122. On August 10, 1936, a temperature of 120°F was recorded in Arkansas. On February 13, 1905, Arkansas recorded a temperature of −29°F. What is the difference between these two temperatures? (*Source: World Almanac and Book of Facts.*)

123. The Standard and Poor's 500, an index measuring the performance of 500 leading stocks, had an annual return of 15.79% in 2006. For 2000, its annual return was −9.10%. Find the difference between these two percents. (*Source*: Legg Mason Wood Walker, Inc.)

124. The low-carb diet craze was responsible for a first-quarter loss in Krispy Kreme doughnut sales in 2004. The company reported a loss of $24.4 million. One year earlier the company had reported a profit of $13.1 million. Express the difference between these two figures as a negative amount. (*Source*: Krispy Kreme Doughnuts.)

125. Andrew McGinnis has $48.35 in his checking account. He uses his debit card to make purchases of $35.99 and $20.00, which overdraws his account. His bank charges his account an overdraft fee of $28.50. He then deposits his paycheck for $66.27 from his part-time job at Arby's. What is the balance in his account?

126. Kayla Koolbeck has $37.50 in her checking account. She uses her debit card to make purchases of $25.99 and $19.34, which overdraws her account. Her bank charges her account an overdraft fee of $25.00. She then deposits her paycheck for $58.66 from her part-time job at Subway. What is the balance in her account?

127. Kyle Evangelista owes \$382.45 on his Visa account. He returns two items costing \$25.10 and \$34.50 for credit. Then he makes purchases of \$45.00 and \$98.17.

(a) How much should his payment be if he wants to pay off the balance on the account?

(b) Instead of paying off the balance, he makes a payment of \$300 and then incurs a finance charge of \$24.66. What is the balance on his account?

128. Adam Gross owes \$237.59 on his MasterCard account. He returns one item costing \$47.25 for credit and then makes two purchases of \$12.39 and \$20.00.

(a) How much should his payment be if he wants to pay off the balance on the account?

(b) Instead of paying off the balance, he makes a payment of \$75.00 and incurs a finance charge of \$32.06. What is the balance on his account?

129. The graph shows profits and losses in thousands of dollars for a private company.

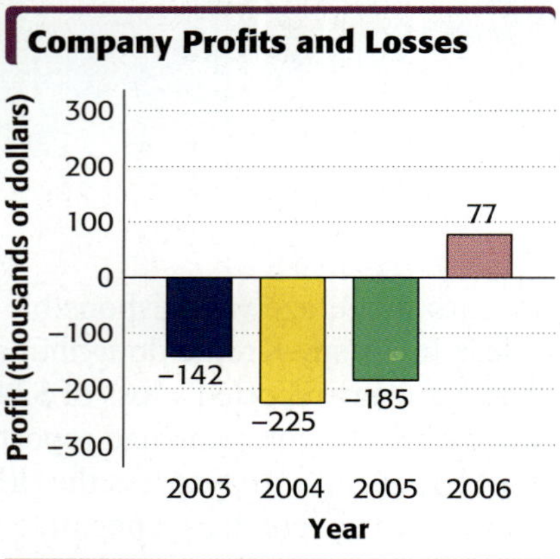

(a) What was the total profit or loss for the years 2003 through 2006?

(b) Find the difference between the profit or loss in 2006 and that in 2005.

(c) Find the difference between the profit or loss in 2004 and that in 2003.

130. The graph shows annual returns in percent for Class A shares of the AIM Charter Fund.

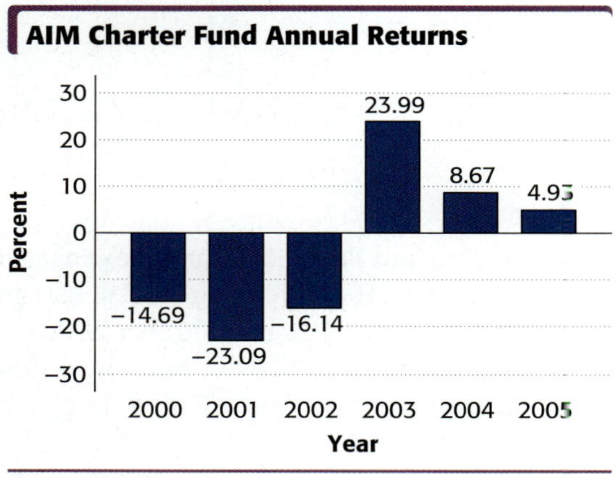

Source: AIM.

(a) Find the sum of the percents for the years shown in the graph.

(b) Find the difference between the returns in 2003 and 2002.

(c) Find the difference between the returns in 2001 and 2000.

The table shows Social Security finances (in billions of dollars). Use this table to work Exercises 131 and 132.

Year	Tax Revenue	Cost of Benefits
2000	538	409
2010*	916	710
2020*	1479	1405
2030*	2041	2542

*Projected

Source: Social Security Board of Trustees.

131. Find the difference between Social Security tax revenue and cost of benefits for each year shown in the table.

132. Interpret your answer for the year 2030.

1.3 Exponents, Roots, and Order of Operations

OBJECTIVES

1. Use exponents.
2. Identify exponents and bases.
3. Find square roots.
4. Use the order of operations.
5. Evaluate algebraic expressions for given values of variables.

Two or more numbers whose product is a third number are **factors** of that third number. For example, 2 and 6 are factors of 12 since $2 \cdot 6 = 12$. Other factors of 12 are $1, 3, 4, 12, -1, -2, -3, -4, -6$, and -12.

OBJECTIVE 1 Use exponents. In algebra, we use *exponents* as a way of writing products of repeated factors. For example, the product $2 \cdot 2 \cdot 2 \cdot 2 \cdot 2$ is written

$$\underbrace{2 \cdot 2 \cdot 2 \cdot 2 \cdot 2}_{5 \text{ factors of } 2} = 2^5.$$

The number 5 shows that 2 is used as a factor 5 times. The number 5 is the **exponent,** and 2 is the **base.**

$$2^5 \leftarrow \text{Exponent}; \quad 2 \leftarrow \text{Base}$$

Read 2^5 as "2 to the fifth power," or "2 to the fifth." Multiplying out the five 2s gives

$$2^5 = 2 \cdot 2 \cdot 2 \cdot 2 \cdot 2 = 32.$$

Exponential Expression

If a is a real number and n is a natural number,

$$a^n = \underbrace{a \cdot a \cdot a \cdot \cdots a}_{n \text{ factors of } a},$$

where n is the **exponent,** a is the **base,** and a^n is an **exponential expression.** Exponents are also called **powers.**

EXAMPLE 1 Using Exponential Notation

Write using exponents.

(a) $4 \cdot 4 \cdot 4$

Here, 4 is used as a factor 3 times, so

$$\underbrace{4 \cdot 4 \cdot 4}_{3 \text{ factors of } 4} = 4^3.$$

Read 4^3 as "4 **cubed.**"

(b) $\frac{3}{5} \cdot \frac{3}{5} = \left(\frac{3}{5}\right)^2$ 2 factors of $\frac{3}{5}$

Read $\left(\frac{3}{5}\right)^2$ as "$\frac{3}{5}$ **squared.**"

(c) $(-6)(-6)(-6)(-6) = (-6)^4$ 4 factors of -6

Read $(-6)^4$ as "-6 to the fourth power," or "-6 to the fourth."

(d) $(0.3)(0.3)(0.3)(0.3)(0.3) = (0.3)^5$

(e) $x \cdot x \cdot x \cdot x \cdot x \cdot x = x^6$

Work Problem 1 at the Side. ▶

1 Write each expression using exponents.

(a) $3 \cdot 3 \cdot 3 \cdot 3 \cdot 3$

(b) $\frac{2}{7} \cdot \frac{2}{7} \cdot \frac{2}{7} \cdot \frac{2}{7}$

(c) $(-10)(-10)(-10)$

(d) $(0.5)(0.5)$

(e) $y \cdot y \cdot y \cdot y \cdot y \cdot y \cdot y \cdot y$

ANSWERS

1. **(a)** 3^5 **(b)** $\left(\frac{2}{7}\right)^4$ **(c)** $(-10)^3$ **(d)** $(0.5)^2$ **(e)** y^8

In Example 1, we used the terms *squared* and *cubed* to refer to powers of 2 and 3, respectively. The term *squared* comes from the figure of a square, which has the same measure for both length and width, as shown in Figure 11(a). Similarly, the term *cubed* comes from the figure of a cube. As shown in Figure 11(b), the length, width, and height of a cube have the same measure.

(a) $3 \cdot 3 = 3$ squared, or 3^2 **(b)** $6 \cdot 6 \cdot 6 = 6$ cubed, or 6^3

Figure 11

EXAMPLE 2 Evaluating Exponential Expressions

Evaluate.

(a) [$5^2 = 5 \cdot 5$, NOT $5 \cdot 2$.] $5^2 = 5 \cdot 5 = 25$ 5 is used as a factor 2 times.

(b) $\left(\frac{2}{3}\right)^3 = \frac{2}{3} \cdot \frac{2}{3} \cdot \frac{2}{3} = \frac{8}{27}$ $\frac{2}{3}$ is used as a factor 3 times.

(c) $2^6 = 2 \cdot 2 \cdot 2 \cdot 2 \cdot 2 \cdot 2 = 64$ **(d)** $(1.2)^3 = (1.2)(1.2)(1.2) = 1.728$

(e) $(-2)^4 = (-2)(-2)(-2)(-2) = 16$

(f) $(-3)^5 = (-3)(-3)(-3)(-3)(-3) = -243$

Parts (e) and (f) of Example 2 suggest the following generalization.

> The product of an *even* number of negative factors is positive.
>
> The product of an *odd* number of negative factors is negative.

◀ *Work Problem* **2** *at the Side.*

> **Calculator Tip** Most calculators have a key labeled x^y or y^x that can be used to raise a number to a power. See your owner's manual for more information.

OBJECTIVE 2 Identify exponents and bases.

EXAMPLE 3 Identifying Exponents and Bases

Identify the exponent and the base. Then evaluate.

(a) 3^6
The exponent is 6, and the base is 3.
$3^6 = 3 \cdot 3 \cdot 3 \cdot 3 \cdot 3 \cdot 3 = 729$

(b) 5^4
The exponent is 4, and the base is 5.
$5^4 = 5 \cdot 5 \cdot 5 \cdot 5 = 625$

Continued on Next Page

2 Evaluate.

(a) 5^3

(b) 3^4

(c) $(-4)^5$

(d) $(-3)^4$

(e) $(0.75)^2$

(f) $\left(\frac{2}{5}\right)^4$

ANSWERS

2. (a) 125 **(b)** 81 **(c)** -1024 **(d)** 81 **(e)** 0.5625 **(f)** $\frac{16}{625}$

(c) $(-2)^6$

The exponent 6 applies to the number -2, so the base is -2.

$$(-2)^6 = (-2)(-2)(-2)(-2)(-2)(-2) = 64 \qquad \text{The base is } -2.$$

(d) -2^6

Since there are no parentheses, the exponent 6 applies *only* to the number 2, not to -2; the base is 2.

$$-2^6 = -(2 \cdot 2 \cdot 2 \cdot 2 \cdot 2 \cdot 2) = -64 \qquad \text{The base is 2.}$$

CAUTION

As shown in Examples 3(c) and (d), it is important to distinguish between $-a^n$ and $(-a)^n$.

$$-a^n = -1\underbrace{(a \cdot a \cdot a \cdots a)}_{n \text{ factors of } a} \qquad \text{The base is } a.$$

$$(-a)^n = \underbrace{(-a)(-a)\cdots(-a)}_{n \text{ factors of } -a} \qquad \text{The base is } -a.$$

Be careful evaluating an exponential expression with a negative sign.

Work Problem **3** at the Side. ▶

OBJECTIVE 3 Find square roots. As we saw in Example 2(a), $5^2 = 5 \cdot 5 = 25$, so 5 squared is 25. The opposite (inverse) of squaring a number is called taking its **square root.** For example, a square root of 25 is 5. Another square root of 25 is -5, since $(-5)^2 = 25$. Thus, 25 has two square roots, 5 and -5.

We write the **positive** or **principal square root** of a number with the symbol $\sqrt{}$, called a **radical sign.** The positive or principal square root of 25 is written $\sqrt{25} = 5$. The negative square root of 25 is written $-\sqrt{25} = -5$. ***Since the square of any nonzero real number is positive, the square root of a negative number, such as $\sqrt{-25}$, is not a real number.***

EXAMPLE 4 Finding Square Roots

Find each square root that is a real number.

(a) $\sqrt{36} = 6$, since 6 is positive and $6^2 = 36$.

(b) $\sqrt{0} = 0$, since $0^2 = 0$.

(c) $\sqrt{\dfrac{9}{16}} = \dfrac{3}{4}$, since $\left(\dfrac{3}{4}\right)^2 = \dfrac{9}{16}$.

(d) $\sqrt{0.16} = 0.4$, since $(0.4)^2 = 0.16$.

(e) $\sqrt{100} = 10$, since $10^2 = 100$.

(f) $-\sqrt{100} = -10$, since the negative sign is outside the radical sign.

(g) $\sqrt{-100}$ is not a real number, since the negative sign is inside the radical sign. No *real number* squared equals -100.

Notice the difference among the square roots in parts (e), (f), and (g). Part (e) is the positive or principal square root of 100, part (f) is the negative square root of 100, and part (g) is the square root of -100, which is not a real number.

Work Problem **4** at the Side. ▶

3 Identify the exponent and the base. Then evaluate.

(a) 7^3

(b) $(-5)^4$

(c) -5^4

(d) $-(0.9)^5$

4 Find each square root that is a real number.

(a) $\sqrt{9}$

(b) $\sqrt{49}$

(c) $-\sqrt{81}$

(d) $\sqrt{\dfrac{121}{81}}$

(e) $\sqrt{0.25}$

(f) $\sqrt{-9}$

(g) $-\sqrt{-169}$

ANSWERS

3. **(a)** 3; 7; 343 **(b)** 4; -5; 625 **(c)** 4; 5; -625 **(d)** 5; 0.9; -0.59049

4. **(a)** 3 **(b)** 7 **(c)** -9 **(d)** $\dfrac{11}{9}$ **(e)** 0.5 **(f)** not a real number **(g)** not a real number

5 Simplify.

(a) $5 \cdot 9 + 2 \cdot 4$

(b) $4 - 12 \div 4 \cdot 2$

CAUTION

The symbol $\sqrt{}$ is used only for the *positive* square root, except that $\sqrt{0} = 0$. The symbol $-\sqrt{}$ is used for the negative square root.

Calculator Tip Most calculators have a square root key, usually labeled $\sqrt{x}$, that allows us to find the square root of a number. On some models, the square root key must be used in conjunction with the key marked INV or 2nd.

OBJECTIVE 4 Use the order of operations. To simplify an expression such as $5 + 2 \cdot 3$, what should we do first—add 5 and 2, or multiply 2 and 3? When an expression involves more than one operation symbol, we use the following **order of operations.**

Order of Operations

1. Work separately above and below any **fraction bar.**
2. If **grouping symbols** such as **parentheses ( ), square brackets [ ],** or **absolute value bars | |** are present, start with the innermost set and work outward.
3. Evaluate all **powers, roots,** and **absolute values.**
4. **Multiply** or **divide** in order from left to right.
5. **Add** or **subtract** in order from left to right.

EXAMPLE 5 Using the Order of Operations

Simplify.

(a) $5 + 2 \cdot 3$

$= 5 + 6$ Multiply.

$= 11$ Add.

(b) $24 \div 3 \cdot 2 + 6$

Multiplications and divisions are done in the order in which they appear from left to right, so divide first.

$24 \div 3 \cdot 2 + 6$

$= 8 \cdot 2 + 6$ Divide.

$= 16 + 6$ Multiply.

$= 22$ Add.

Work Problem **5** *at the Side.*

Calculator Tip Most calculators follow the order of operations given here. You may want to try some of Examples 5–7 to see whether your calculator gives the same answers. Use the parentheses keys to insert parentheses where they are needed. (To work Example 7 with a calculator, put parentheses around the numerator and the denominator.)

ANSWERS

5. (a) 53 **(b)** -2

EXAMPLE 6 **Using the Order of Operations**

Simplify.

(a)

$$\begin{aligned} & 4 \cdot 3^2 + 7 - (2 + 8) && \text{Work inside the parentheses first.} \\ &= 4 \cdot 3^2 + 7 - 10 && \text{Add inside parentheses.} \\ &= 4 \cdot 9 + 7 - 10 && \text{Evaluate the power.} \\ &= 36 + 7 - 10 && \text{Multiply.} \\ &= 43 - 10 && \text{Add.} \\ &= 33 && \text{Subtract.} \end{aligned}$$

$3^2 = 3 \cdot 3$, NOT $3 \cdot 2$.

(b)

$$\begin{aligned} & \frac{1}{2} \cdot 4 + (6 \div 3 - 7) && \text{Work inside the parentheses first.} \\ &= \frac{1}{2} \cdot 4 + (2 - 7) && \text{Divide inside parentheses.} \\ &= \frac{1}{2} \cdot 4 + (-5) && \text{Subtract inside parentheses.} \\ &= 2 + (-5) && \text{Multiply.} \\ &= -3 && \text{Add.} \end{aligned}$$

Work Problem **6** at the Side. ▶

EXAMPLE 7 **Using the Order of Operations**

Simplify.

$$\begin{aligned} & \frac{5 + 2^4}{6\sqrt{9} - 9 \cdot 2} && \text{Work separately above and below the fraction bar.} \\ &= \frac{5 + 16}{6 \cdot 3 - 9 \cdot 2} && \text{Evaluate powers and roots.} \\ &= \frac{5 + 16}{18 - 18} && \text{Multiply.} \\ &= \frac{21}{0} && \text{Add and subtract.} \end{aligned}$$

Because division by 0 is undefined, the given expression is undefined.

Work Problem **7** at the Side. ▶

OBJECTIVE 5 **Evaluate algebraic expressions for given values of variables.** A collection of numbers, variables, operation symbols, and grouping symbols, such as

$$6ab, \quad 5m - 9n, \quad \text{and} \quad -2(x^2 + 4y), \qquad \text{Algebraic expressions}$$

is called an **algebraic expression.** Algebraic expressions have different numerical values for different values of the variables. We can evaluate such expressions by *substituting* given values for the variables.

For example, if movie tickets cost \$9 each, the amount in dollars you pay for x tickets can be represented by the algebraic expression $9x$. We can substitute different numbers of tickets to get the costs to purchase those tickets.

6 Simplify.

(a) $(4 + 2) - 3^2 - (8 - 3)$

(b) $6 + \frac{2}{3}(-9) - \frac{5}{8} \cdot 16$

7 Simplify.

(a) $\dfrac{10 - 6 + 2\sqrt{9}}{11 \cdot 2 - 3(2)^2}$

(b) $\dfrac{-4(8) + 6(3)}{3\sqrt{49} - \frac{1}{2}(42)}$

ANSWERS

6. (a) -8 **(b)** -10
7. (a) 1 **(b)** undefined

EXAMPLE 8 **Evaluating Expressions**

Evaluate each expression if $m = -4, n = 5, p = -6$, and $q = 25$.

(a) $5m - 9n$

$= 5(-4) - 9(5)$ Substitute; $m = -4$ and $n = 5$.

$= -20 - 45$ Multiply.

$= -65$ Subtract.

Use parentheses around substituted values to avoid errors.

(b) $\frac{m + 2n}{4p}$

$= \frac{-4 + 2(5)}{4(-6)}$ Substitute; $m = -4, n = 5$, and $p = -6$.

$= \frac{-4 + 10}{-24}$ Work separately above and below the fraction bar.

$= \frac{6}{-24} = -\frac{1}{4}$ Write in lowest terms; also, $\frac{a}{-b} = -\frac{a}{b}$.

(c) $-3m^3 - n^2(\sqrt{q})$

$= -3(-4)^3 - (5)^2(\sqrt{25})$ Substitute; $m = -4, n = 5$, and $q = 25$.

$= -3(-64) - 25(5)$ Evaluate the powers and the root.

$= 192 - 125$ Multiply.

$= 67$ Subtract.

◀ Work Problem **8** at the Side.

8 Evaluate each expression if $w = 4, x = -12, y = 64$, and $z = -3$.

(a) $5x - 2w$

(b) $-6(x - \sqrt{y})$

(c) $\frac{5x - 3 \cdot \sqrt{y}}{x - 1}$

(d) $w^2 + 2z^3$

EXAMPLE 9 **Evaluating an Expression in an Application**

The amount in billions of dollars that Americans have spent on their pets each year from 1996 to 2008 can be approximated by substituting the year for x in the following expression and then evaluating.

$$1.909x - 3791$$

(*Source:* American Pet Products Manufacturers Association.)

(a) Approximate the amount Americans spent on their pets in 1996.

$1.909x - 3791$

$= 1.909(1996) - 3791$ Let $x = 1996$.

≈ 19.4 Use a calculator; round to the nearest tenth.

≈ means "is approximately equal to."

In 1996, Americans spent about \$19.4 billion on their pets.

◀ Work Problem **9** at the Side.

(b) Give the results found above and in Problem 9 in a table. How has the amount Americans have spent on their pets changed from 1996 to 2008?

Year	Amount Spent on Pets (in billions of dollars)
1996	19.4
2002	30.8
2008	42.3

The amount spent on pets has more than doubled from 1996 to 2008.

9 Use the expression in Example 9 to approximate the amount Americans spent on their pets in 2002 and 2008. Round answers to the nearest tenth.

ANSWERS

8. (a) -68 (b) 120 (c) $\frac{84}{13}$ (d) -38

9. 2002: \$30.8 billion; 2008: \$42.3 billion

1.3 Exercises

FOR EXTRA HELP

MyMathLab | Math XL PRACTICE | WATCH | DOWNLOAD | READ | REVIEW

Decide whether each statement is true *or* false. *If* false, *correct the statement so it is* true.

1. $-4^6 = (-4)^6$

2. $-4^7 = (-4)^7$

3. $\sqrt{16}$ is a positive number.

4. $3 + 5 \cdot 6 = 3 + (5 \cdot 6)$

5. $(-2)^7$ is a negative number.

6. $(-2)^8$ is a positive number.

7. The product of 8 positive factors and 8 negative factors is positive.

8. The product of 3 positive factors and 3 negative factors is positive.

9. In the exponential expression -3^5, -3 is the base.

10. $\sqrt{a}$ is positive for all positive numbers a.

11. Evaluate each exponential expression.
(a) 8^2 **(b)** -8^2 **(c)** $(-8)^2$ **(d)** $-(-8)^2$

12. Evaluate each exponential expression.
(a) 4^3 **(b)** -4^3 **(c)** $(-4)^3$ **(d)** $-(-4)^3$

Write each expression using exponents. See Example 1.

13. $8 \cdot 8 \cdot 8$

14. $10 \cdot 10 \cdot 10 \cdot 10$

15. $\frac{1}{2} \cdot \frac{1}{2}$

16. $\frac{3}{4} \cdot \frac{3}{4} \cdot \frac{3}{4} \cdot \frac{3}{4} \cdot \frac{3}{4}$

17. $(-4)(-4)(-4)(-4)$

18. $(-9)(-9)(-9)$

19. $z \cdot z \cdot z \cdot z \cdot z \cdot z \cdot z$

20. $a \cdot a \cdot a \cdot a \cdot a \cdot a$

Evaluate each expression. See Examples 2 and 3.

21. 4^2

22. 2^4

23. 0.28^3

24. 0.91^3

25. $\left(\frac{1}{5}\right)^3$

26. $\left(\frac{1}{6}\right)^4$

27. $\left(\frac{4}{5}\right)^4$

28. $\left(\frac{7}{10}\right)^3$

29. $(-5)^3$

30. $(-3)^5$

31. $(-2)^8$

32. $(-3)^6$

33. -3^6 34. -4^6 35. -8^4 36. -10^3

Identify the exponent and the base in each expression. Do not evaluate. See Example 3.

37. $(-4.1)^7$ 38. $(-3.4)^9$ 39. -4.1^7 40. -3.4^9

Find each square root. If it is not a real number, say so. See Example 4.

41. $\sqrt{81}$ 42. $\sqrt{64}$ 43. $\sqrt{169}$ 44. $\sqrt{225}$

45. $-\sqrt{400}$ 46. $-\sqrt{900}$ 47. $\sqrt{\dfrac{100}{121}}$ 48. $\sqrt{\dfrac{225}{169}}$

49. $-\sqrt{0.49}$ 50. $-\sqrt{0.64}$ 51. $\sqrt{-36}$ 52. $\sqrt{-121}$

53. Match each square root with the appropriate value or description.

(a) $\sqrt{144}$ A. -12

(b) $\sqrt{-144}$ B. 12

(c) $-\sqrt{144}$ C. Not a real number

54. Explain why $\sqrt{-900}$ is not a real number.

55. If a is a positive number, is $-\sqrt{-a}$ positive, negative, or not a real number?

56. If a is a positive number, is $-\sqrt{a}$ positive, negative, or not a real number?

Simplify each expression. See Examples 5–7.

57. $12 + 3 \cdot 4$ 58. $15 + 5 \cdot 2$ 59. $2[-5 - (-7)]$ 60. $3[-8 - (-2)]$

61. $-12\left(-\frac{3}{4}\right) - (-5)$ 62. $-7\left(-\frac{2}{14}\right) - (-8)$ 63. $6 \cdot 3 - 12 \div 4$ 64. $9 \cdot 4 - 8 \div 2$

65. $10 + 30 \div 2 \cdot 3$ 66. $12 + 24 \div 3 \cdot 2$ 67. $-3(5)^2 - (-2)(-8)$ 68. $-9(2)^2 - (-3)(-2)$

69. $5 - 7 \cdot 3 - (-2)^3$ 70. $-4 - 3 \cdot 5 + 6^2$ 71. $-7(\sqrt{36}) - (-2)(-3)$

72. $-3\left(\sqrt{64}\right) - (-3)(-7)$

73. $-14\left(-\frac{2}{7}\right) \div (2 \cdot 6 - 10)$

74. $-12\left(-\frac{3}{4}\right) - (6 \cdot 5 \div 3)$

75. $6|4 - 5| - 24 \div 3$
(*Hint:* Start inside the absolute value bars.)

76. $-4|2 - 4| + 8 \cdot 2$

77. $|-6 - 5|(-8) + 3^2$

78. $(-6 - 3)|-2 - 3| \div 9$

79. $\dfrac{\left(-5 + \sqrt{4}\right)\left(-2^2\right)}{-5 - 1}$

80. $\dfrac{\left(-9 + \sqrt{16}\right)\left(-3^2\right)}{-4 - 1}$

81. $\dfrac{2(-5) + (-3)(-2)}{-8 + 3^2 - 1}$

82. $\dfrac{3(-4) + (-5)(-8)}{2^3 - 2 - 6}$

Evaluate each expression if $a = -3$, $b = 64$, and $c = 6$. See Example 8.

83. $3a + \sqrt{b}$

84. $-2a - \sqrt{b}$

85. $\sqrt{b} + c - a$

86. $\sqrt{b} - c + a$

87. $4a^3 + 2c$

88. $-3a^4 - 3c$

89. $\dfrac{2c + a^3}{4b + 6a}$

90. $\dfrac{3c + a^2}{2b - 6c}$

Evaluate each expression if $w = 4$, $x = -\frac{3}{4}$, $y = \frac{1}{2}$, and $z = 1.25$. See Example 8.

91. $wy - 8x$

92. $wz - 12y$

93. $xy + y^4$

94. $xy - x^2$

Solve each problem. See Example 9.

Residents of Linn County, Iowa, in the Cedar Rapids Community School District can use the expression

$$(v \times 0.5485 - 4850) \div 1000 \times 31.44$$

to determine their property taxes, where v is assessed home value. (*Source: The Gazette,* August 19, 2000.) Use the expression to calculate the amount of property taxes to the nearest dollar that the owner of a home with each of the following values would pay. Follow the order of operations.

95. \$150,000

96. \$200,000

97. \$250,000

The Blood Alcohol Concentration (BAC) of a person who has been drinking is given by the expression

$$\text{number of oz} \times \%\text{ alcohol} \times 0.075 \div \text{body weight in lb} - \text{hr of drinking} \times 0.015.$$

(*Source:* Lawlor, J., *Auto Math Handbook: Mathematical Calculations, Theory, and Formulas for Automotive Enthusiasts,* HP Books, 1991.)

98. Suppose a policeman stops a 190-lb man who, in 2 hr, has ingested four 12-oz beers (48 oz), each having a 3.2% alcohol content.

(a) Substitute the values in the formula, and write the expression for the man's BAC.

(b) Calculate the man's BAC to the nearest thousandth. Follow the order of operations.

99. Find the BAC to the nearest thousandth for a 135-lb woman who, in 3 hr, has drunk three 12-oz beers (36 oz), each having a 4.0% alcohol content.

100. Calculate the BACs to the nearest thousandth in Exercises 98 and 99 if each person weighs 25 lb more and the rest of the variables stay the same. How does increased weight affect a person's BAC?

101. Predict how decreased weight would affect the BAC of each person in Exercises 98 and 99. Calculate the BACs to the nearest thousandth if each person weighs 25 lb less and the rest of the variables stay the same.

102. An approximation of federal spending on education in billions of dollars from 2001 through 2005 can be obtained using the expression

$$9.0499x - 18{,}071.87,$$

where x represents the year. (*Source:* U.S. Department of the Treasury.)

(a) Use this expression to complete the table. Round answers to the nearest tenth.

Year	Education Spending (in billions of dollars)
2001	37.0
2002	46.0
2003	
2004	
2005	

(b) How has the amount of federal spending on education changed from 2001 to 2005?

103. An approximation of the average price of a theater ticket in the United States from 1977 through 2007 can be obtained by using the expression.

$$0.1399x - 274.4,$$

where x represents the year. (*Source:* National Association of Theater Owners.)

(a) Use the expression to complete the table. Round answers to the nearest cent.

Year	Average Price (in dollars)
1977	
1987	3.58
1997	
2007	

(b) How has the average price of a theater ticket in the United States changed from 1977 to 2007?

1.4 Properties of Real Numbers

OBJECTIVES

1. Use the distributive property.
2. Use the inverse properties.
3. Use the identity properties.
4. Use the commutative and associative properties.
5. Use the multiplication property of 0.

The study of any object is simplified when we know the properties of the object. For example, a property of water is that it freezes when cooled to 0°C. Knowing this helps us to predict the behavior of water.

The study of numbers is no different. The basic properties of real numbers reflect results that occur consistently in work with numbers, so they have been generalized to apply to expressions with variables as well.

OBJECTIVE 1 Use the distributive property. Notice that

$$2(3+5) = 2 \cdot 8 = 16$$

and

$$2 \cdot 3 + 2 \cdot 5 = 6 + 10 = 16,$$

so

$$2(3+5) = 2 \cdot 3 + 2 \cdot 5.$$

This idea is illustrated by the divided rectangle in Figure 12. Similarly,

$$-4[5+(-3)] = -4(2) = -8$$

and

$$-4(5) + (-4)(-3) = -20 + 12 = -8,$$

so

$$-4[5+(-3)] = -4(5) + (-4)(-3).$$

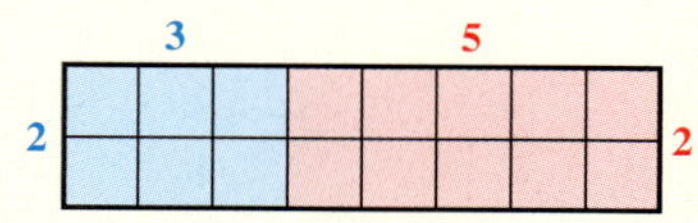

Area of left part is $2 \cdot 3 = 6$.
Area of right part is $2 \cdot 5 = 10$.
Area of total rectangle is $2(3+5) = 16$.

Figure 12

These arithmetic examples are generalized to *all* real numbers as the **distributive property of multiplication with respect to addition,** or simply the **distributive property.**

Distributive Property

For any real numbers a, b, and c,

$$a(b+c) = ab + ac \quad \text{and} \quad (b+c)a = ba + ca.$$

The distributive property can also be written

$$ab + ac = a(b+c) \quad \text{and} \quad ba + ca = (b+c)a.$$

It can be extended to more than two numbers as well.

$$a(b+c+d) = ab + ac + ad$$

This property is important because it provides a way to rewrite a product $a(b+c)$ as a sum $ab+ac$, or a sum as a product.

Note

When we rewrite $a(b+c)$ as $ab+ac$, we sometimes refer to the process as "removing" or "clearing" parentheses.

1 Use the distributive property, where possible, to rewrite each expression.

(a) $8(m+n)$

(b) $-4(p-5)$

(c) $3k+6k$

(d) $-6m+2m$

(e) $2r+3s$

(f) $5(4p-2q+r)$

2 Use the distributive property to evaluate each expression.

(a) $14 \cdot 5 + 14 \cdot 85$

(b) $78 \cdot 33 + 22 \cdot 33$

ANSWERS

1. (a) $8m+8n$ (b) $-4p+20$ (c) $9k$ (d) $-4m$ (e) cannot be rewritten (f) $20p-10q+5r$
2. (a) 1260 (b) 3300

EXAMPLE 1 Using the Distributive Property

Use the distributive property, where possible, to rewrite each expression.

(a) $3(x+y)$ — Use the first form of the property to rewrite the given product as a sum.

$= 3x + 3y$

(b) $-2(5+k)$

$= -2(5) + (-2)(k)$

$= -10 - 2k$

(c) $4x + 8x$ — Use the second form of the property to rewrite the given sum as a product.

$= (4+8)x$

$= 12x$

(d) $3r - 7r$

$= 3r + (-7r)$ — Definition of subtraction

$= [3 + (-7)]r$ — Distributive property

$= -4r$

(e) $5p + 7q$

Because there is no common number or variable here, we cannot use the distributive property to rewrite the expression.

(f) $6(x + 2y - 3z)$

$= 6x + 6(2y) + 6(-3z)$

$= 6x + 12y - 18z$

As illustrated in Example 1(d), the distributive property can also be used for subtraction, so

$$a(b-c) = ab - ac.$$

◀ Work Problem **1** at the Side.

The distributive property can be used to mentally perform calculations.

EXAMPLE 2 Using the Distributive Property for Calculation

Evaluate $38 \cdot 17 + 38 \cdot 3$.

$38 \cdot 17 + 38 \cdot 3$

$= 38(17+3)$ — Distributive property

$= 38(20)$ — Add inside parentheses.

$= 760$ — Multiply.

◀ Work Problem **2** at the Side.

OBJECTIVE 2 Use the inverse properties. In Section 1.1, we saw that the *additive inverse* (or *opposite*) of a number a is $-a$ and that additive inverses have a sum of 0.

$$5 \text{ and } -5, \quad -\frac{1}{2} \text{ and } \frac{1}{2}, \quad -34 \text{ and } 34$$ Additive inverses (sum of 0)

In **Section 1.2,** we saw that the *multiplicative inverse* (or *reciprocal*) of a number a is $\frac{1}{a}$ $(a \neq 0)$ and that multiplicative inverses have a product of 1.

$$5 \text{ and } \frac{1}{5}, \quad -\frac{1}{2} \text{ and } -2, \quad \frac{3}{4} \text{ and } \frac{4}{3} \qquad \text{Multiplicative inverses (product of 1)}$$

This discussion leads to the **inverse properties.**

Inverse Properties

For any real number a,

$$\mathbf{a + (-a) = 0} \quad \text{and} \quad \mathbf{-a + a = 0}$$

$$\mathbf{a \cdot \frac{1}{a} = 1} \quad \text{and} \quad \mathbf{\frac{1}{a} \cdot a = 1 \quad (a \neq 0).}$$

The inverse properties "undo" addition or multiplication. Think of putting on your shoes when you get up in the morning and then taking them off before you go to bed at night. These are inverse operations that undo each other.

Work Problem 3 at the Side. ▶

OBJECTIVE 3 **Use the identity properties.** The numbers 0 and 1 each have a special property. Zero is the only number that can be added to any number to get that number. Adding 0 to any number leaves the identity of the number unchanged. For this reason, 0 is called the **identity element for addition,** or the **additive identity.**

In a similar way, multiplying any number by 1 leaves the identity of the number unchanged, so 1 is the **identity element for multiplication,** or the **multiplicative identity.** The **identity properties** summarize this discussion.

Identity Properties

For any real number a,

$$\mathbf{a + 0 = 0 + a = a}$$

$$\mathbf{a \cdot 1 = 1 \cdot a = a.}$$

The identity properties leave the identity of a real number unchanged. Think of a child wearing a costume on Halloween. The child's appearance is changed, but his or her identity is unchanged.

EXAMPLE 3 **Using the Identity Property $1 \cdot a = a$**

Simplify each expression.

(a) $12m + m$

$= 12m + 1m$ — Identity property; $m = 1 \cdot m$, or $1m$

$= (12 + 1)m$ — Distributive property

$= 13m$ — Add inside parentheses.

(b) $y + y$

$= 1y + 1y$ — Identity property

$= (1 + 1)y$ — Distributive property

$= 2y$ — Add inside parentheses.

Continued on Next Page

3 Complete each statement.

(a) $4 +$ _____ $= 0$

(b) $-7.1 +$ _____ $= 0$

(c) $-9 + 9 =$ _____

(d) $5 \cdot$ _____ $= 1$

(e) $-\frac{3}{4} \cdot$ _____ $= 1$

(f) $7 \cdot \frac{1}{7} =$ _____

ANSWERS

3. (a) -4 **(b)** 7.1 **(c)** 0 **(d)** $\frac{1}{5}$ **(e)** $-\frac{4}{3}$ **(f)** 1

4 Simplify each expression.

(a) $p - 3p$

(b) $r + r + r$

(c) $-(3 + 4p)$

(d) $-(k - 2)$

(c)

$$-(m - 5n)$$

$$= -1(m - 5n) \quad \text{Identity property}$$

$$= -1(m) + (-1)(-5n) \quad \text{Distributive property}$$

$$= -m + 5n \quad \text{Multiply.}$$

Multiply *each* term by -1. Be careful with signs.

◀ Work Problem **4** at the Side.

Expressions such as $12m$ and $5n$ from Example 3 are examples of *terms*. A **term** is a number or the product of a number and one or more variables raised to powers. The numerical factor in a term is called the **numerical coefficient,** or just the **coefficient.** Some examples of terms and their coefficients are shown in the table.

Term	Numerical Coefficient
$-7y$	-7
$34r^3$	34
$-26x^5yz^4$	-26
$-k = -1k$	-1
$r = 1r$	1
$\frac{3x}{8} = \frac{3}{8}x$	$\frac{3}{8}$
$\frac{x}{3} = \frac{1x}{3} = \frac{1}{3}x$	$\frac{1}{3}$

Terms with exactly the same variables raised to exactly the same powers are called **like terms.** Some examples of like terms are

$$5p \text{ and } -21p \qquad -6x^2 \text{ and } 9x^2. \qquad \text{Like terms}$$

Some examples of **unlike terms** are

$$3m \text{ and } 16x \qquad 7y^3 \text{ and } -3y^2. \qquad \text{Unlike terms}$$

OBJECTIVE 4 Use the commutative and associative properties. Simplifying expressions as in Example 3(a) and (b) is called **combining like terms.** ***Only like terms may be combined.*** To combine like terms in an expression such as

$$-2m + 5m + 3 - 6m + 8,$$

we need two more properties. From arithmetic, we know that

$$3 + 9 = 12 \quad \text{and} \quad 9 + 3 = 12$$

$$3 \cdot 9 = 27 \quad \text{and} \quad 9 \cdot 3 = 27.$$

The order of the numbers being added or multiplied does not matter. The same answers result. Also,

$$(5 + 7) + 2 = 12 + 2 = 14$$

$$5 + (7 + 2) = 5 + 9 = 14,$$

and

$$(5 \cdot 7) \cdot 2 = 35 \cdot 2 = 70$$

$$5(7 \cdot 2) = 5 \cdot 14 = 70.$$

The way in which the numbers being added or multiplied are grouped does not matter. The same answers result.

ANSWERS

4. **(a)** $-2p$ **(b)** $3r$ **(c)** $-3 - 4p$ **(d)** $-k + 2$

These arithmetic examples can be extended to algebra.

Commutative and Associative Properties

For any real numbers a, b, and c,

$$\left.\begin{aligned} a + b &= b + a \\ ab &= ba. \end{aligned}\right\} \text{ Commutative properties}$$

Interchange the order of the two terms or factors.

Also,

$$\left.\begin{aligned} a + (b + c) &= (a + b) + c \\ a(bc) &= (ab)c. \end{aligned}\right\} \text{ Associative properties}$$

Shift parentheses among the three terms or factors; order stays the same.

The commutative properties are used to change the *order* of the terms or factors in an expression. Think of *commuting* from home to work and then from work to home. The associative properties are used to *regroup* the terms or factors of an expression. Remember, to *associate* is to be part of a group.

EXAMPLE 4 Using the Commutative and Associative Properties

Simplify $-2m + 5m + 3 - 6m + 8$.

$$\begin{aligned} &-2m + 5m + 3 - 6m + 8 \\ &= (-2m + 5m) + 3 - 6m + 8 && \text{Associative property} \\ &= (-2 + 5)m + 3 - 6m + 8 && \text{Distributive property} \\ &= 3m + 3 - 6m + 8 && \text{Add inside parentheses.} \end{aligned}$$

By the order of operations, the next step would be to add $3m$ and 3, but they are unlike terms. To get $3m$ and $-6m$ together, use the associative and commutative properties.

$$\begin{aligned} &= [3m + (3 - 6m)] + 8 && \text{Associative property} \\ &= [3m + (-6m + 3)] + 8 && \text{Commutative property} \\ &= [(3m + [-6m]) + 3] + 8 && \text{Associative property} \\ &= (-3m + 3) + 8 && \text{Combine like terms.} \\ &= -3m + (3 + 8) && \text{Associative property} \\ &= -3m + 11 && \text{Add.} \end{aligned}$$

In practice, many of the steps are not written down, but you should realize that the commutative and associative properties are used whenever the terms in an expression are rearranged and regrouped to combine like terms.

Work Problem 5 at the Side. ▶

EXAMPLE 5 Using the Properties of Real Numbers

Simplify each expression.

(a) $5y - 8y - 6y + 11y$

$$\begin{aligned} &= (5 - 8 - 6 + 11)y && \text{Distributive property} \\ &= 2y && \text{Combine like terms.} \end{aligned}$$

Continued on Next Page

5 Simplify each expression.

(a) $-3w + 7 - 8w - 2$

(b) $12b - 9 + 4b - 7b + 1$

ANSWERS

5. (a) $-11w + 5$ (b) $9b - 8$

6 Simplify each expression.

(a) $4x - 7x - 10x + 5x$

(b) $9 - 2(a - 3) + 4 - a$

(c) $10 - 3(6 + 2t)$

(d) $7x - (4x - 2)$

(e) $(4m)(2n)$

7 Complete each statement.

(a) $197 \cdot 0 =$ _____

(b) $0\left(-\frac{8}{9}\right) =$ _____

(c) $0 \cdot$ _____ $= 0$

Answers

6. **(a)** $-8x$ **(b)** $19 - 3a$ **(c)** $-8 - 6t$ **(d)** $3x + 2$ **(e)** $8mn$
7. **(a)** 0 **(b)** 0 **(c)** any real number

(b) $3x + 4 - 5(x + 1) - 8$

$= 3x + 4 - 5x - 5 - 8$ Distributive property

$= 3x - 5x + 4 - 5 - 8$ Commutative property

$= -2x - 9$ Combine like terms.

(c) $8 - (3m + 2)$

$= 8 - 1(3m + 2)$ Identity property

$= 8 - 3m - 2$ Distributive property

$= 6 - 3m$ Combine like terms.

(d) $(3x)(5)(y)$

$= [(3x)(5)]y$ Associative property

$= [3(x \cdot 5)]y$ Associative property

$= [3(5x)]y$ Commutative property

$= [(3 \cdot 5)x]y$ Associative property

$= (15x)y$ Multiply.

$= 15(xy)$ Associative property

$= 15xy$

As previously mentioned, many of these steps are not usually written out.

◀ Work Problem **6** at the Side.

CAUTION

Be careful. Notice that the distributive property does not apply in Example 5(d), because there is no addition or subtraction involved.

$$(3x)(5)(y) \neq (3x)(5) \cdot (3x)(y)$$

OBJECTIVE 5 **Use the multiplication property of 0.** The additive identity property gives a special property of 0, namely that $a + 0 = a$ for any real number a. The **multiplication property of 0** gives a special property of 0 that involves multiplication: The product of any real number and 0 is 0.

Multiplication Property of 0

For any real number a,

$$a \cdot 0 = 0 \quad \text{and} \quad 0 \cdot a = 0.$$

◀ Work Problem **7** at the Side.

1.4 Exercises

FOR EXTRA HELP MyMathLab Math XL PRACTICE WATCH DOWNLOAD READ REVIEW

Choose the correct response in Exercises 1–4.

1. The identity element for addition is

A. $-a$ **B.** 0 **C.** 1 **D.** $\frac{1}{a}$.

2. The identity element for multiplication is

A. $-a$ **B.** 0 **C.** 1 **D.** $\frac{1}{a}$.

3. The additive inverse of a is

A. $-a$ **B.** 0 **C.** 1 **D.** $\frac{1}{a}$.

4. The multiplicative inverse of a, where $a \neq 0$, is

A. $-a$ **B.** 0 **C.** 1 **D.** $\frac{1}{a}$.

Complete each statement.

5. The multiplication property of 0 says that the ________ of 0 and any real number is ________.

6. The commutative property is used to change the ________ of two terms or factors.

7. The associative property is used to change the ________ of three terms or factors.

8. Like terms are terms with the ________ variables raised to the ________ powers.

9. When simplifying an expression, only ________ terms can be combined.

10. The coefficient in the term $-8yz^2$ is ________.

Use the properties of real numbers to simplify each expression. See Examples 1 and 3.

11. $2(m + p)$ **12.** $3(a + b)$ **13.** $-5(2d - f)$ **14.** $-2(3m - n)$ **15.** $5k + 3k$

16. $6a + 5a$ **17.** $7r - 9r$ **18.** $4n - 6n$ **19.** $a - 7a$ **20.** $s + 9s$

21. $-8z + 4w$ **22.** $-12k + 3r$ **23.** $-(4b - c)$ **24.** $-(2g - h)$

Use the distributive property to calculate each value mentally. See Example 2.

25. $96 \cdot 19 + 4 \cdot 19$ **26.** $27 \cdot 60 + 27 \cdot 40$ **27.** $58 \cdot \frac{3}{2} - 8 \cdot \frac{3}{2}$

28. $\frac{8}{5} \cdot 17 + \frac{8}{5} \cdot 13$ **29.** $4.31(69) + 4.31(31)$ **30.** $\frac{4}{5}(17) + \frac{4}{5}(23)$

Simplify each expression. See Examples 1 and 3–5.

31. $-12y + 4y + 3 + 2y$ **32.** $-5r - 9r + 8r - 5$ **33.** $-6p + 11p - 4p + 6 + 5$

34. $-8x - 5x + 3x - 12 + 9$

35. $3(k + 2) - 5k + 6 + 3$

36. $5(r - 3) + 6r - 2r + 4$

37. $-2(m + 1) + 3(m - 4)$

38. $6(a - 5) - 4(a + 6)$

39. $0.25(8 + 4p) - 0.5(6 + 2p)$

40. $0.4(10 - 5x) - 0.8(5 + 10x)$

41. $-(2p + 5) + 3(2p + 4) - 2p$

42. $-(7m - 12) - 2(4m + 7) - 8m$

43. $2 + 3(2z - 5) - 3(4z + 6) - 8$

44. $-4 + 4(4k - 3) - 6(2k + 8) + 7$

Complete each statement so that the indicated property is illustrated. Simplify each answer, if possible.

45. $5x + 8x =$ ______________ (distributive property)

46. $9y - 6y =$ ______________ (distributive property)

47. $5(9r) =$ ______________ (associative property)

48. $-4 + (12 + 8) =$ ______________ (associative property)

49. $5x + 9y =$ ______________ (commutative property)

50. $-5(7) =$ ______________ (commutative property)

51. $1 \cdot 7 =$ ______________ (identity property)

52. $-12x + 0 =$ ______________ (identity property)

53. $8(-4 + x) =$ ______________ (distributive property)

54. $3(x - y + z) =$ ______________ (distributive property)

55. Give an "everyday" example of a commutative operation.

56. Give an "everyday" example of inverse operations.

Relating Concepts (Exercises 57–62) For Individual or Group Work

While it may seem that simplifying the expression $3x + 4 + 2x + 7$ *to* $5x + 11$ *is fairly easy, there are several important steps that require mathematical justification. These steps are usually done mentally.* ***Work Exercises 57–62 in order,*** *providing the property that justifies each statement in the given simplification. (These steps could be done in other orders.)*

57. $3x + 4 + 2x + 7 = (3x + 4) + (2x + 7)$ ______________

58. $= 3x + (4 + 2x) + 7$ ______________

59. $= 3x + (2x + 4) + 7$ ______________

60. $= (3x + 2x) + (4 + 7)$ ______________

61. $= (3 + 2)x + (4 + 7)$ ______________

62. $= 5x + 11$ ______________

Chapter 1 Summary

▸ Key Terms

Section	Term	Definition	Example
1.1	**set**	A set is a collection of objects.	
	elements	The elements (**members**) of a set are the numbers or objects that make up the set.	
	finite set	If the number of elements in a set can be listed or counted and the counting process comes to an end, then the set is a finite set.	
	infinite set	If the number of elements in a set cannot be listed or counted, then the set is an infinite set.	
	empty set	The set with no elements is called the empty (**null**) set.	
	variable	A variable is a letter used to represent a number or a set of numbers.	
	set-builder notation	Set-builder notation is used to describe a set of numbers without listing them.	
	number line	A number line is a line with a scale to indicate the set of real numbers.	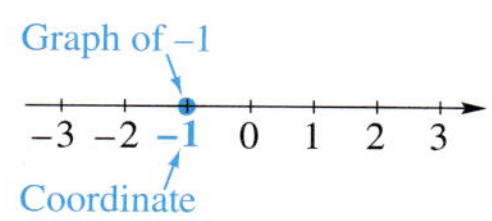
	coordinate	The number that corresponds to a point on the number line is its coordinate.	
	graph	The point on the number line that corresponds to a number is its graph.	
	additive inverse	The additive inverse (**negative, opposite**) of a number a is $-a$.	
	signed numbers	Positive and negative numbers are signed numbers.	
	absolute value	The absolute value of a number is its distance from 0 on a number line.	
	equation	An equation is a mathematical statement that two quantities are equal.	
	inequality	An inequality is a mathematical statement that two quantities are not equal.	
1.2	**sum**	The answer to an addition problem is called the sum.	$2 + 3 = 5$ ← Sum
	difference	The answer to a subtraction problem is called the difference.	$5 - 4 = 1$ ← Difference
	product	The answer to a multiplication problem is called the product.	$2 \cdot 3 = 6$ ← Product
	reciprocals	Two numbers whose product is 1 are reciprocals (**multiplicative inverses**).	
	quotient	The answer to a division problem is called the quotient.	$20 \div 4 = 5$ ← Quotient
1.3	**factors**	Two (or more) numbers whose product is a third number are factors of that third number.	
	exponent	An exponent (**power**) is a number that shows how many times a factor is repeated in a product.	
	base	The base is the number that is a repeated factor in a product.	2^5 ← Exponent; 2 ← Base
	exponential expression	A base with an exponent is called an exponential expression.	
	square root	A square root of a number r is a number that can be squared to obtain r.	
	algebraic expression	A collection of numbers, variables, operation symbols, and grouping symbols is an algebraic expression.	
1.4	**term**	A term is a number or the product of a number and one or more variables.	
	coefficient	A coefficient (**numerical coefficient**) is the numerical factor of a term.	
	like terms	Like terms are terms with the same variables raised to the same powers.	
	combining like terms	Combining like terms is a method of adding or subtracting like terms by using the properties of real numbers.	

▶ New Symbols

$\{a, b\}$	set containing the elements a and b
$\emptyset$	empty (null) set
$\{x \mid x$ has property $P\}$	set-builder notation
$\lvert x \rvert$	absolute value of x
$\neq$	is not equal to
$<$	is less than
$\leq$	is less than or equal to
$>$	is greater than
$\geq$	is greater than or equal to
a^m	m factors of a
$\sqrt{}$	radical sign
$\sqrt{a}$	positive (or principal) square root of a
$\approx$	is approximately equal to

▶ Test Your Word Power

See how well you have learned the vocabulary in this chapter. Answers, with examples, follow the Quick Review.

1. The **empty set** is a set
   - **A.** with 0 as its only element
   - **B.** with an infinite number of elements
   - **C.** with no elements
   - **D.** of ideas.

2. A **variable** is
   - **A.** a symbol used to represent an unknown number
   - **B.** a value that makes an equation true
   - **C.** a solution of an equation
   - **D.** the answer in a division problem.

3. The **absolute value** of a number is
   - **A.** the graph of the number
   - **B.** the reciprocal of the number
   - **C.** the opposite of the number
   - **D.** the distance between 0 and the number on a number line.

4. The **reciprocal** of a nonzero number a is
   - **A.** a
   - **B.** $\frac{1}{a}$
   - **C.** $-a$
   - **D.** 1.

5. A **factor** is
   - **A.** the answer in an addition problem
   - **B.** the answer in a multiplication problem
   - **C.** one of two or more numbers that are added to get another number
   - **D.** any number that divides evenly into a given number.

6. An **exponential expression** is
   - **A.** a number that is a repeated factor in a product
   - **B.** a number or a variable written with an exponent
   - **C.** a number that shows how many times a factor is repeated in a product
   - **D.** an expression that involves addition.

7. A **term** is
   - **A.** a numerical factor
   - **B.** a number or a product of numbers and variables raised to powers
   - **C.** one of several variables with the same exponents
   - **D.** a sum of numbers and variables raised to powers.

8. A **numerical coefficient** is
   - **A.** the numerical factor in a term
   - **B.** the number of terms in an expression
   - **C.** a variable raised to a power
   - **D.** the variable factor in a term.

9. The **identity element** for multiplication is
   - **A.** 0
   - **B.** a
   - **C.** 1
   - **D.** $\frac{1}{a}$.

▶ Quick Review

Concepts	Examples
1.1 Basic Concepts	
Sets of Numbers	
Natural Numbers $\{1, 2, 3, 4, \ldots\}$	$10, 25, 143$
Whole Numbers $\{0, 1, 2, 3, 4, \ldots\}$	$0, 8, 47$
Integers $\{\ldots, -2, -1, 0, 1, 2, \ldots\}$	$-22, -7, 0, 4, 9$
Rational Numbers $\left\{\frac{p}{q} \mid p \text{ and } q \text{ are integers}, q \neq 0\right\}$ (all terminating or repeating decimals)	$-\frac{2}{3}, -0.14, 0, 6, \frac{5}{8}, 0.33333\ldots$
Irrational Numbers $\{x \mid x \text{ is a real number that is not rational}\}$ (all nonterminating, nonrepeating decimals)	$\pi, \sqrt{3}, -\sqrt{22}$
Real Numbers $\{x \mid x \text{ is a rational or an irrational number}\}$	$-3, 0.7, \pi, -\frac{2}{3}$
Absolute Value $\lvert a \rvert = \begin{cases} a & \text{if } a \text{ is positive or } 0 \\ -a & \text{if } a \text{ is negative} \end{cases}$	$\lvert 12 \rvert = 12$ $\lvert -12 \rvert = 12$
1.2 Operations on Real Numbers	
Addition *Same sign:* Add the absolute values. The sum has the same sign as the given numbers.	$-2 + (-7) = -(2 + 7) = -9$
Different signs: Find the absolute values of the numbers, and subtract the lesser absolute value from the greater. The sum has the same sign as the number with the greater absolute value.	$-5 + 8 = 8 - 5 = 3$ $-12 + 4 = -(12 - 4) = -8$
Subtraction For all real numbers a and b, $a - b = a + (-b)$.	$-5 - (-3) = -5 + 3 = -2$
Multiplication and Division *Same sign:* The answer is positive when multiplying or dividing two numbers with the same sign.	$-3(-8) = 24$ $\frac{-15}{-5} = 3$
Different signs: The answer is negative when multiplying or dividing two numbers with different signs.	$-7(5) = -35$ $\frac{-24}{12} = -2$
Division For all real numbers a and b (where $b \neq 0$), $a \div b = \frac{a}{b} = a \cdot \frac{1}{b}$.	$\frac{2}{3} \div \frac{5}{6} = \frac{2}{3} \cdot \frac{6}{5} = \frac{4}{5}$ Multiply by the reciprocal.

Concepts	Examples

1.3 Exponents, Roots, and Order of Operations

Concepts	Examples
The product of an even number of negative factors is positive.	$(-5)^2$ is positive: $(-5)^2 = (-5)(-5) = 25$
The product of an odd number of negative factors is negative.	$(-5)^3$ is negative: $(-5)^3 = (-5)(-5)(-5) = -125$
Order of Operations **1.** Work separately above and below any fraction bar. **2.** If parentheses, brackets, or absolute value bars are present, start with the innermost set and work outward. **3.** Evaluate all exponents, roots, and absolute values. **4.** Multiply or divide in order from left to right. **5.** Add or subtract in order from left to right.	$\frac{12+3}{5\cdot 2} = \frac{15}{10} = \frac{3}{2}$ $(-6)[2^2 - (3+4)] + 3$ $= (-6)[2^2 - 7] + 3$ $= (-6)[4 - 7] + 3$ $= (-6)[-3] + 3$ $= 18 + 3$ $= 21$

1.4 Properties of Real Numbers

Concepts	Examples
Distributive Property $a(b + c) = ab + ac$	$12(4 + 2) = 12 \cdot 4 + 12 \cdot 2$
Inverse Properties $a + (-a) = 0$ and $-a + a = 0$	$5 + (-5) = 0 \quad -12 + 12 = 0$
$a \cdot \frac{1}{a} = 1$ and $\frac{1}{a} \cdot a = 1$	$5 \cdot \frac{1}{5} = 1 \quad -\frac{1}{3}(-3) = 1$
Identity Properties $a + 0 = 0 + a = a$ and $a \cdot 1 = 1 \cdot a = a$	$-32 + 0 = -32 \quad 17.5 \cdot 1 = 17.5$
Commutative Properties $a + b = b + a$ and $ab = ba$	$9 + (-3) = -3 + 9 \quad 6(-4) = (-4)6$
Associative Properties $a + (b + c) = (a + b) + c$ and $a(bc) = (ab)c$	$7 + (5 + 3) = (7 + 5) + 3 \quad -4(6 \cdot 3) = (-4 \cdot 6)3$
Multiplication Property of 0 $a \cdot 0 = 0$ and $0 \cdot a = 0$	$4 \cdot 0 = 0 \quad 0(-3) = 0$

ANSWERS TO TEST YOUR WORD POWER

1. C; *Example:* The set of whole numbers less than 0 is the empty set, written $\emptyset$.
2. A; *Examples:* a, b, c
3. D; *Examples:* $|2| = 2$ and $|-2| = 2$
4. B; *Examples:* 3 is the reciprocal of $\frac{1}{3}$; $-\frac{5}{2}$ is the reciprocal of $-\frac{2}{5}$.
5. D; *Examples:* 2 and 5 are factors of 10 since both divide evenly (without remainder) into 10; other factors of 10 are $-10, -5, -2, -1, 1$, and 10.
6. B; *Examples:* 3^4 and x^{10}
7. B; *Examples:* $6, \frac{x}{2}, -4ab^2$
8. A; *Examples:* The term $8z$ has numerical coefficient 8, and $-10x^3y$ has numerical coefficient -10.
9. C; *Example:* $1 \cdot 5 = 5 \cdot 1 = 5$

Chapter 1 Review Exercises

If you need help with any of these Review Exercises, look in the section indicated in brackets.

[1.1] *Graph the elements of each set on a number line.*

1. $\left\{-4, -1, 2, \frac{9}{4}, 4\right\}$

2. $\left\{-5, -\frac{11}{4}, -0.5, 0, 3, \frac{13}{3}\right\}$

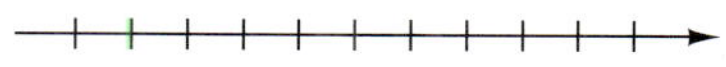

Find the value of each expression.

3. $|-16|$

4. $|23|$

5. $-|-4|$

6. $|-8| - |-3|$

Let set $S = \left\{-9, -\frac{4}{3}, -\sqrt{4}, -0.25, 0, 0.\overline{35}, \frac{5}{3}, \sqrt{7}, \sqrt{-9}, \frac{12}{3}\right\}$. *Simplify the elements of S as necessary, and then list the elements that belong to the specified set.*

7. Whole numbers

8. Integers

9. Rational numbers

10. Real numbers

Write each set by listing its elements.

11. $\{x \mid x$ is a natural number between 3 and 9$\}$

12. $\{y \mid y$ is a whole number less than 4$\}$

Write true *or* false *for each inequality.*

13. $4 \cdot 2 \leq |12 - 4|$

14. $2 + |-2| > 4$

15. $4(3 + 7) > -|40|$

The graph shows the percent change in domestic car sales from January 2004 to January 2005 for various automakers. Use it to work Exercises 16–19.

16. Which automaker had the greatest change in sales? What was that change?

17. Which automaker had the least change in sales? What was that change?

18. *True* or *false:* The absolute value of the percent change for Ford was greater than the absolute value of the percent change for General Motors.

19. *True* or *false:* The percent change for Toyota was more than four times greater than the percent change for Mazda.

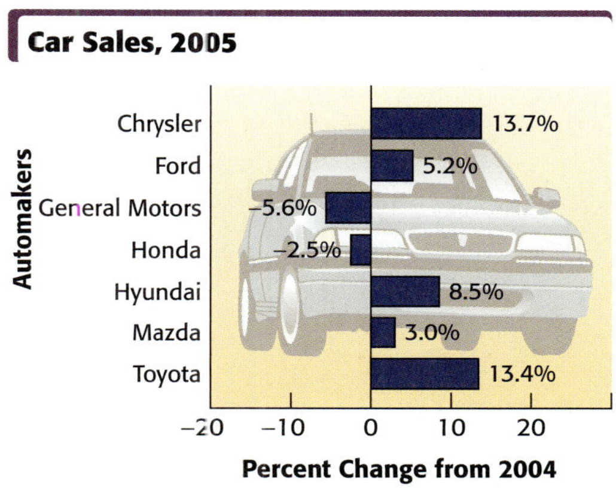

Source: *Chicago Tribune*, February 12, 2006.

[1.2] *Add or subtract as indicated.*

20. $-\frac{5}{8} - \left(-\frac{7}{3}\right)$

21. $-\frac{4}{5} - \left(-\frac{3}{10}\right)$

22. $-5 + (-11) + 20 - 7$

23. $-9.42 + 1.83 - 7.6 - 1.9$

24. $-15 + (-13) + (-11)$

25. $-1 - 3 - (-10) + (-7)$

26. $\frac{3}{4} - \left(\frac{1}{2} - \frac{9}{10}\right)$

27. $-\frac{2}{3} - \left(\frac{1}{6} - \frac{5}{9}\right)$

28. $-|-12| - |-9| + (-4) - |10|$

29. Telescope Peak, altitude 11,049 ft, is next to Death Valley, 282 ft below sea level. Find the difference between these altitudes. (*Source: World Almanac and Book of Facts.*)

Multiply or divide as indicated.

30. $2(-5)(-3)(-3)$

31. $-\frac{3}{7}\left(-\frac{14}{9}\right)$

32. $-4.6(2.48)$

33. $\frac{75}{-5}$

34. $\frac{-2.3754}{-0.74}$

35. Which one of the following is undefined: $\frac{5}{7-7}$ or $\frac{7-7}{5}$?

[1.3] *Evaluate each expression.*

36. 10^4

37. $\left(\frac{3}{7}\right)^3$

38. $(-5)^3$

39. -5^3

40. $(1.7)^2$

Find each square root. If it is not a real number, say so.

41. $\sqrt{400}$

42. $-\sqrt{196}$

43. $\sqrt{\frac{64}{121}}$

44. $-\sqrt{0.81}$

45. $\sqrt{-64}$

Simplify each expression.

46. $-14\left(\frac{3}{7}\right) + 6 \div 3$

47. $-\frac{2}{3}[5(-2) + 8 - 4^3]$

48. $\frac{-5(3^2) + 9(\sqrt{4}) - 5}{6 - 5(-2)}$

Evaluate each expression if $k = -4$, $m = 2$, and $n = 16$.

49. $4k - 7m$

50. $-3\sqrt{n} + m + 5k$

51. $\frac{4m^3 - 3n}{7k^2 - 10}$

52. The following expression for *body mass index* (BMI) can help determine ideal body weight.

$$704 \times (\text{weight in pounds}) \div (\text{height in inches})^2$$

A BMI of 19 to 25 corresponds to a healthy weight. (*Source: Washington Post.*)

(a) Carlos Beltran is 6 ft 1 in. tall and weighs 205 lb. (*Source:* mlb.com) Find his BMI (to the nearest whole number).

(b) Calculate your BMI.

[1.4] *Simplify each expression.*

53. $2q + 19q$

54. $13z - 17z$

55. $-m + 6m$

56. $5p - p$

57. $-2(k + 3)$

58. $6(r + 3)$

59. $9(2m + 3n)$

60. $-(-p + 6q) - (2p - 3q)$

61. $-3y + 6 - 5 + 4y$

62. $2a + 3 - a - 1 - a - 2$

63. $-3(4m - 2) + 2(3m - 1) - 4(3m + 1)$

Complete each statement so that the indicated property is illustrated. Simplify each answer, if possible.

64. $2x - 3x =$ __________________ (distributive property)

65. $-4 \cdot 1 =$ __________________ (identity property)

66. $2(4x) =$ __________________ (associative property)

67. $-3 + 13 =$ __________________ (commutative property)

68. $-3 + 3 =$ __________________ (inverse property)

69. $5(x + z) =$ __________________ (distributive property)

70. $0 + 7 =$ __________________ (identity property)

71. $8 \cdot \frac{1}{8} =$ __________________ (inverse property)

72. $3a - 5a + 6a =$ __________________ (distributive property)

73. $\frac{9}{28} \cdot 0 =$ __________________ (multiplication property of 0)

Mixed Review Exercises*

The table gives U.S. exports and imports with Canada, in millions of dollars.

Year	Exports	Imports
2003	169,924	221,595
2004	189,880	256,360
2005	211,349	287,870

Source: U.S. Census Bureau.

Determine the absolute value of the difference between imports and exports for each year. Is the balance of trade (exports minus imports) in each year positive or negative?

74. 2003

75. 2004

76. 2005

Perform the indicated operations.

77. $\left(-\frac{4}{5}\right)^4$

78. $-\frac{5}{8}(-40)$

79. $-25\left(-\frac{4}{5}\right) + 3^3 - 32 \div \sqrt{4}$

80. $-8 + |-14| + |-3|$

81. $\dfrac{6 \cdot \sqrt{4} - 3 \cdot \sqrt{16}}{-2 \cdot 5 + 7(-3) - 10}$

82. $-\sqrt{25}$

83. $-\frac{10}{21} \div \left(-\frac{5}{14}\right)$

84. $0.8 - 4.9 - 3.2 + 1.14$

85. -3^2

86. $\dfrac{-38}{-19}$

87. $-2(k - 1) + 3k - k$

88. $-\sqrt{-100}$

89. Evaluate $-m(3k^2 + 5m)$ if $k = -4$ and $m = 2$.

90. To evaluate $(3 + 2)^2$, should you work within the parentheses first, or should you square 3 and square 2 and then add?

*The order of exercises in this final group does not correspond to the order in which topics occur in the chapter. This random ordering should help you prepare for the chapter test in yet another way.

Chapter 1 Test

CHAPTER Test Prep VIDEO CD Use the Chapter Test Prep Video CD to see fully worked-out solutions to any of the exercises you want to review.

1. Graph $\{-3, 0.75, \frac{5}{3}, 5, 6.3\}$ on the number line. **1.**

Let $A = \{-\sqrt{6}, -1, -0.5, 0, 3, \sqrt{25}, 7.5, \frac{24}{2}, \sqrt{-4}\}$. *First simplify each element as needed, and then list the elements from A that belong to each set.*

2. Whole numbers **2.** ______

3. Integers **3.** ______

4. Rational numbers **4.** ______

5. Real numbers **5.** ______

Perform the indicated operations.

6. $-6 + 14 + (-11) - (-3)$ **6.** ______

7. $10 - 4 \cdot 3 + 6(-4)$ **7.** ______

8. $7 - 4^2 + 2(6) + (-4)^2$ **8.** ______

9. $\dfrac{10 - 24 + (-6)}{\sqrt{16}(-5)}$ **9.** ______

10. $\dfrac{-2[3 - (-1 - 2) + 2]}{\sqrt{9}(-3) - (-2)}$ **10.** ______

11. $\dfrac{8 \cdot 4 - 3^2 \cdot 5 - 2(-1)}{-3 \cdot 2^3 + 1}$ **11.** ______

The table shows the heights in feet of some selected mountains and the depths in feet (as negative numbers) of some selected ocean trenches.

Mountain	Height	Trench	Depth
Foraker	17,400	Philippine	−32,995
Wilson	14,246	Cayman	−24,721
Pikes Peak	14,110	Java	−23,376

Source: World Almanac and Book of Facts.

12. What is the difference between the height of Mt. Foraker and the depth of the Philippine Trench? **12.** ______

13. What is the difference between the height of Pikes Peak and the depth of the Java Trench? **13.** ______

14. How much deeper is the Cayman Trench than the Java Trench? **14.** ______

15. ______

16. ______

17. ______

18. (a) ______

(b) ______

(c) ______

19. ______

20. ______

21. ______

22. ______

23. ______

24. ______

25. ______

26. ______

27. ______

28. ______

29. ______

30. ______

Find each square root. If it is not a real number, say so.

15. $\sqrt{196}$

16. $-\sqrt{225}$

17. $\sqrt{-16}$

18. For the expression $\sqrt{a}$, under what conditions will its value be **(a)** positive, **(b)** not real, **(c)** 0?

Evaluate each expression if $k = -3$, $m = -3$, and $r = 25$.

19. $\sqrt{r} + 2k - m$

20. $\dfrac{8k + 2m^2}{r - 2}$

21. Simplify $-3(2k - 4) + 4(3k - 5) - 2 + 4k$.

22. How does the subtraction sign affect the terms $-4r$ and 6 when simplifying $(3r + 8) - (-4r + 6)$? What is the simplified form?

Match each statement in Column I with the appropriate property in Column II. Answers may be used more than once.

I	II
23. $6 + (-6) = 0$	**A.** Distributive property
24. $4 + 5 = 5 + 4$	**B.** Inverse property
25. $-2 + (3 + 6) = (-2 + 3) + 6$	**C.** Identity property
26. $5x + 15x = (5 + 15)x$	**D.** Associative property
27. $13 \cdot 0 = 0$	**E.** Commutative property
28. $-9 + 0 = -9$	**F.** Multiplication property of 0
29. $4 \cdot 1 = 4$	
30. $(a + b) + c = (b + a) + c$	

Study Skills

USING YOUR TEXTBOOK

Your textbook is a valuable resource. To find out what it has to offer, we look at some general features that will help in all chapters.

Look in the very front of the book for the Table of Contents. Use it to locate topics in your text. Each chapter is divided into sections, and each section has a number, such as 3.5.

Chapter 3 → 3.5 ← Section 5 within Chapter 3

Your instructor will also use these numbers to assign readings and homework.

Pay special attention to these features as you work through the chapters in your book.

- **Objectives** Each section lists the objectives in the upper corner of the page and again within the section as the corresponding material is presented. An objective tells you what you will be able to do after you complete the section. To check your learning, go back to the list of objectives when you are finished with a section and ask yourself if you can do them all.
- **Margin Exercises** The exercises in the margins in your textbook give you immediate practice and actively involve you in learning the material.
- **Cautions** The yellow boxes titled CAUTION provide warnings about common errors that students often make or trouble spots you will want to avoid.
- **Pointers** These comments in small shaded boxes in examples provide on-the-spot warnings and reminders, point out key steps, and give other helpful tips.
- **Notes** These light purple boxes provide additional explanations or emphasize important ideas in the discussion.
- **Problem-Solving Hints** These green boxes provide helpful problem-solving tips and strategies to use when you are working applications.
- **Calculator Tips** A small, red calculator indicates a Calculator Tip, which provides helpful information about using your calculator. A calculator beside an exercise is a recommendation to use your calculator to work that exercise.

List a page number from Chapter 1 or 2 for each of these features.

✓ A *Caution* appears on page ________.

✓ A *Pointer* appears on page ________.

✓ A *Note* appears on page ________.

✓ A *Problem-Solving Hint* appears on page ________.

✓ A *Calculator Tip* appears on page ________.

OBJECTIVES

1. Become familiar with the features of this text.
2. Locate the Index, Answers, and Solutions sections.

Chapter Features

End of Chapter Features

At the end of most chapters, there are four important sections.

- **Chapter Summary** Turn to the Summary for Chapter 2. It lists the chapter's **Key Terms** (arranged by section), as well as any **New Symbols. Test Your Word Power** checks your understanding of the math vocabulary. The **Quick Review** lists the main concepts in each section of the chapter with corresponding worked-out examples. **Answers to Test Your Word Power** are given right after the Quick Review.
- **Review Exercises** Use these exercises as a way to check your understanding of all the concepts in the chapter. If you have trouble, the red numbers in brackets tell you which section of the chapter to go back to for more explanations. Make sure you do the **Mixed Review Exercises** to practice for tests.
- **Chapter Test** Take the test as a practice exam to be sure you know how to work all types of problems without looking back at the chapter.
- **Cumulative Review Exercises (starting with Chapter 2)** These exercises help you maintain the skills you've learned in all previous chapters. Working on previous skills throughout the course will be a big help on the final exam.

Answers

To find out if you've worked the exercises correctly, your textbook provides many of the answers. **Answers to the margin problems** are at the bottom of each page in the margin area. For homework, you can find the **answers to all of the odd-numbered section exercises** in the **Answers to Selected Exercises** section near the end of your textbook. *All* of the answers are given for the Chapter Review Exercises, Chapter Tests, and Cumulative Review Exercises.

Flag the Answers section with a sticky note or other device, so that you can turn to it quickly.

Solutions

The **Solutions to Selected Exercises** section is located near the end of the text and provides step-by-step, worked out solutions to some exercises that have a twist or are a bit more difficult. Exercises that have a blue square around the exercise number, such as **11.** , have a complete solution in the Solutions section.

Index

Now find the **Index.** All of the topics, vocabulary, and concepts are listed in alphabetical order in the Index. There may be several subheadings listed under the main word or several page numbers listed. Usually, the *first* place that a word appears in the textbook is where it is introduced and defined.

Look up the following in the index. Go to the page or pages listed and find each word. Write down the page that introduces or defines each one.

Conjugates ______________ *Reciprocals* ______________

Pythagorean formula (What two places can you find this in the Index?)

____

2

Linear Equations and Applications

2.1 Linear Equations in One Variable

2.2 Formulas and Percent

2.3 Applications of Linear Equations

2.4 Further Applications of Linear Equations

Summary Exercises on Solving Applied Problems

Despite increasing competition from the Internet and video games, television remains a popular form of entertainment throughout the world. In 2006, 111.4 million American households owned at least one TV set, and average viewing time for all viewers exceeded 30 hours per week. During the 2006–2007 season, favorite prime-time television programs were *American Idol* and *Dancing with the Stars.* (*Source:* Nielsen Media Research.)

In Section 2.2 we discuss the concept of *percent*—one of the most common everyday applications of mathematics—and use it in Exercises 47 and 48 to determine additional information about televisions in U.S. households.

2.1 Linear Equations in One Variable

OBJECTIVES

1. Decide whether a number is a solution of a linear equation.
2. Solve linear equations using the addition and multiplication properties of equality.
3. Solve linear equations using the distributive property.
4. Solve linear equations with fractions or decimals.
5. Identify conditional equations, contradictions, and identities.

In **Chapter 1,** we reviewed *algebraic expressions*. Examples include

$$8x + 9, \quad y - 4, \quad \text{and} \quad \frac{x^3y^8}{z}. \qquad \text{Algebraic expressions}$$

Equations and inequalities compare algebraic expressions, just as a balance scale compares the weights of two quantities. Recall from **Section 1.1** that an ***equation*** is a statement that two algebraic expressions are equal. ***An equation always contains an equals sign, while an expression does not.***

$$\underbrace{3x - 7}_{\text{Left side}} = \underbrace{2}_{\text{Right side}} \qquad\qquad 3x - 7$$

Equation (to solve) — Expression (to simplify or evaluate)

◀ Work Problem 1 at the Side.

A *linear equation in one variable* involves only real numbers and one variable raised to the first power. Examples include

$$x + 1 = -2, \quad x - 3 = 5, \quad \text{and} \quad 2k + 5 = 10. \qquad \text{Linear equations}$$

Linear Equation in One Variable

A **linear equation in one variable** can be written in the form

$$Ax + B = C,$$

where A, B, and C are real numbers, with $A \neq 0$.

A linear equation is a **first-degree equation** since the greatest power on the variable is one. Some examples of equations that are not linear (that is, *nonlinear*) are

$$x^2 + 3y = 5, \quad \frac{8}{x} = -22, \quad \text{and} \quad \sqrt{x} = 6. \qquad \text{Nonlinear equations}$$

OBJECTIVE 1 Decide whether a number is a solution of a linear equation. If the variable in an equation can be replaced by a real number that makes the statement true, then that number is a **solution** of the equation. For example, 8 is a solution of the equation $x - 3 = 5$, since replacing x with 8 gives a true statement, $8 - 3 = 5$. An equation is *solved* by finding its **solution set,** the set of all solutions. The solution set of the equation $x - 3 = 5$ is $\{8\}$.

◀ Work Problem 2 at the Side.

Equivalent equations are equations that have the same solution set. To solve an equation, we usually start with the given equation and replace it with a series of simpler equivalent equations. For example,

$$5x + 2 = 17, \quad 5x = 15, \quad \text{and} \quad x = 3 \qquad \text{Equivalent equations}$$

are all equivalent since each has the solution set $\{3\}$.

1 Decide whether each of the following is an *equation* or an *expression*.

(a) $9x = 10$

(b) $9x + 10$

(c) $3 + 5x - 8x + 9$

(d) $3 + 5x = -8x + 9$

2 Are the given numbers solutions of the given equations?

(a) $3k = 15; 5$

(b) $r + 5 = 4; 1$

(c) $-8m = 12; \frac{3}{2}$

ANSWERS

1. **(a)** equation **(b)** expression **(c)** expression **(d)** equation
2. **(a)** yes **(b)** no **(c)** no

OBJECTIVE 2 Solve linear equations using the addition and multiplication properties of equality. We use two important properties to produce equivalent equations.

Addition and Multiplication Properties of Equality

Addition Property of Equality

For all real numbers A, B, and C, the equations

$$A = B \quad \text{and} \quad A + C = B + C \quad \text{are equivalent.}$$

In words, ***the same number may be added to each side of an equation without changing the solution set.***

Multiplication Property of Equality

For all real numbers A and B, and for $C \neq 0$, the equations

$$A = B \quad \text{and} \quad AC = BC \quad \text{are equivalent.}$$

In words, ***each side of an equation may be multiplied by the same nonzero number without changing the solution set.***

Because subtraction and division are defined in terms of addition and multiplication, respectively, these properties can be extended:

The same number may be subtracted from each side of an equation, and each side of an equation may be divided by the same nonzero number, without changing the solution set.

EXAMPLE 1 Solving a Linear Equation

Solve $4x - 2x - 5 = 4 + 6x + 3$.

The goal is to get x alone on one side of the equation.

$$2x - 5 = 7 + 6x \qquad \text{Combine like terms.}$$

Next, use the addition property to get the terms with x on the same side of the equation and the remaining terms (the numbers) on the other side. One way to do this is to first subtract $6x$ from each side.

$$2x - 5 - 6x = 7 + 6x - 6x \qquad \text{Subtract } 6x.$$

$$-4x - 5 = 7 \qquad \text{Combine like terms.}$$

$$-4x - 5 + 5 = 7 + 5 \qquad \text{Add 5.}$$

$$-4x = 12 \qquad \text{Combine like terms.}$$

$$\frac{-4x}{-4} = \frac{12}{-4} \qquad \text{Divide by } -4.$$

$$x = -3 \qquad \text{Proposed solution}$$

Check by substituting -3 for x in the *original* equation.

Check

$$4x - 2x - 5 = 4 + 6x + 3 \qquad \text{Original equation}$$

$$4(-3) - 2(-3) - 5 \stackrel{?}{=} 4 + 6(-3) + 3 \qquad \text{Let } x = -3.$$

$$-12 + 6 - 5 \stackrel{?}{=} 4 - 18 + 3 \qquad \text{Multiply.}$$

$$-11 = -11 \qquad \text{True}$$

Use parentheses around substituted values to avoid errors.

This is *not* the solution.

The true statement indicates that $\{-3\}$ is the solution set.

Work Problem 3 at the Side. ▶

3 Solve and check.

(a) $3p + 2p + 1 = -24$

(b) $3p = 2p + 4p + 5$

(c) $4x + 8x = 17x - 9 - 1$

(d) $-7 + 3t - 9t = 12t - 5$

ANSWERS

3. (a) $\{-5\}$ (b) $\left\{-\frac{5}{3}\right\}$ (c) $\{2\}$ (d) $\left\{-\frac{1}{9}\right\}$

The steps to solve a linear equation in one variable are as follows.

4 Solve and check.

(a) $5p + 4(3 - 2p) = 2 + p - 10$

(b) $3(z - 2) + 5z = 2$

(c) $-2 + 3(x + 4) = 8x$

(d) $6 - (4 + m) = 8m - 2(3m + 5)$

Solving a Linear Equation in One Variable

Step 1 **Clear fractions.** Eliminate any fractions by multiplying each side by the least common denominator.

Step 2 **Simplify each side separately.** Use the distributive property to clear parentheses and combine like terms as needed.

Step 3 **Isolate the variable terms on one side.** Use the addition property to get all terms with variables on one side of the equation and all numbers on the other.

Step 4 **Isolate the variable.** Use the multiplication property to get an equation with just the variable (with coefficient 1) on one side.

Step 5 **Check.** Substitute the proposed solution into the original equation.

OBJECTIVE 3 Solve linear equations using the distributive property. In Example 1 we did not use Step 1 or the distributive property in Step 2 as given in the box. Many equations require one or both of these steps.

EXAMPLE 2 Solving a Linear Equation

Solve $2(k - 5) + 3k = k + 6$.

Step 1 Since there are no fractions in this equation, Step 1 does not apply.

Step 2 Use the distributive property to simplify and combine like terms on the left side of the equation.

$$2(k - 5) + 3k = k + 6$$

$$2k - 10 + 3k = k + 6 \qquad 2(k - 5) = 2(k) - 2(5) = 2k - 10$$

$$5k - 10 = k + 6 \qquad \text{Combine like terms.}$$

Be sure to distribute over *all* terms within parentheses.

Step 3 Next, use the addition property of equality.

$$5k - 10 - k = k + 6 - k \qquad \text{Subtract } k.$$

$$4k - 10 = 6 \qquad \text{Combine like terms.}$$

$$4k - 10 + 10 = 6 + 10 \qquad \text{Add 10.}$$

$$4k = 16 \qquad \text{Combine like terms.}$$

Step 4 Use the multiplication property of equality to get just k on the left.

$$\frac{4k}{4} = \frac{16}{4} \qquad \text{Divide by 4.}$$

$$k = 4$$

Step 5 Check by substituting 4 for k in the original equation.

Check

$$2(k - 5) + 3k = k + 6 \qquad \text{Original equation}$$

$$2(4 - 5) + 3(4) \stackrel{?}{=} 4 + 6 \qquad \text{Let } k = 4.$$

$$2(-1) + 12 \stackrel{?}{=} 10$$

$$10 = 10 \qquad \text{True}$$

Always check your work.

The solution checks, so the solution set is $\{4\}$.

◀ Work Problem 4 at the Side.

ANSWERS

4. (a) $\{5\}$ (b) $\{1\}$ (c) $\{2\}$ (d) $\{4\}$

CAUTION
Notice in Examples 1 and 2 that the equals signs are aligned in columns. ***Do not use more than one equals sign in a horizontal line of work when solving an equation.***

OBJECTIVE 4 Solve linear equations with fractions or decimals. When fractions or decimals appear as coefficients in equations, our work can be made easier if we multiply each side of the equation by the least common denominator (LCD) of all the fractions. This is an application of the multiplication property of equality, and it produces an equivalent equation with integer coefficients.

EXAMPLE 3 Solving a Linear Equation with Fractions

Solve $\frac{x+7}{6} + \frac{2x-8}{2} = -4$.

Step 1 Start by eliminating the fractions. Multiply each side by the LCD.

$$6\left(\frac{x+7}{6} + \frac{2x-8}{2}\right) = 6(-4) \quad \text{The LCD is 6.}$$

Step 2

$$6\left(\frac{x+7}{6}\right) + 6\left(\frac{2x-8}{2}\right) = 6(-4) \quad \text{Distributive property}$$
$$(x+7) + 3(2x-8) = -24 \quad \text{Multiply.}$$
$$x + 7 + 3(2x) - 3(8) = -24 \quad \text{Distributive property}$$
$$x + 7 + 6x - 24 = -24 \quad \text{Multiply.}$$
$$7x - 17 = -24 \quad \text{Combine like terms.}$$

Step 3

$$7x - 17 + 17 = -24 + 17 \quad \text{Add 17.}$$
$$7x = -7 \quad \text{Combine like terms.}$$

Step 4

$$\frac{7x}{7} = \frac{-7}{7} \quad \text{Divide by 7.}$$
$$x = -1$$

Step 5 Check by substituting -1 for x in the original equation.

Check

$$\frac{x+7}{6} + \frac{2x-8}{2} = -4 \quad \text{Original equation}$$
$$\frac{-1+7}{6} + \frac{2(-1)-8}{2} \stackrel{?}{=} -4 \quad \text{Let } x = -1.$$
$$\frac{6}{6} + \frac{-10}{2} \stackrel{?}{=} -4$$
$$1 - 5 \stackrel{?}{=} -4$$
$$-4 = -4 \quad \text{True}$$

The solution checks, so the solution set is $\{-1\}$.

Work Problem 5 at the Side. ▶

5 Solve and check.

(a) $\frac{2p}{7} - \frac{p}{2} = -3$

(b) $\frac{k+1}{2} + \frac{k+3}{4} = \frac{1}{2}$

ANSWERS

5. (a) $\{14\}$ **(b)** $\{-1\}$

6 Solve and check using the method of Example 4.

$$0.04x + 0.06(20 - x) = 0.05(50)$$

In **Sections 2.2** and **2.3** we solve problems involving interest rates and concentrations of solutions. These problems involve percents that are converted to decimals. The equations that are used to solve such problems involve decimal coefficients. We can clear these decimals by multiplying by a power of 10, such as

$$10^1 = 10, \quad 10^2 = 100, \quad \text{and so on,}$$

that will allow us to obtain integer coefficients.

EXAMPLE 4 Solving a Linear Equation with Decimals

Solve $0.06x + 0.09(15 - x) = 0.07(15)$.

Because each decimal number is given in hundredths, multiply each side of the equation by 100. A number can be multiplied by 100 by moving the decimal point two places to the right.

$$0.06x + 0.09(15 - x) = 0.07(15)$$

$$0.06x + 0.09(15 - x) = 0.07(15) \quad \text{Multiply by 100.}$$

Move decimal points 2 places to the right.

$$6x + 9(15 - x) = 7(15)$$

$$6x + 9(15) - 9(x) = 7(15) \quad \text{Distributive property}$$

$$6x + 135 - 9x = 105 \quad \text{Multiply.}$$

$$-3x + 135 = 105 \quad \text{Combine like terms.}$$

$$-3x + 135 - 135 = 105 - 135 \quad \text{Subtract 135.}$$

$$-3x = -30 \quad \text{Combine like terms.}$$

$$\frac{-3x}{-3} = \frac{-30}{-3} \quad \text{Divide by } -3.$$

$$x = 10$$

Check by substituting 10 for x in the original equation.

Check

$$0.06x + 0.09(15 - x) = 0.07(15) \quad \text{Original equation}$$

$$0.06(10) + 0.09(15 - 10) \stackrel{?}{=} 0.07(15) \quad \text{Let } x = 10.$$

$$0.06(10) + 0.09(5) \stackrel{?}{=} 0.07(15)$$

$$0.6 + 0.45 \stackrel{?}{=} 1.05$$

$$1.05 = 1.05 \quad \text{True}$$

The solution set is $\{10\}$.

◀ Work Problem 6 at the Side.

Examples 3 and 4 illustrate related methods for solving equations with fractions or decimals as coefficients. In both cases, the first step is to eliminate (or "clear") the equation of fractions by multiplying both sides of the equation by the LCD or of decimals by multiplying both sides by a power of 10. Many students prefer these methods because they allow all of the remaining work to be done with integer coefficients.

Some students, however, prefer to solve an equation with decimal coefficients by working with the decimals, which requires fewer steps. The next example shows how to solve the equation from Example 4 without clearing decimals.

ANSWER

6. $\{-65\}$

EXAMPLE 5 Solving a Linear Equation without Clearing Decimals

Solve $0.06x + 0.09(15 - x) = 0.07(15)$.

$$0.06x + 0.09(15 - x) = 0.07(15)$$

$$0.06x + 1.35 - 0.09x = 1.05 \quad \text{Distributive property}$$

Be careful with decimal points.

$$-0.03x + 1.35 = 1.05 \quad \text{Combine like terms.}$$

$$-0.03x + 1.35 - 1.35 = 1.05 - 1.35 \quad \text{Subtract 1.35.}$$

$$-0.03x = -0.3 \quad \text{Combine like terms.}$$

$$\frac{-0.03x}{-0.03} = \frac{-0.3}{-0.03} \quad \text{Divide by } -0.03.$$

$$x = 10$$

As in Example 4, we see that the solution set is $\{10\}$.

Work Problem 7 at the Side. ▶

7 Solve and check using the method of Example 5.

$$0.10(x - 6) + 0.05x = 0.06(50)$$

Either of the methods illustrated in Examples 4 and 5 can be used to solve any equation with decimal coefficients.

Note

Because of space limitations, we will not always show the check when solving an equation. ***To be sure that your solution is correct, you should always check your work.***

OBJECTIVE 5 Identify conditional equations, contradictions, and identities. All of the preceding equations had solution sets containing one element; for example, $2(k - 5) + 3k = k + 6$ has solution set $\{4\}$. This is an example of a *conditional equation,* one which is true only for certain values of the variables. Some linear equations, called *contradictions,* have no solution, while others, called *identities,* have an infinite number of solutions. The table below summarizes these types of equations.

Type of Linear Equation	Number of Solutions	Indication When Solving
Conditional	One	Final line is $x =$ a number. (See Example 6(a).)
Identity	Infinite; solution set {all real numbers}	Final line is true, such as $0 = 0$. (See Example 6(b).)
Contradiction	None; solution set $\emptyset$	Final line is false, such as $-15 = -20$. (See Example 6(c).)

Note

Recall from **Section 1.1** that we use the symbol $\emptyset$ to represent the empty set (or null set), which is the set containing no elements. If an equation has no solution, there are no elements in its solution set, so the solution set is the empty set.

ANSWER

7. $\{24\}$

8 Solve each equation. Decide whether it is a *conditional equation,* an *identity,* or a *contradiction.*

(a) $5(x+2) - 2(x+1) = 3x + 1$

(b) $\frac{x+1}{3} + \frac{2x}{3} = x + \frac{1}{3}$

(c) $5(3x+1) = x + 5$

EXAMPLE 6 Recognizing Conditional Equations, Identities, and Contradictions

Solve each equation. Decide whether it is a *conditional equation,* an *identity,* or a *contradiction.*

(a)

$$5(2x+6) - 2 = 7(x+4)$$

$$10x + 30 - 2 = 7x + 28 \quad \text{Distributive property}$$

$$10x + 28 = 7x + 28 \quad \text{Combine like terms.}$$

$$10x + 28 - 7x - 28 = 7x + 28 - 7x - 28 \quad \text{Subtract } 7x\text{; subtract } 28.$$

$$3x = 0 \quad \text{Combine like terms.}$$

$$\frac{3x}{3} = \frac{0}{3} \quad \text{Divide by 3.}$$

$$x = 0$$

The solution set, $\{0\}$, has only one element, so $5(2x+6) - 2 = 7(x+4)$ is a conditional equation.

(b)

$$5x - 15 = 5(x-3)$$

$$5x - 15 = 5x - 15 \quad \text{Distributive property}$$

$$5x - 15 - 5x + 15 = 5x - 15 - 5x + 15 \quad \text{Subtract } 5x\text{; add } 15.$$

$$0 = 0 \quad \text{True}$$

The final line, the *true* statement $0 = 0$, indicates that the solution set is {all real numbers}, and the equation $5x - 15 = 5(x-3)$ is an identity. (Notice that the first step yielded $5x - 15 = 5x - 15$, which is true for all values of x. We could have identified the equation as an identity at that point.)

(c)

$$5x - 15 = 5(x-4)$$

$$5x - 15 = 5x - 20 \quad \text{Distributive property}$$

$$5x - 15 - 5x = 5x - 20 - 5x \quad \text{Subtract } 5x.$$

$$-15 = -20 \quad \text{False}$$

Since the result, $-15 = -20$, is *false,* the equation has no solution. The solution set is $\emptyset$, so the equation $5x - 15 = 5(x-4)$ is a contradiction.

◀ Work Problem **8** at the Side.

CAUTION

A common error in solving an equation like that in Example 6(a) is to think that the equation has no solution and write the solution set as $\emptyset$. This equation has one solution, the number 0, so it is a conditional equation with solution set $\{0\}$.

ANSWERS

8. **(a)** contradiction; $\emptyset$
(b) identity; {all real numbers}
(c) conditional; $\{0\}$

2.1 Exercises

FOR EXTRA HELP MyMathLab MathXL PRACTICE WATCH DOWNLOAD READ REVIEW

1. Which equations are linear equations in x?

   A. $3x + x - 2 = 0$ **B.** $12 = x^2$

   C. $9x - 4 = 9$ **D.** $\frac{1}{8}x - \frac{1}{x} = 0$

2. Which of the equations in Exercise 1 are nonlinear equations in x? Explain why.

3. Decide whether 6 is a solution of $3(x + 4) = 5x$ by substituting 6 for x. If it is not a solution, explain why.

4. Use substitution to decide whether -2 is a solution of $5(x + 4) - 3(x + 6) = 9(x + 1)$. If it is not a solution, explain why.

5. The equation $4[x + (2 - 3x)] = 2(4 - 4x)$ is an identity. Let x represent the number of letters in your last name. Is this number a solution of this equation? Check your answer.

6. In Example 1, a student looked at the check and thought that $\{-11\}$ should be given as the solution set. Explain why this is not correct.

7. Identify each as an *expression* or an *equation*.

   (a) $5x = 10$

   (b) $5x + 10$

   (c) $5x + 6(x - 3) = 12x + 6$

   (d) $5x + 6(x - 3) - (12x + 6)$

8. Explain why $7x + 10 = 7x + 9$ cannot have a solution. (No work is necessary.)

9. The following work contains a common student error.

   $3x - 2(2x - 3) = 3x + 7$

   $8x - 4x - 6 = 3x + 7$ Distributive property

   $4x - 6 = 3x + 7$ Combine like terms.

   $x = 13$ Subtract $3x$; add 6.

   WHAT WENT WRONG? Give the correct solution.

10. When clearing parentheses in the expression

$$-5m - (2m - 4) + 5$$

on the right side of the equation in Exercise 35 to follow, the $-$ sign before the parenthesis acts like a factor representing what number? Clear parentheses and simplify this expression.

Solve and check each equation. See Examples 1 and 2.

11. $9x + 10 = 1$

12. $7x - 4 = 31$

13. $5x + 2 = 3x - 6$

14. $9p + 1 = 7p - 9$

15. $7x - 5x + 15 = x + 8$

16. $2x + 4 - x = 4x - 5$

17. $12w + 15w - 9 + 5 = -3w + 5 - 9$

18. $-4t + 5t - 8 + 4 = 6t - 4$

19. $3(2t - 4) = 20 - 2t$

20. $2(3 - 2x) = x - 4$

21. $-5(x + 1) + 3x + 2 = 6x + 4$

22. $5(x + 3) + 4x - 5 = 4 - 2x$

23. $2(x + 3) = -4(x + 1)$

24. $4(t - 9) = 8(t + 3)$

25. $3(2w + 1) - 2(w - 2) = 5$

26. $4(x - 2) + 2(x + 3) = 6$

27. $2x + 3(x - 4) = 2(x - 3)$

28. $6x - 3(5x + 2) = 4(1 - x)$

29. $6p - 4(3 - 2p) = 5(p - 4) - 10$

30. $-2k - 3(4 - 2k) = 2(k - 3) + 2$

31. $2[w - (2w + 4) + 3] = 2(w + 1)$

32. $4[2t - (3 - t) + 5] = -(2 + 7t)$

33. $-[2z - (5z + 2)] = 2 + (2z + 7)$

34. $-[6x - (4x + 8)] = 9 + (6x + 3)$

35. $-3m + 6 - 5(m - 1) = -5m - (2m - 4) + 5$

36. $4(k + 2) - 8k - 5 = -3k + 9 - 2(k + 6)$

37. $-3(x + 2) + 4(3x - 8) = 2(4x + 7) + 2(3x - 6)$

38. $-7(2x + 1) + 5(3x + 2) = 6(2x - 4) - (12x + 3)$

39. In order to solve the linear equation

$$\frac{3}{4}x - \frac{1}{3}x = \frac{5}{6}x - 5,$$

we are allowed to multiply each side by the least common denominator of all the fractions in the equation. What is this least common denominator?

40. Suppose that in solving the equation

$$\frac{1}{3}x + \frac{1}{2}x = \frac{1}{6}x,$$

you begin by multiplying each side by 12, rather than the *least* common denominator, 6. Would you get the correct solution anyway? Explain.

41. To solve a linear equation with decimals, we multiply by a power of 10 so that all coefficients are integers. What is the smallest power of 10 that will accomplish this goal in each equation?

(a) $0.05x + 0.12(x + 5000) = 940$ (Exercise 55)

(b) $0.006(x + 2) = 0.007x + 0.009$ (Exercise 61)

42. The expression $0.06(10 - x)(100)$ is equivalent to which of the following?

A. $0.06 - 0.06x$ **B.** $60 - 6x$

C. $6 - 6x$ **D.** $6 - 0.06x$

Solve and check each equation. See Examples 3–5.

43. $\frac{m}{2} + \frac{m}{3} = 10$

44. $\frac{x}{5} - \frac{x}{4} = 2$

45. $\frac{3}{4}x + \frac{5}{2}x = 13$

46. $\frac{8}{3}x - \frac{1}{2}x = -13$

47. $\frac{1}{5}x - 2 = \frac{2}{3}x - \frac{2}{5}x$

48. $\frac{3}{4}x - \frac{1}{3}x = \frac{5}{6}x - 5$

49. $\frac{x - 8}{5} + \frac{8}{5} = -\frac{x}{3}$

50. $\frac{2r - 3}{7} + \frac{3}{7} = -\frac{r}{3}$

51. $\frac{3x - 1}{4} + \frac{x + 3}{6} = 3$

52. $\frac{3x + 2}{7} - \frac{x + 4}{5} = 2$

53. $\frac{4t + 1}{3} = \frac{t + 5}{6} + \frac{t - 3}{6}$

54. $\frac{2x + 5}{5} = \frac{3x + 1}{2} + \frac{-x + 7}{2}$

55. $0.05x + 0.12(x + 5000) = 940$

56. $0.09k + 0.13(k + 300) = 61$

57. $0.02(50) + 0.08r = 0.04(50 + r)$

58. $0.20(14{,}000) + 0.14t = 0.18(14{,}000 + t)$

59. $0.05x + 0.10(200 - x) = 0.45x$

60. $0.08x + 0.12(260 - x) = 0.48x$

61. $0.006(x + 2) = 0.007x + 0.009$

62. $0.004x + 0.006(50 - x) = 0.004(68)$

63. Explain the distinction between a conditional equation, an identity, and a contradiction.

64. A student tried to solve the equation $8x = 7x$ by dividing each side by x, obtaining $8 = 7$. He gave the solution set as $\emptyset$. ***WHAT WENT WRONG?***

65. Suppose you solve a linear equation and obtain, as your final result, an equation in Column I. Match each result with the solution set in Column II for the original equation.

I	II
(a) $7 = 7$	**A.** $\{0\}$
(b) $x = 0$	**B.** {all real numbers}
(c) $7 = 0$	**C.** $\emptyset$

66. Which one of the following linear equations does *not* have {all real numbers} as its solution set?

A. $4x = 5x - x$ **B.** $3(x + 4) = 3x + 12$

C. $4x = 3x$ **D.** $\frac{3}{4}x = 0.75x$

Decide whether each equation is a conditional equation, *an* identity, *or a* contradiction. *Give the solution set. See Example 6.*

67. $-x + 4x - 9 = 3(x - 4) - 5$

68. $-12x + 2x - 11 = -2(5x - 3) + 4$

69. $-11x + 4(x - 3) + 6x = 4x - 12$

70. $3x - 5(x + 4) + 9 = -11 + 15x$

71. $-2(t + 3) - t - 4 = -3(t + 4) + 2$

72. $4(2d + 7) = 2d + 25 + 3(2d + 1)$

73. $7[2 - (3 + 4x)] - 2x = -9 + 2(1 - 15x)$

74. $4[6 - (1 + 2x)] + 10x = 2(10 - 3x) + 8x$

Study Skills

HOMEWORK: HOW, WHY, AND WHEN

It is best for your brain if you keep up with the reading and homework in your math class. The more times you work with the information, the more you learn. So, give yourself every opportunity to read, work problems, and review your mathematics.

Here are two options for reading your math textbook. Read the short descriptions below and decide which will be best for you.

OBJECTIVES

1. **Select an appropriate strategy for homework.**
2. **Use textbook features effectively.**

Preview before Class; Read Carefully after Class

Abby learns best by listening to her teacher explain things. She "gets it" when she sees the instructor work problems on the board. She likes to ask questions in class and put the information in her notes. She has learned that it helps if she has *previewed* the section before the lecture, so she knows generally what to expect in class. *But after the class instruction*, when Abby gets home, she finds that she can better understand the math textbook. She remembers what her teacher said, and she can double-check her notes if she gets confused. So, **Abby carefully reads the section in her text *after* she hears the classroom lecture on the topic.**

Read Carefully before Class

De'Lore, on the other hand, feels he learns well by reading on his own. He prefers to read the section and try working the example problems before coming to class. That way, he already knows what the teacher is going to talk about. Then, he can follow the teacher's examples more easily. It is also easier for him to take notes in class. De'Lore likes to have his questions answered right away, which he can do if he has already read the chapter section. So, **De'Lore carefully reads the section in his text *before* he hears the classroom lecture on the topic.**

Notice that there is no one right way to work with your textbook. You must always figure out what works best for you. Note also that both Abby and De'Lore work with one section at a time. ***The key is that you read the textbook regularly.*** The rest of this activity will give you some ideas of how to make the most of your reading.

Try the following steps as you read your math textbook.

- **Read slowly.** Read only one section—or even part of a section—at a time.
- **Do the sample problems in the margins as you go.** Check them right away. The answers are at the bottom of the page.
- If your mind wanders, **work problems on separate paper and write explanations in your own words.**
- **Make study cards as you read each section.** Pay special attention to the colored boxes in the book. Make cards for new vocabulary, rules, procedures, formulas, and sample problems.
- **NOW**, you are ready to do your homework assignment.

Why Are These Techniques Brain Friendly?

The steps here encourage you to be actively working with the material in your text. You learn best when you are actively doing something.

These methods require you to try several different techniques, not just the same thing over and over.

Also, the techniques allow you to take small breaks in your learning. Those rest periods are crucial for good learning.

> **Which two or three steps will be most helpful for you?**
>
> 1. ______ 2. ______ 3. ______

Homework

Teachers assign homework so you can learn the material and then remember the material through practice. In learning, you get good at what you practice. So, completing homework every day will build your confidence, strengthen your skills, and prepare you for exams.

If you have read each section in your textbook according to the steps above, you will probably encounter few difficulties with the exercises in the homework. Here are some additional suggestions that will help you succeed with the homework.

- **If you have trouble with a problem,** find a similar worked example problem in the section. Pay attention to *every line* of the worked example to see how to get from step to step. Work it yourself too, on separate paper; don't just look at it.
- **If it is hard to remember the steps** to follow for certain procedures, write the steps on a separate card. Then write a short explanation of each step. Keep the card nearby while you do the exercises, but try *not* to look at it.
- **If you aren't sure you are working the assigned exercises correctly,** choose two or three odd-numbered problems that are a similar type and work them. Then check the answers in the answer section of your book and see if you are doing them correctly. If you aren't, go back to the section in the text and review the examples and find out how to correct your errors. Finally, when you are sure you understand, try the assigned problems again.
- **If the problem or a similar problem has a blue screen around the problem number,** such as **11.**, there is a worked-out solution in the selected solutions section at the back of the book. Study this solution.
- **Make sure you do some homework every day,** even if your math class does not meet each day.

> **What are your biggest homework concerns?**
>
> List your two main concerns below. Then write a **solution** for each one.
>
> 1. **Concern:** ______ **Solution:** ______
>
> 2. **Concern:** ______ **Solution:** ______

Why Are These Suggestions Brain Friendly?

Your brain will learn mathematics as you study the worked examples in the text and try doing them yourself on separate paper. Then, when you see similar problems in the homework, you will already have experience to work from.

Giving yourself a practice test by trying to remember the steps (without looking at your study cards or notes) is an excellent way to reinforce what you are learning.

Correcting errors right away is how you learn and reinforce the correct procedures. It is hard to unlearn a mistake, so always check frequently to see that you are on the right track.

2.2 Formulas and Percent

OBJECTIVES

1. Solve a formula for a specified variable.
2. Solve applied problems using formulas.
3. Solve percent problems.
4. Solve problems involving percent increase or decrease.

A **mathematical model** is an equation or inequality that describes a real situation. Models for many applied problems already exist; they are called *formulas*. A **formula** is an equation in which variables are used to describe a relationship. Some formulas that we will be using are

$$d = rt, \quad I = prt, \quad \text{and} \quad P = 2L + 2W. \qquad \text{Formulas}$$

A list of some common formulas used in algebra is given inside the covers of this book.

OBJECTIVE 1 Solve a formula for a specified variable. In some applications, the appropriate formula may be solved for a different variable than the one to be found. For example, the formula $I = prt$ says that interest on a loan or investment equals principal (amount borrowed or invested) times rate (percent) times time at interest (in years). To determine how long it will take for an investment at a stated interest rate to earn a predetermined amount of interest, it would help to first solve the formula for t. This process is called **solving for a specified variable** or **solving a literal equation.**

The steps used in the following examples are very similar to those used in solving linear equations from **Section 2.1.** ***When you are solving for a specified variable, the key is to treat that variable as if it were the only one; treat all other variables like numbers (constants).***

EXAMPLE 1 Solving for a Specified Variable

Solve the formula $I = prt$ for t.

We solve this formula for t by treating I, p, and r as constants (having fixed values) and treating t as the only variable. We first write the formula so that the variable for which we are solving, t, is on the left side. Then we use the properties of the previous section as follows.

$$prt = I \qquad \text{Our goal is to isolate } t.$$

$$(pr)t = I \qquad \text{Associative property}$$

$$\frac{(pr)t}{pr} = \frac{I}{pr} \qquad \text{Divide by } pr.$$

$$t = \frac{I}{pr}$$

The result is a formula for t, time in years.

Work Problem 1 *at the Side.* ▶

1 Solve $I = prt$ for each given variable.

(a) p

(b) r

ANSWERS

1. **(a)** $p = \frac{I}{rt}$ **(b)** $r = \frac{I}{pt}$

2 (a) Solve the formula

$$P = a + b + c$$

for a.

(b) Solve the formula

$$V = \frac{1}{3}\pi r^2 h$$

for h.

To solve an equation for a specified variable, follow these steps.

Solving for a Specified Variable

Step 1 If the equation contains fractions, multiply both sides by the LCD to clear the fractions.

Step 2 Transform so that all terms containing the specified variable are on one side of the equation and all terms without that variable are on the other side.

Step 3 Divide each side by the factor that is the coefficient of the specified variable.

EXAMPLE 2 Solving for a Specified Variable

Solve the formula $P = 2L + 2W$ for W.

This formula gives the relationship between perimeter of a rectangle, P, length of the rectangle, L, and width of the rectangle, W. See Figure 1.

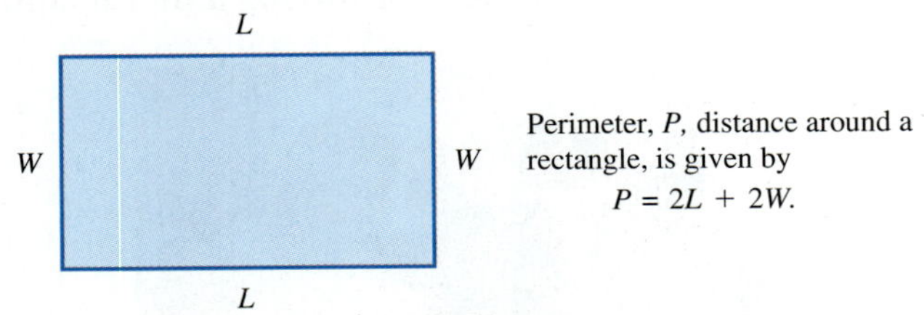

Figure 1

We solve the formula for W by isolating W on one side of the equals sign.

$$P = 2L + 2W$$

Step 1 is not needed here.

Step 2

$$P - 2L = 2L + 2W - 2L \quad \text{Subtract } 2L.$$

$$P - 2L = 2W$$

Step 3

$$\frac{P - 2L}{2} = \frac{2W}{2} \quad \text{Divide by 2.}$$

$$\frac{P - 2L}{2} = W, \quad \text{or} \quad W = \frac{P - 2L}{2}$$

◀ Work Problem 2 at the Side.

CAUTION

In Step 3 of Example 2, we cannot simplify the fraction by dividing 2 into the term $2L$. Based on the order of operations **(Section 1.3),** the fraction bar serves as a grouping symbol. Thus, the subtraction in the numerator must be done before the division.

$$\frac{P - 2L}{2} \neq P - L$$

ANSWERS

2. (a) $a = P - b - c$ (b) $h = \frac{3V}{\pi r^2}$

EXAMPLE 3 **Solving a Formula with Parentheses**

The formula for the perimeter of a rectangle is sometimes written in the equivalent form $P = 2(L + W)$. Solve this form for W.

One way to begin is to use the distributive property on the right side of the equation to get $P = 2L + 2W$, which we would then solve as in Example 2. Another way to begin is to divide by the coefficient 2.

$$P = 2(L + W)$$

$$\frac{P}{2} = L + W \qquad \text{Divide by 2.}$$

$$\frac{P}{2} - L = W, \quad \text{or} \quad W = \frac{P}{2} - L \qquad \text{Subtract } L.$$

We can show that this result is equivalent to our result in Example 2 by rewriting L as $\frac{2}{2}L$.

$$\frac{P}{2} - L = W$$

$$\frac{P}{2} - \frac{2}{2}(L) = W \qquad \tfrac{2}{2} = 1, \text{ so } L = \tfrac{2}{2}(L).$$

$$\frac{P}{2} - \frac{2L}{2} = W$$

$$\frac{P - 2L}{2} = W \qquad \text{Subtract fractions.}$$

The final line agrees with the result in Example 2.

Work Problem **3** *at the Side.* ▶

In Examples 1–3, we solved formulas for specified variables. In Example 4, we solve an equation with two variables for one of these variables. This process will be useful when we work with equations like this one, called *linear equations in two variables,* in **Chapter 4.**

EXAMPLE 4 **Solving an Equation for One of the Variables**

Solve the equation $3x - 4y = 12$ for y.

Our goal is to isolate y on one side of the equation.

$$3x - 4y = 12$$

$$3x - 4y - 3x = 12 - 3x \qquad \text{Subtract } 3x.$$

$$-4y = 12 - 3x$$

$$\frac{-4y}{-4} = \frac{12 - 3x}{-4} \qquad \text{Divide by } -4.$$

$$y = \frac{12 - 3x}{-4}$$

There are other equivalent forms of the final answer that are also correct. For example, since $\frac{a}{-b} = \frac{-a}{b}$ **(Section 1.2),**

Multiply *both* terms of the numerator by -1.

$$y = \frac{12 - 3x}{-4} \text{ can be written as } y = \frac{-(12 - 3x)}{4}, \quad \text{or} \quad y = \frac{3x - 12}{4}.$$

Work Problem **4** *at the Side.* ▶

3 Solve the formula

$$M = \frac{1}{3}(a + b + c)$$

for b.

4 Solve each equation for y.

(a) $2x + 7y = 5$

(b) $5x - 6y = 12$

ANSWERS

3. $b = 3M - a - c$

4. (a) $y = \frac{5 - 2x}{7}$

(b) $y = \frac{12 - 5x}{-6}$, or $y = \frac{5x - 12}{6}$

5 Solve each problem.

(a) A triangle has an area of 36 in.^{2} (square inches) and a base of 12 in. Find its height.

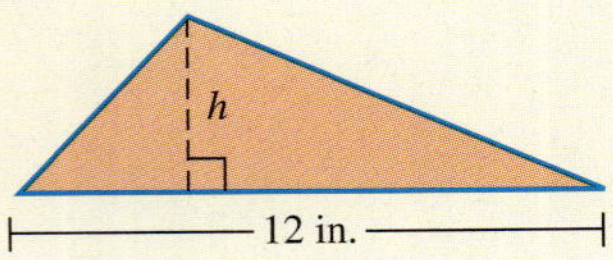

(b) The distance is 500 mi and the time is 20 hr. Find the rate.

(c) In 2006, Sam Hornish, Jr. won the Indianapolis 500 (mile) race with a speed of 157.085 mph. (*Source:* www.indy500.com) Find his time to the nearest thousandth.

ANSWERS

5. **(a)** 6 in. **(b)** 25 mph **(c)** 3.183 hr

OBJECTIVE 2 Solve applied problems using formulas. The distance formula, $d = rt$, relates d, the distance traveled, r, the rate or speed, and t, the travel time.

EXAMPLE 5 Finding Average Speed

Janet Branson found that on average it took her $\frac{3}{4}$ hr each day to drive a distance of 15 mi to work. What was her average speed?

Find the formula for speed (rate) r by solving $d = rt$ for r.

$$d = rt$$

$$\frac{d}{t} = \frac{rt}{t} \qquad \text{Divide by } t.$$

$$\frac{d}{t} = r, \quad \text{or} \quad r = \frac{d}{t}$$

Notice that only Step 3 was needed to solve for r in this example. Now find the speed by substituting the given values of d and t into this formula.

$$r = \frac{15}{\frac{3}{4}} \qquad \text{Let } d = 15,\ t = \tfrac{3}{4}.$$

$$r = 15 \cdot \frac{4}{3} \qquad \text{Multiply by the reciprocal of } \tfrac{3}{4}.$$

$$r = 20$$

Her average speed was 20 mph. (That is, at times she may have traveled a little faster or slower than 20 mph, but overall her speed was 20 mph.)

◀ Work Problem **5** at the Side.

OBJECTIVE 3 Solve percent problems. An important everyday use of mathematics involves the concept of percent. Percent is written with the symbol %. The word **percent** means "per one hundred." One percent means "one per one hundred" or "one one-hundredth."

$$\mathbf{1\% = 0.01 \quad or \quad 1\% = \frac{1}{100}}$$

Solving a Percent Problem

Let a represent a partial amount of b, the base, or whole amount. Then the following equation can be used to solve a percent problem.

$$\frac{\textbf{amount } a}{\textbf{base } b} = \textbf{percent (represented as a decimal)}$$

For example, if a class consists of 50 students and 32 are males, then the percent of males in the class is

$$\frac{\text{amount } a}{\text{base } b} = \frac{32}{50} \qquad \text{Let } a = 32,\ b = 50.$$

$$= 0.64$$

$$= 64\%.$$

EXAMPLE 6 **Solving Percent Problems**

(a) A 50-L mixture of acid and water contains 10 L of acid. What is the percent of acid in the mixture?

The given amount of the mixture is 50 L, and the part that is acid is 10 L. Let x represent the percent of acid in the mixture. Then,

$$x = \frac{10}{50} \begin{array}{l} \leftarrow \text{partial amount} \\ \leftarrow \text{whole amount} \end{array}$$

$$x = 0.20, \quad \text{or} \quad 20\%.$$

(b) If a savings account balance of \$4780 earns 5% interest in one year, how much interest is earned?

Let x represent the amount of interest earned (that is, the part of the whole amount invested). Since $5\% = 0.05$, the equation is

$$\frac{x}{4780} = 0.05 \qquad \frac{\text{amount } a}{\text{base } b} = \text{percent}$$

$$x = 0.05(4780) \qquad \text{Multiply by 4780.}$$

$$x = 239.$$

The interest earned is \$239.

Work Problem **6** at the Side. ▶

6 Solve each problem.

(a) A mixture of gasoline and oil contains 20 oz, of which 1 oz is oil. What percent of the mixture is oil?

(b) An automobile salesman earns a 6% commission on every car he sells. How much does he earn on a car that sells for \$22,000?

EXAMPLE 7 **Interpreting Percents from a Graph**

In 2007, Americans spent about \$41.2 billion on their pets. Use the graph in Figure 2 to determine how much of this amount was spent on pet food.

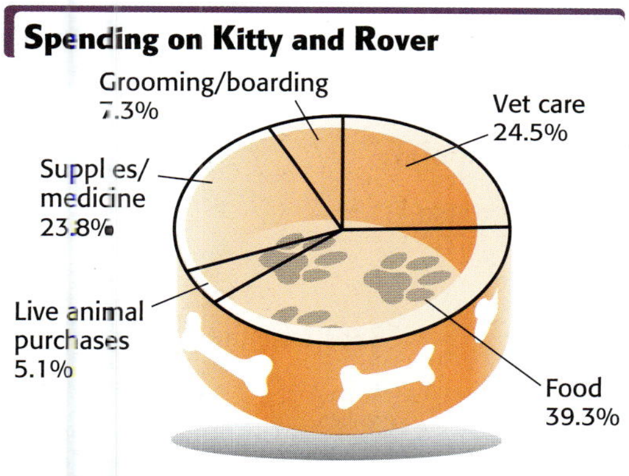

Source: American Pet Products Manufacturers Association Inc.

Figure 2

Dotty

According to the graph, 39.3% was spent on food. Let x represent this amount in billions of dollars.

$$\frac{x}{41.2} = 0.393 \qquad 39.3\% = 0.393$$

$$x = 41.2(0.393) \qquad \text{Multiply by 41.2.}$$

$$x \approx 16.2 \qquad \text{Nearest tenth}$$

Therefore, about \$16.2 billion was spent on pet food.

Work Problem **7** at the Side. ▶

7 Refer to Figure 2. How much was spent on pet supplies/medicine? Round your answer to the nearest tenth.

ANSWERS

6. **(a)** 5% **(b)** \$1320
7. \$9.8 billion

8 Solve each problem.

(a) Cara bought a jacket on sale for \$56. The regular price of the jacket was \$80. What was the percent markdown?

(b) When it was time for Liam to renew the lease on his apartment, the landlord raised his rent from \$650 to \$689 a month. What was the percent increase?

OBJECTIVE 4 Solve problems involving percent increase or decrease. Percent is often used to express a change in some quantity. Buying an item that has been marked up, getting a raise at your job, and the inflation rate are all common applications of **percent increase.** Buying an item on sale, being offered a special discount on tickets to an event, and declining population are common applications of ***percent decrease.*** To solve problems of this type, we use the percent equation from page 76 in the following form:

$$\textbf{percent change} = \frac{\textbf{amount of change}}{\textbf{base}}.$$

Subtract to find this.

EXAMPLE 8 Solving Problems About Percent Increase or Decrease

(a) An electronics store marked up a laptop computer from their cost of \$1200 to a selling price of \$1464. What was the percent markup?

"Markup" is a name for an increase. Let x = the percent increase (as a decimal).

$$\text{percent increase} = \frac{\text{amount of increase}}{\text{base}}$$

$$x = \frac{1464 - 1200}{1200} \quad \text{Substitute the given values.}$$

Subtract to find the *amount* of increase.

Use the original cost.

$$x = \frac{264}{1200}$$

$$x = 0.22 \quad \text{Use a calculator.}$$

The computer was marked up 22%.

(b) The enrollment in a community college declined from 12,750 during one school year to 11,350 the following year. Find the percent decrease to the nearest tenth.

Let x = the percent decrease (as a decimal).

$$\text{percent decrease} = \frac{\text{amount of decrease}}{\text{base}}$$

$$x = \frac{12{,}750 - 11{,}350}{12{,}750} \quad \text{Substitute the given values.}$$

Subtract to find the *amount* of decrease.

Use the original number.

$$x = \frac{1400}{12{,}750}$$

$$x \approx 0.11 \quad \text{Use a calculator.}$$

The college enrollment decreased by about 11%.

CAUTION

When calculating a percent increase or decrease, be sure that you use the original number (*before* the increase or decrease) as the base. A common error is to use the final number (*after* the increase or decrease) in the denominator of the fraction.

Work Problem 8 at the Side.

ANSWERS

8. (a) 30% **(b)** 6%

2.2 Exercises

FOR EXTRA HELP MyMathLab MathXL PRACTICE WATCH DOWNLOAD READ REVIEW

Relating Concepts (Exercises 1–6) For Individual or Group Work

Consider the following equations:

First Equation	*Second Equation*
$\dfrac{7x+8}{3}=12$	$\dfrac{ax+k}{c}=t \quad (c \neq 0)$.

Solving the second equation for x requires the same logic as solving the first equation for x. When solving for x, we treat all other variables as though they were constants. ***Work Exercises 1–6 in order,*** *to see the "parallel logic" of solving for x in the two equations.*

1. **(a)** Clear the first equation of fractions by multiplying each side by 3.

   (b) Clear the second equation of fractions by multiplying each side by c.

2. **(a)** Transform so that the term involving x is the left side of the first equation and the constants are on the right by subtracting 8 from each side.

   (b) Transform so that the term involving x is on the left side of the second equation by subtracting k from each side.

3. **(a)** Simplify the terms in the first equation.

   (b) Simplify the terms in the second equation.

4. **(a)** Divide each side of the first equation by the coefficient of x.

   (b) Divide each side of the second equation by the coefficient of x.

5. Look at your answer for the second equation. What restriction must be placed on the variables? Why is this necessary?

6. Write a short paragraph summarizing what you have learned in this group of exercises.

Solve each formula for the specified variable. See Examples 1–3.

7. $A = LW$ for W (area of a rectangle)

8. $d = rt$ for t (distance)

9. $P = 2L + 2W$ for L (perimeter of a rectangle)

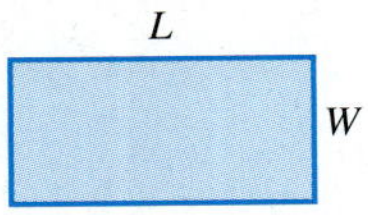

10. $A = bh$ for b (area of a parallelogram)

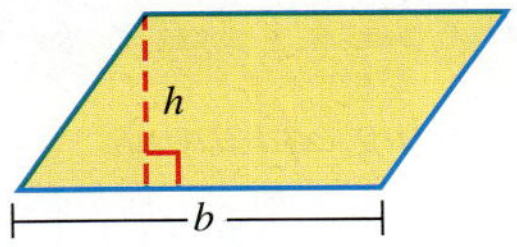

11. $V = LWH$ (volume of a rectangular solid)

(a) for W **(b)** for H

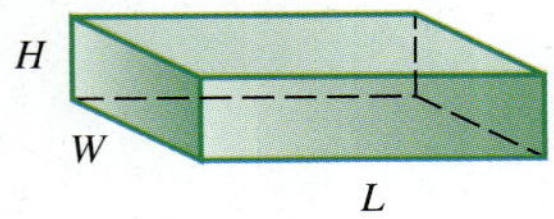

12. $P = a + b + c$ (perimeter of a triangle)

(a) for b **(b)** for c

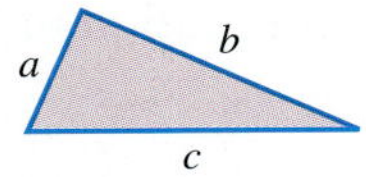

13. $C = 2\pi r$ for r (circumference of a circle)

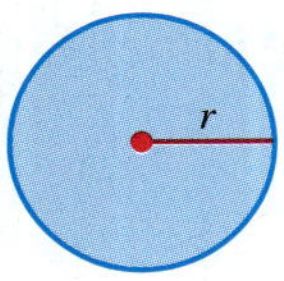

14. $A = \frac{1}{2}bh$ for h (area of a triangle)

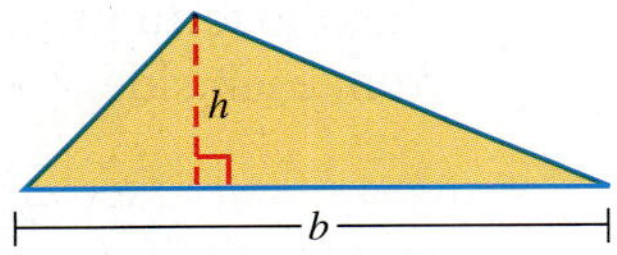

15. $A = \frac{1}{2}h(b + B)$ (area of a trapezoid)

(a) for h **(b)** for B

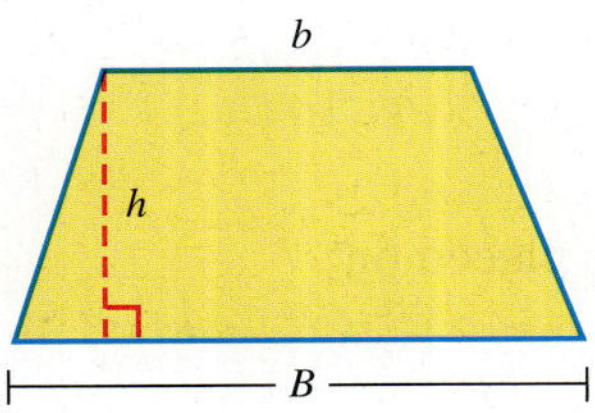

16. $V = \pi r^2 h$ for h
(volume of a right circular cylinder)

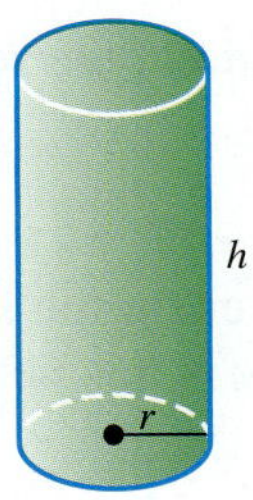

17. $F = \frac{9}{5}C + 32$ for C (Celsius to Fahrenheit)

18. $C = \frac{5}{9}(F - 32)$ for F (Fahrenheit to Celsius)

Solve each equation for y. See Example 4.

19. $4x + 9y = 11$

20. $-7x + 8y = 11$

21. $-3x + 2y = 5$

22. $5x - 3y = 12$

23. $6x - 5y = 7$

Relating Concepts (Exercises 24–26) For Individual or Group Work

The ***surface area*** *of any solid three-dimensional figure is the total area of its surface. For a rectangular solid like that shown in the figure, the surface area A is*

$$A = 2HW + 2LW + 2LH.$$

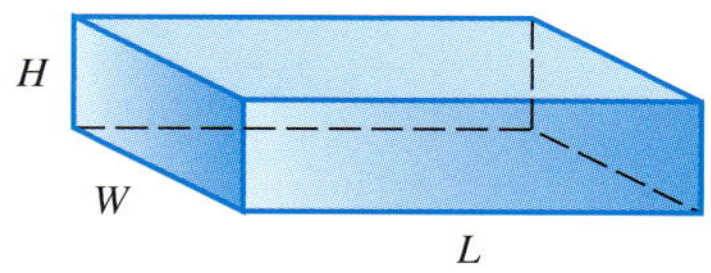

We can solve this formula for L as follows:

$$A = 2HW + 2LW + 2LH$$

$$A - 2HW = 2LW + 2LH \qquad \text{Subtract } 2HW.$$

$$A - 2HW = L(2W + 2H) \qquad \text{Distributive property}$$

$$\frac{A - 2HW}{2W + 2H} = L, \quad \text{or} \quad L = \frac{A - 2HW}{2W + 2H}. \qquad \text{Divide by } 2W + 2H.$$

In the third line, we used the distributive property in reverse to write $2LW + 2LH$ *as* $L(2W + 2H)$. *This is called* ***factoring****, a topic of* ***Chapter 7****.*

Use the distributive property to solve each formula for the specified variable.

24. $k = dF - DF$ for F

25. $Mv = mv - Vm$ for m

26. $A = 2HW + 2LW + 2LH$ for W

Solve each problem. See Example 5.

27. In 2008, Ryan Newman won the Daytona 500 (mile) race with a speed of 152.672 mph. Find his time to the nearest thousandth. (*Source:* www.daytona500.com)

28. In 2007, rain shortened the Indianapolis 500 race to 415 mi. It was won by Dario Franchitti, who averaged 151.774 mph. What was his time to the nearest thousandth? (*Source:* www.indy500.com)

29. As of 2006, the highest temperature ever recorded in Tennessee was 45°C. Find the corresponding Fahrenheit temperature. (*Source:* National Climatic Data Center.)

30. As of 2006, the lowest temperature ever recorded in South Dakota was −58°F. Find the corresponding Celsius temperature. (*Source:* National Climatic Data Center.)

31. The base of the Great Pyramid of Cheops is a square whose perimeter is 920 m. What is the length of each side of this square? (*Source: Atlas of Ancient Archaeology*, 1994.)

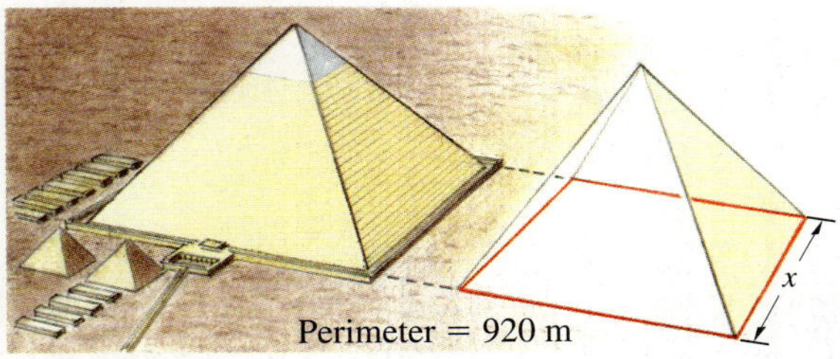

32. Marina City in Chicago is a complex of two residential towers that resemble corncobs. Each tower has a concrete cylindrical core with a 35-ft diameter and is 588 ft tall. Find the volume of the core of one of the towers to the nearest whole number. (*Hint:* Use the π key on your calculator.) (*Source:* www.architechgallery.com; www.aviewoncities.com)

33. The circumference of a circle is 370π in. What is its radius? What is its diameter?

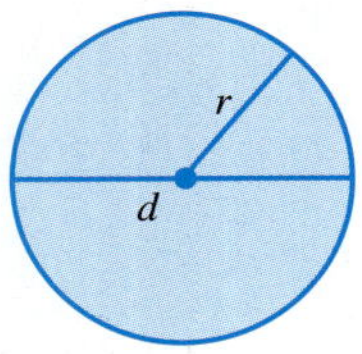

34. The radius of a circle is 2.5 in. What is its diameter? What is its circumference?

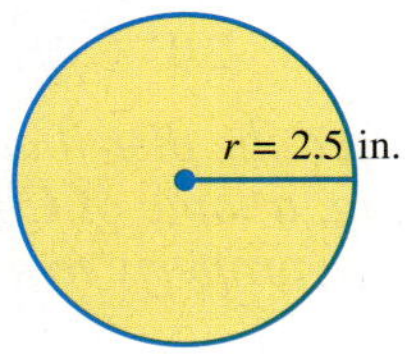

35. A sheet of standard-size copy paper measures 8.5 in. by 11 in. If a ream (500 sheets) of this paper has a volume of 187 in.^{3}, how thick is the ream?

36. Copy paper (Exercise 35) also comes in legal size, which has the same width, but is longer than standard size. If a ream of legal-size copy paper has the same thickness as the standard-size paper and a volume of 238 in.^{3}, what is the length of a sheet of legal paper?

Solve each problem. See Example 6.

37. A mixture of alcohol and water contains a total of 36 oz of liquid. There are 9 oz of pure alcohol in the mixture. What percent of the mixture is water? What percent is alcohol?

38. A mixture of acid and water is 35% acid. If the mixture contains a total of 40 L, how many liters of pure acid are in the mixture? How many liters of pure water are in the mixture?

39. A real estate agent earned \$6900 commission on a property sale of \$230,000. What is her rate of commission?

40. A certificate of deposit for one year pays \$288 simple interest on a principal of \$6400. What is the annual interest rate being paid on this deposit?

When a consumer loan is paid off ahead of schedule, the finance charge is smaller than if the loan were paid off over its scheduled life. By one method, called the ***rule of 78,*** *the amount of unearned interest (finance charge that need not be paid) is given by*

$$u = f \cdot \frac{k(k + 1)}{n(n + 1)},$$

where u is the amount of unearned interest (money saved) when a loan scheduled to run n payments is paid off k payments ahead of schedule. The total scheduled finance charge is f. Use this formula to solve Exercises 41–44.

41. Rhonda Alessi bought a new Ford and agreed to pay it off in 36 monthly payments. The total finance charge is $700. Find the unearned interest if she pays the loan off 4 payments ahead of schedule.

42. Finley Westmoreland bought a car and agreed to pay it off in 36 monthly payments. The total finance charge on the loan was $600. With 12 payments remaining, Finley decided to pay the loan in full. Find the amount of unearned interest.

43. The finance charge on a loan taken out by Vic Denicola is $380.50. If there were 24 equal monthly installments needed to repay the loan, and the loan is paid in full with 8 months remaining, find the amount of unearned interest.

44. Joe Maggiore is scheduled to repay a loan in 24 equal monthly installments. The total finance charge on the loan is $450. With 9 payments remaining, he decides to repay the loan in full. Find the amount of unearned interest.

Exercises 45 and 46 deal with winning percentage in the team standings for Major League Baseball. Winning percentage (Pct.) is commonly expressed as a decimal rounded to the nearest thousandth. To find the winning percentage of a team, divide the number of wins (W) *by the total number of games played* (W + L).

45. The final 2007 standings of the Central Division of the American League are shown in the table. Find the winning percentage of each team.

(a) Detroit **(b)** Minnesota
(c) Chicago **(d)** Kansas City

	W	L	Pct.
Cleveland	96	66	.593
Detroit	88	74	
Minnesota	79	83	
Chicago	72	90	
Kansas City	69	93	

46. Repeat Exercise 45 for the following standings for the Central Division of the National League.

(a) Chicago **(b)** St. Louis
(c) Houston **(d)** Pittsburgh

	W	L	Pct.
Chicago	85	77	
Milwaukee	83	79	.512
St. Louis	78	84	
Houston	73	89	
Cincinnati	72	90	.444
Pittsburgh	68	94	

As mentioned in the chapter introduction, 111.4 *million U.S. households owned at least one TV set in 2006. (Source:* Nielsen Media Research.*) Use this information to solve Exercises 47 and 48. Round answers to the nearest percent.*

47. About 57.9 million U.S. households owned 3 or more TV sets in 2006. What percent of those owning at least one TV set was this?

48. About 93.6 million households that owned at least one TV set in 2006 had a DVD player. What percent of those owning at least one TV set had a DVD player?

An average middle-income family will spend \$242,070 *to raise a child born in 2004 from birth to age 17. The graph shows the percents spent for various categories. Use the graph to answer Exercises 49–52. See Example 7.*

49. To the nearest dollar, how much will be spent to provide housing for the child?

50. To the nearest dollar, how much will be spent for health care?

The Cost of Parenthood

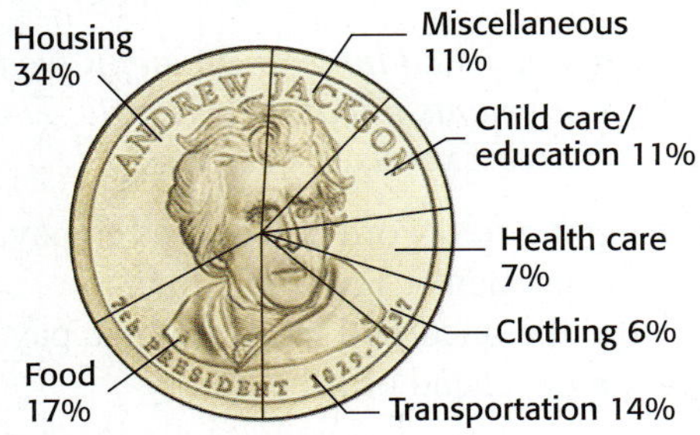

Source: U.S. Department of Agriculture.

51. About \$41,000 will be spent for food. To the nearest percent, what percent of the cost of raising a child from birth to age 17 is this? Does your answer agree with the percent shown in the graph?

52. About \$34,000 will be spent for transportation. To the nearest percent, what percent of the cost of raising a child to age 17 is this? Does your answer agree with the percent shown in the graph?

Solve each problem about percent increase or percent decrease. See Example 8.

53. After 1 yr on the job, Mollie got a raise from \$10.50 per hour to \$11.34 per hour. What was the percent increase in her hourly wage?

54. Sean bought a ticket to a rock concert at a discount. The regular price of the ticket was \$70.00, but he only paid \$59.50. What was the percent discount?

55. Between July 1, 2000, and July 1, 2007, the estimated population of Pittsfield, Massachusetts declined from 134,953 to 129,798. What was the percent decrease to the nearest tenth? (*Source:* U.S. Census Bureau.)

56. Between July 1, 2000, and July 1, 2007, the estimated population of Anchorage, Alaska grew from 320,391 to 362,340. What was the percent increase to the nearest tenth? (*Source:* U.S. Census Bureau.)

57. In April 2008, the audio CD of the Original 2003 Broadway Cast Recording of the musical *Wicked* was available on amazon.com for \$9.97. The list price (full price) of this CD was \$18.98. To the nearest tenth, what was the percent discount? (*Source:* www.amazon.com)

58. In April 2008, the DVD of the movie *Alvin and the Chipmunks* was released. This DVD had a list price of \$29.99 and was for sale on amazon.com at \$15.99. To the nearest tenth, what was the percent discount? (*Source:* www.amazon.com)

2.3 Applications of Linear Equations

OBJECTIVES

1. **Translate from words to mathematical expressions.**
2. **Write equations from given information.**
3. **Distinguish between expressions and equations.**
4. **Use the six steps in solving an applied problem.**
5. **Solve percent problems.**
6. **Solve investment problems.**
7. **Solve mixture problems.**

OBJECTIVE 1 Translate from words to mathematical expressions. Producing a mathematical model of a real situation often involves translating verbal statements into mathematical statements. Although the problems we will be working with are simple ones, the methods we use will also apply to more difficult problems later.

Problem-Solving Hint

Usually there are key words and phrases in a verbal problem that translate into mathematical expressions involving addition, subtraction, multiplication, and division. Translations of some commonly used expressions follow.

TRANSLATING FROM WORDS TO MATHEMATICAL EXPRESSIONS

Verbal Expression	Mathematical Expression (where x and y are numbers)
Addition	
The **sum** of a number and 7	$x + 7$
6 **more than** a number	$x + 6$
3 **plus** a number	$3 + x$
24 **added to** a number	$x + 24$
A number **increased by** 5	$x + 5$
The **sum** of two numbers	$x + y$
Subtraction	
2 **less than** a number	$x - 2$
2 **less** a number	$2 - x$
12 **minus** a number	$12 - x$
A number **decreased by** 12	$x - 12$
A number **subtracted from** 10	$10 - x$
The **difference between** two numbers	$x - y$
Multiplication	
16 **times** a number	$16x$
A number **multiplied by** 6	$6x$
$\frac{2}{3}$ **of** a number (used with fractions and percent)	$\frac{2}{3}x$
$\frac{3}{4}$ **as much as** a number	$\frac{3}{4}x$
Twice (2 times) a number	$2x$
The **product** of two numbers	xy
Division	
The **quotient** of 8 and a number	$\frac{8}{x}$ $(x \neq 0)$
A number **divided by** 13	$\frac{x}{13}$
The **ratio** of two numbers or the **quotient** of two numbers	$\frac{x}{y}$ $(y \neq 0)$

Work Problem 1 at the Side. ▶

1 Translate each verbal expression as a mathematical expression. Use x as the variable.

(a) 9 added to a number

(b) The difference between 7 and a number

(c) Four times a number

(d) The quotient of 7 and a nonzero number

ANSWERS

1. (a) $9 + x$, or $x + 9$ **(b)** $7 - x$ **(c)** $4x$ **(d)** $\frac{7}{x}$ $(x \neq 0)$

CAUTION

Because subtraction and division are not commutative operations, be careful to correctly translate expressions involving them. For example, "2 less than a number" is translated as $x - 2$, *not* $2 - x$. "A number subtracted from 10" is expressed as $10 - x$, *not* $x - 10$.

For division, the number *by which* we are dividing is the denominator, and the number *into which* we are dividing is the numerator. For example, "a number divided by 13" and "13 divided into x" both translate as $\frac{x}{13}$. Similarly, "the quotient of x and y" is translated as $\frac{x}{y}$.

OBJECTIVE 2 Write equations from given information. The symbol for equality, $=$, is often indicated by the word ***is***. In fact, any words that indicate the idea of "sameness" translate to $=$.

EXAMPLE 1 Translating Words into Equations

Translate each verbal sentence into an equation.

Verbal Sentence	Equation
Twice a number, **decreased by** 3, **is** 42.	$2x - 3 = 42$
If the **product of a number and 12** is decreased by 7, the result **is** 105.	$12x - 7 = 105$
The **quotient of a number and the number plus 4 is** 28.	$\frac{x}{x+4} = 28$
The **quotient of a number and 4,** plus the number, **is** 10.	$\frac{x}{4} + x = 10$

◀ Work Problem **2** at the Side.

2 Translate each verbal sentence into an equation. Use x as the variable.

(a) The sum of a number and 6 is 28.

(b) If twice a number is decreased by 3, the result is 17.

(c) The product of a number and 7 is twice the number plus 12.

(d) The quotient of a number and 6, added to twice the number, is 7.

OBJECTIVE 3 Distinguish between expressions and equations. An expression translates as a phrase. An equation includes the $=$ symbol and translates as a sentence.

EXAMPLE 2 Distinguishing between Expressions and Equations

Decide whether each is an *expression* or an *equation*.

(a) $2(3 + x) - 4x + 7$
There is no equals sign, so this is an expression.

(b) $2(3 + x) - 4x + 7 = -1$
Because of the equals sign, this is an equation.

Note that the expression in part (a) simplifies to the expression $-2x + 13$, and the equation in part (b) has solution 7.

◀ Work Problem **3** at the Side.

3 Decide whether each is an *expression* or an *equation*.

(a) $5x - 3(x + 2) = 7$

(b) $5x - 3(x + 2)$

OBJECTIVE 4 Use the six steps in solving an applied problem. While there is no one method that will allow us to solve all types of applied problems, the following six steps are helpful.*

*__Appendix A__ *Strategies for Problem Solving* introduces additional methods and tips for solving applied problems.

ANSWERS

2. (a) $x + 6 = 28$ **(b)** $2x - 3 = 17$
(c) $7x = 2x + 12$ **(d)** $\frac{x}{6} + 2x = 7$

3. (a) equation **(b)** expression

Solving an Applied Problem

Step 1 **Read** the problem, several times if necessary, until you *understand* what is given and what is to be found.

Step 2 **Assign a variable** to represent the unknown value, using diagrams or tables as needed. Write down what the variable represents. Express any other unknown values in terms of the variable.

Step 3 **Write an equation** using the variable expression(s).

Step 4 **Solve** the equation.

Step 5 **State the answer** to the problem. Does it seem reasonable?

Step 6 **Check** the answer in the words of the original problem.

EXAMPLE 3 **Solving a Geometry Problem**

The length of a rectangle is 1 cm more than twice the width. The perimeter of the rectangle is 110 cm. Find the length and the width of the rectangle.

Step 1 **Read** the problem. We must find the length and width of the rectangle. We are given that the length is 1 cm more than twice the width, and the perimeter is 110 cm.

Step 2 **Assign a variable.** Let W = the width; then $1 + 2W$ = the length. Make a sketch, as in Figure 3.

Figure 3

Step 3 **Write an equation.** The perimeter of a rectangle is given by the formula $P = 2L + 2W$.

$$P = 2L + 2W$$
$$110 = 2(1 + 2W) + 2W \quad \text{Let } L = 1 + 2W \text{ and } P = 110.$$

Step 4 **Solve** the equation obtained in Step 3.

$$110 = 2 + 4W + 2W \quad \text{Distributive property}$$
$$110 = 2 + 6W \quad \text{Combine like terms.}$$
$$110 - 2 = 2 + 6W - 2 \quad \text{Subtract 2.}$$
$$108 = 6W$$
$$\frac{108}{6} = \frac{6W}{6} \quad \text{Divide by 6.}$$
$$18 = W$$

We also need to find the length.

Step 5 **State the answer.** The width of the rectangle is 18 cm and the length is $1 + 2W = 1 + 2(18) = 37$ cm.

Step 6 **Check.** The length, 37 cm, is 1 more than twice the width, $2(18)$ cm. The perimeter is $2(37) + 2(18) = 110$ cm, as required.

Work Problem (4) at the Side. ▶

(4) Solve the problem.

The length of a rectangle is 5 cm more than its width. The perimeter is five times the width. What are the dimensions of the rectangle?

ANSWER

4. width: 10 cm; length: 15 cm

5 Solve the problem.

For the 2007 baseball season, the Major League Baseball leaders for RBIs (runs batted in) were Matt Holliday of the Colorado Rockies and Alex Rodriguez of the New York Yankees. These two players had a total of 293 RBIs, and Holliday had 19 fewer RBIs than Rodriguez. How many RBIs did each player have? (*Source: World Almanac and Book of Facts.*)

EXAMPLE 4 Finding Unknown Numerical Quantities

Two outstanding major league pitchers in recent years are Johan Santana and Aaron Harang. In 2006, they combined for a total of 461 strikeouts. Santana had 29 more strikeouts than Harang. How many strikeouts did each pitcher have? (*Source: World Almanac and Book of Facts.*)

Step 1 **Read** the problem. We are asked to find the number of strikeouts each pitcher had.

Step 2 **Assign a variable** to represent the number of strikeouts for one of the men.

Let $s =$ the number of strikeouts for Aaron Harang.

We must also find the number of strikeouts for Johan Santana. Since he had 29 more strikeouts than Harang,

$s + 29 =$ the number of strikeouts for Santana.

Step 3 **Write an equation.** The sum of the numbers of strikeouts is 461, so

Harang's strikeouts	+	Santana's strikeouts	=	Total
↓		↓		↓
s	+	$(s + 29)$	=	461.

Step 4 **Solve** the equation.

$$s + (s + 29) = 461$$

$$2s + 29 = 461 \qquad \text{Combine like terms.}$$

$$2s + 29 - 29 = 461 - 29 \qquad \text{Subtract 29.}$$

$$2s = 432$$

$$\frac{2s}{2} = \frac{432}{2} \qquad \text{Divide by 2.}$$

$$s = 216$$

Don't stop here.

Step 5 **State the answer.** We let s represent the number of strikeouts for Harang, so Harang had 216. Then Santana had

$s + 29 = 216 + 29 = 245$ strikeouts.

Step 6 **Check.** 245 is 29 more than 216, and the sum of 216 and 245 is 461. The conditions of the problem are satisfied, and our answer checks.

CAUTION

A common error in solving applied problems is forgetting to answer all the questions asked in the problem. In Example 4, we were asked for the number of strikeouts *each* player had, so there was an extra step at the end in order to find the number Santana had.

◀ *Work Problem 5 at the Side.*

ANSWER

5. Holliday: 137; Rodriguez: 156

OBJECTIVE 5 Solve percent problems. Recall from **Section 2.2** that percent means "per one hundred," so 5% means 0.05, 14% means 0.14, and so on.

EXAMPLE 5 Solving a Percent Problem

In 2006, total annual health expenditures in the United States were about \$2000 billion (or \$2 trillion). This was an increase of 180% over the total for 1990. What were the approximate total health expenditures in billions of dollars in the United States in 1990? (*Source:* U.S. Centers for Medicare & Medicaid Services.)

Step 1 **Read** the problem. We are given that the total health expenditures increased by 180% from 1990 to 2006, and \$2000 million was spent in 2006. We must find the expenditures in 1990.

Step 2 **Assign a variable.** Let x represent the total health expenditures for 1990.

$$180\% = 180(0.01) = 1.8,$$

so $1.8x$ represents the additional expenditures since 1990.

Step 3 **Write an equation** from the given information.

the expenditures in 1990 + the increase = 2000

$$x + 1.8x = 2000$$

Note the x in $1.8x$.

Step 4 **Solve** the equation.

$$1x + 1.8x = 2000 \quad \text{Identity property}$$
$$2.8x = 2000 \quad \text{Combine like terms.}$$
$$x \approx 714 \quad \text{Divide by 2.8.}$$

Step 5 **State the answer.** Total health expenditures in the United States for 1990 were about \$714 billion.

Step 6 **Check** that the increase, $2000 - 714 = 1286$, is about 180% of 714.

CAUTION

Avoid two common errors that occur in solving problems like the one in Example 5.

1. Do not try to find 180% of 2000 and subtract that amount from 2000. The 180% should be applied to *the amount in 1990, not the amount in 2006.*
2. Do not write the equation as

$$x + 1.8 = 2000. \quad \text{Incorrect}$$

The percent must be multiplied by some number; in this case, the number is the amount spent in 1990, giving $1.8x$.

Work Problem 6 at the Side. ▶

6 Solve each problem.

(a) Mark Schorr bought an LCD high-definition TV that had been marked up 25% over cost. If he paid \$2375 for the TV, what was the store's cost?

(b) Michelle Raymond was paid \$162 for a week's work at her part-time job after 10% deductions for taxes. How much did she earn before the deductions were made?

ANSWERS

6. (a) \$1900 **(b)** \$180

OBJECTIVE 6 Solve investment problems. The investment problems in this chapter deal with *simple interest*. In most real-world applications, *compound interest* (covered in a later chapter) is used.

7 Solve each problem.

(a) A woman invests \$72,000 in two ways—some at 5% and some at 3%. Her total annual interest income is \$3160. Find the amount she invests at each rate.

(b) A man has \$34,000 to invest. He invests some at 5% and the balance at 4%. His total annual interest income is \$1545. Find the amount he invests at each rate.

EXAMPLE 6 Solving an Investment Problem

Mark LeBeau has \$40,000 to invest. He will put part of the money in an account paying 4% interest and the remainder into stocks paying 6% interest. His accountant tells him that the total annual income from these investments should be \$2040. How much should he invest at each rate?

Step 1 **Read** the problem again. We must find the two amounts.

Step 2 **Assign a variable.**

Let $x =$ the amount to invest at 4%;

$40{,}000 - x =$ the amount to invest at 6%.

The formula for interest is $I = prt$. Here the time, t, is 1 year. Make a table to organize the given information.

Principal	Rate (as a decimal)	Interest
x	0.04	$0.04x$
$40{,}000 - x$	0.06	$0.06(40{,}000 - x)$
40,000	XXXXX	2040 ← Total

Multiply principal, rate, and time (here, 1 yr) to find the interest.

Step 3 **Write an equation.** The last column of the table gives the equation.

$$\underset{\text{interest at 4\%}}{0.04x} + \underset{\text{interest at 6\%}}{0.06(40{,}000 - x)} = \underset{\text{total interest}}{2040}$$

Step 4 **Solve** the equation. We do so without clearing decimals.

$$0.04x + 0.06(40{,}000) - 0.06x = 2040 \quad \text{Distributive property}$$
$$0.04x + 2400 - 0.06x = 2040 \quad \text{Multiply.}$$
$$-0.02x + 2400 = 2040 \quad \text{Combine like terms.}$$
$$-0.02x = -360 \quad \text{Subtract 2400.}$$
$$x = 18{,}000 \quad \text{Divide by } -0.02.$$

Step 5 **State the answer.** Mark should invest \$18,000 at 4%. At 6%, he should invest $\$40{,}000 - \$18{,}000 = \$22{,}000$.

Step 6 **Check.** Find the annual interest at each rate and the total.

$$0.04(\$18{,}000) = \$720 \quad \text{and} \quad 0.06(\$22{,}000) = \$1320$$
$$\$720 + \$1320 = \$2040, \quad \text{as required.}$$

◀ Work Problem 7 at the Side.

Problem-Solving Hint

In Example 6, we chose to let the variable represent the amount invested at 4%. Students often ask, "Can I let the variable represent the other unknown?" The answer is yes. The equation will be different, but in the end the two answers will be the same.

ANSWERS

7. **(a)** \$50,000 at 5%; \$22,000 at 3%
(b) \$18,500 at 5%; \$15,500 at 4%

OBJECTIVE 7 Solve mixture problems. Mixture problems involving rates of concentration can be solved with linear equations.

EXAMPLE 7 Solving a Mixture Problem

A chemist must mix 8 L of a 40% acid solution with some 70% solution to get a 50% solution. How much of the 70% solution should be used?

Step 1 **Read** the problem. The problem asks for the amount of 70% solution to be used.

Step 2 **Assign a variable.** Let $x =$ the number of liters of 70% solution to be used. The information in the problem is illustrated in Figure 4.

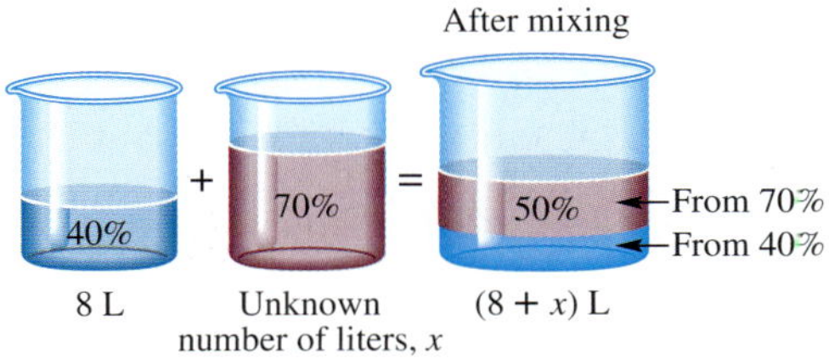

Figure 4

Use the given information to complete the following table.

Number of Liters	Percent (as a decimal)	Liters of Pure Acid
8	0.40	$0.40(8) = 3.2$
x	0.70	$0.70x$
$8 + x$	0.50	$0.50(8 + x)$

Sum must equal

The numbers in the right column were found by multiplying the strengths and the numbers of liters. The number of liters of pure acid in the 40% solution plus the number of liters of pure acid in the 70% solution must equal the number of liters of pure acid in the 50% solution.

Step 3 **Write an equation.**

$$3.2 + 0.70x = 0.50(8 + x)$$

Step 4 **Solve.**

$$3.2 + 0.70x = 4 + 0.50x \quad \text{Distributive property}$$

$$0.20x = 0.8 \quad \text{Subtract 3.2 and } 0.50x.$$

$$x = 4 \quad \text{Divide by 0.20.}$$

Step 5 **State the answer.** The chemist should use 4 L of the 70% solution.

Step 6 **Check.** 8 L of 40% solution plus 4 L of 70% solution is

$$8(0.40) + 4(0.70) = \mathbf{6\ L}$$

of acid. Similarly, 8 + 4 or 12 L of 50% solution has

$$12(0.50) = \mathbf{6\ L}$$

of acid in the mixture. The total amount of pure acid is 6 L both before and after mixing, so the answer checks.

Work Problem 8 at the Side. ▶

8 Solve each problem.

(a) How many liters of a 10% solution should be mixed with 60 L of a 25% solution to get a 15% solution?

(b) How many pounds of candy worth \$8 per lb should be mixed with 100 lb of candy worth \$4 per lb to get a mixture that can be sold for \$7 per lb?

ANSWERS

8. **(a)** 120 L **(b)** 300 lb

9 Solve each problem.

(a) How much pure acid should be added to 6 L of 30% acid to increase the concentration to 50% acid?

(b) How much water must be added to 20 L of 50% antifreeze solution to reduce it to 40% antifreeze?

Answers

9. (a) 2.4 L (b) 5 L

Problem-Solving Hint

When pure water is added to a solution, remember that water is 0% of the chemical (acid, alcohol, etc.). Similarly, pure chemical is 100% chemical.

EXAMPLE 8 Solving a Mixture Problem When One Ingredient Is Pure

The octane rating of gasoline is a measure of its antiknock qualities. For a standard fuel, the octane rating is the percent of isooctane. How many liters of pure isooctane should be mixed with 200 L of 94% isooctane, referred to as 94 octane, to get a mixture that is 98% isooctane?

Step 1 **Read** the problem. The problem asks for the amount of pure isooctane.

Step 2 **Assign a variable.** Let $x =$ the number of liters of pure (100%) isooctane. Complete a table with the given information. Recall that $100\% = 100(0.01) = 1$.

Number of Liters	Percent (as a decimal)	Liters of Pure Isooctane
x	1	x
200	0.94	$0.94(200)$
$x + 200$	0.98	$0.98(x + 200)$

Step 3 **Write an equation.** The equation comes from the last column of the table, as in Example 7.

$$x + 0.94(200) = 0.98(x + 200)$$

Step 4 **Solve.**

$$x + 0.94(200) = 0.98x + 0.98(200) \quad \text{Distributive property}$$
$$x + 188 = 0.98x + 196 \quad \text{Multiply.}$$
$$0.02x = 8 \quad \text{Subtract } 0.98x \text{ and } 188.$$
$$x = 400 \quad \text{Divide by } 0.02.$$

Step 5 **State the answer.** 400 L of isooctane are needed.

Step 6 **Check** by showing that

$$400 + 0.94(200) = 0.98(400 + 200)$$

is true.

◀ Work Problem 9 at the Side.

2.3 Exercises

FOR EXTRA HELP

MyMathLab | Math XL PRACTICE | WATCH | DOWNLOAD | READ | REVIEW

In each of the following, (a) translate as an expression and (b) translate as an equation or inequality. Use x to represent the number.

1. **(a)** 12 more than a number ________
   (b) 12 is more than a number. ________

2. **(a)** 3 less than a number ________
   (b) 3 is less than a number. ________

3. **(a)** 4 less than a number _______
   (b) 4 is less than a number. _______

4. **(a)** 6 greater than a number ________
   (b) 6 is greater than a number. ________

5. Which one of the following is *not* a valid translation of "20% of a number"?

   A. $0.20x$ **B.** $0.2x$ **C.** $\frac{x}{5}$ **D.** $20x$

6. Explain why $24 - x$ is *not* a correct translation of "24 less than a number."

Translate each verbal phrase into a mathematical expression. Use x to represent the unknown number. See Example 1.

7. Twice a number, increased by 18

8. The product of 8 and a number, increased by 14

9. 15 decreased by four times a number

10. 12 less than one-third of a number

11. The product of 10 and 6 less than a number

12. The product of 8 less than a number and 7 more than the number

13. The quotient of five times a number and 9

14. The quotient of 12 and seven times a nonzero number

Use the variable x for the unknown, and write an equation representing the verbal sentence. Then solve the problem. See Example 1.

15. The sum of a number and 6 is -31. Find the number.

16. The sum of a number and -4 is 12. Find the number.

17. If the product of a number and -4 is subtracted from the number, the result is 9 more than the number. Find the number.

18. If the quotient of a number and 6 is added to twice the number, the result is 8 less than the number. Find the number.

19. When $\frac{2}{3}$ of a number is subtracted from 12, the result is 10. Find the number.

20. When 75% of a number is added to 6, the result is 3 more than the number. Find the number.

Decide whether each is an expression *or an* equation. *See Example 2.*

21. $5(x + 3) - 8(2x - 6)$

22. $-7(y + 4) + 13(y - 6)$

23. $5(x + 3) - 8(2x - 6) = 12$

24. $-7(y + 4) + 13(y - 6) = 18$

25. $\frac{t}{2} - \frac{t + 5}{6} - 8$

26. $\frac{t}{2} - \frac{t + 5}{6} = 8$

In Exercises 27 and 28, complete the six suggested problem-solving steps to solve each problem.

27. Two of the leading U.S. research universities are Massachusetts Institute of Technology (MIT) and Stanford University. In a recent year, these two universities secured 230 patents on various inventions. Stanford secured 38 fewer patents than MIT. How many patents did each university secure? (*Source:* Association of University Technology Managers.)

Step 1 **Read** the problem carefully. What are you asked to find?

____

____

Step 2 **Assign a variable.** Let x = the number of patents MIT secured.

Then $x - 38$ = ________________

____________________.

Step 3 **Write an equation.**

________ + ________ = 230

Step 4 **Solve** the equation.

x = ________

Step 5 **State the answer.** MIT secured ________ patents, and Stanford secured ________ patents.

Step 6 **Check.** The number of Stanford patents was ________ fewer than the number of ________________, and the total number of patents was 134 + ________ = ________.

28. In a recent sample of book buyers, 70 more shopped at large chain bookstores than at small chain/independent bookstores. A total sample of 442 book buyers shopped at these two types of stores. How many buyers shopped at each type of bookstore? (*Source:* Book Industry Study Group)

Step 1 **Read** the problem carefully. What are you asked to find?

____

____

Step 2 **Assign a variable.** Let x = the number of book buyers at large chain bookstores.

Then $x - 70$ = ________________

____________________.

Step 3 **Write an equation.**

________ + ________ = 442

Step 4 **Solve** the equation.

x = ________

Step 5 **State the answer.** There were ________ large chain bookstore shoppers and ________ small chain/independent shoppers.

Step 6 **Check.** The number of ________________ was ________ more than the number of ________________, and the total number of these shoppers was 256 + ________ = ________.

Solve each problem. See Examples 3 and 4.

29. The John Hancock Center in Chicago has a rectangular base. The length of the base measures 65 ft less than twice the width. The perimeter of this base is 860 ft. What are the dimensions of the base?

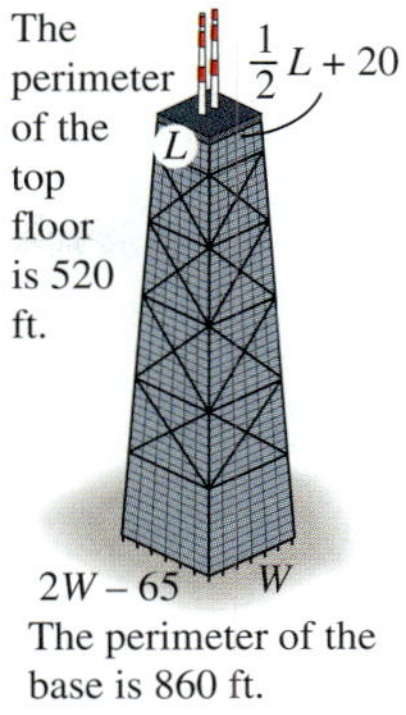

30. The Vietnam Veterans Memorial in Washington, D.C., is in the shape of two sides of an isosceles triangle. If the two walls of equal length were joined by a straight line of 438 ft, the perimeter of the resulting triangle would be 931.5 ft. Find the lengths of the two walls. (*Source:* Pamphlet obtained at Vietnam Veterans Memorial.)

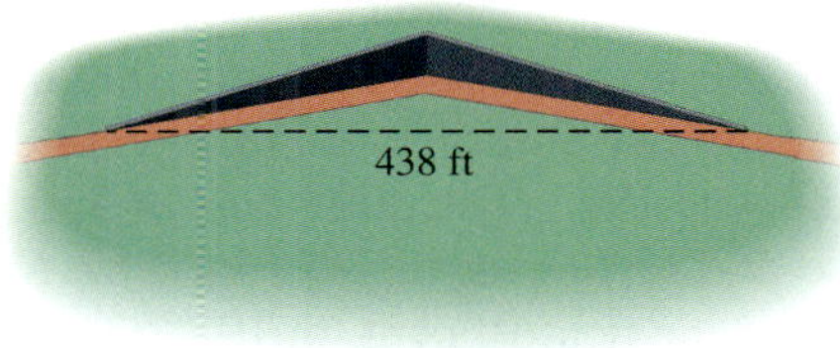

31. The Bermuda Triangle supposedly causes trouble for aircraft pilots. It has a perimeter of 3075 mi. The shortest side measures 75 mi less than the middle side, and the longest side measures 375 mi more than the middle side. Find the lengths of the three sides.

32. The John Hancock Center (Exercise 29) tapers as it rises. The top floor is rectangular and has perimeter 520 ft. The width of the top floor measures 20 ft more than one-half its length. What are the dimensions of the top floor?

33. Galileo Galilei conducted experiments involving Italy's famous Leaning Tower of Pisa to investigate the relationship between an object's speed of fall and its weight. The Leaning Tower is 804 ft shorter than the Eiffel Tower in Paris, France. The two towers have a total height of 1164 ft. How tall is each tower? (*Source: Microsoft Encarta Encyclopedia.*)

34. Two of the longest-running Broadway shows were *Cats*, which played from 1982 through 2000, and *Les Misérables,* which played from 1987 through 2005. Together, there were 14,165 performances of these two shows during their Broadway runs. There were 805 fewer performances of *Les Misérables* than of *Cats*. How many performances were there of each show? (*Source:* The League of American Theatres and Producers.)

35. In 2008, the New York Yankees and the Detroit Tigers had the highest payrolls in Major League Baseball. The Tigers' payroll was \$70.4 million less than the Yankees' payroll, and the two payrolls totaled \$347.8 million. What was the payroll for each team? (*Source:* The Associated Press.)

36. Ted Williams and Rogers Hornsby were two great hitters. Together they got 5584 hits in their careers. Hornsby got 276 more hits than Williams. How many base hits did each get? (*Source:* Neft, D. S. and Cohen, R. M., *The Sports Encyclopedia: Baseball,* St. Martins Griffin; New York, 2006.)

Solve each percent problem. See Example 5.

37. In 2005, the average cost of a traditional Thanksgiving dinner for 10, featuring turkey, stuffing, cranberries, pumpkin pie, and trimmings, was $36.78, an increase of 3.1% over the cost in 2004. What was the cost, to the nearest cent, in 2004? (*Source:* American Farm Bureau.)

38. Refer to Exercise 37. The first year that information on the cost of a traditional Thanksgiving dinner was collected was 1987. The 2005 cost of $36.78 was an increase of 37.5% over the cost in 1987. What was the cost, to the nearest cent, in 1987? (*Source:* American Farm Bureau.)

39. In 2000, the population of Cedar Rapids, Iowa, was 237,230. The 2007 population was estimated at 106.6% of the 2000 population. What was the 2007 population? (*Source:* U.S. Census Bureau.)

40. The consumer price index (CPI) in November 2007 was 210.2. This represented a 4.3% increase from a year earlier. To the nearest tenth, what was the CPI in November 2006? (*Source:* U.S. Bureau of Labor Statistics.)

41. At the end of a day, Erich Bergen found that the total cash register receipts at the motel where he works amounted to $2725. This included the 9% sales tax charged. Find the amount of the tax.

42. Phlash Phelps sold his house for $159,000. He got this amount knowing that he would have to pay a 6% commission to his agent. What amount did he have after the agent was paid?

Solve each investment problem. See Example 6.

43. Jay Jenkins earned $12,000 last year by giving tennis lessons. He invested part at 3% simple interest and the rest at 4%. He earned a total of $440 in interest. How much did he invest at each rate?

Principal	Rate (as a decimal)	Interest
x	0.03	
	0.04	
12,000	XXXXXX	440

44. Stuart Sudak won $60,000 on a slot machine in Las Vegas. He invested part at 2% simple interest and the rest at 3%. He earned a total of $1600 in annual interest. How much was invested at each rate?

Principal	Rate (as a decimal)	Interest
x	0.02	
	XXXXXX	1600

45. Michelle Renda invested some money at 4.5% simple interest and $1000 less than twice this amount at 3%. Her total annual income from the interest was $1020. How much was invested at each rate?

46. Toshira Hashimoto invested some money at 3.5% simple interest, and $5000 more than 3 times this amount at 4%. He earned $1440 in annual interest. How much did he invest at each rate?

47. Vincente and Ricarda Pérez have invested \$27,000 in bonds paying 7%. How much additional money should they invest in a certificate of deposit paying 4% simple interest so that the total annual return on the two investments will be 6%?

48. Carol Hurst received a year-end bonus of \$17,000 from her company and invested the money in an account paying 6.5%. How much additional money should she deposit in an account paying 5% so that the annual return on the two investments will be 6%?

Solve each problem involving rates of concentration and mixtures. See Examples 7 and 8.

49. Ten liters of a 4% acid solution must be mixed with a 10% solution to get a 6% solution. How many liters of the 10% solution are needed?

Liters of Solution	Percent (as a decimal)	Liters of Pure Acid
10	0.04	
x	0.10	
	0.06	

50. How many liters of a 14% alcohol solution must be mixed with 20 L of a 50% solution to get a 30% solution?

Liters of Solution	Percent (as a decimal)	Liters of Pure Alcohol
x	0.14	
	0.50	

51. In a chemistry class, 12 L of a 12% alcohol solution must be mixed with a 20% solution to get a 14% solution. How many liters of the 20% solution are needed?

52. How many liters of a 10% alcohol solution must be mixed with 40 L of a 50% solution to get a 40% solution?

53. How much pure dye must be added to 4 gal of a 25% dye solution to increase the solution to 40%? (*Hint:* Pure dye is 100% dye.)

54. How much water must be added to 6 gal of a 4% insecticide solution to reduce the concentration to 3%? (*Hint:* Water is 0% insecticide.)

55. Randall Albritton wants to mix 50 lb of nuts worth \$2 per lb with some nuts worth \$6 per lb to make a mixture worth \$5 per lb. How many pounds of \$6 nuts must he use?

56. Lee Ann Spahr wants to mix tea worth 2¢ per oz with 100 oz of tea worth 5¢ per oz to make a mixture worth 3¢ per oz. How much 2¢ tea should be used?

57. Why is it impossible to add two mixtures of candy worth \$4 per lb and \$5 per lb to obtain a final mixture worth \$6 per lb?

58. Write an equation based on the following problem, solve the equation, and explain why the problem has no solution.

How much 30% acid should be mixed with 15 L of 50% acid to obtain a mixture that is 60% acid?

Relating Concepts (Exercises 59–63) For Individual or Group Work

Consider each problem.

Problem A

Jack has \$800 invested in two accounts. One pays 5% interest per year and the other pays 10% interest per year. The amount of yearly interest is the same as he would get if the entire \$800 was invested at 8.75%. How much does he have invested at each rate?

Problem B

Jill has 800 L of acid solution. She obtained it by mixing some 5% acid with some 10% acid. Her final mixture of 800 L is 8.75% acid. How much of each of the 5% and 10% solutions did she use to get her final mixture?

In Problem A, let x represent the amount invested at 5% interest, and in Problem B, let y represent the amount of 5% acid used. ***Work Exercises 59–63 in order.***

59. (a) Write an expression in x that represents the amount of money Jack invested at 10% in Problem A.

(b) Write an expression in y that represents the amount of 10% acid solution Jill used in Problem B.

60. (a) Write expressions that represent the amount of interest Jack earns per year at 5% and at 10%.

(b) Write expressions that represent the amount of pure acid in Jill's 5% and 10% acid solutions.

61. (a) The sum of the two expressions in part (a) of Exercise 60 must equal the total amount of interest earned in one year. Write an equation representing this fact.

(b) The sum of the two expressions in part (b) of Exercise 60 must equal the amount of pure acid in the final mixture. Write an equation representing this fact.

62. (a) Solve Problem A.

(b) Solve Problem B.

63. Explain the similarities between the processes used in solving Problems A and B.

Study Skills

TAKING LECTURE NOTES

Study the set of sample math notes below, and read the comments about them. Then try to incorporate the techniques into your own math note taking in class.

OBJECTIVE

1 Identify and apply note taking strategies.

- **Always include the date and title** of the day's lecture topic at the top of every page. **Always begin a new day with a new page.**
- **Skipping lines** makes the notes easier to read.
- **A Caution box alerts you to a common error or point of confusion.**
- **The examples show the correct answers as well as typical mistakes** students make.
- The notes connect familiar concepts (sentences) to new concepts (algebra).
- **A star marks an important reminder.** This is a warning to avoid future mistakes. **Note the underlining,** too, which highlights important information.
- Notice the use of columns, which allows for examples and explanations or related concepts to be close together. Whenever you know you'll be given concepts to compare or contrast, a series of steps to follow, or examples with explanations, **try the column method.**
- **Note the arrows,** which clearly show connections in the material.

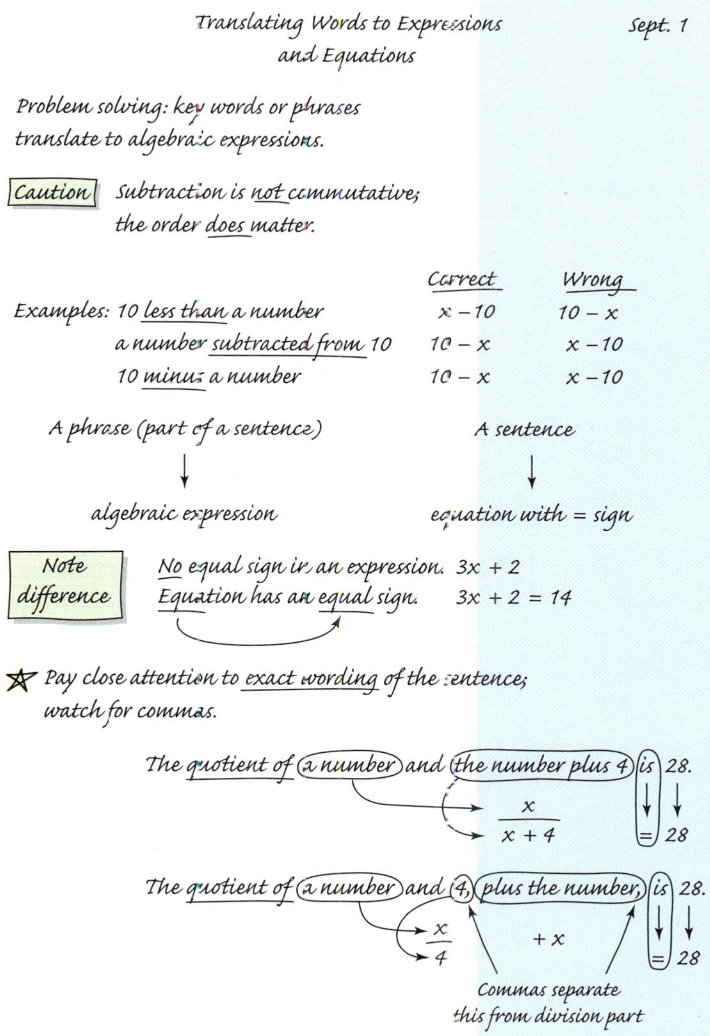

Why Are These Notes Brain Friendly?

The notes are easy to look at, and you are more likely to respond positively to things that are visually pleasing. Other techniques that are visually memorable are the use of spacing (the two columns), stars, underlining, and circling. All of these methods allow you to take note of important concepts and steps.

The notes are also systematic, which means that they use certain techniques regularly. This way, you can easily recognize the topic of the day, the signals that show an important point, and the steps to follow for procedures. When you develop a system that you always use in your notes, they are easy to understand later when you are reviewing for a test.

Now Try This ▶▶▶

Find one or two people in your math class to work with. Compare each other's lecture notes over a period of a week or so. Ask yourself the following questions as you examine the notes.

1. **What are you doing to show the main points** or larger concepts in your notes (such as underlining, boxing, using stars or capital letters, etc.)?
2. **In what ways do you set off the explanations** for worked problems, examples, and subpoints or smaller concepts (such as indenting, using arrows, circling or boxing, etc.)?
3. **What does your instructor do** to show that he or she is moving from one idea to the next (such as saying "Next," "Any questions," "Now," or erasing the board, etc.)?
4. **How do you mark a change in ideas** or topics in your notes (such as skipping lines, using dashes or numbers, etc.)?
5. **What explanations (in words) do you give yourself** in your notes, so that when those new concepts from lecture are fading, you can read your notes and still remember them when doing your homework?
6. **What new ideas did you learn** by examining your classmates' notes?
7. **What new techniques will you try in your own note taking?** List two or three of these techniques that you will use next time you take notes in math class.

2.4 Further Applications of Linear Equations

There are three common applications of linear equations that we did not discuss in **Section 2.3**—money problems, uniform motion problems, and problems involving the angles of a triangle.

OBJECTIVES

1. Solve problems about different denominations of money.
2. Solve problems about uniform motion.
3. Solve problems about angles.

OBJECTIVE 1 Solve problems about different denominations of money. These problems are very similar to the simple interest problems in **Section 2.3.**

Problem-Solving Hint

In problems involving money, use the fact that

number of monetary units of the same kind × **denomination** = **total monetary value**.

For example, 30 dimes have a monetary value of $\$0.10(30) = \$3$. Fifteen five-dollar bills have a value of $\$5(15) = \$75$.

EXAMPLE 1 Solving a Money Denomination Problem

For a bill totaling \$5.65, a cashier received 25 coins consisting of nickels and quarters. How many of each type of coin did the cashier receive?

Step 1 **Read** the problem. The problem asks that we find the number of nickels and the number of quarters the cashier received.

Step 2 **Assign a variable.**

Let x = the number of nickels;
then $25 - x$ = the number of quarters.

We can organize the information in a table.

	Number of Coins	Denomination	Value
Nickels	x	\$0.05	$0.05x$
Quarters	$25 - x$	\$0.25	$0.25(25 - x)$
			5.65 ← Total

Step 3 **Write an equation.** From the last column of the table,

$$0.05x + 0.25(25 - x) = 5.65.$$

Step 4 **Solve.**

$$5x + 25(25 - x) = 565 \quad \text{Multiply by 100.}$$
$$5x + 625 - 25x = 565 \quad \text{Distributive property}$$
$$-20x = -60 \quad \text{Subtract 625; combine like terms.}$$
$$x = 3 \quad \text{Divide by } -20.$$

Move decimal points 2 places to the right.

Step 5 **State the answer.** There are 3 nickels and $25 - 3 = 22$ quarters.

Step 6 **Check.** The cashier has $3 + 22 = 25$ coins, and the value of the coins is $\$0.05(3) + \$0.25(22) = \$5.65$, as required.

Work Problem 1 at the Side. ▶

1 Solve the problem.

At the end of a day, a cashier had 26 coins consisting of dimes and half-dollars. The total value of these coins was \$8.60. How many of each type did he have?

ANSWER

1. 11 dimes, 15 half-dollars

CAUTION

Be sure that your answer is reasonable when working problems like Example 1. Because you are dealing with a number of coins, the correct answer can neither be negative nor a fraction.

OBJECTIVE 2 Solve problems about uniform motion.

Problem-Solving Hint

Uniform motion problems use the distance formula, $d = rt$. ***When rate (or speed) is given in miles per hour, time must be given in hours.*** To solve such problems, ***draw a sketch*** to illustrate what is happening, and ***make a table*** to summarize the given information.

EXAMPLE 2 Solving a Motion Problem (Opposite Directions)

Two cars leave the same place at the same time, one going east and the other west. The eastbound car averages 40 mph, while the westbound car averages 50 mph. In how many hours will they be 300 mi apart?

Step 1 **Read** the problem. We must find the time it takes for the two cars to be 300 mi apart.

Step 2 **Assign a variable.** A sketch shows what is happening in the problem: The cars are going in *opposite* directions. See Figure 5.

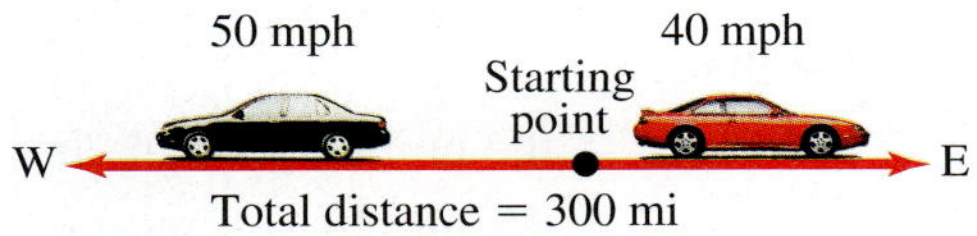

Figure 5

Let x represent the time traveled by each car. Organize the information in a table. ***Fill in each distance by multiplying rate by time*** using the formula $d = rt$. The sum of the two distances is 300.

	Rate	Time	Distance
Eastbound Car	40	x	$40x$
Westbound Car	50	x	$50x$
			300 ← Total

Step 3 **Write an equation.** $40x + 50x = 300$

Step 4 **Solve.**

$$90x = 300 \quad \text{Combine like terms.}$$

$$x = \frac{300}{90} = \frac{10}{3} \quad \text{Divide by 90; lowest terms}$$

Step 5 **State the answer.** The cars travel $\frac{10}{3} = 3\frac{1}{3}$ hr, or 3 hr and 20 min.

Step 6 **Check.** The eastbound car traveled $40\left(\frac{10}{3}\right) = \frac{400}{3}$ mi, and the westbound car traveled $50\left(\frac{10}{3}\right) = \frac{500}{3}$ mi, for a total of $\frac{400}{3} + \frac{500}{3} = \frac{900}{3} = 300$ mi, as required.

CAUTION

It is a common error to write 300 as the distance for *each* car in Example 2. Three hundred miles is the *total* distance traveled.

As in Example 2, in general, the equation for a problem involving motion in *opposite* directions is of the form

partial distance + partial distance = total distance.

Work Problem 2 at the Side. ▶

EXAMPLE 3 **Solving a Motion Problem (Same Direction)**

Geoff can bike to work in $\frac{3}{4}$ hr. By bus, the trip takes $\frac{1}{4}$ hr. If the bus travels 20 mph faster than Geoff rides his bike, how far is it to his workplace?

Step 1 **Read** the problem. We must find the distance between Geoff's home and his workplace.

Step 2 **Assign a variable.** Although the problem asks for a distance, it is easier here to let x be Geoff's speed when he rides his bike to work. Then the speed of the bus is $x + 20$. For the trip by bike,

$$d = rt = x \cdot \frac{3}{4} = \frac{3}{4}x,$$

and by bus,

$$d = rt = (x + 20) \cdot \frac{1}{4} = \frac{1}{4}(x + 20).$$

Summarize this information in a table.

	Rate	Time	Distance
Bike	x	$\frac{3}{4}$	$\frac{3}{4}x$
Bus	$x + 20$	$\frac{1}{4}$	$\frac{1}{4}(x + 20)$

Same

Step 3 **Write an equation.** The key to setting up the correct equation is to realize that the distance in each case is the same. See Figure 6.

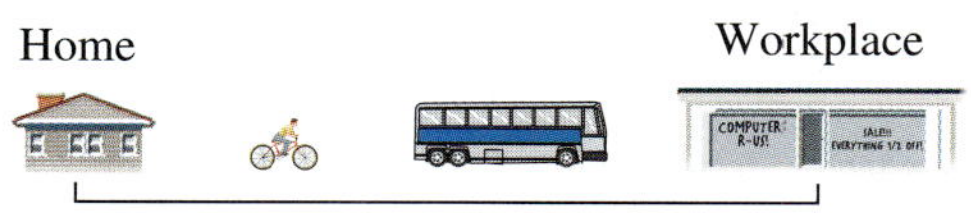

Figure 6

$$\frac{3}{4}x = \frac{1}{4}(x + 20) \quad \text{The distance is the same.}$$

Step 4 **Solve.**

$$4\left(\frac{3}{4}x\right) = 4\left(\frac{1}{4}\right)(x + 20) \quad \text{Multiply by 4.}$$

$$3x = x + 20 \quad \text{Multiply; identity property}$$

$$2x = 20 \quad \text{Subtract } x.$$

$$x = 10 \quad \text{Divide by 2.}$$

Continued on Next Page

2 Solve the problem.

Two cars leave the same location at the same time. One travels north at 60 mph and the other south at 45 mph. In how many hours will they be 420 mi apart?

ANSWER

2. 4 hr

3 Solve the problem.

Elayn begins jogging at 5:00 A.M., averaging 3 mph. Clay leaves at 5:30 A.M., following her, averaging 5 mph. How long will it take him to catch up to her? (*Hint:* 30 min $= \frac{1}{2}$ hr.)

4 Solve the problem.

One angle in a triangle is 15° larger than a second angle. The third angle is 25° larger than twice the second angle. Find the measure of each angle.

Step 5 **State the answer.** The required distance is given by

$$d = \frac{3}{4}x = \frac{3}{4}(10) = \frac{30}{4} = 7.5 \text{ mi.}$$

Step 6 **Check** by finding the distance using

$$d = \frac{1}{4}(x + 20) = \frac{1}{4}(10 + 20) = \frac{30}{4} = 7.5 \text{ mi.}$$

The same result

As in Example 3, the equation for a problem involving motion in the same direction is often of the form

one distance = other distance.

Problem-Solving Hint

In Example 3 it was easier to let the variable represent a quantity other than the one that we were asked to find. This is the case in some problems. It takes practice to learn when this approach is best, and practice means working lots of problems.

◀ *Work Problem* **3** *at the Side.*

OBJECTIVE 3 Solve problems about angles. An important result of Euclidean geometry (the geometry of the Greek mathematician Euclid) is that ***the sum of the angle measures of any triangle is 180°.*** This property is used in the next example.

EXAMPLE 4 Finding Angle Measures

Find the value of x, and determine the measure of each angle in Figure 7.

Step 1 **Read** the problem. We are asked to find the measure of each angle.

Step 2 **Assign a variable.** Let x represent the measure of one angle.

Step 3 **Write an equation.** The sum of the three measures shown in the figure must be 180°.

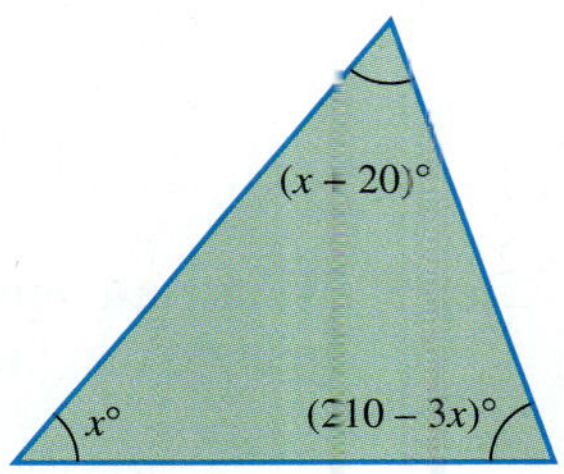

Figure 7

$$x + (x + 20) + (210 - 3x) = 180$$

Step 4 **Solve.**

$$-x + 230 = 180 \quad \text{Combine like terms.}$$
$$-x = -50 \quad \text{Subtract 230.}$$
$$x = 50 \quad \text{Divide by } -1.$$

Step 5 **State the answer.** One angle measures 50°, another measures $x + 20 = 50 + 20 = 70°$, and the third measures $210 - 3x = 210 - 3(50) = 60°$.

Step 6 **Check.** Since $50° + 70° + 60° = 180°$, the answer is correct.

◀ *Work Problem* **4** *at the Side.*

ANSWERS

3. $\frac{3}{4}$ hr, or 45 min
4. 35°, 50°, and 95°

2.4 Exercises

FOR EXTRA HELP

MyMathLab | Math XL PRACTICE | WATCH | DOWNLOAD | READ | REVIEW

Solve each problem.

1. What amount of money is found in a piggy bank containing 38 nickels and 26 dimes?

2. The distance between Cape Town, South Africa, and Miami is 7700 mi. If a jet averages 480 mph between the two cities, what is its travel time in hours?

3. Tri Phong traveled from Denver to Pittsburgh, a distance of 1320 mi, in 24 hr. What was his rate in miles per hour?

4. A square has perimeter 40 in. What would be the perimeter of an equilateral triangle whose sides each measure the same length as the side of the square?

Solve each problem. See Example 1.

5. Otis Taylor has a box of coins that he uses when playing poker with his friends. The box currently contains 44 coins, consisting of pennies, dimes, and quarters. The number of pennies is equal to the number of dimes, and the total value is \$4.37. How many of each denomination of coin does he have in the box?

Number of Coins	Denomination	Value
x	0.01	$0.01x$
x		
	0.25	
		4.37 ← Total

6. Nana Nantambu found some coins while looking under her sofa pillows. There were equal numbers of nickels and quarters, and twice as many half-dollars as quarters. If she found \$2.60 in all, how many of each denomination of coin did she find?

Number of Coins	Denomination	Value
x	0.05	$0.05x$
x		
$2x$	0.50	
		2.60 ← Total

7. In Canada, \$1 and \$2 bills have been replaced by coins. The \$1 coins are called "loonies" because they have a picture of a loon (a well-known Canadian bird) on the reverse, and the \$2 coins are called "toonies." When Marissa returned home to San Francisco from a trip to Vancouver, she found that she had acquired 37 of these coins, with a total value of 51 Canadian dollars. How many coins of each denomination did she have?

8. Luke Corey works at an ice cream shop. At the end of his shift, he counted the bills in his cash drawer and found 119 bills with a total value of \$347. If all of the bills are \$5 bills and \$1 bills, how many of each denomination were in his cash drawer?

9. Dave Bowers collects U.S. gold coins. He has a collection of 53 coins. Some are \$10 coins, and the rest are \$20 coins. If the face value of the coins is \$780, how many of each denomination does he have?

10. In the 19th century, the United States minted two-cent and three-cent pieces. Frances Steib has three times as many three-cent pieces as two-cent pieces, and the face value of these coins is \$2.42. How many of each denomination does she have?

11. In 2008, general admission to the Field Museum in Chicago cost \$14 for adults and \$11 for children and seniors. If \$24,726 was collected from the sale of 2010 general admission tickets, how many adult tickets were sold? (*Source:* www.fieldmuseum.org)

12. For a high school production of *West Side Story,* student tickets cost \$5 each while nonstudent tickets cost \$8. If 480 tickets were sold for the Saturday night show and a total of \$2895 was collected, how many tickets of each type were sold?

In Exercises 13–16, find the rate based on the information provided. Round your answers to the nearest hundredth. All events were at the 2004 Summer Olympics in Athens, Greece. (Source: World Almanac and Book of Facts.)

	Event	Participant	Distance	Time
13.	100-m hurdles, women	Joanna Hayes, USA	100 m	12.37 sec
14.	400-m hurdles, women	Fani Halkia, Greece	400 m	52.82 sec
15.	400-m hurdles, men	Felix Sánchez, DO	400 m	47.63 sec
16.	400-m run, men	Jeremy Wariner, USA	400 m	44.00 sec

Solve each problem. See Examples 2 and 3.

17. Two steamers leave a port on a river at the same time, traveling in opposite directions. Each is traveling 22 mph. How long will it take for them to be 110 mi apart?

	Rate	Time	Distance
First Steamer		t	
Second Steamer	22		
			110

18. A train leaves Dayton, Ohio, and travels north at 85 km per hr. Another train leaves at the same time and travels south at 95 km per hr. How long will it take before they are 315 km apart?

	Rate	Time	Distance
First Train	85	t	
Second Train			
			315

19. Agents Mulder and Scully are driving to Georgia to investigate "Big Blue," a giant aquatic reptile reported to inhabit one of the local lakes. Mulder leaves Washington at 8:30 A.M. and averages 65 mph. His partner, Scully, leaves at 9:00 A.M., following the same path and averaging 68 mph. At what time will Scully catch up with Mulder?

20. Lois and Clark are covering separate stories and have to travel in opposite directions. Lois leaves the *Daily Planet* at 8:00 A.M. and travels at 35 mph. Clark leaves at 8:15 A.M. and travels at 40 mph. At what time will they be 140 mi apart?

21. It took Charmaine 3.6 hr to drive to her mother's house on Saturday morning for a weekend visit. On her return trip on Sunday night, traffic was heavier, so the trip took her 4 hr. Her average speed on Sunday was 5 mph slower than on Saturday. What was her average speed on Sunday?

22. Sarah Kueffer commutes to her office in Redwood City, California, by train. When she walks to the train station, it takes her 40 min. When she rides her bike, it takes her 12 min. Her average walking speed is 7 mph less than her average biking speed. Find the distance from Sarah's house to the train station.

23. Johnny leaves Memphis to visit his cousin, Anne Hoffman, in the town of Hornsby, Tennessee, 80 mi away. He travels at an average speed of 50 mph. One-half hour later, Anne leaves to visit Johnny, traveling at an average speed of 60 mph. How long after Anne leaves will it be before they meet?

24. On an automobile trip, Heather Dowdell maintained a steady speed for the first two hours. Rush-hour traffic slowed her speed by 25 mph for the last part of the trip. The entire trip, a distance of 125 mi, took $2\frac{1}{2}$ hr. What was her speed during the first part of the trip?

Find the measure of each angle in the triangles shown. See Example 4.

25.

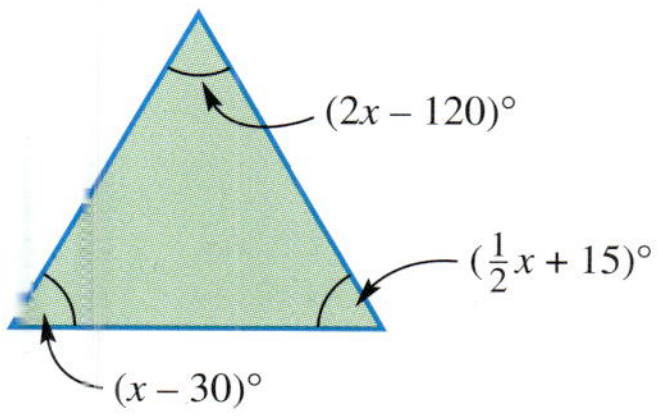

26.

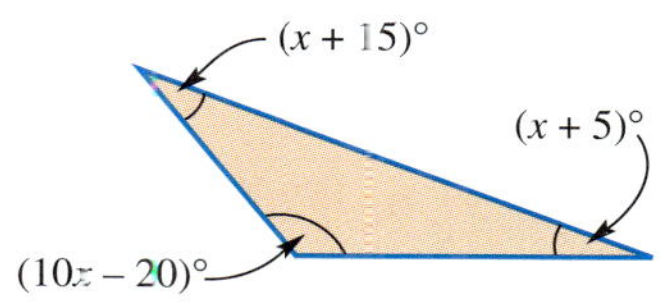

27.

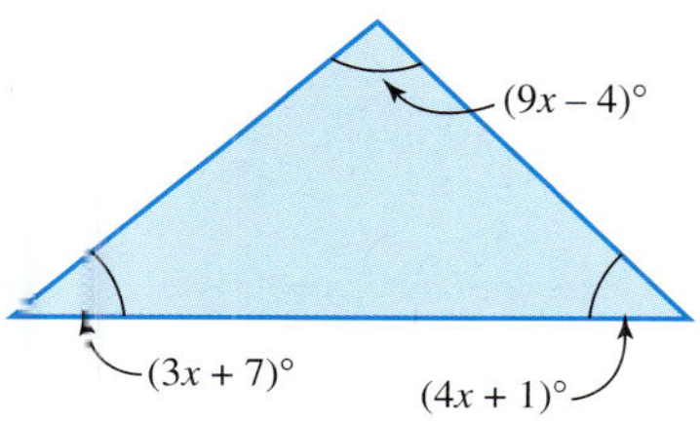

28.

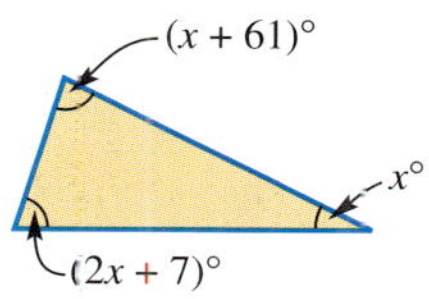

Relating Concepts (Exercises 29–32) For Individual or Group Work

Consider the following two figures. ***Work Exercises 29–32 in order.***

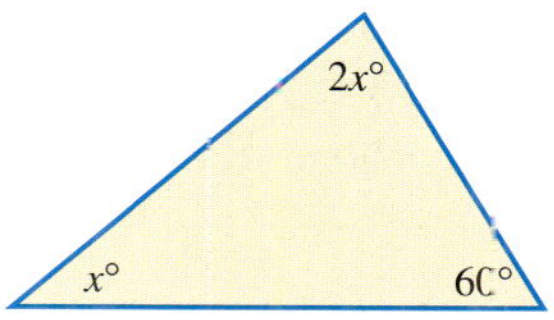

Figure A

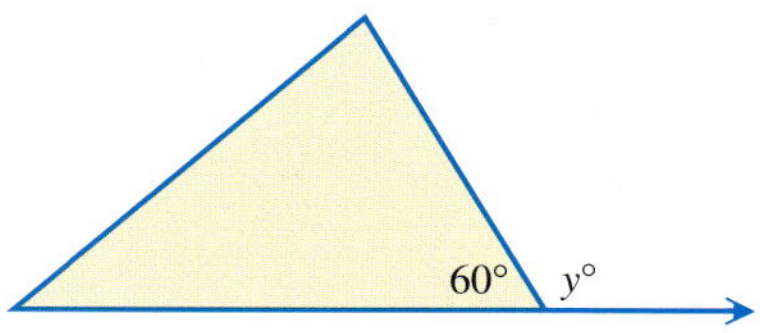

Figure B

29. Solve for the measures of the unknown angles in Figure A.

30. Solve for the measure of the unknown angle marked $y°$ in Figure B.

31. Add the measures of the two angles you found in Exercise 29. How does the sum compare to the measure of the angle you found in Exercise 30?

32. From Exercises 29–31, make a conjecture (an educated guess) about the relationship among the angles marked ①, ②, and ③ in the figure shown here.

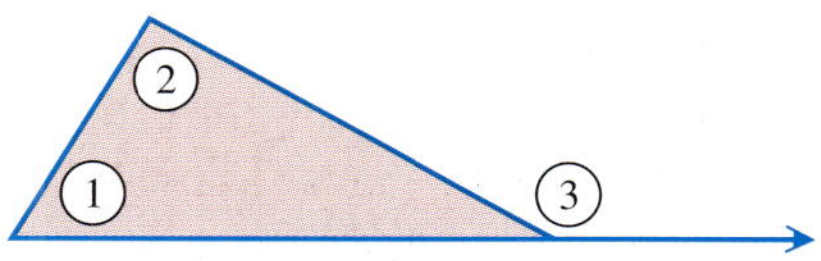

In Exercises 33 and 34, the angles marked with variable expressions are called ***vertical angles.*** *It is shown in geometry that vertical angles have equal measures. Find the measure of each angle.*

33.

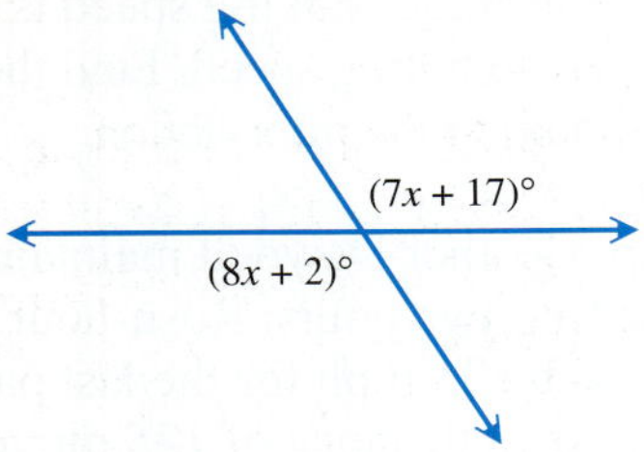

34.

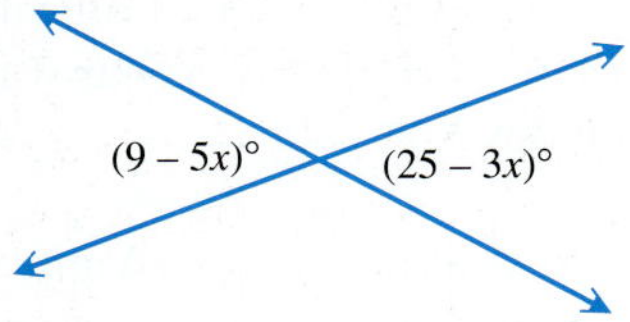

35. Two angles whose sum is equal to 90° are called **complementary angles.** Find the measures of the complementary angles shown in the figure.

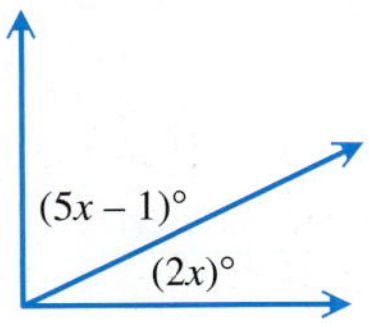

36. Two angles whose sum is equal to 180° are called **supplementary angles.** Find the measures of the supplementary angles shown in the figure.

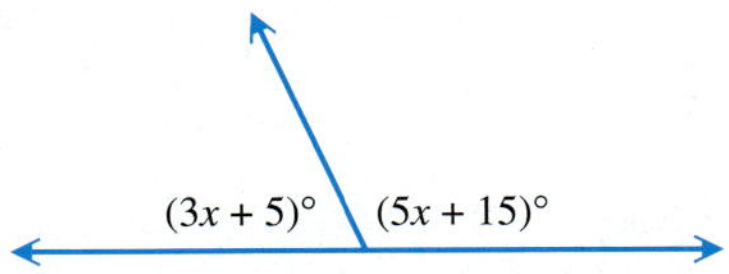

Another type of application often studied in algebra courses involves *consecutive integers*. **Consecutive integers** are integers that follow each other in counting order, such as 8, 9, and 10. Suppose we wish to solve the following problem:

Find three consecutive integers such that the sum of the first and third, increased by 3, *is* 50 *more than the second.*

Let x represent the first of the unknown integers. Then $x + 1 =$ the second, and $x + 2 =$ the third. The equation to solve would be

Sum of the first and third	increased by 3	is	50 more than the second.
↓	↓	↓	↓
$x + (x + 2)$	$+ 3$	$=$	$(x + 1) + 50.$

$$2x + 5 = x + 51$$

$$x = 46$$

The solution of this equation is 46, meaning that the first integer is $x = 46$, the second is $46 + 1 = 47$, and the third is $46 + 2 = 48$. The three integers are 46, 47, and 48. Check by substituting these numbers back into the words of the original problem.

Solve each problem involving consecutive integers.

37. Find three consecutive integers such that the sum of the first and twice the second is 22 more than twice the third.

38. Find four consecutive integers such that the sum of the first three is 62 more than the fourth.

39. If I add my current age to the age I will be next year on this date, the sum is 95 yr. How old will I be 10 years from today?

40. Two pages facing each other in this book have 365 as the sum of their page numbers. What are the two page numbers?

Summary Exercises on Solving Applied Problems

The applications that follow are of the various types introduced in this chapter. Use the strategies you have developed to solve each problem.

1. The length of a rectangle is 3 in. more than its width. If the length were decreased by 2 in. and the width were increased by 1 in., the perimeter of the resulting rectangle would be 24 in. Find the dimensions of the original rectangle.

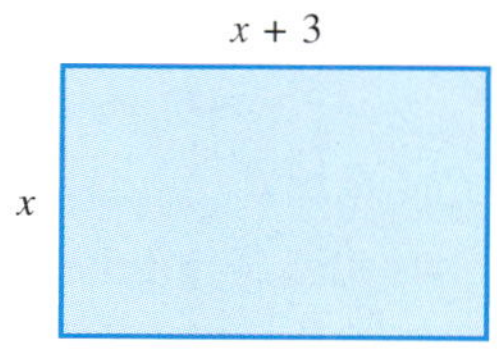

2. The perimeter of a triangle is 34 in. The middle side is twice as long as the shortest side. The longest side is 2 in. less than three times the shortest side. Find the lengths of the three sides.

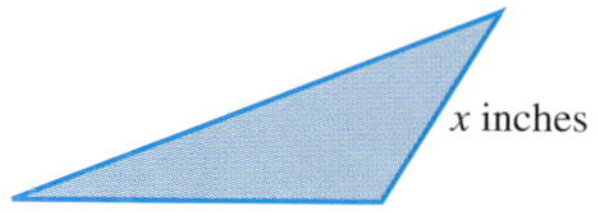

3. After a discount of 37%, the sale price for a *Harry Potter* Paperback Boxed Set (Books 1–6) by J. K. Rowling was \$35.87. What was the regular price of the set of books to the nearest cent? (*Source:* www.amazon.com)

4. An electronics store offered a DVD recorder for \$255. This was the sale price, after the regular price had been discounted 40%. What was the regular price?

5. Ceci Fleming invested an amount of money at 4% annual simple interest and twice that amount at 5%. The total annual interest is \$77. How much did she invest at each rate?

6. Meredith Ruhberg invested an amount of money at 3% annual simple interest, and \$3000 more than that amount at 4%. The total annual interest is \$960. How much did she invest at each rate?

7. Kobe Bryant of the Los Angeles Lakers was the leading scorer in the NBA for both the 2005–2006 and 2006–2007 seasons. He scored a total of 5262 points during these two seasons and scored 402 fewer points in 2006–2007 than in 2005–2006. How many points did he score in each season? (*Source: World Almanac and Book of Facts.*)

8. Two of the top-grossing American movies of the early 21st century were *Shrek 2* and *Spider-Man.* In the United States, *Shrek 2* grossed \$29.5 million more than *Spider-Man,* and together the two films brought in \$844.9 million. How much did each movie gross? (*Source:* www.movieweb.com)

9. Joshua Rogers has a sheet of tin 12 cm by 16 cm. He plans to make a box by cutting equal squares out of each of the four corners and folding up the remaining edges. How large a square should he cut so that the finished box will have a length that is 5 cm less than twice the width?)

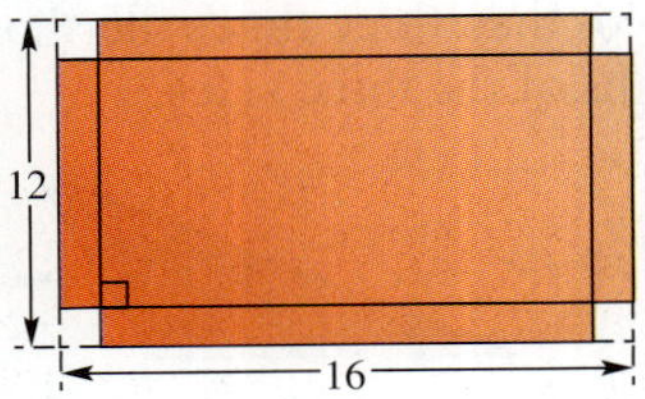

10. Atlanta and Cincinnati are 440 mi apart. John leaves Cincinnati, driving toward Atlanta at an average speed of 60 mph. Pat leaves Atlanta at the same time, driving toward Cincinnati in her antique auto, averaging 28 mph. How long will it take them to meet?

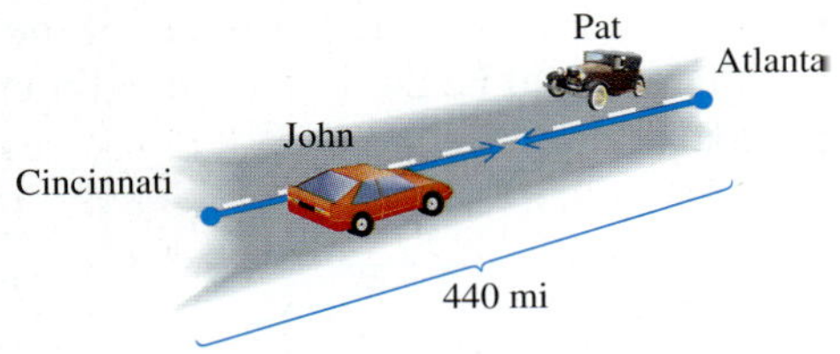

11. A pharmacist has 20 L of a 10% drug solution. How many liters of 5% solution must be added to get a mixture that is 8% drug?

12. A certain metal is 20% tin. How many kilograms of this metal must be mixed with 80 kg of a metal that is 70% tin to get a metal that is 50% tin?

13. A cashier has a total of 126 bills in fives and tens. The total value of the money is $840. How many of each type of bill does he have?

14. A newspaper recycling collection bin is in the shape of a box, 1.5 ft wide and 5 ft long. If the volume of the bin is 75 ft^3, find the height.

15. The sum of the least and greatest of three consecutive integers is 45 more than the middle integer. What are the three integers?

16. If the lesser of two consecutive odd integers is doubled, the result is 7 more than the greater of the two integers. Find the two integers.

17. Find the measure of each angle.

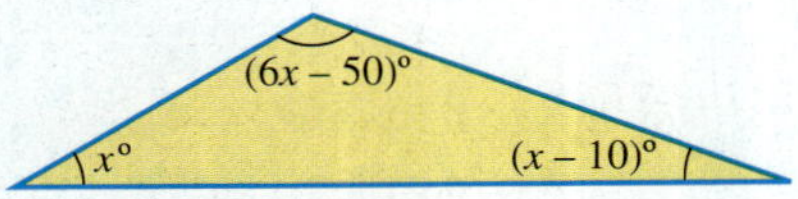

18. Find the measure of each marked angle.

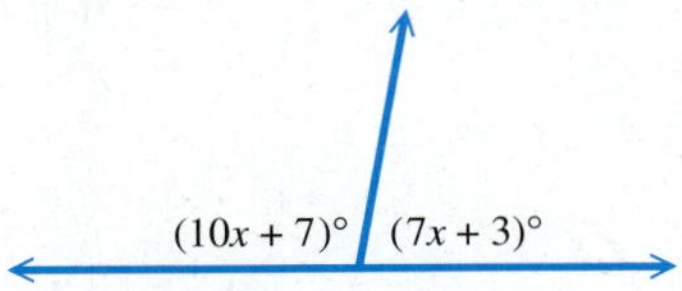

Study Skills

REVIEWING A CHAPTER

This activity is really about preparing for tests. Some of the suggestions are ones that you will learn to use a little later in the term, but it is a good idea to start trying them out now.

- **Make a study card for each key term and concept.** Include a definition, an example, a sketch, and a page reference. Include a symbol or formula if there is one. See the *Using Study Cards* activity for a quick look at some sample study cards.

- **Go back to the section** to find more explanations or information about any new vocabulary, formulas, or symbols.

- **Use the Chapter Summary** to practice each type of problem. Do not expect the Summary to substitute for reading and working through the whole chapter. First, take the "Test Your Word Power" quiz to check your understanding of new vocabulary. The answers are at the end of the Quick Review. Then read the Quick Review. **Pay special attention to the headings.** Study the explanations and examples given for each concept. **Try to think about the whole chapter.**

- **Reread your lecture notes** to see what your instructor has emphasized in class. Then review that material in your text.

- **Do the Review Exercises.**
  - ✓ Check your answers after you're done with each section of exercises.
  - ✓ If you get stuck on a problem, first check the Chapter Summary. If that doesn't clear up your confusion, then check the section and your lecture notes.
  - ✓ Pay attention to direction words for the problems, such as *simplify, round, solve,* and *estimate*.
  - ✓ Make study cards for especially difficult problems.

- **Do the Mixed Review Exercises.** This is a good check to see if you can still do the problems when they are in mixed-up order. **Check your answers carefully** in the answer section at the back of your book. Are your answers exact and complete? Make sure you are labeling answers correctly, using the right units. For example, does your answer need to include \$, cm^2, ft, and so on?

OBJECTIVES

1. **Use the Chapter Summary to practice every type of problem.**
2. **Create study cards for vocabulary.**
3. **Practice by doing review and mixed review exercises.**
4. **Take the Chapter Test as a practice test.**

Use These Chapter Reviewing Techniques

Why Are These Activities Brain Friendly?

You have already become familiar with the features of your textbook. This activity requires you to make good use of them. Your *brain needs repetition* in order to learn. By following the steps outlined here, you will be reinforcing the concepts, procedures, and skills you need to use for tests (and for the next chapters).

This combination of techniques also provides repetition in different ways. A thorough review of each chapter will help you be sure that you understand the concepts *completely and accurately*. Also, taking the Chapter Test will *simulate the testing situation*, which gives you practice in test taking conditions.

- ▶ **Take the Chapter Test as if it were a real test.** If your instructor has skipped sections in the chapter, figure out which problems to skip on the test before you start.
  - ✓ **Time yourself** just as you would for a real test.
  - ✓ **Use a calculator or notes** just as you would be permitted to (or not) on a real test.
  - ✓ **Take the test in one sitting,** just like a real test is given in one sitting.
  - ✓ **Show all your work.** Practice showing your work just the way your instructor has asked you to show it.
  - ✓ **Practice neatness.** Can someone else follow your steps?
  - ✓ **Check your answers** in the back of the book.

Notice that reviewing a chapter will take some time. You cannot fully learn the material by rushing through a review in one night. But if you use the suggestions over a few days or evenings, you will notice that you understand the material more thoroughly and remember it longer.

Now Try This ▶▶▶

Follow the reviewing techniques listed above for your next test. For each technique, write a comment *about how it worked for you in the spaces below.*

1. **Make a study card for each vocabulary word and concept.**
2. **Go back to the section** to find more explanations or information.
3. **Take the Word Power Quiz and use the Quick Review** to review each concept in the chapter.
4. **Study your lecture notes** to see what your instructor has emphasized in class.
5. **Do the Review Exercises,** following the specific suggestions on the previous page.
6. **Do the Mixed Review Exercises.**
7. **Take the Chapter Test** as if it were a real test.

Chapter 2 ▶▶▶ Summary

▶ Key Terms

	Term	Definition
2.1	**linear (first-degree) equation in one variable**	A linear equation in one variable can be written in the form $Ax + B = C$, where A, B, and C are real numbers, with $A \neq 0$.
	solution	A solution of an equation is a number that makes the equation true when substituted for the variable.
	solution set	The solution set of an equation is the set of all its solutions.
	equivalent equations	Equivalent equations are equations that have the same solution set.
	conditional equation	An equation that is true only for certain value(s) of the variable is called a conditional equation.
	contradiction	An equation that has no solution (that is, its solution set is $\emptyset$) is called a contradiction.
	identity	An equation that is satisfied by every valid replacement of the variable is called an identity.
2.2	**mathematical model**	A mathematical model is an equation or inequality that describes a real situation.
	formula	A formula is an equation in which variables are used to describe a relationship.
	percent	One percent (1%) means "one per hundred."
2.4	**vertical angles**	Angles ① and ② shown in the figure are called vertical angles. They have equal measures.
	complementary angles	Two angles whose sum is 90° are called complementary angles.
	supplementary angles	Two angles whose sum is 180° are called supplementary angles.
	consecutive integers	Two integers that differ by one are called consecutive integers.

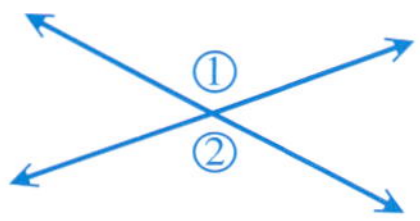

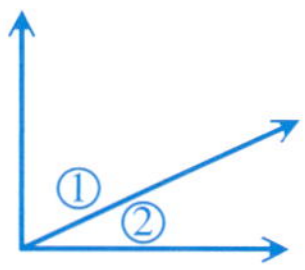

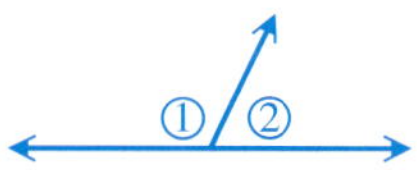

▶ Test Your Word Power

See how well you have learned the vocabulary in this chapter. Answers, with examples, follow the Quick Review.

1. An **algebraic expression** is
- **A.** an expression that uses any of the four basic operations or the operations of raising to powers or taking roots on any collection of variables and numbers
- **B.** an expression that contains fractions
- **C.** an equation that uses any of the four basic operations or the operation of taking roots on any collection of variables and numbers
- **D.** an equation in algebra.

2. An **equation** is
- **A.** an algebraic expression
- **B.** an expression that contains fractions
- **C.** an expression that uses any of the four basic operations or the operations of raising to powers or taking roots on any collection of variables and numbers
- **D.** a statement that two algebraic expressions are equal.

3. A **solution set** is the set of numbers that
- **A.** make an expression undefined
- **B.** make an equation false
- **C.** make an equation true
- **D.** make an expression equal to 0.

▸ Quick Review

Concepts	Examples

2.1 Linear Equations in One Variable

Addition and Multiplication Properties of Equality
The same number may be added to (or subtracted from) each side of an equation to obtain an equivalent equation. Similarly, the same nonzero number may be multiplied by or divided into each side of an equation to obtain an equivalent equation.

Solving a Linear Equation in One Variable

Step 1 Clear fractions.

Step 2 Simplify each side separately.

Step 3 Isolate the variable terms on one side.

Step 4 Isolate the variable.

Step 5 Check.

Solve the equation.

$$\begin{aligned} 4(8 - 3t) &= 32 - 8(t + 2) \\ 32 - 12t &= 32 - 8t - 16 && \text{Distributive property} \\ 32 - 12t &= 16 - 8t \\ 32 - 12t + 12t &= 16 - 8t + 12t && \text{Add } 12t. \\ 32 &= 16 + 4t \\ 32 - 16 &= 16 + 4t - 16 && \text{Subtract 16.} \\ 16 &= 4t \\ \frac{16}{4} &= \frac{4t}{4} && \text{Divide by 4.} \\ 4 &= t \end{aligned}$$

The solution set is $\{4\}$. This can be checked by substituting 4 for t in the original equation.

2.2 Formulas

Solving a Formula for a Specified Variable (Solving a Literal Equation)

Step 1 If the equation contains fractions, multiply both sides by the LCD to clear the fractions.

Step 2 Transform so that all terms with the specified variable are on one side and all terms without that variable are on the other side.

Step 3 Divide each side by the factor that is the coefficient of the specified variable.

Solve for h: $A = \frac{1}{2}bh$.

$$\begin{aligned} A &= \frac{1}{2}bh \\ 2A &= 2\left(\frac{1}{2}bh\right) && \text{Multiply by 2.} \\ 2A &= bh \\ \frac{2A}{b} &= h, \quad \text{or} \quad h = \frac{2A}{b} && \text{Divide by } b. \end{aligned}$$

2.3 Applications of Linear Equations

Solving an Applied Problem

Step 1 Read the problem.

Step 2 Assign a variable.

How many liters of 30% alcohol solution and 80% alcohol solution must be mixed to obtain 100 L of 50% alcohol solution?

Let x = number of liters of 30% solution needed;
then $100 - x$ = number of liters of 80% solution needed.

Liters of Solution	Percent (as a decimal)	Liters of Pure Alcohol
x	0.30	$0.30x$
$100 - x$	0.80	$0.80(100 - x)$
100	0.50	$0.50(100)$

Concepts	Examples

2.3 Applications of Linear Equations *(continued)*

Step 3 Write an equation.

The equation is

$$0.30x + 0.80(100 - x) = 0.50(100).$$

Step 4 Solve the equation.

The solution of the equation is 60.

Step 5 State the answer.

60 L of 30% solution and $100 - 60 = 40$ L of 80% solution are needed.

Step 6 Check.

$$0.30(60) + 0.80(100 - 60) = 50 \text{ is true.}$$

2.4 Further Applications of Linear Equations

To solve a uniform motion problem, draw a sketch and make a table. Use the formula

$$d = rt.$$

Two cars start from towns 400 mi apart and travel toward each other. They meet after 4 hr. Find the speed of each car if one travels 20 mph faster than the other.

Let x = speed of the slower car in miles per hour;
then $x + 20$ = speed of the faster car.

Use the information in the problem and $d = rt$ to complete the table.

	Rate	Time	Distance	
Slower Car	x	4	$4x$	
Faster Car	$x + 20$	4	$4(x + 20)$	
			400	← Total

A sketch shows that the sum of the distances, $4x$ and $4(x + 20)$, must be 400.

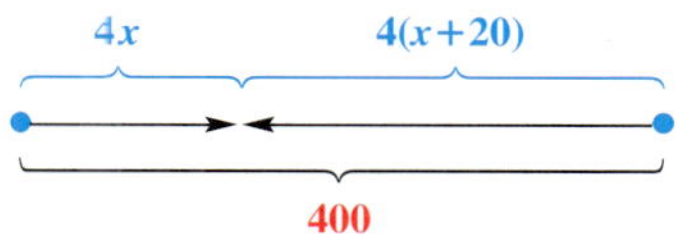

The equation is

$$4x + 4(x + 20) = 400.$$

Solving this equation gives $x = 40$. The slower car travels 40 mph, and the faster car travels $40 + 20 = 60$ mph.

Problems involving denominations of money and mixture problems are solved using methods similar to the one used for the mixture problem shown in the example above for **Section 2.3.**

ANSWERS TO TEST YOUR WORD POWER

1. A; *Examples:* $\frac{3y - 1}{2}, 6 + \sqrt{2x}, 4a^3b - c$
2. D; *Examples:* $2a + 3 = 7$; $3y = -8$, $x^2 = 4$
3. C; *Example:* $\{8\}$ is the solution set of $2x + 5 = 21$.

Math in the Media

LEARN MATH, LOSE WEIGHT

The 2004 movie *Mean Girls* stars Lindsay Lohan as Cady Heron, a teenage girl who has been home-schooled until her senior year in high school. A scene in the school cafeteria features her sitting with the Plastics (the "mean girls" of the title). Regina George, played by Rachel McAdams, is reading a candy bar wrapper.

REGINA: *120 calories and 48 calories from fat. What percent is that? I'm only eating food with less than 30% calories from fat.*

CADY: *It's 40%. (Responding to a quizzical look from Regina.) Well, 48 over 120 equals x over 100, and then you cross-multiply and get the value of x.*

REGINA: *Whatever . . . I'm getting cheese fries.*

1. Show that Cady's answer is correct. Let x represent the percent, set up the equation, and solve it. Show all steps.

Use Cady's method to find the percent calories from fat for each of the following candy bars. Round to the nearest percent.

2. 2.05 oz Milky Way: 260 calories, with 90 calories from fat

3. 1.61 oz Almond Joy: 220 calories, with 120 calories from fat

4. 1.85 oz PayDay: 250 calories, with 120 calories from fat

5. 2.1 oz Butterfinger: 270 calories, with 100 calories from fat

6. 1.76 oz Snickers: 230 calories, with 100 calories from fat

Chapter 2 Review Exercises

[2.1] *Solve each equation.*

1. $-(8 + 3x) + 5 = 2x + 6$

2. $-(r + 5) - (2 + 7r) + 8r = 3r - 8$

3. $\dfrac{m - 2}{4} + \dfrac{m + 2}{2} = 8$

4. $\dfrac{2q + 1}{3} - \dfrac{q - 1}{4} = 0$

5. $5(2x - 3) = 6(x - 1) + 4x$

6. $-3x + 2(4x + 5) = 10$

7. $\dfrac{1}{2}x - \dfrac{3}{8}x = \dfrac{1}{4}x + 2$

8. $0.05x + 0.03(1200 - x) = 42$

9. Which equation has $\{0\}$ as its solution set?

A. $x - 7 = 7$ **B.** $9x = 10x$

C. $x + 4 = -4$ **D.** $8x - 8 = 8$

10. Give the steps you would use to solve the equation $-2x + 5 = 7$.

Decide whether each equation is a conditional equation, *an* identity, *or a* contradiction. *Give the solution set.*

11. $7r - 3(2r - 5) + 5 + 3r = 4r + 20$

12. $8p - 4p - (p - 7) + 9p + 13 = 12p$

13. $-2r + 6(r - 1) + 3r - (4 - r) = -(r + 5) - 5$

[2.2] *Solve each formula for the indicated variable.*

14. $V = LWH$ for L

15. $A = \dfrac{1}{2}h(b + B)$ for b

16. $4x + 7y = 9$ for y

Solve each problem.

17. A rectangular solid has a volume of 180 ft^3. Its length is 9 ft and its width is 4 ft. Find its height.

18. The number of students attending college in the United States in 2005 was 15.3 million. In 2006, this number had increased to 17.5 million. To the nearest tenth, what was the percent increase? (*Source:* U.S. Census Bureau.)

19. Find the simple interest rate that Francis Castellucio is earning, if a principal of $30,000 earns $6600 interest in 4 yr.

20. If the Fahrenheit temperature is 77°, what is the corresponding Celsius temperature?

21. For 2005, total U.S. government spending was about $2500 billion (or $2.5 trillion). The circle graph shows how the spending was divided.

(a) About how much was spent on Social Security?

(b) About how much did the U.S. government spend on education and social services in 2005?

2005 U.S. Government Spending

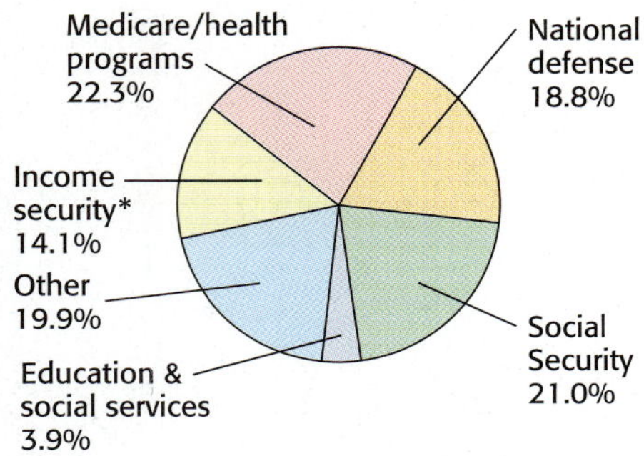

*Includes pensions for government workers, unemployment compensation, food stamps, and other such programs.

Source: U.S. Office of Management and Budget.

22. The drum that Wade purchased has a circumference of 200π mm. Find the measure of its radius.

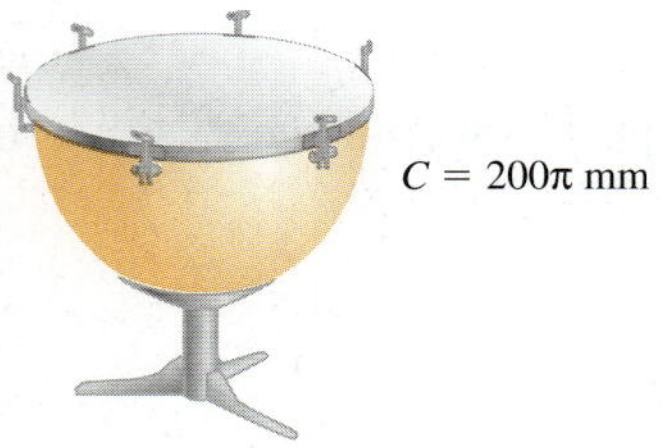

[2.3] *Write each phrase as a mathematical expression, using x as the variable.*

23. One-fifth of a number, subtracted from 14

24. The product of 6 and a number, divided by 3 more than the number

Solve each problem.

25. The length of a rectangle is 3 m less than twice the width. The perimeter of the rectangle is 42 m. Find the length and width of the rectangle.

26. In a triangle with two sides of equal length, the third side measures 15 in. less than the sum of the two equal sides. The perimeter of the triangle is 53 in. Find the lengths of the three sides.

27. A candy clerk has three times as many kilograms of chocolate creams as peanut clusters. The clerk has 48 kg of the two candies altogether. How many kilograms of peanut clusters does the clerk have?

28. How many liters of a 20% solution of a chemical should be mixed with 15 L of a 50% solution to get a 30% mixture?

29. How much water should be added to 30 L of a 40% acid solution to reduce it to a 30% solution?

Liters of Solution	Percent (as a decimal)	Liters of Pure Acid
	0.40	
x	0	
	0.30	

30. Anna Mae Wood invested some money at 6% and \$4000 less than this amount at 4%. Find the amount invested at each rate if her total annual interest income is \$840.

Principal	Rate (as a decimal)	Interest
x	0.06	
	0.04	
XXXXX	XXXXX	

[2.4]

31. Which choice is the best *estimate* for the average speed of a trip of 405 mi that lasted 8.2 hr?

A. 50 mph

B. 30 mph

C. 60 mph

D. 40 mph

32. (a) A driver averaged 53 mph and took 10 hr to travel from Memphis to Chicago. What is the distance between Memphis and Chicago?

(b) A small plane traveled from Warsaw to Rome, averaging 164 mph. The trip took 2 hr. What is the distance from Warsaw to Rome?

33. A passenger train and a freight train leave a town at the same time and go in opposite directions. They travel at 60 mph and 75 mph, respectively. How long will it take for them to be 297 mi apart?

	Rate	Time	Distance
Passenger Train	60	x	
Freight Train	75	x	
XXXXX	XXX	XXX	

34. Two cars leave towns 230 km apart at the same time, traveling directly toward one another. One car travels 15 km per hr slower than the other. They pass one another 2 hr later. What are their speeds?

	Rate	Time	Distance
Faster Car	x	2	
Slower Car	$x - 15$	2	
XXXXX	XXX	XXX	

35. An automobile averaged 45 mph for the first part of a trip and 50 mph for the second part. If the entire trip took 4 hr and covered 195 mi, for how long was the rate 45 mph?

36. An 85-mi trip to the beach took the Valenzuela family 2 hr. During the second hour, a rainstorm caused them to average 7 mph less than they traveled during the first hour. Find their average rate for the first hour.

Mixed Review Exercises

Solve each equation.

37. $(7 - 2k) + 3(5 - 3k) = k + 8$

38. $\frac{4x + 2}{4} + \frac{3x - 1}{8} = \frac{x + 6}{16}$

39. $-5(6p + 4) - 2p = -32p + 14$

40. $0.08x + 0.04(x + 200) = 188$

41. $5(2r - 3) + 7(2 - r) = 3(r + 2) - 7$

42. $Ax + By = C$ for x

Solve each problem.

43. A square is such that if each side were increased by 4 in., the perimeter would be 8 in. less than twice the perimeter of the original square. Find the length of a side of the original square.

44. Two cars start from the same point and travel in opposite directions. The car traveling west leaves 1 hr later than the car traveling east. The eastbound car travels 40 mph, and the westbound car travels 60 mph. When they are 240 mi apart, how long had each car traveled?

45. The two most-visited sites in the National Park System in 2006 were the Blue Ridge Parkway in North Carolina and Virginia, and the Golden Gate National Recreation Area in California. There were a total of 32.44 million visits to the two sites, with 5.46 million more visits to the Blue Ridge Parkway than to the Golden Gate Recreation Area. How many visits were there to each site? (*Source:* National Park Service.)

46. Some money is invested at 3% simple annual interest and \$600 more than that amount is invested at 5%. After 1 year, a total of \$126 interest was earned. How much was invested at each rate?

Principal	Rate (as a decimal)	Interest
x	0.03	
$x + 600$	0.05	
XXXXXX	XXXXXX	126 ← Total

Chapter 2 Test

Use the Chapter Test Prep Video CD to see fully worked-out solutions to any of the exercises you want to review

Solve each equation.

1. $3(2x - 2) - 4(x + 6) = 4x + 8$ — 1. ______

2. $0.08x + 0.06(x + 9) = 1.24$ — 2. ______

3. $\frac{x + 6}{10} + \frac{x - 4}{15} = 1$ — 3. ______

4. Decide whether each equation is a *conditional equation,* an *identity,* or a *contradiction*. Give its solution set.

   (a) $3x - (2 - x) + 4x + 2 = 8x + 3$ — 4. (a) ______

   (b) $\frac{x}{3} + 7 = \frac{5x}{6} - 2 - \frac{x}{2} + 9$ — (b) ______

   (c) $-4(2x - 6) = 5x + 24 - 7x$ — (c) ______

5. Solve for v: $S = -16t^2 + vt$. — 5. ______

6. Solve for y: $-3x + 2y = 6$. — 6. ______

Solve each problem.

7. The 2007 Daytona 500 (mile) race was won by Kevin Harvik, who averaged 149.335 mph. What was Harvik's time, to the nearest thousandth? (*Source:* www.daytona500.com) — 7. ______

8. A certificate of deposit pays \$1733.75 in simple interest for 1 year on a principal of \$36,500. What is the rate of interest? — 8. ______

9. Of the 36,826 offices, stations, and branches of the U.S. Postal Service in 2006, only 27,318 were actually classified as post offices. What percent to the nearest tenth were classified as post offices? (*Source:* U.S. Postal Service.) — 9. ______

10. ______________________

11. ______________________

12. ______________________

13. ______________________

14. ______________________

15. ______________________

10. Tyler McGinnis invested some money at 3% simple interest and some at 5% simple interest. The total amount of his investments was \$32,000, and the interest he earned during the first year was \$1320. How much did he invest at each rate?

11. Two cars leave from the same point at the same time, traveling in opposite directions. One travels 15 mph slower than the other. After 6 hr, they are 630 mi apart. Find the rate of each car.

12. Find the measure of each angle.

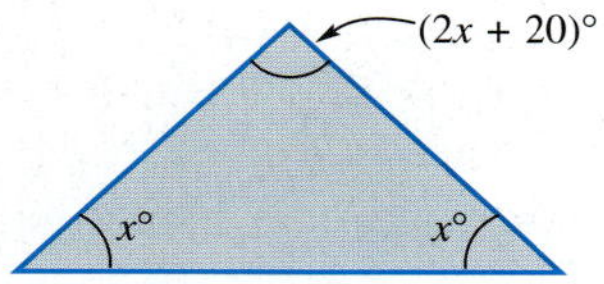

The formula

$$A = \frac{24f}{b(p + 1)}$$

gives the approximate annual interest rate for a consumer loan paid off with monthly payments. Here f is the finance charge on the loan, p is the number of payments, and b is the original amount of the loan. Use this formula to solve Problems 13 and 14.

13. Find the approximate annual interest rate for an installment loan to be repaid in 24 monthly installments. The finance charge on the loan is \$200, and the original loan balance is \$1920.

14. Find the approximate annual interest rate for an automobile loan to be repaid in 36 monthly installments. The finance charge on the loan is \$740, and the amount financed is \$3600. (Round to the nearest hundredth of a percent.)

15. The circle graph shows the percents of various occupations in a representative sample of stockholders. Based on the figure, in a group of 5000 stockholders, how many would you expect to be white-collar workers?

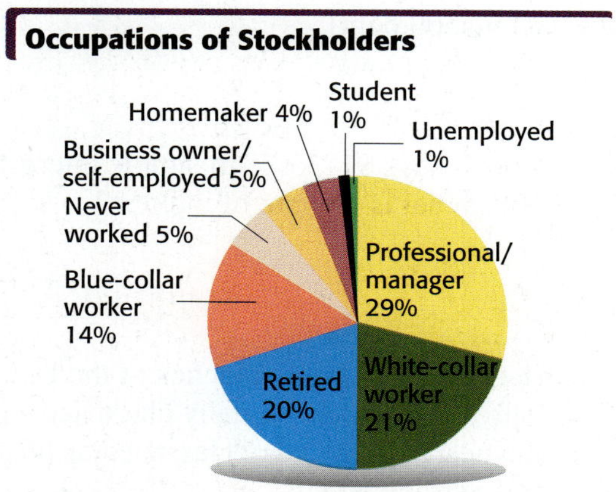

Source: Study by Peter D. Hart Research Associates for the Nasdaq Stock Market.

Study Skills

PREPARING FOR TESTS

Many activities besides studying can improve your test scores. You may not realize that eating the right foods and getting enough exercise and sleep can also improve your scores. Your brain (and therefore your ability to think) is affected by the condition of your whole body. So, part of your preparation for tests includes keeping yourself in good physical shape as well as spending time on the actual course material. Try these suggestions and see the difference.

OBJECTIVES

1. **Restate the importance of sleep and good nutrition as it affects learning.**
2. **Explain the affect of anxiety and stress on learning.**

Performance Health Tips to Improve Your Test Score	Explanation
Get seven to eight hours of sleep the night before the exam. (It's helpful to get that much sleep *every* night.)	***Fatigue and exhaustion reduce efficiency.*** They also cause poor memory and recall. If you didn't sleep much the night before a test, 20 minutes of relaxation or meditation can help. (Also see the comments below about eating carbohydrates to help you sleep.)
Eat a small, high-energy meal about two hours before the test. Start the meal with a small amount of protein such as fish, chicken, or nonfat yogurt. Include carbohydrates if you like, but no high-fat foods.	Just 3 to 4 ounces of protein increases the amount of a chemical in the brain called tyrosine, which ***improves your alertness, accuracy, and motivation.*** High-fat foods dull your mind and slow down your brain.
Drink plenty of water. Don't wait until you feel thirsty; your body is already dehydrated by the time you feel it.	Research suggests that staying well hydrated improves the electrochemical communications in your brain.
Keep up with the material as it is covered in class.	***Cramming doesn't work;*** you cannot learn that quickly. ***Studying every day*** using these study skills techniques is the way to give your brain the time it needs.

Techniques to Prevent Anxiety	Explanation
Practice slow, deep breathing for five minutes each day. Then do a minute or two of deep breathing right before the test. Also, if you feel your anxiety building during the test, stop for a minute, close your eyes, and do some deep breathing.	When test anxiety hits, you breathe more quickly and shallowly, which causes hyperventilation. Symptoms may be confusion, inability to concentrate, shaking, dizziness, and more. ***Slow, deep breathing will calm you and prevent panic.***
Do 15 to 20 minutes of moderate exercise (like walking) shortly before the test. Daily exercise is even better.	***Exercise reduces stress*** and will help prevent "blanking out" on a test. Exercise also increases your alertness, clear thinking, and energy.

(*continued*)

Techniques to Prevent Anxiety	Explanation
To help you sleep the night before the test, or any time you need to calm down, ***eat high carbohydrate foods*** such as popcorn, bread, rice, crackers, muffins, bagels, pasta, corn, baked potatoes (not fries or chips), and cereals.	Carbohydrates increase the level of a chemical in the brain called serotonin, which has a ***calming effect on the mind.*** It reduces feelings of tension and stress and improves your ability to concentrate. You only need to eat a small amount, like half a bagel, to get this effect.
Before the test, ***go easy on caffeinated beverages*** such as coffee, tea, and soft drinks. Do not eat candy bars or other sugary snacks.	Extra caffeine can ***make you jittery,*** "hyper," and shaky for the test. It can increase the tendency to panic. Too much sugar causes negative emotional reactions in some people.

Now Try This ▶▶▶

What will you do to improve your next test score? List the three or four tips you think will help you the most.

1. ____________________
2. ____________________
3. ____________________
4. ____________________

What changes will you have to make in order to try the tips you chose?

____

____

____

____

____

____

See *Tips for Taking Math Tests* and *Preparing for Your Math Final Exam* for more ideas about managing anxiety. (Check the Table of Contents to find their locations.)

Cumulative Review Exercises ▶▶▶ Chapters 1–2

Let $A = \{-8, -\frac{2}{3}, -\sqrt{6}, 0, \frac{4}{5}, 9, \sqrt{36}\}$. Simplify the elements of A as necessary, and then list the elements that belong to each set listed in Exercises 1–6.

1. Natural numbers

2. Whole numbers

3. Integers

4. Rational numbers

5. Irrational numbers

6. Real numbers

Add or subtract, as indicated.

7. $-\frac{4}{3} - \left(-\frac{2}{7}\right)$

8. $|-4.2| + |5.6| - |-1.9|$

9. $(-2)^4 + (-2)^3$

10. $\sqrt{25} - \frac{\sqrt{100}}{2}$

Evaluate each expression.

11. $(-3)^5$

12. $\left(\frac{6}{7}\right)^3$

13. 4^5

14. -4^6

15. Which one of the following is not a real number: $-\sqrt{49}$ or $\sqrt{-49}$?

16. Which one of the following is undefined: $\frac{4-4}{4+4}$ or $\frac{4+4}{4-4}$?

Evaluate each expression if $a = 2$, $b = -3$, and $c = 4$.

17. $-3a + 2b - c$

18. $-2b^2 - c^2$

19. $-8(a^2 + b^3)$

20. $\frac{3a^3 - b}{4 + 3c}$

Simplify each expression.

21. $-7r + 5 - 13r + 12$

22. $-(3k + 8) - 2(4k - 7) + 3(8k + 12)$

Identify the property of real numbers illustrated in each equation.

23. $(a + b) + 8 = 8 + (a + b)$

24. $5x + 13x = (5 + 13)x$

25. $-13 + 13 = 0$

Solve each equation.

26. $-4x + 7(2x + 3) = 7x + 36$

27. $-\frac{3}{5}x + \frac{2}{3}x = 2$

28. $0.06x + 0.03(100 + x) = 4.35$

29. $P = a + b + c$ for c

30. $4(2x - 6) + 3(x - 2) = 11x + 1$

31. $\frac{2}{3}x + \frac{5}{8}x = \frac{31}{24}x$

Solve each problem.

32. How much pure alcohol should be added to 7 L of 10% alcohol to increase the concentration to 30% alcohol?

33. A coin collection contains 29 coins. It consists of pennies, nickels, and quarters. The number of quarters is 4 fewer than the number of nickels, and the face value of the collection is $2.69. How many of each denomination are there in the collection?

34. Kathy Manley invested some money at 5% simple annual interest and $2000 more than that amount at 6%. Her annual interest from the two investments totaled $670. How much did she invest at each rate?

35. Jack and Jill are running in the Fresh Water Fun Run. Jack runs at 7 mph and Jill runs at 5 mph. If they start at the same time, how long will it be before Jack is $\frac{1}{4}$ mi ahead of Jill?

36. Clark's rule, a formula used in reducing drug dosage according to weight from the recommended adult dosage to a child dosage, is

$$\frac{\text{weight of child in pounds}}{150} \times \text{adult dose} = \text{child's dose}.$$

Find a child's dosage if the child weighs 55 lb and the recommended adult dosage is 120 mg.

37. The body mass index, or BMI, of a person is given by the formula

$$\text{BMI} = \frac{704 \times (\text{weight in pounds})}{(\text{height in inches})^2}.$$

Jimmy Rollins, the National League Most Valuable Player for 2007, is listed as being 5 ft, 8 in. tall and weighing 160 lb. What is his BMI (to the nearest tenth)? (*Source: Reader's Digest*, October 1993; www.mlb.com)

38. Since 1975, the number of daily newspapers has steadily declined. According to the table,

(a) by how much did the number of daily newspapers decrease between 1975 and 2005?

(b) by what *percent* to the nearest tenth did the number of daily newspapers decrease from 1975 to 2005?

Year	Number of Daily Newspapers
1975	1756
1980	1745
1985	1676
1990	1611
1995	1533
2000	1480
2005	1452

Source: Editor & Publisher Co.

3

Linear Inequalities and Absolute Value

3.1 Linear Inequalities in One Variable

3.2 Set Operations and Compound Inequalities

3.3 Absolute Value Equations and Inequalities

Summary Exercises on Solving Linear and Absolute Value Equations and Inequalities

The cost of a college education has risen rapidly in the last 15 years. Average higher education tuition and fees for students at public 2-year institutions increased 135% from the 1990–1991 school year to the 2005–2006 school year. (*Source:* National Center for Education Statistics, U.S. Department of Education.)

In Exercises 65–68 in Section 3.2, we use *set operations* and *compound inequalities* to work problems related to college student expenses.

3.1 Linear Inequalities in One Variable

OBJECTIVES

1. Graph intervals on a number line.
2. Solve linear inequalities using the addition property.
3. Solve linear inequalities using the multiplication property.
4. Solve linear inequalities with three parts.
5. Solve applied problems using linear inequalities.

Solving inequalities is closely related to solving equations. In this section we introduce properties for solving inequalities.

Inequalities are algebraic expressions related by

$<$ "is less than,"	$\leq$ "is less than or equal to,"
$>$ "is greater than,"	$\geq$ "is greater than or equal to."

We solve an inequality by finding all real number solutions for it. For example, the solution set of $x \leq 2$ includes *all* real numbers that are less than or equal to 2, not just the integers less than or equal to 2. For example, $-2.5, -1.7, -1, \frac{1}{2}, \sqrt{2}, \frac{7}{4}$, and 2 are real numbers less than or equal to 2 and are therefore solutions of $x \leq 2$.

OBJECTIVE 1 Graph intervals on a number line. A good way to show the solution set of an inequality is by graphing. We graph all the real numbers satisfying $x \leq 2$ by placing a square bracket at 2 on a number line and drawing an arrow extending from the bracket to the left (to represent the fact that all numbers less than 2 are also part of the graph). The graph is shown in Figure 1.

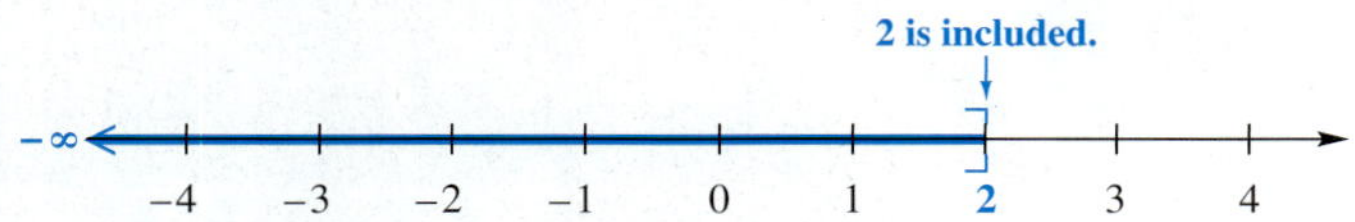

Figure 1

The set of numbers less than or equal to 2 is an example of an **interval** on the number line. To write intervals, we use **interval notation.** For example, the interval of all numbers less than or equal to 2 is written $(-\infty, 2]$. The negative infinity symbol $-\infty$ does not indicate a number. It is used to show that the interval includes all real numbers less than 2. As on the number line, the square bracket indicates that 2 is included in the solution set. ***A parenthesis is always used next to the infinity symbol. The set of real numbers is written in interval notation as*** $(-\infty, \infty)$.

EXAMPLE 1 Graphing Intervals Written in Interval Notation on Number Lines

Write each inequality in interval notation and graph it.

(a) $x > -5$

The statement $x > -5$ says that x can represent any number greater than -5, but x cannot equal -5. This interval is written $(-5, \infty)$. We show this solution set on a number line by placing a parenthesis at -5 and drawing an arrow to the right, as in Figure 2. The parenthesis at -5 shows that -5 is *not* part of the graph.

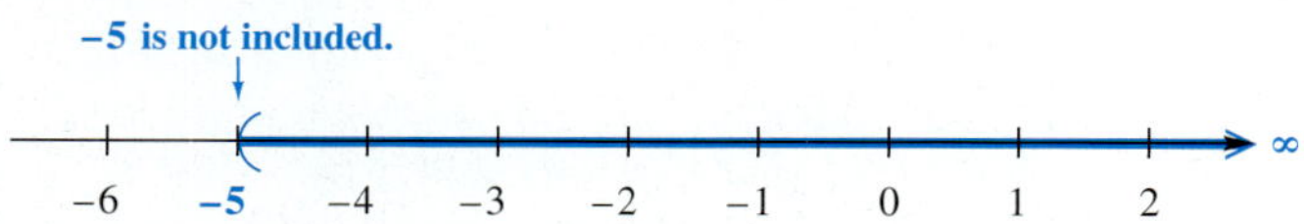

Figure 2

Continued on Next Page

(b) $-1 \leq x < 3$

This statement is read "-1 is less than or equal to x *and* x is less than 3." Thus, we want the set of numbers that are *between* -1 and 3, with -1 included and 3 excluded. In interval notation, we write the solution set as $[-1, 3)$, using a square bracket at -1 because it is part of the graph and a parenthesis at 3 because it is not part of the graph. See Figure 3.

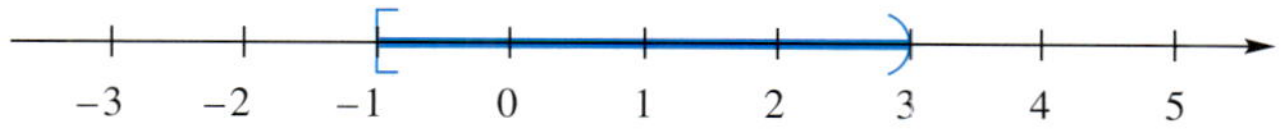

Figure 3

Work Problem **1** *at the Side.* ▶

1 Write each inequality in interval notation and graph it.

(a) $x < -1$

(b) $x \geq -3$

(c) $-4 \leq x < 2$

(d) $0 < x < 3.5$

We now summarize the various types of intervals.

Type of Interval	Set-Builder Notation	Interval Notation	Graph
Open interval	$\{x \mid a < x < b\}$	(a, b)	
Closed interval	$\{x \mid a \leq x \leq b\}$	$[a, b]$	
Half-open (or half-closed) interval	$\{x \mid a \leq x < b\}$	$[a, b)$	
	$\{x \mid a < x \leq b\}$	$(a, b]$	
Disjoint interval*	$\{x \mid x < a \text{ or } x > b\}$	$(-\infty, a) \cup (b, \infty)$	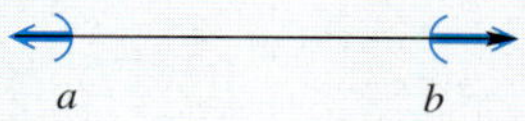
Infinite interval	$\{x \mid x > a\}$	(a, ∞)	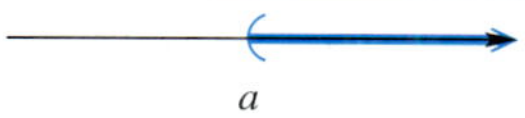
	$\{x \mid x \geq a\}$	$[a, \infty)$	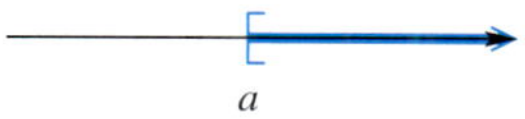
	$\{x \mid x < a\}$	$(-\infty, a)$	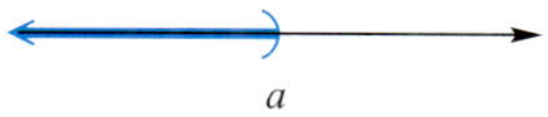
	$\{x \mid x \leq a\}$	$(-\infty, a]$	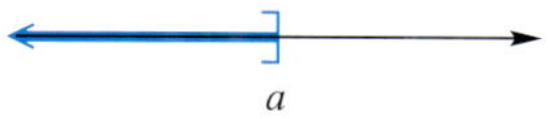
	$\{x \mid x \text{ is a real number}\}$	$(-\infty, \infty)$	

An **inequality** is a statement with algebraic expressions related by $<$, $\leq$, $>$, or $\geq$.

Linear Inequality

A **linear inequality in one variable** can be written in the form

$$Ax + B < C,$$

where A, B, and C are real numbers, with $A \neq 0$.

(While we give definitions and rules only for $<$, they are also valid for $>$, $\leq$, and $\geq$.)

*We will work with disjoint intervals in **Section 3.2** when we study *set operations* and *compound inequalities*.

Answers

1. (a) $(-\infty, -1)$

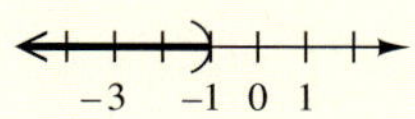

(b) $[-3, \infty)$

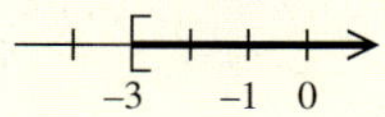

(c) $[-4, 2)$

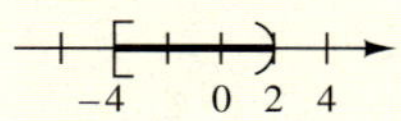

(d) $(0, 3.5)$

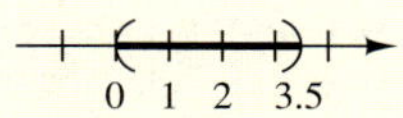

Examples of linear inequalities include

$$x + 5 < 2, \quad t - 3 \geq 5, \quad \text{and} \quad 2k + 5 \leq 10. \qquad \text{Linear inequalities}$$

OBJECTIVE 2 Solve linear inequalities using the addition property. We solve an inequality by finding all numbers that make the inequality true. Usually, an inequality has an infinite number of solutions. These solutions, like solutions of equations, are found by producing a series of simpler equivalent inequalities. **Equivalent inequalities** are inequalities with the same solution set. We use the *addition property of inequality* to produce equivalent inequalities.

Addition Property of Inequality

For all real numbers A, B, and C, the inequalities

$$A < B \quad \text{and} \quad A + C < B + C$$

are equivalent.

In words, adding the same number to each side of an inequality does not change the solution set.

EXAMPLE 2 Using the Addition Property of Inequality

Solve $x - 7 < -12$, and graph the solution set.

$$x - 7 < -12$$
$$x - 7 + 7 < -12 + 7 \qquad \text{Add 7.}$$
$$x < -5$$

Check Substitute -5 for x in the *equation* $x - 7 = -12$.

$$x - 7 = -12$$
$$-5 - 7 \stackrel{?}{=} -12 \qquad \text{Let } x = -5.$$
$$-12 = -12 \qquad \text{True}$$

The result, a true statement, shows that -5 is the boundary point. Now we test a number on each side of -5 to verify that numbers *less than* -5 make the *inequality* true. We choose -4 and -6.

$$x - 7 < -12$$

$-4 - 7 \stackrel{?}{<} -12$ Let $x = -4$.	$-6 - 7 \stackrel{?}{<} -12$ Let $x = -6$.
$-11 < -12$ False	$-13 < -12$ True
-4 is *not* in the solution set.	-6 is in the solution set.

Thus, $(-\infty, -5)$, the infinite interval graphed in Figure 4, is the solution set.

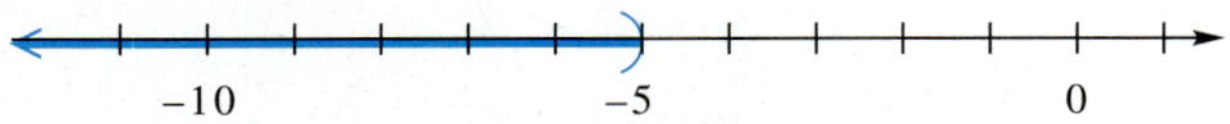

Figure 4

As with equations, the addition property of inequality can be used to *subtract* the same number from each side of an inequality. For example, to solve the inequality $x + 4 > 10$, we subtract 4 from each side to get $x > 6$.

◀ Work Problem 2 at the Side.

2 Solve each inequality, check your solutions, and graph the solution set.

(a) $x - 3 < -9$

(b) $p + 6 < 8$

ANSWERS

2. (a) $(-\infty, -6)$

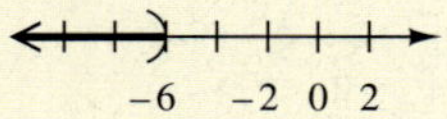

(b) $(-\infty, 2)$

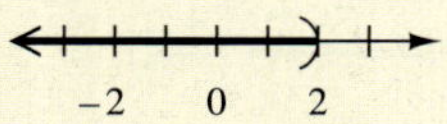

EXAMPLE 3 **Using the Addition Property of Inequality**

Solve $14 + 2m \leq 3m$, and graph the solution set.

$$14 + 2m \leq 3m$$

$$14 + 2m - 2m \leq 3m - 2m \qquad \text{Subtract } 2m.$$

Pay careful attention here.

$$14 \leq m \qquad \text{Combine like terms.}$$

$$m \geq 14 \qquad \text{Rewrite.}$$

The inequality $14 \leq m$ (14 is less than or equal to m) can also be written $m \geq 14$ (m is greater than or equal to 14). ***Notice that in each case, the inequality symbol points to the lesser number, 14.***

Check

$$14 + 2m = 3m$$

$$14 + 2(14) \stackrel{?}{=} 3(14) \qquad \text{Let } m = 14.$$

$$42 = 42 \qquad \text{True}$$

So 14 satisfies the equality part of $\leq$. Choose 10 and 15 as test points.

$$14 + 2m < 3m$$

$14 + 2(10) \stackrel{?}{<} 3(10)$ Let $m = 10$.	$14 + 2(15) \stackrel{?}{<} 3(15)$ Let $m = 15$.
$34 < 30$ False	$44 < 45$ True
10 is not in the solution set.	15 is in the solution set.

The check confirms that $[14, \infty)$ is the solution set. See Figure 5.

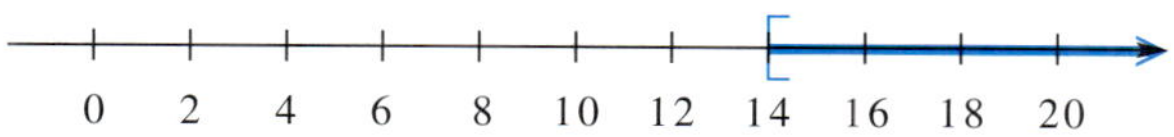

Figure 5

Work Problem **3** *at the Side.* ▶

CAUTION
To avoid errors, rewrite an inequality such as $14 \leq m$ as $m \geq 14$ so that the variable is on the left, as in Example 3.

OBJECTIVE 3 **Solve linear inequalities using the multiplication property.** Solving an inequality such as $3x \leq 15$ requires dividing each side by 3 using the *multiplication property of inequality*. Consider

$$-2 < 5,$$

a true statement. Multiply each side by 8.

$$-2(8) < 5(8) \qquad \text{Multiply by 8.}$$

$$-16 < 40 \qquad \text{True}$$

The result is true. Start again with $-2 < 5$, and multiply each side by -8.

$$-2(-8) < 5(-8) \qquad \text{Multiply by } -8.$$

$$16 < -40 \qquad \text{False}$$

The result, $16 < -40$, is false. To make it true, we must change the direction of the inequality symbol to get

$$16 > -40. \qquad \text{True}$$

Work Problem **4** *at the Side.* ▶

3 Solve $2k - 5 \geq 1 + k$, check, and graph the solution set.

4 Multiply both sides of each inequality by -5. Then insert the correct symbol, either $<$ or $>$, in the first blank, and fill in the other blanks in part (b).

(a) $7 < 8$

-35 ______ -40

(b) $-1 > -4$

5 ______ ______

ANSWERS

3. $[6, \infty)$

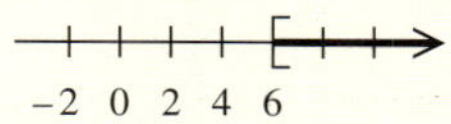

4. (a) $>$ **(b)** $<$; 20

5 Solve, check, and graph the solution set of each inequality.

(a) $2x < -10$

(b) $-7k \geq 8$

(c) $-9m < -81$

As these examples suggest, multiplying each side of an inequality by a *negative* number reverses the direction of the inequality symbol. The same is true for dividing by a negative number since division is defined in terms of multiplication.

Multiplication Property of Inequality

For all real numbers A, B, and C, with $C \neq 0$,

(a) the inequalities

$$A < B \quad \text{and} \quad AC < BC \quad \text{are equivalent } \textbf{if } C > 0;$$

(b) the inequalities

$$A < B \quad \text{and} \quad AC > BC \quad \text{are equivalent } \textbf{if } C < 0.$$

In words, each side of an inequality may be multiplied (or divided) by a *positive* number without changing the direction of the inequality symbol. ***Multiplying (or dividing) by a negative number requires that we reverse the direction of the inequality symbol.***

EXAMPLE 4 **Using the Multiplication Property of Inequality**

Solve each inequality, and graph the solution set.

(a) $5m \leq -30$

Use the multiplication property to divide each side by 5. ***Since 5 is positive, do not reverse the direction of inequality symbol.***

$$5m \leq -30$$

$$\frac{5m}{5} \leq \frac{-30}{5} \qquad \text{Divide by 5.}$$

$$m \leq -6$$

Check that the solution set is the interval $(-\infty, -6]$, graphed in Figure 6.

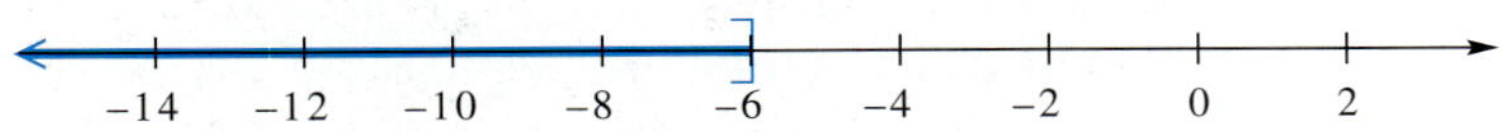

Figure 6

(b) $-4k \leq 32$

Divide each side by -4. ***Since -4 is negative, reverse the direction of inequality symbol.***

$$-4k \leq 32$$

$$\frac{-4k}{-4} \geq \frac{32}{-4} \qquad \text{Divide by } -4\text{; reverse the direction of the symbol.}$$

$$k \geq -8$$

Reverse the inequality symbol when dividing by a negative number.

Check the solution set. Figure 7 shows the graph of the solution set, $[-8, \infty)$.

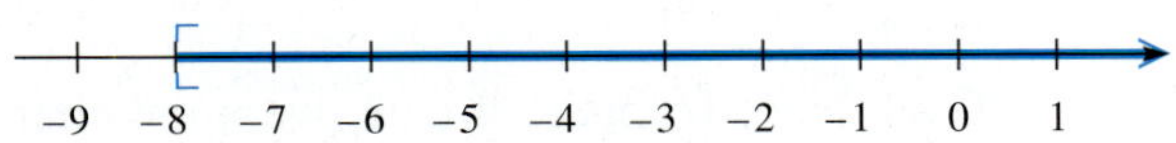

Figure 7

◀ Work Problem 5 at the Side.

ANSWERS

5. (a) $(-\infty, -5)$

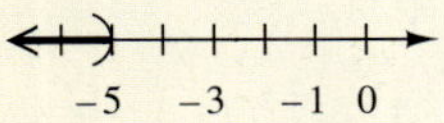

(b) $\left(-\infty, -\frac{8}{7}\right]$

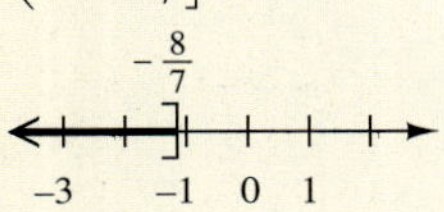

(c) $(9, \infty)$

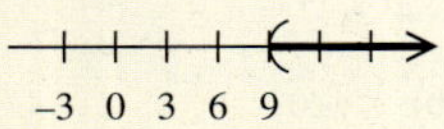

Solving a Linear Inequality

Step 1 **Simplify each side separately.** Clear parentheses, fractions, and decimals using the distributive property as needed, and combine like terms.

Step 2 **Isolate the variable terms on one side.** Use the addition property of inequality to get all terms with variables on one side of the inequality and all numbers on the other side.

Step 3 **Isolate the variable.** Use the multiplication property of inequality to change the inequality to the form $x < k$ or $x > k$.

CAUTION

Reverse the direction of the inequality symbol only when multiplying or dividing each side of an inequality by a negative number.

EXAMPLE 5 Solving a Linear Inequality

Solve $-3(x + 4) + 2 \geq 7 - x$, and graph the solution set.

Step 1 $-3(x + 4) + 2 \geq 7 - x$

$-3x - 12 + 2 \geq 7 - x$ Distributive property

$-3x - 10 \geq 7 - x$ Combine like terms.

Step 2 $-3x - 10 + x \geq 7 - x + x$ Add x.

$-2x - 10 \geq 7$ Combine like terms.

$-2x - 10 + 10 \geq 7 + 10$ Add 10.

$-2x \geq 17$ Combine like terms.

Step 3 $\dfrac{-2x}{-2} \leq \dfrac{17}{-2}$ Divide by -2; change $\geq$ to $\leq$.

$x \leq -\dfrac{17}{2}$

Be sure to reverse the inequality symbol.

Figure 8 shows the graph of the solution set, $\left(-\infty, -\frac{17}{2}\right]$.

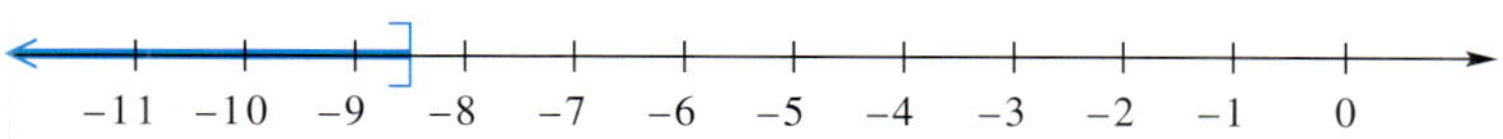

Figure 8

Work Problem **6** at the Side. ▶

Note

In Step 2 of Example 5, if we add $3x$ (instead of x) to both sides, we get

$-3x - 10 + 3x \geq 7 - x + 3x$ Add $3x$.

$-10 \geq 2x + 7$

$-10 - 7 \geq 2x + 7 - 7$ Subtract 7.

$-17 \geq 2x$

$-\dfrac{17}{2} \geq x$, or $x \leq -\dfrac{17}{2}$. Divide by 2; rewrite.

The result "$-\frac{17}{2}$ is greater than or equal to x" means the same thing as "x is less than or equal to $-\frac{17}{2}$." Thus, the solution set is the same.

6 Solve, check, and graph the solution set of each inequality.

(a) $x + 4(2x - 1) \geq x + 2$

(b) $m - 2(m - 4) \leq 3m$

ANSWERS

6. (a) $\left[\frac{3}{4}, \infty\right)$

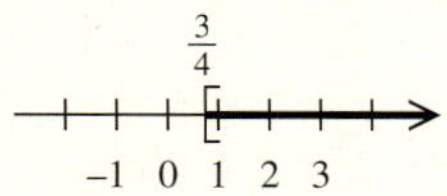

(b) $[2, \infty)$

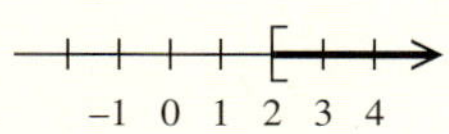

7 Solve, check, and graph the solution set of each inequality.

(a) $5 - 3(m - 1) \leq 2(m + 3) + 1$

(b) $\frac{1}{4}(m + 3) + 2 \leq \frac{3}{4}(m + 8)$

8 Rewrite each three-part inequality using the order in which the numbers appear on the number line.

(a) $1 > x > -1$

(b) $16 \geq p \geq 11$

(c) $-2 > t \geq -8$

EXAMPLE 6 Solving a Linear Inequality with Fractions

Solve $-\frac{2}{3}(r - 3) - \frac{1}{2} < \frac{1}{2}(5 - r)$, and graph the solution set.

Step 1 To clear fractions, multiply by the least common denominator, 6.

$$-\frac{2}{3}(r - 3) - \frac{1}{2} < \frac{1}{2}(5 - r)$$

$$6\left[-\frac{2}{3}(r - 3) - \frac{1}{2}\right] < 6\left[\frac{1}{2}(5 - r)\right] \quad \text{Multiply by 6, the LCD.}$$

Be careful here.

$$6\left[-\frac{2}{3}(r - 3)\right] - 6\left(\frac{1}{2}\right) < 6\left[\frac{1}{2}(5 - r)\right] \quad \text{Distributive property}$$

$$-4(r - 3) - 3 < 3(5 - r) \quad \text{Multiply.}$$

$$-4r + 12 - 3 < 15 - 3r \quad \text{Distributive property}$$

$$-4r + 9 < 15 - 3r$$

Step 2

$$-4r + 9 + 3r < 15 - 3r + 3r \quad \text{Add } 3r.$$

$$-r + 9 < 15$$

$$-r + 9 - 9 < 15 - 9 \quad \text{Subtract 9.}$$

$$-r < 6$$

Step 3

$$-1(-r) > -1(6) \quad \text{Multiply by } -1\text{; change } < \text{ to } >.$$

$$r > -6$$

Check that the solution set is $(-6, \infty)$. See Figure 9.

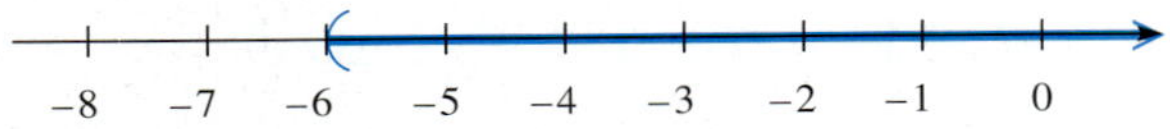

Figure 9

◀ *Work Problem* **7** *at the Side.*

OBJECTIVE 4 Solve linear inequalities with three parts. For some applications, it is necessary to work with a **three-part inequality** such as

$$3 < x + 2 < 8,$$

where $x + 2$ is *between* 3 and 8. We solve this inequality as follows.

$$3 - 2 < x + 2 - 2 < 8 - 2 \quad \text{Subtract 2 from } each \text{ part.}$$

$$1 < x < 6$$

Thus, x must be between 1 and 6 so that $x + 2$ will be between 3 and 8. The solution set, the open interval $(1, 6)$, is graphed in Figure 10.

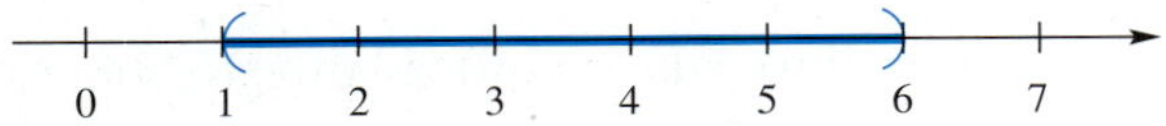

Figure 10

CAUTION

In three-part inequalities, the order of the parts is important. For example, do *not* write $8 < x + 2 < 3$, since this implies that $8 < 3$, a false statement. ***We write three-part inequalities so that the symbols point in the same direction, and both point toward the lesser number.***

◀ *Work Problem* **8** *at the Side.*

ANSWERS

7. (a) $\left[\frac{1}{5}, \infty\right)$

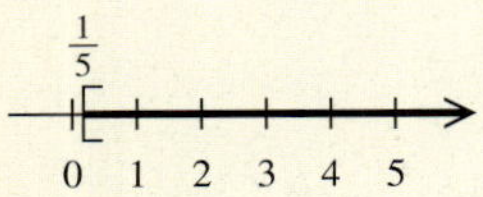

(b) $\left[-\frac{13}{2}, \infty\right)$

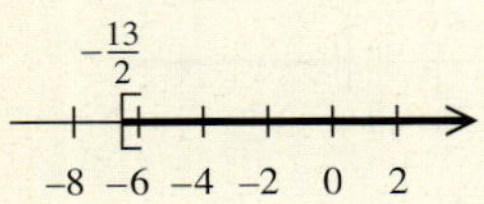

8. (a) $-1 < x < 1$
(b) $11 \leq p \leq 16$
(c) $-8 \leq t < -2$

EXAMPLE 7 Solving a Three-Part Inequality

Solve $-2 \le -3k - 1 \le 5$, and graph the solution set.

Begin by adding 1 to each of the three parts to isolate the variable term in the middle.

$$-2 + 1 \le -3k - 1 + 1 \le 5 + 1 \qquad \text{Add 1 to each part.}$$

$$-1 \le -3k \le 6$$

$$\frac{-1}{-3} \ge \frac{-3k}{-3} \ge \frac{6}{-3} \qquad \text{Divide each part by } -3\text{; reverse the direction of the inequality symbols.}$$

$$\frac{1}{3} \ge k \ge -2$$

$$-2 \le k \le \frac{1}{3} \qquad \text{Rewrite in order based on the number line.}$$

Check that the solution set is the closed interval $[-2, \frac{1}{3}]$. See Figure 11.

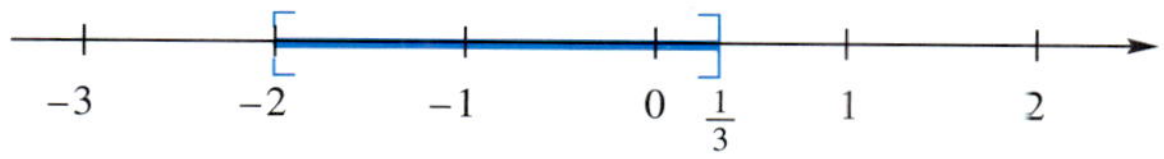

Figure 11

Work Problem 9 at the Side. ▶

Examples of the types of solution sets to be expected from solving linear equations and linear inequalities are shown below.

SOLUTIONS OF LINEAR EQUATIONS AND INEQUALITIES

Equation or Inequality	Typical Solution Set	Graph of Solution Set
Linear equation $5x + 4 = 14$	$\{2\}$	
Linear inequality $5x + 4 < 14$	$(-\infty, 2)$	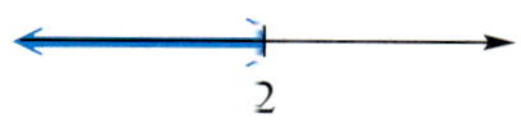
$5x + 4 > 14$	$(2, \infty)$	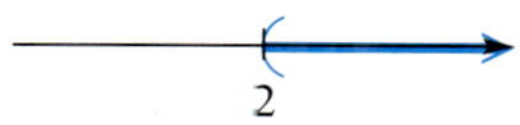
Three-part inequality $-1 \le 5x + 4 \le 14$	$[-1, 2]$	

OBJECTIVE 5 Solve applied problems using linear inequalities. Besides the familiar "is less than" and "is greater than," other expressions also indicate inequalities, as shown in the table below.

Word Expression	Interpretation
a **exceeds** b	$a > b$
a **is at least** b	$a \ge b$
a **is no less than** b	$a \ge b$
a **is at most** b	$a \le b$
a **is no more than** b	$a \le b$

9 Solve, check, and graph the solution set of each inequality.

(a) $-3 \le x - 1 \le 7$

(b) $5 < 3x - 4 < 9$

ANSWERS

9. (a) $[-2, 8]$

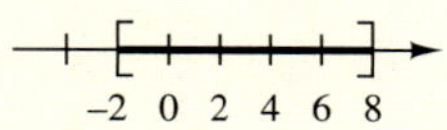

(b) $\left(3, \frac{13}{3}\right)$

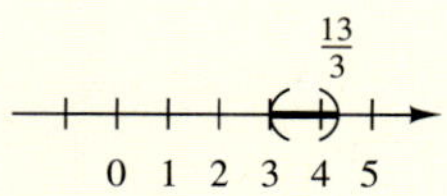

10 Solve the problem.

A rental company charges $10 to rent a leaf blower, plus $7.50 per hr. Marge Ruhberg can spend no more than $40 to blow leaves from her driveway and pool deck. What is the *maximum* amount of time she can use the rented leaf blower?

In Examples 8 and 9, we use the six problem-solving steps from **Section 2.3,** changing Step 3 from "Write an equation" to "Write an inequality."

EXAMPLE 8 Using a Linear Inequality to Solve a Rental Problem

A rental company charges $20 to rent a chain saw, plus $9 per hr. Tom Ruhberg can spend no more than $65 to clear some logs from his yard. What is the *maximum* amount of time he can use the rented saw?

Step 1 **Read** the problem again.

Step 2 **Assign a variable.** Let $h =$ the number of hours he can rent the saw.

Step 3 **Write an inequality.** He must pay $20, plus 9h$, to rent the saw for h hours, and this amount must be *no more than* $65.

$$\underbrace{20 + 9h}_{\text{Cost of renting}} \quad \underbrace{\leq}_{\text{is no more than}} \quad \underbrace{65}_{\text{65 dollars.}}$$

Step 4 **Solve.**

$$9h \leq 45 \quad \text{Subtract 20.}$$

$$h \leq 5 \quad \text{Divide by 9.}$$

Step 5 **State the answer.** He can use the saw for a maximum of 5 hr. (He may use it for less time, as indicated by the inequality $h \leq 5$.)

Step 6 **Check.** If Tom uses the saw for **5** hr, he will spend

$$20 + 9(5) = 65 \text{ dollars}, \quad \text{the maximum amount.}$$

◀ *Work Problem* **10** *at the Side.*

11 Solve the problem.

Alex Lose has grades of 92, 90, and 84 on his first three history tests. What grade must he make on his fourth test in order to keep an average of at least 90?

EXAMPLE 9 Finding an Average Test Score

Emma Saska has scores of 88, 86, and 90 on her first three algebra tests. An average score of at least 90 will earn an A in the class. What possible scores on her fourth test will earn her an A average?

Let x represent the score on the fourth test. Her average score must be at least 90. To find the average of four numbers, add them and then divide by 4.

$$\underbrace{\frac{88 + 86 + 90 + x}{4}}_{\text{Average}} \quad \underbrace{\geq}_{\text{is at least}} \quad \underbrace{90}_{90.}$$

$$\frac{264 + x}{4} \geq 90 \quad \text{Add the scores.}$$

$$264 + x \geq 360 \quad \text{Multiply by 4.}$$

$$x \geq 96 \quad \text{Subtract 264.}$$

She must score **96** or more on her fourth test.

Check $\frac{88 + 86 + 90 + 96}{4} = \frac{360}{4} = 90,$ the minimum score.

A score of 96 or more will give an average of at least 90, as required.

◀ *Work Problem* **11** *at the Side.*

ANSWERS

10. 4 hr

11. at least 94

3.1 Exercises

FOR EXTRA HELP MyMathLab Math XL PRACTICE WATCH DOWNLOAD READ REVIEW

Match each inequality in Column I with the correct graph or interval in Column II.

I	II
1. $x \leq 3$	**A.** 
2. $x > 3$	**B.** 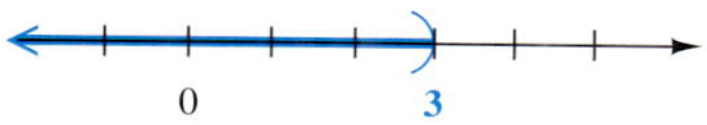
3. $x < 3$	**C.** $(3, \infty)$
4. $x \geq 3$	**D.** $(-\infty, 3]$
5. $-3 \leq x \leq 3$	**E.** $(-3, 3)$
6. $-3 < x < 3$	**F.** $[-3, 3]$

7. Refer to the graph, and write an inequality or a three-part inequality for each description.

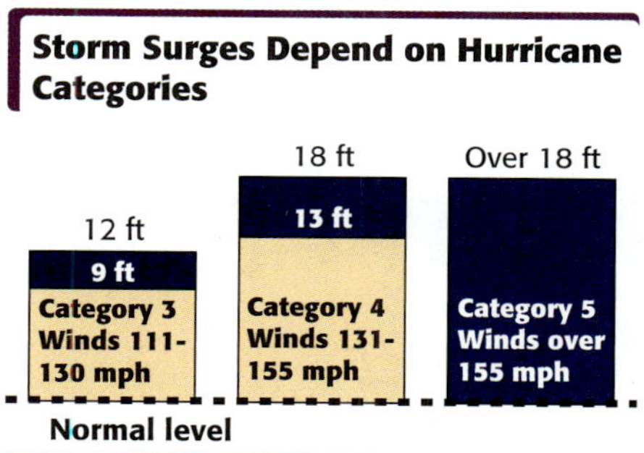

Source: National Oceanic and Atmospheric Administration.

(a) The wind speed s (in miles per hour) of a Category 4 hurricane

(b) The wind speed s (in miles per hour) of a Category 5 hurricane

(c) The storm surge x (in feet) from a Category 3 hurricane

(d) The storm surge x (in feet) from a Category 5 hurricane

8. Dr. Paul Donohue writes a syndicated column in which readers question him on a variety of health topics. Reader C. J. wrote, "Many people say they can weigh more because they have a large frame. How is frame size determined?" Here is Dr. Donohue's response:

> *"For a man, a wrist circumference between* 6.75 *and* 7.25 in. *[inclusive] indicates a medium frame. Anything above is a large frame and anything below, a small frame."*

Using x to represent wrist circumference in inches, write an inequality or a three-part inequality that represents wrist circumference for a male with the indicated frame size.

(a) Small frame

(b) Medium frame

(c) Large frame

(*Source: The Gazette,* Cedar Rapids, Iowa, October 4, 2004.)

9. A student solved the following inequality as shown.

$$4x \geq -64$$
$$\frac{4x}{4} \leq \frac{-64}{4}$$
$$x \leq -16$$

Solution set: $(-\infty, -16]$

WHAT WENT WRONG? Give the correct solution set.

10. Explain how you will determine whether to use parentheses or brackets when graphing the solution set of an inequality.

Solve each inequality, giving solution sets in both interval and graph forms. Check your answers. See Examples 1–6.

11. $x - 4 \leq 3$

12. $t - 3 \leq 1$

13. $4x + 1 \geq 21$

14. $5t + 2 \geq 52$

15. $5x > -25$

16. $7x < -28$

17. $-4x < 16$

18. $-2m > 10$

19. $-\frac{3}{4}r \geq 30$

20. $-\frac{2}{3}x \leq 12$

21. $-1.3m \geq -5.2$

22. $-2.5x \leq -1.25$

23. $\frac{3k - 1}{4} > 5$

24. $\frac{5z - 6}{8} < 8$

25. $\frac{2k - 5}{-4} > 5$

26. $\frac{3z - 2}{-5} < 6$

27. $3k + 1 < -20$

28. $5z + 6 > -29$

29. $x + 4(2x - 1) \geq x$

30. $m - 2(m - 4) \leq 3m$

31. $-(4 + r) + 2 - 3r < -14$

32. $-(9 + k) - 5 + 4k \geq 4$

33. $-3(z - 6) > 2z - 2$

34. $-2(x + 4) \leq 6x + 16$

35. $\frac{2}{3}(3k - 1) \geq \frac{3}{2}(2k - 3)$

36. $\frac{7}{5}(10m - 1) < \frac{2}{3}(6m + 5)$

37. $-\frac{1}{4}(p + 6) + \frac{3}{2}(2p - 5) < 10$

38. $\frac{3}{5}(k - 2) - \frac{1}{4}(2k - 7) \leq 3$

Relating Concepts (Exercises 39–43) For Individual or Group Work

Work Exercises 39–43 in order.

39. Solve the linear equation

$$5(x + 3) - 2(x - 4) = 2(x + 7),$$

and graph the solution set on a number line.

40. Solve the linear inequality

$$5(x + 3) - 2(x - 4) > 2(x + 7),$$

and graph the solution set on a number line.

41. Solve the linear inequality

$$5(x + 3) - 2(x - 4) < 2(x + 7),$$

and graph the solution set on a number line.

42. Graph all the solution sets of the equation and inequalities in Exercises 39–41 on the same number line. What set do you obtain?

43. Based on the results of Exercises 39–41, complete the following using a conjecture (educated guess): The solution set of

$$-3(x + 2) = 3x + 12$$

is $\{-3\}$, and the solution set of

$$-3(x + 2) < 3x + 12$$

is $(-3, \infty)$. Therefore the solution set of

$$-3(x + 2) > 3x + 12$$

is ______________.

44. Which is the graph of $-2 < x$?

A. 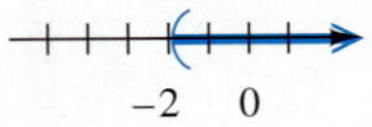

B. 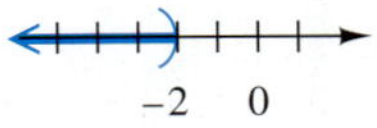

C. 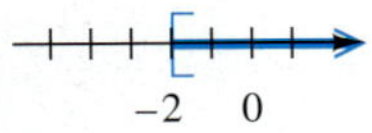

D. 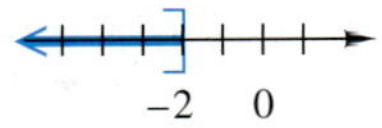

Solve each inequality, giving solution sets in both interval and graph forms. Check your answers. See Example 7.

45. $-4 < x - 5 < 6$

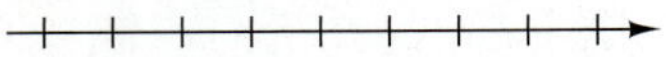

46. $-1 < x + 1 < 8$

47. $-9 \le k + 5 \le 15$

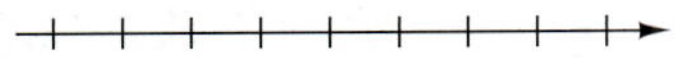

48. $-4 \le m + 3 \le 10$

49. $-6 \le 2(z + 2) \le 16$

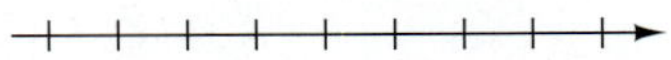

50. $-15 < 3(p + 2) < 24$

51. $-16 < 3t + 2 < -10$

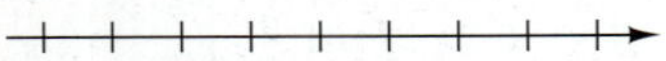

52. $-19 < 3x - 5 \le 1$

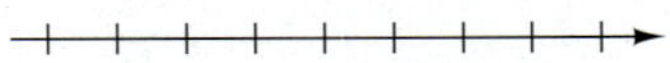

53. $4 < -9x + 5 \le 8$

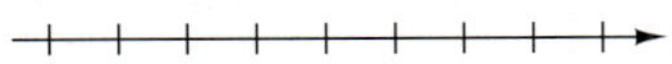

54. $4 < -2x + 3 \le 8$

55. $-1 \le \frac{2x - 5}{6} \le 5$

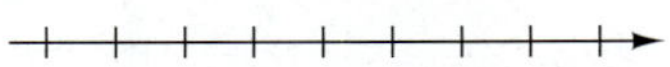

56. $-3 < \frac{3m + 1}{4} \le 3$

The weather forecast by time of day for the U.S. Olympic Track and Field Trials, in Sacramento, California, is shown in the figure. Use this graph to work Exercises 57–60.

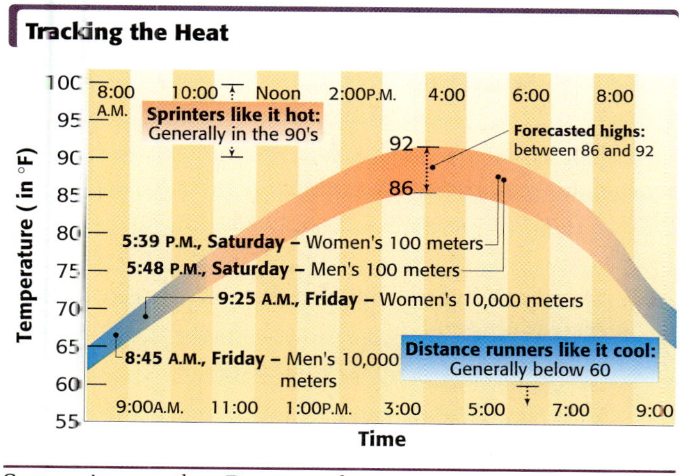

Source: Accuweather, Bee research.

57. Sprinters prefer Fahrenheit temperatures in the 90's. Using the upper boundary of the forecast, in what time period is the temperature expected to be at least 90°F?

58. Distance runners prefer cool temperatures. During what time period are temperatures predicted to be no more than 70°F? Use the lower forecast boundary.

59. What range of temperatures is predicted for the Women's 100-m event?

60. What range of temperatures is forecast for the Men's 10,000-m event?

Solve each problem. See Examples 8 and 9.

61. Amber is signing up for cell phone service. She is trying to decide between Plan A, which costs \$54.99 a month with a free phone included, and Plan B, which costs \$49.99 a month, but would require her to buy a phone for \$129. Under either plan, Amber does not expect to go over the included number of monthly minutes. After how many months would Plan B be a better deal?

62. Stuart and Tracy Sudak need to rent a truck to move their belongings to their new apartment. They can rent a truck of the size they need from U-Haul for \$29.95 a day plus 28 cents per mile or from Budget Truck Rentals for \$34.95 a day plus 25 cents per mile. After how many miles would the Budget rental be a better deal than the U-Haul one?

63. Bonnie Boehme earned scores of 90 and 82 on her first two tests in English Literature. What score must she make on her third test to keep an average of 84 or greater?

64. Scott Barnett scored 92 and 96 on his first two tests in Methods in Teaching Mathematics. What score must he make on his third test to keep an average of 90 or greater?

A product will produce a profit only when the revenue (R) from selling the product exceeds the cost (C) of producing it. Find the least whole number of units x that must be sold for each business to show a profit for the item described.

65. Peripheral Visions, Inc. finds that the cost to produce x studio-quality DVDs is

$$C = 20x + 100,$$

while the revenue produced from them is $R = 24x$ (C and R in dollars).

66. Speedy Delivery finds that the cost to make x deliveries is

$$C = 3x + 2300,$$

while the revenue produced from them is $R = 5.50x$ (C and R in dollars).

67. A BMI (body mass index) between 19 and 25 is considered healthy. Use the formula

$$\text{BMI} = \frac{704 \times (\text{weight in pounds})}{(\text{height in inches})^2}$$

to find the weight range w, to the nearest pound, that gives a healthy BMI for each height. (*Source: Washington Post*.)

(a) 72 in. **(b)** Your height in inches

68. To achieve the maximum benefit from exercising, the heart rate in beats per minute should be in the target heart rate zone (THR). For a person aged A, the formula is

$$0.7(220 - A) \le THR \le 0.85(220 - A).$$

Find the THR to the nearest whole number for each age. (*Source:* Hockey, Robert V., *Physical Fitness: The Pathway to Healthful Living*, Times Mirror/Mosby College Publishing, 1989.)

(a) 35 **(b)** Your age

Find the unknown numbers in each description.

69. Six times a number is between -12 and 12.

70. Half a number is between -3 and 2.

71. When 1 is added to twice a number, the result is greater than or equal to 7.

72. If 8 is subtracted from a number, then the result is at least 5.

73. One third of a number is added to 6, giving a result of at least 3.

74. Three times a number, minus 5, is no more than 7.

Study Skills

USING STUDY CARDS

You may have used "flash cards" in other classes. In math, "study cards" can be helpful. The main things to remember in math besides terms and definitions are *sets of steps to follow* to solve problems (and how to know which set of steps to follow) and *concepts about how math works* (principles). So, the cards may look different but will be just as useful.

In this two-part activity, you will find four types of study cards to use in math. Look carefully at what kinds of information to put on them and where to put it. Then use them the way you would any flash card:

- to quickly review when you have a few minutes,
- for daily reviews,
- to review before a quiz or test.

Remember, the most helpful thing about study cards is making them. After each card description you will find an assignment to try, marked **Now Try This**.

OBJECTIVES

1 **Create study cards for new terms.**

2 **Create study cards for new procedures.**

New Vocabulary Cards

To make a new vocabulary card, put the word (spelled correctly) and the page number where it is found on the front of the card. On the back, write:

- the definition (in your own words if possible),
- an example, including any exceptions or other special information,
- any related words, and
- a sample problem (if appropriate).

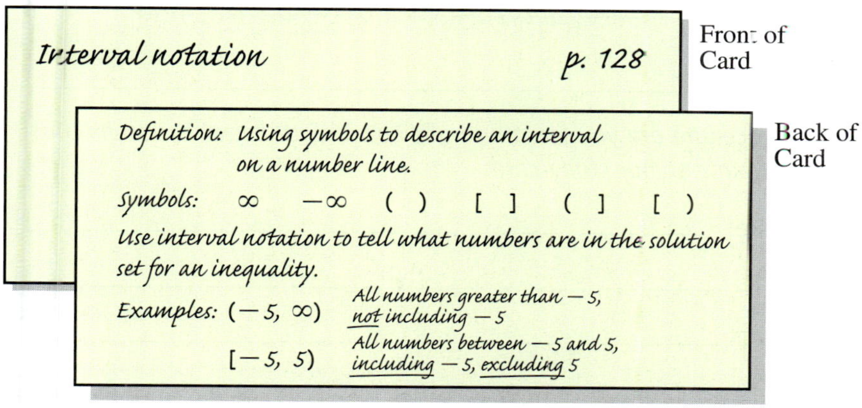

◀◀◀ Now Try This

List 4 new vocabulary words/concepts you need to learn right now. Make a card for each one.

__________ __________ __________ __________

Procedure ("Steps") Cards

To make a procedure (steps) card, write the name of the procedure at the top on the front of the card. Then write each step *in words*. If you need to learn abbreviations for some words, include them along with the whole words written out. On the back of the card, put an example of the procedure, showing each step you need to take. You can review by looking at the front and practicing a new worked example, or by looking at the back and remembering the procedure and its steps.

Front of Card

Solving a Linear Inequality

1. *Simplify each side separately. (Clear parentheses and combine like terms.)*
2. *Isolate variable terms on one side. (Add or subtract the same number from both sides.)*
3. *Isolate the variable. (Divide both sides by the same number; if dividing by a negative number, reverse direction of inequality.)*

Back of Card

Solve $-3(x+4)+2 \geq 7-x$ *and graph the solution set.*

$-3(x+4)+2 \geq 7-x$	*Clear parentheses.*
$-3x-12+2 \geq 7-x$	*Combine like terms.*
$-3x-10 \geq 7-x$	*Both sides are simplified.*
$-3x-10+x \geq 7-x+x$	*Add x to both sides.*
$-2x-10 \geq 7$	*Variable term sill not isolated.*
$-2x-10+10 \geq 7+10$	*Add 10 to both sides.*
$\frac{-2x}{-2} \geq \frac{17}{-2}$	*Divide both sides by −2; dividing by negative, reverse direction of inequality symbol.*
$x \leq -\frac{17}{2}$	$-\frac{17}{2} = -8\frac{1}{2}$

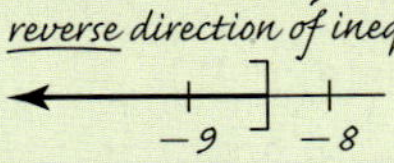

Now Try This ▶▶▶

What procedure are you learning right now? Write below the steps that you will put on your study card.

Procedure: ______________________

Step 1 ______________________

Step 2 ______________________

Step 3 ______________________

Step 4 ______________________

Step 5 ______________________

3.2 Set Operations and Compound Inequalities

OBJECTIVES

1. Find the intersection of two sets.
2. Solve compound inequalities with the word *and*.
3. Find the union of two sets.
4. Solve compound inequalities with the word *or*.

The table shows symptoms of an overactive thyroid and an underactive thyroid.

Underactive Thyroid	Overactive Thyroid
Sleepiness, s	Insomnia, i
Dry hands, d	Moist hands, m
Intolerance of cold, c	Intolerance of heat, h
Goiter, g	Goiter, g

Source: The Merck Manual of Diagnosis and Therapy, 16th Edition, Merck Research Laboratories, 1992.

Let N be the set of symptoms for an underactive thyroid, and let O be the set of symptoms for an overactive thyroid. Suppose we are interested in the set of symptoms that are found in *both* sets N *and* O. In this section, we discuss the use of the words *and* and *or* as they relate to sets and inequalities.

OBJECTIVE 1 Find the intersection of two sets. The intersection of two sets is defined using the word *and*.

Intersection of Sets

For any two sets A and B, the **intersection** of A and B, symbolized $A \cap B$, is defined as follows:

$$A \cap B = \{x \mid x \text{ is an element of } A \textbf{ and } x \text{ is an element of } B\}.$$

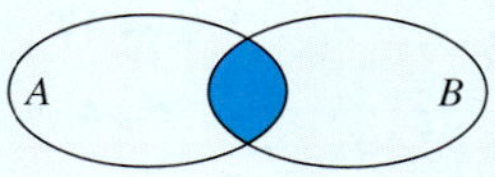

EXAMPLE 1 Finding the Intersection of Two Sets

Let $A = \{1, 2, 3, 4\}$ and $B = \{2, 4, 6\}$. Find $A \cap B$.

The set $A \cap B$ contains those elements that belong to both A *and* B.

$$\begin{aligned} A \cap B &= \{1, 2, 3, 4\} \cap \{2, 4, 6\} \\ &= \{2, 4\} \end{aligned}$$

Work Problem 1 at the Side. ▶

1 List the elements in each set.

(a) $A \cap B$, if $A = \{3, 4, 5, 6\}$ and $B = \{5, 6, 7\}$

(b) $N \cap O$ (Refer to the thyroid table.)

A **compound inequality** consists of two inequalities linked by a connective word such as *and* or *or*. Examples of compound inequalities are

$$x + 1 \leq 9 \quad \textbf{and} \quad x - 2 \geq 3 \qquad \text{Compound inequalities}$$

and

$$2x > 4 \quad \textbf{or} \quad 3x - 6 < 5.$$

OBJECTIVE 2 Solve compound inequalities with the word *and*.

Solving a Compound Inequality with *and*

Step 1 Solve each inequality individually.

Step 2 Since the inequalities are joined with *and*, the solution set of the compound inequality will include all numbers that satisfy both inequalities in Step 1 (the intersection of the solution sets).

ANSWERS

1. (a) $\{5, 6\}$ (b) $\{g\}$

2 Solve each compound inequality, and graph the solution set.

(a) $x < 10$ and $x > 2$

(b) $x + 3 \leq 1$ and $x - 4 \geq -12$

EXAMPLE 2 Solving a Compound Inequality with *and*

Solve the compound inequality $x + 1 \leq 9$ and $x - 2 \geq 3$.

Step 1 Solve each inequality individually.

$$x + 1 \leq 9 \quad \text{and} \quad x - 2 \geq 3$$
$$x + 1 - 1 \leq 9 - 1 \quad \text{and} \quad x - 2 + 2 \geq 3 + 2$$
$$x \leq 8 \quad \text{and} \quad x \geq 5$$

Step 2 Because the inequalities are joined with the word *and,* the solution set will include all numbers that satisfy *both* inequalities in Step 1 at the same time. Thus, the compound inequality is true whenever $x \leq 8$ and $x \geq 5$ are both true. The top graph in Figure 12 shows $x \leq 8$, and the bottom graph shows $x \geq 5$.

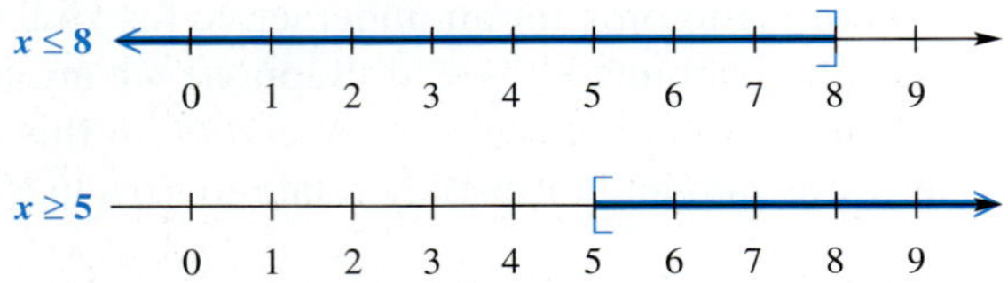

Figure 12

Find the intersection of the two graphs in Figure 12 to get the solution set of the compound inequality. The intersection of the two graphs in Figure 13 shows that the solution set is the closed interval [5, 8].

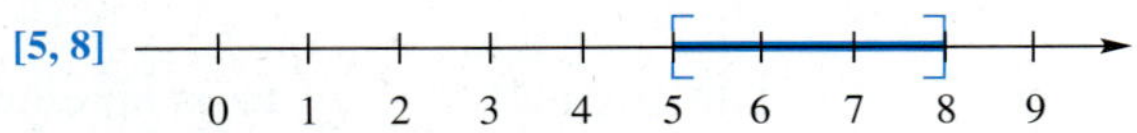

Figure 13

◀ *Work Problem* **2** *at the Side.*

3 Solve

$2x \geq x - 1$ and $3x \geq 3 + 2x$,

and graph the solution set.

EXAMPLE 3 Solving a Compound Inequality with *and*

Solve the compound inequality $-3x - 2 > 5$ and $5x - 1 \leq -21$.

Step 1 Solve each inequality individually.

$$-3x - 2 > 5 \quad \text{and} \quad 5x - 1 \leq -21$$
$$-3x > 7 \quad \text{and} \quad 5x \leq -20$$
$$x < -\frac{7}{3} \quad \text{and} \quad x \leq -4$$

Remember to reverse the inequality symbol.

The graphs of $x < -\frac{7}{3}$ and $x \leq -4$ are shown in Figure 14.

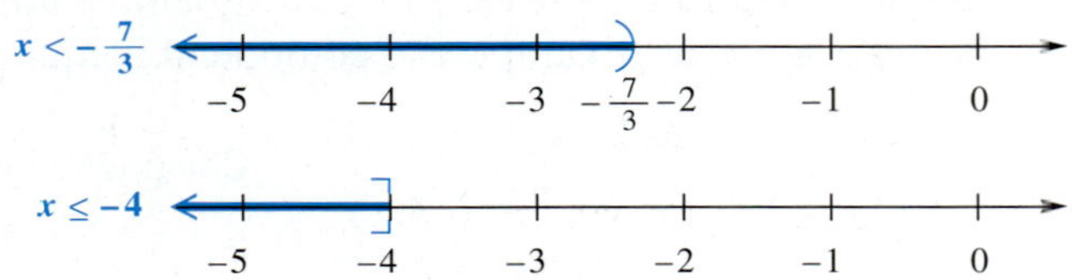

Figure 14

Step 2 Now find all values of x that satisfy both conditions; that is, the real numbers that are less than $-\frac{7}{3}$ and also less than or equal to -4. See Figure 15. The solution set is the infinite interval $(-\infty, -4]$.

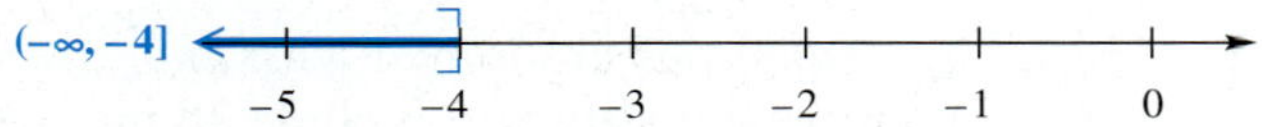

Figure 15

◀ *Work Problem* **3** *at the Side.*

ANSWERS

2. (a) $(2, 10)$

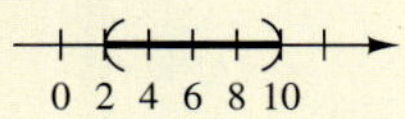

(b) $[-8, -2]$

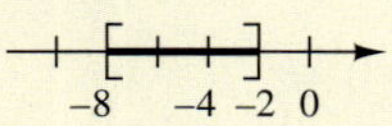

3. $[3, \infty)$

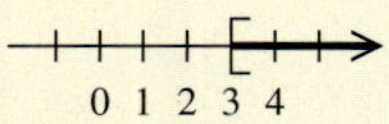

EXAMPLE 4 **Solving a Compound Inequality with *and***

Solve $x + 2 < 5$ and $x - 10 > 2$.

First solve each inequality individually.

$$x + 2 < 5 \quad \text{and} \quad x - 10 > 2$$
$$x < 3 \quad \text{and} \quad x > 12$$

The graphs of $x < 3$ and $x > 12$ are shown in Figure 16.

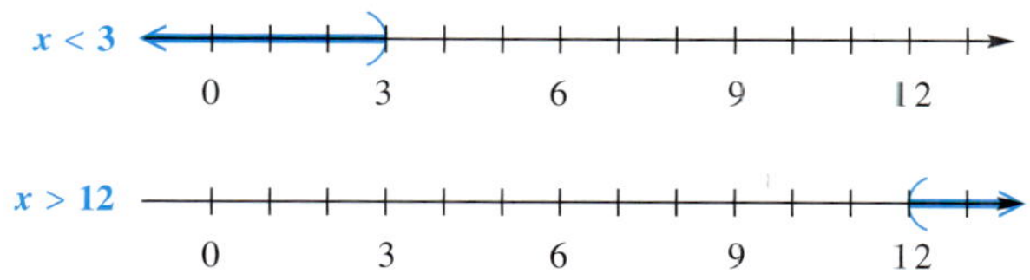

Figure 16

There is no number that is both less than 3 *and* greater than 12, so the given compound inequality has no solution. The solution set is $\emptyset$. See Figure 17.

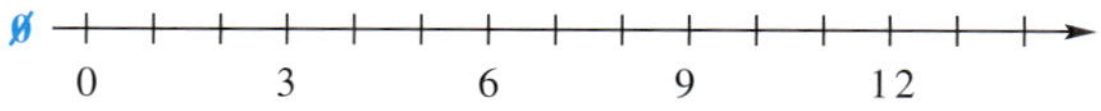

Figure 17

Work Problem **4** *at the Side.* ▶

OBJECTIVE 3 **Find the union of two sets.** The union of two sets is defined using the word *or*.

Union of Sets

For any two sets A and B, the **union** of A and B, symbolized $A \cup B$, is defined as follows:

$$A \cup B = \{x \mid x \text{ is an element of } A \textbf{ or } x \text{ is an element of } B\}.$$

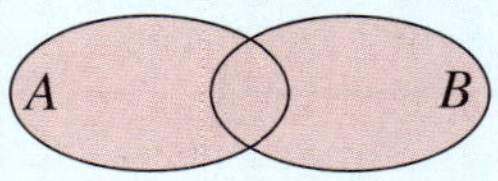

EXAMPLE 5 **Finding the Union of Two Sets**

Let $A = \{1, 2, 3, 4\}$ and $B = \{2, 4, 6\}$. Find $A \cup B$.

Begin by listing all the elements of set A: 1, 2, 3, 4. Then list any additional elements from set B. In this case the elements 2 and 4 are already listed, so the only additional element is 6. Therefore,

$$A \cup B = \{1, 2, 3, 4\} \cup \{2, 4, 6\}$$
$$= \{1, 2, 3, 4, 6\}.$$

The union consists of all elements in either A *or* B (or both).

Note

Although the elements 2 and 4 appeared in both sets A and B, they are written only once in $A \cup B$.

Work Problem **5** *at the Side.* ▶

4 Solve.

(a) $x < 5$ and $x > 5$

(b) $x + 2 > 3$ and $2x + 1 < -3$

5 List the elements in each set.

(a) $A \cup B$, if $A = \{3, 4, 5, 6\}$ and $B = \{5, 6, 7\}$

(b) $N \cup O$ from the thyroid table at the beginning of this section

ANSWERS

4. (a) $\emptyset$ **(b)** $\emptyset$

5. (a) $\{3, 4, 5, 6, 7\}$ **(b)** $\{s, d, c, g, i, m, h\}$

6 Solve. Give each solution set in both interval and graph forms.

(a) $x + 2 > 3$ or $2x + 1 < -3$

(b) $x - 1 > 2$ or $3x + 5 < 2x + 6$

OBJECTIVE 4 **Solve compound inequalities with the word *or*.** Use the following steps.

Solving a Compound Inequality with *or*

Step 1 Solve each inequality individually.

Step 2 Since the inequalities are joined with *or,* the solution set includes all numbers that satisfy either one of the two inequalities in Step 1 (the union of the solution sets).

EXAMPLE 6 **Solving a Compound Inequality with *or***

Solve $6x - 4 < 2x$ or $-3x \leq -9$.

Step 1 Solve each inequality individually.

$$6x - 4 < 2x \quad \text{or} \quad -3x \leq -9$$
$$4x < 4$$
$$x < 1 \quad \text{or} \quad x \geq 3$$

Remember to reverse the inequality symbol.

The graphs of these two inequalities are shown in Figure 18.

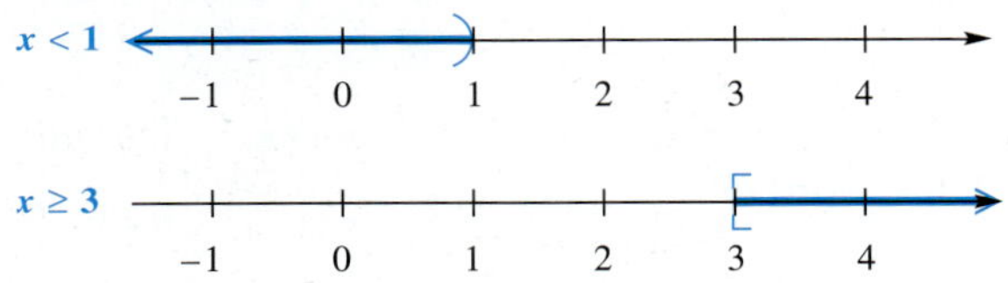

Figure 18

Step 2 Since the inequalities are joined with *or,* we find the union of the two solution sets, as shown in Figure 19. The solution set is the disjoint interval

$$(-\infty, 1) \cup [3, \infty).$$

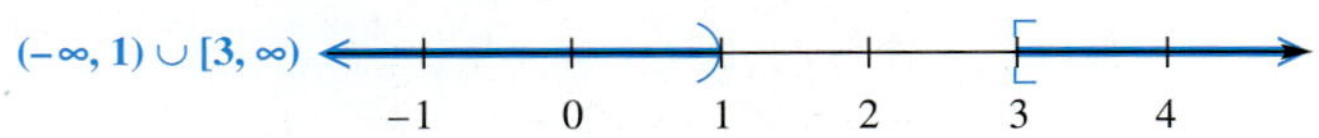

Figure 19

CAUTION
When inequalities are used to write the solution set in Example 6, it *must* be written as

$$x < 1 \quad \text{or} \quad x \geq 3,$$

which keeps the numbers 1 and 3 in their order on the number line. Writing $3 \leq x < 1$ would imply that $3 \leq 1$, which is ***FALSE.*** There is no other way to write the solution set of such a union.

◀ *Work Problem* **6** *at the Side.*

ANSWERS

6. (a) $(-\infty, -2) \cup (1, \infty)$

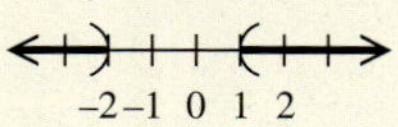

(b) $(-\infty, 1) \cup (3, \infty)$

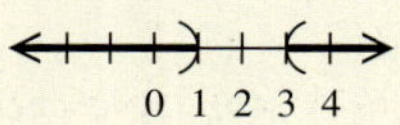

EXAMPLE 7 **Solving a Compound Inequality with *or***

Solve $-4x + 1 \geq 9$ or $5x + 3 \leq -12$.

First, solve each inequality individually.

$$\begin{aligned} -4x + 1 &\geq 9 \quad \text{or} \quad 5x + 3 \leq -12 \\ -4x &\geq 8 \quad \text{or} \quad 5x \leq -15 \\ x &\leq -2 \quad \text{or} \quad x \leq -3 \end{aligned}$$

The graphs of these two inequalities are shown in Figure 20.

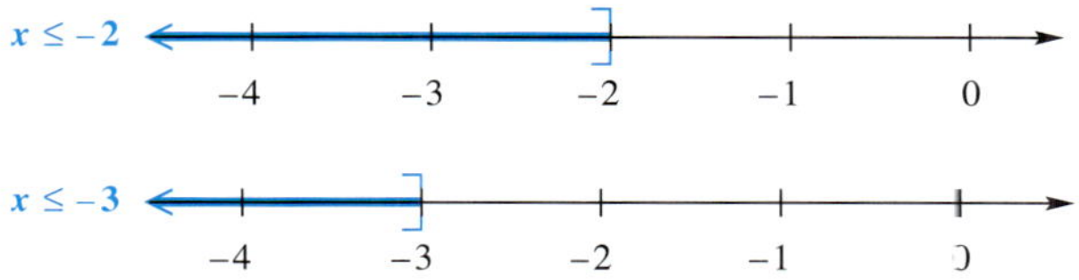

Figure 20

By taking the union, we obtain the interval $(-\infty, -2]$. See Figure 21.

Figure 21

Work Problem 7 at the Side. ▶

EXAMPLE 8 **Solving a Compound Inequality with *or***

Solve $-2x + 5 \geq 11$ or $4x - 7 \geq -27$.

Solve each inequality individually.

$$\begin{aligned} -2x + 5 &\geq 11 \quad \text{or} \quad 4x - 7 \geq -27 \\ -2x &\geq 6 \quad \text{or} \quad 4x \geq -20 \\ x &\leq -3 \quad \text{or} \quad x \geq -5 \end{aligned}$$

The graphs of these two inequalities are shown in Figure 22.

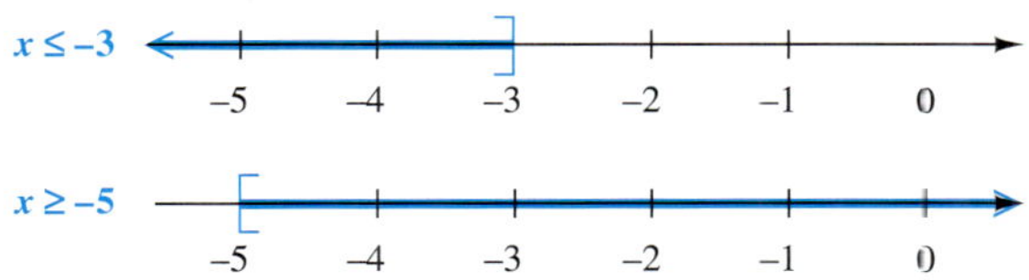

Figure 22

By taking the union, we obtain every real number as a solution, since every real number satisfies at least one of the two inequalities. The set of all real numbers is written in interval notation as $(-\infty, \infty)$ and graphed as in Figure 23.

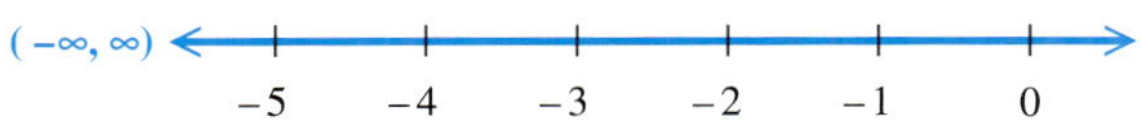

Figure 23

Work Problem 8 at the Side. ▶

7 Solve. Give each solution set in both interval and graph forms.

(a) $2x + 1 \leq 9$ or $2x + 3 \leq 5$

(b) $3x - 4 > 2$ or $-2x + 5 < 3$

8 Solve

$3x - 2 \leq 13$ or $x + 5 \geq 7$.

Give the solution set in both interval and graph forms.

ANSWERS

7. (a) $(-\infty, 4]$

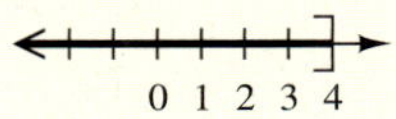

(b) $(1, \infty)$

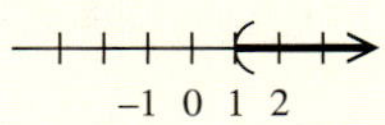

8. $(-\infty, \infty)$

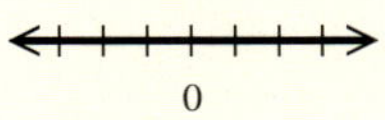

9 From Example 9, list the elements that satisfy each set.

(a) The set of films with admissions greater than 130,000,000 and gross income less than $800,000,000

(b) The set of films with admissions greater than 130,000,000 or gross income less than $800,000,000

ANSWERS

9. (a) $\emptyset$ **(b)** {*Gone with the Wind, Star Wars, The Sound of Music, E.T., The Ten Commandments*}

EXAMPLE 9 **Applying Intersection and Union**

The five highest-grossing domestic films (adjusted for inflation) as of July, 2005, are listed in the table.

FIVE ALL-TIME HIGHEST GROSSING DOMESTIC FILMS

Film	Admissions	Gross Income
Gone with the Wind	202,044,569	$1,293,085,000
Star Wars	178,119,595	$1,139,965,000
The Sound of Music	142,415,376	$911,458,000
E.T.	141,925,359	$908,322,298
The Ten Commandments	131,000,000	$838,400,000

Source: Exhibitor Relations Co., Inc.

List the elements of the following sets.

(a) The set of top-five films with admissions greater than 180,000,000 *and* gross income greater than $1,000,000,000

The only film that satisfies both conditions is *Gone with the Wind,* so the set is

$$\{\textit{Gone with the Wind}\}.$$

(b) The set of top-five films with admissions less than 170,000,000 *or* gross income greater than $1,000,000,000

Here, a film that satisfies at least one of the conditions is in the set. This set includes all five films:

{*Gone with the Wind, Star Wars, The Sound of Music, E.T., The Ten Commandments*}.

◀ Work Problem **9** at the Side.

3.2 Exercises

FOR EXTRA HELP — MyMathLab — Math XL PRACTICE — WATCH — DOWNLOAD — READ — REVIEW

Decide whether each statement is true *or* false. *If it is* false, *explain why.*

1. The union of the solution sets of $2x + 1 = 3$, $2x + 1 > 3$, and $2x + 1 < 3$ is $(-\infty, \infty)$.

2. The intersection of the sets $\{x \mid x \geq 5\}$ and $\{x \mid x \leq 5\}$ is $\emptyset$.

3. The union of the sets $(-\infty, 6)$ and $(6, \infty)$ is $\{6\}$.

4. The intersection of the sets $[6, \infty)$ and $(-\infty, 6]$ is $\{6\}$.

Let $A = \{1, 2, 3, 4, 5, 6\}$, $B = \{1, 3, 5\}$, $C = \{1, 6\}$, *and* $D = \{4\}$. *Specify each set. See Examples 1 and 5.*

5. $A \cap D$ **6.** $B \cap C$ **7.** $B \cap \emptyset$ **8.** $A \cap \emptyset$

9. $A \cup B$ **10.** $B \cup D$ **11.** $B \cup C$ **12.** $C \cup B$

Two sets are specified by graphs. Graph the intersection of the two sets.

13.
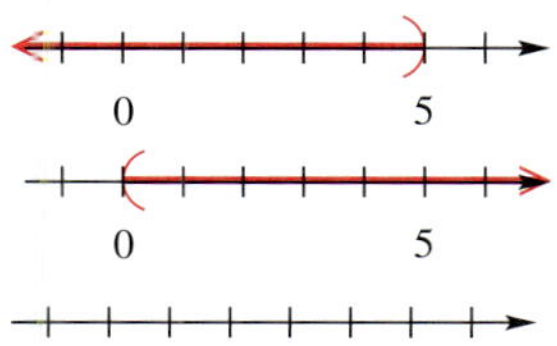

14.
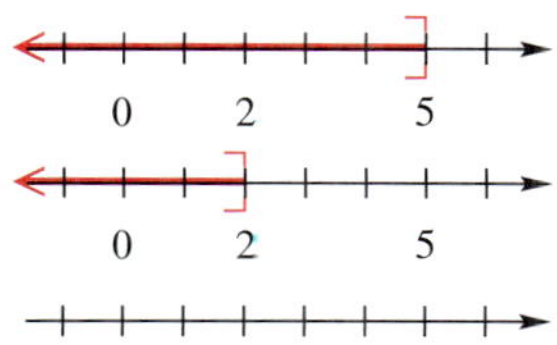

15.
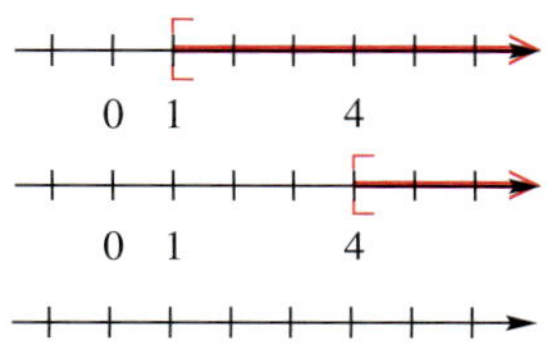

Two sets are specified by graphs. Graph the union of the two sets.

16.
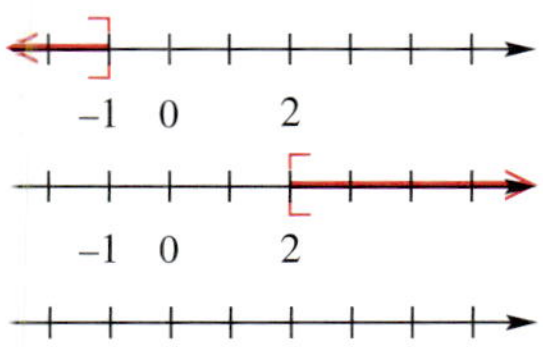

17.
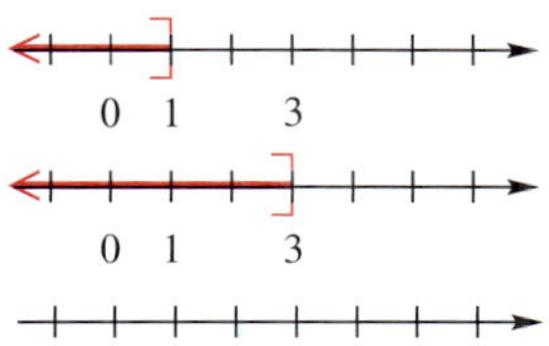

18.
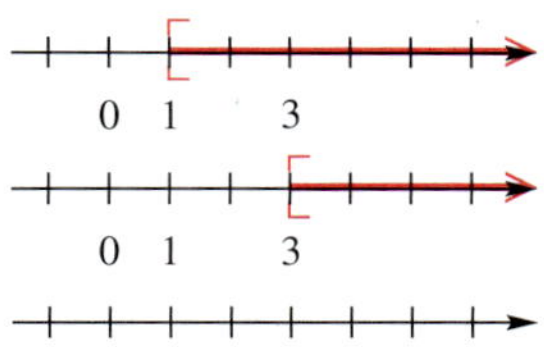

19. Give an example of intersection applied to a real-life situation.

20. A compound inequality uses one of the words *and* or *or*. Explain how you will determine whether to use *intersection* or *union* when graphing the solution set.

For each compound inequality, give the solution set in both interval and graph forms. See Examples 2–4.

21. $x < 2$ and $x > -3$

22. $x < 5$ and $x > 0$

23. $x \leq 2$ and $x \leq 5$

24. $x \geq 3$ and $x \geq 6$

25. $x \leq 3$ and $x \geq 6$

26. $x \leq -1$ and $x \geq 3$

27. $x - 3 \leq 6$ and $x + 2 \geq 7$

28. $x + 5 \leq 11$ and $x - 3 \geq -1$

29. $3x - 4 \leq 8$ and $4x - 1 \leq 15$

30. $7x + 6 \leq 48$ and $-4x \geq -24$

For each compound inequality, give the solution set in both interval and graph forms. See Examples 6–8.

31. $x \leq 1$ or $x \leq 8$

32. $x \geq 1$ or $x \geq 8$

33. $x \geq -2$ or $x \geq 5$

34. $x \leq -2$ or $x \leq 6$

35. $x + 3 \geq 1$ or $x - 8 \leq -4$

36. $x + 6 \geq 11$ or $x - 4 \leq 3$

37. $x + 2 > 7$ or $1 - x > 6$

38. $x + 1 > 3$ or $x + 4 < 2$

39. $x + 1 > 3$ or $-4x + 1 \geq 5$

40. $3x < x + 12$ or $x + 1 > 10$

41. $4x - 8 > 0$ or $4x - 1 < 7$

42. $3x < x + 12$ or $3x - 8 > 10$

Express each set in the simplest interval form.

43. $(-\infty, -1] \cap [-4, \infty)$

44. $[-1, \infty) \cap (-\infty, 9]$

45. $(-\infty, -6] \cap [-9, \infty)$

46. $(5, 11] \cap [6, \infty)$

47. $(-\infty, 3) \cup (-\infty, -2)$

48. $[-9, 1] \cup (-\infty, -3)$

49. $[3, 6] \cup (4, 9)$

50. $[-1, 2] \cup (0, 5)$

For each compound inequality, state whether intersection or union should be used. Then give the solution set in both interval and graph forms. See Examples 2–4 and 6–8.

51. $x < -1$ and $x > -5$

52. $x > -1$ and $x < 7$

53. $x < 4$ or $x < -2$

54. $x < 5$ or $x < -3$

55. $x + 1 \geq 5$ and $x - 2 \leq 10$

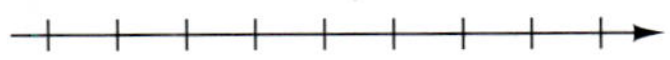

56. $2x - 6 \leq -18$ and $2x \geq -18$

57. $-3x \leq -6$ or $-3x \geq 0$

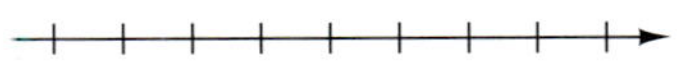

58. $-8x \leq -24$ or $-5x \geq 15$

Relating Concepts (Exercises 59–64) For Individual or Group Work

The figures represent the backyards of neighbors Luigi, Mario, Than, and Joe. Find the area and the perimeter of each yard. Suppose that each resident has 150 ft *of fencing and enough sod to cover* 1400 ft^2 *of lawn.*

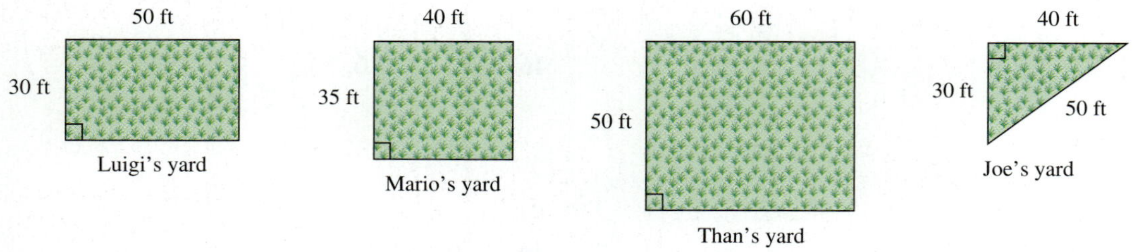

Give the name or names of the residents whose yards satisfy each description.

59. The yard can be fenced *and* the yard can be sodded.

60. The yard can be fenced *and* the yard cannot be sodded.

61. The yard cannot be fenced *and* the yard can be sodded.

62. The yard cannot be fenced *and* the yard cannot be sodded.

63. The yard can be fenced *or* the yard can be sodded.

64. The yard cannot be fenced *or* the yard can be sodded.

Average expenses for full-time college students at 2-year institutions during the 2005–2006 academic year are shown in the table.

COLLEGE EXPENSES (IN DOLLARS), 2-YEAR INSTITUTIONS

Type of Expense	Public Schools	Private Schools
Tuition and fees	1935	12,450
Board rates	2306	4726
Dormitory charges	2251	3994

Source: National Center for Education Statistics, U.S. Department of Education.

Use the table to list the elements of each set. See Example 9.

65. The set of expenses that are less than \$2500 for public schools *and* are greater than \$5000 for private schools

66. The set of expenses that are greater than \$2300 for public schools *and* are less than \$12,000 for private schools

67. The set of expenses that are less than \$2300 for public schools *or* are greater than \$10,000 for private schools

68. The set of expenses that are greater than \$12,000 *or* are less than \$2000

Study Skills

USING STUDY CARDS REVISITED

OBJECTIVES

1 **Create study cards for difficult problems.**

2 **Create study cards of quiz problems.**

This is the second part of the Study Cards activity. As you get further into a chapter, you can choose particular problems that will serve as a good test review. Here are two more types of study cards that will help you.

Tough Problems Card

When you are doing your homework and find yourself saying, "This is really hard," or "I'm having trouble with this," make a tough problem study card. On the front, write out the procedure to work the type of problem *in words*. If there are special notes (like what *not* to do), include them. On the back, work at least one example. Make sure you label what you are doing.

Front of Card

Solving a Linear Inequality with Fractions
First step: Clear the inequality of fractions.
— *Find a common denominator.*
— *Multiply each term by the common denominator.*

Back of Card

Solve $\frac{3}{4}(m-3)+2 \le \frac{1}{2}(m+8)$ *and graph the solution set*

$\frac{4}{1}\left[\frac{3}{4}(m-3)\right]+4(2) \le \frac{4}{1}\left[\frac{1}{2}(m+8)\right]$ — *Common denom. is 4. Multiply every term by 4.*

$3(m-3)+8 \le 2(m+8)$
$3m-9+8 \le 2m+16$ — *Simplify each side.*
$3m-1 \le 2m+16$

$3m-1-2m \le 2m+16-2m$ — *Subtract 2m.*
$m-1 \le 16$
$m-1+1 \le 16+1$
$m \le 17$

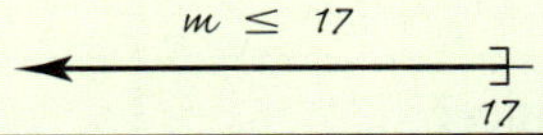

Now Try This

Choose three types of difficult problems, and work them out on study cards. Be sure to put the words for solving the problem on one side and the worked problem on the other side.

Practice Quiz Cards

Quiz study cards cover each type of problem you learn. They are useful when you prepare for a test. To make a quiz card, put the problem with the direction words (like *solve, simplify, estimate*) on the front of a card, and work the problem on the back. If you like, include the page number from the text. When you review, work the problem on a separate paper and check it by looking at the back of your quiz card.

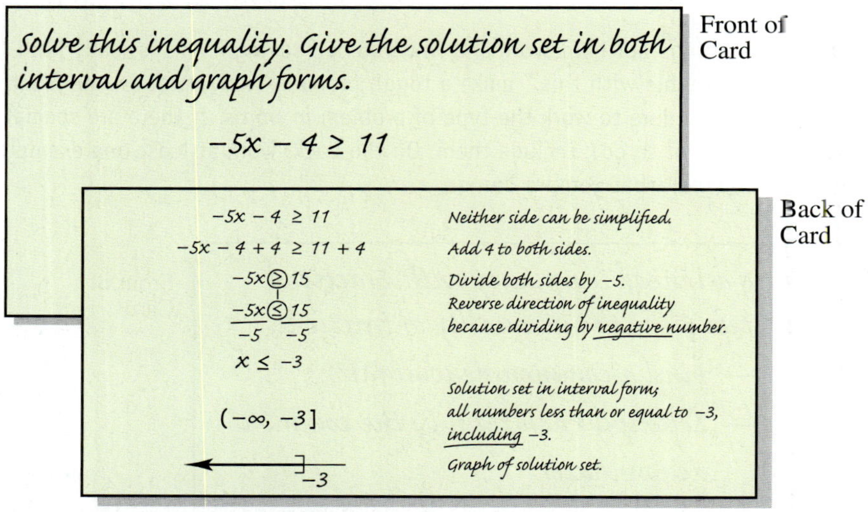

Now Try This ▶▶▶

Choose three problems from different sections of this chapter, and work them on study cards. Be sure you don't just choose the easiest problems.

Why Are Study Cards Brain Friendly?

First, **making the study cards is an active technique.** You have to make decisions about what is most important and how to put it on a card. This kind of thinking is more involved than just memorizing, and as a result, you will understand the concepts better and remember them longer.

Second, **the cards are visually appealing** (if you write neatly and try some color). You will remember a visual image longer and may even be able to "picture in your mind" how your cards look. This will help you during tests.

Third, because **study cards are small and portable,** you can review them easily whenever you have a few minutes. Even while you're waiting for a bus or have a few minutes between classes, you can take out your cards and read over them. After a while, the information will become automatic and easier to remember.

3.3 Absolute Value Equations and Inequalities

In a production line, quality is controlled by randomly choosing items from the line and checking to see how selected measurements vary from the optimum measure. These differences are sometimes positive and sometimes negative, so they are expressed with absolute value. For example, a machine that fills quart milk cartons might be set to release 1 qt (32 oz) plus or minus 2 oz per carton. Then the number of ounces in each carton should satisfy the *absolute value inequality* $|x - 32| \leq 2$, where x is the number of ounces.

OBJECTIVES

1 Use the distance definition of absolute value.

2 Solve equations of the form $|ax + b| = k$, for $k > 0$.

3 Solve inequalities of the form $|ax + b| < k$ and of the form $|ax + b| > k$, for $k > 0$.

4 Solve absolute value equations that involve rewriting.

5 Solve equations of the form $|ax + b| = |cx + d|$.

6 Solve special cases of absolute value equations and inequalities.

OBJECTIVE 1 Use the distance definition of absolute value. In **Section 1.1,** we saw that the absolute value of a number x, written $|x|$, represents the distance from x to 0 on the number line. For example, the solutions of $|x| = 4$ are 4 and -4, as shown in Figure 24.

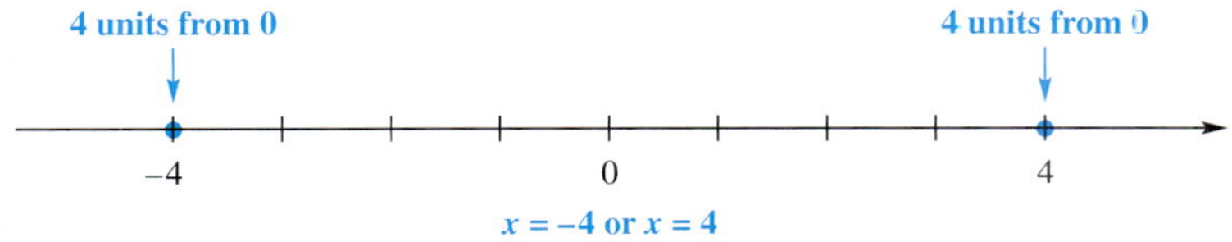

Figure 24

Because absolute value represents distance from 0, it is reasonable to interpret the solutions of $|x| > 4$ to be all numbers that are *more* than 4 units from 0. The set $(-\infty, -4) \cup (4, \infty)$ fits this description. Figure 25 shows the graph of the solution set of $|x| > 4$. Because the graph consists of two separate intervals, the solution set is described using *or* as $x < -4$ or $x > 4$.

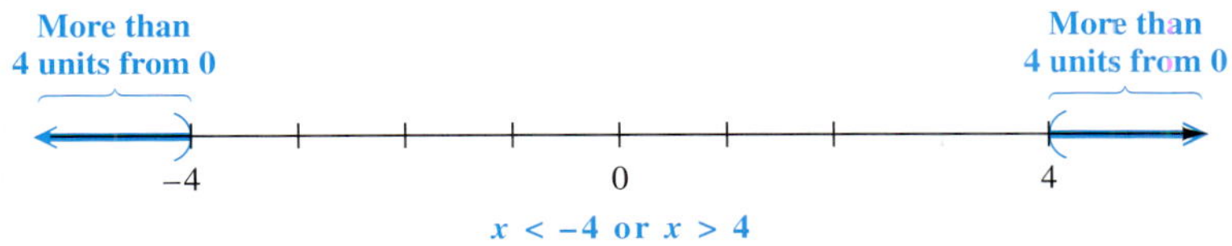

Figure 25

The solution set of $|x| < 4$ consists of all numbers that are *less* than 4 units from 0 on the number line. Another way of thinking of this is to think of all numbers *between* -4 and 4. This set of numbers is given by $(-4, 4)$, as shown in Figure 26. Here, the graph shows that $-4 < x < 4$, which means $x > -4$ *and* $x < 4$.

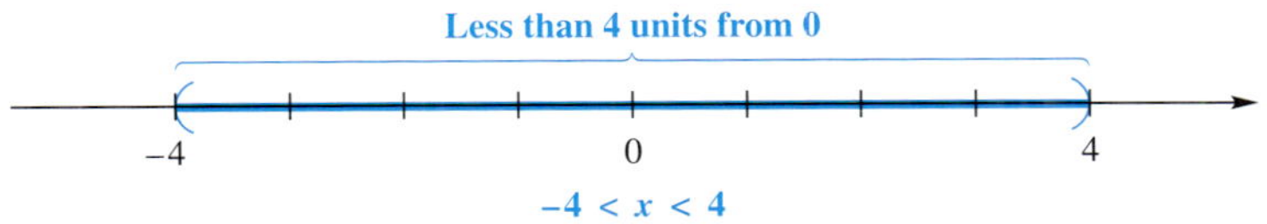

Figure 26

Work Problem **1** *at the Side.* ▶

The equation and inequalities just described are examples of **absolute value equations and inequalities.** They involve the absolute value of a variable expression and generally take the form

$$|ax + b| = k, \quad |ax + b| > k, \quad \text{or} \quad |ax + b| < k,$$

where k is a positive number. From Figures 24–26, we see that

$|x| = 4$ has the same solution set as $x = -4$ **or** $x = 4$,

$|x| > 4$ has the same solution set as $x < -4$ **or** $x > 4$,

$|x| < 4$ has the same solution set as $x > -4$ **and** $x < 4$.

1 Graph the solution set of each equation or inequality.

(a) $|x| = 3$

(b) $|x| > 3$

(c) $|x| < 3$

ANSWERS

1. (a)

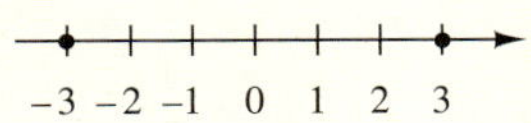

(b)

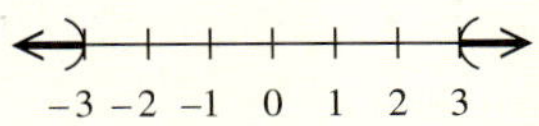

(c)

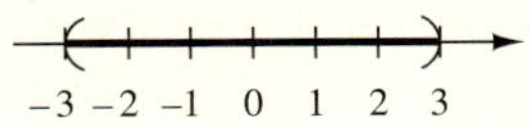

Thus, we can solve an absolute value equation or inequality by solving the appropriate compound equation or inequality.

Solving Absolute Value Equations and Inequalities

Let k be a positive real number, and p and q be real numbers.

Case 1 To solve $|ax + b| = k$, solve the compound equation

$$ax + b = k \quad \text{or} \quad ax + b = -k.$$

The solution set is usually of the form $\{p, q\}$, which includes two numbers.

Case 2 To solve $|ax + b| > k$, solve the compound inequality

$$ax + b > k \quad \text{or} \quad ax + b < -k.$$

The solution set is of the form $(-\infty, p) \cup (q, \infty)$, which is a disjoint interval.

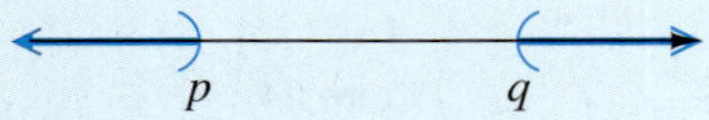

Case 3 To solve $|ax + b| < k$, solve the three-part inequality

$$-k < ax + b < k.$$

The solution set is of the form (p, q), an open interval.

OBJECTIVE 2 Solve equations of the form $|ax + b| = k$, for $k > 0$. *Remember that because absolute value refers to distance from the origin, an absolute value equation will have two parts.*

EXAMPLE 1 Solving an Absolute Value Equation

Solve $|2x + 1| = 7$.

For $|2x + 1|$ to equal 7, $2x + 1$ must be 7 units from 0 on the number line. This can happen only when $2x + 1 = 7$ or $2x + 1 = -7$. This is Case 1 in the preceding box. Solve this compound equation as follows.

$$\begin{aligned} 2x + 1 &= 7 \quad \text{or} \quad 2x + 1 = -7 \\ 2x &= 6 \quad \text{or} \quad 2x = -8 \\ x &= 3 \quad \text{or} \quad x = -4 \end{aligned}$$

Check by substituting 3 and then -4 in the original absolute value equation to verify that the solution set is $\{-4, 3\}$. The graph is shown in Figure 27.

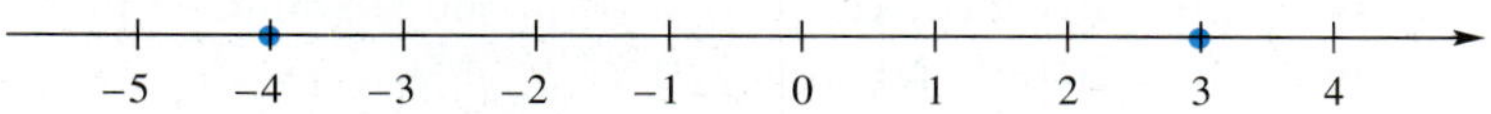

Figure 27

◀ *Work Problem 2 at the Side.*

2 Solve, check, and graph the solution set of each equation.

(a) $|x + 2| = 3$

(b) $|3x - 4| = 11$

ANSWERS

2. (a) $\{-5, 1\}$

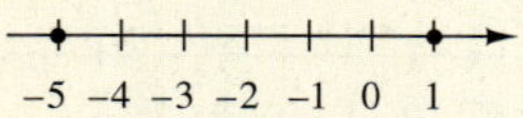

(b) $\left\{-\frac{7}{3}, 5\right\}$

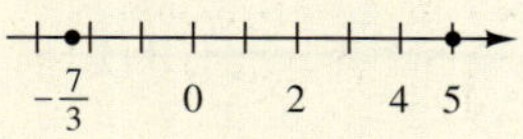

> **Note**
> Some people prefer to write the compound statements in Cases 1 and 2 of the box on the previous page as the equivalent forms
> $$ax + b = k \quad \text{or} \quad -(ax + b) = k$$
> and
> $$ax + b > k \quad \text{or} \quad -(ax + b) > k.$$
> These forms produce the same results.

OBJECTIVE 3 **Solve inequalities of the form $|ax + b| < k$ and of the form $|ax + b| > k$, for $k > 0$.**

EXAMPLE 2 **Solving an Absolute Value Inequality with >**

Solve $|2x + 1| > 7$.

By Case 2 in the box on the preceding page, this absolute value inequality is rewritten as

$$2x + 1 > 7 \quad \text{or} \quad 2x + 1 < -7,$$

because $2x + 1$ must represent a number that is *more* than 7 units from 0 on either side of the number line. Now, solve the compound inequality.

$$\begin{aligned} 2x + 1 &> 7 \quad \text{or} \quad 2x + 1 < -7 \\ 2x &> 6 \quad \text{or} \quad 2x < -8 \\ x &> 3 \quad \text{or} \quad x < -4 \end{aligned}$$

Check these solutions. The solution set is $(-\infty, -4) \cup (3, \infty)$, a disjoint interval. See the graph in Figure 28.

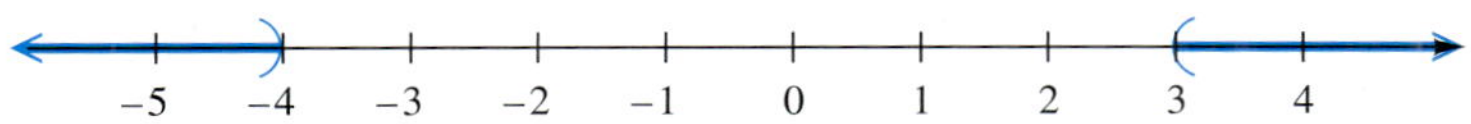

Figure 28

Work Problem **3** at the Side. ▶

EXAMPLE 3 **Solving an Absolute Value Inequality with <**

Solve $|2x + 1| < 7$.

The expression $2x + 1$ must represent a number that is less than 7 units from 0 on either side of the number line. Another way of thinking of this is to realize that $2x + 1$ must be between -7 and 7. As Case 3 in the box on the preceding page shows, this is written as the three-part inequality

$$-7 < 2x + 1 < 7.$$

$$-8 < 2x < 6 \qquad \text{Subtract 1 from each part.}$$

$$-4 < x < 3 \qquad \text{Divide each part by 2.}$$

Check that the solution set is $(-4, 3)$, so the graph consists of the open interval shown in Figure 29.

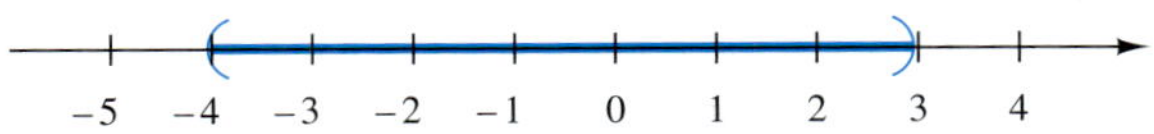

Figure 29

Work Problem **4** at the Side. ▶

3 Solve, check, and graph the solution set of each inequality.

(a) $|x + 2| > 3$

(b) $|3x - 4| \geq 11$

4 Solve, check, and graph the solution set of each inequality.

(a) $|x + 2| < 3$

(b) $|3x - 4| \leq 11$

ANSWERS

3. (a) $(-\infty, -5) \cup (1, \infty)$

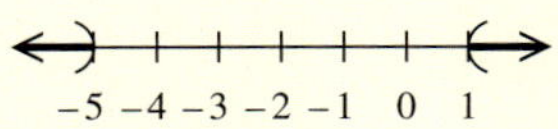

(b) $\left(-\infty, -\frac{7}{3}\right] \cup [5, \infty)$

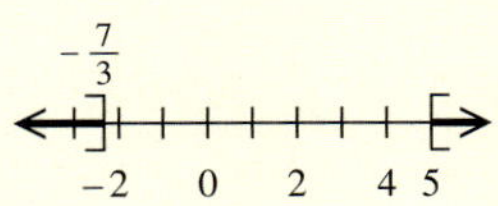

4. (a) $(-5, 1)$

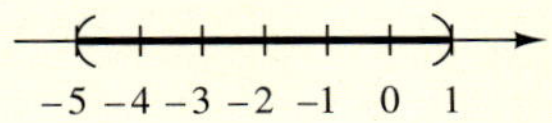

(b) $\left[-\frac{7}{3}, 5\right]$

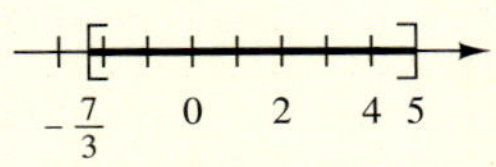

5 Solve $|5x + 2| - 9 = -7$. Check your solutions.

Look back at Figures 27, 28, and 29, with the graphs of $|2x + 1| = 7$, $|2x + 1| > 7$, and $|2x + 1| < 7$. If we find the union of the three sets, we get the set of all real numbers. This is because, for any value of x, $|2x + 1|$ will satisfy one and only one of the following: It is either equal to 7, greater than 7, or less than 7.

CAUTION

When solving absolute value equations and inequalities of the types in Examples 1, 2, and 3, remember the following.

1. The methods described apply when the constant is alone on one side of the equation or inequality and is *positive*.
2. Absolute value equations and absolute value inequalities of the form $|ax + b| > k$ translate into "or" compound statements.
3. Absolute value inequalities of the form $|ax + b| < k$ translate into "and" compound statements, which may be written as three-part inequalities.
4. An "or" statement *cannot* be written in three parts. It would be incorrect to use $-7 > 2x + 1 > 7$ in Example 2, because this would imply that $-7 > 7$, which is *false*.

OBJECTIVE 4 **Solve absolute value equations that involve rewriting.** Sometimes an absolute value equation or inequality requires some rewriting before it can be set up as a compound statement.

EXAMPLE 4 **Solving an Absolute Value Equation That Requires Rewriting**

Solve $|x + 3| + 5 = 12$.

First, get the absolute value alone on one side of the equals sign.

$$|x + 3| + 5 = 12$$

$$|x + 3| + 5 - 5 = 12 - 5 \quad \text{Subtract 5.}$$

$$|x + 3| = 7$$

Now use the method shown in Example 1.

$$x + 3 = 7 \quad \text{or} \quad x + 3 = -7$$

$$x = 4 \quad \text{or} \quad x = -10$$

Check these solutions by substituting each one in the original equation.

Check $|x + 3| + 5 = 12$

$\|4 + 3\| + 5 \stackrel{?}{=} 12$	Let $x = 4$.	$\|-10 + 3\| + 5 \stackrel{?}{=} 12$	Let $x = -10$.
$\|7\| + 5 \stackrel{?}{=} 12$		$\|-7\| + 5 \stackrel{?}{=} 12$	
$12 = 12$	True	$12 = 12$	True

The check confirms that the solution set is $\{-10, 4\}$.

◀ *Work Problem* **5** *at the Side.*

CAUTION

When solving an equation like the one in Example 4, do *not* simply drop the absolute value bars.

ANSWER

5. $\left\{-\frac{4}{5}, 0\right\}$

We use a method similar to that used in Example 4 to solve an absolute value *inequality* that requires rewriting:

$\lvert x+3 \rvert + 5 \geq 12$	$\lvert x+3 \rvert + 5 \leq 12$
$\lvert x+3 \rvert \geq 7$	$\lvert x+3 \rvert \leq 7$
$x+3 \geq 7 \quad \text{or} \quad x+3 \leq -7$	$-7 \leq x+3 \leq 7$
$x \geq 4 \quad \text{or} \quad x \leq -10.$	$-10 \leq x \leq 4.$
Solution set: $(-\infty, -10] \cup [4, \infty)$	Solution set: $[-10, 4]$

Work Problem 6 at the Side. ▶

OBJECTIVE 5 Solve equations of the form $\lvert ax + b \rvert = \lvert cx + d \rvert$. By definition, for two expressions to have the same absolute value, they must either be equal or be negatives of each other.

> **Solving $\lvert ax + b \rvert = \lvert cx + d \rvert$**
>
> To solve an absolute value equation of the form
>
> $$\lvert ax + b \rvert = \lvert cx + d \rvert,$$
>
> solve the compound equation
>
> $$ax + b = cx + d \quad \text{or} \quad ax + b = -(cx + d).$$

EXAMPLE 5 Solving an Equation with Two Absolute Values

Solve $\lvert z + 6 \rvert = \lvert 2z - 3 \rvert$.

This equation is satisfied either if $z + 6$ and $2z - 3$ are equal to each other, or if $z + 6$ and $2z - 3$ are negatives of each other. Thus,

$$z + 6 = 2z - 3 \quad \text{or} \quad z + 6 = -(2z - 3)$$
$$z + 9 = 2z \quad \text{or} \quad z + 6 = -2z - 3$$
$$9 = z \quad \text{or} \quad 3z = -3$$
$$z = 9 \quad \text{or} \quad z = -1.$$

Check that the solution set is $\{9, -1\}$.

Work Problem 7 at the Side. ▶

OBJECTIVE 6 Solve special cases of absolute value equations and inequalities. When a typical absolute value equation or inequality involves a *negative* constant or 0 alone on one side, we use the properties of absolute value to solve. Keep the following in mind.

> **Special Cases of Absolute Value**
>
> *Case 1* The absolute value of an expression can never be negative—that is, $\lvert a \rvert \geq 0$ for all real numbers a.
>
> *Case 2* The absolute value of an expression equals 0 only when the expression is equal to 0.

6 Solve each inequality, and graph the solution set.

(a) $\lvert x + 2 \rvert - 3 > 2$

(b) $\lvert 3x + 2 \rvert + 4 \leq 15$

7 Solve each equation.

(a) $\lvert k - 1 \rvert = \lvert 5k + 7 \rvert$

(b) $\lvert 4r - 1 \rvert = \lvert 3r + 5 \rvert$

ANSWERS

6. (a) $(-\infty, -7) \cup (3, \infty)$

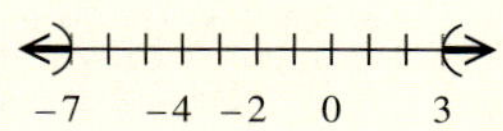

(b) $\left[-\frac{13}{3}, 3\right]$

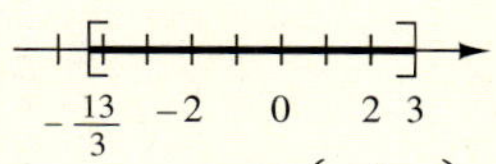

7. (a) $\{-1, -2\}$ **(b)** $\left\{-\frac{4}{7}, 6\right\}$

8 Solve each equation.

(a) $|6x + 7| = -5$

(b) $\left|\frac{1}{4}x - 3\right| = 0$

9 Solve.

(a) $|x| > -1$

(b) $|x| < -5$

(c) $|x + 2| \leq 0$

(d) $|t - 10| - 2 \leq -3$

ANSWERS

8. **(a)** $\emptyset$ **(b)** $\{12\}$
9. **(a)** $(-\infty, \infty)$ **(b)** $\emptyset$ **(c)** $\{-2\}$ **(d)** $\emptyset$

EXAMPLE 6 Solving Special Cases of Absolute Value Equations

Solve each equation.

(a) $|5r - 3| = -4$

See Case 1 in the box on the preceding page. ***The absolute value of an expression can never be negative,*** so there are no solutions for this equation. The solution set is $\emptyset$.

(b) $|7x - 3| = 0$

See Case 2 in the box on the preceding page. The expression $7x - 3$ will equal 0 *only* if

$$7x - 3 = 0$$

$$7x = 3 \qquad \text{Add 3.}$$

$$x = \frac{3}{7}. \qquad \text{Divide by 7.}$$

Thus, the solution set of the original equation is $\{\frac{3}{7}\}$, with just one element. Check this solution by substituting it in the original equation.

◀ Work Problem **8** at the Side.

EXAMPLE 7 Solving Special Cases of Absolute Value Inequalities

Solve each inequality.

(a) $|x| \geq -4$

The absolute value of a number is always greater than or equal to 0. Thus, $|x| \geq -4$ is true for *all* real numbers. The solution set is $(-\infty, \infty)$.

(b) $|x + 6| - 3 < -5$

$$|x + 6| < -2 \qquad \text{Add 3 to each side.}$$

There is no number whose absolute value is less than -2, so this inequality has no solution. The solution set is $\emptyset$.

(c) $|x - 7| + 4 \leq 4$

$$|x - 7| \leq 0 \qquad \text{Subtract 4 from each side.}$$

The value of $|x - 7|$ will never be less than 0. However, $|x - 7|$ will *equal* 0 when $x = 7$. Therefore, the solution set is $\{7\}$.

◀ Work Problem **9** at the Side.

3.3 Exercises

FOR EXTRA HELP

MyMathLab | Math XL PRACTICE | WATCH | DOWNLOAD | READ | REVIEW

Match each absolute value equation or inequality in Column I with the graph of its solution set in Column II.

1.

I	II
$\|x\| = 5$	**A.** 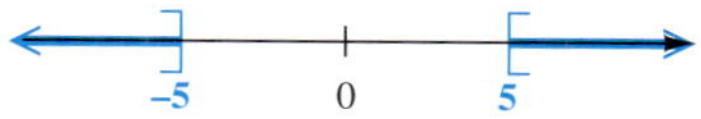
$\|x\| < 5$	**B.** 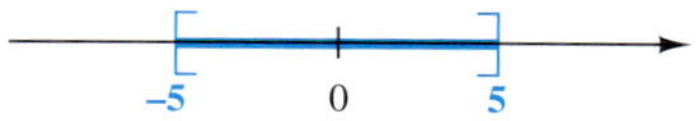
$\|x\| > 5$	**C.** 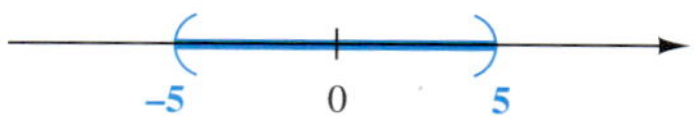
$\|x\| \le 5$	**D.** 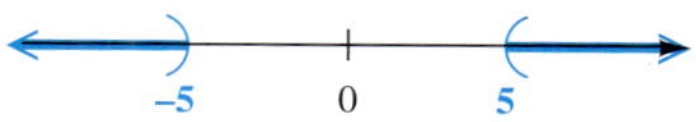
$\|x\| \ge 5$	**E.** 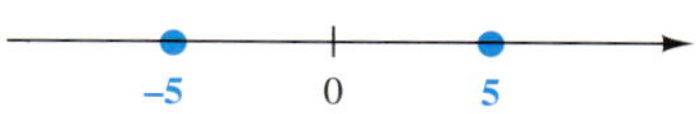

2.

I	II
$\|x\| = 9$	**A.** 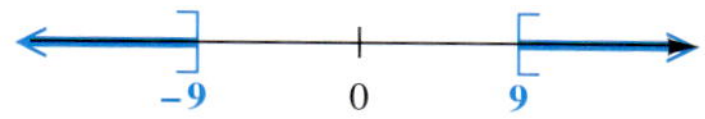
$\|x\| > 9$	**B.** 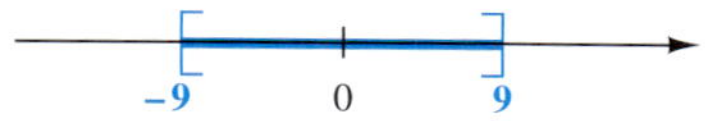
$\|x\| \ge 9$	**C.** 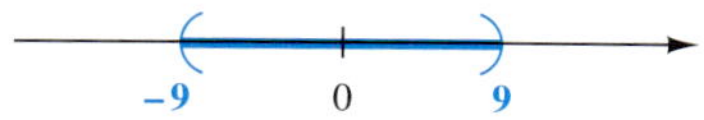
$\|x\| < 9$	**D.** 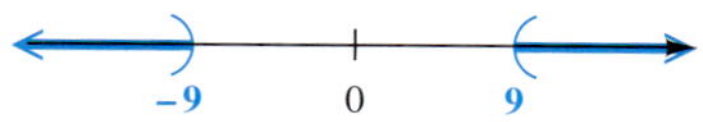
$\|x\| \le 9$	**E.** 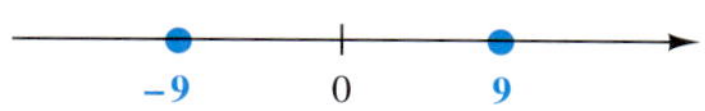

3. How many solutions will $|ax + b| = k$ have if
(a) $k = 0$; **(b)** $k > 0$; **(c)** $k < 0$?

4. Explain when to use *and* and when to use *or* if you are solving an absolute value equation or inequality of the form $|ax + b| = k$, $|ax + b| < k$, or $|ax + b| > k$, where k is a positive number.

Solve each equation. See Example 1.

5. $|x| = 12$ **6.** $|x| = 14$ **7.** $|4x| = 20$ **8.** $|5x| = 30$

9. $|x - 3| = 9$ **10.** $|p - 5| = 13$ **11.** $|2x + 1| = 9$ **12.** $|2x + 3| = 19$

13. $|4r - 5| = 17$ **14.** $|5t - 1| = 21$ **15.** $|2x + 5| = 14$ **16.** $|2x - 9| = 18$

17. $\left|\frac{1}{2}x + 3\right| = 2$ **18.** $\left|\frac{2}{3}q - 1\right| = 5$ **19.** $\left|1 - \frac{3}{4}k\right| = 7$ **20.** $\left|2 - \frac{5}{2}m\right| = 14$

Solve each inequality, and graph the solution set. See Example 2.

21. $|x| > 3$

22. $|x| > 2$

23. $|k| \geq 4$

24. $|r| \geq 1$

25. $|t + 2| > 8$

26. $|r + 5| > 20$

27. $|3x - 1| \geq 8$

28. $|4x + 1| \geq 21$

29. $|3 - x| > 5$

30. $|5 - x| > 3$

31. The graph of the solution set of $|2x + 1| = 9$ is given here.

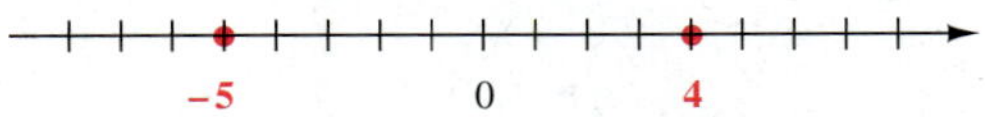

Without actually doing the algebraic work, graph the solution set of each inequality, referring to the graph above.

(a) $|2x + 1| < 9$

0

(b) $|2x + 1| > 9$

0

32. The graph of the solution set of $|3x - 4| < 5$ is given here.

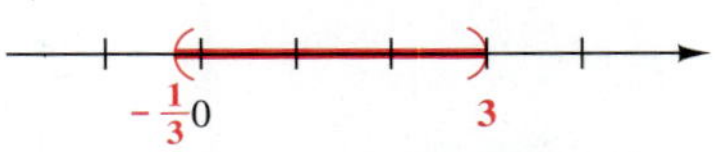

Without actually doing the algebraic work, graph the solution set of the equation and the inequality, referring to the graph above.

(a) $|3x - 4| = 5$

0

(b) $|3x - 4| > 5$

0

Solve each inequality, and graph the solution set. See Example 3. (Hint: Compare your answers to those in Exercises 21–30.)

33. $|x| \leq 3$

34. $|x| \leq 2$

35. $|k| < 4$

36. $|r| < 1$

37. $|t + 2| \leq 8$

38. $|r + 5| \leq 20$

39. $|3x - 1| < 8$

40. $|4x + 1| < 21$

41. $|3 - x| \leq 5$

42. $|5 - x| \leq 3$

Exercises 43–50 represent a sampling of the various types of absolute value equations and inequalities. Decide which method of solution applies, find the solution set, and graph. See Examples 1–3.

43. $|-4 + k| > 6$

44. $|-3 + t| > 5$

45. $|7 + 2z| = 5$

46. $|9 - 3p| = 3$

47. $|3r - 1| \leq 11$

48. $|2s - 6| \leq 6$

49. $|-3x - 8| \le 4$

50. $|-2x - 6| \le 5$

Solve each equation or inequality. Give the solution set using set notation for equations and interval notation for inequalities. See Example 4.

51. $|x| - 1 = 4$

52. $|x| + 3 = 10$

53. $|x + 4| + 1 = 2$

54. $|x + 5| - 2 = 12$

55. $|2x + 1| + 3 > 8$

56. $|6x - 1| - 2 > 6$

57. $|x + 5| - 6 \le -1$

58. $|r - 2| - 3 \le 4$

Solve each equation. See Example 5.

59. $|3x + 1| = |2x + 4|$

60. $|7x + 12| = |x - 8|$

61. $\left|m - \frac{1}{2}\right| = \left|\frac{1}{2}m - 2\right|$

62. $\left|\frac{2}{3}r - 2\right| = \left|\frac{1}{3}r + 3\right|$

63. $|6x| = |9x + 1|$

64. $|13x| = |2x + 1|$

65. $|2p - 6| = |2p + 11|$

66. $|3x - 1| = |3x + 9|$

Solve each equation or inequality. See Examples 6 and 7.

67. $|x| \geq -10$

68. $|x| \geq -15$

69. $|12t - 3| = -8$

70. $|13w + 1| = -3$

71. $|4x + 1| = 0$

72. $|6r - 2| = 0$

73. $|2q - 1| < -6$

74. $|8n + 4| < -4$

75. $|x + 5| > -9$

76. $|x + 9| > -3$

77. $|7x + 3| \leq 0$

78. $|4x - 1| \leq 0$

79. $|5x - 2| \geq 0$

80. $|4 + 7x| \geq 0$

81. $|10z + 7| > 0$

82. $|4x + 1| > 0$

83. $|x - 2| + 3 \geq 2$

84. $|k - 4| + 5 \geq 4$

85. The 2007 recommended daily intake (RDI) of calcium for females aged 19–50 is 1000 mg/day. Actual vitamin needs vary from person to person. Write an absolute value inequality to express the RDI plus or minus 100 mg and solve it. (*Source:* Food and Nutrition Board, National Academy of Sciences Institute of Medicine.)

86. The average clotting time of blood is 7.45 sec with a variation of plus or minus 3.6 sec. Write this statement as an absolute value inequality and solve it.

Relating Concepts (Exercises 87–90) For Individual or Group Work

The 10 tallest buildings in Kansas City, Missouri, are listed along with their heights.

Building	Height (in feet)
One Kansas City Place	632
Town Pavilion	591
Hyatt Regency Crown Center	504
Kansas City Power and Light	481
Fidelity Bank and Trust Building	454
City Hall	443
1201 Walnut	427
Federal Office Building	413
Commerce Tower	407
City Center Square	404

Source: *World Almanac and Book of Facts.*

Use this information to ***work Exercises 87–90 in order.***

87. To find the average of a group of numbers, we add the numbers and then divide by the number of items added. Use a calculator to find the average of the heights.

88. Let k represent the average height of these buildings. If a height x satisfies the inequality

$$|x - k| < t,$$

then the height is said to be within t feet of the average. Using your result from Exercise 87, list the buildings that are within 50 ft of the average.

89. Repeat Exercise 88, but find the buildings that are within 75 ft of the average.

90. **(a)** Write an absolute value inequality that describes the height of a building that is *not* within 75 ft of the average.

(b) Solve the inequality you wrote in part (a).

(c) Use the result of part (b) to find the buildings that are not within 75 ft of the average.

(d) Confirm that your answer to part (c) makes sense by comparing it with your answer to Exercise 89.

Summary Exercises on Solving Linear and Absolute Value Equations and Inequalities

This section of miscellaneous equations and inequalities provides practice in solving all the types introduced in **Chapters 2 and 3.** You might wish to refer to the boxes in these chapters that summarize the various methods of solution.

Solve each equation or inequality. Give the solution set using set notation for equations and interval notation for inequalities.

1. $4z + 1 = 49$

2. $|m - 1| = 6$

3. $6q - 9 = 12 + 3q$

4. $3p + 7 = 9 + 8p$

5. $|a + 3| = -4$

6. $2m + 1 \leq m$

7. $8r + 2 \geq 5r$

8. $4(a - 11) + 3a = 20a - 31$

9. $2q - 1 = -7$

10. $|3a - 7| - 4 = 0$

11. $6z - 5 \leq 3z + 10$

12. $|5z - 8| + 9 \geq 7$

13. $9x - 3(x + 1) = 8x - 7$

14. $|x| \geq 8$

15. $9x - 5 \geq 9x + 3$

16. $13p - 5 > 13p - 8$

17. $|q| < 5.5$

18. $4z - 1 = 12 + z$

19. $\frac{2}{3}x + 8 = \frac{1}{4}x$

20. $-\frac{5}{8}x \geq -20$

21. $\frac{1}{4}p < -6$

22. $7z - 3 + 2z = 9z - 8z$

23. $\frac{3}{5}q - \frac{1}{10} = 2$

24. $|r - 1| < 7$

25. $r + 9 + 7r = 4(3 + 2r) - 3$

26. $6 - 3(2 - p) < 2(1 + p) + 3$

27. $|2p - 3| > 11$

28. $\frac{x}{4} - \frac{2x}{3} = -10$

29. $|5a + 1| \le 0$

30. $5z - (3 + z) \ge 2(3z + 1)$

31. $-2 \le 3x - 1 \le 8$

32. $-1 \le 6 - x \le 5$

33. $|7z - 1| = |5z + 3|$

34. $|p + 2| = |p + 4|$

35. $|1 - 3x| \ge 4$

36. $\frac{1}{2} \le \frac{2}{3}r \le \frac{5}{4}$

37. $-(m + 4) + 2 = 3m + 8$

38. $\frac{p}{6} - \frac{3p}{5} = p - 86$

39. $-6 \le \frac{3}{2} - x \le 6$

40. $|5 - x| < 4$

41. $|x - 1| \ge -6$

42. $|2r - 5| = |r + 4|$

43. $8q - (1 - q) = 3(1 + 3q) - 4$

44. $8x - (x + 3) = -(2x + 1) - 12$

45. $|r - 5| = |r + 9|$

46. $|r + 2| < -3$

47. $2x + 1 > 5$ or $3x + 4 < 1$

48. $1 - 2x \ge 5$ and $7 + 3x \ge -2$

Study Skills

MANAGING YOUR TIME

OBJECTIVES

1. Create a semester schedule.
2. Create a "to do" list.

Many college students find themselves juggling a difficult schedule and multiple responsibilities. Perhaps you are going to school, working part time, and managing family demands. Here are some tips to help you develop good time management skills and habits.

- **Read the syllabus for each class.** Check on class policies, such as attendance, late homework, and make-up tests. Find out how you are graded. Keep the syllabus in your notebook.
- **Make a semester or quarter calendar.** Put test dates and major due dates for *all* your classes on the same calendar. That way you will see which weeks are the really busy ones. Try using a different color pen for each class. A sample semester calendar is given on the next page.
- **Make a weekly schedule.** After you fill in your classes and other regular responsibilities (such as work, picking up kids from school, etc.), block off some study periods during the day that you can guarantee you will use for studying. Aim for 2 hours of study for each 1 hour you are in class.
- **Make "to-do" lists.** Then use them by crossing off the tasks as you complete them. You might even number them in the order they need to be done (most important ones first).
- **Break big assignments into smaller chunks.** They won't seem so big that way. Make deadlines for each small part so that you stay on schedule.
- **Give yourself short breaks when studying.** Do not try to study for hours at a time. Your body needs rest between periods of learning. Try to give yourself a 10 minute break each hour or so. You will learn more and remember it longer.
- **If you get off schedule, just try to get back on schedule tomorrow.** We all slip from time to time. Make a new "to-do" list and start doing the most important things first.
- **Get help when you need it.** Talk with your instructor during office hours. Also, most colleges have some kind of learning center, tutoring center, or counseling office. If you feel lost or overwhelmed, ask for help. Someone can help you prioritize and decide what to spend your time on right away.

What two or three of the suggestions above will you try this week? How do you think they will help you?

1. ____________________
2. ____________________
3. ____________________

Why Are These Techniques Brain Friendly?

We are creatures of habit. We enjoy a little routine. For example, if you choose the same study time and place each day, you will find that you quickly settle in to your reading or homework.

You function better when you are calm. Too much rushing around at the last minute to get your homework and studying done makes it more difficult for you to learn and remember. So, a little planning can really pay off.

Building rest into your schedule is also good for you. Many people benefit from short "power" naps.

We've suggested using color on your calendars. Messy and hard to read calendars will not be helpful, and you probably won't look at them often.

We provide this sample semester calendar as a place to record tests, due dates, study time, appointments, and other regular responsibilities.

SEMESTER CALENDAR

WEEK	MON	TUES	WED	THUR	FRI	SAT	SUN
1							
2							
3							
4							
5							
6							
7							
8							
9							
10							
11							
12							
13							
14							
15							
16							

Chapter 3 Summary

▸ Key Terms

	Term	Definition
3.1	**interval**	An interval is a portion of a number line.
	interval notation	The notation used to indicate an interval on the number line is called interval notation.
	inequality	An inequality is a statement with algebraic expressions related by $<$, $\leq$, $>$, or $\geq$.
	linear inequality in one variable	A linear inequality in the variable x can be written in the form $Ax + B < C$, where A, B, and C are real numbers, with $A \neq 0$. (Other inequality symbols may be used.)
	equivalent inequalities	Equivalent inequalities are inequalities with the same solution set.
3.2	**intersection**	The intersection of two sets A and B is the set of elements that belong to both A and B.
	compound inequality	A compound inequality is formed by joining two inequalities with a connective word such as *and* or *or*.
	union	The union of two sets A and B is the set of elements that belong to either A or B (or both).
3.3	**absolute value equation; absolute value inequality**	Absolute value equations and inequalities are equations and inequalities that involve the absolute value of a variable expression.

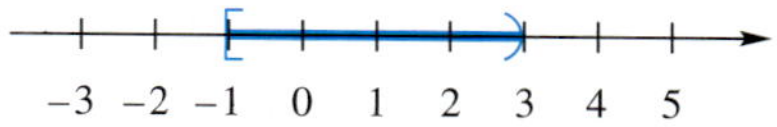

The interval [−1, 3)

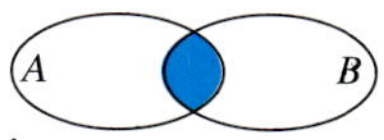

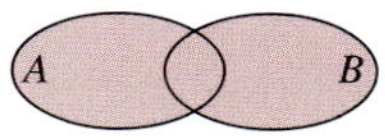

▸ New Symbols

∞	infinity	$\cap$	set intersection
$-\infty$	negative infinity	$\cup$	set union
$(-\infty, \infty)$	the set of real numbers		

▸ Test Your Word Power

See how well you have learned the vocabulary in this chapter. Answers, with examples, follow the Quick Review.

1. An **inequality** is
   - **A.** a statement that two algebraic expressions are equal
   - **B.** a point on a number line
   - **C.** an equation with no solutions
   - **D.** a statement with algebraic expressions related by $<$, $\leq$, $>$, or $\geq$.

2. **Interval notation** is
   - **A.** a portion of a number line
   - **B.** a special notation for describing a point on a number line
   - **C.** a way to use symbols to describe an interval on a number line
   - **D.** a notation to describe unequal quantities.

3. The **intersection** of two sets A and B is the set of elements that belong
   - **A.** to both A and B
   - **B.** to either A or B, or both
   - **C.** to either A or B, but not both
   - **D.** to just A.

4. The **union** of two sets A and B is the set of elements that belong
   - **A.** to both A and B
   - **B.** to either A or B, or both
   - **C.** to either A or B, but not both
   - **D.** to just B.

▸ Quick Review

Concepts	Examples

3.1 Linear Inequalities in One Variable

Solving Linear Inequalities in One Variable

Step 1 Simplify each side of the inequality by clearing parentheses, fractions, and decimals, as needed, and combining like terms.

Step 2 Use the addition property of inequality to get all terms with variables on one side and all terms without variables on the other side.

Step 3 Use the multiplication property of inequality to write the inequality in the form $x < k$ or $x > k$.

If an inequality is multiplied or divided by a negative number, the direction of the inequality symbol must be reversed.

To solve a three-part inequality, work with all three parts at the same time.

Solve $3(x + 2) - 5x \leq 12$.

$$3x + 6 - 5x \leq 12 \quad \text{Distributive property}$$
$$-2x + 6 \leq 12$$
$$-2x + 6 - 6 \leq 12 - 6 \quad \text{Subtract 6.}$$
$$-2x \leq 6$$
$$\frac{-2x}{-2} \geq \frac{6}{-2} \quad \text{Divide by } -2; \text{ change } \leq \text{ to } \geq.$$
$$x \geq -3$$

The solution set $[-3, \infty)$ is graphed below.

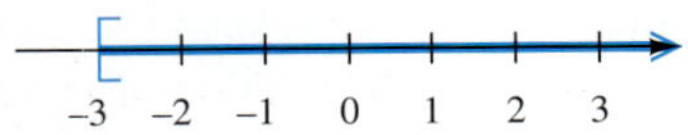

Solve $-4 < 2x + 3 \leq 7$.

$$-4 - 3 < 2x + 3 - 3 \leq 7 - 3 \quad \text{Subtract 3.}$$
$$-7 < 2x \leq 4$$
$$\frac{-7}{2} < \frac{2x}{2} \leq \frac{4}{2} \quad \text{Divide by 2.}$$
$$-\frac{7}{2} < x \leq 2$$

The solution set $(-\frac{7}{2}, 2]$ is graphed below.

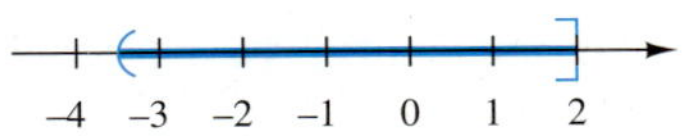

3.2 Set Operations and Compound Inequalities

Solving a Compound Inequality

Step 1 Solve each inequality in the compound inequality individually.

Step 2 If the inequalities are joined with *and,* the solution set is the intersection of the two individual solution sets.

If the inequalities are joined with *or,* the solution set is the union of the two individual solution sets.

Solve $x + 1 > 2$ and $2x < 6$.

$$x + 1 > 2 \quad \text{and} \quad 2x < 6$$
$$x > 1 \quad \text{and} \quad x < 3$$

The solution set is $(1, 3)$.

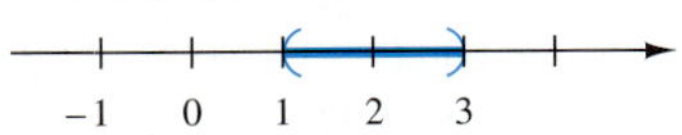

Solve $x \geq 4$ or $x \leq 0$.
The solution set is $(-\infty, 0] \cup [4, \infty)$.

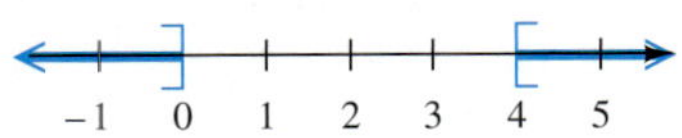

Concepts	Examples

3.3 Absolute Value Equations and Inequalities

Let k be a positive number.

To solve $|ax + b| = k$, solve the compound equation

$$ax + b = k \quad \text{or} \quad ax + b = -k.$$

Solve $|x - 7| = 3$.

$$x - 7 = 3 \quad \text{or} \quad x - 7 = -3$$
$$x = 10 \quad \text{or} \quad x = 4$$

The solution set is $\{4, 10\}$.

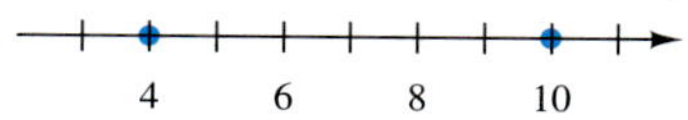

To solve $|ax + b| > k$, solve the compound inequality

$$ax + b > k \quad \text{or} \quad ax + b < -k.$$

Solve $|x - 7| > 3$.

$$x - 7 > 3 \quad \text{or} \quad x - 7 < -3$$
$$x > 10 \quad \text{or} \quad x < 4$$

The solution set is $(-\infty, 4) \cup (10, \infty)$.

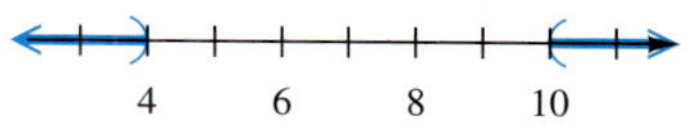

To solve $|ax + b| < k$, solve the compound inequality

$$-k < ax + b < k.$$

Solve $|x - 7| < 3$.

$$-3 < x - 7 < 3$$
$$4 < x < 10 \qquad \text{Add 7 to each part.}$$

The solution set is $(4, 10)$.

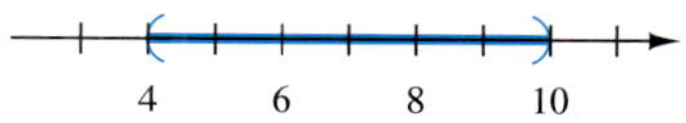

To solve an absolute value equation of the form

$$|ax + b| = |cx + d|,$$

solve the compound equation

$$ax + b = cx + d \quad \text{or} \quad ax + b = -(cx + d).$$

Solve $|x + 2| = |2x - 6|$.

$$x + 2 = 2x - 6 \quad \text{or} \quad x + 2 = -(2x - 6)$$
$$x = 8 \qquad\qquad x + 2 = -2x + 6$$
$$3x = 4$$
$$x = \frac{4}{3}$$

The solution set is $\{\frac{4}{3}, 8\}$.

ANSWERS TO TEST YOUR WORD POWER

1. D; *Examples:* $x < 5$, $7 + 2k \geq 11$, $-5 < 2z - 1 \leq 3$
2. C; *Examples:* $(-\infty, 5]$, $(1, \infty)$, $[-3, 3)$
3. A; *Example:* If $A = \{2, 4, 6, 8\}$ and $B = \{1, 2, 3\}$, $A \cap B = \{2\}$.
4. B; *Example:* Using the preceding sets A and B, $A \cup B = \{1, 2, 3, 4, 6, 8\}$.

Math in the Media

AN APPLICATION OF MATHEMATICS ANY STUDENT CAN APPRECIATE

The May 19, 2008 issue of *USA Today* included an article entitled "Great Education Debate: Reforming the Grade System." Pros and cons of a new movement to make 50% (rather than 0%) the minimum score for teachers to assign on graded assignments were discussed. Mo Denis, president of the Nevada Parent Teacher Association, is not sure how his kids would react if there was a minimum-F policy at their school.

"I don't know if that would motivate them more or less," says Denis, who is a state assemblyman from Las Vegas. ***"I wonder if they would just figure out what they needed to do to pass the class. They seem to know how to do that particular kind of math really well."***

Suppose that one intermediate algebra teacher bases final grades on points earned for activities as given in the Graded Classwork table on the left. To determine final grades, the teacher strictly adheres to the point ranges given in the Grade Distribution table on the right.

GRADED CLASSWORK

Activity	Points Available
Homework and vocabulary	45
Daily activities (scaled)	55
Lab participation and completion	100
Major exams (3 at 100 points)	300
Final Exam	150
Total points	650

GRADE DISTRIBUTION

Grade	Points Required
A	585–650
B	520–584
C	455–519
IP*	$<$ 455 and active
F	$<$ 455 and inactive

*In Progress

Notice that exams account for 450 of the possible 650 points. The remaining 200 points should be fairly easy to earn by keeping up with the day-to-day course requirements.

Assumption: You earn a "baseline" number of points based on the following criteria:

(1) You earn *all* of the homework and vocabulary points. **(2)** You earn a minimum of 50 points based on daily activities. **(3)** You earn a minimum of 90 lab participation and completion points.

1. Assume that you earn the baseline number of points. Let x represent the test points to be earned. Write and solve linear inequalities to find the minimum number of points that you need in test scores to earn grades no lower than A, B, and C. What "test average" is each minimum score? Round *up* to the nearest whole percent.

2. Write a compound inequality to find the range of points that you need in test scores to earn a B average. Solve the inequality. What range of "test averages" are those minimum scores? Round *up* to the nearest whole percent.

3. Suppose that Mark earns only 15 points in homework and vocabulary, 40 points in daily activities, and 50 points in lab participation. Write and solve linear inequalities to find the minimum number of points that Mark needs in test scores to earn grades no lower than A, B, and C. What "test average" is each minimum score? Round *up* to the nearest whole percent.

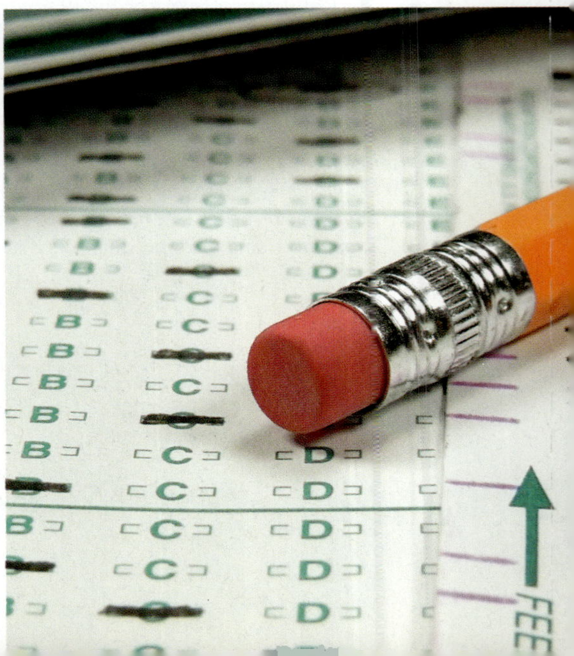

Chapter 3 Review Exercises

[3.1] *Solve each inequality. Give the solution set in both interval and graph forms.*

1. $-\frac{2}{3}x < 6$

2. $-5x - 4 \geq 11$

3. $\frac{5a + 3}{-4} < -3$

4. $\frac{9x + 5}{-3} > 3$

5. $5 - (6 - 4t) \geq 2t - 7$

6. $-6 \leq 2k \leq 24$

7. $8 \leq 3x - 1 < 14$

8. $-4 < 3 - 2z < 9$

Solve each problem.

9. The perimeter of a rectangular playground must be no greater than 120 m. The width of the playground must be 22 m. Find the possible lengths of the playground.

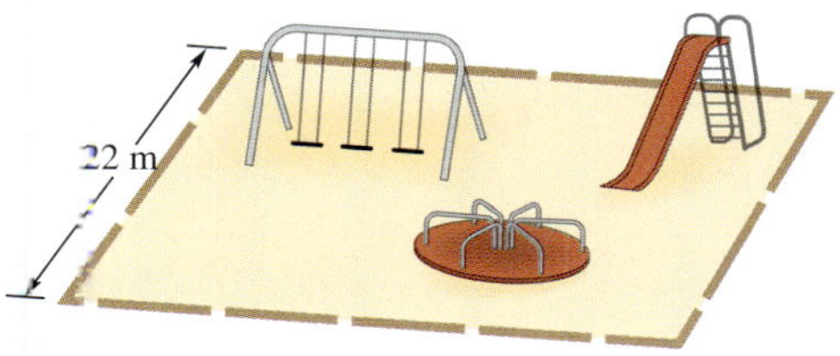

10. A group of college students wants to buy tickets to attend a performance of Monty Python's *Spamalot* at the Cadillac Palace Theatre in Chicago. The best price they can find is a group rate of \$89 per ticket. If they have \$2000 available to spend on tickets and they qualify for a \$50 group discount, how many tickets can they purchase?

11. To pass algebra, a student must have an average of at least 70 on five tests. On the first four tests, a student has scores of 75, 79, 64, and 71. What possible scores on the fifth test would guarantee a passing score in the class?

12. While solving the inequality

$$10x + 2(x - 4) < 12x - 13,$$

a student did all the work correctly and obtained the statement $-8 < -13$. The student did not know what to do at this point, because the variable "disappeared." How would you explain to the student the interpretation of this result?

[3.2] *Let* $A = \{a, b, c, d\}$, $B = \{a, c, e, f\}$, *and* $C = \{a, e, f, g\}$. *Find each set.*

13. $A \cap B$

14. $A \cap C$

15. $B \cup C$

16. $A \cup C$

Solve each compound inequality. Give the solution set in both interval and graph forms.

17. $x > 4$ and $x < 7$

18. $x + 4 > 12$ and $x - 2 < 12$

19. $x > 5$ or $x \leq -3$

20. $x \geq -2$ or $x < 2$

21. $x - 4 > 6$ and $x + 3 \leq 10$

22. $-5x + 1 \geq 11$ or $3x + 5 \geq 26$

Express each union or intersection in simplest interval form.

23. $(-3, \infty) \cap (-\infty, 4)$

24. $(-\infty, 6) \cap (-\infty, 2)$

25. $(4, \infty) \cup (9, \infty)$

26. $(1, 2) \cup (1, \infty)$

27. The numbers of civilian workers (to the nearest thousand) for several states in a recent year are shown in the table.

NUMBER OF WORKERS

State	Female	Male
Illinois	2,979,000	3,407,000
Maine	334,000	362,000
North Carolina	1,973,000	2,270,000
Oregon	840,000	1,010,000
Utah	538,000	668,000
Wisconsin	1,456,000	1,615,000

Source: U.S. Bureau of Labor Statistics.

List the elements of each set.

(a) The set of states with less than 1 million female workers *and* more than 1 million male workers

(b) The set of states with less than 1 million female workers *or* more than 2 million male workers

(c) The set of states with a total of more than 7 million civilian workers

[3.3] *Solve each absolute value equation.*

28. $|x| = 7$

29. $|x + 2| = 9$

30. $|3k - 7| = 8$

31. $|z - 4| = -12$

32. $|2k - 7| + 4 = 11$

33. $|4a + 2| - 7 = -3$

34. $|3p + 1| = |p + 2|$

35. $|2m - 1| = |2m + 3|$

Solve each absolute value inequality. Give the solution set in both interval and graph forms.

36. $|x| < 12$

37. $|-x + 6| \leq 7$

38. $|2p + 5| \leq 1$

39. $|x + 1| \geq -3$

40. $|5r - 1| > 9$

41. $|3x + 6| \geq 0$

Mixed Review Exercises

Solve.

42. $(7 - 2x) + 3(5 - 3x) \geq x + 8$

43. $x < 5$ and $x \geq -4$

44. $\frac{3}{4}(a - 2) - \frac{1}{3}(5 - 2a) < -2$

45. To qualify for a company pension plan, an employee must average at least \$1000 per month in earnings. During the first four months of the year, an employee made \$900, \$1200, \$1040, and \$760. What possible amounts earned during the fifth month will qualify the employee?

46. $-5r \geq -10$

47. $|7x - 2| > 9$

48. $|2x - 10| = 20$

49. $|m + 3| \leq 13$

50. $x \geq -2$ or $x < 4$

51. $|m - 1| = |2m + 3|$

In Exercises 52 and 53, sketch the graph of each solution set.

52. $x > 6$ and $x < 8$

53. $-5x + 1 \geq 6$ or $3x + 5 \geq 26$

54. If $k < 0$, what is the solution set of

(a) $|5x + 3| < k$, **(b)** $|5x + 3| > k$, **(c)** $|5x + 3| = k$?

Chapter 3 Test

Use the Chapter Test Prep Video CD to see fully worked-out solutions to any of the exercises you want to review

1. What is the special rule that must be remembered when multiplying or dividing each side of an inequality by a negative number?

Solve each inequality. Give the solution set in both interval and graph forms.

2. $4 - 6(x + 3) \leq -2 - 3(x + 6) + 3x$

3. $-\frac{4}{7}x > -16$

4. $-6 \leq \frac{4}{3}x - 2 \leq 2$

5. Which one of the following inequalities is equivalent to $x < -3$?

A. $-3x < 9$ **B.** $-3x > -9$ **C.** $-3x > 9$ **D.** $-3x < -9$

6. The graph shows the percentage of the U.S. population that was foreign born for selected years. During which years was the percentage
(a) at least 7%, **(b)** less than 6%, **(c)** between 10% and 12%?

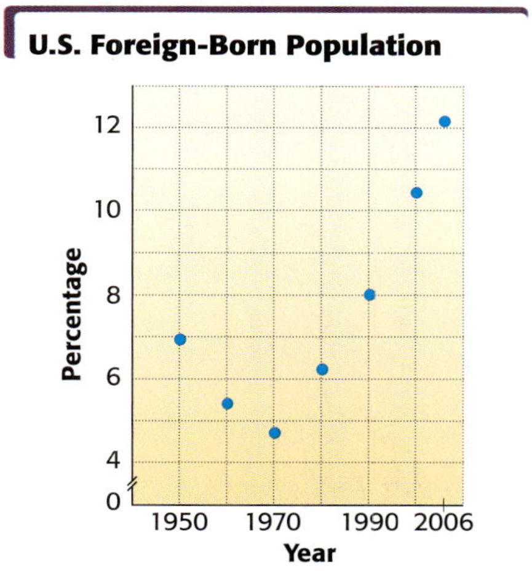

Source: U.S. Census Bureau.

Solve each problem.

7. Justin Sudak must have an average of at least 80 on the four tests in a course to get a B. He had scores of 83, 76, and 79 on the first three tests. What possible scores on the fourth test would guarantee him a B in the course?

8. A product will break even or produce a profit only if the revenue R (in dollars) from selling the product is at least the cost C (in dollars) of producing it. Suppose that the cost to produce x units of carpet is $C = 50x + 5000$, while the revenue is $R = 60x$. For what values of x is R at least equal to C? Give the answer using interval notation.

9. (a) ______________ (b) ______________

9. Let $A = \{1, 2, 5, 7\}$ and $B = \{1, 5, 9, 12\}$. Find

(a) $A \cap B$, **(b)** $A \cup B$.

10. ______________

10. Solve $x \leq 2$ and $x \geq 2$.

Solve each compound or absolute value inequality. For Exercises 11–15, give the solution set in both interval and graph forms.

11. $3k \geq 6$ and $k - 4 < 5$

12. $-4x \leq -24$ or $4x - 2 < 10$

13. $|4x + 3| \leq 7$

14. $|5 - 6x| > 12$

15. $|-3x + 4| - 4 < -1$

16. ______________

16. $|7 - x| \leq -1$

Solve each absolute value equation.

17. ______________

17. $|3k - 2| + 1 = 8$

18. ______________

18. $|3 - 5x| = |2x + 8|$

19. ______________

19. $|4x + 3| + 5 = 4$

20. ______________

20. If $k < 0$, what is the solution set of $|8x - 5| < k$? of $|8x - 5| > k$? of $|8x - 5| = k$?

Study Skills

TIPS FOR TAKING MATH TESTS

OBJECTIVES

1. **Apply suggestions to tests and quizzes.**
2. **Develop a set of "best practices" to apply while testing.**

Techniques to Improve Your Test Score	Comments
Come prepared with a pencil, eraser, and calculator, if allowed. If you are easily distracted, sit in the corner farthest from the door.	***Working in pencil lets you erase,*** keeping your work neat and readable.
Scan the entire test, note the point value of different problems, and plan your time accordingly. Allow at least five minutes to check your work at the end of the testing time.	If you have 50 minutes to do 20 problems, $50 \div 20 = 2.5$ minutes per problem. ***Spend less time on easy ones,*** more time on problems with higher point values.
Read directions carefully, and circle any significant words. When you finish a problem, read the directions again to make sure you did what was asked.	***Pay attention to announcements*** written on the board or made by your instructor. Ask if you don't understand. You don't want to get problems wrong because you misread the directions.
Show your work. Most math teachers give partial credit if some of the steps in your work are correct, even if the final answer is wrong. ***Write neatly.*** If you like to scribble when first working or checking a problem, do it on scratch paper.	***If your teacher can't read your writing, you won't get credit for it.*** If you need more space to work, ask if you can use extra pieces of paper that you hand in with your test paper.
Check that the answer to an application problem is reasonable and makes sense. Read the problem again to make sure you've answered the question.	***Use common sense.*** Can the father really be seven years old? Would a month's rent be $32,140? Label your answer: $, years, inches, etc.
To check for careless errors, rework the problem without looking at your previous work. Cover your work with a piece of scratch paper, and pretend you are doing the problem for the first time. Then compare the two answers.	If you just "look over" your work, your mind can make the same mistake again without noticing it. ***Reworking the problem from the beginning forces you to rethink it.*** If possible, use a different method to solve the problem the second time.

Techniques to Reduce Anxiety	Comments
Do not try to review up until the last minute before the test. Instead, go for a walk, do some deep breathing, and arrive just in time for the test. Ignore other students.	Listening to anxious classmates before the test ***may cause you to panic.*** Moderate exercise and deep breathing will calm your mind.

(continued)

Why Are These Suggestions Brain Friendly?

Several of the suggestions address anxiety. Reducing anxiety allows you to think clearly.

Remember that your subconscious continues to work on a difficult problem even if you skip it and go on to the next one. Your mind will often come through for you if you are open to the idea.

Some of the suggestions ask you to use your common sense. Follow the directions, show your work, write neatly, and pay attention to whether your answers really make sense.

Techniques to Reduce Anxiety	Comments
Do a "knowledge dump" as soon as you get the test. Write important notes to yourself in a corner of the test paper, such as formulas or common errors you want to watch out for.	***Writing down tips and things that you've memorized lets you relax.*** You won't have to worry about forgetting those things and can refer to them as needed.
Do the easy problems first in order to build confidence. If you feel your anxiety starting to build, *immediately* stop for a minute, close your eyes, and take several slow, deep breaths.	***Greater confidence helps you get the easier problems correct.*** Anxiety causes shallow breathing, which leads to confusion and reduced concentration. Deep breathing calms you
As you work on more difficult problems, you may have negative thoughts such as, "I can't do it," or "Who cares about this test anyway?" In your mind, think "STOP" and take several deep, slow breaths. Or, replace the negative thoughts with positive ones.	***Try writing one of these positive statements on the top of your test paper.*** • I know I can do it. • I can do this one step at a time. • I've studied hard, and I'll do the best I can. • This test is a positive challenge for me to show what I've learned.
Read the harder problems twice. Write down *anything* that might help solve the problem: a formula, a diagram, etc. If you still can't get it, circle the problem and ***come back to it later***. Do *not* erase any of the things you wrote down.	If you know even a little bit about the problem, write it down. ***The answer may come to you as you work on it, or you may get partial credit.*** Don't spend too long on any one problem. Your subconscious mind will work on the tough problem while you go on with the test.
If you still can't solve a difficult problem when you come back to it the second time, ***make a guess and do not change it.*** In this situation, your first guess is your best bet. On any problems, do not change the answer just because you're a little unsure. ***Change it only if you find an obvious mistake.***	If you are thinking about changing an answer, be sure you have a good reason for changing it. If you cannot find a specific error, leave your first answer alone. ***When the test is returned, check to see if changing answers helped or hurt you.***
Ignore students who finish early. Use the entire test time. You do not get extra credit for finishing early. Use the extra time to rework problems and correct careless errors.	Students who leave early are often the ones who didn't study or who are too anxious to continue working. If they bother you, ***sit as far from the door as possible.***

Cumulative Review Exercises ▶▶▶ Chapters 1–3

1. Write $\frac{108}{144}$ in lowest terms.

2. Is the statement *true* or *false?*

$$\frac{8(7) - 5(6 + 2)}{3 \cdot 5 + 1} \geq 1$$

Perform the indicated operations.

3. $\frac{5}{6} + \frac{1}{4} - \frac{7}{15}$

4. $\frac{9}{8} \cdot \frac{16}{3} \div \frac{5}{8}$

5. $9 - (-4) + (-2)$

6. $\frac{-4(9)(-2)}{-3^2}$

7. $|-7 - 1|(-4) + (-4)$

Evaluate each exponential expression.

8. $(-5)^3$

9. $\left(\frac{3}{2}\right)^4$

Evaluate each expression if $x = 2$, $y = -3$, and $z = 4$.

10. $-2y + 4(x - 3z)$

11. $\frac{3x^2 - y^2}{4z}$

Name each property illustrated.

12. $7(k + m) = 7k + 7m$

13. $3 + (5 + 2) = 3 + (2 + 5)$

14. Simplify $-4(k + 2) + 3(2k - 1)$.

Solve each equation, and check the solution.

15. $4 - 5(a + 2) = 3(a + 1) - 1$

16. $\frac{2}{3}x + \frac{3}{4}x = -17$

17. $\frac{2x + 3}{5} = \frac{x - 4}{2}$

18. $|3m - 5| = |m + 2|$

19. $3x + 4y = 24$ for y

20. $A = P(1 + ni)$ for n

Solve each inequality. Give the solution set in both interval and graph forms.

21. $3 - 2(x + 7) \leq -x + 3$

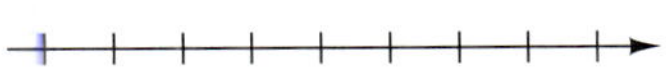

22. $-4 < 5 - 3x \leq 0$

23. $2x + 1 > 5$ or $2 - x > 2$

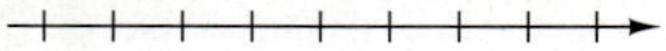

24. $|-7k + 3| \geq 4$

Solve each problem.

25. Luke Roth invested some money at 7% interest and the same amount at 10%. His total interest for the year was $150 less than one-tenth of the total amount he invested. How much did he invest at each rate?

26. A dietician must use three foods, A, B, and C, in a diet. He must include twice as many grams of food A as food C, and 5 g of food B. The three foods must total at most 24 g. What is the largest amount of food C that the dietician can use?

27. Zach Schneider got scores of 88 and 78 on his first two tests. What score must he make on his third test to keep an average of 80 or greater?

28. Two cars are 400 mi apart. Both start at the same time and travel toward one another. They meet 4 hr later. If the speed of one car is 20 mph faster than the other, what is the speed of each car?

29. Since 1995, the number of daily newspapers has steadily declined.

Year	Number of Daily Newspapers
1995	1533
1997	1509
1999	1483
2001	1468
2003	1456
2005	1452

Source: Statistical Abstract of the United States.

According to the table,

(a) by how much did the number of daily newspapers decrease between 1995 and 2005?

(b) by what *percent* (to the nearest tenth) did the number of daily newspapers decrease from 1995 to 2005?

30. For a woven hanging, Janette Krauss needs three pieces of yarn, which she will cut from a 40 cm piece. The longest piece is to be 3 times as long as the middle-sized piece, and the shortest piece is to be 5 cm shorter than the middle-sized piece. What lengths should she cut?

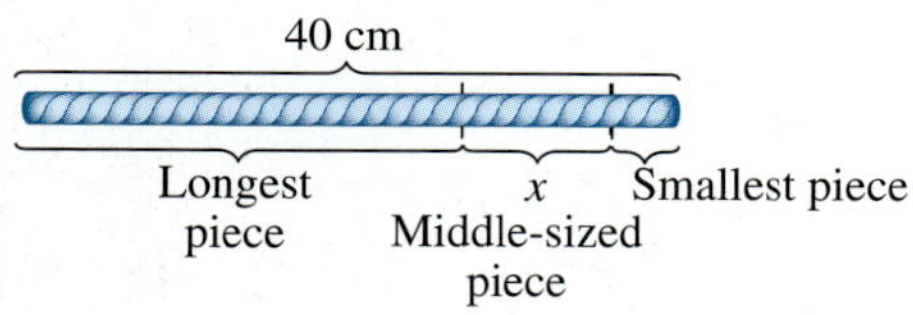

4

Graphs, Linear Equations, and Functions

4.1 The Rectangular Coordinate System

4.2 Slope of a Line

4.3 Linear Equations in Two Variables

Summary Exercises on Slopes and Equations of Lines

4.4 Linear Inequalities in Two Variables

4.5 Introduction to Functions

The two most common measures of temperature are Fahrenheit (F) and Celsius (C). We often see signs displaying one or both of these types of temperature. It is fairly common knowledge that water freezes at 32°F, or 0°C, and boils at 212°F, or 100°C. Because there is a *linear* relationship between the Fahrenheit and Celsius temperature scales, using these two equivalences, we can derive the familiar formulas for converting from one scale to the other, as seen in Exercises 71–78 of Section 4.3.

Graphs are widely used in the media because they present a great deal of information in a concise form. In this chapter, we see how information such as the relationship between the two temperature scales can be depicted by graphs.

4.1 The Rectangular Coordinate System

OBJECTIVES

1. Interpret a line graph.
2. Plot ordered pairs.
3. Find ordered pairs that satisfy a given equation.
4. Graph lines.
5. Find *x*- and *y*-intercepts.
6. Recognize equations of horizontal and vertical lines and lines passing through the origin.
7. Use the midpoint formula.

OBJECTIVE 1 Interpret a line graph. The line graph in Figure 1 shows personal spending (in billions of dollars) on medical care in the United States from 2000 through 2005. About how much was spent on medical care in 2005?

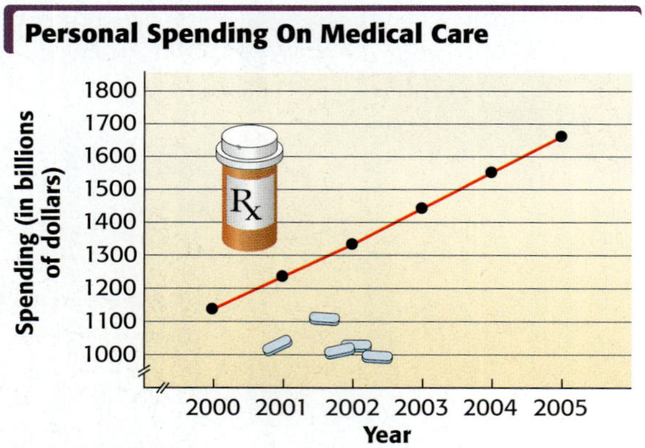

Source: U.S. Centers for Medicare and Medicaid Services.

Figure 1

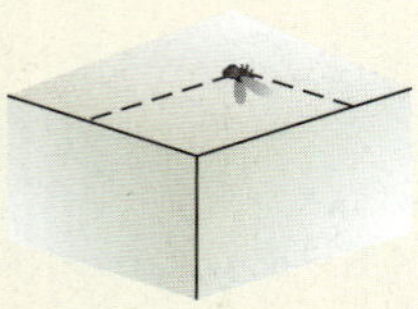

Locating a fly on a ceiling

The line graph in Figure 1 presents information based on a method for locating a point in a plane developed by René Descartes, a 17th-century French mathematician. Legend has it that Descartes, who was lying in bed ill, was watching a fly crawl about on the ceiling near a corner of the room. It occurred to him that the location of the fly on the ceiling could be described by determining its distances from the two adjacent walls. See the figure in the margin. In this chapter we use this insight to plot points and graph linear equations in two variables whose graphs are straight lines.

OBJECTIVE 2 Plot ordered pairs. Each of the pairs of numbers $(3, 1)$, $(-5, 6)$, and $(4, -1)$ is an example of an **ordered pair**—that is, a pair of numbers written within parentheses in which the order of the numbers is important. We graph an ordered pair using two perpendicular number lines that intersect at their 0 points, as shown in Figure 2. The common 0 point is called the **origin.** The position of any point in this plane is determined by referring to the horizontal number line, the ***x*-axis,** and the vertical number line, the ***y*-axis.** The first number in the ordered pair indicates the position relative to the *x*-axis, and the second number indicates the position relative to the *y*-axis. The *x*-axis and the *y*-axis make up a **rectangular** (or **Cartesian,** for Descartes) **coordinate system.**

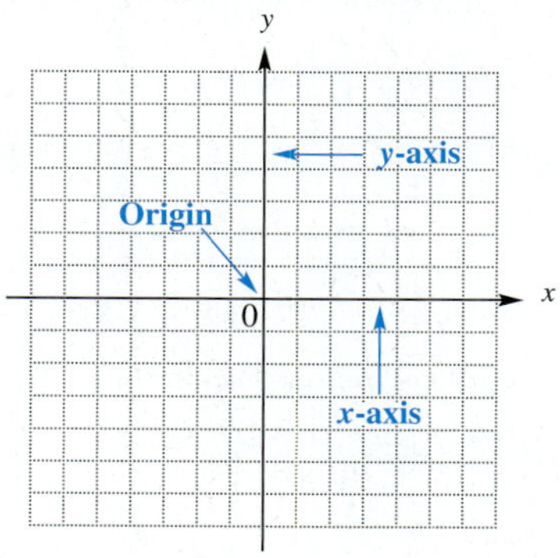

Figure 2

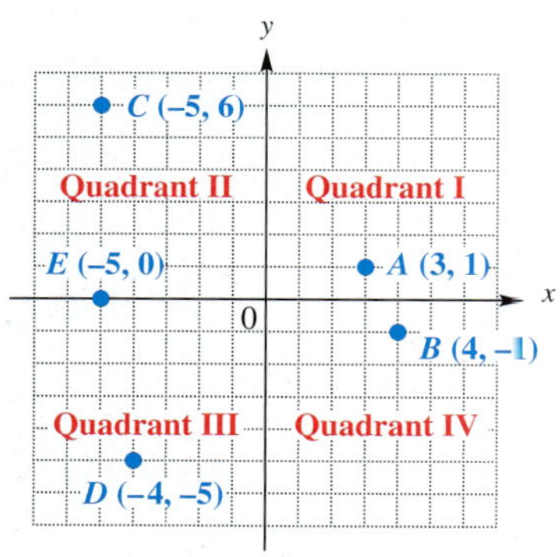

Figure 3

To locate, or **plot,** the point on the graph that corresponds to the ordered pair (3, 1), we move three units from 0 to the right along the x-axis, and then one unit up parallel to the y-axis. The point corresponding to the ordered pair (3, 1) is labeled A in Figure 3 on the preceding page. Additional points are labeled B–E. The phrase "the point corresponding to the ordered pair (3, 1)" is often abbreviated as "the point (3, 1)." The numbers in the ordered pairs are called **components** and are the **coordinates** of the corresponding point.

We can relate this method of locating ordered pairs to the line graph in Figure 1. We move along the horizontal axis to a year, then up parallel to the vertical axis to find medical spending for that year. Thus, we can write the ordered pair (2005, 1660) to indicate that in 2005, personal spending on medical care was about $1660 billion.

CAUTION

The parentheses used to represent an ordered pair are also used to represent an open interval (introduced in **Section 3.1**). The context of the discussion tells whether ordered pairs or open intervals are being represented.

The four regions of the graph, shown in Figure 3, are called **quadrants I, II, III,** and **IV,** reading counterclockwise from the upper-right quadrant. The points on the x-axis and y-axis do not belong to any quadrant. For example, point E in Figure 3 belongs to no quadrant.

Work Problem **1** *at the Side.* ▶

1 Plot each point. Name the quadrant (if any) in which each point is located.

(a) $(-4, 2)$

(b) $(3, -2)$

(c) $(-5, -6)$

(d) $(4, 6)$

(e) $(-3, 0)$

(f) $(0, -5)$

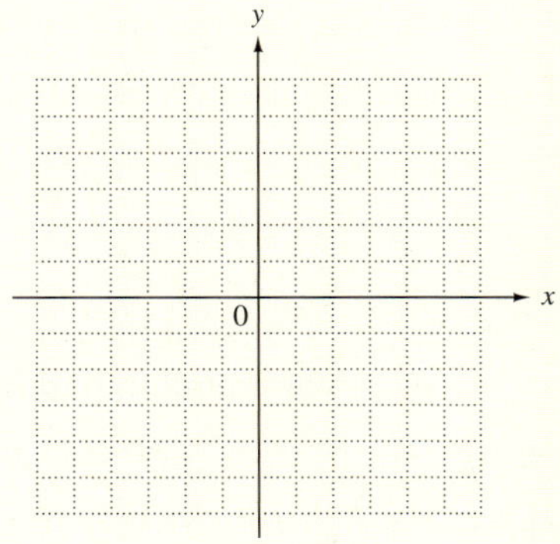

OBJECTIVE 3 Find ordered pairs that satisfy a given equation. Each solution to an equation with two variables, such as

$$2x + 3y = 6,$$

will include two numbers, one for each variable. To keep track of which number goes with which variable, we write the solutions as ordered pairs. ***(If x and y are used as the variables, the x-value is given first.)*** For example, we can show that $(6, -2)$ is a solution of $2x + 3y = 6$ by substitution.

$$2x + 3y = 6$$

$$2(6) + 3(-2) \stackrel{?}{=} 6 \quad \text{Let } x = 6, y = -2.$$

Use parentheses to avoid errors.

$$12 - 6 \stackrel{?}{=} 6 \quad \text{Multiply.}$$

$$6 = 6 \quad \text{True}$$

Because the pair of numbers $(6, -2)$ makes the equation true, it is a solution. On the other hand, (5, 1) is *not* a solution of the equation $2x + 3y = 6$ because

$$\begin{aligned} &2x + 3y \\ &= 2(5) + 3(1) \\ &= 10 + 3 \\ &= 13, \quad \textit{not} \quad 6. \end{aligned}$$

To find ordered pairs that satisfy an equation, select any number for one of the variables, substitute it into the equation for that variable, and then solve for the other variable. Two other ordered pairs satisfying $2x + 3y = 6$ are (0, 2) and (3, 0).

ANSWERS

1.

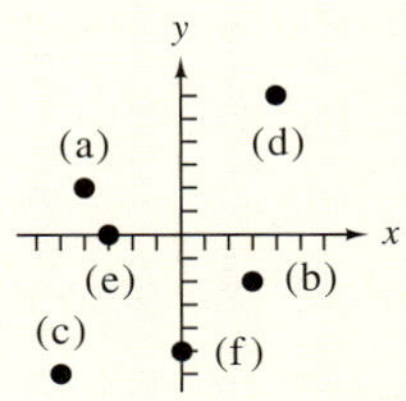

(a) II **(b)** IV **(c)** III **(d)** I **(e)** no quadrant **(f)** no quadrant

Since any real number could be selected for one variable and would lead to a real number for the other variable, linear equations in two variables have an infinite number of solutions.

EXAMPLE 1 **Completing Ordered Pairs and Making a Table**

In parts (a) and (b), complete each ordered pair for $2x + 3y = 6$. Then, in part (c), write the results as a table of ordered pairs.

(a) $(-3, \quad)$
Replace x with -3 in the equation to find y.

$$2x + 3y = 6$$
$$2(-3) + 3y = 6 \quad \text{Let } x = -3.$$
$$-6 + 3y = 6 \quad \text{Multiply.}$$
$$3y = 12 \quad \text{Add 6.}$$
$$y = 4 \quad \text{Divide by 3.}$$

The ordered pair is $(-3, 4)$.

(b) $(\quad, -4)$
Replace y with -4 in the equation to find x.

$$2x + 3y = 6$$
$$2x + 3(-4) = 6 \quad \text{Let } y = -4.$$
$$2x - 12 = 6 \quad \text{Multiply.}$$
$$2x = 18 \quad \text{Add 12.}$$
$$x = 9 \quad \text{Divide by 2.}$$

The ordered pair is $(9, -4)$.

(c) We write a table of these ordered pairs as shown.

x	y	
-3	4	← Represents the ordered pair $(-3, 4)$
9	-4	← Represents the ordered pair $(9, -4)$

◀ Work Problem 2 at the Side.

2 (a) Complete each ordered pair for $3x - 4y = 12$.

$(0, \quad)$

$(\quad, 0)$

$(\quad, -2)$

$(-4, \quad)$

(b) Make a table of the ordered pairs you found in part (a).

OBJECTIVE 4 Graph lines. The **graph of an equation** is the set of points corresponding to all ordered pairs that satisfy the equation. It gives a "picture" of the equation. Most equations in two variables are satisfied by an infinite number of ordered pairs, so their graphs include an infinite number of points.

To graph an equation, we plot a number of ordered pairs that satisfy the equation until we have enough points to suggest the shape of the graph. For example, to graph $2x + 3y = 6$, we plot the ordered pairs found in Objective 3 and Example 1. These points, shown in a table of values and plotted in Figure 4(a), appear to lie on a straight line. If all the ordered pairs that satisfy the equation $2x + 3y = 6$ were graphed, they would form the straight line shown in Figure 4(b).

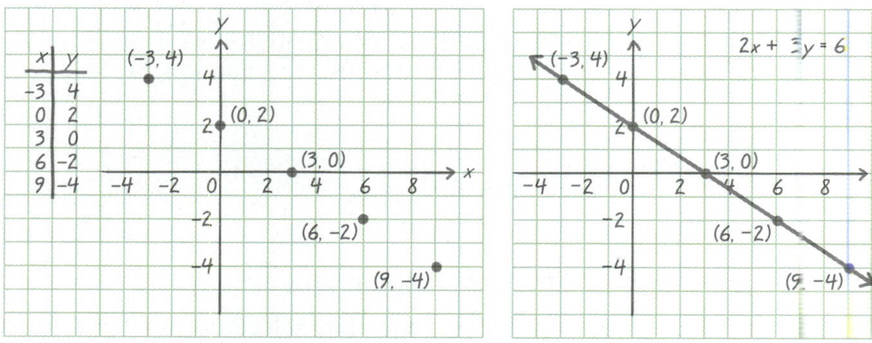

(a) (b)

Figure 4

ANSWERS

2. (a) $(0, -3), (4, 0), \left(\frac{4}{3}, -2\right), (-4, -6)$

(b)

x	y
0	-3
4	0
$\frac{4}{3}$	-2
-4	-6

Work Problem **3** *at the Side.* ▶

The equation $2x + 3y = 6$ is called a **first-degree equation** because it has no term with a variable to a power greater than one.

The graph of any first-degree equation in two variables is a straight line.

Since first-degree equations with two variables have straight-line graphs, they are called *linear equations in two variables.*

Linear Equation in Two Variables

A **linear equation in two variables** is an equation that can be written in the form

$$Ax + By = C,$$

where A, B, and C are real numbers, and A and B are not both 0. This form is called **standard form.**

OBJECTIVE 5 **Find x- and y-intercepts.** A straight line is determined if any two different points on the line are known, so finding two different points is enough to graph the line. Two useful points for graphing are the x- and y-intercepts. The ***x*-intercept** is the point (if any) where the line intersects the x-axis; likewise, the ***y*-intercept** is the point (if any) where the line intersects the y-axis.* See Figure 5.

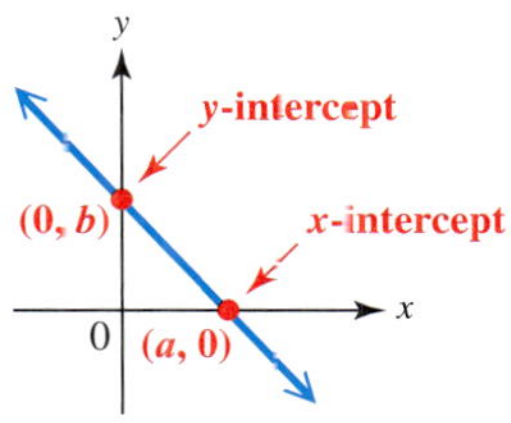

Figure 5

The y-value of the point where a line intersects the x-axis is 0. Similarly, the x-value of the point where a line intersects the y-axis is 0. This suggests a method for finding the x- and y-intercepts.

Finding Intercepts

When graphing the equation of a line, find the intercepts as follows:

Let $y = 0$ to find the x-intercept.

Let $x = 0$ to find the y-intercept.

EXAMPLE 2 **Finding Intercepts**

Find the x- and y-intercepts of $4x - y = -3$, and graph the equation.

First find the intercepts. Let $y = 0$ and then let $x = 0$.

***x*-intercept**		***y*-intercept**	
$4x - y = -3$		$4x - y = -3$	
$4x - 0 = -3$	Let $y = 0$.	$4(0) - y = -3$	Let $x = 0$.
$4x = -3$		$-y = -3$	
$x = -\frac{3}{4}$	x-intercept is $(-\frac{3}{4}, 0)$.	$y = 3$	y-intercept is $(0, 3)$.

Continued on Next Page

* Some texts define an intercept as a number, not a point.

3 Graph $3x - 4y = 12$. Use the points from Problem 2 in the margin on the previous page.

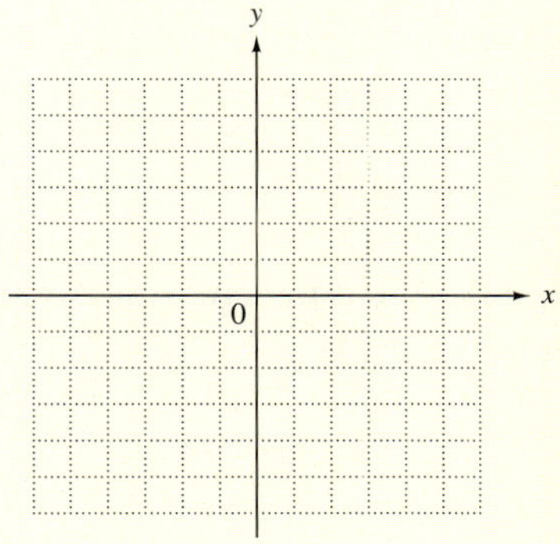

ANSWER

3.

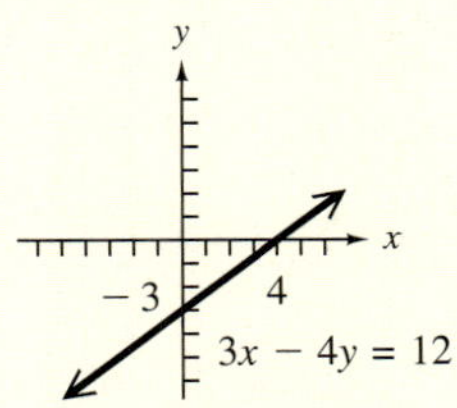

4 Find the intercepts, and graph $2x - y = 4$.

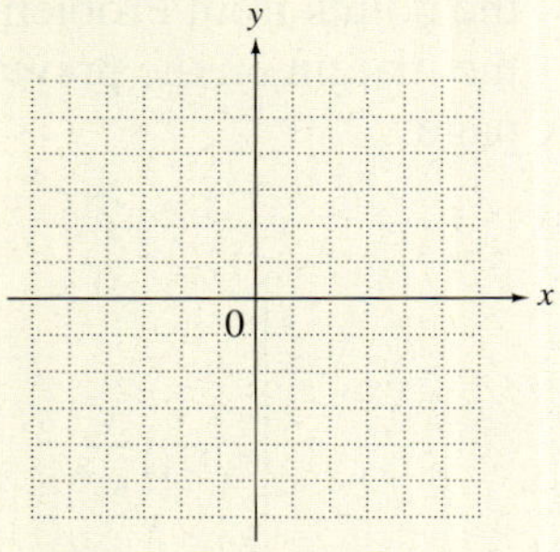

5 Find the intercepts, and graph $y + 4 = 0$.

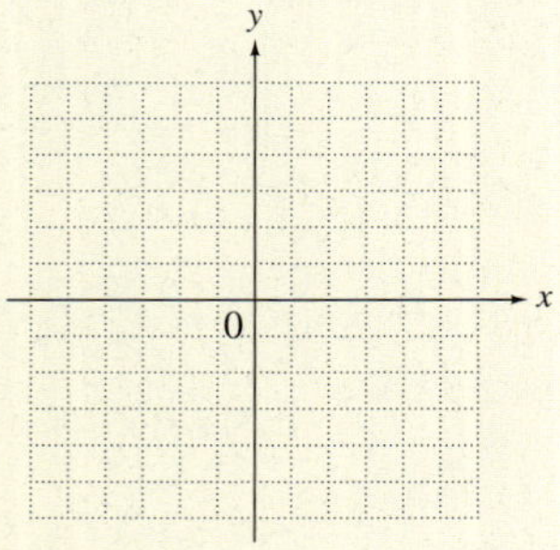

6 Find all intercepts, and graph the line $x = 2$.

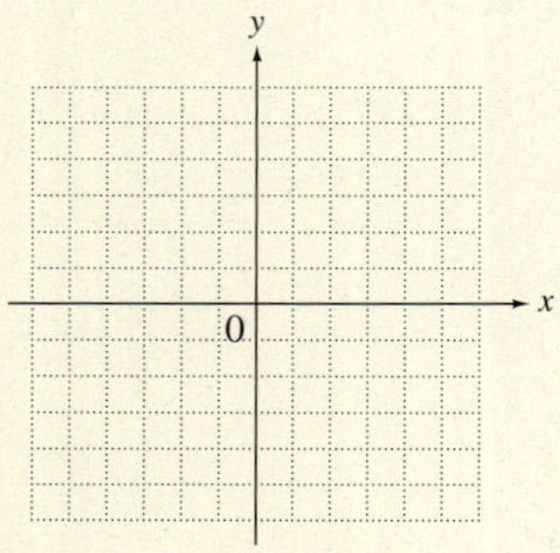

ANSWERS

4. x-intercept is $(2, 0)$; y-intercept is $(0, -4)$.

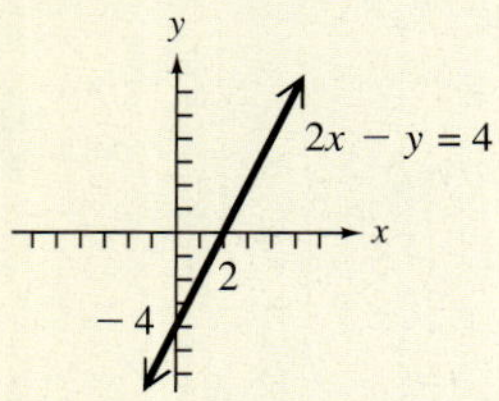

5. no x-intercept; y-intercept is $(0, -4)$.

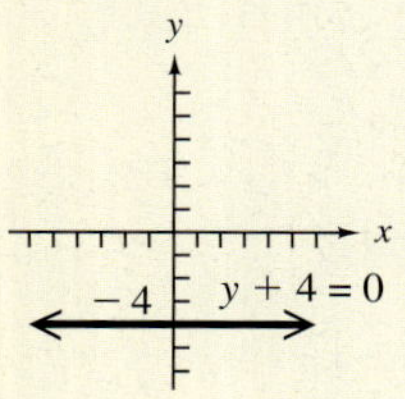

6. no y-intercept; x-intercept is $(2, 0)$.

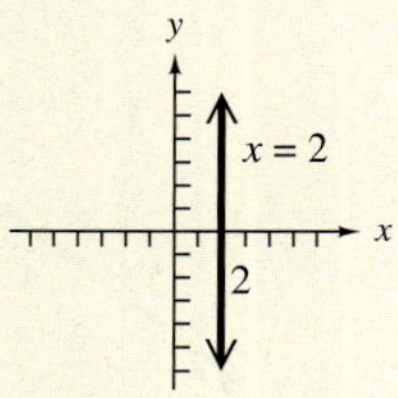

The intercepts are the two points $(-\frac{3}{4}, 0)$ and $(0, 3)$. We show these ordered pairs in the table next to Figure 6, along with the point $(-2, -5)$, and use these points to draw the graph. Verify by substitution in the equation $4x - y = -3$ that $(-2, -5)$ also lies on the graph.

x	y
$-\frac{3}{4}$	0
0	3
-2	-5

Use a third point as a check.

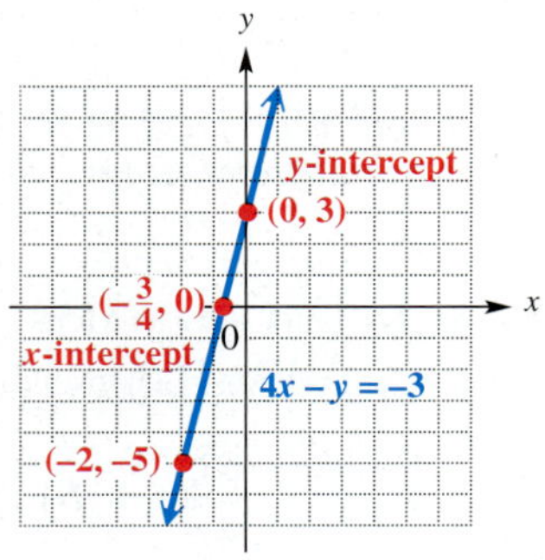

Figure 6

◀ *Work Problem* **4** *at the Side.*

OBJECTIVE 6 Recognize equations of horizontal and vertical lines and lines passing through the origin. A graph can fail to have an x-intercept or a y-intercept, which is why the phrase "if any" was added when discussing intercepts.

EXAMPLE 3 Graphing Horizontal and Vertical Lines

Graph each equation.

(a) $y = 2$

Since y is always 2, there is no value of x corresponding to $y = 0$, so the graph has no x-intercept. The y-intercept is $(0, 2)$. Plot any two other points with y-coordinate 2—for example, $(-1, 2)$ and $(3, 2)$. The graph in Figure 7, shown with a table of ordered pairs, is a horizontal line.

x	y
-1	2
0	2
3	2

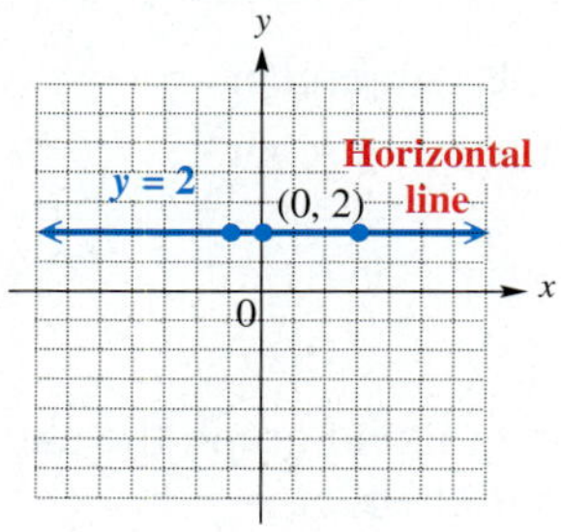

Figure 7

x	y
-1	-4
-1	0
-1	5

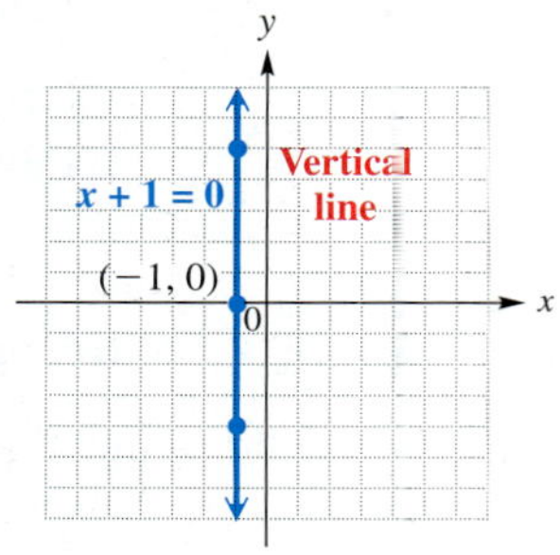

Figure 8

(b) $x + 1 = 0$

This equation can be rewritten as $x = -1$. Since x is always -1, there is no value of y that makes $x = 0$, so the graph has no y-intercept. The x-intercept is $(-1, 0)$. Plot any two other points with x-coordinate -1—for example, $(-1, -4)$ and $(-1, 5)$. The only way a straight line can have no y-intercept is if it is vertical, as in Figure 8.

◀ *Work Problems* **5** *and* **6** *at the Side.*

CAUTION

To avoid confusing equations of horizontal and vertical lines remember that

1. An equation with only the variable x will always intersect the *x-axis* and thus will be *vertical*.
2. An equation with only the variable y will always intersect the *y-axis* and thus will be *horizontal*.

Some lines have both the x- and y-intercepts at the origin.

EXAMPLE 4 Graphing a Line That Passes through the Origin

Graph $x + 2y = 0$.

Find the intercepts.

x-intercept		y-intercept	
$x + 2y = 0$		$x + 2y = 0$	
$x + 2(0) = 0$	Let $y = 0$.	$0 + 2y = 0$	Let $x = 0$.
$x + 0 = 0$		$y = 0$	y-intercept is (0, 0).
$x = 0$	x-intercept is (0, 0).		

Both intercepts are the same ordered pair, (0, 0), which means that the graph passes through the origin. To find another point to graph the line, choose any nonzero number for x, say $x = 4$, and solve for y.

$$x + 2y = 0$$
$$4 + 2y = 0 \quad \text{Let } x = 4.$$
$$2y = -4 \quad \text{Subtract 4.}$$
$$y = -2 \quad \text{Divide by 2.}$$

This gives the ordered pair $(4, -2)$. To find this additional point, we could have chosen any number (except 0) for y instead of x. The points (0, 0) and $(4, -2)$ lead to the graph shown in Figure 9. As a check, verify that $(-2, 1)$ also lies on the line.

x	y
−2	1
0	0
4	−2

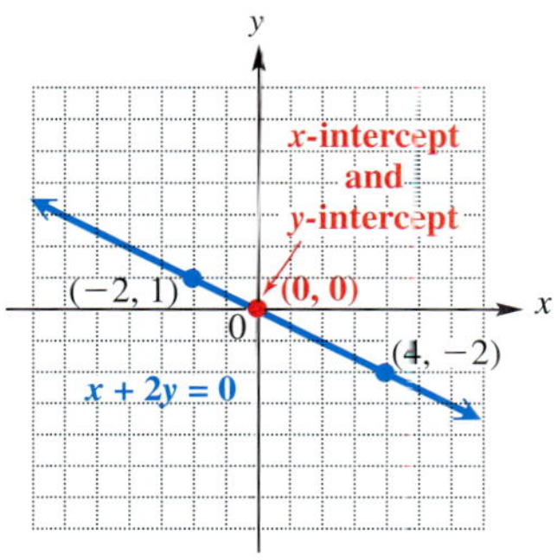

Figure 9

Work Problem 7 at the Side. ▶

7 Find the intercepts, and graph the line $3x - y = 0$.

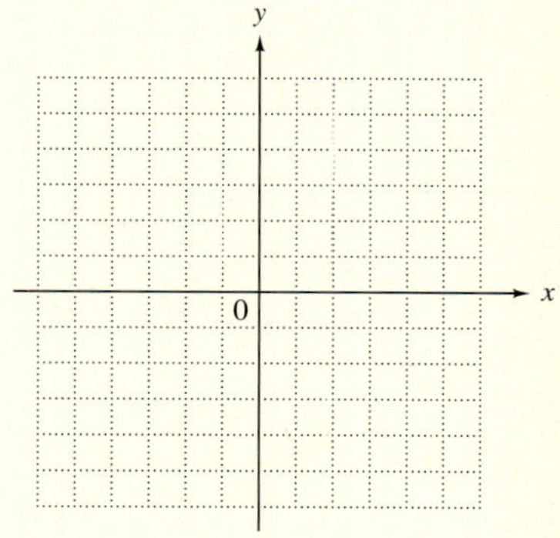

ANSWER

7. Both intercepts are (0, 0).

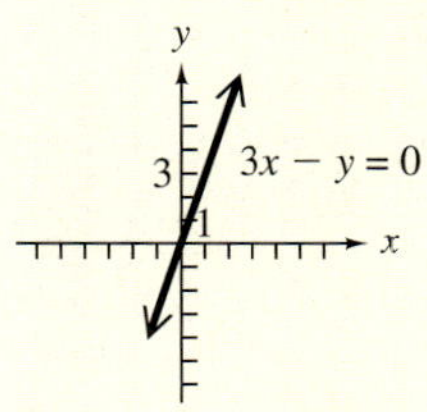

8 Find the coordinates of the midpoint of line segment PQ with endpoints $P(-5, 8)$ and $Q(2, 4)$.

OBJECTIVE 7 Use the midpoint formula. If the coordinates of the endpoints of a line segment are known, then the coordinates of the *midpoint* of the segment can be found. Figure 10 shows that segment PQ has endpoints $P(-8, 4)$ and $Q(3, -2)$. R is the point with the same x-coordinate as P and the same y-coordinate as Q. So the coordinates of R are $(-8, -2)$.

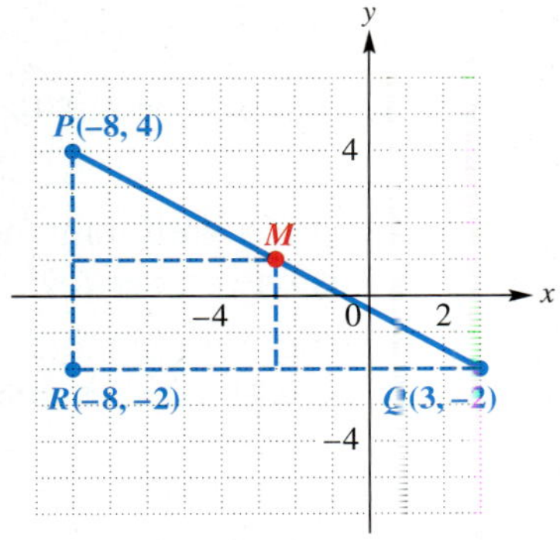

Figure 10

The x-coordinate of the midpoint M of PQ is the same as the x-coordinate of the midpoint of RQ. Since RQ is horizontal, the x-coordinate of its midpoint is the *average* of the x-coordinates of its endpoints:

$$\frac{1}{2}(-8 + 3) = -2.5.$$

The y-coordinate of M is the average of the y-coordinates of the midpoint of PR:

$$\frac{1}{2}(4 + (-2)) = 1.$$

The midpoint of PQ is $M(-2.5, 1)$. This discussion leads to the *midpoint formula.*

Midpoint Formula

If the endpoints of a line segment PQ are (x_1, y_1) and (x_2, y_2), its midpoint M is

$$\left(\frac{x_1 + x_2}{2}, \frac{y_1 + y_2}{2}\right).$$

The small numbers 1 and 2 in these ordered pairs are called **subscripts.** Read (x_1, y_1) as "x-sub-one, y-sub-one."

EXAMPLE 5 Finding the Coordinates of a Midpoint

Find the coordinates of the midpoint of line segment PQ with endpoints $P(4, -3)$ and $Q(6, -1)$.

Use the midpoint formula with $x_1 = 4$, $x_2 = 6$, $y_1 = -3$, and $y_2 = -1$:

$$\left(\frac{4 + 6}{2}, \frac{-3 + (-1)}{2}\right) = \left(\frac{10}{2}, \frac{-4}{2}\right) = (5, -2). \longleftarrow \text{Midpoint}$$

◀ Work Problem 8 at the Side.

Note

When finding the coordinates of the midpoint of a line segment, remember that you are finding the *average* of the x-coordinates and the *average* of the y-coordinates of the endpoints of the segment. In both cases, add the corresponding coordinates and divide the sum by 2.

Answer

8. $(-1.5, 6)$

4.1 Exercises

In Exercises 1 and 2, answer each part by locating ordered pairs on the graphs. See Example 1.

1. The graph indicates U.S. federal government tax revenues in billions of dollars.

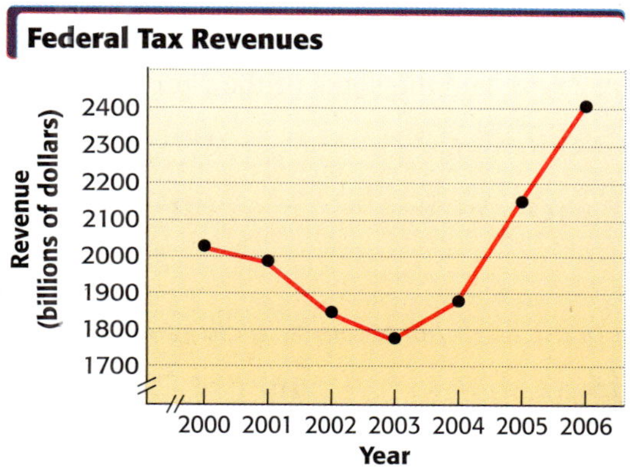

Source: Office of Management and Budget.

(a) If the ordered pair (x, y) represents a point on the graph, what does x represent? What does y represent?

(b) Estimate revenue in 2006.

(c) Write an ordered pair (x, y) that gives approximate federal tax revenues in 2006.

(d) What does the ordered pair (2004, 1880) mean in the context of this graph?

2. The graph indicates personal spending in billions of dollars on medical care in the United States.

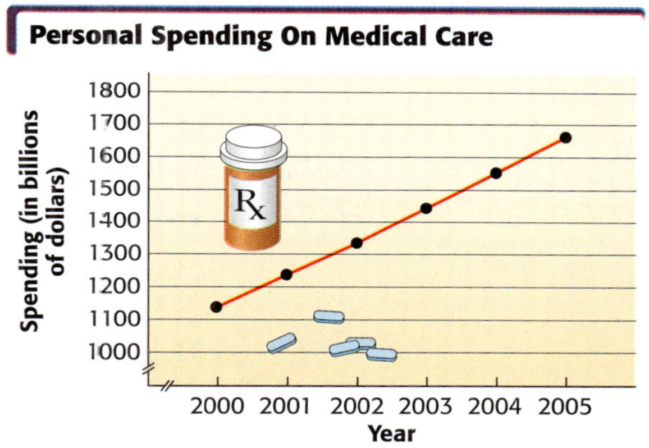

Source: U.S. Centers for Medicare and Medicaid Services.

(a) If (x, y) represents a point on the graph, what does x represent? What does y represent?

(b) What was spending in 2003?

(c) In what year was spending about \$1140 billion?

Fill in each blank with the correct response.

3. The point with coordinates (0, 0) is called the ______ of a rectangular coordinate system.

4. For any value of x, the point $(x, 0)$ lies on the ______-axis.

5. To find the x-intercept of a line, we let ______ equal 0 and solve for ______.

6. The equation ______ $(x$ or $y)$ $= 4$ has a horizontal line as its graph.

7. To graph a straight line, we must find a minimum of ______ points.

8. The point (______, 4) is on the graph of $2x - 3y = 0$.

9. The equation of the x-axis is ______.

10. The equation of the y-axis is ______.

Name the quadrant, if any, in which each point is located.

11. **(a)** $(1, 6)$ **(b)** $(-4, -2)$ **(c)** $(-3, 6)$ **(d)** $(7, -5)$ **(e)** $(-3, 0)$

12. **(a)** $(-2, -10)$ **(b)** $(4, 8)$ **(c)** $(-9, 12)$ **(d)** $(3, -9)$ **(e)** $(0, -8)$

13. Use the given information to determine the possible quadrants in which the point (x, y) must lie. (*Hint:* Consider the signs of the coordinates in each quadrant, and the signs of their product and quotient.)

(a) $xy > 0$ **(b)** $xy < 0$

(c) $\frac{x}{y} < 0$ **(d)** $\frac{x}{y} > 0$

14. What must be true about the coordinates of any point that lies on an axis?

Locate each point on the rectangular coordinate system.

15. $(2, 3)$ **16.** $(-1, 2)$ **17.** $(-3, -2)$ **18.** $(1, -4)$

19. $(0, 5)$ **20.** $(-2, -4)$ **21.** $(-2, 4)$ **22.** $(3, 0)$

23. $(-2, 0)$ **24.** $(3, -3)$

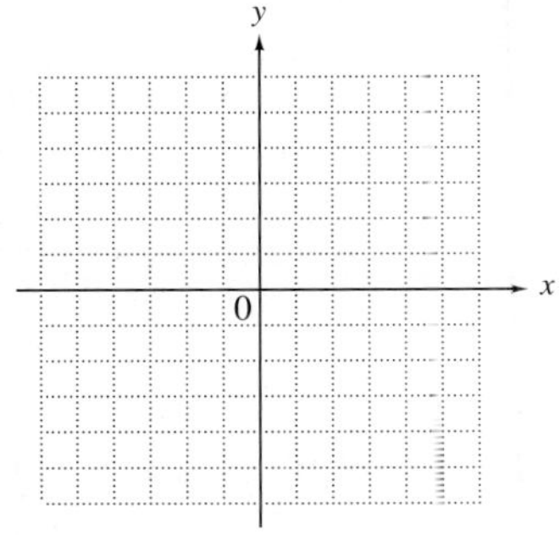

In each exercise, complete the given table for the equation, and then graph the equation. See Example 1 and Figure 4.

25. $x - y = 3$

x	y
0	
	0
5	
2	

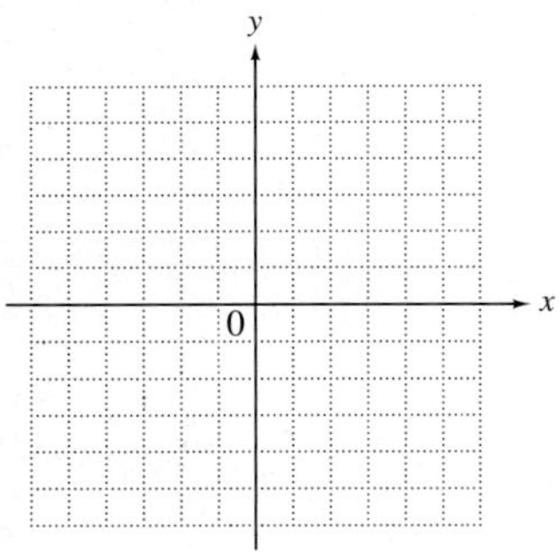

26. $x - y = 5$

x	y
0	
	0
1	
3	

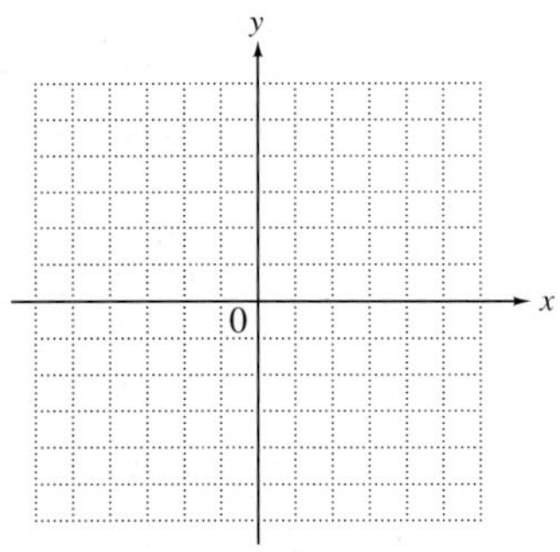

27. $x + 2y = 5$

x	y
0	
	0
2	
	2

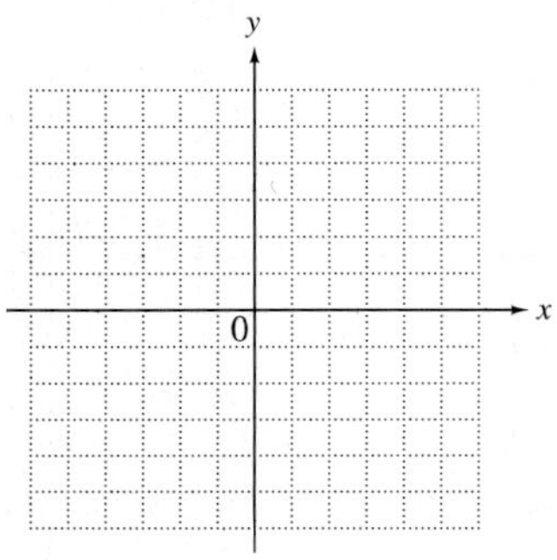

28. $x + 3y = -5$

x	y
0	
	0
1	
	−1

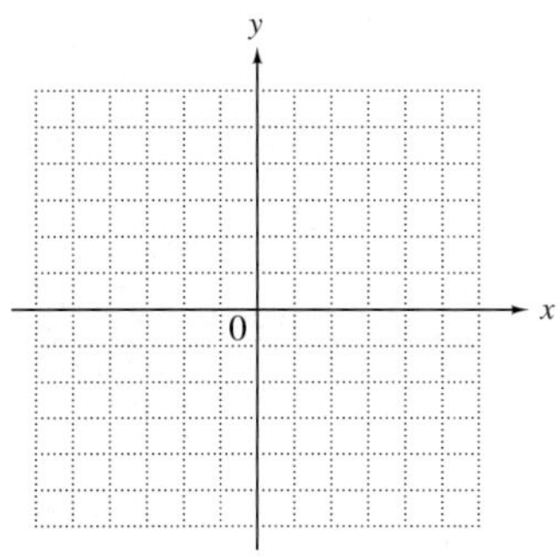

29. $4x - 5y = 20$

x	y
0	
	0
2	
	-3

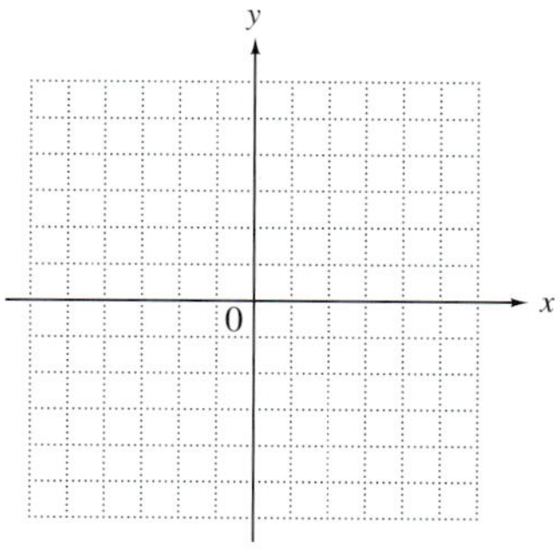

30. $6x - 5y = 30$

x	y
0	
	0
3	
	-2

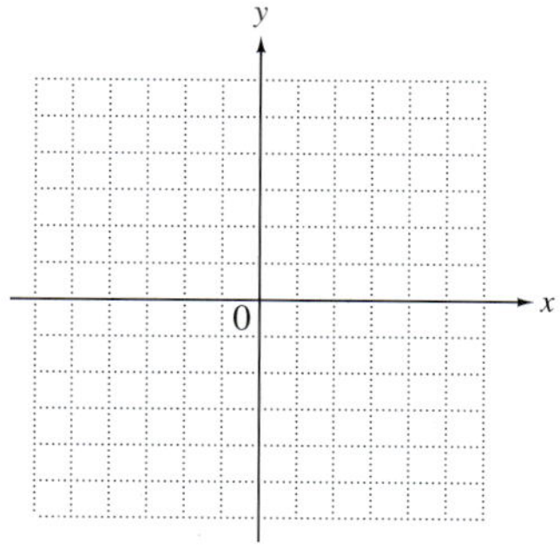

Find the x- and y-intercepts. Then graph each equation. See Examples 2–4.

31. $2x + 3y = 12$

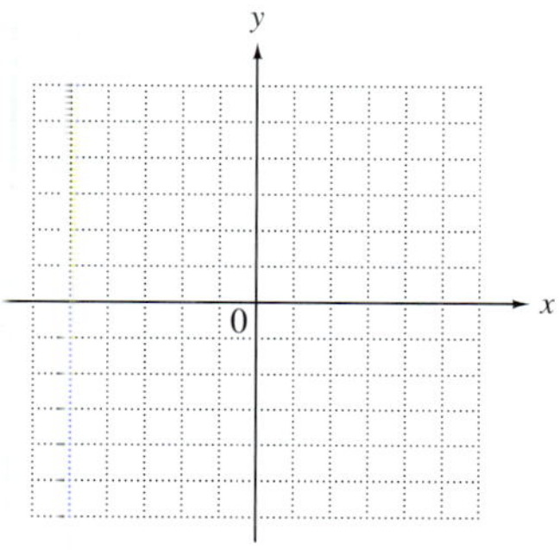

32. $5x + 2y = 10$

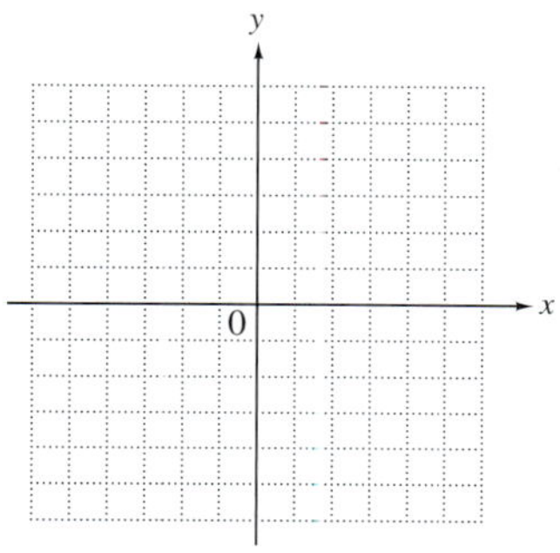

33. $x - 3y = 6$

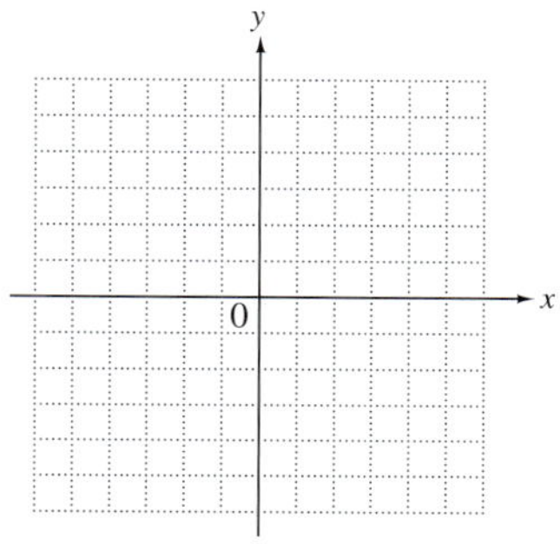

34. $x - 2y = -4$

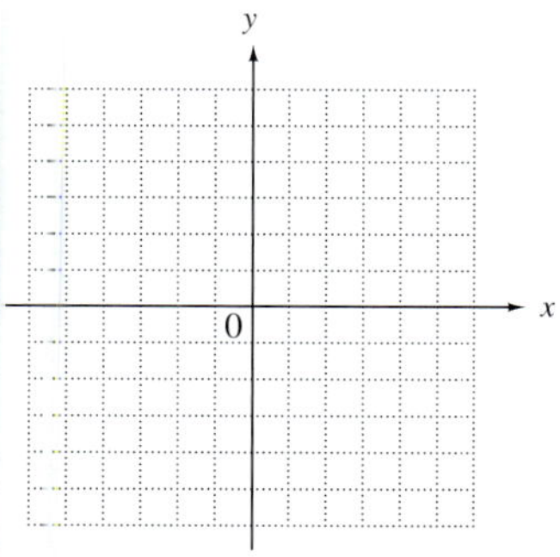

35. $3x - 7y = 9$

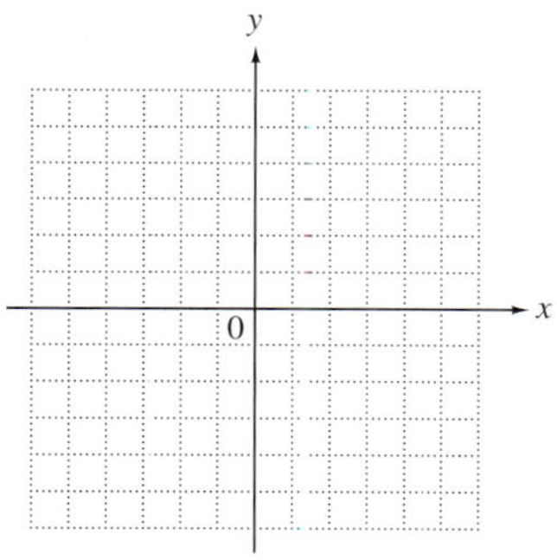

36. $5x + 6y = -10$

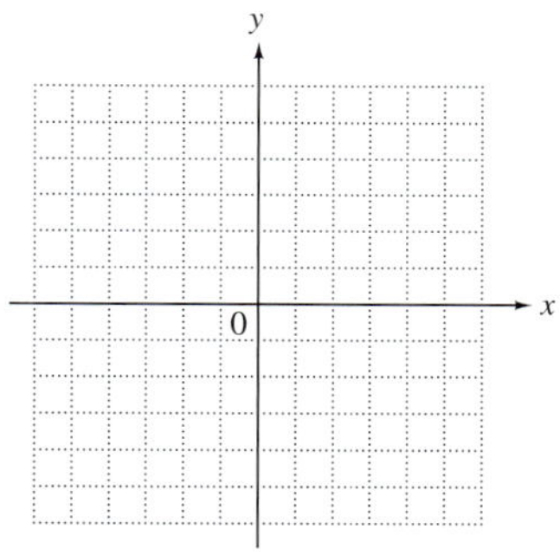

37. $y = 5$

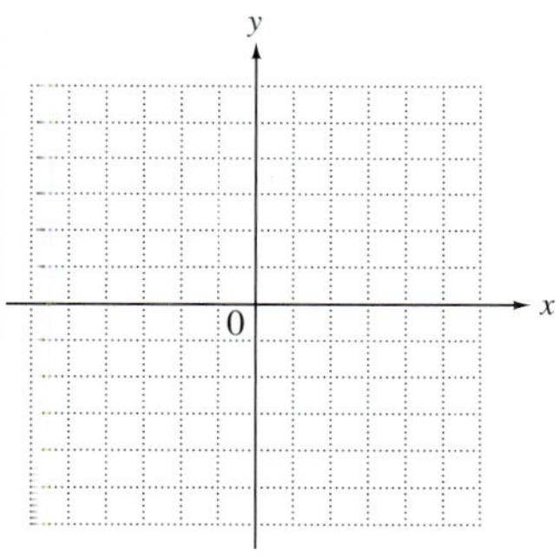

38. $y = -3$

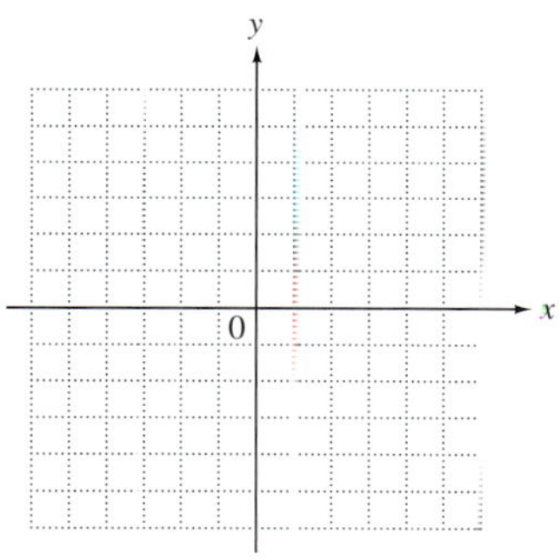

39. $x = 5$

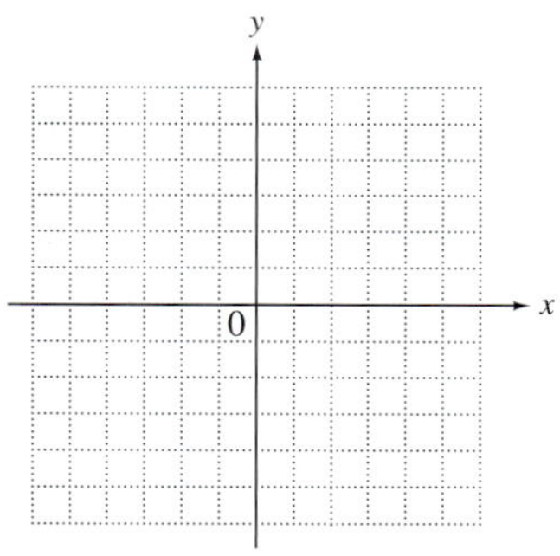

40. $x = -3$

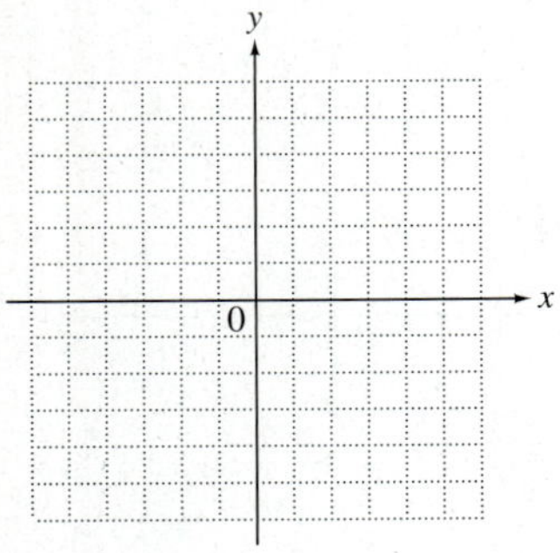

41. $x + 5y = 0$

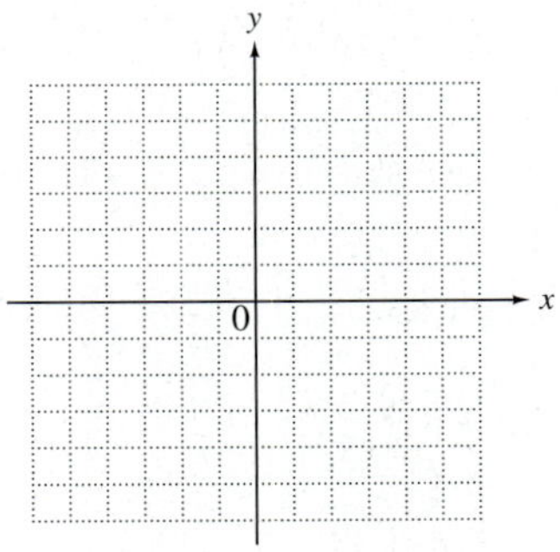

42. $x - 3y = 0$

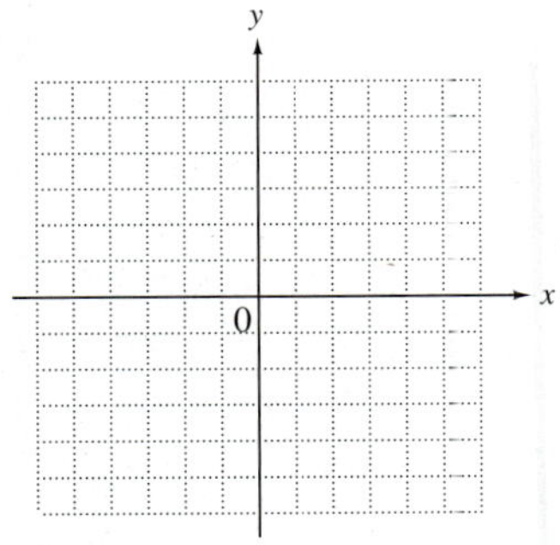

43. $2x = 3y$

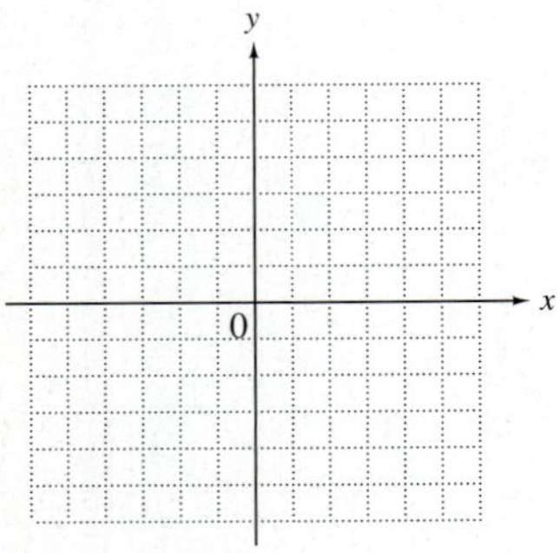

44. $3x = -4y$

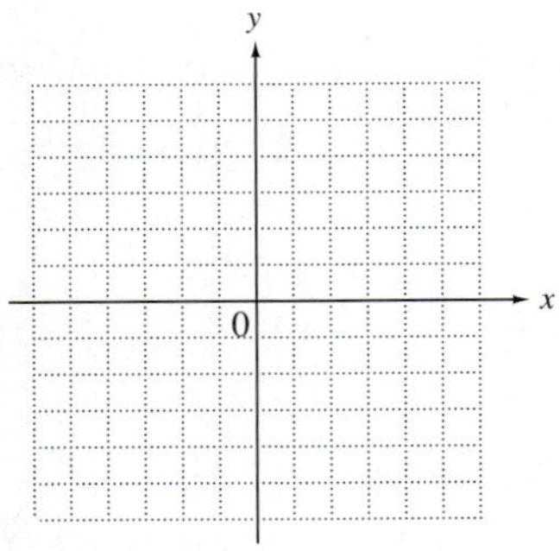

Find the midpoint of each segment with the given endpoints. See Example 5.

45. $(-8, 4)$ and $(-2, -6)$

46. $(5, 2)$ and $(-1, 8)$

47. $(3, -6)$ and $(6, 3)$

48. $(-10, 4)$ and $(7, 1)$

49. $(-9, 3)$ and $(9, 8)$

50. $(4, -3)$ and $(-1, 3)$

51. $(2.5, 3.1)$ and $(1.7, -1.3)$

52. $(6.2, 5.8)$ and $(1.4, -0.6)$

53. $\left(\frac{1}{2}, \frac{1}{3}\right)$ and $\left(\frac{3}{2}, \frac{5}{3}\right)$

54. $\left(\frac{21}{4}, \frac{2}{5}\right)$ and $\left(\frac{7}{4}, \frac{3}{5}\right)$

55. $\left(-\frac{1}{3}, \frac{2}{7}\right)$ and $\left(-\frac{1}{2}, \frac{1}{14}\right)$

56. $\left(\frac{3}{5}, -\frac{1}{3}\right)$ and $\left(\frac{1}{2}, -\frac{7}{2}\right)$

4.2 Slope of a Line

Slope (steepness) is used in many practical ways. The slope of a highway (sometimes called the *grade*) is often given as a percent. For example, a 10% (or $\frac{10}{100} = \frac{1}{10}$) slope means the highway rises 1 unit for every 10 horizontal units. Stairs and roofs have slopes too, as shown in Figure 11.

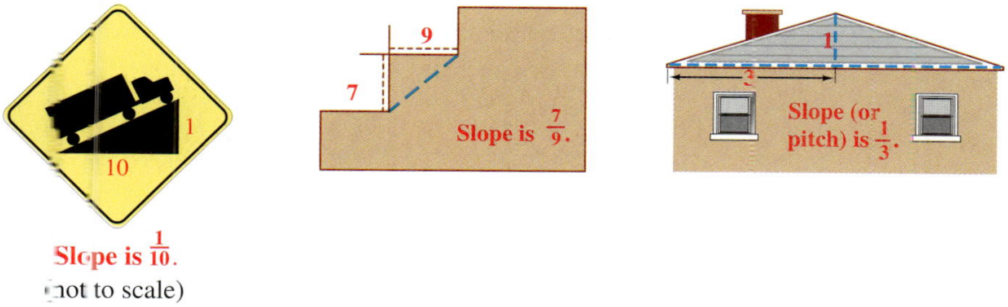

Figure 11

In each example mentioned, slope is the ratio of vertical change, or **rise,** to horizontal change, or **run.** A simple way to remember this is to think "slope is rise over run."

OBJECTIVE 1 Find the slope of a line, given two points on the line. To obtain a formal definition of the slope of a line, we designate two different points on the line. To differentiate between the points, we write them as (x_1, y_1) and (x_2, y_2). See Figure 12. As in our work with midpoints in **Section 4.1,** we use subscripts to write the coordinates of two points when working with slope.

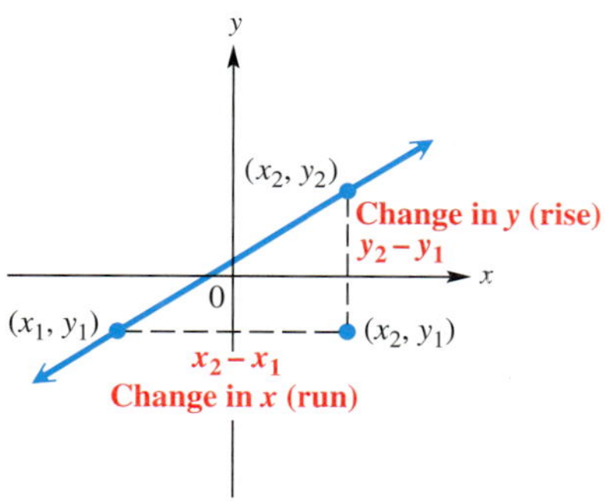

Figure 12

As we move along the line in Figure 12 from (x_1, y_1) to (x_2, y_2), the y-value changes (vertically) from y_1 to y_2, an amount equal to $y_2 - y_1$. As y changes from y_1 to y_2, the value of x changes (horizontally) from x_1 to x_2 by the amount $x_2 - x_1$. The ratio of the change in y to the change in x (the rise over the run) is called the *slope* of the line, with the letter m traditionally used for slope.

Slope Formula

The **slope** of the line through the distinct points (x_1, y_1) and (x_2, y_2) is

$$m = \frac{\textbf{rise}}{\textbf{run}} = \frac{\textbf{change in } y}{\textbf{change in } x} = \frac{y_2 - y_1}{x_2 - x_1} \quad (x_1 \neq x_2).$$

Work Problem 1 at the Side. ▶

OBJECTIVES

1. **Find the slope of a line, given two points on the line.**
2. **Find the slope of a line, given an equation of the line.**
3. **Graph a line, given its slope and a point on the line.**
4. **Use slopes to determine whether two lines are parallel, perpendicular, or neither.**
5. **Solve problems involving average rate of change.**

1 Use the information given for the walkway in the figure to find the following.

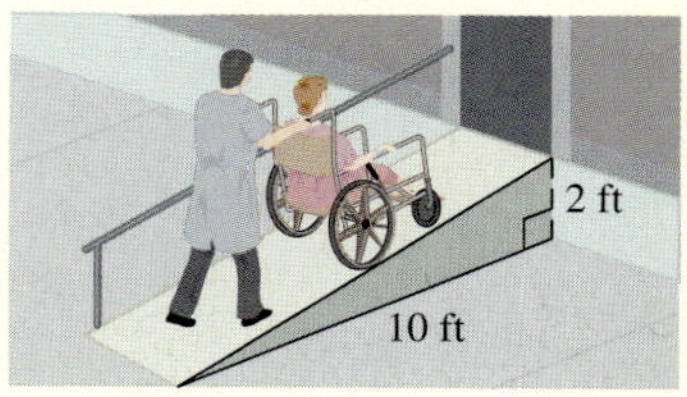

(a) The rise

(b) The run

(c) The slope

ANSWERS

1. (a) 2 ft **(b)** 10 ft **(c)** $\frac{2}{10}$, or $\frac{1}{5}$

2 Find the slope of the line through each pair of points.

(a) $(-2, 7), (4, -3)$

(b) $(1, 2), (8, 5)$

(c) $(8, -4), (3, -2)$

EXAMPLE 1 Finding the Slope of a Line

Find the slope of the line through the points $(2, -1)$ and $(-5, 3)$.

If $(2, -1) = (x_1, y_1)$ and $(-5, 3) = (x_2, y_2)$, then

$$m = \frac{y_2 - y_1}{x_2 - x_1} = \frac{3 - (-1)}{-5 - 2} = \frac{4}{-7} = -\frac{4}{7}.$$

The slope is $-\frac{4}{7}$. See Figure 13.

On the other hand, if the ordered pairs are interchanged so that $(2, -1) = (x_2, y_2)$ and $(-5, 3) = (x_1, y_1)$, the slope is the same.

$$m = \frac{y_2 - y_1}{x_2 - x_1} = \frac{-1 - 3}{2 - (-5)} = \frac{-4}{7} = -\frac{4}{7}$$

y-values are in the numerator, *x*-values in the denominator.

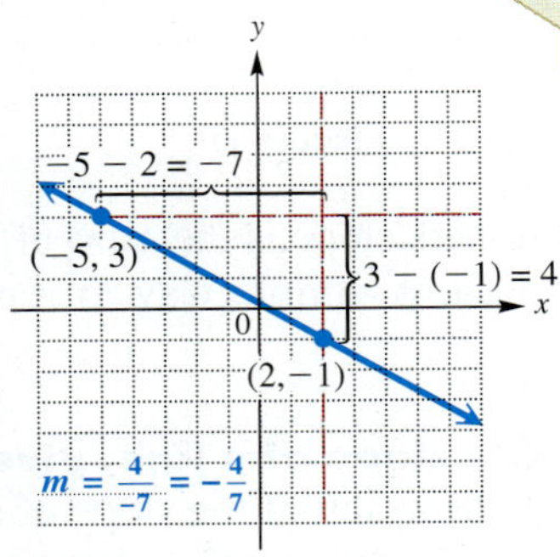

Figure 13

Example 1 suggests the following important ideas regarding slope:

1. The slope is the same no matter which point we consider first.
2. Using similar triangles from geometry, we can show that the slope is the same no matter which two different points on the line we choose.

CAUTION

In calculating slope, be careful to subtract the y-values and the x-values in the same order.

Correct	**Incorrect**
$\frac{y_2 - y_1}{x_2 - x_1}$ or $\frac{y_1 - y_2}{x_1 - x_2}$	~~$\frac{y_2 - y_1}{x_1 - x_2}$ or $\frac{y_1 - y_2}{x_2 - x_1}$~~

Also, remember that the change in y is the numerator and the change in x is the denominator.

◀ Work Problem **2** at the Side.

OBJECTIVE 2 Find the slope of a line, given an equation of the line. When an equation of a line is given, one way to find the slope is to use the definition of slope by first finding two different points on the line.

ANSWERS

2. (a) $-\frac{5}{3}$ (b) $\frac{3}{7}$ (c) $-\frac{2}{5}$

EXAMPLE 2 **Finding the Slope of a Line**

Find the slope of the line $4x - y = -8$.

The intercepts can be used as the two different points needed to find the slope. Let $y = 0$ to find that the x-intercept is $(-2, 0)$. Then let $x = 0$ to find that the y-intercept is $(0, 8)$. Use these two points in the slope formula. The slope is

$$m = \frac{\text{rise}}{\text{run}} = \frac{8 - 0}{0 - (-2)} = \frac{8}{2} = 4.$$

Work Problem **3** at the Side. ▶

EXAMPLE 3 **Finding the Slopes of Horizontal and Vertical Lines**

Find the slope of each line.

(a) $y = 2$

The graph of $y = 2$ is a horizontal line. (See Figure 14.) To find the slope, select two different points on the line, such as $(3, 2)$ and $(-1, 2)$, and use the slope formula.

$$m = \frac{\text{rise}}{\text{run}} = \frac{2 - 2}{3 - (-1)} = \frac{0}{4} = 0$$

In this case, the *rise* is 0, so the slope is 0.

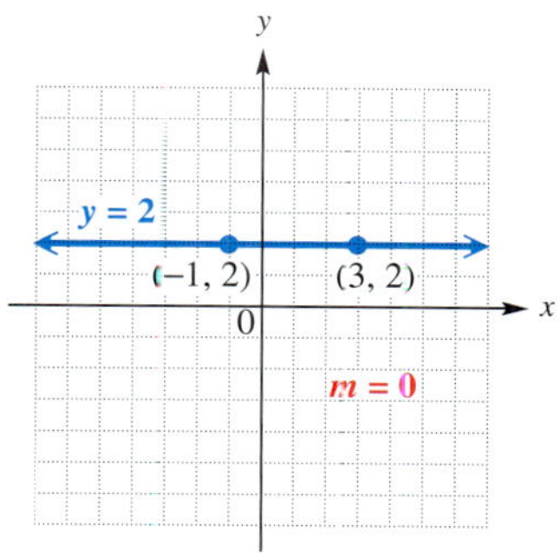

Figure 14

(b) $x + 1 = 0$

The graph of $x + 1 = 0$, or $x = -1$, is a vertical line. (See Figure 15.) Two points that satisfy the equation $x = -1$ are $(-1, 5)$ and $(-1, -4)$. If we use these two points to try to find the slope, we obtain

$$m = \frac{\text{rise}}{\text{run}} = \frac{-4 - 5}{-1 - (-1)} = \frac{-9}{0}.$$

Since division by 0 is undefined, the slope is undefined. This is why the definition of slope includes the restriction that $x_1 \neq x_2$.

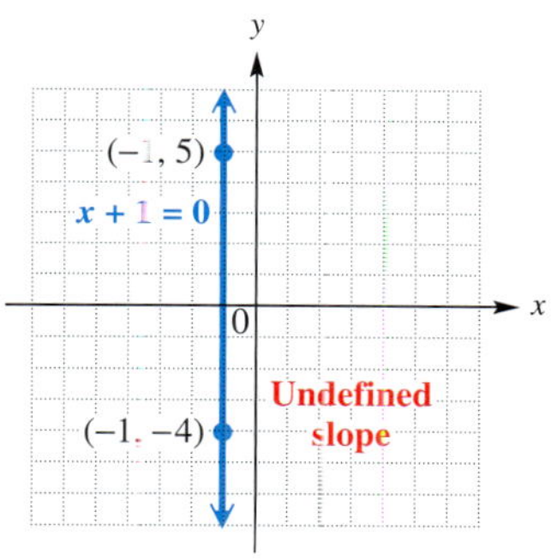

Figure 15

Generalizing from Example 3, we can make the following statements about horizontal and vertical lines.

Slopes of Horizontal and Vertical Lines

The slope of a horizontal line is 0.

The slope of a vertical line is undefined.

Work Problem **4** at the Side. ▶

3 Find the slope of each line.

(a) $2x + y = 6$

(b) $3x - 4y = 12$

4 Find the slope of each line.

(a) $x = -6$

(b) $y + 5 = 0$

Answers

3. **(a)** -2 **(b)** $\frac{3}{4}$

4. **(a)** undefined **(b)** 0

5 Find the slope of the graph of $2x - 5y = 8$.

The slope of a line can also be found directly from its equation. Look again at the equation $4x - y = -8$ from Example 2. Solve this equation for y.

$$4x - y = -8 \quad \text{Equation from Example 2}$$
$$-y = -4x - 8 \quad \text{Subtract } 4x.$$
$$y = 4x + 8 \quad \text{Multiply by } -1.$$

Notice that the slope, 4, found using the slope formula in Example 2, is the same number as the coefficient of x in the equation $y = 4x + 8$. We will see in the next section that this always happens, *as long as the equation is solved for y.*

EXAMPLE 4 Finding the Slope from an Equation

Find the slope of the graph of $3x - 5y = 8$.

Solve the equation for y.

$$3x - 5y = 8$$
$$-5y = -3x + 8 \quad \text{Subtract } 3x.$$
$$y = \frac{3}{5}x - \frac{8}{5} \quad \text{Divide by } -5.$$

The slope is given by the coefficient of x, so the slope is $\frac{3}{5}$.

◀ Work Problem **5** at the Side.

OBJECTIVE 3 Graph a line, given its slope and a point on the line. Example 5 shows how to graph a straight line by using the slope and one point on the line.

EXAMPLE 5 Using the Slope and a Point to Graph Lines

Graph each line.

(a) With slope $\frac{2}{3}$ through the point $(-1, 4)$

First locate and plot the point $P(-1, 4)$ on a graph as shown in Figure 16. Then use the slope to find a second point. From the slope formula,

$$m = \frac{\text{change in } y}{\text{change in } x} = \frac{2}{3},$$

so move 2 units *up* and then 3 units to the *right* to locate another point on the graph, $R(2, 6)$. The line through $P(-1, 4)$ and $R(2, 6)$ is the required graph.

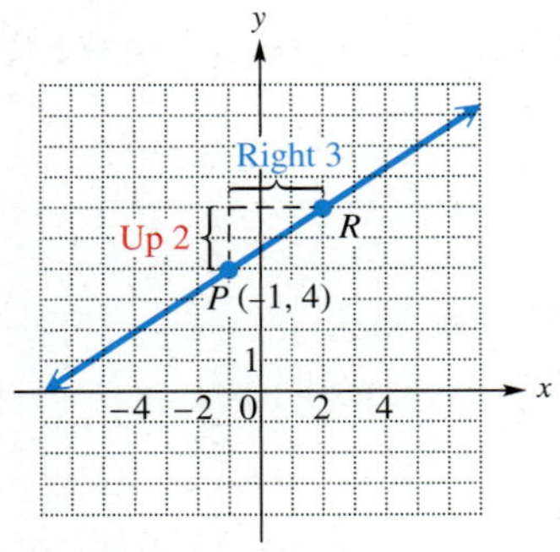

Figure 16

Continued on Next Page

ANSWER

5. $\frac{2}{5}$

(b) Through (3, 1) with slope -4

Start by locating the point $P(3, 1)$ on a graph. Find a second point R on the line by writing -4 as $\frac{-4}{1}$ and using the slope formula.

$$m = \frac{\text{change in } y}{\text{change in } x} = \frac{-4}{1}$$

Move 4 units *down* from (3, 1), and then move 1 unit to the *right*. Draw a line through this second point $R(4, -3)$ and $P(3, 1)$, as shown in Figure 17.

The slope also could be written as

$$m = \frac{\text{change in } y}{\text{change in } x} = \frac{4}{-1}.$$

In this case, the second point R is located 4 units *up* and 1 unit to the *left*. Verify that this approach produces the line in Figure 17.

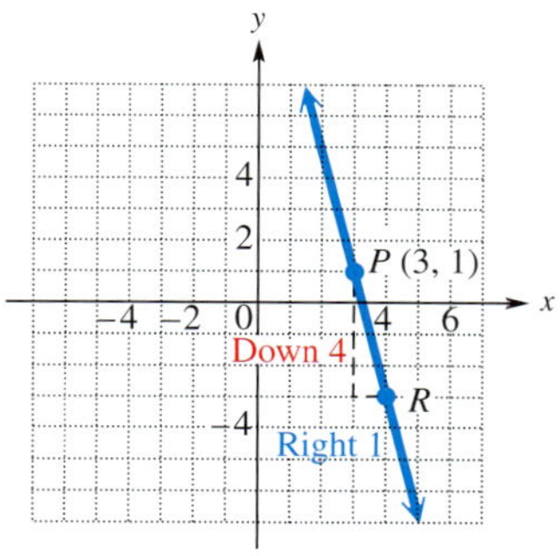

Figure 17

Work Problem **6** at the Side. ▶

In Example 5(a), the slope of the line is the *positive* number $\frac{2}{3}$. The graph of the line in Figure 16 goes up (rises) from left to right. The line in Example 5(b) has a *negative* slope, -4. As Figure 17 shows, its graph goes down (falls) from left to right. These facts suggest the following generalization.

Orientation of a Line in the Plane

A positive slope indicates that the line goes *up* (rises) from left to right.

A negative slope indicates that the line goes *down* (falls) from left to right.

Figure 18 shows lines of positive, 0, negative, and undefined slopes.

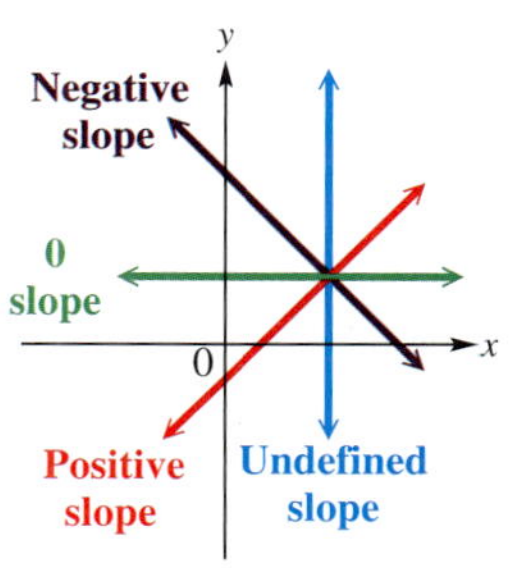

Figure 18

6 Graph each line.

(a) Through $(1, -3)$; $m = -\frac{3}{4}$

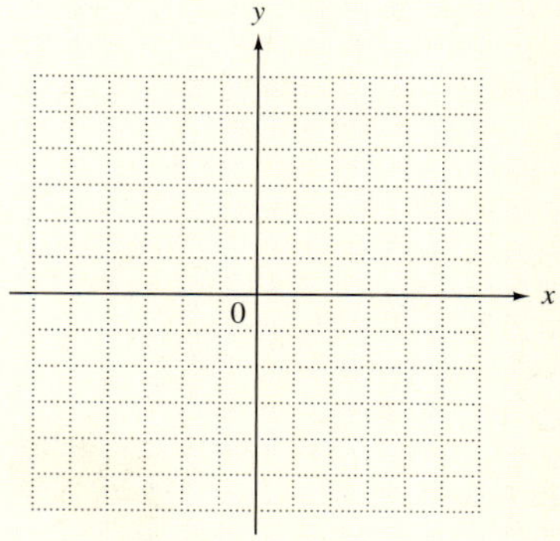

(b) Through $(-1, -4)$; $m = 2$

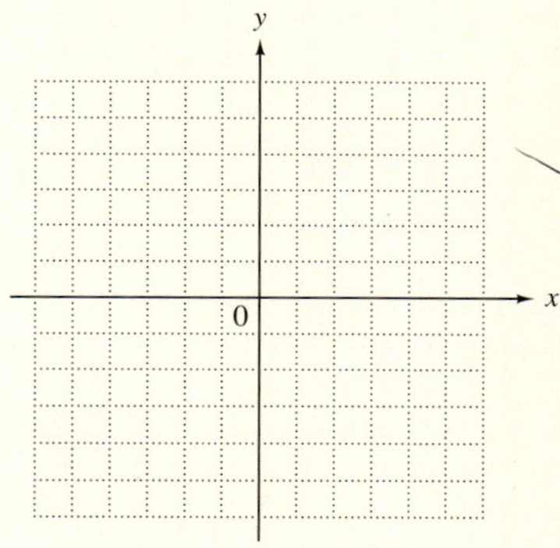

ANSWERS

6. (a)

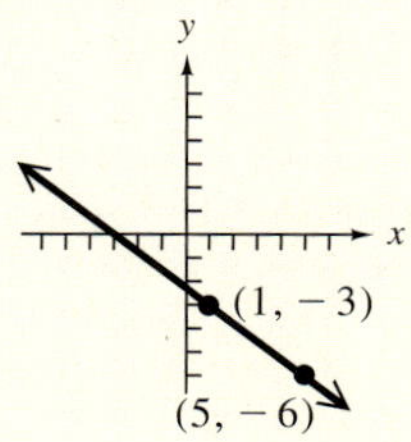

(b)

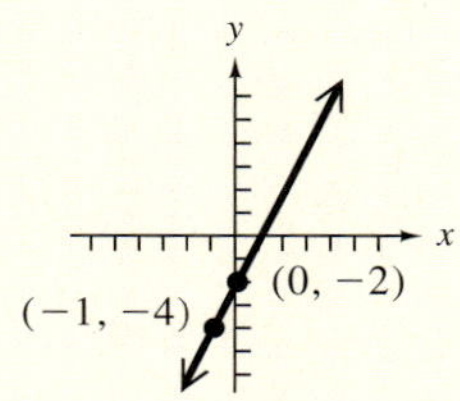

OBJECTIVE 4 Use slopes to determine whether two lines are parallel, perpendicular, or neither. The slopes of a pair of parallel or perpendicular lines are related in a special way. The slope of a line measures the steepness of the line. Since parallel lines have equal steepness, their slopes must be equal; also, lines with the same slope are parallel.

Slopes of Parallel Lines

Two nonvertical lines with the same slope are parallel.

Two nonvertical parallel lines have the same slope.

EXAMPLE 6 Determining Whether Two Lines Are Parallel

Are the lines L_1, through $(-2, 1)$ and $(4, 5)$, and L_2, through $(3, 0)$ and $(0, -2)$, parallel?

$$\text{The slope of } L_1 \text{ is } \quad m_1 = \frac{5-1}{4-(-2)} = \frac{4}{6} = \frac{2}{3}.$$

$$\text{The slope of } L_2 \text{ is } \quad m_2 = \frac{-2-0}{0-3} = \frac{-2}{-3} = \frac{2}{3}.$$

Because the slopes are equal, the two lines are parallel.

To see how the slopes of perpendicular lines are related, consider a nonvertical line with slope $\frac{a}{b}$. If this line is rotated 90°, the vertical change and the horizontal change are reversed and the slope is $-\frac{b}{a}$, since the horizontal change is now negative. See Figure 19. Thus, the slopes of perpendicular lines have product -1 and are negative reciprocals of each other. For example, if the slopes of two lines are $\frac{3}{4}$ and $-\frac{4}{3}$, then the lines are perpendicular because $\frac{3}{4}\left(-\frac{4}{3}\right) = -1$.

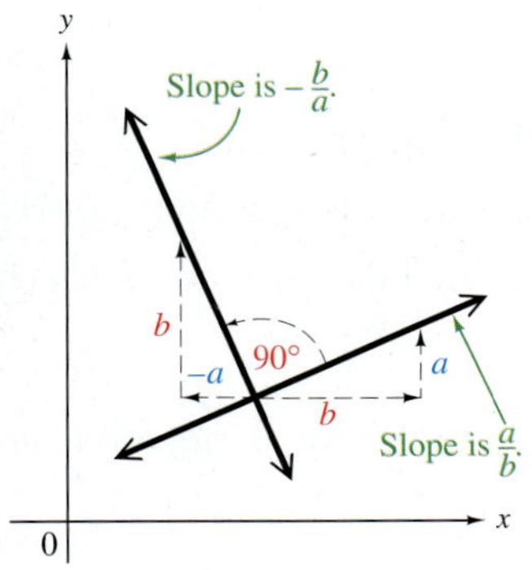

Figure 19

Slopes of Perpendicular Lines

If neither is vertical, perpendicular lines have slopes that are negative reciprocals; that is, their product is -1. Also, lines with slopes that are negative reciprocals are perpendicular.

A line with 0 slope is perpendicular to a line with undefined slope.

EXAMPLE 7 Determining Whether Two Lines Are Perpendicular

Are the lines with equations $2y = 3x - 6$ and $2x + 3y = -6$ perpendicular?

Find the slope of each line by first solving each equation for y.

$$2y = 3x - 6$$
$$y = \frac{3}{2}x - 3 \quad \text{Divide by 2.}$$
($\frac{3}{2}$ ← Slope)

$$2x + 3y = -6$$
$$3y = -2x - 6 \quad \text{Subtract } 2x.$$
$$y = -\frac{2}{3}x - 2 \quad \text{Divide by 3.}$$
($-\frac{2}{3}$ ← Slope)

Since the product of the slopes is $\frac{3}{2}\left(-\frac{2}{3}\right) = -1$, the lines are perpendicular.

EXAMPLE 8 Determining Whether Two Lines Are Parallel, Perpendicular, or Neither

Determine whether the lines with equations $2x - 5y = 8$ and $2x + 5y = 8$ are *parallel, perpendicular,* or *neither.*

Find the slope of each line by first solving each equation for y.

$$2x - 5y = 8$$
$$-5y = -2x + 8$$
$$y = \frac{2}{5}x - \frac{8}{5}$$
($\frac{2}{5}$ ← Slope)

$$2x + 5y = 8$$
$$5y = -2x + 8$$
$$y = -\frac{2}{5}x + \frac{8}{5}$$
($-\frac{2}{5}$ ← Slope)

The slopes, $\frac{2}{5}$ and $-\frac{2}{5}$, are not equal, and they are not negative reciprocals because their product is $-\frac{4}{25}$, not -1. Thus, the two lines are neither parallel nor perpendicular.

Work Problem 7 at the Side. ▶

OBJECTIVE 5 Solve problems involving average rate of change.

We know that the slope of a line is the ratio of the vertical change in y to the corresponding horizontal change in x. Thus, the slope formula applied to any two points on the line gives the **average rate of change** in y per unit change in x, where the value of y depends on the value of x.

For example, if the height of a boy increased from 60 to 68 in. between the ages of 12 and 16, as shown in the graph below, then the boy's average growth rate (or average change in height) from ages 12 to 16 was

$$\frac{\text{Change in height } y \rightarrow 68 - 60}{\text{Change in age } x \rightarrow 16 - 12} = \frac{8}{4} = 2 \text{ in. per year.}$$

The boy may actually have grown more than 2 in. during some years and less than 2 in. during other years. If we plotted ordered pairs (age, height) for those years, the points would be a little above or below the line shown in the graph. If we drew a line connecting any two of those points and calculated the average rate of change, it would likely be slightly different than that found above. However using the height data for ages 12 and 16, the boy's *average* change in height was 2 in. per year over these years.

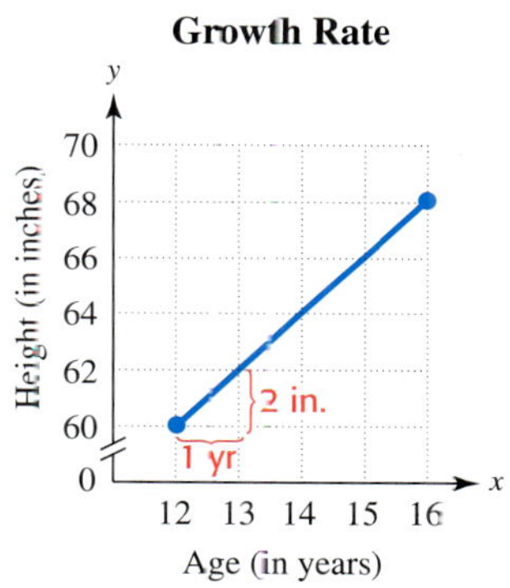

7 Write *parallel, perpendicular,* or *neither* for each pair of two distinct lines.

(a) The line through $(-1, 2)$ and $(3, 5)$ and the line through $(4, 7)$ and $(8, 10)$

(b) The line through $(5, -9)$ and $(3, 7)$ and the line through $(0, 2)$ and $(8, 3)$

(c) $2x - y = 4$ and $2x + y = 6$

(d) $3x + 5y = 6$ and $5x - 3y = 2$

ANSWERS

7. **(a)** parallel **(b)** perpendicular **(c)** neither **(d)** perpendicular

8 Americans spent an average of 886 hr in 2003 watching cable and satellite TV. (*Source:* Veronis Suhler Stevenson.)

(a) Using this data for 2003 and the data for 2000 from the graph in Figure 20, find the average rate of change to the nearest tenth of an hour from 2000 to 2003.

(b) How does the average rate of change from part (a) compare to the average rate of change from 2000 to 2005 found in Example 9?

EXAMPLE 9 Interpreting Slope as Average Rate of Change

The graph in Figure 20 approximates the average number of hours per year spent watching cable and satellite TV for each person in the United States during the years 2000 through 2005. Find the average rate of change in number of hours per year.

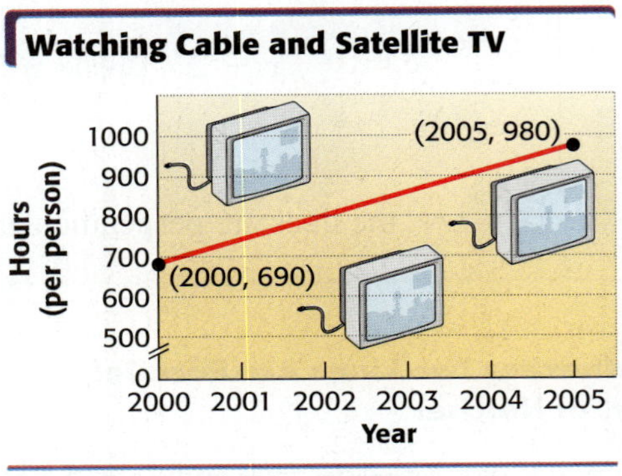

Source: Veronis Suhler Stevenson.

Figure 20

To determine the average rate of change, we need two pairs of data. From the graph, if $x = 2000$, then $y = 690$ and if $x = 2005$, then $y = 980$. Thus, we have the ordered pairs (2000, 690) and (2005, 980). By the slope formula,

$$\text{average rate of change} = \frac{980 - 690}{2005 - 2000} = \frac{290}{5} = 58$$

A positive slope indicates an increase.

This means that the average time per person spent watching cable and satellite TV *increased* by 58 hr per year from 2000 through 2005.

◀ Work Problem **8** at the Side.

9 In 2000, 942.5 million compact discs were sold in the United States. In 2006, 614.9 million CDs were sold. Find the average rate of change in CDs sold per year. (*Source:* Recording Industry Association of America.)

EXAMPLE 10 Interpreting Slope as Average Rate of Change

During the year 2000, the average person in the United States spent 866 hr watching broadcast TV. In 2004, the average number of hours per person spent watching broadcast TV was 678. Find the average rate of change in number of hours per year. (*Source:* Veronis Suhler Stevenson.)

To use the slope formula, we need two ordered pairs. Here, we let one ordered pair be (2000, 866) and the other be (2004, 678).

$$\text{average rate of change} = \frac{678 - 866}{2004 - 2000} = \frac{-188}{4} = -47$$

A negative slope indicates a decrease.

The graph in Figure 21 confirms that the line through the ordered pairs falls from left to right and, therefore, has negative slope. Thus, the average time per person spent watching broadcast TV *decreased* by 47 hr per year from 2000 through 2004.

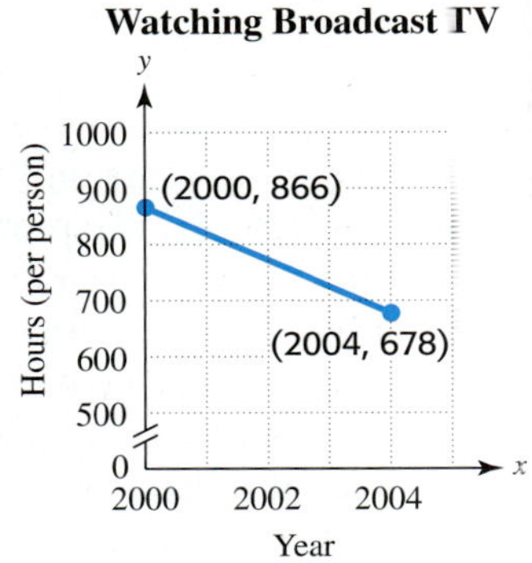

Figure 21

◀ Work Problem **9** at the Side.

ANSWERS

8. (a) 65.3 hr per yr
(b) It is greater than the average rate of change for 2000–2005.

9. −54.6 million CDs per yr

4.2 Exercises

1. A ski slope drops 30 ft for every horizontal 100 ft. Which of the following express its slope? (There are several correct choices.)

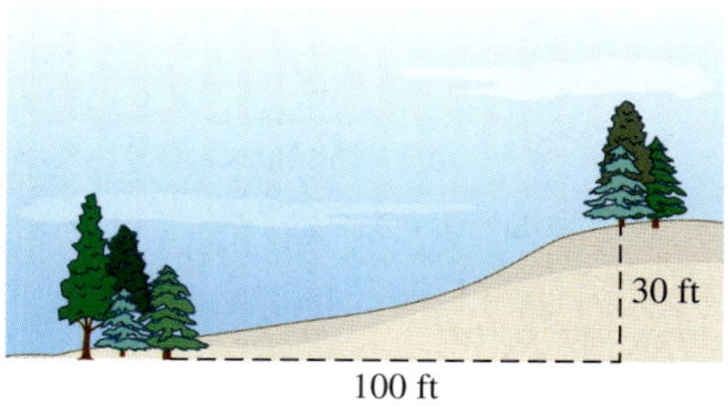

A. 0.3 B. $\frac{3}{10}$ C. $3\frac{1}{3}$

D. $\frac{30}{100}$ E. $\frac{10}{3}$ F. 30

2. A hill has slope 0.05. How many feet in the vertical direction correspond to a run of 50 ft?

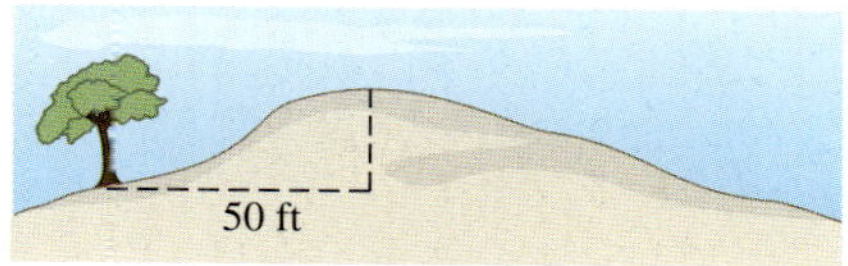

NOT TO SCALE

Use the given figure to determine the slope of the line segment described, by counting the number of units of "rise," the number of units of "run," and then finding the quotient.

3. AB 4. BC 5. CD 6. DE

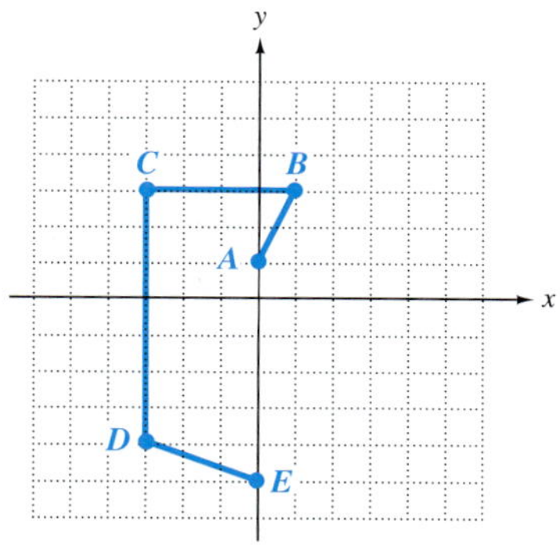

Evaluate each expression for m, applying the slope formula.

7. $m = \frac{6 - 2}{5 - 3}$ 8. $m = \frac{5 - 7}{4 - 2}$ 9. $m = \frac{-4 - (-4)}{-3 - (-5)}$

10. $m = \frac{-5 - (-5)}{3 - 2}$ 11. $m = \frac{-6 - 0}{-3 - (-3)}$ 12. $m = \frac{7 - (-2)}{-3 - (-3)}$

Find the slope of the line through each pair of points using the slope formula. See Example 1.

13. $(-2, -3)$ and $(-1, 5)$ 14. $(-4, 3)$ and $(-3, -4)$ 15. $(-4, 1)$ and $(2, 6)$

16. $(-3, -3)$ and $(5, 6)$ 17. $(2, 4)$ and $(-4, 4)$ 18. $(-6, 3)$ and $(2, 3)$

Find the slope of each line.

19.

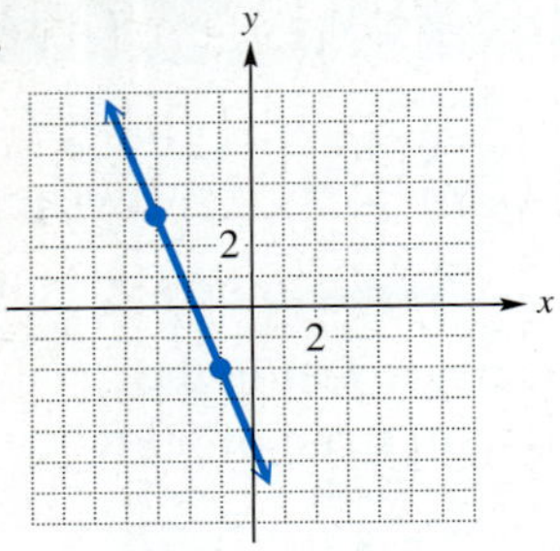

20.

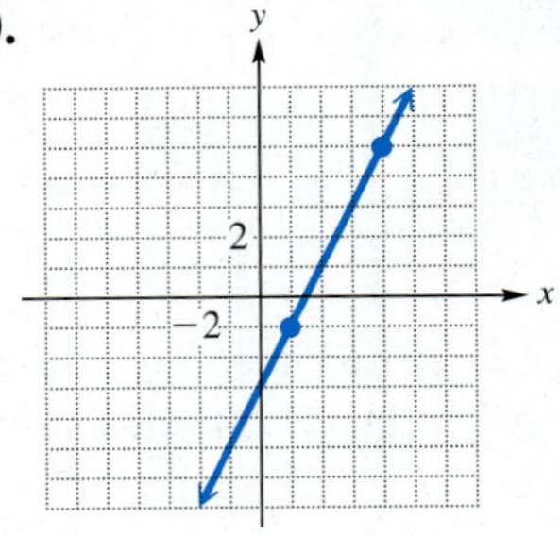

21.

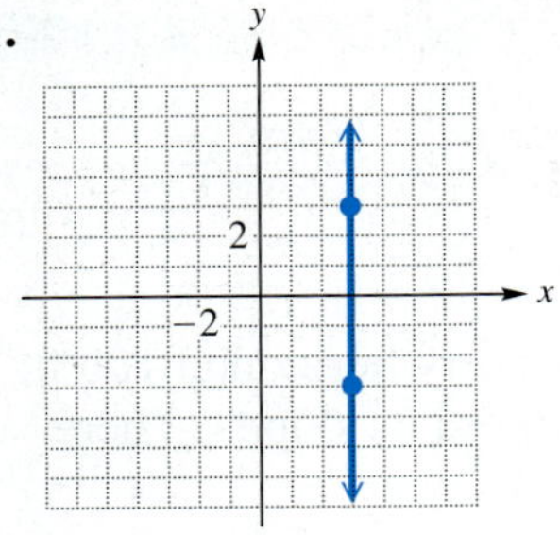

22.

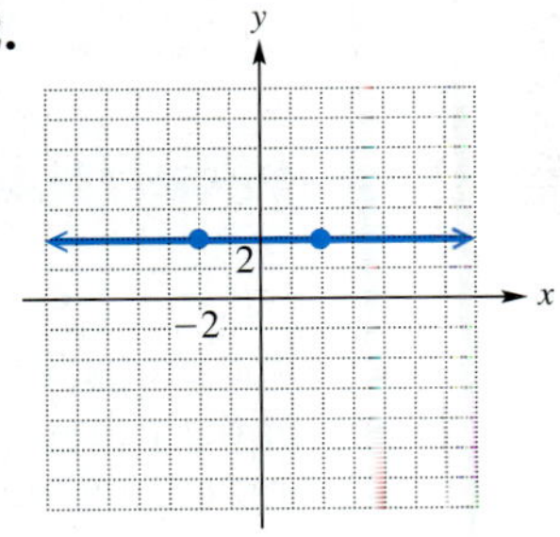

Based on the figure shown here, determine which line satisfies the given description.

23. The line has positive slope.

24. The line has negative slope.

25. The line has slope 0.

26. The line has undefined slope.

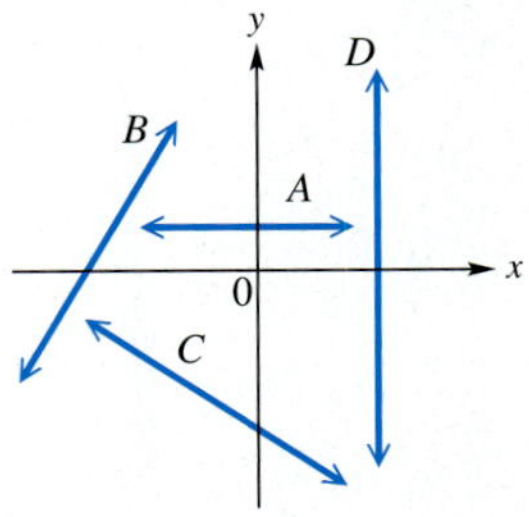

Find the slope of each line, and sketch the graph. See Examples 2–4.

27. $x + 2y = 4$

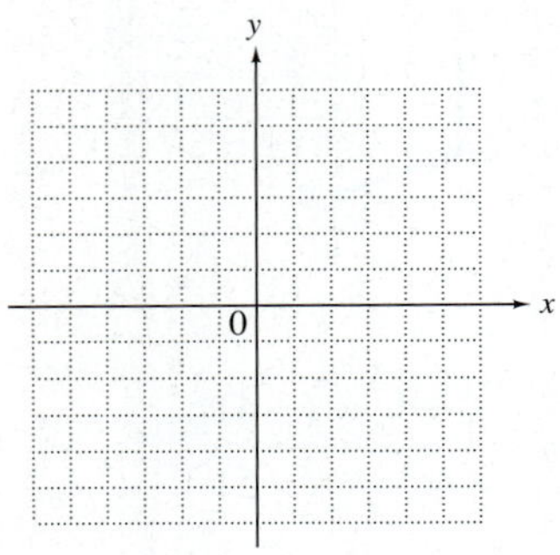

28. $x + 3y = -6$

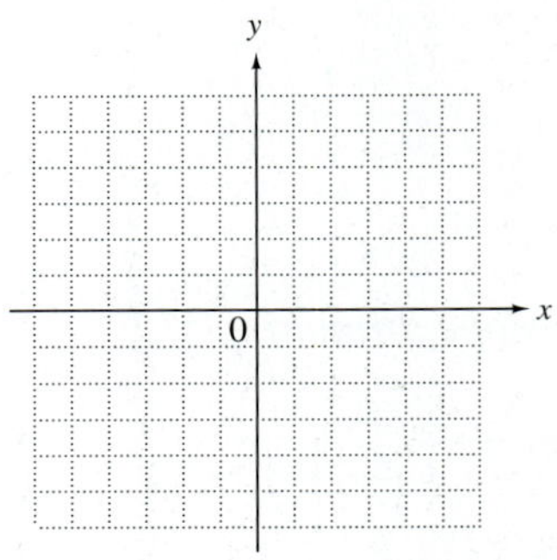

29. $-x + y = 4$

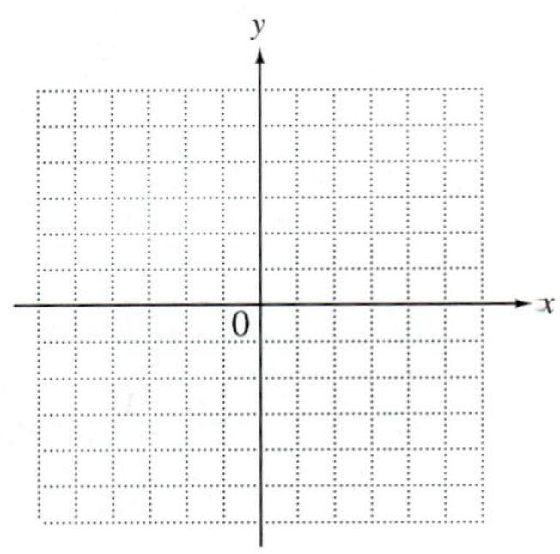

30. $-x + y = 6$

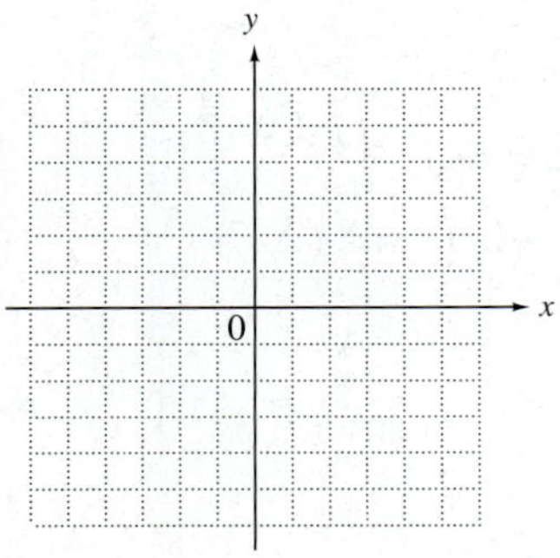

31. $6x + 5y = 30$

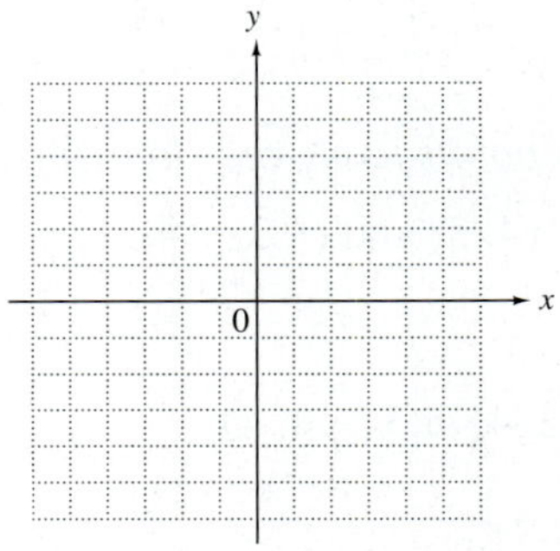

32. $3x + 4y = 12$

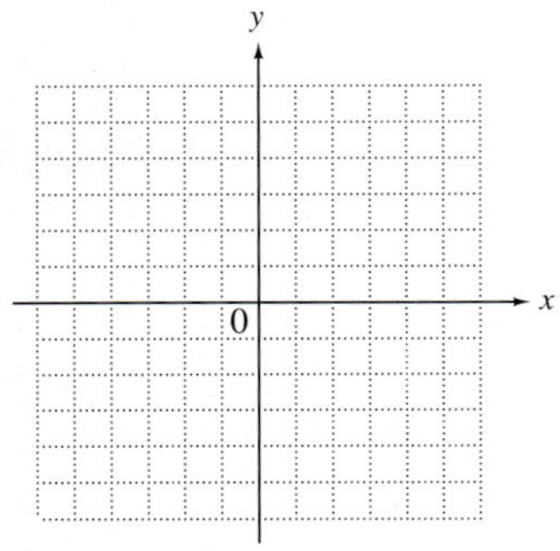

33. $x + 2 = 0$

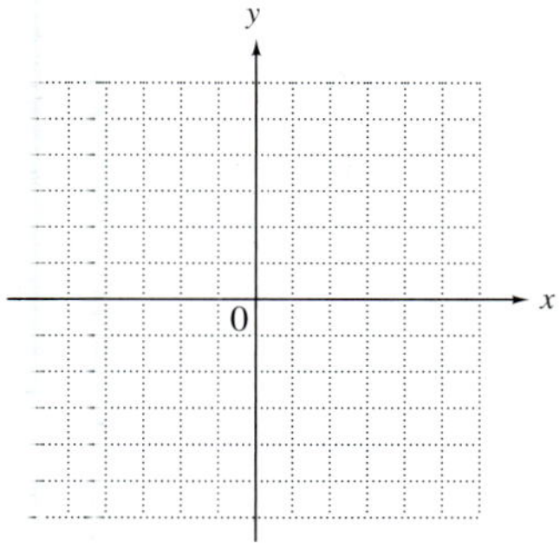

34. $x - 4 = 0$

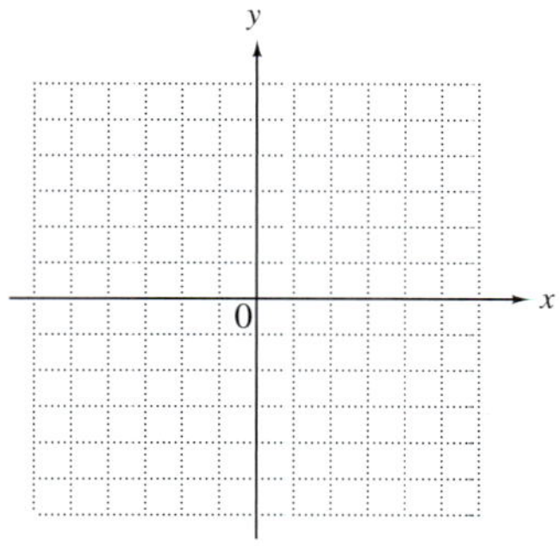

35. $y = 4x$

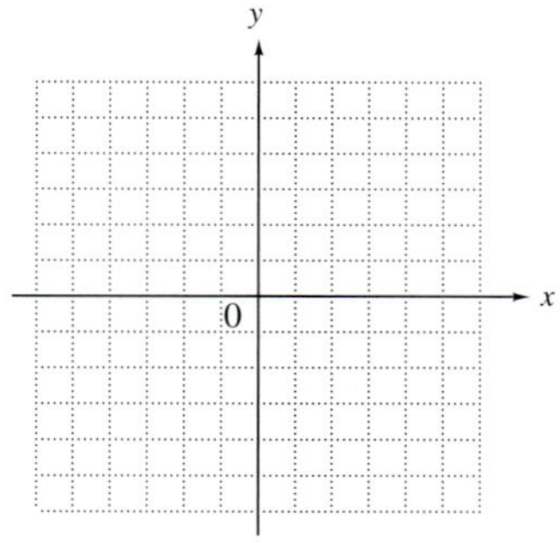

36. $y = -3x$

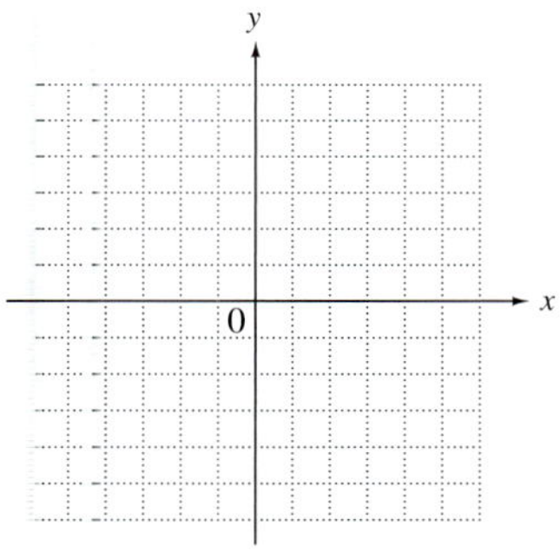

37. $y - 3 = 0$

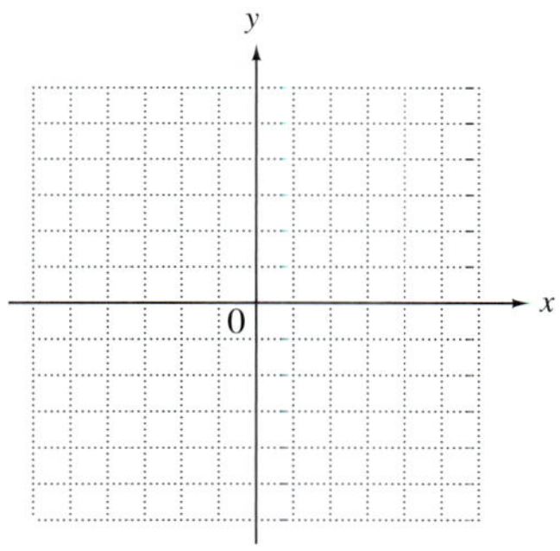

38. $y + 5 = 0$

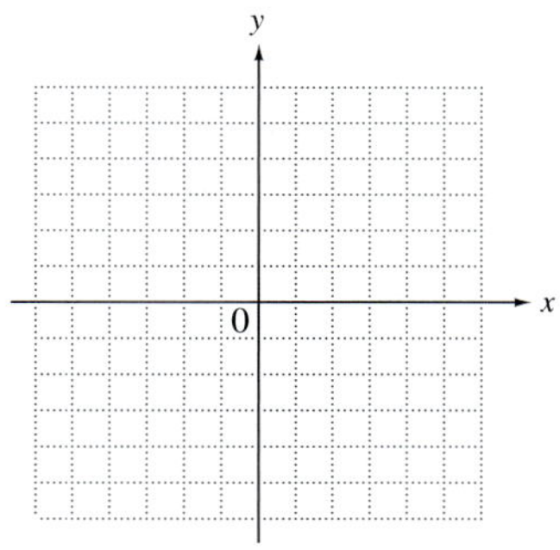

Use the method shown in Example 5 to graph each line.

39. Through $(-4, 2)$; $m = \frac{1}{2}$

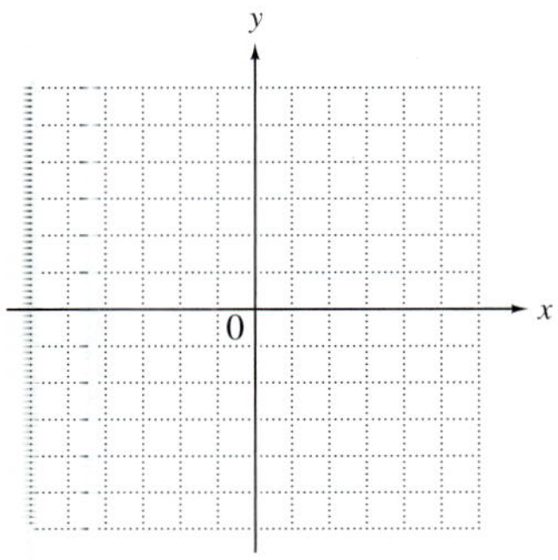

40. Through $(-2, -3)$; $m = \frac{5}{4}$

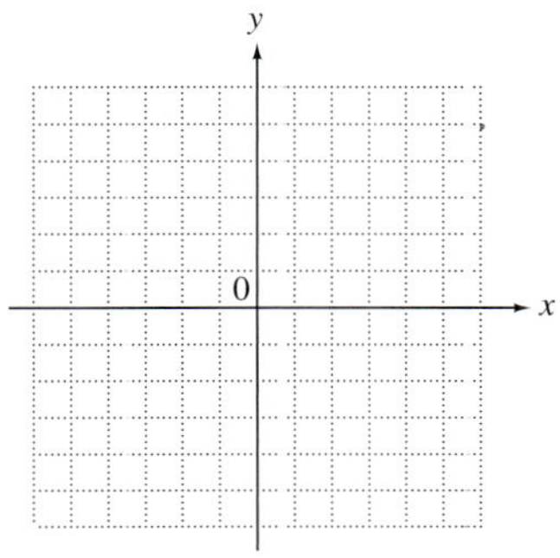

41. Through $(0, -2)$; $m = -\frac{2}{3}$

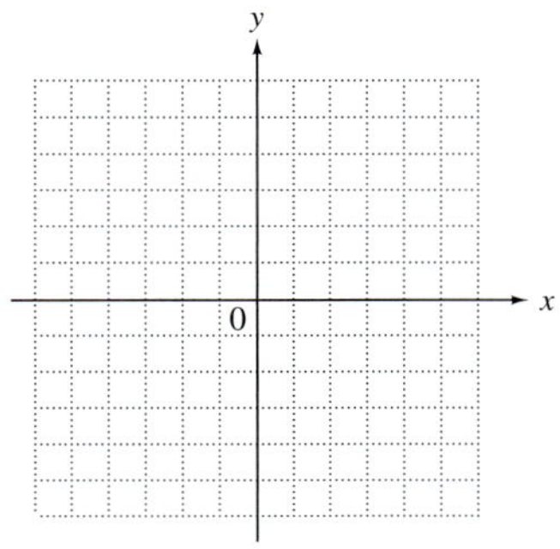

42. Through $(0, -4)$; $m = -\frac{3}{2}$

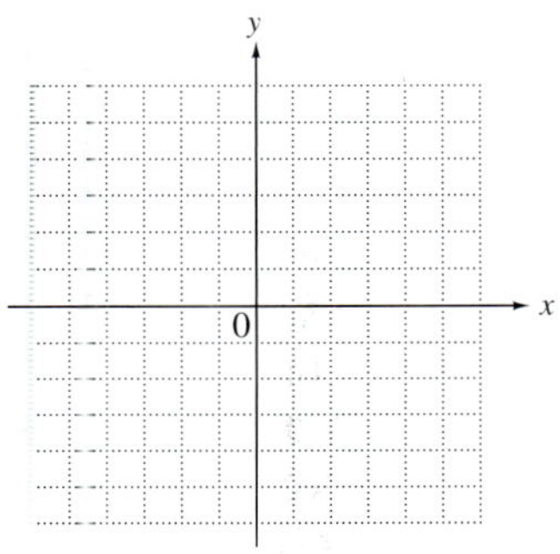

43. Through $(-1, -2)$; $m = 3$

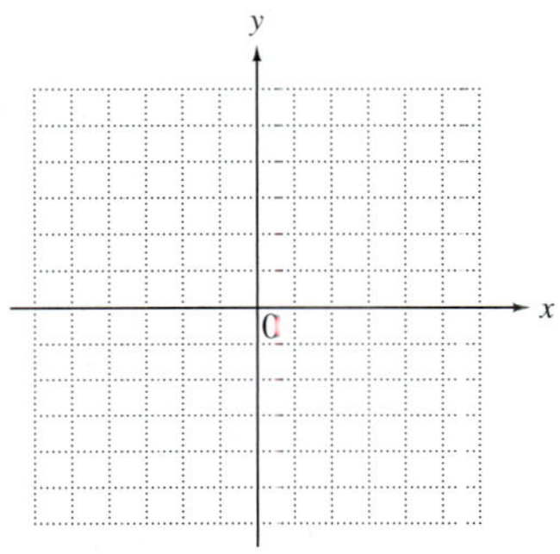

44. Through $(-2, -4)$; $m = 4$

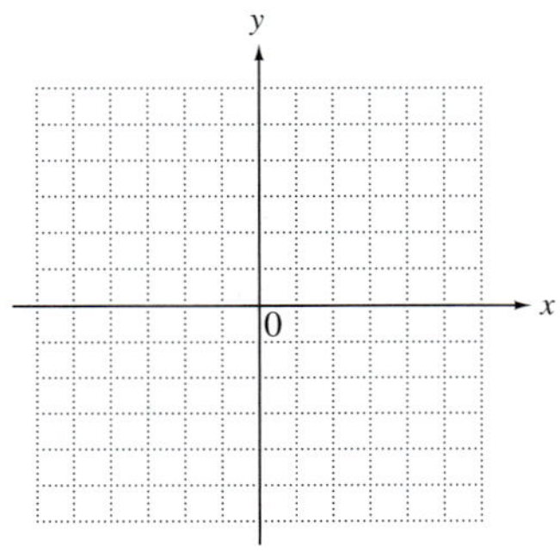

Decide whether the lines in each pair are parallel, perpendicular, *or* neither. *See Examples 6–8.*

45. The line through $(4, 6)$ and $(-8, 7)$ and the line through $(-5, 5)$ and $(7, 4)$

46. The line through $(15, 9)$ and $(12, -7)$ and the line through $(8, -4)$ and $(5, -20)$

47. $2x + 5y = -7$ and $5x - 2y = 1$

48. $x + 4y = 7$ and $4x - y = 3$

49. $2x + y = 6$ and $x - y = 4$

50. $4x - 3y = 6$ and $3x - 4y = 2$

51. $3x = y$ and $2y - 6x = 5$

52. $x = 6$ and $6 - x = 8$

53. $2x + 5y = -8$ and $6 + 2x = 5y$

54. $4x + y = 0$ and $5x - 8 = 2y$

55. $4x - 3y = 8$ and $4y + 3x = 12$

56. $2x = y + 3$ and $2y + x = 3$

Use the concept of slope to solve each problem.

57. The upper deck at U.S. Cellular Field (formerly Comiskey Park) in Chicago has produced, among other complaints, displeasure with its steepness. It is 160 ft from home plate to the front of the upper deck and 250 ft from home plate to the back. The top of the upper deck is 63 ft above the bottom. What is its slope? (Consider the slope as a positive number.)

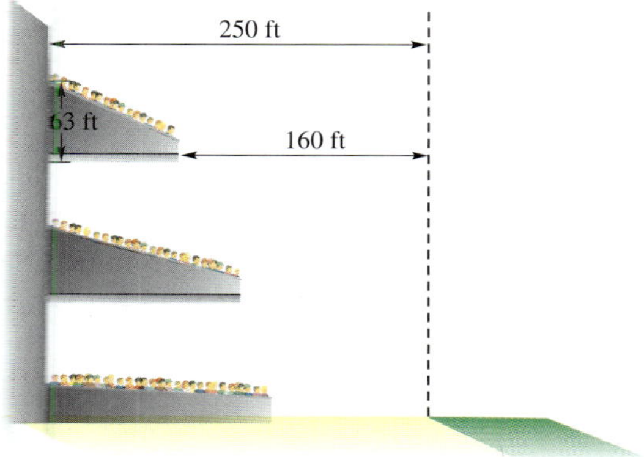

58. When designing the TD Bank North Garden arena in Boston, architects designed the ramps leading up to the entrances so that circus elephants would be able to march up the ramps. The maximum grade (or slope) that an elephant will walk on is 13%. Suppose that such a ramp were constructed with a horizontal run of 150 ft. What would be the maximum vertical rise the architects could use?

Find and interpret the average rate of change illustrated in each graph.

59.

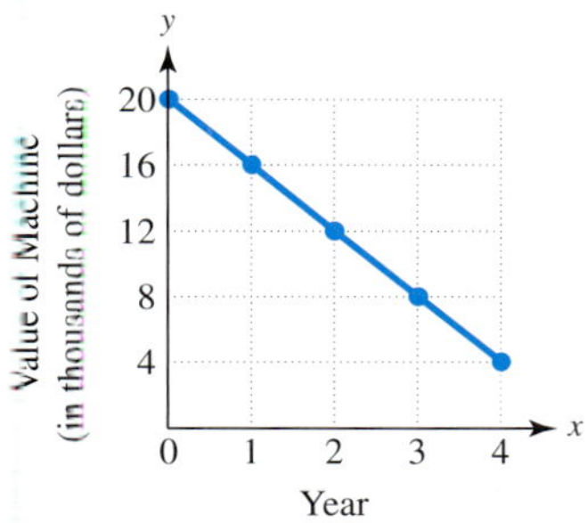

60.

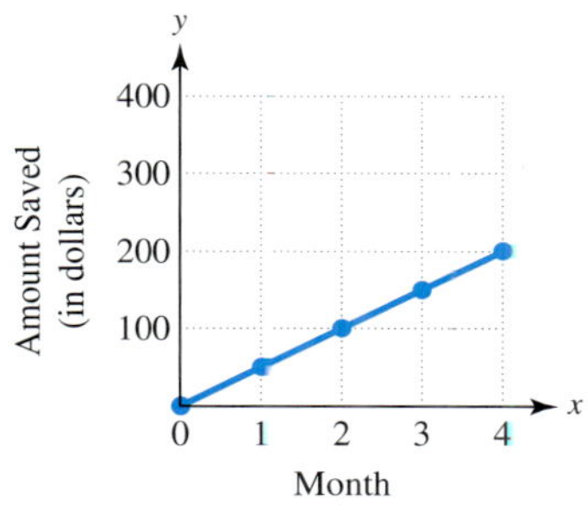

61.

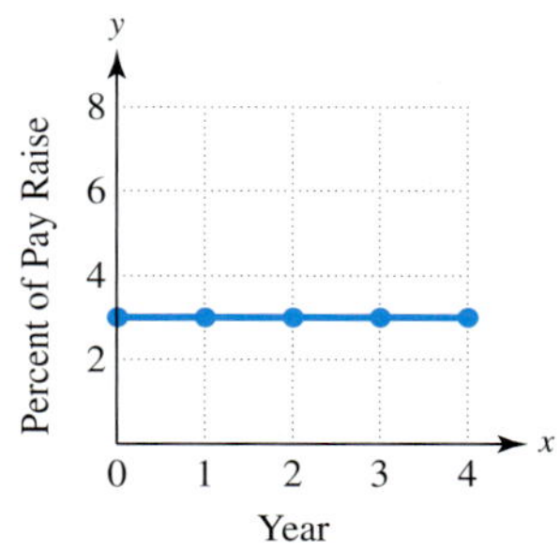

Use the idea of average rate of change to solve each problem. Round answers to the nearest thousandth. See Examples 9 and 10.

62. The graph provides a good approximation of the number of mobile homes (in thousands) placed in use in the United States during 2000–2006.

(a) Use the given ordered pairs to find the average rate of change in the number of mobile homes per year during this period.

(b) Interpret what a negative slope means in this situation.

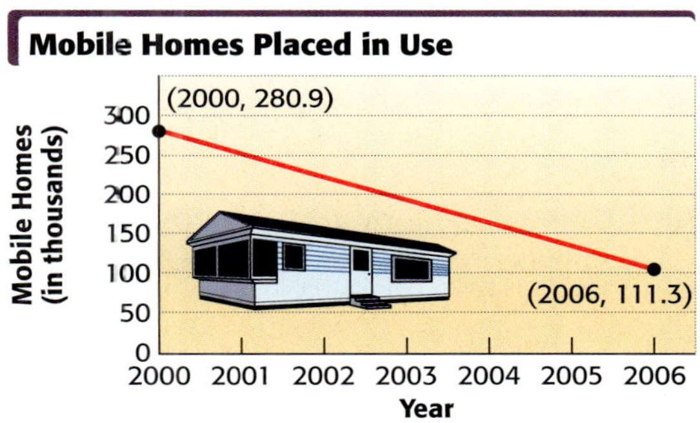

Source: U.S. Census Bureau.

63. Personal spending on recreation in the United States (in billions of dollars) in recent years is closely approximated by the graph.

(a) Use the given ordered pairs to determine the average rate of change in these expenditures per year.

(b) Explain how a positive slope is interpreted in this situation.

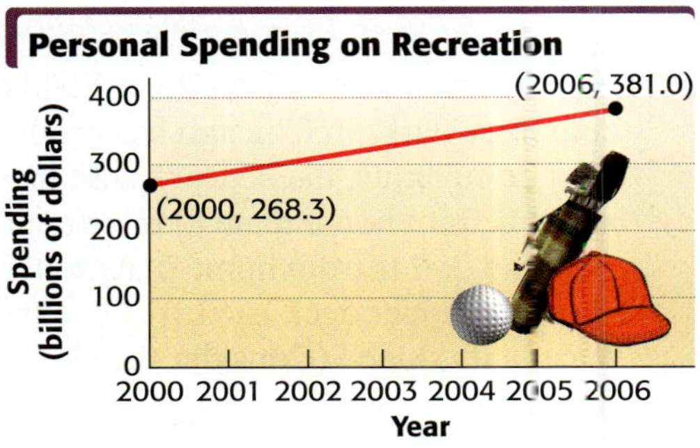

Source: U.S. Department of Commerce.

64. The total amount spent on plasma TVs in the United States changed from \$1590 million in 2003 to \$5705 million in 2006. Find and interpret the average rate of change in sales, in millions of dollars per year. Round your answer to the nearest hundredth. (*Source:* Consumer Electronics Association.)

65. The total amount spent on analog TVs in the United States changed from \$5836 million in 2003 to \$1424 million in 2006. Find and interpret the average rate of change in sales, in millions of dollars per year. Round your answer to the nearest hundredth. (*Source:* Consumer Electronics Association.)

Relating Concepts (Exercises 66–71) For Individual or Group Work

In these exercises we investigate a method of determining whether three points lie on the same straight line. (Such points are said to be ***collinear.****) The points we consider are* $A(3, 1)$, $B(6, 2)$, *and* $C(9, 3)$. ***Work Exercises 66–71 in order.***

66. Find the slope of segment AB.

67. Find the slope of segment BC.

68. Find the slope of segment AC.

69. If slope of AB = slope of BC = slope of AC, then A, B, and C are collinear. Use the results of Exercises 66–68 to show that this statement is satisfied

70. Use the slope formula to determine whether the points $(1, -2)$, $(3, -1)$, and $(5, 0)$ are collinear.

71. Repeat Exercise 70 for the points $(0, 6)$, $(4, -5)$, and $(-2, 12)$.

4.3 Linear Equations in Two Variables

OBJECTIVE 1 Write an equation of a line, given its slope and *y*-intercept. In **Section 4.2** we found the slope of a line from the equation of the line by solving the equation for y. For example, we found that the slope of the line with equation

$$y = 4x + 8$$

is 4, the coefficient of x. What does the number 8 represent?

To find out, suppose a line has slope m and y-intercept $(0, b)$. We can find an equation of this line by choosing another point (x, y) on the line, as shown in Figure 22. Using the slope formula,

$$m = \frac{y - b}{x - 0}$$

$$m = \frac{y - b}{x}$$

$$mx = y - b \quad \text{Multiply by } x.$$

$$mx + b = y \quad \text{Add } b.$$

$$y = mx + b. \quad \text{Rewrite.}$$

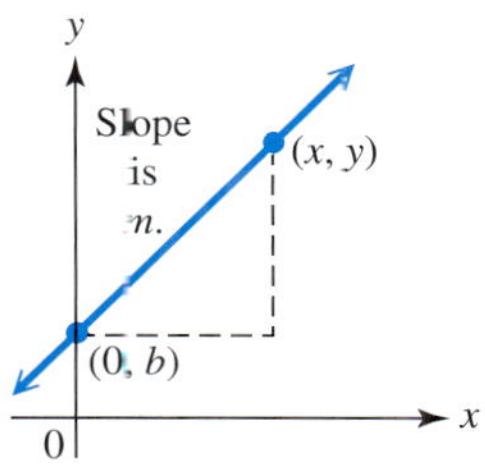

Figure 22

This last equation is called the *slope-intercept form* of the equation of a line, because we can identify the slope m and y-intercept $(0, b)$ at a glance. Thus, in the line with equation $y = 4x + 8$, the number 8 indicates that the y-intercept is $(0, 8)$.

Slope-Intercept Form

The **slope-intercept form** of the equation of a line with slope m and y-intercept $(0, b)$ is

$$y = mx + b.$$

↑ Slope ↑ y-intercept is $(0, b)$.

EXAMPLE 1 Writing an Equation of a Line

Write an equation of the line with slope $-\frac{4}{5}$ and y-intercept $(0, -2)$.

Here $m = -\frac{4}{5}$ and $b = -2$. Substitute these values into the slope-intercept form.

$$y = mx + b \quad \text{Slope-intercept form}$$

$$y = -\frac{4}{5}x - 2 \quad m = -\frac{4}{5};\ b = -2$$

Work Problem 1 at the Side. ▶

Note

The slope-intercept form of a linear equation is the most useful. Every linear equation (of a nonvertical line) has a *unique* (one and only one) slope-intercept form. In **Section 4.5** we study *linear functions,* which are defined using slope-intercept form. Also, this is the form we use when graphing a line with a graphing calculator.

OBJECTIVES

1. Write an equation of a line, given its slope and y-intercept.
2. Graph a line, using its slope and y-intercept.
3. Write an equation of a line, given its slope and a point on the line.
4. Write an equation of a line, given two points on the line.
5. Write an equation of a line parallel or perpendicular to a given line.
6. Write an equation of a line that models real data.

1 Write an equation in slope-intercept form for each line with the given slope and y-intercept.

(a) Slope 2; y-intercept $(0, -3)$

(b) Slope $-\frac{2}{3}$; y-intercept $(0, 0)$

ANSWERS

1. (a) $y = 2x - 3$ **(b)** $y = -\frac{2}{3}x$

2 Graph each line, using its slope and y-intercept.

(a) $y = 2x + 3$

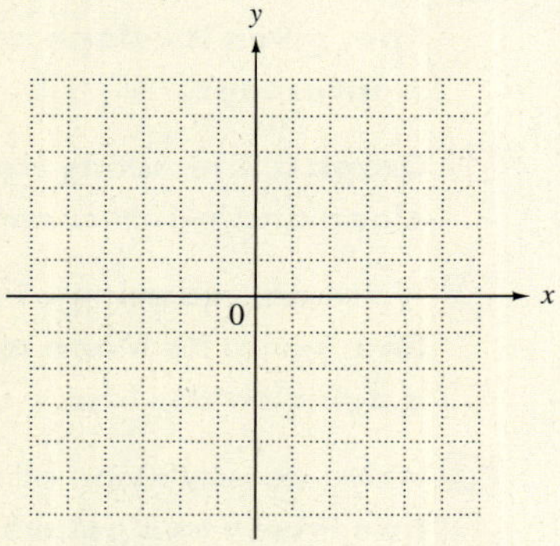

(b) $3x + 4y = 8$

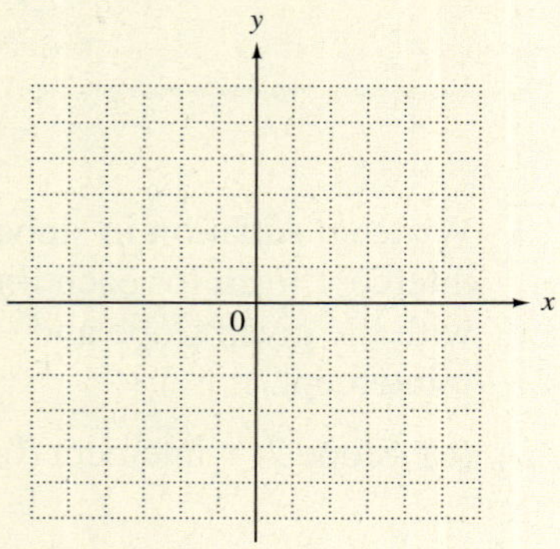

OBJECTIVE 2 Graph a line, using its slope and y-intercept.

EXAMPLE 2 Graphing Lines Using Slope and y-Intercept

Graph each line, using its slope and y-intercept.

(a) $y = 3x - 6$

Here $m = 3$ and $b = -6$. Plot the y-intercept $(0, -6)$. The slope 3 can be interpreted as

$$m = \frac{\text{rise}}{\text{run}} = \frac{\text{change in } y}{\text{change in } x} = \frac{3}{1}.$$

From $(0, -6)$, move 3 units *up* and 1 unit to the *right,* and plot a second point at $(1, -3)$. Join the two points with a straight line to obtain the graph in Figure 23.

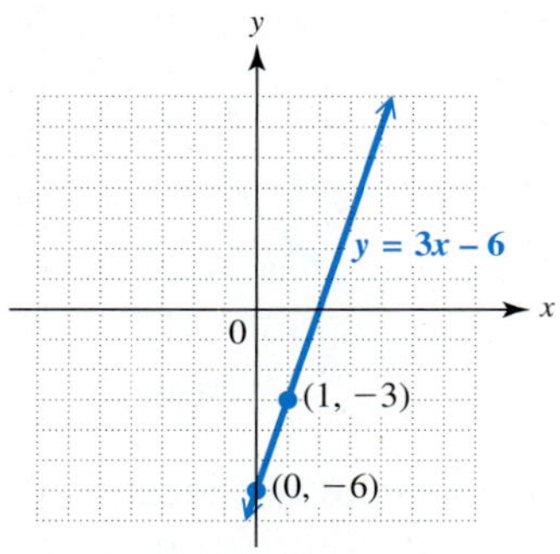

Figure 23

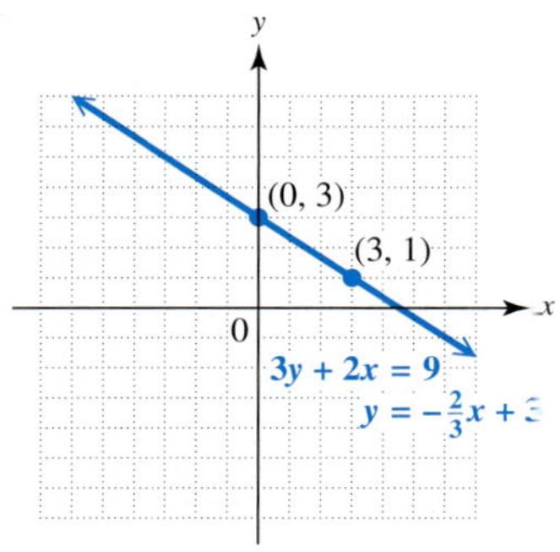

Figure 24

(b) $3y + 2x = 9$

Write the equation in slope-intercept form by solving for y.

$$3y + 2x = 9$$

$$3y = -2x + 9 \quad \text{Subtract } 2x.$$

$$y = -\frac{2}{3}x + 3 \quad \text{Slope-intercept form}$$

Slope ⟶ $-\frac{2}{3}$; y-intercept is $(0, 3)$.

To graph this equation, plot the y-intercept $(0, 3)$. The slope can be interpreted as either $\frac{-2}{3}$ or $\frac{2}{-3}$. Using $\frac{-2}{3}$, begin at $(0, 3)$ and move 2 units *down* and 3 units to the *right* to locate the point $(3, 1)$. The line through these two points is the required graph. See Figure 24. (Verify that the point obtained using $\frac{2}{-3}$ as the slope is also on this line.)

◀ Work Problem **2** at the Side.

OBJECTIVE 3 Write an equation of a line, given its slope and a point on the line. Let m represent the slope of a line and (x_1, y_1) represent a given point on the line. Let (x, y) represent any other point on the line. See Figure 25. Then by the slope formula,

$$m = \frac{y - y_1}{x - x_1}$$

$$m(x - x_1) = y - y_1 \quad \text{Multiply by } x - x_1.$$

$$y - y_1 = m(x - x_1). \quad \text{Rewrite.}$$

This last equation is the *point-slope form* of the equation of a line.

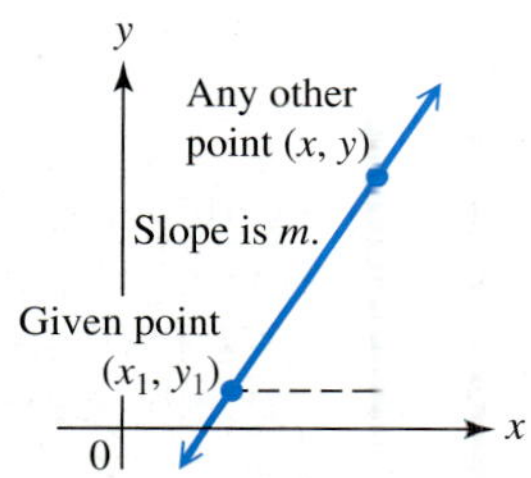

Figure 25

ANSWERS

2. (a)

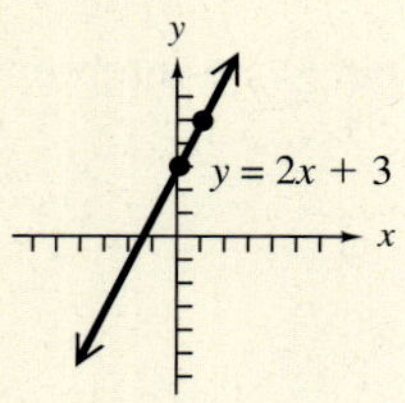

(b)

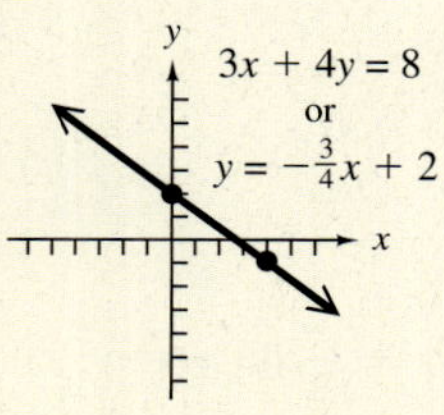

Point-Slope Form

The **point-slope form** of the equation of a line with slope m passing through the point (x_1, y_1) is

$$y - y_1 = m(x - x_1).$$

(Slope: m; Given point: (x_1, y_1))

To use this form to write the equation of a line, we need to know the coordinates of a point (x_1, y_1) and the slope m of the line.

EXAMPLE 3 **Finding the Equation of a Line, Given the Slope and a Point**

Find an equation of the line with slope $\frac{1}{3}$ passing through the point $(-2, 5)$.

Method 1 Use the point-slope form of the equation of a line, with $(x_1, y_1) = (-2, 5)$ and $m = \frac{1}{3}$.

$$y - y_1 = m(x - x_1) \qquad \text{Point-slope form}$$

$$y - 5 = \frac{1}{3}[x - (-2)] \qquad y_1 = 5,\ m = \tfrac{1}{3},\ x_1 = -2$$

$$y - 5 = \frac{1}{3}(x + 2)$$

$$3y - 15 = x + 2 \qquad \text{Multiply by 3.}$$

$$3y = x + 17 \qquad \text{Add 15.}$$

$$y = \frac{1}{3}x + \frac{17}{3} \qquad \text{Divide by 3; slope-intercept form}$$

Method 2 An alternative method for finding this equation uses the slope-intercept form, with $(x, y) = (-2, 5)$ and $m = \frac{1}{3}$.

$$y = mx + b \qquad \text{Slope-intercept form}$$

$$5 = \frac{1}{3}(-2) + b \qquad \text{Substitute for } y, m, \text{ and } x.$$

Solve for b.

$$5 = -\frac{2}{3} + b \qquad \text{Multiply.}$$

$$b = \frac{17}{3} \qquad 5 = \tfrac{15}{3};\ \text{add } \tfrac{2}{3};\ \text{rewrite.}$$

Knowing that $m = \frac{1}{3}$ and $b = \frac{17}{3}$ gives the equation $y = \frac{1}{3}x + \frac{17}{3}$.

In **Section 4.1,** we defined *standard form* for a linear equation as

$$Ax + By = C,$$

where A, B, and C are real numbers. In most cases, A, B, and C are rational numbers. For consistency, in this book we will give answers so that A, B, and C are integers with greatest common factor 1, and $A \geq 0$. For example, the equation in Example 3 is written in standard form as $x - 3y = -17$.

3 **(a)** Write an equation in standard form for the line through $(-2, 7)$ with $m = 3$.

(b) Write an equation in slope-intercept form for the line through $(1, 3)$ with $m = -\frac{5}{4}$.

4 Write an equation in standard form for each line.

(a) Through $(-1, 2)$ and $(5, 7)$

(b) Through $(-2, 6)$ and $(1, 4)$

Answers

3. **(a)** $3x - y = -13$ **(b)** $y = -\frac{5}{4}x + \frac{17}{4}$

4. **(a)** $5x - 6y = -17$ **(b)** $2x + 3y = 14$

Note

The definition of "standard form" is not standard from one text to another. Any linear equation can be written in many different (all equally correct) forms. For example, the equation $2x + 3y = 8$ can be written as

$$2x = 8 - 3y, \quad 3y = 8 - 2x, \quad x + \tfrac{3}{2}y = 4, \quad 4x + 6y = 16,$$

and so on. In addition to writing it in standard form $Ax + By = C$ with $A \geq 0$, let us agree that the form $2x + 3y = 8$ is preferred over any multiples of each side, such as $4x + 6y = 16$. (To write $4x + 6y = 16$ this way, divide each side by 2.)

◀ Work Problem **3** at the Side.

OBJECTIVE 4 **Write an equation of a line, given two points on the line.** To find an equation of a line when two points on the line are known, first use the slope formula to find the slope of the line. Then use the slope with either of the given points and the point-slope form of the equation of a line.

EXAMPLE 4 **Finding an Equation of a Line, Given Two Points**

Find an equation of the line through the points $(-4, 3)$ and $(5, -7)$. Write the equation in standard form.

First find the slope by using the slope formula.

$$m = \frac{-7 - 3}{5 - (-4)} = -\frac{10}{9}$$

Use either $(-4, 3)$ or $(5, -7)$ as (x_1, y_1) in the point-slope form of the equation of a line. If we choose $(-4, 3)$, then $-4 = x_1$ and $3 = y_1$.

$$y - y_1 = m(x - x_1) \qquad \text{Point-slope form}$$

$$y - 3 = -\frac{10}{9}[x - (-4)] \qquad y_1 = 3,\ m = -\tfrac{10}{9},\ x_1 = -4$$

$$y - 3 = -\frac{10}{9}(x + 4)$$

$$9y - 27 = -10x - 40 \qquad \text{Multiply by 9; distributive property}$$

$$10x + 9y = -13 \qquad \text{Standard form}$$

Verify that if $(5, -7)$ were used, the same equation would result.

◀ Work Problem **4** at the Side.

A horizontal line has slope 0. From the point-slope form, the equation of a horizontal line through the point (a, b) is

$$y - y_1 = m(x - x_1) \qquad \text{Point-slope form}$$

$$y - b = 0(x - a) \qquad y_1 = b,\ m = 0,\ x_1 = a$$

$$y - b = 0 \qquad \text{Multiplication property of 0}$$

$$y = b. \qquad \text{Add } b.$$

Notice that the point-slope form does not apply to a vertical line, since the slope of a vertical line is undefined. A vertical line through the point (a, b) has equation $x = a$.

In summary, horizontal and vertical lines have the following special equations.

> **Equations of Horizontal and Vertical Lines**
>
> The horizontal line through the point (a, b) has equation $y = b$.
>
> The vertical line through the point (a, b) has equation $x = a$.

Work Problem **5** *at the Side.* ▶

OBJECTIVE 5 Write an equation of a line parallel or perpendicular to a given line. As mentioned in the previous section, parallel lines have the same slope and perpendicular lines have slopes with product -1.

EXAMPLE 5 Finding Equations of Parallel or Perpendicular Lines

Find the equation in slope-intercept form of the line passing through the point $(-4, 5)$ and **(a)** parallel to the line $2x + 3y = 6$; **(b)** perpendicular to the line $2x + 3y = 6$.

(a) We find the slope of the line $2x + 3y = 6$ by solving for y.

$$2x + 3y = 6$$

$$3y = -2x + 6 \quad \text{Subtract } 2x.$$

$$y = -\frac{2}{3}x + 2 \quad \text{Divide by 3.}$$

(Slope: $-\frac{2}{3}$)

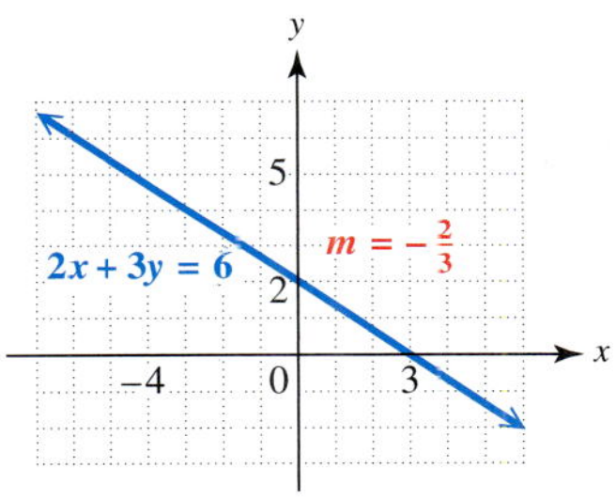

Figure 26

The slope is given by the coefficient of x, so $m = -\frac{2}{3}$. See Figure 26. Since parallel lines have the same slope, the required equation of the line through $(-4, 5)$ and parallel to $2x + 3y = 6$ must also have slope $-\frac{2}{3}$. To find this equation, we use the point-slope form, with $(x_1, y_1) = (-4, 5)$ and $m = -\frac{2}{3}$.

$$y - 5 = -\frac{2}{3}[x - (-4)] \quad y_1 = 5, m = -\tfrac{2}{3}, x_1 = -4$$

$$y - 5 = -\frac{2}{3}(x + 4)$$

$$y - 5 = -\frac{2}{3}x - \frac{8}{3} \quad \text{Distributive property}$$

$$y = -\frac{2}{3}x - \frac{8}{3} + \frac{15}{3} \quad \text{Add } 5 = \tfrac{15}{3}.$$

$$y = -\frac{2}{3}x + \frac{7}{3} \quad \text{Combine like terms.}$$

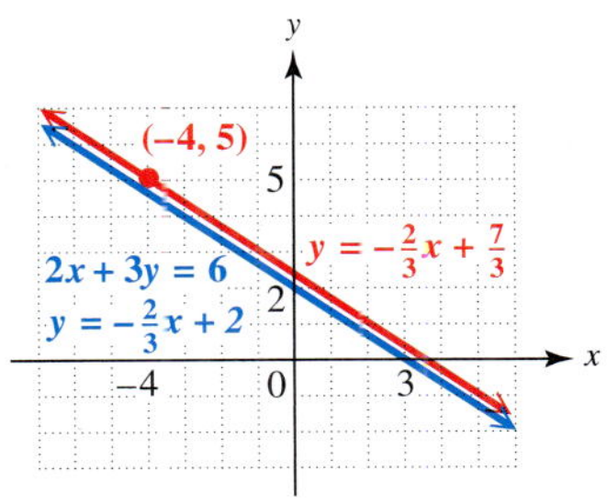

Figure 27

We did not clear fractions after the substitution step here because we want the equation in slope-intercept form—that is, solved for y. Both lines are shown in Figure 27.

Continued on Next Page

5 Write an equation for each line.

(a) Through $(8, -2)$; $m = 0$

(b) The vertical line through $(3, 5)$

ANSWERS

5. (a) $y = -2$ **(b)** $x = 3$

6 Write an equation in slope-intercept form of the line passing through the point $(-8, 3)$ and

(a) parallel to the line $2x - 3y = 10$.

(b) perpendicular to the line $2x - 3y = 10$.

(b) The given line $2x + 3y = 6$ can be written in slope-intercept form as

$$y = -\frac{2}{3}x + 2,$$

so the line has slope $-\frac{2}{3}$. To be perpendicular to the line $2x + 3y = 6$, a line must have a slope that is the negative reciprocal of $-\frac{2}{3}$, which is $\frac{3}{2}$. We use the point $(-4, 5)$ and slope $\frac{3}{2}$ in the point-slope form to get the equation of the perpendicular line shown in Figure 28.

$$y - 5 = \frac{3}{2}[x - (-4)] \quad y_1 = 5, m = \tfrac{3}{2}, x_1 = -4$$

$$y - 5 = \frac{3}{2}(x + 4)$$

$$y - 5 = \frac{3}{2}x + 6 \quad \text{Distributive property}$$

$$y = \frac{3}{2}x + 11 \quad \text{Add 5.}$$

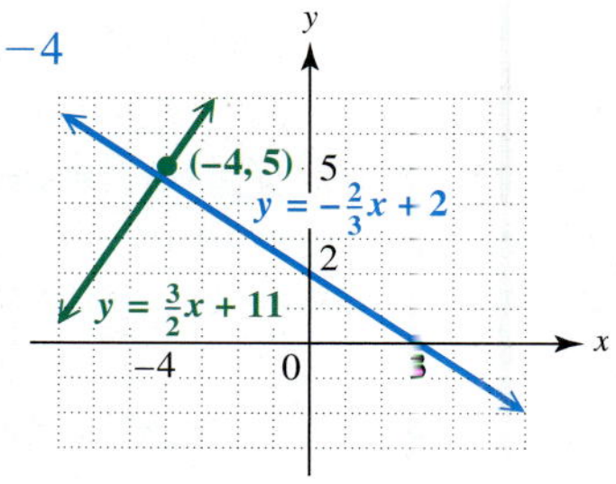

Figure 28

◀ *Work Problem* **6** *at the Side.*

A summary of the various forms of linear equations follows.

FORMS OF LINEAR EQUATIONS

Equation	Description	When to Use
$y = mx + b$	**Slope-Intercept Form** Slope is m. y-intercept is $(0, b)$.	The slope and y-intercept can be easily identified and used to quickly graph the equation.
$y - y_1 = m(x - x_1)$	**Point-Slope Form** Slope is m. Line passes through (x_1, y_1).	This form is ideal for finding the equation of a line if the slope and a point on the line or two points on the line are known.
$Ax + By = C$	**Standard Form** Slope is $-\frac{A}{B}$ $(B \neq 0)$. x-intercept is $(\frac{C}{A}, 0)$ $(A \neq 0)$. y-intercept is $(0, \frac{C}{B})$ $(B \neq 0)$.	The x- and y-intercepts can be found quickly and used to graph the equation. The slope must be calculated.
$y = b$	**Horizontal Line** Slope is 0. y-intercept is $(0, b)$.	If the graph intersects only the y-axis, then y is the only variable in the equation.
$x = a$	**Vertical Line** Slope is undefined. x-intercept is $(a, 0)$.	If the graph intersects only the x-axis, then x is the only variable in the equation.

OBJECTIVE 6 Write an equation of a line that models real data. We can use the information presented in this section to write equations of lines that mathematically describe, or *model,* real data if the given set of data changes at a fairly constant rate. In this case, the data fit a linear pattern, and the rate of change is the slope of the line.

ANSWERS

6. (a) $y = \frac{2}{3}x + \frac{25}{3}$ (b) $y = -\frac{3}{2}x - 9$

EXAMPLE 6 Determining a Linear Equation to Describe Real Data

Suppose it is time to fill your car with gasoline. At your local station, 89-octane gas is selling for \$4.50 per gal.

(a) Write an equation that describes the cost y to buy x gallons of gas.

Experience has taught you that the total price you pay is determined by the number of gallons you buy multiplied by the price per gallon (in this case, \$4.50). As you pump the gas, two sets of numbers flash by: the number of gallons pumped and the price for that number of gallons.

The table uses ordered pairs to illustrate this situation.

Number of Gallons Pumped	Price of This Number of Gallons
0	$0(\$4.50) = \$\ 0.00$
1	$1(\$4.50) = \$\ 4.50$
2	$2(\$4.50) = \$\ 9.00$
3	$3(\$4.50) = \$13.50$
4	$4(\$4.50) = \$18.00$

If we let x denote the number of gallons pumped, then the total price y in dollars can be found by using the linear equation

Total price → y; Number of gallons → x

$$y = 4.50x.$$

Theoretically, there are infinitely many ordered pairs (x, y) that satisfy this equation, but here we are limited to nonnegative values for x, since we cannot have a negative number of gallons. There is also a practical maximum value for x in this situation, which varies from one car to another. What determines this maximum value?

(b) You can also get a car wash at the gas station if you pay an additional \$3.00. Write an equation that defines the price for gas and a car wash.

Since an additional \$3.00 will be charged, you pay $4.50x + 3.00$ dollars for x gallons of gas and a car wash, described by

$$y = 4.5x + 3. \quad \text{Delete unnecessary 0s.}$$

(c) Interpret the ordered pairs (5, 25.5) and (10, 48) in relation to the equation from part (b).

The ordered pair (5, 25.5) indicates that the price of 5 gal of gas and a car wash is \$25.50. Similarly, (10, 48) indicates that the price of 10 gal of gas and a car wash is \$48.00.

Work Problem 7 at the Side. ▶

Note

In Example 6(a), the ordered pair (0, 0) satisfied the equation, so the linear equation has the form $y = mx$, where $b = 0$. If a situation involves an initial charge b plus a charge per unit m as in Example 6(b), the equation has the form $y = mx + b$, where $b \neq 0$.

7 **(a)** Suppose it costs \$0.10 per minute to make a long-distance call. Write an equation to describe the cost y to make an x-minute call.

(b) Suppose there is a flat rate of \$0.20 plus a charge of \$0.10 per minute to make a call. Write an equation that gives the cost y for a call of x minutes.

(c) Interpret the ordered pair (15, 1.7) in relation to the equation from part (b).

ANSWERS

7. (a) $y = 0.1x$ (Note: $0.10x = 0.1x$)
(b) $y = 0.1x + 0.2$
(c) The ordered pair (15, 1.7) indicates that the price of a 15-minute call is \$1.70.

8 The percentage of the U.S. population 25 yr and older with at least a high school diploma is shown in the table for selected years.

Year	Percent
1950	34.3
1960	41.1
1970	52.3
1980	66.5
1990	77.6
2000	84.1
2005	85.2

Source: U.S. Census Bureau.

(a) Let $x = 0$ represent 1950, $x = 10$ represent 1960, and so on. Use the data for 1950 and 2000 to find an equation that models the data.

(b) Use the equation from part (a) to approximate the percentage, to the nearest tenth, of the U.S. population 25 yr and older who were at least high school graduates in 1995.

EXAMPLE 7 Finding an Equation of a Line That Models Data

Average annual tuition and fees for in-state students at public four-year colleges are shown in the table for selected years and graphed as ordered pairs of points in the **scatter diagram** in Figure 29, where $x = 0$ represents 1990, $x = 4$ represents 1994, and so on, and y represents the cost in dollars.

Year	Cost (in dollars)
1990	2035
1994	2820
1996	3151
1998	3486
2000	3774
2002	4273
2004	4920

Source: U.S. National Center for Education Statistics.

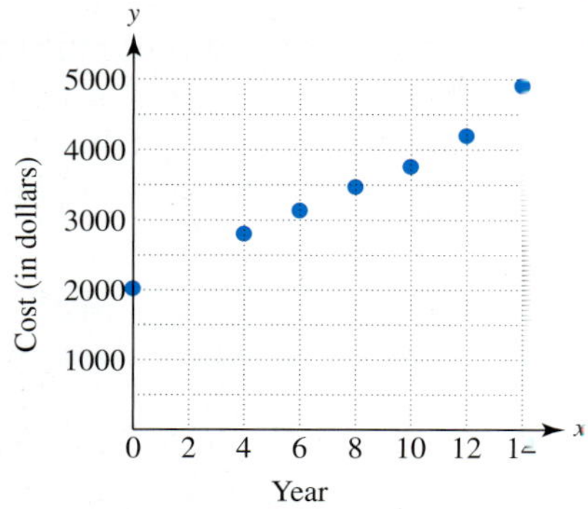

Figure 29

(a) Find an equation that models the data.

Since the points in Figure 29 lie approximately on a straight line, we can write a linear equation that models the relationship between year x and cost y. We choose two data points, (0, 2035) and (12, 4273), to find the slope of the line.

$$m = \frac{4273 - 2035}{12 - 0} = \frac{2238}{12} = 186.5$$

Start with the x- and y-values of the same point.

The slope 186.5 indicates that the cost of tuition and fees increased by about \$186.50 per year from 1990 to 2002. We use this slope, the y-intercept (0, 2035), and the slope-intercept form to write an equation of the line. Thus,

$$y = 186.5x + 2035.$$

(b) Use the equation from part (a) to approximate the cost of tuition and fees at public four-year colleges in 2006.

The value $x = 16$ corresponds to the year 2006.

$$y = 186.5x + 2035$$
$$y = 186.5(16) + 2035 \quad \text{Substitute 16 for } x.$$
$$y = 5019$$

According to the model, average tuition and fees for in-state students at public four-year colleges in 2006 were about \$5019.

Note
In Example 7, if we had chosen different data points, we would have found a slightly different equation. However, all such equations should yield similar results, since the data points are approximately linear.

◀ *Work Problem* **8** *at the Side.*

ANSWERS

8. **(a)** $y = 0.996x + 34.3$ **(b)** 79.1%

4.3 Exercises

FOR EXTRA HELP — MyMathLab — Math XL PRACTICE — WATCH — DOWNLOAD — READ — REVIEW

1. The following equations all represent the same line. Which one is in standard form as defined in the text?

A. $3x - 2y = 5$ **B.** $2y = 3x - 5$

C. $\frac{3}{5}x - \frac{2}{5}y = 1$ **D.** $3x = 2y + 5$

2. Which equation is in point-slope form?

A. $y = 6x + 2$ **B.** $4x + y = 9$

C. $y - 3 = 2(x - 1)$ **D.** $2y = 3x - 7$

3. Which equation in Exercise 2 is in slope-intercept form?

4. Write the equation $y + 2 = -3(x - 4)$ in slope-intercept form.

5. Write the equation from Exercise 4 in standard form.

6. Write the equation $10x - 7y = 70$ in slope-intercept form.

Match each equation with the graph that it most closely resembles. (Hint: Determining the signs of m and b will help you make your decision.)

7. $y = 2x + 3$ **8.** $y = -2x + 3$ **9.** $y = -2x - 3$ **10.** $y = 2x - 3$

11. $y = 2x$ **12.** $y = -2x$ **13.** $y = 3$ **14.** $y = -3$

A.
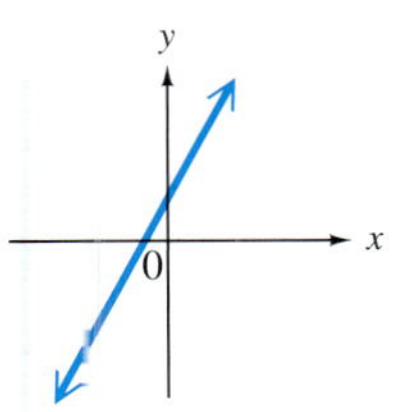

B.
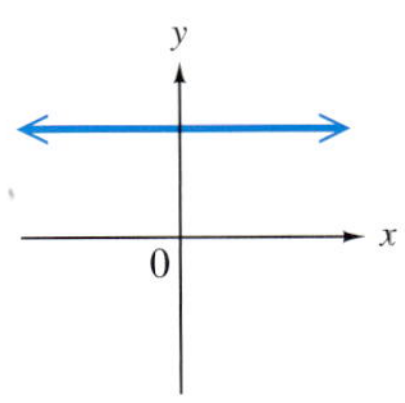

C.
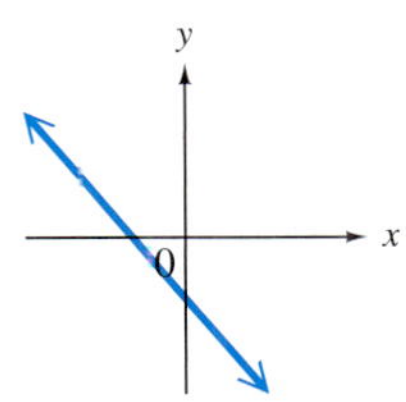

D.
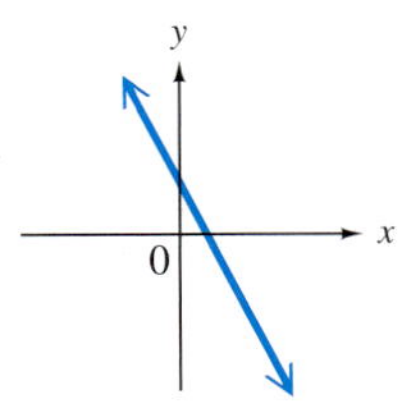

E.
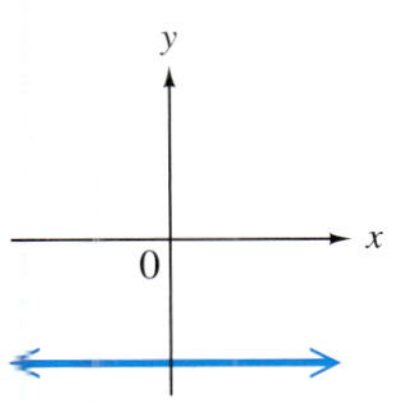

F.
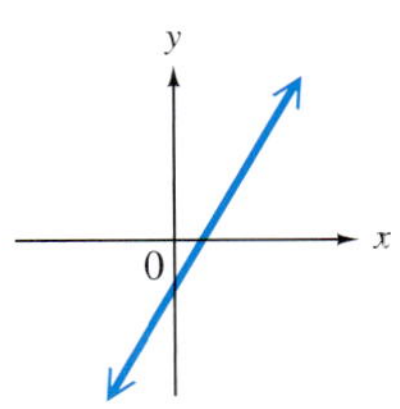

G.
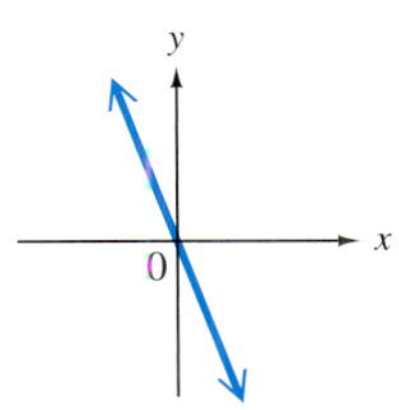

H.
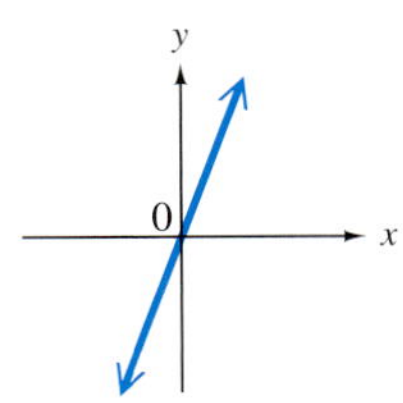

Write the equation in slope-intercept form of the line satisfying the given conditions. See Example 1.

15. $m = 5; b = 15$ **16.** $m = -2; b = 12$ **17.** $m = -\frac{2}{3}$; y-intercept $\left(0, \frac{4}{5}\right)$

18. $m = -\frac{5}{8}$; y-intercept $\left(0, -\frac{1}{3}\right)$

19. Slope $\frac{2}{5}$; y-intercept $(0, 5)$

20. Slope $-\frac{3}{4}$; y-intercept $(0, 7)$

For each equation, ***(a)*** *write it in slope-intercept form,* ***(b)*** *give the slope of the line,* ***(c)*** *give the y-intercept, and* ***(d)*** *graph the line. See Example 2.*

21. $-x + y = 2$

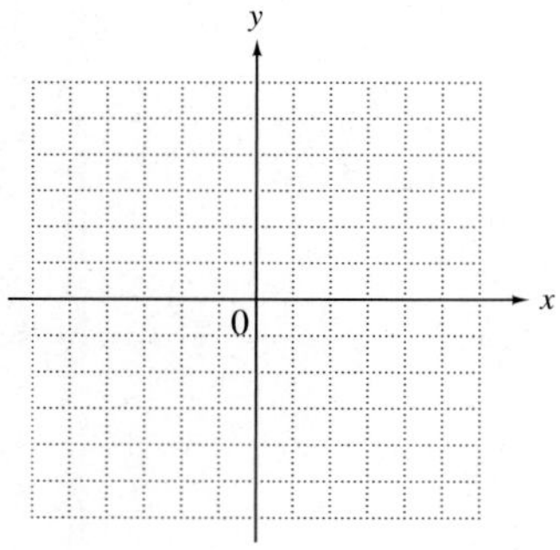

22. $-x + y = 5$

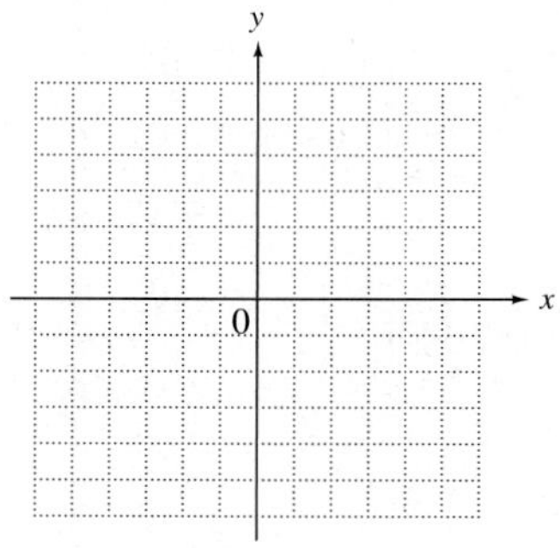

23. $4x - 5y = 20$

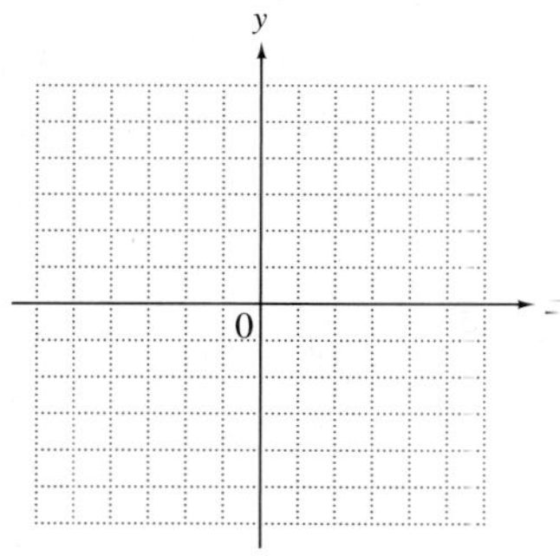

24. $7x - 3y = 3$

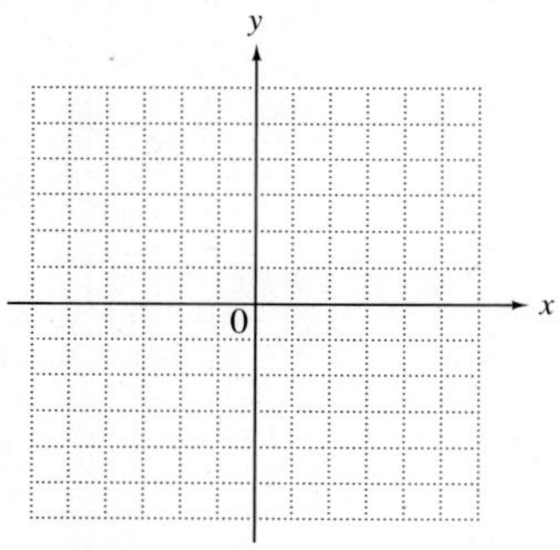

25. $x + 2y = -4$

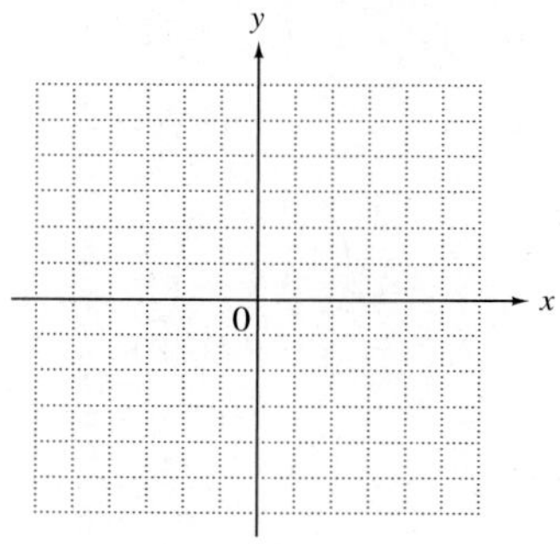

26. $x + 3y = -9$

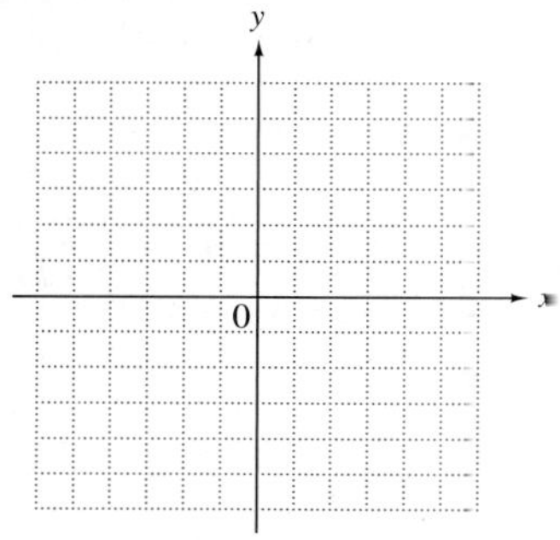

Write the equation in standard form of the line satisfying the given conditions. See Example 3.

27. Through $(-2, 4)$; $m = -\frac{3}{4}$

28. Through $(-1, 6)$; $m = -\frac{5}{6}$

29. Through $(5, 8)$; $m = -2$

30. Through $(12, 10)$; $m = 1$

31. Through $(-5, 4)$; $m = \frac{1}{2}$

32. Through $(7, -2)$; $m = \frac{1}{4}$

33. Through $(-4, 12)$; horizontal

34. Through $(1, 5)$; horizontal

Write an equation that satisfies the given conditions.

35. Through $(9, 10)$; undefined slope

36. Through $(-2, 8)$; 0 slope

37. Through $(0.5, 0.2)$; horizontal

38. Through $\left(\frac{5}{8}, \frac{2}{9}\right)$; vertical

Write the equation in standard form of the line through the given points. See Example 4.

39. $(3, 4)$ and $(5, 8)$

40. $(5, -2)$ and $(-3, 14)$

41. $(6, 1)$ and $(-2, 5)$

42. $(-2, 5)$ and $(-8, 1)$

43. $\left(-\frac{2}{5}, \frac{2}{5}\right)$ and $\left(\frac{4}{3}, \frac{2}{3}\right)$

44. $\left(\frac{3}{4}, \frac{8}{3}\right)$ and $\left(\frac{2}{5}, \frac{2}{3}\right)$

45. $(2, 5)$ and $(1, 5)$

46. $(-2, 2)$ and $(4, 2)$

47. $(7, 6)$ and $(7, -8)$

48. $(13, 5)$ and $(13, -1)$

Write the equation in slope-intercept form of the line satisfying the given conditions. See Example 5.

49. Through $(7, 2)$; parallel to $3x - y = 8$

50. Through $(4, 1)$; parallel to $2x + 5y = 10$

51. Through $(-2, -2)$; parallel to $-x + 2y = 10$

52. Through $(-1, 3)$; parallel to $-x + 3y = 12$

53. Through (8, 5); perpendicular to $2x - y = 7$

54. Through (2, −7); perpendicular to $5x + 2y = 18$

55. Through (−2, 7); perpendicular to $x = 9$

56. Through (8, 4); perpendicular to $x = -3$

Write an equation in the form $y = mx$ for each situation. Then give the three ordered pairs associated with the equation for x-values of 0, 5, *and* 10. *See Example 6(a).*

57. x represents the number of hours traveling at 45 mph, and y represents the distance traveled (in miles).

58. x represents the number of compact discs sold at $16 each, and y represents the total cost of the discs (in dollars).

59. x represents the number of gallons of gas sold at $5.00 per gal, and y represents the total cost of the gasoline (in dollars).

60. x represents the number of days a DVD movie is rented at $2.50 per day, and y represents the total charge for the rental (in dollars).

For each situation, ***(a)*** *write an equation in the form $y = mx + b$,* ***(b)*** *find and interpret the ordered pair associated with the equation for $x = 5$, and* ***(c)*** *answer the question. See Examples 6(b) and 6(c).*

61. A membership to the Midwest Athletic Club costs $99 plus $41 per month. (*Source:* Midwest Athletic Club.) Let x represent the number of months and y represent the cost. How much does the first year's membership cost?

62. For a family membership, the athletic club in Exercise 61 charges a membership fee of $159 plus $60 for each additional family member after the first. Let x represent the number of additional family members and y represent the cost. What is the membership fee for a four-person family?

63. A cell phone plan includes 900 anytime minutes for $60 per month, plus a one-time activation fee of $36. A Nokia 6085 cell phone is included at no additional charge. (*Source:* AT&T.) Let x represent the number of months of service and y represent the cost. If you sign a 2-yr contract, how much will this cell phone plan cost? (Assume that you never use more than the allotted number of minutes.)

64. Another cell phone plan includes 450 anytime minutes for $40 per month, plus $50 for a Nokia 6555 cell phone and $36 for a one-time activation fee. (*Source:* AT&T.) Let x represent the number of months of service and y represent the cost. If you sign a 1-yr contract, how much will this cell phone plan cost? (Assume that you never use more than the allotted number of minutes.)

65. A rental car costs \$50 plus \$0.20 per mile. Let x represent the number of miles driven and y represent the total charge to the renter. How many miles was the car driven if the renter paid \$84.60?

66. There is a \$30 fee to rent a chain saw, plus \$6 per day. Let x represent the number of days the saw is rented and y represent the charge to the user in dollars. If the total charge is \$138, for how many days is the saw rented?

Solve each problem. In part (a), give equations in slope-intercept form. (Round the slope to the nearest tenth.) See Example 7. (Source for Exercises 67 and 68: Consumer Electronics Association.)

67. Total sales of digital cameras in the United States (in millions of dollars) are shown in the graph, where the year 2003 corresponds to $x = 0$.

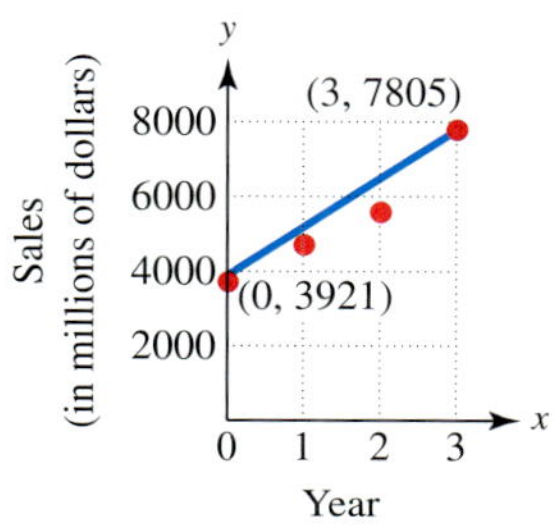

(a) Use the ordered pairs from the graph to write an equation that models the data. What does the slope tell us in the context of this problem?

(b) Use the equation from part (a) to approximate the sales of digital cameras in the United States in 2007.

68. Total sales of fax machines in the United States (in millions of dollars) are shown in the graph, where the year 2003 corresponds to $x = 0$.

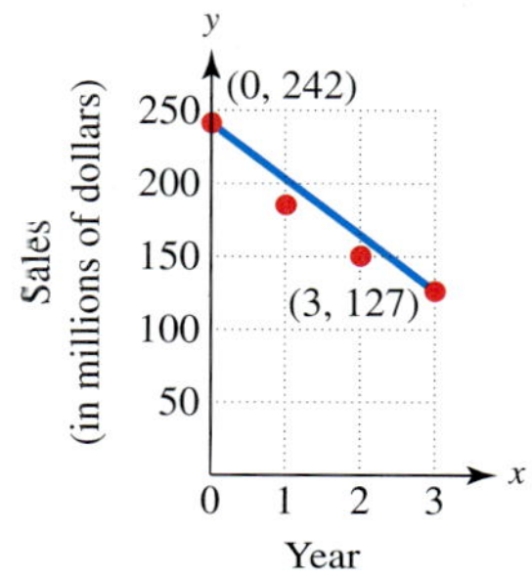

(a) Use the ordered pairs from the graph to write an equation that models the data. What does the slope tell us in the context of this problem?

(b) Use the equation from part (a) to approximate the sales of fax machines in the United States in 2007.

69. The number of pieces of first class mail delivered in the United States is shown in the bar graph.

(a) Use the information given for the years 2003 and 2007, letting $x = 3$ represent 2003 and $x = 7$ represent 2007, and letting y represent the number of pieces of mail (in millions), to write an equation that models the data.

(b) Use the equation to approximate the number of pieces of first class mail delivered in 2005. How does this result compare to the actual value, 98,071 million?

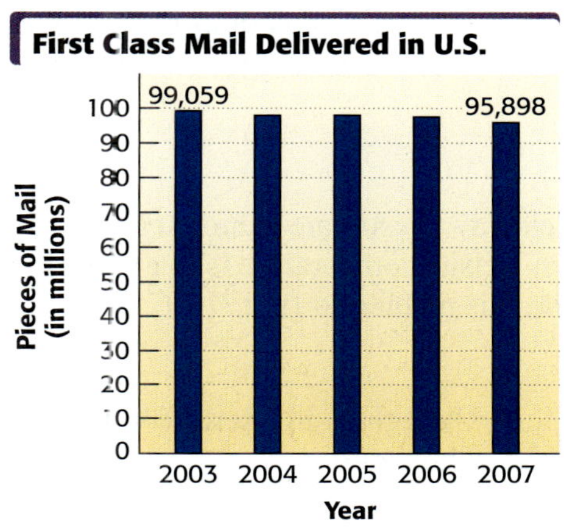

Source: U.S. Postal Service.

70. Median household income of all Americans is shown in the bar graph.

(a) Use the information given for the years 2000 and 2005, letting $x = 0$ represent 2000, $x = 5$ represent 2005, and y represent the median income, to write an equation that models median household income.

(b) Use the equation to approximate the median income for 2003. How does your result compare to the actual value, \$52,680?

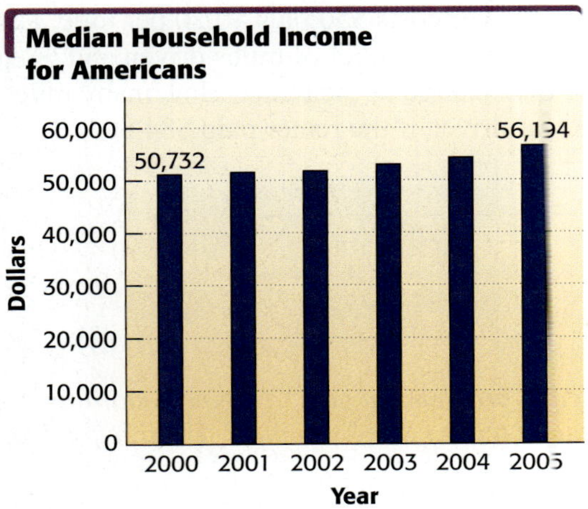

Source: U.S. Census Bureau.

Relating Concepts (Exercises 71–78) For Individual or Group Work

*In **Section 2.2** we learned how formulas can be applied to problem solving. In Exercises 71–78, we will see how the formula that relates Celsius and Fahrenheit temperatures is derived.*
Work Exercises 71–78 in order.

71. There is a linear relationship between Celsius and Fahrenheit temperatures. When $C = 0°$, $F =$ ________°, and when $C = 100°$, $F =$ ________°.

72. Think of ordered pairs of temperatures (C, F), where C and F represent corresponding Celsius and Fahrenheit temperatures. The equation that relates the two scales has a straight-line graph that contains the two points determined in Exercise 71.

(a) What are these two points?

(b) Find the slope of the line described in part (a).

73. Now think of the point-slope form of the equation in terms of C and F, where C replaces x and F replaces y. Use the slope you found in Exercise 72(b) and one of the two points determined earlier, and find the equation that gives F in terms of C.

74. To obtain another form of the formula, use the equation you found in Exercise 73 and solve for C in terms of F. For what temperature does $F = C$?

75. A quick way to estimate Fahrenheit temperature for a given Celsius temperature is to double C and add 30. Use this method to find F if $C = 15$.

76. Use the equation found in Exercise 73 to find F if $C = 15$. How does the answer compare with your answer to Exercise 75?

77. Use the method given in Exercise 75 to estimate the Fahrenheit temperature given $C = 30$. Then use the equation from Exercise 73 to find F when $C = 30$. How do the temperatures compare?

78. Explain why the method given in Exercise 75 to estimate Fahrenheit temperature gives a good approximation of $F = \frac{9}{5}C + 32$.

Summary Exercises on Slopes and Equations of Lines

Find the slope of each line.

1. Through $(3, -3)$ and $(8, -6)$

2. Through $(4, -5)$ and $(-1, -5)$

3. $3x - 7y = 21$

4. $x - 4 = 0$

*For each line described, find an equation of the line **(a)** in slope-intercept form and **(b)** in standard form.*

5. Through $(4, -2)$ with slope -3

6. Through $(-3, 6)$ with slope $\frac{2}{3}$

7. Through the points $(-2, 6)$ and $(4, 1)$

8. Through the points $(4, -8)$ and $(-4, 12)$

9. Through $(-2, 5)$ and parallel to the graph of $3x - y = 4$

10. Through the origin and perpendicular to the graph of $2x - 5y = 6$

11. Through $(5, -8)$ and parallel to the graph of $y = 4$

12. Through $\left(\frac{3}{4}, -\frac{7}{9}\right)$ and perpendicular to the graph of $x = \frac{2}{3}$

13. Through $(-4, 2)$ and parallel to the line through $(3, 9)$ and $(6, 11)$

14. Through $(4, -2)$ and perpendicular to the line through $(3, 7)$ and $(5, 6)$

15. Through $(-4, 12)$ and the midpoint of the segment with endpoints $(5, 8)$ and $(-3, 2)$

16. Through $(0, 3)$ and the midpoint of the segment with endpoints $(2, 8)$ and $(-4, 12)$

17. Through $(2, -1)$ and parallel to the graph of $y = \frac{1}{5}x + \frac{7}{4}$

18. Through $(0, -6)$ and perpendicular to the graph of $y = \frac{4}{3}x + \frac{3}{8}$

19. Through the points $(0.3, 1.5)$ and $(0.4, 1.7)$

20. Through $(2.5, 1.75)$ and parallel to the graph of $y = 0.5x + 3.25$

Match the description in Column I with the correct equation in Column II.

I	II
21. Slope -0.5, $b = -2$	**A.** $y = -\frac{1}{2}x$
22. x-intercept $(4, 0)$, y-intercept $(0, 2)$	**B.** $y = -\frac{1}{2}x - 2$
23. Passes through $(4, -2)$ and $(0, 0)$	**C.** $x - 2y = 2$
24. $m = \frac{1}{2}$, passes through $(-2, -2)$	**D.** $x + 2y = 4$
25. $m = \frac{1}{2}$, passes through the origin	**E.** $x = 2y$

4.4 Linear Inequalities in Two Variables

OBJECTIVES

1. Graph linear inequalities in two variables.
2. Graph the intersection of two linear inequalities.
3. Graph the union of two linear inequalities.

OBJECTIVE 1 Graph linear inequalities in two variables. In **Section 3.1** we graphed linear inequalities in one variable on the number line. We now graph linear inequalities in two variables on a rectangular coordinate system.

Linear Inequality in Two Variables

An inequality that can be written as

$$Ax + By < C \quad \text{or} \quad Ax + By > C,$$

where A, B, and C are real numbers and A and B are not both 0, is a **linear inequality in two variables.**

The symbols $\leq$ and $\geq$ may replace $<$ and $>$ in the definition.

Consider the graph in Figure 30. The graph of the line $x + y = 5$ divides the points in the rectangular coordinate system into three sets:

1. Those points that lie on the line itself and satisfy the equation $x + y = 5$ [like (0, 5), (2, 3), and (5, 0)];
2. Those that lie in the half-plane above the line and satisfy the inequality $x + y > 5$ [like (5, 3) and (2, 4)];
3. Those that lie in the half-plane below the line and satisfy the inequality $x + y < 5$ [like (0, 0) and $(-3, -1)$].

The graph of the line $x + y = 5$ is called the **boundary line** for the inequalities $x + y > 5$ and $x + y < 5$. Graphs of linear inequalities in two variables are *regions* in the real number plane that may or may not include boundary lines.

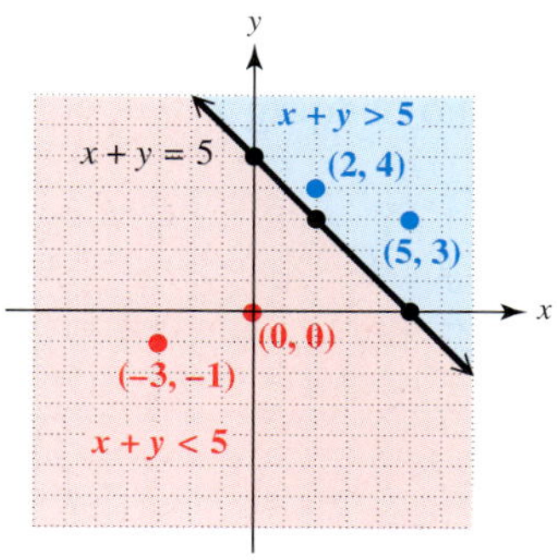

Figure 30

To graph a linear inequality in two variables, follow these steps.

Graphing a Linear Inequality

Step 1 **Draw the graph of the straight line that is the boundary.** Make the line solid if the inequality involves $\leq$ or $\geq$; make the line dashed if the inequality involves $<$ or $>$.

Step 2 **Choose a test point.** Choose any point not on the line, and substitute the coordinates of this point in the inequality.

Step 3 **Shade the appropriate region.** Shade the region that includes the test point if it satisfies the original inequality; otherwise, shade the region on the other side of the boundary line.

1 Graph each inequality.

(a) $x + y \leq 4$

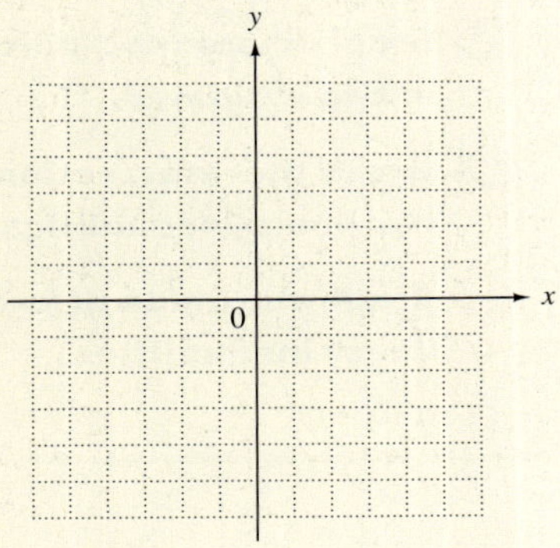

(b) $3x + y \geq 6$

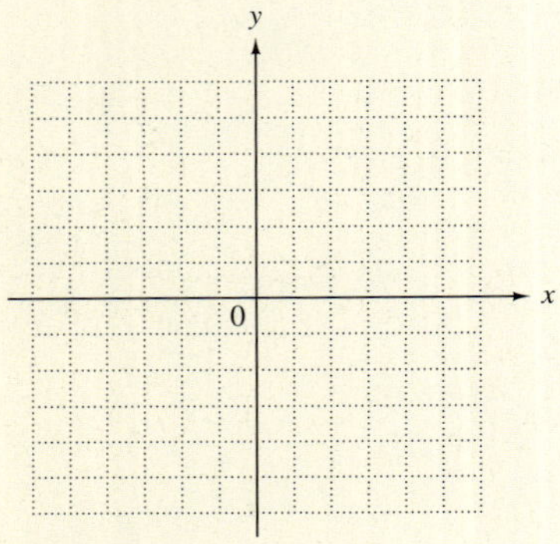

EXAMPLE 1 Graphing a Linear Inequality

Graph $3x + 2y \geq 6$.

Step 1 First graph the line $3x + 2y = 6$. The graph of this line, the boundary of the graph of the inequality, is shown in Figure 31.

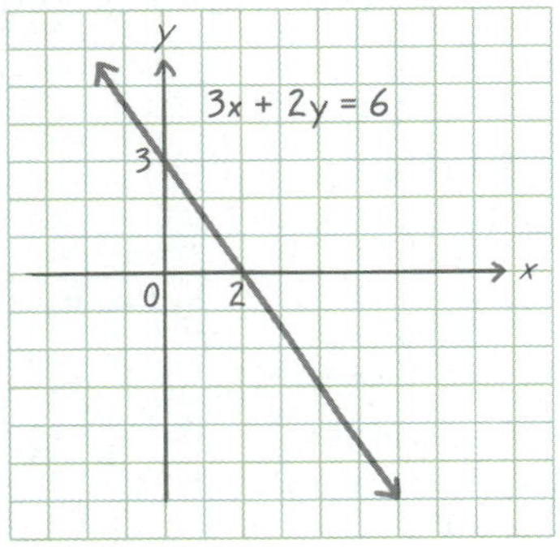

Figure 31

Step 2 The graph of the inequality $3x + 2y \geq 6$ includes the points of the boundary line $3x + 2y = 6$ and either the points *above* the line $3x + 2y = 6$ or the points *below* that line. To decide which, select any point not on the line $3x + 2y = 6$ as a test point. The origin, (0, 0), is often a good choice. Substitute the values from the test point (0, 0) for x and y in the inequality.

$$3x + 2y > 6$$

$$3(0) + 2(0) \overset{?}{>} 6 \quad \text{Let } x = 0 \text{ and } y = 0.$$

$$0 > 6 \quad \text{False}$$

Step 3 Because the result is false, (0, 0) does *not* satisfy the inequality, and so the solution set includes all points on the other side of the line. This region is shaded in Figure 32.

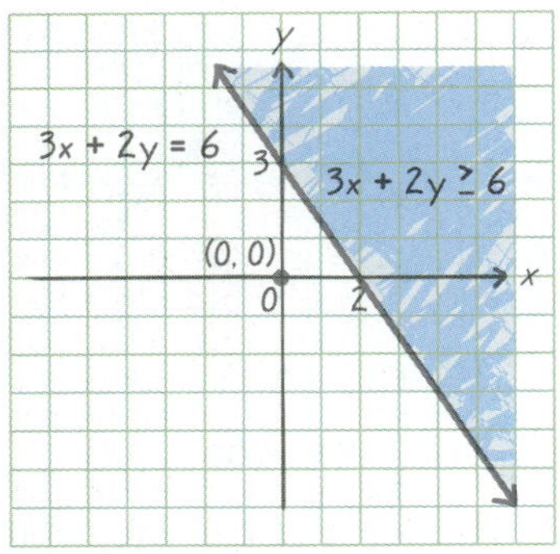

Figure 32

◀ *Work Problem* **1** *at the Side.*

If the inequality is written in the form $y > mx + b$ or $y < mx + b$, the inequality symbol indicates which half-plane to shade.

If $y > mx + b$, shade **above** the boundary line.

If $y < mx + b$, shade **below** the boundary line.

This method works only if the inequality is solved for y.

ANSWERS

1. (a)

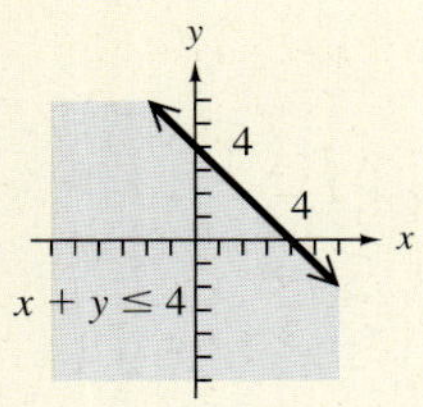

(b)

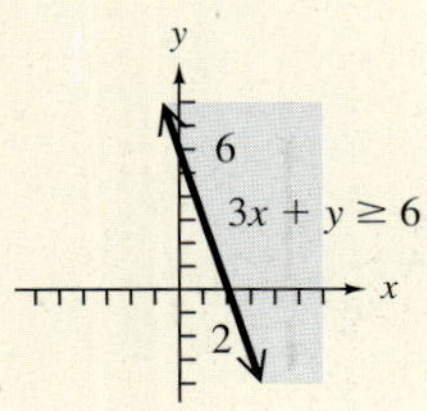

EXAMPLE 2 Graphing a Linear Inequality with Boundary Passing through the Origin

Graph $3x - 4y > 0$.

First graph the boundary line. If $x = 0$, then $y = 0$. Thus, this line passes through the origin. Two other points on the line are $(4, 3)$ and $(-4, -3)$. The points of the boundary line do not belong to the inequality $3x - 4y > 0$ (because inequality symbol does not include equality). For this reason, the line is dashed. Now solve the inequality for y.

$$3x - 4y > 0$$

$$-4y > -3x \quad \text{Subtract } 3x.$$

$$y < \frac{3}{4}x \quad \text{Divide by } -4\text{; change } > \text{ to } <.$$

Because of the *is less than* symbol, shade *below* the line. As a check, choose a test point not on the line. Because the origin is on the line, we must choose a different point, such as $(2, -1)$. Substitute for x and y in the original inequality.

$$3x - 4y > 0$$

$$3(2) - 4(-1) \overset{?}{>} 0 \quad \text{Let } x = 2 \text{ and } y = -1.$$

$$6 + 4 \overset{?}{>} 0$$

$$10 > 0 \quad \text{True}$$

This result agrees with the decision to shade below the line. The solution set, graphed in Figure 33, includes only those points in the shaded half-plane (not those on the line).

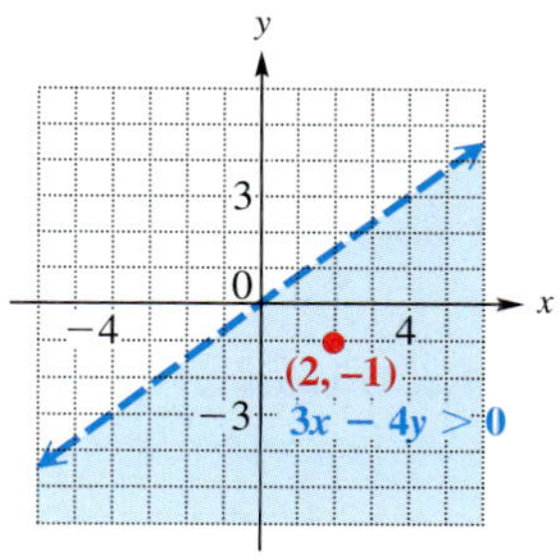

Figure 33

Work Problem 2 at the Side. ▶

2 Graph each inequality.

(a) $x + y > 0$

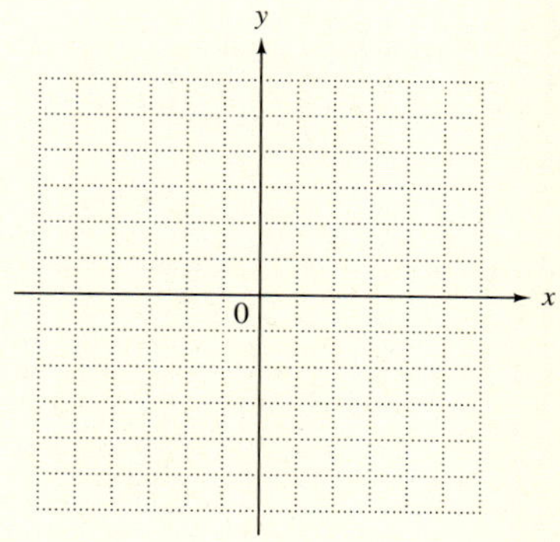

(b) $3x - 2y > 0$

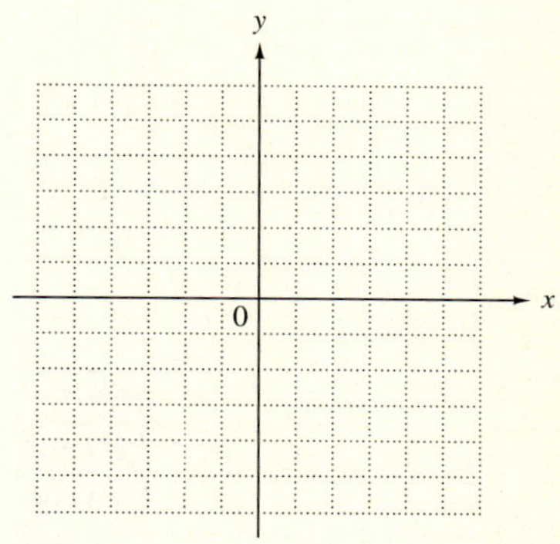

OBJECTIVE 2 Graph the intersection of two linear inequalities.

In **Section 3.2** we discussed how the words *and* and *or* are used with compound inequalities. In that section, the inequalities had one variable. Those ideas can be extended to include inequalities in two variables.

A pair of inequalities joined with the word *and* is interpreted as the intersection of the solution sets of the inequalities. ***The graph of the intersection of two or more inequalities is the region of the plane where all points satisfy all of the inequalities at the same time.***

ANSWERS

2. (a)

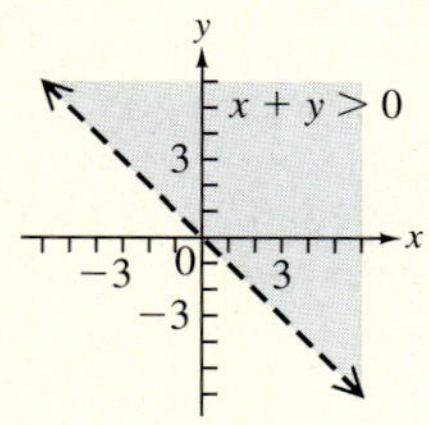

(b)

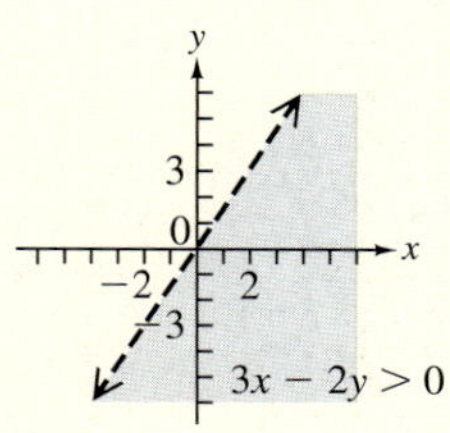

3 Graph $x - y \leq 4$ and $x \geq -2$.

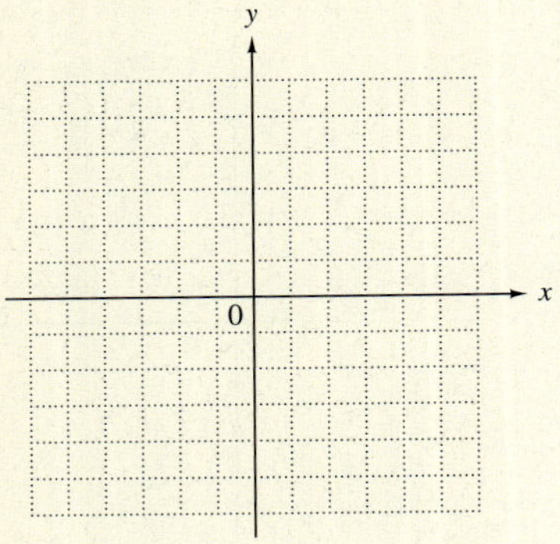

4 Graph $7x - 3y < 21$ or $x > 2$.

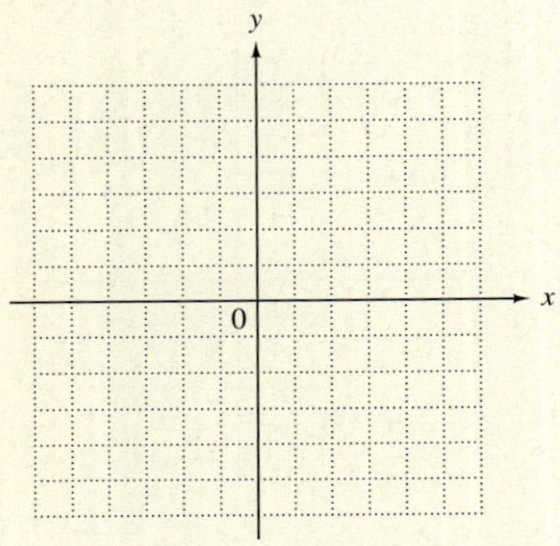

EXAMPLE 3 Graphing the Intersection of Two Inequalities

Graph $2x + 4y \geq 5$ and $x \geq 1$.

To begin, we graph each of the two inequalities $2x + 4y \geq 5$ and $x \geq 1$ separately. The graph of $2x + 4y \geq 5$ is shown in Figure 34(a), and the graph of $x \geq 1$ is shown in Figure 34(b).

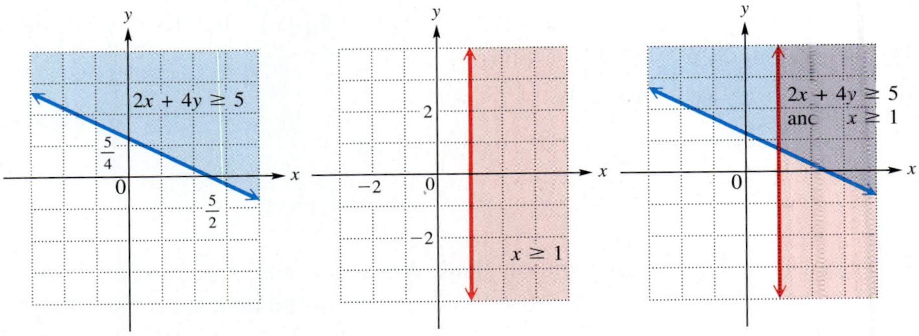

(a) (b) (c)

Figure 34

In practice, the two graphs in Figures 34(a) and 34(b) are graphed on the same axes. Then we use heavy shading to identify the intersection of the graphs, as shown in Figure 34(c). To check, we can use a test point from each of the four regions formed by the intersection of the boundary lines. Verify that only ordered pairs in the heavily shaded region satisfy both inequalities.

◀ Work Problem **3** at the Side.

OBJECTIVE 3 Graph the union of two linear inequalities. When two inequalities are joined by the word *or,* we must find the union of the graphs of the inequalities. ***The graph of the union of two inequalities includes all of the points that satisfy either inequality.***

EXAMPLE 4 Graphing the Union of Two Inequalities

Graph $2x + 4y \geq 5$ or $x \geq 1$.

The graphs of the two inequalities are shown in Figures 34(a) and 34(b) in Example 3. The graph of the union is shown in Figure 35.

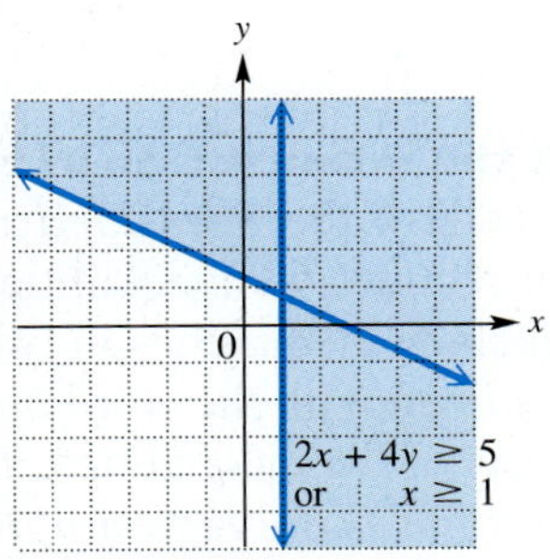

Figure 35

◀ Work Problem **4** at the Side.

ANSWERS

3.

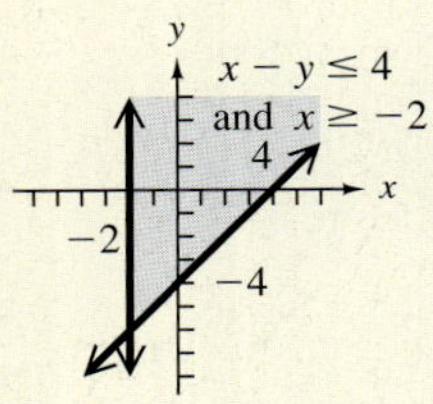

4.

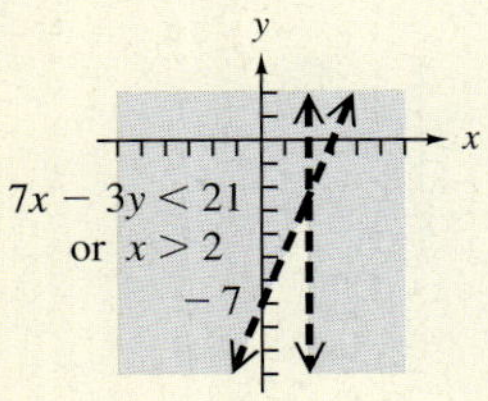

4.4 Exercises

FOR EXTRA HELP MyMathLab Math XL PRACTICE WATCH DOWNLOAD READ REVIEW

In each statement, fill in the first blank with one of the words solid *or* dashed. *Fill in the second blank with one of the words* above *or* below.

1. The boundary of the graph of $y \le -x + 2$ will be a ________ line, and the shading will be ________ the line.

2. The boundary of the graph of $y < -x + 2$ will be a ________ line, and the shading will be ________ the line.

3. The boundary of the graph of $y > -x + 2$ will be a ________ line, and the shading will be ________ the line.

4. The boundary of the graph of $y \ge -x + 2$ will be a ________ line, and the shading will be ________ the line.

5. How is the boundary line $Ax + By = C$ used in graphing either $Ax + By < C$ or $Ax + By > C$?

6. Describe the two methods discussed in the text for deciding which region is the solution set of a linear inequality in two variables.

Graph each linear inequality. See Examples 1 and 2.

7. $x + y \le 2$

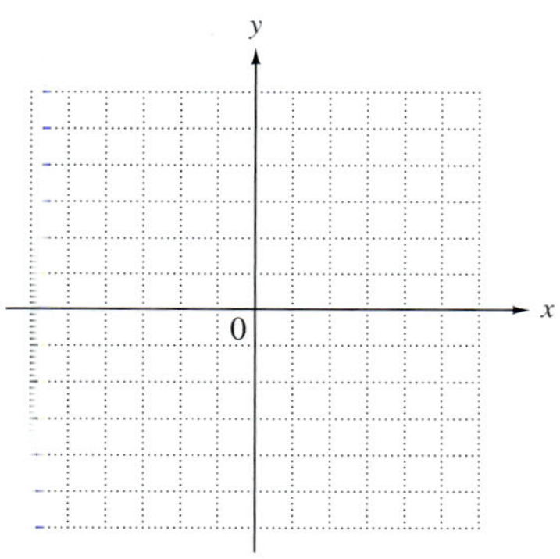

8. $x + y \le -3$

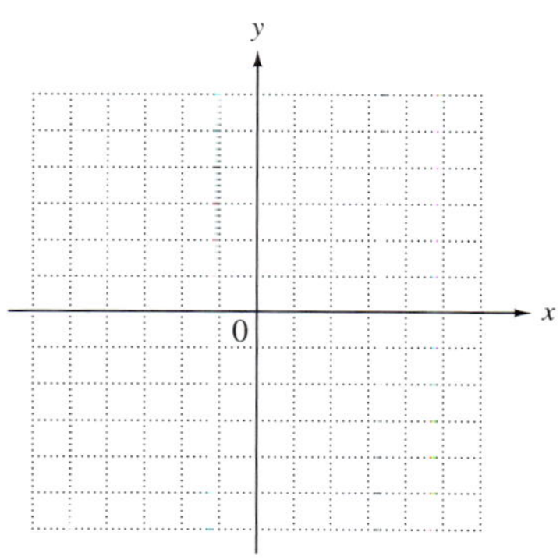

9. $4x - y < 4$

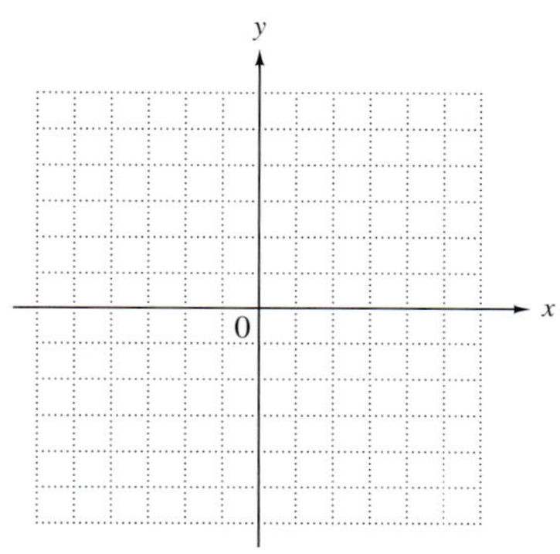

10. $3x - y < 3$

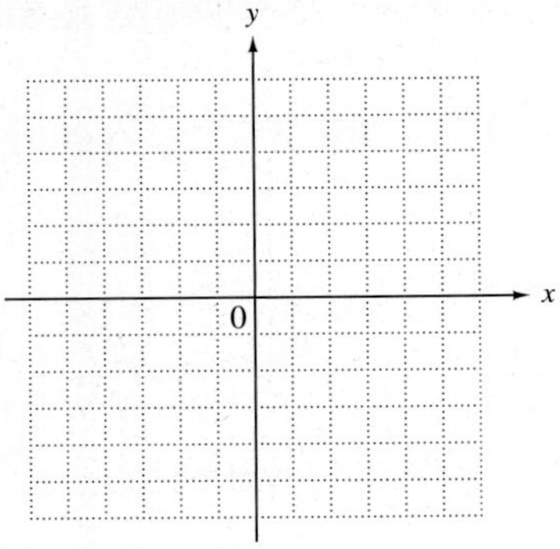

11. $x + 3y \geq -2$

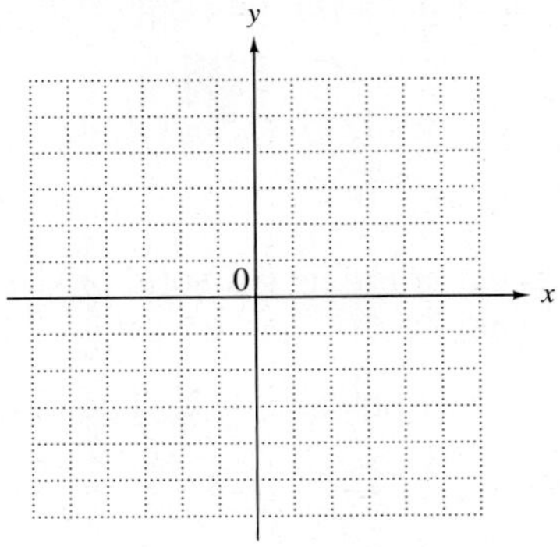

12. $x + 4y \geq -3$

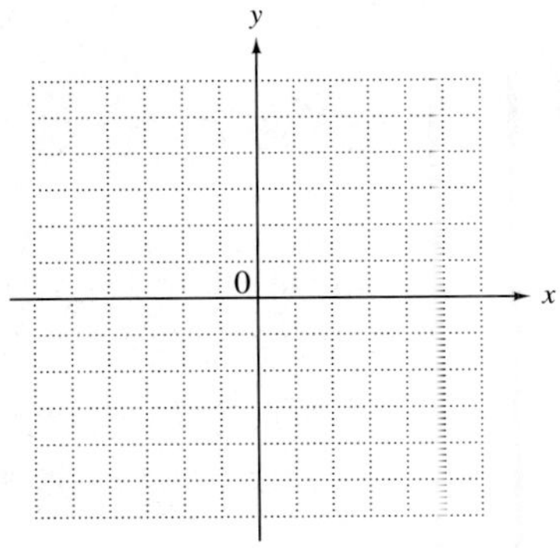

13. $x + y > 0$

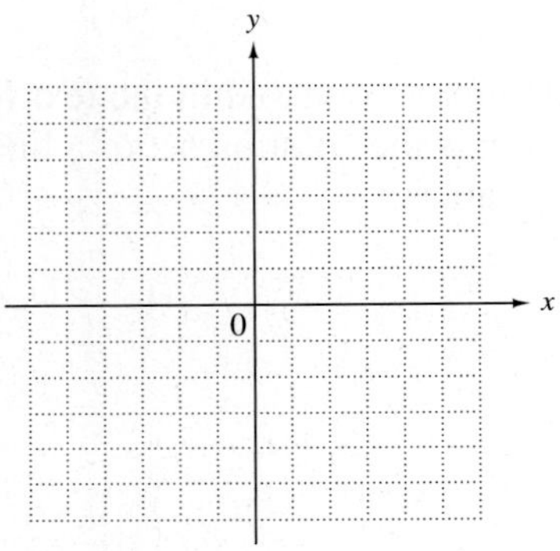

14. $x + 2y > 0$

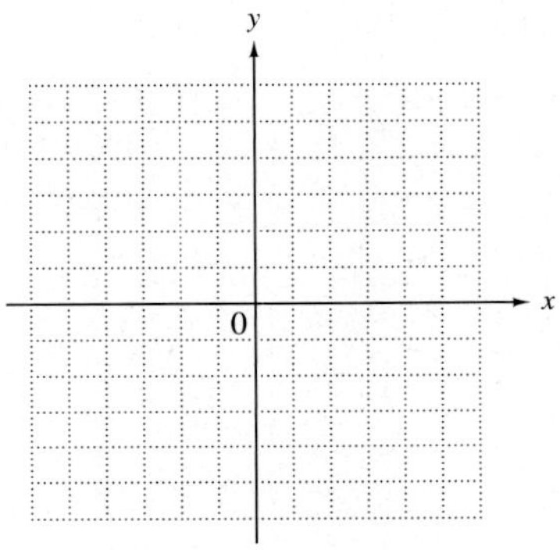

15. $x - 3y \leq 0$

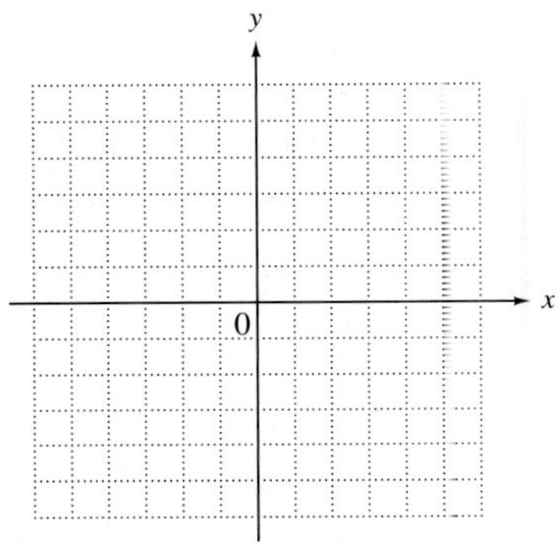

16. $x - 5y \leq 0$

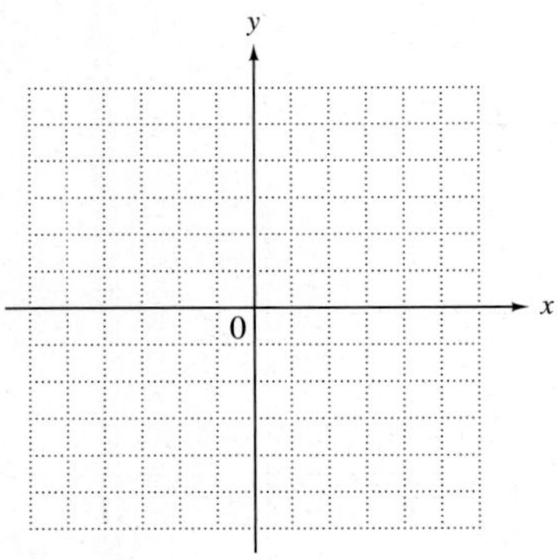

17. $y < x$

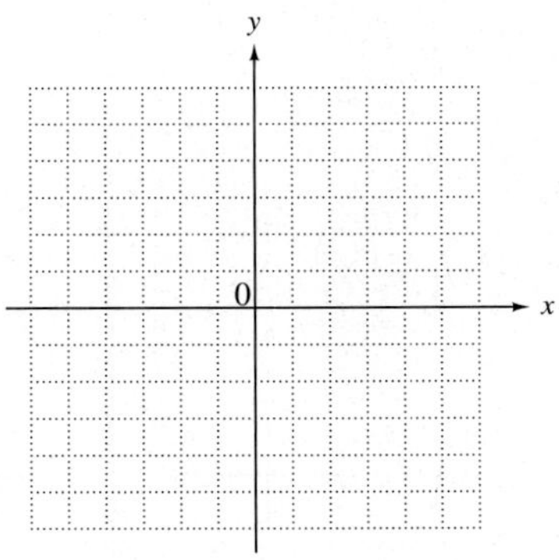

18. $y \leq 4x$

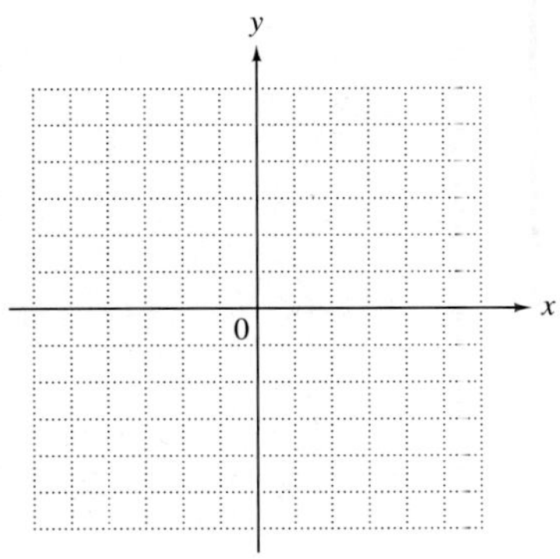

Graph the intersection of each pair of inequalities. See Example 3.

19. $x + y \leq 1$ and $x \geq 1$

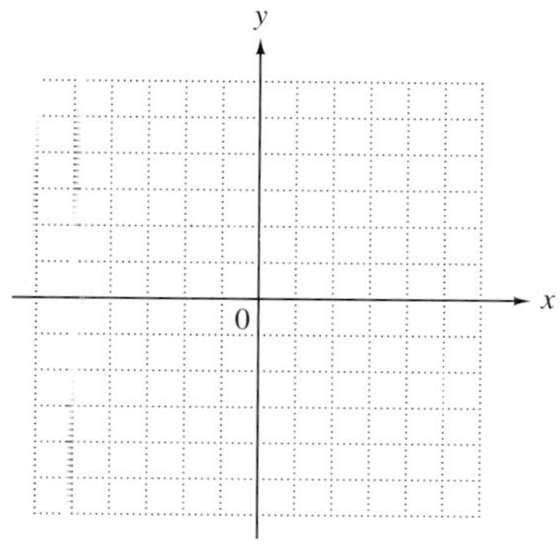

20. $x - y \geq 2$ and $x \geq 3$

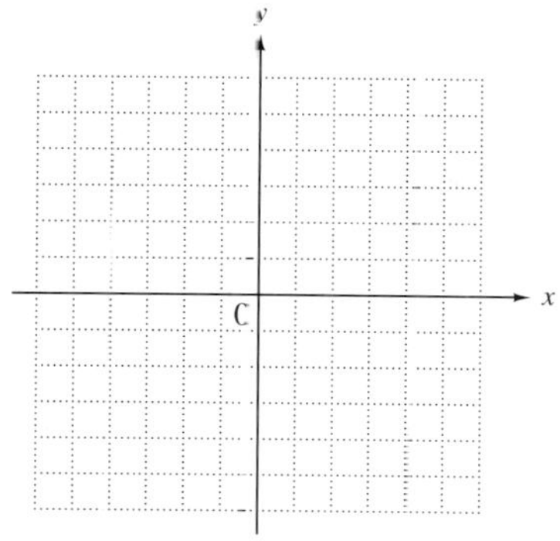

21. $2x - y \geq 2$ and $y < 4$

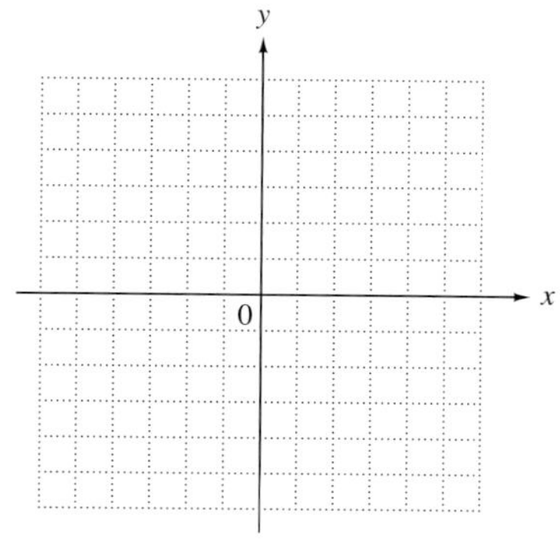

22. $3x - y \geq 3$ and $y < 3$

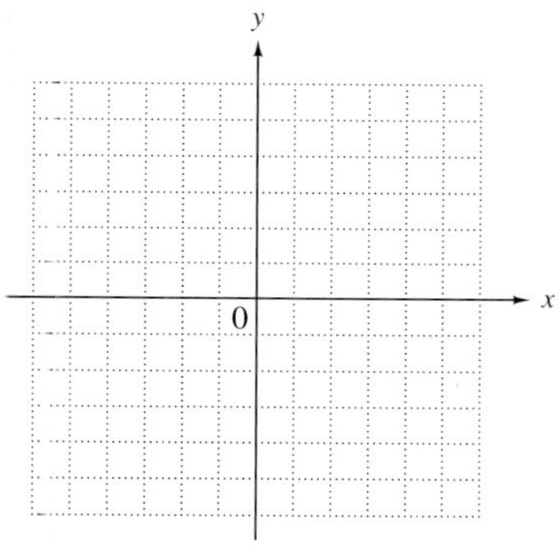

23. $x + y > -5$ and $y < -2$

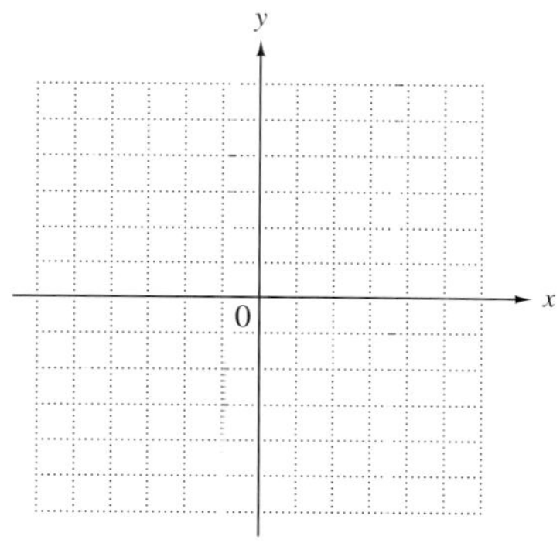

24. $6x - 4y < 10$ and $y > 2$

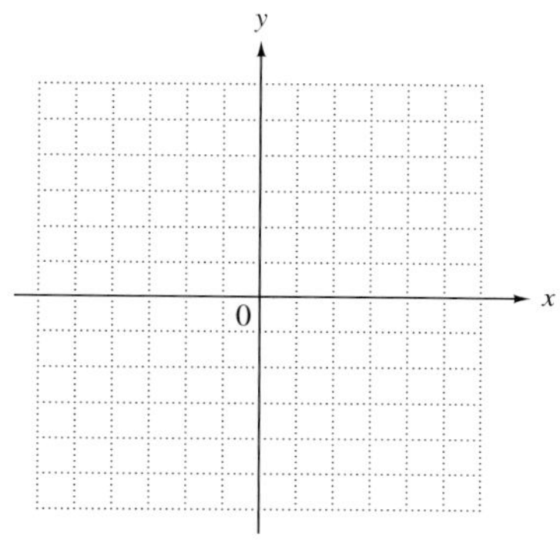

Use the method described in **Section 3.3** *to write each inequality as a compound inequality, and graph its solution set in the rectangular coordinate plane.*

25. $|x| \geq 3$

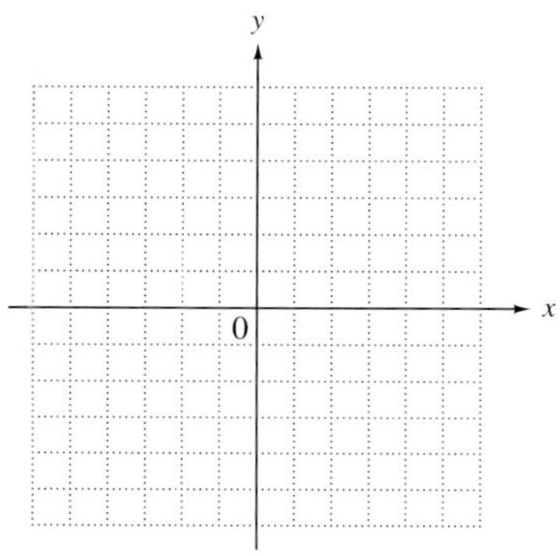

26. $|y| < 5$

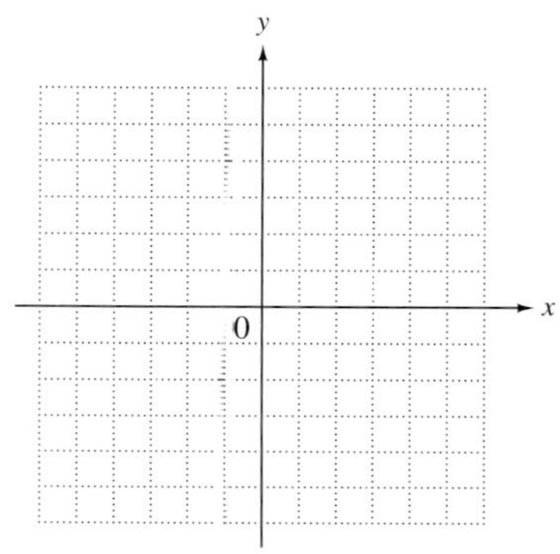

27. $|y + 1| < 2$

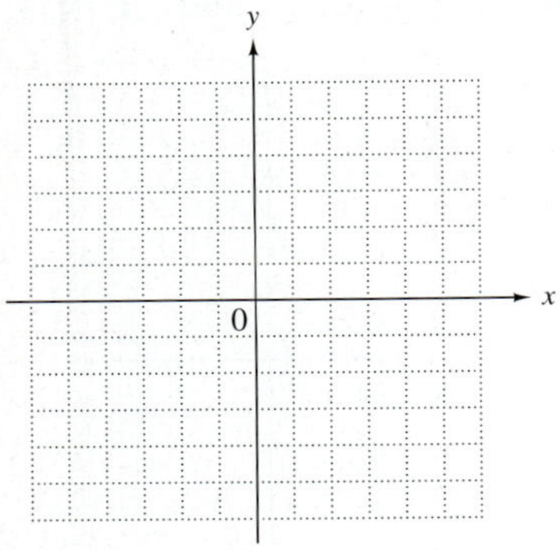

28. $|x - 2| \geq 1$

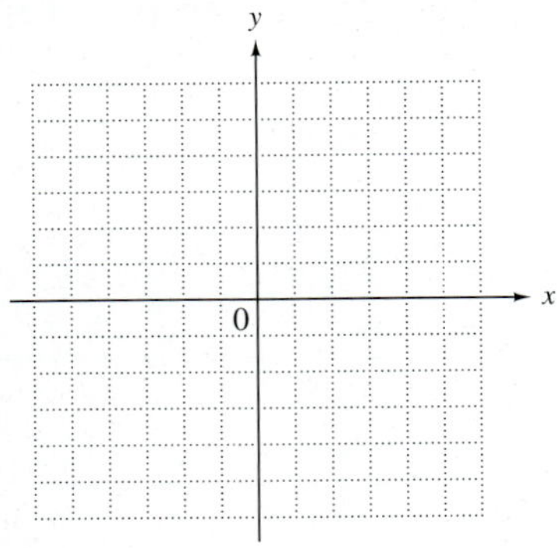

Graph the union of each pair of inequalities. See Example 4.

29. $x - y \geq 1$ or $y \geq 2$

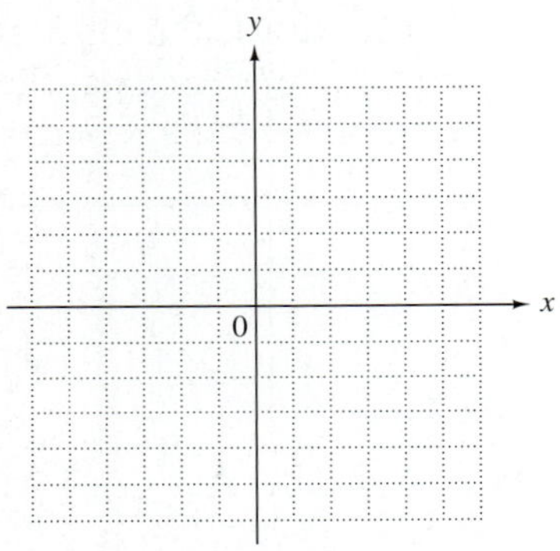

30. $x + y \leq 2$ or $y \geq 3$

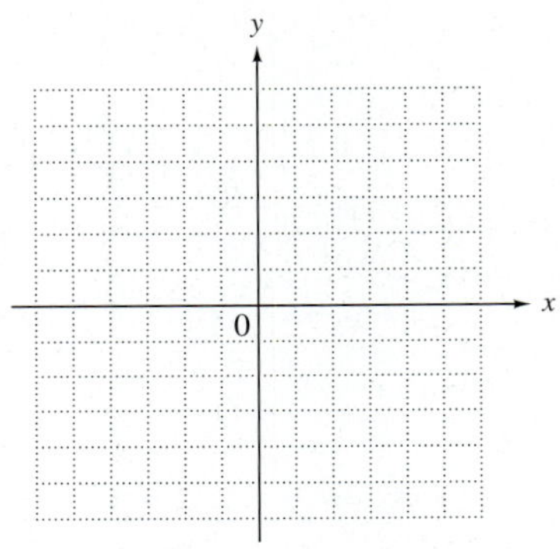

31. $x - 2 > y$ or $x < 1$

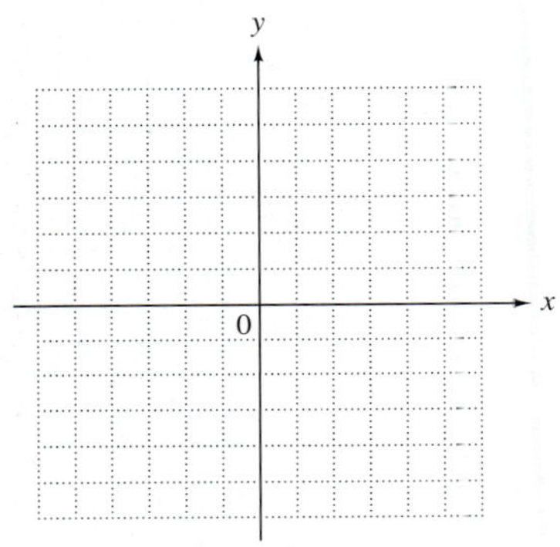

32. $x + 3 < y$ or $x > 3$

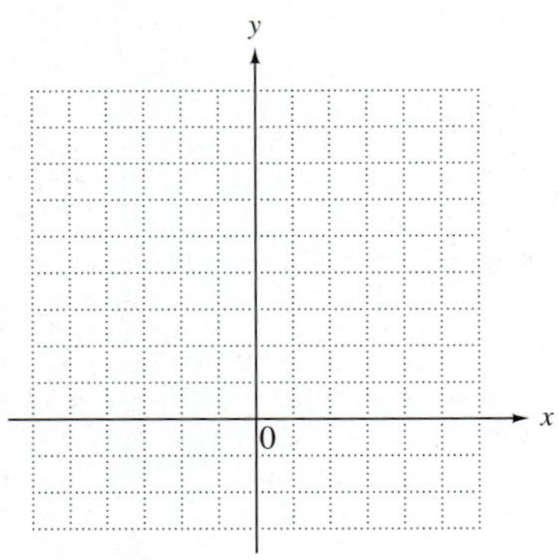

33. $3x + 2y < 6$ or $x - 2y > 2$

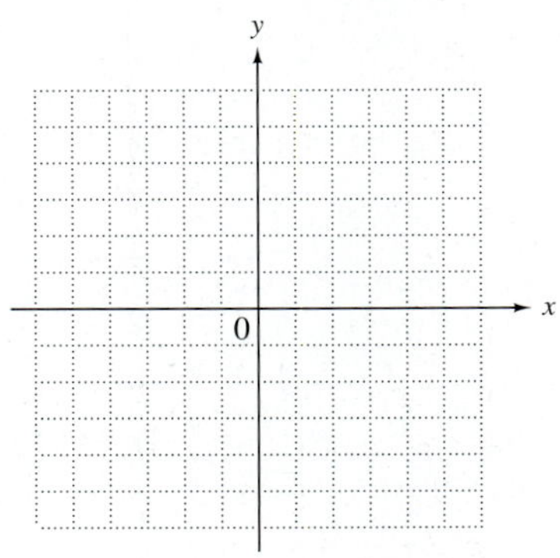

34. $x - y \geq 1$ or $x + y \leq 4$

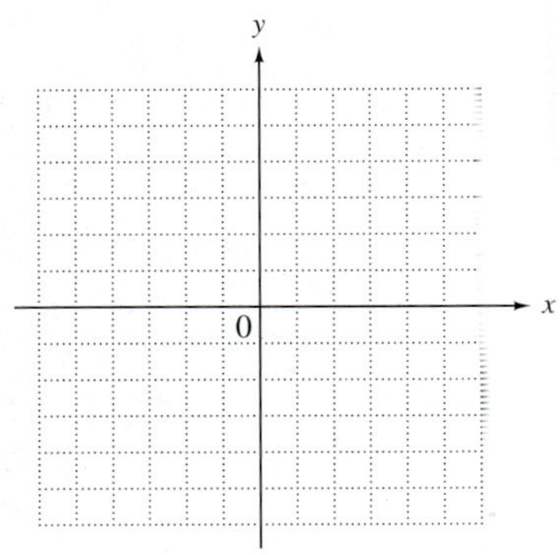

4.5 Introduction to Functions

OBJECTIVES

1. Define and identify relations and functions.
2. Find domain and range.
3. Identify functions defined by graphs and equations.
4. Use function notation.
5. Graph linear and constant functions.

We often describe one quantity in terms of another. Consider the following.

- The amount of your paycheck if you are paid hourly depends on the number of hours you worked.
- The cost at the gas station depends on the number of gallons of gas you pumped into your car.
- The distance traveled by a car moving at a constant speed depends on the time traveled.

We can use ordered pairs to represent these corresponding quantities. For example, we indicate the relationship between the amount of your paycheck and hours worked by writing ordered pairs in which the first number represents hours worked and the second number represents paycheck amount in dollars. Then the ordered pair (5, 40) indicates that when you work 5 hr, your paycheck is $40. Similarly, the ordered pairs (10, 80) and (20, 160) show that working 10 hr results in an $80 paycheck and working 20 hr results in a $160 paycheck.

Work Problem 1 *at the Side.* ▶

1 What would the ordered pair (40, 320) in the correspondence between number of hours worked and paycheck amount (in dollars) indicate?

Since the amount of your paycheck *depends* on the number of hours worked, your paycheck amount is called the *dependent variable,* and the number of hours worked is called the *independent variable*. Generalizing, if the value of the variable y depends on the value of the variable x, then y is the **dependent variable** and x is the **independent variable.**

Independent variable → (x, y) ← Dependent variable

OBJECTIVE 1 Define and identify relations and functions. Since we can write related quantities using ordered pairs, a set of ordered pairs such as

$$\{(5, 40), (10, 80), (20, 160), (40, 320)\}$$

is called a *relation*.

Relation

A **relation** is any set of ordered pairs.

A special kind of relation, called a *function,* is very important in mathematics and its applications.

Function

A **function** is a relation in which, for each value of the first component of the ordered pairs, there is *exactly one value* of the second component.

ANSWER

1. It indicates that when you work 40 hr, your paycheck is $320.

2 Determine whether each relation defines a function.

(a) $\{(0, 3), (-1, 2), (-1, 3)\}$

(b) $\{(2, -2), (4, -4), (6, -6)\}$

(c) $\{(-1, 5), (0, 5)\}$

EXAMPLE 1 Determining Whether Relations Are Functions

Tell whether each relation defines a function.

$$F = \{(1, 2), (-2, 4), (3, -1)\}$$
$$G = \{(-2, -1), (-1, 0), (0, 1), (1, 2), (2, 2)\}$$
$$H = \{(-4, 1), (-2, 1), (-2, 0)\}$$

Relations F and G are functions, because for each different x-value there is exactly one y-value. Notice that in G, the last two ordered pairs have the same y-value (1 is paired with 2, and 2 is paired with 2). This does not violate the definition of function, since the first components (x-values) are different and each is paired with only one second component (y-value)

In relation H, however, the last two ordered pairs have the *same* x-value paired with *two different* y-values (-2 is paired with both 1 and 0), so H is a relation but not a function. ***In a function, no two ordered pairs can have the same first component and different second components.***

Different y-values

$$H = \{(-4, 1), (\mathbf{-2}, \mathbf{1}), (\mathbf{-2}, \mathbf{0})\} \quad \text{Not a function}$$

Same x-value

◀ Work Problem **2** at the Side.

In a function, there is **exactly one** ***value of the dependent variable, the second component, for each value of the independent variable, the first component. This is what makes functions so important in applications.***

Relations and functions can also be expressed as a correspondence or *mapping* from one set to another, as shown in Figure 36 for function F and relation H from Example 1. The arrow from 1 to 2 indicates that the ordered pair (1, 2) belongs to F—each first component is paired with exactly one second component. In the mapping for set H, which is not a function, the first component -2 is paired with two different second components, 1 and 0.

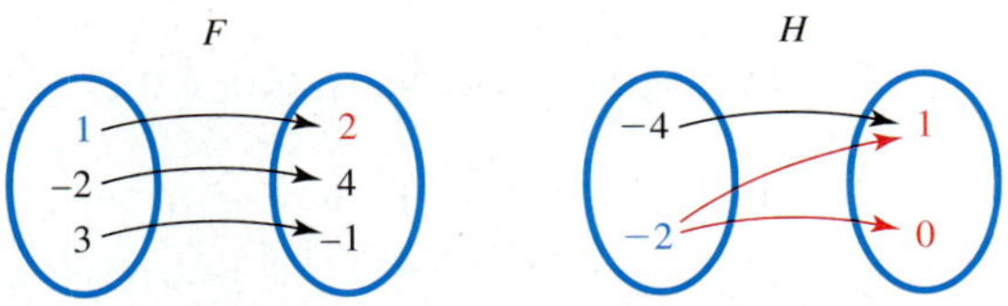

F is a function. *H* is not a function.

Figure 36

Since relations and functions are sets of ordered pairs, we can represent them using tables and graphs. A table and graph for function F is shown in Figure 37.

x	y
1	2
−2	4
3	−1

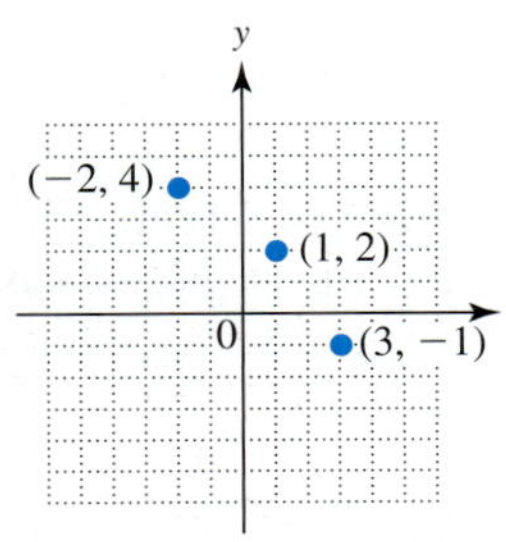

Graph of *F*

Figure 37

ANSWERS

2. **(a)** not a function **(b)** function **(c)** function

Finally, we can describe a relation or function using a rule that tells how to determine the dependent variable for a specific value of the independent variable. The rule may be given in words, such as "the dependent variable is twice the independent variable." Usually, however, the rule is given as an equation:

$$y = 2x.$$

Dependent variable (y) ↑ ↑ Independent variable (x)

An equation is the most efficient way to define a relation or function.

Note

Another way to think of a function relationship is to think of the independent variable as an input and the dependent variable as an output. This is illustrated by the input-output (function) machine in the margin for the function defined by $y = 2x$.

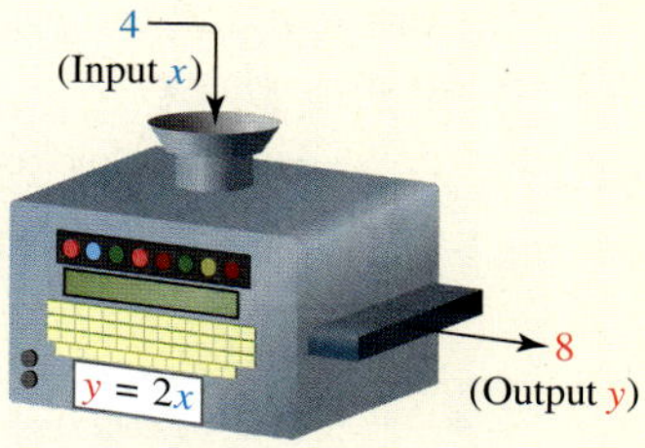

Function machine

OBJECTIVE 2 **Find domain and range.** For every relation, there are two important sets of elements called the *domain* and *range.*

Domain and Range

In a relation, the set of all values of the independent variable (x) is the **domain.** The set of all values of the dependent variable (y) is the **range.**

EXAMPLE 2 **Finding Domains and Ranges of Relations**

Give the domain and range of each relation. Tell whether the relation defines a function.

(a) $\{(3, -1), (4, 2), (4, 5), (6, 8)\}$

The domain, the set of x-values, is $\{3, 4, 6\}$; the range, the set of y-values, is $\{-1, 2, 5, 8\}$. This relation is not a function because the same x-value 4 is paired with two different y-values, 2 and 5.

(b)

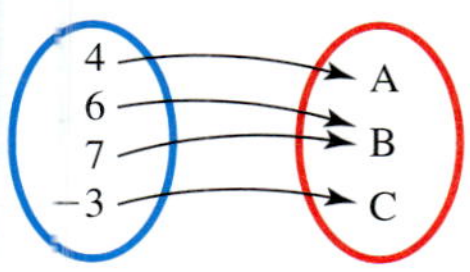

The domain of this relation is

$$\{4, 6, 7, -3\}.$$

The range is

$$\{A, B, C\}.$$

This mapping defines a function—each x-value corresponds to exactly one y-value.

(c)

x	y
−5	2
0	2
5	2

This is a table of ordered pairs, so the domain is the set of x-values, $\{-5, 0, 5\}$, and the range is the set of y-values, $\{2\}$. The table defines a function because each different x-value corresponds to exactly one y-value (even though it is the same y-value).

Work Problem 3 at the Side. ▶

3 Give the domain and range of each relation. Does the relation define a function?

(a) $\{(4, 0), (4, 1), (4, 2)\}$

(b)

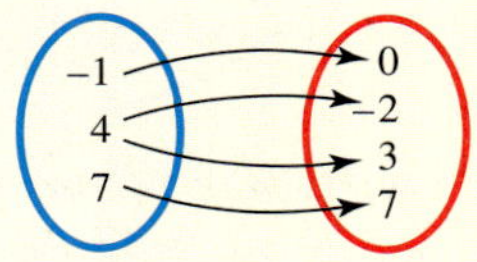

(c)

Year	Cell Phone Subscribers (in thousands)
2002	140,766
2003	158,722
2004	182,140
2005	207,896
2006	233,041

Source: CTIA-The Wireless Association.

Answers

3. **(a)** domain: $\{4\}$; range: $\{0, 1, 2\}$; No, the relation does not define a function.
(b) domain: $\{-1, 4, 7\}$; range: $\{0, -2, 3, 7\}$; No, the relation does not define a function.
(c) domain: $\{2002, 2003, 2004, 2005, 2006\}$; range: $\{140{,}766, 158{,}722, 182{,}140, 207{,}896, 233{,}041\}$; Yes, the relation defines a function.

4 Give the domain and range of each relation.

(a)

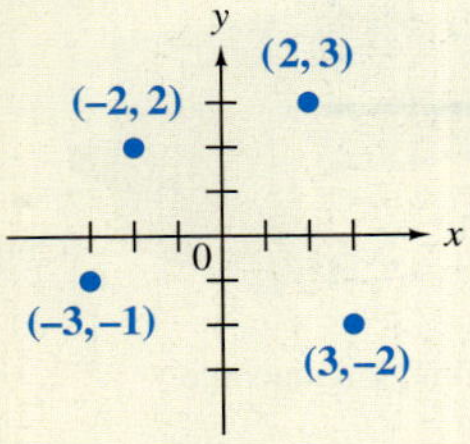

(b)

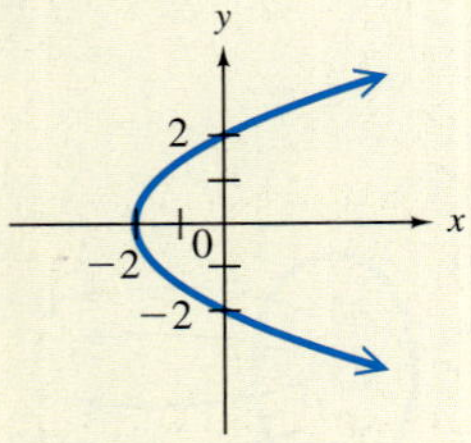

(c)

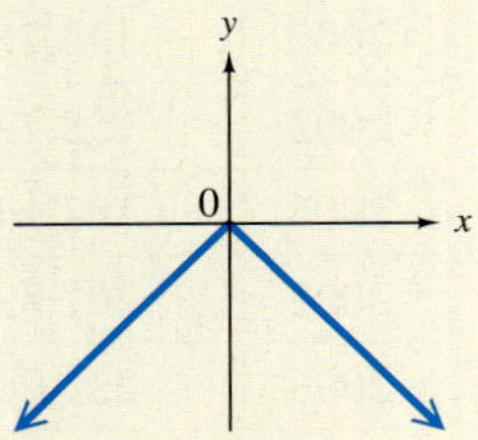

The graph of a relation gives a picture of the relation, which can be used to determine its domain and range.

EXAMPLE 3 Finding Domains and Ranges from Graphs

Give the domain and range of each relation.

(a)

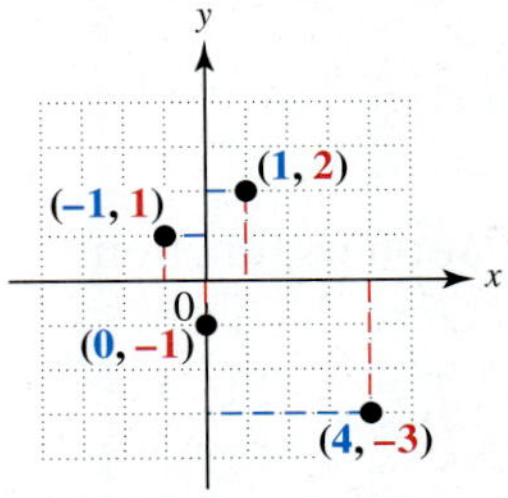

The domain is the set of x-values,

$$\{-1, 0, 1, 4\}.$$

The range is the set of y-values,

$$\{-3, -1, 1, 2\}.$$

(b)

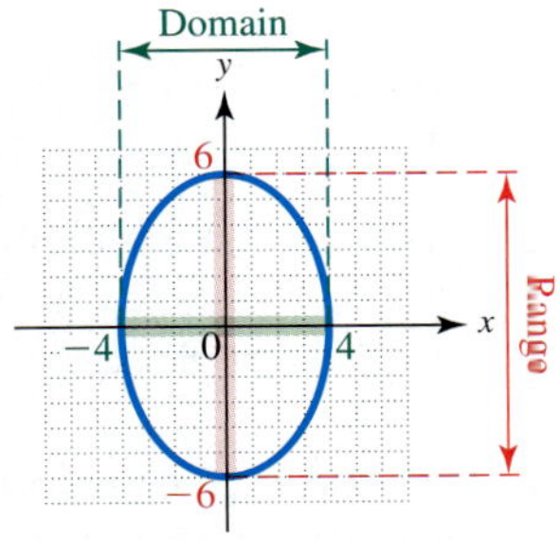

The x-values of the points on the graph include all numbers between −4 and 4, inclusive. The y-values include all numbers between −6 and 6, inclusive. Using interval notation,

the domain is $[-4, 4]$;

the range is $[-6, 6]$

(c)

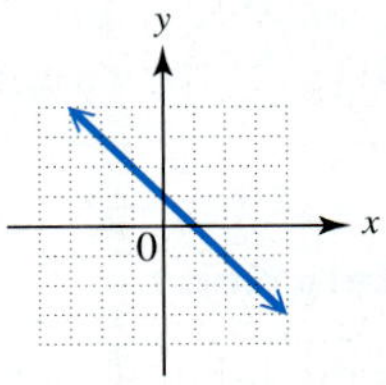

The arrowheads indicate that the line extends indefinitely left and right, as well as up and down. Therefore, both the domain and the range include all real numbers, written $(-\infty, \infty)$.

(d)

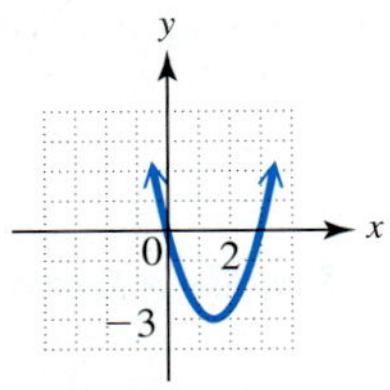

The arrowheads indicate that the graph extends indefinitely left and right, as well as upward. The domain is $(-\infty, \infty)$. Because there is a least y-value, −3, the range includes all numbers greater than or equal to −3, written $[-3, \infty)$.

◀ Work Problem **4** at the Side.

Since relations are often defined by equations, such as $y = 2x + 3$ and $y^2 = x$, we must sometimes determine the domain of a relation from its equation. We assume the following agreement on the domain of a relation.

Agreement on Domain

The domain of a relation is assumed to be all real numbers that produce real numbers when substituted for the independent variable.

To illustrate this agreement, since any real number can be used as a replacement for x in $y = 2x + 3$, the domain of this function is the set of all real numbers. The function defined by $y = \frac{1}{x}$ has all real numbers except 0 as domain, since y is undefined if $x = 0$. ***In general, the domain of a function defined by an algebraic expression is all real numbers, except those numbers that lead to division by 0 or an even root of a negative number.***

ANSWERS

4. (a) domain: $\{-3, -2, 2, 3\}$; range: $\{-2, -1, 2, 3\}$
(b) domain: $[-2, \infty)$; range: $(-\infty, \infty)$
(c) domain: $(-\infty, \infty)$; range: $(-\infty, 0]$

OBJECTIVE 3 Identify functions defined by graphs and equations. Since each value of x in a function corresponds to only one value of y, any vertical line drawn through the graph of a function must intersect the graph in at most one point. This is the *vertical line test* for a function.

> **Vertical Line Test**
>
> If every vertical line intersects the graph of a relation in no more than one point, then the relation represents a function.

For example, the graph shown in Figure 38(a) is not the graph of a function since a vertical line intersects the graph in more than one point. The graph in Figure 38(b) does represent a function.

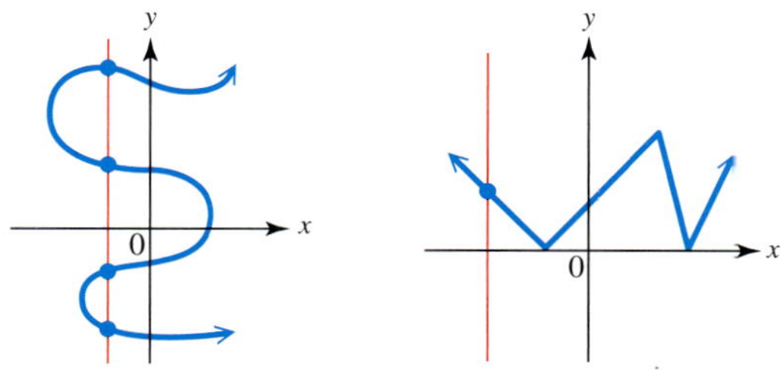

Not a function—the same x-value corresponds to four different y-values. (a)

Function—each x-value corresponds to only one y-value. (b)

Figure 38

EXAMPLE 4 Using the Vertical Line Test

Use the vertical line test to determine whether each relation graphed in Example 3 is a function.

(a)

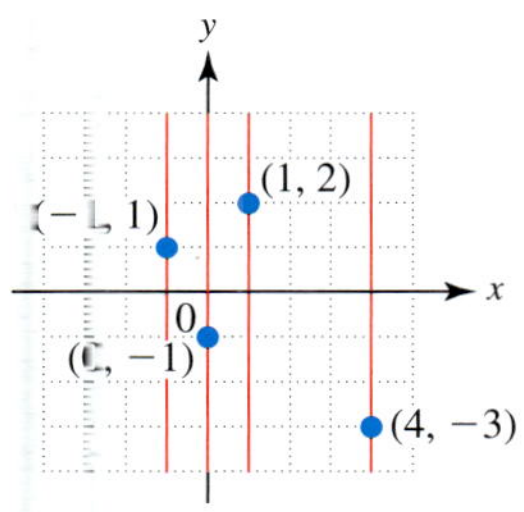

Function

(b)

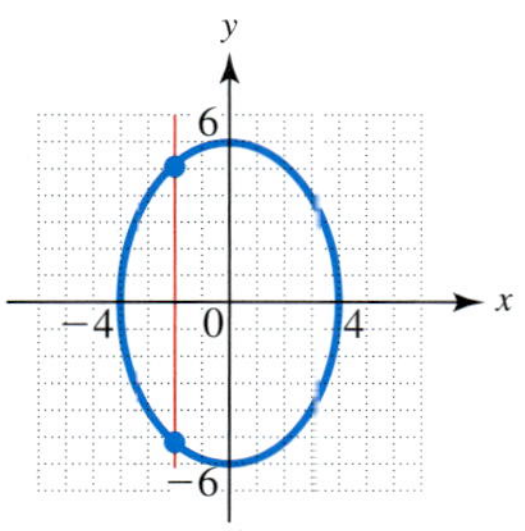

Not a function

(c)

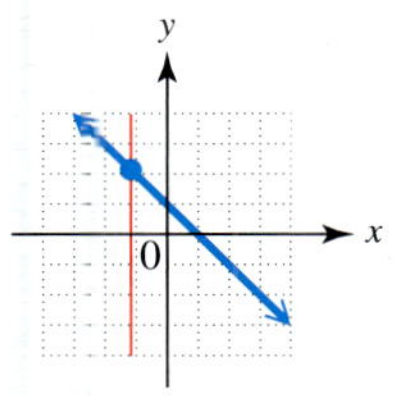

Function

(d)

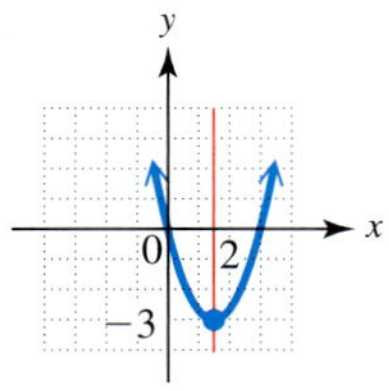

Function

The graphs in (a), (c), and (d) represent functions. The graph of the relation in (b) fails the vertical line test, since the same x-value corresponds to two different y-values; therefore, it is not the graph of a function.

Work Problem 5 at the Side. ▶

5 Use the vertical line test to decide which graphs represent functions.

A.

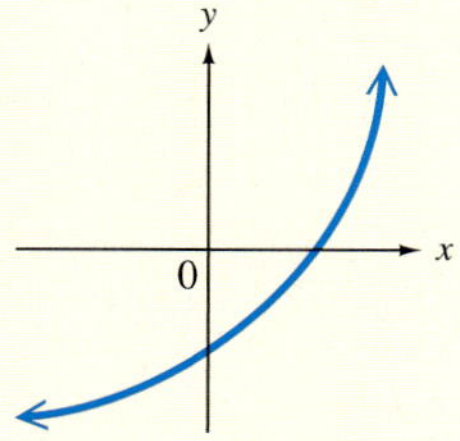

B.

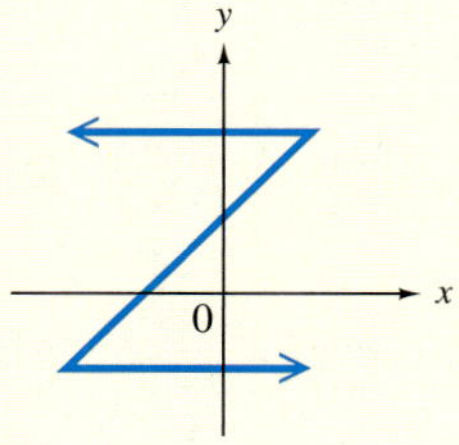

C.

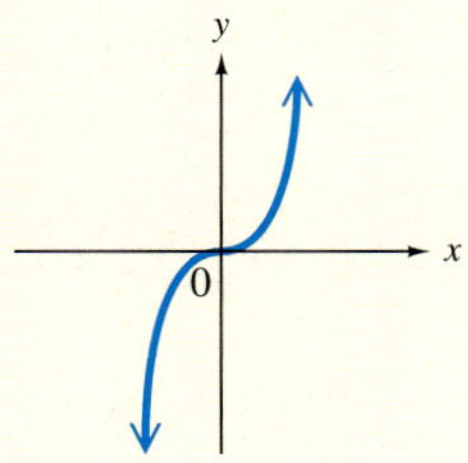

ANSWER

5. A and C are graphs of functions.

Note

Graphs that do not represent functions are still relations. ***Remember that all equations and graphs represent relations and that all relations have a domain and range.***

It can be more difficult to decide whether a relation defined by an equation or an inequality is a function. The next example gives some hints that may help.

EXAMPLE 5 Identifying Functions from Their Equations

Decide whether each relation defines a function and give the domain.

(a) $y = x + 4$

In the defining equation, $y = x + 4$, y is always found by adding 4 to x. Thus, each value of x corresponds to just one value of y and the relation defines a function; x can be any real number, so the domain is $(-\infty, \infty)$.

(b) $y = \sqrt{2x - 1}$

For any choice of x in the domain, there is exactly one corresponding value for y (the radical is a nonnegative number), so this equation defines a function. Since the equation involves a square root, the quantity under the radical sign cannot be negative. Thus,

$$2x - 1 \geq 0$$
$$2x \geq 1$$
$$x \geq \frac{1}{2},$$

and the domain of the function is $[\frac{1}{2}, \infty)$.

(c) $y^2 = x$

The ordered pairs (16, 4) and (16, −4) both satisfy this equation. Since one value of x, 16, corresponds to two values of y, 4 and −4, this equation does not define a function. Because x is equal to the square of y, the values of x must always be nonnegative. The domain of the relation is $[0, \infty)$.

(d) $y \leq x - 1$

By definition, y is a function of x if every value of x leads to exactly one value of y. Here a particular value of x, say 1, corresponds to many values of y. The ordered pairs (1, 0), (1, −1), (1, −2), (1, −3), and so on, all satisfy the inequality. Thus, this relation does not define a function. Any number can be used for x, so the domain is the set of real numbers, $(-\infty, \infty)$

(e) $y = \frac{5}{x - 1}$

Given any value of x in the domain, we find y by subtracting 1, then dividing the result into 5. This process produces exactly one value of y for each value in the domain, so this equation defines a function. The domain includes all real numbers except those that make the denominator 0. We find these numbers by setting the denominator equal to 0 and solving for x.

$$x - 1 = 0$$
$$x = 1$$

The domain includes all real numbers *except* 1, written $(-\infty, 1) \cup (1, \infty)$.

◀ Work Problem 6 at the Side.

6 Decide whether each relation defines a function, and give the domain.

(a) $y = 6x + 12$

(b) $y \leq 4x$

(c) $y = -\sqrt{3x - 2}$

(d) $y^2 = 25x$

Answers

6. **(a)** yes; $(-\infty, \infty)$ **(b)** no; $(-\infty, \infty)$
(c) yes; $\left[\frac{2}{3}, \infty\right)$ **(d)** no; $[0, \infty)$

In summary, three variations of the definition of function are given here.

Variations of the Definition of Function

1. A **function** is a relation in which, for each value of the first component of the ordered pairs, there is exactly one value of the second component.
2. A **function** is a set of ordered pairs in which no first component is repeated.
3. A **function** is a rule or correspondence that assigns exactly one range value to each distinct domain value.

OBJECTIVE 4 Use function notation. When a function f is defined with a rule or an equation using x and y for the independent and dependent variables, we say "y is a function of x" to emphasize that y *depends on* x. We use the notation

$$y = f(x),$$

called **function notation,** to express this and read $f(x)$ as "f of x." (In this special notation the parentheses do not indicate multiplication.) The letter f stands for *function*. For example, if $y = 9x - 5$, we can name this function f and write

$$f(x) = 9x - 5.$$

Note that ***f(x) is just another name for the dependent variable y.*** For example, if $y = f(x) = 9x - 5$ and $x = 2$, then we find y, or $f(2)$, by replacing x with 2.

$$\begin{aligned} y &= f(2) \\ &= 9 \cdot 2 - 5 \\ &= 18 - 5 \\ &= 13. \end{aligned}$$

For function f, the statement "if $x = 2$, then $y = 13$" is represented by the ordered pair $(2, 13)$ and is abbreviated with function notation as

$$f(2) = 13.$$

Read $f(2)$ as "f of 2" or "f at 2." Also,

$$f(0) = 9 \cdot 0 - 5 = -5 \quad \text{and} \quad f(-3) = 9(-3) - 5 = -32.$$

These ideas can be illustrated as follows.

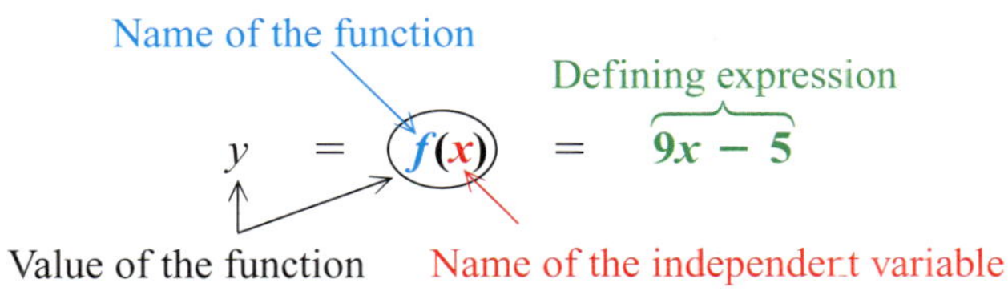

CAUTION

The symbol $f(x)$ *does not* indicate "f times x," but represents the y-value for the indicated x-value. As just shown, $f(2)$ is the y-value that corresponds to the x-value 2.

7 Find $f(-3)$, $f(p)$, and $f(m+1)$.

(a) $f(x) = 6x - 2$

(b) $f(x) = \dfrac{-3x+5}{2}$

(c) $f(x) = \dfrac{1}{6}x - 1$

EXAMPLE 6 Using Function Notation

Let $f(x) = -x^2 + 5x - 3$. Find the following.

(a) $f(2)$

Do not read this as "f times 2." Read it as "f of 2."

$$f(x) = -x^2 + 5x - 3 \quad \text{The base in } -x^2 \text{ is } x, \text{ not } (-x).$$
$$f(2) = -2^2 + 5 \cdot 2 - 3 \quad \text{Replace } x \text{ with 2.}$$
$$f(2) = -4 + 10 - 3 \quad \text{Apply the exponent; multiply.}$$
$$f(2) = 3 \quad \text{Add and subtract.}$$

Since $f(2) = 3$, the ordered pair $(2, 3)$ belongs to f.

(b) $f(q)$

$$f(x) = -x^2 + 5x - 3$$
$$f(q) = -q^2 + 5q - 3 \quad \text{Replace } x \text{ with } q.$$

The replacement of one variable with another is important in later courses.

Sometimes letters other than f, such as g, h, or capital letters F, G, and H, are used to name functions.

EXAMPLE 7 Using Function Notation

Let $g(x) = 2x + 3$. Find and simplify $g(a+1)$.

$$g(x) = 2x + 3$$
$$g(a+1) = 2(a+1) + 3 \quad \text{Replace } x \text{ with } a+1.$$
$$g(a+1) = 2a + 2 + 3$$
$$g(a+1) = 2a + 5$$

◀ *Work Problem* **7** *at the Side.*

Functions can be evaluated in a variety of ways, as shown in Example 8.

EXAMPLE 8 Using Function Notation

For each function, find $f(3)$.

(a)
$$f(x) = 3x - 7$$
$$f(3) = 3(3) - 7$$
$$f(3) = 9 - 7$$
$$f(3) = 2$$

(b) $f = \{(-3, 5), (0, 3), (3, 1), (5, -1)\}$

We want $f(3)$, the y-value of the ordered pair where $x = 3$. As indicated by the ordered pair $(3, 1)$, when $x = 3$, $y = 1$, so $f(3) = 1$.

(c)

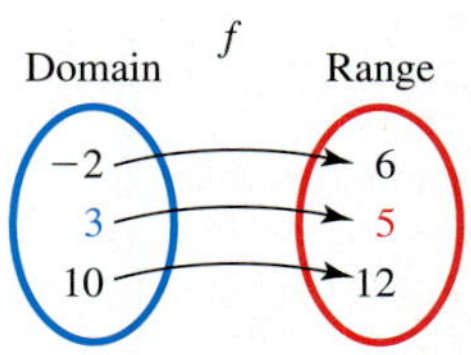

The domain element 3 is paired with 5 in the range, so $f(3) = 5$.

Continued on Next Page

Answers

7. (a) -20; $6p - 2$; $6m + 4$
(b) 7; $\dfrac{-3p+5}{2}$; $\dfrac{-3m+2}{2}$
(c) $-\dfrac{3}{2}$; $\dfrac{1}{6}p - 1$; $\dfrac{1}{6}(m+1) - 1$, or $\dfrac{1}{6}m - \dfrac{5}{6}$

(d)

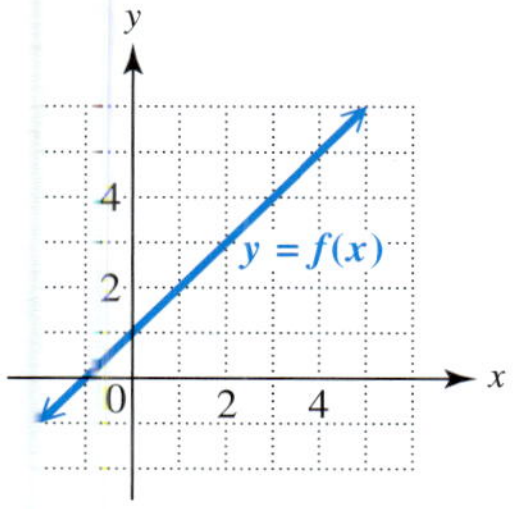

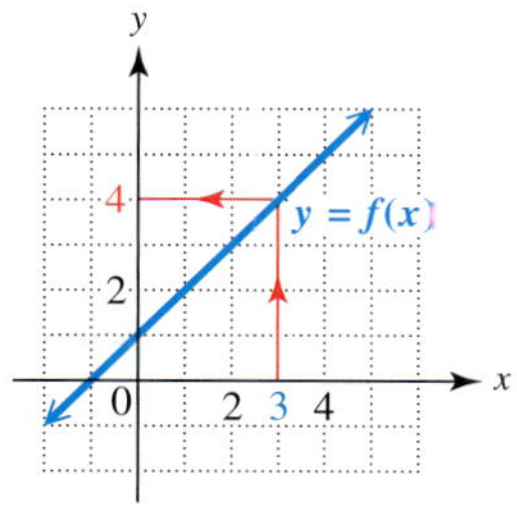

Figure 39

To evaluate $f(3)$, find 3 on the x-axis. See Figure 39. Then move up until the graph of f is reached. Moving horizontally to the y-axis gives 4 for the corresponding y-value. Thus, $f(3) = 4$.

Work Problem **8** *at the Side.* ▶

If a function f is defined by an equation with x and y instead of with function notation, use the following steps to find $f(x)$.

Writing an Equation Using Function Notation

Step 1 Solve the equation for y.

Step 2 Replace y with $f(x)$.

EXAMPLE 9 **Writing Equations Using Function Notation**

Rewrite each equation using function notation $f(x)$. Then find $f(-2)$ and $f(a)$.

(a) $y = x^2 + 1$

This equation is already solved for y. Since $y = f(x)$,

$$f(x) = x^2 + 1.$$

To find $f(-2)$, let $x = -2$.

$$f(-2) = (-2)^2 + 1$$
$$f(-2) = 4 + 1$$
$$f(-2) = 5$$

Find $f(a)$ by letting $x = a$: $\quad f(a) = a^2 + 1.$

(b) $x - 4y = 5$

First solve $x - 4y = 5$ for y. Then replace y with $f(x)$.

$$x - 4y = 5$$
$$x - 5 = 4y \qquad \text{Add } 4y\text{; subtract 5.}$$
$$y = \frac{x-5}{4}, \quad \text{so} \quad f(x) = \frac{1}{4}x - \frac{5}{4}$$

Now find $f(-2)$ and $f(a)$.

$$f(-2) = \frac{1}{4}(-2) - \frac{5}{4} = -\frac{7}{4} \qquad \text{Let } x = -2.$$

$$f(a) = \frac{1}{4}a - \frac{5}{4} \qquad \text{Let } x = a.$$

Work Problem **9** *at the Side.* ▶

8 For each function, find $f(-2)$.

(a) $f(x) = -4x - 8$

(b) $f = \{(0, 5), (-1, 3), (-2, 1)\}$

(c)

x	$f(x)$
-4	16
-2	4
0	0
2	4
4	16

9 Rewrite each equation using function notation $f(x)$. Then find $f(-1)$.

(a) $y = \sqrt{x + 2}$

(b) $x^2 - 4y = 3$

Answers

8. **(a)** 0 **(b)** 1 **(c)** 4

9. **(a)** $f(x) = \sqrt{x+2}$; 1

(b) $f(x) = \frac{x^2 - 3}{4}$, or $f(x) = \frac{1}{4}x^2 - \frac{3}{4}$; $-\frac{1}{2}$

10 Graph each linear function. Give the domain and range.

(a) $f(x) = \frac{3}{4}x - 2$

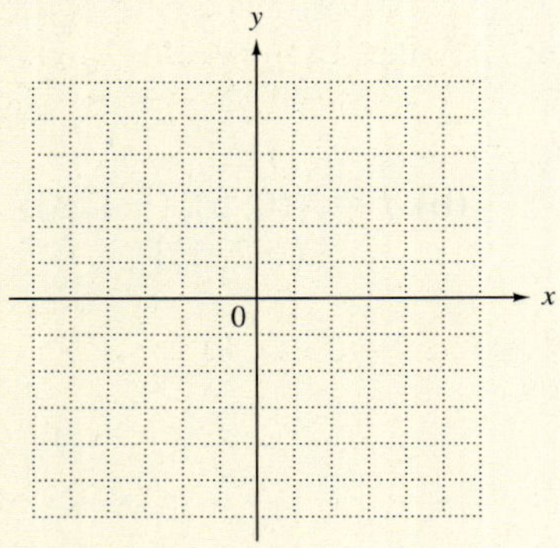

(b) $g(x) = 3$

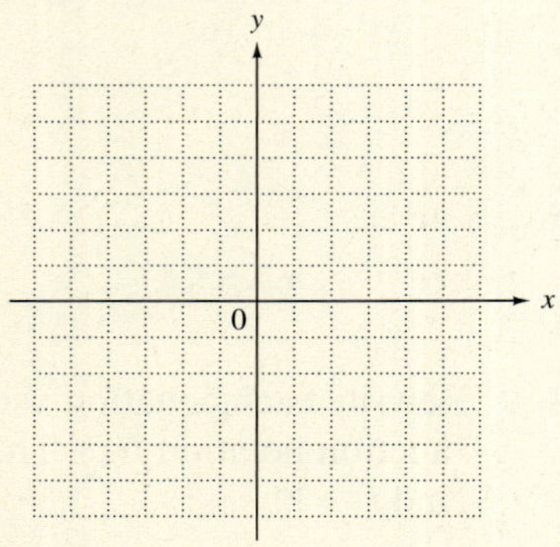

ANSWERS

10. (a)

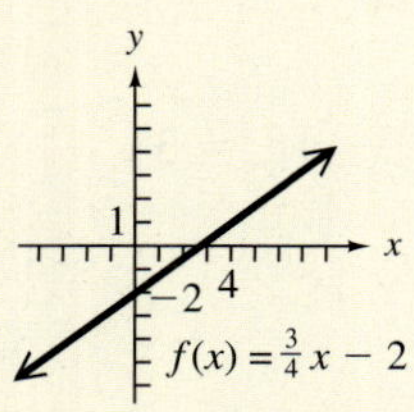

domain: $(-\infty, \infty)$; range: $(-\infty, \infty)$

(b)

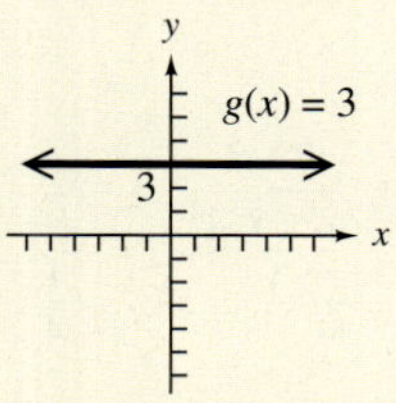

domain: $(-\infty, \infty)$; range: $\{3\}$

OBJECTIVE 5 Graph linear and constant functions. Our first two-dimensional graphing was of straight lines. Linear equations (except for vertical lines with equations $x = a$) define *linear functions*.

> **Linear Function**
>
> A function that can be defined by
>
> $$f(x) = ax + b,$$
>
> for real numbers a and b, is a **linear function.** The value of a is the slope m of the graph of the function.

A linear function defined by $f(x) = b$ (whose graph is a horizontal line) is sometimes called a **constant function.** The domain of any linear function is $(-\infty, \infty)$. The range of a nonconstant linear function is $(-\infty, \infty)$, while the range of the constant function defined by $f(x) = b$ is $\{b\}$.

EXAMPLE 10 Graphing Linear and Constant Functions

Graph each function. Give the domain and range.

(a) $f(x) = \frac{1}{4}x - \frac{5}{4}$

Recall from **Section 4.3** that m is the slope of the line and $(0, b)$ is the y-intercept. In Example 9(b), we wrote the equation $x - 4y = 5$ as the linear function defined by

$$f(x) = \frac{1}{4}x - \frac{5}{4}.$$

Slope ↑ $\frac{1}{4}$; y-intercept is $(0, -\frac{5}{4})$.

To graph this function, plot the y-intercept $(0, -\frac{5}{4})$ and use the definition of slope as $\frac{\text{rise}}{\text{run}}$ to find a second point on the line. Since the slope is $\frac{1}{4}$, move 1 unit up from $(0, -\frac{5}{4})$ and 4 units to the right to the point $(4, -\frac{1}{4})$. Draw the straight line through these points to obtain the graph shown in Figure 40. The domain and range are both $(-\infty, \infty)$.

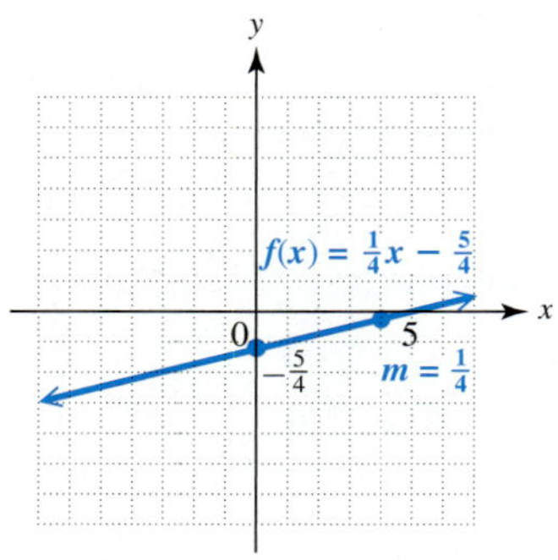

Figure 40

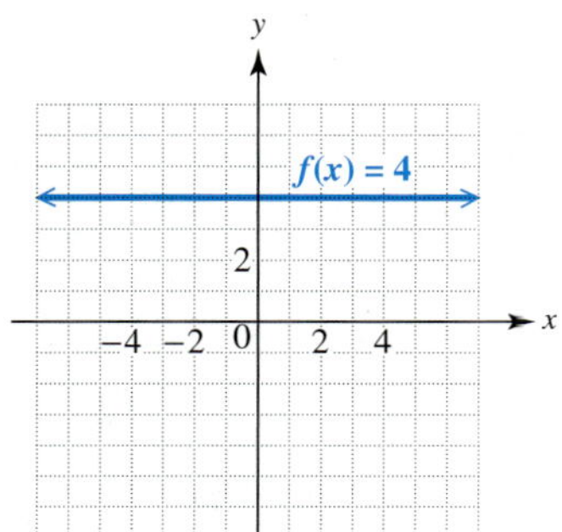

Figure 41

(b) $f(x) = 4$

This is a constant function. Its graph is the horizontal line containing all points with y-coordinate equal to 4. See Figure 41. The domain is $(-\infty, \infty)$ and the range is $\{4\}$.

◀ Work Problem **10** at the Side.

4.5 Exercises

1. In an ordered pair of a relation, is the first element the independent or the dependent variable?

2. Give an example of a relation that is not a function, having domain $\{-3, 2, 6\}$ and range $\{4, 6\}$. (There are many possible correct answers.)

3. Explain what is meant by each term.
   (a) Relation **(b)** Domain of a relation
   (c) Range of a relation **(d)** Function

4. Describe the use of the vertical line test.

Decide whether each relation is a function, and give the domain and the range. Use the vertical line test in Exercises 17–22. See Examples 1–4.

5. $\{(5, 1), (3, 2), (4, 9), (7, 3)\}$

6. $\{(8, 0), (5, 4), (9, 3), (3, 9)\}$

7. $\{(2, 4), (0, 2), (2, 6)\}$

8. $\{(9, -2), (-3, 5), (9, 1)\}$

9. $\{(-3, 1), (4, 1), (-2, 7)\}$

10. $\{(-12, 5), (-10, 3), (8, 3)\}$

11. $\{(1, 1), (1, -1), (0, 0), (2, 4), (2, -4)\}$

12. $\{(2, 5), (3, 7), (4, 9), (5, 11)\}$

13.

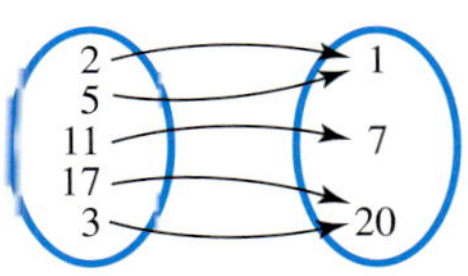

14.

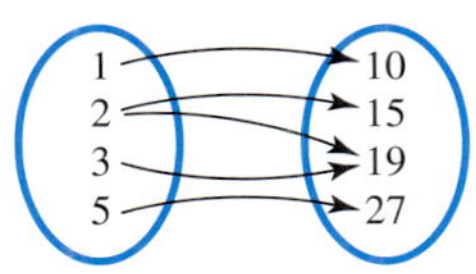

15.

x	y
1	5
1	2
1	−1
1	−4

16.

x	y
4	−3
2	−3
0	−3
−2	−3

17.
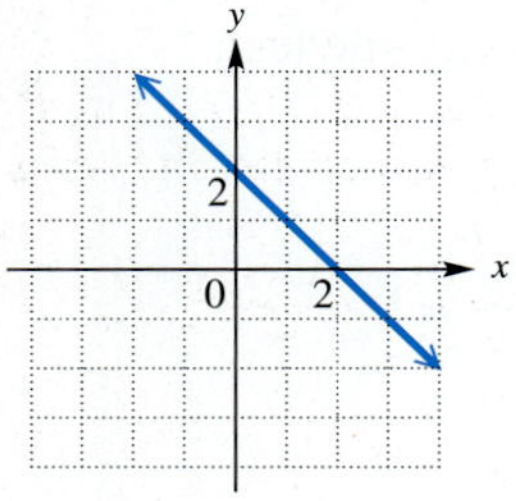

18.
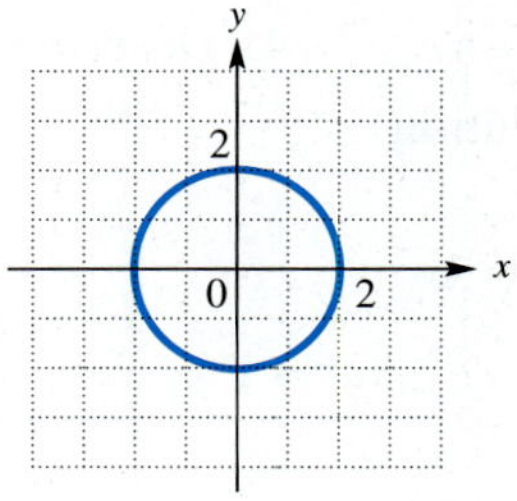

19.
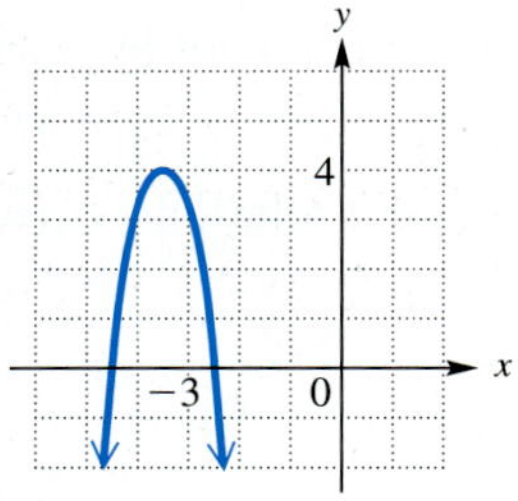

20.
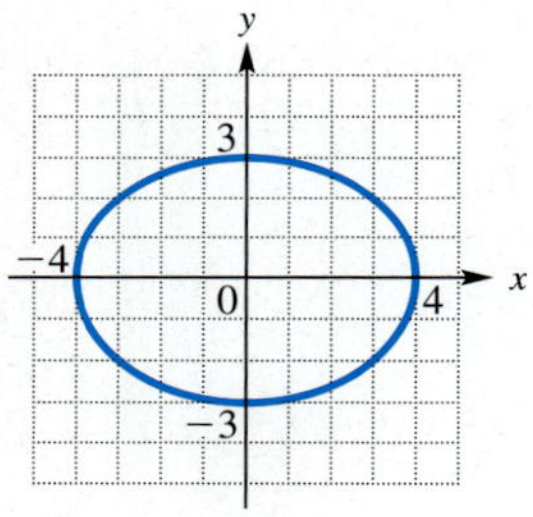

21.
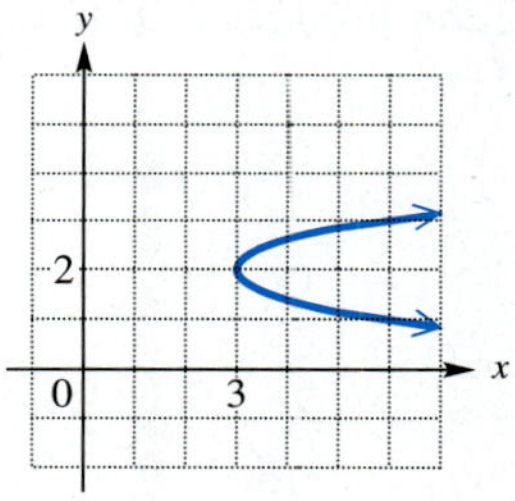

22.
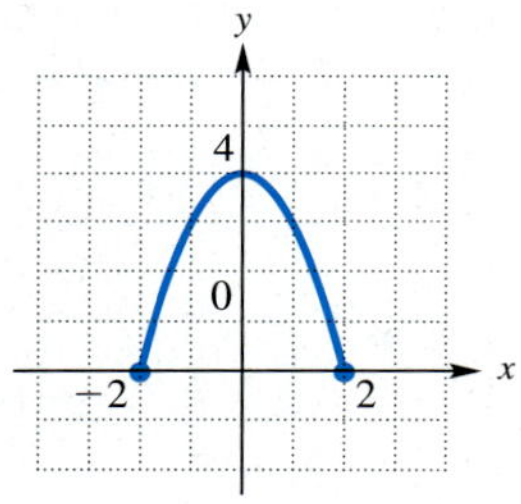

Decide whether each relation defines y as a function of x. Give the domain. See Example 5.

23. $y = x^2$ **24.** $y = x^3$ **25.** $x = y^6$ **26.** $x = y^4$

27. $y = 2x - 6$ **28.** $y = -6x + 8$ **29.** $x + y < 4$ **30.** $x - y < 3$

31. $y = \sqrt{x}$

32. $y = -\sqrt{x}$

33. $xy = 1$

34. $xy = -3$

35. $y = \sqrt{4x + 2}$

36. $y = \sqrt{9 - 2x}$

37. $y = \dfrac{2}{x - 9}$

38. $y = \dfrac{-7}{x - 16}$

39. Refer to the graph to answer the questions.

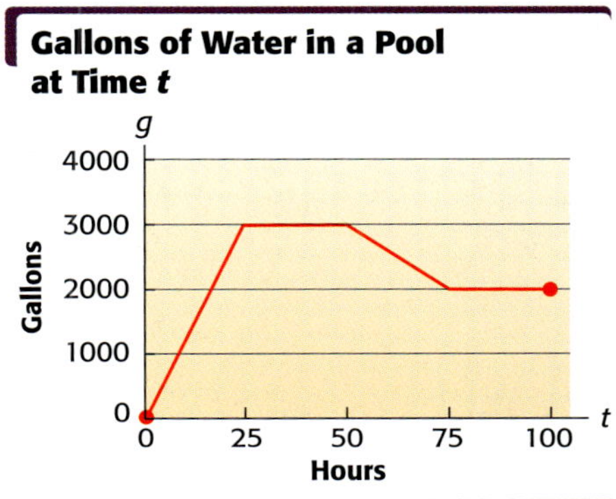

(a) What numbers are possible values of the dependent variable?

(b) For how long is the water level increasing? decreasing?

(c) How many gallons are in the pool after 90 hr?

(d) Call this function g. What is $g(0)$? What does it mean in this example?

40. The graph shows the daily megawatts of electricity used on a record-breaking summer day in Sacramento, California.

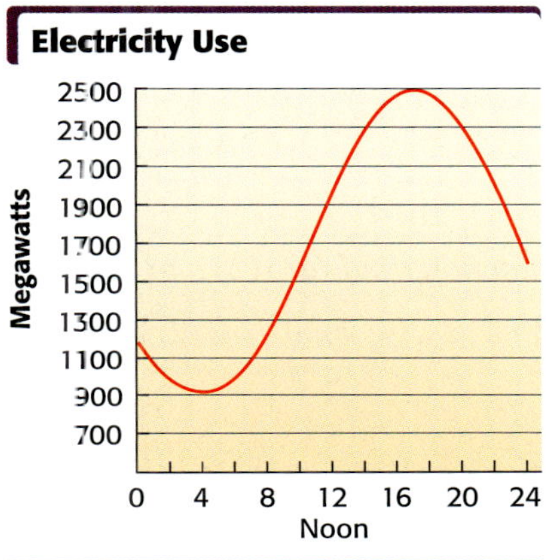

Source: Sacramento Municipal Utility District.

(a) Is this the graph of a function?

(b) What is the domain?

(c) Estimate the number of megawatts used at 8 A.M.

(d) At what time was the most electricity used? the least electricity?

41. Give an example of a function from everyday life. (*Hint:* Fill in the blanks: _______ depends on _______, so _______ is a function of _______.)

42. Choose the correct response: The notation $f(3)$ means

A. the variable f times 3 or $3f$

B. the value of the dependent variable when the independent variable is 3

C. the value of the independent variable when the dependent variable is 3

D. f equals 3.

Let $f(x) = -3x + 4$ and $g(x) = -x^2 + 4x + 1$. Find the following. See Examples 6 and 7.

43. $f(0)$

44. $f(-3)$

45. $g(-2)$

46. $g(10)$

47. $f(p)$

48. $g(k)$

49. $f(-x)$

50. $g(-x)$

51. $f(x + 2)$

52. $g\left(-\frac{1}{x}\right)$

53. $g\left(\frac{p}{3}\right)$

54. $f(3t - 2)$

*For each function, find **(a)** $f(2)$ and **(b)** $f(-1)$. See Example 8.*

55. $f = \{(-1, 3), (4, 7), (0, 6), (2, 2)\}$

56. $f = \{(2, 5), (3, 9), (-1, 11), (5, 3)\}$

57.

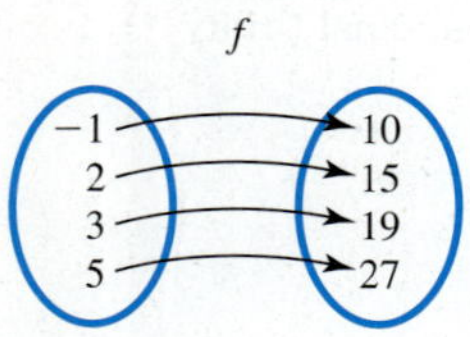

58.

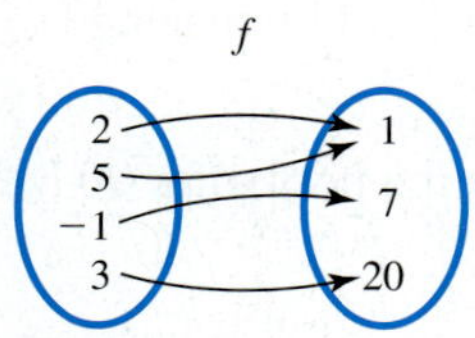

59.

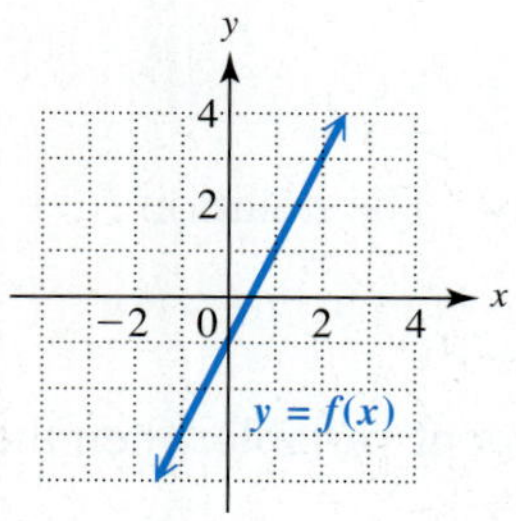

60.

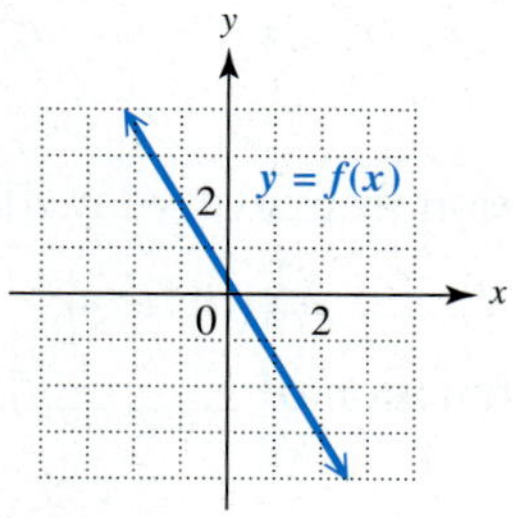

61. Fill in each blank with the correct response.

The equation $2x + y = 4$ has a straight _______ as its graph. One point that lies on the graph is (3, _______). If we solve the equation for y and use function notation, we have a _______ function defined by $f(x) =$ _______. For this function, $f(3) =$ _______, meaning that the point (_______, _______) lies on the graph of the function.

62. Which of the following defines a linear function?

A. $y = \frac{2}{5}x - 3$ **B.** $y = \frac{1}{x}$

C. $y = x^2$ **D.** $y = \sqrt{x}$

An equation that defines y as a function f of x is given. ***(a)*** *Solve for y in terms of x, and replace y with the function notation f(x).* ***(b)*** *Find f(3). See Example 9.*

63. $x + 3y = 12$ **64.** $x - 4y = 8$ **65.** $y + 2x^2 = 3$

66. $y - 3x^2 = 2$ **67.** $4x - 3y = 8$ **68.** $-2x + 5y = 9$

Graph each linear or constant function. Give the domain and range. See Example 10.

69. $f(x) = -2x + 5$ **70.** $g(x) = 4x - 1$ **71.** $h(x) = \frac{1}{2}x + 2$

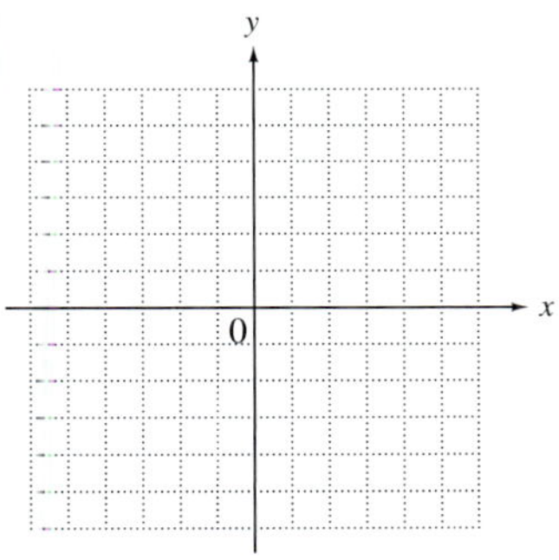

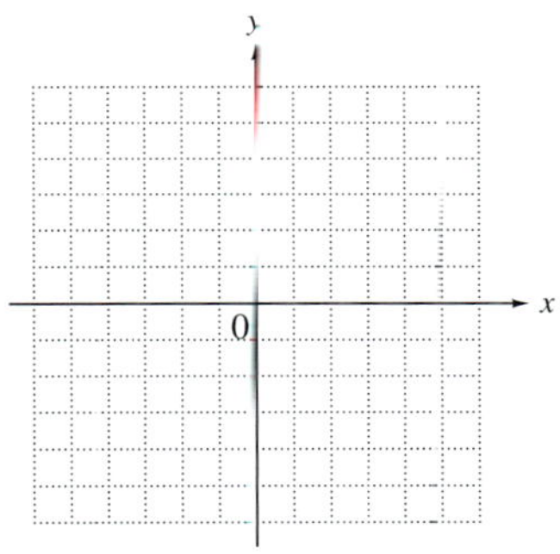

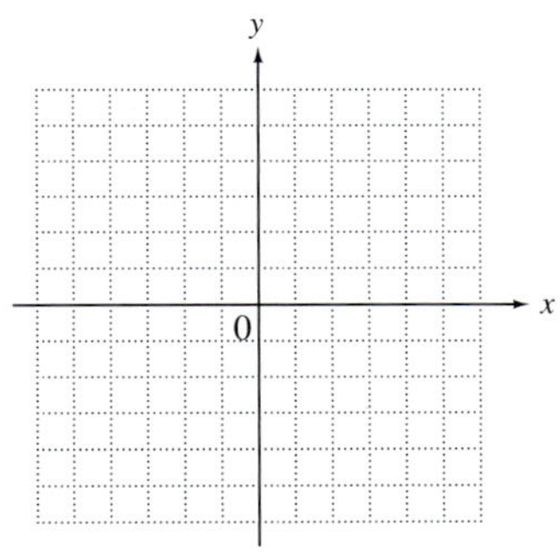

72. $F(x) = -\frac{1}{4}x + 1$ **73.** $g(x) = -4$ **74.** $f(x) = 5$

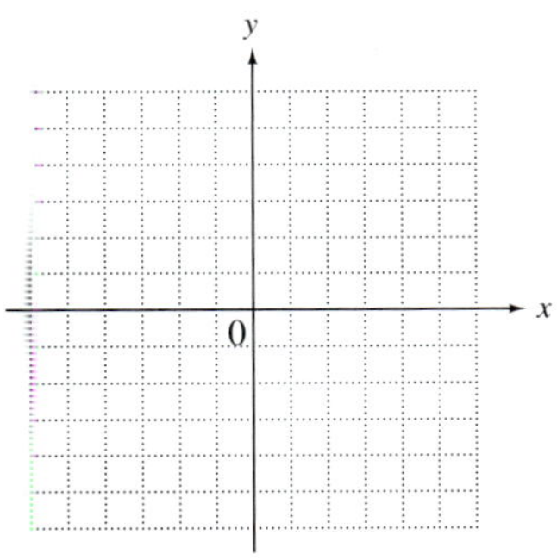

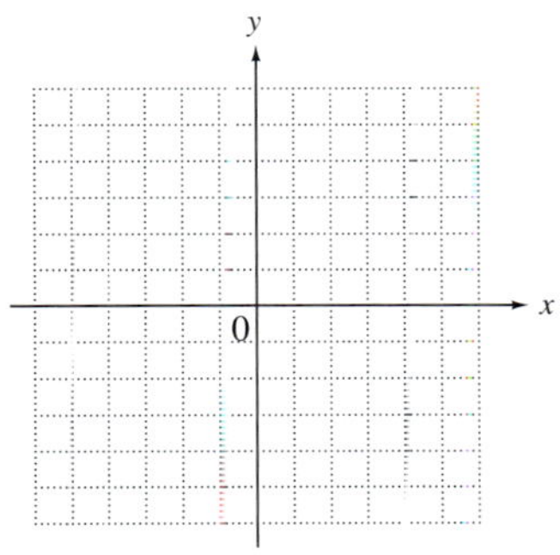

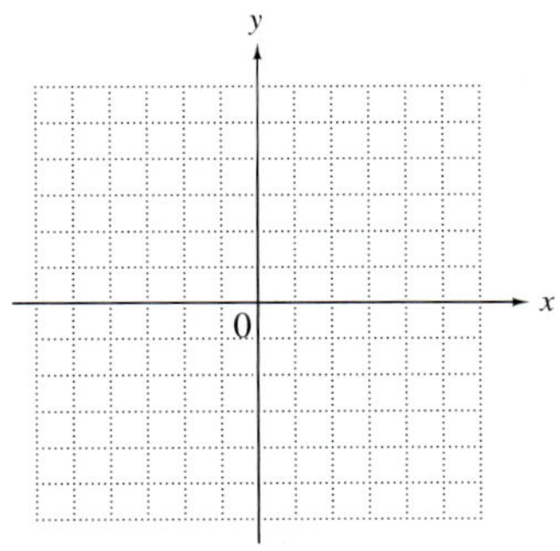

Solve each problem.

75. Suppose that a taxicab driver charges \$2.50 per mi.

(a) Fill in the table with the correct response for the price $f(x)$ he charges for a trip of x miles.

x	$f(x)$
0	
1	
2	
3	

(b) The linear function that gives a rule for the amount charged is $f(x) =$ _______.

(c) Graph this function for the domain $\{0, 1, 2, 3\}$.

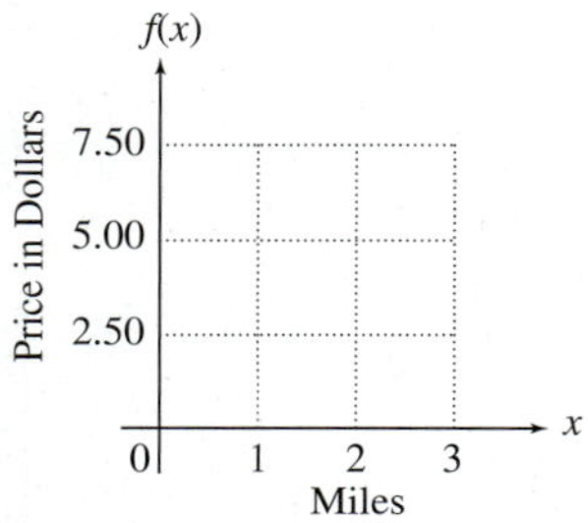

76. Suppose that a package weighing x pounds costs $f(x)$ dollars to ship to a given location, where $f(x) = 3.75x$.

(a) What is the value of $f(3)$?

(b) In your own words, describe what 3 and the value $f(3)$ mean in part (a), using the terms *independent variable* and *dependent variable.*

(c) How much would it cost to mail a 5-lb package? Write the answer using function notation.

Forensic scientists use the lengths of certain bones to calculate the height of a person. Two bones often used are the tibia (t), the bone from the ankle to the knee, and the femur (r), the bone from the knee to the hip socket. A person's height (h) is determined from the lengths of these bones using functions defined by the following formulas. All measurements are in centimeters.

Functions for men: $h(r) = 69.09 + 2.24r$ or $h(t) = 81.69 + 2.39t$

Functions for women: $h(r) = 61.41 + 2.32r$ or $h(t) = 72.57 + 2.53t$

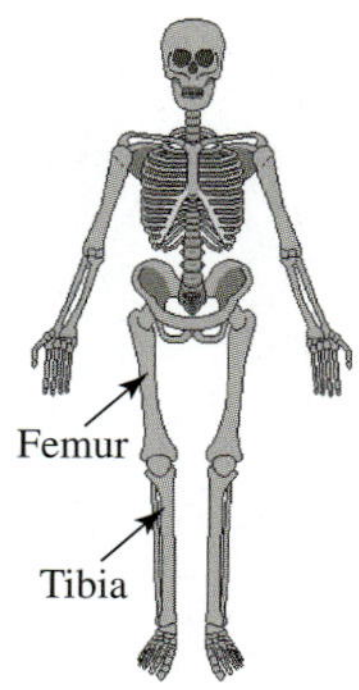

77. Find the height of a man with a femur measuring 56 cm.

78. Find the height of a man with a tibia measuring 40 cm.

79. Find the height of a woman with a femur measuring 50 cm.

80. Find the height of a woman with a tibia measuring 36 cm.

Chapter 4 Summary

Key Terms

Section	Term	Definition
4.1	**ordered pair**	An ordered pair is a pair of numbers written in parentheses in which the order of the numbers is important.
	origin	When two number lines intersect at a right angle, the origin is the common 0 point, with coordinates (0, 0).
	x-axis	The horizontal number line in a rectangular coordinate system is called the x-axis.
	y-axis	The vertical number line in a rectangular coordinate system is called the y-axis.
	rectangular (Cartesian) coordinate system	Two number lines that intersect at a right angle at their 0 points form a rectangular coordinate system, also called the Cartesian coordinate system.
	plot	To plot an ordered pair is to locate it on a rectangular coordinate system.
	components	The two numbers in an ordered pair are the components of the ordered pair.
	coordinate	Each number in an ordered pair represents a coordinate of the corresponding point.
	quadrant	A quadrant is one of the four regions in the plane determined by a rectangular coordinate system.
	graph of an equation	The graph of an equation is the set of points corresponding to all ordered pairs that satisfy the equation.
	first-degree equation	A first-degree equation has no term with a variable to a power greater than one.
	linear equation in two variables	A first-degree equation with two variables is a linear equation in two variables.
	x-intercept	The point where a line intersects the x-axis is the x-intercept.
	y-intercept	The point where a line intersects the y-axis is the y-intercept.
4.2	**rise**	The rise of a line is the vertical change between two points on the line.
	run	The run of a line is the horizontal change between two points on the line.
	slope	The ratio of the change in y compared to the change in x (rise/run) along a line is the slope of the line.
4.4	**linear inequality in two variables**	A linear inequality in two variables is a first-degree inequality with two variables.
	boundary line	In the graph of a linear inequality, the boundary line separates the region that satisfies the inequality from the region that does not satisfy the inequality.
4.5	**dependent variable**	If the quantity y depends on x, then y is called the dependent variable in a relation between x and y.
	independent variable	If y depends on x, then x is the independent variable in a relation between x and y.
	relation	A relation is a set of ordered pairs of real numbers.
	function	A function is a set of ordered pairs in which each value of the first component, x, corresponds to exactly one value of the second component, y.
	domain	The domain of a relation is the set of first components (x-values) of the ordered pairs of the relation.
	range	The range of a relation is the set of second components (y-values) of the ordered pairs of the relation.
	function notation	The function notation $f(x)$ is another way to represent the dependent variable y for the function f.

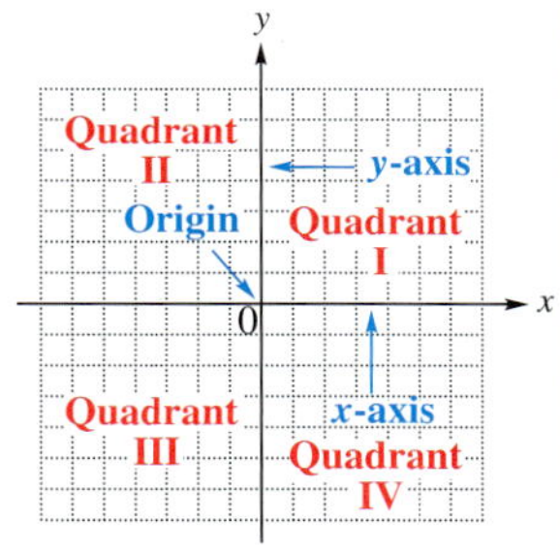

Rectangular coordinate system

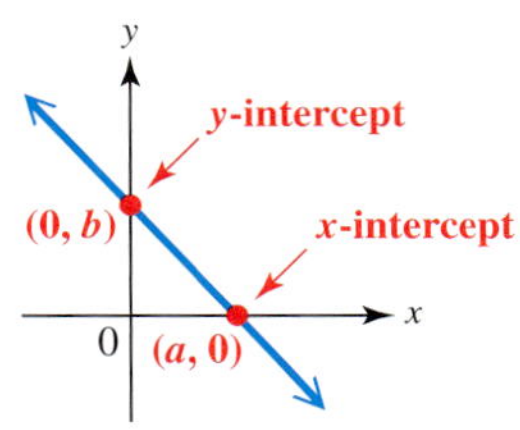

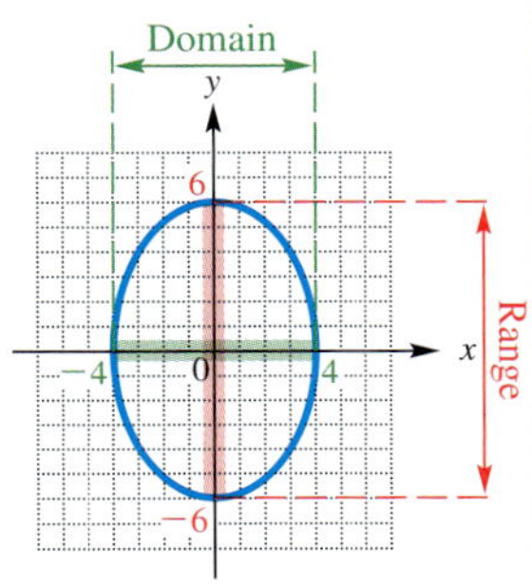

Graph of a relation

▸ Key Terms

linear function A function that is defined by $f(x) = mx + b$ is a linear function.

constant function A constant function is a linear function of the form $f(x) = b$, for a real number b.

▸ New Symbols

(a, b)	ordered pair	m	slope
x_1	a specific value of the variable x (read "x-sub-one")	$f(x)$	function f of x (read "f of x")

▸ Test Your Word Power

See how well you have learned the vocabulary in this chapter. Answers, with examples, follow the Quick Review.

1. An **ordered pair** is a pair of numbers written
   - **A.** in numerical order between brackets
   - **B.** between parentheses or brackets
   - **C.** between parentheses in which order is important
   - **D.** between parentheses in which order does not matter.

2. The **coordinates** of a point are
   - **A.** the numbers in the corresponding ordered pair
   - **B.** the solution of an equation
   - **C.** the values of the x-and y-intercepts
   - **D.** the graph of the point.

3. A **linear equation in two variables** is an equation that can be written in the form
   - **A.** $Ax + By < C$
   - **B.** $ax = b$
   - **C.** $y = x^2$
   - **D.** $Ax + By = C$.

4. An **intercept** is
   - **A.** the point where the x-axis and y-axis intersect
   - **B.** a pair of numbers written between parentheses in which order matters
   - **C.** one of the four regions determined by a rectangular coordinate system
   - **D.** the point where a graph intersects the x-axis or the y-axis.

5. The **slope** of a line is
   - **A.** the measure of the run over the rise of the line
   - **B.** the distance between two points on the line
   - **C.** the ratio of the change in y to the change in x along the line
   - **D.** the horizontal change compared to the vertical change of two points on the line.

6. In a relationship between two variables x and y, the **independent variable** is
   - **A.** x, if x depends on y
   - **B.** x, if y depends on x
   - **C.** either x or y
   - **D.** the larger of x and y.

7. In a relationship between two variables x and y, the **dependent variable** is
   - **A.** y, if y depends on x
   - **B.** y, if x depends on y
   - **C.** either x or y
   - **D.** the smaller of x and y.

8. A **relation** is
   - **A.** a set of ordered pairs
   - **B.** the ratio of the change in y to the change in x along a line
   - **C.** the set of all possible values of the independent variable
   - **D.** all the second components of a set of ordered pairs.

9. A **function** is
   - **A.** the numbers in an ordered pair
   - **B.** a set of ordered pairs in which each x-value corresponds to exactly one y-value
   - **C.** a pair of numbers written between parentheses in which order matters
   - **D.** the set of all ordered pairs that satisfy an equation.

10. The **domain** of a function is
    - **A.** the set of all possible values of the dependent variable y
    - **B.** a set of ordered pairs
    - **C.** the difference between the x-values
    - **D.** the set of all possible values of the independent variable x.

11. The **range** of a function is
    - **A.** the set of all possible values of the dependent variable y
    - **B.** a set of ordered pairs
    - **C.** the difference between the y-values
    - **D.** the set of all possible values of the independent variable x.

▶ Quick Review

Concepts	Examples

4.1 The Rectangular Coordinate System

Finding Intercepts
To find the x-intercept, let $y = 0$ and solve for x.

To find the y-intercept, let $x = 0$ and solve for y.

Find the intercepts of the graph of $2x + 3y = 12$.

x-intercept	y-intercept
$2x + 3(0) = 12$	$2(0) + 3y = 12$
$2x = 12$	$3y = 12$
$x = 6$	$y = 4$
The x-intercept is $(6, 0)$.	The y-intercept is $(0, 4)$.

Midpoint Formula
If the endpoints of a line segment PQ are $P(x_1, y_1)$ and $Q(x_2, y_2)$, then its midpoint M is

$$\left(\frac{x_1 + x_2}{2}, \frac{y_1 + y_2}{2}\right).$$

The midpoint of the segment with endpoints $(4, -7)$ and $(-10, -13)$ is

$$\left(\frac{4 + (-10)}{2}, \frac{-7 + (-13)}{2}\right) = (-3, -10).$$

4.2 Slope of a Line

If $x_1 \neq x_2$, then the slope m is given by

$$m = \frac{\textbf{rise}}{\textbf{run}} = \frac{\textbf{change in } y}{\textbf{change in } x} = \frac{y_2 - y_1}{x_2 - x_1}.$$

Find the slope of the graph of $2x + 3y = 12$.
Use the intercepts $(6, 0)$ and $(0, 4)$ and the slope formula.

$$m = \frac{4 - 0}{0 - 6} = \frac{4}{-6} = -\frac{2}{3} \quad x_1 = 6, y_1 = 0, x_2 = 0, y_2 = 4$$

A horizontal line has 0 slope.

The graph of the line $x = 3$ has undefined slope.

A vertical line has undefined slope.

The graph of the line $y = -5$ has slope $m = 0$.

Parallel lines have equal slopes.

The lines $y = 2x + 3$ and $4x - 2y = 6$ are **parallel**; both have $m = 2$.

$y = 2x + 3$	$4x - 2y = 6$
$m = 2$	$-2y = -4x + 6$
	$y = 2x - 3$
	$m = 2$

The slopes of perpendicular lines are negative reciprocals (with a product of -1).

The lines $y = 3x - 1$ and $x + 3y = 4$ are **perpendicular**; their slopes are negative reciprocals.

$y = 3x - 1$	$x + 3y = 4$
$m = 3$	$3y = -x + 4$
	$y = -\frac{1}{3}x + \frac{4}{3}$
	$m = -\frac{1}{3}$

4.3 Linear Equations in Two Variables

Slope-Intercept Form
$y = mx + b$

$y = 2x + 3$ $m = 2$, y-intercept is $(0, 3)$.

Point-Slope Form
$y - y_1 = m(x - x_1)$

$y - 3 = 4(x - 5)$ $(5, 3)$ is on the line, $m = 4$.

Standard Form
$Ax + By = C$, where A, B, and C are real numbers, and A and B are not both 0.

$2x - 5y = 8$ Standard form

(continued)

Concepts	Examples

4.3 Linear Equations in Two Variables (continued)

Horizontal Line

$y = b$

$y = 4$ Horizontal line

Vertical Line

$x = a$

$x = -1$ Vertical line

4.4 Linear Inequalities in Two Variables

Graphing a Linear Inequality

Step 1 Draw the graph of the line that is the boundary. Make the line solid if the inequality involves $\leq$ or $\geq$; make the line dashed if the inequality involves $<$ or $>$.

Step 2 Choose any point not on the line as a test point. Substitute the coordinates in the inequality.

Step 3 Shade the region that includes the test point if the test point satisfies the original inequality; otherwise, shade the region on the other side of the boundary line.

Graph $2x - 3y \leq 6$.

Draw the graph of $2x - 3y = 6$. Use a solid line because of the inclusion of equality in the symbol $\leq$.

Choose (0, 0), for example.

$2(0) - 3(0) = 0$, and $0 \leq 6$ True

Shade the side of the line that includes (0, 0).

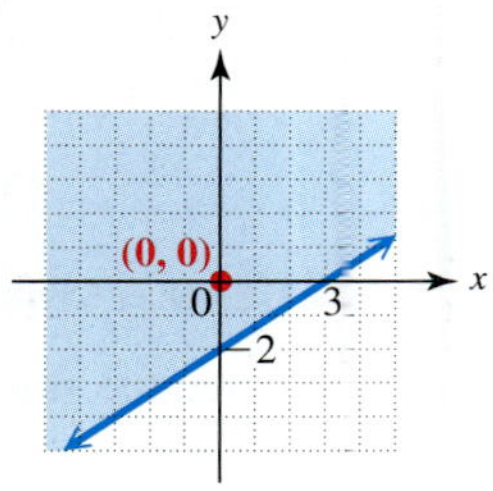

4.5 Introduction to Functions

A **function** is a set of ordered pairs such that, for each first component, there is one and only one second component. The set of first components is called the **domain,** and the set of second components is called the **range.**

$y = f(x) = x^2$ defines a function f with domain $(-\infty, \infty)$ and range $[0, \infty)$.

To evaluate a function using function notation (that is, $f(x)$ notation) for a given value of x, substitute the value wherever x appears.

If $f(x) = x^2 - 7x + 12$, then

$$f(1) = 1^2 - 7(1) + 12 = 6.$$

To write the equation that defines a function f in function notation, follow these steps.

Step 1 Solve the equation for y.

Step 2 Replace y with $f(x)$.

Write $2x + 3y = 12$ in function notation for function f.

$$3y = -2x + 12 \quad \text{Subtract } 2x.$$

$$y = -\frac{2}{3}x + 4 \quad \text{Divide by 3.}$$

$$f(x) = -\frac{2}{3}x + 4$$

ANSWERS TO TEST YOUR WORD POWER

1. C; *Examples:* (0, 3), (3, 8), (4, 0)
2. A; *Example:* The point associated with the ordered pair (1, 2) has x-coordinate 1 and y-coordinate 2.
3. D; *Examples:* $3x + 2y = 6$, $x = y - 7$, $4x = y$
4. D; *Example:* In Figure 4(b) of **Section 4.1,** the x-intercept is (3, 0) and the y-intercept is (0, 2).
5. C; *Example:* The line through (3, 6) and (5, 4) has slope $\frac{4 - 6}{5 - 3} = \frac{-2}{2} = -1$.
6. B; *Example:* See Answer 7, which follows.
7. A; *Example:* When borrowing money, the amount you borrow (independent variable) determines the size of your payments (dependent variable).
8. A; *Example:* The set $\{(2, 0), (4, 3), (6, 6), (8, 9)\}$ defines a relation.
9. B; *Example:* The relation given in Answer 8 is a function since each x-value corresponds to exactly one y-value.
10. D; *Example:* In the function in Answer 8, the domain is the set of x-values, $\{2, 4, 6, 8\}$.
11. A; *Example:* In the function in Answer 8, the range is the set of y-values, $\{0, 3, 6, 9\}$.

Chapter 4 Review Exercises

[4.1] *Complete the table of ordered pairs for each equation, and then graph the equation.*

1. $3x + 2y = 6$

x	y
0	
	0
	−2

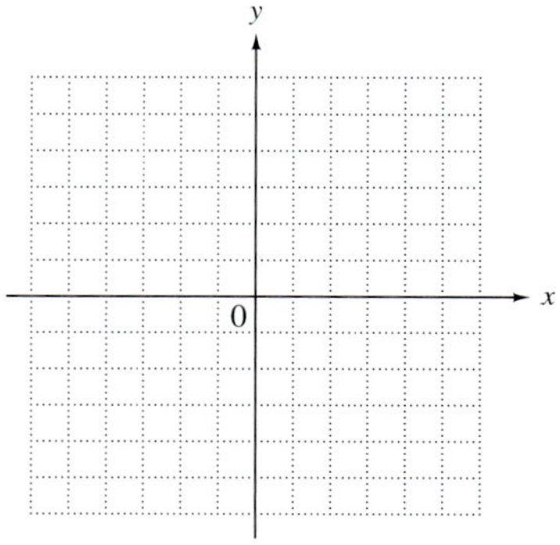

2. $x - y = 6$

x	y
2	
	−3
1	
	−2

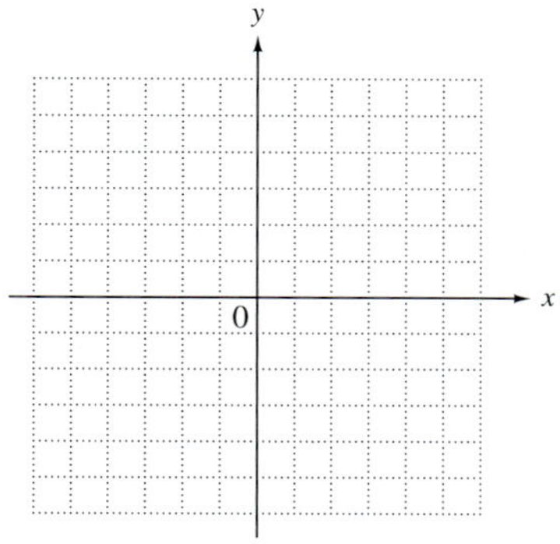

Find the x- and y-intercepts, and then graph each equation.

3. $4x + 3y = 12$

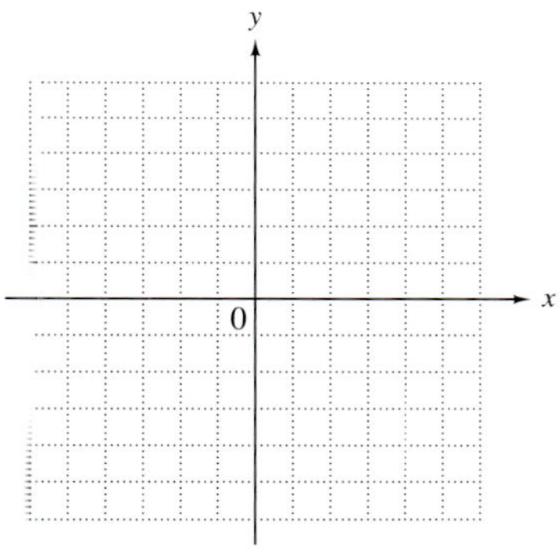

4. $5x + 7y = 15$

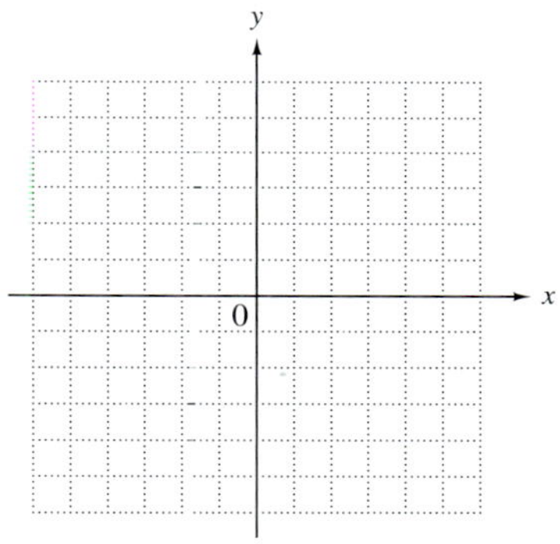

Use the midpoint formula to find the midpoint of each segment with the given endpoints.

5. $(-8, -12)$ and $(8, 16)$

6. $(0, -5)$ and $(-9, 8)$

7. $(3.8, 8.6)$ and $(1.4, 15.2)$

8. $(15.5, -6.3)$ and $(-6.5, -12.7)$

[4.2] *Find the slope of each line.*

9. Through $(-1, 2)$ and $(4, -6)$

10. $y = 2x + 3$

11. $-3x + 4y = 5$

12. $y = 4$

13. A line parallel to $3y = -2x + 5$

14. A line perpendicular to $3x - y = 6$

Tell whether the line has positive, negative, 0, *or* undefined *slope.*

15.

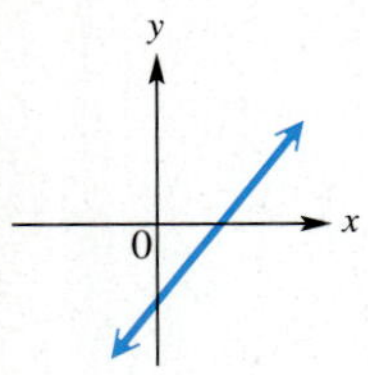

16.

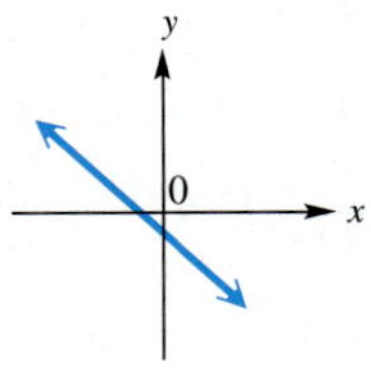

17.

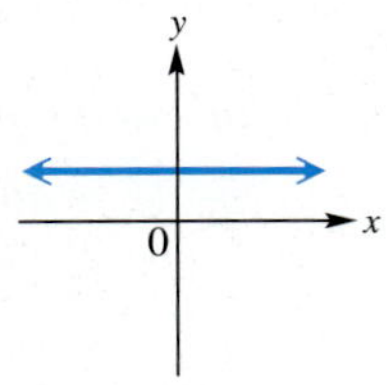

18.

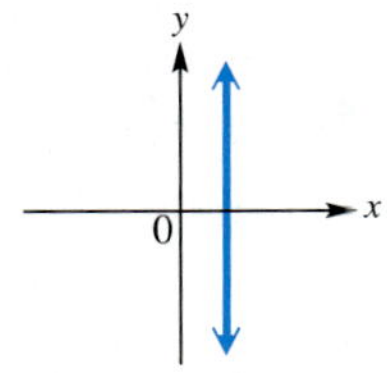

19. If the pitch of a roof is $\frac{1}{4}$, how many feet in the horizontal direction correspond to a rise of 3 ft?

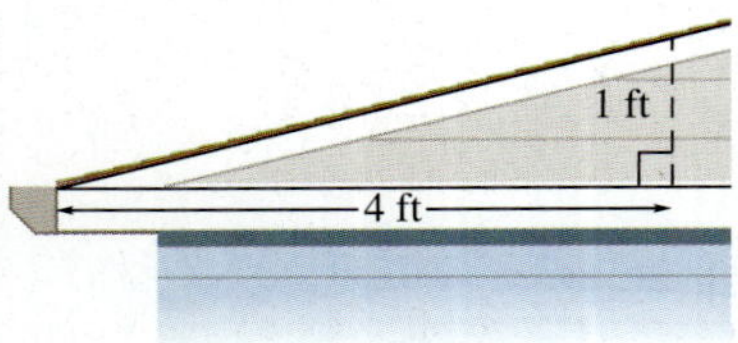

20. Family income in the United States has steadily increased for many years (primarily due to inflation). In 1980 the median family income was about \$21,000 per yr. In 2005 it was about \$56,200 per yr. Find the average rate of change of median family income over that period. (*Source:* U.S. Census Bureau.)

[4.3] *Write an equation in slope-intercept form for each line.*

21. Slope $\frac{3}{5}$; y-intercept $(0, -8)$

22. Slope $-\frac{1}{3}$; y-intercept $(0, 5)$

23. Slope 0; y-intercept $(0, 12)$

Write an equation for each line.

24. Undefined slope; through $(2, 7)$

25. Horizontal; through $(-1, 4)$

26. Vertical; through $(0.3, 0.6)$

Write an equation for each line ***(a)*** *in slope-intercept form and* ***(b)*** *in standard form.*

27. Through $(2, -5)$ and $(1, 4)$

28. Through $(-3, -1)$ and $(2, 6)$

29. Parallel to $4x - y = 3$ and through $(6, -2)$

30. Perpendicular to $2x - 5y = 7$ and through $(0, 1)$

31. The Midwest Athletic Club (**Section 4.3,** Exercises 61 and 62) offers two special membership plans. (*Source:* Midwest Athletic Club.) For each plan, write a linear equation in slope-intercept form and give the cost y in dollars of a 1-yr membership. Let x represent the number of months.

(a) Executive VIP/Gold membership: \$159 fee plus \$57 per month

(b) Executive Regular/Silver membership: \$159 fee plus \$47 per month

[4.4] *Graph each inequality.*

32. $3x - 2y \leq 12$

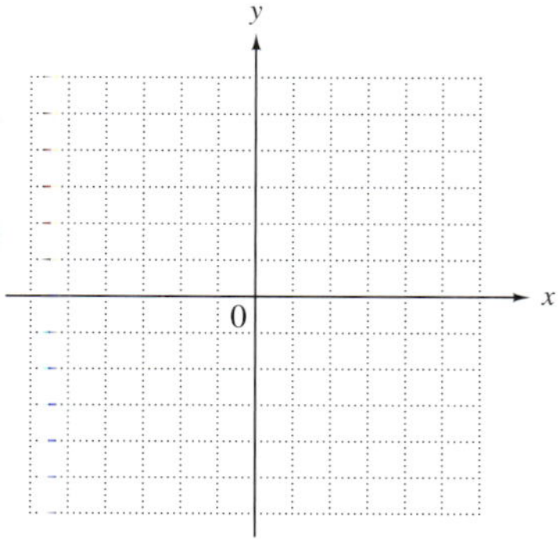

33. $5x - y > 6$

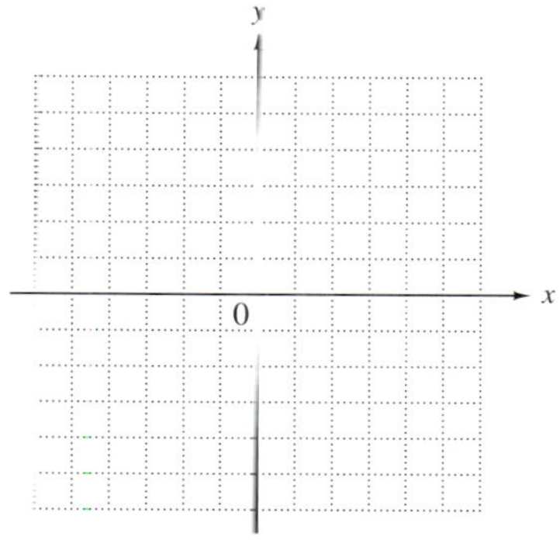

34. $x \geq 2$ or $y \geq 2$

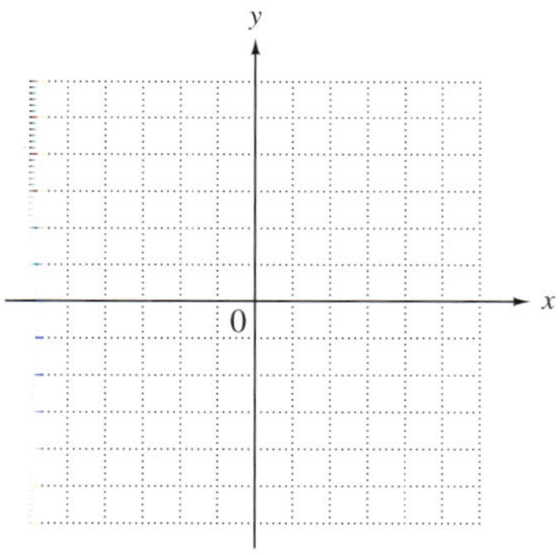

35. $2x + y \leq 1$ and $x \geq 2y$

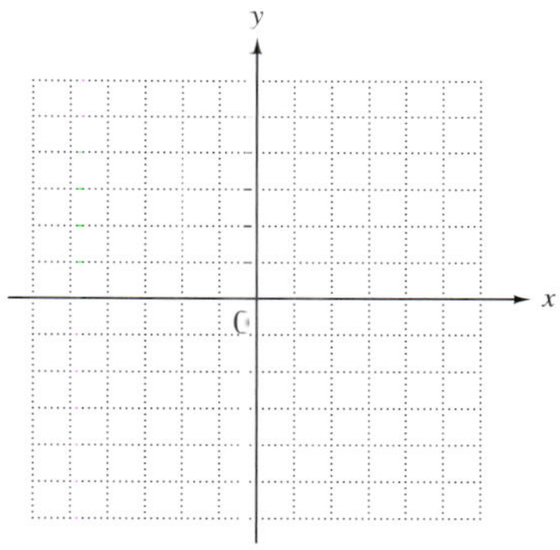

[4.5] *Give the domain and range of each relation. Identify any functions.*

36. $\{(-4, 2), (-4, -2), (1, 5), (1, -5)\}$

37.

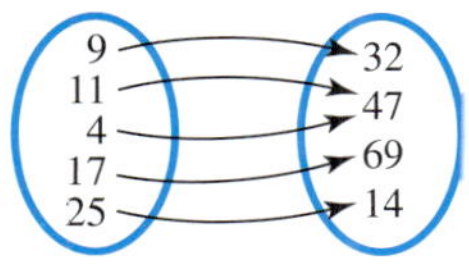

38.

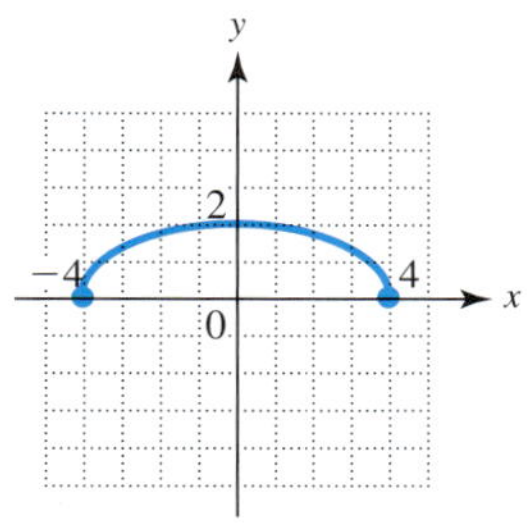

Determine whether each relation defines y as a function of x. Identify any linear functions. Give the domain in each case.

39. $y = 3x - 3$

40. $y < x + 2$

41. $y = |x - 4|$

42. $y = \sqrt{4x + 7}$

43. $x = y^2$

44. $y = \dfrac{7}{x - 36}$

45. The table shows life expectancy at birth in the United States for selected years.

Year	Life Expectancy at Birth (in years)
1943	63.3
1953	68.8
1963	69.9
1973	71.4
1983	74.6
1993	75.5
2003	77.6

Source: Centers for Disease Control and Prevention.

(a) Does the table define a function?

(b) What are the domain and range?

(c) Call this function f. Give two ordered pairs that belong to f.

(d) Find $f(2003)$. What does it mean?

(e) If $f(x) = 75.5$, what does x equal?

Given $f(x) = -2x^2 + 3x - 6$, *find each of the following.*

46. $f(0)$

47. $f(3)$

48. $f(p)$

49. $f(-k)$

50. The equation $2x^2 - y = 0$ defines y as a function of x. Rewrite it using $f(x)$ notation, and find $f(3)$.

51. The linear equation $2x + 5y = 7$ defines a function. If $y = f(x)$, which one of the following defines the same function?

A. $f(x) = -\frac{2}{5}x + \frac{7}{5}$

B. $f(x) = -\frac{2}{5}x - \frac{7}{5}$

C. $f(x) = \frac{2}{5}x - \frac{7}{5}$

D. $f(x) = \frac{2}{5}x + \frac{7}{5}$

52. Describe the graph of a constant function.

Chapter 4 Test

Use the Chapter Test Prep Video CD to see fully worked-out solutions to any of the exercises you want to review

1. Find the slope of the line through (6, 4) and (−4, −1).

1. ________

For each line, find the slope and the x- and y-intercepts.

2. $3x - 2y = 13$

2. ________

3. $y = 5$

3. ________

4. Describe the graph of a line with undefined slope in a rectangular coordinate system.

4. ________

Find the x- and y-intercepts, and graph each equation.

5. $4x - 3y = -12$

5. ________

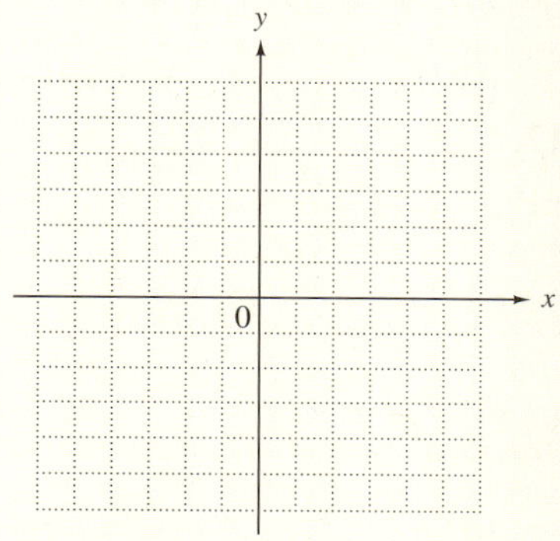

6. $y - 2 = 0$

6. ________

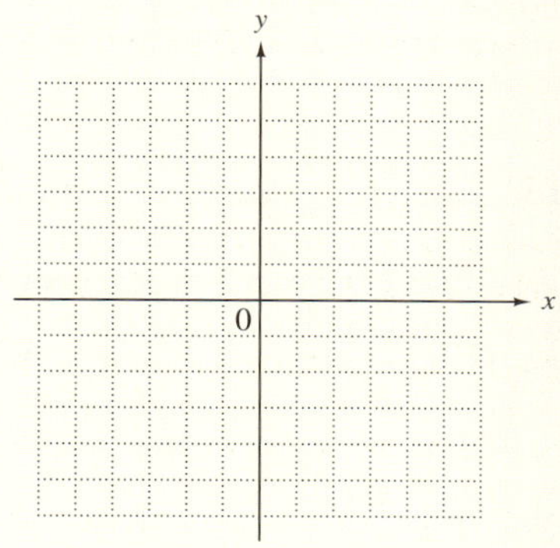

7. $y = -2x$

7. ________

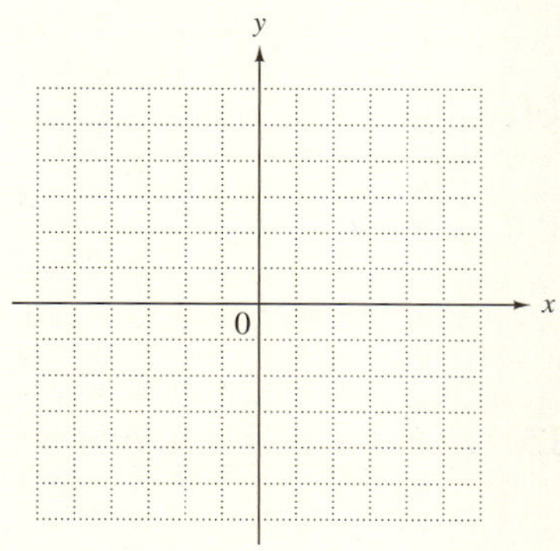

8.

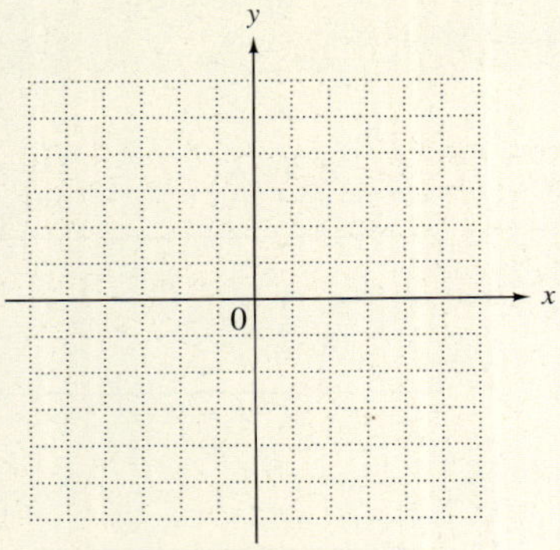

9. ______________

10. ______________

11. (a) ______________

(b) ______________

12. ______________

____

13. ______________

____

14. ______________

15. ______________

8. Graph $3x - 2y > 6$.

Write the equation of each line in standard form.

9. Through $(-3, 14)$ and $(-6, 9)$ **10.** Through $(4, -1)$; $m = -5$

11. Write the equation in slope-intercept form for the line through $(-7, 2)$ and

(a) parallel to $3x + 5y = 6$. **(b)** perpendicular to $y = 2x$.

12. Which of the following is the graph of a function? Give its domain and range.

A.

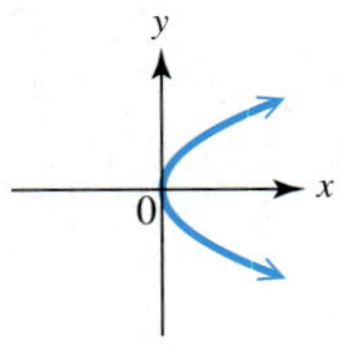

B.

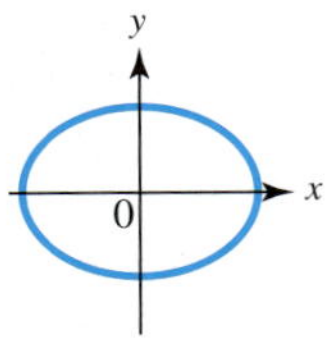

C.

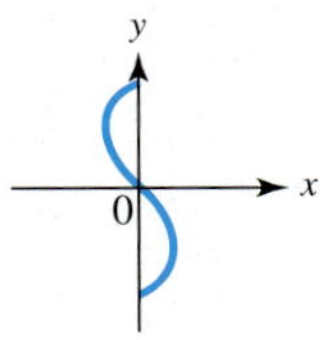

D.

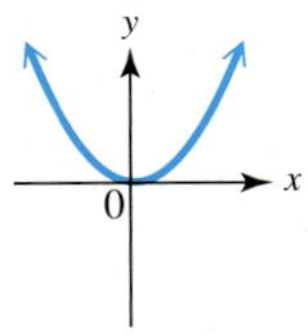

13. Which of the following does not define a function? Give its domain and range.

A. $\{(0, 1), (-2, 3), (4, 8)\}$ **B.** $y = 2x - 6$

C. $y = \sqrt{x + 2}$ **D.**

x	y
0	1
3	2
0	2
6	3

14. If $f(x) = -x^2 + 2x - 1$, find $f(1)$ and $f(a)$.

15. In 1980, there were 119,000 farms in Iowa. As of 2005, there were 89,000. Find and interpret the average rate of change in the number of farms per year. (*Source:* U.S. Department of Agriculture.)

Study Skills

ANALYZING YOUR TEST RESULTS

OBJECTIVES

1. Determine the reason for errors.
2. Develop a plan to avoid test taking errors.
3. Review material to correct misunderstandings.

After taking a test, many students heave a big sigh of relief and try to forget it ever happened. Don't fall into this trap. ***An exam is a learning opportunity.*** It gives you clues about *what your instructor thinks is important*, what *concepts and skills are valued* in mathematics, and *if you are on the right track.*

After the test is returned, do the following:

- **Jot down problems that caused you trouble.** Find out how to solve them by checking your textbook or notes, or asking your instructor. You might see those same problems again on a final exam.

- **Find out what you got wrong and why you had points deducted.** Write down the problem so you can learn how to do it correctly. Sometimes you only have a short time in class to review your test. If you need more time, ask your instructor if you can look at the test in his or her office.

Find Out Why You Made the Errors You Made

Here is a list of typical reasons you might make errors on math tests.

1. You read the directions wrong.
2. You read the question wrong or skipped over something.
3. You made a computation error (maybe even a minor one).
4. Your answer is not accurate.
5. Your answer is not complete.
6. You labeled your answer wrong. For example, you labeled it "ft" and it should have been "ft^2."
7. You didn't show your work.
8.* You didn't understand the concept.
9.* You were unable to go from words (in an application) to setting up the problem.
10.* You were unable to apply a procedure to a new situation.
11. You were anxious and made errors even when you knew the material.

The first seven errors are test taking errors. They are easy to correct if you carefully read test questions and directions, proofread or rework the problems, show all your work, and double-check units and labels every time.

The three starred errors (*) are test preparation errors. Remember that you need to practice the kinds of problems that you will see on tests. So, for example, if application problems are difficult for you, you must do more application problems. If you have practiced the various techniques outlined in the Study Skills, you will be less likely to make these kinds of errors on tests.

The last item isn't really an error. **Anxiety can play a big part in your test results.** Go back to the *Preparing for Tests* Study Skill, and read the suggestions for managing anxiety. Five minutes of brisk walking before your test can help you relax. Also, practicing a relaxation technique while you do your homework will make it more likely that you will benefit from using it during a test. Deep breathing is one such technique. When you are anxious, you tend to breathe more shallowly, which can make you feel confused and easily distracted.

Make a Plan for the Next Test

Make this plan based on your results from a test. You might review the Chapter Summary and work the problems in the Chapter Review Exercises or the Chapter Test. Perhaps ask your instructor for more help.

Now Try This ▶▶▶

Below is a record sheet to use to track your progress in test taking. Use it to find out if you tend to make particular kinds of errors on tests. Then you can work specifically on correcting them. For each category of errors, check the appropriate box when you made one of the errors.

Test Taking Errors

Test	Read directions wrong	Read question wrong	Computation error	Not exact or accurate	Not complete	Labeled wrong	Didn't show work
1							
2							
3							
4							
5							

What will you do to avoid these kinds of errors on your next test?

____

____

Test Preparation Errors

Test	Didn't understand concept	Didn't set up problem correctly	Couldn't apply concept to new situation
1			
2			
3			
4			
5			

What will you do to avoid these kinds of errors on your next test?

____

____

Anxiety

Test	Felt anxious *before* the exam	Felt anxious *during* the exam	Blanked out on questions	Got questions wrong that I knew how to do
1				
2				
3				
4				
5				

What will you do to reduce your anxiety before the next test? ________

____

____

Cumulative Review Exercises ▸▸▸ Chapters 1–4

Decide whether each statement is always true, sometimes true, *or* never true. *If the statement is* sometimes true, *give examples where it is true and where it is false.*

1. The absolute value of a negative number equals the additive inverse of the number.

2. The quotient of two integers with nonzero denominator is a rational number.

3. The sum of two negative numbers is positive.

4. The sum of a positive number and a negative number is 0.

Perform each operation.

5. $-|-2| - 4 + |-3| + 7$

6. $(-0.8)^2$

7. $\sqrt{-64}$

8. $-\frac{2}{3}\left(-\frac{12}{5}\right)$

Simplify.

9. $-(-4m + 3)$

10. $3x^2 - 4x + 4 + 9x - x^2$

11. $\dfrac{3\sqrt{16} - (-1)7}{4 + (-6)}$

12. Write $-3 < x \le 5$ in interval notation.

13. Is $\sqrt{\dfrac{-2 + 4}{-5}}$ a real number?

Evaluate if $p = -4$, $q = -2$, *and* $r = 5$.

14. $-3(2q - 3p)$

15. $|p|^3 - |q^3|$

16. $\dfrac{\sqrt{r}}{-p + 2q}$

Solve

17. $2z - 5 + 3z = 4 - (z + 2)$

18. $\dfrac{3a - 1}{5} + \dfrac{a + 2}{2} = -\dfrac{3}{10}$

19. $V = \frac{1}{3}\pi r^2 h$ for h

20. Two planes leave the Dallas-Fort Worth airport at the same time. One travels east at 550 mph, and the other travels west at 500 mph. Assuming no wind, how long will it take for the planes to be 2100 mi apart?

21. Ms. Bell must take at least 30 units of a certain medication each day. She can get the medication from white pills or yellow pills, each of which contains 3 units of the drug. To provide other benefits, she needs to take twice as many of the yellow pills as white pills. Find the least number of white pills that will satisfy these requirements.

22. If each side of a square were increased by 4 in., the perimeter would be 8 in. less than twice the perimeter of the original square. Find the length of a side of the original square.

23. How are the solution sets of a linear equation and the two associated inequalities related?

Solve.

24. $3 - 2(m + 3) < 4m$

25. $2k + 4 < 10$ and $3k - 1 > 5$

26. $2k + 4 > 10$ or $3k - 1 < 5$

27. $|5x + 3| = 13$

28. $|x + 2| < 9$

29. $|2x - 5| \geq 9$

30. Complete the ordered pairs (0,), (, 0), and (2,) for the equation $3x - 4y = 12$.

31. Graph $-4x + 2y = 8$, and give the intercepts.

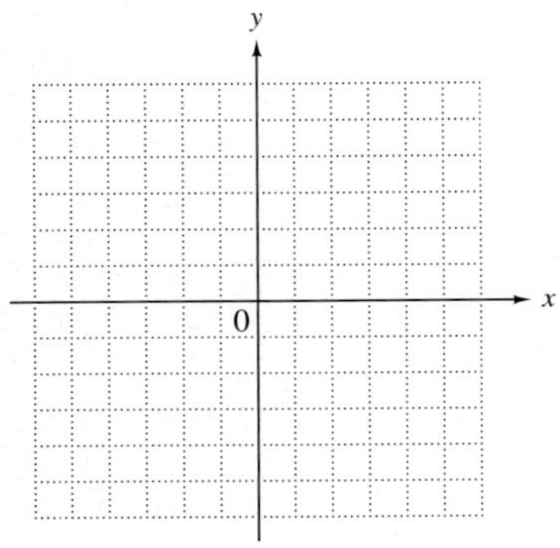

Find the slope of each line.

32. Through $(-5, 8)$ and $(-1, 2)$

33. Parallel to $y = -\frac{1}{2}x + 5$

34. Perpendicular to $4x - 3y = 12$

Write an equation in slope-intercept form for each line.

35. Slope $-\frac{3}{4}$; y-intercept $(0, -1)$

36. Horizontal; through $(2, -2)$

37. Through $(4, -3)$ and $(1, 1)$

38. For the function defined by $f(x) = -4x + 10$,

(a) what is the domain?

(b) what is $f(-3)$?

Use the graph to answer Exercises 39 and 40.

39. Find and interpret the average rate of change in the number of motor scooters sold in the United States from 1997 to 2004.

40. Write an equation in slope-intercept form that models the number y of motor scooters sold (in thousands) in year x, where $x = 0$ represents 1997.

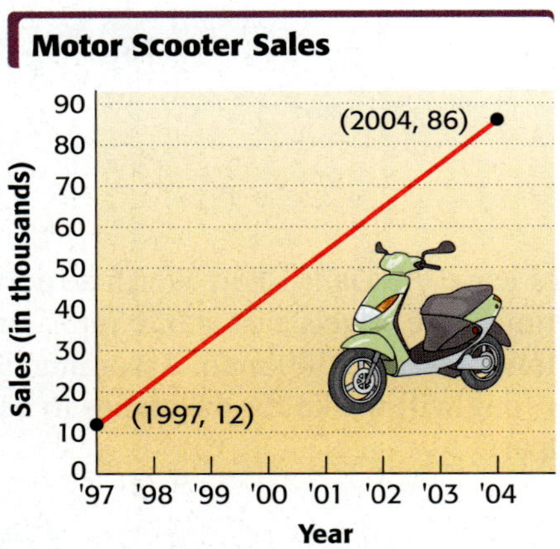

Source: Motorcycle Industry Council.

5

Systems of Linear Equations

5.1 Systems of Linear Equations in Two Variables

5.2 Systems of Linear Equations in Three Variables

5.3 Applications of Systems of Linear Equations

5.4 Solving Systems of Linear Equations by Matrix Methods

In the early 1970s, the NBC television network aired *The Bill Cosby Show*, in which the popular comedian played Chet Kincaid, a Los Angeles high school physical education teacher. In the episode "Let x Equal a Lousy Weekend," Chet must substitute for the algebra teacher. He and the entire class are stumped by the following problem:

How many pounds of candy that sells for \$0.75 *per lb must be mixed with candy that sells for* \$1.25 *per lb to obtain* 9 *lb of a mixture that should sell for* \$0.96 *per lb?*

The smartest student in the class eventually helps Chet solve this problem. In Exercise 31 of Section 5.3, we ask you to use a *system of linear equations*, the topic of this chapter, to do so.

5.1 Systems of Linear Equations in Two Variables

OBJECTIVES

1. Solve linear systems by graphing.
2. Decide whether an ordered pair is a solution of a linear system.
3. Solve linear systems (with two equations and two variables) by substitution.
4. Solve linear systems (with two equations and two variables) by elimination.
5. Solve special systems.

In recent years, the sale of digital cameras has increased, while that of conventional cameras has decreased. These trends can be seen in the graph in Figure 1. The two straight-line graphs intersect at the point in time when the two types of cameras had the *same* sales.

Say Cheese!

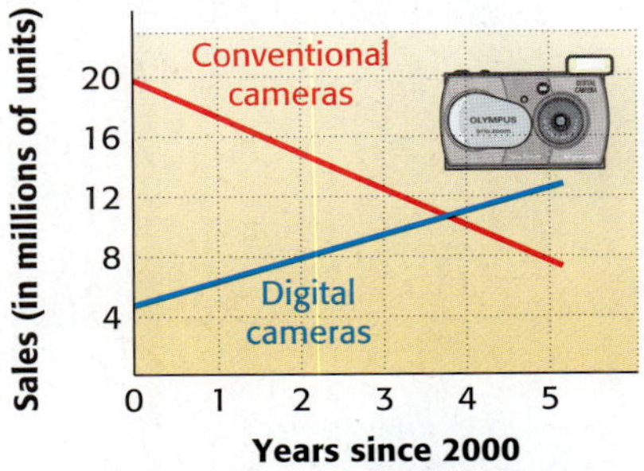

Source: Consumer Electronics Association.

Figure 1

$$\begin{aligned} 2.5x + y &= 19.4 \\ -1.7x + y &= 4.4 \end{aligned}$$ Linear system of equations

(Here, $x = 0$ represents 2000, $x = 1$ represents 2001, and so on; y represents sales in millions of units.)

As shown beside Figure 1, we can use a linear equation to model the graph of digital camera sales (the blue equation) and another linear equation to model the graph of conventional camera sales (the red equation). Such a set of equations is called a **system of equations**—in this case, a **linear system of equations.** The point where the graphs in Figure 1 intersect is a solution of each of the individual equations. It is also the solution of the linear system of equations.

OBJECTIVE 1 Solve linear systems by graphing. The **solution set of a system of equations** contains all ordered pairs that satisfy all the equations of the system *at the same time*. Another example of a linear system is

$$\begin{aligned} x + y &= 5 \\ 2x - y &= 4. \end{aligned}$$ Linear system of equations

One way to find the solution set of a linear system of equations is to graph each equation and find the point where the graphs intersect.

EXAMPLE 1 Solving a System by Graphing

Solve the system of equations by graphing.

$$x + y = 5 \quad (1)$$
$$2x - y = 4 \quad (2)$$

When we graph these linear equations as shown in Figure 2 on the next page, the graph suggests that the point of intersection is the ordered pair (3, 2).

Continued on Next Page

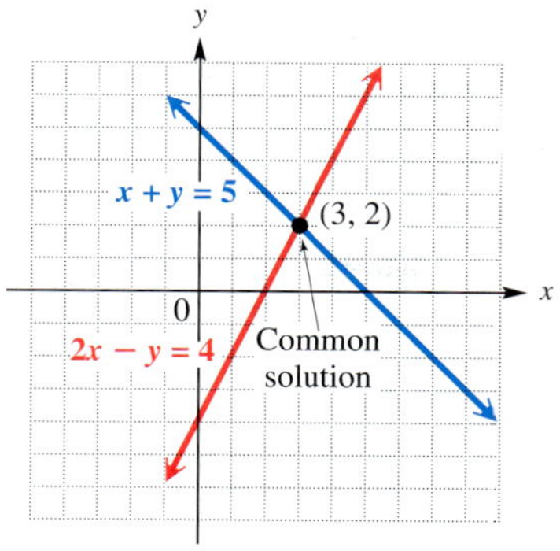

Figure 2

To be sure that (3, 2) is a solution of *both* equations, we check by substituting 3 for x and 2 for y in each equation.

Check

$x + y = 5$ (1)	$2x - y = 4$ (2)
$3 + 2 \stackrel{?}{=} 5$	$2(3) - 2 \stackrel{?}{=} 4$
$5 = 5$ True	$6 - 2 \stackrel{?}{=} 4$
	$4 = 4$ True

Since (3, 2) makes both equations true, $\{(3, 2)\}$ is the solution set of the system.

Work Problem **1** *at the Side.* ▶

Calculator Tip A graphing calculator can be used to solve a system. Each equation must be solved for y before being entered in the calculator. The point of intersection of the graphs, which is the solution of the system, can then be displayed. Consult your owner's manual for details.

OBJECTIVE 2 Decide whether an ordered pair is a solution of a linear system. To decide if an ordered pair is a solution of a system, we substitute the ordered pair in both equations of the system, just as we did when we checked the solution in Example 1.

EXAMPLE 2 Deciding Whether an Ordered Pair Is a Solution

Decide whether the given ordered pair is a solution of the given system.

(a) $\begin{aligned} x + y &= 6 \\ 4x - y &= 14 \end{aligned}$; (4, 2)

Replace x with 4 and y with 2 in each equation of the system.

$x + y = 6$	$4x - y = 14$
$4 + 2 \stackrel{?}{=} 6$	$4(4) - 2 \stackrel{?}{=} 14$
$6 = 6$ True	$16 - 2 \stackrel{?}{=} 14$
	$14 = 14$ True

Since (4, 2) makes both equations true, (4, 2) is a solution of the system.

Continued on Next Page

1 Solve each system of equations by graphing.

(a) $\begin{aligned} x - y &= 3 \quad (1) \\ 2x - y &= 4 \quad (2) \end{aligned}$

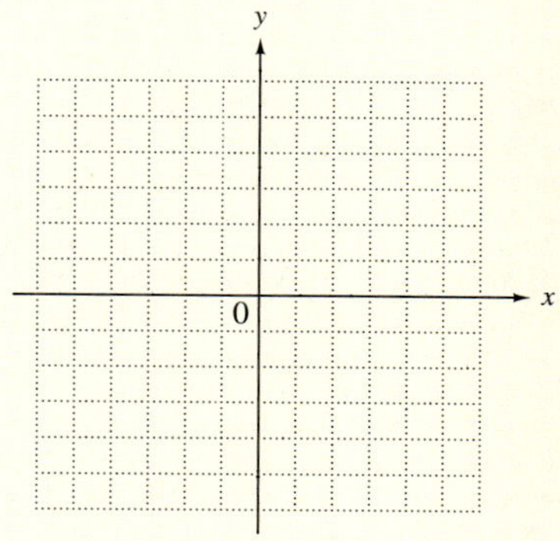

(b) $\begin{aligned} 2x + y &= -5 \quad (1) \\ -x + 3y &= 6 \quad (2) \end{aligned}$

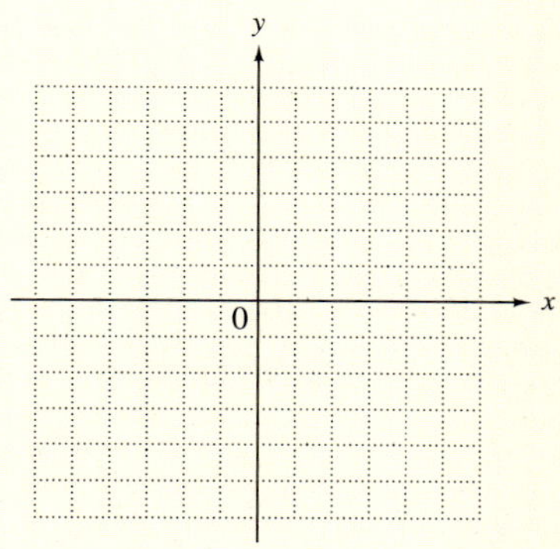

ANSWERS

1. (a) $\{(1, -2)\}$

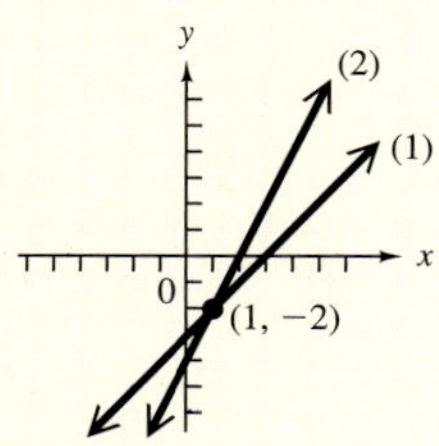

(b) $\{(-3, 1)\}$

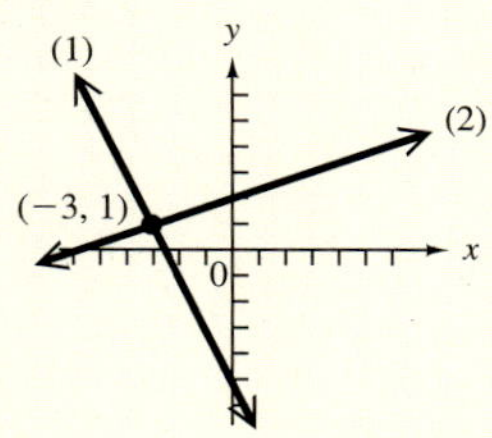

2 Are the given ordered pairs solutions of the given systems?

(a) $\begin{aligned} 2x + y &= -6 \\ x + 3y &= 2 \end{aligned}$; $(-4, 2)$

(b) $\begin{aligned} 9x - y &= -4 \\ 4x + 3y &= 11 \end{aligned}$; $(-1, 5)$

(b) $\begin{aligned} 3x + 2y &= 11 \\ x + 5y &= 36 \end{aligned}$; $(-1, 7)$

Replace x with -1 and y with 7 in each equation of the system.

$3x + 2y = 11$	$x + 5y = 36$
$3(-1) + 2(7) \stackrel{?}{=} 11$	$-1 + 5(7) \stackrel{?}{=} 36$
$-3 + 14 \stackrel{?}{=} 11$	$-1 + 35 \stackrel{?}{=} 36$
$11 = 11$ True	$34 = 36$ False

The ordered pair $(-1, 7)$ is not a solution of the system, since it does not make *both* equations true.

◀ *Work Problem* **2** *at the Side.*

Since the graph of a linear equation is a straight line, there are three possibilities for the solution set of a linear system in two variables.

Graphs of Linear Systems in Two Variables

Case 1 **The two graphs intersect in a single point.** The coordinates of this point give the only solution of the system. Since the system has a solution, it is **consistent.** The equations are *not* equivalent, so they are **independent.** See Figure 3(a).

Case 2 **The graphs are parallel lines.** There is no solution common to both equations, so the solution set is $\emptyset$ and the system is **inconsistent.** Since the equations are *not* equivalent, they are **independent.** See Figure 3(b).

Case 3 **The graphs are the same line.** Since any solution of one equation of the system is a solution of the other, the solution set is an infinite set of ordered pairs representing the points on the line. This type of system is **consistent** because there is a solution. The equations are equivalent, so they are **dependent.** See Figure 3(c).

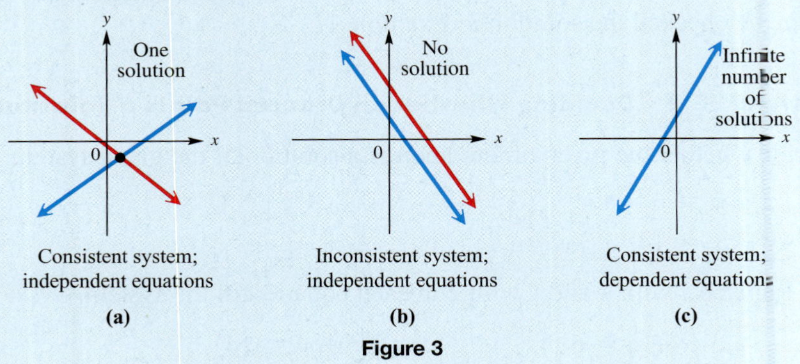

Figure 3

OBJECTIVE 3 Solve linear systems (with two equations and two variables) by substitution. Since it can be difficult to read exact coordinates, especially if they are not integers, from a graph, we usually use algebraic methods to solve systems. One such method, the **substitution method,** is most useful for solving linear systems in which one equation is solved or can be easily solved for one variable in terms of the other.

Answers

2. (a) yes (b) no

EXAMPLE 3 Solving a System by Substitution

Solve the system.

$$2x - y = 6 \quad (1)$$
$$x = y + 2 \quad (2)$$

Since equation (2) is solved for x, substitute $y + 2$ for x in equation (1).

$$\begin{aligned} 2x - y &= 6 && (1) \\ 2(y + 2) - y &= 6 && \text{Let } x = y + 2. \\ 2y + 4 - y &= 6 && \text{Distributive property} \\ y + 4 &= 6 && \text{Combine like terms.} \\ y &= 2 && \text{Subtract 4.} \end{aligned}$$

Be sure to use parentheses here.

We found y. Now find x by substituting 2 for y in equation (2).

$$x = y + 2 = 2 + 2 = 4$$

Write the x-value first in the ordered pair.

Thus, $x = 4$ and $y = 2$, giving the ordered pair (4, 2). Check this solution in both equations of the original system.

Check

$2x - y = 6 \quad (1)$	$x = y + 2 \quad (2)$
$2(4) - 2 \stackrel{?}{=} 6$	$4 \stackrel{?}{=} 2 + 2$
$8 - 2 \stackrel{?}{=} 6$	$4 = 4$ True
$6 = 6$ True	

Since (4, 2) makes both equations true, the solution set is $\{(4, 2)\}$.

CAUTION

Be careful! Even though we found y first in Example 3, ***the x-coordinate is always written first in the ordered-pair solution of a system.***

Work Problem 3 at the Side. ▶

The substitution method is summarized as follows.

Solving a Linear System by Substitution

Step 1 **Solve one of the equations for either variable.** If one of the variable terms has coefficient 1 or -1, choose it, since the substitution method is usually easier this way.

Step 2 **Substitute** for that variable in the other equation. The result should be an equation with just one variable.

Step 3 **Solve** the equation from Step 2.

Step 4 **Find the other value.** Substitute the result from Step 3 into the equation from Step 1 to find the value of the other variable.

Step 5 **Check** the ordered-pair solution in *both* of the *original* equations. Then write the solution set.

---

3 Solve by substitution.

(a) $7x - 2y = -2$
$y = 3x$

(b) $5x - 3y = -6$
$x = 2 - y$

ANSWERS

3. (a) $\{(-2, -6)\}$ **(b)** $\{(0, 2)\}$

4 Solve by substitution.

(a) $3x - y = 10$
$2x + 5y = 1$

(b) $4x - 5y = -11$
$x + 2y = 7$

ANSWERS

4. (a) $\{(3, -1)\}$ **(b)** $\{(1, 3)\}$

EXAMPLE 4 Solving a System by Substitution

Solve the system.

$$3x + 2y = 13 \quad (1)$$
$$4x - y = -1 \quad (2)$$

Step 1 First solve one of the equations for x or y. Since the coefficient of y in equation (2) is -1, it is easiest to solve for y in equation (2).

$$4x - y = -1 \quad (2)$$
$$-y = -1 - 4x \quad \text{Subtract } 4x.$$
$$\mathbf{y = 1 + 4x} \quad \text{Multiply by } -1.$$

Step 2 Substitute $1 + 4x$ for y in equation (1).

$$3x + 2y = 13 \quad (1)$$
$$3x + 2(1 + 4x) = 13 \quad \text{Let } y = 1 + 4x.$$

Step 3 Solve for x.

$$3x + 2 + 8x = 13 \quad \text{Distributive property}$$
$$11x = 11 \quad \text{Combine like terms; subtract 2.}$$
$$\mathbf{x = 1} \quad \text{Divide by 11.}$$

Step 4 Now solve for y. From Step 1, $y = 1 + 4x$, so if $x = 1$, then

$$y = 1 + 4(1) = 5. \quad \text{Let } x = 1.$$

Step 5 Check the solution $(1, 5)$ in both equations (1) and (2).

Check

$3x + 2y = 13$ (1)	$4x - y = -1$ (2)
$3(1) + 2(5) \stackrel{?}{=} 13$	$4(1) - 5 \stackrel{?}{=} -1$
$3 + 10 \stackrel{?}{=} 13$	$4 - 5 \stackrel{?}{=} -1$
$13 = 13$ True	$-1 = -1$ True

The solution set is $\{(1, 5)\}$.

◀ *Work Problem* **4** *at the Side.*

EXAMPLE 5 Solving a System with Fractional Coefficients

Solve the system.

$$\frac{2}{3}x - \frac{1}{2}y = \frac{7}{6} \quad (1)$$
$$3x - y = 6 \quad (2)$$

This system will be easier to solve if we clear the fractions in equation (1).

$$\mathbf{6}\left(\frac{2}{3}x - \frac{1}{2}y\right) = \mathbf{6}\left(\frac{7}{6}\right) \quad \text{Multiply (1) by the LCD, 6.}$$
$$\mathbf{6} \cdot \frac{2}{3}x - \mathbf{6} \cdot \frac{1}{2}y = \mathbf{6} \cdot \frac{7}{6} \quad \text{Distributive property}$$
$$4x - 3y = 7 \quad (3)$$

Remember to multiply *each* term by 6.

Now the system consists of equations (2) and (3).

$$3x - y = 6 \quad (2)$$
$$4x - 3y = 7 \quad (3)$$

This equation is equivalent to equation (1).

Continued on Next Page

To use the substitution method, we solve equation (2) for y.

$$\begin{aligned} 3x - y &= 6 && \text{(2)} \\ -y &= 6 - 3x && \text{Subtract } 3x. \\ y &= 3x - 6 && \text{Multiply by } -1\text{; rewrite.} \end{aligned}$$

Substitute $3x - 6$ for y in equation (3).

$$\begin{aligned} 4x - 3y &= 7 && \text{(3)} \\ 4x - 3(3x - 6) &= 7 && \text{Let } y = 3x - 6. \\ 4x - 9x + 18 &= 7 && \text{Distributive property} \\ -5x + 18 &= 7 && \text{Combine like terms.} \\ -5x &= -11 && \text{Subtract 18.} \\ x &= \frac{11}{5} && \text{Divide by } -5. \end{aligned}$$

Be careful with signs.

Since $y = 3x - 6$ and $x = \frac{11}{5}$,

$$y = 3\left(\frac{11}{5}\right) - 6 = \frac{33}{5} - \frac{30}{5} = \frac{3}{5}. \qquad 6 = \frac{30}{5}$$

A check verifies that the solution set is $\left\{\left(\frac{11}{5}, \frac{3}{5}\right)\right\}$.

Work Problem **5** *at the Side.* ▶

5 Solve by substitution.

(a) $-2x + 5y = 22$

$\frac{1}{2}x + \frac{1}{4}y = \frac{1}{2}$

(b) $\frac{1}{5}x + \frac{2}{3}y = -\frac{8}{5}$

$3x - y = 9$

> **Note**
>
> If an equation in a system contains decimal coefficients, it is best to first clear the decimals by multiplying by 10, 100, or 1000, depending on the number of decimal places. Then solve the system. For example, we multiply *each side* of the equation
>
> $$0.5x + 0.75y = 3.25$$
>
> by 100 to get the equivalent equation
>
> $$50x + 75y = 325.$$

OBJECTIVE 4 Solve linear systems (with two equations and two variables) by elimination. Another algebraic method, the **elimination method,** involves combining the two equations in a system so that one variable is eliminated. This is done using the following logic:

If $a = b$ and $c = d$, then $a + c = b + d$.

EXAMPLE 6 Solving a System by Elimination

Solve the system.

$$\begin{aligned} 2x + 3y &= -6 && \text{(1)} \\ 4x - 3y &= 6 && \text{(2)} \end{aligned}$$

Notice that adding the equations together will eliminate the variable y.

$$\begin{aligned} 2x + 3y &= -6 && \text{(1)} \\ 4x - 3y &= 6 && \text{(2)} \\ \hline 6x \quad &= 0 && \text{Add.} \\ x &= 0 && \text{Solve for } x. \end{aligned}$$

Continued on Next Page

ANSWERS

5. (a) $\{(-1, 4)\}$ (b) $\{(2, -3)\}$

6 Solve by elimination.

(a) $3x - y = -7$
$2x + y = -3$

(b) $-2x + 3y = -10$
$2x + 2y = 5$

To find y, substitute 0 for x in either equation (1) or equation (2).

$$2x + 3y = -6 \quad (1)$$
$$2(0) + 3y = -6 \quad \text{Let } x = 0.$$
$$0 + 3y = -6 \quad \text{Multiply.}$$
$$3y = -6 \quad \text{Add.}$$
$$y = -2 \quad \text{Divide by 3.}$$

The solution of the system is $(0, -2)$. Check by substituting 0 for x and -2 for y in both equations of the original system. The solution set is $\{(0, -2)\}$.

◀ *Work Problem* **6** *at the Side.*

By adding the equations in Example 6, we eliminated the variable y because the coefficients of the y-terms were opposites. In many cases the coefficients will *not* be opposites, and we must transform one or both equations so that the coefficients of one pair of variable terms are opposites.

Solving a Linear System by Elimination

Step 1 **Write both equations in standard form $Ax + By = C$.**

Step 2 **Make the coefficients of one pair of variable terms opposites.** Multiply one or both equations by appropriate numbers so that the sum of the coefficients of either the x- or y-terms is 0.

Step 3 **Add** the new equations to eliminate a variable. The sum should be an equation with just one variable.

Step 4 **Solve** the equation from Step 3 for the remaining variable.

Step 5 **Find the other value.** Substitute the result from Step 4 into either of the original equations and solve for the other variable.

Step 6 **Check** the ordered-pair solution in *both* of the *original* equations. Then write the solution set.

EXAMPLE 7 **Solving a System by Elimination**

Solve the system.

$$5x - 2y = 4 \quad (1)$$
$$2x + 3y = 13 \quad (2)$$

Step 1 Both equations are in standard form.

Step 2 Suppose that you wish to eliminate the variable x. One way to do this is to multiply equation (1) by 2 and equation (2) by -5.

$$10x - 4y = 8 \quad \text{2 times each side of equation (1)}$$
$$-10x - 15y = -65 \quad -5 \text{ times each side of equation (2)}$$

Step 3 Now add.

$$\begin{array}{r} 10x - 4y = 8 \\ \underline{-10x - 15y = -65} \\ -19y = -57 \end{array} \quad \text{Add.}$$

Step 4 Solve for y. $\quad y = 3 \quad$ Divide by -19.

Continued on Next Page

Answers

6. (a) $\{(-2, 1)\}$ (b) $\left\{\left(\frac{7}{2}, -1\right)\right\}$

Step 5 To find x, substitute 3 for y in either equation (1) or (2).

$$\begin{aligned} 2x + 3y &= 13 && (2) \\ 2x + 3(3) &= 13 && \text{Let } y = 3. \\ 2x + 9 &= 13 && \text{Multiply.} \\ 2x &= 4 && \text{Subtract 9.} \\ x &= 2 && \text{Divide by 2.} \end{aligned}$$

Step 6 To check, substitute 2 for x and 3 for y in equations (1) and (2).

Check

$$\begin{aligned} 5x - 2y &= 4 && (1) \\ 5(2) - 2(3) &\stackrel{?}{=} 4 \\ 10 - 6 &\stackrel{?}{=} 4 \\ 4 &= 4 && \text{True} \end{aligned}
\qquad
\begin{aligned} 2x + 3y &= 13 && (2) \\ 2(2) + 3(3) &\stackrel{?}{=} 13 \\ 4 + 9 &\stackrel{?}{=} 13 \\ 13 &= 13 && \text{True} \end{aligned}$$

The solution set is $\{(2, 3)\}$.

Work Problem 7 at the Side. ▶

OBJECTIVE 5 Solve special systems. As we saw in Figures 3(b) and (c), some systems have no solution or an infinite number of solutions.

EXAMPLE 8 Solving a System of Dependent Equations

Solve the system.

$$\begin{aligned} 2x - y &= 3 && (1) \\ 6x - 3y &= 9 && (2) \end{aligned}$$

We multiply equation (1) by -3, and then add the result to equation (2).

$$\begin{aligned} -6x + 3y &= -9 && -3 \text{ times each side of equation (1)} \\ 6x - 3y &= 9 && (2) \\ \hline 0 &= 0 && \text{True} \end{aligned}$$

Adding these equations gives the true statement $0 = 0$. In the original system, we could get equation (2) from equation (1) by multiplying equation (1) by 3. Because of this, equations (1) and (2) are equivalent and have the same graph, as shown in Figure 4. The equations are dependent.

The solution set is the set of all points on the line with equation $2x - y = 3$, written in set-builder notation **(Section 1.1)** as

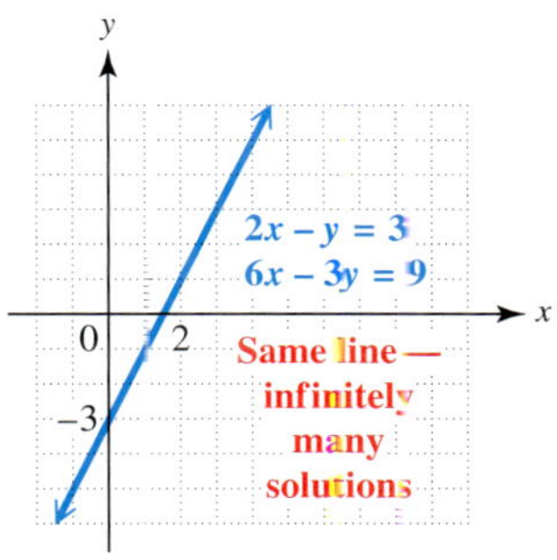

Figure 4

$$\{(x, y) \mid 2x - y = 3\}$$

and read "the set of all ordered pairs (x, y), such that $2x - y = 3$."

Note

When a system has dependent equations and an infinite number of solutions, as in Example 8, either equation of the system or an equivalent equation could be used to write the solution set. ***In this book, we use the equation in standard form with coefficients that are integers having greatest common factor 1 and positive coefficient of x.***

Work Problem 8 at the Side. ▶

7 Solve by elimination.

(a) $x + 3y = 8$
$2x - 5y = -17$

(b) $6x - 2y = -21$
$-3x + 4y = 36$

(c) $2x + 3y = 19$
$3x - 7y = -6$

8 Solve the system. Then graph both equations.

$$\begin{aligned} 2x + y &= 6 && (1) \\ -8x - 4y &= -24 && (2) \end{aligned}$$

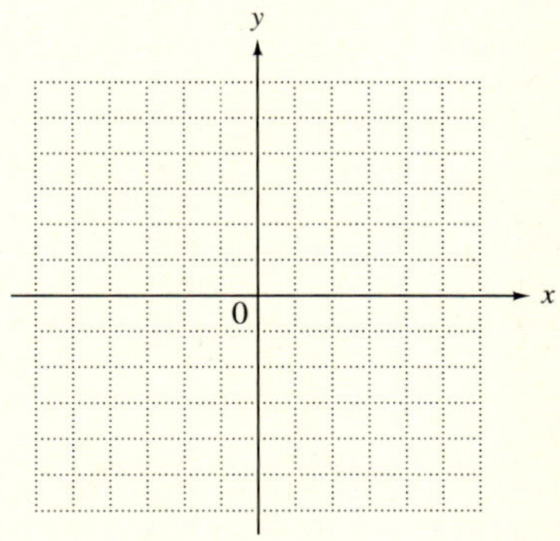

ANSWERS

7. (a) $\{(-1, 3)\}$ **(b)** $\left\{\left(-\frac{2}{3}, \frac{17}{2}\right)\right\}$
(c) $\{(5, 3)\}$

8. $\{(x, y) \mid 2x + y = 6\}$

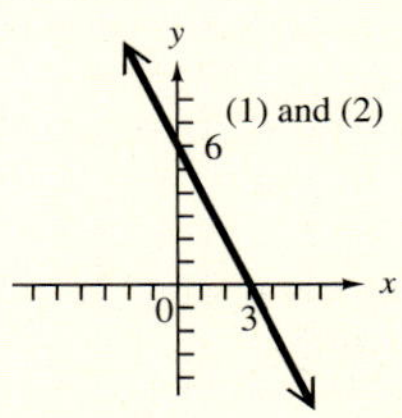

EXAMPLE 9 Solving an Inconsistent System

Solve the system.

$$x + 3y = 4 \quad (1)$$
$$-2x - 6y = 3 \quad (2)$$

Multiply equation (1) by 2, and then add the result to equation (2).

$$\begin{aligned} 2x + 6y &= 8 && \text{Equation (1) multiplied by 2} \\ -2x - 6y &= 3 && (2) \\ \hline 0 &= 11 && \text{False} \end{aligned}$$

The result of the addition step is a false statement, which indicates that the system is inconsistent. As shown in Figure 5, the graphs of the equations of the system are parallel lines. There are no ordered pairs that satisfy both equations, so there is no solution for the system. The solution set is $\emptyset$.

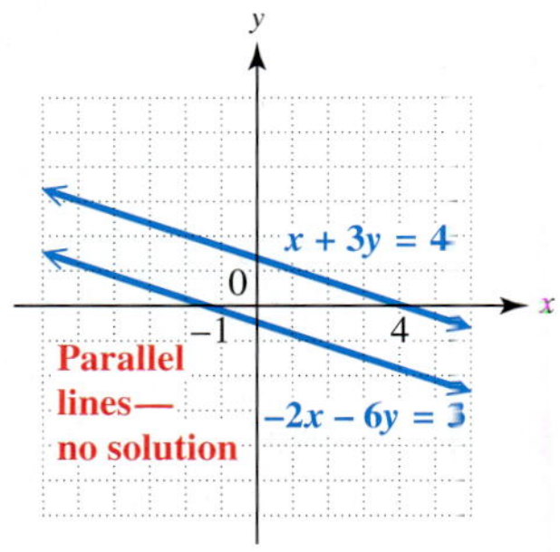

Figure 5

◀ Work Problem 9 at the Side.

The results of Examples 8 and 9 are generalized as follows.

Special Cases of Linear Systems

If both variables are eliminated when a system of linear equations is solved, then

1. there are infinitely many solutions if the resulting statement is *true*;
2. there is no solution if the resulting statement is *false*.

Slopes and y-intercepts can be used to decide if the graphs of a system of equations are parallel lines or if they coincide. In Example 8, writing each equation in slope-intercept form shows that both lines have slope 2 and y-intercept $(0, -3)$, so the graphs are the same line and the system has an infinite number of solutions.

◀ Work Problem 10 at the Side.

In Example 9, both equations have slope $-\frac{1}{3}$, but the y-intercepts are $(0, \frac{4}{3})$ and $(0, -\frac{1}{2})$, showing that the graphs are two distinct parallel lines. Thus, the system has no solution.

◀ Work Problem 11 at the Side.

9 Solve the system. Then graph both equations.

$$4x - 3y = 8 \quad (1)$$
$$8x - 6y = 14 \quad (2)$$

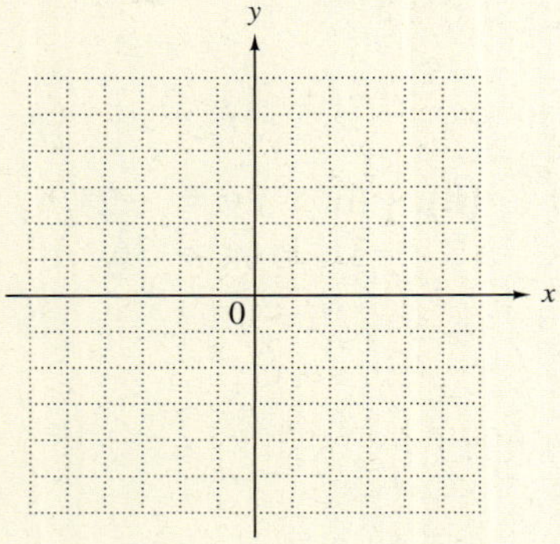

10 Write the equations of Example 8

$$2x - y = 3 \quad (1)$$
$$6x - 3y = 9 \quad (2)$$

in slope-intercept form. Use function notation.

11 Write the equations of Example 9

$$x + 3y = 4 \quad (1)$$
$$-2x - 6y = 3 \quad (2)$$

in slope-intercept form. Use function notation.

ANSWERS

9. $\emptyset$

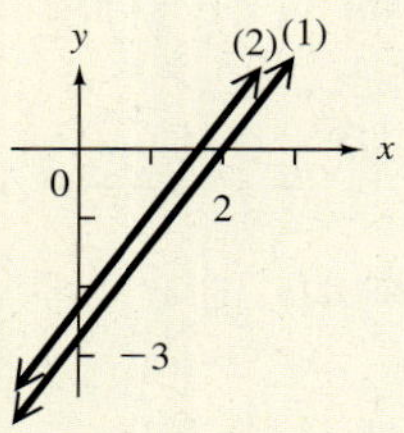

10. Both equations are $f(x) = 2x - 3$.

11. $f(x) = -\frac{1}{3}x + \frac{4}{3}$; $f(x) = -\frac{1}{3}x - \frac{1}{2}$

5.1 Exercises

FOR EXTRA HELP

MyMathLab | Math XL PRACTICE | WATCH | DOWNLOAD | READ | REVIEW

Fill in the blanks with the correct responses.

1. If $(3, -6)$ is a solution of a linear system in two variables, then substituting ________ for x and ________ for y leads to true statements in *both* equations.

2. A solution of a system of independent linear equations in two variables is a(n) ______________.

3. Which ordered pair could possibly be a solution of the graphed system of equations? Why?

A. $(3, 3)$
B. $(-3, 3)$
C. $(-3, -3)$
D. $(3, -3)$

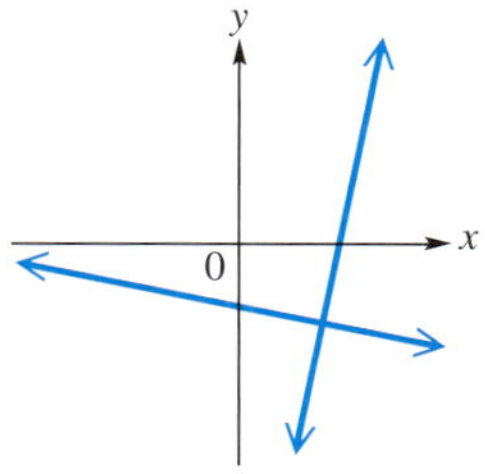

4. Which ordered pair could possibly be a solution of the graphed system of equations? Why?

A. $(3, 0)$
B. $(-3, 0)$
C. $(0, 3)$
D. $(0, -3)$

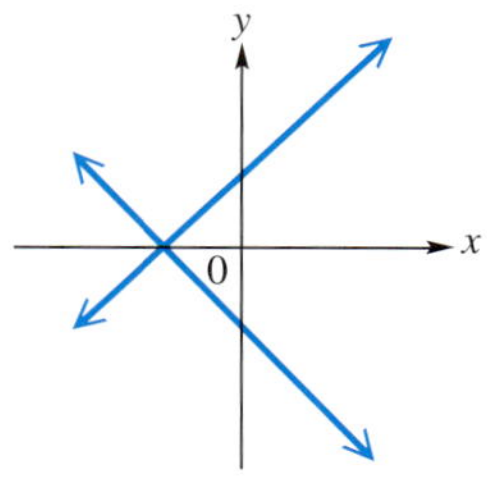

Match each system in Exercises 5–8 with the correct graph in A–D.

5. $x - y = 6$
$x - y = 0$

6. $x + y = -6$
$x - y = 0$

7. $x + y = 0$
$x - y = -6$

8. $x + y = 0$
$x - y = 6$

A.

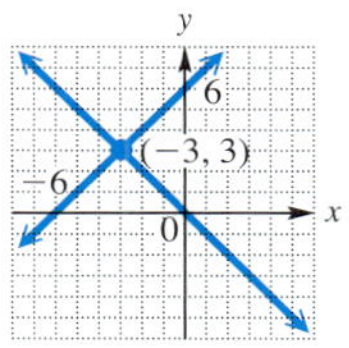

B.

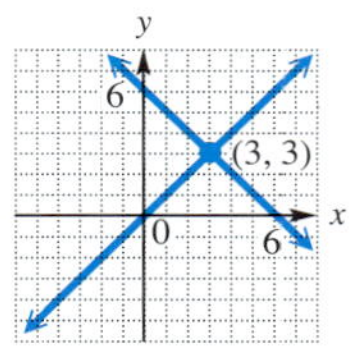

C.

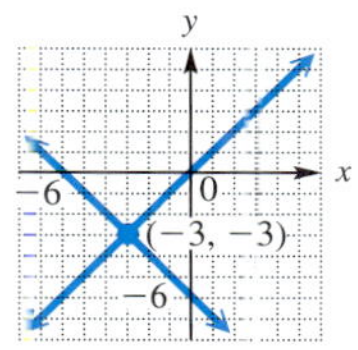

D.

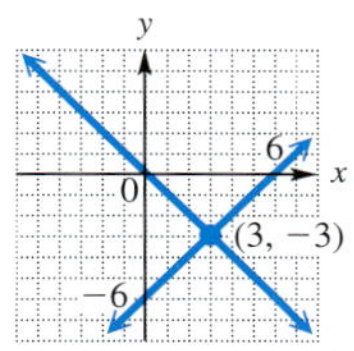

Solve each system by graphing. See Example 1.

9. $x + y = -5$
$-2x + y = 1$

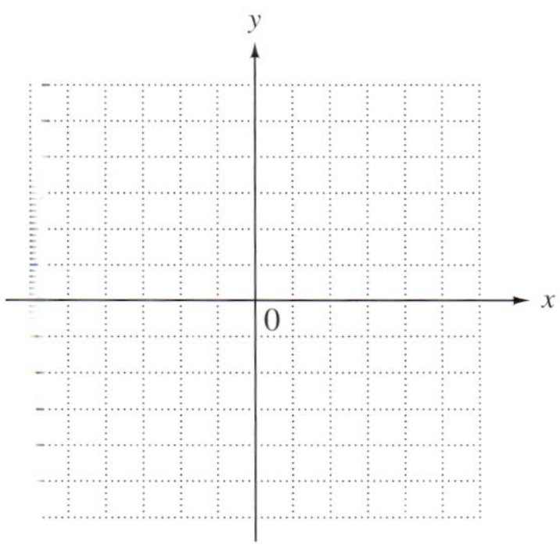

10. $x + y = 4$
$2x - y = 2$

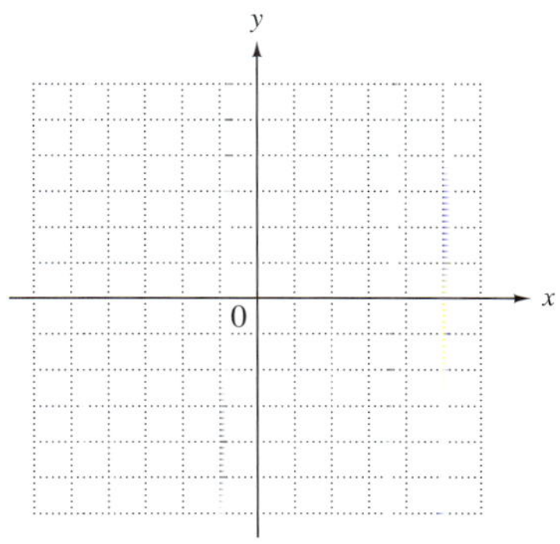

11. $x - 4y = -4$
$3x + y = 1$

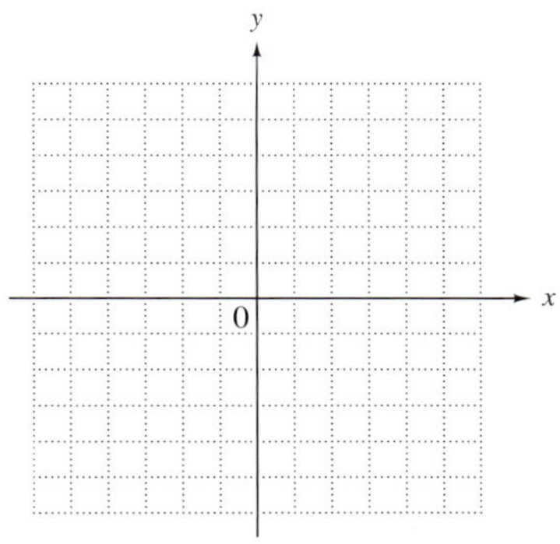

12. $\begin{aligned} 6x - y &= 2 \\ x - 2y &= 4 \end{aligned}$

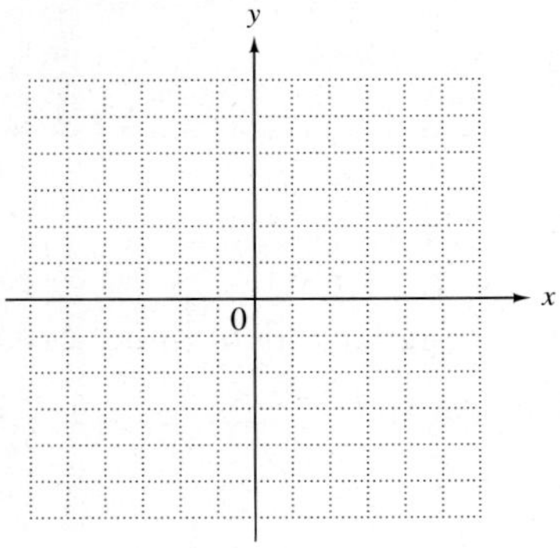

13. $\begin{aligned} 2x + 3y &= -6 \\ x - 3y &= -3 \end{aligned}$

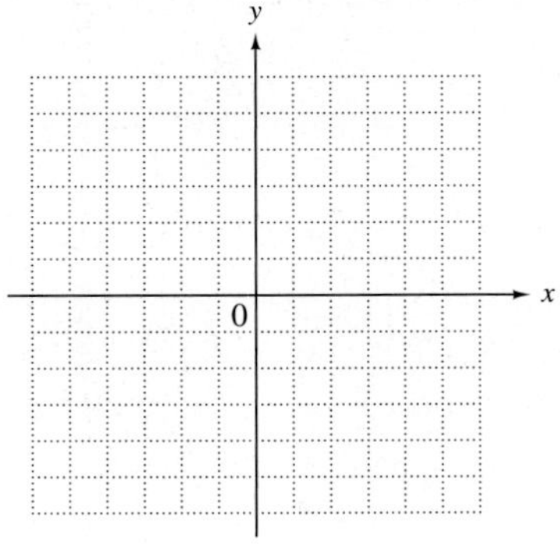

14. $\begin{aligned} 3x + 4y &= 12 \\ x - 4y &= 4 \end{aligned}$

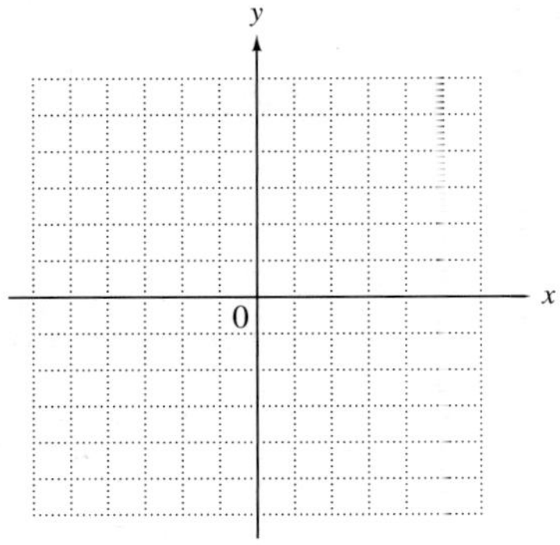

Decide whether the given ordered pair is a solution of the given system. See Example 2.

15. $\begin{aligned} x + y &= 6 \\ x - y &= 4 \end{aligned}$; $(5, 1)$

16. $\begin{aligned} x - y &= 17 \\ x + y &= -1 \end{aligned}$; $(8, -9)$

17. $\begin{aligned} 2x - y &= 8 \\ 3x + 2y &= 20 \end{aligned}$; $(5, 2)$

18. $\begin{aligned} 3x - 5y &= -12 \\ x - y &= 1 \end{aligned}$; $(-1, 2)$

19. $\begin{aligned} 4x + 3y &= -1 \\ -2x + 5y &= 3 \end{aligned}$; $(-1, 1)$

20. $\begin{aligned} 3x - 5y &= 7 \\ 2x + 3y &= 30 \end{aligned}$; $(9, 4)$

Solve each system by substitution. If the system is inconsistent or has dependent equations, say so. See Examples 3–5, 8, and 9.

21. $\begin{aligned} 4x + y &= 6 \\ y &= 2x \end{aligned}$

22. $\begin{aligned} 2x - y &= 6 \\ y &= 5x \end{aligned}$

23. $\begin{aligned} -x - 4y &= -14 \\ y &= 2x - 1 \end{aligned}$

24. $\begin{aligned} -3x - 5y &= -17 \\ y &= 4x + 8 \end{aligned}$

25. $\begin{aligned} 3x - 4y &= -22 \\ -3x + y &= 0 \end{aligned}$

26. $\begin{aligned} -3x + y &= -5 \\ x + 2y &= 0 \end{aligned}$

27. $\begin{aligned} 5x - 4y &= 9 \\ 3 - 2y &= -x \end{aligned}$

28. $\begin{aligned} 6x - y &= -9 \\ 4 + 7x &= -y \end{aligned}$

29. $\begin{aligned} x &= 3y + 5 \\ x &= \frac{3}{2}y \end{aligned}$

30. $\begin{aligned} x &= 6y - 2 \\ x &= \frac{3}{4}y \end{aligned}$

31. $\begin{aligned} \frac{1}{2}x + \frac{1}{3}y &= 3 \\ -3x + y &= 0 \end{aligned}$

32. $\begin{aligned} \frac{1}{4}x - \frac{1}{5}y &= 9 \\ 5x - y &= 0 \end{aligned}$

33. $y = 2x$
$4x - 2y = 0$

34. $x = 3y$
$3x - 9y = 0$

35. $5x - 25y = 5$
$x = 5y$

36. $8x + 2y = 4$
$y = -4x$

Solve each system by elimination. If the system is inconsistent or has dependent equations, say so. See Examples 6–9.

37. $-2x + 3y = -16$
$2x - 5y = 24$

38. $6x + 5y = -7$
$-6x - 11y = 1$

39. $2x - 5y = 11$
$3x + y = 8$

40. $-2x + 3y = 1$
$-4x + y = -3$

41. $3x + 4y = -6$
$5x + 3y = 1$

42. $4x + 3y = 1$
$3x + 2y = 2$

43. $3x + 3y = 0$
$4x + 2y = 3$

44. $8x + 4y = 0$
$4x - 2y = 2$

45. $7x + 2y = 6$
$-14x - 4y = -12$

46. $x - 4y = 2$
$4x - 16y = 8$

47. $\frac{x}{2} + \frac{y}{3} = -\frac{1}{3}$
$\frac{x}{2} + 2y = -7$

48. $\frac{x}{5} + y = \frac{6}{5}$
$\frac{x}{10} + \frac{y}{3} = \frac{5}{6}$

49. $5x - 5y = 3$
$x - y = 12$

50. $2x - 3y = 7$
$-4x + 6y = 14$

Write each equation in slope-intercept form, and then tell how many solutions the system has. Do not actually solve.

51. $3x + 7y = 4$
$6x + 14y = 3$

52. $-x + 2y = 8$
$4x - 8y = 1$

53. $2x = -3y + 1$
$6x = -9y + 3$

54. $5x = -2y + 1$
$10x = -4y + 2$

55. Assuming you want to minimize the amount of work required, tell whether you would use the substitution or elimination method to solve each system. Explain your answers. *Do not actually solve.*

(a) $6x - y = 5$
$y = 11x$

(b) $3x + y = -7$
$x - y = -5$

(c) $3x - 2y = 0$
$9x + 8y = 7$

Solve each system by the method of your choice. (For Exercises 56–58, see your answers for Exercise 55.)

56. $6x - y = 5$
$y = 11x$

57. $3x + y = -7$
$x - y = -5$

58. $3x - 2y = 0$
$9x + 8y = 7$

59. $2x + 3y = 10$
$-3x + y = 18$

60. $3x - 5y = 7$
$2x + 3y = 30$

61. $\frac{1}{2}x - \frac{1}{8}y = -\frac{1}{4}$
$-4x + y = 2$

62. $\frac{1}{6}x + \frac{1}{3}y = 8$
$\frac{1}{4}x + \frac{1}{2}y = 12$

63. $0.3x + 0.2y = 0.4$
$0.5x + 0.4y = 0.7$

64. $0.2x + 0.5y = 6$
$0.4x + y = 9$

Answer the questions in Exercises 65–68 by observing the graphs provided.

65. The figure shows graphs that represent supply and demand for a certain brand of low-fat frozen yogurt at various prices per half-gallon (in dollars).

(a) At what price does supply equal demand?

(b) For how many half-gallons does supply equal demand?

(c) What are the supply and demand at a price of \$2 per half-gallon?

66. La Bronda Jones compared the monthly payments she would incur for two types of mortgages: fixed-rate and variable-rate. Her observations led to the following graphs.

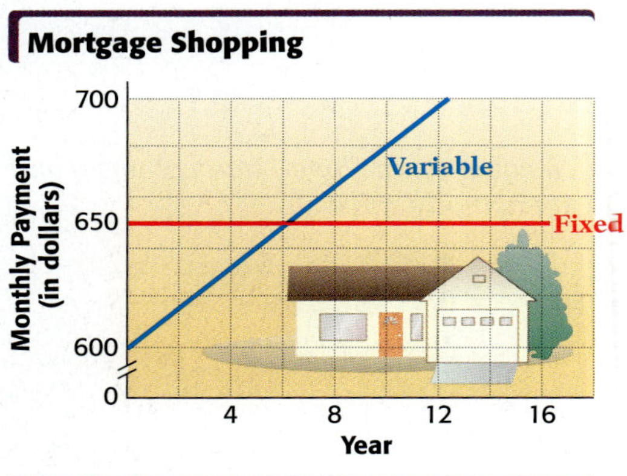

(a) For which years would the monthly payment be more for the fixed-rate mortgage than for the variable-rate mortgage?

(b) In what year would the payments be the same, and what would those payments be?

67. If the rates of growth between 1990 and 2000 continue, the populations of Houston, Phoenix, Dallas, and Philadelphia will follow the trends indicated in the graph.

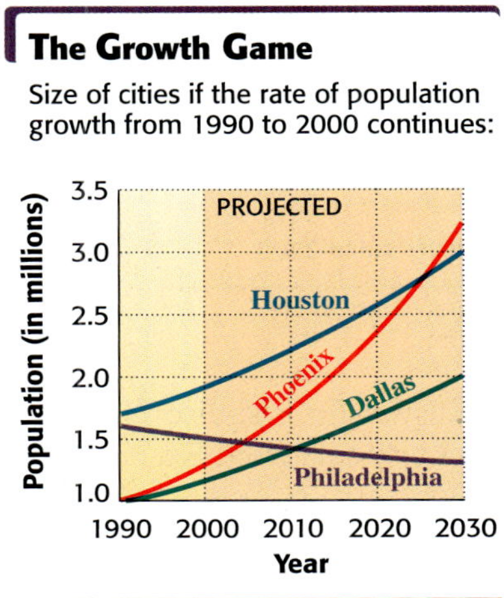

Source: U.S. Census Bureau, *Chronicle* research.

(a) Which cities will experience population growth?

(b) Which city will experience population decline?

(c) Rank the city populations from least to greatest for the year 2020.

(d) In which year will the population of Dallas equal that of Philadephia? About what will this population be?

(e) Write as an ordered pair (year, population in millions) the point at which Houston and Phoenix will have the same population.

68. The graph shows network share (the percentage of TV sets in use) for the early evening news programs for three major broadcast networks from 1994 through 2004.

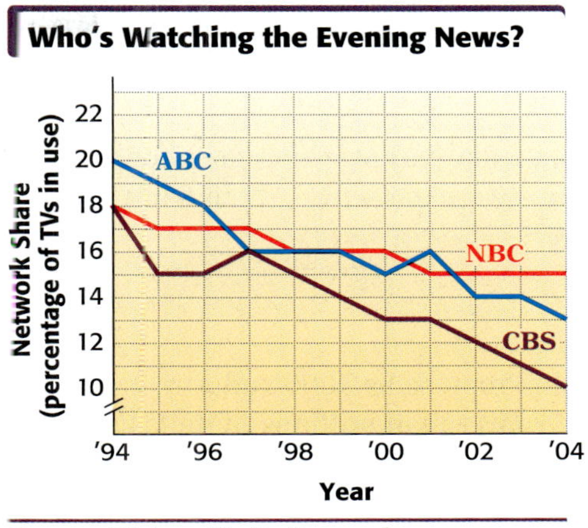

Source: Nielsen Media Research.

(a) Between what years did the ABC early evening news dominate?

(b) During what year did ABC's dominance end? Which network equaled ABC's share that year? What was that share?

(c) During what years did ABC and NBC have equal network share? What was the share for each of those years?

(d) Find the first year on the graph in which two networks had equal share. Which networks were these? Write their share as an ordered pair of the form (year, share).

(e) Describe the general trend in viewership for the three major networks during the years shown.

Use the graph given in Figure 1 at the beginning of this section (repeated here) to work Exercises 69–72.

69. For which years during the period 2000–2004 were sales of digital cameras less than sales of conventional cameras?

70. Estimate the year in which sales for the two types of cameras were the same. About what was this sales figure?

Say Cheese!

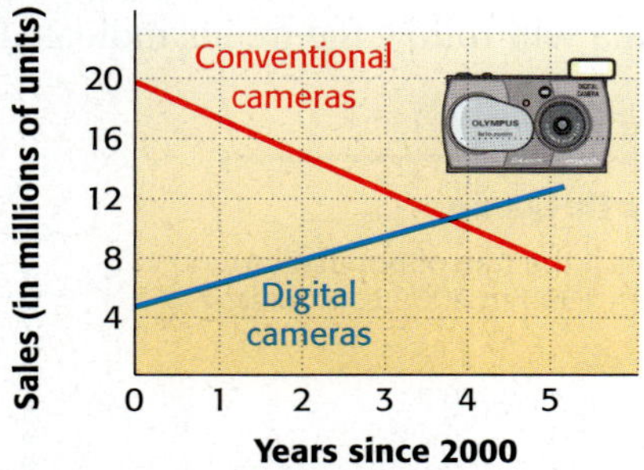

Source: Consumer Electronics Association.

71. If $x = 0$ represents 2000 and $x = 4$ represents 2004, sales (y) in millions of units can be modeled by the linear equations in the following system.

$$2.5x + y = 19.4 \quad \text{Conventional cameras}$$
$$-1.7x + y = 4.4 \quad \text{Digital cameras}$$

Solve this system. Express values as decimals rounded to the nearest tenth. Write the solution as an ordered pair of the form (year, sales).

72. Interpret your answer for Exercise 71. How does it compare to your estimate from Exercise 70?

Relating Concepts (Exercises 73–76) For Individual or Group Work

Work Exercises 73–76 in order *to see the connections between systems of linear equations and the graphs of linear functions.*

73. Use elimination or substitution to solve the system.

$$3x + y = 6 \quad (1)$$
$$-2x + 3y = 7 \quad (2)$$

74. For equation (1) in the system of Exercise 73, solve for y and rename it $f(x)$. What special kind of function is f?

75. For equation (2) in the system of Exercise 73, solve for y and rename it $g(x)$. What special kind of function is g?

76. Use the result of Exercise 73 to fill in the blanks with the appropriate responses:

Because the graphs of f and g are straight lines that are neither parallel nor coincide, they intersect in exactly _______ point. The coordinates of the point are (_______, _______). Using function notation, this is given by f(_______) = _______ and g(_______) = _______.

5.2 Systems of Linear Equations in Three Variables

OBJECTIVES

1. Understand the geometry of systems of three equations in three variables.
2. Solve linear systems (with three equations and three variables) by elimination.
3. Solve linear systems (with three equations and three variables) in which some of the equations have missing terms.
4. Solve special systems.

A solution of an equation in three variables, such as

$$2x + 3y - z = 4, \quad \text{Linear equation in three variables}$$

is called an **ordered triple** and is written (x, y, z). For example, the ordered triple $(0, 1, -1)$ is a solution of the equation, because

$$2(0) + 3(1) - (-1) = 4$$

is a true statement. Verify that another solution of this equation is $(10, -3, 7)$.

We now extend the term *linear equation* to equations of the form

$$Ax + By + Cz + \dots + Dw = K,$$

where not all the coefficients $A, B, C, \dots, D$ equal 0. For example,

$$2x + 3y - 5z = 7 \quad \text{and} \quad x - 2y - z + 3u - 2w = 8$$

are linear equations, the first with three variables and the second with five.

OBJECTIVE 1 Understand the geometry of systems of three equations in three variables. Consider the solution of a system such as

$$\begin{aligned} 4x + 8y + z &= 2 \\ x + 7y - 3z &= -14 \\ 2x - 3y + 2z &= 3. \end{aligned} \quad \text{System of linear equations in three variables}$$

Theoretically, a system of this type can be solved by graphing. However, the graph of a linear equation with three variables is a *plane*, not a line. Since the graph of each equation of the system is a plane, which requires three-dimensional graphing, the graphing method is not practical for solving such systems. However, it does illustrate the number of solutions possible for these systems, as shown in Figure 6.

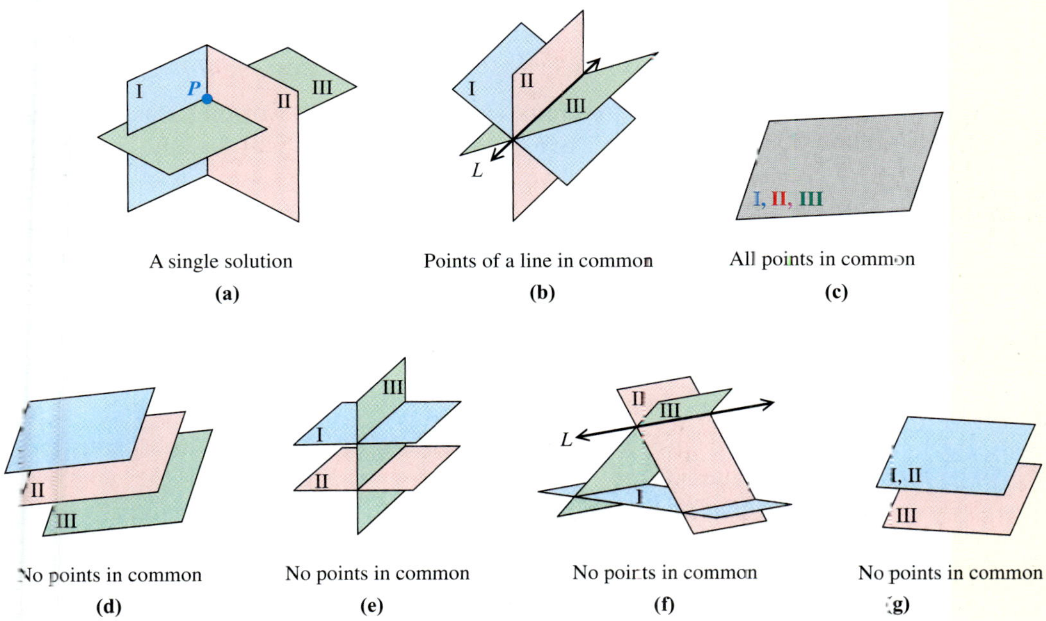

Figure 6

Figure 6 on the preceding page illustrates the following cases.

Graphs of Linear Systems in Three Variables

Case 1 **The three planes may meet at a single, common point** that forms the solution set of the system. See Figure 6(a).

Case 2 **The three planes may have the points of a line in common** so that the infinite set of points that satisfy the equation of the line forms the solution set of the system. See Figure 6(b).

Case 3 **The three planes may coincide** so that the solution set of the system is the set of all points on a plane. See Figure 6(c)

Case 4 **The planes may have no points common to all three** so that there is no solution of the system; the solution set is $\emptyset$. See Figures 6(d)–(g).

OBJECTIVE 2 Solve linear systems (with three equations and three variables) by elimination. Is it possible to solve a system of three equations in three variables such as the one that follows?

$$\begin{aligned} 4x + 8y + z &= 2 \\ x + 7y - 3z &= -14 \\ 2x - 3y + 2z &= 3 \end{aligned}$$

As mentioned, graphing to find the solution set of such a system is impractical. We solve these systems with an extension of the elimination method from **Section 5.1,** summarized as follows.

Solving a Linear System in Three Variables*

Step 1 **Select a variable and an equation.** A good choice for the variable, which we call the *focus variable,* is one that has coefficient 1 or -1. Then select an equation, usually the one that contains the focus variable, as the *working equation*.

Step 2 **Eliminate the focus variable.** Use the working equation and one of the other two equations of the original system. The result is an equation in two variables.

Step 3 **Eliminate the focus variable again.** Use the working equation and the remaining equation of the original system. The result is another equation in two variables.

Step 4 **Write the equations in two variables that result from Steps 2 and 3 as a system, and solve it.** Doing this gives the values of two of the variables.

Step 5 **Find the value of the remaining variable.** Substitute the values of the two variables found in Step 4 into the working equation to obtain the value of the focus variable.

Step 6 **Check** the ordered-triple solution in *each* of the *original* equations of the system. Then write the solution set.

*The authors wish to thank Christine Heinecke Lehmann of Purdue University North Central for her suggestions here.

EXAMPLE 1 Solving a System in Three Variables

Solve the system.

$$\begin{aligned} 4x + 8y + z &= 2 && (1) \\ x + 7y - 3z &= -14 && (2) \\ 2x - 3y + 2z &= 3 && (3) \end{aligned}$$

Step 1 Since z in equation (1) has coefficient 1, we choose z as the focus variable and (1) as the working equation. (Another option would be to choose x as the focus variable, since it also has coefficient 1, and use (2) as the working equation.)

$$4x + 8y + \underset{\text{Focus variable}}{z} = 2 \qquad (1) \leftarrow \text{Working equation}$$

Step 2 Multiply working equation (1) by 3 and add the result to equation (2) to eliminate focus variable z.

$$\begin{aligned} 12x + 24y + 3z &= 6 && \text{Multiply each side of (1) by 3.} \\ x + 7y - 3z &= -14 && (2) \\ \hline 13x + 31y &= -8 && \text{Add.} \quad (4) \end{aligned}$$

Step 3 Multiply working equation (1) by -2 and add the result to remaining equation (3) to again eliminate focus variable z.

$$\begin{aligned} -8x - 16y - 2z &= -4 && \text{Multiply each side of (1) by } -2. \\ 2x - 3y + 2z &= 3 && (3) \\ \hline -6x - 19y &= -1 && \text{Add.} \quad (5) \end{aligned}$$

Step 4 Write the equations in two variables that result in Steps 2 and 3 as a system.

Make sure these equations have the same variables.

$$\begin{aligned} 13x + 31y &= -8 && (4) \quad \text{The result from Step 2} \\ -6x - 19y &= -1 && (5) \quad \text{The result from Step 3} \end{aligned}$$

Now solve this system. We choose to eliminate x.

$$\begin{aligned} 78x + 186y &= -48 && \text{Multiply each side of (4) by 6.} \\ -78x - 247y &= -13 && \text{Multiply each side of (5) by 13.} \\ \hline -61y &= -61 && \text{Add.} \\ y &= 1 && \text{Divide by } -61. \end{aligned}$$

Substitute 1 for y in either equation (4) or (5) to find x.

$$\begin{aligned} -6x - 19y &= -1 && (5) \\ -6x - 19(1) &= -1 && \text{Let } y = 1. \\ -6x - 19 &= -1 && \text{Multiply.} \\ -6x &= 18 && \text{Add 19.} \\ x &= -3 && \text{Divide by } -6. \end{aligned}$$

Step 5 Now substitute the two values we found in Step 4 in working equation (1) to find the value of the remaining variable, focus variable z.

$$\begin{aligned} 4x + 8y + z &= 2 && (1) \\ 4(-3) + 8(1) + z &= 2 && \text{Let } x = -3 \text{ and } y = 1. \\ -4 + z &= 2 && \text{Multiply; add.} \\ z &= 6 && \text{Add 4.} \end{aligned}$$

Continued on Next Page

Step 6 It appears that the ordered triple $(-3, 1, 6)$ is the only solution of the system. We must check that the solution satisfies all three original equations of the system. For equation (1),

Check

$$4x + 8y + z = 2 \quad (1)$$
$$4(-3) + 8(1) + 6 \stackrel{?}{=} 2$$
$$-12 + 8 + 6 \stackrel{?}{=} 2$$
$$2 = 2. \quad \text{True}$$

◀ Work Problem 1 at the Side.

Because $(-3, 1, 6)$ also satisfies equations (2) and (3), the solution set is $\{(-3, 1, 6)\}$.

◀ Work Problem 2 at the Side.

OBJECTIVE 3 Solve linear systems (with three equations and three variables) in which some of the equations have missing terms. If a linear system has an equation missing a term or terms, one elimination step can be omitted.

EXAMPLE 2 Solving a System of Equations with Missing Terms

Solve the system.

$$6x - 12y = -5 \quad (1) \quad \text{Missing } z$$
$$8y + z = 0 \quad (2) \quad \text{Missing } x$$
$$9x - z = 12 \quad (3) \quad \text{Missing } y$$

Since equation (3) is missing the variable y, one way to begin is to eliminate y again using equations (1) and (2).

$$12x - 24y = -10 \quad \text{Multiply each side of (1) by 2.}$$
$$24y + 3z = 0 \quad \text{Multiply each side of (2) by 3.}$$
$$12x + 3z = -10 \quad \text{Add. (4)}$$

Leave space for the missing terms.

Use the resulting equation (4) in x and z, together with equation (3), $9x - z = 12$, to eliminate z. Multiply equation (3) by 3.

$$27x - 3z = 36 \quad \text{Multiply each side of (3) by 3.}$$
$$12x + 3z = -10 \quad (4)$$
$$39x = 26 \quad \text{Add.}$$
$$x = \frac{26}{39} = \frac{2}{3} \quad \text{Divide by 39; lowest terms}$$

We can find z by substituting this value for x into equation (3).

$$9x - z = 12 \quad (3)$$
$$9\left(\frac{2}{3}\right) - z = 12 \quad \text{Let } x = \tfrac{2}{3}.$$
$$6 - z = 12 \quad \text{Multiply.}$$
$$z = -6 \quad \text{Subtract 6; multiply by } -1.$$

Continued on Next Page

1 Check that the solution $(-3, 1, 6)$ satisfies equations (2) and (3) of Example 1.

(a) $x + 7y - 3z = -14 \quad (2)$

Does the solution satisfy equation (2)?

(b) $2x - 3y + 2z = 3 \quad (3)$

Does the solution satisfy equation (3)?

2 Solve each system.

(a)
$$x + y + z = 2$$
$$x - y + 2z = 2$$
$$-x + 2y - z = 1$$

(b)
$$2x + y + z = 9$$
$$-x - y + z = 1$$
$$3x - y + z = 9$$

Answers

1. (a) yes (b) yes
2. (a) $\{(-1, 1, 2)\}$ (b) $\{(2, 1, 4)\}$

We can find y by substituting -6 for z in equation (2).

$$\begin{aligned} 8y + z &= 0 && (2) \\ 8y - 6 &= 0 && \text{Let } z = -6. \\ 8y &= 6 && \text{Add 6.} \\ y &= \frac{6}{8} = \frac{3}{4} && \text{Divide by 8; lowest terms} \end{aligned}$$

Thus, $x = \frac{2}{3}$, $y = \frac{3}{4}$, and $z = -6$. Check these values in each of the original equations of the system to verify that the solution set of the system is $\left\{\left(\frac{2}{3}, \frac{3}{4}, -6\right)\right\}$.

Note

Another way to solve the system in Example 2 is to begin by eliminating the variable z from equations (2) and (3). The resulting equation together with equation (1) forms a system of two equations in the variables x and y. Try working Example 2 this way to see that the same solution results.

There are often multiple ways to solve a system of equations. Some ways may involve more work than others.

Work Problem 3 at the Side. ▶

OBJECTIVE 4 Solve special systems. Linear systems with three variables may be inconsistent or may include dependent equations.

EXAMPLE 3 Solving an Inconsistent System with Three Variables

Solve the system.

$$\begin{aligned} 2x - 4y + 6z &= 5 && (1) \\ -x + 3y - 2z &= -1 && (2) \\ x - 2y + 3z &= 1 && (3) \end{aligned}$$

Use as the working equation, with focus variable x.

Eliminate x by adding equations (2) and (3) to get the equation

$$y + z = 0.$$

Eliminate x again, using equations (1) and (3).

$$\begin{aligned} -2x + 4y - 6z &= -2 && \text{Multiply each side of (3) by } -2. \\ 2x - 4y + 6z &= 5 && (1) \\ \hline 0 &= 3 && \text{False} \end{aligned}$$

This false statement indicates that equations (1) and (3) have no common solution. Thus, the system is inconsistent and the solution set is $\emptyset$. The graph of this system would show these two planes parallel to one another.

Note

If a false statement results when adding as in Example 3, it is not necessary to go any further with the solution. Since two of the three planes are parallel, it is not possible for the three planes to have any common points.

Work Problem 4 at the Side. ▶

3 Solve each system.

(a) $\begin{aligned} x - y &= 6 \\ 2y + 5z &= 1 \\ 3x - 4z &= 8 \end{aligned}$

(b) $\begin{aligned} 5x - y &= 26 \\ 4y + 3z &= -4 \\ x + z &= 5 \end{aligned}$

4 Solve each system.

(a) $\begin{aligned} 3x - 5y + 2z &= 1 \\ 5x + 8y - z &= 4 \\ -6x + 10y - 4z &= 5 \end{aligned}$

(b) $\begin{aligned} 7x - 9y + 2z &= 0 \\ y + z &= 0 \\ 8x - z &= 0 \end{aligned}$

ANSWERS

3. (a) $\{(4, -2, 1)\}$ **(b)** $\{(5, -1, 0)\}$

4. (a) $\emptyset$ **(b)** $\{(0, 0, 0)\}$

5 Solve the system.

$$\begin{aligned} x - y + z &= 4 \\ -3x + 3y - 3z &= -12 \\ 2x - 2y + 2z &= 8 \end{aligned}$$

6 Solve the system.

$$\begin{aligned} 2x + 3y - z &= 8 \\ \frac{1}{2}x + \frac{3}{4}y - \frac{1}{4}z &= 2 \\ x + \frac{3}{2}y - \frac{1}{2}z &= -6 \end{aligned}$$

EXAMPLE 4 **Solving a System of Dependent Equations with Three Variables**

Solve the system.

$$\begin{aligned} 2x - 3y + 4z &= 8 && (1) \\ -x + \frac{3}{2}y - 2z &= -4 && (2) \\ 6x - 9y + 12z &= 24 && (3) \end{aligned}$$

Multiplying each side of equation (1) by 3 gives equation (3). Multiplying each side of equation (2) by -6 also gives equation (3). Because of this, the equations are dependent. All three equations have the same graph, as illustrated in Figure 6(c). The solution set is written

$$\{(x, y, z) \mid 2x - 3y + 4z = 8\}. \qquad \text{Set-builder notation}$$

Although any one of the three equations could be used to write the solution set, we use the equation with coefficients that are integers with greatest common factor 1, as we did in **Section 5.1.**

◀ *Work Problem* **5** *at the Side.*

EXAMPLE 5 **Solving Another Special System**

Solve the system.

$$\begin{aligned} 2x - y + 3z &= 6 && (1) \\ x - \frac{1}{2}y + \frac{3}{2}z &= 3 && (2) \\ 4x - 2y + 6z &= 1 && (3) \end{aligned}$$

Multiplying each side of equation (2) by 2 gives equation (1). Thus, these two equations are dependent.

Equations (1) and (3) are not equivalent, however. Multiplying equation (3) by $\frac{1}{2}$ gives

$$2x - y + 3z = \frac{1}{2}, \qquad (3) \text{ multiplied by } \tfrac{1}{2}$$

which is *not* equivalent to equation (1). Instead, we obtain two equations with the same coefficients, but with different constant terms. The graphs of equations (1) and (3) have no points in common (that is, the planes are parallel). Thus, the system is inconsistent and the solution set is $\emptyset$, as illustrated in Figure 6(g).

◀ *Work Problem* **6** *at the Side.*

ANSWERS

5. $\{(x, y, z) \mid x - y + z = 4\}$
6. $\emptyset$

5.2 Exercises

1. Explain what the following statement means: The solution set of the system

$$\begin{aligned} 2x + y + z &= 3 \\ 3x - y + z &= -2 \\ 4x - y + 2z &= 0 \end{aligned}$$

is $\{(-1, 2, 3)\}$.

2. The two equations

$$\begin{aligned} x + y + z &= 6 \\ 2x - y + z &= 3 \end{aligned}$$

have a common solution of $(1, 2, 3)$. Which equation would complete a system of three linear equations in three variables having solution set $\{(1, 2, 3)\}$?

A. $3x + 2y - z = 1$ **B.** $3x + 2y - z = 4$

C. $3x + 2y - z = 5$ **D.** $3x + 2y - z = 6$

Solve each system of equations. See Example 1.

3. $\begin{aligned} 2x - 5y + 3z &= -1 \\ x + 4y - 2z &= 9 \\ x - 2y - 4z &= -5 \end{aligned}$

4. $\begin{aligned} x + 3y - 6z &= 7 \\ 2x - y + z &= 1 \\ x + 2y + 2z &= -1 \end{aligned}$

5. $\begin{aligned} 3x + 2y + z &= 8 \\ 2x - 3y + 2z &= -16 \\ x + 4y - z &= 20 \end{aligned}$

6. $\begin{aligned} -3x + y - z &= -10 \\ -4x + 2y + 3z &= -1 \\ 2x + 3y - 2z &= -5 \end{aligned}$

7. $\begin{aligned} x + 2y + z &= 4 \\ 2x + y - z &= -1 \\ x - y - z &= -2 \end{aligned}$

8. $\begin{aligned} x - 2y + 5z &= -7 \\ -2x - 3y + 4z &= -14 \\ -3x + 5y - z &= -7 \end{aligned}$

9. $\begin{aligned} -x + 2y + 6z &= 2 \\ 3x + 2y + 6z &= 6 \\ x + 4y - 3z &= 1 \end{aligned}$

10. $\begin{aligned} 2x + y + 2z &= 1 \\ x + 2y + z &= 2 \\ x - y - z &= 0 \end{aligned}$

11. $$\begin{aligned} 2x + 5y + 2z &= 0 \\ 4x - 7y - 3z &= 1 \\ 3x - 8y - 2z &= -6 \end{aligned}$$

12. $$\begin{aligned} 5x - 2y + 3z &= -9 \\ 4x + 3y + 5z &= 4 \\ 2x + 4y - 2z &= 14 \end{aligned}$$

13. $$\begin{aligned} x + 2y + 3z &= 1 \\ -x - y + 3z &= 2 \\ -6x + y + z &= -2 \end{aligned}$$

14. $$\begin{aligned} x + y - z &= -2 \\ 2x - y + z &= -5 \\ -x + 2y - 3z &= -4 \end{aligned}$$

Solve each system of equations. See Example 2.

15. $$\begin{aligned} 2x - 3y + 2z &= -1 \\ x + 2y + z &= 17 \\ 2y - z &= 7 \end{aligned}$$

16. $$\begin{aligned} 2x - y + 3z &= 6 \\ x + 2y - z &= 8 \\ 2y + z &= 1 \end{aligned}$$

17. $$\begin{aligned} 4x + 2y - 3z &= 6 \\ x - 4y + z &= -4 \\ -x + 2z &= 2 \end{aligned}$$

18. $$\begin{aligned} 2x + 3y - 4z &= 4 \\ x - 6y + z &= -16 \\ -x + 3z &= 8 \end{aligned}$$

19. $$\begin{aligned} -5x + 2y + z &= 5 \\ -3x - 2y - z &= 3 \\ -x + 6y &= 1 \end{aligned}$$

20. $$\begin{aligned} x + y - z &= 0 \\ 2y - z &= 1 \\ 2x + 3y - 4z &= -4 \end{aligned}$$

21. $$\begin{aligned} 2x + y &= 6 \\ 3y - 2z &= -4 \\ 3x - 5z &= -7 \end{aligned}$$

22. $$\begin{aligned} 4x - 8y &= -7 \\ 4y + z &= 7 \\ -8x + z &= -4 \end{aligned}$$

23. Using your immediate surroundings, give an example of three planes that

(a) intersect in a single point;

(b) do not intersect;

(c) intersect in infinitely many points.

24. Suppose that a system has infinitely many ordered triple solutions of the form (x, y, z) such that

$$x + y + 2z = 1.$$

Give three specific ordered triples that are solutions of the system.

Solve each system of equations. If the system is inconsistent or has dependent equations, say so. See Examples 1–5.

25.
$$\begin{aligned} 2x + 2y - 6z &= 5 \\ -3x + y - z &= -2 \\ -x - y + 3z &= 4 \end{aligned}$$

26.
$$\begin{aligned} -2x + 5y + z &= -3 \\ 5x + 14y - z &= -11 \\ 7x + 9y - 2z &= -5 \end{aligned}$$

27.
$$\begin{aligned} -5x + 5y - 20z &= -40 \\ x - y + 4z &= 8 \\ 3x - 3y + 12z &= 24 \end{aligned}$$

28.
$$\begin{aligned} x + 4y - z &= 3 \\ -2x - 8y + 2z &= -6 \\ 3x + 12y - 3z &= 9 \end{aligned}$$

29.
$$\begin{aligned} 2x + y - z &= 6 \\ 4x + 2y - 2z &= 12 \\ -x - \frac{1}{2}y + \frac{1}{2}z &= -3 \end{aligned}$$

30.
$$\begin{aligned} 2x - 8y + 2z &= -10 \\ -x + 4y - z &= 5 \\ \frac{1}{8}x - \frac{1}{2}y + \frac{1}{8}z &= -\frac{5}{8} \end{aligned}$$

31.
$$\begin{aligned} x + y - 2z &= 0 \\ 3x - y + z &= 0 \\ 4x + 2y - z &= 0 \end{aligned}$$

32.
$$\begin{aligned} 2x + 3y - z &= 0 \\ x - 4y + 2z &= 0 \\ 3x - 5y - z &= 0 \end{aligned}$$

33.
$$\begin{aligned} x - 2y + \frac{1}{3}z &= 4 \\ 3x - 6y + z &= 12 \\ -6x + 12y - 2z &= -3 \end{aligned}$$

34.
$$\begin{aligned} 4x + y - 2z &= 3 \\ x + \frac{1}{4}y - \frac{1}{2}z &= \frac{3}{4} \\ 2x + \frac{1}{2}y - z &= 1 \end{aligned}$$

35.
$$\begin{aligned} x + 5y - 2z &= -1 \\ -2x + 8y + z &= -4 \\ 3x - y + 5z &= 19 \end{aligned}$$

36.
$$\begin{aligned} x + 3y + z &= 2 \\ 4x + y + 2z &= -4 \\ 5x + 2y + 3z &= -2 \end{aligned}$$

Relating Concepts (Exercises 37–44) For Individual or Group Work

Suppose that on a distant planet a function of the form

$$f(x) = ax^2 + bx + c \quad (a \neq 0)$$

describes the height in feet of a projectile x seconds after it has been projected upward. ***Work Exercises 37–44 in order,*** *to see how this can be related to a system of three equations in three variables a, b, and c.*

37. After 1 sec, the height of a certain projectile is 128 ft. Thus, $f(1) = 128$. Use this information to write one equation in the variables a, b, and c. (*Hint:* Substitute 1 for x and 128 for $f(x)$.)

38. After 1.5 sec, the height is 140 ft. Write a second equation in a, b, and c.

39. After 3 sec, the height is 80 ft. Write a third equation in a, b, and c.

40. Write a system of three equations in a, b, and c, based on your answers in Exercises 37–39. Solve the system.

41. What is the function f for this particular projectile?

42. In the function f written in Exercise 41, the ______ of the projectile is a function of the ______ elapsed after it was projected.

43. What was the initial height of the projectile? (*Hint:* Find $f(0)$.)

44. The projectile reaches its maximum height in 1.625 sec. Find its maximum height.

5.3 Applications of Systems of Linear Equations

OBJECTIVES

1. Solve geometry problems using two variables.
2. Solve money problems using two variables.
3. Solve mixture problems using two variables.
4. Solve distance-rate-time problems using two variables.
5. Solve problems with three variables using a system of three equations.

Many applied problems involve more than one unknown quantity. Although some problems with two unknowns can be solved using just one variable (as in **Chapter 2**), an alternative method of solution uses two variables. To solve a problem in this way using two unknowns, we must write two equations that relate the unknown quantities. The system formed by the pair of equations can then be solved using the methods of this chapter.

Problems that can be solved by writing a system of equations have been of interest historically. The following problem, which is given in the exercises for this section, first appeared in a Hindu work that dates back to about A.D. 850.

> *The mixed price of 9 citrons [a lemonlike fruit shown in the photo] and 7 fragrant wood apples is 107; again, the mixed price of 7 citrons and 9 fragrant wood apples is 101. O you arithmetician, tell me quickly the price of a citron and the price of a wood apple here, having distinctly separated those prices well.*

The following steps, based on the six-step problem-solving method first introduced in **Section 2.3,** give a strategy for solving applied problems using more than one variable.

Solving an Applied Problem by Writing a System of Equations

Step 1 **Read** the problem, several times if necessary, until you understand what is given and what is to be found.

Step 2 **Assign variables** to represent the unknown values, using diagrams or tables as needed. *Write down* what each variable represents.

Step 3 **Write a system of equations** that relates the unknowns.

Step 4 **Solve** the system of equations.

Step 5 **State the answer** to the problem. Does it seem reasonable?

Step 6 **Check** the answer in the words of the original problem.

OBJECTIVE 1 Solve geometry problems using two variables. Problems about the perimeter of a geometric figure often involve two unknowns and can be solved using systems of equations.

1 Solve the problem.

The length of the foundation of a rectangular house is to be 6 m more than its width. Find the length and width of the house if the perimeter must be 48 m.

EXAMPLE 1 Finding the Dimensions of a Soccer Field

Unlike football, where the dimensions of a playing field cannot vary, a rectangular soccer field may have a width between 50 and 100 yd and a length between 100 and 130 yd. Suppose that one particular field has a perimeter of 320 yd. Its length measures 40 yd more than its width. What are the dimensions of this field? (*Source: Microsoft Encarta Encyclopedia.*)

Step 1 **Read** the problem again. We must find the dimensions of the field.

Step 2 **Assign variables.** Let $L =$ the length and $W =$ the width. Figure 7 shows a soccer field with these variables as labels.

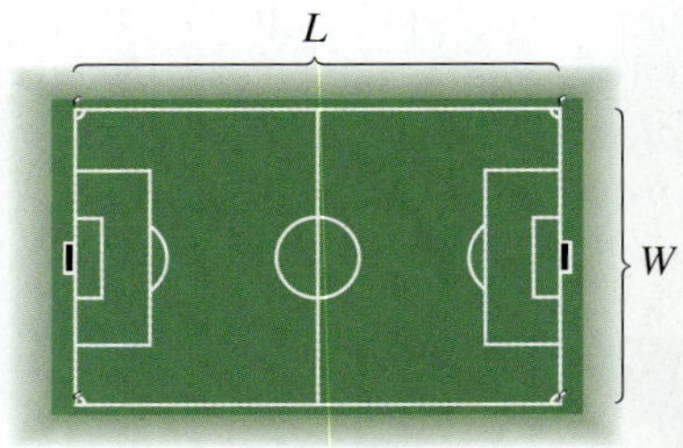

Figure 7

Step 3 **Write a system of equations.** Because the perimeter is 320 yd, we find one equation by using the perimeter formula:

$$2L + 2W = 320.$$

Because the length is 40 yd more than the width, we have

$$L = W + 40.$$

The system is, therefore,

$$\begin{aligned} 2L + 2W &= 320 && (1) \\ L &= W + 40. && (2) \end{aligned}$$

Step 4 **Solve** the system of equations. Since equation (2) is solved for L, we can use the substitution method. We substitute $W + 40$ for L in equation (1), and solve for W.

$$\begin{aligned} 2L + 2W &= 320 && (1) \\ 2(W + 40) + 2W &= 320 && \text{Let } L = W + 40. \\ 2W + 80 + 2W &= 320 && \text{Distributive property} \\ 4W + 80 &= 320 && \text{Combine like terms.} \\ 4W &= 240 && \text{Subtract 80.} \\ W &= 60 && \text{Divide by 4.} \end{aligned}$$

Be sure to use parentheses around $W + 40$.

Let $W = 60$ in the equation $L = W + 40$ to find L.

$$L = 60 + 40 = 100$$

Step 5 **State the answer.** The length is **100** yd, and the width is **60** yd. Both dimensions are within the ranges given in the problem.

Step 6 **Check.** The perimeter is $2(100) + 2(60) = 320$ yd, and the length, 100 yd, is indeed 40 yd more than the width, since $100 - 40 = 60$. The answer is correct.

◀ *Work Problem* **1** *at the Side.*

ANSWER

1. length: 15 m; width: 9 m

OBJECTIVE 2 Solve money problems using two variables.

EXAMPLE 2 Solving a Problem about Ticket Prices

For the 2005–2006 National Hockey League and National Basketball Association seasons, two hockey tickets and one basketball ticket purchased at their average prices would have cost \$128.30. One hockey ticket and two basketball tickets would have cost \$133.03. What were the average ticket prices for the two sports? (*Source:* Team Marketing Report, Chicago.)

Step 1 **Read** the problem again. There are two unknowns.

Step 2 **Assign variables.**

Let h = the average price for a hockey ticket

and b = the average price for a basketball ticket.

Step 3 **Write a system of equations.** Because two hockey tickets and one basketball ticket cost a total of \$128.30, one equation for the system is

$$2h + b = 128.30.$$

By similar reasoning, the second equation is

$$h + 2b = 133.03.$$

Therefore, the system is

$$\begin{aligned} 2h + b &= 128.30 & \text{(1)} \\ h + 2b &= 133.03. & \text{(2)} \end{aligned}$$

Step 4 **Solve** the system of equations. To eliminate h, multiply equation (2) by -2 and add.

$$\begin{aligned} 2h + b &= 128.30 & &\text{(1)} \\ -2h - 4b &= -266.06 & &\text{Multiply each side of (2) by } -2. \\ \hline -3b &= -137.76 & &\text{Add.} \\ \mathbf{b} &= \mathbf{45.92} & &\text{Divide by } -3. \end{aligned}$$

To find the value of h, let $b = 45.92$ in equation (2).

$$\begin{aligned} h + 2b &= 133.03 & &\text{(2)} \\ h + 2(45.92) &= 133.03 & &\text{Let } b = 45.92. \\ h + 91.84 &= 133.03 & &\text{Multiply.} \\ \mathbf{h} &= \mathbf{41.19} & &\text{Subtract 91.84.} \end{aligned}$$

Step 5 **State the answer.** The average price for one basketball ticket was \$45.92. For one hockey ticket, the average price was \$41.19.

Step 6 **Check** that these values satisfy the conditions stated in the problem.

Work Problem **2** *at the Side.* ▶

2 Solve the problem.

For recent Major League Baseball and National Football League seasons, based on average ticket prices, three baseball tickets and two football tickets would have cost \$181.41, while two baseball tickets and one football ticket would have cost \$101.29. What were the average ticket prices for the two sports? (*Source:* Team Marketing Report, Chicago.)

ANSWER

2. baseball: \$21.17; football: \$58.95

OBJECTIVE 3 Solve mixture problems using two variables. We solved mixture problems in **Section 2.3** using one variable. For many mixture problems we can use more than one variable and a system of equations.

EXAMPLE 3 Solving a Mixture Problem

How many ounces each of 5% hydrochloric acid and 20% hydrochloric acid must be combined to get 10 oz of solution that is 12.5% hydrochloric acid?

Step 1 **Read** the problem. Two solutions of different strengths are being mixed together to get a specific amount of a solution with an "in-between" strength.

Step 2 **Assign variables.**

Let $x =$ the number of ounces of 5% solution

and $y =$ the number of ounces of 20% solution.

Use a table to summarize the information from the problem. We multiply the amount of each solution (given in the first column) by its concentration of acid (given in the second column) to get the amount of acid in that solution (given in the third column).

Ounces of Solution	Percent (as a decimal)	Ounces of Pure Acid
x	$5\% = 0.05$	$0.05x$
y	$20\% = 0.20$	$0.20y$
10	$12.5\% = 0.125$	$(0.125)10$

Gives equation (1) (first column); Gives equation (2) (third column)

Figure 8 also illustrates what is happening in the problem.

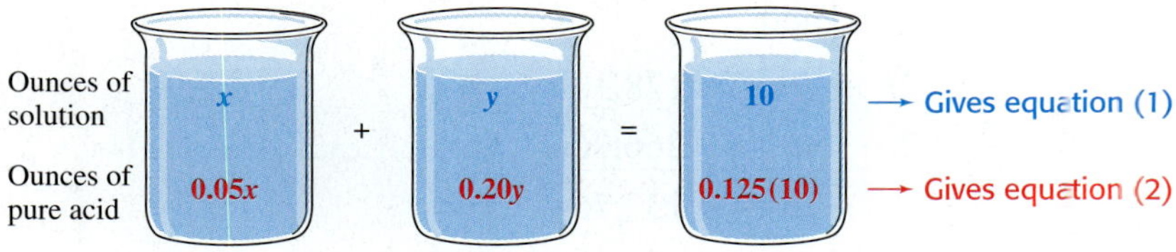

Figure 8

Step 3 **Write a system of equations.** When the x ounces of 5% solution and the y ounces of 20% solution are combined, the total number of ounces is 10, so

$$x + y = 10. \quad (1)$$

The ounces of acid in the 5% solution $(0.05x)$ plus the ounces of acid in the 20% solution $(0.20y)$ should equal the total ounces of acid in the mixture, which is $(0.125)10$, or 1.25. That is,

$$0.05x + 0.20y = 1.25. \quad (2)$$

Notice that these equations can be quickly determined by reading down in the table or across in Figure 8.

Continued on Next Page

Step 4 **Solve** the system of equations (1) and (2). Eliminate x by first multiplying equation (2) by 100 to clear it of decimals and then multiplying equation (1) by -5.

$$\begin{aligned} 5x + 20y &= 125 && \text{Multiply each side of (2) by 100.} \\ -5x - 5y &= -50 && \text{Multiply each side of (1) by } -5. \\ \hline 15y &= 75 && \text{Add.} \\ y &= 5 && \text{Divide by 15.} \end{aligned}$$

Because $y = 5$ and $x + y = 10$, the value of x is also 5.

Step 5 **State the answer.** The desired mixture will require 5 oz of the 5% solution and 5 oz of the 20% solution.

Step 6 **Check.**

Total amount of solution: $x + y = 5 \text{ oz} + 5 \text{ oz}$
$= \mathbf{10} \text{ oz}$, as required.

Total amount of acid: 5% of 5 oz + 20% of 5 oz
$= 0.05(5) + 0.20(5)$
$= \mathbf{1.25} \text{ oz}$

Percent of acid in solution:

$$\frac{\text{Total acid} \rightarrow \mathbf{1.25}}{\text{Total solution} \rightarrow \mathbf{10}} = 0.125, \quad \text{or} \quad 12.5\%, \quad \text{as required.}$$

Work Problem **3** *at the Side.* ▶

OBJECTIVE 4 Solve distance-rate-time problems using two variables. Motion problems require the distance formula, $d = rt$, where d is distance, r is rate (or speed), and t is time. These applications often lead to systems of equations.

EXAMPLE 4 Solving a Motion Problem

A car travels 250 km in the same time that a truck travels 225 km. If the speed of the car is 8 km per hr faster than the speed of the truck, find both speeds.

Step 1 **Read** the problem again. Given the distances traveled, we need to find the speed of each vehicle.

Step 2 **Assign variables.**

Let x = the speed of the car
and y = the speed of the truck.

Fill in the given information for each vehicle (distances) and the assigned variables for the unknown speeds (rates) in a table.

	d	r	t
Car	250	x	$\frac{250}{x}$
Truck	225	y	$\frac{225}{y}$

The times must be equal.

To get the expressions for time, solve the distance formula, $d = rt$, for t. Since $\frac{d}{r} = t$, the two times can be written as $\frac{250}{x}$ and $\frac{225}{y}$.

Continued on Next Page

3 Solve each problem.

(a) A grocer has some \$4 per lb coffee and some \$8 per lb coffee, which he will mix to make 50 lb of \$5.60 per lb coffee. How many pounds of each should be used?

(b) Some 40% ethyl alcohol solution is to be mixed with some 80% solution to get 200 L of a 50% solution. How many liters of each should be used?

ANSWERS

3. **(a)** 30 lb of \$4; 20 lb of \$8
(b) 150 L of 40%; 50 L of 80%

4 Solve the problem.

A train travels 600 mi in the same time that a truck travels 520 mi. Find the speed of each vehicle if the train's average speed is 8 mph faster than the truck's.

Step 3 **Write a system of equations.** The problem states that the car travels 8 km per hr faster than the truck. Since the two speeds are x and y,

$$x = y + 8. \qquad (1)$$

Both vehicles travel for the same time, so from the table,

$$\text{Time for car} \rightarrow \frac{250}{x} = \frac{225}{y}. \leftarrow \text{Time for truck}$$

This is not a linear equation. Multiplying each side by xy gives

$$xy \cdot \frac{250}{x} = \frac{225}{y} \cdot xy \qquad \text{Multiply by the LCD, } xy.$$

$$\frac{250xy}{x} = \frac{225xy}{y}$$

$$250y = 225x, \qquad (2)$$

which is linear. The system is

$$x = y + 8 \qquad (1)$$
$$250y = 225x. \qquad (2)$$

Step 4 **Solve** the system of equations by substitution. Replace x with $y + 8$ in equation (2).

$$250y = 225x \qquad (2)$$
$$250y = 225(y + 8) \qquad \text{Let } x = y + 8.$$
$$250y = 225y + 1800 \qquad \text{Distributive property}$$
$$25y = 1800 \qquad \text{Subtract } 225y.$$
$$y = 72 \qquad \text{Divide by 25.}$$

Be sure to use parentheses around $y + 8$.

Because $x = y + 8$, the value of x is

$$72 + 8 = 80.$$

Step 5 **State the answer.** The car's speed is 80 km per hr, and the truck's speed is 72 km per hr.

Step 6 **Check.** This is especially important since one of the equations had variable denominators.

$$\text{Car: } t = \frac{d}{r} = \frac{250}{80} = 3.125$$
$$\text{Truck: } t = \frac{d}{r} = \frac{225}{72} = 3.125$$

Times are equal.

Since $80 - 72 = 8$, the conditions of the problem are satisfied.

◀ Work Problem 4 at the Side.

OBJECTIVE 5 Solve problems with three variables using a system of three equations.

Problem-Solving Hint

If an application requires finding *three* unknown quantities, we can use a system of *three* equations to solve it. We extend the method used for two unknowns.

ANSWER

4. train: 60 mph; truck: 52 mph

EXAMPLE 5 Solving a Problem Involving Prices

At Panera Bread, a loaf of honey wheat bread costs \$2.95, a loaf of sunflower bread costs \$2.99, and a loaf of French bread costs \$5.79. On a recent day, three times as many loaves of honey wheat bread were sold as sunflower bread. The number of loaves of French bread sold was 5 less than the number of loaves of honey wheat bread sold. Total receipts for these breads were \$87.89. How many loaves of each type of bread were sold? (*Source:* Panera Bread menu.)

Step 1 **Read** the problem again. There are three unknowns in this problem.

Step 2 **Assign variables** to represent the three unknowns.

Let $x =$ the number of loaves of honey wheat bread,
$y =$ the number of loaves of sunflower bread,
and $z =$ the number of loaves of French bread.

Step 3 **Write a system of three equations.** Since three times as many loaves of honey wheat bread were sold as sunflower bread,

$x = 3y$, or $x - 3y = 0$. Subtract $3y$. (1)

Also,

Number of loaves of French	equals	5 less than the number of loaves of honey wheat.
↓	↓	↓
z	$=$	$x - 5$,

$-x + z = -5$ Subtract x.

$x - z = 5$. Multiply by -1. (2)

Multiplying the cost of a loaf of each kind of bread by the number of loaves of that kind sold and adding gives the total receipts.

$$2.95x + 2.99y + 5.79z = 87.89$$

Multiply each side of this equation by 100 to clear it of decimals.

$$295x + 299y + 579z = 8789 \quad (3)$$

Step 4 **Solve** the system of three equations,

$$x - 3y = 0 \quad (1)$$
$$x - z = 5 \quad (2)$$
$$295x + 299y + 579z = 8789, \quad (3)$$

using the method shown in **Section 5.2.**

Work Problem 5 at the Side. ▶

Thus, we find that $x = 12$, $y = 4$, and $z = 7$.

Step 5 **State the answer.** The solution set is $\{(12, 4, 7)\}$, meaning that 12 loaves of honey wheat bread, 4 loaves of sunflower bread, and 7 loaves of French bread were sold.

Step 6 **Check.** Since $12 = 3 \cdot 4$, the number of loaves of honey wheat bread is three times the number of loaves of sunflower bread. Also, $12 - 7 = 5$, so the number of loaves of French bread is 5 less than the number of loaves of honey wheat bread. Multiply the appropriate cost per loaf by the number of loaves sold and add the results to check that total receipts were \$87.89.

Work Problem 6 at the Side. ▶

5 Solve the system of equations from Example 5.

$$x - 3y = 0 \quad (1)$$
$$x - z = 5 \quad (2)$$
$$295x + 299y + 579z = 8789 \quad (3)$$

6 Solve the problem.

A department store display features three kinds of perfume: Felice, Vivid, and Joy. There are 10 more bottles of Felice than Vivid, and 3 fewer bottles of Joy than Vivid. Each bottle of Felice costs \$8, Vivid costs \$15, and Joy costs \$32. The total value of all the perfume is \$589. How many bottles of each are there?

ANSWERS

5. $\{(12, 4, 7)\}$
6. 21 bottles of Felice; 11 of Vivid; 8 of Joy

7 Solve the problem.

A paper mill makes newsprint, bond, and copy machine paper. Each ton of newsprint requires 3 tons of recycled paper and 1 ton of wood pulp. Each ton of bond requires 2 tons of recycled paper, 4 tons of wood pulp, and 3 tons of rags. A ton of copy machine paper requires 2 tons of recycled paper, 3 tons of wood pulp, and 2 tons of rags. The mill has 4200 tons of recycled paper, 5800 tons of wood pulp, and 3900 tons of rags. How much of each kind of paper can be made from these supplies?

EXAMPLE 6 **Solving a Business Production Problem**

A company produces three color television sets, models X, Y, and Z. Each model X set requires 2 hr of electronics work, 2 hr of assembly time, and 1 hr of finishing time. Each model Y requires 1, 3, and 1 hr of electronics, assembly, and finishing time, respectively. Each model Z requires 3, 2, and 2 hr of the same work, respectively. There are 100 hr available for electronics, 100 hr available for assembly, and 65 hr available for finishing per week. How many of each model should be produced each week if all available time must be used?

Step 1 **Read** the problem again. There are three unknowns.

Step 2 **Assign variables.**

Let $x =$ the number of model X produced per week,
$y =$ the number of model Y produced per week,
and $z =$ the number of model Z produced per week.

We organize the information in a table.

	Each Model X	Each Model Y	Each Model Z	Totals	
Hours of Electronics Work	2	1	3	100	→ Gives equation (1)
Hours of Assembly Time	2	3	2	100	→ Gives equation (2)
Hours of Finishing Time	1	1	2	65	→ Gives equation (3)

Step 3 **Write a system of three equations.** The x model X sets require $2x$ hours of electronics, the y model Y sets require $1y$ (or y) hours of electronics, and the z model Z sets require $3z$ hours of electronics. Since 100 hr are available for electronics,

$$2x + y + 3z = 100. \qquad (1)$$

Similarly, from the fact that 100 hr are available for assembly,

$$2x + 3y + 2z = 100, \qquad (2)$$

and the fact that 65 hr are available for finishing leads to the equation

$$x + y + 2z = 65. \qquad (3)$$

Notice that by reading *across* the table, we can quickly determine the coefficients and constants in the equations of the system

Step 4 **Solve** the system of equations (1), (2), and (3), namely,

$$\begin{aligned} 2x + y + 3z &= 100 \qquad (1)\\ 2x + 3y + 2z &= 100 \qquad (2)\\ x + y + 2z &= 65 \qquad (3) \end{aligned}$$

to find $x = 15$, $y = 10$, and $z = 20$.

Step 5 **State the answer.** The company should produce 15 model X, 10 model Y, and 20 model Z sets per week.

Step 6 **Check** that these values satisfy the conditions of the problem

◀ *Work Problem* **7** *at the Side.*

ANSWER

7. 400 tons of newsprint; 900 tons of bond; 600 tons of copy machine paper

5.3 Exercises

Solve each problem. See Example 1.

1. During the 2007 Major League Baseball regular season, the Cleveland Indians played 162 games. They won 30 more games than they lost. What was their win-loss record that year?

2. Refer to Exercise 1. During the same 162-game season, the Chicago White Sox lost 18 more games than they won. What was the team's win-loss record?

2007 MLB Final Standings
American League Central

Team	W	L
Cleveland	___	___
Detroit	88	74
Minnesota	79	83
Chicago	___	___
Kansas City	69	93

Source: www.mlb.com

3. Venus and Serena measured a tennis court and found that it was 42 ft longer than it was wide and had a perimeter of 228 ft. What were the length and the width of the tennis court?

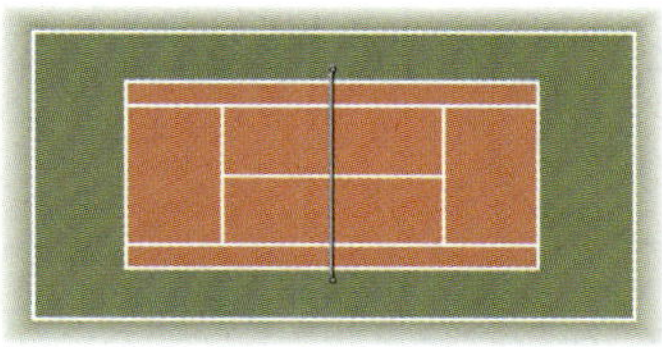

4. Wilt and Oscar found that the width of their basketball court was 44 ft less than the length. If the perimeter was 288 ft, what were the length and the width of their court?

5. The two biggest Fortune 500 companies in 2005 were Wal-Mart and ExxonMobil. ExxonMobil's revenue was \$24 billion more than that of Wal-Mart. Total revenue for the two companies was \$656 billion. What was the revenue for each company? (*Source:* Fortune 500.)

6. In 2007, U.S. exports to Canada were \$112 billion more than exports to Mexico. Together, exports to these two countries totaled \$386 billion. How much were exports to each country? (*Source:* U.S. Census Bureau.)

In Exercises 7 and 8, find the measures of the angles marked x and y. Remember that (1) *the sum of the measures of the angles of a triangle is* 180°, (2) *supplementary angles have a sum of* 180°, *and* (3) *vertical angles have equal measures.*

7.

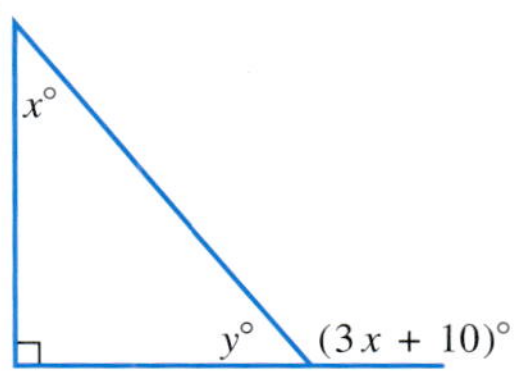

8.

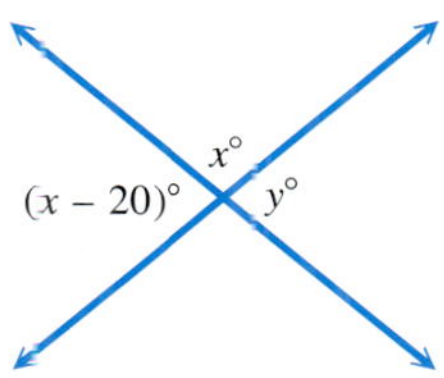

The Fan Cost Index (FCI) represents the cost of four average-price tickets, four small soft drinks, two small beers, four hot dogs, parking for one car, two game programs, and two souvenir caps to a sporting event. (Source: www.teammarketing.com*)*

Use the concept of FCI in Exercises 9 and 10. See Example 2.

9. For the 2005–2006 season, the FCI prices for the National Hockey League (NHL) and the National Basketball Association (NBA) totaled \$514.69. The hockey FCI was \$20.05 less than that of basketball. What were the FCIs for these sports?

10. In 2005, the FCI prices for Major League Baseball (MLB) and the National Football League (NFL) totaled \$501.01. The football FCI was \$158.63 more than that of baseball. What were the FCIs for these sports?

Solve each problem. See Example 2.

11. Andrew McGinnis works at Arby's. During one particular day he sold 15 Junior Roast Beef sandwiches and 10 Big Montana sandwiches, totaling \$75.25. Another day he sold 30 Junior Roast Beef sandwiches and 5 Big Montana sandwiches, totaling \$84.65. How much did each type of sandwich cost? (*Source:* Arby's menu.)

12. London and New York are among the most expensive cities worldwide for business travelers. On the basis of average costs per day for each city (which include the costs of business-class lodging and three meals), 2 days in London and 3 days in New York costs \$2099, while 4 days in London and 2 days in New York costs \$2586. What is the average cost per day for each city? (*Source:* Runzheimer International.)

The formulas $p = br$ (percentage = base × rate) and $I = prt$ (simple interest = principal × rate × time) are used in the applications in Exercises 17–24. To prepare to use these formulas, answer the questions in Exercises 13 and 14.

13. If a container of liquid contains 60 oz of solution, what is the number of ounces of pure acid if the given solution contains the following acid concentrations?

(a) 10% **(b)** 25% **(c)** 40% **(d)** 50%

14. If \$5000 is invested in an account paying simple annual interest, how much interest will be earned during the first year at the following rates?

(a) 2% **(b)** 3% **(c)** 4% **(d)** 3.5%

15. If a pound of turkey costs \$1.29, how much will x pounds cost?

16. If a ticket to the movie *The Final Season* costs \$8.50 and y tickets are sold, how much is collected from the sale?

Solve each problem. See Example 3.

17. How many gallons each of 25% alcohol and 35% alcohol should be mixed to get 20 gal of 32% alcohol?

Gallons of Solution	Percent (as a Decimal)	Gallons of Pure Alcohol
x	25% = 0.25	
y	35% = 0.35	
20	32% =	

18. How many liters each of 15% acid and 33% acid should be mixed to get 120 L of 21% acid?

Liters of Solution	Percent (as a Decimal)	Liters of Pure Acid
x	15% = 0.15	
y	33% =	
120	21% =	

19. Pure acid is to be added to a 10% acid solution to obtain 54 L of a 20% acid solution. What amounts of each should be used? (*Hint:* Pure acid is 100% acid.)

20. A truck radiator holds 36 L of fluid. How much pure antifreeze must be added to a mixture that is 4% antifreeze to fill the radiator with a mixture that is 20% antifreeze?

21. A party mix is made by adding nuts that sell for \$2.50 per kg to a cereal mixture that sells for \$1 per kg. How much of each should be added to get 30 kg of a mix that will sell for \$1.70 per kg?

	Number of Kilograms	Price per Kilogram	Value
Nuts	x	2.50	
Cereal	y	1.00	
Mixture		1.70	

22. A popular fruit drink is made by mixing fruit juices. Such a drink with 50% juice is to be mixed with another drink that is 30% juice to get 200 L of a drink that is 45% juice. How much of each should be used?

	Liters of Drink	Percent (as a Decimal)	Liters of Pure Juice
50% Juice	x	0.50	
30% Juice	y	0.30	
Mixture		0.45	

23. A total of \$3000 is invested, part at 2% simple interest and part at 4%. If the total annual return from the two investments is \$100, how much is invested at each rate?

Principal	Rate (as a Decimal)	Interest
x	0.02	$0.02x$
y	0.04	$0.04y$
3000	XXXXX	100

24. An investor must invest a total of \$15,000 in two accounts, one paying 4% annual simple interest, and the other 3%. If he wants to earn \$550 annual interest, how much should he invest at each rate?

Principal	Rate (as a Decimal)	Interest
x	0.04	
y	0.03	
15,000	XXXXX	

The formula $d = rt$ (distance = rate × time) is used in the applications in Exercises 27–30. To prepare to use this formula, answer the questions in Exercises 25 and 26.

25. If the speed of a boat in still water is 10 mph, and the speed of the current of a river is x mph, what is the speed of the boat

(a) going upstream (that is, against the current, which slows the boat down);

(b) going downstream (that is, with the current, which speeds the boat up)?

26. If the speed of a killer whale is 25 mph and the whale swims for y hours, how many miles does the whale travel?

Solve each problem. See Example 4.

27. A motor scooter travels 20 mi in the same time that a bicycle covers 8 mi. If the speed of the scooter is 5 mph more than twice the speed of the bicycle, find both speeds.

28. A train travels 150 km in the same time that a plane covers 400 km. If the speed of the plane is 20 km per hr less than 3 times the speed of the train, find both speeds.

29. In his motorboat, Bill Ruhberg travels upstream at top speed to his favorite fishing spot, a distance of 36 mi, in 2 hr. Returning, he finds that the trip downstream, still at top speed, takes only 1.5 hr. Find the speed of Bill's boat and the speed of the current. Let $x =$ the speed of the boat in still water and $y =$ the speed of the current.

	r	t	d
Upstream	$x - y$	2	
Downstream	$x + y$		

30. Traveling for 3 hr into a steady headwind, a plane flies 1650 mi. The pilot determines that flying *with* the same wind for 2 hr, he could make a trip of 1300 mi. Find the speed of the plane and the speed of the wind.

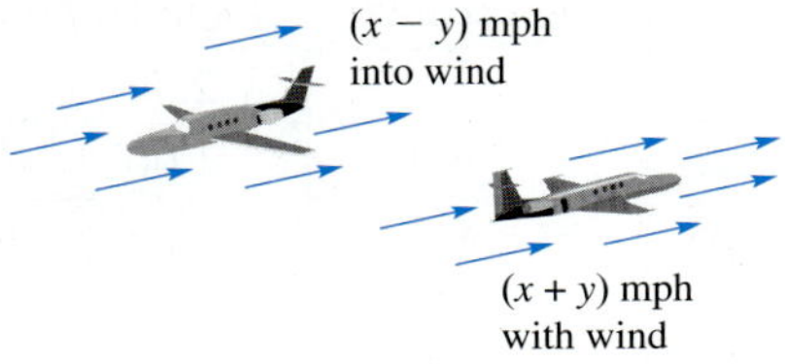

Solve each problem. See Examples 1–6.

31. (See the Chapter Introduction.) How many pounds of candy that sells for \$0.75 per lb must be mixed with candy that sells for \$1.25 per lb to obtain 9 lb of a mixture that should sell for \$0.96 per lb?

32. The average cost of tuition and fees at a 4-yr public college or university during the 2007–2008 school year was \$3824 more than at a 2-yr community college. Suppose a student plans to attend a local community college for 2 yr and then transfer to a public 4-yr university for 2 yr. The student expected to pay \$17,092 tuition and fees for the 4 yr, assuming the 2007–2008 rates were locked in. What was the cost of tuition and fees during the 2007–2008 year at each type of school? (*Source:* The College Board.)

33. Tickets to a production of *Cats* at Shelton State Community College cost \$5 for general admission or \$4 with a student ID. If 184 people paid to see a performance and \$812 was collected, how many of each type of ticket were sold?

34. At a business meeting at Panera Bread, the bill (without tax) for two cappuccinos and three house lattes was \$14.55. At another table, the bill for one cappuccino and two house lattes was \$8.77. How much did each type of beverage cost? (*Source:* Panera Bread menu.)

35. The mixed price of 9 citrons and 7 fragrant wood apples is 107; again, the mixed price of 7 citrons and 9 fragrant wood apples is 101. O you arithmetician, tell me quickly the price of a citron and the price of a wood apple here, having distinctly separated those prices well. (*Source:* Hindu work, A.D. 850.) (*Hint:* "Mixed price" refers to the price of a mixture of the two fruits.)

36. Braving blizzard conditions on the planet Hoth, Luke Skywalker sets out at top speed in his snow speeder for a rebel base 4800 mi away. He travels into a steady headwind and makes the trip in 3 hr. Returning, he finds that the trip back, still at top speed but now with a tailwind, takes only 2 hr. Find the top speed of Luke's snow speeder and the speed of the wind.

	r	t	d
Into Headwind			
With Tailwind			

Solve each problem involving three unknowns. See Examples 5 and 6. (In Exercises 37–40, remember that the sum of the measures of the angles of a triangle is 180°.)

37. In the figure, $z = x + 10$ and $x + y = 100$. Determine a third equation involving x, y, and z, and then find the measures of the three angles.

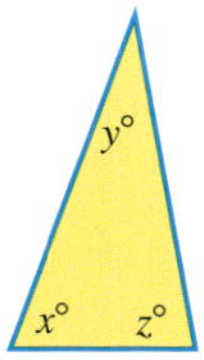

38. In the figure, x is 10 less than y and x is 20 less than z. Write a system of equations and find the measures of the three angles.

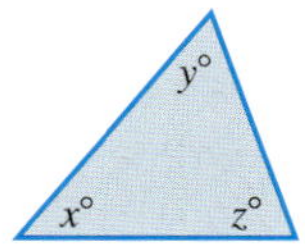

39. In a certain triangle, the measure of the second angle is 10° more than three times the first. The third angle measure is equal to the sum of the measures of the other two. Find the measures of the three angles.

40. The measure of the largest angle of a triangle is 12° less than the sum of the measures of the other two. The smallest angle measures 58° less than the largest. Find the measures of the angles.

41. The perimeter of a triangle is 70 cm. The longest side is 4 cm less than the sum of the other two sides. Twice the shortest side is 9 cm less than the longest side. Find the length of each side of the triangle.

42. The perimeter of a triangle is 56 in. The longest side measures 4 in. less than the sum of the other two sides. Three times the shortest side is 4 in. more than the longest side. Find the lengths of the three sides.

43. In a random sample of 100 Americans of voting age, 10 more Americans identify themselves as Independents than Republicans. Six fewer Americans identify themselves as Republicans than Democrats. Assuming that all of those sampled are Republican, Democrat, or Independent, how many of those in the sample identify themselves with each political affiliation? (*Source:* The Gallup Organization.)

44. In the 2004 Summer Olympics in Athens, Greece, the United States earned 6 more gold medals than bronze. The number of silver medals earned was 19 less than twice the number of bronze medals. The United Stated earned a total of 103 medals. How many of each kind of medal did the United States earn? (*Source: World Almanac and Book of Facts.*)

45. Tickets for one show on the Harlem Globetrotters' 2006 "Unstoppable" Tour cost \$14, \$20, or, for VIP seats, \$50. Five times as many \$14 tickets were sold as VIP tickets. The number of \$14 tickets was 15 more than the sum of the number of \$20 tickets and the number of VIP tickets. Sales of all three kinds of tickets totaled \$11,700. How many of each kind of ticket were sold? (*Source*: www.ticketmaster.com)

46. Three kinds of tickets are available for a *Third Day* concert: "up close," "in the middle," and "far out." "Up close" tickets cost \$10 more than "in the middle" tickets, while "in the middle" tickets cost \$10 more than "far out" tickets. Twice the cost of an "up close" ticket is \$20 more than 3 times the cost of a "far out" ticket. Find the price of each kind of ticket.

47. A wholesaler supplies college T-shirts to three college bookstores: A, B, and C. The wholesaler recently shipped a total of 800 T-shirts to the three bookstores. In order to meet student demand at the three colleges, twice as many T-shirts were shipped to bookstore B as to bookstore A, and the number shipped to bookstore C was 40 less than the sum of the numbers shipped to the other two bookstores. How many T-shirts were shipped to each bookstore?

48. An office supply store sells three models of computer desks: A, B, and C. In January, the store sold a total of 85 computer desks. The number of model B desks was five more than the number of model C desks, and the number of model A desks was four more than twice the number of model C desks. How many of each model did the store sell in January?

49. During the 2005–2006 National Hockey League regular season, the Calgary Flames played 82 games. Their wins and losses totaled 71. They tied 14 fewer games than they lost. How many wins, losses, and ties did they have?

50. (Refer to Exercise 49.) During the same 82-game season, the Minnesota Wild had a total of 44 losses and ties. They had two more wins than losses. How many wins, losses, and ties did they have?

2005-2006 NHL Final Standings

Team	W	L	T	Pts
Calgary	—	—	—	103
Colorado	43	30	9	95
Edmonton	41	28	13	95
Vancouver	42	32	8	92
Minnesota	—	—	—	84

Source: www.sportzdomain.com

5.4 Solving Systems of Linear Equations by Matrix Methods

OBJECTIVES

1. **Define a matrix.**
2. **Write the augmented matrix for a system.**
3. **Use row operations to solve a system with two equations.**
4. **Use row operations to solve a system with three equations.**
5. **Use row operations to solve special systems.**

OBJECTIVE 1 Define a matrix. An ordered array of numbers such as

$$\text{Rows}\ \rightarrow \overset{\text{Columns}}{\overset{\downarrow\ \ \downarrow\ \ \downarrow}{\begin{bmatrix} 2 & 3 & 5 \\ 7 & 1 & 2 \end{bmatrix}}} \qquad \text{Matrix}$$

is called a **matrix.** The numbers are called **elements** of the matrix. *Matrices* (the plural of *matrix*) are named according to the number of **rows** and **columns** they contain. The rows are read horizontally, and the columns are read vertically. For example, the first row in the preceding matrix is 2 3 5 and the first column is $\begin{matrix} 2 \\ 7 \end{matrix}$. This matrix is a 2×3 (read "two by three") matrix because it has 2 rows and 3 columns. The number of rows followed by the number of columns gives the **dimensions** of the matrix.

$$\begin{bmatrix} -1 & 0 \\ 1 & -2 \end{bmatrix} \quad \begin{matrix} 2 \times 2 \\ \text{matrix} \end{matrix} \qquad\qquad \begin{bmatrix} 8 & -1 & -3 \\ 2 & 1 & 6 \\ 0 & 5 & -3 \\ 5 & 9 & 7 \end{bmatrix} \quad \begin{matrix} 4 \times 3 \\ \text{matrix} \end{matrix}$$

A **square matrix** is one that has the same number of rows as columns. The 2×2 matrix is a square matrix.

Calculator Tip Figure 9 shows how a graphing calculator displays the preceding two matrices. Work with matrices is made much easier by using technology when available. Consult your owner's manual for details.

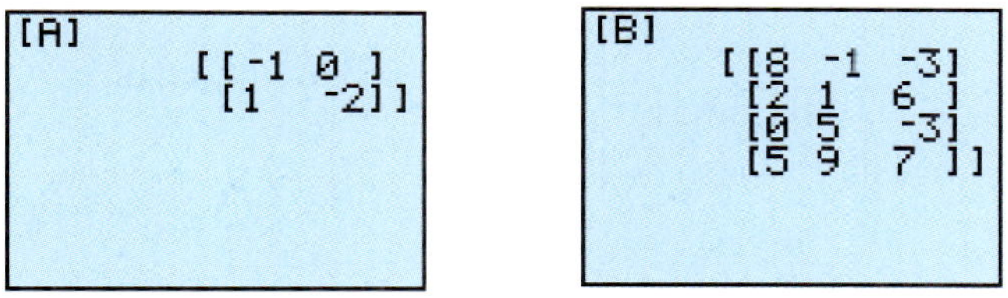

Figure 9

In this section, we discuss a method of solving linear systems that uses matrices. The advantage of this new method is that it can be done by a graphing calculator or a computer, allowing large systems of equations to be solved easily.

OBJECTIVE 2 Write the augmented matrix for a system. To solve a linear system using matrices, we begin by writing an *augmented matrix* for the system. An **augmented matrix** has a vertical bar that separates the columns of the matrix into two groups. For example, to solve the system

$$\begin{aligned} x - 3y &= 1 \\ 2x + y &= -5, \end{aligned}$$

we start with the augmented matrix

$$\left[\begin{array}{cc|c} 1 & -3 & 1 \\ 2 & 1 & -5 \end{array}\right]. \qquad \text{Augmented matrix}$$

System of equations:

$$\begin{aligned} x - 3y &= 1 \\ 2x + y &= -5 \end{aligned}$$

Augmented matrix:

$$\left[\begin{array}{cc|c} 1 & -3 & 1 \\ 2 & 1 & -5 \end{array}\right]$$

Coefficients of the variables — The bar separates the coefficients from the constants. — Constants

Notice that we place the coefficients of the variables to the left of the bar, and the constants to the right. ***The matrix is just a shorthand way of writing the system of equations, so the rows of the augmented matrix can be treated the same as the equations of a system of equations.***

We know that exchanging the position of two equations in a system does not change the system. Also, multiplying any equation in a system by a nonzero number does not change the system. Comparable changes to the augmented matrix of a system of equations produce new matrices that correspond to systems with the same solutions as the original system.

The following **row operations** produce new matrices that lead to systems having the same solutions as the original system.

Matrix Row Operations

1. Any two rows of the matrix may be interchanged.
2. The numbers in any row may be multiplied by any nonzero real number.
3. Any row may be transformed by adding to the numbers of the row the product of a real number and the corresponding numbers of another row.

Examples of these row operations follow.

Row operation 1:

$$\begin{bmatrix} 2 & 3 & 9 \\ 4 & 8 & -3 \\ 1 & 0 & 7 \end{bmatrix} \text{ becomes } \begin{bmatrix} 1 & 0 & 7 \\ 4 & 8 & -3 \\ 2 & 3 & 9 \end{bmatrix}.$$ Interchange row 1 and row 3.

Row operation 2:

$$\begin{bmatrix} 2 & 3 & 9 \\ 4 & 8 & -3 \\ 1 & 0 & 7 \end{bmatrix} \text{ becomes } \begin{bmatrix} 6 & 9 & 27 \\ 4 & 8 & -3 \\ 1 & 0 & 7 \end{bmatrix}.$$ Multiply the numbers in row 1 by 3.

Row operation 3:

$$\begin{bmatrix} 2 & 3 & 9 \\ 4 & 8 & -3 \\ 1 & 0 & 7 \end{bmatrix} \text{ becomes } \begin{bmatrix} 0 & 3 & -5 \\ 4 & 8 & -3 \\ 1 & 0 & 7 \end{bmatrix}.$$ Multiply the numbers in row 3 by -2; add them to the corresponding numbers in row 1

The third row operation corresponds to the way we eliminated a variable from a pair of equations in the previous sections.

OBJECTIVE 3 Use row operations to solve a system with two equations. Row operations can be used to rewrite a matrix. The goal is a matrix in the form

$$\left[\begin{array}{cc|c} 1 & a & b \\ 0 & 1 & c \end{array}\right] \quad \text{or} \quad \left[\begin{array}{ccc|c} 1 & a & b & c \\ 0 & 1 & d & e \\ 0 & 0 & 1 & f \end{array}\right]$$

for systems with two or three equations, respectively. Notice that there are 1s down the diagonal from upper left to lower right and 0s below the 1s. A matrix written this way is said to be in **row echelon form.** When these matrices are rewritten as systems of equations, the value of one variable is known, and the rest can be found by substitution. The following examples illustrate this method.

EXAMPLE 1 **Using Row Operations to Solve a System with Two Variables**

Use row operations to solve the system.

$$x - 3y = 1$$
$$2x + y = -5$$

We start with the augmented matrix of the system.

$$\left[\begin{array}{cc|c} 1 & -3 & 1 \\ 2 & 1 & -5 \end{array}\right] \quad \text{Augmented matrix}$$

Now we use the various row operations to change this matrix into one that leads to a system that is easier to solve.

It is best to work by columns. We start with the first column and make sure that there is a 1 in the first row, first column position. There is already a 1 in this position. Next, we get 0 in every position below the first. To get a 0 in row two, column one, we use the third row operation and add to the numbers in row two the result of multiplying each number in row one by -2. (We abbreviate this as $-2R_1 + R_2$.) Row one remains unchanged.

$$\left[\begin{array}{cc|c} 1 & -3 & 1 \\ 2 + 1(-2) & 1 + -3(-2) & -5 + 1(-2) \end{array}\right]$$

Original number from row two ↑   ↑ -2 times number from row one

$$\left[\begin{array}{cc|c} 1 & -3 & 1 \\ 0 & 7 & -7 \end{array}\right] \quad -2R_1 + R_2$$

The matrix now has a 1 in the first position of column one, with 0 in every position below the first.

Now we go to column two. An entry of 1 is needed in row two, column two. We get this 1 by using the second row operation, multiplying each number of row two by $\frac{1}{7}$.

Stop here—this matrix is in row echelon form.

$$\left[\begin{array}{cc|c} 1 & -3 & 1 \\ 0 & 1 & -1 \end{array}\right] \quad \tfrac{1}{7}R_2$$

This augmented matrix leads to the system of equations

$$\begin{aligned} 1x - 3y &= 1 \\ 0x + 1y &= -1, \end{aligned} \quad \text{or} \quad \begin{aligned} x - 3y &= 1 \\ y &= -1. \end{aligned}$$

From the second equation, $y = -1$. We substitute -1 for y in the first equation to get

$$\begin{aligned} x - 3y &= 1 & & \\ x - 3(-1) &= 1 & & \text{Let } y = -1. \\ x + 3 &= 1 & & \text{Multiply.} \\ x &= -2. & & \text{Subtract 3.} \end{aligned}$$

Write the values of x and y in the correct order.

The solution set of the system is $\{(-2, -1)\}$. Check this solution by substitution in both equations of the system.

Work Problem **1** *at the Side.* ▶

1 Use row operations to solve the system.

$$x - 2y = 9$$
$$3x + y = 13$$

ANSWER

1. $\{(5, -2)\}$

Calculator Tip If the augmented matrix of the system in Example 1 is entered as matrix A in a graphing calculator (Figure 10(a)) and the row echelon form of the matrix is found (Figure 10(b)), the system becomes

$$x + \frac{1}{2}y = -\frac{5}{2}$$
$$y = -1.$$

While this system looks different from the one we obtained in Example 1, it is equivalent, since its solution set is also $\{(-2, -1)\}$.

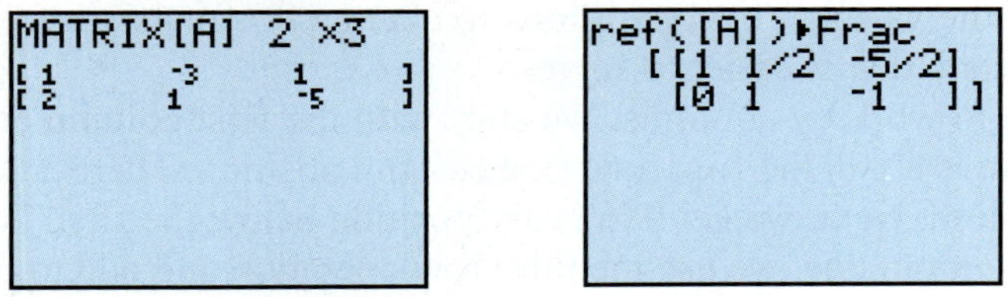

(a) (b)

Figure 10

OBJECTIVE 4 Use row operations to solve a system with three equations. As before, we use row operations to get 1s down the diagonal from left to right and all 0s below each 1.

EXAMPLE 2 Using Row Operations to Solve a System with Three Variables

Use row operations to solve the system.

$$\begin{aligned} x - y + 5z &= -6 \\ 3x + 3y - z &= 10 \\ x + 3y + 2z &= 5 \end{aligned}$$

Start by writing the augmented matrix of the system.

$$\left[\begin{array}{ccc|c} 1 & -1 & 5 & -6 \\ 3 & 3 & -1 & 10 \\ 1 & 3 & 2 & 5 \end{array}\right] \quad \text{Augmented matrix}$$

This matrix already has 1 in row one, column one. Next get 0s in the rest of column one. First, add to row two the results of multiplying each number of row one by -3. This gives the matrix

$$\left[\begin{array}{ccc|c} 1 & -1 & 5 & -6 \\ 0 & 6 & -16 & 28 \\ 1 & 3 & 2 & 5 \end{array}\right]. \quad -3R_1 + R_2$$

Now add to the numbers in row three the results of multiplying each number of row one by -1.

$$\left[\begin{array}{ccc|c} 1 & -1 & 5 & -6 \\ 0 & 6 & -16 & 28 \\ 0 & 4 & -3 & 11 \end{array}\right] \quad -1R_1 + R_3$$

Continued on Next Page

Get 1 in row two, column two by multiplying each number in row two by $\frac{1}{6}$.

$$\left[\begin{array}{ccc|c} 1 & -1 & 5 & -6 \\ 0 & 1 & -\frac{8}{3} & \frac{14}{3} \\ 0 & 4 & -3 & 11 \end{array}\right] \quad \frac{1}{6}R_2$$

Introduce 0 in row three, column two by adding to row three the results of multiplying each number in row two by -4.

$$\left[\begin{array}{ccc|c} 1 & -1 & 5 & -6 \\ 0 & 1 & -\frac{8}{3} & \frac{14}{3} \\ 0 & 0 & \frac{23}{3} & -\frac{23}{3} \end{array}\right] \quad -4R_2 + R_3$$

Finally, obtain 1 in row three, column three by multiplying each number in row three by $\frac{3}{23}$.

$$\left[\begin{array}{ccc|c} 1 & -1 & 5 & -6 \\ 0 & 1 & -\frac{8}{3} & \frac{14}{3} \\ 0 & 0 & 1 & -1 \end{array}\right] \quad \frac{3}{23}R_3$$

This final matrix gives the system of equations

$$x - y + 5z = -6$$
$$y - \frac{8}{3}z = \frac{14}{3}$$
$$z = -1.$$

Substitute -1 for z in the second equation, $y - \frac{8}{3}z = \frac{14}{3}$, to get $y = 2$. Substitute 2 for y and -1 for z in the first equation, $x - y + 5z = -6$, to get $x = 1$. The solution set is $\{(1, 2, -1)\}$. Check by substitution in the original system.

Work Problem **2** *at the Side.* ▶

OBJECTIVE 5 **Use row operations to solve special systems.**

EXAMPLE 3 **Recognizing Inconsistent Systems or Dependent Equations**

Use row operations to solve each system.

(a) $2x - 3y = 8$
$-6x + 9y = 4$

$$\left[\begin{array}{cc|c} 2 & -3 & 8 \\ -6 & 9 & 4 \end{array}\right] \quad \text{Write the augmented matrix.}$$

$$\left[\begin{array}{cc|c} 1 & -\frac{3}{2} & 4 \\ -6 & 9 & 4 \end{array}\right] \quad \frac{1}{2}R_1$$

$$\left[\begin{array}{cc|c} 1 & -\frac{3}{2} & 4 \\ \mathbf{0} & \mathbf{0} & \mathbf{28} \end{array}\right] \quad 6R_1 + R_2$$

The corresponding system of equations is

$$x - \frac{3}{2}y = 4$$
$$0 = 28, \quad \text{False}$$

which has no solution and is inconsistent. The solution set is $\emptyset$.

Continued on Next Page

2 Use row operations to solve the system.

$$2x - y + z = 7$$
$$x - 3y - z = 7$$
$$-x + y - 5z = -9$$

ANSWER

2. $\{(2, -2, 1)\}$

3 Use row operations to solve each system.

(a) $\begin{aligned} x - y &= 2 \\ -2x + 2y &= 2 \end{aligned}$

(b) $\begin{aligned} x - y &= 2 \\ -2x + 2y &= -4 \end{aligned}$

Answers

3. (a) $\emptyset$ **(b)** $\{(x, y) | x - y = 2\}$

(b) $\begin{aligned} -10x + 12y &= 30 \\ 5x - 6y &= -15 \end{aligned}$

$$\left[\begin{array}{cc|c} -10 & 12 & 30 \\ 5 & -6 & -15 \end{array}\right] \quad \text{Write the augmented matrix.}$$

$$\left[\begin{array}{cc|c} 1 & -\frac{6}{5} & -3 \\ 5 & -6 & -15 \end{array}\right] \quad -\tfrac{1}{10}R_1$$

$$\left[\begin{array}{cc|c} 1 & -\frac{6}{5} & -3 \\ 0 & 0 & 0 \end{array}\right] \quad -5R_1 + R_2$$

The corresponding system is

$$x - \frac{6}{5}y = -3$$

$$0 = 0, \qquad \text{True}$$

which has dependent equations. Using the second equation of the original system, we write the solution set as

$$\{(x, y) | 5x - 6y = -15\}.$$

◀ *Work Problem* **3** *at the Side.*

5.4 Exercises

1. Consider the matrix $\begin{bmatrix} -2 & 3 & 1 \\ 0 & 5 & -3 \\ 1 & 4 & 8 \end{bmatrix}$, and answer the following.

(a) What are the elements of the second row?

(b) What are the elements of the third column?

(c) Is this a square matrix? Explain.

(d) Give the matrix obtained by interchanging the first and third rows.

(e) Give the matrix obtained by multiplying the first row by $-\frac{1}{2}$.

(f) Give the matrix obtained by multiplying the third row by 3 and adding it to the first row.

2. Give the dimensions of each matrix.

(a) $\begin{bmatrix} 3 & -7 \\ 4 & 5 \\ -1 & 0 \end{bmatrix}$ **(b)** $\begin{bmatrix} 4 & 9 & 0 \\ -1 & 2 & -4 \end{bmatrix}$ **(c)** $\begin{bmatrix} 6 & 3 \\ -2 & 5 \\ 4 & 10 \\ 1 & -11 \end{bmatrix}$

Complete the steps in the matrix solution of each system by filling in the blanks. Give the final system and the solution set. See Example 1.

3. $\begin{aligned} 4x + 8y &= 44 \\ 2x - y &= -3 \end{aligned}$

$\left[\begin{array}{cc|c} 4 & 8 & 44 \\ 2 & -1 & -3 \end{array}\right]$

$\left[\begin{array}{cc|c} 1 & \underline{\quad} & \underline{\quad} \\ 2 & -1 & -3 \end{array}\right]$ $\frac{1}{4}R_1$

$\left[\begin{array}{cc|c} 1 & 2 & 11 \\ 0 & \underline{\quad} & \underline{\quad} \end{array}\right]$ $-2R_1 + R_2$

$\left[\begin{array}{cc|c} 1 & 2 & 11 \\ 0 & 1 & \underline{\quad} \end{array}\right]$ $-\frac{1}{5}R_2$

4. $\begin{aligned} 2x - 5y &= -1 \\ 3x + y &= 7 \end{aligned}$

$\left[\begin{array}{cc|c} 2 & -5 & -1 \\ 3 & 1 & 7 \end{array}\right]$

$\left[\begin{array}{cc|c} 1 & -\frac{5}{2} & \underline{\quad} \\ 3 & 1 & 7 \end{array}\right]$ $\frac{1}{2}R_1$

$\left[\begin{array}{cc|c} 1 & -\frac{5}{2} & -\frac{1}{2} \\ 0 & \underline{\quad} & \underline{\quad} \end{array}\right]$ $-3R_1 + R_2$

$\left[\begin{array}{cc|c} 1 & -\frac{5}{2} & -\frac{1}{2} \\ 0 & 1 & \underline{\quad} \end{array}\right]$ $\frac{2}{17}R_2$

Use row operations to solve each system. See Examples 1 and 3.

5. $\begin{aligned} x + y &= 5 \\ x - y &= 3 \end{aligned}$

6. $\begin{aligned} x + 2y &= 7 \\ x - y &= -2 \end{aligned}$

7. $\begin{aligned} 2x + 4y &= 6 \\ 3x - y &= 2 \end{aligned}$

8. $\begin{aligned} 4x + 5y &= -7 \\ x - y &= 5 \end{aligned}$

9. $\begin{aligned} 3x + 4y &= 13 \\ 2x - 3y &= -14 \end{aligned}$

10. $\begin{aligned} 5x + 2y &= 8 \\ 3x - y &= 7 \end{aligned}$

11. $\begin{aligned} -4x + 12y &= 36 \\ x - 3y &= 9 \end{aligned}$

12. $\begin{aligned} 2x - 4y &= 8 \\ -3x + 6y &= 5 \end{aligned}$

13. $\begin{aligned} 2x + y &= 4 \\ 4x + 2y &= 8 \end{aligned}$

14. $\begin{aligned} -3x - 4y &= 1 \\ 6x + 8y &= -2 \end{aligned}$

15. $\begin{aligned} \frac{1}{2}x + \frac{1}{3}y &= 0 \\ \frac{2}{3}x + \frac{3}{4}y &= 0 \end{aligned}$

16. $\begin{aligned} 1.2x + 0.3y &= 0 \\ 2.9x - 0.6y &= 0 \end{aligned}$

Complete the steps in the matrix solution of each system by filling in the blanks. Give the final system and the solution set. See Example 2.

17. $\begin{aligned} x + y - z &= -3 \\ 2x + y + z &= 4 \\ 5x - y + 2z &= 23 \end{aligned}$

$$\left[\begin{array}{ccc|c} 1 & 1 & -1 & -3 \\ 2 & 1 & 1 & 4 \\ 5 & -1 & 2 & 23 \end{array}\right]$$

$$\left[\begin{array}{ccc|c} 1 & 1 & -1 & -3 \\ 0 & \underline{\quad} & \underline{\quad} & \underline{\quad} \\ 0 & \underline{\quad} & \underline{\quad} & \underline{\quad} \end{array}\right] \begin{array}{l} -2R_1 + R_2 \\ -5R_1 + R_3 \end{array}$$

$$\left[\begin{array}{ccc|c} 1 & 1 & -1 & -3 \\ 0 & 1 & \underline{\quad} & \underline{\quad} \\ 0 & -6 & 7 & 38 \end{array}\right] \quad -1R_2$$

$$\left[\begin{array}{ccc|c} 1 & 1 & -1 & -3 \\ 0 & 1 & -3 & -10 \\ 0 & 0 & \underline{\quad} & \underline{\quad} \end{array}\right] \quad 6R_2 + R_3$$

$$\left[\begin{array}{ccc|c} 1 & 1 & -1 & -3 \\ 0 & 1 & -3 & -10 \\ 0 & 0 & 1 & \underline{\quad} \end{array}\right] \quad -\tfrac{1}{11}R_3$$

18. $\begin{aligned} 2x + y + 2z &= 11 \\ 2x - y - z &= -3 \\ 3x + 2y + z &= 9 \end{aligned}$

$$\left[\begin{array}{ccc|c} 2 & 1 & 2 & 11 \\ 2 & -1 & -1 & -3 \\ 3 & 2 & 1 & 9 \end{array}\right]$$

$$\left[\begin{array}{ccc|c} 1 & \underline{\quad} & \underline{\quad} & \underline{\quad} \\ 2 & -1 & -1 & -3 \\ 3 & 2 & 1 & 9 \end{array}\right] \quad \tfrac{1}{2}R_1$$

$$\left[\begin{array}{ccc|c} 1 & \frac{1}{2} & 1 & \frac{11}{2} \\ 0 & \underline{\quad} & \underline{\quad} & \underline{\quad} \\ 0 & \underline{\quad} & \underline{\quad} & \underline{\quad} \end{array}\right] \begin{array}{l} -2R_1 + R_2 \\ -3R_1 + R_3 \end{array}$$

$$\left[\begin{array}{ccc|c} 1 & \frac{1}{2} & 1 & \frac{11}{2} \\ 0 & 1 & \underline{\quad} & \underline{\quad} \\ 0 & \frac{1}{2} & -2 & -\frac{15}{2} \end{array}\right] \quad -\tfrac{1}{2}R_2$$

$$\left[\begin{array}{ccc|c} 1 & \frac{1}{2} & 1 & \frac{11}{2} \\ 0 & 1 & \frac{3}{2} & 7 \\ 0 & 0 & \underline{\quad} & \underline{\quad} \end{array}\right] \quad -\tfrac{1}{2}R_2 + R_3$$

$$\left[\begin{array}{ccc|c} 1 & \frac{1}{2} & 1 & \frac{11}{2} \\ 0 & 1 & \frac{3}{2} & 7 \\ 0 & 0 & 1 & \underline{\quad} \end{array}\right] \quad -\tfrac{4}{11}R_3$$

Use row operations to solve each system. See Examples 2 and 3.

19. $\begin{aligned} x + y - 3z &= 1 \\ 2x - y + z &= 9 \\ 3x + y - 4z &= 8 \end{aligned}$

20. $\begin{aligned} 2x + 4y - 3z &= -18 \\ 3x + y - z &= -5 \\ x - 2y + 4z &= 14 \end{aligned}$

21. $\begin{aligned} x + y - z &= 6 \\ 2x - y + z &= -9 \\ x - 2y + 3z &= 1 \end{aligned}$

22. $\begin{aligned} x + 3y - 6z &= 7 \\ 2x - y + 2z &= 0 \\ x + y + 2z &= -1 \end{aligned}$

23. $\begin{aligned} x - y &= 1 \\ y - z &= 6 \\ x + z &= -1 \end{aligned}$

24. $\begin{aligned} x + y &= 1 \\ 2x - z &= 0 \\ y + 2z &= -2 \end{aligned}$

25. $\begin{aligned} 4x + 8y + 4z &= 9 \\ x + 3y + 4z &= 10 \\ 5x + 10y + 5z &= 12 \end{aligned}$

26. $\begin{aligned} x + 2y + 3z &= -2 \\ 2x + 4y + 6z &= -5 \\ x - y + 2z &= 5 \end{aligned}$

27. $\begin{aligned} x - 2y + z &= 4 \\ 3x - 6y + 3z &= 12 \\ -2x + 4y - 2z &= -8 \end{aligned}$

28. $\begin{aligned} x + 3y + z &= 1 \\ 2x + 6y + 2z &= 2 \\ 3x + 9y + 3z &= 3 \end{aligned}$

29. $\begin{aligned} 5x + 3y - z &= 0 \\ 2x - 3y + z &= 0 \\ x + 4y - 2z &= 0 \end{aligned}$

30. $\begin{aligned} 4x + 5y - z &= 0 \\ 7x - 5y + z &= 0 \\ x + 3y - 2z &= 0 \end{aligned}$

Chapter 5 Summary

Key Terms

5.1	**system of equations**	Two or more equations that are to be solved at the same time form a system of equations.
	linear system	A linear system is a system of equations that contains only linear equations.
	solution set of a system	All ordered pairs that satisfy all the equations of a system at the same time make up the solution set of the system.
	consistent system	A system is consistent if it has a solution.
	independent equations	Independent equations are equations whose graphs are different lines.
	inconsistent system	A system is inconsistent if it has no solution.
	dependent equations	Dependent equations are equations whose graphs are the same line.
5.4	**matrix**	A matrix is a rectangular array of numbers, consisting of horizontal **rows** and vertical **columns.**
	elements of a matrix	The numbers in a matrix are its elements.
	square matrix	A square matrix is a matrix that has the same number of rows as columns.
	augmented matrix	An augmented matrix is a matrix that has a vertical bar that separates the columns of the matrix into two groups.
	row echelon form	If a matrix is written with 1s down the diagonal from upper left to lower right and 0s below the 1s, it is said to be in row echelon form.

New Symbols

(x, y, z) ordered triple

$\begin{bmatrix} a & b & c \\ d & e & f \end{bmatrix}$ matrix with 2 rows, 3 columns (2×3)

Test Your Word Power

See how well you have learned the vocabulary in this chapter. Answers, with examples, follow the Quick Review.

1. A **system of equations** consists of
   - **A.** at least two equations with different variables
   - **B.** two or more equations that have an infinite number of solutions
   - **C.** two or more equations that are to be solved at the same time
   - **D.** two or more inequalities that are to be solved.

2. The **solution set of a system of equations in two variables** is
   - **A.** all ordered pairs that satisfy one equation of the system
   - **B.** all ordered pairs that satisfy all the equations of the system at the same time
   - **C.** any ordered pair that satisfies one or more equations of the system
   - **D.** the set of values that make all the equations of the system false.

3. An **inconsistent system** is a system of equations
   - **A.** with one solution
   - **B.** with no solution
   - **C.** with an infinite number of solutions
   - **D.** that have the same graph.

4. **Dependent equations**
   - **A.** have different graphs
   - **B.** have no solution
   - **C.** have one solution
   - **D.** are different forms of the same equation.

5. A **matrix** is
   - **A.** an ordered pair of numbers
   - **B.** an array of numbers with the same number of rows and columns
   - **C.** a pair of numbers written between brackets
   - **D.** a rectangular array of numbers.

▶ Quick Review

Concepts	Examples
5.1 Systems of Linear Equations in Two Variables	
Solving a Linear System by Substitution	Solve by substitution. $$\begin{aligned} 4x - y &= 7 \quad (1) \\ 3x + 2y &= 30 \quad (2) \end{aligned}$$
Step 1 Solve one of the equations for either variable.	Solve for y in equation (1). $$y = 4x - 7$$
Step 2 Substitute for that variable in the other equation. The result should be an equation with just one variable. ***Step 3*** Solve the equation from Step 2.	Substitute $4x - 7$ for y in equation (2), and solve for x. $$\begin{aligned} 3x + 2y &= 30 && (2) \\ 3x + 2(4x - 7) &= 30 && \text{Let } y = 4x - 7. \\ 3x + 8x - 14 &= 30 && \text{Distributive property} \\ 11x - 14 &= 30 && \text{Combine like terms.} \\ 11x &= 44 && \text{Add 14.} \\ x &= 4 && \text{Divide by 11.} \end{aligned}$$
Step 4 Find the value of the other variable by substituting the result from Step 3 into the equation from Step 1.	Substitute 4 for x in the equation $y = 4x - 7$ to find that $y = 9$.
Step 5 Check the ordered-pair solution in *both* of the *original* equations. Then write the solution set.	Check to see that $\{(4, 9)\}$ is the solution set.
Solving a Linear System by Elimination	Solve by elimination.
Step 1 Write both equations in standard form.	$$\begin{aligned} 5x + y &= 2 \quad (1) \\ 2x - 3y &= 11 \quad (2) \end{aligned}$$
Step 2 Make the coefficients of one pair of variable terms opposites.	To eliminate y, multiply equation (1) by 3, and add the result to equation (2).
Step 3 Add the new equations. The sum should be an equation with just one variable. ***Step 4*** Solve the equation from Step 3.	$$\begin{aligned} 15x + 3y &= 6 && \text{3 times equation (1)} \\ 2x - 3y &= 11 && (2) \\ \hline 17x &= 17 && \text{Add.} \\ x &= 1 && \text{Divide by 17.} \end{aligned}$$
Step 5 Find the value of the other variable by substituting the result from Step 4 into either of the original equations.	Let $x = 1$ in equation (1), and solve for y. $$\begin{aligned} 5(1) + y &= 2 \\ y &= -3 \end{aligned}$$
Step 6 Check the ordered-pair solution in *both* of the *original* equations. Then write the solution set.	Check to verify that $\{(1, -3)\}$ is the solution set.
If the result of the addition step (Step 3) is a false statement, such as $0 = 4$, the graphs are parallel lines and *there is no solution. The solution set is* $\emptyset$.	$$\begin{aligned} x - 2y &= 6 \\ -x + 2y &= -2 \\ \hline 0 &= 4 \end{aligned}$$ Solution set: $\emptyset$
If the result is a true statement, such as $0 = 0$, the graphs are the same line, and an *infinite number of ordered pairs are solutions. The solution set is written in set-builder notation as* $\{(x, y) \mid \_\_\_\_\_\_\}$, *where a form of the equation is written in the blank.*	$$\begin{aligned} x - 2y &= 6 \\ -x + 2y &= -6 \\ \hline 0 &= 0 \end{aligned}$$ Solution set: $\{(x, y) \mid x - 2y = 6\}$

Concepts	Examples

5.2 Systems of Linear Equations in Three Variables

Solving a Linear System in Three Variables

Step 1 Select a focus variable, preferably one with coefficient 1 or -1, and a working equation.

Solve the system.

$$\begin{aligned} x + 2y - z &= 6 \quad (1) \\ x + y + z &= 6 \quad (2) \\ 2x + y - z &= 7 \quad (3) \end{aligned}$$

We choose z as the focus variable and (2) as the working equation.

Step 2 Eliminate the focus variable, using the working equation and one of the equations of the system.

Add equations (1) and (2) to get

$$2x + 3y = 12. \quad (4)$$

Step 3 Eliminate the focus variable again, using the working equation and the remaining equation of the system.

Add equations (2) and (3) to get

$$3x + 2y = 13. \quad (5)$$

Step 4 Solve the system of two equations in two variables formed by the equations from Steps 2 and 3.

Use equations (4) and (5) to eliminate x.

$$\begin{aligned} -6x - 9y &= -36 && \text{Multiply (4) by } -3. \\ 6x + 4y &= 26 && \text{Multiply (5) by 2.} \\ \hline -5y &= -10 && \text{Add.} \\ y &= 2 && \text{Divide by } -5. \end{aligned}$$

To find x, substitute 2 for y in equation (4).

$$\begin{aligned} 2x + 3(2) &= 12 && \text{Let } y = 2 \text{ in (4).} \\ 2x + 6 &= 12 \\ 2x &= 6 \\ x &= 3 \end{aligned}$$

Step 5 Find the value of the remaining variable.

Substitute 3 for x and 2 for y in working equation (2).

$$\begin{aligned} x + y + z &= 6 \quad (2) \\ 3 + 2 + z &= 6 \\ z &= 1 \end{aligned}$$

Step 6 Check the ordered-triple solution in each of the original equations of the system. Then write the solution set.

A check of the solution (3, 2, 1) confirms that the solution set is $\{(3, 2, 1)\}$.

5.3 Applications of Systems of Linear Equations

Use the six-step problem-solving method.

Step 1 Read the problem carefully.

Step 2 Assign variables.

Step 3 Write a system of equations that relates the unknowns.

Step 4 Solve the system.

Step 5 State the answer.

Step 6 Check.

The perimeter of a rectangle is 18 ft. The length is 3 ft more than twice the width. What are the dimensions of the rectangle?

Let x represent the length and y represent the width. From the perimeter formula, one equation is $2x + 2y = 18$. From the problem, another equation is $x = 3 + 2y$. Solve the system

$$\begin{aligned} 2x + 2y &= 18 \\ x &= 3 + 2y \end{aligned}$$

to get $x = 7$ and $y = 2$. The length is 7 ft, and the width is 2 ft. Since the perimeter is

$$2(7) + 2(2) = 18, \quad \text{and} \quad 3 + 2(2) = 7,$$

the solution checks.

Concepts	Examples

5.4 Solving Systems of Linear Equations by Matrix Methods

Matrix Row Operations

1. Any two rows of the matrix may be interchanged.

$$\begin{bmatrix} 1 & 5 & 7 \\ 3 & 9 & -2 \\ 0 & 6 & 4 \end{bmatrix} \text{ becomes } \begin{bmatrix} 3 & 9 & -2 \\ 1 & 5 & 7 \\ 0 & 6 & 4 \end{bmatrix} \quad \text{Interchange } R_1 \text{ and } R_2.$$

2. The numbers in any row may be multiplied by any nonzero real number.

$$\begin{bmatrix} 1 & 5 & 7 \\ 3 & 9 & -2 \\ 0 & 6 & 4 \end{bmatrix} \text{ becomes } \begin{bmatrix} 1 & 5 & 7 \\ 1 & 3 & -\frac{2}{3} \\ 0 & 6 & 4 \end{bmatrix} \quad \tfrac{1}{3}R_2$$

3. Any row may be transformed by adding to the numbers of the row the product of a real number and the numbers of another row.

$$\begin{bmatrix} 1 & 5 & 7 \\ 3 & 9 & -2 \\ 0 & 6 & 4 \end{bmatrix} \text{ becomes } \begin{bmatrix} 1 & 5 & 7 \\ 0 & -6 & -23 \\ 0 & 6 & 4 \end{bmatrix} \quad -3R_1 - R_2$$

A system can be solved by matrix methods. Write the augmented matrix, and use row operations to obtain a matrix in row echelon form.

Solve using row operations.

$$\begin{aligned} x + 3y &= 7 \\ 2x + y &= 4 \end{aligned}$$

$$\left[\begin{array}{cc|c} 1 & 3 & 7 \\ 2 & 1 & 4 \end{array}\right] \quad \text{Augmented matrix}$$

$$\left[\begin{array}{cc|c} 1 & 3 & 7 \\ 0 & -5 & -10 \end{array}\right] \quad -2R_1 + R_2$$

$$\left[\begin{array}{cc|c} 1 & 3 & 7 \\ 0 & 1 & 2 \end{array}\right] \quad -\tfrac{1}{5}R_2$$

$$\begin{aligned} x + 3y &= 7 \\ y &= 2 \end{aligned}$$

When $y = 2$, $x + 3(2) = 7$, so $x = 1$. The solution set is $\{(1, 2)\}$.

ANSWERS TO TEST YOUR WORD POWER

1. C; *Example:* $\begin{aligned} 3x - y &= 3 \\ 2x + y &= 7 \end{aligned}$
2. B; *Example:* The ordered pair (2, 3) satisfies both equations of the system in Answer 1, so $\{(2, 3)\}$ is the solution set of the system.
3. B; *Example:* The equations of two parallel lines form an inconsistent system. Their graphs never intersect, so the system has no solution.
4. D; *Example:* The equations $4x - y = 8$ and $8x - 2y = 16$ are dependent because their graphs are the same line.
5. D; *Examples:* $\begin{bmatrix} 3 & -1 & 0 \\ 4 & 2 & 1 \end{bmatrix}, \begin{bmatrix} 1 & 2 \\ 4 & 3 \end{bmatrix}$

Chapter 5 Review Exercises

[5.1] 1. Solve by graphing:

$$x + 3y = 8$$
$$2x - y = 2.$$

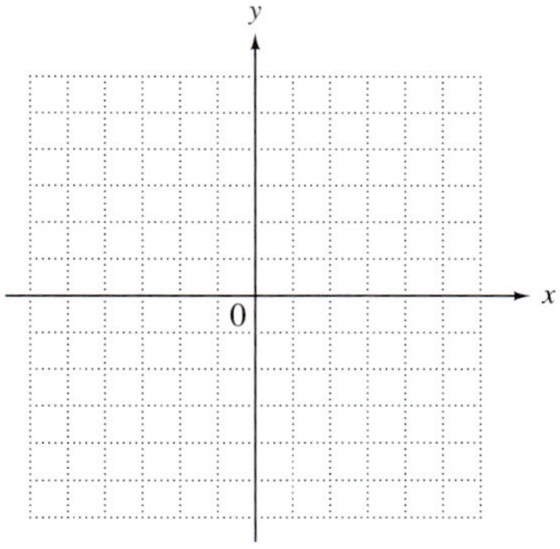

2. Which one of the following ordered pairs is not a solution of the equation $3x + 2y = 6$?

A. (2, 0) B. (0, 3)
C. (4, −3) D. (3, −2)

3. The graph shows the trends during the years 1975 through 2002 relating to bachelor's degrees awarded in the United States.

(a) Between what years shown on the horizontal axis did the number of degrees for men and women reach equal numbers?

(b) When the number of degrees for men and women reached equal numbers, what was that number (approximately)?

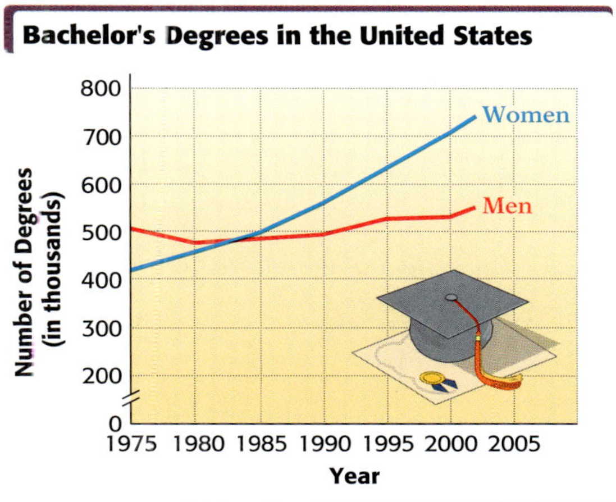

Source: U.S. National Center for Education Statistics, *Digest of Education Statistics*, annual.

Solve each system using the substitution method.

4. $3x + y = -4$
$x = \frac{2}{3}y$

5. $9x - y = -4$
$y = x + 4$

6. $-5x + 2y = -2$
$x + 6y = 26$

Solve each system using the elimination method. If a system is inconsistent or has dependent equations, say so.

7. $6x + 5y = 4$
$-4x + 2y = 8$

8. $\frac{x}{6} + \frac{y}{6} = -\frac{1}{2}$
$x - y = -9$

9. $4x + 5y = 9$
$3x + 7y = -1$

10. $-3x + y = 6$
$2y = 12 + 6x$

11. $5x - 4y = 2$
$-10x + 8y = 7$

12. $3x + 3y = 0$
$-2x - y = 0$

Suppose that two linear equations are graphed on the same set of coordinate axes. Sketch what the graph might look like if the system has the given description.

13. The system has a single solution.

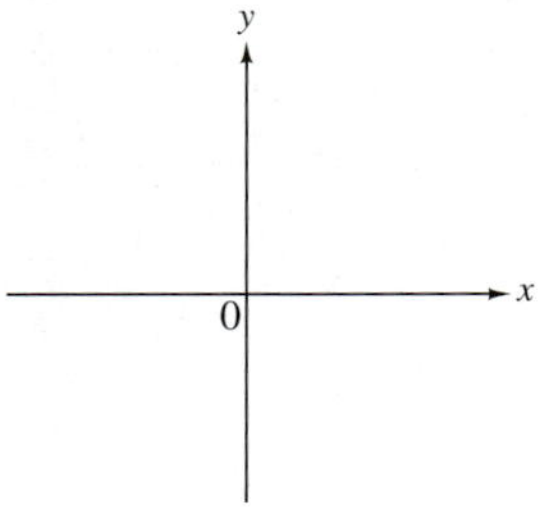

14. The system has no solution.

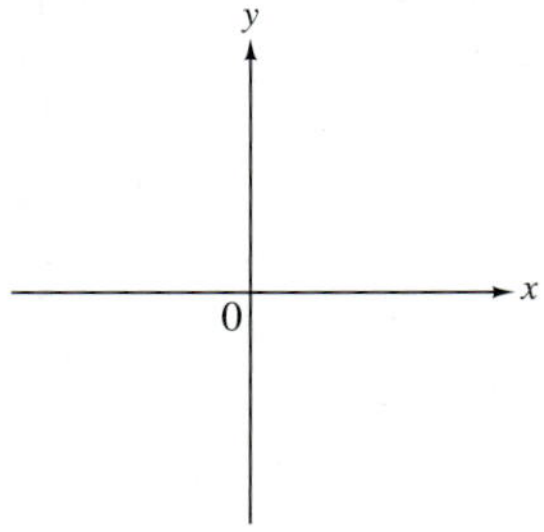

15. The system has infinitely many solutions.

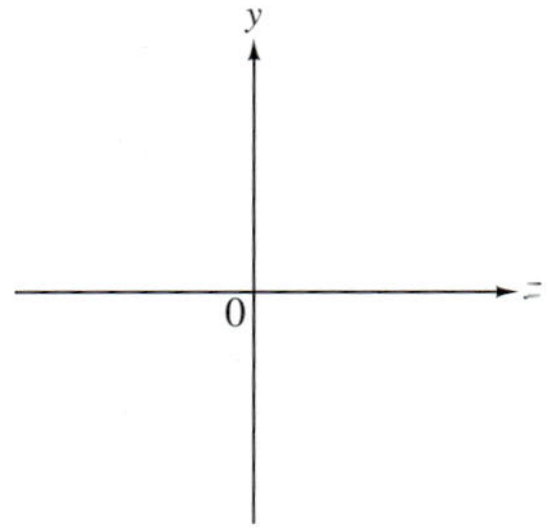

16. Without doing any algebraic work, explain why the following system has $\emptyset$ as its solution set. Base your answer only on your knowledge of the graphs of the two lines.

$$\begin{aligned} y &= 3x + 2 \\ y &= 3x - 4 \end{aligned}$$

[5.2] *Solve each system of equations. If a system is inconsistent or has dependent equations, say so.*

17.
$$\begin{aligned} 2x + 3y - z &= -16 \\ x + 2y + 2z &= -3 \\ -3x + y + z &= -5 \end{aligned}$$

18.
$$\begin{aligned} 3x - y - z &= -8 \\ 4x + 2y + 3z &= 15 \\ -6x + 2y + 2z &= 10 \end{aligned}$$

19.
$$\begin{aligned} 4x - y &= 2 \\ 3y + z &= 9 \\ x + 2z &= 7 \end{aligned}$$

[5.3] *Solve each problem using a system of equations.*

20. A regulation National Hockey League ice rink has perimeter 570 ft. The length is 30 ft longer than twice the width. What are the dimensions of an NHL ice rink? (*Source: Microsoft Encarta Encyclopedia.*)

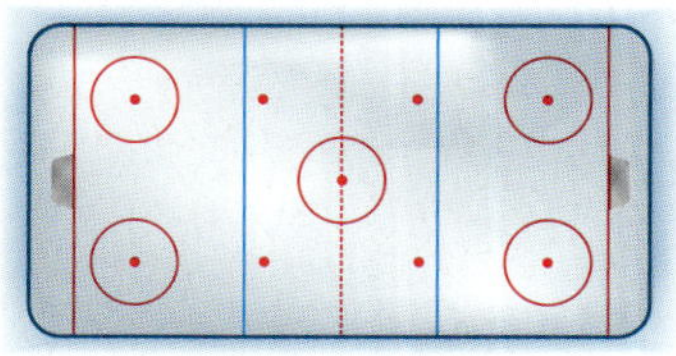

21. In 2004, the Boston Red Sox and the Chicago Cubs had the most expensive average ticket prices in Major League Baseball. Four Red Sox tickets and four Cubs tickets purchased at their average prices cost \$276.88, while two Red Sox tickets and six Cubs tickets cost \$252.24. Find the average ticket price for a Red Sox ticket and a Cubs ticket. (*Source:* AP.)

22. A plane flies 560 mi in 1.75 hr traveling with the wind. The return trip later against the same wind takes the plane 2 hr. Find the speed of the plane and the speed of the wind.

	r	*t*	*d*
With Wind	$x + y$	1.75	
Against Wind		2	

23. Sweet's Candy Store is offering a special mix for Valentine's Day. Ms. Sweet will mix some \$2 per lb nuts with some \$1 per lb chocolate candy to get 100 lb of mix, which she will sell at \$1.30 per lb. How many pounds of each should she use?

	Number of Pounds	Price per Pound	Value
Nuts	x		
Chocolate	y		
Mixture	100		

24. A biologist wants to grow two types of algae, green and brown. She has 15 kg of nutrient X and 26 kg of nutrient Y. A vat of green algae needs 2 kg of nutrient X and 3 kg of nutrient Y, while a vat of brown algae needs 1 kg of nutrient X and 2 kg of nutrient Y. How many vats of each type of algae should she grow in order to use all the nutrients?

25. The sum of the measures of the angles of a triangle is 180°. The largest angle measures 10° less than the sum of the other two. The measure of the middle-sized angle is the average of the other two. Find the measures of the three angles.

26. How many liters each of 8%, 10%, and 20% hydrogen peroxide should be mixed together to get 8 L of 12.5% solution, if the amount of 8% solution used must be 2 L more than the amount of 20% solution used?

27. In the great baseball year of 1961, Yankee teammates Mickey Mantle, Roger Maris, and John Blanchard combined for 136 home runs. Mantle hit 7 fewer than Maris. Maris hit 40 more than Blanchard. What were the home run totals for each player? (*Source:* Neft, David S. and Richard M. Cohen, *The Sports Encyclopedia: Baseball 2003.*)

[5.4] *Solve each system using row operations.*

28. $$\begin{aligned} 2x + 5y &= -4 \\ 4x - y &= 14 \end{aligned}$$

29. $$\begin{aligned} 6x + 3y &= 9 \\ -7x + 2y &= 17 \end{aligned}$$

30. $$\begin{aligned} x + 2y - z &= 1 \\ 3x + 4y + 2z &= -2 \\ -2x - y + z &= -1 \end{aligned}$$

Mixed Review Exercises

31. Which system, A or B, would be easier to solve using the substitution method? Why?

A. $$\begin{aligned} 5x - 3y &= 7 \\ 2x + 8y &= 3 \end{aligned}$$

B. $$\begin{aligned} 7x + 2y &= 4 \\ y &= -3x + 1 \end{aligned}$$

Solve by any method.

32. $\frac{2}{3}x + \frac{1}{6}y = \frac{19}{2}$
$\frac{1}{3}x - \frac{2}{9}y = 2$

33. $2x - 5y = 8$
$3x + 4y = 10$

34. $x = 7y + 10$
$2x + 3y = 3$

35. $x + 4y = 17$
$-3x + 2y = -9$

36. $-7x + 3y = 12$
$5x + 2y = 8$

37. $2x + 5y - z = 12$
$-x + y - 4z = -10$
$-8x - 20y + 4z = 31$

38. To make a 10% acid solution for chemistry class, Xavier wants to mix some 5% solution with 10 L of 20% solution. How many liters of 5% solution should he use?

39. In the 2006 Winter Olympics in Turino, Italy, the top medal-winning countries were Germany, the United States, and Canada, with a combined total of 78 medals. Germany won four more medals than the United States, while Canada won one fewer medal than the United States. How many medals did each country win? (*Source:* www.nbcolympics.com)

Relating Concepts (Exercises 40–44) For Individual or Group Work

Thus far in this text we have studied only linear *equations. In later chapters we will study the graphs of other kinds of equations. One such graph is a* circle, *which has an equation of the form*

$$x^2 + y^2 + ax + by + c = 0. \quad \text{Equation of a circle}$$

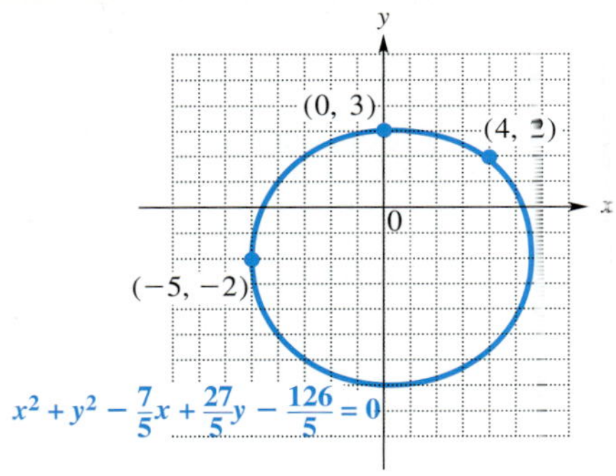

It is a fact from geometry that given three noncollinear *points (that is, points that do not all lie on the same straight line), there will be a circle that contains them. For example, the points* (4, 2), (−5, −2), *and* (0, 3) *lie on the circle whose equation is shown in the figure.* ***Work Exercises 40–44 in order,*** *to find an equation of the circle passing through the points* (2, 1), (−1, 0), *and* (3, 3).

40. Let $x = 2$ and $y = 1$ in the equation $x^2 + y^2 + ax + by + c = 0$ to find an equation in a, b, and c.

41. Let $x = -1$ and $y = 0$ to find a second equation in a, b, and c.

42. Let $x = 3$ and $y = 3$ to find a third equation in a, b, and c.

43. Solve the system of equations formed by your answers in Exercises 40–42 to find the values of a, b, and c. What is the equation of the circle?

44. Explain why the relation whose graph is a circle is not a function.

Chapter 5 Test

CHAPTER Test Prep VIDEO CD Use the Chapter Test Prep Video CD to see fully worked-out solutions to any of the exercises you want to review

The graph shows the numbers of people in the United States of two different races living with AIDS during the period 1995–2005. Use the graph to answer the questions in Exercises 1 and 2.

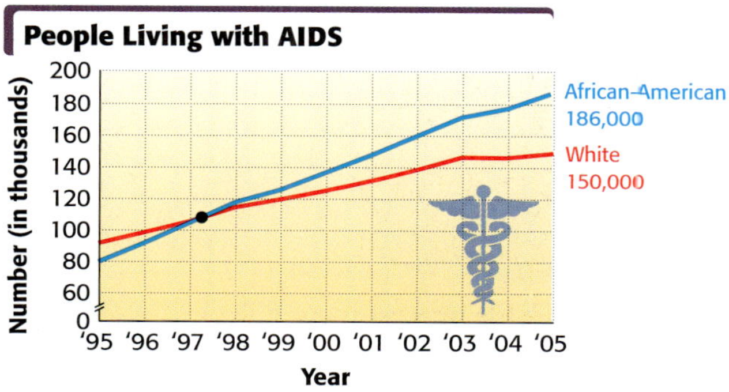

Source: U.S. Centers for Disease Control and Prevention.

1. In what year were the numbers the same? 1. ______

2. How many people of each race were living with AIDS that year? 2. ______

3. Use a graph to solve the system. 3. ______

$$x + y = 7$$
$$x - y = 5$$

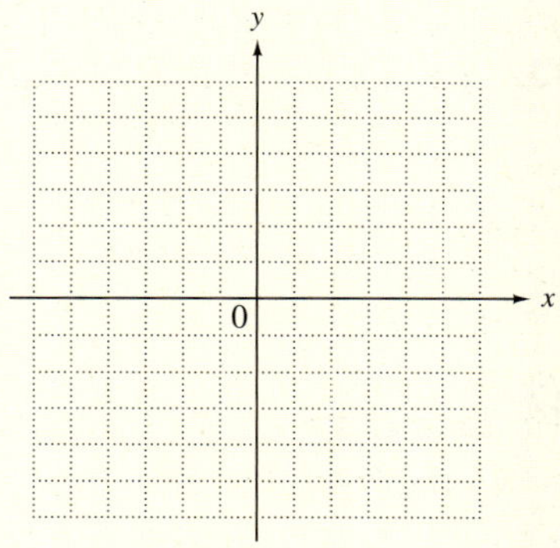

Solve each system by substitution or elimination. If a system is inconsistent or has dependent equations, say so.

4. $2x - 3y = 24$
$y = -\frac{2}{3}x$

5. $12x - 5y = 8$
$3x = \frac{5}{4}y + 2$

4. ______

5. ______

6. $3x - y = -8$
$2x + 6y = 3$

7. $3x + y = 12$
$2x - y = 3$

6. ______

7. ______

8. ____________________

9. ____________________

8. $\begin{aligned} -5x + 2y &= -4 \\ 6x + 3y &= -6 \end{aligned}$

9. $\begin{aligned} 3x + 4y &= 8 \\ 8y &= 7 - 6x \end{aligned}$

10. ____________________

11. ____________________

10. $\begin{aligned} 3x + 5y + 3z &= 2 \\ 6x + 5y + z &= 0 \\ 3x + 10y - 2z &= 6 \end{aligned}$

11. $\begin{aligned} 4x + y + z &= 11 \\ x - y - z &= 4 \\ y + 2z &= 0 \end{aligned}$

Solve each problem using a system of equations.

12. ____________________

12. Julia Roberts is one of the biggest box-office stars in Hollywood. As of December 2007, her two top-grossing domestic films, *Ocean's Eleven* and *Runaway Bride,* together earned \$335.5 million. If *Runaway Bride* grossed \$31.3 million less than *Ocean's Eleven,* how much did each film gross? (*Source:* www.the-numbers.com)

13. ____________________

13. Two cars start from points 420 mi apart and travel toward each other. They meet after 3.5 hr. Find the average speed of each car if one travels 30 mph slower than the other.

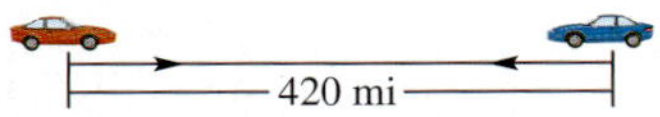

14. ____________________

14. A chemist needs 12 L of a 40% alcohol solution. She must mix a 20% solution and a 50% solution. How many liters of each will be required to obtain what she needs?

15. ____________________

15. A local electronics store will sell 7 AC adaptors and 2 rechargeable flashlights for \$86, or 3 AC adaptors and 4 rechargeable flashlights for \$84. What is the price of a single AC adaptor and a single rechargeable flashlight?

16. ____________________

16. The owner of a tea shop wants to mix three kinds of tea to make 100 oz of a mixture that will sell for \$0.83 per oz. He uses Orange Pekoe, which sells for \$0.80 per oz, Irish Breakfast, for \$0.85 per oz, and Earl Grey, for \$0.95 per oz. If he wants to use twice as much Orange Pekoe as Irish Breakfast, how much of each kind of tea should he use?

Solve each system using row operations.

17. ____________________

18. ____________________

17. $\begin{aligned} 3x + 2y &= 4 \\ 5x + 5y &= 9 \end{aligned}$

18. $\begin{aligned} x + 3y + 2z &= 11 \\ 3x + 7y + 4z &= 23 \\ 5x + 3y - 5z &= -14 \end{aligned}$

Cumulative Review Exercises Chapters 1–5

Evaluate.

1. $(-3)^4$
2. -3^4
3. $-(-3)^4$
4. $\sqrt{0.49}$
5. $-\sqrt{0.49}$
6. $\sqrt{-0.49}$

Evaluate if $x = -4$, $y = 3$, and $z = 6$.

7. $|2x| + y^2 - z^3$
8. $-5(x^3 - y^3)$
9. $\dfrac{2x^2 - x + z}{y^2 - z}$

Solve each equation.

10. $7(2x + 3) - 4(2x + 1) = 2(x + 1)$
11. $0.04x + 0.06(x - 1) = 1.04$
12. $ax + by = c$ for x
13. $|6x - 8| = 4$

Solve each inequality.

14. $\frac{2}{3}x + \frac{5}{12}x \leq 20$
15. $|3x + 2| \leq 4$
16. $|12t + 7| \geq 0$

17. A survey measured public recognition of the most popular contemporary advertising slogans. Complete the results shown in the table if 2500 people were surveyed.

Slogan (product or company)	Percent Recognition (nearest tenth of a percent)	Actual Number Who Recognized Slogan (nearest whole number)
Please Don't Squeeze the . . . (Charmin)	80.4%	
The Breakfast of Champions (Wheaties)	72.5%	
The King of Beers (Budweiser)		1570
Like a Good Neighbor (State Farm)		1430

(Other slogans included "You're in Good Hands" (Allstate), "Snap, Crackle, Pop" (Rice Krispies), and "The Un-Cola" (7-Up).)
Source: Department of Integrated Marketing Communications, Northwestern University.

Solve each problem.

18. A jar contains only pennies, nickels, and dimes. The number of dimes is 1 more than the number of nickels, and the number of pennies is 6 more than the number of nickels. How many of each denomination can be found in the jar, if the total value is \$4.80?

19. Two angles of a triangle have the same measure. The measure of the third angle is 4° less than twice the measure of each of the equal angles. Find the measures of the three angles.

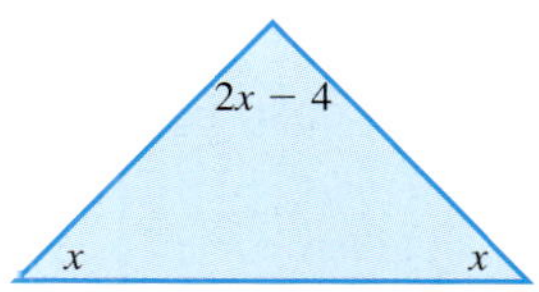

Measures are in degrees.

In Exercises 20–25, point A has coordinates $(-2, 6)$ *and point B has coordinates* $(4, -2)$.

20. What is the equation of the horizontal line through A?

21. What is the equation of the vertical line through B?

22. What is the slope of AB?

23. What is the slope of a line perpendicular to line AB?

24. What is the standard form of the equation of line AB?

25. Write the equation of the line in the form of a linear function.

26. Graph the linear function whose graph has slope $\frac{2}{3}$ and passes through the point $(-1, -3)$.

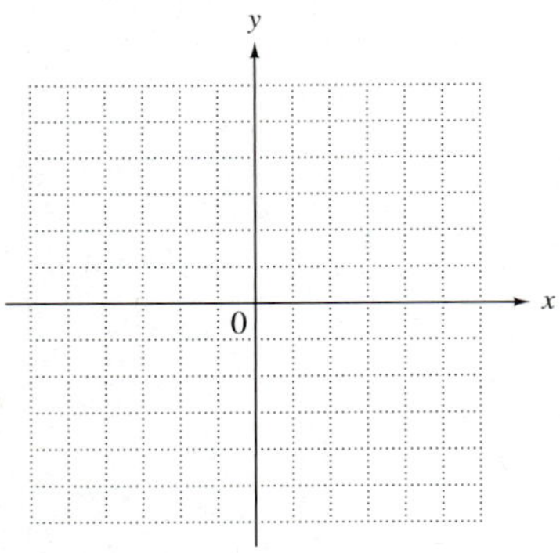

27. Graph the inequality $-3x - 2y \le 6$.

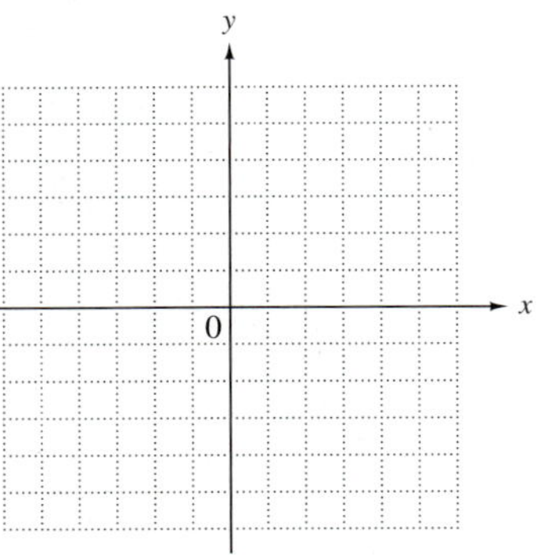

Solve by any method.

28.
$$\begin{aligned} -2x + 3y &= -15 \\ 4x - y &= 15 \end{aligned}$$

29.
$$\begin{aligned} x + y + z &= 10 \\ x - y - z &= 0 \\ -x + y - z &= -4 \end{aligned}$$

Solve each problem using a system of equations.

30. A grocer plans to mix candy that sells for \$1.20 per lb with candy that sells for \$2.40 per lb to get a mixture that he plans to sell for \$1.65 per lb. How much of the \$1.20 and \$2.40 candy should he use if he wants 80 lb of the mix?

31. A small company took out three loans totaling \$25,000. The company was able to borrow some of the money at 8%. It borrowed \$2000 more than $\frac{1}{2}$ the amount of the 8% loan at 10%, and the rest at 9%. The total annual interest was \$2220. How much did the company borrow at each rate?

The graph shows a company's costs to produce computer parts and the revenue from the sale of those parts.

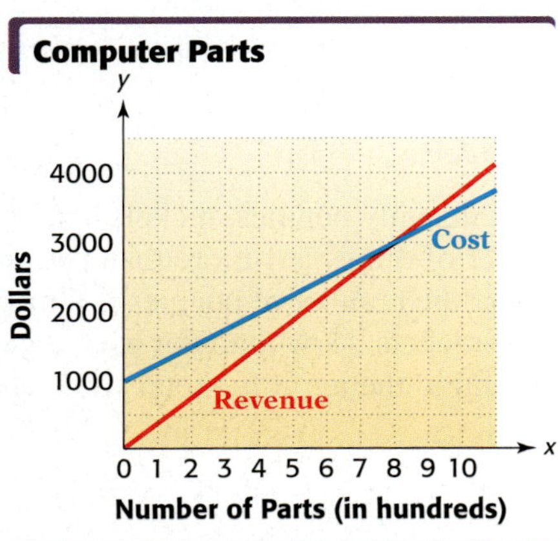

32. At what production level does the cost equal the revenue? What is the revenue at that point?

33. Profit is revenue less cost. Estimate the profit on the sale of 1100 parts.

6

Exponents, Polynomials, and Polynomial Functions

6.1 Integer Exponents and Scientific Notation

6.2 Adding and Subtracting Polynomials

6.3 Polynomial Functions

6.4 Multiplying Polynomials

6.5 Dividing Polynomials

One of the most popular tourist sites in the world is the magnificent Taj Mahal (or "the Taj"), at Agra, India, which attracts about 2.4 million visitors a year. The Taj, a domed white marble tomb built by Emperor Shah Jahan in memory of his wife, was completed in the year 1648. (*Source:* www.greatbuildings.com, www.forbestraveler.com)

We introduced the concept of function in Section 4.5 and extend our work to include *polynomial functions* in this chapter. In Exercise 11 of Section 6.3, we use a polynomial function to model the amount Americans spend on foreign travel.

6.1 Integer Exponents and Scientific Notation

OBJECTIVES

1. Use the product rule for exponents.
2. Define 0 and negative exponents.
3. Use the quotient rule for exponents.
4. Use the power rules for exponents.
5. Simplify exponential expressions.
6. Use the rules for exponents with scientific notation.

Recall from **Section 1.3** that we use exponents to write products of repeated factors. For example,

$$2^5 \quad \text{is defined as} \quad 2 \cdot 2 \cdot 2 \cdot 2 \cdot 2 = 32.$$

The number 5, the **exponent,** shows that the **base** 2 appears as a factor 5 times. The quantity 2^5 is called an **exponential,** or a **power.** We read 2^5 as "2 to the fifth power" or "2 to the fifth."

OBJECTIVE 1 Use the product rule for exponents. The product $2^5 \cdot 2^3$ can be simplified as follows.

$$5 + 3 = 8$$

$$2^5 \cdot 2^3 = (2 \cdot 2 \cdot 2 \cdot 2 \cdot 2)(2 \cdot 2 \cdot 2) = 2^8$$

This result, that products of exponential expressions with the *same base* are found by adding exponents, is generalized as the **product rule for exponents.**

Product Rule for Exponents

If m and n are natural numbers and a is any real number, then

$$a^m \cdot a^n = a^{m+n}.$$

In words, when multiplying powers of like bases, keep the same base and add the exponents.

To see that the product rule is true, use the definition of an exponent.

$$a^m = \underbrace{a \cdot a \cdot a \cdots a}_{a \text{ appears as a factor } m \text{ times.}} \qquad a^n = \underbrace{a \cdot a \cdot a \cdots a}_{a \text{ appears as a factor } n \text{ times.}}$$

From this,

$$a^m \cdot a^n = \underbrace{a \cdot a \cdot a \cdots a}_{m \text{ factors}} \cdot \underbrace{a \cdot a \cdot a \cdots a}_{n \text{ factors}}$$

$$= \underbrace{a \cdot a \cdot a \cdots a}_{(m+n) \text{ factors}}$$

$$a^m \cdot a^n = a^{m+n}.$$

EXAMPLE 1 Using the Product Rule for Exponents

Apply the product rule for exponents, if possible, in each case.

(a) $3^4 \cdot 3^7 = 3^{4+7} = 3^{11}$ — Do not multiply the bases. Keep the same base.

(b) $5^3 \cdot 5 = 5^3 \cdot 5^1 = 5^{3+1} = 5^4$

(c) $y^3 \cdot y^8 \cdot y^2 = y^{3+8+2} = y^{13}$

(d) $(5y^2)(-3y^4)$
$= 5(-3)y^2y^4$
$= -15y^{2+4}$
$= -15y^6$

(e) $(7p^3q)(2p^5q^2)$
$= 7(2)p^3p^5q^1q^2$
$= 14p^8q^3$

(f) $x^2 \cdot y^4$ The bases are not the same; the product rule does not apply.

◀ Work Problem 1 at the Side.

1 Apply the product rule for exponents, if possible, in each case.

(a) $m^8 \cdot m^6$

(b) $r^7 \cdot r$

(c) $k^4k^3k^6$

(d) $m^5 \cdot p^4$

(e) $(-4a^3)(6a^2)$

(f) $(-5p^4)(-9p^5)$

ANSWERS

1. (a) m^{14} (b) r^8 (c) k^{13} (d) The product rule does not apply. (e) $-24a^5$ (f) $45p^9$

CAUTION

Be careful in problems like Example 1(a) not to multiply the bases. Notice that $3^4 \cdot 3^7 = 3^{11}$, *not* 9^{11}. ***Keep the same base and add the exponents.***

OBJECTIVE 2 Define 0 and negative exponents. Suppose we multiply 4^2 by 4^0. By the product rule, extended to whole numbers,

$$4^2 \cdot 4^0 = 4^{2+0} = 4^2.$$

For the product rule to hold true, 4^0 must equal 1, and so we define a^0 this way for any nonzero real number a.

Zero Exponent

If a is any nonzero real number, then

$$\mathbf{a^0 = 1.}$$

The expression 0^0 is undefined.*

EXAMPLE 2 Using 0 as an Exponent

Evaluate.

(a) $6^0 = 1$

(b) $(-6)^0 = 1$

The base is 6, not −6.

Here the base is −6.

(c) $-6^0 = -(6^0) = -1$

(d) $-(-6)^0 = -1$

(e) $5^0 + 12^0$
$= 1 + 1$
$= 2$

(f) $(8k)^0 = 1, \quad k \neq 0$

Work Problem **2** at the Side. ▶

To define a negative exponent, we extend the product rule. For example,

$$8^2 \cdot 8^{-2} = 8^{2+(-2)} = 8^0 = 1.$$

Here 8^{-2} is the reciprocal of 8^2. But $\frac{1}{8^2}$ is the reciprocal of 8^2, and a number can have only one reciprocal. Therefore, $8^{-2} = \frac{1}{8^2}$. We can generalize this result as follows.

Negative Exponent

For any natural number n and any nonzero real number a,

$$\mathbf{a^{-n} = \frac{1}{a^n}.}$$

With this definition, the expression a^n is meaningful for any integer exponent n and any nonzero real number a.

*In advanced treatments, 0^0 is called an *indeterminate form*.

2 Evaluate.

(a) 29^0

(b) $(-29)^0$

(c) $-(-29)^0$

(d) -29^0

(e) $8^0 - 15^0$

(f) $(-15p^5)^0, \quad p \neq 0$

ANSWERS

2. (a) 1 **(b)** 1 **(c)** −1 **(d)** −1 **(e)** 0 **(f)** 1

CAUTION

A negative exponent does not indicate that an expression represents a negative number. Negative exponents lead to reciprocals.

Expression	Example	
a^{-m}	$3^{-2} = \frac{1}{3^2} = \frac{1}{9}$	**Not negative**
$-a^{-m}$	$-3^{-2} = -\frac{1}{3^2} = -\frac{1}{9}$	**Negative**

EXAMPLE 3 Using Negative Exponents

In parts (a)–(f), write with only positive exponents.

(a) $2^{-3} = \frac{1}{2^3}$

(b) $6^{-1} = \frac{1}{6^1} = \frac{1}{6}$

(c) $(5z)^{-3} = \frac{1}{(5z)^3}, \quad z \neq 0$

Base is $5z$.

(d) $5z^{-3} = 5\left(\frac{1}{z^3}\right) = \frac{5}{z^3}, \quad z \neq 0$

Base is z.

(e) $-m^{-2} = -\frac{1}{m^2}, \quad m \neq 0$

(What is the base here?)

(f) $(-m)^{-2} = \frac{1}{(-m)^2}, \quad m \neq 0$

(What is the base here?)

In parts (g) and (h), evaluate.

(g) $3^{-1} + 4^{-1}$

$= \frac{1}{3} + \frac{1}{4}$

$= \frac{4}{12} + \frac{3}{12} \qquad \frac{1}{3} \cdot \frac{4}{4} = \frac{4}{12}; \frac{1}{4} \cdot \frac{3}{3} = \frac{3}{12}$

$= \frac{7}{12}$

(h) $5^{-1} - 2^{-1}$

$= \frac{1}{5} - \frac{1}{2}$

$= \frac{2}{10} - \frac{5}{10}$

$= -\frac{3}{10}$

CAUTION

In Example 3(g), note that $3^{-1} + 4^{-1} \neq (3 + 4)^{-1}$. The expression on the left is equal to $\frac{7}{12}$, as shown in the solution, while the expression on the right is $7^{-1} = \frac{1}{7}$. Similar reasoning can be applied to part (h).

Work Problem 3 at the Side.

EXAMPLE 4 Using Negative Exponents

Evaluate.

(a) $\frac{1}{2^{-3}} = \frac{1}{\frac{1}{2^3}} = 1 \div \frac{1}{2^3} = 1 \cdot \frac{2^3}{1} = 2^3 = 8$

Multiply by the reciprocal of the divisor.

Continued on Next Page

3 In parts (a)–(f), write with only positive exponents. In parts (g) and (h), evaluate.

(a) 6^{-3}

(b) 8^{-1}

(c) $(2x)^{-4}, \quad x \neq 0$

(d) $7r^{-6}, \quad r \neq 0$

(e) $-q^{-4}, \quad q \neq 0$

(f) $(-q)^{-4}, \quad q \neq 0$

(g) $3^{-1} + 5^{-1}$

(h) $4^{-1} - 2^{-1}$

ANSWERS

3. (a) $\frac{1}{6^3}$ (b) $\frac{1}{8}$ (c) $\frac{1}{(2x)^4}$ (d) $\frac{7}{r^6}$ (e) $-\frac{1}{q^4}$ (f) $\frac{1}{(-q)^4}$ (g) $\frac{8}{15}$ (h) $-\frac{1}{4}$

(b) $\dfrac{2^{-3}}{3^{-2}} = \dfrac{\frac{1}{2^3}}{\frac{1}{3^2}} = \dfrac{1}{2^3} \div \dfrac{1}{3^2} = \dfrac{1}{2^3} \cdot \dfrac{3^2}{1} = \dfrac{3^2}{2^3} = \dfrac{9}{8}$

Example 4 suggests the following generalizations.

Special Rules for Negative Exponents

If $a \neq 0$ and $b \neq 0$, then $\dfrac{1}{a^{-n}} = a^n$ and $\dfrac{a^{-n}}{b^{-m}} = \dfrac{b^m}{a^n}$.

Work Problem 4 at the Side. ▶

OBJECTIVE 3 Use the quotient rule for exponents. A quotient, such as $\frac{a^8}{a^3}$, can be simplified in much the same way as a product. (In all quotients of this type, assume that the denominator is not 0.) Using the definition of an exponent,

$$\frac{a^8}{a^3} = \frac{a \cdot a \cdot a \cdot a \cdot a \cdot a \cdot a \cdot a}{a \cdot a \cdot a} = a \cdot a \cdot a \cdot a \cdot a = a^5.$$

Notice that $8 - 3 = 5$. In the same way,

$$\frac{a^3}{a^8} = \frac{a \cdot a \cdot a}{a \cdot a \cdot a \cdot a \cdot a \cdot a \cdot a \cdot a} = \frac{1}{a^5} = a^{-5}.$$

Here, $3 - 8 = -5$. These examples suggest the **quotient rule for exponents.**

Quotient Rule for Exponents

If a is any nonzero real number and m and n are integers, then

$$\frac{a^m}{a^n} = a^{m-n}.$$

In words, when dividing powers of like bases, keep the same base and subtract the exponent of the denominator from the exponent of the numerator.

EXAMPLE 5 Using the Quotient Rule for Exponents

Apply the quotient rule for exponents, if possible, and write each result using only positive exponents.

Numerator exponent

Denominator exponent

(a) $\dfrac{3^7}{3^2} = 3^{7-2} = 3^5$

Minus sign

(b) $\dfrac{p^6}{p^2} = p^{6-2} = p^4, \quad p \neq 0$

(c) $\dfrac{k^7}{k^{12}} = k^{7-12} = k^{-5} = \dfrac{1}{k^5}, \quad k \neq 0$

(d) $\dfrac{2^7}{2^{-3}} = 2^{7-(-3)} = 2^{7+3} = 2^{10}$

Use parentheses to avoid errors.

Continued on Next Page

4 Evaluate.

(a) $\dfrac{1}{4^{-3}}$

(b) $\dfrac{3^{-3}}{9^{-1}}$

Answers

4. (a) 64 **(b)** $\frac{1}{3}$

5 Apply the quotient rule for exponents, if possible, and write each result using only positive exponents. Assume that all variables are nonzero.

(a) $\frac{4^8}{4^6}$ (b) $\frac{x^{12}}{x^3}$

(c) $\frac{r^5}{r^8}$ (d) $\frac{2^8}{2^{-4}}$

(e) $\frac{6^{-3}}{6^4}$ (f) $\frac{8}{8^{-1}}$

(g) $\frac{t^{-4}}{t^{-6}}$ (h) $\frac{x^3}{y^5}$

6 Use one or more power rules to simplify each expression.

(a) $(r^5)^4$

(b) $\left(\frac{3}{4}\right)^3$

(c) $(9x)^3$

(d) $(5r^6)^3$

(e) $\left(\frac{-3n^4}{m}\right)^3, \quad m \neq 0$

Answers

5. (a) 4^2 (b) x^9 (c) $\frac{1}{r^3}$ (d) 2^{12} (e) $\frac{1}{6^7}$ (f) 8^2 (g) t^2 (h) The quotient rule does not apply.

6. (a) r^{20} (b) $\frac{27}{64}$ (c) $729x^3$ (d) $125r^{18}$ (e) $\frac{-27n^{12}}{m^3}$

(e) $\frac{8^{-2}}{8^5} = 8^{-2-5} = 8^{-7} = \frac{1}{8^7}$

(f) $\frac{6}{6^{-1}} = \frac{6^1}{6^{-1}} = 6^{1-(-1)} = 6^2$

(g) $\frac{z^{-5}}{z^{-8}} = z^{-5-(-8)} = z^3, \quad z \neq 0$

Be careful with signs.

(h) $\frac{a^3}{b^4}, \quad b \neq 0$ This expression cannot be simplified further.

The quotient rule does not apply because the bases are different.

◀ Work Problem **5** at the Side.

OBJECTIVE 4 Use the power rules for exponents. The expression $(3^4)^2$ can be simplified as

$$(3^4)^2 = 3^4 \cdot 3^4 = 3^{4+4} = 3^8,$$

where $4 \cdot 2 = 8$. This example suggests the first **power rule for exponents.** The other two power rules can be demonstrated with similar examples.

Power Rules for Exponents

If a and b are real numbers and m and n are integers, then

(a) $(a^m)^n = a^{mn}$, **(b)** $(ab)^m = a^m b^m$,

and **(c)** $\left(\frac{a}{b}\right)^m = \frac{a^m}{b^m} \quad (b \neq 0).$

In words,

(a) to raise a power to a power, multiply exponents;

(b) to raise a product to a power, raise each factor to that power; and

(c) to raise a quotient to a power, raise the numerator and the denominator to that power.

EXAMPLE 6 Using the Power Rules for Exponents

Use one or more power rules to simplify each expression.

(a) $(p^8)^3 = p^{8 \cdot 3} = p^{24}$

(b) $\left(\frac{2}{3}\right)^4 = \frac{2^4}{3^4} = \frac{16}{81}$

(c) $(3y)^4 = 3^4y^4 = 81y^4$

(d) $(6p^7)^2 = 6^2p^{7 \cdot 2} = 6^2p^{14} = 36p^{14}$

(e) $\left(\frac{-2m^5}{z}\right)^3 = \frac{(-2)^3m^{5 \cdot 3}}{z^3} = \frac{(-2)^3m^{15}}{z^3} = \frac{-8m^{15}}{z^3}, \quad z \neq 0$

◀ Work Problem **6** at the Side.

The reciprocal of a^n is $\frac{1}{a^n} = \left(\frac{1}{a}\right)^n$. Also, by definition, a^n and a^{-n} are reciprocals since

$$a^n \cdot a^{-n} = a^n \cdot \frac{1}{a^n} = 1.$$

Thus, since both are reciprocals of a^n,

$$a^{-n} = \left(\frac{1}{a}\right)^n.$$

Some examples of this result are

$$6^{-3} = \left(\frac{1}{6}\right)^3 \quad \text{and} \quad \left(\frac{1}{3}\right)^{-2} = 3^2.$$

This discussion can be generalized as follows.

More Special Rules for Negative Exponents

If $a \neq 0$ and $b \neq 0$ and n is an integer, then

$$a^{-n} = \left(\frac{1}{a}\right)^n \quad \text{and} \quad \left(\frac{a}{b}\right)^{-n} = \left(\frac{b}{a}\right)^n.$$

In words, any nonzero number raised to the negative nth power is equal to the reciprocal of that number raised to the nth power

EXAMPLE 7 Using Negative Exponents with Fractions

Write each expression with only positive exponents and then evaluate.

(a) $\left(\frac{3}{7}\right)^{-2}$

$= \left(\frac{7}{3}\right)^2$ ← Change the fraction to its reciprocal and change the sign of the exponent.

$= \frac{49}{9}$

(b) $\left(\frac{4}{5}\right)^{-3}$

$= \left(\frac{5}{4}\right)^3$

$= \frac{125}{64}$

Work Problem **7** *at the Side.* ▶

7 Write each expression with only positive exponents and then evaluate.

(a) $\left(\frac{3}{4}\right)^{-3}$

(b) $\left(\frac{5}{6}\right)^{-2}$

The definitions and rules of this section are summarized here.

Definitions and Rules for Exponents

For all integers m and n and all real numbers a and b, the following rules apply.

Product Rule $a^m \cdot a^n = a^{m+n}$

Quotient Rule $\frac{a^m}{a^n} = a^{m-n} \quad (a \neq 0)$

Zero Exponent $a^0 = 1 \quad (a \neq 0)$

Negative Exponent $a^{-n} = \frac{1}{a^n} \quad (a \neq 0)$

(continued)

ANSWERS

7. **(a)** $\left(\frac{4}{3}\right)^3; \frac{64}{27}$ **(b)** $\left(\frac{6}{5}\right)^2; \frac{36}{25}$

Definitions and Rules for Exponents *(continued)*

Power Rules $(a^m)^n = a^{mn}$

$(ab)^m = a^m b^m$

$\left(\frac{a}{b}\right)^m = \frac{a^m}{b^m} \quad (b \neq 0)$

Special Rules $\frac{1}{a^{-n}} = a^n \quad (a \neq 0)$

$\frac{a^{-n}}{b^{-m}} = \frac{b^m}{a^n} \quad (a, b \neq 0)$

$a^{-n} = \left(\frac{1}{a}\right)^n \quad (a \neq 0)$

$\left(\frac{a}{b}\right)^{-n} = \left(\frac{b}{a}\right)^n \quad (a, b \neq 0)$

OBJECTIVE 5 Simplify exponential expressions.

EXAMPLE 8 Using the Definitions and Rules for Exponents

Simplify each expression so that no negative exponents appear in the final result. Assume that all variables represent nonzero real numbers.

(a) $3^2 \cdot 3^{-5}$

$= 3^{2+(-5)}$

$= 3^{-3}$

$= \frac{1}{3^3}$, or $\frac{1}{27}$

(b) $x^{-3} \cdot x^{-4} \cdot x^2$

$= x^{-3+(-4)+2}$

$= x^{-5}$

$= \frac{1}{x^5}$

(c) $(4^{-2})^{-5}$

$= 4^{(-2)(-5)}$

$= 4^{10}$

(d) $(x^{-4})^6$

$= x^{(-4)6}$

$= x^{-24}$

$= \frac{1}{x^{24}}$

(e) $\frac{x^{-4}y^2}{x^2y^{-5}}$

$= \frac{x^{-4}}{x^2} \cdot \frac{y^2}{y^{-5}}$

$= x^{-4-2} \cdot y^{2-(-5)}$

$= x^{-6}y^7$

$= \frac{y^7}{x^6}$

(f) $(2^3x^{-2})^{-2}$

$= (2^3)^{-2} \cdot (x^{-2})^{-2}$

$= 2^{-6}x^4$

$= \frac{x^4}{2^6}$, or $\frac{x^4}{64}$

Continued on Next Page

(g) $\left(\frac{3x^2}{y}\right)^2\left(\frac{4x^3}{y^{-2}}\right)^{-1}$

$= \frac{3^2(x^2)^2}{y^2} \cdot \frac{y^{-2}}{4x^3}$ Combination of rules

$= \frac{9x^4}{y^2} \cdot \frac{y^{-2}}{4x^3}$ Power rule (a)

$= \frac{9}{4}x^{4-3}y^{-2-2}$ Quotient rule

$= \frac{9x}{4y^4}$ $a^{-n} = \left(\frac{1}{a}\right)^n$

(h) $\left(\frac{-4m^5n^4}{24mn^{-7}}\right)^{-2}$

$= \left(\frac{m^{5-1}n^{4-(-7)}}{-6}\right)^{-2}$ Quotient rule; divide coefficients.

$= \left(\frac{m^4n^{11}}{-6}\right)^{-2}$ Subtract exponents.

$= \frac{(m^4)^{-2}(n^{11})^{-2}}{(-6)^{-2}}$ Power rules (b) and (c)

$= \frac{m^{-8}n^{-22}}{(-6)^{-2}}$ Power rule (a)

$= \frac{(-6)^2}{m^8n^{22}}$ $\left(\frac{a}{b}\right)^{-n} = \left(\frac{b}{a}\right)^n$

The sign on -6 does *not* change in this step.

$= \frac{36}{m^8n^{22}}$ $(-6)^2 = 36$

Work Problem **8** at the Side. ▶

Note

There is often more than one way to simplify expressions like those in Example 8. For instance, we could simplify Example 8(e) as follows.

$$\frac{x^{-4}y^2}{x^2y^{-5}} = \frac{y^5y^2}{x^4x^2} = \frac{y^7}{x^6} \qquad \text{Use } \frac{a^{-n}}{b^{-m}} = \frac{b^m}{a^n}\text{; product rule}$$

OBJECTIVE 6 **Use the rules for exponents with scientific notation.** The number of one-celled organisms that will sustain a whale for a few hours is 400,000,000,000,000, and the shortest wavelength of visible light is approximately 0.0000004 m. It is often simpler to write these numbers using *scientific notation.*

In scientific notation, a number is written with the decimal point after the first nonzero digit and multiplied by a power of 10.

Scientific Notation

A number is written in **scientific notation** when it is expressed in the form

$$a \times 10^n, \qquad \text{where } 1 \le |a| < 10, \text{ and } n \text{ is an integer.}$$

8 Simplify each expression so that no negative exponents appear in the final result. Assume that all variables represent nonzero real numbers.

(a) $5^4 \cdot 5^{-6}$

(b) $x^{-4} \cdot x^{-6} \cdot x^8$

(c) $(5^{-3})^{-2}$

(d) $(y^{-2})^7$

(e) $\frac{a^{-3}b^5}{a^4b^{-2}}$

(f) $(3^2k^{-4})^{-1}$

(g) $\left(\frac{2y}{x^3}\right)^2\left(\frac{4y}{x}\right)^{-1}$

(h) $\left(\frac{-28a^3b^{-5}}{7a^{-7}b^3}\right)^{-3}$

Answers

8. (a) $\frac{1}{5^2}$, or $\frac{1}{25}$ **(b)** $\frac{1}{x^2}$ **(c)** 5^6 **(d)** $\frac{1}{y^{14}}$ **(e)** $\frac{b^7}{a^7}$ **(f)** $\frac{k^4}{3^2}$, or $\frac{k^4}{9}$ **(g)** $\frac{y}{x^5}$ **(h)** $-\frac{b^{24}}{64a^{30}}$

9 Write each number in scientific notation.

(a) 400,000

(b) 29,800,000

(c) -6083

(d) 0.00172

(e) 0.0000000503

(f) -0.0031

ANSWERS

9. (a) 4×10^5 **(b)** 2.98×10^7 **(c)** -6.083×10^3 **(d)** 1.72×10^{-3} **(e)** 5.03×10^{-8} **(f)** -3.1×10^{-3}

For example, in scientific notation,

$$8000 = 8 \times 1000 = 8 \times 10^3.$$

It is customary to use $\times$ rather than $\cdot$.

The following numbers are *not* in scientific notation.

0.230×10^4 — 0.230 is less than 1.

46.5×10^{-3} — 46.5 is greater than 10.

To write a number in scientific notation, use the following steps. (If the number is negative, ignore the negative sign, go through these steps, and then attach a negative sign to the result.)

Converting to Scientific Notation

Step 1 **Position the decimal point.** Place a caret, ^, to the right of the first nonzero digit, where the decimal point will be placed.

Step 2 **Determine the numeral for the exponent.** Count the number of digits from the decimal point to the caret. This number gives the absolute value of the exponent on 10.

Step 3 **Determine the sign for the exponent.** Decide whether multiplying by 10^n should make the result of Step 1 greater or less. The exponent should be positive to make the result greater; it should be negative to make the result less.

It is helpful to remember that for $n \geq 1$, $10^{-n} < 1$ and $10^n \geq 10$.

EXAMPLE 9 Writing Numbers in Scientific Notation

Write each number in scientific notation.

(a) 820,000

Step 1 Place a caret to the right of the 8 (the first nonzero digit) to mark the new location of the decimal point.

$$8_\wedge 20{,}000$$

Step 2 Count from the decimal point, which is understood to be after the last 0, to the caret.

$$8.20{,}000. \leftarrow \text{Decimal point}$$

Count 5 places.

Step 3 Since the number 8.2 is to be made greater, the exponent on 10 is positive.

$$820{,}000 = 8.2 \times 10^5$$

(b) 0.0000072
Count from left to right.

$$0.000007.2$$

6 places

Since the number 7.2 is to be made less, the exponent on 10 is negative.

$$0.0000072 = 7.2 \times 10^{-6}$$

(c) $-0.0000462 = -4.62 \times 10^{-5}$

Count 5 places.

◀ Work Problem 9 at the Side.

Converting from Scientific Notation

Multiplying a number by a positive power of 10 makes the number greater, so move the decimal point to the right n places if n is positive in 10^n.

Multiplying by a negative power of 10 makes a number less, so move the decimal point to the left $|n|$ places if n is negative.

If n is 0, leave the decimal point where it is.

EXAMPLE 10 Converting from Scientific Notation to Standard Notation

Write each number in standard notation.

(a) 6.93×10^7

$$6.9300000. \quad \text{(7 places)}$$ Attach 0s as necessary.

We moved the decimal point 7 places to the right. (We had to attach five 0s.)

$$6.93 \times 10^7 = 69{,}300{,}000$$

(b) 4.7×10^{-6}

$$.0000004.7 \quad \text{(6 places)}$$

We moved the decimal point 6 places to the left.

$$4.7 \times 10^{-6} = 0.0000047$$ Attach a leading zero.

(c) $-1.083 \times 10^0 = -1.083 \times 1 = -1.083$

Work Problem **10** at the Side. ▶

EXAMPLE 11 Using Scientific Notation in Computation

Evaluate.

$$\frac{1{,}920{,}000 \times 0.0015}{0.000032 \times 45{,}000}$$

$= \dfrac{1.92 \times 10^6 \times 1.5 \times 10^{-3}}{3.2 \times 10^{-5} \times 4.5 \times 10^4}$ Express all numbers in scientific notation.

$= \dfrac{1.92 \times 1.5 \times 10^6 \times 10^{-3}}{3.2 \times 4.5 \times 10^{-5} \times 10^4}$ Commutative property

$= \dfrac{1.92 \times 1.5 \times 10^3}{3.2 \times 4.5 \times 10^{-1}}$ Product rule

$= \dfrac{1.92 \times 1.5}{3.2 \times 4.5} \times 10^4$ Quotient rule

$= 0.2 \times 10^4$ Simplify.

Don't stop here.

$= (2 \times 10^{-1}) \times 10^4$ Write 0.2 in scientific notation.

$= 2 \times 10^3$ Product rule

$= 2000$ Standard notation

Work Problem **11** at the Side. ▶

10 Write each number in standard notation.

(a) 4.98×10^5

(b) 6.8×10^{-7}

(c) -5.372×10^0

11 Evaluate.

$$\frac{200{,}000 \times 0.0003}{0.06 \times 4{,}000{,}000}$$

Answers

10. **(a)** 498,000 **(b)** 0.00000068 **(c)** −5.372

11. 2.5×10^{-4}, or 0.00025

12 The distance to the sun is 9.3×10^7 mi. How long would it take a rocket, traveling at 3.2×10^3 mph, to reach the sun? (*Hint:* $t = \frac{d}{r}$.)

Calculator Tip To enter numbers in scientific notation, you can use the EE or EXP key on a scientific calculator.

1.025 EE 4 or 1.025 EXP 4 means 1.025×10^4.

For instance, to work Example 11 using a popular model calculator with an EE key, enter the following symbols.

1.92 EE 6 × 1.5 EE 3 +/− ÷ (3.2 EE 5 +/− × 4.5 EE 4) =

The EXP key is used in exactly the same way. Notice that the negative exponent −3 is entered by pressing 3, then +/−. ***(Keystrokes vary among different models of calculators,*** so you should refer to your owner's manual if this sequence does not apply to your particular model.)

Calculators use the letter E to display numbers in scientific notation. For example,

3.62E5 means 3.62×10^5,

while

3.62E⁻4 means 3.62×10^{-4}.

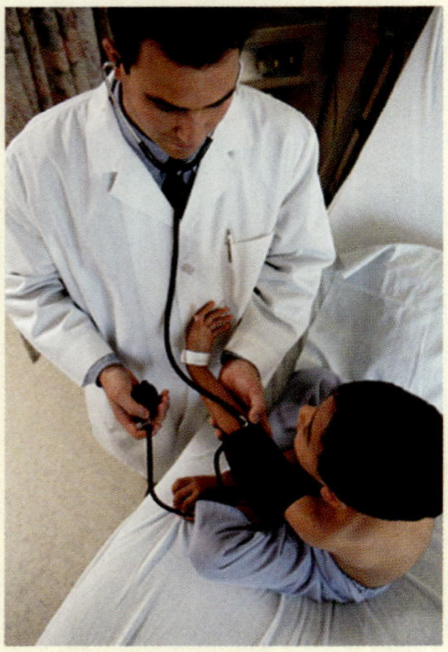

EXAMPLE 12 **Using Scientific Notation to Solve Problems**

In 1990, the national health care expenditure in the United States was \$714.0 billion. By 2005, this figure had risen by a factor of 2.8; that is, it almost tripled in 15 years. (*Source:* U.S. Centers for Medicare & Medicaid Services.)

(a) Write the 1990 health care expenditure using scientific notation.

714.0 billion

$= 714.0 \times 10^9$ — 1 billion $= 10^9$

$= (7.140 \times 10^2) \times 10^9$ — Write 714.0 in scientific notation.

$= 7.140 \times 10^{11}$ — Product rule

In 1990, the expenditure was $\$7.140 \times 10^{11}$.

(b) What was the expenditure in 2005?
Multiply the result in part (a) by 2.8.

$(7.140 \times 10^{11}) \times 2.8$

$= (2.8 \times 7.140) \times 10^{11}$ — Commutative and associative properties

$= 19.992 \times 10^{11}$ — Round to three decimal places.

$= (1.9992 \times 10^1) \times 10^{11}$ — Write 19.992 in scientific notation.

$= 1.9992 \times 10^{12}$ — Product rule

The 2005 expenditure was about \$1,999,200,000,000 (almost \$2 trillion).

◀ Work Problem **12** at the Side.

Answer

12. approximately 2.9×10^4 hr

6.1 Exercises

FOR EXTRA HELP — MyMathLab — Math XL PRACTICE — WATCH — DOWNLOAD — READ — REVIEW

Decide whether each expression has been simplified correctly. If not, correct it.

1. $(ab)^2 = ab^2$

2. $(5x)^3 = 5^3x^3$

3. $\left(\dfrac{4}{a}\right)^3 = \dfrac{4^3}{a} \quad (a \neq 0)$

4. $y^2 \cdot y^6 = y^{12}$

5. $x^3 \cdot x^4 = x^7$

6. $xy^0 = 0 \quad (y \neq 0)$

Apply the product rule for exponents, if possible, in each case. See Example 1.

7. $13^4 \cdot 13^8$

8. $9^6 \cdot 9^4$

9. $8^9 \cdot 8$

10. $12 \cdot 12^6$

11. $x^3 \cdot x^5 \cdot x^9$

12. $y^4 \cdot y^5 \cdot y^6$

13. $(-3w^5)(9w^3)$

14. $(-5x^2)(3x^4)$

15. $(2x^2y^5)(9xy^3)$

16. $(8s^4t)(3s^3t^5)$

17. $r^2 \cdot s^4$

18. $p^3 \cdot q^2$

In Exercises 19 and 20, match the expression in Column I with its equivalent expression in Column II. Choices may be used once, more than once, or not at all. See Example 2.*

19.

	I	II
	(a) 9^0	**A.** 0
	(b) -9^0	**B.** 1
	(c) $(-9)^0$	**C.** -1
	(d) $-(-9)^0$	**D.** 9
		E. -9

20.

	I	II
	(a) $2x^0$	**A.** 0
	(b) $-2x^0$	**B.** 1
	(c) $(2x)^0$	**C.** -1
	(d) $(-2x)^0$	**D.** 2
	(*Note*: $x \neq 0$)	**E.** -2

Evaluate. Assume that all variables represent nonzero numbers. See Example 2.

21. 17^0

22. 24^0

23. -5^0

24. -14^0

25. $(-15)^0$

26. $(-20)^0$

27. $-4^0 - m^0$

28. $-8^0 - k^0$

---

*The authors thank Mitchel Levy of Broward Community College for his suggestions for Exercises 19, 20, 29, 30, 55, and 56.

In Exercises 29 and 30, match the expression in Column I with its equivalent expression in Column II. Choices may be used once, more than once, or not at all. See Example 3.

I	II
29. (a) 4^{-2}	**A.** 16
(b) -4^{-2}	**B.** $\frac{1}{16}$
(c) $(-4)^{-2}$	**C.** -16
(d) $-(-4)^{-2}$	**D.** $-\frac{1}{16}$

I	II
30. (a) 5^{-3}	**A.** 125
(b) -5^{-3}	**B.** -125
(c) $(-5)^{-3}$	**C.** $\frac{1}{125}$
(d) $-(-5)^{-3}$	**D.** $-\frac{1}{125}$

Write each expression with only positive exponents. Assume that all variables represent nonzero numbers. In Exercises 43–46, simplify each expression. See Example 3.

31. 5^{-4} **32.** 7^{-2} **33.** 8^{-1} **34.** 12^{-1}

35. $(4x)^{-2}$ **36.** $(5t)^{-3}$ **37.** $4x^{-2}$ **38.** $5t^{-3}$

39. $-a^{-3}$ **40.** $-b^{-4}$ **41.** $(-a)^{-4}$ **42.** $(-b)^{-6}$

43. $5^{-1} + 6^{-1}$ **44.** $2^{-1} + 8^{-1}$ **45.** $8^{-1} - 3^{-1}$ **46.** $6^{-1} - 4^{-1}$

Evaluate each expression. See Examples 4 and 7.

47. $\frac{1}{4^{-2}}$ **48.** $\frac{1}{3^{-3}}$ **49.** $\frac{2^{-2}}{3^{-3}}$ **50.** $\frac{3^{-3}}{2^{-2}}$

51. $\left(\frac{2}{3}\right)^{-3}$ **52.** $\left(\frac{3}{2}\right)^{-3}$ **53.** $\left(\frac{4}{5}\right)^{-2}$ **54.** $\left(\frac{5}{4}\right)^{-2}$

In Exercises 55 and 56, match the expression in Column I with its equivalent expression in Column II. Choices may be used once, more than once, or not at all.

I	II
55. (a) $\left(\frac{1}{3}\right)^{-1}$	**A.** $\frac{1}{3}$
(b) $\left(-\frac{1}{3}\right)^{-1}$	**B.** 3
(c) $-\left(\frac{1}{3}\right)^{-1}$	**C.** $-\frac{1}{3}$
(d) $-\left(-\frac{1}{3}\right)^{-1}$	**D.** -3

I	II
56. (a) $\left(\frac{2}{5}\right)^{-2}$	**A.** $\frac{25}{4}$
(b) $\left(-\frac{2}{5}\right)^{-2}$	**B.** $-\frac{25}{4}$
(c) $-\left(\frac{2}{5}\right)^{-2}$	**C.** $\frac{4}{25}$
(d) $-\left(-\frac{2}{5}\right)^{-2}$	**D.** $-\frac{4}{25}$

Apply the quotient rule for exponents, if applicable, and write each result using only positive exponents. Assume that all variables represent nonzero numbers. See Example 5.

57. $\frac{4^8}{4^6}$ **58.** $\frac{5^9}{5^7}$ **59.** $\frac{x^{12}}{x^8}$ **60.** $\frac{y^{14}}{y^{10}}$

61. $\frac{r^7}{r^{10}}$ **62.** $\frac{y^8}{y^{12}}$ **63.** $\frac{6^4}{6^{-2}}$ **64.** $\frac{7^5}{7^{-3}}$

65. $\frac{6^{-3}}{6^7}$ **66.** $\frac{5^{-4}}{5^2}$ **67.** $\frac{7}{7^{-1}}$ **68.** $\frac{8}{8^{-1}}$

69. $\frac{r^{-3}}{r^{-6}}$ **70.** $\frac{s^{-4}}{s^{-8}}$ **71.** $\frac{x^3}{y^2}$ **72.** $\frac{y^5}{t^3}$

Use one or more power rules to simplify each expression. Assume that all variables represent nonzero numbers. See Example 6.

73. $(x^3)^5$ **74.** $(y^5)^4$ **75.** $\left(\frac{3}{5}\right)^3$ **76.** $\left(\frac{4}{3}\right)^2$

77. $(4t)^3$ **78.** $(5t)^4$ **79.** $(-6x^2)^3$ **80.** $(-2x^5)^5$

81. $\left(\frac{-4m^2}{t}\right)^3$ **82.** $\left(\frac{-5n^4}{r^2}\right)^3$ **83.** $\left(\frac{-s^3}{t^5}\right)^4$ **84.** $\left(\frac{-2a^4}{b^5}\right)^6$

Simplify each expression so that no negative exponents appear in the final result. Assume that all variables represent nonzero numbers. See Example 8.

85. $3^5 \cdot 3^{-6}$ **86.** $4^4 \cdot 4^{-6}$ **87.** $a^{-3}a^2a^{-4}$ **88.** $k^{-5}k^{-3}k^4$

89. $(k^2)^{-3}k^4$ **90.** $(x^3)^{-4}x^5$ **91.** $-4r^{-2}(r^4)^2$ **92.** $-2m^{-1}(m^3)^2$

93. $(5a^{-1})^4(a^2)^{-3}$ **94.** $(3p^{-4})^2(p^3)^{-1}$ **95.** $(z^{-4}x^3)^{-1}$ **96.** $(y^{-2}z^4)^{-3}$

97. $\dfrac{(p^{-2})^3}{5p^4}$ **98.** $\dfrac{(m^4)^{-1}}{9m^3}$ **99.** $\dfrac{4a^5(a^{-1})^3}{(a^{-2})^{-2}}$ **100.** $\dfrac{12k^{-2}(k^{-3})^{-4}}{6k^5}$

101. $\dfrac{(2k)^2m^{-5}}{(km)^{-3}}$ **102.** $\dfrac{(3rs)^{-2}}{3^2r^2s^{-4}}$ **103.** $\left(\dfrac{3k^{-2}}{k^4}\right)^{-1}\cdot\dfrac{2}{k}$ **104.** $\left(\dfrac{7m^{-2}}{m^{-3}}\right)^{-2}\cdot\dfrac{m^3}{4}$

105. $\left(\dfrac{2p}{q^2}\right)^3\left(\dfrac{3p^4}{q^{-4}}\right)^{-1}$ **106.** $\left(\dfrac{5z^3}{2a^2}\right)^{-3}\left(\dfrac{8a^{-1}}{15z^{-2}}\right)^{-3}$ **107.** $\left(\dfrac{3a^{-4}b^6}{15a^2b^{-4}}\right)^{-2}$ **108.** $\left(\dfrac{9r^3s^{-5}}{-18r^{-8}s^{-2}}\right)^{-3}$

Write each number in scientific notation. See Example 9.

109. 530 **110.** 1600 **111.** 0.830 **112.** 0.0072

113. 0.00000692 **114.** 0.875 **115.** −38,500 **116.** −976,000,000

Write each number in standard notation. See Example 10.

117. 7.2×10^4 **118.** 8.91×10^2 **119.** 2.54×10^{-3} **120.** 5.42×10^{-4}

121. -6×10^4 **122.** -9×10^3 **123.** 1.2×10^{-5} **124.** 2.7×10^{-6}

Evaluate. See Example 11.

125. $\dfrac{3 \times 10^{-2}}{12 \times 10^3}$ **126.** $\dfrac{5 \times 10^{-3}}{25 \times 10^2}$ **127.** $\dfrac{0.05 \times 1600}{0.0004}$ **128.** $\dfrac{0.003 \times 40{,}000}{0.00012}$

Solve each problem. See Example 12.

129. The U.S. budget first passed **\$1,000,000,000** in 1917. Seventy years later in 1987 it exceeded **\$1,000,000,000,000** for the first time. President George W. Bush's budget request for fiscal 2009 was about **\$3,100,000,000,000.** If stacked in dollar bills, this amount would stretch **210,385** mi, almost 90% of the distance to the moon. Write the four boldfaced numbers in scientific notation. (*Source:* www.gpoaccess.gov, *The New York Times*.)

130. By area, the largest of the fifty United States is Alaska, with land area of about **365,482,000** acres, while the smallest is Rhode Island, with land area of about **677,000** acres. The total land area of the United States is about **2,271,343,000** acres. Write these three numbers in scientific notation. (*Source:* General Services Administration.)

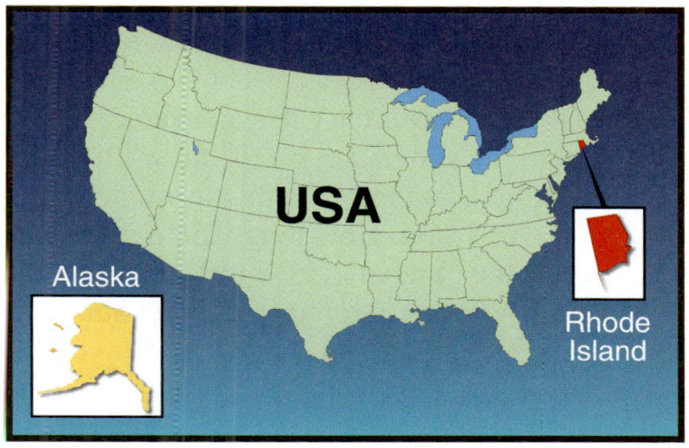

131. In May 2008, the population of the United States was 304.1 million. (*Source:* U.S. Census Bureau.)

(a) Write the May 2008 population using scientific notation.

(b) Write \$1 trillion, that is, \$1,000,000,000,000, using scientific notation.

(c) Using your answers from parts (a) and (b), calculate how much each person in the United States in the year 2008 would have had to contribute in order to make someone a trillionaire. Write this amount in standard notation to the nearest dollar.

132. In the early years of the Powerball Lottery, a player would choose five numbers from 1 through 49 and one number from 1 through 42. It can be shown that there are about 8.009×10^7 different ways to do this. Suppose that a group of 2000 persons decided to purchase tickets for all these numbers and each ticket cost \$1.00. How much should each person have expected to pay? (*Source:* www.powerball.com)

POWERBALL

133. In 2006, the population of Japan was 1.275×10^6, which was 38.64 times the population of Monaco. What was the population of Monaco? (*Source:* U.S. Census Bureau.)

134. The speed of light is approximately 3×10^{10} cm per sec. How long does it take light to travel 9×10^{12} cm?

135. The average distance from Earth to the sun is 9.3×10^7 mi. How long would it take a rocket, traveling at 2.9×10^3 mph, to reach the sun?

136. A *light-year* is the distance that light travels in one year. Find the number of miles in a light-year if light travels 1.86×10^5 mi per sec.

137. (a) The planet Mercury has an average distance from the sun of 3.6×10^7 mi, while the average distance of Venus from the sun is 6.7×10^7 mi. How long would it take a spacecraft traveling at 1.55×10^3 mph to travel from Venus to Mercury? (Give your answer in hours, in standard notation.)

(b) Use the information from part (a) to find the number of days it would take the spacecraft to travel from Venus to Mercury. Round your answer to the nearest whole number of days.

138. When the distance between the centers of the moon and Earth is 4.60×10^8 m, an object on the line joining the centers of the moon and Earth exerts the same gravitational force on each when it is 4.14×10^8 m from the center of Earth. How far is the object from the center of the moon at that point?

139. In some cases, $-a^n$ and $(-a)^n$ do give the same result for $a > 0$. Using $a = 2$ and $n = 2, 3, 4,$ and 5, draw a conclusion as to when they are equal and when they are opposites.

140. Your friend evaluated $4^5 \cdot 4^2$ as 16^7. ***WHAT WENT WRONG?*** Give the correct answer.

141. In your own words, describe how to rewrite a fraction raised to a negative power as a fraction raised to a positive power.

142. Explain in your own words how to raise a power to a power.

6.2 Adding and Subtracting Polynomials

OBJECTIVES

1. Know the basic definitions for polynomials.
2. Find the degree of a polynomial.
3. Add and subtract polynomials.

OBJECTIVE 1 Know the basic definitions for polynomials. Just as whole numbers are the basis of arithmetic, *polynomials* are fundamental in algebra. To understand polynomials, we review several words from **Section 1.4.** A **term** is a number, a variable, or the product or quotient of a number and one or more variables raised to powers. Examples of terms include

$$4x, \quad \frac{1}{2}m^5\left(\text{or } \frac{m^5}{2}\right), \quad -7z^9, \quad 6x^2z, \quad \frac{5}{3x^2}, \quad \text{and} \quad 9. \qquad \text{Terms}$$

The number in the product is called the **numerical coefficient,** or just the **coefficient.*** In the term $8x^3$, the coefficient is 8. In the term $-4p^5$, it is -4. The coefficient of the term k is understood to be 1 (since k can be written as $1k$.) The coefficient of $-r$ is -1. In the term $\frac{x}{3}$, the coefficient is $\frac{1}{3}$ since $\frac{x}{3} = \frac{1x}{3} = \frac{1}{3}x$.

Work Problem 1 at the Side. ▶

1 Identify the numerical coefficient of each term.

(a) $-9m^5$

(b) $12y^2x$

(c) x

(d) $-y$

(e) $\frac{z}{4}$

Recall that any combination of variables or constants (numerical values) joined by the basic operations of addition, subtraction, multiplication, and division (except by 0), or raising to powers or taking roots is called an **algebraic expression.** One kind of algebraic expression is a *polynomial.*

Polynomial

A **polynomial** is a term or a finite sum of terms in which all variables have whole number exponents and no variables appear in denominators or under radicals.

Examples of polynomials include

$$3x - 5, \quad 4m^3 - 5m^2p + 8, \quad \text{and} \quad -5t^2s^3. \qquad \text{Polynomials}$$

Even though the expression $3x - 5$ involves subtraction, it is a sum of terms since it could be written as $3x + (-5)$.

Some examples of expressions that are not polynomials are

$$x^{-1} + 3x^{-2}, \quad \sqrt{9 - x}, \quad \text{and} \quad \frac{1}{x}. \qquad \text{Not polynomials}$$

The first of these is not a polynomial because it has negative integer exponents, the second because it involves a variable under a radical, and the third because it contains a variable in the denominator.

Most of the polynomials used in this book contain only one variable. A polynomial containing only the variable x is called a **polynomial in x.** A polynomial in one variable is written in **descending powers** of the variable if the exponents on the variable decrease from left to right. For example,

$$x^5 - 6x^2 + 12x - 5$$

is a polynomial in descending powers of x. The term -5 in this polynomial can be thought of as $-5x^0$, since $-5x^0 = -5(1) = -5$.

*More generally, any factor in a term is the coefficient of the product of the remaining factors. For example, $3x^2$ is the coefficient of y in the term $3x^2y$, and $3y$ is the coefficient of x^2 in $3x^2y$.

ANSWERS

1. (a) -9 (b) 12 (c) 1 (d) -1 (e) $\frac{1}{4}$

2 Write each polynomial in descending powers of the variable.

(a) $-4 + 9y + y^3$

(b) $-3z^4 + 2z^3 + z^5 - 6z$

(c) $-12m^{10} + 8m^9 + 10m^{12}$

3 Identify each polynomial as a *trinomial, binomial, monomial,* or *none of these.*

(a) $12m^4 - 6m^2$

(b) $-6y^3 + 2y^2 - 8y$

(c) $3a^5$

(d) $-2k^{10} + 2k^9 - 8k^5 + 2k$

4 Give the degree of each polynomial.

(a) $9y^4 + 8y^3 - 6$

(b) $-12m^7 + 11m^3 + m^9$

(c) $-2k$

(d) 10

(e) $3mn^2 + 2m^3n$

Answers

2. **(a)** $y^3 + 9y - 4$ **(b)** $z^5 - 3z^4 + 2z^3 - 6z$ **(c)** $10m^{12} - 12m^{10} + 8m^9$
3. **(a)** binomial **(b)** trinomial **(c)** monomial **(d)** none of these
4. **(a)** 4 **(b)** 9 **(c)** 1 **(d)** 0 **(e)** 4

EXAMPLE 1 Writing Polynomials in Descending Powers

Write each polynomial in descending powers of the variable.

(a) $y - 6y^3 + 8y^5 - 9y^4 + 12$ is written as $8y^5 - 9y^4 - 6y^3 + y + 12$.

(b) $-2 + m + 6m^2 - 4m^3$ is written as $-4m^3 + 6m^2 + m - 2$.

◀ *Work Problem* **2** *at the Side.*

Some polynomials with a specific number of terms are so common that they are given special names. A polynomial with exactly three terms is a **trinomial,** and a polynomial with exactly two terms is a **binomial.** A single-term polynomial is a **monomial.** The table that follows gives examples.

Type of Polynomial	Examples
Monomial	$5x$, $7m^9$, -8, x^2y^2
Binomial	$3x^2 - 6$, $11y + 8$, $5a^2b + 3a$
Trinomial	$y^2 + 11y + 6$, $8p^3 - 7p + 2m$, $-3 + 2k^5 + 9z^4$
None of these	$p^3 - 5p^2 + 2p - 5$, $-9z^3 + 5c^3 + 2m^5 + 11r^2 - 7r$

◀ *Work Problem* **3** *at the Side.*

OBJECTIVE 2 Find the degree of a polynomial. The **degree of a term** with one variable is the exponent on the variable. For example, the degree of $2x^3$ is 3, the degree of $-x^4$ is 4, and the degree of $17x$ (that is, $17x^1$) is 1. The degree of a term in more than one variable is defined to be the sum of the exponents on the variables. For example, the degree of $5x^3y^7$ is 10, because $3 + 7 = 10$.

The greatest degree of any term in a polynomial is called the **degree of the polynomial.** In most cases, we will be interested in finding the degree of a polynomial in one variable. For example,

$$4x^3 - 2x^2 - 3x + 7$$

has degree 3, because the greatest degree of any term is 3 (the degree of $4x^3$).

The table shows several polynomials and their degrees.

Polynomial	Degree
$9x^2 - 5x + 8$	2
$17m^9 + 18m^{14} - 9m^3$	14
$5x$	1, because $5x = 5x^1$
-2	0, because $-2 = -2x^0$ (Any nonzero constant has degree 0.)
$5a^2b^5$	7, because $2 + 5 = 7$
$13xy^4 + x^3y^9 + 7xy$	12, because the degrees of the terms are 5, 12, and 2; 12 is the greatest.

Note

The number 0 has no degree, since 0 times a variable to any power is 0.

◀ *Work Problem* **4** *at the Side.*

OBJECTIVE 3 Add and subtract polynomials. We use the distributive property to simplify polynomials by combining terms. For example,

$$x^3 + 4x^2 + 5x^2 - 1$$
$$= x^3 + (4 + 5)x^2 - 1 \quad \text{Distributive property}$$
$$= x^3 + 9x^2 - 1.$$

The terms in the polynomial $4x + 5x^2$ cannot be combined. ***Only terms containing exactly the same variables to the same powers may be combined.*** As mentioned in **Section 1.4,** such terms are called **like terms.**

CAUTION
Remember that only like terms can be combined.

EXAMPLE 2 Combining Like Terms

Combine like terms.

(a) $-5y^3 + 8y^3 - y^3$
$= (-5 + 8 - 1)y^3$ Distributive property
$= 2y^3$

(b) $6x + 5y - 9x + 2y$
$= 6x - 9x + 5y + 2y$ Commutative property
$= -3x + 7y$ Combine like terms.

Since $-3x$ and $7y$ are unlike terms, no further simplification is possible.

(c) $5x^2y - 6xy^2 + 9x^2y + 13xy^2$
$= 5x^2y + 9x^2y - 6xy^2 + 13xy^2$
$= 14x^2y + 7xy^2$

Work Problem **5** at the Side. ▶

Adding Polynomials
To add two polynomials, combine like terms.

Polynomials can be added horizontally or vertically.

EXAMPLE 3 Adding Polynomials

Add: $(3a^5 - 9a^3 + 4a^2) + (-8a^5 + 8a^3 + 2)$.

Use the commutative and associative properties to rearrange the polynomials so that like terms are together. Then use the distributive property to combine like terms.

$$(3a^5 - 9a^3 + 4a^2) + (-8a^5 + 8a^3 + 2)$$
$$= 3a^5 - 8a^5 - 9a^3 + 8a^3 + 4a^2 + 2$$
$$= -5a^5 - a^3 + 4a^2 + 2 \quad \text{Combine like terms.}$$

Add these same two polynomials vertically by placing like terms in columns.

$$\begin{array}{r} 3a^5 - 9a^3 + 4a^2 \phantom{{}+2} \\ \underline{-8a^5 + 8a^3 \phantom{{}+4a^2} + 2} \\ -5a^5 - a^3 + 4a^2 + 2 \end{array}$$

Work Problem **6** at the Side. ▶

5 Combine like terms.

(a) $11x + 12x - 7x - 3x$

(b) $11p^5 + 4p^5 - 6p^3 + 8p^3$

(c) $2y^2z^4 + 3y^4 + 5y^4 - 9y^4z^2$

6 Add, using both the horizontal and vertical methods.

(a) $(12y^2 - 7y + 9)$
$+ (-4y^2 - 11y + 5)$

(b)
$$\begin{array}{r} -6r^5 + 2r^3 - r^2 \\ \underline{8r^5 - 2r^3 + 5r^2} \end{array}$$

ANSWERS

5. (a) $13x$ **(b)** $15p^5 + 2p^3$
(c) $2y^2z^4 + 8y^4 - 9y^4z^2$
6. (a) $8y^2 - 18y + 14$ **(b)** $2r^5 + 4r^2$

In **Section 1.2,** we defined subtraction of real numbers as

$$a - b = a + (-b).$$

That is, we add the first number (minuend) and the negative (or opposite) of the second (subtrahend). We can give a similar definition for subtraction of polynomials by defining the **negative of a polynomial** as that polynomial with the sign of every coefficient changed.

Subtracting Polynomials

To subtract two polynomials, add the first polynomial and the negative of the *second* polynomial.

EXAMPLE 4 Subtracting Polynomials

Subtract: $(-6m^2 - 8m + 5) - (-5m^2 + 7m - 8)$.

Change every sign in the second polynomial and add.

$$\begin{aligned}
&(-6m^2 - 8m + 5) - (-5m^2 + 7m - 8) \\
&= -6m^2 - 8m + 5 + 5m^2 - 7m + 8 && \text{Definition of subtraction} \\
&= -6m^2 + 5m^2 - 8m - 7m + 5 + 8 && \text{Rearrange terms.} \\
&= -m^2 - 15m + 13 && \text{Combine like terms.}
\end{aligned}$$

Check by adding the sum, $-m^2 - 15m + 13$, to the second polynomial. The result should be the first polynomial.

To subtract these two polynomials vertically, write the first polynomial above the second, lining up like terms in columns.

$$\begin{array}{r} -6m^2 - 8m + 5 \\ \underline{-5m^2 + 7m - 8} \end{array}$$

Change all the signs in the second polynomial, and add.

$$\begin{array}{rl} -6m^2 - 8m + 5 & \\ \underline{+5m^2 - 7m + 8} & \text{Change all signs.} \\ -m^2 - 15m + 13 & \text{Add in columns.} \end{array}$$

◀ Work Problem 7 at the Side.

7 Subtract, using both the horizontal and vertical methods.

(a) $(6y^3 - 9y^2 + 8)$
$- (2y^3 + y^2 + 5)$

(b) $\begin{array}{r} 6y^3 - 2y^2 + 5y \\ \underline{-2y^3 + 8y^2 - 11y} \end{array}$

Answers

7. (a) $4y^3 - 10y^2 + 3$
(b) $8y^3 - 10y^2 + 16y$

6.2 Exercises

FOR EXTRA HELP — MyMathLab — Math XL PRACTICE — WATCH — DOWNLOAD — READ — REVIEW

We defined a polynomial written in descending powers in the text. Sometimes we write a polynomial in **ascending powers,** *with the degree of the terms increasing from left to right. Decide whether each polynomial is written in* descending *powers,* ascending *powers, or* neither. *See Example 1.*

1. $2r^3 + x - 3x^2$ **2.** $3x^5 + x^4 - 2x^3 + x$ **3.** $4p^3 - 8p^5 + p^7$

4. $q^2 + 3q^4 - 2q + 1$ **5.** $-m^3 + 5m^2 + 3m + 10$ **6.** $4 - x + 3x^2$

Give the numerical coefficient and the degree of each term.

7. $7z$ **8.** $3r$ **9.** $-15p^2$ **10.** $-27k^3$

11. x^4 **12.** y^6 **13.** $-mn^5$ **14.** $-a^5b$

Identify each polynomial as a monomial, binomial, trinomial, *or* none of these. *Give the degree of each.*

Polynomial	Type	Degree
15. 24		
17. $7m - 21$		
19. $2r^3 + 3r^2 + 5r$		
21. $-5p^4q - 3p^3q^2 + 2pq^3 - q^4$		

Polynomial	Type	Degree
16. 5		
18. $-x^2 + 3x^5$		
20. $5z^2 - 5z + 7$		
22. $8s^3t - 3s^2t^2 + 2st^3 + 9$		

Combine like terms. See Example 2.

23. $5z^4 + 3z^4$ **24.** $8r^5 - 2r^5$ **25.** $-m^3 + 2m^3 + 6m^3$ **26.** $3p^4 + 5p^4 - 2p^4$

27. $x + x + x + x + x$ **28.** $z - z - z + z$ **29.** $m^4 - 3m^2 + m$ **30.** $5a^5 + 2a^4 - 9a^3$

31. $y^3 + 7y - 4y^2$ **32.** $2c^2 - 4 + 8 - c^2$ **33.** $2k + 3k^2 + 5k^2 - 7$ **34.** $4x^2 + 2x - 6x^2 - 6$

35. $n^4 - 2n^3 + n^2 - 3n^4 + n^3$ **36.** $2q^3 + 3q^2 - 4q - q^3 + 5q^2$

Add or subtract as indicated. See Examples 3 and 4.

37. Add.
$$\begin{array}{r} -12p^2 + 4p - 1 \\ 3p^2 + 7p - 8 \\ \hline \end{array}$$

38. Add.
$$\begin{array}{r} -6y^3 + 8y + 5 \\ 9y^3 + 4y - 6 \\ \hline \end{array}$$

39. Subtract.
$$\begin{array}{r} 12a + 15 \\ 7a - 3 \\ \hline \end{array}$$

40. Subtract.
$$\begin{array}{r} -3b + 6 \\ 2b - 8 \\ \hline \end{array}$$

41. Subtract.
$$\begin{array}{r} 6m^2 - 11m + 5 \\ -8m^2 + 2m - 1 \\ \hline \end{array}$$

42. Subtract.
$$\begin{array}{r} -4z^2 + 2z - 1 \\ 3z^2 - 5z + 2 \\ \hline \end{array}$$

43. Add.
$$\begin{array}{r} 12z^2 - 11z + 8 \\ 5z^2 + 16z - 2 \\ -4z^2 + 5z - 9 \\ \hline \end{array}$$

44. Add.
$$\begin{array}{r} -6m^3 + 2m^2 + 5m \\ 8m^3 + 4m^2 - 6m \\ -3m^3 + 2m^2 - 7m \\ \hline \end{array}$$

45. Add.
$$\begin{array}{r} 6y^3 - 9y^2 \qquad + 8 \\ 4y^3 + 2y^2 + 5y \qquad \\ \hline \end{array}$$

46. Add.
$$\begin{array}{r} -7r^8 + 2r^6 - r^5 \qquad \\ 3r^6 \qquad + 5 \\ \hline \end{array}$$

47. Subtract.
$$\begin{array}{r} -5a^4 \qquad + 8a^2 - 9 \\ 6a^3 - a^2 + 2 \\ \hline \end{array}$$

48. Subtract.
$$\begin{array}{r} -2m^3 + 8m^2 \qquad \\ m^4 - m^3 \qquad\qquad + 2m \\ \hline \end{array}$$

49. $(3r + 8) - (2r - 5)$

50. $(2d + 7) - (3d - 1)$

51. $(5x^2 + 7x - 4) + (3x^2 - 6x + 2)$

52. $(4k^3 + k^2 + k) + (2k^3 - 4k^2 - 3k)$

53. $(2a^2 + 3a - 1) - (4a^2 + 5a + 6)$

54. $(q^4 - 2q^2 + 10) - (3q^4 + 5q^2 - 5)$

55. $(z^5 + 3z^2 + 2z) - (4z^5 + 2z^2 - 5z)$

56. $(5t^3 - 3t^2 + 2t) - (4t^3 + 2t^2 + 3t)$

6.3 Polynomial Functions

OBJECTIVE 1 Recognize and evaluate polynomial functions. In **Chapter 4** we studied linear (first-degree polynomial) functions, defined as $f(x) = ax + b$. Now we consider more general polynomial functions.

Polynomial Function

A **polynomial function of degree *n*** is defined by

$$f(x) = a_nx^n + a_{n-1}x^{n-1} + \cdots + a_1x + a_0,$$

for real numbers $a_n, a_{n-1}, \ldots, a_1$, and a_0, where $a_n \neq 0$ and n is a whole number.

Another way of describing a polynomial function is to say that it is a function defined by a polynomial in one variable, consisting of one or more terms. It is usually written in descending powers of the variable, and its degree is the degree of the polynomial that defines it.

Suppose that the polynomial $3x^2 - 5x + 7$ defines function f. Then

$$f(x) = 3x^2 - 5x + 7.$$

If $x = -2$, then $f(x) = 3x^2 - 5x + 7$ takes on the value

$f(-2) = 3(-2)^2 - 5(-2) + 7$ Let $x = -2$.

$f(-2) = 3 \cdot 4 + 10 + 7$

$f(-2) = 12 + 10 + 7$

$f(-2) = 29.$

Thus, $f(-2) = 29$ and the ordered pair $(-2, 29)$ belongs to f.

EXAMPLE 1 Evaluating Polynomial Functions

Let $f(x) = 4x^3 - x^2 + 5$. Find each value.

(a) $f(3)$

$f(x) = 4x^3 - x^2 + 5$

$f(3) = 4 \cdot 3^3 - 3^2 + 5$ Substitute 3 for x.

$= 4 \cdot 27 - 9 + 5$ Order of operations

$= 108 - 9 + 5$ Multiply.

$= 104$ Subtract; add.

Read this as "f of 3," not "f times 3."

Thus, $f(3) = 104$ and the ordered pair $(3, 104)$ belongs to f.

(b) $f(-4)$

$f(x) = 4x^3 - x^2 + 5$

$f(-4) = 4 \cdot (-4)^3 - (-4)^2 + 5$ Let $x = -4$; use parentheses.

$= 4 \cdot (-64) - 16 + 5$ Be careful with signs.

$= -256 - 16 + 5$ Multiply.

$= -267$ Subtract; add.

So, $f(-4) = -267$. The ordered pair $(-4, -267)$ belongs to f.

Work Problem 1 at the Side. ▶

OBJECTIVES

1. Recognize and evaluate polynomial functions.
2. Use a polynomial function to model data.
3. Add and subtract polynomial functions.
4. Graph basic polynomial functions.

1 Let $f(x) = -x^2 + 5x - 11$. Find each value.

(a) $f(1)$

(b) $f(-4)$

(c) $f(0)$

ANSWERS

1. (a) -7 (b) -47 (c) -11

2 Use the function in Example 2 to approximate the number of high school students in the United States in 2000.

While f is the most common letter used to represent functions, recall that other letters such as g and h are also used. The capital letter P is often used for polynomial functions. Note that the function defined as

$$P(x) = 4x^3 - x^2 + 5$$

yields the same ordered pairs as the function f in Example 1.

OBJECTIVE 2 Use a polynomial function to model data. Polynomial functions can be used to approximate data. They are usually valid for small intervals, and they allow us to predict (with caution) what might happen for values just outside the intervals. These intervals are often periods of years, as shown in Example 2.

EXAMPLE 2 Using a Polynomial Model to Approximate Data

The number of high school students in the United States over the period 1990 to 2005 can be modeled by the polynomial function

$$P(x) = -0.01085x^2 + 0.4572x + 12.70,$$

where $x = 0$ corresponds to the year 1990, $x = 1$ corresponds to 1991, and so on, and $P(x)$ is in millions. (*Source:* U.S. Census Bureau.) Use this function to approximate the number of high school students in 2005.

Since $x = 15$ corresponds to the year 2005, we must find $P(15)$.

$$P(x) = -0.01085x^2 + 0.4572x + 12.70$$

$$P(15) = -0.01085(15)^2 + 0.4572(15) + 12.70 \quad \text{Let } x = 15$$

$$P(15) \approx 17.1 \quad \text{Use a calculator to evaluate.}$$

Thus, in 2005, there were about 17.1 million high school students in the United States.

◀ *Work Problem* **2** *at the Side.*

OBJECTIVE 3 Add and subtract polynomial functions. The operations of addition, subtraction, multiplication, and division are also defined for functions. For example, businesses use the equation "profit equals revenue minus cost," written using function notation as

$$\underset{\substack{\text{Profit}\\\text{function}}}{P(x)} = \underset{\substack{\text{Revenue}\\\text{function}}}{R(x)} - \underset{\substack{\text{Cost}\\\text{function}}}{C(x)},$$

where x is the number of items produced and sold. Thus, the profit function is found by subtracting the cost function from the revenue function.

We define the following **operations on functions.**

Adding and Subtracting Functions

If $f(x)$ and $g(x)$ define functions, then

$$(f + g)(x) = f(x) + g(x) \quad \text{Sum function}$$

and

$$(f - g)(x) = f(x) - g(x). \quad \text{Difference function}$$

In each case, the domain of the new function is the intersection of the domains of $f(x)$ and $g(x)$.

ANSWER

2. 16.2 million

EXAMPLE 3 **Adding and Subtracting Functions**

Find each of the following for the polynomial functions defined by

$$f(x) = x^2 - 3x + 7 \quad \text{and} \quad g(x) = -3x^2 - 7x + 7.$$

(a) $(f - g)(x)$ — This notation does *not* indicate the distributive property.

$= f(x) + g(x)$	Use the definition.
$= (x^2 - 3x + 7) + (-3x^2 - 7x + 7)$	Substitute.
$= -2x^2 - 10x + 14$	Add the polynomials.

(b) $(f - g)(x)$

$= f(x) - g(x)$	Use the definition.
$= (x^2 - 3x + 7) - (-3x^2 - 7x + 7)$	Substitute.
$= (x^2 - 3x + 7) + (3x^2 + 7x - 7)$	Change to addition.
$= 4x^2 + 4x$	Add.

Work Problem **3** *at the Side.* ▶

EXAMPLE 4 **Adding and Subtracting Functions**

Find each of the following for the polynomial functions defined by

$$f(x) = 10x^2 - 2x \quad \text{and} \quad g(x) = 2x.$$

(a) $(f - g)(2)$

$(f + g)(2) = f(2) + g(2)$	Use the definition.
$= \underbrace{[10(2)^2 - 2(2)]}_{f(x) = 10x^2 - 2x} + \underbrace{2(2)}_{g(x) = 2x}$ (This is a key step.)	Substitute.
$= [40 - 4] + 4$	Order of operations
$= 40$	Subtract; add.

Alternatively, we could first find $(f + g)(x)$.

$(f + g)(x)$	
$= f(x) + g(x)$	Use the definition.
$= (10x^2 - 2x) + 2x$	Substitute.
$= 10x^2$	Combine like terms.

Then,

$(f + g)(2)$	
$= 10(2)^2$	Substitute.
$= 40.$	The result is the same.

(b) $(f - g)(x)$ and $(f - g)(1)$

$(f - g)(x)$	
$= f(x) - g(x)$	Use the definition.
$= (10x^2 - 2x) - 2x$	Substitute.
$= 10x^2 - 4x$	Combine like terms.

Then,

$(f - g)(1)$	
$= 10(1)^2 - 4(1)$	Substitute.
$= 6.$	

Confirm that $f(1) - g(1)$ gives the same result.

Work Problem **4** *at the Side.* ▶

3 Let

$$f(x) = 3x^2 + 8x - 6$$

and $g(x) = -4x^2 + 4x - 8.$

Find each function.

(a) $(f + g)(x)$

(b) $(f - g)(x)$

4 For

$$f(x) = 18x^2 - 24x$$

and $g(x) = 3x,$

find each of the following.

(a) $(f + g)(x)$ and $(f + g)(-1)$

(b) $(f - g)(x)$ and $(f - g)(1)$

ANSWERS

3. **(a)** $-x^2 + 12x - 14$
(b) $7x^2 + 4x + 2$

4. **(a)** $18x^2 - 21x$; 39
(b) $18x^2 - 27x$; -9

OBJECTIVE 4 Graph basic polynomial functions. Functions were introduced in **Section 4.5.** Recall that each input (or x-value) of a function results in one output (or y-value).

The simplest polynomial function is the **identity function,** defined by $f(x) = x$. The domain (set of x-values) of this function is all real numbers, $(-\infty, \infty)$, and it pairs each real number with itself. Therefore, the range (set of y-values) is also $(-\infty, \infty)$. Its graph is a straight line, as first seen in **Chapter 4.** (Notice that a *linear function* is a specific kind of polynomial function.) Figure 1 shows its graph and a table of selected ordered pairs.

x	$f(x) = x$
−2	−2
−1	−1
0	0
1	1
2	2

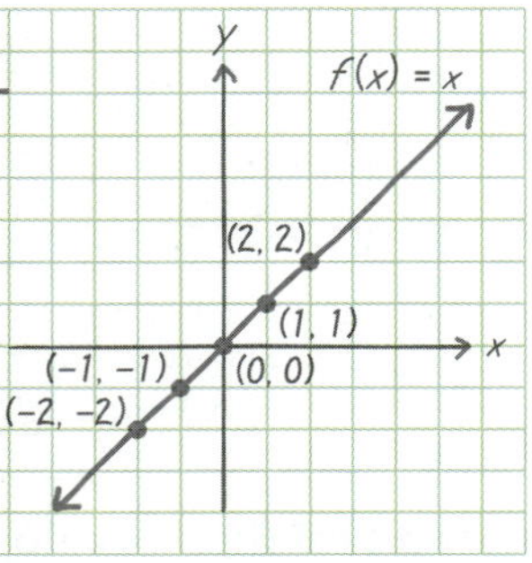

Figure 1

Another polynomial function, defined by $f(x) = x^2$, is the **squaring function.** For this function, every real number is paired with its square. The input can be any real number, so the domain is $(-\infty, \infty)$. Since the square of any real number is nonnegative, the range is $[0, \infty)$. Its graph is a **parabola.** Figure 2 shows the graph and a table of selected ordered pairs.

x	$f(x) = x^2$
−2	4
−1	1
0	0
1	1
2	4

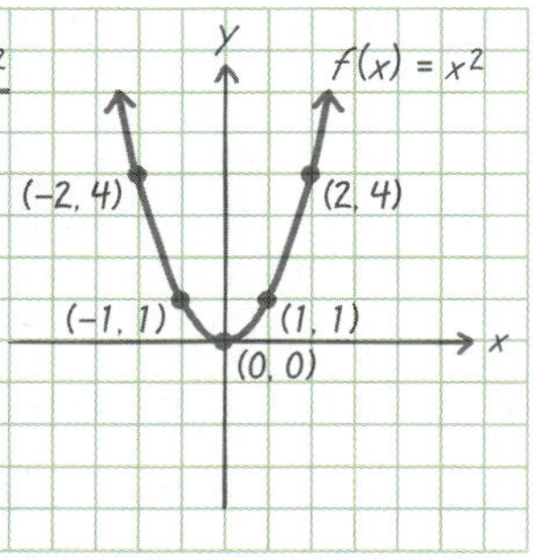

Figure 2

The **cubing function** is defined by $f(x) = x^3$. Every real number is paired with its cube. The domain and the range are both $(-\infty, \infty)$. Its graph is neither a line nor a parabola. See Figure 3 and the table of ordered pairs. (Polynomial functions of degree 3 and greater are studied in detail in more advanced courses.)

x	$f(x) = x^3$
−2	−8
−1	−1
0	0
1	1
2	8

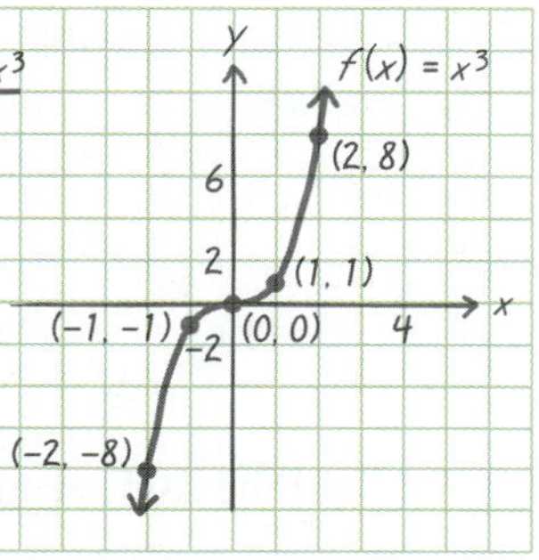

Figure 3

EXAMPLE 5 Graphing Variations of the Identity, Squaring, and Cubing Functions

Graph each function by creating a table of ordered pairs. Give the domain and the range of each function by observing the graphs.

(a) $f(x) = 2x$

To find each range value, multiply the domain value by 2. Plot the points and join them with a straight line. See Figure 4. Both the domain and the range are $(-\infty, \infty)$.

x	$f(x) = 2x$
-2	-4
-1	-2
0	0
1	2
2	4

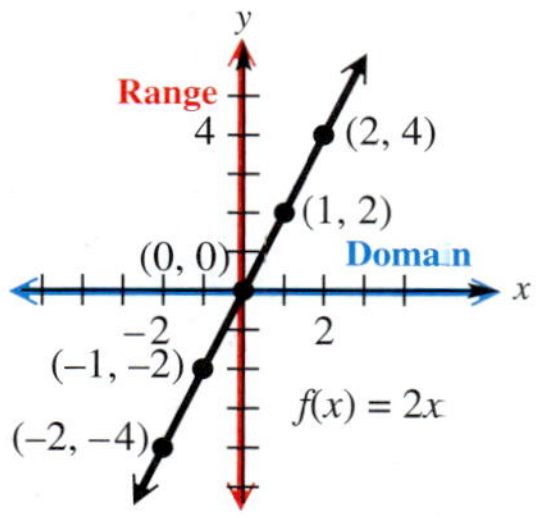

Figure 4

(b) $f(x) = -x^2$

For each input x, square it and then take its opposite. Plotting and joining the points gives a parabola that opens down. See the table and Figure 5. The domain is $(-\infty, \infty)$, and the range is $(-\infty, 0]$.

x	$f(x) = -x^2$
-2	-4
-1	-1
0	0
1	-1
2	-4

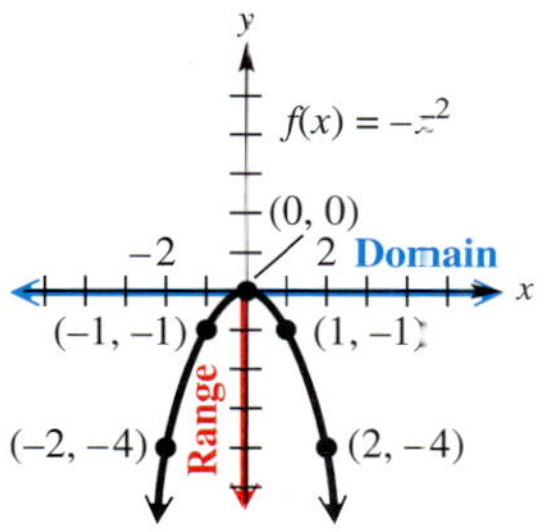

Figure 5

(c) $f(x) = x^3 - 2$

For this function, cube the input and then subtract 2 from the result. The graph is that of the cubing function *shifted* 2 units down. See the table and Figure 6. The domain and the range are both $(-\infty, \infty)$.

x	$f(x) = x^3 - 2$
-2	-10
-1	-3
0	-2
1	-1
2	6

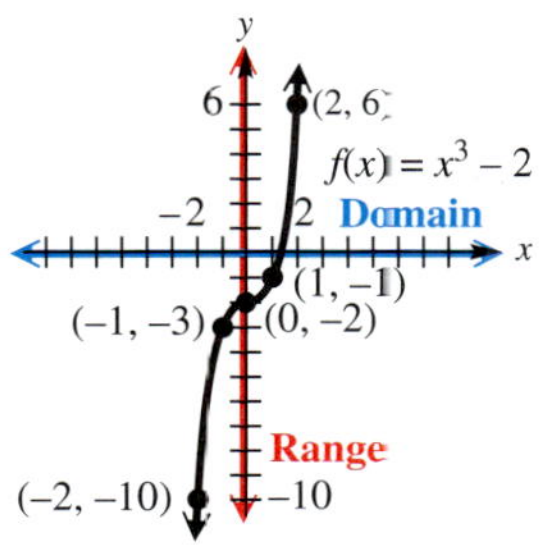

Figure 6

Work Problem **5** *at the Side.* ▶

5 Graph $f(x) = -2x^2$. Give the domain and the range.

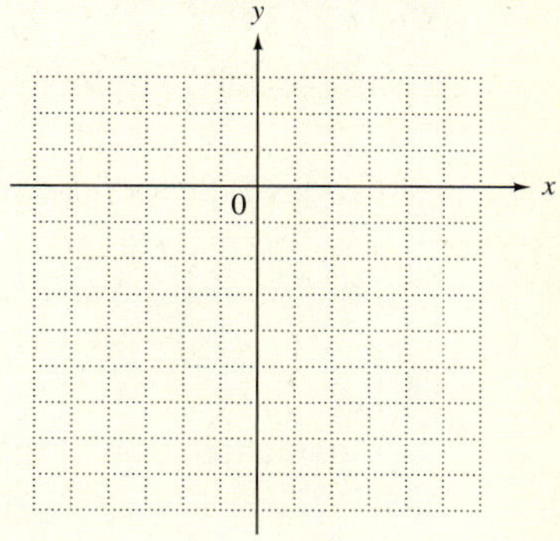

ANSWER

5.

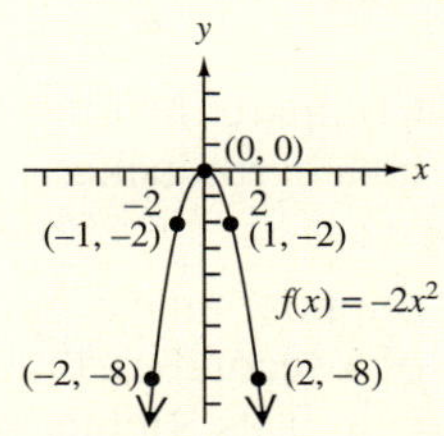

domain: $(-\infty, \infty)$; range: $(-\infty, 0]$

Math in the Media

HOW CAN 8.0 BE ONE THOUSAND TIMES 5.0?

Charles F. Richter devised a scale in 1935 to compare the intensities, or relative power, of earthquakes. The **intensity** of an earthquake (often mentioned in newspaper reports) is measured relative to the intensity of a standard **zero-level** earthquake of intensity I_0. The relationship is equivalent to $I = I_0 \times 10^R$, where R is the **Richter scale** measure. For example, if an earthquake has magnitude 5.0 on the Richter scale, then its intensity is calculated as $I = I_0 \times 10^{5.0} = I_0 \times 100{,}000$, which is 100,000 times as intense as a zero-level earthquake.

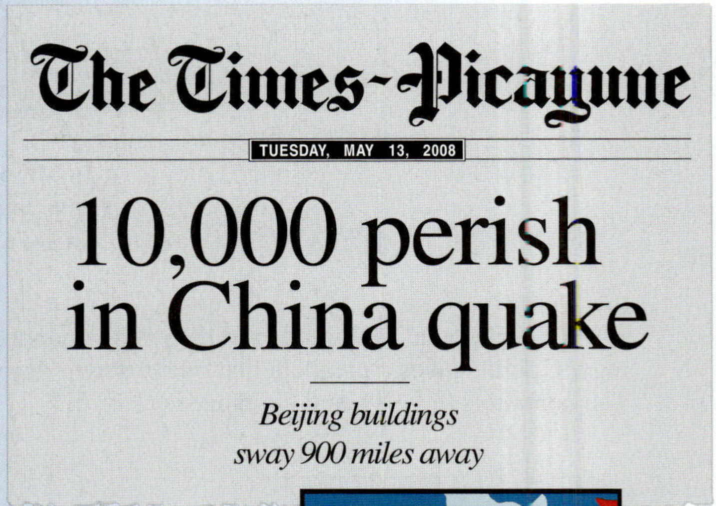

The Times-Picayune

TUESDAY, MAY 13, 2008

10,000 perish in China quake

Beijing buildings sway 900 miles away

To compare an earthquake that measures 8.1 on the Richter scale to one that measures 5.2, find the ratio of the intensities:

$$\frac{\text{intensity } 8.1}{\text{intensity } 5.2} = \frac{I_0 \times 10^{8.1}}{I_0 \times 10^{5.2}} = \frac{10^{8.1}}{10^{5.2}} = 10^{8.1-5.2}$$

$$= 10^{2.9} \approx 794. \quad \text{(Use a calculator.)}$$

Therefore, an earthquake that measures 8.1 on the Richter scale is almost 800 times as intense as one that measures 5.2.

The table gives Richter scale measurements for selected earthquakes in the United States.

	Earthquake	Richter Scale Measurement
2001	Washington: Olympia, Seattle, Tacoma	6.8
2002	Alaska: Slana, Mentasta Lake, Fairbanks	7.9
2005	Montana: Dillon, Silver Star, Twin Bridges	5.6
2006	Hawaiian Islands (tsunami)	6.7
2007	Utah: Huntington	4.2

Source: www.ngdc.noaa.gov

1. Compare the intensity of the 2002 Alaska earthquake to that of the 2007 Utah earthquake.

2. Compare the intensity of the 2001 Washington earthquake to that of the 2007 Utah earthquake.

3. Compare the intensity of the 2006 Hawaiian Islands tsunami earthquake to the 2005 Montana earthquake.

4. Suppose an earthquake measures a value of x on the Richter scale. How would the intensity of a second earthquake compare if its Richter scale measure is $x + 3.0$? How would it compare if its Richter scale measure is $x - 1.0$?

6.3 Exercises

*For each polynomial function, find **(a)** $f(-1)$ and **(b)** $f(2)$. See Example 1.*

1. $f(x) = 6x - 4$
2. $f(x) = -2x + 5$
3. $f(x) = x^2 - 3x + 4$
4. $f(x) = 3x^2 + x - 5$
5. $f(x) = 5x^4 - 3x^2 + 6$
6. $f(x) = -4x^4 + 2x^2 - 1$
7. $f(x) = -x^2 + 2x^3 - 8$
8. $f(x) = -x^2 - x^3 + 11x$

Solve each problem. See Example 2.

9. The number of airports in the United States during the period from 1980 through 2005 can be approximated by the polynomial function defined by

$$f(x) = -2.3425x^2 + 248.04x + 15{,}160,$$

where $x = 0$ represents 1980, $x = 1$ represents 1981, and so on. Use this function to approximate the number of airports in each given year. (*Source*: U.S. Bureau of Transportation Statistics.)

(a) 1980 **(b)** 1995 **(c)** 2005

10. The percent of births in the United States that were to unmarried women during the period from 1990 through 2005 can be approximated by the polynomial function defined by

$$f(x) = 0.0097x^3 - 0.2364x^2 + 2.057x + 26.60,$$

where $x = 0$ represents 1990, $x = 1$ represents 1991, and so on. Use this function to approximate the percent (to the nearest tenth) of births to unmarried women in each given year. (*Source:* National Center for Health Statistics.)

(a) 1990 **(b)** 1998 **(c)** 2005

11. The amount spent by Americans on foreign travel during the period from 1985 through 2006 can be modeled by the polynomial function defined by

$$P(x) = -0.00189x^3 + 0.1193x^2 + 2.027x + 28.19,$$

where $x = 0$ represents 1985, $x = 1$ represents 1986, and so on, and $P(x)$ is in billions of dollars. Use this function to approximate the amount spent by Americans on foreign travel in each given year. Round answers to the nearest tenth. (*Source:* U.S. Department of Commerce.)

(a) 1985

(b) 2000

(c) 2006

12. Imports of Fair Trade Certified™ coffee into the United States during the period from 2000 through 2006 can be modeled by the polynomial function defined by

$$P(x) = 1667x^2 + 22.78x + 4300,$$

where $x = 0$ corresponds to the year 2000, $x = 1$ corresponds to 2001, and so on, and $P(x)$ is in thousands of pounds. Use this function to approximate the amount of Fair Trade coffee imported into the United States in each given year. (*Source:* TransFair USA.)

(a) 2000

(b) 2003

(c) 2006

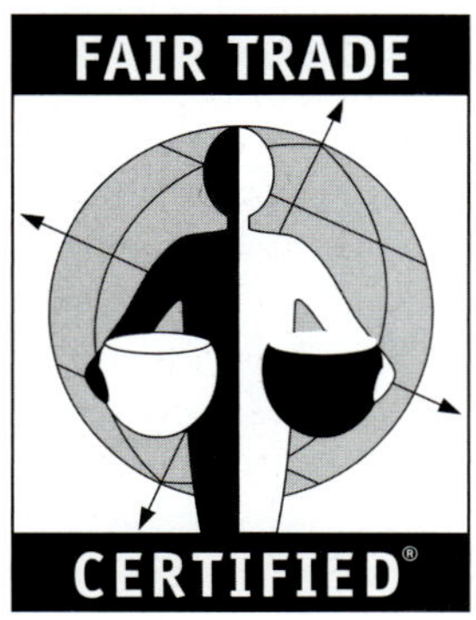

For each pair of functions, find **(a)** $(f + g)(x)$ *and* **(b)** $(f - g)(x)$. *See Example 3.*

13. $f(x) = 5x - 10,\ g(x) = 3x + 7$

14. $f(x) = -4x + 1,\ g(x) = 6x + 2$

15. $f(x) = 4x^2 + 8x - 3,\ g(x) = -5x^2 + 4x - 9$

16. $f(x) = 3x^2 - 9x + 10,\ g(x) = -4x^2 + 2x + 12$

Let $f(x) = x^2 - 9$, $g(x) = 2x$, *and* $h(x) = x - 3$. *Find each of the following. See Example 4.*

17. $(f + g)(x)$

18. $(f - g)(x)$

19. $(f + g)(3)$

20. $(f - g)(-3)$

21. $(f - h)(x)$

22. $(f + h)(x)$

23. $(f - h)(-3)$

24. $(f + h)(-2)$

25. $(g + h)(-10)$

26. $(g - h)(10)$

27. $(g - h)(-3)$

28. $(g + h)\left(\frac{1}{3}\right)$

Graph each function by creating a table of ordered pairs. Give the domain and the range. See Example 5.

29. $f(x) = -2x + 1$

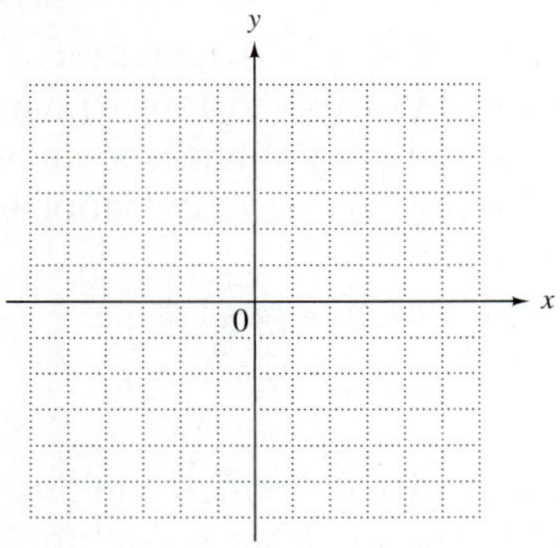

30. $f(x) = 3x + 2$

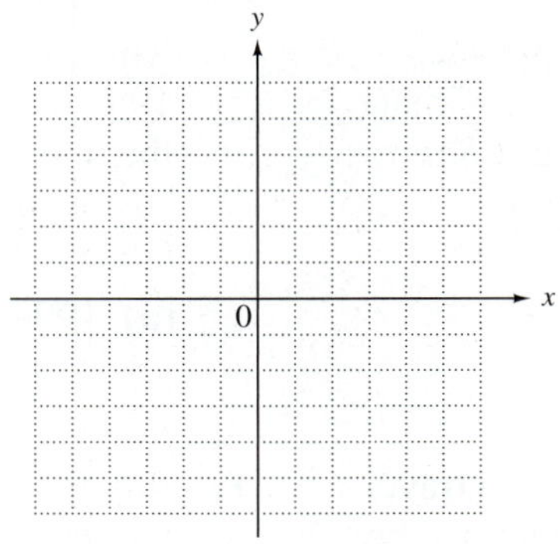

31. $f(x) = -3x^2$

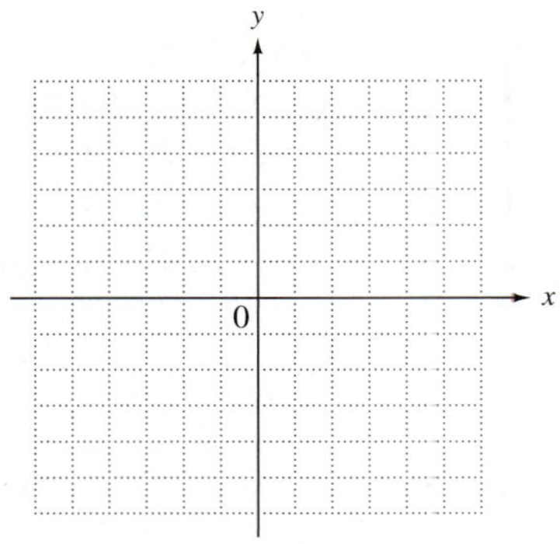

32. $f(x) = \frac{1}{2}x^2$

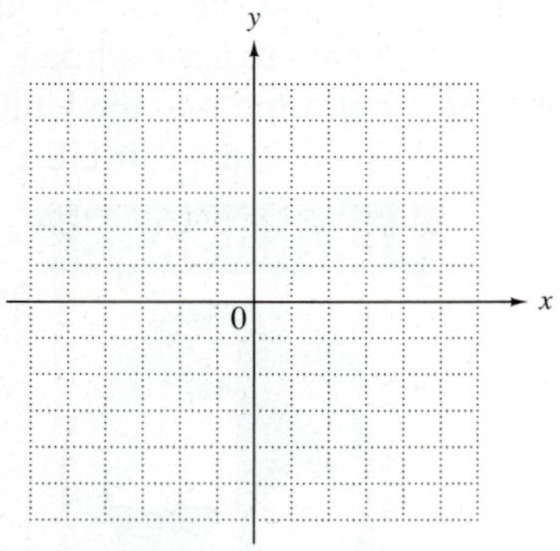

33. $f(x) = x^3 + 1$

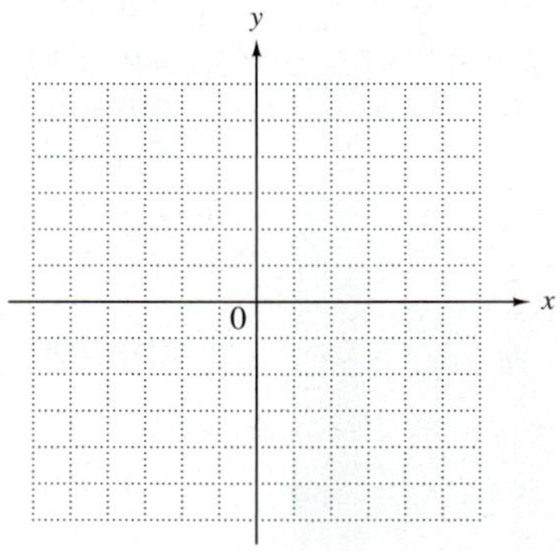

34. $f(x) = -x^3 + 2$

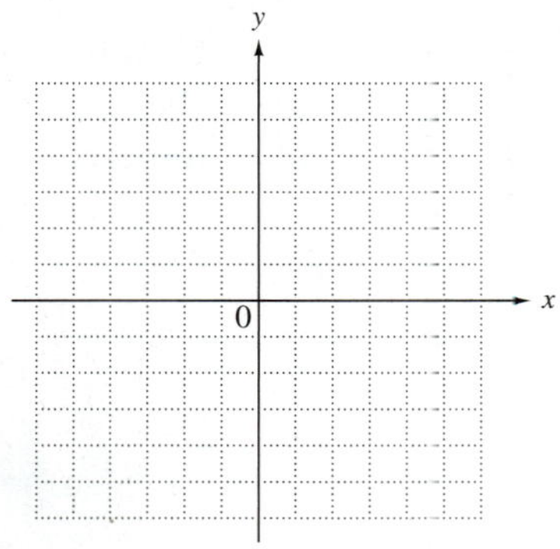

6.4 Multiplying Polynomials

OBJECTIVES

1. Multiply terms.
2. Multiply any two polynomials.
3. Multiply binomials.
4. Find the product of the sum and difference of two terms.
5. Find the square of a binomial.
6. Multiply polynomial functions.

OBJECTIVE 1 Multiply terms. Recall that the product of the two terms $3x^4$ and $5x^3$ is found as follows:

$$(3x^4)(5x^3)$$
$$= 3 \cdot 5 \cdot x^4 \cdot x^3 \quad \text{Commutative and associative properties}$$
$$= 15x^{4+3} \quad \text{Product rule for exponents}$$
$$= 15x^7.$$

EXAMPLE 1 Multiplying Monomials

Find each product.

(a) $-4a^3(3a^5)$
$= -4(3)a^3 \cdot a^5$
$= -12a^8$

(b) $2m^2z^4(8m^3z^2)$
$= 2(8)m^2 \cdot m^3 \cdot z^4 \cdot z^2$
$= 16m^5z^6$

Work Problem 1 at the Side. ▶

OBJECTIVE 2 Multiply any two polynomials.

EXAMPLE 2 Multiplying Polynomials

Find each product.

(a) $-2(8x^3 - 9x^2)$
$= -2(8x^3) - 2(-9x^2)$ Distributive property
$= -16x^3 + 18x^2$ Multiply.

(b) $5x^2(-4x^2 + 3x - 2)$
$= 5x^2(-4x^2) + 5x^2(3x) + 5x^2(-2)$
$= -20x^4 + 15x^3 - 10x^2$

(c) $(3x - 4)(2x^2 + x)$

$(3x - 4)(2x^2 + x)$ Distributive property; multiply each term of $2x^2 + x$ by $3x - 4$.

Treat $3x - 4$ as a single expression.

$= (3x - 4)(2x^2) + (3x - 4)(x)$

Now use the distributive property two more times.

$= 3x(2x^2) + (-4)(2x^2) + (3x)(x) + (-4)(x)$
$= 6x^3 - 8x^2 + 3x^2 - 4x$
$= 6x^3 - 5x^2 - 4x$ Combine like terms.

(d) $2x^2(x + 1)(x - 3)$
$= 2x^2[(x + 1)(x) + (x + 1)(-3)]$ Distributive property
$= 2x^2[x^2 + x - 3x - 3]$ Distributive property
$= 2x^2(x^2 - 2x - 3)$ Combine like terms.
$= 2x^4 - 4x^3 - 6x^2$ Distributive property

Work Problem 2 at the Side. ▶

1 Find each product.

(a) $-6m^5(2m^4)$

(b) $8k^3y(9ky^3)$

2 Find each product.

(a) $-2r(9r - 5)$

(b) $3p^2(5p^3 + 2p^2 - 7)$

(c) $(4a - 5)(3a + 6)$

(d) $3x^3(x + 4)(x - 6)$

ANSWERS

1. (a) $-12m^9$ (b) $72k^4y^4$
2. (a) $-18r^2 + 10r$
(b) $15p^5 + 6p^4 - 21p^2$
(c) $12a^2 + 9a - 30$
(d) $3x^5 - 6x^4 - 72x^3$

3 Find each product.

(a) $\begin{array}{r} 2m - 5 \\ \underline{3m + 4} \end{array}$

(b) $\begin{array}{r} 5a^3 - 6a^2 + 2a - 3 \\ \underline{2a - 5} \end{array}$

EXAMPLE 3 **Multiplying Polynomials Vertically**

Find each product.

(a) $(5a - 2b)(3a + b)$

$$\begin{array}{rrrl} & 5a & -2b & \\ & 3a & +\ b & \\ \hline & 5ab & -2b^2 & \longleftarrow \text{Multiply } b(5a - 2b). \\ 15a^2 & -6ab & & \longleftarrow \text{Multiply } 3a(5a - 2b). \\ \hline 15a^2 & -\ ab & -2b^2 & \text{Combine like terms.} \end{array}$$

(b) $(3m^3 - 2m^2 + 4)(3m - 5)$

$$\begin{array}{rrrrrl} & 3m^3 & -2m^2 & & +\ 4 & \\ & & & 3m & -\ 5 & \\ \hline & -15m^3 & +10m^2 & & -20 & \text{Multiply } -5(3m^3 - 2m^2 + 4). \\ 9m^4 & -\ 6m^3 & & +12m & & \text{Multiply } 3m(3m^3 - 2m^2 + 4). \\ \hline 9m^4 & -21m^3 & +10m^2 & +12m & -20 & \text{Combine like terms.} \end{array}$$

Be sure to write like terms in columns.

◀ Work Problem **3** at the Side.

Note

We can also use a rectangle to model polynomial multiplication. For example, to find the product

$$(5a - 2b)(3a + b)$$

from Example 3(a), label a rectangle with each term as shown here. Then put the product of each pair of monomials in the appropriate box.

	$3a$	b
$5a$		
$-2b$		

	$3a$	b
$5a$	$15a^2$	$5ab$
$-2b$	$-6ab$	$-2b^2$

The product of the original binomials is the sum of these four monomial products.

$$\begin{aligned} &(5a - 2b)(3a + b) \\ &= 15a^2 + 5ab - 6ab - 2b^2 \\ &= 15a^2 - ab - 2b^2 \end{aligned}$$

OBJECTIVE 3 **Multiply binomials.** When working with polynomials, the product of two binomials occurs repeatedly. There is a shortcut method for finding these products. Recall that a binomial has just two terms, such as $3x - 4$ or $2x + 3$. We can find the product of these binomials using the distributive property as follows.

$$\begin{aligned} &(3x - 4)(2x + 3) \\ &= 3x(2x + 3) - 4(2x + 3) \\ &= 3x(2x) + 3x(3) - 4(2x) - 4(3) \\ &= 6x^2 + 9x - 8x - 12 \end{aligned}$$

Answers

3. (a) $6m^2 - 7m - 20$
(b) $10a^4 - 37a^3 + 34a^2 - 16a + 15$

Before combining like terms to find the simplest form of the answer, we check the origin of each of the four terms in the sum $6x^2 + 9x - 8x - 12$. First, $6x^2$ is the product of the two *first* terms.

$(3x - 4)(2x + 3) \qquad 3x(2x) = \mathbf{6x^2} \qquad$ First terms

To get $9x$, the *outer* terms are multiplied.

$(3x - 4)(2x + 3) \qquad 3x(3) = \mathbf{9x} \qquad$ Outer terms

The term $-8x$ comes from the *inner* terms.

$(3x - 4)(2x + 3) \qquad -4(2x) = \mathbf{-8x} \qquad$ Inner terms

Finally, -12 comes from the *last* terms.

$(3x - 4)(2x + 3) \qquad -4(3) = \mathbf{-12} \qquad$ Last terms

The product is found by combining these four results.

$$\begin{aligned}&(3x - 4)(2x + 3)\\ &= 6x^2 + 9x - 8x - 12\\ &= 6x^2 + x - 12\end{aligned}$$

To keep track of the order of multiplying these terms, we use the initials FOIL (**F**irst, **O**uter, **I**nner, **L**ast). All the steps of the FOIL method can be done as follows. Try to do as many of these steps as possible mentally.

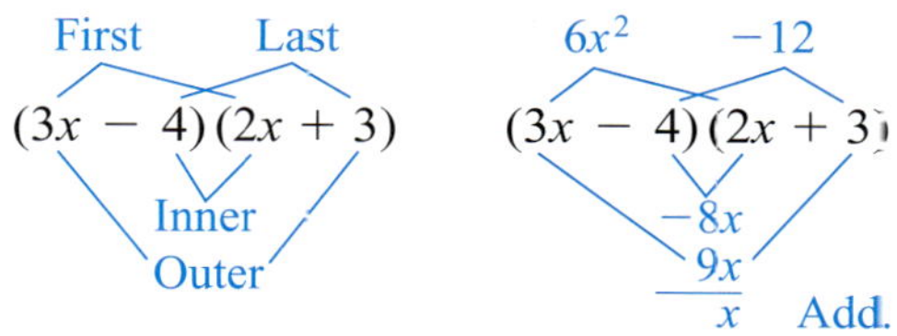

> **CAUTION**
> The FOIL method is an extension of the distributive property, and the acronym ***"FOIL" applies only to multiplying two binomials.***

EXAMPLE 4 **Using the FOIL Method**

Use the FOIL method to find each product.

(a) $(4m - 5)(3m + 1)$

First terms $\qquad (4m - 5)(3m + 1) \qquad 4m(3m) = \mathbf{12m^2}$

Outer terms $\qquad (4m - 5)(3m + 1) \qquad 4m(1) = \mathbf{4m}$

Inner terms $\qquad (4m - 5)(3m + 1) \qquad -5(3m) = \mathbf{-15m}$

Last terms $\qquad (4m - 5)(3m + 1) \qquad -5(1) = \mathbf{-5}$

Simplify by combining the four terms.

$$\begin{aligned}&(4m - 5)(3m + 1)\\ &\qquad\ \text{F}\qquad\ \text{O}\qquad\ \ \text{I}\qquad\ \text{L}\\ &= 12m^2 + 4m - 15m - 5\\ &= 12m^2 - 11m - 5\end{aligned}$$

Continued on Next Page

The procedure can be written in compact form as follows.

$12m^2$ -5

$(4m - 5)(3m + 1)$

$-15m$

$4m$

$-11m$ Add.

Combine these four results to get $12m^2 - 11m - 5$.

(b) $(6a - 5b)(3a + 4b)$

First   Outer   Inner   Last

$$= 18a^2 + 24ab - 15ab - 20b^2$$
$$= 18a^2 + 9ab - 20b^2$$

(c) $(2k + 3z)(5k - 3z)$

$$= 10k^2 + 9kz - 9z^2 \qquad \text{FOIL}$$

◀ Work Problem 4 at the Side.

4 Use the FOIL method to find each product.

(a) $(3z + 2)(z + 1)$

(b) $(5r - 3)(2r - 5)$

(c) $(4p + 5q)(3p - 2q)$

OBJECTIVE 4 Find the product of the sum and difference of two terms. Some types of binomial products occur frequently. For example, the product of the sum and difference of the same two terms, x and y, is

$$(x + y)(x - y)$$
$$= x^2 - xy + xy - y^2 \qquad \text{FOIL}$$
$$= x^2 - y^2. \qquad \text{Combine like terms.}$$

Product of the Sum and Difference of Two Terms

The **product of the sum and difference of the two terms x and y** is the difference of the squares of the terms.

$$(x + y)(x - y) = x^2 - y^2$$

EXAMPLE 5 Multiplying the Sum and Difference of Two Terms

Find each product.

(a) $(p + 7)(p - 7)$
$$= p^2 - 7^2$$
$$= p^2 - 49$$

(b) $(2r + 5)(2r - 5)$
$$= (2r)^2 - 5^2$$
$$= 2^2r^2 - 25$$
$$= 4r^2 - 25$$

(c) $(6m + 5n)(6m - 5n)$
$$= (6m)^2 - (5n)^2$$
$$= 36m^2 - 25n^2$$

(d) $2x^3(x + 3)(x - 3)$
$$= 2x^3(x^2 - 9)$$
$$= 2x^5 - 18x^3$$

◀ Work Problem 5 at the Side.

5 Find each product.

(a) $(m + 5)(m - 5)$

(b) $(x - 4y)(x + 4y)$

(c) $(7m - 2n)(7m + 2n)$

(d) $4y^2(y + 7)(y - 7)$

OBJECTIVE 5 Find the square of a binomial. Another special binomial product is the *square of a binomial*. To find the square of a sum $x + y$, or $(x + y)^2$, multiply $x + y$ by itself.

$$(x + y)(x + y)$$
$$= x^2 + xy + xy + y^2 \qquad \text{FOIL}$$
$$= x^2 + 2xy + y^2 \qquad \text{Combine like terms.}$$

A similar result is true for the square of a difference.

ANSWERS

4. (a) $3z^2 + 5z + 2$ (b) $10r^2 - 31r + 15$ (c) $12p^2 + 7pq - 10q^2$
5. (a) $m^2 - 25$ (b) $x^2 - 16y^2$ (c) $49m^2 - 4n^2$ (d) $4y^4 - 196y^2$

Square of a Binomial

The **square of a binomial** is the sum of the square of the first term, twice the product of the two terms, and the square of the last term.

$$(x + y)^2 = x^2 + 2xy + y^2$$
$$(x - y)^2 = x^2 - 2xy + y^2$$

EXAMPLE 6 Squaring Binomials

Find each product.

(a) $(m + 7)^2$

$= m^2 + 2 \cdot m \cdot 7 + 7^2$ $\quad (x + y)^2 = x^2 + 2xy + y^2$

$= m^2 + 14m + 49$

(b) $(p - 5)^2$

$= p^2 - 2 \cdot p \cdot 5 + 5^2$ $\quad (x - y)^2 = x^2 - 2xy + y^2$

$= p^2 - 10p + 25$

(c) $(2p + 3v)^2$

$= (2p)^2 + 2(2p)(3v) + (3v)^2$

$= 4p^2 + 12pv + 9v^2$

(d) $(3r - 5s)^2$

$= (3r)^2 - 2(3r)(5s) + (5s)^2$

$= 9r^2 - 30rs + 25s^2$

CAUTION

As the products in the formula for the square of a binomial show,

$$(x + y)^2 \neq x^2 + y^2.$$

More generally,

$$(x + y)^n \neq x^n + y^n \qquad (n \neq 1).$$

Work Problem **6** *at the Side.* ▶

We can use the patterns for the special products with more complicated products, as the following example shows.

EXAMPLE 7 Multiplying More Complicated Binomials

Use special products to find each product.

(a) $[(3p - 2) + 5q][(3p - 2) - 5q]$

$= (3p - 2)^2 - (5q)^2$ $\quad$ Product of sum and difference of terms

$= 9p^2 - 12p + 4 - 25q^2$ $\quad$ Square both quantities.

(b) $[(2z + r) + 1]^2$

$= (2z + r)^2 + 2(2z + r)(1) + 1^2$ $\quad$ Square of a binomial

$= 4z^2 + 4zr + r^2 + 4z + 2r + 1$ $\quad$ Square again; use the distributive property.

Continued on Next Page

6 Find each product.

(a) $(a + 2)^2$

(b) $(2m - 5)^2$

(c) $(y + 6z)^2$

(d) $(3k - 2n)^2$

ANSWERS

6. **(a)** $a^2 + 4a + 4$
(b) $4m^2 - 20m + 25$
(c) $y^2 + 12yz + 36z^2$
(d) $9k^2 - 12kn + 4n^2$

(c) $(x + y)^3$

$= (x + y)^2(x + y)$

$= (x^2 + 2xy + y^2)(x + y)$ Square $x + y$.

$= x^3 + 2x^2y + xy^2 + x^2y + 2xy^2 + y^3$ Distributive property

$= x^3 + 3x^2y + 3xy^2 + y^3$ Combine like terms.

This does not equal $x^3 + y^3$.

(d) $(2a + b)^4$

$= (2a + b)^2(2a + b)^2$

$= (4a^2 + 4ab + b^2)(4a^2 + 4ab + b^2)$ Square $2a + b$.

$= 16a^4 + 16a^3b + 4a^2b^2 + 16a^3b + 16a^2b^2 + 4ab^3 + 4a^2b^2 + 4ab^3 + b^4$

$= 16a^4 + 32a^3b + 24a^2b^2 + 8ab^3 + b^4$

◀ Work Problem **7** at the Side.

OBJECTIVE 6 Multiply polynomial functions. In Section 6.3, we added and subtracted functions. Functions can also be multiplied.

Multiplying Functions

If $f(x)$ and $g(x)$ define functions, then

$$(fg)(x) = f(x) \cdot g(x). \quad \text{Product function}$$

The domain of the product function is the intersection of the domains of $f(x)$ and $g(x)$.

EXAMPLE 8 Multiplying Polynomial Functions

For $f(x) = 3x + 4$ and $g(x) = 2x^2 + x$, find $(fg)(x)$ and $(fg)(-1)$.

$(fg)(x)$

$= f(x) \cdot g(x)$ Use the definition.

$= (3x + 4)(2x^2 + x)$ Substitute.

$= 6x^3 + 3x^2 + 8x^2 + 4x$ FOIL

$= 6x^3 + 11x^2 + 4x$ Combine like terms.

Then $(fg)(-1)$

$= 6(-1)^3 + 11(-1)^2 + 4(-1)$ Let $x = -1$.

$= -6 + 11 - 4$ Be careful with signs.

$= 1.$

(Another way to find $(fg)(-1)$ is to find $f(-1)$ and $g(-1)$ and then multiply the results. Verify this by showing that $f(-1) \cdot g(-1)$ equals 1. This follows from the definition.)

◀ Work Problem **8** at the Side.

CAUTION

Write the product $f(x) \cdot g(x)$ as $(fg)(x)$, *not* $f(g(x))$, which has a different mathematical meaning, as discussed in **Section 12.1.**

---

7 Find each product.

(a) $[(m - 2n) - 3] \cdot [(m - 2n) + 3]$

(b) $[(k - 5h) + 2]^2$

(c) $(p + 2q)^3$

(d) $(x + 2)^4$

8 For

$f(x) = 2x + 7$

and $g(x) = x^2 - 4$,

find $(fg)(x)$ and $(fg)(2)$.

Answers

7. (a) $m^2 - 4mn + 4n^2 - 9$
(b) $k^2 - 10kh + 25h^2 + 4k - 20h + 4$
(c) $p^3 + 6p^2q + 12pq^2 + 8q^3$
(d) $x^4 + 8x^3 + 24x^2 + 32x + 16$
8. $2x^3 + 7x^2 - 8x - 28$; 0

6.4 Exercises

FOR EXTRA HELP MyMathLab Math XL PRACTICE WATCH DOWNLOAD READ REVIEW

Find each product. See Examples 1–3.

1. $-8m^3(3m^2)$
2. $4p^2(-5p^4)$
3. $3x(-2x + 5)$
4. $5y(-6y - 1)$
5. $-q^3(2 + 3q)$
6. $-3a^4(4 - a)$
7. $6k^2(3k^2 + 2k + 1)$
8. $5r^3(2r^2 - 3r - 4)$
9. $(2m + 3)(3m^2 - 4m - 1)$
10. $(4z - 2)(z^2 + 3z + 5)$
11. $4x^3(x - 3)(x + 2)$
12. $2y^5(y - 8)(y + 2)$
13. $(2y + 3)(3y - 4)$
14. $(5m - 3)(2m + 6)$
15. $\begin{array}{r} 5m - 3n \\ \underline{5m + 3n} \end{array}$
16. $\begin{array}{r} 2k + 6q \\ \underline{2k - 6q} \end{array}$
17. $\begin{array}{r} -b^2 + 3b + 3 \\ \underline{2b + 4} \end{array}$
18. $\begin{array}{r} -r^2 - 4r + 8 \\ \underline{3r - 2} \end{array}$
19. $\begin{array}{r} 2z^3 - 5z^2 + 8z - 1 \\ \underline{4z + 3} \end{array}$
20. $\begin{array}{r} 3z^4 - 2z^3 + z - 5 \\ \underline{2z - 5} \end{array}$
21. $\begin{array}{r} 2p^2 + 3p - 6 \\ \underline{3p^2 - 4p - 1} \end{array}$
22. $\begin{array}{r} 5y^2 - 2y + 4 \\ \underline{2y^2 + y + 3} \end{array}$

Use the FOIL method to find each product. See Example 4.

23. $(m + 5)(m - 8)$

24. $(p - 6)(p + 4)$

25. $(4k + 3)(3k - 2)$

26. $(5w + 2)(2w + 5)$

27. $(z - w)(3z + 4w)$

28. $(s + t)(2s - 5t)$

29. $(6c - d)(2c + 3d)$

30. $(2m - n)(3m + 5n)$

31. $(0.2x + 1.3)(0.5x - 0.1)$

32. $(0.5y - 0.4)(0.1y + 2.1)$

33. $\left(3r + \frac{1}{4}y\right)(r - 2y)$

34. $\left(5w - \frac{2}{3}z\right)(w + 5z)$

Find each product. See Example 5.

35. $(2p - 3)(2p + 3)$

36. $(3x - 8)(3x + 8)$

37. $(5m - 1)(5m + 1)$

38. $(6y + 3)(6y - 3)$

39. $(3a + 2c)(3a - 2c)$

40. $(5r - 4s)(5r + 4s)$

41. $\left(4x - \frac{2}{3}\right)\left(4x + \frac{2}{3}\right)$

42. $\left(3t + \frac{5}{4}\right)\left(3t - \frac{5}{4}\right)$

43. $(4m + 7n^2)(4m - 7n^2)$

44. $(2k^2 + 6h)(2k^2 - 6h)$

45. $5y^3(y + 2)(y - 2)$

46. $3x^3(x - 4)(x + 4)$

Find each square. See Example 6.

47. $(y - 5)^2$

48. $(a - 3)^2$

49. $(2p + 7)^2$

50. $(3z + 8)^2$ **51.** $(4n - 3m)^2$ **52.** $(5r - 7s)^2$

53. $\left(k - \frac{5}{7}p\right)^2$ **54.** $\left(q - \frac{3}{4}r\right)^2$

Find each product. See Example 7.

55. $[(5x + 1) + 6y]^2$ **56.** $[(3m - 2) + p]^2$ **57.** $[(2a + b) - 3][(2a + b) + 3]$

58. $[(m + p) + 5][(m + p) - 5]$ **59.** $[(2h - k) + j][(2h - k) - j]$ **60.** $[(3m - y) + z][(3m - y) - z]$

61. $(x + 2)^3$ **62.** $(z - 3)^3$ **63.** $(5r - s)^3$

64. $(x + 3y)^3$ **65.** $(q - 2)^4$ **66.** $(m - p)^4$

Find the area of each figure. Express it as a polynomial in descending powers of the variable x. Refer to the formulas on the inside covers of this book if necessary.

67.

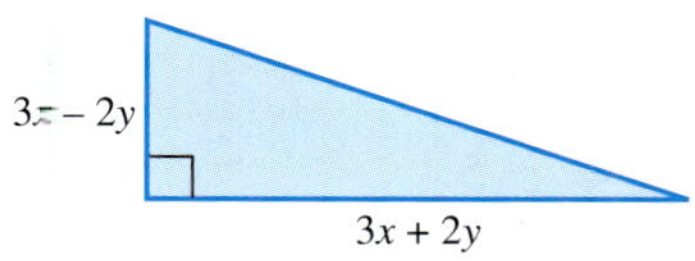

68.

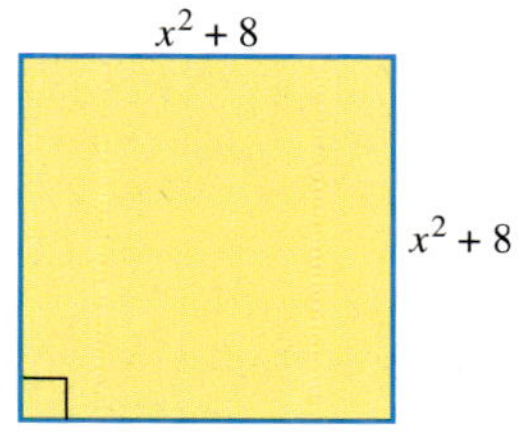

69.

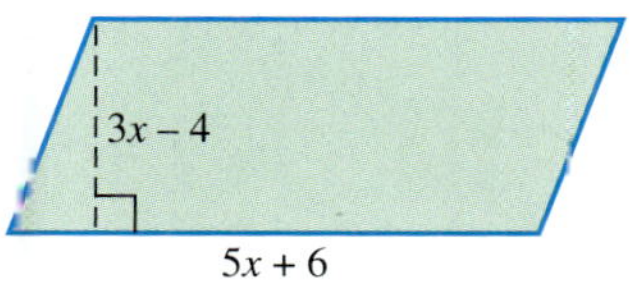

70.

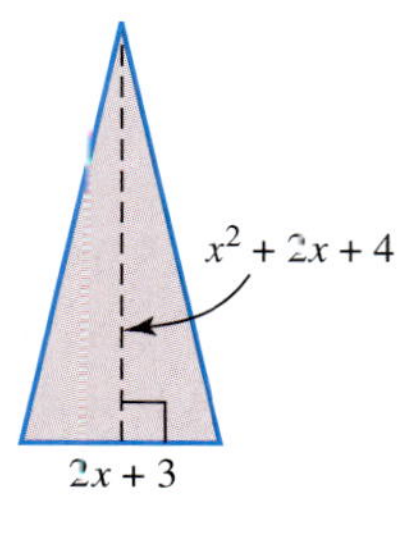

Relating Concepts (Exercises 71–78) For Individual or Group Work

Consider the figure. ***Work Exercises 71–78 in order.***

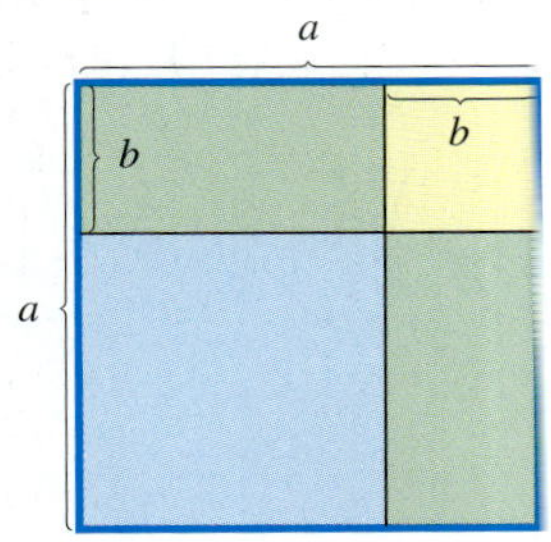

71. What is the length of each side of the blue square in terms of a and b?

72. What is the formula for the area of a square? Use the formula to write an expression, in the form of a product, for the area of the blue square.

73. Each green rectangle has an area of ______________. Therefore, the total area in green is represented by the polynomial ________.

74. The yellow square has an area of ________.

75. The area of the entire colored region is represented by ________, because each side of the entire colored region has length ________.

76. The area of the blue square is equal to the area of the entire colored region minus the total area of the green squares minus the area of the yellow square. Write this as a simplified polynomial in a and b.

77. What must be true about the expressions for the area of the blue square you found in Exercises 72 and 76?

78. Write a statement of equality based on your answer in Exercise 77. How does this reinforce one of the main ideas of this section?

For each pair of functions, find the product $(fg)(x)$. See Example 8.

79. $f(x) = 2x,\ g(x) = 5x - 1$

80. $f(x) = 3x,\ g(x) = 6x - 8$

81. $f(x) = x + 1,\ g(x) = 2x - 3$

82. $f(x) = x - 7,\ g(x) = 4x + 5$

83. $f(x) = 2x - 3,\ g(x) = 4x^2 + 6x + 9$

84. $f(x) = 3x + 4,\ g(x) = 9x^2 - 12x + 16$

Let $f(x) = x^2 - 9$, $g(x) = 2x$, and $h(x) = x - 3$. Find each of the following. See Example 8.

85. $(fg)(2)$

86. $(fh)(1)$

87. $(fh)(-1)$

88. $(gh)(-3)$

89. $(fg)(-2)$

Show that each statement is false by replacing x with 2 and y with 3. Then, rewrite each statement with the correct product.

90. $(x + y)^2 = x^2 + y^2$

91. $(x + y)^3 = x^3 + y^3$

92. $(x + y)^4 = x^4 + y^4$

6.5 Dividing Polynomials

OBJECTIVES

1. Divide a polynomial by a monomial.
2. Divide a polynomial by a polynomial of two or more terms.
3. Divide polynomial functions.

OBJECTIVE 1 Divide a polynomial by a monomial. Recall that a monomial is a single term, such as $8x$, $-9m^4$, or $11y^2$.

Dividing by a Monomial

To divide a polynomial by a monomial, divide each term in the polynomial by the monomial, and then write each quotient in lowest terms.

EXAMPLE 1 Dividing a Polynomial by a Monomial

Divide.

(a) $\dfrac{15x^2 - 12x + 6}{3}$

$= \dfrac{15x^2}{3} - \dfrac{12x}{3} + \dfrac{6}{3}$ Divide each term by 3.

$= 5x^2 - 4x + 2$ Write in lowest terms.

Check $3(5x^2 - 4x + 2) = 15x^2 - 12x + 6$

Divisor (3), Quotient ($5x^2 - 4x + 2$), Original polynomial ($15x^2 - 12x + 6$)

(b) $\dfrac{5m^3 - 9m^2 + 10m}{5m^2}$

$= \dfrac{5m^3}{5m^2} - \dfrac{9m^2}{5m^2} + \dfrac{10m}{5m^2}$ Divide each term by $5m^2$.

$= m - \dfrac{9}{5} + \dfrac{2}{m}$ Write in lowest terms.

This result, $m - \frac{9}{5} + \frac{2}{m}$, is not a polynomial. (Why?) The quotient of two polynomials need not be a polynomial.

(c) $\dfrac{8xy^2 - 9x^2y + 6x^2y^2}{x^2y^2}$

$= \dfrac{8xy^2}{x^2y^2} - \dfrac{9x^2y}{x^2y^2} + \dfrac{6x^2y^2}{x^2y^2}$

$= \dfrac{8}{x} - \dfrac{9}{y} + 6$

Work Problem 1 at the Side. ▶

1 Divide.

(a) $\dfrac{12p + 30}{6}$

(b) $\dfrac{9y^3 - 4y^2 + 8y}{2y^2}$

(c) $\dfrac{8a^2b^2 - 20ab^3}{4a^3b}$

OBJECTIVE 2 Divide a polynomial by a polynomial of two or more terms. This process is similar to that for dividing whole numbers.

EXAMPLE 2 Dividing a Polynomial by a Polynomial

Divide $\dfrac{2m^2 + m - 10}{m - 2}$.

$m - 2\overline{)2m^2 + m - 10}$ Write both polynomials in descending powers.

Continued on Next Page

ANSWERS

1. (a) $2p + 5$ (b) $\dfrac{9y}{2} - 2 + \dfrac{4}{y}$ (c) $\dfrac{2b}{a} - \dfrac{5b^2}{a^2}$

2 Divide.

(a) $\dfrac{2r^2 + r - 21}{r - 3}$

(b) $\dfrac{2k^2 + 17k + 30}{2k + 5}$

Answers

2. (a) $2r + 7$ (b) $k + 6$

Divide the first term of the dividend $\mathbf{2m^2} + m - 10$ by the first term of the divisor $\boldsymbol{m} - 2$. Since $\dfrac{2m^2}{m} = 2m$, place this result above the division line.

$$\begin{array}{r} \mathbf{2m} \\ m - 2\overline{)2m^2 + m - 10} \end{array} \quad \longleftarrow \text{Result of } \tfrac{2m^2}{m}$$

Multiply $m - 2$ and $2m$, and write the result below $2m^2 + m - 10$.

$$\begin{array}{r} 2m \phantom{{}+ m - 10} \\ m - 2\overline{)2m^2 + m - 10} \\ \underline{2m^2 - 4m} \phantom{{}- 10} \end{array} \quad \longleftarrow 2m(m - 2) = 2m^2 - 4m$$

Now subtract by mentally changing the signs on $2m^2 - 4m$ and adding.

To subtract, add the opposite.

$$\begin{array}{r} 2m \phantom{{}+ m - 10} \\ m - 2\overline{)2m^2 + m - 10} \\ \underline{2m^2 - 4m} \phantom{{}- 10} \\ 5m \phantom{{}- 10} \end{array} \quad \longleftarrow \text{Subtract. The difference is } 5m.$$

Bring down -10 and continue by dividing $5m$ by m.

$$\begin{array}{r} 2m + \mathbf{5} \\ m - 2\overline{)2m^2 + m - 10} \\ \underline{2m^2 - 4m} \phantom{{}- 10} \\ 5m - 10 \\ \underline{5m - 10} \\ 0 \end{array} \quad \begin{array}{l} \longleftarrow \tfrac{5m}{m} = 5 \\ \\ \\ \longleftarrow \text{Bring down } -10. \\ \longleftarrow 5(m - 2) = 5m - 10 \\ \longleftarrow \text{Subtract. The difference is } 0. \end{array}$$

Finally, $(2m^2 + m - 10) \div (m - 2) = 2m + 5$. Check by multiplying $m - 2$ and $2m + 5$. The result should be $2m^2 + m - 10$.

◀ *Work Problem* **2** *at the Side.*

EXAMPLE 3 Dividing a Polynomial with a Missing Term

Divide $3x^3 - 2x + 5$ by $x - 3$.

Make sure that $3x^3 - 2x + 5$ is in descending powers of the variable. Add a term with 0 coefficient as a placeholder for the missing x^2-term.

$$x - 3\overline{)3x^3 + \mathbf{0x^2} - 2x + 5} \quad \longleftarrow \text{Missing term}$$

Start with $\dfrac{3x^3}{x} = 3x^2$.

$$\begin{array}{r} 3x^2 \phantom{{}- 2x + 5} \\ x - 3\overline{)3x^3 + 0x^2 - 2x + 5} \\ \underline{3x^3 - 9x^2} \phantom{{}- 2x + 5} \end{array} \quad \begin{array}{l} \longleftarrow \tfrac{3x^3}{x} = 3x^2 \\ \\ \longleftarrow 3x^2(x - 3) \end{array}$$

Subtract by mentally changing the signs on $3x^3 - 9x^2$ and adding.

$$\begin{array}{r} 3x^2 \phantom{{}- 2x + 5} \\ x - 3\overline{)3x^3 + 0x^2 - 2x + 5} \\ \underline{3x^3 - 9x^2} \phantom{{}- 2x + 5} \\ 9x^2 \phantom{{}- 2x + 5} \end{array} \quad \longleftarrow \text{Subtract.}$$

Bring down the next term.

$$\begin{array}{r} 3x^2 \phantom{{}- 2x + 5} \\ x - 3\overline{)3x^3 + 0x^2 - 2x + 5} \\ \underline{3x^3 - 9x^2} \phantom{{}- 2x + 5} \\ 9x^2 - 2x \phantom{{}+ 5} \end{array} \quad \longleftarrow \text{Bring down } -2x.$$

Continued on Next Page

In the next step, $\frac{9x^2}{x} = 9x$.

$$\begin{array}{r l}
 3x^2 + 9x & \leftarrow \frac{9x^2}{x} = 9x \\
x - 3\overline{)3x^3 + 0x^2 - 2x + 5} & \\
3x^3 - 9x^2 & \\
9x^2 - 2x & \\
9x^2 - 27x & \leftarrow 9x(x-3) \\
25x + 5 & \leftarrow \text{Subtract; bring down 5.}
\end{array}$$

Finally, $\frac{25x}{x} = 25$.

$$\begin{array}{r l}
 3x^2 + 9x + 25 & \leftarrow \frac{25x}{x} = 25 \\
x - 3\overline{)3x^3 + 0x^2 - 2x + 5} & \\
3x^3 - 9x^2 & \\
9x^2 - 2x & \\
9x^2 - 27x & \\
25x + 5 & \\
25x - 75 & \leftarrow 25(x-3) \\
80 & \leftarrow \text{Remainder}
\end{array}$$

Write the remainder, 80, as the numerator of the fraction $\frac{80}{x-3}$. The answer is

$$3x^2 + 9x + 25 + \frac{80}{x-3}.$$

Be sure to add $\frac{\text{remainder}}{\text{divisor}}$.
Don't forget the + sign.

Check Multiply $x - 3$ (the divisor) and $3x^2 + 9x + 25$ (the quotient), and then add 80 (the remainder). The result should be $3x^3 - 2x + 5$.

CAUTION
Remember to add $\frac{\text{remainder}}{\text{divisor}}$ to the quotient when writing the answer.

Work Problem 3 at the Side. ▶

EXAMPLE 4 Dividing by a Polynomial with a Missing Term

Divide $6r^4 + 9r^3 + 2r^2 - 8r + 7$ by $3r^2 - 2$.

Write $3r^2 - 2$ as $3r^2 + 0r - 2$ and divide as usual.

$$\begin{array}{r l}
2r^2 + 3r + 2 & \\
3r^2 + 0r - 2\overline{)6r^4 + 9r^3 + 2r^2 - 8r + 7} & \\
6r^4 + 0r^3 - 4r^2 & \\
9r^3 + 6r^2 - 8r & \\
9r^3 + 0r^2 - 6r & \\
6r^2 - 2r + 7 & \\
6r^2 + 0r - 4 & \\
-2r + 11 & \leftarrow \text{Remainder}
\end{array}$$

Missing term (points to $0r$)

Stop when the degree of the remainder is less than the degree of the divisor.

Since the degree of the remainder, $-2r + 11$, is less than the degree of the divisor, $3r^2 - 2$, the process is now finished. The answer is written

$$2r^2 + 3r + 2 + \frac{-2r + 11}{3r^2 - 2}.$$

Work Problem 4 at the Side. ▶

3 Divide.

$$\frac{3k^3 + 9k - 14}{k - 2}$$

4 Divide.

(a) $\frac{3r^5 - 15r^4 - 2r^3 + 19r^2 - 7}{3r^2 - 2}$

(b) $\frac{4x^4 - 7x^2 + x + 5}{2x^2 - x}$

ANSWERS

3. $3k^2 + 6k + 21 + \frac{28}{k - 2}$

4. (a) $r^3 - 5r^2 + 3 + \frac{-1}{3r^2 - 2}$
(b) $2x^2 + x - 3 + \frac{-2x + 5}{2x^2 - x}$

5 Divide $2p^3 + 7p^2 + 9p + 4$ by $2p + 2$.

6 For
$$f(x) = 2x^2 + 17x + 30$$
and $g(x) = 2x + 5$,
find $\left(\frac{f}{g}\right)(x)$ and $\left(\frac{f}{g}\right)(-1)$.

ANSWERS

5. $p^2 + \frac{5}{2}p + 2$

6. $x + 6, \quad x \neq -\frac{5}{2}; 5$

EXAMPLE 5 Finding a Quotient with a Fractional Coefficient

Divide $2p^3 + 5p^2 + p - 2$ by $2p + 2$.

$\frac{3p^2}{2p} = \frac{3}{2}p$

$$\begin{array}{r} p^2 + \frac{3}{2}p - 1 \\ 2p + 2\overline{)2p^3 + 5p^2 + p - 2} \\ \underline{2p^3 + 2p^2} \qquad\qquad \\ 3p^2 + p \qquad \\ \underline{3p^2 + 3p} \qquad \\ -2p - 2 \\ \underline{-2p - 2} \\ 0 \end{array}$$

Since the remainder is 0, the quotient is $p^2 + \frac{3}{2}p - 1$.

◀ Work Problem 5 at the Side.

CAUTION
When dividing a polynomial by a polynomial of two or more terms:

1. Be sure the terms in both polynomials are in descending powers.
2. Write any missing terms with 0 placeholders.

OBJECTIVE 3 Divide polynomial functions.

Dividing Functions
If $f(x)$ and $g(x)$ define functions, then

$$\left(\frac{f}{g}\right)(x) = \frac{f(x)}{g(x)}. \quad \text{Quotient function}$$

The domain of the quotient function is the intersection of the domains of $f(x)$ and $g(x)$, excluding any values of x for which $g(x) = 0$.

EXAMPLE 6 Dividing Polynomial Functions

For $f(x) = 2x^2 + x - 10$ and $g(x) = x - 2$, find $\left(\frac{f}{g}\right)(x)$ and $\left(\frac{f}{g}\right)(-3)$.

$$\left(\frac{f}{g}\right)(x) = \frac{f(x)}{g(x)} = \frac{2x^2 + x - 10}{x - 2}$$

This quotient, found in Example 2, with x replacing m, is $2x + 5$, so

$$\left(\frac{f}{g}\right)(x) = 2x + 5, \quad x \neq 2.$$

Then $$\left(\frac{f}{g}\right)(-3) = 2(-3) + 5 = -1. \quad \text{Let } x = -3.$$

(Which is easier to find here: $\left(\frac{f}{g}\right)(-3)$ or $\frac{f(-3)}{g(-3)}$?)

◀ Work Problem 6 at the Side.

6.5 Exercises

Divide. See Example 1.

1. $\dfrac{15x^3 - 10x^2 + 5}{5}$

2. $\dfrac{27m^4 - 18m^3 + 9m}{9}$

3. $\dfrac{9y^2 + 12y - 15}{3y}$

4. $\dfrac{80r^2 - 40r + 10}{10r}$

5. $\dfrac{15m^3 + 25m^2 + 30m}{5m^2}$

6. $\dfrac{64x^3 - 72x^2 + 12x}{8x^3}$

7. $\dfrac{14m^2n^2 - 21mn^3 + 28m^2n}{14m^2n}$

8. $\dfrac{24h^2k + 56hk^2 - 28hk}{16h^2k^2}$

Divide. See Examples 2–5.

9. $\dfrac{y^2 + 3y - 18}{y + 6}$

10. $\dfrac{q^2 + 4q - 32}{q - 4}$

11. $\dfrac{3t^2 + 17t + 10}{3t + 2}$

12. $\dfrac{2k^2 - 3k - 20}{2k + 5}$

13. $\dfrac{p^2 + 2p + 20}{p + 6}$

14. $\dfrac{x^2 + 11x + 16}{x + 8}$

15. $(2z^3 - 5z^2 + 6z - 15) \div (2z - 5)$

16. $(3p^3 + p^2 + 18p + 6) \div (3p + 1)$

17. $(4x^3 + 9x^2 - 10x + 3) \div (4x + 1)$

18. $(10z^3 - 26z^2 + 17z - 13) \div (5z - 3)$

19. $\dfrac{14x + 6x^3 - 15 - 19x^2}{3x^2 - 2x + 4}$

20. $\dfrac{37m - 18m^2 - 13 + 8m^3}{2m^2 - 3m + 6}$

21. $(3x^3 - x + 4) \div (x - 2)$

22. $(4x^3 - 3x - 2) \div (x + 1)$

23. $(2x^3 - 11x^2 + 28) \div (x - 5)$

24. $(3x^3 - 4x + 2) \div (x - 1)$

25. $\dfrac{4k^4 + 6k^3 + 3k - 1}{2k^2 + 1}$

26. $\dfrac{6y^4 + 9y^3 + 10y^2 + 6y + 4}{3y^2 + 2}$

27. $(9z^4 - 13z^3 + 23z^2 - 10z + 8) \div (z^2 - z + 2)$

28. $(2q^4 + 5q^3 - 11q^2 + 11q - 20) \div (2q^2 - q + 2)$

29. $\left(2x^2 - \dfrac{7}{3}x - 1\right) \div (3x + 1)$

30. $\left(m^2 + \dfrac{7}{2}m + 3\right) \div (2m + 3)$

31. $\left(3a^2 - \dfrac{23}{4}a - 5\right) \div (4a + 3)$

32. $\left(3q^2 + \dfrac{19}{5}q - 3\right) \div (5q - 2)$

For each pair of functions, find the quotient $\left(\frac{f}{g}\right)(x)$ *and give any x-values that are not in the domain of the quotient function. See Example 6.*

33. $f(x) = 10x^2 - 2x, g(x) = 2x$

34. $f(x) = 18x^2 - 24x, g(x) = 3x$

35. $f(x) = 2x^2 - x - 3, g(x) = x + 1$

36. $f(x) = 4x^2 - 23x - 35, g(x) = x - 7$

37. $f(x) = 8x^3 - 27, g(x) = 2x - 3$

38. $f(x) = 27x^3 + 64, g(x) = 3x + 4$

Let $f(x) = x^2 - 9$, $g(x) = 2x$, *and* $h(x) = x - 3$. *Find each of the following. See Example 6.*

39. $\left(\dfrac{f}{g}\right)(x)$

40. $\left(\dfrac{f}{h}\right)(x)$

41. $\left(\dfrac{f}{g}\right)(2)$

42. $\left(\dfrac{f}{h}\right)(1)$

43. $\left(\dfrac{h}{g}\right)(x)$

44. $\left(\dfrac{f}{h}\right)(-3)$

45. $\left(\dfrac{h}{g}\right)(3)$

46. $\left(\dfrac{f}{g}\right)(-1)$

Solve each problem.

47. The volume of a box is $(2p^3 + 15p^2 + 28p)$ cubic feet. The height is p feet and the length is $(p + 4)$ feet. Find an expression for the width.

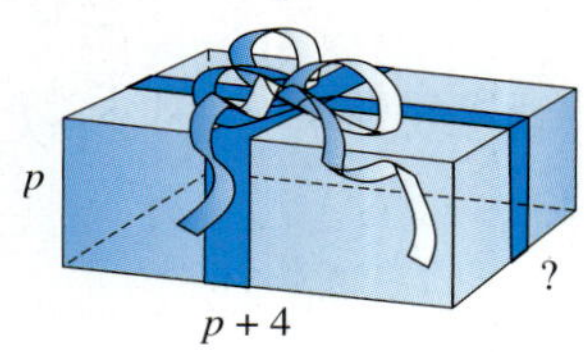

48. Suppose a car goes $(2m^3 + 15m^2 + 13m - 63)$ kilometers in $(2m + 9)$ hours. Find an expression for its rate.

Chapter 6 Summary

Key Terms

6.2

term A term is a number, a variable, or the product or quotient of a number and one or more variables raised to powers.

coefficient (numerical coefficient) A coefficient is a factor in a term (usually used for the numerical factor).

algebraic expression An algebraic expression is any combination of variables or constants (numerical values) joined by the basic operations of addition, subtraction, multiplication, and division (except by 0), or raising to powers or taking roots.

polynomial A polynomial is a term or a finite sum of terms in which all variables have whole number exponents and no variables appear in denominators.

polynomial in x A polynomial in x is a polynomial containing only the variable x.

descending powers A polynomial in one variable is written in descending powers if the exponents on the variable in the terms decrease from left to right.

trinomial A trinomial is a polynomial with exactly three terms.

binomial A binomial is a polynomial with exactly two terms.

monomial A monomial is a polynomial with exactly one term.

degree of a term The degree of a term with one variable is the exponent on that variable.

degree of a polynomial The degree of a polynomial is the greatest degree of any of the terms in the polynomial.

negative of a polynomial The negative of a polynomial is obtained by changing the sign of every coefficient in the polynomial.

6.3

polynomial function of degree n A function defined by $f(x) = a_n x^n + a_{n-1}x^{n-1} + \cdots + a_1 x + a_0$, where $a_n \neq 0$ and n is a whole number, is a polynomial function of degree n.

identity function The simplest polynomial function is the identity function, defined by $f(x) = x$.

squaring function The polynomial function defined by $f(x) = x^2$ is called the squaring function.

cubing function The polynomial function defined by $f(x) = x^3$ is called the cubing function.

Test Your Word Power

See how well you have learned the vocabulary in this chapter. Answers, with examples, follow the Quick Review.

1. A **polynomial** is an algebraic expression made up of
- **A.** a term or a finite product of terms with positive coefficients and exponents
- **B.** the sum of two or more terms with whole number coefficients and exponents
- **C.** the product of two or more terms with positive exponents
- **D.** a term or a finite sum of terms with real coefficients and whole number exponents.

2. A **monomial** is a polynomial with
- **A.** only one term
- **B.** exactly two terms
- **C.** exactly three terms
- **D.** more than three terms.

3. A **binomial** is a polynomial with
- **A.** only one term
- **B.** exactly two terms
- **C.** exactly three terms
- **D.** more than three terms.

4. A **trinomial** is a polynomial with
- **A.** only one term
- **B.** exactly two terms
- **C.** exactly three terms
- **D.** more than three terms.

5. **FOIL** is a method for
- **A.** adding two binomials
- **B.** adding two trinomials
- **C.** multiplying two binomials
- **D.** multiplying two trinomials.

▸ Quick Review

Concepts	Examples

6.1 Integer Exponents and Scientific Notation

Definitions and Rules for Exponents

Apply the rules for exponents.

Rule	Definition	Examples
Product Rule	$a^m \cdot a^n = a^{m+n}$	$3^4 \cdot 3^2 = 3^6$
Quotient Rule	$\frac{a^m}{a^n} = a^{m-n} \quad (a \neq 0)$	$\frac{2^5}{2^3} = 2^2$
Zero Exponent	$a^0 = 1 \quad (a \neq 0)$	$27^0 = 1, \quad (-5)^0 = 1$
Negative Exponent	$a^{-n} = \frac{1}{a^n} \quad (a \neq 0)$	$5^{-2} = \frac{1}{5^2}$
Power Rules	$(a^m)^n = a^{mn}, \quad (ab)^m = a^m b^m$	$(6^3)^4 = 6^{12}, \quad (5p)^4 = 5^4p^4$
	$\left(\frac{a}{b}\right)^n = \frac{a^n}{b^n} \quad (b \neq 0)$	$\left(\frac{2}{3}\right)^5 = \frac{2^5}{3^5}$
Special Rules	$\frac{1}{a^{-n}} = a^n \quad (a \neq 0)$	$\frac{1}{x^{-3}} = x^3$
	$\frac{a^{-n}}{b^{-m}} = \frac{b^m}{a^n} \quad (a, b \neq 0)$	$\frac{r^{-3}}{t^{-4}} = \frac{t^4}{r^3}$
	$a^{-n} = \left(\frac{1}{a}\right)^n \quad (a \neq 0)$	$4^{-3} = \left(\frac{1}{4}\right)^3$
	$\left(\frac{a}{b}\right)^{-n} = \left(\frac{b}{a}\right)^n \quad (a, b \neq 0)$	$\left(\frac{4}{7}\right)^{-2} = \left(\frac{7}{4}\right)^2$

Scientific Notation

A number is in scientific notation when it is expressed in the form

$$a \times 10^n,$$

where $1 \leq |a| < 10$, and n is an integer.

Write 23,500,000,000 in scientific notation.

$$23{,}500{,}000{,}000 = 2.35 \times 10^{10}$$

Write 4.3×10^{-6} in standard notation.

$$4.3 \times 10^{-6} = 0.0000043$$

6.2 Adding and Subtracting Polynomials

Add or subtract polynomials by combining like terms.

$$(5x^4 + 3x^2) - (7x^4 + x^2 - x)$$
$$= -2x^4 + 2x^2 + x$$

6.3 Polynomial Functions

The graph of $f(x) = x$ is a line, and the graph of $f(x) = x^2$ is a parabola. The graph of $f(x) = x^3$ is neither of these. They define the identity, squaring, and cubing functions, respectively.

Graph the identity, squaring, and cubing functions.

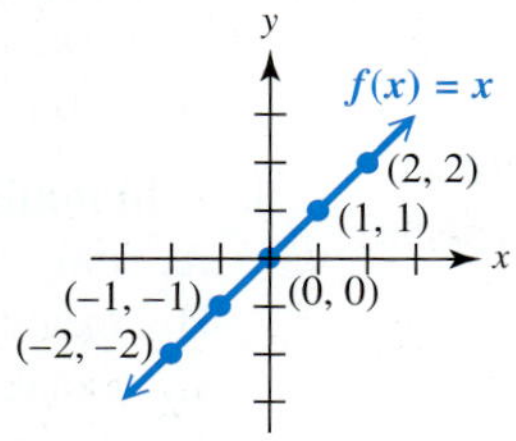

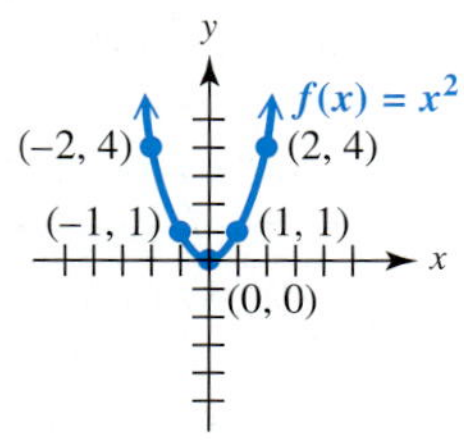

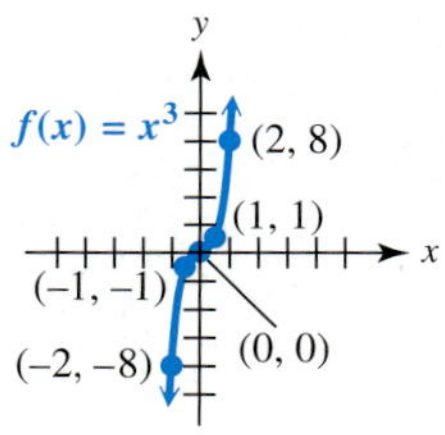

Concepts	Examples
6.4 Multiplying Polynomials	
To multiply two polynomials, multiply each term of one by each term of the other.	$(x^3 + 3x)(4x^2 - 5x + 2)$ $= 4x^5 + 12x^3 - 5x^4 - 15x^2 + 2x^3 + 6x$ $= 4x^5 - 5x^4 + 14x^3 - 15x^2 + 6x$
To multiply two binomials, use the **FOIL** method. Multiply the **First** terms, the **Outer** terms, the **Inner** terms, and the **Last** terms. Then add these products.	$(2x + 3)(x - 7)$ $= 2x(x) + 2x(-7) + 3x + 3(-7)$ FOIL $= 2x^2 - 14x + 3x - 21$ $= 2x^2 - 11x - 21$
Special Products $(x + y)(x - y) = x^2 - y^2$ $(x + y)^2 = x^2 + 2xy + y^2$ $(x - y)^2 = x^2 - 2xy + y^2$	$(3m + 8)(3m - 8)$ $= 9m^2 - 64$ $(5a + 3b)^2$ $= 25a^2 + 30ab + 9b^2$ $(2k - 1)^2$ $= 4k^2 - 4k + 1$
6.5 Dividing Polynomials	
Dividing by a Monomial To divide a polynomial by a monomial, divide each term in the polynomial by the monomial, and then write each fraction in lowest terms.	$\frac{2x^3 - 4x^2 + 6x - 8}{2x}$ $= \frac{2x^3}{2x} - \frac{4x^2}{2x} + \frac{6x}{2x} - \frac{8}{2x}$ $= x^2 - 2x + 3 - \frac{4}{x}$
Dividing by a Polynomial Use the "long division" process.	Divide $\frac{m^3 - m^2 + 2m + 5}{m + 1}$. $$\begin{array}{r} m^2 - 2m + 4 \\ m + 1\overline{)m^3 - m^2 + 2m + 5} \\ \underline{m^3 + m^2} \qquad\qquad \\ -2m^2 + 2m \qquad \\ \underline{-2m^2 - 2m} \qquad \\ 4m + 5 \\ \underline{4m + 4} \\ 1 \end{array}$$ 1 ← Remainder The answer is $m^2 - 2m + 4 + \frac{1}{m + 1}$.

ANSWERS TO TEST YOUR WORD POWER

1. D; *Example:* $5x^3 + 2x^2 - 7$
2. A; *Examples:* $-4, 2t^3, 15a^2b$
3. B; *Example:* $3t^3 + 5t$
4. C; *Example:* $2a^2 - 3ab + b^2$
5. C; *Example:* $(m + 4)(m - 3)$

F O I L

$= m(m) + m(-3) + 4m + 4(-3)$

$= m^2 + m - 12$

Math in the Media

PRIME NUMBERS IN PRIME TIME

The 1997 movie *Contact*, based on the Carl Sagan novel of the same name, portrays Jodie Foster as scientist Ellie Arroway. After years of searching, Ellie makes contact with intelligent life in outer space. Her contact is verified after receiving radio signals that indicate **prime numbers:**

$$2, 3, 5, 7, 11, 13, \quad \text{and so on.}$$

Her superiors, evidently not familiar with prime numbers, are not convinced and ask her why the aliens just don't speak English? Her response (accompanied by quizzical looks from the bosses) follows.

> *Well, maybe because 70% of the planet speaks other languages. Mathematics is the only universal language, Senator. It's no coincidence they're using primes . . . prime numbers—that would be integers that are divisible only by themselves and 1.*

Integers greater than 1 that are not prime are called **composite numbers,** because they are composed of prime factors in one and only one way. In **Chapter 7,** we will extend this idea to factoring polynomials into their prime factors.

1. A prime number is a positive integer greater than 1 whose only factors are 1 and itself. List the first fifteen prime numbers.

2. A recurring feature on the NBC *Today Show* is *Where in the World is Matt Lauer?* Co-host Matt Lauer travels to exotic places during one week, always giving a hint as to where he will be on the following day. In his travels in 2008, Matt gave this clue regarding his next destination: **It's an anagram of a synonym of a homophone of an even prime number.** Where was Matt the next day?

3. The 1996 film *The Mirror Has Two Faces* stars Jeff Bridges as mathematician Gregory Larkin, who has written a book on the Twin Prime Conjecture. Search the Internet to discover the statement of this famous unproved conjecture.

4. Watch the episode "Prime Suspect" from the first season of the CBS television series *NUMB3RS*. The story is based on the premise that a mathematician was very close to proving a famous unsolved problem involving prime numbers, the Riemann Hypothesis, and his daughter was kidnapped as a result. Why did the criminals kidnap the child?

Chapter 6 Review Exercises

[6.1] *Simplify. Write answers with only positive exponents. Assume that all variables represent positive real numbers.*

1. 4^3

2. $\left(\frac{1}{3}\right)^4$

3. $(-5)^3$

4. $\frac{2}{(-3)^{-2}}$

5. $\left(\frac{2}{3}\right)^{-4}$

6. $\left(\frac{5}{4}\right)^{-2}$

7. $5^{-2} + 6^{-1}$

8. $-3^0 + 3^0$

9. $(-3x^4y^3)(4x^{-2}y^5)$

10. $\frac{6m^{-4}n^3}{-3mn^2}$

11. $\frac{(5p^{-2}q)(4p^5q^{-3})}{2p^{-5}q^5}$

12. $\frac{x^{-2}y^{-4}}{x^{-4}y^{-2}}$

13. $(3^{-4})^2$

14. $(x^{-4})^{-2}$

15. $(xy^{-3})^{-2}$

16. $(z^{-3})^3z^{-6}$

17. $(5m^{-3})^2(m^4)^{-3}$

18. $\left(\frac{5z^{-3}}{z^{-1}}\right)\left(\frac{5}{z^2}\right)^{-1}$

19. $\left(\frac{6m^{-4}}{m^{-9}}\right)^{-1}\left(\frac{m^{-2}}{16}\right)$

20. $\left(\frac{3r^5}{5r^{-3}}\right)^{-2}\left(\frac{9r^{-1}}{2r^{-5}}\right)^3$

21. $\left(\frac{3w^{-2}z^4}{-6wz^{-5}}\right)^{-2}$

Write each number in scientific notation.

22. 13,450

23. 0.0000000765

24. 0.138

25. In July 2007, the total population of the United States was estimated at **299,400,000.** Of this amount, about **74,000** Americans were centenarians, that is, age **100** or older. Write the three boldfaced numbers using scientific notation. (*Source:* U.S. Census Bureau.)

Write each number in standard notation.

26. 1.21×10^6

27. 5.8×10^{-3}

Use scientific notation to compute. Give answers in both scientific notation and standard notation.

28. $\dfrac{16 \times 10^4}{8 \times 10^8}$

29. $\dfrac{6 \times 10^{-2}}{4 \times 10^{-5}}$

30. $\dfrac{0.0000000164}{0.0004}$

31. $\dfrac{0.0009 \times 12{,}000{,}000}{400{,}000}$

[6.2] *Give the numerical coefficient of each term.*

32. $14p^5$

33. $-z$

34. $0.045x^4$

35. $504p^3r^5$

For each polynomial, **(a)** *write in descending powers,* **(b)** *identify as* monomial, binomial, trinomial, *or* none of these, *and* **(c)** *give the degree.*

36. $9k + 11k^3 - 3k^2$

37. $14m^6 + 9m^7$

38. $-7q^5r^3$

39. Give an example of a polynomial in the variable x such that it has degree 5, is lacking a third-degree term, and is in descending powers of the variable.

Add or subtract as indicated.

40. Add.

$$\begin{array}{r} 3x^2 - 5x + 6 \\ \underline{-4x^2 + 2x - 5} \end{array}$$

41. Subtract.

$$\begin{array}{r} -5y^3 \qquad\quad + 8y - 3 \\ \underline{\qquad\quad 4y^2 + 2y + 9} \end{array}$$

42. $(4a^3 - 9a + 15) - (-2a^3 + 4a^2 + 7a)$

43. $(3y^2 + 2y - 1) + (5y^2 - 11y + 6)$

44. Find the perimeter of the triangle.

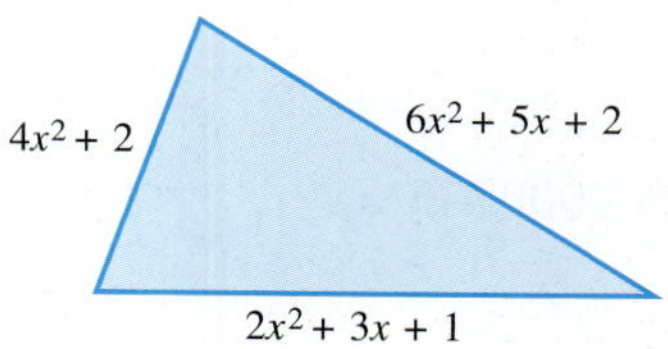

[6.3]

45. For the polynomial function defined by $f(x) = -2x^2 - 5x + 7$, find each value.

(a) $f(-2)$ **(b)** $f(3)$

46. For $f(x) = 2x + 3$ and $g(x) = 5x^2 - 3x + 2$, find each of the following.

(a) $(f + g)(x)$ **(b)** $(f - g)(x)$ **(c)** $(f + g)(-1)$ **(d)** $(f - g)(-1)$

47. The number of twin births in the United States during the period 1990 through 2005 can be modeled by the polynomial function defined by

$$f(x) = -23.334x^3 + 645.75x^2 - 1861.4x + 95{,}072,$$

where $x = 0$ corresponds to 1990, $x = 1$ corresponds to 1991, and so on. Use this model to approximate the number of twin births in each given year. (*Source:* National Center for Health Statistics.)

(a) 1990 **(b)** 2000 **(c)** 2005

Graph each polynomial function defined as follows.

48. $f(x) = -2x + 5$

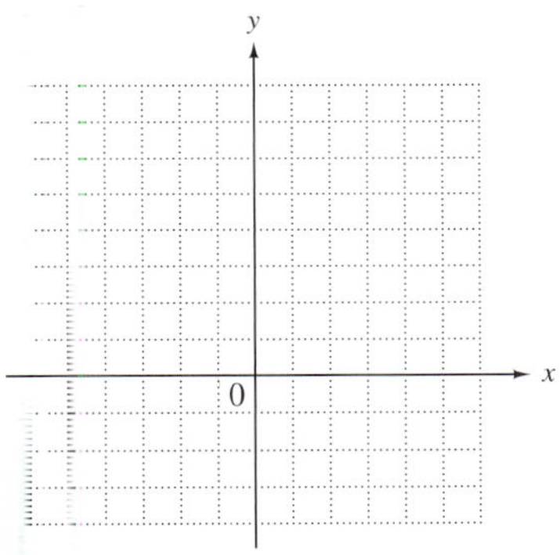

49. $f(x) = x^2 - 6$

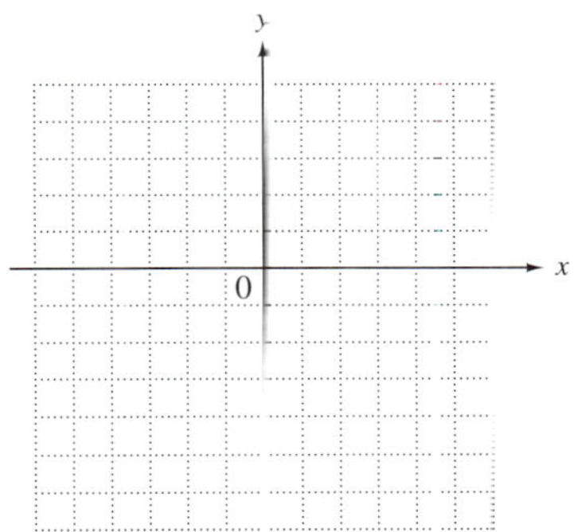

50. $f(x) = -x^3 + 1$

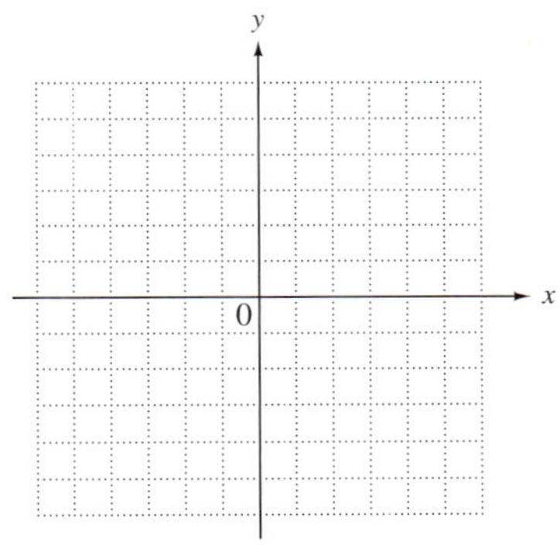

[6.4] *Find each product.*

51. $-6k(2k^2 + 7)$

52. $(7y - 8)(2y + 3)$

53. $(3w - 2t)(2w - 3t)$

54. $(2p^2 + 6p)(5p^2 - 4)$

55. $(3z^3 - 2z^2 + 4z - 1)(3z - 2)$

56. $(6r^2 - 1)(6r^2 + 1)$

57. $\left(z + \frac{3}{5}\right)\left(z - \frac{3}{5}\right)$

58. $(4m + 3)^2$

59. $(2x + 5)^3$

[6.5] *Divide.*

60. $\dfrac{4y^3 - 12y^2 + 5y}{4y}$

61. $\dfrac{2p^3 + 9p^2 + 27}{2p - 3}$

62. $\dfrac{5p^4 + 15p^3 - 33p^2 - 9p + 18}{5p^2 - 3}$

Mixed Review Exercises

63. Match each expression (a)–(i) in Column I with its equivalent expression A–I in Column II. Choices may be used once, more than once, or not at all.

I

(a) 4^{-2}

(b) -4^2

(c) 4^0

(d) $(-4)^0$

(e) $(-4)^{-2}$

(f) -4^0

(g) $-4^0 + 4^0$

(h) $-4^0 - 4^0$

(i) $4^{-2} + 4^{-1}$

II

A. $\frac{1}{16}$

B. 0

C. 1

D. $-\frac{1}{16}$

E. -1

F. $\frac{5}{16}$

G. -16

H. -2

I. none of these

64. In 2008, the estimated population of Luxembourg was 4.86×10^5. The population density was 487 people per mi^2. Based on this information, what is the area of Luxembourg to the nearest square mile? (*Source: The 2008 World Factbook.*)

Perform the indicated operations, and then simplify. Write answers with only positive exponents. Assume that all variables represent nonzero real numbers.

65. $(4x + 1)(2x - 3)$

66. $\dfrac{6^{-1}y^3(y^2)^{-2}}{6y^{-4}(y^{-1})}$

67. $(y^6)^{-5}(2y^{-3})^{-4}$

68. $(2x - 9)^2$

69. $\dfrac{20y^3x^3 + 15y^4x + 25yx^4}{10yx^2}$

70. $7p^5(3p^4 + p^3 + 2p^2)$

71. $\dfrac{(-z^{-2})^3}{5(z^{-3})^{-1}}$

72. $\dfrac{x^3 + 7x^2 + 7x - 12}{x + 5}$

73. $(-5 + 11w) + (6 + 5w) + (-15 - 8w^2)$

74. $(2k - 1) - (3k^2 - 2k + 6)$

Chapter 6 Test

Use the Chapter Test Prep Video CD to see fully worked-out solutions to any of the exercises you want to review

1. Match each expression in Column I with its equivalent expression in Column II. Choices may be used once, more than once, or not at all.

I	II
(a) 7^{-2}	**A.** 1
(b) 7^0	**B.** $\frac{1}{9}$
(c) -7^0	**C.** $\frac{1}{49}$
(d) $(-7)^0$	**D.** -1
(e) -7^2	**E.** -49
(f) $7^{-1} + 2^{-1}$	**F.** $\frac{9}{14}$
(g) $(7 + 2)^{-1}$	**G.** $\frac{2}{7}$
(h) $\frac{7^{-1}}{2^{-1}}$	**H.** 0
(i) $(-7)^{-2}$	**I.** none of these

Simplify. Write answers with only positive exponents. Assume that all variables represent nonzero real numbers.

2. $(3x^{-2}y^3)^{-2}(4x^3y^{-4})$

3. $\frac{36r^{-4}(r^2)^{-3}}{6r^4}$

4. $\left(\frac{4p^2}{q^4}\right)^3\left(\frac{6p^8}{q^{-8}}\right)^{-2}$

5. $(-2x^4y^{-3})^0(-4x^{-3}y^{-8})^2$

6. Write 9.1×10^{-7} in standard notation.

7. Use scientific notation to simplify $\frac{2{,}500{,}000 \times 0.00003}{0.05 \times 5{,}000{,}000}$. Write the answer in both scientific notation and standard notation.

8. If $f(x) = -2x^2 + 5x - 6$ and $g(x) = 7x - 3$, find each of the following.

(a) $f(4)$ **(b)** $(f + g)(x)$ **(c)** $(f - g)(x)$ **(d)** $(f - g)(-2)$

9. Graph the function defined by $f(x) = -2x^2 + 3$.

1. (a) ____ (b) ____ (c) ____
(d) ____ (e) ____ (f) ____
(g) ____ (h) ____ (i) ____

2. ________

3. ________

4. ________

5. ________

6. ________

7. ________

8. (a) ________
(b) ________
(c) ________
(d) ________

9.

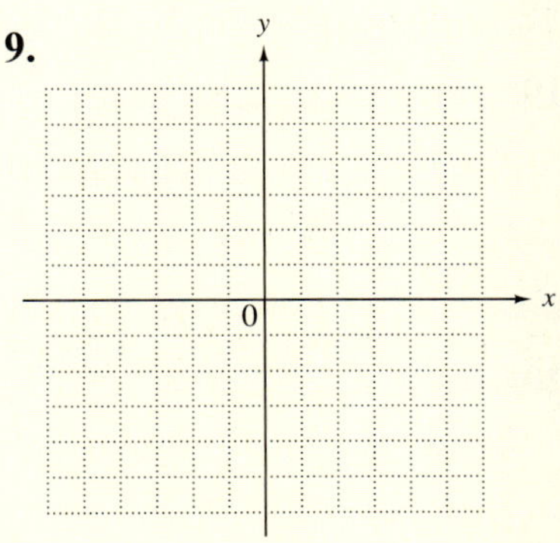

10. (a) ____________
(b) ____________
(c) ____________

10. The number of medical doctors, in thousands, in the United States during the period from 1980 through 2005 can be modeled by the polynomial function defined by

$$f(x) = 0.139x^2 + 14.16x + 465.9,$$

where $x = 0$ corresponds to 1980, $x = 1$ corresponds to 1981, and so on. Use this model to approximate the number of doctors to the nearest thousand in each given year. (*Source:* American Medical Association.)

(a) 1980 **(b)** 1995 **(c)** 2005

Perform the indicated operations.

11. ____________

11. $(4x^3 - 3x^2 + 2x - 5) - (3x^3 + 11x + 8) + (x^2 - x)$

12. ____________

13. ____________

12. $(5x - 3)(2x + 1)$

13. $(2m - 5)(3m^2 + 4m - 5)$

14. ____________

15. ____________

14. $(6x + y)(6x - y)$

15. $(3k + q)^2$

16. ____________

17. ____________

16. $[2y + (3z - x)][2y - (3z - x)]$

17. $\dfrac{16p^3 - 32p^2 + 24p}{4p^2}$

18. ____________

18. $(x^3 + 3x^2 - 6) \div (x - 2)$

19. (a) ____________
(b) ____________

19. If $f(x) = x^2 + 3x + 2$ and $g(x) = x + 1$, find each of the following.

(a) $(fg)(x)$ **(b)** $(fg)(-2)$

20. (a) ____________
(b) ____________

20. Use $f(x)$ and $g(x)$ from Problem 19 to find each of the following.

(a) $\left(\dfrac{f}{g}\right)(x)$ **(b)** $\left(\dfrac{f}{g}\right)(-2)$

Cumulative Review Exercises — Chapters 1–6

Match each number in Column I with the choice or choices of sets of numbers in Column II to which the number belongs.

I		II	
1. 24	**2.** 0	**A.** Natural numbers	**B.** Whole numbers
3. 2.16	**4.** $-\sqrt{36}$	**C.** Integers	**D.** Rational numbers
5. $\sqrt{13}$	**6.** $-\frac{4}{5}$	**E.** Irrational numbers	**F.** Real numbers

Evaluate.

7. $9 \cdot 4 - 16 \div 4$

8. $-|8 - 13| - |-4| + |-9|$

Solve.

9. $-5(8 - 2z) + 4(7 - z) = 7(8 + z) - 3$

10. $3(x + 2) - 5(x + 2) = -2x - 4$

11. $2(m + 5) - 3m + 1 > 5$

12. $|3x - 1| = 2$

13. $|3z + 1| \geq 7$

14. A survey polled teens about the most important inventions of the twentieth century. Complete the results shown in the table if 1500 teens were surveyed.

Most Important Invention	Percent	Actual Number
Personal computer		480
Pacemaker	26%	
Wireless communication	18%	
Television		150

Source: Lemelson-MIT Program.

15. Find the measure of each angle of the triangle.

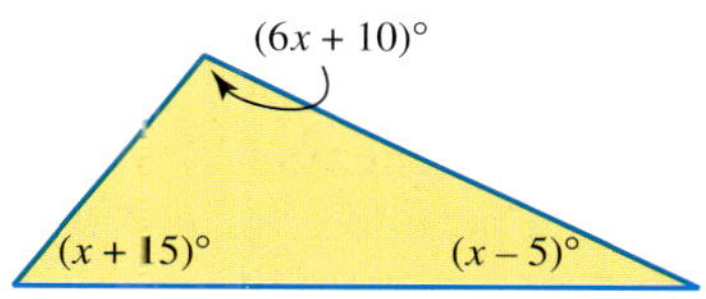

16. Find the slope of the line through $(-4, 5)$ and $(2, -3)$. Then write an equation of the line in standard form.

Graph each equation or inequality.

17. $-3x + 4y = 12$

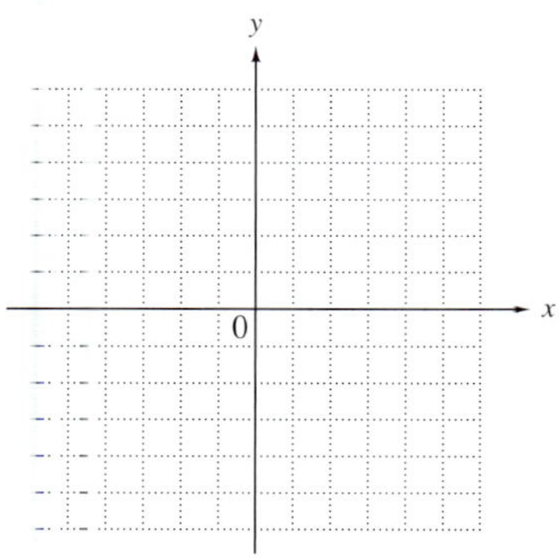

18. $y \leq 2x - 6$

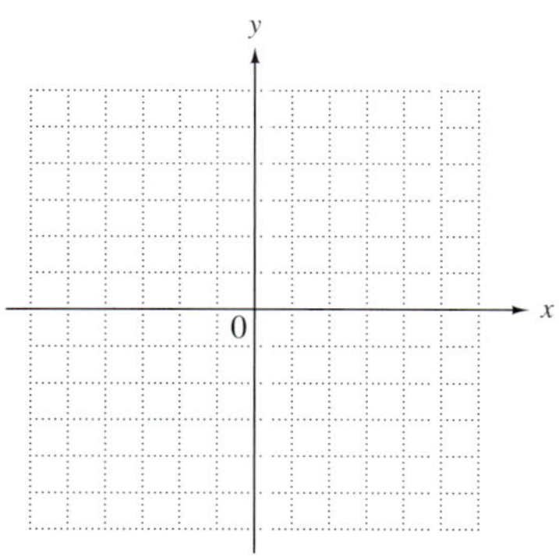

19. $3x + 2y < 0$

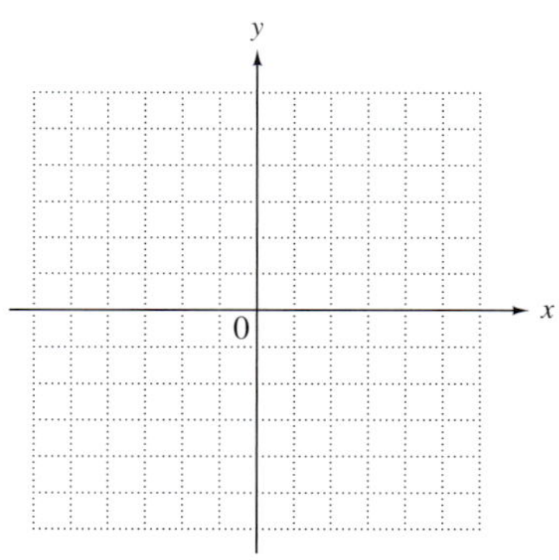

20. The graph shows the number of pounds of shrimp caught in the United States (in thousands of pounds) in selected years.

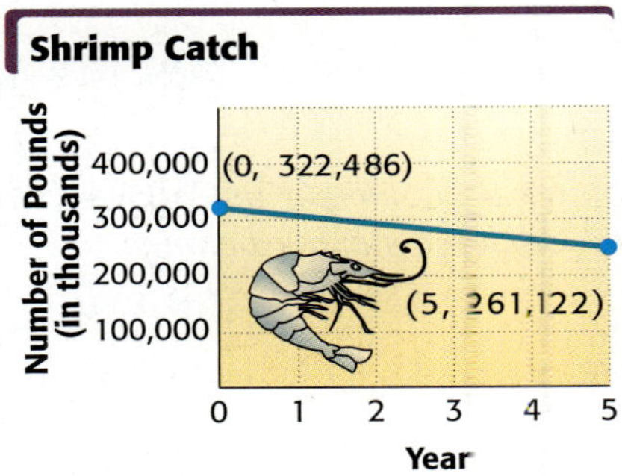

Source: National Oceanic and Atmospheric Administration.

(a) Use the information given in the graph to find and interpret the average rate of change in the number of pounds of shrimp caught per year.

(b) If $x = 0$ represents the year 2000, $x = 1$ represents 2001, and so on, use your answer from part (a) to write an equation of the line in slope-intercept form that models the annual amount of shrimp caught (in thousands of pounds) for the years 2000 through 2005.

(c) Use the equation from part (b) to approximate the amount of shrimp caught in 2003.

21. Give the domain and range of the relation $\{(-4, -2), (-1, 0), (2, 0), (5, 2)\}$. Does this relation define a function?

Solve each system.

22. $\begin{aligned} 3x - 4y &= 1 \\ 2x + 3y &= 12 \end{aligned}$

23. $\begin{aligned} 3x - 2y &= 4 \\ -6x + 4y &= 7 \end{aligned}$

24. $\begin{aligned} x + 3y - 6z &= 7 \\ 2x - y + z &= 1 \\ x + 2y + 2z &= -1 \end{aligned}$

25. The Star-Spangled Banner that flew over Fort McHenry during the War of 1812 had a perimeter of 144 ft. Its length measured 12 ft more than its width. Use a system of equations to find the dimensions of this flag, which is displayed in the Smithsonian Institution's Museum of American History in Washington, D.C. (*Source:* National Park Service brochure.)

Simplify. Write answers with only positive exponents. Assume that all variables represent positive real numbers.

26. $\left(\frac{2m^3n}{p^2}\right)^3$

27. $\frac{x^{-6}y^3z^{-1}}{x^7y^{-4}z}$

28. $(2m^{-2}n^3)^{-3}$

Perform the indicated operations.

29. $2(3x^2 - 8x + 1) - 4(x^2 - 3x - 9)$

30. $(3x + 2y)(5x - y)$

31. $(8m + 5n)(8m - 5n)$

32. $\frac{m^3 - 3m^2 - 5m + 4}{m - 1}$

7

Factoring

7.1 Greatest Common Factors; Factoring by Grouping

7.2 Factoring Trinomials

7.3 Special Factoring

7.4 A General Approach to Factoring

7.5 Solving Equations by Factoring

Factoring is used to solve *quadratic equations*, which have many useful applications. An important one is to express the distance a falling or projected object travels in a specific time. Such equations are used in astronomy and the space program to describe the motion of objects in space.

In Section 7.5, we use the concepts of this chapter to explore how to find the heights of objects after they are projected or dropped.

7.1 Greatest Common Factors; Factoring by Grouping

OBJECTIVES

1 Factor out the greatest common factor.

2 Factor by grouping.

Writing a polynomial as the product of two or more simpler polynomials is called **factoring** the polynomial. For example, the product of $3x$ and $5x - 2$ is $15x^2 - 6x$, and $15x^2 - 6x$ can be factored as the product $3x(5x - 2)$.

$$3x(5x - 2) = 15x^2 - 6x \quad \text{Multiplying}$$
$$15x^2 - 6x = 3x(5x - 2) \quad \text{Factoring}$$

Notice that both multiplying and factoring use the distributive property, but in opposite directions. ***Factoring "undoes," or reverses, multiplying.***

OBJECTIVE 1 Factor out the greatest common factor. The first step in factoring a polynomial is to find the *greatest common factor*. The **greatest common factor (GCF)** is the largest term that is a factor of all terms in the polynomial. For example, the greatest common factor for $8x + 12$ is 4, since 4 is the largest factor that *divides into* both $8x$ and 12.

$$8x + 12$$
$$= 4(2x) + 4(3) \quad \text{Distributive property}$$
$$= 4(2x + 3)$$

As a check, multiply 4 and $2x + 3$ to obtain $8x + 12$. Using the distributive property this way is called **factoring out the greatest common factor.**

EXAMPLE 1 Factoring Out the Greatest Common Factor

Factor out the greatest common factor.

(a) $9z - 18$
Since 9 is the GCF, factor 9 from each term.

$$9z - 18$$
$$= 9 \cdot z - 9 \cdot 2$$
$$= 9(z - 2)$$

Check Multiply $9(z - 2)$ to obtain $9z - 18$. ← Original polynomial

(b) $56m + 35p$
$= 7(8m + 5p)$

(c) $2y + 5$ There is no common factor other than 1.

(d) $12 + 24z$

$= 12 \cdot 1 + 12 \cdot 2z$ Identity property

$= 12(1 + 2z)$ 12 is the GCF.

Remember to write the 1.

Check $12(1 + 2z)$

$= 12(1) + 12(2z)$ Distributive property

$= 12 + 24z$ Original polynomial

CAUTION
Always check answers by multiplying.

◀ Work Problem 1 at the Side.

1 Factor out the greatest common factor.

(a) $7k + 28$

(b) $32m + 24$

(c) $8a - 9$

(d) $5z + 5$

ANSWERS

1. **(a)** $7(k + 4)$ **(b)** $8(4m + 3)$
(c) There is no common factor other than 1.
(d) $5(z + 1)$

EXAMPLE 2 **Factoring Out the Greatest Common Factor**

Factor out the greatest common factor.

(a) $9x^2 + 12x^3$

The numerical part of the GCF is 3, the largest number that divides into both 9 and 12. The least exponent that appears on x is 2. The GCF is $3x^2$.

$$9x^2 + 12x^3$$
$$= 3x^2(3) + 3x^2(4x) \qquad \text{GCF} = 3x^2$$
$$= 3x^2(3 + 4x)$$

(b) $32p^4 - 24p^3 + 40p^5$

$$= 8p^3(4p) + 8p^3(-3) + 8p^3(5p^2) \qquad \text{GCF} = 8p^3$$
$$= 8p^3(4p - 3 + 5p^2)$$

(c) $3k^4 - 15k^7 + 24k^9$

$$= 3k^4(1 - 5k^3 + 8k^5)$$

Remember the 1.

(d) $24m^3n^2 - 18m^2n + 6m^4n^3$

$$= 6m^2n(4mn) + 6m^2n(-3) + 6m^2n(m^2n^2)$$
$$= 6m^2n(4mn - 3 + m^2n^2)$$

(e) $25x^2y^3 + 30y^5 - 15x^4y^7$

$$= 5y^3(5x^2 + 6y^2 - 3x^4y^4)$$

In each case, remember to check the factored form by multiplying.

Work Problem **2** at the Side. ▶

EXAMPLE 3 **Factoring Out a Binomial Factor**

Factor out the greatest common factor.

(a) $(x + 5)(x + 6) + (x + 5)(2x + 5)$ — The greatest common factor is $x + 5$.

$= (x + 5)[(x + 6) + (2x + 5)]$ — Factor out $x + 5$.

$= (x + 5)(3x + 11)$ — Combine like terms.

(b) $z^2(m - n) + x^2(m - n)$

$$= (m - n)(z^2 + x^2)$$

(c) $p(r + 2s)^2 - q(r + 2s)^3$

$= (r + 2s)^2[p - q(r + 2s)]$ — Factor out the common factor.

$= (r + 2s)^2(p - qr - 2qs)$ — Be careful with signs.

(d) $(p - 5)(p + 2) - (p - 5)(3p + 4)$

$= (p - 5)[(p + 2) - (3p + 4)]$ — Factor out $p - 5$.

$= (p - 5)[p + 2 - 3p - 4]$ — Distributive property

$= (p - 5)[-2p - 2]$ — Combine like terms.

$= (p - 5)[-2(p + 1)]$ — Look for a common factor.

$= -2(p - 5)(p + 1)$ — Commutative property

Work Problem **3** at the Side. ▶

2 Factor out the greatest common factor.

(a) $16y^4 + 8y^3$

(b) $14p^2 - 9p^3 + 6p^4$

(c) $15z^2 + 45z^5 - 60z^6$

(d) $4x^2z - 2xz + 8z^2$

(e) $12y^5x^2 + 8y^3x^3$

(f) $5m^4x^3 + 15m^5x^6 - 20m^4x^6$

3 Factor out the greatest common factor.

(a) $(a + 2)(a - 3) + (a + 2)(a + 6)$

(b) $(y - 1)(y + 3) - (y - 1)(y + 4)$

(c) $k^2(a + 5b) + m^2(a + 5b)^2$

(d) $r^2(y + 6) + r^2(y + 3)$

ANSWERS

2. (a) $8y^3(2y + 1)$ (b) $p^2(14 - 9p + 6p^2)$
(c) $15z^2(1 + 3z^3 - 4z^4)$
(d) $2z(2x^2 - x + 4z)$
(e) $4y^3x^2(3y^2 + 2x)$
(f) $5m^4x^3(1 + 3mx^3 - 4x^3)$
3. (a) $(a + 2)(2a + 3)$
(b) $(y - 1)(-1)$, or $-y + 1$
(c) $(a + 5b)(k^2 + m^2a + 5m^2b)$
(d) $r^2(2y + 9)$

When the coefficient of the term of greatest degree is negative, we sometimes prefer to factor out the -1 that is understood along with the GCF.

EXAMPLE 4 Factoring Out a Negative Common Factor

Factor $-a^3 + 3a^2 - 5a$ in two ways.

First, a could be used as the common factor, giving

$$\begin{aligned} &-a^3 + 3a^2 - 5a \\ &= a(-a^2) + a(3a) + a(-5) && \text{Factor out } a. \\ &= a(-a^2 + 3a - 5). \end{aligned}$$

Because of the leading negative sign, $-a$ could be used as the common factor.

$$\begin{aligned} &-a^3 + 3a^2 - 5a \\ &= -a(a^2) + (-a)(-3a) + (-a)(5) && \text{Factor out } -a. \\ &= -a(a^2 - 3a + 5) \end{aligned}$$

Sometimes there may be a reason to prefer one of these forms over the other, but either is correct.

◀ Work Problem 4 at the Side.

4 Factor each polynomial in two ways.

(a) $-k^2 + 3k$

(b) $-6r^3 - 5r^2 + 14r$

Note

The answer section in this book will usually give the factored form where the common factor has a positive coefficient.

OBJECTIVE 2 Factor by grouping. Sometimes the *individual terms* of a polynomial have a greatest common factor of 1, but we can still factor the polynomial by using a process called **factoring by grouping.** ***We usually factor by grouping when a polynomial has more than three terms.***

EXAMPLE 5 Factoring by Grouping

Factor $ax - ay + bx - by$.

Group the terms as follows:

Terms with common factor a ↓ ↓   Terms with common factor b ↓ ↓

$$(ax - ay) + (bx - by).$$

Then factor $ax - ay$ as $a(x - y)$ and factor $bx - by$ as $b(x - y)$.

$$\begin{aligned} &ax - ay + bx - by \\ &= (ax - ay) + (bx - by) && \text{Group the terms.} \\ &= a(x - y) + b(x - y) && \text{Factor each group.} \\ &= (x - y)(a + b) && \text{The common factor is } x - y. \end{aligned}$$

Check
$$\begin{aligned} &(x - y)(a + b) \\ &= xa + xb - ya - yb && \text{Multiply using the FOIL method.} \\ &= ax + bx - ay - by && \text{Commutative property} \\ &= ax - ay + bx - by && \text{Original polynomial} \end{aligned}$$

◀ Work Problem 5 at the Side.

5 Factor $6p - 6q + rp - rq$.

ANSWERS

4. **(a)** $k(-k + 3)$; $-k(k - 3)$
(b) $r(-6r^2 - 5r + 14)$; $-r(6r^2 + 5r - 14)$

5. $(p - q)(6 + r)$

EXAMPLE 6 Factoring by Grouping

Factor $3x - 3y - ax + ay$.

$$3x - 3y - ax + ay$$

Pay close attention here.

$$= (3x - 3y) + (-ax + ay) \quad \text{Group the terms.}$$

$$= 3(x - y) + a(-x + y) \quad \text{Factor out 3; factor out } a.$$

The factors $(x - y)$ and $(-x + y)$ are opposites. If we factor out $-a$ instead of a in the second group of terms, we get

$$(3x - 3y) + (-ax + ay)$$

$$= 3(x - y) - a(x - y) \quad \text{Be careful with signs.}$$

$$= (x - y)(3 - a). \quad \text{Factor out } x - y.$$

Check $(x - y)(3 - a)$

$$= 3x - ax - 3y + ay \quad \text{Multiply using the FOIL method.}$$

$$= 3x - 3y - ax + ay \quad \text{Original polynomial}$$

Work Problem 6 at the Side. ▶

Note

In Example 6, a different grouping would lead to a different factored form, $(a - 3)(y - x)$. Verify by multiplying that this form is also correct.

Use the following steps to factor by grouping.

Factoring by Grouping

Step 1 **Group terms.** Collect the terms into groups so that each group has a common factor.

Step 2 **Factor within the groups.** Factor out the common factor in each group.

Step 3 **Factor the entire polynomial.** If each group now has a common factor, factor it out. If not, try a different grouping.

Always check the factored form by multiplying.

EXAMPLE 7 Factoring by Grouping

Factor $6ax + 12bx + a + 2b$.

$$6ax + 12bx + a + 2b$$

$$= (6ax + 12bx) + (a + 2b) \quad \text{Group the terms.}$$

Now factor $6x$ from the first group, and use the identity property of multiplication to introduce the factor 1 in the second group.

$$= 6x(a + 2b) + 1(a + 2b) \quad \text{Remember to write the 1.}$$

$$= (a + 2b)(6x + 1) \quad \text{Factor out } a + 2b.$$

Check by multiplying.

Work Problem 7 at the Side. ▶

6 Factor $xy - 2y - 4x + 8$.

7 Factor $2xy + 3y + 2x + 3$.

ANSWERS

6. $(x - 2)(y - 4)$
7. $(2x + 3)(y + 1)$

8 Factor.

(a) $mn + 6 + 2n + 3m$

(b) $10x^2y^2 - 18 + 15y^2 - 12x^2$

9 Factor.

(a) $12wy + 4wz - 24xy - 8xz$

(b) $6bxy + 3xyz + 6bxz + 3xz^2$

EXAMPLE 8 **Rearranging Terms before Factoring by Grouping**

Factor $p^2q^2 - 10 - 2q^2 + 5p^2$.

Neither the first two terms nor the last two terms have a common factor except 1. We can rearrange the terms and group them as follows.

$(p^2q^2 - 2q^2) + (5p^2 - 10)$ Rearrange and group the terms.

Don't stop here. $= q^2(p^2 - 2) + 5(p^2 - 2)$ Factor out the common factors.

$= (p^2 - 2)(q^2 + 5)$ Factor out $p^2 - 2$.

Check $(p^2 - 2)(q^2 + 5)$

$= p^2q^2 + 5p^2 - 2q^2 - 10$ FOIL

$= p^2q^2 - 10 - 2q^2 + 5p^2$ Original polynomial

CAUTION

In Example 8, do not stop at the step

$$q^2(p^2 - 2) + 5(p^2 - 2).$$

This expression is *not in factored form* because it is a *sum* of two terms, $q^2(p^2 - 2)$ and $5(p^2 - 2)$, not a *product*.

◀ Work Problem **8** at the Side.

EXAMPLE 9 **Factoring Out a Common Factor before Factoring by Grouping**

Factor $10ax - 5ay + 10bx - 5by$.

Start by factoring out the greatest common factor, 5, from each term.

$10ax - 5ay + 10bx - 5by$

$= 5(2ax - ay + 2bx - by)$ Factor out the GCF, 5.

$= 5[(2ax - ay) + (2bx - by)]$ Group the terms inside the brackets.

$= 5[a(2x - y) + b(2x - y)]$ Factor out the common factors.

$= 5[(2x - y)(a + b)]$ Factor out $2x - y$.

$= 5(2x - y)(a + b)$ Write without the brackets.

Check $5(2x - y)(a + b)$

$= 5(2ax + 2bx - ay - by)$ FOIL

$= 10ax + 10bx - 5ay - 5by$ Distributive property

$= 10ax - 5ay + 10bx - 5by$ Original polynomial

◀ Work Problem **9** at the side.

Answers

8. **(a)** $(m + 2)(n + 3)$
(b) $(2x^2 + 3)(5y^2 - 6)$

9. **(a)** $4(3y + z)(w - 2x)$
(b) $3x(2b + z)(y + z)$

7.1 Exercises

Find the greatest common factor for each list of terms.

1. $9m^3, 3m^2, 15m$

2. $4a^2, 6a, 2a^3$

3. $16xy^3, 24x^2y^2, 8x^2y$

4. $10m^2n^2, 25mn^3, 50m^2n$

5. $6m(r+t)^2, 3p(r+t)^4$

6. $7z^2(m+n)^4, 9z^3(m+n)^5$

7. Which one of the following has the greatest common factor of $6x^3y^4 - 12x^5y^2 + 24x^4y^8$ as one of the factors?

A. $6x^3y^2(y^2 - 2x^2 + 4xy^6)$

B. $6xy(x^2y^3 - 2x^4y + 4x^3y^7)$

C. $2x^3y^2(3y^2 - 6x^2 + 12xy^6)$

D. $6x^2y^2(xy^2 - 2x^3 + 4x^2y^6)$

8. When directed to factor the polynomial $4x^2y^5 - 8xy^3$ completely, a student responded with $2xy^3(2xy^2 - 4)$. When the teacher did not give him full credit, he complained because when his factors are multiplied, the product is the original polynomial. ***WHAT WENT WRONG?*** Give the correct answer.

Factor out the greatest common factor. See Examples 1–4.

9. $10x - 30$

10. $15y - 60$

11. $8s - 16t$

12. $35p + 70q$

13. $6 + 12r$

14. $9 + 18m$

15. $8k^3 + 24k$

16. $9z^4 + 27z$

17. $3xy - 5xy^2$

18. $5h^2j - 7hj$

19. $-4p^3q^4 - 2p^2q^5$

20. $-3z^5w^2 - 18z^3w^4$

21. $21x^5 + 35x^4 - 14x^3$

22. $18k^3 - 36k^4 + 48k^5$

23. $36p^4 + 9p^2 - 27p^3$

24. $42z^6 + 7z^3 - 14z^4$

25. $15a^2c^3 - 25ac^2 + 5a^2c$

26. $15y^3z^3 + 27y^2z^4 - 36yz^5$

27. $-27m^3p^5 + 5r^4s^3 - 8x^5z^4$

28. $-50r^4t^2 + 81x^3y^3 - 49p^2q^4$

29. $(m-4)(m+2) + (m-4)(m+3)$

30. $(z-5)(z+7) + (z-5)(z+9)$

31. $(2z - 1)(z + 6) - (2z - 1)(z - 5)$

32. $(3x + 2)(x - 4) - (3x + 2)(x + 8)$

33. $5(2 - x)^3 - (2 - x)^4 + 4(2 - x)^2$

34. $3(5 - x)^4 + 2(5 - x)^3 - (5 - x)^2$

Factor each polynomial twice. First use a common factor with a positive coefficient, and then use a common factor with a negative coefficient. See Example 4.

35. $-r^3 + 3r^2 + 5r$

36. $-t^4 + 8t^3 - 12t$

37. $-12s^5 + 48s^4$

38. $-16y^4 + 64y^3$

39. $-2x^5 + 6x^3 + 4x^2$

40. $-5a^3 + 10a^4 - 15a^5$

Factor by grouping. See Examples 5–9.

41. $mx + 3qx + my + 3qy$

42. $2k + 2h + jk + jh$

43. $10m + 2n + 5mk + nk$

44. $3ma + 3mb + 2ab + 2b^2$

45. $4 - 2q - 6p + 3pq$

46. $20 + 5m + 12n + 3mn$

47. $p^2 - 4zq + pq - 4pz$

48. $r^2 - 9tw + 3rw - 3rt$

49. $7ab + 35bc + a + 5c$

50. $6kn + 2mn + 3k + m$

51. $m^3 + 4m^2 - 6m - 24$

52. $2a^3 + a^2 - 14a - 7$

53. $-3a^3 - 3ab^2 + 2a^2b + 2b^3$

54. $-16m^3 + 4m^2p^2 - 4mp + p^3$

55. $4 + xy - 2y - 2x$

56. $10ab - 21 - 6b + 35a$

57. $8 + 9y^4 - 6y^3 - 12y$

58. $x^3y^2 - 3 - 3y^2 + x^3$

59. $2mx + 6qx + 2my + 6qy$

60. $12 - 6q - 18p + 9pq$

61. $2x^3y^2 + x^2y^2 - 14xy^2 - 7y^2$

Factor out the variable that is raised to the lesser exponent.

62. $k^{-2} - 2k^{-4}$

63. $3m^{-5} + m^{-3}$

64. $8q^{-2} - 5q^{-3}$

65. $3p^{-3} + 2p^{-2}$

7.2 Factoring Trinomials

OBJECTIVES

1. **Factor trinomials when the coefficient of the squared term is 1.**
2. **Factor trinomials when the coefficient of the squared term is not 1.**
3. **Use an alternative method for factoring trinomials.**
4. **Factor by substitution.**

OBJECTIVE 1 Factor trinomials when the coefficient of the squared term is 1. We begin by finding the product of $x + 3$ and $x - 5$.

$$
\begin{aligned}
&(x + 3)(x - 5) \\
&= x^2 - 5x + 3x - 15 \\
&= x^2 - 2x - 15
\end{aligned}
$$

By this result, the factored form of $x^2 - 2x - 15$ is $(x + 3)(x - 5)$.

Multiplying →

Factored form → $(x + 3)(x - 5) = x^2 - 2x - 15$ ← Product

← Factoring

Since multiplying and factoring are operations that "undo" each other, factoring trinomials involves using the FOIL method backwards. As shown here, the x^2-term comes from multiplying x and x, and -15 comes from multiplying 3 and -5.

Product of x and x is x^2.

$$(x + 3)(x - 5) = x^2 - 2x - 15$$

Product of 3 and -5 is -15.

We find the $-2x$ in $x^2 - 2x - 15$ by multiplying the outer terms, multiplying the inner terms, and adding.

Outer terms: $x(-5) = -5x$

$$(x + 3)(x - 5)$$

Inner terms: $3 \cdot x = 3x$

Add to get $-2x$.

Based on this example, follow these steps to factor a trinomial $x^2 + bx + c$, where 1 is the coefficient of the squared term. (A procedure for factoring a trinomial when the coefficient of the squared term is *not* 1 follows later in this section.)

Factoring $x^2 + bx + c$

Step 1 **Find pairs whose product is *c*.** Find all pairs of integers whose product is c, the third term of the trinomial.

Step 2 **Find pairs whose sum is *b*.** Choose the pair whose sum is b, the coefficient of the middle term.

If there are no such integers, the polynomial cannot be factored.

A polynomial that cannot be factored with integer coefficients is a **prime polynomial.** Some examples of prime polynomials are

$x^2 + x + 2$, $x^2 - x - 1$, and $2x^2 + x + 7$. Prime polynomials

1 Factor each trinomial.

(a) $p^2 + 6p + 5$

(b) $a^2 + 9a + 20$

(c) $k^2 - k - 6$

(d) $b^2 - 7b + 10$

(e) $y^2 - 8y + 6$

2 Factor each trinomial.

(a) $x^2 + 2nx - 8n^2$

(b) $x^2 - 7xz + 9z^2$

Answers

1. (a) $(p + 1)(p + 5)$ (b) $(a + 5)(a + 4)$ (c) $(k - 3)(k + 2)$ (d) $(b - 5)(b - 2)$ (e) prime
2. (a) $(x - 2n)(x + 4n)$ (b) prime

EXAMPLE 1 Factoring Trinomials in $x^2 + bx + c$ Form

Factor each trinomial.

(a) $y^2 + 2y - 35$

Step 1 Find pairs of integers whose product is c, -35.	***Step 2*** Write sums of the pairs of integers from Step 1, looking for a sum of b, 2.
$35(-1)$	$35 + (-1) = 34$
$-35(1)$	$-35 + 1 = -34$
$7(-5)$	$7 + (-5) = 2$ ← Coefficient of the middle term
$-7(5)$	$-7 + 5 = -2$

The integers 7 and -5 have the necessary product and sum, so

$$y^2 + 2y - 35 \quad \text{factors as} \quad (y + 7)(y - 5).$$

Multiply to check

(b) $r^2 + 8r + 12$

Look for two integers with a product of 12 and a sum of 8. Of all pairs of integers having a product of 12, only the pair 6 and 2 has a sum of 8.

$$r^2 + 8r + 12 \quad \text{factors as} \quad (r + 6)(r + 2).$$

Because of the commutative property, it would be equally correct to write $(r + 2)(r + 6)$. ***Check by using FOIL to multiply the factored form.***

EXAMPLE 2 Recognizing a Prime Polynomial

Factor $m^2 + 6m + 7$.

Look for two integers whose product is 7 and whose sum is 6. Only 7 and 1 and -7 and -1 give a product of 7. Neither pair has a sum of 6, so $m^2 + 6m + 7$ cannot be factored with integer coefficients and is prime.

◀ *Work Problem* **1** *at the Side.*

EXAMPLE 3 Factoring a Trinomial in Two Variables

Factor $x^2 + 6ax - 16a^2$.

Look at this trinomial as a trinomial in the form $x^2 + bx + c$, where $b = 6a$ and $c = -16a^2$.

Step 1 Find pairs of expressions whose product is $-16a^2$.	***Step 2*** Write sums of the pairs of expressions from Step 1, looking for a sum of $6a$.
$16a(-a)$	$16a + (-a) = 15a$
$-16a(a)$	$-16a + a = -15a$
$8a(-2a)$	$8a + (-2a) = 6a$
$-8a(2a)$	$-8a + 2a = -6a$
$-4a(4a)$	$-4a + 4a = 0$

The expressions $8a$ and $-2a$ have the necessary product and sum, so

$$x^2 + 6ax - 16a^2 \quad \text{factors as} \quad (x + 8a)(x - 2a).$$

Check $(x + 8a)(x - 2a)$

$= x^2 - 2ax + 8ax - 16a^2$ FOIL

$= x^2 + 6ax - 16a^2$ Original polynomial

◀ *Work Problem* **2** *at the Side.*

EXAMPLE 4 **Factoring a Trinomial with a Common Factor**

Factor $16y^3 - 32y^2 - 48y$.

$$16y^3 - 32y^2 - 48y$$
$$= 16y(y^2 - 2y - 3) \quad \text{Factor out the GCF, } 16y.$$

To factor $y^2 - 2y - 3$, look for two integers whose product is -3 and whose sum is -2. The necessary integers are -3 and 1.

$$= 16y(y - 3)(y + 1)$$

Remember to include the GCF, $16y$.

CAUTION

When factoring, always look for a common factor first. Remember to write the common factor as part of the answer.

Work Problem 3 at the Side. ▶

OBJECTIVE 2 **Factor trinomials when the coefficient of the squared term is not 1.** We can use a generalization of the method shown in Objective 1 to factor a trinomial of the form $ax^2 + bx + c$, where $a \neq 1$. To factor $3x^2 + 7x + 2$, for example, we first identify the values of a, b, and c.

$$\begin{array}{ccccc} ax^2 & + & bx & + & c \\ \downarrow & & \downarrow & & \downarrow \\ 3x^2 & + & 7x & + & 2, \end{array} \quad \text{so} \quad a = 3, \quad b = 7, \quad c = 2.$$

The product ac is $3 \cdot 2 = 6$, so we must find two integers having a product of 6 and a sum of 7 (since the middle term has coefficient $b = 7$). The necessary integers are 1 and 6, so we write $7x$ as $1x + 6x$, or $x + 6x$. Thus,

$$3x^2 + 7x + 2$$
$$= 3x^2 + \underbrace{x + 6x}_{x + 6x = 7x} + 2$$
$$= (3x^2 + x) + (6x + 2) \quad \text{Group the terms.}$$
$$= x(3x + 1) + 2(3x + 1) \quad \text{Factor each group.}$$
$$= (3x + 1)(x + 2). \quad \text{Factor out the common factor.}$$

Check by multiplying.

EXAMPLE 5 **Factoring a Trinomial in $ax^2 + bx + c$ Form**

Factor $12r^2 - 5r - 2$.

Since $a = 12$, $b = -5$, and $c = -2$, the product ac is -24. The two integers whose product is -24 and whose sum is b, -5, are 3 and -8.

$$12r^2 - 5r - 2$$
$$= 12r^2 + 3r - 8r - 2 \quad \text{Write } -5r \text{ as } 3r - 8r.$$
$$= 3r(4r + 1) - 2(4r + 1) \quad \text{Factor by grouping.}$$
$$= (4r + 1)(3r - 2) \quad \text{Factor out the common factor.}$$

Check by multiplying.

Work Problem 4 at the Side. ▶

3 Factor $5m^4 - 5m^3 - 100m^2$.

4 Factor each trinomial.

(a) $3y^2 - 11y - 4$

(b) $6k^2 - 19k + 10$

ANSWERS

3. $5m^2(m - 5)(m + 4)$
4. (a) $(y - 4)(3y + 1)$ (b) $(2k - 5)(3k - 2)$

OBJECTIVE 3 Use an alternative method for factoring trinomials. This method involves trying repeated combinations and using FOIL.

EXAMPLE 6 Factoring Trinomials in $ax^2 + bx + c$ Form

Factor each trinomial.

(a) $3x^2 + 7x + 2$

To factor this trinomial we use an alternative method. The goal is to find the correct numbers to fill in the blanks.

$$3x^2 + 7x + 2 = (\_\_\_x + \_\_\_)(\_\_\_x + \_\_\_)$$

Addition signs are used, since all the signs in the trinomial indicate addition. The first two expressions have a product of $3x^2$, so they must be $3x$ and x.

$$3x^2 + 7x + 2 = (3x + \_\_\_)(x + \_\_\_)$$

The product of the two last terms must be 2, so the numbers must be 2 and 1. There is a choice. The 2 could be placed with the $3x$ or with the x. Only one of these choices will give the correct middle term, $7x$. We use the FOIL method to check each one.

$(3x + 2)(x + 1)$	$(3x + 1)(x + 2)$
Outer: $3x$; Inner: $2x$	Outer: $6x$; Inner: x
$3x + 2x = 5x$	$6x + x = 7x$
Wrong middle term	Correct middle term

Therefore, $3x^2 + 7x + 2$ factors as $(3x + 1)(x + 2)$. (Compare to the answer obtained using factoring by grouping on the preceding page.)

(b) $12r^2 - 5r - 2$

We note that the trinomial has no common factor (except 1). This means that neither of its factors can have a common factor. We should keep this in mind as we choose factors. We try 4 and 3 for the two first terms.

$$12r^2 - 5r - 2 = (4r\_\_\_)(3r\_\_\_)$$

The factors of -2 are -2 and 1 or 2 and -1. We try both possibilities.

$(4r - 2)(3r + 1)$	$(4r - 1)(3r + 2)$
Wrong: $4r - 2$ has a common factor of 2. This cannot be correct, since 2 is not a factor of $12r^2 - 5r - 2$.	Outer: $8r$; Inner: $-3r$
	$8r - 3r = 5r$
	Wrong middle term

The middle term on the right is $5r$, instead of the $-5r$ that is needed. We get $-5r$ by interchanging the signs of the second terms in the factors.

$$(4r + 1)(3r - 2)$$

Outer: $-8r$; Inner: $3r$

$-8r + 3r = -5r$

Correct middle term

Thus, $12r^2 - 5r - 2$ factors as $(4r + 1)(3r - 2)$. (Compare to Example 5.)

◀ Work Problem 5 at the Side.

5 Factor each trinomial.

(a) $10x^2 + 17x + 3$

(b) $16y^2 - 34y - 15$

(c) $8t^2 - 13t + 5$

ANSWERS

5. (a) $(5x + 1)(2x + 3)$
(b) $(8y + 3)(2y - 5)$
(c) $(8t - 5)(t - 1)$

We summarize the method used in Example 6 as follows.

Factoring $ax^2 + bx + c$

Step 1 **Find pairs whose product is a.** Write all pairs of integer factors of a, the coefficient of the second-degree term.

Step 2 **Find pairs whose product is c.** Write all pairs of integer factors of c, the last term.

Step 3 **Choose inner and outer terms.** Use FOIL and various combinations of the factors from Steps 1 and 2 until the necessary middle term is found.

If no such combinations exist, the trinomial is prime.

EXAMPLE 7 **Factoring a Trinomial in Two Variables**

Factor $18m^2 - 19mx - 12x^2$.

There is no common factor (except 1). Follow the steps to factor the trinomial. There are many possible factors of both 18 and -12. Try 6 and 3 for 18 and -3 and 4 for -12.

$(6m - 3x)(3m + 4x)$	$(6m + 4x)(3m - 3x)$
Wrong: common factor	Wrong: common factors

Since 6 and 3 do not work as factors of 18, try 9 and 2 instead, with 3 and -4 as factors of -12.

$(9m + 3x)(2m - 4x)$	$(9m - 4x)(2m + 3x)$ (outer: $27mx$; inner: $-8mx$)
Wrong: common factors	$27mx + (-8mx) = 19mx$ Wrong middle term

The result on the right differs from the correct middle term only in sign, so interchange the signs in the factors.

$$18m^2 - 19mx - 12x^2 \quad \text{factors as} \quad (9m + 4x)(2m - 3x)$$ Check by multiplying.

Work Problem **6** at the Side. ▶

EXAMPLE 8 **Factoring $ax^2 + bx + c$, with $a < 0$**

Factor $-3x^2 + 16x + 12$.

While we could factor this trinomial directly, it is helpful to first factor out -1, so that the coefficient of the x^2-term is positive.

$$\begin{aligned} &-3x^2 + 16x + 12 \\ &= -1(3x^2 - 16x - 12) && \text{Factor out } -1. \\ &= -1(3x + 2)(x - 6) && \text{Factor the trinomial.} \\ &= -(3x + 2)(x - 6) \end{aligned}$$

This factored form can be written in other ways. Two of them are

$$(-3x - 2)(x - 6) \quad \text{and} \quad (3x + 2)(-x + 6).$$

Verify that these both give the original trinomial when multiplied.

Work Problem **7** at the Side. ▶

6 Factor each trinomial.

(a) $7p^2 + 15pq + 2q^2$

(b) $6m^2 + 7mn - 5n^2$

(c) $12z^2 - 5zy - 2y^2$

(d) $8m^2 + 18mx - 5x^2$

7 Factor each trinomial.

(a) $-6r^2 + 13r + 5$

(b) $-8x^2 + 10x - 3$

ANSWERS

6. **(a)** $(7p + q)(p + 2q)$
(b) $(3m + 5n)(2m - n)$
(c) $(3z - 2y)(4z + y)$
(d) $(4m - x)(2m + 5x)$

7. **(a)** $-(2r - 5)(3r + 1)$
(b) $-(4x - 3)(2x - 1)$

8 Factor each trinomial.

(a) $2m^3 - 4m^2 - 6m$

(b) $12r^4 + 6r^3 - 90r^2$

(c) $30y^5 - 55y^4 - 50y^3$

9 Factor each polynomial.

(a) $6(a - 1)^2 + (a - 1) - 2$

(b) $8(z + 5)^2 - 2(z + 5) - 3$

(c) $15(m - 4)^2 - 11(m - 4) + 2$

10 Factor each trinomial.

(a) $y^4 + y^2 - 6$

(b) $2p^4 + 7p^2 - 15$

(c) $6r^4 - 13r^2 + 5$

ANSWERS

8. (a) $2m(m + 1)(m - 3)$
(b) $6r^2(r + 3)(2r - 5)$
(c) $5y^3(2y - 5)(3y + 2)$

9. (a) $(2a - 3)(3a - 1)$
(b) $(4z + 17)(2z + 11)$
(c) $(3m - 13)(5m - 22)$

10. (a) $(y^2 - 2)(y^2 + 3)$
(b) $(2p^2 - 3)(p^2 + 5)$
(c) $(3r^2 - 5)(2r^2 - 1)$

EXAMPLE 9 Factoring a Trinomial with a Common Factor

Factor $16y^3 + 24y^2 - 16y$.

$$
\begin{aligned}
&16y^3 + 24y^2 - 16y \\
&= 8y(2y^2 + 3y - 2) && \text{GCF} = 8y \\
&= 8y(2y - 1)(y + 2) && \text{Factor the trinomial.}
\end{aligned}
$$

Remember the common factor.

◀ Work Problem **8** at the Side.

OBJECTIVE 4 Factor by substitution.

EXAMPLE 10 Factoring a Polynomial Using Substitution

Factor $2(x + 3)^2 + 5(x + 3) - 12$.

Since the binomial $x + 3$ appears to powers 2 and 1, we let the substitution variable represent $x + 3$. We may choose any letter we wish except x. We choose t to represent $x + 3$.

$$
\begin{aligned}
&2(x + 3)^2 + 5(x + 3) - 12 \\
&= 2t^2 + 5t - 12 && \text{Let } t = x + 3. \\
&= (2t - 3)(t + 4) && \text{Factor.} \\
&= [2(x + 3) - 3][(x + 3) + 4] && \text{Replace } t \text{ with } x + 3. \\
&= (2x + 6 - 3)(x + 7) && \text{Simplify.} \\
&= (2x + 3)(x + 7)
\end{aligned}
$$

CAUTION

Remember to make the final substitution of $x + 3$ for t in Example 10.

◀ Work Problem **9** at the Side.

EXAMPLE 11 Factoring a Trinomial in $ax^4 + bx^2 + c$ Form

Factor $6y^4 + 7y^2 - 20$.

The variable y appears to powers in which the larger exponent is twice the smaller exponent. We can let a substitution variable equal the smaller power. Here, we let $t = y^2$.

$$
\begin{aligned}
&6y^4 + 7y^2 - 20 \\
&= 6(y^2)^2 + 7y^2 - 20 && y^4 = (y^2)^2 \\
&= 6t^2 + 7t - 20 && \text{Substitute } t \text{ for } y^2. \\
&= (3t - 4)(2t + 5) && \text{Factor.} \\
&= (3y^2 - 4)(2y^2 + 5) && t = y^2
\end{aligned}
$$

Don't stop here. Replace t with y^2.

◀ Work Problem **10** at the Side.

Note

Some students feel comfortable factoring polynomials like the one in Example 11 directly, without using the substitution method.

7.2 Exercises

FOR EXTRA HELP MyMathLab Math XL PRACTICE WATCH DOWNLOAD READ REVIEW

1. Which one of the following is *not* a valid way of starting the process of factoring $12x^2 + 29x + 10$?
   A. $(12x \quad)(x \quad)$ B. $(4x \quad)(3x \quad)$
   C. $(6x \quad)(2x \quad)$ D. $(8x \quad)(4x \quad)$

2. Which one of the following is the completely factored form of $2x^6 - 5x^5 - 3x^4$?
   A. $x^4(2x + 1)(x - 3)$ B. $x^4(2x - 1)(x + 3)$
   C. $(2x^5 + x^4)(x - 3)$ D. $x^3(2x^2 + x)(x - 3)$

3. Which one of the following is the completely factored form of $4x^2 - 4x - 24$?
   A. $4(x - 2)(x + 3)$ B. $4(x + 2)(x + 3)$
   C. $4(x + 2)(x - 3)$ D. $4(x - 2)(x - 3)$

4. Which one of the following is *not* a factored form of $-x^2 + 16x - 60$?
   A. $(x - 10)(-x + 6)$ B. $(-x - 10)(x + 6)$
   C. $(-x + 10)(x - 6)$ D. $-1(x - 10)(x - 6)$

Factor each trinomial. See Examples 1–9.

5. $y^2 - 7y - 30$
6. $z^2 + 2z - 24$
7. $p^2 - p - 56$
8. $k^2 - 11k + 30$
9. $m^2 - 11m + 60$
10. $p^2 - 12p - 27$
11. $a^2 - 2ab - 35b^2$
12. $z^2 + 8zw + 15w^2$
13. $y^2 - 3yq - 15q^2$
14. $k^2 - 11hk + 28h^2$
15. $x^2 - 11xy + 18y^2$
16. $p^2 - 5pq - 18q^2$
17. $-6m^2 - 13m + 15$
18. $-15y^2 + 17y + 18$
19. $10x^2 + 3x - 18$
20. $8k^2 + 34k + 35$
21. $20k^2 + 47k + 24$
22. $27z^2 + 42z - 5$
23. $15a^2 - 22ab + 8b^2$
24. $15p^2 + 24pq + 8q^2$
25. $36m^2 - 60m + 25$
26. $25r^2 - 90r + 81$
27. $40x^2 + xy + 6y^2$
28. $14c^2 - 17cd - 6d^2$
29. $6x^2z^2 + 5xz - 4$
30. $8m^2n^2 - 10mn + 3$
31. $24x^2 + 42x + 15$
32. $36x^2 + 18x - 4$

33. $-15a^2 - 70a + 120$

34. $-12a^2 - 10a + 42$

35. $11x^3 - 110x^2 + 264x$

36. $9k^3 + 36k^2 - 189k$

37. $2x^3y^3 - 48x^2y^4 + 288xy^5$

38. $6m^3n^2 - 24m^2n^3 - 30mn^4$

Factor each trinomial. See Example 10.

39. $10(k+1)^2 - 7(k+1) + 1$

40. $4(m-5)^2 - 4(m-5) - 15$

41. $3(m+p)^2 - 7(m+p) - 20$

42. $4(x-y)^2 - 23(x-y) - 6$

43. $a^2(a+b)^2 - ab(a+b) - 6b^2$

44. $m^2(m-p)^2 + mp(m-p) - 2p^2$

Factor each trinomial. See Example 11.

45. $2x^4 - 9x^2 - 18$

46. $6z^4 + z^2 - 1$

47. $16x^4 + 16x^2 + 3$

48. $9r^4 + 9r^2 + 2$

49. $12p^6 - 32p^3r + 5r^2$

50. $2y^6 + 7xy^3 + 6x^2$

Relating Concepts (Exercises 51–56) For Individual or Group Work

If the terms of a polynomial have no common factor, then none of the terms of its factors can have a common factor, as seen in Examples 6 and 7. ***Work Exercises 51–56 in order.***

51. Is 2 a factor of the composite number 45?

52. List all positive integer factors of 45. Is 2 a factor of any of these factors?

53. Is 5 a factor of $10x^2 + 29x + 10$?

54. Factor $10x^2 + 29x + 10$. Is 5 a factor of either of its factors?

55. Suppose that k is an odd integer and you are asked to factor $2x^2 + kx + 8$. Why is $2x + 4$ not a possible choice for a factor in factoring this polynomial?

56. The polynomial $12y^2 - 11y - 15$ can be factored using the methods of this section. Explain why $3y + 15$ cannot be one of its factors.

7.3 Special Factoring

OBJECTIVE 1 Factor a difference of squares. The special products from **Section 6.4** are used in reverse when factoring. The product of the sum and difference of two terms leads to a **difference of squares.**

Difference of Squares

$$x^2 - y^2 = (x + y)(x - y)$$

EXAMPLE 1 Factoring Differences of Squares

Factor each polynomial.

(a) $t^2 - 36$

$= t^2 - 6^2$ $\quad 36 = 6^2$

$= (t + 6)(t - 6)$ $\quad$ Factor the difference of squares.

(b) $4a^2 - 64$

$= 4(a^2 - 16)$ $\quad$ Factor out the common factor, 4.

$= 4(a + 4)(a - 4)$ $\quad$ Factor the difference of squares.

(c) $16m^2 - 49p^2 = (4m)^2 - (7p)^2 = (4m + 7p)(4m - 7p)$

(with $x^2 - y^2 = (x + y)(x - y)$, $x \to 4m$, $y \to 7p$)

(d) $81k^2 - (a + 2)^2 = (9k)^2 - (a + 2)^2 = (9k + a + 2)(9k - [a + 2])$

$= (9k + a + 2)(9k - a - 2)$

(with $x^2 - y^2 = (x + y)(x - y)$, $x \to 9k$, $y \to a + 2$)

We could have used the method of substitution here.

(e) $x^4 - 81$

$= (x^2 + 9)(x^2 - 9)$ $\quad$ Factor the difference of squares.

$= (x^2 + 9)(x + 3)(x - 3)$ $\quad$ Factor the difference of squares again.

Work Problem 1 at the Side. ▶

CAUTION

Assuming that the greatest common factor is 1, it is not possible to factor (with real numbers) a sum of squares, such as $x^2 + 9$ in Example 1(e). In particular, $x^2 + y^2 \neq (x + y)^2$, as shown next.

OBJECTIVE 2 Factor a perfect square trinomial. Two other special products from **Section 6.4** lead to the following rules for factoring.

Perfect Square Trinomial

$$x^2 + 2xy + y^2 = (x + y)^2$$
$$x^2 - 2xy + y^2 = (x - y)^2$$

OBJECTIVES

1. Factor a difference of squares.
2. Factor a perfect square trinomial.
3. Factor a difference of cubes.
4. Factor a sum of cubes.

1 Factor each polynomial.

(a) $p^2 - 100$

(b) $2x^2 - 18$

(c) $9a^2 - 16b^2$

(d) $(m + 3)^2 - 49z^2$

(e) $y^4 - 16$

ANSWERS

1. **(a)** $(p + 10)(p - 10)$
   (b) $2(x + 3)(x - 3)$
   (c) $(3a - 4b)(3a + 4b)$
   (d) $(m + 3 + 7z)(m + 3 - 7z)$
   (e) $(y^2 + 4)(y + 2)(y - 2)$

Because the trinomial $x^2 + 2xy + y^2$ is the square of $x + y$, it is called a **perfect square trinomial.** In this pattern, both the first and the last terms of the trinomial must be perfect squares. In the factored form $(x + y)^2$, twice the product of the first and the last terms must give the middle term of the trinomial. It is important to understand these patterns in terms of words, since they occur with many different symbols (other than x and y).

$4m^2 + 20m + 25$	$p^2 - 8p + 64$
Perfect square trinomial; $4m^2 = (2m)^2$, $25 = 5^2$, and $2(2m)(5) = 20m$.	Not a perfect square trinomial; middle term would have to be $16p$ or $-16p$.

◀ Work Problem **2** at the Side.

2 Identify any perfect square trinomials.

(a) $z^2 + 12z + 36$

(b) $2x^2 - 4x + 4$

(c) $9a^2 + 12ab + 16b^2$

EXAMPLE 2 Factoring Perfect Square Trinomials

Factor each polynomial.

(a) $144p^2 - 120p + 25$

Here, $144p^2 = (12p)^2$ and $25 = 5^2$. The sign on the middle term is $-$, so if $144p^2 - 120p + 25$ is a perfect square trinomial, the factored form will have to be

$$(12p - 5)^2.$$

Take twice the product of the two terms to see if this is correct.

$$2(12p)(-5) = -120p$$

This is the middle term of the given trinomial, so

$$144p^2 - 120p + 25 \quad \text{factors as} \quad (12p - 5)^2.$$

(b) $4m^2 + 20mn + 49n^2$

If this is a perfect square trinomial, it will equal $(2m + 7n)^2$. By the pattern described earlier, if multiplied out, this squared binomial has a middle term of $2(2m)(7n) = 28mn$, which *does not equal* $20mn$. Verify that this trinomial cannot be factored by the methods of the previous section either. It is prime.

(c) $(r + 5)^2 + 6(r + 5) + 9$

$= [(r + 5) + 3]^2$ $\quad$ $2(r + 5)(3) = 6(r + 5)$, the middle term.

$= (r + 8)^2$

(d) $m^2 - 8m + 16 - p^2$

Since there are four terms, we use factoring by grouping. The first three terms form a perfect square trinomial. Group them together, and factor as follows.

$(m^2 - 8m + 16) - p^2$

$= (m - 4)^2 - p^2$ $\quad$ Factor the perfect square trinomial.

$= (m - 4 + p)(m - 4 - p)$ $\quad$ Factor the difference of squares.

◀ Work Problem **3** at the Side.

3 Factor each polynomial.

(a) $49z^2 - 14zk + k^2$

(b) $9a^2 + 48ab + 64b^2$

(c) $(k + m)^2 - 12(k + m) + 36$

(d) $x^2 - 2x + 1 - y^2$

Perfect square trinomials, of course, can be factored using the general methods shown earlier for other trinomials. The patterns given here provide "shortcuts."

ANSWERS

2. **(a)** perfect square trinomial
(b) not a perfect square trinomial
(c) not a perfect square trinomial
3. **(a)** $(7z - k)^2$ **(b)** $(3a + 8b)^2$
(b) $[(k + m) - 6]^2$, or $(k + m - 6)^2$
(c) $(x - 1 + y)(x - 1 - y)$

OBJECTIVE 3 Factor a difference of cubes. A **difference of cubes**, such as $x^3 - y^3$, can be factored as follows.

Difference of Cubes

$$x^3 - y^3 = (x - y)(x^2 + xy + y^2)$$

We could check this pattern by finding the product of $x - y$ and $x^2 + xy + y^2$.

EXAMPLE 3 Factoring Differences of Cubes

Factor each polynomial.

$$x^3 - y^3 = (x - y)(x^2 + x \cdot y + y^2)$$

(a) $m^3 - 8 = m^3 - 2^3 = (m - 2)(m^2 + m \cdot 2 + 2^2)$

$= (m - 2)(m^2 + 2m + 4)$

Check $(m - 2)(m^2 + 2m + 4)$

$= m^3 + 2m^2 + 4m - 2m^2 - 4m - 8$

$= m^3 - 8$

(b) $27x^3 - 8y^3$

$= (3x)^3 - (2y)^3$

$= (3x - 2y)[(3x)^2 + (3x)(2y) + (2y)^2]$

$= (3x - 2y)(9x^2 + 6xy + 4y^2)$

(c) $1000k^3 - 27n^3$

$= (10k)^3 - (3n)^3$

$= (10k - 3n)[(10k)^2 + (10k)(3n) + (3n)^2]$

$= (10k - 3n)(100k^2 + 30kn + 9n^2)$

Work Problem 4 at the Side. ▶

4 Factor each polynomial.

(a) $x^3 - 1000$

(b) $8k^3 - y^3$

(c) $27a^3 - 64b^3$

OBJECTIVE 4 Factor a sum of cubes. While the binomial $x^2 + y^2$ (a sum of *squares*) cannot be factored with real numbers, a **sum of cubes**, such as $x^3 + y^3$, is factored as follows.

Sum of Cubes

$$x^3 + y^3 = (x + y)(x^2 - xy + y^2)$$

To verify this result, find the product of $x + y$ and $x^2 - xy + y^2$.

Note

The sign of the second term in the binomial factor of a sum or difference of cubes is always the *same* as the sign in the original polynomial. In the trinomial factor, the sign of the middle term is the *opposite* of the sign of the second term in the binomial factor; the last term is ***always positive***. To remember the signs in the sum and difference of cubes formulas, use the memory aid ***soap*** (*same, opposite, always positive*).

Answers

4. **(a)** $(x - 10)(x^2 + 10x + 100)$
(b) $(2k - y)(4k^2 + 2ky + y^2)$
(c) $(3a - 4b)(9a^2 + 12ab + 16b^2)$

5 Factor each polynomial.

(a) $8p^3 + 125$

(b) $27m^3 + 125n^3$

(c) $2x^3 + 2000$

(d) $(a - 4)^3 + b^3$

EXAMPLE 4 Factoring Sums of Cubes

Factor each polynomial.

(a) $r^3 + 27$

$= r^3 + 3^3$

$= (r + 3)(r^2 - 3r + 3^2)$

$= (r + 3)(r^2 - 3r + 9)$

(b) $27z^3 + 125$

$= (3z)^3 + 5^3$

$= (3z + 5)[(3z)^2 - (3z)(5) + 5^2]$

$= (3z + 5)(9z^2 - 15z + 25)$

(c) $125t^3 + 216s^6$

$= (5t)^3 + (6s^2)^3$

$= (5t + 6s^2)[(5t)^2 - (5t)(6s^2) + (6s^2)^2]$

$= (5t + 6s^2)(25t^2 - 30ts^2 + 36s^4)$

(d) $3x^3 + 192$

$= 3(x^3 + 64)$ Factor out the common factor.

$= 3(x + 4)(x^2 - 4x + 16)$ Factor the sum of cubes.

Remember the common factor.

(e) $(x + 2)^3 + t^3$

$= [(x + 2) + t][(x + 2)^2 - (x + 2)t + t^2]$

$= (x + 2 + t)(x^2 + 4x + 4 - xt - 2t + t^2)$

CAUTION
A common error when factoring $x^3 + y^3$ or $x^3 - y^3$ is to think that the xy-term has a coefficient of 2. Since there is no coefficient of 2, the trinomials $x^2 + xy + y^2$ and $x^2 - xy + y^2$ cannot be factored further.

◀ Work Problem **5** at the Side.

The special types of factoring are summarized here. ***These should be memorized.***

Special Types of Factoring

Difference of Squares	$x^2 - y^2 = (x + y)(x - y)$
Perfect Square Trinomial	$x^2 + 2xy + y^2 = (x + y)^2$
	$x^2 - 2xy + y^2 = (x - y)^2$
Difference of Cubes	$x^3 - y^3 = (x - y)(x^2 + xy + y^2)$
Sum of Cubes	$x^3 + y^3 = (x + y)(x^2 - xy + y^2)$

ANSWERS

5. **(a)** $(2p + 5)(4p^2 - 10p + 25)$
(b) $(3m + 5n)(9m^2 - 15mn + 25n^2)$
(c) $2(x + 10)(x^2 - 10x + 100)$
(d) $(a - 4 + b) \cdot (a^2 - 8a + 16 - ab + 4b + b^2)$

7.3 Exercises

FOR EXTRA HELP

MyMathLab | Math XL PRACTICE | WATCH | DOWNLOAD | READ | REVIEW

1. Which of the following binomials are differences of squares?

   A. $64 - m^2$ **B.** $2x^2 - 25$
   C. $k^2 + 9$ **D.** $4z^4 - 49$

2. Which of the following binomials are sums or differences of cubes?

   A. $64 + y^3$ **B.** $125 - p^6$
   C. $9x^3 + 125$ **D.** $(x + y)^3 - 1$

3. Which of the following trinomials are perfect squares?

   A. $x^2 - 8x - 16$ **B.** $4m^2 + 20m + 25$
   C. $9z^4 + 30z^2 + 25$ **D.** $25a^2 - 45a + 81$

4. Insert the correct signs in the blanks.

   (a) $8 + t^3 = (2 \_\_ t)(4 \_\_ 2t \_\_ t^2)$

   (b) $z^3 - 1 = (z \_\_ 1)(z^2 \_\_ z \_\_ 1)$

Factor each polynomial. See Examples 1–4.

5. $p^2 - 16$
6. $k^2 - 9$
7. $25x^2 - 4$
8. $36m^2 - 25$
9. $18a^2 - 98b^2$
10. $32c^2 - 98d^2$
11. $64m^4 - 4y^4$
12. $243x^4 - 3t^4$
13. $(y + z)^2 - 81$
14. $(h - k)^2 - 9$
15. $16 - (x + 3y)^2$
16. $64 - (r + 2t)^2$
17. $p^4 - 256$
18. $a^4 - 625$
19. $k^2 - 6k + 9$
20. $x^2 + 10x + 25$
21. $4z^2 + 4zw + w^2$
22. $9y^2 + 6yz + z^2$
23. $16m^2 - 8m + 1 - n^2$
24. $25c^2 - 20c + 4 - d^2$
25. $4r^2 - 12r + 9 - s^2$
26. $9a^2 - 24a + 16 - b^2$
27. $x^2 - y^2 + 2y - 1$
28. $-k^2 - h^2 + 2kh + 4$
29. $98m^2 + 84mn + 18n^2$
30. $80z^2 - 40zw + 5w^2$
31. $(p + q)^2 + 2(p + q) + 1$
32. $(x + y)^2 + 6(x + y) + 9$
33. $(a - b)^2 + 8(a - b) + 16$
34. $(m - n)^2 + 4(m - n) + 4$

35. $y^3 - 64$

36. $t^3 - 216$

37. $r^3 + 343$

38. $m^3 + 512$

39. $8x^3 - y^3$

40. $z^3 - 125p^3$

41. $64g^3 + 27h^3$

42. $27a^3 + 8b^3$

43. $24n^3 + 81p^3$

44. $250x^3 - 16y^3$

45. $(y + z)^3 - 64$

46. $(p - q)^3 + 125$

47. $m^6 - 125$

48. $x^6 + 729$

49. $125y^6 + z^3$

Relating Concepts (Exercises 50–55) For Individual or Group Work

The binomial $x^6 - y^6$ may be considered either as a difference of squares or a difference of cubes. ***Work Exercises 50–55 in order.***

50. Factor $x^6 - y^6$ by first factoring as a difference of squares. Then factor further by considering one of the factors as a sum of cubes and the other factor as a difference of cubes.

51. Based on your answer in Exercise 50, fill in the blank with the correct factors so that $x^6 - y^6$ is factored completely:

$x^6 - y^6 = (x - y)(x + y)$ ______________.

52. Factor $x^6 - y^6$ by first factoring as a difference of cubes. Then factor further by considering one of the factors as a difference of squares.

53. Based on your answer in Exercise 52, fill in the blank with the correct factor so that $x^6 - y^6$ is factored:

$x^6 - y^6 = (x - y)(x + y)$ ______________.

54. Notice that the factor you wrote in the blank in Exercise 53 is a fourth-degree polynomial, while the two factors you wrote in the blank in Exercise 51 are both second-degree polynomials. What must be true about the product of the two factors you wrote in the blank in Exercise 51? Verify this.

55. If you have a choice of factoring as a difference of squares or a difference of cubes, how should you start to more easily obtain the factored form of the polynomial? Base the answer on your results in Exercises 50–54 and the methods of factoring explained in this section.

7.4 A General Approach to Factoring

A polynomial is completely factored when **(1)** it is written as a *product* of prime polynomials with integer coefficients, and **(2)** none of the polynomial factors can be factored further.

Factoring a Polynomial

Step 1 **Factor out any common factor.**

Step 2 **If the polynomial is a binomial,** check to see if it is the difference of squares, the difference of cubes, or the sum of cubes.

If the polynomial is a trinomial, check to see if it is a perfect square trinomial. If it is not, factor as in **Section 7.2.**

If the polynomial has more than three terms, try to factor by grouping.

Step 3 ***Check the factored form by multiplying.***

OBJECTIVES

1. **Factor out any common factor.**
2. **Factor binomials.**
3. **Factor trinomials.**
4. **Factor polynomials with more than three terms.**

OBJECTIVE 1 Factor out any common factor. ***This step is always the same, regardless of the number of terms in the polynomial.***

EXAMPLE 1 Factoring Out a Common Factor

Factor each polynomial.

(a) $9p - 45$

$= 9(p + 5)$ GCF = 9

(b) $8m^2p^2 + 4mp$

$= 4mp(2mp + 1)$

(c) $5x(a + b) - y(a + b)$

$= (a + b)(5x - y)$ Factor out $a + b$.

Work Problem 1 at the Side.

1 Factor each polynomial.

(a) $8x - 80$

(b) $2x^3 + 10x^2 - 2x$

(c) $12m(p - q) - 7n(p - q)$

OBJECTIVE 2 Factor binomials. Use one of the following rules.

Factoring a Binomial

For a **binomial** (two terms), check for the following patterns.

Difference of squares	$x^2 - y^2 = (x + y)(x - y)$
Difference of cubes	$x^3 - y^3 = (x - y)(x^2 + xy + y^2)$
Sum of cubes	$x^3 + y^3 = (x + y)(x^2 - xy + y^2)$

EXAMPLE 2 Factoring Binomials

Factor each binomial if possible.

(a) $64m^2 - 9n^2$

$= (8m)^2 - (3n)^2$ Difference of squares

$= (8m + 3n)(8m - 3n)$

(b) $8p^3 - 27$ $8p^3 = (2p)^3$; $27 = 3^3$

$= (2p - 3)[(2p)^2 + (2p)(3) + 3^2]$

$= (2p - 3)(4p^2 + 6p + 9)$

(c) $1000m^3 + 1$

$= (10m)^3 + 1^3$ Sum of cubes

$= (10m + 1)[(10m)^2 - (10m)(1) + 1^2]$

$= (10m + 1)(100m^2 - 10m + 1)$

(d) $25m^2 + 121$ is prime.

It is the *sum* of squares.

Work Problem 2 at the Side.

2 Factor each binomial if possible.

(a) $36x^2 - y^2$

(b) $4t^2 + 1$

(c) $125x^3 - 27y^3$

(d) $x^3 + 343y^3$

ANSWERS

1. (a) $8(x - 10)$
(b) $2x(x^2 + 5x - 1)$
(c) $(p - q)(12m - 7n)$

2. (a) $(6x + y)(6x - y)$
(b) prime
(c) $(5x - 3y)(25x^2 + 15xy + 9y^2)$
(d) $(x + 7y)(x^2 - 7xy + 49y^2)$

3 Factor each trinomial.

(a) $16m^2 + 56m + 49$

(b) $r^2 + 18r + 72$

(c) $8t^2 - 13t + 5$

(d) $6x^2 - 3x - 63$

4 Factor each polynomial.

(a) $p^3 - 2pq^2 + p^2q - 2q^3$

(b) $9x^2 + 24x + 16 - y^2$

(c) $64a^3 + 16a^2 + b^3 - b^2$

Answers

3. **(a)** $(4m + 7)^2$
(b) $(r + 6)(r + 12)$
(c) $(8t - 5)(t - 1)$
(d) $3(2x - 7)(x + 3)$

4. **(a)** $(p + q)(p^2 - 2q^2)$
(b) $(3x + 4 + y)(3x + 4 - y)$
(c) $(4a + b)(16a^2 - 4ab + b^2 + 4a - b)$

Note

The binomial $25m^2 + 625$ is a sum of squares. It *can* be factored, however, as $25(m^2 + 25)$ because it has a common factor, 25.

OBJECTIVE 3 Factor trinomials. Consider the following

Factoring a Trinomial

For a **trinomial** (three terms), decide whether it is a perfect square trinomial of the form

$$x^2 + 2xy + y^2 = (x + y)^2 \quad \text{or} \quad x^2 - 2xy + y^2 = (x - y)^2.$$

If not, use the general factoring methods of **Section 7.2.**

EXAMPLE 3 Factoring Trinomials

Factor each trinomial.

(a) $p^2 + 10p + 25$
$= (p + 5)^2 \qquad 2(p)(5) = 10p$

(b) $49z^2 - 42z + 9$
$= (7z - 3)^2 \qquad 2(7z)(3) = 42z$

(c) $y^2 - 5y - 6$
$= (y - 6)(y + 1)$ — The numbers -6 and 1 have a product of -6 and a sum of -5.

(d) $2k^2 - k - 6$
$= (2k + 3)(k - 2)$

(e) $28z^2 + 6z - 10$
$= 2(14z^2 + 3z - 5)$
$= 2(7z + 5)(2z - 1)$

◀ Work Problem 3 at the Side.

OBJECTIVE 4 Factor polynomials with more than three terms. Consider factoring by grouping.

EXAMPLE 4 Factoring Polynomials with More than Three Terms

Factor each polynomial.

(a) $20k^3 + 4k^2 - 45k - 9$

$= (20k^3 + 4k^2) - (45k + 9)$ Group the terms.

$= 4k^2(5k + 1) - 9(5k + 1)$ Factor each group.

$= (5k + 1)(4k^2 - 9)$ $5k + 1$ is a common factor.

$= (5k + 1)(2k + 3)(2k - 3)$ Difference of squares

(b) $4a^2 + 4a + 1 - b^2$

$= (4a^2 + 4a + 1) - b^2$ Group the first three terms.

$= (2a + 1)^2 - b^2$ Perfect square trinomial

$= (2a + 1 + b)(2a + 1 - b)$ Difference of squares

(c) $8m^3 + 4m^2 - n^3 - n^2$

$= (8m^3 - n^3) + (4m^2 - n^2)$ Rearrange and group the terms.

$= (2m - n)(4m^2 + 2mn + n^2) + (2m - n)(2m + n)$ Factor each group.

$= (2m - n)(4m^2 + 2mn + n^2 + 2m + n)$ Factor out $2m - n$.

◀ Work Problem 4 at the Side.

7.4 Exercises

FOR EXTRA HELP

MyMathLab · MathXL PRACTICE · WATCH · DOWNLOAD · READ · REVIEW

Factor each polynomial completely. We have randomly included all the different types of factoring exercises here to give you practice in applying factoring strategies. See Examples 1–4.

1. $100a^2 - 9b^2$

2. $10r^2 + 13r - 3$

3. $18p^5 - 24p^3 + 12p^6$

4. $15x^2 - 20x$

5. $x^2 + 2x - 35$

6. $9 - a^2 + 2ab - b^2$

7. $225p^2 + 256$

8. $x^3 + 1000$

9. $6b^2 - 17b - 3$

10. $k^2 - 6k + 16$

11. $18m^3n + 3m^2n^2 - 6mn^3$

12. $6t^2 + 19tu - 77u^2$

13. $2p^2 + 11pq + 15q^2$

14. $9m^2 - 45m + 18m^3$

15. $4k^2 + 28kr + 49r^2$

16. $54m^3 - 2000$

17. $mn - 2n + 5m - 10$

18. $9m^2 - 30mn + 25n^2 - p^2$

19. $x^3 + 3x^2 - 9x - 27$

20. $56k^3 - 875$

21. $9r^2 + 100$

22. $8p^3 - 125$

23. $6k^2 - k - 1$

24. $27m^2 + 144mn + 192n^2$

25. $x^4 - 625$

26. $125m^6 + 216$

27. $ab + 6b + ac + 6c$

28. $p^3 + 64$

29. $4y^2 - 8y$

30. $6a^4 - 11a^2 - 10$

31. $14z^2 - 3zk - 2k^2$

32. $12z^3 - 6z^2 + 18z$

33. $256b^2 - 400c^2$

34. $z^2 - zp + 20p^2$

35. $1000z^3 + 512$

36. $64m^2 - 25n^2$

37. $10r^2 + 23rs - 5s^2$

38. $12k^2 - 17kq - 5q^2$

39. $32x^2 + 16x^3 - 24x^5$

40. $48k^4 - 243$

41. $14x^2 - 25xq - 25q^2$

42. $5p^2 - 10p$

43. $y^2 + 3y - 10$

44. $b^2 - 7ba - 18a^2$

45. $2a^3 + 6a^2 - 4a$

46. $12m^2rx + 4mnrx + 40n^2rx$

47. $18p^2 + 53pr - 35r^2$

48. $21a^2 - 5ab - 4b^2$

49. $(x - 2y)^2 - 4$

50. $(3m - n)^2 - 25$

51. $(5r + 2s)^2 - 6(5r + 2s) + 9$

52. $(p + 8q)^2 - 10(p + 8q) + 25$

53. $z^4 - 9z^2 + 20$

54. $21m^4 - 32m^2 - 5$

55. $4(p + 2) + m(p + 2)$

56. $kq - 9q + kr - 9r$

57. $50p^2 - 162$

58. $25x^2 - 20xy + 4y^2$

59. $16a^2 + 8ab + b^2$

60. $40p - 32r$

7.5 Solving Equations by Factoring

We developed methods for solving linear, or first-degree, equations in **Chapter 2.** Solving higher degree polynomial equations requires other methods, one of which involves factoring.

OBJECTIVES

1. Learn and use the zero-factor property.
2. Solve applied problems that require the zero-factor property.

OBJECTIVE 1 Learn and use the zero-factor property. Solving equations by factoring depends on a special property of the number 0, called the **zero-factor property.**

Zero-Factor Property

If two numbers have a product of 0, then at least one of the numbers must be 0. That is,

$$\text{if } ab = 0, \quad \text{then either} \quad a = 0 \quad \text{or} \quad b = 0.$$

To prove the zero-factor property, we first assume $a \neq 0$. (If a does equal 0, then the property is proved already.) If $a \neq 0$, then $\frac{1}{a}$ exists, and each side of $ab = 0$ can be multiplied by $\frac{1}{a}$.

$$ab = 0$$

$$\frac{1}{a} \cdot ab = \frac{1}{a} \cdot 0$$

$$b = 0$$

Thus, if $a \neq 0$, then $b = 0$, and the property is proved.

CAUTION

If $ab = 0$, then $a = 0$ or $b = 0$. However, if $ab = 6$, for example, it is not necessarily true that $a = 6$ or $b = 6$; in fact, it is very likely that *neither* $a = 6$ *nor* $b = 6$. ***The zero-factor property works only for a product equal to 0.***

EXAMPLE 1 Using the Zero-Factor Property to Solve an Equation

Solve $(x + 6)(2x - 3) = 0$.

Here the product of $x + 6$ and $2x - 3$ is 0. By the zero-factor property, this can be true only if

$$x + 6 = 0 \quad \text{or} \quad 2x - 3 = 0.$$

Solve these two equations.

$$x = -6 \quad \text{or} \quad 2x = 3$$

$$x = \frac{3}{2}$$

The solutions are $x = -6$ or $x = \frac{3}{2}$. Check the two solutions -6 and $\frac{3}{2}$ by substitution in the *original* equation.

Continued on Next Page

Check If $x = -6$, then

$$(x + 6)(2x - 3) = 0$$
$$(-6 + 6)[2(-6) - 3] \stackrel{?}{=} 0$$
$$0(-15) = 0. \quad \text{True}$$

If $x = \frac{3}{2}$, then

$$(x + 6)(2x - 3) = 0$$
$$\left(\frac{3}{2} + 6\right)\left(2 \cdot \frac{3}{2} - 3\right) \stackrel{?}{=} 0$$
$$\frac{15}{2}(0) = 0. \quad \text{True}$$

Because both solutions check, the solution set is $\{-6, \frac{3}{2}\}$.

◀ Work Problem 1 at the Side.

1 Solve each equation.

(a) $(3x + 5)(x + 1) = 0$

(b) $(3x + 11)(5x - 2) = 0$

Since the product $(x + 6)(2x - 3)$ equals $2x^2 + 9x - 18$, the equation of Example 1 has a second-degree term and is an example of a *quadratic equation*. ***A quadratic equation has degree 2.***

Quadratic Equation

An equation that can be written in the form

$$ax^2 + bx + c = 0,$$

where a, b, and c are real numbers, with $a \neq 0$, is a **quadratic equation.** This form is called **standard form.**

Quadratic equations are discussed in more detail in **Chapter 10.** The steps for solving a quadratic equation by factoring are summarized here.

Solving a Quadratic Equation by Factoring

Step 1 **Write in standard form.** Rewrite the equation if necessary so that one side is 0.

Step 2 **Factor** the polynomial.

Step 3 **Use the zero-factor property.** Set each variable factor equal to 0.

Step 4 **Find the solution(s).** Solve each equation formed in Step 3.

Step 5 **Check** each solution in the *original* equation.

EXAMPLE 2 **Solving Quadratic Equations by Factoring**

Solve each equation.

(a) $2x^2 + 3x = 2$

Step 1 $2x^2 + 3x = 2$

$2x^2 + 3x - 2 = 0$ Standard form

Step 2 $(x + 2)(2x - 1) = 0$ Factor.

Step 3 $x + 2 = 0$ or $2x - 1 = 0$ Zero-factor property

Step 4 $x = -2$ or $2x = 1$ Solve each equation.

$x = \frac{1}{2}$

Continued on Next Page

Answers

1. (a) $\left\{-\frac{5}{3}, -1\right\}$ (b) $\left\{-\frac{11}{3}, \frac{2}{5}\right\}$

Step 5 Check each solution in the original equation.

Check If $x = -2$, then

$$2x^2 + 3x = 2$$
$$2(-2)^2 + 3(-2) \stackrel{?}{=} 2$$
$$2(4) - 6 \stackrel{?}{=} 2$$
$$8 - 6 \stackrel{?}{=} 2$$
$$2 = 2. \quad \text{True}$$

If $x = \frac{1}{2}$, then

$$2x^2 + 3x = 2$$
$$2\left(\frac{1}{2}\right)^2 + 3\left(\frac{1}{2}\right) \stackrel{?}{=} 2$$
$$2\left(\frac{1}{4}\right) + \frac{3}{2} \stackrel{?}{=} 2$$
$$\frac{1}{2} + \frac{3}{2} \stackrel{?}{=} 2$$
$$2 = 2. \quad \text{True}$$

Because both solutions check, the solution set is $\left\{-2, \frac{1}{2}\right\}$.

(b)

$$4x^2 = 4x - 1$$
$$4x^2 - 4x + 1 = 0 \quad \text{Standard form}$$
$$(2x - 1)^2 = 0 \quad \text{Factor.}$$
$$2x - 1 = 0 \quad \text{Zero-factor property}$$
$$2x = 1 \quad \text{Add 1.}$$
$$x = \frac{1}{2} \quad \text{Divide by 2.}$$

We could factor as $(2x - 1)(2x - 1)$. The same solution results.

There is only one solution, called a **double solution,** because the trinomial is a perfect square. The solution set is $\left\{\frac{1}{2}\right\}$.

Work Problem **2** at the Side. ▶

EXAMPLE 3 **Solving a Quadratic Equation with a Missing Constant Term**

Solve $4z^2 - 20z = 0$.

This quadratic equation has a missing constant term. Comparing it with the standard form $ax^2 + bx + c = 0$ shows that $c = 0$.

$$4z^2 - 20z = 0$$
$$4z(z - 5) = 0 \quad \text{Factor out the GCF.}$$
$$4z = 0 \quad \text{or} \quad z - 5 = 0 \quad \text{Zero-factor property}$$
$$z = 0 \quad \text{or} \quad z = 5 \quad \text{Solve each equation.}$$

Set each *variable* factor equal to 0.

Check If $z = 0$, then

$$4z^2 - 20z = 0$$
$$4(0)^2 - 20(0) \stackrel{?}{=} 0$$
$$0 - 0 = 0. \quad \text{True}$$

If $z = 5$, then

$$4z^2 - 20z = 0$$
$$4(5)^2 - 20(5) \stackrel{?}{=} 0$$
$$100 - 100 = 0. \quad \text{True}$$

The solution set is $\{0, 5\}$.

Work Problem **3** at the Side. ▶

CAUTION
Remember to include 0 as a solution of the equation in Example 3.

2 Solve each equation.

(a) $3x^2 - x = 4$

(b) $25x^2 = -20x - 4$

3 Solve $x^2 + 12x = 0$.

ANSWERS

2. (a) $\left\{-1, \frac{4}{3}\right\}$ **(b)** $\left\{-\frac{2}{5}\right\}$

3. $\{-12, 0\}$

4 Solve $5x^2 - 80 = 0$.

5 Solve

$(x + 6)(x - 2) = -8 + x$.

EXAMPLE 4 Solving a Quadratic Equation with a Missing Linear Term

Solve $3m^2 - 108 = 0$.

$$3m^2 - 108 = 0$$

$$3(m^2 - 36) = 0 \quad \text{Factor out 3}$$

The factor 3 does *not* lead to a solution.

$$3(m + 6)(m - 6) = 0 \quad \text{Factor } m^2 - 36.$$

$$m + 6 = 0 \quad \text{or} \quad m - 6 = 0 \quad \text{Zero-factor property}$$

$$m = -6 \quad \text{or} \quad m = 6$$

Check that the solution set is $\{-6, 6\}$.

◀ Work Problem **4** at the Side.

CAUTION

The factor 3 in Example 4 is not a *variable* factor, so it does not lead to a solution of the equation. In Example 3, however, the factor $4z$ is a variable factor and leads to the solution 0.

EXAMPLE 5 Solving an Equation That Requires Rewriting

Solve $(2q + 1)(q + 1) = 2(1 - q) + 6$.

$$(2q + 1)(q + 1) = 2(1 - q) + 6$$

$$2q^2 + 3q + 1 = 2 - 2q + 6 \quad \text{Multiply on each side.}$$

$$2q^2 + 3q + 1 = 8 - 2q \quad \text{Add on the right.}$$

$$2q^2 + 5q - 7 = 0 \quad \text{Standard form}$$

$$(2q + 7)(q - 1) = 0 \quad \text{Factor.}$$

$$2q + 7 = 0 \quad \text{or} \quad q - 1 = 0 \quad \text{Zero-factor property}$$

$$2q = -7 \quad \text{or} \quad q = 1 \quad \text{Solve each equation.}$$

$$q = -\frac{7}{2}$$

Check $\quad (2q + 1)(q + 1) = 2(1 - q) + 6$

$$\left[2\left(-\frac{7}{2}\right) + 1\right]\left(-\frac{7}{2} + 1\right) \stackrel{?}{=} 2\left[1 - \left(-\frac{7}{2}\right)\right] + 6 \quad \text{Let } q = -\tfrac{7}{2}.$$

$$(-7 + 1)\left(-\frac{5}{2}\right) \stackrel{?}{=} 2\left(\frac{9}{2}\right) + 6 \quad \text{Simplify; } 1 = \tfrac{2}{2}.$$

$$(-6)\left(-\frac{5}{2}\right) \stackrel{?}{=} 9 + 6$$

$$15 = 15 \quad \text{True}$$

Check that 1 is a solution. The solution set is $\left\{-\frac{7}{2}, 1\right\}$.

◀ Work Problem **5** at the Side.

The zero-factor property can be extended to solve certain polynomial equations of degree 3 or higher, as shown in the next example.

Answers

4. $\{-4, 4\}$
5. $\{-4, 1\}$

EXAMPLE 6 Solving an Equation of Degree 3

Solve $-x^3 + x^2 = -6x$.

Start by adding $6x$ to each side to get 0 on the right side.

$$-x^3 + x^2 + 6x = 0$$

$$x^3 - x^2 - 6x = 0 \quad \text{Multiply by } -1.$$

$$x(x^2 - x - 6) = 0 \quad \text{Factor out } x.$$

$$x(x + 2)(x - 3) = 0 \quad \text{Factor the trinomial.}$$

Use the zero-factor property, extended to include the three variable factors.

Remember to set x equal to 0.

$$x = 0 \quad \text{or} \quad x + 2 = 0 \quad \text{or} \quad x - 3 = 0$$

$$x = -2 \qquad\qquad x = 3$$

Check that the solution set is $\{-2, 0, 3\}$.

Work Problem **6** *at the Side.* ▶

6 Solve $3x^3 + x^2 = 4x$.

OBJECTIVE 2 Solve applied problems that require the zero-factor property. An application may lead to a quadratic equation. We continue to use the six-step problem-solving method introduced in **Section 2.3.**

EXAMPLE 7 Using a Quadratic Equation in an Application

A piece of sheet metal is in the shape of a parallelogram. The longer sides of the parallelogram are each 8 m longer than the distance between them. The area of the parallelogram is 48 m^2. Find the length of the longer sides and the distance between them.

Step 1 **Read** the problem again. There will be two answers.

Step 2 **Assign a variable.**

Let x = the distance between the longer sides;

$x + 8$ = the length of each longer side. (See Figure 1.)

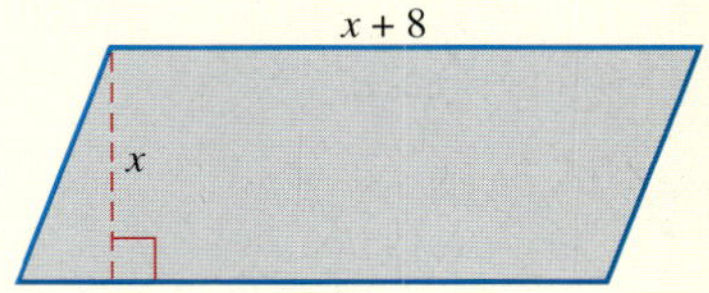

Figure 1

Step 3 **Write an equation.** The area of a parallelogram is given by $A = bh$, where b is the length of the longer side and h is the distance between the longer sides. Here $b = x + 8$ and $h = x$.

$$A = bh$$

$$48 = (x + 8)x \quad \text{Let } A = 48,\ b = x + 8,\ h = x.$$

Step 4 **Solve.**

$$48 = x^2 + 8x \quad \text{Distributive property}$$

$$0 = x^2 + 8x - 48 \quad \text{Standard form}$$

$$0 = (x + 12)(x - 4) \quad \text{Factor.}$$

$$x + 12 = 0 \quad \text{or} \quad x - 4 = 0 \quad \text{Zero-factor property}$$

$$x = -12 \quad \text{or} \quad x = 4$$

Step 5 **State the answer.** ***A distance cannot be negative, so reject −12 as a solution.*** The only possible solution is 4, so the distance between the longer sides is 4 m. The length of the longer sides is $4 + 8 = 12$ m.

Step 6 **Check.** The length of the longer sides is 8 m more than the distance between them, and the area is $4 \cdot 12 = 48$ m^2, so the answer checks.

Work Problem **7** *at the Side.* ▶

7 Solve the problem.

Carl is planning to build a rectangular deck along the back of his house. He wants the area of the deck to be 60 m^2, and the width to be 1 m less than half the length. What length and width should he use?

ANSWERS

6. $\left\{-\frac{4}{3}, 0, 1\right\}$

7. length: 12 m; width: 5 m

8 Solve the problem.
How long will it take the rocket in Example 8 to reach a height of 256 ft?

CAUTION
When an application leads to a quadratic equation, a solution of the equation may not satisfy the physical requirements of the application, as in Example 7. Reject such solutions.

A function defined by a quadratic polynomial is called a ***quadratic function.*** (See **Chapter 10.**) The next example uses such a function.

EXAMPLE 8 Using a Quadratic Function in an Application

Quadratic functions are used to describe the height a falling object or a projected object reaches in a specific time. For example, if a small rocket is launched vertically upward from ground level with an initial velocity of 128 ft per sec, then its height in feet after t seconds is a function defined by

$$h(t) = -16t^2 + 128t,$$

if air resistance is neglected. After how many seconds will the rocket be 220 ft above the ground?

We must let $h(t) = 220$ and solve for t.

$$\begin{aligned} \mathbf{220} &= -16t^2 + 128t && \text{Let } h(t) = 220. \\ 16t^2 - 128t + 220 &= 0 && \text{Standard form} \\ 4t^2 - 32t + 55 &= 0 && \text{Divide by 4.} \\ (2t - 5)(2t - 11) &= 0 && \text{Factor.} \\ 2t - 5 = 0 \quad \text{or} \quad 2t - 11 &= 0 && \text{Zero-factor property} \\ t = 2.5 \quad \text{or} \quad t &= 5.5 \end{aligned}$$

The rocket will reach a height of 220 ft twice: on its way up at 2.5 sec and again on its way down at 5.5 sec.

◀ *Work Problem* **8** *at the Side.*

ANSWER

8. 4 sec

7.5 Exercises

FOR EXTRA HELP

MyMathLab | Math XL PRACTICE | WATCH | DOWNLOAD | READ | REVIEW

1. Explain in your own words how the zero-factor property is used in solving a quadratic equation.

2. One of the following equations is *not* in proper form for using the zero-factor property. Which one is it? Tell why it is not in proper form.

   A. $(x + 2)(x - 6) = 0$

   B. $x(3x - 7) = 0$

   C. $3t(t + 8)(t - 9) = 0$

   D. $y(y - 3) + 6(y - 3) = 0$

Solve each equation using the zero-factor property. See Examples 1–5.

3. $(x + 10)(x - 5) = 0$

4. $(x + 7)(x + 3) = 0$

5. $(3k + 8)(2k - 5) = 0$

6. $(2q + 5)(3q - 4) = 0$

7. $m^2 - 3m - 10 = 0$

8. $x^2 + x - 12 = 0$

9. $z^2 + 9z + 18 = 0$

10. $x^2 - 18x + 80 = 0$

11. $2x^2 = 7x + 4$

12. $2x^2 = 3 - x$

13. $15k^2 - 7k = 4$

14. $12x^2 + 4x = 5$

15. $16x^2 + 24x = -9$

16. $49x^2 + 14x = -1$

17. $2a^2 - 8a = 0$

18. $4x^2 + 16x = 0$

19. $6m^2 - 36m = 0$

20. $3m^2 - 27m = 0$

21. $4p^2 - 16 = 0$

22. $9x^2 - 81 = 0$

23. $-3m^2 + 27 = 0$

24. $-2x^2 + 8 = 0$

25. $(x - 3)(x + 5) = -7$

26. $(x + 8)(x - 2) = -21$

27. $(2x + 1)(x - 3) = 6x + 3$

28. $(3x + 2)(x - 3) = 7x - 1$

29. $(5x + 1)(x + 3) = -2(5x + 1)$

30. $(3x + 1)(x - 3) = 2 + 3(x + 5)$

31. $(x + 3)(x - 6) = (2x + 2)(x - 6)$

32. $(2x + 1)(x + 5) = (x + 11)(x + 3)$

Solve each equation. See Example 6.

33. $2x^3 - 9x^2 - 5x = 0$

34. $6x^3 - 13x^2 - 5x = 0$

35. $x^3 - 2x^2 = 3x$

36. $z^3 - 6z^2 = -8z$

37. $9t^3 = 16t$

38. $25x^3 = 64x$

39. $2r^3 + 5r^2 - 2r - 5 = 0$

40. $2p^3 + p^2 - 98p - 49 = 0$

41. A student tried to solve the equation in Exercise 37 by first dividing each side by t, obtaining $9t^2 = 16$. She then solved the resulting equation by the zero-factor property to get the solution set $\{-\frac{4}{3}, \frac{4}{3}\}$. ***WHAT WENT WRONG?*** Give the correct solution set.

42. Without actually solving each equation, determine which one of the following has 0 in its solution set.

A. $4x^2 - 25 = 0$ **B.** $x^2 + 2x - 3 = 0$

C. $6x^2 + 9x + 1 = 0$ **D.** $x^3 + 4x^2 = 3x$

Solve each problem. See Examples 7 and 8.

43. A garden has an area of 320 ft^2. Its length is 4 ft more than its width. What are the dimensions of the garden?

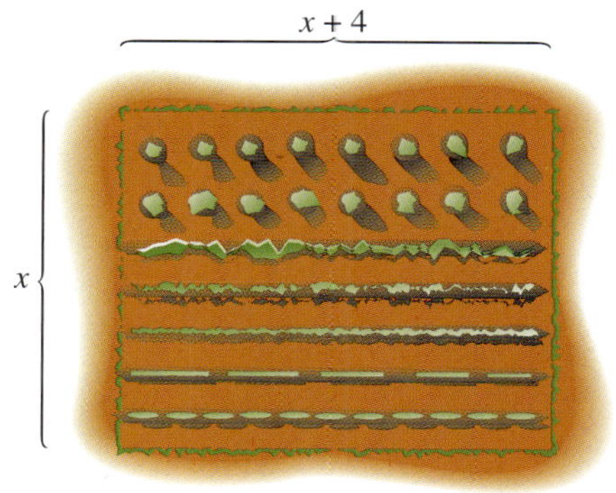

44. A square mirror has sides measuring 2 ft less than the sides of a square painting. If the difference between their areas is 32 ft^2, find the lengths of the sides of the mirror and the painting.

45. The base of a parallelogram is 7 ft more than the height. If the area of the parallelogram is 60 ft^2, what are the measures of the base and the height?

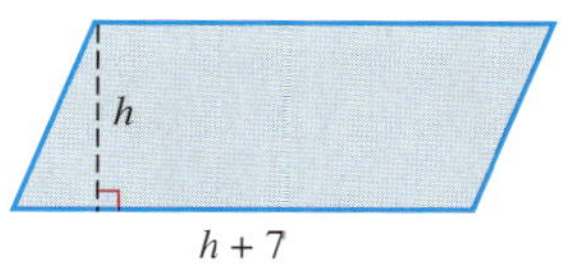

46. A sign has the shape of a triangle. The length of the base is 3 m less than the height. If the area is 44 m^2, what are the measures of the base and the height?

47. A farmer has 300 ft of fencing and wants to enclose a rectangular area of 5000 ft^2. What dimensions should she use?

48. A rectangular landfill has an area of 30,000 ft^2. Its length is 200 ft more than its width. What are the dimensions of the landfill?

49. Find two consecutive integers such that the sum of their squares is 61.

50. Find two consecutive integers such that their product is 72.

51. A box with no top is to be constructed from a piece of cardboard whose length measures 6 in. more than its width. The box is to be formed by cutting squares that measure 2 in. on each side from the four corners and then folding up the sides. If the volume of the box will be 110 in.^{3}, what are the dimensions of the piece of cardboard?

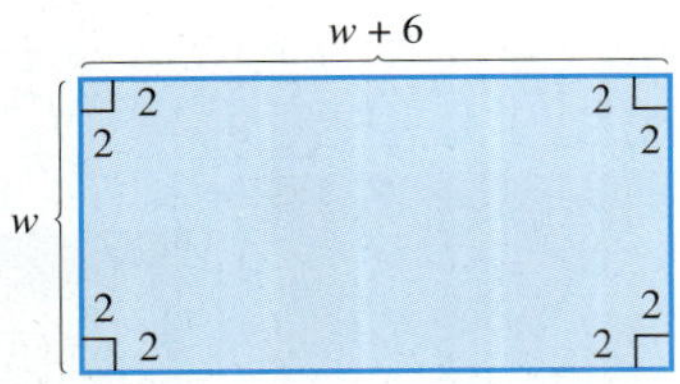

52. The surface area of the box with open top shown in the figure is 161 in.^{2}. Find the dimensions of the base. (*Hint:* The surface area is a function defined by $S(x) = x^2 + 16x$.)

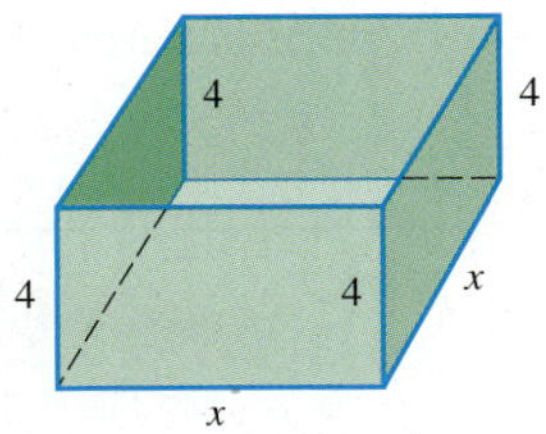

53. Refer to Example 8. After how many seconds will the rocket be 240 ft above the ground? 112 ft above the ground?

54. If an object is projected upward with an initial velocity of 64 ft per sec from a height of 80 ft, then its height in feet t seconds after it is projected is a function defined by

$$f(t) = -16t^2 + 64t + 80.$$

How long after it is projected will it hit the ground? (*Hint:* When it hits the ground, its height is 0 ft.)

55. If a rock is dropped from a building 576 ft high, then its distance in feet from the ground t seconds later is a function defined by

$$f(t) = -16t^2 + 576.$$

How long after it is dropped will it hit the ground?

56. If a baseball is dropped from a helicopter 625 ft above the ground, then its distance in feet from the ground t seconds later is a function defined by

$$f(t) = -16t^2 + 625.$$

How long after it is dropped will it hit the ground?

Chapter 7 Summary

Key Terms

7.1 greatest common factor The product of the largest common numerical factor and each variable factor of least degree common to every term in a polynomial is the greatest common factor of the terms of the polynomial.

7.2 prime polynomial A polynomial that cannot be factored with integer coefficients is a prime polynomial.

7.5 quadratic equation An equation that can be written in the form $ax^2 + bx + c = 0$, where a, b, and c are real numbers, with $a \neq 0$, is a quadratic equation. This form is called **standard form.**

Test Your Word Power

See how well you have learned the vocabulary in this chapter. Answers, with examples, follow the Quick Review.

1. **Factoring** is
   A. a method of multiplying polynomials
   B. the process of writing a polynomial as a product
   C. the answer in a multiplication problem
   D. a way to add the terms of a polynomial.

2. A **difference of squares** is a binomial
   A. that can be factored as the difference of two cubes
   B. that cannot be factored
   C. that is squared
   D. that can be factored as the product of the sum and difference of two terms.

3. A **perfect square trinomial** is a trinomial
   A. that can be factored as the square of a binomial
   B. that cannot be factored
   C. that is multiplied by a binomial
   D. where all terms are perfect squares.

4. A **quadratic equation** is a polynomial equation of
   A. degree one
   B. degree two
   C. degree three
   D. degree four.

5. The **zero-factor property** is used to
   A. factor a perfect square trinomial
   B. factor by grouping
   C. solve a polynomial equation of degree 2 or more
   D. solve a linear equation.

Quick Review

Concepts	Examples
7.1 Greatest Common Factors; Factoring by Grouping	
The Greatest Common Factor The product of the largest common numerical factor and each common variable raised to the least exponent that appears on that variable in any term is the greatest common factor of the terms of the polynomial.	Factor $4x^2y - 50xy^2$. $4x^2y - 50xy^2 = 2xy(2x - 25y)$ The greatest common factor is $2xy$.
Factoring by Grouping Group the terms so that each group has a common factor. Factor out the common factor in each group. If the groups now have a common factor, factor it out. If not, try a different grouping. ***Always check the factored form by multiplying.***	Factor by grouping. $5a - 5b - ax + bx$ $= (5a - 5b) + (-ax + bx)$ Group the terms. $= 5(a - b) - x(a - b)$ Factor out 5 and $-x$. $= (a - b)(5 - x)$ Factor out $a - b$.

Concepts	Examples
7.2 Factoring Trinomials	
To factor a trinomial, choose factors of the first term and factors of the last term. Then, place them in a pair of parentheses of this form: ( )( ). Try various combinations of the factors until the correct middle term of the trinomial is found.	Factor $15x^2 + 14x - 8$. The factors of 15 are 5 and 3, and 15 and 1. The factors of -8 are -4 and 2, 4 and -2, -1 and 8, and 1 and -8. Various combinations of these factors lead to $15x^2 + 14x - 8 = (5x - 2)(3x + 4)$. Check by multiplying.
7.3 Special Factoring	
Difference of Squares $x^2 - y^2 = (x + y)(x - y)$	$4m^2 - 25n^2 = (2m)^2 - (5n)^2 = (2m + 5n)(2m - 5n)$
Perfect Square Trinomials $x^2 + 2xy + y^2 = (x + y)^2$; $x^2 - 2xy + y^2 = (x - y)^2$	$9y^2 + 6y + 1 = (3y + 1)^2$; $16p^2 - 56p + 49 = (4p - 7)^2$
Difference of Cubes $x^3 - y^3 = (x - y)(x^2 + xy + y^2)$	$8 - 27a^3 = (2 - 3a)(4 + 6a + 9a^2)$
Sum of Cubes $x^3 + y^3 = (x + y)(x^2 - xy + y^2)$	$64z^3 + 1 = (4z + 1)(16z^2 - 4z + 1)$
7.4 A General Approach to Factoring	
See pages 409–410 for guidelines and examples.	
7.5 Solving Equations by Factoring	
Step 1 Rewrite the equation if necessary so that one side is 0.	Solve. $2x^2 + 5x = 3$; $2x^2 + 5x - 3 = 0$ Standard form
Step 2 Factor the polynomial.	$(x + 3)(2x - 1) = 0$ Factor.
Step 3 Set each factor equal to 0.	$x + 3 = 0$ or $2x - 1 = 0$ Zero-factor property
Step 4 Solve each equation from Step 3.	$x = -3$ or $2x = 1$, $x = \frac{1}{2}$ Solve each equation.
Step 5 Check each solution.	A check verifies that the solution set is $\{-3, \frac{1}{2}\}$.

ANSWERS TO TEST YOUR WORD POWER

1. B; *Example:* $x^2 - 5x - 14$ factors as $(x - 7)(x + 2)$.
2. D; *Example:* $b^2 - 49$ is the difference of the squares b^2 and 7^2. It can be factored as $(b + 7)(b - 7)$.
3. A; *Example:* $a^2 + 2a + 1$ is a perfect square trinomial; its factored form is $(a + 1)^2$.
4. B; *Examples:* $x^2 - 3x + 2 = 0, x^2 - 9 = 0, 2m^2 = 6m + 8$
5. C; *Example:* Use the zero-factor property to write $(x + 4)(x - 2) = 0$ as $x + 4 = 0$ or $x - 2 = 0$, and then solve each linear equation to find the solution set $\{-4, 2\}$.

Chapter 7 Review Exercises

[7.1] *Factor out the greatest common factor.*

1. $21y^2 + 35y$

2. $12q^2b + 8qb^2 - 20q^3b^2$

3. $(x + 3)(4x - 1) - (x + 3)(3x + 2)$

4. $(z + 1)(z - 4) + (z + 1)(2z + 3)$

Factor by grouping.

5. $4m + nq + mn + 4q$

6. $x^2 + 5y + 5x + xy$

7. $2m + 6 - am - 3a$

8. $2am - 2bm - ap + bp$

[7.2] *Factor completely.*

9. $3p^2 - p - 4$

10. $12r^2 - 5r - 3$

11. $10m^2 + 37m + 30$

12. $10k^2 - 11kh + 3h^2$

13. $9x^2 + 4xy - 2y^2$

14. $24x - 2x^2 - 2x^3$

15. $2k^4 - 5k^2 - 3$

16. $p^2(p + 2)^2 + p(p + 2)^2 - 6(p + 2)^2$

[7.3] *Factor completely.*

17. $16x^2 - 25$

18. $9t^2 - 49$

19. $x^2 + 14x + 49$

20. $9k^2 - 12k + 4$

21. $r^3 + 27$

22. $125x^3 - 1$

23. $m^6 - 1$

24. $x^8 - 1$

25. $x^2 + 6x + 9 - 25y^2$

[7.5] *Solve each equation.*

26. $(x + 1)(5x + 2) = 0$

27. $p^2 - 5p + 6 = 0$

28. $6z^2 = 5z + 50$

29. $6r^2 + 7r = 3$

30. $-4m^2 + 36 = 0$

31. $6x^2 + 9x = 0$

32. $(2x + 1)(x - 2) = -3$

33. $x^2 - 8x + 16 = 0$

34. $2x^3 - x^2 - 28x = 0$

Solve each problem.

35. A triangular wall brace creates the shape of a right triangle. One of the perpendicular sides is 1 ft longer than twice the other. The area enclosed by the triangle is 10.5 ft^2. Find the shorter of the perpendicular sides.

The area is 10.5 ft^2.

36. A rectangular parking lot has a length 20 ft more than its width. Its area is 2400 ft^2. What are the dimensions of the lot?

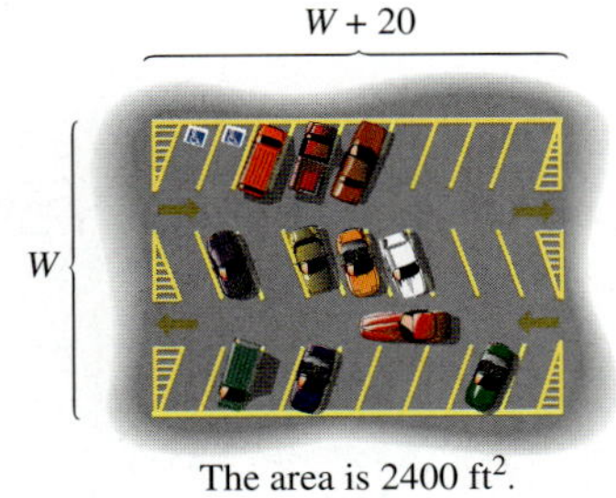

The area is 2400 ft^2.

A rock is projected directly upward from ground level. After t seconds, its height in feet is given by $f(t) = -16t^2 + 256t$ (if air resistance is neglected).

37. When will the rock return to the ground?

38. After how many seconds will it be 240 ft above the ground?

39. Why does the question in Exercise 38 have two answers?

Mixed Review Exercises

Factor completely.

40. $30a + am - am^2$

41. $8 - a^3$

42. $81k^2 - 16$

43. $9x^2 + 13xy - 3y^2$

44. $15y^3 + 20y^2$

45. $25z^2 - 30zm + 9m^2$

Solve.

46. $5x^2 - 17x - 12 = 0$

47. $x^3 - x = 0$

48. $3m^2 - 9m = 0$

49. When Europeans arrived in America, many native Americans of the Northeast lived in *longhouses* that sheltered several related families. The rectangular floor area of a typical Huron longhouse was about 2750 ft^2. The length was 85 ft greater than the width. What were the dimensions of the floor?

50. The length of a rectangular picture frame is 2 in. longer than its width. The area enclosed by the frame is 48 in.^{2}. What is the width?

Chapter 7 Test

Use the Chapter Test Prep Video CD to see fully worked-out solutions to any of the exercises you want to review

Factor.

1. $11z^2 - 44z$ 1. ______

2. $10x^2y^5 - 5x^2y^3 - 25x^5y^3$ 2. ______

3. $3x - by + bx + 3y$ 3. ______

4. $-2x^2 - x + 36$ 4. ______

5. $6x^2 + 11x - 35$ 5. ______

6. $4p^2 + 3pq - q^2$ 6. ______

7. $16a^2 + 40ab + 25b^2$ 7. ______

8. $x^2 + 2x + 1 - 4z^2$ 8. ______

9. $a^3 + 2a^2 - ab^2 - 2b^2$ 9. ______

10. $9k^2 - 121j^2$ 10. ______

11. $y^3 - 216$ 11. ______

12. $6k^4 - k^2 - 35$ 12. ______

13. ______________________

13. $27x^6 + 1$

14. ______________________

14. $-x^2 + x + 30$

15. ______________________

15. $(t^2 + 3)^2 + 4(t^2 + 3) - 5$

16. ______________________

16. Explain why $(x^2 + 2y)p + 3(x^2 + 2y)$ is not in factored form. Then factor the polynomial.

17. ______________________

17. Which one of the following is *not* a factored form of $-x^2 - x + 12$?

A. $(3 - x)(x + 4)$ **B.** $-(x - 3)(x + 4)$

C. $(-x + 3)(x + 4)$ **D.** $(x - 3)(-x + 4)$

Solve each equation.

18. ______________________

18. $3x^2 + 8x = -4$

19. ______________________

19. $3x^2 - 5x = 0$

20. ______________________

20. $5m(m - 1) = 2(1 - m)$

Solve each problem.

21. ______________________

21. The area of the rectangle shown is 40 in.^{2}. Find the length and the width of the rectangle.

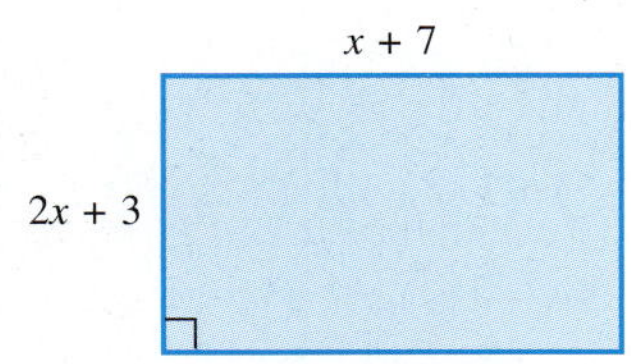

The area is 40 in.^{2}.

22. ______________________

22. A ball is projected upward from ground level. After t seconds, its height in feet is a function defined by $f(t) = -16t^2 + 96t$. After how many seconds will it reach a height of 128 ft?

Cumulative Review Exercises Chapters 1–7

Simplify each expression.

1. $-2(m - 3)$

2. $-(-4m + 3)$

3. $3x^2 - 4x + 4 + 9x - x^2$

Evaluate if $p = -4$, $q = -2$, and $r = 5$.

4. $-3(2q - 3p)$

5. $8r^2 + q^2$

6. $\dfrac{\sqrt{r}}{-p + 2q}$

7. $\dfrac{rp + 6r^2}{p^2 + q - 1}$

Solve.

8. $2z - 5 + 3z = 4 - (z + 2)$

9. $\dfrac{3a - 1}{5} + \dfrac{a - 2}{2} = -\dfrac{3}{10}$

10. $-\dfrac{4}{3}d \geq -5$

11. $3 - 2(m + 3) < 4m$

12. $2k + 4 < 10$ and $3k - 1 > 5$

13. $2k + 4 > 10$ or $3k - 1 < 5$

14. $|5x - 3| - 10 = 3$

15. $|x + 2| < 9$

16. $|2y - 5| \geq 9$

17. $V = lwh$ for h

18. Two planes leave the Dallas-Fort Worth airport at the same time. One travels east at 550 mph, and the other travels west at 500 mph. Assuming no wind, how long will it take for the planes to be 2100 mi apart?

	r	t	d
Eastbound plane	550	x	
Westbound plane	500	x	

19. Graph $4x + 2y = -8$.

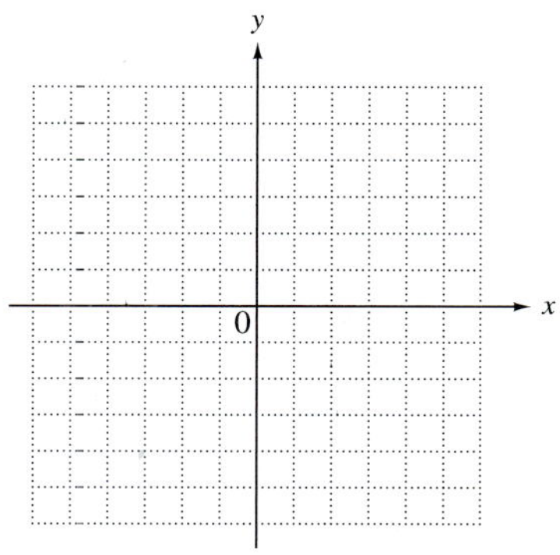

20. Find the slope of the line through the points $(-4, 8)$ and $(-2, 6)$.

21. What is the slope of the line shown here?

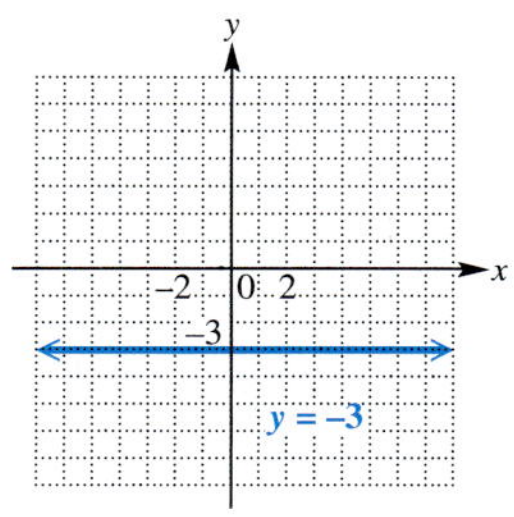

Use the function defined by $f(x) = 2x + 7$ to find the following.

22. $f(-4)$

23. The x-intercept of its graph

24. The y-intercept of its graph

Solve each system.

25. $3x - 2y = -7$
$2x + 3y = 17$

26. $2x + 3y - 6z = 5$
$8x - y + 3z = 7$
$3x + 4y - 3z = 7$

Perform the indicated operations. Assume variables represent nonzero real numbers.

27. $(3x^2y^{-1})^{-2}(2x^{-3}y)^{-1}$

28. $\dfrac{5m^{-2}y^3}{3m^{-3}y^{-1}}$

Perform the indicated operations.

29. $(3x^3 + 4x^2 - 7) - (2x^3 - 8x^2 + 3x)$

30. $(7x + 3y)^2$

31. $(2p + 3)(5p^2 - 4p - 8)$

Factor.

32. $16w^2 + 50wz - 21z^2$

33. $4x^2 - 4x + 1 - y^2$

34. $4y^2 - 36y + 81$

35. $100x^4 - 81$

36. $8p^3 + 27$

Solve.

37. $(p + 4)(2p + 3)(p - 1) = 0$

38. $9q^2 = 6q - 1$

39. A sign is to have the shape of a triangle with a height 3 ft greater than the length of the base. How long should the base be if the area is to be 14 ft^2?

40. A game board has the shape of a rectangle. The longer sides are each 2 in. longer than the distance between them. The area of the board is 288 in.^{2}. Find the length of the longer sides and the distance between them.

8

Rational Expressions and Functions

8.1 Rational Expressions and Functions; Multiplying and Dividing

8.2 Adding and Subtracting Rational Expressions

8.3 Complex Fractions

8.4 Equations with Rational Expressions and Graphs

Summary Exercises on Rational Expressions and Equations

8.5 Applications of Rational Expressions

8.6 Variation

Americans have been car crazy ever since the first automobiles hit the road early in the twentieth century. Nowhere is this more apparent than in Michigan, long the home of automobile manufacturing in the United States. To coincide with the 100th anniversary of the Model T, the 100th birthday of General Motors Corporation, and the first Autopalooza, a summer auto tourism festival expected to draw over one million visitors, Michigan officially designated 2008 as the Year of the Car. (*Source: The Gazette*, May 25, 2008.)

In Exercises 67 and 68 of Section 8.2, we use a *rational expression* to determine the cost of restoring a vintage automobile.

8.1 Rational Expressions and Functions; Multiplying and Dividing

OBJECTIVES

1. Define rational expressions.
2. Define rational functions and describe their domains.
3. Write rational expressions in lowest terms.
4. Multiply rational expressions.
5. Find reciprocals for rational expressions.
6. Divide rational expressions.

OBJECTIVE 1 Define rational expressions. In arithmetic, a rational number is the quotient of two integers, with the denominator not 0. In algebra, a **rational expression** or *algebraic fraction* is the quotient of two polynomials, again with the denominator not 0. For example,

$$\frac{x}{y}, \quad \frac{-a}{4}, \quad \frac{m+4}{m-2}, \quad \frac{8x^2-2x+5}{4x^2+5x}, \quad \text{and} \quad x^5\left(\text{or } \frac{x^5}{1}\right) \qquad \text{Rational expressions}$$

are rational expressions. Rational expressions are the elements of the set

$$\left\{\frac{P}{Q}\,\middle|\, \textbf{P and Q are polynomials, with } Q \neq 0\right\}.$$

OBJECTIVE 2 Define rational functions and describe their domains. A function that is defined by a rational expression is called a **rational function** and has the form

$$f(x) = \frac{P(x)}{Q(x)}, \quad \text{where } Q(x) \neq 0.$$

The domain of a rational function includes all real numbers except those that make $Q(x)$, that is, the denominator, equal to 0. For example, the domain of

$$f(x) = \frac{2}{\underbrace{x-5}_{\text{Cannot equal 0}}}$$

includes all real numbers except 5, because 5 would make the denominator equal to 0. Figure 1 shows a graph of the function defined by

$$f(x) = \frac{2}{x-5}.$$

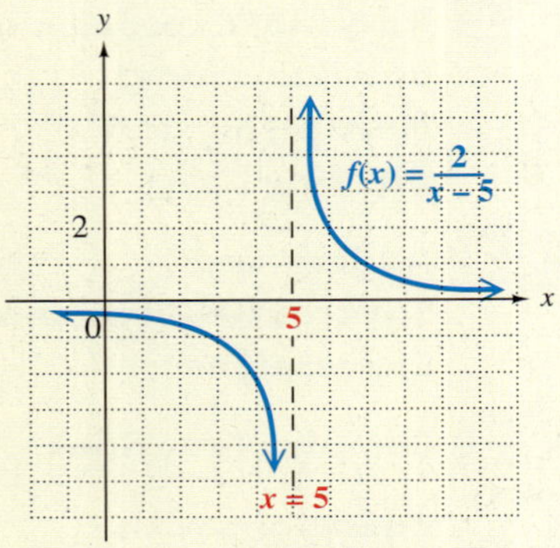

Figure 1

Notice that the graph does not exist when $x = 5$. It does not intersect the dashed vertical line whose equation is $x = 5$. This line is an **asymptote.** We will discuss graphs of rational functions in more detail in **Section 8.4.**

EXAMPLE 1 Finding Numbers That Are Not in the Domains of Rational Functions

Find all numbers that are not in the domain of each rational function. Then give the domain using set notation.

(a) $f(x) = \frac{3}{7x - 14}$

The only values that cannot be used are those that make the denominator 0. To find these values, set the denominator equal to 0 and solve the resulting equation.

$$7x - 14 = 0$$

$$7x = 14 \qquad \text{Add 14.}$$

$$x = 2 \qquad \text{Divide by 7.}$$

The number 2 cannot be used as a replacement for x. The domain of f includes all real numbers except 2, written using set notation as $\{x \mid x \neq 2\}$.

Continued on Next Page

(b) $g(x) = \dfrac{3 + x}{x^2 - 4x + 3}$ Values that make the denominator 0 must be excluded.

$$x^2 - 4x + 3 = 0 \quad \text{Set the denominator equal to 0.}$$
$$(x - 1)(x - 3) = 0 \quad \text{Factor.}$$
$$x - 1 = 0 \quad \text{or} \quad x - 3 = 0 \quad \text{Zero-factor property}$$
$$x = 1 \quad \text{or} \quad x = 3 \quad \text{Solve each equation.}$$

The domain of g includes all real numbers except 1 and 3, written $\{x \mid x \neq 1, 3\}$.

(c) $h(x) = \dfrac{8x + 2}{3}$

The denominator, 3, can never be 0, so the domain of h includes all real numbers, written $(-\infty, \infty)$.

(d) $f(x) = \dfrac{2}{x^2 + 4}$

Setting $x^2 + 4$ equal to 0 leads to $x^2 = -4$. There is no real number whose square is -4. Therefore, any real number can be used as a replacement for x. As in part (c), the domain of f consists of all real numbers $(-\infty, \infty)$.

Work Problem **1** at the Side. ▶

OBJECTIVE 3 Write rational expressions in lowest terms. In arithmetic, we write the fraction $\frac{15}{20}$ in lowest terms by dividing the numerator and denominator by 5 to get $\frac{3}{4}$. We write rational expressions in lowest terms in a similar way, using the **fundamental property of rational numbers.**

Fundamental Property of Rational Numbers

If $\frac{a}{b}$ is a rational number and if c is any nonzero real number, then

$$\frac{a}{b} = \frac{ac}{bc}.$$

In words, the numerator and denominator of a rational number may either be multiplied or divided by the same nonzero number without changing the value of the rational number.

Because $\frac{c}{c}$ is equivalent to 1, the fundamental property is based on the identity property of multiplication.

A rational expression is a quotient of two polynomials. Since the value of a polynomial is a real number for every value of the variable for which it is defined, any statement that applies to rational numbers will also apply to rational expressions.

We use the following steps to write rational expressions in lowest terms.

Writing a Rational Expression in Lowest Terms

Step 1 **Factor** both numerator and denominator to find their greatest common factor (GCF).

Step 2 **Apply the fundamental property.**

1 Find all numbers that are not in the domain of each rational function. Then give the domain using set notation.

(a) $f(x) = \dfrac{x + 4}{x - 6}$

(b) $f(x) = \dfrac{x + 6}{x^2 - x - 6}$

(c) $f(x) = \dfrac{3 + 2x}{5}$

(d) $f(x) = \dfrac{2}{x^2 + 1}$

Answers

1. **(a)** 6; $\{x \mid x \neq 6\}$ **(b)** $-2, 3$; $\{x \mid x \neq -2, 3\}$ **(c)** none; The domain consists of all real numbers $(-\infty, \infty)$. **(d)** none; The domain consists of all real numbers $(-\infty, \infty)$.

EXAMPLE 2 Writing Rational Expressions in Lowest Terms

Write each rational expression in lowest terms.

(a) $\frac{8k}{16} = \frac{k \cdot 8}{2 \cdot 8} = \frac{k}{2} \cdot 1 = \frac{k}{2}$ Factor; apply the fundamental property.

(b) $\frac{8 + k}{16}$ Be careful. The numerator cannot be factored.

This expression cannot be simplified further and is in lowest terms.

(c) $\frac{a^2 - a - 6}{a^2 + 5a + 6}$

$= \frac{(a - 3)(a + 2)}{(a + 3)(a + 2)}$ Factor the numerator and the denominator.

$= \frac{a - 3}{a + 3} \cdot 1$ Fundamental property

$= \frac{a - 3}{a + 3}$ Lowest terms

(d) $\frac{y^2 - 4}{2y + 4}$

$= \frac{(y + 2)(y - 2)}{2(y + 2)}$ Factor the difference of squares in the numerator; factor the denominator.

$= \frac{y - 2}{2}$ Lowest terms

(e) $\frac{x^3 - 27}{x - 3}$

$= \frac{(x - 3)(x^2 + 3x + 9)}{x - 3}$ Factor the difference of cubes.

$= x^2 + 3x + 9$ Lowest terms

(f) $\frac{pr + qr + ps + qs}{pr + qr - ps - qs}$

$= \frac{(pr + qr) + (ps + qs)}{(pr + qr) - (ps + qs)}$ Group the terms.

$= \frac{r(p + q) + s(p + q)}{r(p + q) - s(p + q)}$ Factor within the groups.

$= \frac{(p + q)(r + s)}{(p + q)(r - s)}$ Factor by grouping.

$= \frac{r + s}{r - s}$ Lowest terms

CAUTION

Be careful! ***When using the fundamental property of rational numbers, only common factors may be divided.*** For example,

$$\frac{y-2}{2} \neq y \quad \text{and} \quad \frac{y-2}{2} \neq y-1$$

because the 2 in $y - 2$ is not a *factor* of the numerator. ***Remember to factor before writing a fraction in lowest terms.***

Work Problem 2 at the Side. ▶

In the rational expression from Example 2(c),

$$\frac{a^2 - a - 6}{a^2 + 5a + 6}, \quad \text{or} \quad \frac{(a-3)(a+2)}{(a+3)(a+2)},$$

a can take any value except -3 or -2 since these values make the denominator 0. In the simplified rational expression

$$\frac{a-3}{a+3},$$

a cannot equal -3. Because of this,

$$\frac{a^2 - a - 6}{a^2 + 5a + 6} = \frac{a-3}{a+3}$$

for all values of a except -3 or -2. From now on such statements of equality will be made with the understanding that they apply only for those real numbers that make neither denominator equal 0. We will no longer state such restrictions.

EXAMPLE 3 Writing Rational Expressions in Lowest Terms

Write each rational expression in lowest terms.

(a) $\dfrac{m-3}{3-m}$

Here, the numerator and denominator are opposites. This expression can be written in lowest terms by writing the denominator as $-1(m-3)$.

$$\frac{m-3}{3-m} = \frac{m-3}{-1(m-3)} = \frac{1}{-1} = -1$$

The numerator could have been rewritten instead to get the same result.

(b) $\dfrac{r^2 - 16}{4 - r}$

$= \dfrac{(r+4)(r-4)}{4-r}$ Factor the difference of squares in the numerator.

$= \dfrac{(r+4)(r-4)}{-1(r-4)}$ Write $4 - r$ as $-1(r-4)$.

$= \dfrac{r+4}{-1}$ Fundamental property

$= -(r+4)$ or $-r-4$ Lowest terms

2 Write each rational expression in lowest terms.

(a) $\dfrac{y^2 + 2y - 3}{y^2 - 3y + 2}$

(b) $\dfrac{3y + 9}{y^2 - 9}$

(c) $\dfrac{y + 2}{y^2 + 4}$

(d) $\dfrac{1 + p^3}{1 + p}$

(e) $\dfrac{3x + 3y + rx + ry}{5x + 5y - rx - ry}$

ANSWERS

2. **(a)** $\dfrac{y+3}{y-2}$ **(b)** $\dfrac{3}{y-3}$
(c) already in lowest terms
(d) $1 - p + p^2$ **(e)** $\dfrac{3+r}{5-r}$

3 Write each rational expression in lowest terms.

(a) $\frac{y-2}{2-y}$

(b) $\frac{8-b}{8+b}$

(c) $\frac{p-2}{4-p^2}$

As shown in Example 3, the quotient $\frac{a}{-a}$ $(a \neq 0)$ can be simplified as

$$\frac{a}{-a} = \frac{a}{-1(a)} = \frac{1}{-1} = -1.$$

Quotient of Opposites

In general, if the numerator and the denominator of a rational expression are opposites, the expression equals -1.

Based on this result, the following are true:

$$\frac{q-7}{7-q} = -1 \quad \text{and} \quad \frac{-5a+2b}{5a-2b} = -1.$$

Numerator and denominator in each expression are opposites.

However, the following expression cannot be simplifed further.

$$\frac{r-2}{r+2}$$ Numerator and denominator are *not* opposites.

◀ Work Problem **3** at the Side.

OBJECTIVE 4 Multiply rational expressions. To multiply rational expressions, follow these steps. (In practice, we usually simplify before multiplying.)

Multiplying Rational Expressions

Step 1 **Factor** all numerators and denominators as completely as possible.

Step 2 **Apply the fundamental property.**

Step 3 **Multiply** remaining factors in the numerator and remaining factors in the denominator. Leave the denominator in factored form.

Step 4 **Check** to be sure the product is in lowest terms.

EXAMPLE 4 Multiplying Rational Expressions

Multiply.

(a) $\frac{5p-5}{p} \cdot \frac{3p^2}{10p-10}$

$= \frac{5(p-1)}{p} \cdot \frac{3p \cdot p}{2 \cdot 5(p-1)}$ Factor.

$= \frac{5(p-1)}{5(p-1)} \cdot \frac{p}{p} \cdot \frac{3p}{2}$ Commutative property

$= \frac{1}{1} \cdot \frac{1}{1} \cdot \frac{1}{1} \cdot \frac{3p}{2}$ Fundamental property

$= \frac{3p}{2}$ Lowest terms

Continued on Next Page

Answers

3. (a) -1 (b) already in lowest terms
(c) $\frac{-1}{2+p}$

(b) $\dfrac{k^2+2k-15}{k^2-4k+3} \cdot \dfrac{k^2-k}{k^2+k-20}$

$= \dfrac{(k+5)(k-3)}{(k-3)(k-1)} \cdot \dfrac{k(k-1)}{(k+5)(k-4)}$ Factor.

$= \dfrac{k}{k-4}$ Lowest terms

(c) $(p-4) \cdot \dfrac{3}{5p-20}$

$= \dfrac{p-4}{1} \cdot \dfrac{3}{5p-20}$ Write $p-4$ as $\frac{p-4}{1}$.

$= \dfrac{p-4}{1} \cdot \dfrac{3}{5(p-4)}$ Factor.

$= \dfrac{3}{5}$ Multiply; lowest terms

(d) $\dfrac{x^2+2x}{x+1} \cdot \dfrac{x^2-1}{x^3+x^2}$

$= \dfrac{x(x+2)}{x+1} \cdot \dfrac{(x+1)(x-1)}{x^2(x+1)}$ Factor.

$= \dfrac{(x+2)(x-1)}{x(x+1)}$ Multiply; lowest terms.

(e) $\dfrac{x-6}{x^2-12x+36} \cdot \dfrac{x^2-3x-18}{x^2+7x+12}$

$= \dfrac{x-6}{(x-6)^2} \cdot \dfrac{(x+3)(x-6)}{(x+3)(x+4)}$ Factor.

$= \dfrac{1}{x+4}$ Lowest terms

Remember to include 1 in the numerator when all other factors are eliminated.

Work Problem 4 at the Side. ▶

4 Multiply.

(a) $\dfrac{2r+4}{5r} \cdot \dfrac{3r}{5r+10}$

(b) $\dfrac{c^2+2c}{c^2-4} \cdot \dfrac{c^2-4c+4}{c^2-c}$

(c) $\dfrac{m^2-16}{m+2} \cdot \dfrac{1}{m+4}$

(d) $\dfrac{x-3}{x^2+2x-15} \cdot \dfrac{x^2-25}{x^2+3x-40}$

OBJECTIVE 5 **Find reciprocals for rational expressions.** The rational numbers $\frac{a}{b}$ and $\frac{c}{d}$ are reciprocals of each other if they have a product of 1. The **reciprocal** of a rational expression is defined in the same way: ***Two rational expressions are reciprocals of each other if they have a product of 1. Recall that 0 has no reciprocal.*** The table shows several rational expressions and their reciprocals.

Rational Expression	Reciprocal
$\dfrac{5}{k}$	$\dfrac{k}{5}$
$\dfrac{m^2-9m}{2}$	$\dfrac{2}{m^2-9m}$
$\dfrac{0}{4}$	undefined

ANSWERS

4. (a) $\dfrac{6}{25}$ (b) $\dfrac{c-2}{c-1}$ (c) $\dfrac{m-4}{m+2}$ (d) $\dfrac{1}{x+8}$

The examples in the table on the previous page suggest the following.

Finding the Reciprocal

To find the reciprocal of a nonzero rational expression, interchange the numerator and denominator of the expression.

◀ Work Problem 5 at the Side.

OBJECTIVE 6 Divide rational expressions. Dividing rational expressions is like dividing rational numbers.

Dividing Rational Expressions

To divide two rational expressions, multiply the first (the dividend) by the reciprocal of the second (the divisor).

EXAMPLE 5 Dividing Rational Expressions

Divide.

(a) $\frac{2z}{9} \div \frac{5z^2}{18}$

$= \frac{2z}{9} \cdot \frac{18}{5z^2}$ Multiply by the reciprocal of the divisor.

$= \frac{2z}{9} \cdot \frac{2 \cdot 9}{5z^2}$ Factor.

$= \frac{4}{5z}$ Multiply; lowest terms

(b) $\frac{8k - 16}{3k} \div \frac{3k - 6}{4k^2}$

$= \frac{8k - 16}{3k} \cdot \frac{4k^2}{3k - 6}$ Multiply by the reciprocal.

$= \frac{8(k - 2)}{3k} \cdot \frac{4k^2}{3(k - 2)}$ Factor.

$= \frac{32k}{9}$ Multiply; lowest terms

(c) $\frac{5m^2 + 17m - 12}{3m^2 + 7m - 20} \div \frac{5m^2 + 2m - 3}{15m^2 - 34m + 15}$

$= \frac{5m^2 + 17m - 12}{3m^2 + 7m - 20} \cdot \frac{15m^2 - 34m + 15}{5m^2 + 2m - 3}$ Definition of division

$= \frac{(5m - 3)(m + 4)}{(m + 4)(3m - 5)} \cdot \frac{(3m - 5)(5m - 3)}{(5m - 3)(m + 1)}$ Factor.

$= \frac{5m - 3}{m + 1}$ Lowest terms

◀ Work Problem 6 at the Side.

5 Find each reciprocal.

(a) $\frac{-3}{r}$

(b) $\frac{7}{y + 8}$

(c) $\frac{a^2 + 7a}{2a - 1}$

(d) $\frac{0}{-5}$

6 Divide.

(a) $\frac{16k^2}{5} \div \frac{3k}{10}$

(b) $\frac{5p + 2}{6} \div \frac{15p + 6}{5}$

(c) $\frac{y^2 - 2y - 3}{y^2 + 4y + 4} \div \frac{y^2 - 1}{y^2 + y - 2}$

Answers

5. **(a)** $\frac{r}{-3}$ **(b)** $\frac{y + 8}{7}$ **(c)** $\frac{2a - 1}{a^2 + 7a}$
(d) There is no reciprocal.

6. **(a)** $\frac{32k}{3}$ **(b)** $\frac{5}{18}$ **(c)** $\frac{y - 3}{y + 2}$

8.1 Exercises

FOR EXTRA HELP MyMathLab Math XL PRACTICE WATCH DOWNLOAD READ REVIEW

Rational expressions can often be written in lowest terms in seemingly different ways. For example,

$$\frac{y-3}{-5} \quad \text{and} \quad \frac{-y+3}{5}$$

look different, but we get the second expression by multiplying the first by -1 in both the numerator and denominator. To practice recognizing equivalent rational expressions, match the expressions in Exercises 1–6 with their equivalents in Choices A–F.

1. $\frac{x-3}{x+4}$ **2.** $\frac{x+3}{x-4}$ **3.** $\frac{x-3}{x-4}$ **4.** $\frac{x-3}{x+4}$ **5.** $\frac{3-x}{x+4}$ **6.** $\frac{x+3}{4-x}$

A. $\frac{-x-3}{4-x}$ **B.** $\frac{-x-3}{-x-4}$ **C.** $\frac{3-x}{-x-4}$ **D.** $\frac{-x+3}{-x+4}$ **E.** $\frac{x-3}{-x-4}$ **F.** $\frac{-x-3}{x-4}$

7. In Example 1(a), we showed that the domain of the rational function defined by $f(x) = \frac{3}{7x-14}$ does not include 2. Explain in your own words why this is so. In general, how do we find the value or values excluded from the domain of a rational function?

8. The domain of the rational function defined by $g(x) = \frac{x+1}{x^2+3}$ includes all real numbers. Explain.

Find all numbers that are not in the domain of each function. Then give the domain using set notation. See Example 1.

9. $f(x) = \frac{x}{x-7}$ **10.** $f(x) = \frac{x}{x+3}$ **11.** $f(x) = \frac{6x-5}{7x+1}$ **12.** $f(x) = \frac{8x-3}{2x+7}$

13. $f(x) = \frac{12x+3}{x}$ **14.** $f(x) = \frac{9x+8}{x}$ **15.** $f(x) = \frac{3x+1}{2x^2+x-6}$ **16.** $f(x) = \frac{2x+4}{3x^2+11x-42}$

17. $f(x) = \frac{x+2}{14}$ **18.** $f(x) = \frac{x-9}{26}$ **19.** $f(x) = \frac{2x^2-3x+4}{3x^2+8}$ **20.** $f(x) = \frac{9x^2-8x+3}{4x^2+1}$

21. **(a)** Identify the two *terms* in the numerator and the two *terms* in the denominator of the rational expression $\frac{x^2 + 4x}{x + 4}$.

(b) Describe the steps you would use to write this rational expression in lowest terms. (*Hint:* It simplifies to x.)

22. Only one of the following rational expressions can be simplified. Which one is it?

A. $\frac{x^2 + 2}{x^2}$ **B.** $\frac{x^2 + 2}{2}$

C. $\frac{x^2 + y^2}{y^2}$ **D.** $\frac{x^2 - 5x}{x}$

23. Only one of the following rational expressions is *not* equivalent to $\frac{x - 3}{4 - x}$. Which one is it?

A. $\frac{3 - x}{x - 4}$ **B.** $\frac{x + 3}{4 + x}$

C. $-\frac{3 - x}{4 - x}$ **D.** $-\frac{x - 3}{x - 4}$

24. Which two of the following rational expressions equal -1?

A. $\frac{2x + 3}{2x - 3}$ **B.** $\frac{2x - 3}{3 - 2x}$

C. $\frac{2x + 3}{3 + 2x}$ **D.** $\frac{2x + 3}{-2x - 3}$

Write each rational expression in lowest terms. See Example 2.

25. $\frac{x^2(x + 1)}{x(x + 1)}$

26. $\frac{y^3(y - 4)}{y^2(y - 4)}$

27. $\frac{(x + 4)(x - 3)}{(x + 5)(x + 4)}$

28. $\frac{(2x + 7)(x - 1)}{(2x + 3)(2x + 7)}$

29. $\frac{4x(x + 3)}{8x^2(x - 3)}$

30. $\frac{5y^2(y + 8)}{15y(y - 8)}$

31. $\frac{3x + 7}{3}$

32. $\frac{4x - 9}{4}$

33. $\frac{6m + 18}{7m + 21}$

34. $\frac{5r - 20}{3r - 12}$

35. $\frac{3z^2 + z}{18z + 6}$

36. $\frac{2x^2 - 5x}{16x - 40}$

37. $\frac{2t + 6}{t^2 - 9}$

38. $\frac{5s - 25}{s^2 - 25}$

39. $\frac{x^2 + 2x - 15}{x^2 + 6x + 5}$

40. $\dfrac{y^2 - 5y - 14}{y^2 + y - 2}$

41. $\dfrac{8x^2 - 10x - 3}{8x^2 - 6x - 9}$

42. $\dfrac{12x^2 - 4x - 5}{8x^2 - 6x - 5}$

43. $\dfrac{a^3 + b^3}{a + b}$

44. $\dfrac{r^3 - s^3}{r - s}$

45. $\dfrac{2c^2 + 2cd - 60d^2}{2c^2 - 12cd + 10d^2}$

46. $\dfrac{3s^2 - 9st - 54t^2}{3s^2 - 6st - 72t^2}$

47. $\dfrac{ac - ad + bc - bd}{ac - ad - bc + bd}$

48. $\dfrac{2xy + 2xw + y + w}{2xy + y - 2xw - w}$

Write each rational expression in lowest terms. See Example 3.

49. $\dfrac{7 - b}{b - 7}$

50. $\dfrac{r - 13}{13 - r}$

51. $\dfrac{x^2 - y^2}{y - x}$

52. $\dfrac{m^2 - n^2}{n - m}$

53. $\dfrac{(a - 3)(x + y)}{(3 - a)(x - y)}$

54. $\dfrac{(8 - p)(x + 2)}{(p - 8)(x - 2)}$

55. $\dfrac{5k - 10}{20 - 10k}$

56. $\dfrac{7x - 21}{63 - 21x}$

57. $\dfrac{a^2 - b^2}{a^2 + b^2}$

58. $\dfrac{p^2 + q^2}{p^2 - q^2}$

Multiply or divide as indicated. See Examples 4 and 5.

59. $\dfrac{(x + 2)(x + 1)}{(x + 3)(x - 2)} \cdot \dfrac{(x + 3)(x + 4)}{(x + 2)(x + 1)}$

60. $\dfrac{(x + 3)(x - 4)}{(x - 4)(x + 2)} \cdot \dfrac{(x + 5)(x - 6)}{(x + 3)(x - 6)}$

61. $\dfrac{(2x + 3)(x - 4)}{(x + 8)(x - 4)} \div \dfrac{(x - 4)(x + 2)}{(x - 4)(x + 8)}$

62. $\dfrac{(6x + 5)(x - 3)}{(x + 9)(x - 1)} \div \dfrac{(x - 3)(2x + 7)}{(x - 1)(x + 9)}$

63. $\dfrac{7t + 7}{-6} \div \dfrac{4t + 4}{15}$

64. $\dfrac{8z - 16}{-20} \div \dfrac{3z - 6}{40}$

65. $\dfrac{4x}{8x + 4} \cdot \dfrac{14x + 7}{6}$

66. $\dfrac{12x - 20}{5x} \cdot \dfrac{6}{9x - 15}$

67. $\dfrac{p^2 - 25}{4p} \cdot \dfrac{2}{5 - p}$

68. $\dfrac{a^2 - 1}{4a} \cdot \dfrac{2}{1 - a}$

69. $\dfrac{m^2 - 49}{m + 1} \div \dfrac{7 - m}{m}$

70. $\dfrac{k^2 - 4}{3k^2} \div \dfrac{2 - k}{11k}$

71. $\dfrac{12x - 10y}{3x + 2y} \cdot \dfrac{6x + 4y}{10y - 12x}$

72. $\dfrac{9s - 12t}{2s + 2t} \cdot \dfrac{3s + 3t}{4t - 3s}$

73. $\dfrac{x^2 - 25}{x^2 + x - 20} \cdot \dfrac{x^2 + 7x + 12}{x^2 - 2x - 15}$

74. $\dfrac{t^2 - 49}{t^2 + 4t - 21} \cdot \dfrac{t^2 + 8t + 15}{t^2 - 2t - 35}$

75. $\dfrac{6x^2 + 5xy - 6y^2}{12x^2 - 11xy + 2y^2} \div \dfrac{4x^2 - 12xy + 9y^2}{8x^2 - 14xy + 3y^2}$

76. $\dfrac{8a^2 - 6ab - 9b^2}{6a^2 - 5ab - 6b^2} \div \dfrac{4a^2 + 11ab + 6b^2}{9a^2 + 12ab + 4b^2}$

77. $\dfrac{3k^2 + 17kp + 10p^2}{6k^2 + 13kp - 5p^2} \div \dfrac{6k^2 + kp - 2p^2}{6k^2 - 5kp + p^2}$

78. $\dfrac{16c^2 + 24cd + 9d^2}{16c^2 - 16cd + 3d^2} \div \dfrac{16c^2 - 9d^2}{16c^2 - 24cd + 9d^2}$

79. $\left(\dfrac{6k^2 - 13k - 5}{k^2 + 7k} \div \dfrac{2k - 5}{k^3 + 6k^2 - 7k}\right) \cdot \dfrac{k^2 - 5k + 6}{3k^2 - 8k - 3}$

80. $\left(\dfrac{2x^3 + 3x^2 - 2x}{3x - 15} \div \dfrac{2x^3 - x^2}{x^2 - 3x - 10}\right) \cdot \dfrac{5x^2 - 10x}{3x^2 + 12x + 12}$

Study Skills

MAKING A MIND MAP

OBJECTIVES

1. **Create mind maps for appropriate concepts.**
2. **Visually show how concepts relate to each other using arrows or lines.**

Mind mapping is a visual way to show information that you have learned. It is an excellent way to review. Mapping is flexible and can be personalized, which is helpful for your memory. You remember information better if it is pleasing to look at, colorful, and shows connections between ideas. Take advantage of that by creating maps that are

- easy to read,
- use color in a systematic way, and
- clearly show you how different concepts are related (using arrows or dotted lines, for example).

Directions for Making a Mind Map

Here are some general directions for making a mind map. After you read them, work on completing the map that has been started for you on the next page. It is from **Section 8.1.**

- To begin a mind map, write the concept in the center of a piece of paper and either circle it or draw a box around it.
- Make a line out from the center concept, and draw a box large enough to write the definition of the concept.
- Think of the other aspects (subpoints) of the concept that you have learned, such as procedures to follow or formulas. Make a separate line and box connecting each subpoint to the center.
- From each of the new boxes, add the information you've learned. You can continue making new lines and boxes or circles, or you can list items below the new information.
- Use color to highlight the major points. For example, everything related to one subpoint might be the same color. That way you can easily see related ideas.
- Use arrows, underlining, or small drawings to help you remember.

Why Is Mapping Brain Friendly?

Remember that you learn better when you are actively thinking about and working with information. Making a map requires you to think hard about *how to place the information, how to show connections* between parts of the map, and *how color will be useful*. It also takes a lot of thinking to fill in all related details and *show how those details connect* to the larger concept. It is time well spent.

Try This Mind Map Using Sections 8.1 and 8.2

On a separate sheet of paper, make a map that summarizes *Multiplying and Dividing Rational Expressions* (**Section 8.1**). Follow the directions and use the starter map below.

- Notice that *dividing rational expressions requires two steps first*, then you follow the same steps as you do when multiplying. Notice the placement of those two steps and the arrow leading you to the next steps.
- Notice under "multiplying" that the written steps are on the left, and the example is on the right.
- **Your job is to complete the map by writing an example for dividing rational expressions. Then make another map for adding and subtracting rational expressions when you get to Section 8.2.**

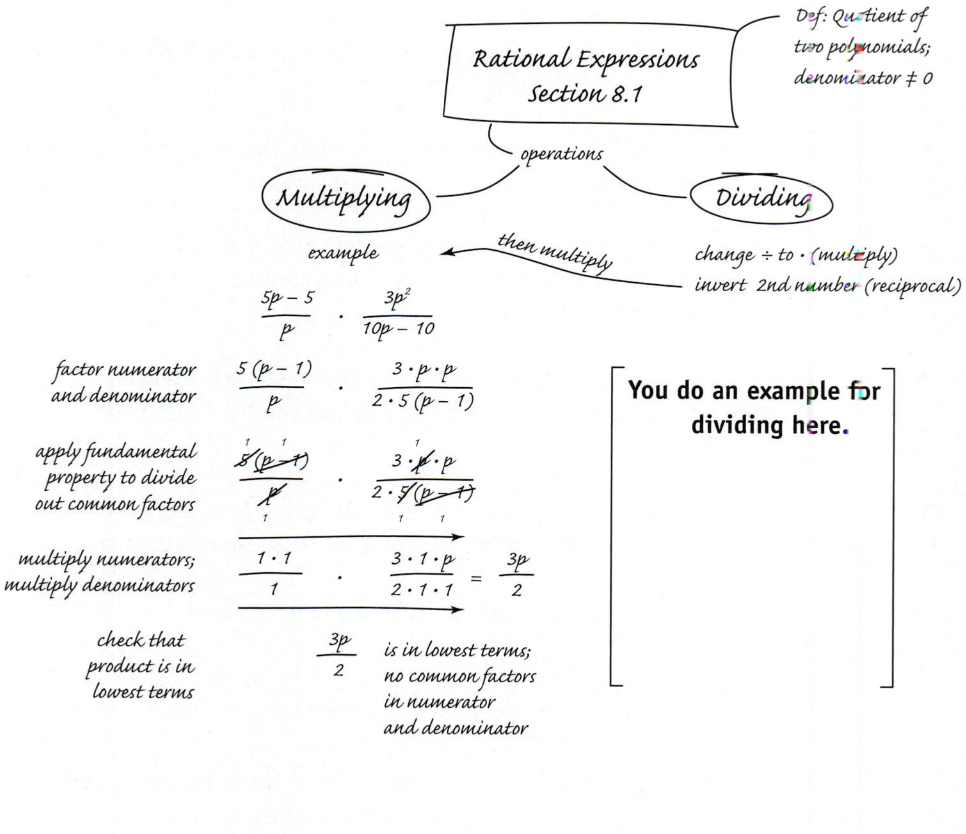

8.2 Adding and Subtracting Rational Expressions

OBJECTIVE 1 Add and subtract rational expressions with the same denominator. We do so as we would with rational numbers.

Adding or Subtracting Rational Expressions

Step 1 **If the denominators are the same,** add or subtract the numerators. Place the result over the common denominator.

If the denominators are different, first find the least common denominator. Write all rational expressions with this LCD, and then add or subtract the numerators. Place the result over the common denominator.

Step 2 **Simplify.** Write all answers in lowest terms.

EXAMPLE 1 Adding and Subtracting Rational Expressions with the Same Denominator

Add or subtract as indicated.

(a) $\frac{3y}{5} + \frac{x}{5} = \frac{3y + x}{5}$ ← Add the numerators. ← Keep the common denominator.

(b) $\frac{7}{2r^2} - \frac{11}{2r^2}$

$= \frac{7 - 11}{2r^2}$ Subtract the numerators; keep the common denominator.

$= \frac{-4}{2r^2}$

$= -\frac{2}{r^2}$ Lowest terms

(c) $\frac{m}{m^2 - p^2} + \frac{p}{m^2 - p^2}$

$= \frac{m + p}{m^2 - p^2}$ Add the numerators; keep the common denominator.

$= \frac{m + p}{(m + p)(m - p)}$ Factor.

$= \frac{1}{m - p}$ Lowest terms (Remember to write 1 in the numerator.)

(d) $\frac{4}{x^2 + 2x - 8} + \frac{x}{x^2 + 2x - 8}$

$= \frac{4 + x}{x^2 + 2x - 8}$ Add.

$= \frac{4 + x}{(x - 2)(x + 4)}$ Factor.

$= \frac{1}{x - 2}$ Lowest terms

Work Problem 1 at the Side. ▶

OBJECTIVES

1. Add and subtract rational expressions with the same denominator.
2. Find a least common denominator.
3. Add and subtract rational expressions with different denominators.

1 Add or subtract.

(a) $\frac{3m}{8} + \frac{5n}{8}$

(b) $\frac{7}{3a} + \frac{10}{3a}$

(c) $\frac{2}{y^2} - \frac{5}{y^2}$

(d) $\frac{a}{a + b} + \frac{b}{a + b}$

(e) $\frac{2y - 1}{y^2 + y - 2} - \frac{y}{y^2 + y - 2}$

ANSWERS

1. (a) $\frac{3m + 5n}{8}$ (b) $\frac{17}{3a}$ (c) $-\frac{3}{y^2}$ (d) 1 (e) $\frac{1}{y + 2}$

OBJECTIVE 2 Find a least common denominator. We add or subtract rational expressions with different denominators by first writing them with a common denominator, usually the **least common denominator (LCD).**

Finding the Least Common Denominator

Step 1 **Factor** each denominator.

Step 2 **Find the least common denominator.** The LCD is the product of all different factors from each denominator, with each factor raised to the *greatest* power that occurs in any denominator.

EXAMPLE 2 Finding Least Common Denominators

Assume that the given expressions are denominators of fractions. Find the LCD for each group.

(a) $5xy^2, \quad 2x^3y$

Each denominator is already factored.

$$5xy^2 = 5 \cdot x \cdot y^2$$
$$2x^3y = 2 \cdot x^3 \cdot y$$

Greatest exponent on x is 3.

$$\text{LCD} = 5 \cdot 2 \cdot x^3 \cdot y^2 \leftarrow \text{Greatest exponent on } y \text{ is 2.}$$
$$= 10x^3y^2$$

(b) $k - 3, \quad k$

Each denominator is already factored. The LCD, an expression divisible by *both* $k - 3$ and k, is

$$k(k - 3).$$

Don't forget the factor k.

It is usually best to leave a least common denominator in factored form.

(c) $y^2 - 2y - 8, \quad y^2 + 3y + 2$

Factor the denominators.

$$\left.\begin{aligned} y^2 - 2y - 8 &= (y - 4)(y + 2) \\ y^2 + 3y + 2 &= (y + 2)(y + 1) \end{aligned}\right\} \text{Factor.}$$

The LCD, divisible by both polynomials, is $(y - 4)(y + 2)(y + 1)$.

(d) $8z - 24, \quad 5z^2 - 15z$

$$\left.\begin{aligned} 8z - 24 &= 8(z - 3) \\ 5z^2 - 15z &= 5z(z - 3) \end{aligned}\right\} \text{Factor.}$$

The LCD is $8 \cdot 5z \cdot (z - 3) = 40z(z - 3)$.

(e) $m^2 + 5m + 6, \quad m^2 + 4m + 4, \quad 2m + 6$

$$\left.\begin{aligned} m^2 + 5m + 6 &= (m + 3)(m + 2) \\ m^2 + 4m + 4 &= (m + 2)^2 \\ 2m + 6 &= 2(m + 3) \end{aligned}\right\} \text{Factor.}$$

The LCD is $2(m + 3)(m + 2)^2$.

◀ *Work Problem* **2** *at the Side.*

2 Find the LCD for each group of denominators.

(a) $5k^3s, \quad 10ks^4$

(b) $3 - x, \quad 9 - x^2$

(c) $z, \quad z + 6$

(d) $2y^2 - 3y - 2,$
$2y^2 + 3y + 1$

(e) $x^2 - 2x + 1,$
$x^2 - 4x + 3,$
$4x - 4$

ANSWERS

2. **(a)** $10k^3s^4$ **(b)** $(3 + x)(3 - x)$
(c) $z(z + 6)$ **(d)** $(y - 2)(2y + 1)(y + 1)$
(e) $4(x - 3)(x - 1)^2$

OBJECTIVE 3 Add and subtract rational expressions with different denominators. Before adding or subtracting two rational expressions, we write each expression with the least common denominator by multiplying its numerator and denominator by the factors needed to get the LCD. This procedure is valid because we are multiplying each rational expression by a form of 1, the identity element for multiplication.

Consider the sum $\frac{7}{15} + \frac{5}{12}$. The LCD for 15 and 12 is 60. Multiply $\frac{7}{15}$ by $\frac{4}{4}$ (a form of 1) and multiply $\frac{5}{12}$ by $\frac{5}{5}$ so that each fraction has denominator 60. Then add the numerators.

$$\frac{7}{15} + \frac{5}{12}$$

$$= \frac{7 \cdot 4}{15 \cdot 4} + \frac{5 \cdot 5}{12 \cdot 5} \quad \text{Fundamental property}$$

$$= \frac{28}{60} + \frac{25}{60}$$

$$= \frac{28 + 25}{60} \quad \text{Add the numerators; keep the common denominator.}$$

$$= \frac{53}{60}$$

EXAMPLE 3 Adding and Subtracting Rational Expressions with Different Denominators

Add or subtract as indicated.

(a) $\frac{5}{2p} + \frac{3}{8p}$ The LCD for $2p$ and $8p$ is $8p$.

$$= \frac{5 \cdot 4}{2p \cdot 4} + \frac{3}{8p} \quad \text{Fundamental property}$$

$$= \frac{20}{8p} + \frac{3}{8p}$$

$$= \frac{20 + 3}{8p} \quad \text{Add the numerators; keep the common denominator.}$$

$$= \frac{23}{8p}$$

(b) $\frac{6}{r} - \frac{5}{r - 3}$ The LCD is $r(r - 3)$.

$$= \frac{6(r-3)}{r(r-3)} - \frac{r \cdot 5}{r(r-3)} \quad \text{Fundamental property}$$

$$= \frac{6r - 18}{r(r-3)} - \frac{5r}{r(r-3)} \quad \text{Distributive and commutative properties}$$

$$= \frac{6r - 18 - 5r}{r(r-3)} \quad \text{Subtract the numerators.}$$

$$= \frac{r - 18}{r(r-3)} \quad \text{Combine terms in the numerator.}$$

Work Problem 3 at the Side. ▶

3 Add or subtract.

(a) $\frac{6}{7} + \frac{1}{5}$

(b) $\frac{8}{3k} - \frac{2}{9k}$

(c) $\frac{2}{y} - \frac{1}{y + 4}$

ANSWERS

3. (a) $\frac{37}{35}$ **(b)** $\frac{22}{9k}$ **(c)** $\frac{y + 8}{y(y + 4)}$

4 Subtract.

(a) $\dfrac{5x+7}{2x+7} - \dfrac{-x-14}{2x+7}$

(b) $\dfrac{2}{r-2} - \dfrac{r}{r-1}$

CAUTION

Sign errors can easily occur when a rational expression with two or more terms in the numerator is being subtracted. In this case, ***the subtraction sign must be distributed to every term in the numerator of the fraction that follows it.*** Study Example 4 carefully to see how this is done.

EXAMPLE 4 Using the Distributive Property When Subtracting Rational Expressions

Subtract.

(a) $\dfrac{7x}{3x+1} - \dfrac{x-2}{3x+1}$

The denominators are the same for both rational expressions. The subtraction sign must be applied to *both* terms in the numerator of the second rational expression. Notice the careful use of the distributive property here.

$$\frac{7x}{3x+1} - \frac{x-2}{3x+1}$$

Use parentheses to avoid errors.

$$= \frac{7x - (x-2)}{3x+1}$$ Subtract the numerators; keep the common denominator.

Be careful with signs.

$$= \frac{7x - x + 2}{3x+1}$$ Distributive property

$$= \frac{6x+2}{3x+1}$$ Combine terms in the numerator.

$$= \frac{2(3x+1)}{3x+1}$$ Factor the numerator.

$$= 2$$ Lowest terms

(b) $\dfrac{1}{q-1} - \dfrac{1}{q+1}$ The LCD is $(q-1)(q+1)$.

$$= \frac{1(q+1)}{(q-1)(q+1)} - \frac{1(q-1)}{(q+1)(q-1)}$$ Fundamental property

$$= \frac{(q+1)-(q-1)}{(q-1)(q+1)}$$ Subtract.

$$= \frac{q+1-q+1}{(q-1)(q+1)}$$ Be careful with signs. Distributive property

$$= \frac{2}{(q-1)(q+1)}$$ Combine terms in the numerator.

◄ Work Problem **4** at the Side.

In some problems, rational expressions to be added or subtracted have denominators that are opposites of each other, such as

$$\frac{y}{y-2} + \frac{8}{2-y}.$$ Denominators are opposites.

The next example illustrates how to proceed in such a problem.

ANSWERS

4. (a) 3 **(b)** $\dfrac{-r^2+4r-2}{(r-2)(r-1)}$

EXAMPLE 5 **Adding Rational Expressions with Denominators That Are Opposites**

Add.

$$\frac{y}{y-2}+\frac{8}{2-y}$$

$$=\frac{y}{y-2}+\frac{8(-1)}{(2-y)(-1)}$$ Multiply the second expression by $\frac{-1}{-1}$.

$$=\frac{y}{y-2}+\frac{-8}{y-2}$$ The LCD is $y-2$.

$$=\frac{y-8}{y-2}$$ Add the numerators.

We could use $2-y$ as the common denominator and rewrite the first expression.

$$\frac{y}{y-2}+\frac{8}{2-y}$$

$$=\frac{y(-1)}{(y-2)(-1)}+\frac{8}{2-y}$$ Multiply the first expression by $\frac{-1}{-1}$.

$$=\frac{-y+8}{2-y}, \text{ or equivalently } \frac{8-y}{2-y}$$ The LCD is $2-y$.

Work Problem 5 at the Side. ▶

EXAMPLE 6 **Adding and Subtracting Three Rational Expressions**

Add and subtract as indicated.

$$\frac{3}{x-2}+\frac{5}{x}-\frac{6}{x^2-2x}$$

$$=\frac{3}{x-2}+\frac{5}{x}-\frac{6}{x(x-2)}$$ Factor the third denominator.

$$=\frac{3x}{x(x-2)}+\frac{5(x-2)}{x(x-2)}-\frac{6}{x(x-2)}$$ The LCD is $x(x-2)$; fundamental property

$$=\frac{3x+5(x-2)-6}{x(x-2)}$$ Add and subtract the numerators.

$$=\frac{3x+5x-10-6}{x(x-2)}$$ Distributive property

$$=\frac{8x-16}{x(x-2)}$$ Combine terms in the numerator.

$$=\frac{8(x-2)}{x(x-2)}$$ Factor the numerator.

$$=\frac{8}{x}$$ Lowest terms

Work Problem 6 at the Side. ▶

5 Add or subtract as indicated.

(a) $\frac{8}{x-4}+\frac{2}{4-x}$

(b) $\frac{9}{2x-9}-\frac{4}{9-2x}$

6 Add and subtract as indicated.

$$\frac{4}{x-5}+\frac{-2}{x}-\frac{10}{x^2-5x}$$

ANSWERS

5. (a) $\frac{6}{x-4}$, or $\frac{-6}{4-x}$ **(b)** $\frac{13}{2x-9}$, or $\frac{-13}{9-2x}$

6. $\frac{2}{x-5}$

7 Subtract.

$$\frac{-a}{a^2 + 3a - 4} - \frac{4a}{a^2 + 7a + 12}$$

EXAMPLE 7 **Subtracting Rational Expressions**

Subtract.

$$\frac{m + 4}{m^2 - 2m - 3} - \frac{2m - 3}{m^2 - 5m + 6}$$

$$= \frac{m + 4}{(m - 3)(m + 1)} - \frac{2m - 3}{(m - 3)(m - 2)}$$ Factor each denominator.

$$= \frac{(m + 4)(m - 2)}{(m - 3)(m + 1)(m - 2)} - \frac{(2m - 3)(m + 1)}{(m - 3)(m - 2)(m + 1)}$$ Fundamental property

The LCD is $(m - 3)(m + 1)(m - 2)$.

$$= \frac{(m + 4)(m - 2) - (2m - 3)(m + 1)}{(m - 3)(m + 1)(m - 2)}$$ Subtract the numerators.

Note the careful use of parentheses.

$$= \frac{m^2 + 2m - 8 - (2m^2 - m - 3)}{(m - 3)(m + 1)(m - 2)}$$ Multiply in the numerator.

Be careful with signs.

$$= \frac{m^2 + 2m - 8 - 2m^2 + m + 3}{(m - 3)(m + 1)(m - 2)}$$ Distributive property

$$= \frac{-m^2 + 3m - 5}{(m - 3)(m + 1)(m - 2)}$$ Combine terms in the numerator.

If we try to factor the numerator, we find that this rational expression is in lowest terms.

◀ Work Problem **7** at the Side.

8 Add.

$$\frac{4}{p^2 - 6p + 9} + \frac{1}{p^2 + 2p - 15}$$

EXAMPLE 8 **Adding Rational Expressions**

Add.

$$\frac{5}{x^2 + 10x + 25} + \frac{2}{x^2 + 7x + 10}$$

$$= \frac{5}{(x + 5)^2} + \frac{2}{(x + 5)(x + 2)}$$ Factor each denominator

$$= \frac{5(x + 2)}{(x + 5)^2(x + 2)} + \frac{2(x + 5)}{(x + 5)^2(x + 2)}$$ The LCD is $(x + 5)^2(x + 2)$; fundamental property

$$= \frac{5(x + 2) + 2(x + 5)}{(x + 5)^2(x + 2)}$$ Add the numerators.

$$= \frac{5x + 10 + 2x + 10}{(x + 5)^2(x + 2)}$$ Distributive property

$$= \frac{7x + 20}{(x + 5)^2(x + 2)}$$ Combine terms in the numerator.

◀ Work Problem **8** at the Side.

ANSWERS

7. $\dfrac{-5a^2 + a}{(a + 4)(a - 1)(a + 3)}$

8. $\dfrac{5p + 17}{(p - 3)^2(p + 5)}$

8.2 Exercises

FOR EXTRA HELP MyMathLab | Math XL PRACTICE | WATCH | DOWNLOAD | READ | REVIEW

1. Write an explanation for adding or subtracting rational expressions that have a common denominator.

2. Write an explanation for adding or subtracting rational expressions that have different denominators.

Add or subtract as indicated. Write all answers in lowest terms. See Example 1.

3. $\dfrac{7}{t} + \dfrac{2}{t}$

4. $\dfrac{5}{r} + \dfrac{9}{r}$

5. $\dfrac{11}{5x} - \dfrac{1}{5x}$

6. $\dfrac{7}{4y} - \dfrac{3}{4y}$

7. $\dfrac{5x + 4}{6x + 5} + \dfrac{x + 1}{6x + 5}$

8. $\dfrac{6y + 12}{4y + 3} + \dfrac{2y - 6}{4y + 3}$

9. $\dfrac{x^2}{x + 5} - \dfrac{25}{x + 5}$

10. $\dfrac{y^2}{y + 6} - \dfrac{36}{y + 6}$

11. $\dfrac{4}{p^2 + 7p + 12} + \dfrac{p}{p^2 + 7p + 12}$

12. $\dfrac{5}{x^2 + x - 20} + \dfrac{x}{x^2 + x - 20}$

13. $\dfrac{a^3}{a^2 + ab + b^2} - \dfrac{b^3}{a^2 + ab + b^2}$

14. $\dfrac{p^3}{p^2 - pq + q^2} + \dfrac{q^3}{p^2 - pq + q^2}$

Assume that the expressions given are denominators of fractions. Find the least common denominator (LCD) for each group. See Example 2.

15. $18x^2y^3$, $24x^4y^5$

16. $24a^3b^4$, $18a^5b^2$

17. $z - 2$, z

18. $k + 3$, k

19. $2y + 8$, $y - 4$

20. $3r - 21$, $r - 7$

21. $x^2 - 81$, $x^2 + 18x + 81$

22. $y^2 - 16$, $y^2 - 8y + 16$

23. $m + n$, $m - n$, $m^2 - n^2$

24. $r + s$, $r - s$, $r^2 - s^2$

25. $x^2 - 3x - 4$, $x + x^2$

26. $y^2 - 8y + 12$, $y^2 - 6y$

27. $2t^2 + 7t - 15$, $t^2 + 3t - 10$

28. $s^2 - 3s - 4$, $3s^2 + s - 2$

29. $2y + 6$, $y^2 - 9$, y

30. $9x + 18$, $x^2 - 4$, x

31. One student added two rational expressions and obtained the answer $\frac{3}{5-y}$. Another student obtained the answer $\frac{-3}{y-5}$ for the same problem. Is it possible that both answers are correct? Explain.

32. Consider the following incorrect work. ***WHAT WENT WRONG?***

$$\frac{x}{x+2} - \frac{4x-1}{x+2} = \frac{x-4x-1}{x+2} = \frac{-3x-1}{x+2}$$

Add or subtract as indicated. Write all answers in lowest terms. See Examples 3–8.

33. $\frac{8}{t} + \frac{7}{3t}$

34. $\frac{5}{x} + \frac{9}{4x}$

35. $\frac{5}{12x^2y} - \frac{11}{6xy}$

36. $\frac{7}{18a^3b^2} - \frac{2}{9ab}$

37. $\frac{1}{x-1} - \frac{1}{x}$

38. $\frac{3}{x-3} - \frac{1}{x}$

39. $\frac{3a}{a+1} + \frac{2a}{a-3}$

40. $\frac{2x}{x+4} + \frac{3x}{x-7}$

41. $\frac{17y+3}{9y+7} - \frac{-10y-18}{9y+7}$

42. $\frac{7x+8}{3x+2} - \frac{x+4}{3x+2}$

43. $\frac{2}{4-x} + \frac{5}{x-4}$

44. $\frac{3}{2-t} + \frac{1}{t-2}$

45. $\frac{w}{w-z} - \frac{z}{z-w}$

46. $\frac{a}{a-b} - \frac{b}{b-a}$

47. $\frac{5}{12+4x} - \frac{7}{9+3x}$

48. $\frac{3}{10x+15} - \frac{8}{12x+18}$

49. $\frac{4x}{x-1} - \frac{2}{x+1} - \frac{4}{x^2-1}$

50. $\frac{4}{x+3} - \frac{x}{x-3} - \frac{18}{x^2-9}$

51. $\dfrac{15}{y^2 + 3y} + \dfrac{2}{y} + \dfrac{5}{y + 3}$

52. $\dfrac{7}{t - 2} - \dfrac{6}{t^2 - 2t} - \dfrac{3}{t}$

53. $\dfrac{5}{x - 2} + \dfrac{1}{x} + \dfrac{2}{x^2 - 2x}$

54. $\dfrac{5x}{x - 3} + \dfrac{2}{x} + \dfrac{6}{x^2 - 3x}$

55. $\dfrac{3x}{x + 1} + \dfrac{4}{x - 1} - \dfrac{6}{x^2 - 1}$

56. $\dfrac{5x}{x + 3} + \dfrac{x + 2}{x} - \dfrac{6}{x^2 + 3x}$

57. $\dfrac{4}{x + 1} + \dfrac{1}{x^2 - x + 1} - \dfrac{12}{x^3 + 1}$

58. $\dfrac{5}{x + 2} + \dfrac{2}{x^2 - 2x + 4} - \dfrac{60}{x^3 + 8}$

59. $\dfrac{2x - 4}{x + 3} + \dfrac{3}{x} - \dfrac{6}{x^2 + 3x}$

60. $\dfrac{4x + 1}{x + 5} - \dfrac{2}{x} + \dfrac{10}{x^2 + 5x}$

61. $\dfrac{3}{x^2 - 5x + 6} - \dfrac{2}{x^2 - 4x + 4}$

62. $\dfrac{2}{m^2 - 4m + 4} + \dfrac{3}{m^2 + m - 6}$

63. $\dfrac{3}{x^2 - 4x + 4} + \dfrac{7}{x^2 + 5x + 6}$

64. $\dfrac{5}{x^2 + 6x + 9} - \dfrac{2}{x^2 + 4x + 3}$

65. $\dfrac{5x}{x^2 - xy - 2y^2} - \dfrac{3x}{x^2 + 5xy - 6y^2}$

66. $\dfrac{6x}{6x^2 + 5xy - 4y^2} - \dfrac{2y}{9x^2 - 16y^2}$

*A **concours d'elegance** is a competition in which a maximum of 100 points is awarded to a car based on its general attractiveness. The function defined by the rational expression*

$$c(x) = \frac{1010}{49(101 - x)} - \frac{10}{49}$$

approximates the cost, in thousands of dollars, of restoring a car so that it will win x points.

Use this information to work Exercises 67 and 68.

67. Simplify the expression for $c(x)$ by performing the indicated subtraction.

68. Use the simplified expression to determine how much it would cost to win 95 points.

Relating Concepts (Exercises 69–74) For Individual or Group Work

In Example 6 we showed that

$$\frac{3}{x-2} + \frac{5}{x} - \frac{6}{x^2 - 2x} \quad \textit{simplifies to} \quad \frac{8}{x}.$$

Algebra is, in a sense, a generalized form of arithmetic. ***Work Exercises 69–74 in order,*** *to see how the algebra in this example is related to the arithmetic of common fractions.*

69. Perform the following operations, and express your answer in lowest terms.

$$\frac{3}{7} + \frac{5}{9} - \frac{6}{63}$$

70. Substitute 9 for x in the given problem from Example 6. Compare this problem to the one given in Exercise 69. What do you notice?

71. Now substitute 9 for x in the answer given in Example 6. Do your results agree with the result you obtained in Exercise 69?

72. Replace x in the problem from Example 6 with the number of letters in your last name, assuming that this number is not 2. If your last name has two letters, let $x = 3$. Now predict the answer to your problem. Verify that your prediction is correct.

73. Why will $x = 2$ not work for the problem from Example 6?

74. What other value of x is not allowed in the problem given from Example 6?

8.3 Complex Fractions

A **complex fraction** is an expression having a fraction in the numerator, denominator, or both. Examples of complex fractions include

$$\frac{1+\frac{1}{x}}{2}, \quad \frac{\frac{4}{y}}{6-\frac{3}{y}}, \quad \text{and} \quad \frac{\frac{m^2-9}{m+1}}{\frac{m+3}{m^2-1}}. \qquad \text{Complex fractions}$$

OBJECTIVES

1. Simplify complex fractions by simplifying the numerator and denominator (Method 1).
2. Simplify complex fractions by multiplying by a common denominator (Method 2).
3. Compare the two methods of simplifying complex fractions.
4. Simplify rational expressions with negative exponents.

OBJECTIVE 1 Simplify complex fractions by simplifying the numerator and denominator (Method 1). There are two different methods for simplifying complex fractions.

Simplifying a Complex Fraction: Method 1

Step 1 Simplify the numerator and denominator separately.

Step 2 Divide by multiplying the numerator by the reciprocal of the denominator.

Step 3 Simplify the resulting fraction, if possible.

In Step 2, we are treating the complex fraction as a quotient of two rational expressions and dividing. Before performing this step, be sure that both the numerator and denominator are single fractions.

EXAMPLE 1 Simplifying Complex Fractions by Method 1

Use Method 1 to simplify each complex fraction.

(a) $\dfrac{\frac{x+1}{x}}{\frac{x-1}{2x}}$ Both the numerator and the denominator are already simplified. (Step 1)

$= \dfrac{x+1}{x} \div \dfrac{x-1}{2x}$ Write as a division problem.

$= \dfrac{x+1}{x} \cdot \dfrac{2x}{x-1}$ Multiply by the reciprocal of $\frac{x-1}{2x}$. (Step 2)

$= \dfrac{2x(x+1)}{x(x-1)}$ Multiply.

$= \dfrac{2(x+1)}{x-1}$ Simplify. (Step 3)

Continued on Next Page

1 Use Method 1 to simplify each complex fraction.

(a) $\dfrac{\dfrac{a+2}{5a}}{\dfrac{a-3}{7a}}$

(b) $\dfrac{2+\dfrac{1}{k}}{2-\dfrac{1}{k}}$

(c) $\dfrac{\dfrac{r^2-4}{4}}{1+\dfrac{2}{r}}$

(b) $\dfrac{2+\dfrac{1}{y}}{3-\dfrac{2}{y}}$

$= \dfrac{\dfrac{2y}{y}+\dfrac{1}{y}}{\dfrac{3y}{y}-\dfrac{2}{y}}$ Simplify the numerator and denominator. (Step 1)

$= \dfrac{\dfrac{2y+1}{y}}{\dfrac{3y-2}{y}}$ $\dfrac{\frac{2y+1}{y}}{\frac{3y-2}{y}}$ means $\frac{2y+1}{y} \div \frac{3y-2}{y}$.

$= \dfrac{2y+1}{y} \cdot \dfrac{y}{3y-2}$ Multiply by the reciprocal of $\frac{3y-2}{y}$. (Step 2)

$= \dfrac{2y+1}{3y-2}$ Multiply and simplify. (Step 3)

◀ *Work Problem* **1** *at the Side.*

OBJECTIVE 2 Simplify complex fractions by multiplying by a common denominator (Method 2). The second method for simplifying complex fractions uses the identity property of multiplication.

Simplifying a Complex Fraction: Method 2

Step 1 Multiply the numerator and denominator of the complex fraction by the least common denominator of the fractions in the numerator and the fractions in the denominator of the complex fraction.

Step 2 Simplify the resulting fraction, if possible.

EXAMPLE 2 Simplifying Complex Fractions by Method 2

Use Method 2 to simplify each complex fraction.

(a) $\dfrac{2+\dfrac{1}{y}}{3-\dfrac{2}{y}}$ This is the same fraction as in Example 1(b) above. Compare the solution methods.

$= \dfrac{\left(2+\dfrac{1}{y}\right)\cdot y}{\left(3-\dfrac{2}{y}\right)\cdot y}$ Multiply the numerator and denominator by the LCD, y. (Step 1)

$= \dfrac{2\cdot y+\dfrac{1}{y}\cdot y}{3\cdot y-\dfrac{2}{y}\cdot y}$ Distributive property

$= \dfrac{2y+1}{3y-2}$ Simplify. (Step 2)

Continued on Next Page

Answers

1. (a) $\dfrac{7(a+2)}{5(a-3)}$ (b) $\dfrac{2k+1}{2k-1}$ (c) $\dfrac{r(r-2)}{4}$

(b) $\dfrac{2p - \dfrac{5}{p-1}}{3p - \dfrac{2}{p}}$

$= \dfrac{\left(2p + \dfrac{5}{p-1}\right) \cdot p(p-1)}{\left(3p - \dfrac{2}{p}\right) \cdot p(p-1)}$ Multiply the numerator and denominator by the LCD, $p(p-1)$.

$= \dfrac{2p[p(p-1)] + \dfrac{5}{p-1} \cdot p(p-1)}{3p[p(p-1)] - \dfrac{2}{p} \cdot p(p-1)}$ Distributive property

$= \dfrac{2p[p(p-1)] + 5p}{3p[p(p-1)] - 2(p-1)}$ Multiply.

$= \dfrac{2p^3 - 2p^2 + 5p}{3p^3 - 3p^2 - 2p + 2}$ Multiply; lowest terms

Work Problem **2** *at the Side.* ▶

OBJECTIVE 3 Compare the two methods of simplifying complex fractions. Choosing whether to use Method 1 or Method 2 to simplify a complex fraction is usually a matter of preference. Some students prefer one method over the other, while other students feel comfortable with both methods and rely on practice with many examples to determine which method they will use on a particular problem. In the next example, we illustrate how to simplify a complex fraction using both methods so that you can observe the processes and decide for yourself the pros and cons of each method.

EXAMPLE 3 Simplifying Complex Fractions Using Both Methods

Use both Method 1 and Method 2 to simplify each complex fraction.

Method 1

(a) $\dfrac{\dfrac{2}{x-3}}{\dfrac{5}{x^2-9}}$

$= \dfrac{\dfrac{2}{x-3}}{\dfrac{5}{(x-3)(x+3)}}$

$= \dfrac{2}{x-3} \div \dfrac{5}{(x-3)(x+3)}$

$= \dfrac{2}{x-3} \cdot \dfrac{(x-3)(x+3)}{5}$

$= \dfrac{2(x-3)}{5}$

Method 2

(a) $\dfrac{\dfrac{2}{x-3}}{\dfrac{5}{x^2-9}}$

$= \dfrac{\dfrac{2}{x-3}}{\dfrac{5}{(x-3)(x+3)}}$

$= \dfrac{\dfrac{2}{x-3} \cdot (x-3)(x+3)}{\dfrac{5}{(x-3)(x+3)} \cdot (x-3)(x+3)}$

$= \dfrac{2(x+3)}{5}$

Continued on Next Page

2 Use Method 2 to simplify each complex fraction.

(a) $\dfrac{\dfrac{5}{y} + 6}{\dfrac{8}{3y} - 1}$

(b) $\dfrac{\dfrac{1}{y} + \dfrac{1}{y-1}}{\dfrac{1}{y} - \dfrac{2}{y-1}}$

ANSWERS

2. (a) $\dfrac{15 + 18y}{8 - 3y}$ **(b)** $\dfrac{2y-1}{-y-1}$, or $\dfrac{1-2y}{y+1}$

Method 1

(b) $\dfrac{\frac{1}{x}+\frac{1}{y}}{\frac{1}{x^2}-\frac{1}{y^2}}$

$= \dfrac{\frac{y}{xy}+\frac{x}{xy}}{\frac{y^2}{x^2y^2}-\frac{x^2}{x^2y^2}}$

$= \dfrac{\frac{y+x}{xy}}{\frac{y^2-x^2}{x^2y^2}}$

$= \dfrac{y+x}{xy} \div \dfrac{y^2-x^2}{x^2y^2}$

$= \dfrac{y+x}{xy} \cdot \dfrac{x^2y^2}{(y-x)(y+x)}$

$= \dfrac{xy}{y-x}$

Method 2

(b) $\dfrac{\frac{1}{x}+\frac{1}{y}}{\frac{1}{x^2}-\frac{1}{y^2}}$

$= \dfrac{\left(\frac{1}{x}+\frac{1}{y}\right)\cdot x^2y^2}{\left(\frac{1}{x^2}-\frac{1}{y^2}\right)\cdot x^2y^2}$

$= \dfrac{\left(\frac{1}{x}\right)x^2y^2+\left(\frac{1}{y}\right)x^2y^2}{\left(\frac{1}{x^2}\right)x^2y^2-\left(\frac{1}{y^2}\right)x^2y^2}$

$= \dfrac{xy^2+x^2y}{y^2-x^2}$

$= \dfrac{xy(y+x)}{(y+x)(y-x)}$

$= \dfrac{xy}{y-x}$

◀ Work Problem 3 at the Side.

3 Use both methods to simplify each complex fraction.

(a) $\dfrac{\frac{5}{y+2}}{\frac{-3}{y^2-4}}$

(b) $\dfrac{\frac{1}{a}-\frac{1}{b}}{\frac{1}{a^2}-\frac{1}{b^2}}$

OBJECTIVE 4 Simplify rational expressions with negative exponents. To simplify such expressions, we begin by rewriting the expressions with only positive exponents.

EXAMPLE 4 Simplifying a Rational Expression with Negative Exponents

Simplify, using only positive exponents in the answer.

$$\frac{m^{-1}+p^{-2}}{2m^{-2}-p^{-1}}$$

$$= \frac{\frac{1}{m}+\frac{1}{p^2}}{\frac{2}{m^2}-\frac{1}{p}} \quad \text{Write with positive exponents.}$$

The base of $2m^{-2}$ is m, *not* $2m$: $2m^{-2} = \frac{2}{m^2}$.

$$= \frac{m^2p^2\left(\frac{1}{m}+\frac{1}{p^2}\right)}{m^2p^2\left(\frac{2}{m^2}-\frac{1}{p}\right)} \quad \text{Simplify by Method 2, multiplying the numerator and denominator by the LCD, } m^2p^2.$$

$$= \frac{m^2p^2\cdot\frac{1}{m}+m^2p^2\cdot\frac{1}{p^2}}{m^2p^2\cdot\frac{2}{m^2}-m^2p^2\cdot\frac{1}{p}} \quad \text{Distributive property}$$

$$= \frac{mp^2+m^2}{2p^2-m^2p} \quad \text{Lowest terms}$$

◀ Work Problem 4 at the Side.

4 Simplify each expression, using only positive exponents in the answer.

(a) $\dfrac{r^{-2}-s^{-1}}{4r^{-1}+s^{-2}}$

(b) $\dfrac{b^{-4}}{b^{-5}+2}$

Answers

3. (Both methods give the same answers.)
(a) $\dfrac{5(y-2)}{-3}$ (b) $\dfrac{ab}{b+a}$

4. (a) $\dfrac{s^2-r^2s}{4rs^2+r^2}$ (b) $\dfrac{b}{1+2b^5}$

8.3 Exercises

FOR EXTRA HELP MyMathLab Math XL PRACTICE WATCH DOWNLOAD READ REVIEW

1. Explain in your own words Method 1 for simplifying complex fractions.

2. Method 2 for simplifying complex fractions says that we can multiply both the numerator and the denominator of the complex fraction by the same nonzero expression. What property of real numbers from **Section 1.4** justifies this method?

Use either method to simplify each complex fraction. See Examples 1–3.

3. $\dfrac{\frac{12}{x-1}}{\frac{6}{x}}$

4. $\dfrac{\frac{24}{t+4}}{\frac{6}{t}}$

5. $\dfrac{\frac{k+1}{2k}}{\frac{3k-1}{4k}}$

6. $\dfrac{\frac{1-r}{4r}}{\frac{-1-r}{8r}}$

7. $\dfrac{\frac{4z^2x^4}{9}}{\frac{12x^2z^5}{15}}$

8. $\dfrac{\frac{3y^2x^3}{8}}{\frac{9y^3x^4}{16}}$

9. $\dfrac{\frac{1}{x}+1}{-\frac{1}{x}+1}$

10. $\dfrac{\frac{2}{k}-1}{\frac{2}{k}+1}$

11. $\dfrac{\frac{3}{x}+\frac{3}{y}}{\frac{3}{x}-\frac{3}{y}}$

12. $\dfrac{\frac{4}{r}-\frac{4}{s}}{\frac{4}{r}+\frac{4}{s}}$

13. $\dfrac{\frac{8x-24y}{10}}{\frac{x-3y}{5x}}$

14. $\dfrac{\frac{10x-5y}{12}}{\frac{2x-y}{6y}}$

15. $\dfrac{\frac{x^2-16y^2}{xy}}{\frac{1}{y}-\frac{4}{x}}$

16. $\dfrac{\frac{2}{s}-\frac{3}{t}}{\frac{4t^2-9s^2}{st}}$

17. $\dfrac{y-\frac{y-3}{3}}{\frac{4}{9}+\frac{2}{3y}}$

18. $\dfrac{p - \dfrac{p+2}{4}}{\dfrac{3}{4} - \dfrac{5}{2p}}$

19. $\dfrac{\dfrac{x+2}{x} + \dfrac{1}{x+2}}{\dfrac{5}{x} + \dfrac{x}{x+2}}$

20. $\dfrac{\dfrac{y+3}{y} - \dfrac{4}{y-1}}{\dfrac{y}{y-1} + \dfrac{1}{y}}$

Relating Concepts (Exercises 21–26) For Individual or Group Work

Simplifying a complex fraction by Method 1 is a good way to review the methods of adding, subtracting, multiplying, and dividing rational expressions. Method 2 gives a good review of the fundamental property of rational expressions. Refer to the following complex fraction, and ***work Exercises 21–26 in order.***

$$\frac{\dfrac{4}{m} + \dfrac{m+2}{m-1}}{\dfrac{m+2}{m} - \dfrac{2}{m-1}}$$

21. Add the fractions in the numerator.

22. Subtract as indicated in the denominator.

23. Divide your answer from Exercise 21 by your answer from Exercise 22.

24. Go back to the original complex fraction and find the least common denominator of all denominators.

25. Multiply the numerator and denominator of the complex fraction by your answer from Exercise 24.

26. Your answers for Exercises 23 and 25 should be the same. Write an explanation comparing the two methods. Which method do you prefer? Explain why.

Simplify each expression, using only positive exponents in the answer. See Example 4.

27. $\dfrac{1}{x^{-2} + y^{-2}}$

28. $\dfrac{1}{p^{-2} - q^{-2}}$

29. $\dfrac{x^{-2} + y^{-2}}{x^{-1} + y^{-1}}$

30. $\dfrac{x^{-1} - y^{-1}}{x^{-2} - y^{-2}}$

31. $\dfrac{x^{-1} + 2y^{-1}}{2y + 4x}$

32. $\dfrac{a^{-2} - 4b^{-2}}{3b - 6a}$

8.4 Equations with Rational Expressions and Graphs

In **Section 8.1,** we defined the domain of a rational function as the set of all possible values of the variable. (We can also refer to this as "the domain of the variable.") Any value that makes the denominator 0 is excluded.

OBJECTIVES

1. Determine the domain of the variable in a rational equation.
2. Solve rational equations.
3. Recognize the graph of a rational function.

OBJECTIVE 1 Determine the domain of the variable in a rational equation. The **domain of the variable in a rational equation** is the intersection (overlap) of the domains of the rational expressions in the equation.

EXAMPLE 1 Determining Domains in Rational Equations

Find the domain of the variable in each equation.

(a) $\frac{2}{x} - \frac{3}{2} = \frac{7}{2x}$

The domains of the three rational terms of the equation are, in order, $\{x \mid x \neq 0\}$, $(-\infty, \infty)$, and $\{x \mid x \neq 0\}$. The intersection of these three domains is all real numbers except 0, which may be written $\{x \mid x \neq 0\}$.

(b) $\frac{2}{x-3} - \frac{3}{x+3} = \frac{12}{x^2-9}$

The domains of these three terms are, respectively, $\{x \mid x \neq 3\}$, $\{x \mid x \neq -3\}$, and $\{x \mid x \neq \pm 3\}$. ($\pm$ is read "positive or negative," or "plus or minus.") The domain of the variable is the intersection of the three domains, all real numbers except 3 and -3, written $\{x \mid x \neq \pm 3\}$.

Work Problem 1 at the Side. ▶

1 Find the domain of the variable in each equation.

(a) $\frac{3}{x} + \frac{1}{2} = \frac{5}{6x}$

(b) $\frac{4}{x-5} - \frac{2}{x+5} = \frac{1}{x^2-25}$

OBJECTIVE 2 Solve rational equations. The easiest way to solve most equations involving rational expressions is to multiply all terms in the equation by the least common denominator. This step will clear the equation of all denominators. ***We can do this only with equations, not expressions.***

CAUTION
When each side of an equation is multiplied by a *variable* expression, the resulting "solutions" may not satisfy the original equation. ***You must either determine and observe the domain or check all proposed solutions in the original equation. It is wise to do both.***

EXAMPLE 2 Solving a Rational Equation

Solve $\frac{2}{x} - \frac{3}{2} = \frac{7}{2x}$.

The domain, which excludes 0, was found in Example 1(a).

$$2x\left(\frac{2}{x} - \frac{3}{2}\right) = 2x\left(\frac{7}{2x}\right) \quad \text{Multiply by the LCD, } 2x.$$

$$2x\left(\frac{2}{x}\right) - 2x\left(\frac{3}{2}\right) = 2x\left(\frac{7}{2x}\right) \quad \text{Distributive property}$$

$$4 - 3x = 7 \quad \text{Multiply.}$$

$$-3x = 3 \quad \text{Subtract 4.}$$

$$\text{Proposed solution} \longrightarrow x = -1 \quad \text{Divide by } -3.$$

Continued on Next Page

ANSWERS
1. (a) $\{x \mid x \neq 0\}$ (b) $\{x \mid x \neq \pm 5\}$

Check Replace x with -1 in the original equation.

$$\frac{2}{x} - \frac{3}{2} = \frac{7}{2x} \quad \text{Original equation}$$

$$\frac{2}{-1} - \frac{3}{2} \stackrel{?}{=} \frac{7}{2(-1)} \quad \text{Let } x = -1.$$

$$-2 - \frac{3}{2} \stackrel{?}{=} -\frac{7}{2}$$

$$-\frac{7}{2} = -\frac{7}{2} \quad \text{True}$$

The solution set is $\{-1\}$.

◀ *Work Problem* **2** *at the Side.*

EXAMPLE 3 **Solving a Rational Equation with No Solution**

Solve $\frac{2}{x-3} - \frac{3}{x+3} = \frac{12}{x^2-9}$.

Using the result from Example 1(b), we know that the domain excludes 3 and -3, since these values make one or more of the denominators in the equation equal 0. Multiply each side by the LCD, $(x+3)(x-3)$.

$$(x+3)(x-3)\left(\frac{2}{x-3} - \frac{3}{x+3}\right) = (x+3)(x-3)\left(\frac{12}{x^2-9}\right)$$

$$(x+3)(x-3)\left(\frac{2}{x-3}\right) - (x+3)(x-3)\left(\frac{3}{x+3}\right)$$

$$= (x+3)(x-3)\left(\frac{12}{x^2-9}\right) \quad \text{Distributive property}$$

$$2(x+3) - 3(x-3) = 12 \quad \text{Multiply.}$$

$$2x + 6 - 3x + 9 = 12 \quad \text{Distributive property}$$

$$-x + 15 = 12 \quad \text{Combine like terms.}$$

$$-x = -3 \quad \text{Subtract 15.}$$

$$\text{Proposed solution} \longrightarrow x = 3 \quad \text{Divide by } -1.$$

Since 3 is not in the domain, it cannot be a solution of the equation. Substituting 3 in the original equation shows why.

Check

$$\frac{2}{x-3} - \frac{3}{x+3} = \frac{12}{x^2-9} \quad \text{Original equation}$$

$$\frac{2}{3-3} - \frac{3}{3+3} \stackrel{?}{=} \frac{12}{3^2-9} \quad \text{Let } x = 3.$$

$$\frac{2}{0} - \frac{3}{6} \stackrel{?}{=} \frac{12}{0}$$

Since division by 0 is undefined, the given equation has no solution, and the solution set is $\emptyset$.

◀ *Work Problem* **3** *at the Side.*

2 Solve $-\frac{3}{20} + \frac{2}{x} = \frac{5}{4x}$.

3 Solve each equation.

(a) $\frac{3}{x+1} = \frac{1}{x-1} - \frac{2}{x^2-1}$

(b) $\frac{1}{x-3} + \frac{1}{x+3} = \frac{6}{x^2-9}$

ANSWERS

2. $\{5\}$

3. (a) $\emptyset$ **(b)** $\emptyset$

EXAMPLE 4 **Solving a Rational Equation**

Solve $\frac{3}{p^2 + p - 2} - \frac{1}{p^2 - 1} = \frac{7}{2(p^2 + 3p + 2)}$.

Factor each denominator to find the LCD, $2(p - 1)(p + 2)(p + 1)$. The domain excludes 1, -2, and -1. Multiply each side by the LCD.

$$2(p-1)(p+2)(p+1)\left(\frac{3}{(p+2)(p-1)} - \frac{1}{(p+1)(p-1)}\right)$$
$$= 2(p-1)(p+2)(p+1)\left(\frac{7}{2(p+2)(p+1)}\right)$$

$2 \cdot 3(p + 1) - 2(p + 2) = 7(p - 1)$ Distributive property

$6p + 6 - 2p - 4 = 7p - 7$ Distributive property

$4p + 2 = 7p - 7$ Combine like terms.

$9 = 3p$ Subtract $4p$; add 7.

Proposed solution ⟶ $3 = p$ Divide by 3.

Note that 3 is in the domain. Substitute 3 for p in the original equation to check that the solution set is $\{3\}$.

Work Problem 4 at the Side. ▶

EXAMPLE 5 **Solving a Rational Equation That Leads to a Quadratic Equation**

Solve $\frac{2}{3x + 1} = \frac{1}{x} - \frac{6x}{3x + 1}$.

Since the denominator $3x + 1$ cannot equal 0, $-\frac{1}{3}$ is excluded from the domain, as is 0. Multiply each side by the LCD, $x(3x + 1)$.

$$x(3x+1)\left(\frac{2}{3x+1}\right) = x(3x+1)\left(\frac{1}{x} - \frac{6x}{3x+1}\right)$$

$$x(3x+1)\left(\frac{2}{3x+1}\right) = x(3x+1)\left(\frac{1}{x}\right) - x(3x+1)\left(\frac{6x}{3x+1}\right)$$

Distributive property

$$2x = 3x + 1 - 6x^2$$

Write this quadratic equation in standard form with 0 on the right side.

$6x^2 - 3x + 2x - 1 = 0$

$6x^2 - x - 1 = 0$ Standard form

$(3x + 1)(2x - 1) = 0$ Factor.

$3x + 1 = 0$ or $2x - 1 = 0$ Zero-factor property

$x = -\frac{1}{3}$ or $x = \frac{1}{2}$ Proposed solutions

Because $-\frac{1}{3}$ is not in the domain, it is not a solution. Check that the solution set is $\{\frac{1}{2}\}$.

Work Problem 5 at the Side. ▶

4 Solve
$\frac{4}{x^2 + x - 6} - \frac{1}{x^2 - 4} = \frac{2}{x^2 + 5x + 6}$.

5 Solve
$\frac{1}{x + 4} + \frac{x}{x - 4} = \frac{-8}{x^2 - 16}$.

ANSWERS

4. $\{-9\}$
5. $\{-1\}$

OBJECTIVE 3 Recognize the graph of a rational function. As mentioned in **Section 8.1,** a function defined by a quotient of polynomials is a **rational function.** Because one or more values of x may be excluded from the domain of most rational functions, their graphs are often **discontinuous.** That is, there will be one or more breaks in the graph. For example, we use point plotting and observing the domain to graph the simple rational function defined by

$$f(x) = \frac{1}{x}.$$

The domain of this function includes all real numbers except 0. Thus, there will be no point on the graph with $x = 0$. The vertical line with equation $x = 0$ is called a **vertical asymptote** of the graph. The horizontal line with equation $y = 0$ is called a **horizontal asymptote.** We show some typical ordered pairs in the table for both negative and positive x-values.

x	-3	-2	-1	$-\frac{1}{2}$	$-\frac{1}{4}$	$-\frac{1}{10}$	$\frac{1}{10}$	$\frac{1}{4}$	$\frac{1}{2}$	1	2	3
y	$-\frac{1}{3}$	$-\frac{1}{2}$	-1	-2	-4	-10	10	4	2	1	$\frac{1}{2}$	$\frac{1}{3}$

Notice that the closer positive values of x are to 0, the larger y is. Similarly, the closer negative values of x are to 0, the smaller (more negative) y is. Using this observation, excluding 0 from the domain, and plotting the points in the table, we obtain the graph in Figure 2.

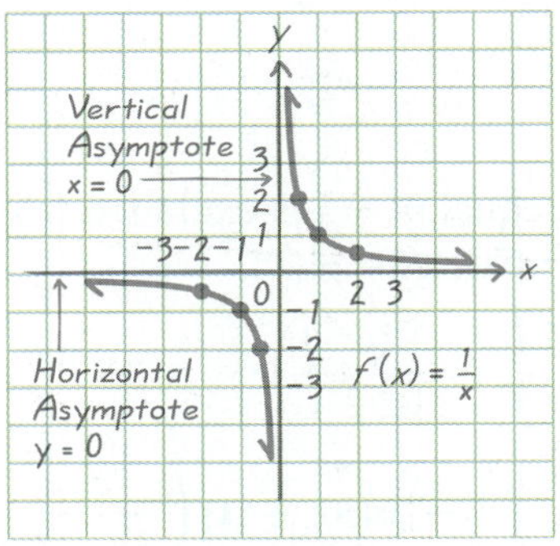

Figure 2

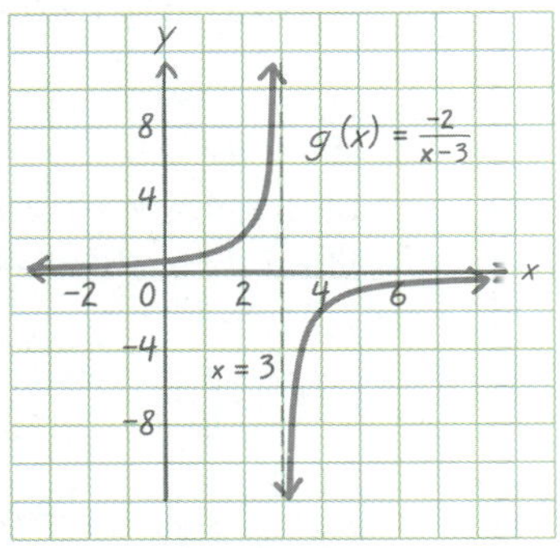

Figure 3

The graph of

$$g(x) = \frac{-2}{x-3}$$

is shown in Figure 3. Some ordered pairs are shown in the table.

x	-2	-1	0	1	2	2.5	2.75	3.25	3.5	4	5	6
y	$\frac{2}{5}$	$\frac{1}{2}$	$\frac{2}{3}$	1	2	4	8	-8	-4	-2	-1	$-\frac{2}{3}$

There is no point on the graph for $x = 3$ because 3 is excluded from the domain. The dashed line $x = 3$ represents the vertical asymptote and is not part of the graph. As suggested by the points from the table, the graph gets closer to the vertical asymptote as the x-values get closer to 3. Again, $y = 0$ is a horizontal asymptote.

◀ Work Problem 6 at the Side.

6 Graph each rational function, and give the equations of the vertical and horizontal asymptotes.

(a) $f(x) = -\frac{1}{x}$

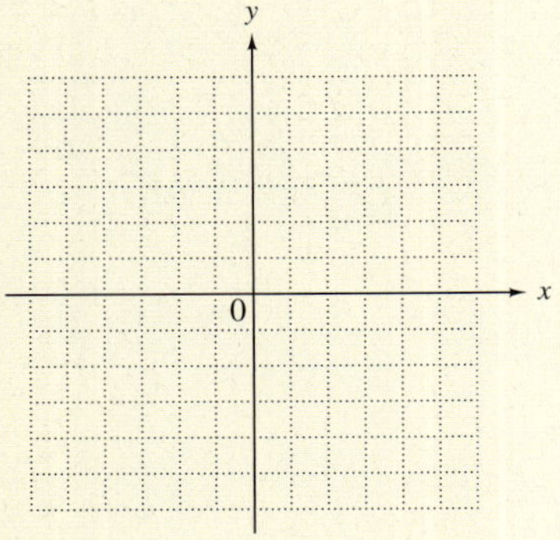

(b) $f(x) = \frac{2}{x+3}$

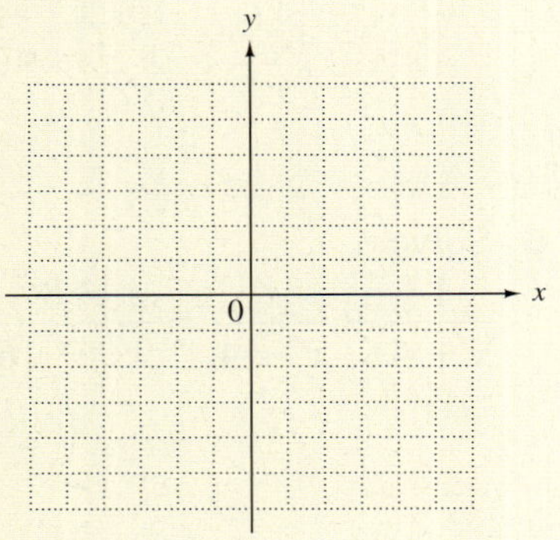

ANSWERS

6. (a) vertical asymptote: $x = 0$; horizontal asymptote: $y = 0$

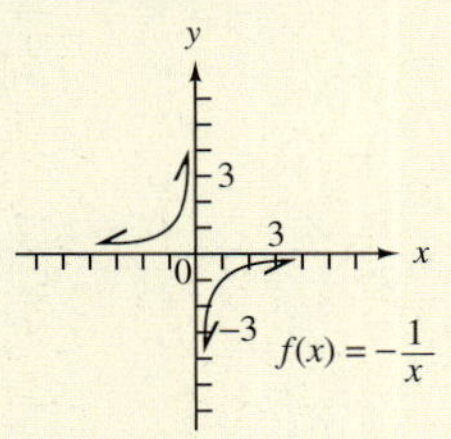

(b) vertical asymptote: $x = -3$; horizontal asymptote: $y = 0$

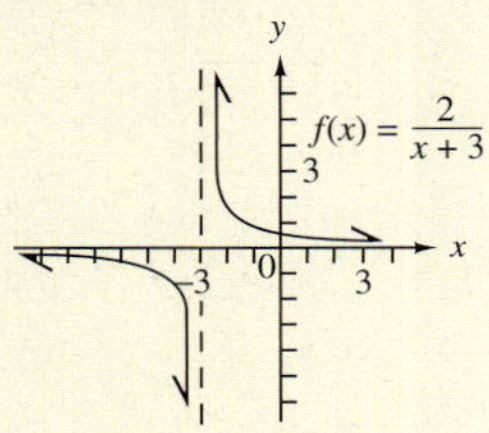

8.4 Exercises

FOR EXTRA HELP

MyMathLab — Math XL PRACTICE — WATCH — DOWNLOAD — READ — REVIEW

As explained in this section, any values that would cause a denominator to equal 0 must be excluded from the domain and consequently as solutions of an equation that has variable expressions in the denominators. **(a)** *Without actually solving the equation, list all possible numbers that would have to be rejected if they appeared as potential solutions.* **(b)** *Then give the domain using set notation. See Example 1.*

1. $\frac{1}{x-1} - \frac{1}{x-2} = 0$

2. $\frac{3}{x+4} - \frac{2}{x-9} = 0$

3. $\frac{5}{3x+5} - \frac{1}{x} = \frac{1}{2x+3}$

4. $\frac{6}{4x+7} - \frac{3}{x} = \frac{5}{6x-13}$

5. $\frac{1}{3x} + \frac{1}{2x} = \frac{x}{3}$

6. $\frac{5}{6x} - \frac{8}{2x} = \frac{x}{4}$

7. $\frac{3x+1}{x-4} = \frac{6x+5}{2x-7}$

8. $\frac{4x-1}{2x+3} = \frac{12x-25}{6x-2}$

9. $\frac{2}{x^2-x} + \frac{1}{x+3} = \frac{4}{x-2}$

10. Suppose that in solving the following equation, all of your algebraic steps are correct. Is it possible that your proposed solution would have to be rejected? Explain.

$$\frac{x+7}{4} - \frac{x+3}{3} = \frac{x}{12}$$

Solve each equation. See Examples 2–5.

11. $\frac{-5}{2x} + \frac{3}{4x} = \frac{-7}{4}$

12. $\frac{6}{5x} - \frac{2}{3x} = \frac{-8}{45}$

13. $x - \frac{24}{x} = -2$

14. $p + \frac{15}{p} = -8$

15. $\frac{x - 4}{x + 6} = \frac{2x + 3}{2x - 1}$

16. $\frac{5x - 8}{x + 2} = \frac{5x - 1}{x + 3}$

17. $\frac{3x + 1}{x - 4} = \frac{6x + 5}{2x - 7}$

18. $\frac{4x - 1}{2x + 3} = \frac{12x - 25}{6x - 2}$

19. $\frac{1}{y - 1} + \frac{5}{12} = \frac{-2}{3y - 3}$

20. $\frac{4}{m + 2} - \frac{11}{9} = \frac{1}{3m + 6}$

21. $\frac{-2}{3t - 6} - \frac{1}{36} = \frac{-3}{4t - 8}$

22. $\frac{3}{4m + 2} = \frac{17}{2} - \frac{7}{2m + 1}$

23. $\frac{3}{k + 2} - \frac{2}{k^2 - 4} = \frac{1}{k - 2}$

24. $\frac{3}{x - 2} + \frac{21}{x^2 - 4} = \frac{14}{x + 2}$

25. $\frac{1}{y + 2} + \frac{3}{y + 7} = \frac{5}{y^2 + 9y + 14}$

26. $\frac{1}{t + 3} + \frac{4}{t + 5} = \frac{2}{t^2 + 8t + 15}$

27. $\frac{9}{x} + \frac{4}{6x - 3} = \frac{2}{6x - 3}$

28. $\frac{5}{n} + \frac{4}{6 - 3n} = \frac{2n}{6 - 3n}$

29. $\frac{6}{w + 3} + \frac{-7}{w - 5} = \frac{-48}{w^2 - 2w - 15}$

30. $\frac{2}{r - 5} + \frac{3}{2r + 1} = \frac{22}{2r^2 - 9r - 5}$

31. $\frac{x}{x - 3} + \frac{4}{x + 3} = \frac{18}{x^2 - 9}$

32. $\frac{2x}{x - 3} + \frac{4}{x + 3} = \frac{-24}{x^2 - 9}$

33. $\frac{6}{x - 4} + \frac{5}{x} = \frac{-20}{x^2 - 4x}$

34. $\frac{7}{x-4}+\frac{3}{x}=\frac{-12}{x^2-4x}$

35. $\frac{2}{4x+7}+\frac{x}{3}=\frac{6}{12x+21}$

36. $\frac{5x+14}{x^2-9}=\frac{-2x^2-5x+2}{x^2-9}+\frac{2x+4}{x-3}$

37. $\frac{4x-7}{4x^2-9}=\frac{-2x^2+5x-4}{4x^2-9}+\frac{x+1}{2x+3}$

38. What is wrong with the following problem? "Solve $\frac{2x+1}{3x-4}+\frac{1}{2x+3}$."

Graph each rational function. Give the equations of the vertical and horizontal asymptotes. See Objective 3 and Figures 2 and 3.

39. $f(x)=\frac{2}{x}$

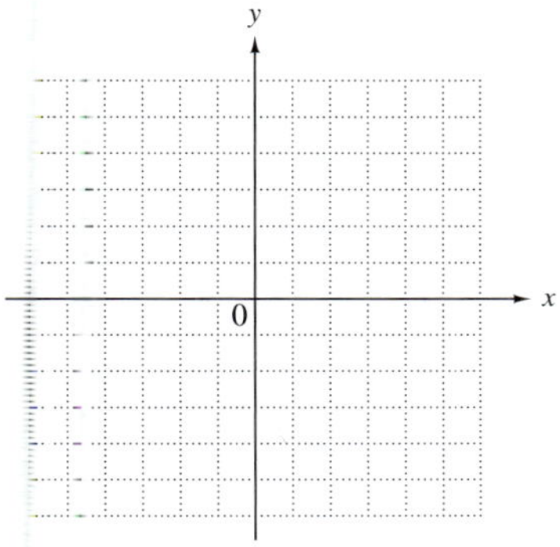

40. $f(x)=\frac{3}{x}$

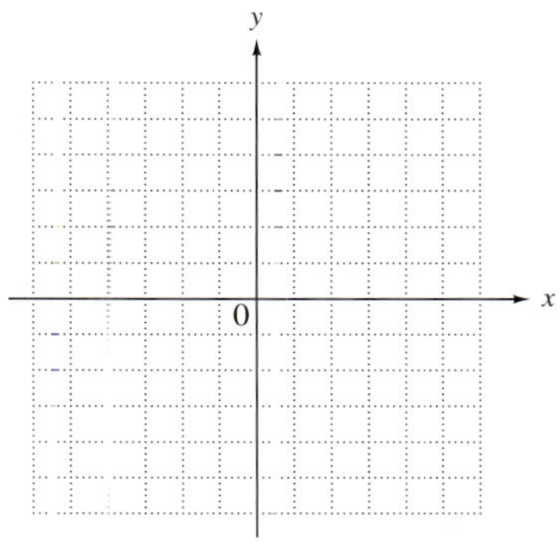

41. $f(x)=\frac{1}{x-2}$

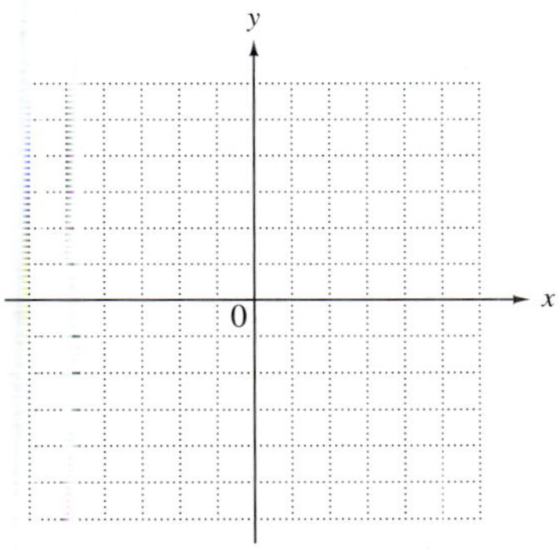

42. $f(x)=\frac{1}{x+2}$

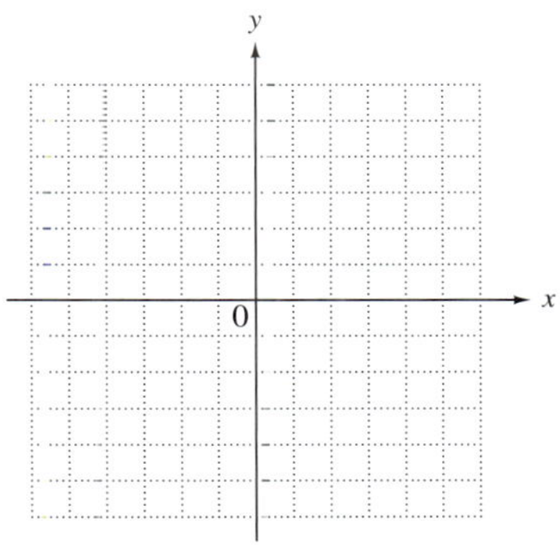

Solve each problem.

43. The average number of vehicles waiting in line to enter a parking area is modeled by the rational function defined by

$$w(x) = \frac{x^2}{2(1 - x)},$$

where x is a quantity between 0 and 1 known as the **traffic intensity.** (*Source*: Mannering, F. and W. Kilareski, *Principles of Highway Engineering and Traffic Control,* John Wiley and Sons, 1990.) To the nearest tenth, find the average number of vehicles waiting for each traffic intensity.

(a) 0.1

(b) 0.8

(c) 0.9

(d) What happens to waiting time as traffic intensity increases?

44. The force required to keep a 2000-lb car going 30 mph from skidding on a curve, where r is the radius of the curve in feet, is given by

$$F(r) = \frac{225{,}000}{r}.$$

(a) What radius must a curve have if a force of 450 lb is needed to keep the car from skidding?

(b) As the radius of the curve is lengthened, how is the force affected?

Relating Concepts (Exercises 45–48) For Individual or Group Work

An equation of the form

$$\frac{A}{x + B} + \frac{x}{x - B} = \frac{C}{x^2 - B^2}$$

will have one rejected solution if the relationship $C = -2AB$ *holds true. (This can be proved using methods not covered in intermediate algebra.) For example, if* $A = 1$ *and* $B = 2$, *then* $C = -2AB = -2(1)(2) = -4$, *and the equation becomes*

$$\frac{1}{x + 2} + \frac{x}{x - 2} = \frac{-4}{x^2 - 4}.$$

This equation has solution set $\{-1\}$*; the potential solution* -2 *must be rejected. To further understand this idea,* ***work Exercises 45–48 in order.***

45. Show that the second equation does indeed have solution set $\{-1\}$ and -2 must be rejected.

46. Let $A = 2$ and let $B = 1$. What is the corresponding value of C? Solve the equation determined by A, B, and C. What is the solution set? What value must be rejected?

47. Let $A = 4$ and let $B = -3$. What is the corresponding value of C? Solve the equation determined by A, B, and C. What is the solution set? What value must be rejected?

48. Choose two numbers of your own, letting one be A and the other be B. Repeat the process described in Exercises 46 and 47.

Summary Exercises on Rational Expressions and Equations

A common student error is to confuse an equation, *such as* $\frac{x}{2} + \frac{x}{3} = -5$, *with an expression, such as* $\frac{x}{2} + \frac{x}{3}$. *Look for the equals sign to distinguish between them. Equations are solved for a numerical answer, while problems involving operations result in simplified expressions.*

Solving an Equation

Solve: $\frac{x}{2} + \frac{x}{3} = -5.$

Multiply each side by the LCD, 6.

$$6\left(\frac{x}{2} + \frac{x}{3}\right) = 6(-5)$$

$$6\left(\frac{x}{2}\right) + 6\left(\frac{x}{3}\right) = 6(-5)$$

$$3x + 2x = -30$$

$$5x = -30$$

$$x = -6$$

Check that the solution set is $\{-6\}$.

Simplifying an Expression Involving an Operation

Add: $\frac{x}{2} + \frac{x}{3}.$

Write both fractions with the LCD, 6.

$$\frac{x}{2} + \frac{x}{3}$$

$$= \frac{x \cdot 3}{2 \cdot 3} + \frac{x \cdot 2}{3 \cdot 2}$$

$$= \frac{3x}{6} + \frac{2x}{6}$$

$$= \frac{3x + 2x}{6}$$

$$= \frac{5x}{6}$$

Identify each exercise as an expression *or an* equation. *Then simplify the expression by performing the indicated operation, or solve the given equation, as appropriate.*

1. $\frac{x}{2} - \frac{x}{4} = 5$

2. $\frac{4x - 20}{x^2 - 25} \cdot \frac{(x + 5)^2}{10}$

3. $\frac{6}{7x} - \frac{4}{x}$

4. $\dfrac{\frac{1}{x} + \frac{1}{y}}{\frac{1}{x} - \frac{1}{y}}$

5. $\frac{5}{7t} = \frac{52}{7} - \frac{3}{t}$

6. $\frac{x - 5}{3} + \frac{1}{3} = \frac{x - 2}{5}$

7. $\frac{7}{6x} + \frac{5}{8x}$

8. $\frac{4}{x} - \frac{8}{x + 1} = 0$

9. $\dfrac{\frac{6}{x + 1} - \frac{1}{x}}{\frac{2}{x} - \frac{4}{x + 1}}$

10. $\frac{8}{r + 2} - \frac{7}{4r + 8}$

11. $\frac{x}{x + y} + \frac{2y}{x - y}$

12. $\frac{3p^2 - 6p}{p + 5} \div \frac{p^2 - 4}{8p + 40}$

13. $\dfrac{x-2}{9} \cdot \dfrac{5}{8-4x}$

14. $\dfrac{a-4}{3} + \dfrac{11}{6} = \dfrac{a+1}{2}$

15. $\dfrac{b^2+b-6}{b^2+2b-8} \cdot \dfrac{b^2+8b+16}{3b+12}$

16. $\dfrac{10z^2-5z}{3z^3-6z^2} \div \dfrac{2z^2+5z-3}{z^2+z-6}$

17. $\dfrac{5}{x^2-2x} - \dfrac{3}{x^2-4}$

18. $\dfrac{6}{t+1} + \dfrac{4}{5t+5} = \dfrac{34}{15}$

19. $\dfrac{\frac{5}{x} - \frac{3}{y}}{\frac{9x^2-25y^2}{x^2y}}$

20. $\dfrac{-2}{a^2+2a-3} - \dfrac{5}{3-3a} = \dfrac{4}{3a+9}$

21. $\dfrac{4y^2-13y+3}{2y^2-9y+9} \div \dfrac{4y^2+11y-3}{6y^2-5y-6}$

22. $\dfrac{8}{3k+9} - \dfrac{8}{15} = \dfrac{2}{5k+15}$

23. $\dfrac{3r}{r-2} = 1 + \dfrac{6}{r-2}$

24. $\dfrac{6z^2-5z-6}{6z^2+5z-6} \cdot \dfrac{12z^2-17z+6}{12z^2-z-6}$

25. $\dfrac{-1}{3-x} - \dfrac{2}{x-3}$

26. $\dfrac{\frac{t}{4} - \frac{1}{t}}{1 + \frac{t+4}{t}}$

27. $\dfrac{2}{y+1} - \dfrac{3}{y^2-y-2} = \dfrac{3}{y-2}$

28. $\dfrac{7}{2x^2-8x} + \dfrac{3}{x^2-16}$

29. $\dfrac{3}{y-3} - \dfrac{3}{y^2-5y+6} = \dfrac{2}{y-2}$

30. $\dfrac{2k + \frac{5}{k-1}}{3k - \frac{2}{k}}$

8.5 Applications of Rational Expressions

OBJECTIVE 1 Find the value of an unknown variable in a formula. Formulas may contain rational expressions, such as $t = \frac{d}{r}$ and $\frac{1}{f} = \frac{1}{p} + \frac{1}{q}$.

EXAMPLE 1 Finding the Value of a Variable in a Formula

In physics, the focal length, f, of a lens is given by the formula

$$\frac{1}{f} = \frac{1}{p} + \frac{1}{q},$$

where p is the distance from the object to the lens and q is the distance from the lens to the image. See Figure 4. Find q if $p = 20$ cm and $f = 10$ cm.

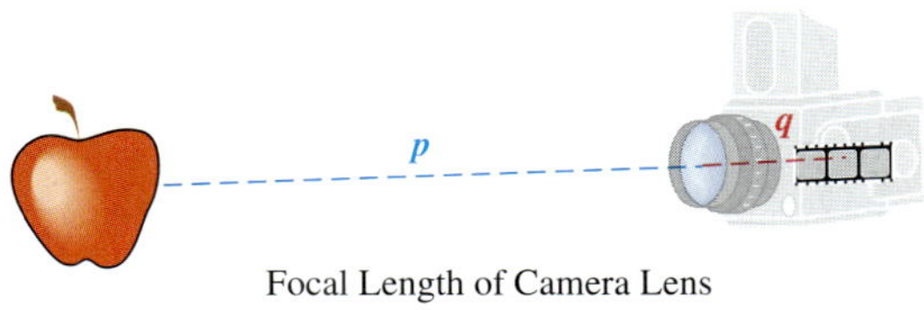

Focal Length of Camera Lens

Figure 4

Replace f with 10 and p with 20.

$$\frac{1}{f} = \frac{1}{p} + \frac{1}{q} \qquad \text{Solve this equation for } q.$$

$$\frac{1}{10} = \frac{1}{20} + \frac{1}{q} \qquad \text{Let } f = 10,\ p = 20.$$

$$20q \cdot \frac{1}{10} = 20q\left(\frac{1}{20} + \frac{1}{q}\right) \qquad \text{Multiply by the LCD, } 20q.$$

$$20q \cdot \frac{1}{10} = 20q\left(\frac{1}{20}\right) + 20q\left(\frac{1}{q}\right) \qquad \text{Distributive property.}$$

$$2q = q + 20 \qquad \text{Multiply.}$$

$$q = 20 \qquad \text{Subtract } q.$$

The distance from the lens to the image is 20 cm.

Work Problem 1 at the Side. ▶

OBJECTIVE 2 Solve a formula for a specified variable. The goal in solving for a specified variable is to isolate it on one side of the equals sign.

EXAMPLE 2 Solving a Formula for a Specified Variable

Solve $\frac{1}{f} = \frac{1}{p} + \frac{1}{q}$ for p.

$$\frac{1}{f} = \frac{1}{p} + \frac{1}{q}$$

$$fpq \cdot \frac{1}{f} = fpq\left(\frac{1}{p} + \frac{1}{q}\right) \qquad \text{Multiply by the LCD, } fpq.$$

$$pq = fq + fp \qquad \text{Distributive property}$$

Continued on Next Page

OBJECTIVES

1. Find the value of an unknown variable in a formula.
2. Solve a formula for a specified variable.
3. Solve applications using proportions.
4. Solve applications about distance, rate, and time.
5. Solve applications about work rates.

1 Use the formula given in Example 1 to answer each part.

(a) Find p if $f = 15$ and $q = 25$.

(b) Find f if $p = 6$ and $q = 9$.

(c) Find q if $f = 12$ and $p = 16$.

ANSWERS

1. (a) $\frac{75}{2}$ **(b)** $\frac{18}{5}$ **(c)** 48

2 Solve $\frac{3}{p} + \frac{3}{q} = \frac{5}{r}$ for q.

Transform the equation so that the terms with p (the specified variable) are on the same side. One way to do this is to subtract fp from each side.

$$pq = fq + fp$$

$$pq - fp = fq \quad \text{Subtract } fp.$$

$$p(q - f) = fq \quad \text{Factor out } p.$$

This is a key step.

$$p = \frac{fq}{q - f} \quad \text{Divide by } q - f.$$

◀ Work Problem **2** at the Side.

EXAMPLE 3 **Solving a Formula for a Specified Variable**

Solve $I = \frac{nE}{R + nr}$ for n.

$$I = \frac{nE}{R + nr}$$

$$(R + nr)I = (R + nr)\frac{nE}{R + nr} \quad \text{Multiply by } R + nr.$$

$$RI + nrI = nE$$

$$RI = nE - nrI \quad \text{Subtract } nrI.$$

$$RI = n(E - rI) \quad \text{Factor out } n.$$

$$\frac{RI}{E - rI} = n \quad \text{Divide by } E - rI.$$

3 Solve $A = \frac{Rr}{R + r}$ for R.

CAUTION

Refer to the steps in Examples 2 and 3 that factor out the desired variable. ***The variable for which you are solving must be a factor on only one side of the equation,*** so that each side can be divided by the remaining factor in the last step.

◀ Work Problem **3** at the Side.

We can now solve problems that translate into equations with rational expressions. To do so, we continue to use the six-step problem-solving method from **Section 2.3.**

OBJECTIVE 3 **Solve applications using proportions.** A **ratio** is a comparison of two quantities. The ratio of a to b may be written in any of the following ways:

$$a \text{ to } b, \quad a : b, \quad \text{or} \quad \frac{a}{b}. \qquad \text{Ratio of } a \text{ to } b$$

Ratios are usually written as quotients in algebra. A **proportion** is a statement that two ratios are equal, such as

$$\frac{a}{b} = \frac{c}{d}. \qquad \text{Proportion}$$

Proportions are a useful and important type of rational equation.

Answers

2. $q = \frac{3rp}{5p - 3r}$, or $q = \frac{-3rp}{3r - 5p}$

3. $R = \frac{-Ar}{A - r}$, or $R = \frac{Ar}{r - A}$

EXAMPLE 4 **Solving a Proportion**

In 2005, about 15 of every 100 Americans had no health insurance coverage. The population at that time was about 296 million. How many million Americans had no health insurance? (*Source*: U.S. Census Bureau.)

Step 1 **Read** the problem.

Step 2 **Assign a variable.** Let $x =$ the number (in millions) who had no health insurance.

Step 3 **Write an equation.** To get an equation, set up a proportion. The ratio 15 to 100 should equal the ratio x to 296.

$$\frac{15}{100} = \frac{x}{296} \quad \text{Write a proportion.}$$

Step 4 **Solve.**

$$29{,}600\left(\frac{15}{100}\right) = 29{,}600\left(\frac{x}{296}\right) \quad \text{Multiply by a common denominator.}$$

$$4440 = 100x \quad \text{Simplify.}$$

$$x = 44.4 \quad \text{Divide by 100.}$$

Step 5 **State the answer.** There were about 44.4 million Americans with no health insurance in 2005.

Step 6 **Check** that the ratio of 44.4 million to 296 million equals $\frac{15}{100}$.

Work Problem 4 at the Side. ▶

EXAMPLE 5 **Solving a Proportion Involving Rates**

Marissa's car uses 10 gal of gas to travel 210 mi. She has 5 gal of gas in the car, and she still needs to drive 640 mi. If we assume the car continues to use gas at the same rate, how many more gallons will she need?

Step 1 **Read** the problem.

Step 2 **Assign a variable.** Let $x =$ the additional number of gallons of gas.

Step 3 **Write an equation.** To get an equation, set up a proportion

$$\frac{\text{gallons} \rightarrow 10}{\text{miles} \rightarrow 210} = \frac{5 + x \leftarrow \text{gallons}}{640 \leftarrow \text{miles}}$$

Step 4 **Solve.** We could multiply by the LCD $10 \cdot 21 \cdot 64$. Instead we use an alternative method that involves *cross products:* For $\frac{a}{b} = \frac{c}{d}$ to be true, then the cross products ad and bc must be equal. Thus,

$$10 \cdot 640 = 210(5 + x) \quad \text{If } \tfrac{a}{b} = \tfrac{c}{d}, \text{ then } ad = bc.$$

$$6400 = 1050 + 210x \quad \text{Multiply; distributive property}$$

$$5350 = 210x \quad \text{Subtract 1050.}$$

$$25.5 \approx x. \quad \text{Divide by 210.}$$

Step 5 **State the answer.** Marissa will need about 25.5 more gallons of gas.

Step 6 **Check.** The 25.5 gal plus the 5 gal equals 30.5 gal.

$$\frac{30.5}{640} \approx \mathbf{0.048} \quad \text{and} \quad \frac{10}{210} \approx \mathbf{0.048}$$

Since the ratios are equal, the answer is correct.

Work Problem 5 at the Side. ▶

4 Solve the problem.

In 2006, approximately 11.7% (that is, 11.7 of every 100) of the 73,740,000 children under 18 yr of age in the United States had no health insurance. How many such children were uninsured? (*Source*: U.S. Census Bureau.)

5 Solve the problem.

Lauren's car uses 15 gal of gasoline to drive 495 mi. She has 6 gal of gasoline in the car, and she wants to know how much more gasoline she will need to drive 600 mi. If we assume that the car continues to use gasoline at the same rate, how many more gallons will she need? (Round your answer to the nearest tenth.)

ANSWERS

4. 8,627,580
5. 12.2 more gallons

OBJECTIVE 4 Solve applications about distance, rate, and time. The next examples use the distance formula $d = rt$ that was first introduced in **Section 2.2.** A familiar example of a rate is speed, which is the ratio of distance to time, or $r = \frac{d}{t}$.

EXAMPLE 6 Solving a Problem about Distance, Rate, and Time

A paddle wheeler goes 10 mi against the current in a river in the same time that it goes 15 mi with the current. If the speed of the current is 3 mph, find the speed of the boat in still water.

Step 1 **Read** the problem. We must find the speed of the boat in still water.

Step 2 **Assign a variable.**

Let $x =$ the speed of the boat in still water.

When the boat is traveling *against* the current, the current slows the boat down, and the speed of the boat is the difference between its speed in still water and the speed of the current. So, the speed against the current is $(x - 3)$ mph.

When the boat is traveling *with* the current, the current speeds the boat up, and the speed of the boat is the sum of its speed in still water and the speed of the current, that is, $(x + 3)$ mph.

Thus, $x - 3 =$ the speed of the boat *against* the current,

and $x + 3 =$ the speed of the boat *with* the current.

Because the time is the same going against the current as with the current, find time in terms of distance and rate (speed) for each situation. Start with the distance formula,

$$d = rt,$$

and divide each side by r to get $t = \frac{d}{r}$. Against the current, the distance is 10 mi and the rate is $(x - 3)$ mph, giving

$$t = \frac{d}{r} = \frac{10}{x - 3}. \quad \text{Time against the current}$$

With the current, the distance is 15 mi and the rate is $(x + 3)$ mph, so

$$t = \frac{d}{r} = \frac{15}{x + 3}. \quad \text{Time with the current}$$

This information is summarized in the following table.

	Distance	Rate	Time
Against Current	10	$x - 3$	$\frac{10}{x - 3}$
With Current	15	$x + 3$	$\frac{15}{x + 3}$

Times are equal.

Step 3 **Write an equation.** Because the times are equal,

$$\frac{10}{x - 3} = \frac{15}{x + 3}.$$

Continued on Next Page

Step 4 **Solve.** $\dfrac{10}{x-3} = \dfrac{15}{x+3}$

$$(x-3)(x-3)\left(\frac{10}{x-3}\right) = (x+3)(x-3)\left(\frac{15}{x+3}\right) \quad \text{Multiply by the LCD.}$$

$$10(x+3) = 15(x-3) \quad \text{Multiply.}$$

$$10x + 30 = 15x - 45 \quad \text{Distributive property}$$

$$30 = 5x - 45 \quad \text{Subtract } 10x.$$

$$75 = 5x \quad \text{Add 45.}$$

$$15 = x \quad \text{Divide by 5.}$$

Step 5 **State the answer.** The speed of the boat in still water is 15 mph.

Step 6 **Check** the answer: $\dfrac{10}{15-3} = \dfrac{15}{15+3}$ is true.

Work Problem **6** at the Side. ▶

6 Solve the problem.

A plane travels 100 mi against the wind in the same time that it takes to travel 120 mi with the wind. The wind speed is 20 mph.

(a) Complete this table.

	d	r	t
Against Wind	100	$x - 20$	
With Wind	120	$x + 20$	

(b) Find the speed of the plane in still air.

EXAMPLE 7 Solving a Problem about Distance, Rate, and Time

At O'Hare International Airport in Chicago, Cheryl and Bill are walking to the gate (at the same speed) to catch their flight to Denver. Bill steps onto the moving sidewalk and continues to walk while Cheryl uses the stationary sidewalk. If the sidewalk moves at 1 m per sec and Bill saves 50 sec covering the 300-m distance, what is their walking speed?

Step 1 **Read** the problem. We must find their walking speed.

Step 2 **Assign a variable.** Let x represent their walking speed in meters per second. Thus Cheryl travels at x meters per second and Bill travels at $(x + 1)$ meters per second. Express their times in terms of the known distances and the variable rates. As in Example 6, start with $d = rt$ and divide each side by r to get $t = \frac{d}{r}$. For Cheryl, the distance is 300 m and the rate is x, so Cheryl's time is

$$t = \frac{d}{r} = \frac{300}{x}. \quad \text{Cheryl's time}$$

Bill travels 300 m at a rate of $x + 1$, so his time is

$$t = \frac{d}{r} = \frac{300}{x+1}. \quad \text{Bill's time}$$

This information is summarized in the following table.

	Distance	Rate	Time
Cheryl	300	x	$\frac{300}{x}$
Bill	300	$x + 1$	$\frac{300}{x+1}$

Step 3 **Write an equation** using the times from the table.

Bill's time is Cheryl's time less 50 seconds.

$$\frac{300}{x+1} = \frac{300}{x} - 50$$

Continued on Next Page

Answers

6. (a) $\dfrac{100}{x-20}$; $\dfrac{120}{x+20}$ (b) 220 mph

7 Solve the problem.

Kathy Manley drove 300 mi north from San Antonio, mostly on the freeway. She usually averaged 55 mph, but an accident slowed her speed through Dallas to 15 mph. If her trip took 6 hr, how many miles did she drive at reduced speed?

	d	r	t
Normal Speed	$300 - x$	55	
Reduced Speed	x	15	

Step 4 **Solve.**

$$\frac{300}{x+1} = \frac{300}{x} - 50$$

$$x(x+1)\left(\frac{300}{x+1}\right) = x(x+1)\left(\frac{300}{x} - 50\right) \quad \text{Multiply by the LCD, } x(x+1).$$

$$x(x+1)\left(\frac{300}{x+1}\right) = x(x+1)\left(\frac{300}{x}\right) - x(x+1)(50) \quad \text{Distributive property}$$

$$300x = 300(x+1) - 50x(x+1) \quad \text{Multiply.}$$

$$300x = 300x + 300 - 50x^2 - 50x \quad \text{Distributive property}$$

$$50x^2 + 50x - 300 = 0 \quad \text{Standard form}$$

$$x^2 + x - 6 = 0 \quad \text{Divide by 50.}$$

$$(x+3)(x-2) = 0 \quad \text{Factor.}$$

$$x + 3 = 0 \quad \text{or} \quad x - 2 = 0 \quad \text{Zero-factor property}$$

$$x = -3 \quad \text{or} \quad x = 2 \quad \text{Solve each equation.}$$

Discard the negative answer, since speed cannot be negative.

Step 5 **State the answer.** Their walking speed is 2 m per sec.

Step 6 **Check** the answer in the words of the original problem.

◀ Work Problem **7** at the Side.

OBJECTIVE 5 Solve applications about work rates. Problems about work are closely related to distance problems.

Problem-Solving Hint

People work at different rates. If the letters r, t, and A represent the rate at which the work is done, the time required, and the amount of work accomplished, respectively, then $A = rt$. Notice the similarity to the distance formula, $d = rt$.

Amount of work can be measured in terms of jobs accomplished. Thus, if 1 job is completed, $A = 1$, and the formula gives the rate as

$$1 = rt$$

$$r = \frac{1}{t}.$$

To solve a work problem, we begin by using the following fact to express all rates of work.

Rate of Work

If a job can be accomplished in t units of time, then the rate of work is

$$\frac{1}{t} \textbf{ job per unit of time.}$$

See if you can identify the six problem-solving steps in the next example.

Answer

7. $11\frac{1}{4}$ mi

EXAMPLE 8 Solving a Problem about Work

Letitia and Kareem are working on a neighborhood cleanup. Kareem can clean up all the trash in the area in 7 hr, while Letitia can do the same job in 5 hr. How long will it take them if they work together?

Let $x =$ the number of hours it will take the two people working together. Just as we made a table for the distance formula, $d = rt$, make a table here for $A = rt$, with $A = 1$. Since $A = 1$, the rate for each person will be $\frac{1}{t}$, where t is the time it takes the person to complete the job alone. For example, since Kareem can clean up all the trash in 7 hr, his rate is $\frac{1}{7}$ of the job per hour. Similarly, Letitia's rate is $\frac{1}{5}$ of the job per hour.

	Rate	Time Working Together	Fractional Part of the Job Done
Kareem	$\frac{1}{7}$	x	$\frac{1}{7}x$
Letitia	$\frac{1}{5}$	x	$\frac{1}{5}x$

Since together they complete 1 job, the sum of the fractional parts accomplished by them should equal 1.

$$\underbrace{\frac{1}{7}x}_{\text{Part done by Kareem}} + \underbrace{\frac{1}{5}x}_{\text{part done by Letitia}} \underset{\text{is}}{=} \underbrace{1}_{\text{1 whole job.}}$$

$$35\left(\frac{1}{7}x + \frac{1}{5}x\right) = 35 \cdot 1 \quad \text{Multiply by the LCD, 35.}$$

$$5x + 7x = 35 \quad \text{Distributive property}$$

$$12x = 35 \quad \text{Combine like terms.}$$

$$x = \frac{35}{12} \quad \text{Divide by 12.}$$

Working together, Kareem and Letitia can do the entire job in $\frac{35}{12}$ hr, or 2 hr, 55 min. Check this result in the original problem.

Work Problem 8 at the Side. ▶

There is another way to approach problems about work. For instance, in Example 8, x represents the number of hours it will take the two people working together to complete the entire job. In one hour, $\frac{1}{x}$ of the entire job will be completed. Kareem completes $\frac{1}{7}$ of the job in one hour, and Letitia completes $\frac{1}{5}$ of the job, so the sum of their rates should equal $\frac{1}{x}$. Thus,

$$\frac{1}{7} + \frac{1}{5} = \frac{1}{x}.$$

Multiplying each side of this equation by $35x$ gives $5x + 7x = 35$. This is the same equation we got in Example 8 in the third line from the bottom. Thus, the solution of the equation is the same using either approach.

8 Solve each problem.

(a) Stan needs 45 min to do the dishes, while Deb can do them in 30 min. How long will it take them if they work together?

	Rate	Time Working Together	Fractional Part of the Job Done
Stan	$\frac{1}{45}$	x	
Deb	$\frac{1}{30}$	x	

(b) Suppose it takes Stan 35 min to do the dishes, and together they can do them in 15 min. How long will it take Deb to do them alone?

ANSWERS

8. (a) 18 min **(b)** $26\frac{1}{4}$ min

Math in the Media

A "BIG LEAGUE" APPLICATION OF RATIONAL EXPRESSIONS AND EQUATIONS

In the 1994 movie *Little Big League*, young Billy Heywood (Luke Edwards) inherits the Minnesota Twins baseball team and becomes manager. He leads the team to the Division Championship and then to the playoffs. But before the final playoff game, the biggest game of the year, he can't keep his mind on his job because a homework problem is giving him trouble:

> *If Joe can paint a house in 3 hours and Sam can paint the same house in 5 hours, how long does it take for them to do it together?*

With the help of one of his players, he is able to solve the problem.

1. Use the method described in Example 8 of **Section 8.5** to solve this problem.

2. Before the player was able to solve the problem correctly, Billy got "help" from some of the other players. The incorrect answers they gave him were

(a) 15 hr **(b)** 8 hr **(c)** 4 hr.

Explain the faulty reasoning behind each of these incorrect answers.

3. The player who gave Billy the correct answer solved the problem as follows:

> *Using the simple formula a times b over a plus b, we get our answer of one and seven-eighths.*

Show that if it takes one person a hours to complete one job and another b hours to complete the same job, then the expression stated by the player,

$$\frac{a \cdot b}{a + b}$$

actually does give the number of hours it would take them to do the job together. (*Hint:* Refer to Example 8, and use a and b rather than 7 and 5. Then solve the resulting formula for x.)

8.5 Exercises

FOR EXTRA HELP MyMathLab Math XL PRACTICE WATCH DOWNLOAD READ REVIEW

In Exercises 1–4, a familiar formula is given. Give the letter of the choice that is an equivalent form of the given formula.

1. $p = br$ (percent)

A. $b = \frac{p}{r}$ **B.** $r = \frac{b}{p}$

C. $b = \frac{r}{p}$ **D.** $p = \frac{r}{b}$

2. $V = LWH$ (geometry)

A. $H = \frac{LW}{V}$ **B.** $L = \frac{V}{WH}$

C. $L = \frac{WH}{V}$ **D.** $W = \frac{H}{VL}$

3. $m = \frac{F}{a}$ (physics)

A. $a = mF$ **B.** $F = \frac{m}{a}$

C. $F = \frac{a}{m}$ **D.** $F = ma$

4. $I = \frac{E}{R}$ (electricity)

A. $R = \frac{I}{E}$ **B.** $R = IE$

C. $E = \frac{I}{R}$ **D.** $E = RI$

Solve each problem. See Example 1.

5. A gas law in chemistry says that

$$\frac{PV}{T} = \frac{pv}{t}.$$

Suppose that $T = 300$, $t = 350$, $V = 9$, $P = 50$, and $v = 8$. Find p.

6. In work with electric circuits, the formula

$$\frac{1}{a} = \frac{1}{b} + \frac{1}{c}$$

occurs. Find b if $a = 8$ and $c = 12$.

7. A formula from anthropology says that

$$c = \frac{100b}{L}.$$

Find L if $c = 80$ and $b = 5$.

8. The gravitational force between two masses is given by

$$F = \frac{GMm}{d^2}.$$

Find M to the nearest thousandth if $F = 10$, $G = 6.67 \times 10^{-11}$, $m = 1$, and $d = 3 \times 10^{-6}$.

Solve each formula for the specified variable. See Examples 2 and 3.

9. $F = \frac{GMm}{d^2}$ for G (physics)

10. $F = \frac{GMm}{d^2}$ for M (physics)

11. $\frac{1}{a} = \frac{1}{b} + \frac{1}{c}$ for a (electricity)

12. $\frac{1}{a} = \frac{1}{b} + \frac{1}{c}$ for b (electricity)

13. $\frac{PV}{T} = \frac{pv}{t}$ for v (chemistry)

14. $\frac{PV}{T} = \frac{pv}{t}$ for T (chemistry)

15. $I = \frac{nE}{R + nr}$ for r (engineering)

16. $a = \frac{V - v}{t}$ for V (physics)

17. $A = \frac{1}{2}h(b + B)$ for b (mathematics)

18. $S = \frac{n}{2}(a + \ell)d$ for n (mathematics)

19. $\frac{E}{e} = \frac{R + r}{r}$ for r (engineering)

20. $y = \frac{x + z}{a - x}$ for x

21. To solve the equation $m = \frac{ab}{a - b}$ for a, what is the first step?

22. Suppose you are asked to solve the equation

$$rp - rq = p + q$$

for r. What is the first step?

Solve each problem mentally. Use proportions in Exercises 23 and 24.

23. In a mathematics class, 3 of every 4 students are girls. If there are 28 students in the class, how many are girls? How many are boys?

24. In a certain southern state, sales tax on a purchase of \$1.50 is \$0.12. What is the sales tax on a purchase of \$9.00?

25. If Marin can mow her yard in 2 hr, what is her rate (in job per hour)?

26. A van traveling from Atlanta to Detroit averages 50 mph and takes 14 hr to make the trip. What is the driving distance from Atlanta to Detroit?

Use a proportion to solve each problem. Give answers to the nearest tenth if an approximation is needed. See Examples 4 and 5.

27. On a map of the United States, the distance between Seattle and Durango is 4.125 in. The two cities are actually 1238 miles apart. On this same map, what would be the distance between Chicago and El Paso, two cities that are actually 1606 mi apart? (*Source:* Universal Map Atlas.)

28. On a map of the United States, the distance between Reno and Phoenix is 2.5 in. The two cities are actually 768 miles apart. On this same map, what would be the distance between St. Louis and Jacksonville, two cities that are actually 919 mi apart? (*Source:* Universal Map Atlas.)

29. On a world globe, the distance between New York and Cairo, two cities that are actually 5619 mi apart, is 8.5 in. On this same globe, how far apart are Madrid and Rio de Janeiro, two cities that are actually 5045 mi apart? (*Source:* Author's globe, *World Almanac and Book of Facts.*)

30. On a world globe, the distance between San Francisco and Melbourne, two cities that are actually 7856 mi apart, is 11.375 in. On this same globe, how far apart are Mexico City and Singapore, two cities that are actually 10,327 mi apart? (*Source:* Author's globe, *World Almanac and Book of Facts.*)

Solve each problem. See Examples 4 and 5.

31. On May 23, 2008, the Boston Red Sox were in first place in the East Division of the American League, having won 31 of their first 50 regular season games. If the team continued to win the same fraction of its games, how many games would the Red Sox win for the complete 162-game season? Round your answer to the nearest whole number. (*Source:* www.mlb.com)

32. During 2004–2005, the ratio of teachers to students in public elementary and secondary schools was approximately 1 to 16. If a public school had 846 students, how many teachers would be at the school if this ratio was valid for that school? Round your answer to the nearest whole number. (*Source*: U.S. National Center for Education Statistics.)

33. Biologists tagged 500 fish in a lake on January 1. On February 1 they returned and collected a random sample of 400 fish, 8 of which had been previously tagged. Approximately how many fish does the lake have based on this experiment?

34. Suppose that in the experiment of Exercise 33, 10 of the previously tagged fish were collected on February 1. What would be the estimate of the fish population?

35. Bruce Johnston's Shelby Cobra uses 5 gal of gasoline to drive 156 mi. He has 3 gal of gasoline in the car, and he wants to know how much more gasoline he will need to drive 300 mi. If we assume that the car continues to use gasoline at the same rate, how many more gallons will he need?

36. Mike Love's T-bird uses 6 gal of gasoline to drive 141 miles. He has 4 gal of gasoline in the car, and he wants to know how much more gasoline he will need to drive 275 mi. If we assume that the car continues to use gasoline at the same rate, how many more gallons will he need?

Nurses use proportions to determine the amount of a drug to administer when the dose of the drug is measured in milligrams but the drug is packaged in a diluted form in milliliters. (*Source:* Hoyles, Celia, Richard Noss, and Stefano Pozzi, "Proportional Reasoning in Nursing Practice," *Journal for Research in Mathematics Education,* January 2001.) *For example, to find the number of milliliters of fluid needed to administer* 300 mg *of a drug that comes packaged as* 120 mg *in* 2 mL *of fluid, a nurse sets up the proportion*

$$\frac{120 \text{ mg}}{2 \text{ mL}} = \frac{300 \text{ mg}}{x \text{ mL}},$$

where x represents the amount to administer in milliliters. Use this method to find the correct dose for each prescription.

37. 120 mg of Amakacine packaged as 100 mg in 2-mL vials

38. 1.5 mg of morphine packaged as 20 mg ampules diluted in 10 mL of fluid

In geometry, it is shown that two triangles with corresponding angle measures equal, called **similar triangles,** *have corresponding sides proportional. For example, in the figure, angle A = angle D, angle B = angle E, and angle C = angle F, so the triangles are similar. Then the following ratios of corresponding sides are equal.*

$$\frac{4}{6} = \frac{6}{9} = \frac{2x + 1}{2x + 5}$$

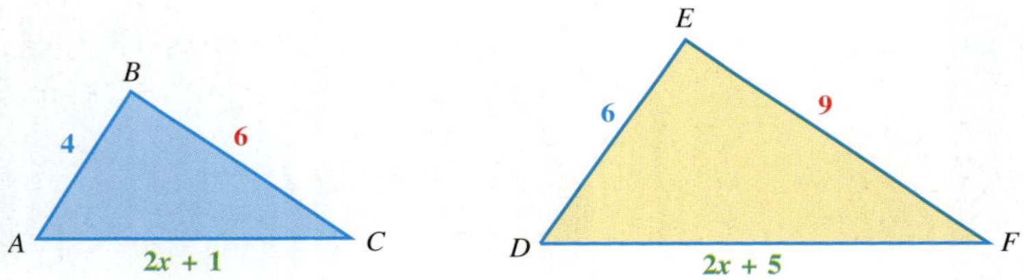

39. Solve for x using the given proportion to find the lengths of the third sides of the triangles.

40. Suppose the following triangles are similar. Find y and the lengths of the two longest sides of each triangle.

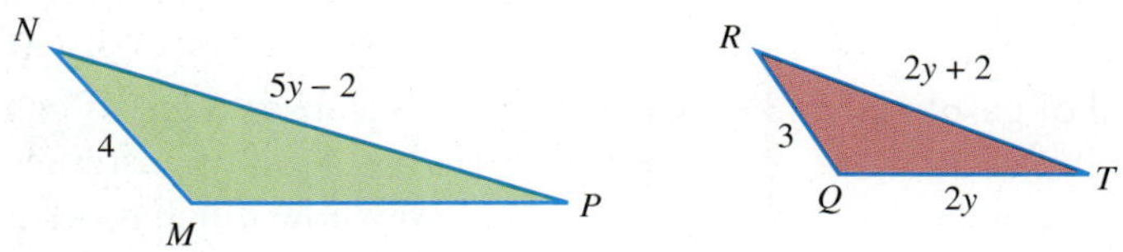

Solve each problem. See Examples 6 and 7.

41. Kellen's boat goes 12 mph. Find the rate of the current of the river if she can go 6 mi upstream in the same amount of time she can go 10 mi downstream.

	Distance	Rate	Time
Downstream	10	$12 + x$	
Upstream	6	$12 - x$	

42. Kasey can travel 8 mi upstream in the same time it takes her to go 12 mi downstream. Her boat goes 15 mph in still water. What is the rate of the current?

	Distance	Rate	Time
Downstream			
Upstream			

43. On his drive from Montpelier, Vermont, to Columbia, South Carolina, Dylan Davis averaged 51 mph. If he had been able to average 60 mph, he would have reached his destination 3 hr earlier. What is the driving distance between Montpelier and Columbia?

44. Leah drove from her apartment to her parents' house for the weekend. Driving to their house on Saturday morning, she was able to average 60 mph because traffic was light. However, returning on Sunday night, she was able to average only 45 mph on the same route, because traffic was heavy. The drive on Sunday took her 1.5 hr longer than the drive on Saturday. What is the distance between Leah's apartment and her parents' house?

45. A private plane traveled from San Francisco to a secret rendezvous. It averaged 200 mph. On the return trip, the average speed was 300 mph. If the total traveling time was 4 hr, how far from San Francisco was the secret rendezvous?

46. Johnny averages 30 mph when he drives on the old highway to his favorite fishing hole, and he averages 50 mph when most of his route is on the interstate. If both routes are the same length, and he saves 2 hr by traveling on the interstate, how far away is the fishing hole?

47. On the first part of a trip to Carmel traveling on the freeway, Marge averaged 60 mph. On the rest of the trip, which was 10 mi longer than the first part, she averaged 50 mph. Find the total distance to Carmel if the second part of the trip took 30 min more than the first part.

48. While on vacation, Jim and Annie decided to drive all day. During the first part of their trip on the highway, they averaged 60 mph. When they got to Houston, traffic caused them to average only 30 mph. The distance they drove in Houston was 100 mi less than their distance on the highway. What was their total driving distance if they spent 50 min more on the highway than they did in Houston?

Solve each problem. See Example 8.

49. Butch and Peggy want to pick up the mess that their grandson, Grant, has made in his playroom. Butch could do it in 15 min working alone. Peggy, working alone, could clean it in 12 min. How long will it take them if they work together?

	Rate	Time Working Together	Fractional Part of the Job Done
Butch	$\frac{1}{15}$	x	
Peggy	$\frac{1}{12}$	x	

50. Lou can groom Jay Beckenstein's dogs in 8 hr, but it takes his business partner, Janet, only 5 hr to groom the same dogs. How long will it take them to groom Jay's dogs if they work together?

	Rate	Time Working Together	Fractional Part of the Job Done
Lou	$\frac{1}{8}$	x	
Janet	$\frac{1}{5}$	x	

51. Jerry and Kuba are laying a hardwood floor. Working alone, Jerry can do the job in 20 hr. If the two of them work together, they can complete the job in 12 hr. How long would it take Kuba to lay the floor working alone?

52. Mrs. Disher is a high school mathematics teacher. She can grade a set of chapter tests in 5 hr working alone. If her student teacher Mr. Howes helps her, it will take 3 hr to grade the tests. How long would it take Mr. Howes to grade the tests if he worked alone?

53. If a vat of acid can be filled by an inlet pipe in 10 hr and emptied by an outlet pipe in 20 hr, how long will it take to fill the vat if both pipes are open?

54. A winery has a vat to hold Chardonnay. An inlet pipe can fill the vat in 9 hr, while an outlet pipe can empty it in 12 hr. How long will it take to fill the vat if both the outlet and the inlet pipes are open?

55. Suppose that Hortense and Mort can clean their entire house in 7 hr, while their toddler, Mimi, just by being around, can completely mess it up in only 2 hr. If Hortense and Mort clean the house while Mimi is at her grandma's, and then start cleaning up after Mimi the minute she gets home, how long does it take from the time Mimi gets home until the whole place is a shambles?

56. An inlet pipe can fill an artificial lily pond in 60 min, while an outlet pipe can empty it in 80 min. Through an error, both pipes are left open. How long will it take for the pond to fill?

8.6 Variation

Certain types of functions are very common, especially in business and the physical sciences. These are functions where y depends on a multiple of x, or y depends on a number divided by x. In such situations, y is said to *vary directly as x* (in the first case) or *vary inversely as x* (in the second case). For example, by the distance formula, the distance traveled varies directly as the rate (or speed) and the time. The simple interest formula and the formulas for area and volume are other familiar examples of *direct variation*.

By contrast, the force required to keep a car from skidding on a curve varies inversely as the radius of the curve. Another example of *inverse variation* is how travel time is inversely proportional to rate or speed.

OBJECTIVES

1. **Write an equation expressing direct variation.**
2. **Find the constant of variation, and solve direct variation problems.**
3. **Solve inverse variation problems.**
4. **Solve joint variation problems.**
5. **Solve combined variation problems.**

OBJECTIVE 1 Write an equation expressing direct variation. The circumference of a circle is given by the formula $C = 2\pi r$, where r is the radius of the circle. See the figure. Circumference is always a constant multiple of the radius. (C is always found by multiplying r by the constant 2π.) Thus,

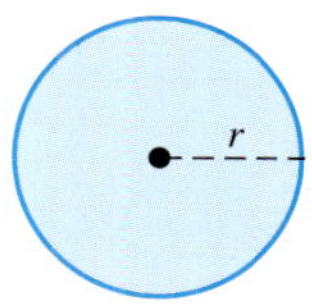

$C = 2\pi r$

As the *radius increases*, the *circumference increases*.

The reverse is also true.

As the *radius decreases*, the *circumference decreases*.

Because of this, the circumference is said to *vary directly* as the radius.

> **Direct Variation**
>
> **y varies directly as x** if there exists a real number k such that
>
> $$y = kx.$$
>
> Also, y is said to be **proportional to** x. The number k is called the **constant of variation.** In direct variation, for $k > 0$, as the value of x increases, the value of y also increases. Similarly, as x decreases, y decreases.

OBJECTIVE 2 Find the constant of variation, and solve direct variation problems. ***The direct variation equation $y = kx$ defines a linear function, where the constant of variation k is the slope of the line.*** For example, we wrote the equation

$$y = 4.50x$$

to describe the cost y to buy x gallons of gas in Example 6 of **Section 4.3.** The cost varies directly as, or is proportional to, the number of gallons of gas purchased. That is,

As the *number* of gallons of gas *increases, cost increases*.

The reverse is also true.

As the *number* of gallons of gas *decreases, cost decreases*.

The constant of variation k is 4.50, the cost of 1 gallon of gas.

1 Find the constant of variation, and write a direct variation equation.

(a) Ginny Michaud is paid a daily wage. One month she worked 17 days and earned \$1334.50.

(b) Distance varies directly as time (at a constant speed). A car travels 100 mi at a constant speed in 2 hr.

2 The charge (in dollars) to customers for electricity (in kilowatt-hours) varies directly as the number of kilowatt-hours used. It costs \$52 to use 800 kilowatt-hours. Find the cost to use 1000 kilowatt-hours.

ANSWERS

1. **(a)** $k = 78.50$; Let E represent her earnings for d days. Then $E = 78.50d$.
   (b) $k = 50$; Let d represent the distance traveled in h hours. Then $d = 50h$.
2. \$65

EXAMPLE 1 Finding the Constant of Variation and the Variation Equation

Stella Frolick is paid an hourly wage. One week she worked 43 hr and was paid \$795.50. How much does she earn per hour?

Let h represent the number of hours she works and P represent her corresponding pay. Then, P **varies directly as** h, so

$$P = kh.$$

Here k represents Stella's hourly wage. Since $P = 795.50$ when $h = 43$,

$$795.50 = 43k$$

$$k = 18.50. \quad \text{Use a calculator.}$$

This is the constant of variation.

Her hourly wage is \$18.50, and P and h are related by

$$P = 18.50h.$$

◀ Work Problem **1** at the Side.

EXAMPLE 2 Solving a Direct Variation Problem

Hooke's law for an elastic spring states that the distance a spring stretches is proportional to the force applied. If a force of 150 newtons* stretches a certain spring 8 cm, how much will a force of 400 newtons stretch the spring?

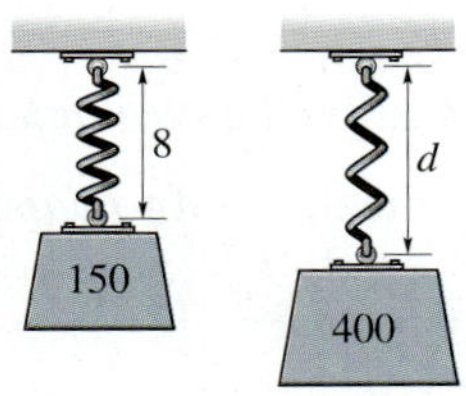

Figure 5

See Figure 5. If d is the distance the spring stretches and f is the force applied, then $d = kf$ for some constant k. Since a force of 150 newtons stretches the spring 8 cm, we can use these values to find k.

$$d = kf \quad \text{Variation equation}$$

$$8 = k \cdot 150 \quad \text{Let } d = 8 \text{ and } f = 150.$$

$$k = \frac{8}{150} \quad \text{Solve for } k.$$

$$k = \frac{4}{75} \quad \text{Lowest terms}$$

Substitute $\frac{4}{75}$ for k in the variation equation $d = kf$ to get

$$d = \frac{4}{75}f.$$

For a force of 400 newtons,

$$d = \frac{4}{75}(400) = \frac{64}{3}. \quad \text{Let } f = 400.$$

The spring will stretch $\frac{64}{3}$ cm, or $21\frac{1}{3}$ cm, if a force of 400 newtons is applied.

◀ Work Problem **2** at the Side.

*A newton is a unit of measure of force used in physics.

In summary, use the following steps to solve a variation problem.

Solving a Variation Problem

Step 1 Write the variation equation.

Step 2 Substitute the initial values and solve for k.

Step 3 Rewrite the variation equation with the value of k from Step 2.

Step 4 Substitute the remaining values, solve for the unknown, and find the required answer.

The direct variation equation $y = kx$ is a linear equation. However, other kinds of variation involve other types of equations. For example, one variable can be proportional to a power of another variable.

Direct Variation as a Power

y* varies directly as the *n*th power of *x if there exists a real number k such that

$$y = kx^n.$$

An example of direct variation as a power is the formula for the area of a circle, $A = \pi r^2$. Here, π is the constant of variation, and the area varies directly as the square of the radius.

EXAMPLE 3 Solving a Direct Variation Problem

The distance a body falls from rest varies directly as the square of the time it falls (disregarding air resistance). If a skydiver falls 64 ft in 2 sec, how far will she fall in 8 sec?

Step 1 If d represents the distance the skydiver falls and t the time it takes to fall, then d is a function of t, and

$$d = kt^2$$

for some constant k.

Step 2 To find the value of k, use the fact that the skydiver falls 64 ft in 2 sec.

$$d = kt^2 \quad \text{Variation equation}$$

$$64 = k(2)^2 \quad \text{Let } d = 64 \text{ and } t = 2.$$

$$k = 16 \quad \text{Find } k.$$

Step 3 Using 16 for k, the variation equation becomes

$$d = 16t^2.$$

Step 4 Let $t = 8$ to find the number of feet the skydiver will fall in 8 sec.

$$d = 16(8)^2 = 1024 \quad \text{Let } t = 8.$$

The skydiver will fall 1024 ft in 8 sec.

Work Problem 3 at the Side. ▶

3 The area of a circle varies directly as the square of its radius. A circle with radius 3 in. has area 28.278 in.^{2}.

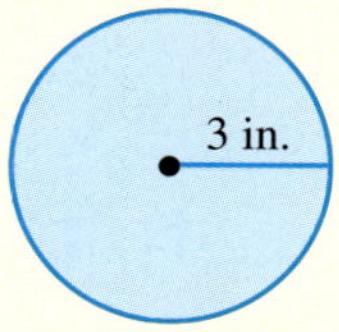

(a) Write a variation equation and give the value of k.

(b) What is the area of a circle with radius 4.1 in.?

Answers

3. (a) $A = kr^2$; 3.142
(b) 52.817 in.^{2} (to the nearest thousandth)

OBJECTIVE 3 Solve inverse variation problems. In direct variation, where $k > 0$, as x increases, y increases. Similarly, as x decreases, y decreases. Another type of variation is *inverse variation*. ***With inverse variation, where k > 0, as one variable increases, the other variable decreases.***

For example, in a closed space, volume decreases as pressure increases, as illustrated by a trash compactor. See Figure 6. As the compactor presses down, the pressure on the trash increases; in turn, the trash occupies a smaller space.

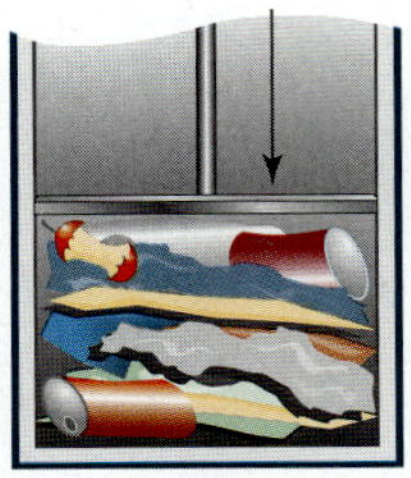

Figure 6

> **Inverse Variation**
>
> **y varies inversely as x** if there exists a real number k such that
>
> $$y = \frac{k}{x}.$$
>
> Also, **y varies inversely as the nth power of x** if there exists a real number k such that
>
> $$y = \frac{k}{x^n}.$$

The inverse variation equation also defines a function. Since x is in the denominator, these functions are rational functions, as seen in **Section 8.1.** Another example of inverse variation comes from the distance formula. In its usual form, the formula is

$$d = rt.$$

Dividing each side by r gives

$$t = \frac{d}{r}.$$

Here, t (time) varies inversely as r (rate or speed), with d (distance) serving as the constant of variation. For example, if the distance between Chicago and Des Moines is 300 mi, then

$$t = \frac{300}{r},$$

and the values of r and t might be any of the following.

$$\left.\begin{aligned} r &= 50, t = 6 \\ r &= 60, t = 5 \\ r &= 75, t = 4 \end{aligned}\right\} \text{As } r \text{ increases, } t \text{ decreases.} \qquad \left.\begin{aligned} r &= 30, t = 10 \\ r &= 25, t = 12 \\ r &= 20, t = 15 \end{aligned}\right\} \text{As } r \text{ decreases, } t \text{ increases.}$$

If we *increase* the rate (speed) we drive, time *decreases*. If we *decrease* the rate (speed) we drive, time *increases*.

EXAMPLE 4 Solving an Inverse Variation Problem

The weight of an object above Earth varies inversely as the square of its distance from the center of Earth. A space shuttle in an elliptical orbit has a maximum distance from the center of Earth **(apogee)** of 6700 mi. Its minimum distance from the center of Earth **(perigee)** is 4090 mi. See Figure 7. If an astronaut in the shuttle weighs 57 lb at its apogee, what does the astronaut weigh at its perigee?

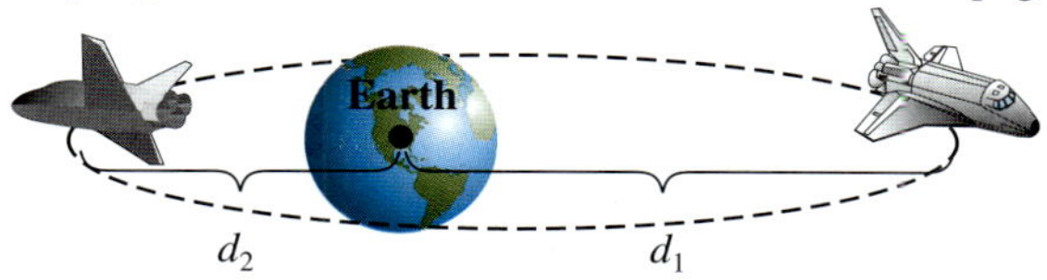

Figure 7

If w is the weight and d is the distance from the center of Earth, then

$$w = \frac{k}{d^2}$$

for some constant k. At the apogee the astronaut weighs 57 lb, and the distance from the center of Earth is 6700 mi. Use these values to find k.

$$57 = \frac{k}{(6700)^2} \qquad \text{Let } w = 57 \text{ and } d = 6700.$$

$$k = 57(6700)^2 \qquad \text{Solve for } k.$$

Then the weight at the perigee with $d = 4090$ mi is

$$w = \frac{57(6700)^2}{(4090)^2} \approx 153 \text{ lb.} \qquad \text{Use a calculator.}$$

Work Problem **4** *at the Side.* ▶

OBJECTIVE 4 Solve joint variation problems. It is possible for one variable to depend on several others. If one variable varies directly as the *product* of several other variables (perhaps raised to powers), the first variable is said to *vary jointly* as the others.

Joint Variation

y varies jointly as x and z if there exists a real number k such that

$$y = kxz.$$

CAUTION

Note that *and* in the expression "y varies jointly as x *and* z" translates as the product

$$y = kxz.$$

The word *and* does not indicate addition here.

4 If the temperature is constant, the volume of a gas varies inversely as the pressure. For a certain gas, the volume is 10 cm^3 when the pressure is 6 kg per cm^2.

(a) Find the variation equation.

(b) Find the volume when the pressure is 12 kg per cm^2.

ANSWERS

4. **(a)** $V = \frac{60}{P}$ **(b)** 5 cm^3

5 The volume of a rectangular box of a given height is proportional to its width and length. A box with width 2 ft and length 4 ft has volume 12 ft^3. Find the volume of a box with the same height that is 3 ft wide and 5 ft long.

EXAMPLE 5 **Solving a Joint Variation Problem**

The interest on a loan or an investment is given by the formula $I = prt$. Here, for a given principal p, the interest earned I varies jointly as the interest rate r and the time t that the principal is left at interest. If an investment earns \$100 interest at 5% for 2 yr, how much interest will the same principal earn at 4.5% for 3 yr?

We use the formula $I = prt$, where p is the constant of variation because it is the same for both investments. For the first investment,

$$I = prt$$

$$100 = p(0.05)(2) \quad \text{Let } I = 100,\ r = 0.05, \text{ and } t = 2.$$

$$100 = 0.1p$$

$$p = 1000. \quad \text{Divide by 0.1.}$$

Now we find I when $p = 1000$, $r = 0.045$, and $t = 3$.

$$I = 1000(0.045)(3) = 135 \quad \text{Let } p = 1000,\ r = 0.045, \text{ and } t = 3.$$

The interest will be \$135.

◀ Work Problem **5** at the Side.

6 The maximum load that a cylindrical column with a circular cross section can hold varies directly as the fourth power of the diameter of the cross section and inversely as the square of the height. A 9-m column 1 m in diameter will support 8 metric tons. How many metric tons can be supported by a column 12 m high and $\frac{2}{3}$ m in diameter?

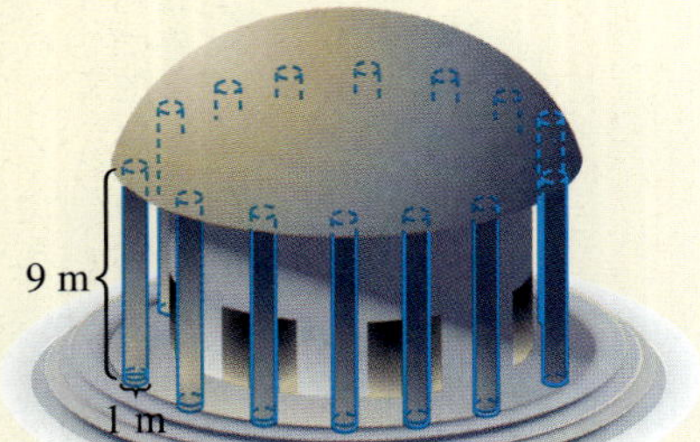

Load = 8 metric tons

OBJECTIVE 5 **Solve combined variation problems.** There are many combinations of direct and inverse variation, called **combined variation.**

EXAMPLE 6 **Solving a Combined Variation Problem**

Body mass index, or BMI, is used by physicians to assess a person's level of fatness. A BMI from 19 through 25 is considered desirable. BMI varies directly as an individual's weight in pounds and inversely as the square of the individual's height in inches. A man who weighs 118 lb and is 64 in. tall has a BMI of 20. (The BMI is rounded to the nearest whole number.) Find the BMI of a man who weighs 165 lb with a height of 70 in.

Let B represent the BMI, w the weight, and h the height. Then

$$B = \frac{kw}{h^2}. \quad \begin{array}{l} \leftarrow \text{BMI varies directly as the weight.} \\ \leftarrow \text{BMI varies inversely as the square of the height} \end{array}$$

To find k, let $B = 20$, $w = 118$, and $h = 64$.

$$20 = \frac{k(118)}{64^2}$$

$$k = \frac{20(64^2)}{118} \quad \text{Multiply by } 64^2; \text{ divide by 118.}$$

$$k \approx 694 \quad \text{Use a calculator.}$$

Now find B when $k = 694$, $w = 165$, and $h = 70$.

$$B = \frac{694(165)}{70^2} \approx 23 \quad \text{Nearest whole number}$$

The man's BMI is 23.

◀ Work Problem **6** at the Side.

ANSWERS

5. 22.5 ft^3

6. $\frac{8}{9}$ metric ton

8.6 Exercises

FOR EXTRA HELP MyMathLab MathXL PRACTICE WATCH DOWNLOAD READ REVIEW

Determine whether each equation represents direct, inverse, joint, *or* combined *variation.*

1. $y = \frac{3}{x}$

2. $y = \frac{8}{x}$

3. $y = 10x^2$

4. $y = 2x^3$

5. $y = 3xz^4$

6. $y = 6x^3z^2$

7. $y = \frac{4x}{wz}$

8. $y = \frac{6x}{st}$

Solve each problem. See Examples 2–5.

9. If x varies directly as y, and $x = 9$ when $y = 3$, find x when $y = 12$.

10. If x varies directly as y, and $x = 10$ when $y = 7$, find y when $x = 50$.

11. If z varies inversely as w, and $z = 10$ when $w = 0.5$, find z when $w = 8$.

12. If t varies inversely as s, and $t = 3$ when $s = 5$, find s when $t = 5$.

13. p varies jointly as q and r^2, and $p = 200$ when $q = 2$ and $r = 3$. Find p when $q = 5$ and $r = 2$.

14. f varies jointly as g^2 and h, and $f = 50$ when $g = 4$ and $h = 2$. Find f when $g = 3$ and $h = 6$.

15. For $k > 0$, if y varies directly as x, when x increases, y __________, and when x decreases, y __________.

16. For $k > 0$, if y varies inversely as x, when x increases, y __________, and when x decreases, y __________.

17. Explain the difference between inverse variation and direct variation.

18. What is meant by the constant of variation in a direct variation problem? If you were to graph the linear equation $y = kx$ for some constant k, what role would the value of k play in the graph?

Solve each problem involving variation. See Examples 1–6.

19. Matt bought 8 gal of gasoline and paid \$36.79. To the nearest tenth of a cent, what is the price of gasoline per gallon?

20. Nora gives horseback rides at Shadow Mountain Ranch. A 2.5-hr ride costs \$50.00. What is the price per hour?

21. The volume of a can of tomatoes is proportional to the height of the can. If the volume of the can is 300 cm^3 when its height is 10.62 cm, find the volume of a can with height 15.92 cm.

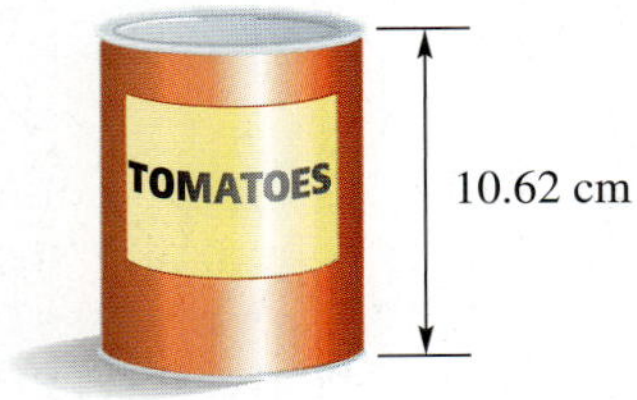

22. The weight of an object on Earth is directly proportional to the weight of that same object on the moon. A 200-lb astronaut would weigh 32 lb on the moon. How much would a 50-lb dog weigh on the moon?

23. For a body falling freely from rest (disregarding air resistance), the distance the body falls varies directly as the square of the time. If an object is dropped from the top of a tower 576 ft high and hits the ground in 6 sec, how far did it fall in the first 4 sec?

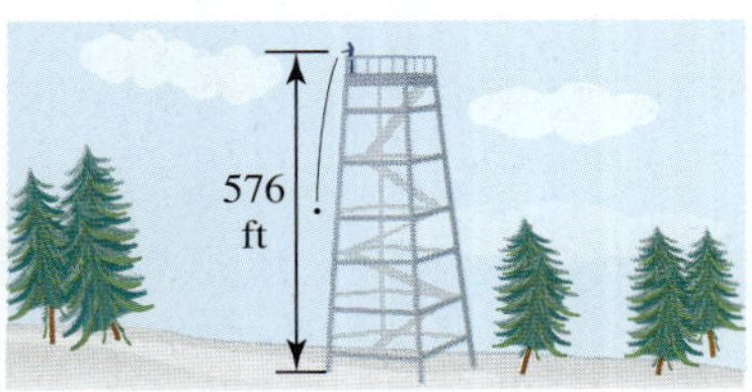

24. The amount of water emptied by a pipe varies directly as the square of the diameter of the pipe. For a certain constant water flow, a pipe emptying into a canal will allow 200 gal of water to escape in an hour. The diameter of the pipe is 6 in. How much water would a 12-in. pipe empty into the canal in an hour, assuming the same water flow?

25. The current in a simple electrical circuit is inversely proportional to the resistance. If the current is 20 amperes (an **ampere** is a unit for measuring current) when the resistance is 5 ohms, find the current when the resistance is 7.5 ohms.

26. The frequency (number of vibrations per second) of a vibrating guitar string varies inversely as its length. That is, a longer string vibrates fewer times in a second than a shorter string. Suppose a guitar string 0.65 m long vibrates 4.3 times per sec. What frequency would a string 0.5 m long have?

27. The amount of light (measured in foot-candles) produced by a light source varies inversely as the square of the distance from the source. If the illumination produced 1 m from a light source is 768 foot-candles, find the illumination produced 6 m from the same source.

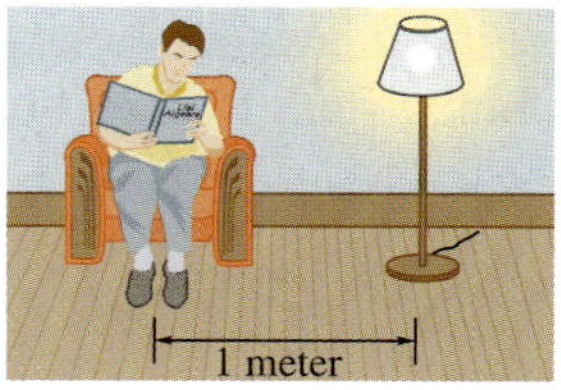

28. The force with which Earth attracts an object above Earth's surface varies inversely with the square of the distance of the object from the center of Earth. If an object 4000 mi from the center of Earth is attracted with a force of 160 lb, find the force of attraction if the object were 6000 mi from the center of Earth.

29. For a given interest rate, simple interest varies jointly as principal and time. If $2000 left in an account for 4 yr earned interest of $280, how much interest would be earned in 6 yr?

30. The collision impact of an automobile varies jointly as its weight and the square of its speed. Suppose a 2000-lb car traveling at 55 mph has a collision impact of 6.1. What is the collision impact of the same car at 65 mph?

31. The force needed to keep a car from skidding on a curve varies inversely as the radius of the curve and jointly as the weight of the car and the square of the speed. If 242 lb of force keep a 2000-lb car from skidding on a curve of radius 500 ft at 30 mph, what force (to the nearest tenth) would keep the same car from skidding on a curve of radius 750 ft at 50 mph?

32. Almost 70% of the new single-family homes sold in the United States in 2006 used natural gas as the primary heating fuel. (*Source:* U.S. Census Bureau.) The volume of gas varies inversely as the pressure and directly as the temperature. (Temperature must be measured in *Kelvin* (K), a unit of measurement used in physics.) If a certain gas occupies a volume of 1.3 L at 300 K and a pressure of 18 newtons per cm^2, find the volume at 340 K and a pressure of 24 newtons per cm^2.

33. The number of long-distance phone calls between two cities in a certain time period varies jointly as the populations of the cities, p_1 and p_2, and inversely as the distance between them. If 80,000 calls are made between two cities 400 mi apart, with populations of 70,000 and 100,000, how many calls are made between cities with populations of 50,000 and 75,000 that are 250 mi apart?

34. A body mass index from 27 through 29 carries a slight risk of weight-related health problems, while one of 30 or more indicates a great increase in risk. Use your own height and weight and the information in Example 6 to determine whether you are at risk.

Exercises 35 and 36 describe weight-estimation formulas that fishermen have used over the years. Girth *is the distance around the body of the fish. Give answers to the nearest tenth. (Source: Sacramento Bee,* November 9, 2000.)

35. The weight of a bass varies jointly as its girth and the square of its length. A prize-winning bass weighed in at 22.7 lb and measured 36 in. long with 21 in. girth. How much would a bass 28 in. long with 18 in. girth weigh?

36. The weight of a trout varies jointly as its length and the square of its girth. One angler caught a trout that weighed 10.5 lb and measured 26 in. long with 18 in. girth. Find the weight of a trout that is 22 in. long with 15 in. girth.

Relating Concepts (Exercises 37–42) For Individual or Group Work

A routine activity such as pumping gasoline can be related to many of the concepts studied in this chapter. Suppose that premium unleaded costs \$4.45 *per gal.* ***Work Exercises 37–42 in order.***

37. 0 gal of gasoline cost \$0.00, while 1 gal costs \$4.45. Represent these two pieces of information as ordered pairs of the form (gallons, price).

38. Use the information from Exercise 37 to find the slope of the line on which the two points lie.

39. Write the slope-intercept form of the equation of the line on which the two points lie.

40. Using function notation, if $f(x) = ax + b$ represents the line from Exercise 39, what are the values of a and b?

41. How does the value of a from Exercise 40 relate to gasoline in this situation? With relationship to the line, what do we call this number?

42. Why does the equation from Exercise 40 satisfy the conditions for direct variation? In the context of variation, what do we call the value of a?

Chapter 8 Summary

▶ Key Terms

8.1	**rational expression**	A rational expression is the quotient of two polynomials with denominator not 0.
	rational function	A rational function is a function that is defined by a rational expression in the form $f(x) = \frac{P(x)}{Q(x)}$, where $Q(x) \neq 0$.
8.2	**least common denominator (LCD)**	The least common denominator in a group of denominators is the product of all different factors from each denominator, with each factor raised to the greatest power that occurs in any denominator.
8.3	**complex fraction**	A complex fraction is an expression having a fraction in the numerator, denominator, or both.
8.4	**domain of the variable in a rational equation**	The domain of the variable in a rational equation is the intersection (overlap) of the domains of the rational expressions in the equation.
	discontinuous	A graph of a function is discontinuous if there are one or more breaks in the graph.
	vertical asymptote	A rational function in simplest form $f(x) = \frac{P(x)}{x - a}$ has the line $x = a$ as a vertical asymptote; the graph approaches the line on each side but does not intersect it.
	horizontal asymptote	A horizontal line that a graph approaches as $\lvert x \rvert$ gets larger and larger without bound is called a horizontal asymptote.
8.5	**ratio**	A ratio is a comparison of two quantities using a quotient.
	proportion	A proportion is a statement that two ratios are equal.
8.6	**varies directly**	y varies directly as x if there exists a real number k such that $y = kx$.
	varies inversely	y varies inversely as x if there exists a real number k such that $y = \frac{k}{x}$.
	constant of variation	In the equations for direct and inverse variation, k is the constant of variation.

▶ Test Your Word Power

See how well you have learned the vocabulary in this chapter. Answers, with examples, follow the Quick Review.

1. A **rational expression** is
   - **A.** an algebraic expression made up of a term or the sum of a finite number of terms with real coefficients and integer exponents
   - **B.** a polynomial equation of degree 2
   - **C.** an expression with one or more fractions in the numerator, denominator, or both
   - **D.** the quotient of two polynomials with denominator not zero.

2. In a given set of fractions, the **least common denominator** is
   - **A.** the smallest denominator of all the denominators
   - **B.** the smallest expression that is divisible by all the denominators
   - **C.** the largest integer that evenly divides the numerator and denominator of all the fractions
   - **D.** the largest denominator of all the denominators.

3. A **complex fraction** is
   - **A.** an algebraic expression made up of a term or the sum of a finite number of terms with real coefficients and integer exponents
   - **B.** a polynomial equation of degree 2
   - **C.** an expression with one or more fractions in the numerator, denominator, or both
   - **D.** the quotient of two polynomials with denominator not zero.

4. A **ratio**
   - **A.** compares two quantities using a quotient
   - **B.** says that two quotients are equal
   - **C.** is a product of two quantities
   - **D.** is a difference between two quantities.

5. A **proportion**
   - **A.** compares two quantities using a quotient
   - **B.** says that two ratios are equal
   - **C.** is a product of two quantities
   - **D.** is a difference between two quantities.

▶ Quick Review

Concepts	Examples

8.1 Rational Expressions and Functions; Multiplying and Dividing

Fundamental Property of Rational Numbers

If $\frac{a}{b}$ is a rational number and if c is any nonzero real number, then

$$\frac{a}{b} = \frac{ac}{bc}.$$

$$\frac{3}{4} = \frac{3 \cdot 5}{4 \cdot 5} = \frac{15}{20}$$

Writing a Rational Expression in Lowest Terms

Step 1 Factor the numerator and the denominator completely.

Step 2 Apply the fundamental property.

Write in lowest terms.

$$\frac{2x+8}{x^2-16}$$
$$= \frac{2(x+4)}{(x-4)(x+4)} \quad \text{Factor.}$$
$$= \frac{2}{x-4} \quad \text{Lowest terms}$$

Multiplying Rational Expressions

Step 1 Factor numerators and denominators.

Step 2 Apply the fundamental property.

Step 3 Multiply the remaining factors in the numerator and in the denominator.

Step 4 Check that the product is in lowest terms.

Multiply.

$$\frac{x^2+2x+1}{x^2-1} \cdot \frac{5}{3x+3}$$
$$= \frac{(x+1)^2}{(x-1)(x+1)} \cdot \frac{5}{3(x+1)} \quad \text{Factor.}$$
$$= \frac{5}{3(x-1)} \quad \text{Multiply; lowest terms}$$

Dividing Rational Expressions

Multiply the first rational expression (the dividend) by the reciprocal of the second (the divisor).

Divide.

$$\frac{2x+5}{x-3} \div \frac{2x^2+3x-5}{x^2-9}$$
$$= \frac{2x+5}{x-3} \cdot \frac{x^2-9}{2x^2+3x-5} \quad \text{Multiply by the reciprocal.}$$
$$= \frac{2x+5}{x-3} \cdot \frac{(x+3)(x-3)}{(2x+5)(x-1)} \quad \text{Factor.}$$
$$= \frac{x+3}{x-1} \quad \text{Multiply; lowest terms}$$

8.2 Adding and Subtracting Rational Expressions

Adding or Subtracting Rational Expressions

Step 1 If the denominators are the same, add or subtract the numerators. Place the result over the common denominator.

If the denominators are different, write all rational expressions with the LCD. Then add or subtract the numerators, and place the result over the common denominator.

Step 2 Make sure that the answer is in lowest terms.

Subtract.

$$\frac{1}{x+6} - \frac{3}{x+2}$$
$$= \frac{x+2}{(x+6)(x+2)} - \frac{3(x+6)}{(x+6)(x+2)}$$
$$= \frac{x+2-3(x+6)}{(x+6)(x+2)}$$
$$= \frac{x+2-3x-18}{(x+6)(x+2)}$$
$$= \frac{-2x-16}{(x+6)(x+2)}$$

Concepts	Examples

8.3 Complex Fractions

Simplifying a Complex Fraction

Simplify the complex fraction.

Method 1 Simplify the numerator and denominator separately, as much as possible. Then multiply the numerator by the reciprocal of the denominator. Write the answer in lowest terms.

Method 1

$$\frac{\frac{1}{x^2}-\frac{1}{y^2}}{\frac{1}{x}+\frac{1}{y}} = \frac{\frac{y^2}{x^2y^2}-\frac{x^2}{x^2y^2}}{\frac{y}{xy}+\frac{x}{xy}}$$

$$= \frac{\frac{y^2-x^2}{x^2y^2}}{\frac{y+x}{xy}} = \frac{y^2-x^2}{x^2y^2} \div \frac{y+x}{xy}$$

$$= \frac{(y+x)(y-x)}{x^2y^2} \cdot \frac{xy}{y+x}$$

$$= \frac{y-x}{xy}$$

Method 2 Multiply the numerator and denominator of the complex fraction by the least common denominator of all fractions appearing in the complex fraction. Then simplify the result.

Method 2

$$\frac{\frac{1}{x^2}-\frac{1}{y^2}}{\frac{1}{x}+\frac{1}{y}} = \frac{x^2y^2\left(\frac{1}{x^2}-\frac{1}{y^2}\right)}{x^2y^2\left(\frac{1}{x}+\frac{1}{y}\right)}$$

$$= \frac{y^2-x^2}{xy^2+x^2y}$$

$$= \frac{(y-x)(y+x)}{xy(y+x)}$$

$$= \frac{y-x}{xy}$$

8.4 Equations with Rational Expressions and Graphs

Solving an Equation with Rational Expressions

To solve an equation involving rational expressions, first determine the domain of the variable. Then multiply all the terms in the equation by the least common denominator. Solve the resulting equation. ***Each proposed solution must be checked to see that it is in the domain of the variable in the equation.***

Solve. $\frac{1}{x} + x = \frac{26}{5}$ Note that 0 is excluded from the domain.

$5 - 5x^2 = 26x$ Multiply by $5x$.

$5x^2 - 26x + 5 = 0$ Subtract $26x$.

$(5x - 1)(x - 5) = 0$ Factor.

$5x - 1 = 0$ or $x - 5 = 0$ Zero-factor property

$x = \frac{1}{5}$ or $x = 5$ Solve each equation.

Both check. The solution set is $\left\{\frac{1}{5}, 5\right\}$.

The graph of a rational function of the type covered in this section may have one or more breaks. At such points, the graph will approach an asymptote.

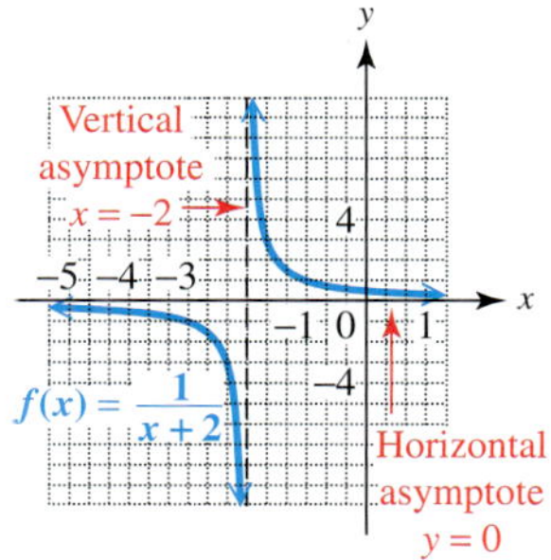

Concepts	Examples

8.5 Applications of Rational Expressions

To solve a motion problem, use the formula

$$d = rt$$

or one of its equivalents,

$$t = \frac{d}{r} \quad \text{or} \quad r = \frac{d}{t}.$$

Solve.

A canal has a current of 2 mph. Find the speed of Amy's boat in still water if it goes 11 mi downstream in the same time that it goes 8 mi upstream.

Let x represent the speed of the boat in still water.

	Distance	Rate	Time
Downstream	11	$x + 2$	$\frac{11}{x+2}$
Upstream	8	$x - 2$	$\frac{8}{x-2}$

Because the times are the same, the equation is

$$\frac{11}{x+2} = \frac{8}{x-2}. \qquad \text{Use } t = \tfrac{d}{r}.$$

$$11(x-2) = 8(x+2) \qquad \text{Multiply by the LCD.}$$

$$11x - 22 = 8x + 16 \qquad \text{Distributive property}$$

$$3x = 38 \qquad \text{Subtract } 8x \text{ and add } 22.$$

$$x = 12\frac{2}{3} \qquad \text{Divide by 3.}$$

The speed in still water is $12\frac{2}{3}$ mph.

To solve a work problem, use the fact that if a complete job is done in t units of time, the rate of work is $\frac{1}{t}$ job per unit of time.

8.6 Variation

If there is some constant k such that

$y = kx^n$, then y varies directly as x^n.

$y = \frac{k}{x^n}$, then y varies inversely as x^n.

$y = kxz$, then y varies jointly as x and z.

The area of a circle **varies directly as** the square of the radius.

$A = kr^2$ Here, $k = \pi$.

Pressure **varies inversely as** volume.

$$p = \frac{k}{V}$$

For a given principal, interest **varies jointly as** interest rate and time.

$I = krt$ k is the given principal.

ANSWERS TO TEST YOUR WORD POWER

1. D; *Examples:* $-\frac{3}{4y^2}, \frac{5x^3}{x+2}, \frac{a+3}{a^2-4a-5}$

2. B; *Example:* The LCD of $\frac{1}{x}, \frac{2}{3}$, and $\frac{5}{x+1}$ is $3x(x+1)$.

3. C; *Examples:* $\dfrac{\frac{2}{3}}{\frac{4}{7}}, \dfrac{x - \frac{1}{x}}{x + \frac{1}{y}}, \dfrac{\frac{2}{a+1}}{a^2-1}$

4. A; *Example:* $\frac{7\text{ in.}}{12\text{ in.}}$ compares two quantities.

5. B; *Example:* The proportion $\frac{2}{3} = \frac{8}{12}$ states that the two ratios are equal.

Chapter 8 Review Exercises

[8.1] *(a) Find all real numbers that are excluded from the domain. (b) Give the domain using set notation.*

1. $f(x) = \dfrac{-7}{3x + 18}$

2. $f(x) = \dfrac{5x + 17}{x^2 - 7x + 10}$

3. $f(x) = \dfrac{9}{x^2 - 18x + 81}$

Write in lowest terms.

4. $\dfrac{12x^2 + 6x}{24x + 12}$

5. $\dfrac{25m^2 - n^2}{25m^2 - 10mn + n^2}$

6. $\dfrac{r - 2}{4 - r^2}$

7. What is meant by the reciprocal of a rational expression?

Multiply or divide. Write the answer in lowest terms.

8. $\dfrac{(2y + 3)^2}{5y} \cdot \dfrac{15y^3}{4y^2 - 9}$

9. $\dfrac{w^2 - 16}{w} \cdot \dfrac{3}{4 - w}$

10. $\dfrac{z^2 - z - 6}{z - 6} \div \dfrac{z^2 + 2z - 15}{z^2 - 6z}$

11. $\dfrac{m^3 - n^3}{m^2 - n^2} \div \dfrac{m^2 + mn + n^2}{m + n}$

[8.2] *Assume that each expression is the denominator of a rational expression. Find the least common denominator for each group.*

12. $32b^3$, $24b^5$

13. $9r^2$, $3r + 1$

14. $6x^2 + 13x - 5$, $9x^2 + 9x - 4$

Add or subtract as indicated.

15. $\dfrac{8}{z} - \dfrac{3}{2z^2}$

16. $\dfrac{5y + 13}{y + 1} - \dfrac{1 - 7y}{y + 1}$

17. $\dfrac{6}{5a + 10} + \dfrac{7}{6a + 12}$

18. $\dfrac{3r}{10r^2 - 3rs - s^2} + \dfrac{2r}{2r^2 + rs - s^2}$

[8.3] *Simplify each complex fraction.*

19. $\dfrac{\dfrac{3}{t}+2}{\dfrac{4}{t}-7}$

20. $\dfrac{\dfrac{2}{m-3n}}{\dfrac{1}{3n-m}}$

21. $\dfrac{\dfrac{3}{p}-\dfrac{2}{q}}{\dfrac{9q^2-4p^2}{qp}}$

22. $\dfrac{x^{-2}-y^{-2}}{x^{-1}-y^{-1}}$

[8.4] *Solve each equation.*

23. $\dfrac{1}{t+4}+\dfrac{1}{2}=\dfrac{3}{2t+8}$

24. $\dfrac{-5m}{m+1}+\dfrac{m}{3m+3}=\dfrac{56}{6m+6}$

25. $\dfrac{2}{k-1}-\dfrac{4k+1}{k^2-1}=\dfrac{-1}{k+1}$

26. $\dfrac{5}{x+2}+\dfrac{3}{x+3}=\dfrac{x}{x^2+5x+6}$

27. After solving the equation

$$\frac{3}{x-3}-\frac{2}{x-2}=\frac{3}{x^2-5x+6},$$

a student got $x = 3$ as her final step. She could not understand why the answer in the back of the book was "∅," because she checked her algebra several times and was sure that all her algebraic work was correct. Was she wrong or was the answer in the back of the book wrong? Explain.

28. Explain the difference between simplifying the expression

$$\frac{4}{x}+\frac{1}{2}-\frac{1}{3}$$

and solving the equation

$$\frac{4}{x}+\frac{1}{2}=\frac{1}{3}.$$

29. Which is the graph of a rational function? Give the equations of its vertical and horizontal asymptotes?

A.

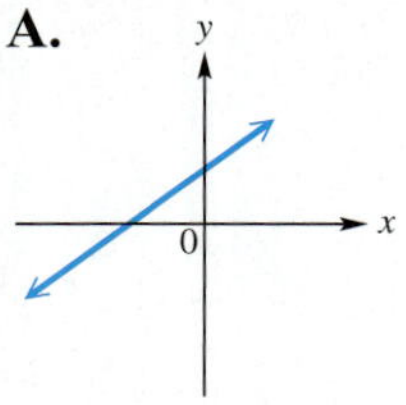

B.

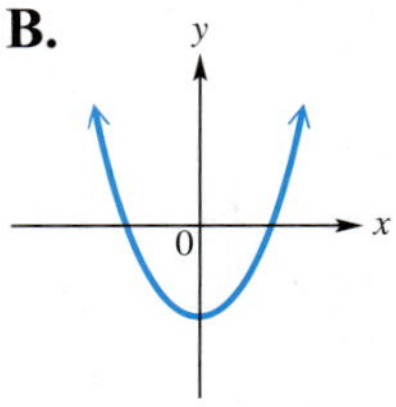

C.

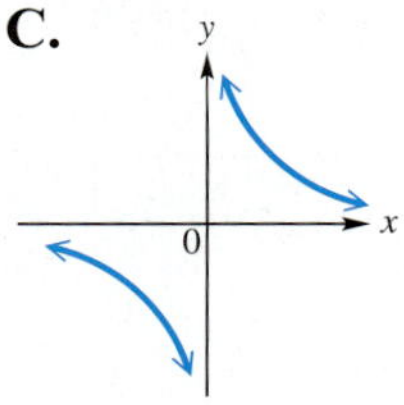

D.

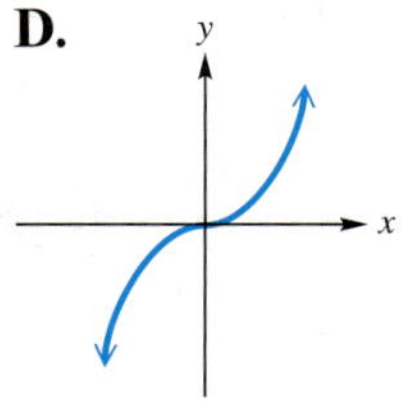

[8.5]

30. According to a law from physics, $\dfrac{1}{A}=\dfrac{1}{B}+\dfrac{1}{C}$. Find A if $B = 30$ and $C = 10$.

Solve each formula for the specified variable.

31. $V = \frac{1}{3}\pi r^2 h$ for h (mathematics)

32. $\mu = \frac{Mv}{M + m}$ for M (electronics)

Solve each problem.

33. A river has a current of 4 km per hr. Find the speed of Lynn McTernan's boat in still water if it goes 40 km downstream in the same time that it takes to go 24 km upstream.

	d	r	t
Upstream	24	$x - 4$	
Downstream	40		

34. A sink can be filled by a cold-water tap in 8 min, and filled by the hot-water tap in 12 min. How long would it take to fill the sink with both taps open?

[8.6] *Solve each variation problem.*

35. In which one of the following does y vary inversely as x?

A. $y = 2x$ **B.** $y = \frac{x}{3}$ **C.** $y = \frac{3}{x}$ **D.** $y = x^2$

36. If m varies inversely as p^2, and $m = 20$ when $p = 2$, find m when $p = 5$.

37. For the subject in a photograph to appear in the same perspective in the photograph as in real life, the viewing distance must be properly related to the amount of enlargement. For a particular camera, the viewing distance varies directly as the amount of enlargement. A picture taken with this camera that is enlarged 5 times should be viewed from a distance of 250 mm. Suppose a print 8.6 times the size of the negative is made. From what distance should it be viewed?

38. The volume of a rectangular box of a given height is proportional to its width and length. A box with width 4 ft and length 8 ft has volume 64 ft^3. Find the volume of a box with the same height that is 3 ft wide and 6 ft long.

Mixed Review Exercises

Write in lowest terms.

39. $\frac{x + 2y}{x^2 - 4y^2}$

40. $\frac{x^2 + 2x - 15}{x^2 - x - 6}$

Perform the indicated operations.

41. $\frac{2}{m} + \frac{5}{3m^2}$

42. $\frac{k^2 - 6k + 9}{1 - 216k^3} \cdot \frac{6k^2 + 17k - 3}{9 - k^2}$

43. $\dfrac{\dfrac{-3}{x} + \dfrac{x}{2}}{1 + \dfrac{x+1}{x}}$

44. $\dfrac{9x^2 + 46x + 5}{3x^2 - 2x - 1} \div \dfrac{x^2 + 11x + 30}{x^3 + 5x^2 - 6x}$

45. $\dfrac{\dfrac{3}{x} - 5}{6 + \dfrac{1}{x}}$

46. $\dfrac{9}{3 - x} - \dfrac{2}{x - 3}$

47. $\dfrac{4y + 16}{30} \div \dfrac{2y + 8}{5}$

48. $\dfrac{t^{-2} + s^{-2}}{t^{-1} - s^{-1}}$

49. $\dfrac{4a}{a^2 - ab - 2b^2} - \dfrac{6b - a}{a^2 + 4ab + 3b^2}$

50. $\dfrac{a}{b} + \dfrac{b}{c} + \dfrac{c}{d}$

Solve each equation.

51. $\dfrac{x + 3}{x^2 - 5x + 4} - \dfrac{1}{x} = \dfrac{2}{x^2 - 4x}$

52. $A = \dfrac{Rr}{R + r}$ for r

53. $1 - \dfrac{5}{r} = \dfrac{-4}{r^2}$

54. $\dfrac{3x}{x - 4} + \dfrac{2}{x} = \dfrac{48}{x^2 - 4x}$

Solve each problem.

55. Anna and Matthew Sudak need to sort a pile of bottles at the recycling center. Working alone, Anna could do the entire job in 9 hr, while Matthew could do the entire job in 6 hr. How long will it take them if they work together?

56. Rebecca Song is a college student who lives in an off-campus apartment. Some days she rides her bike to campus, while other days she walks. When she rides her bike, she gets to her first classroom building 36 min faster than when she walks. If her average walking speed is 3 mph and her average biking speed is 12 mph, how far is it from her apartment to the classroom building?

57. The frequency (number of vibrations per second) of a vibrating guitar string varies inversely as its length. That is, a longer string vibrates fewer times in a second than a shorter string. Suppose a guitar string 0.65 m long vibrates 4.3 times per sec. What frequency would a string 0.5 m long have?

58. The area of a triangle varies jointly as the lengths of the base and height. A triangle with base 10 ft and height 4 ft has area 20 ft^2. Find the area of a triangle with base 3 ft and height 8 ft.

Chapter 8 Test

Use the Chapter Test Prep Video CD to see fully worked-out solutions to any of the exercises you want to review

1. Find all real numbers excluded from the domain of
$f(x) = \frac{x+3}{3x^2 + 2x - 8}$. Then give the domain using set notation.

2. Write $\frac{6x^2 - 13x - 5}{9x^3 - x}$ in lowest terms.

Multiply or divide.

3. $\frac{(x+3)^2}{4} \cdot \frac{6}{2x+6}$

4. $\frac{y^2 - 16}{y^2 - 25} \cdot \frac{y^2 + 2y - 15}{y^2 - 7y + 12}$

5. $\frac{x^2 - 9}{x^3 + 3x^2} \div \frac{x^2 + x - 12}{x^3 + 9x^2 + 20x}$

6. Find the least common denominator for the following group of denominators: $t^2 + t - 6, \quad t^2 + 3t, \quad t^2$.

Add or subtract as indicated.

7. $\frac{7}{6t^2} - \frac{1}{3t}$

8. $\frac{9}{x-7} + \frac{4}{x+7}$

9. $\frac{6}{x+4} + \frac{1}{x+2} - \frac{3x}{x^2 + 6x + 8}$

Simplify each complex fraction.

10. $\dfrac{\frac{12}{r+4}}{\frac{11}{6r+24}}$

11. $\dfrac{\frac{1}{a} - \frac{1}{b}}{\frac{a}{b} - \frac{b}{a}}$

12. $\dfrac{2x^{-2} + y^{-2}}{x^{-1} - y^{-1}}$

13. Identify each of the following as an *expression* to be simplified or an *equation* to be solved. Then simplify the one that is an expression, and solve the one that is an equation.

(a) $\frac{2x}{3} + \frac{x}{4} - \frac{11}{2}$

(b) $\frac{2x}{3} + \frac{x}{4} = \frac{11}{2}$

Solve each equation.

14. $\frac{1}{x} - \frac{4}{3x} = \frac{1}{x-2}$

15. $\frac{y}{y+2} - \frac{1}{y-2} = \frac{8}{y^2 - 4}$

1. ____________________
2. ____________________
3. ____________________
4. ____________________
5. ____________________
6. ____________________
7. ____________________
8. ____________________
9. ____________________
10. ____________________
11. ____________________
12. ____________________
13. (a) ____________________
(b) ____________________
14. ____________________
15. ____________________

16. ______________

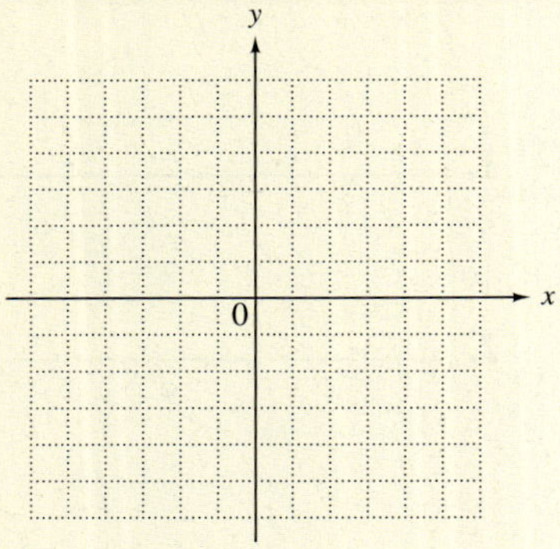

16. Sketch the graph of the function defined by $f(x) = \frac{-2}{x+1}$. Give the equations of its vertical and horizontal asymptotes.

Solve each problem.

17. ______________

17. Wayne can do a job in 9 hr, while Sandra can do the same job in 5 hr. How long would it take them to do the job if they worked together?

18. ______________

18. The rate of the current in a stream is 3 mph. Danielle Lalezhar's boat can go 36 mi downstream in the same time that it takes to go 24 mi upstream. Find the rate of her boat in still water.

19. ______________

19. Biologists collected a sample of 600 fish from Lake Linda on May 1 and tagged each of them. When they returned on June 1, a new sample of 800 fish was collected, and 10 of these had been previously tagged. Use this experiment to determine the approximate fish population of Lake Linda.

20. (a) ______________

(b) ______________

20. In biology, the function defined by

$$g(x) = \frac{5x}{2+x}$$

gives the growth rate g of a population for x units of available food. (*Source:* Smith, J. Maynard, *Models in Ecology,* Cambridge University Press, 1974.)

(a) What amount of food (in appropriate units) would produce a growth rate of 3 units of growth per unit of food?

(b) What is the growth rate if no food is available?

21. ______________

21. The current in a simple electrical circuit is inversely proportional to the resistance. If the current is 80 amps when the resistance is 30 ohms, find the current when the resistance is 12 ohms.

22. ______________

22. The force of the wind blowing on a vertical surface varies jointly as the area of the surface and the square of the velocity. If a wind blowing at 40 mph exerts a force of 50 lb on a surface of 500 ft^2, how much force will a wind of 80 mph place on a surface of 2 ft^2?

Cumulative Review Exercises ▶▶▶ Chapters 1–8

Solve each equation or inequality.

1. $7(2x + 3) - 4(2x + 1) = 2(x + 1)$

2. $|6x - 8| - 4 = 0$

3. $\frac{2}{3}x + \frac{5}{12}x \leq 20$

Solve each problem.

4. Otis Taylor invested some money at 4% interest and twice as much at 3% interest. His interest for the first year was $400. How much did he invest at each rate?

5. A triangle has an area of 42 m^2. The base is 14 m long. Find the height of the triangle.

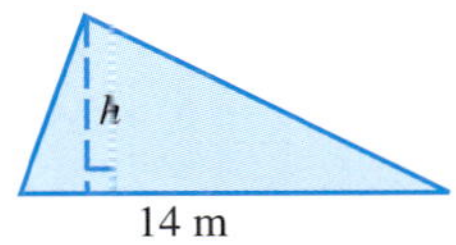

Find the slope of each line.

6. (a) Through $(-5, 8)$ and $(-1, 2)$

(b) Perpendicular to $4x + 3y = 12$, through $(5, 2)$

7. Write an equation of each line in Exercise 6 in the form $y = mx + b$.

Graph.

8. $-4x + 2y = 8$

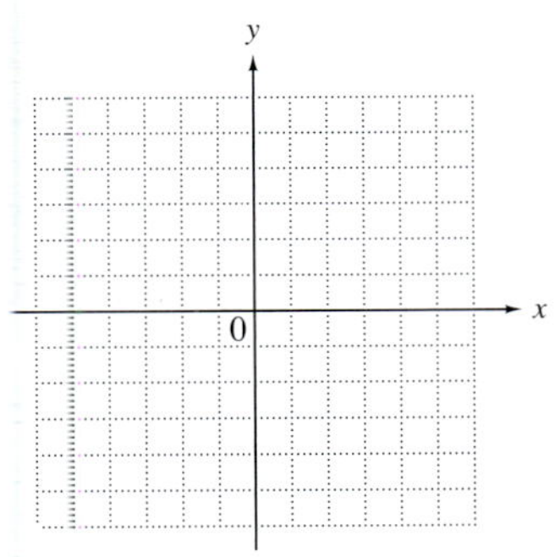

9. $2x + 5y > 10$

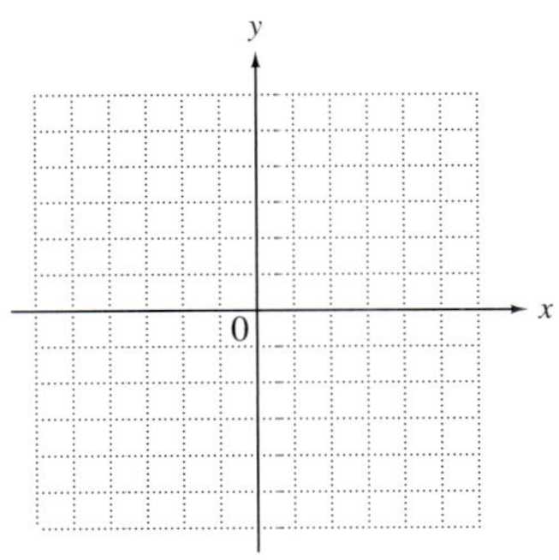

10. $x - y \geq 3$ and $3x + 4y \leq 12$

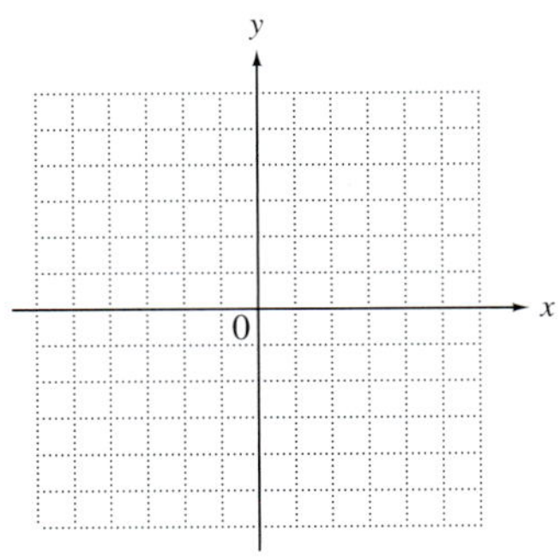

11. Consider the equation $5x - 3y = 8$.

(a) Write y as a function f of x, using function notation $f(x)$.

(b) Find $f(1)$.

12. If $f(x) = 3x + 6$, what is $f(x + 3)$?

Solve each system.

13. $4x - y = -7$
$5x + 2y = 1$

14. $x + y - 2z = -1$
$2x - y + z = -6$
$3x + 2y - 3z = -3$

15. $x + 2y + z = 5$
$x - y + z = 3$
$2x + 4y + 2z = 11$

Perform the indicated operations.

16. $(3y^2 - 2y + 6) - (-y^2 + 5y + 12)$

17. $(3x^3 + 13x^2 - 17x - 7) \div (3x + 1)$

18. $(4f + 3)(3f - 1)$

19. $(7t^3 + 8)(7t^3 - 8)$

20. $\left(\frac{1}{4}x + 5\right)^2$

21. For the polynomial functions defined by

$$f(x) = x^2 + 2x - 3 \quad \text{and} \quad g(x) = 2x^3 - 3x^2 + 4x - 1,$$

find each of the following.

(a) $(f + g)(x)$ **(b)** $(g - f)(x)$ **(c)** $(f + g)(-1)$

Factor each polynomial completely.

22. $2x^2 - 13x - 45$

23. $100t^4 - 25$

24. $8p^3 + 125$

Perform the indicated operations. Express the answer in lowest terms.

25. $\frac{2a^2}{a + b} \cdot \frac{a - b}{4a}$

26. $\frac{x + 4}{x - 2} + \frac{2x - 10}{x - 2}$

27. $\frac{2x}{2x - 1} + \frac{4}{2x + 1} + \frac{8}{4x^2 - 1}$

Solve.

28. $3x^2 + 4x = 7$

29. $\frac{-3x}{x + 1} + \frac{4x + 1}{x} = \frac{-3}{x^2 + x}$

30. $\frac{1}{f} = \frac{1}{p} + \frac{1}{q}$ for q

9

Roots, Radicals, and Root Functions

Tom Skilling is the chief meteorologist for the *Chicago Tribune*. He writes a column titled "Ask Tom Why," in which readers question him on a variety of topics. In the Saturday, August 17, 2002, issue, reader Ted Fleischaker wrote,

> *I cannot remember the formula to calculate the distance to the horizon. I have a stunning view from my 14th-floor condo, 150 ft above the ground. How far can I see?*

Skilling's answer in Section 9.3, Exercise 127, provides a formula for finding the distance to the horizon. The formula includes a *square root*, one of the topics of this chapter.

9.1 Radical Expressions and Graphs

9.2 Rational Exponents

9.3 Simplifying Radical Expressions

9.4 Adding and Subtracting Radical Expressions

9.5 Multiplying and Dividing Radical Expressions

Summary Exercises on Operations with Radicals and Rational Exponents

9.6 Solving Equations with Radicals

9.7 Complex Numbers

9.1 Radical Expressions and Graphs

OBJECTIVES

1. Find roots of numbers.
2. Find principal roots.
3. Graph functions defined by radical expressions.
4. Find *n*th roots of *n*th powers.
5. Use a calculator to find roots.

OBJECTIVE 1 Find roots of numbers. Recall from **Section 1.3** that $6^2 = 36$; that is, 6 *squared* is 36. The opposite (or inverse) of *squaring* a number is taking its *square root*. Thus,

> It is customary to write $\sqrt{\ }$, rather than $\sqrt[2]{\ }$.

$$\sqrt{36} = 6, \quad \text{because} \quad 6^2 = 36.$$

We now extend our discussion of roots to *cube roots* $\sqrt[3]{\ }$, *fourth roots* $\sqrt[4]{\ }$, and higher roots. In general, $\sqrt[n]{a}$ is a number whose nth power equals a. That is,

$$\sqrt[n]{a} = b \quad \textbf{means} \quad b^n = a.$$

The number a is the **radicand,** n is the **index,** or **order,** and the expression $\sqrt[n]{a}$ is a **radical.**

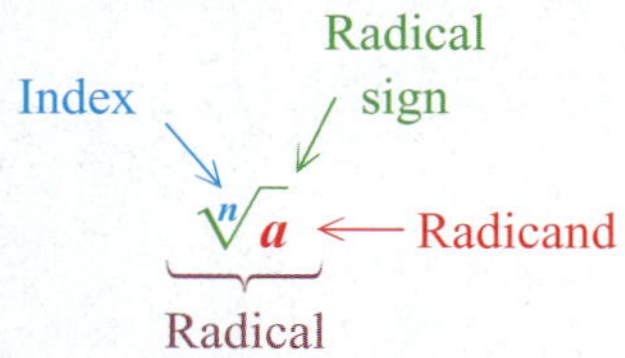

1 Simplify.

(a) $\sqrt[3]{27}$

(b) $\sqrt[3]{1000}$

(c) $\sqrt[4]{256}$

(d) $\sqrt[5]{243}$

(e) $\sqrt[4]{\frac{16}{81}}$

(f) $\sqrt[3]{0.064}$

EXAMPLE 1 Simplifying Higher Roots

Simplify.

(a) $\sqrt[3]{64} = 4$, because $4^3 = 64$.

(b) $\sqrt[3]{125} = 5$, because $5^3 = 125$.

(c) $\sqrt[4]{16} = 2$, because $2^4 = 16$.

(d) $\sqrt[5]{32} = 2$, because $2^5 = 32$.

(e) $\sqrt[3]{\frac{8}{27}} = \frac{2}{3}$, because $\left(\frac{2}{3}\right)^3 = \frac{8}{27}$.

(f) $\sqrt[4]{0.0016} = 0.2$, because $(0.2)^4 = 0.0016$.

◀ Work Problem 1 at the Side.

OBJECTIVE 2 Find principal roots. If n is even, positive numbers have two nth roots. For example, both 4 and -4 are square roots of 16, and 2 and -2 are fourth roots of 16. The notation $\sqrt[n]{a}$ represents the positive root, called the **principal root,** and $-\sqrt[n]{a}$ represents the negative root.

> ***n*th Root**
>
> ***Case 1*** If n is ***even*** and a is ***positive or 0,*** then
>
> $\sqrt[n]{a}$ represents the **principal *n*th root** of a, and
>
> $-\sqrt[n]{a}$ represents the **negative *n*th root** of a.
>
> ***Case 2*** If n is ***even*** and a is ***negative,*** then
>
> $\sqrt[n]{a}$ is not a real number.
>
> ***Case 3*** If n is ***odd,*** then
>
> there is exactly one nth root of a, written $\sqrt[n]{a}$.

If n is even, then the two nth roots of a are often written together as $\pm\sqrt[n]{a}$, with $\pm$ read "positive or negative," or "plus or minus."

ANSWERS

1. (a) 3 (b) 10 (c) 4 (d) 3 (e) $\frac{2}{3}$ (f) 0.4

EXAMPLE 2 Finding Roots

Find each root.

(a) $\sqrt{100} = 10$

While 100 has two square roots, $\sqrt{100}$ represents the principal square root, which is 10.

(b) $-\sqrt{100} = -10$

Here, we want the negative square root, -10.

(c) $\sqrt[4]{81} = 3$ Principal 4th root

(d) $-\sqrt[4]{81} = -3$ Negative 4th root

Parts (a)–(d) illustrate Case 1 in the preceding box.

(e) $\sqrt[4]{-81}$

The index is *even* and the radicand is *negative,* so $\sqrt[4]{-81}$ is not a real number. This is Case 2 in the preceding box.

(f) $\sqrt[3]{8} = 2$, because $2^3 = 8$.

(g) $\sqrt[3]{-8} = -2$, because $(-2)^3 = -8$.

Parts (f) and (g) illustrate Case 3 in the box. The index is *odd,* so each radical represents exactly one *n*th root (regardless of whether the radicand is positive, negative, or 0).

Work Problem 2 at the Side. ▶

OBJECTIVE 3 Graph functions defined by radical expressions.

A **radical expression** is an algebraic expression that contains radicals.

$$3 - \sqrt{x}, \quad \sqrt[3]{x}, \quad \text{and} \quad \sqrt{2x - 1} \qquad \text{Radical expressions}$$

In earlier chapters we graphed functions defined by polynomial and rational expressions. Now we examine the graphs of functions defined by the radical expressions $f(x) = \sqrt{x}$ and $f(x) = \sqrt[3]{x}$.

Figure 1 shows the graph of the **square root function** defined by $f(x) = \sqrt{x}$, together with a table of selected points. Only nonnegative values can be used for x, so the domain is $[0, \infty)$. Because $\sqrt{x}$ is the principal square root of x, it always has a nonnegative value, so the range is also $[0, \infty)$.

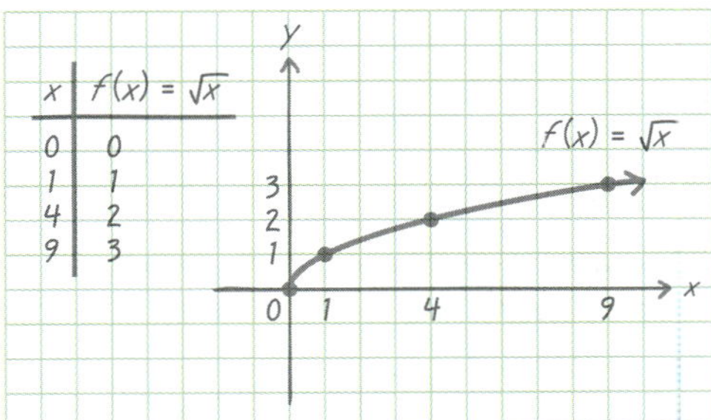

Figure 1

Figure 2 on the next page shows the graph of the **cube root function** defined by $f(x) = \sqrt[3]{x}$, together with a table of selected points. Since any real number (positive, negative, or 0) can be used for x in the cube root function, $\sqrt[3]{x}$ can be positive, negative, or 0. Thus, both the domain and the range of the cube root function are $(-\infty, \infty)$.

2 Find each root.

(a) $\sqrt{36}$

(b) $-\sqrt{36}$

(c) $\sqrt[4]{16}$

(d) $-\sqrt[4]{16}$

(e) $\sqrt[4]{-16}$

(f) $\sqrt[5]{1024}$

(g) $\sqrt[5]{-1024}$

ANSWERS

2. (a) 6 **(b)** -6 **(c)** 2 **(d)** -2
(e) not a real number
(f) 4 **(g)** -4

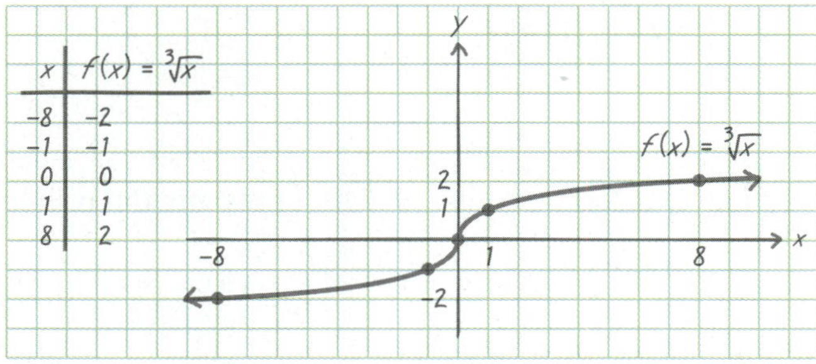

Figure 2

3 Graph each function by creating a table of values. Give the domain and range.

(a) $f(x) = \sqrt{x} + 2$

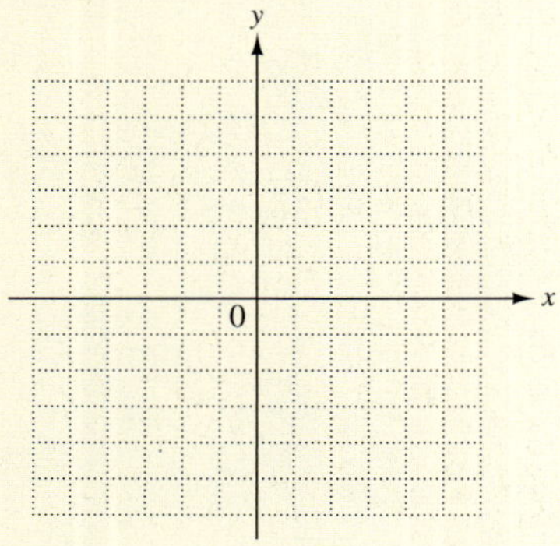

(b) $f(x) = \sqrt[3]{x - 1}$

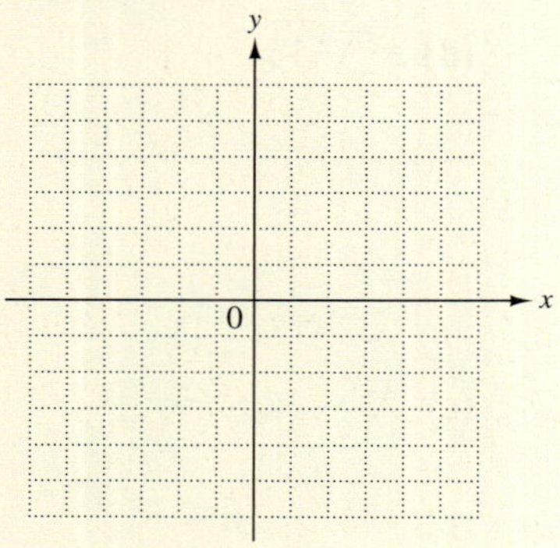

EXAMPLE 3 Graphing Functions Defined with Radicals

Graph each function by creating a table of values. Give the domain and the range.

(a) $f(x) = \sqrt{x - 3}$

A table of values is shown. The x-values were chosen so that the function values are all integers. For the radicand to be nonnegative, we must have $x - 3 \geq 0$, or $x \geq 3$. Therefore, the domain is $[3, \infty)$. Again, function values are positive or 0, so the range is $[0, \infty)$. See the graph in Figure 3.

x	$f(x) = \sqrt{x - 3}$
3	$\sqrt{3 - 3} = 0$
4	$\sqrt{4 - 3} = 1$
7	$\sqrt{7 - 3} = 2$

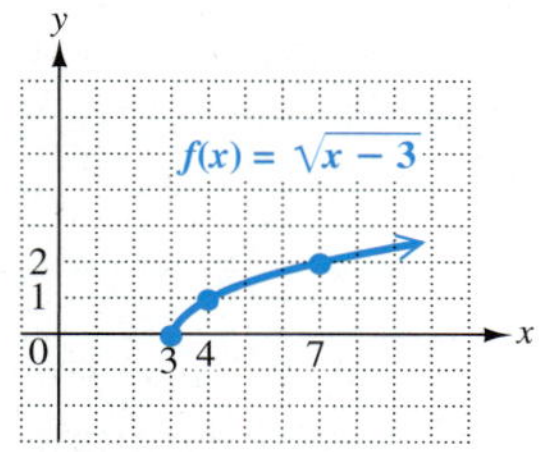

Figure 3

(b) $f(x) = \sqrt[3]{x} + 2$

See the table and Figure 4. Both the domain and the range are $(-\infty, \infty)$.

x	$f(x) = \sqrt[3]{x} + 2$
−8	$\sqrt[3]{-8} + 2 = 0$
−1	$\sqrt[3]{-1} + 2 = 1$
0	$\sqrt[3]{0} + 2 = 2$
1	$\sqrt[3]{1} + 2 = 3$
8	$\sqrt[3]{8} + 2 = 4$

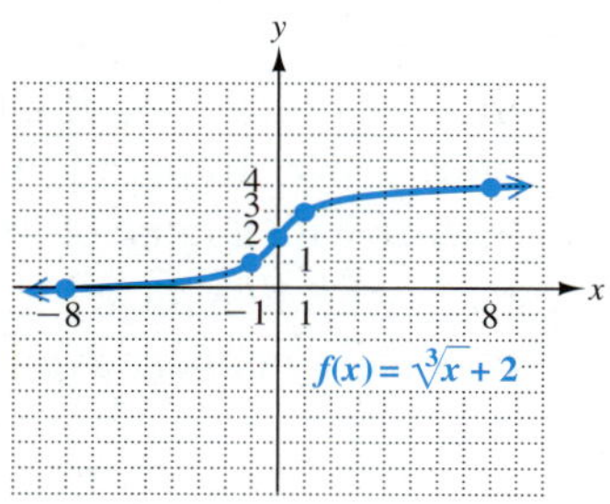

Figure 4

◀ Work Problem **3** at the Side.

OBJECTIVE 4 Find *n*th roots of *n*th powers. What does $\sqrt{a^2}$ equal? Your first answer might be a, but this is not necessarily true. For example, consider the following:

If $a = 6$, then $\sqrt{a^2} = \sqrt{6^2} = \sqrt{36} = 6$.

If $a = -6$, then $\sqrt{a^2} = \sqrt{(-6)^2} = \sqrt{36} = 6$. ← Instead of −6, we get 6, the *absolute value* of −6.

Since the symbol $\sqrt{a^2}$ represents the *nonnegative* square root, we write $\sqrt{a^2}$ with absolute value bars, as $|a|$, because a may be a negative number.

ANSWERS

3. (a) domain: $[0, \infty)$; range: $[2, \infty)$

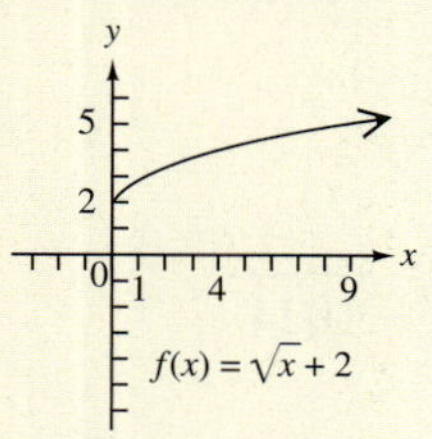

(b) domain: $(-\infty, \infty)$; range: $(-\infty, \infty)$

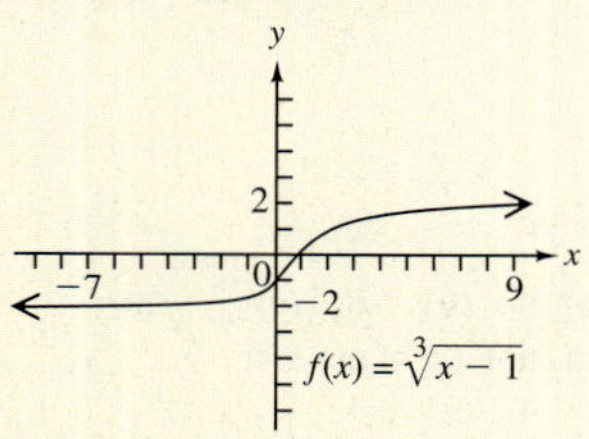

$\sqrt{a^2}$

For any real number a, $\sqrt{a^2} = |a|$.

In words, the principal square root of a^2 is the absolute value of a.

EXAMPLE 4 Simplifying Square Roots Using Absolute Value

Find each square root that is a real number.

(a) $\sqrt{7^2} = |7| = 7$ **(b)** $\sqrt{(-7)^2} = |-7| = 7$

(c) $\sqrt{k^2} = |k|$ **(d)** $\sqrt{(-k)^2} = |-k| = |k|$

Work Problem 4 at the Side. ▶

We can generalize this idea to any nth root.

$\sqrt[n]{a^n}$

If n is an *even* positive integer, then $\sqrt[n]{a^n} = |a|$.

If n is an *odd* positive integer, then $\sqrt[n]{a^n} = a$.

In words, use absolute value when n is even; absolute value is not necessary when n is odd.

EXAMPLE 5 Simplifying Higher Roots Using Absolute Value

Simplify each root.

(a) $\sqrt[6]{(-3)^6} = |-3| = 3$ n is even; use absolute value.

(b) $\sqrt[5]{(-4)^5} = -4$ n is odd.

(c) $-\sqrt[4]{(-9)^4} = -|-9| = -9$ n is even; use absolute value.

(d) $-\sqrt{m^4} = -|m^2| = -m^2$ For all m, $|m^2| = m^2$.

No absolute value bars are needed here because m^2 is nonnegative for any real number value of m.

(e) $\sqrt[3]{a^{12}} = a^4$, because $a^{12} = (a^4)^3$.

(f) $\sqrt[4]{x^{12}} = |x^3|$

We use absolute value bars to guarantee that the result is not negative (because x^3 can be either positive or negative, depending on x). If desired, $|x^3|$ can be written as $x^2 \cdot |x|$.

Work Problem 5 at the Side. ▶

OBJECTIVE 5 Use a calculator to find roots. Radical expressions often represent irrational numbers. To find approximations of such radicals, we usually use a calculator. For example,

$$\sqrt{15} \approx 3.872983346, \quad \sqrt[3]{10} \approx 2.15443469, \quad \text{and} \quad \sqrt[4]{2} \approx 1.189207115,$$

where the symbol $\approx$ means "is approximately equal to." In this book, we often give approximations rounded to three decimal places. Thus,

$$\sqrt{15} \approx 3.873, \quad \sqrt[3]{10} \approx 2.154, \quad \text{and} \quad \sqrt[4]{2} \approx 1.189.$$

4 Find each square root that is a real number.

(a) $\sqrt{15^2}$

(b) $\sqrt{(-12)^2}$

(c) $\sqrt{r^2}$

(d) $\sqrt{(-r)^2}$

5 Simplify.

(a) $\sqrt[4]{(-5)^4}$

(b) $\sqrt[5]{(-7)^5}$

(c) $-\sqrt[6]{(-3)^6}$

(d) $-\sqrt[4]{m^8}$

(e) $\sqrt[3]{x^{24}}$

(f) $\sqrt[6]{y^{18}}$

ANSWERS

4. **(a)** 15 **(b)** 12 **(c)** $|r|$ **(d)** $|r|$

5. **(a)** 5 **(b)** -7 **(c)** -3 **(d)** $-m^2$ **(e)** x^8 **(f)** $|y^3|$

Calculator Tip The methods for finding approximations differ among makes and models, and you should always consult your owner's manual for keystroke instructions. Be aware that graphing calculators often differ from scientific calculators in the order in which keystrokes are made.

Figure 5 shows how the preceding approximations are displayed on a TI-83/84 Plus graphing calculator. In Figure 5(a), eight or nine decimal places are shown, while in Figure 5(b), the number of decimal places is fixed at three.

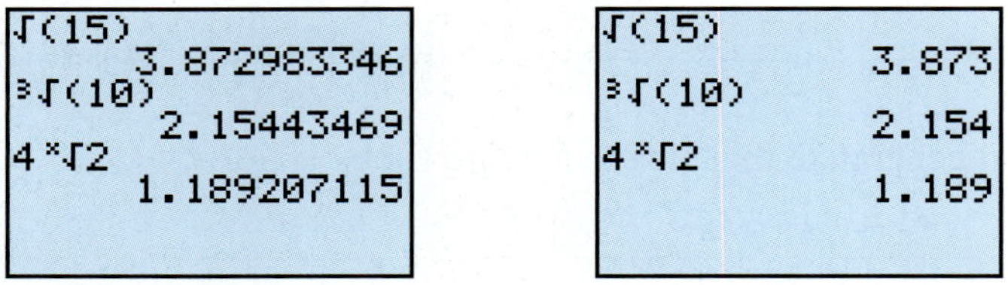

(a) (b)

Figure 5

There is a simple way to check that a calculator approximation is "in the ballpark." Because 16 is a little larger than 15, $\sqrt{16} = 4$ should be a little larger than $\sqrt{15}$. Thus, 3.873 is a reasonable approximation for $\sqrt{15}$.

EXAMPLE 6 Finding Approximations for Roots

Use a calculator to verify that each approximation is correct.

(a) $\sqrt{39} \approx 6.245$ **(b)** $-\sqrt{72} \approx -8.485$

(c) $\sqrt[3]{93} \approx 4.531$ **(d)** $\sqrt[4]{39} \approx 2.499$

◀ *Work Problem* **6** *at the Side.*

EXAMPLE 7 Using Roots to Calculate Resonant Frequency

In electronics, the resonant frequency f of a circuit may be found by the formula

$$f = \frac{1}{2\pi\sqrt{LC}},$$

where f is in cycles per second, L is in henrys, and C is in farads.* Find the resonant frequency f if $L = 5 \times 10^{-4}$ henry and $C = 3 \times 10^{-10}$ farad. Give your answer to the nearest thousand.

Find the value of f when $L = 5 \times 10^{-4}$ and $C = 3 \times 10^{-10}$.

$$f = \frac{1}{2\pi\sqrt{LC}} \quad \text{Given formula}$$

$$f = \frac{1}{2\pi\sqrt{(5 \times 10^{-4})(3 \times 10^{-10})}} \quad \text{Substitute for } L \text{ and } C.$$

$$f \approx 411{,}000 \quad \text{Use a calculator.}$$

The resonant frequency f is approximately 411,000 cycles per sec.

◀ *Work Problem* **7** *at the Side.*

*Henrys and farads are units of measure in electronics.

6 Use a calculator to approximate each radical to three decimal places.

(a) $\sqrt{17}$

(b) $-\sqrt{362}$

(c) $\sqrt[3]{9482}$

(d) $\sqrt[4]{6825}$

7 Use the formula in Example 7 to approximate f to the nearest thousand if

$L = 6 \times 10^{-5}$

and $C = 4 \times 10^{-9}$.

ANSWERS

6. (a) 4.123 **(b)** −19.026 **(c)** 21.166 **(d)** 9.089

7. 325,000 cycles per sec

9.1 Exercises

FOR EXTRA HELP — MyMathLab — Math XL PRACTICE — WATCH — DOWNLOAD — READ — REVIEW

Match each expression from Column I with the equivalent choice from Column II. Answers may be used once, more than once, or not at all. See Examples 1 and 2.

I		II	
1. $-\sqrt{16}$	**2.** $\sqrt{-16}$	**A.** 3	**B.** -2
3. $\sqrt[3]{-27}$	**4.** $\sqrt[5]{-32}$	**C.** 2	**D.** -3
5. $\sqrt[4]{16}$	**6.** $-\sqrt[3]{64}$	**E.** -4	**F.** Not a real number

Choose the closest approximation of each square root.

7. $\sqrt{123.5}$

A. 9 **B.** 10 **C.** 11 **D.** 12

8. $\sqrt{67.8}$

A. 7 **B.** 8 **C.** 9 **D.** 10

Refer to the figure to answer the questions in Exercises 9–10.

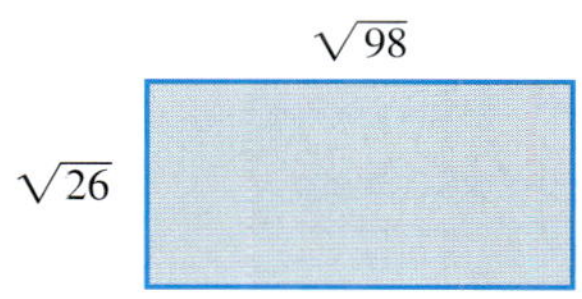

9. Which one of the following is the best estimate of its area?

A. 2500 **B.** 250 **C.** 50 **D.** 100

10. Which one of the following is the best estimate of its perimeter?

A. 15 **B.** 250 **C.** 100 **D.** 30

11. Consider the expression $-\sqrt{-a}$. Decide whether it is positive, negative, 0, or not a real number if

(a) $a > 0$, **(b)** $a < 0$, **(c)** $a = 0$.

12. If n is odd, under what conditions is $\sqrt[n]{a}$

(a) positive, **(b)** negative, **(c)** 0?

Find each root that is a real number. Use a calculator as necessary. See Examples 1 and 2.

13. $-\sqrt{81}$ **14.** $-\sqrt{121}$ **15.** $\sqrt[3]{216}$ **16.** $\sqrt[3]{343}$

17. $\sqrt[3]{-64}$ **18.** $\sqrt[3]{-125}$ **19.** $-\sqrt[3]{512}$ **20.** $-\sqrt[3]{1000}$

21. $\sqrt[4]{1296}$ **22.** $\sqrt[4]{625}$ **23.** $-\sqrt[4]{16}$ **24.** $-\sqrt[4]{256}$

25. $\sqrt[4]{-625}$

26. $\sqrt[4]{-256}$

27. $\sqrt[6]{729}$

28. $\sqrt[6]{64}$

29. $\sqrt[6]{-64}$

30. $\sqrt[6]{-1}$

31. $\sqrt{\frac{64}{81}}$

32. $\sqrt{\frac{100}{9}}$

33. $\sqrt{0.49}$

34. $\sqrt{0.81}$

35. $\sqrt[3]{\frac{64}{27}}$

36. $\sqrt[4]{\frac{81}{16}}$

37. $-\sqrt[6]{\frac{1}{64}}$

38. $-\sqrt[5]{\frac{1}{32}}$

39. $\sqrt[3]{0.001}$

40. $\sqrt[3]{0.125}$

Graph each function and give its domain and range. See Example 3.

41. $f(x) = \sqrt{x+3}$

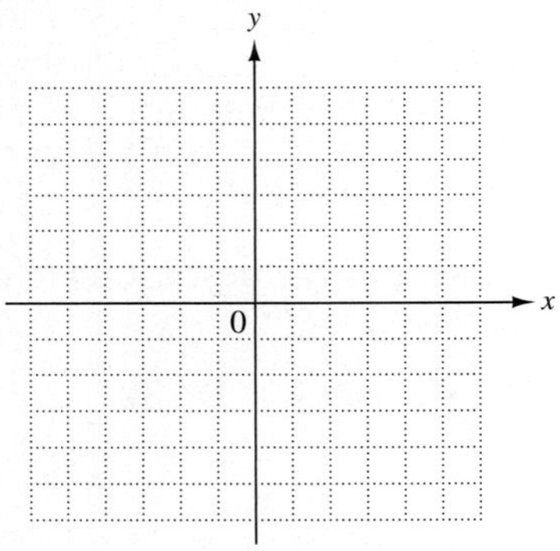

42. $f(x) = \sqrt{x-5}$

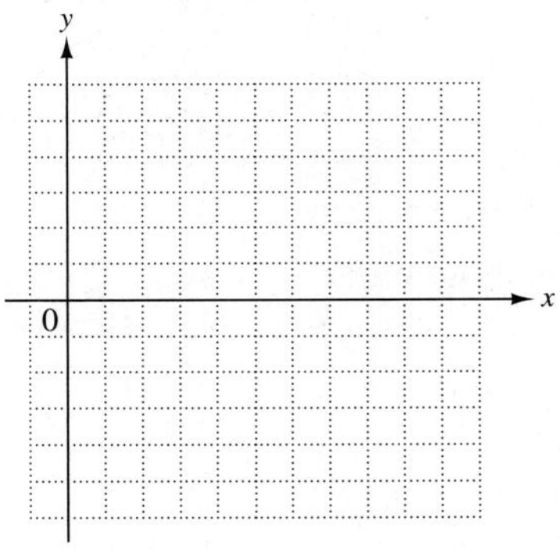

43. $f(x) = \sqrt{x} - 2$

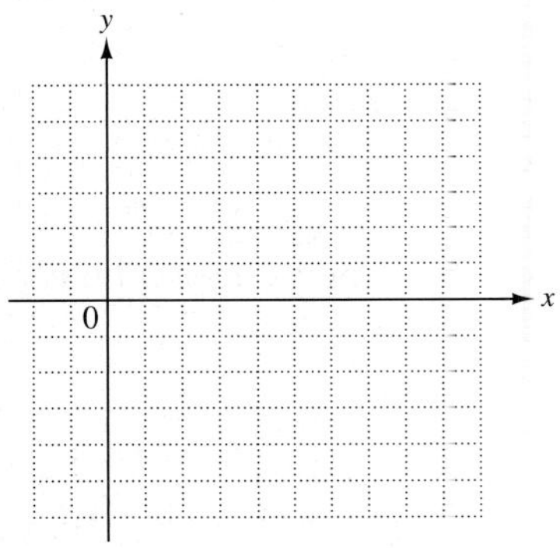

44. $f(x) = \sqrt{x} + 4$

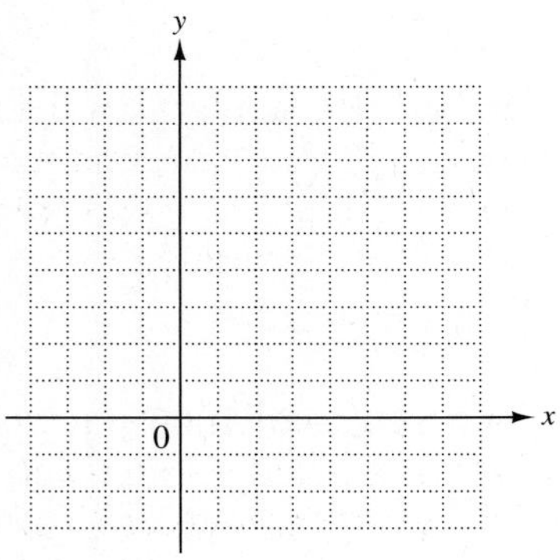

45. $f(x) = \sqrt[3]{x} - 3$

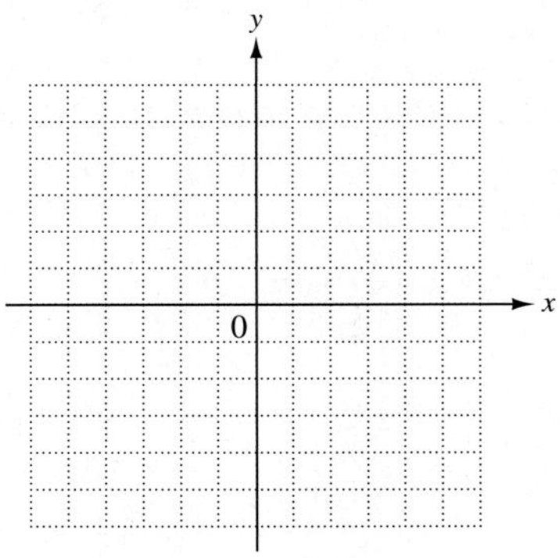

46. $f(x) = \sqrt[3]{x} + 1$

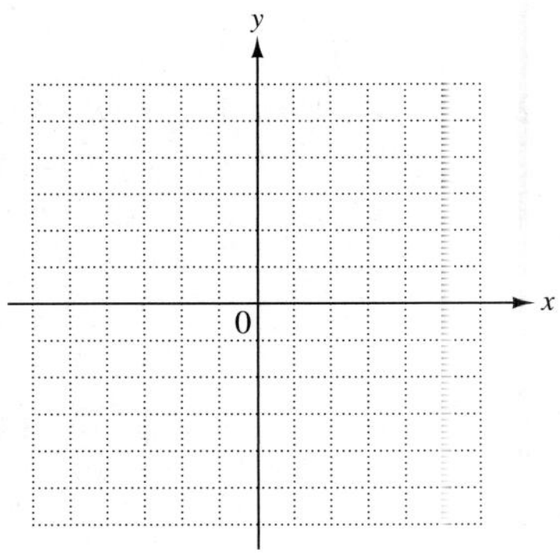

Simplify each root. See Examples 4 and 5.

47. $\sqrt{12^2}$ **48.** $\sqrt{19^2}$ **49.** $\sqrt{(-10)^2}$ **50.** $\sqrt{(-13)^2}$

51. $\sqrt[6]{(-2)^6}$ **52.** $\sqrt[6]{(-4)^6}$ **53.** $\sqrt[5]{(-9)^5}$ **54.** $\sqrt[5]{(-8)^5}$

55. $-\sqrt[6]{(-5)^6}$ **56.** $-\sqrt[6]{(-7)^6}$ **57.** $\sqrt{x^2}$ **58.** $-\sqrt{x^2}$

59. $\sqrt{(-z)^2}$ **60.** $\sqrt{(-q)^2}$ **61.** $\sqrt[3]{x^3}$ **62.** $-\sqrt[3]{x^3}$

63. $\sqrt[3]{x^{15}}$ **64.** $\sqrt[3]{m^9}$ **65.** $\sqrt[6]{x^{30}}$ **66.** $\sqrt[4]{k^{20}}$

Use a calculator to find a decimal approximation for each radical. Round answers to three decimal places. See Example 6.

67. $\sqrt{9483}$ **68.** $\sqrt{6825}$ **69.** $\sqrt{284.361}$ **70.** $\sqrt{846.104}$

71. $-\sqrt{82}$ **72.** $-\sqrt{91}$ **73.** $\sqrt[3]{423}$ **74.** $\sqrt[3]{555}$

75. $\sqrt[4]{100}$ **76.** $\sqrt[4]{250}$ **77.** $\sqrt[5]{23.8}$ **78.** $\sqrt[5]{98.4}$

Solve each problem. See Example 7.

79. Use the formula in Example 7 to calculate the resonant frequency of a circuit to the nearest thousand if $L = 7.237 \times 10^{-5}$ henry and $C = 2.5 \times 10^{-10}$ farad.

80. The threshold weight T for a person is the weight above which the risk of death increases greatly. The threshold weight in pounds for men aged 40–49 is related to height in inches by the formula

$$h = 12.3\sqrt[3]{T}.$$

What height corresponds to a threshold weight of 216 lb for a 43-yr-old man? Round your answer to the nearest inch, and then to the nearest tenth of a foot.

81. According to an article in *The World Scanner Report,* the distance D, in miles, to the horizon from an observer's point of view over water or "flat" earth is given by

$$D = \sqrt{2H},$$

where H is the height of the point of view, in feet. If a person whose eyes are 6 ft above ground level is standing at the top of a hill 44 ft above "flat" earth, approximately how far to the horizon will she be able to see?

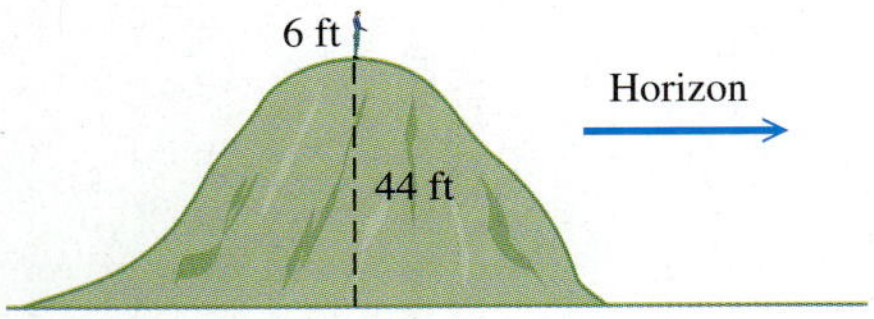

82. The time for one complete swing of a simple pendulum is

$$t = 2\pi\sqrt{\frac{L}{g}},$$

where t is time in seconds, L is the length of the pendulum in feet, and g, the force due to gravity, is about 32 ft per sec^2. Find the time of a complete swing of a 2-ft pendulum to the nearest tenth of a second.

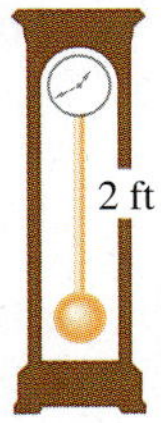

83. **Heron's formula** gives a method of finding the area of a triangle if the lengths of its sides are known. Suppose that a, b, and c are the lengths of the sides. Let s denote one-half of the perimeter of the triangle (called the **semiperimeter**); that is,

$$s = \frac{1}{2}(a + b + c).$$

Then the area of the triangle is

$$A = \sqrt{s(s - a)(s - b)(s - c)}.$$

Find the area of the Bermuda Triangle, if the "sides" of this triangle measure approximately 850 mi, 925 mi, and 1300 mi. Give your answer to the nearest thousand square miles.

84. The Vietnam Veterans' Memorial in Washington, D.C., is in the shape of an unenclosed isosceles triangle with equal sides of length 246.75 ft. If the triangle were enclosed, the third side would have length 438.14 ft. Use Heron's formula from the previous exercise to find the area of this enclosure to the nearest hundred square feet. (*Source:* Information pamphlet obtained at the Vietnam Veterans' Memorial.)

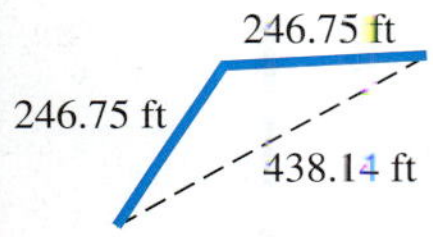

Not to scale

The coefficient of self-induction L (in henrys), the energy P stored in an electronic circuit (in joules), and the current I (in amps) are related by the formula

$$I = \sqrt{\frac{2P}{L}}.$$

Round your answers in Exercises 85 and 86 to the nearest thousandth.

85. Find I if $P = 120$ and $L = 80$.

86. Find I if $P = 100$ and $L = 40$.

9.2 Rational Exponents

OBJECTIVE 1 Use exponential notation for *n*th roots. We now look at exponents that are rational numbers of the form $\frac{1}{n}$, or $1/n$, where n is a natural number.

Consider the product $(3^{1/2})^2 = 3^{1/2} \cdot 3^{1/2}$. Using the rules of exponents from **Section 6.1,** extended to rational exponents, we can simplify this product as follows:

$$(3^{1/2})^2 = 3^{1/2} \cdot 3^{1/2}$$
$$= 3^{1/2+1/2} \quad \text{Product rule: } a^m \cdot a^n = a^{m+n}$$
$$= 3^1 \quad \text{Add exponents.}$$
$$= 3.$$

Also, by definition,

$$\left(\sqrt{3}\right)^2 = \sqrt{3} \cdot \sqrt{3} = 3.$$

Since both $(3^{1/2})^2$ and $\left(\sqrt{3}\right)^2$ are equal to 3, it seems reasonable to define

$$3^{1/2} = \sqrt{3}.$$

This suggests the following generalization.

> **$a^{1/n}$**
>
> If $\sqrt[n]{a}$ is a real number, then $\quad a^{1/n} = \sqrt[n]{a}.$

EXAMPLE 1 Evaluating Exponentials of the Form $a^{1/n}$

Evaluate each exponential.

(a) $64^{1/3} = \sqrt[3]{64} = 4$ (The denominator is the index.)

(b) $100^{1/2} = \sqrt{100} = 10$ (The denominator is the index. $\sqrt{\ }$ means $\sqrt[2]{\ }$.)

(c) $-256^{1/4} = -\sqrt[4]{256} = -4$

(d) $(-256)^{1/4} = \sqrt[4]{-256}$ is not a real number, because the radicand, -256, is negative and the index is even.

(e) $(-32)^{1/5} = \sqrt[5]{-32} = -2$

(f) $\left(\frac{1}{8}\right)^{1/3} = \sqrt[3]{\frac{1}{8}} = \frac{1}{2}$

CAUTION

Notice the difference between parts (c) and (d) in Example 1. The radical in part (c) is the ***negative fourth root of a positive number,*** while the radical in part (d) is the ***principal fourth root of a negative number, which is not a real number.***

Work Problem 1 at the Side. ▶

OBJECTIVES

1. Use exponential notation for *n*th roots.
2. Define and use expressions of the form $a^{m/n}$.
3. Convert between radicals and rational exponents.
4. Use the rules for exponents with rational exponents.

1 Evaluate each exponential.

(a) $8^{1/3}$

(b) $9^{1/2}$

(c) $-81^{1/4}$

(d) $(-81)^{1/4}$

(e) $(-64)^{1/3}$

(f) $\left(\frac{1}{32}\right)^{1/5}$

ANSWERS

1. (a) 2 (b) 3 (c) -3 (d) not a real number (e) -4 (f) $\frac{1}{2}$

OBJECTIVE 2 **Define and use expressions of the form $a^{m/n}$.**

We know that $8^{1/3} = \sqrt[3]{8}$. How should we define a number like $8^{2/3}$? For past rules of exponents to be valid,

$$8^{2/3} = 8^{(1/3)2} = (8^{1/3})^2.$$

Since $8^{1/3} = \sqrt[3]{8}$,

$$8^{2/3} = \left(\sqrt[3]{8}\right)^2 = 2^2 = 4.$$

Generalizing from this example, we define $a^{m/n}$ as follows.

> **$a^{m/n}$**
>
> If m and n are positive integers with m/n in lowest terms, then
>
> $$a^{m/n} = (a^{1/n})^m,$$
>
> provided that $a^{1/n}$ is a real number. If $a^{1/n}$ is not a real number, then $a^{m/n}$ is not a real number.

EXAMPLE 2 **Evaluating Exponentials of the Form $a^{m/n}$**

Evaluate each exponential.

Think: $36^{1/2} = \sqrt{36} = 6$

(a) $36^{3/2} = (36^{1/2})^3 = 6^3 = 216$

Think: $125^{1/3} = \sqrt[3]{125} = 5$

(b) $125^{2/3} = (125^{1/3})^2 = 5^2 = 25$

Be careful. The base is 4.

(c) $-4^{5/2} = -(4^{5/2}) = -(4^{1/2})^5 = -(2)^5 = -32$

Because the base here is 4, the negative sign is *not* affected by the exponent.

(d) $(-27)^{2/3} = [(-27)^{1/3}]^2 = (-3)^2 = 9$

Notice how the $-$ sign is used in parts (c) and (d). In part (c), we first evaluate the exponential and then find its negative. In part (d), the $-$ sign is part of the base, -27.

(e) $(-100)^{3/2} = [(-100)^{1/2}]^3$, which is not a real number, since $(-100)^{1/2}$, or $\sqrt{-100}$, is not a real number.

◀ *Work Problem* **2** *at the Side.*

2 Evaluate each exponential.

(a) $25^{3/2}$

(b) $27^{2/3}$

(c) $-16^{3/2}$

(d) $(-64)^{2/3}$

(e) $(-36)^{3/2}$

Recall from **Section 6.1** that for any natural number n,

$$a^{-n} = \frac{1}{a^n} \quad (a \neq 0).$$

When a rational exponent is negative, we apply this interpretation of negative exponents.

> **$a^{-m/n}$**
>
> If $a^{m/n}$ is a real number, then
>
> $$a^{-m/n} = \frac{1}{a^{m/n}} \quad (a \neq 0).$$

ANSWERS

2. **(a)** 125 **(b)** 9 **(c)** -64 **(d)** 16 **(e)** not a real number

EXAMPLE 3 **Evaluating Exponentials with Negative Rational Exponents**

Evaluate each exponential.

(a) $16^{-3/4} = \frac{1}{16^{3/4}} = \frac{1}{(16^{1/4})^3} = \frac{1}{(\sqrt[4]{16})^3} = \frac{1}{2^3} = \frac{1}{8}$

The denominator of 3/4 is the index and the numerator is the exponent.

(b) $25^{-3/2} = \frac{1}{25^{3/2}} = \frac{1}{(25^{1/2})^3} = \frac{1}{(\sqrt{25})^3} = \frac{1}{5^3} = \frac{1}{125}$

(c) $\left(\frac{8}{27}\right)^{-2/3} = \frac{1}{\left(\frac{8}{27}\right)^{2/3}} = \frac{1}{\left(\sqrt[3]{\frac{8}{27}}\right)^2} = \frac{1}{\left(\frac{2}{3}\right)^2} = \frac{1}{\frac{4}{9}} = \frac{9}{4}$

$\frac{1}{\frac{4}{9}} = 1 \div \frac{4}{9} = 1 \cdot \frac{9}{4}$

We could also use the rule $\left(\frac{b}{a}\right)^{-m} = \left(\frac{a}{b}\right)^m$ here, as follows:

$$\left(\frac{8}{27}\right)^{-2/3} = \left(\frac{27}{8}\right)^{2/3} = \left(\sqrt[3]{\frac{27}{8}}\right)^2 = \left(\frac{3}{2}\right)^2 = \frac{9}{4}.$$

Take the reciprocal only of the base, *not* the exponent.

Work Problem **3** at the Side. ▶

3 Evaluate each exponential.

(a) $36^{-3/2}$

(b) $32^{-4/5}$

(c) $\left(\frac{4}{9}\right)^{-5/2}$

CAUTION

Be careful to distinguish between exponential expressions such as

$16^{-1/4}$, which equals $\frac{1}{2}$; $-16^{1/4}$, which equals -2; and

$-16^{-1/4}$, which equals $-\frac{1}{2}$.

A negative exponent does not necessarily lead to a negative result. Negative exponents lead to reciprocals, which may be positive.

We obtain an alternative definition of $a^{m/n}$ by applying the power rule a little differently than earlier. If all indicated roots are real numbers, then

$$a^{m/n} = a^{m(1/n)} = (a^m)^{1/n}, \qquad \text{so} \qquad a^{m/n} = (a^m)^{1/n}.$$

$a^{m/n}$

If all indicated roots are real numbers, then

$$\mathbf{a^{m/n} = (a^{1/n})^m = (a^m)^{1/n}.}$$

We can now evaluate an expression such as $27^{2/3}$ in two ways:

$27^{2/3} = (27^{1/3})^2 = 3^2 = 9$

or $27^{2/3} = (27^2)^{1/3} = 729^{1/3} = 9.$

The result is the same.

In most cases, it is easier to use $(a^{1/n})^m$.

ANSWERS

3. (a) $\frac{1}{216}$ (b) $\frac{1}{16}$ (c) $\frac{243}{32}$

This rule can also be expressed with radicals as follows.

Radical Form of $a^{m/n}$

If all indicated roots are real numbers, then

$$a^{m/n} = \sqrt[n]{a^m} = \left(\sqrt[n]{a}\right)^m.$$

In words, raise a to the mth power and then take the nth root, or take the nth root of a and then raise to the mth power.

For example,

$$8^{2/3} = \sqrt[3]{8^2} = \sqrt[3]{64} = 4, \quad \text{and} \quad 8^{2/3} = \left(\sqrt[3]{8}\right)^2 = 2^2 = 4,$$

so

$$8^{2/3} = \sqrt[3]{8^2} = \left(\sqrt[3]{8}\right)^2.$$

OBJECTIVE 3 Convert between radicals and rational exponents. Using the definition of rational exponents, we can simplify many problems involving radicals by converting the radicals to numbers with rational exponents. After simplifying, we convert the answer back to radical form.

EXAMPLE 4 Converting between Rational Exponents and Radicals

In (a)–(f), write each exponential as a radical. Assume that all variables represent positive real numbers. Use the definition that takes the root first.

(a) $13^{1/2} = \sqrt{13}$ **(b)** $6^{3/4} = \left(\sqrt[4]{6}\right)^3$ **(c)** $9m^{5/8} = 9\left(\sqrt[8]{m}\right)^5$

(d) $6x^{2/3} - (4x)^{3/5} = 6\left(\sqrt[3]{x}\right)^2 - \left(\sqrt[5]{4x}\right)^3$

(e) $r^{-2/3} = \dfrac{1}{r^{2/3}} = \dfrac{1}{\left(\sqrt[3]{r}\right)^2}$

(f) $(a^2 + b^2)^{1/2} = \sqrt{a^2 + b^2}$ ← $\sqrt{a^2 + b^2} \neq a + b$

In (g)–(i), write each radical as an exponential. Simplify. Assume that all variables represent positive real numbers.

(g) $\sqrt{10} = 10^{1/2}$

(h) $\sqrt[4]{3^8} = 3^{8/4} = 3^2 = 9$

(i) $\sqrt[6]{z^6} = z^{6/6} = z^1 = z$, since z is positive.

◀ Work Problem 4 at the Side.

Note

In Example 4(i), it was not necessary to use absolute value bars, since the directions specifically stated that the variable represents a positive real number. Because the absolute value of the positive real number z is z itself, the answer is simply z.

◀ Work Problem 5 at the Side.

4 Write each exponential as a radical. Assume that all variables represent positive real numbers. Use the definition that takes the root first.

(a) $19^{1/2}$

(b) $5^{2/3}$

(c) $4k^{3/5}$

(d) $5x^{3/5} - (2x)^{3/5}$

(e) $x^{-5/7}$

(f) $(m^3 + n^3)^{1/3}$

5 Write each radical as an exponential and simplify. Assume that all variables represent positive real numbers.

(a) $\sqrt{37}$

(b) $\sqrt[4]{9^8}$

(c) $\sqrt[4]{t^4}$

Answers

4. (a) $\sqrt{19}$ (b) $\left(\sqrt[3]{5}\right)^2$ (c) $4\left(\sqrt[5]{k}\right)^3$ (d) $5\left(\sqrt[5]{x}\right)^3 - \left(\sqrt[5]{2x}\right)^3$ (e) $\dfrac{1}{\left(\sqrt[7]{x}\right)^5}$ (f) $\sqrt[3]{m^3 + n^3}$

5. (a) $37^{1/2}$ (b) 9^2, or 81 (c) t

OBJECTIVE 4 Use the rules for exponents with rational exponents. The definition of rational exponents allows us to apply the rules for exponents first introduced in **Section 6.1.**

Rules for Rational Exponents

Let r and s be rational numbers. For all real numbers a and b for which the indicated expressions exist:

$$a^r \cdot a^s = a^{r+s} \qquad a^{-r} = \frac{1}{a^r} \qquad \frac{a^r}{a^s} = a^{r-s} \qquad \left(\frac{a}{b}\right)^{-r} = \frac{b^r}{a^r}$$

$$(a^r)^s = a^{rs} \qquad (ab)^r = a^r b^r \qquad \left(\frac{a}{b}\right)^r = \frac{a^r}{b^r} \qquad a^{-r} = \left(\frac{1}{a}\right)^r.$$

EXAMPLE 5 Applying Rules for Rational Exponents

Write with only positive exponents. Assume that all variables represent positive real numbers.

(a) $2^{1/2} \cdot 2^{1/4}$

$= 2^{1/2+1/4}$ Product rule

$= 2^{3/4}$ Add exponents.

(b) $\dfrac{5^{2/3}}{5^{7/3}}$

$= 5^{2/3-7/3}$ Quotient rule

$= 5^{-5/3}$ Subtract exponents.

$= \dfrac{1}{5^{5/3}}$ $a^{-r} = \frac{1}{a^r}$

(c) $\dfrac{(x^{1/2}y^{2/3})^4}{y}$

$= \dfrac{(x^{1/2})^4(y^{2/3})^4}{y}$ Power rule

$= \dfrac{x^2y^{8/3}}{y^1}$ Power rule

$= x^2y^{8/3-1}$ Quotient rule

$= x^2y^{5/3}$ $\frac{8}{3} - 1 = \frac{8}{3} - \frac{3}{3} = \frac{5}{3}$

(d) $\left(\dfrac{x^4y^{-6}}{x^{-2}y^{1/3}}\right)^{-2/3}$

$= \dfrac{(x^4)^{-2/3}(y^{-6})^{-2/3}}{(x^{-2})^{-2/3}(y^{1/3})^{-2/3}}$ Power rule

$= \dfrac{x^{-8/3}y^4}{x^{4/3}y^{-2/9}}$ Power rule

$= x^{-8/3-4/3}y^{4-(-2/9)}$ Quotient rule

$= x^{-4}y^{38/9}$ Use parentheses to avoid errors. $4 - \left(-\frac{2}{9}\right) = \frac{36}{9} + \frac{2}{9} = \frac{38}{9}$

$= \dfrac{y^{38/9}}{x^4}$ Definition of negative exponent

Continued on Next Page

6 Write with only positive exponents. Assume that all variables represent positive real numbers.

(a) $11^{3/4} \cdot 11^{5/4}$

(b) $\dfrac{7^{3/4}}{7^{7/4}}$

(c) $\dfrac{9^{2/3}(x^{1/3})^4}{9^{-1/3}}$

(d) $\left(\dfrac{a^3b^{-4}}{a^{-2}b^{1/5}}\right)^{-1/2}$

(e) $a^{2/3}(a^{7/3} + a^{1/3})$

7 Write all radicals as exponentials, and then apply the rules for rational exponents. Give answers in exponential form. Assume that all variables represent positive real numbers.

(a) $\sqrt[5]{m^3} \cdot \sqrt{m}$

(b) $\dfrac{\sqrt[3]{p^5}}{\sqrt{p^3}}$

(c) $\sqrt[4]{\sqrt[3]{x}}$

Answers

6. (a) 11^2, or 121 **(b)** $\frac{1}{7}$ **(c)** $9x^{4/3}$ **(d)** $\frac{b^{21/10}}{a^{5/2}}$ **(e)** $a^3 + a$

7. (a) $m^{11/10}$ **(b)** $p^{1/6}$ **(c)** $x^{1/12}$

The same result is obtained if we simplify within the parentheses first.

$$\left(\frac{x^4y^{-6}}{x^{-2}y^{1/3}}\right)^{-2/3}$$

$= (x^{4-(-2)}y^{-6-1/3})^{-2/3}$ Quotient rule

$= (x^6y^{-19/3})^{-2/3}$ $-6 - \frac{1}{3} = -\frac{18}{3} - \frac{1}{3} = -\frac{19}{3}$

$= (x^6)^{-2/3}(y^{-19/3})^{-2/3}$ Power rule

$= x^{-4}y^{38/9}$ Power rule

$= \dfrac{y^{38/9}}{x^4}$ Definition of negative exponent

(e) $m^{3/4}(m^{5/4} - m^{1/4})$

$= m^{3/4}(m^{5/4}) - m^{3/4}(m^{1/4})$ Distributive property

$= m^{3/4+5/4} - m^{3/4+1/4}$ Product rule

$= m^{8/4} - m^{4/4}$

$= m^2 - m$

Do not make the common mistake of multiplying exponents in the first step.

◀ Work Problem **6** at the Side.

CAUTION

Use the rules of exponents in problems like those in Example 5. Do not convert the expressions to radical form.

EXAMPLE 6 Applying Rules for Rational Exponents

Write all radicals as exponentials, and then apply the rules for rational exponents. Give answers in exponential form. Assume that all variables represent positive real numbers.

(a) $\sqrt[3]{x^2} \cdot \sqrt[4]{x}$

$= x^{2/3} \cdot x^{1/4}$ Convert to rational exponents.

$= x^{2/3+1/4}$ Product rule

$= x^{8/12+3/12}$ Write exponents with a common denominator.

$= x^{11/12}$

(b) $\dfrac{\sqrt{x^3}}{\sqrt[3]{x^2}}$

$= \dfrac{x^{3/2}}{x^{2/3}}$ Convert to rational exponents.

$= x^{3/2-2/3}$ Quotient rule

$= x^{5/6}$ $\frac{3}{2} - \frac{2}{3} = \frac{9}{6} - \frac{4}{6} = \frac{5}{6}$

(c) $\sqrt{\sqrt[4]{z}}$

$= \sqrt{z^{1/4}}$ Convert the inside radical to rational exponents.

$= (z^{1/4})^{1/2}$ Convert to rational exponents.

$= z^{1/8}$ Power rule

◀ Work Problem **7** at the Side.

9.2 Exercises

Match each expression from Column I with the equivalent choice from Column II.

I		II	
1. $2^{1/2}$	2. $(-27)^{1/3}$	A. -4	B. 8
3. $-16^{1/2}$	4. $(-16)^{1/2}$	C. $\sqrt{2}$	D. $-\sqrt{6}$
5. $(-32)^{1/5}$	6. $(-32)^{2/5}$	E. -3	F. $\sqrt{6}$
7. $4^{3/2}$	8. $6^{2/4}$	G. 4	H. -2
9. $-6^{2/4}$	10. $36^{0.5}$	I. 6	J. Not a real number

Evaluate each exponential. See Examples 1–3.

11. $169^{1/2}$
12. $121^{1/2}$
13. $729^{1/3}$
14. $512^{1/3}$
15. $16^{1/4}$
16. $625^{1/4}$
17. $\left(\frac{64}{81}\right)^{1/2}$
18. $\left(\frac{8}{27}\right)^{1/3}$
19. $(-27)^{1/3}$
20. $(-32)^{1/5}$
21. $(-144)^{1/2}$
22. $(-36)^{1/2}$
23. $100^{3/2}$
24. $64^{3/2}$
25. $81^{3/4}$
26. $216^{2/3}$
27. $-16^{5/2}$
28. $-32^{3/5}$
29. $(-8)^{4/3}$
30. $(-243)^{2/5}$
31. $32^{-3/5}$
32. $27^{-4/3}$
33. $64^{-3/2}$
34. $81^{-3/2}$
35. $\left(\frac{125}{27}\right)^{-2/3}$
36. $\left(\frac{64}{125}\right)^{-2/3}$

Write with radicals. Assume that all variables represent positive real numbers. Use the definition that takes the root first. See Example 4.

37. $12^{1/2}$

38. $3^{1/2}$

39. $8^{3/4}$

40. $7^{2/3}$

41. $(9q)^{5/8} - (2x)^{2/3}$

42. $(3p)^{3/4} + (4x)^{1/3}$

43. $(2m)^{-3/2}$

44. $(5y)^{-3/5}$

45. $(2y + x)^{2/3}$

46. $(r + 2z)^{3/2}$

47. $(3m^4 + 2k^2)^{-2/3}$

48. $(5x^2 + 3z^3)^{-5/6}$

49. Show that, in general, $\sqrt{a^2 + b^2} \neq a + b$ by replacing a with 3 and b with 4.

50. Suppose someone claims that $\sqrt[n]{a^n + b^n}$ must equal $a + b$, since when $a = 1$ and $b = 0$, a true statement results:

$$\sqrt[n]{a^n + b^n} = \sqrt[n]{1^n + 0^n} = \sqrt[n]{1^n} = 1 = 1 + 0 = a + b.$$

Explain why this is faulty reasoning.

Simplify by first converting to rational exponents. Assume that all variables represent positive real numbers. See Example 4.

51. $\sqrt{2^{12}}$

52. $\sqrt{5^{10}}$

53. $\sqrt[3]{4^9}$

54. $\sqrt[4]{6^8}$

55. $\sqrt{x^{20}}$

56. $\sqrt{r^{50}}$

57. $\sqrt[3]{x} \cdot \sqrt{x}$

58. $\sqrt[4]{y} \cdot \sqrt[5]{y^2}$

59. $\dfrac{\sqrt[3]{t^4}}{\sqrt[5]{t^4}}$

60. $\dfrac{\sqrt[4]{w^3}}{\sqrt[6]{w}}$

Simplify each expression. Write all answers with positive exponents. Assume that all variables represent positive real numbers. See Example 5.

61. $3^{1/2} \cdot 3^{3/2}$

62. $6^{4/3} \cdot 6^{2/3}$

63. $\dfrac{64^{5/3}}{64^{4/3}}$

64. $\dfrac{125^{7/3}}{125^{5/3}}$

65. $y^{7/3} \cdot y^{-4/3}$

66. $r^{-8/9} \cdot r^{17/9}$

67. $x^{2/3} \cdot x^{-1/4}$

68. $x^{2/5} \cdot x^{-1/3}$

69. $\dfrac{k^{1/3}}{k^{2/3} \cdot k^{-1}}$

70. $\dfrac{z^{3/4}}{z^{5/4} \cdot z^{-2}}$

71. $\dfrac{(x^{1/4}y^{2/5})^{20}}{x^2}$

72. $\dfrac{(r^{1/5}s^{2/3})^{15}}{r^2}$

73. $\dfrac{(x^{2/3})^2}{(x^2)^{7/3}}$

74. $\dfrac{(p^3)^{1/4}}{(p^{5/4})^2}$

75. $\dfrac{m^{3/4}n^{-1/4}}{(m^2n)^{1/2}}$

76. $\dfrac{(a^2b^5)^{-1/4}}{(a^{-3}b^2)^{1/6}}$

77. $\dfrac{p^{1/5}p^{7/10}p^{1/2}}{(p^3)^{-1/5}}$

78. $\dfrac{z^{1/3}z^{-2/3}z^{1/6}}{(z^{-1/6})^3}$

79. $\left(\dfrac{b^{-3/2}}{c^{-5/3}}\right)^2(b^{-1/4}c^{-1/3})^{-1}$

80. $\left(\dfrac{m^{-2/3}}{a^{-3/4}}\right)^4(m^{-3/8}a^{1/4})^{-2}$

81. $\left(\dfrac{p^{-1/4}q^{-3/2}}{3^{-1}p^{-2}q^{-2/3}}\right)^{-2}$

82. $\left(\dfrac{2^{-2}w^{-3/4}x^{-5/8}}{w^{3/4}x^{-1/2}}\right)^{-3}$

83. $p^{2/3}(p^{1/3} + 2p^{4/3})$

84. $z^{5/8}(3z^{5/8} + 5z^{11/8})$

85. $k^{1/4}(k^{3/2} - k^{1/2})$

86. $r^{3/5}(r^{1/2} + r^{3/4})$

87. $6a^{7/4}(a^{-7/4} + 3a^{-3/4})$

88. $4m^{5/3}(m^{-2/3} - 4m^{-5/3})$

89. $5m^{-2/3}(m^{2/3} + m^{-7/3})$

Write radicals as exponentials, and then apply the rules for rational exponents. Give answers in exponential form. Assume that all radicands represent positive real numbers. See Example 6.

90. $\sqrt[5]{x^3} \cdot \sqrt[4]{x}$

91. $\sqrt[6]{y^5} \cdot \sqrt[3]{y^2}$

92. $\dfrac{\sqrt{x^5}}{\sqrt{x^8}}$

93. $\dfrac{\sqrt[3]{k^5}}{\sqrt[3]{k^7}}$

94. $\sqrt{y} \cdot \sqrt[3]{yz}$

95. $\sqrt[3]{xz} \cdot \sqrt{z}$

96. $\sqrt[4]{\sqrt[3]{m}}$ **97.** $\sqrt[3]{\sqrt{k}}$ **98.** $\sqrt{\sqrt[3]{\sqrt[4]{x}}}$

99. $\sqrt[3]{\sqrt[5]{\sqrt{y}}}$ **100.** $\sqrt{y^{5/4}}$ **101.** $\sqrt[3]{x^{5/9}}$

Solve each problem.

102. Meteorologists can determine the duration of a storm by using the function defined by

$$T(D) = 0.07D^{3/2},$$

where D is the diameter of the storm in miles and T is the time in hours. Find the duration of a storm with a diameter of 16 mi. Round your answer to the nearest tenth of an hour.

103. The threshold weight T, in pounds, for a person is the weight above which the risk of death increases greatly. The threshold weight in pounds for men aged 40–49 is related to height in inches by the function defined by

$$h(T) = (1860.867T)^{1/3}.$$

What height corresponds to a threshold weight of 200 lb for a 46-yr-old man? Round your answer to the nearest inch, and then to the nearest tenth of a foot.

*The **windchill factor** is a measure of the cooling effect that the wind has on a person's skin. It calculates the equivalent cooling temperature if there were no wind. The National Weather Service uses the formula*

$$\text{Windchill temperature} = 35.74 + 0.6215T - 35.75V^{4/25} + 0.4275TV^{4/25},$$

where T is the temperature in °F and V is the wind speed in miles per hour, to calculate windchill. The chart gives the windchill factor for various wind speeds and temperatures at which frostbite is a risk, and how quickly it may occur.

Temperature (°F)

Wind speed (mph) / Calm	40	30	20	10	0	−10	−20	−30	−40
5	36	25	13	1	−11	−22	−34	−46	−57
10	34	21	9	−4	−16	−28	−41	−53	−66
15	32	19	6	−7	−19	−32	−45	−58	−71
20	30	17	4	−9	−22	−35	−48	−61	−74
25	29	16	3	−11	−24	−37	−51	−64	−78
30	28	15	1	−12	−26	−39	−53	−67	−80
35	28	14	0	−14	−27	−41	−55	−69	−82
40	27	13	−1	−15	−29	−43	−57	−71	−84

Frostbites times: 30 minutes 10 minutes 5 minutes

Source: National Oceanic and Atmospheric Administration, National Weather Service.

Use the formula to determine the windchill to the nearest tenth of a degree, given the following conditions. Compare your answers with the appropriate entries in the table.

104. 30°F, 15-mph wind

105. 10°F, 30-mph wind

9.3 Simplifying Radical Expressions

OBJECTIVES

1. Use the product rule for radicals.
2. Use the quotient rule for radicals.
3. Simplify radicals.
4. Simplify products and quotients of radicals with different indexes.
5. Use the Pythagorean formula.
6. Use the distance formula.

OBJECTIVE 1 Use the product rule for radicals. Is the product of two nth-root radicals equal to the nth root of the product of the radicands? For example, are $\sqrt{36 \cdot 4}$ and $\sqrt{36} \cdot \sqrt{4}$ equal?

$$\sqrt{36 \cdot 4} = \sqrt{144} = 12$$
$$\sqrt{36} \cdot \sqrt{4} = 6 \cdot 2 = 12$$
The result is the same.

This is an example of the **product rule for radicals.**

Product Rule for Radicals

If $\sqrt[n]{a}$ and $\sqrt[n]{b}$ are real numbers and n is a natural number, then

$$\sqrt[n]{a} \cdot \sqrt[n]{b} = \sqrt[n]{ab}.$$

In words, the product of two nth roots is the nth root of the product.

We justify the product rule using the rules for rational exponents. Since $\sqrt[n]{a} = a^{1/n}$ and $\sqrt[n]{b} = b^{1/n}$,

$$\sqrt[n]{a} \cdot \sqrt[n]{b} = a^{1/n} \cdot b^{1/n} = (ab)^{1/n} = \sqrt[n]{ab}.$$

CAUTION
Use the product rule only when the radicals have the same index.

EXAMPLE 1 Using the Product Rule

Multiply. Assume that all variables represent positive real numbers.

(a) $\sqrt{5} \cdot \sqrt{7} = \sqrt{5 \cdot 7} = \sqrt{35}$

(b) $\sqrt{2} \cdot \sqrt{19} = \sqrt{2 \cdot 19} = \sqrt{38}$

(c) $\sqrt{11} \cdot \sqrt{p} = \sqrt{11p}$

(d) $\sqrt{7} \cdot \sqrt{11xyz} = \sqrt{77xyz}$

Work Problem 1 at the Side.

EXAMPLE 2 Using the Product Rule

Multiply. Assume that all variables represent positive real numbers.

(a) $\sqrt[3]{3} \cdot \sqrt[3]{12} = \sqrt[3]{3 \cdot 12} = \sqrt[3]{36}$

(b) $\sqrt[4]{8y} \cdot \sqrt[4]{3r^2} = \sqrt[4]{24yr^2}$

(c) $\sqrt[6]{10m^4} \cdot \sqrt[6]{5m} = \sqrt[6]{50m^5}$

(d) $\sqrt[4]{2} \cdot \sqrt[5]{2}$ cannot be simplified using the product rule for radicals, because the indexes (4 and 5) are different.

Work Problem 2 at the Side.

1 Multiply. Assume that all variables represent positive real numbers.

(a) $\sqrt{5} \cdot \sqrt{13}$

(b) $\sqrt{10y} \cdot \sqrt{3k}$

2 Multiply. Assume that all variables represent positive real numbers.

(a) $\sqrt[3]{2} \cdot \sqrt[3]{7}$

(b) $\sqrt[6]{8r^2} \cdot \sqrt[6]{2r^3}$

(c) $\sqrt[5]{9y^2x} \cdot \sqrt[5]{8xy^2}$

(d) $\sqrt{7} \cdot \sqrt[3]{5}$

Answers

1. (a) $\sqrt{65}$ (b) $\sqrt{30yk}$
2. (a) $\sqrt[3]{14}$ (b) $\sqrt[6]{16r^5}$ (c) $\sqrt[5]{72y^4x^2}$ (d) cannot be simplified using the produ

OBJECTIVE 2 Use the quotient rule for radicals. The **quotient rule for radicals** is similar to the product rule.

Quotient Rule for Radicals

If $\sqrt[n]{a}$ and $\sqrt[n]{b}$ are real numbers, $b \neq 0$, and n is a natural number, then

$$\sqrt[n]{\frac{a}{b}} = \frac{\sqrt[n]{a}}{\sqrt[n]{b}}.$$

In words, the nth root of a quotient is the quotient of the nth roots.

EXAMPLE 3 Using the Quotient Rule

Simplify. Assume that all variables represent positive real numbers.

(a) $\sqrt{\frac{16}{25}} = \frac{\sqrt{16}}{\sqrt{25}} = \frac{4}{5}$

(b) $\sqrt{\frac{7}{36}} = \frac{\sqrt{7}}{\sqrt{36}} = \frac{\sqrt{7}}{6}$

(c) $\sqrt[3]{-\frac{8}{125}} = \sqrt[3]{\frac{-8}{125}} = \frac{\sqrt[3]{-8}}{\sqrt[3]{125}} = \frac{-2}{5} = -\frac{2}{5}$ $\quad \frac{-a}{b} = -\frac{a}{b}$

(d) $\sqrt[3]{\frac{7}{216}} = \frac{\sqrt[3]{7}}{\sqrt[3]{216}} = \frac{\sqrt[3]{7}}{6}$

(e) $\sqrt[5]{\frac{x}{32}} = \frac{\sqrt[5]{x}}{\sqrt[5]{32}} = \frac{\sqrt[5]{x}}{2}$

Think: $\sqrt[3]{m^6} = m^{6/3} = m^2$

(f) $-\sqrt[3]{\frac{m^6}{125}} = -\frac{\sqrt[3]{m^6}}{\sqrt[3]{125}} = -\frac{m^2}{5}$

◀ Work Problem 3 at the Side.

OBJECTIVE 3 Simplify radicals. We use the product and quotient rules to simplify radicals. A radical is **simplified** if the following four conditions are met.

Conditions for a Simplified Radical

1. The radicand has no factor raised to a power greater than or equal to the index.
2. The radicand has no fractions.
3. No denominator has a radical.
4. Exponents in the radicand and the index of the radical have greatest common factor 1.

3 Simplify. Assume that all variables represent positive real numbers.

(a) $\sqrt{\frac{100}{81}}$

(b) $\sqrt{\frac{11}{25}}$

(c) $\sqrt[3]{-\frac{125}{216}}$

(d) $\sqrt{\frac{y^8}{16}}$

(e) $-\sqrt[3]{\frac{x^2}{r^{12}}}$

ANSWERS

3. (a) $\frac{10}{9}$ (b) $\frac{\sqrt{11}}{5}$ (c) $-\frac{5}{6}$ (d) $\frac{y^4}{4}$ (e) $-\frac{\sqrt[3]{x^2}}{r^4}$

EXAMPLE 4 Simplifying Roots of Numbers

Simplify.

(a) $\sqrt{24}$

Check to see whether 24 is divisible by a perfect square (the square of a natural number) such as 4, 9, 16, ... The largest perfect square that divides into 24 is 4.

$$\sqrt{24}$$
$$= \sqrt{4 \cdot 6} \quad \text{Factor; 4 is a perfect square.}$$
$$= \sqrt{4} \cdot \sqrt{6} \quad \text{Product rule}$$
$$= 2\sqrt{6} \quad \sqrt{4} = 2$$

(b) $\sqrt{108}$

As shown on the left, the number 108 is divisible by the perfect square 36. If this perfect square is not immediately clear, try factoring 108 into its prime factors, as shown on the right.

$$\sqrt{108} = \sqrt{36 \cdot 3} = \sqrt{36} \cdot \sqrt{3} = 6\sqrt{3}$$

$$\sqrt{108}$$
$$= \sqrt{2^2 \cdot 3^3}$$
$$= \sqrt{2^2 \cdot 3^2 \cdot 3}$$
$$= \sqrt{2^2} \cdot \sqrt{3^2} \cdot \sqrt{3} \quad \text{Product rule}$$
$$= 2 \cdot 3 \cdot \sqrt{3} \quad \sqrt{2^2} = 2,\ \sqrt{3^2} = 3$$
$$= 6\sqrt{3} \quad \text{Multiply.}$$

(c) $\sqrt{10}$ No perfect square (other than 1) divides into 10, so $\sqrt{10}$ cannot be simplified further.

(d) $\sqrt[3]{16}$

The largest perfect *cube* that divides into 16 is 8, so factor 16 as $8 \cdot 2$.

Remember to write the index.

$$\sqrt[3]{16}$$
$$= \sqrt[3]{8 \cdot 2} \quad \text{8 is a perfect cube.}$$
$$= \sqrt[3]{8} \cdot \sqrt[3]{2} \quad \text{Product rule}$$
$$= 2\sqrt[3]{2} \quad \sqrt[3]{8} = 2$$

(e) $-\sqrt[4]{162}$

Remember the negative sign in each line.

$$= -\sqrt[4]{81 \cdot 2} \quad \text{81 is a perfect 4th power.}$$
$$= -\sqrt[4]{81} \cdot \sqrt[4]{2} \quad \text{Product rule}$$
$$= -3\sqrt[4]{2} \quad \sqrt[4]{81} = 3$$

Work Problem **4** *at the Side.* ▶

CAUTION

Be careful with which factors belong outside the radical sign and which belong inside. Note in Example 4(b) how $2 \cdot 3$ is written outside because $\sqrt{2^2} = 2$ and $\sqrt{3^2} = 3$. The remaining 3 is left inside the radical.

4 Simplify.

(a) $\sqrt{32}$

(b) $\sqrt{45}$

(c) $\sqrt{300}$

(d) $\sqrt{35}$

(e) $-\sqrt[3]{54}$

(f) $\sqrt[4]{243}$

ANSWERS

4. (a) $4\sqrt{2}$ **(b)** $3\sqrt{5}$ **(c)** $10\sqrt{3}$ **(d)** cannot be simplified further **(e)** $-3\sqrt[3]{2}$ **(f)** $3\sqrt[4]{3}$

5 Simplify. Assume that all variables represent positive real numbers.

(a) $\sqrt{25p^7}$

(b) $\sqrt{72y^3x}$

(c) $\sqrt[3]{-27y^7x^5z^6}$

(d) $-\sqrt[4]{32a^5b^7}$

EXAMPLE 5 Simplifying Radicals Involving Variables

Simplify. Assume that all variables represent positive real numbers.

(a) $\sqrt{16m^3}$

$= \sqrt{16m^2 \cdot m}$ Factor.

$= \sqrt{16m^2} \cdot \sqrt{m}$ Product rule

$= 4m\sqrt{m}$

Absolute value bars are not needed around the m in color because of the assumption that all the variables represent *positive* real numbers.

(b) $\sqrt{200k^7q^8}$

$= \sqrt{10^2 \cdot 2 \cdot (k^3)^2 \cdot k \cdot (q^4)^2}$ Factor.

$= 10k^3q^4\sqrt{2k}$ Remove perfect square factors

(c) $\sqrt[3]{-8x^4y^5}$

$= \sqrt[3]{(-8x^3y^3)(xy^2)}$ Choose $-8x^3y^3$ as the perfect cube that divides into $-8x^4y^5$.

$= \sqrt[3]{-8x^3y^3} \cdot \sqrt[3]{xy^2}$ Product rule

$= -2xy\sqrt[3]{xy^2}$

(d) $-\sqrt[4]{32y^9}$

$= -\sqrt[4]{(16y^8)(2y)}$ $16y^8$ is the largest 4th power that divides into $32y^9$.

$= -\sqrt[4]{16y^8} \cdot \sqrt[4]{2y}$ Product rule

$= -2y^2\sqrt[4]{2y}$

◀ Work Problem **5** at the Side.

Note

From Example 5 we see that if a variable is raised to a power with an exponent divisible by 2, it is a perfect square. If it is raised to a power with an exponent divisible by 3, it is a perfect cube. ***In general, if it is raised to a power with an exponent divisible by n, it is a perfect nth power.***

The conditions for a simplified radical given earlier state that an exponent in the radicand and the index of the radical should have greatest common factor 1. The next example applies this condition.

EXAMPLE 6 Simplifying Radicals by Using Lesser Indexes

Simplify. Assume that all variables represent positive real numbers.

(a) $\sqrt[9]{5^6}$

We can write this radical using rational exponents and then write the exponent in lowest terms. We then express the answer as a radical.

$$\sqrt[9]{5^6} = 5^{6/9} = 5^{2/3} = \sqrt[3]{5^2}, \quad \text{or} \quad \sqrt[3]{25}$$

(b) $\sqrt[4]{p^2} = p^{2/4} = p^{1/2} = \sqrt{p}$ (Recall the assumption that $p > 0$.)

Answers

5. (a) $5p^3\sqrt{p}$ (b) $6y\sqrt{2yx}$ (c) $-3y^2xz^2\sqrt[3]{yx^2}$ (d) $-2ab\sqrt[4]{2ab^3}$

These examples suggest the following rule.

> If m is an integer, n and k are natural numbers, and all indicated roots exist, then
> $$\sqrt[kn]{a^{km}} = \sqrt[n]{a^m}.$$

Work Problem **6** *at the Side.* ▶

OBJECTIVE 4 Simplify products and quotients of radicals with different indexes. Since the product and quotient rules for radicals apply only when they have the same index, we multiply and divide radicals with different indexes by using rational exponents.

EXAMPLE 7 Multiplying Radicals with Different Indexes

Simplify $\sqrt{7} \cdot \sqrt[3]{2}$.

Because the different indexes, 2 and 3, have a least common index of 6, we use rational exponents to write each radical as a sixth root.

$$\sqrt{7} = 7^{1/2} = 7^{3/6} = \sqrt[6]{7^3} = \sqrt[6]{343}$$
$$\sqrt[3]{2} = 2^{1/3} = 2^{2/6} = \sqrt[6]{2^2} = \sqrt[6]{4}$$

Therefore,

$$\sqrt{7} \cdot \sqrt[3]{2}$$
$$= \sqrt[6]{343} \cdot \sqrt[6]{4} \quad \text{Substitute; } \sqrt{7} = \sqrt[6]{343},\ \sqrt[3]{2} = \sqrt[6]{4}$$
$$= \sqrt[6]{1372}. \quad \text{Product rule}$$

Work Problem **7** *at the Side.* ▶

OBJECTIVE 5 Use the Pythagorean formula. The **Pythagorean formula** relates the lengths of the three sides of a right triangle.

Pythagorean Formula

If a and b are the lengths of the shorter sides of a right triangle and c is the length of the longest side, then

$$a^2 + b^2 = c^2.$$

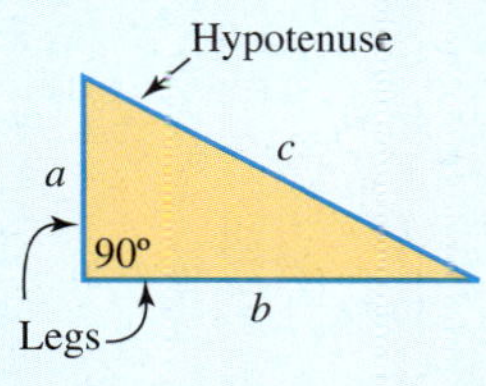

The two shorter sides are the **legs** of the triangle, and the longest side is the **hypotenuse.** The hypotenuse is the side opposite the right angle. Thus,

$$\textbf{leg}^2 + \textbf{leg}^2 = \textbf{hypotenuse}^2.$$

In **Section 10.1,** we will see that an equation such as $x^2 = 7$ has two solutions: $\sqrt{7}$ (the principal, or positive, square root of 7) and $-\sqrt{7}$. Similarly, $c^2 = 52$ has two solutions, $\pm\sqrt{52}$, or $\pm 2\sqrt{13}$. In applications we often choose only the positive square root.

6 Simplify. Assume that all variables represent positive real numbers.

(a) $\sqrt[12]{2^3}$

(b) $\sqrt[6]{t^2}$

7 Simplify $\sqrt{5} \cdot \sqrt[3]{4}$.

ANSWERS

6. (a) $\sqrt[4]{2}$ **(b)** $\sqrt[3]{t}$

7. $\sqrt[6]{2000}$

8 Find the length of the unknown side in each triangle.

(a)

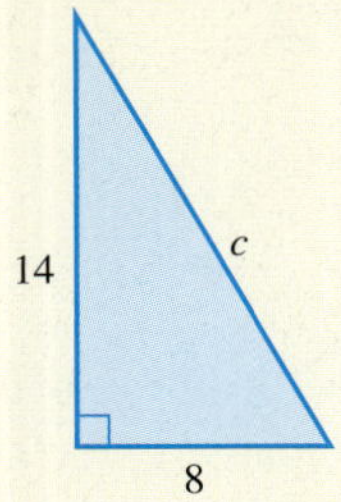

(b)

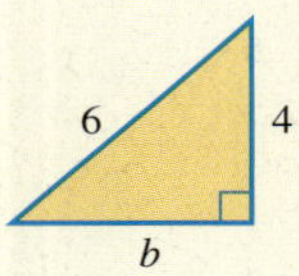

(*Hint:* Write the Pythagorean formula as $b^2 = c^2 - a^2$ here.)

EXAMPLE 8 Using the Pythagorean Formula

Use the Pythagorean formula to find the length of the hypotenuse in the triangle in Figure 6.

To find the length of the hypotenuse c, let $a = 4$ and $b = 6$. Then, use the formula.

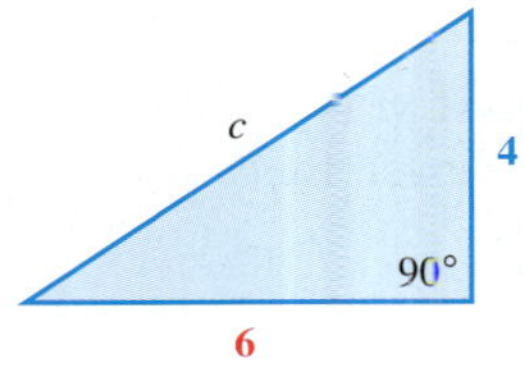

Figure 6

$$a^2 + b^2 = c^2$$

Substitute carefully.

$$4^2 + 6^2 = c^2 \quad \text{Let } a = 4 \text{ and } b = 6.$$

$$16 + 36 = c^2 \quad \text{Apply the exponents.}$$

$$c^2 = 52 \quad \text{Add; interchange sides.}$$

$$c = \sqrt{52} \quad \text{Choose the principal root.}$$

$$c = \sqrt{4 \cdot 13} \quad \text{Factor.}$$

$$c = \sqrt{4} \cdot \sqrt{13} \quad \text{Product rule}$$

$$c = 2\sqrt{13} \quad \text{Simplify.}$$

The length of the hypotenuse is $2\sqrt{13}$.

◀ Work Problem **8** at the Side.

CAUTION

When substituting in the Pythagorean formula $a^2 + b^2 = c^2$, be sure that the lengths of the legs are substituted for a and b and the length of the hypotenuse is substituted for c.

OBJECTIVE 6 Use the distance formula. The *distance formula* allows us to find the distance between two points in the coordinate plane, or the length of the line segment joining those two points.

Figure 7 shows the points $(3, -4)$ and $(-5, 3)$. The vertical line through $(-5, 3)$ and the horizontal line through $(3, -4)$ intersect at the point $(-5, -4)$. Thus, the point $(-5, -4)$ becomes the vertex of the right angle in a right triangle. By the Pythagorean formula, the sum of the squares of the lengths of the two legs a and b of the right triangle in Figure 7 is equal to the square of the length of the hypotenuse, d:

$$\boldsymbol{a^2 + b^2 = d^2}, \qquad \text{or} \qquad \boldsymbol{d^2 = a^2 + b^2}.$$

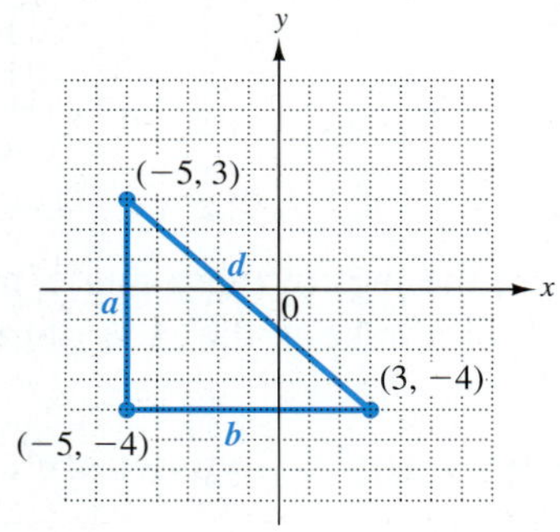

Figure 7

The length a is the difference between the y-coordinates of the endpoints. Since the x-coordinate of both of these points in Figure 7 is -5, the side is vertical, and we can find a by finding the difference between the y-coordinates. We subtract -4 from 3 to get a positive value for a.

$$a = 3 - (-4) = 7$$

Answers

8. (a) $2\sqrt{65}$ **(b)** $2\sqrt{5}$

Similarly, we find b by subtracting -5 from 3.

$$b = 3 - (-5) = 8$$

Substituting these values into the formula, we have

$$d^2 = a^2 + b^2$$

$$d^2 = 7^2 + 8^2 \quad \text{Let } a = 7 \text{ and } b = 8.$$

$$d^2 = 49 + 64 \quad \text{Apply the exponents.}$$

$$d^2 = 113 \quad \text{Add.}$$

$$d = \sqrt{113}. \quad \text{Choose the principal root.}$$

We choose the principal root since distance cannot be negative. Therefore, the distance between $(-5, 3)$ and $(3, -4)$ is $\sqrt{113}$.

Note

It is customary to leave the distance in radical form. Do not use a calculator to get an approximation, unless you are specifically directed to do so.

This result can be generalized. Figure 8 shows the two points (x_1, y_1) and (x_2, y_2). The distance a between (x_1, y_1) and (x_2, y_1) is

$$a = x_2 - x_1,$$

and the distance b between (x_2, y_2) and (x_2, y_1) is

$$b = y_2 - y_1.$$

From the Pythagorean formula,

$$d^2 = a^2 + b^2$$

$$d^2 = (x_2 - x_1)^2 + (y_2 - y_1)^2.$$

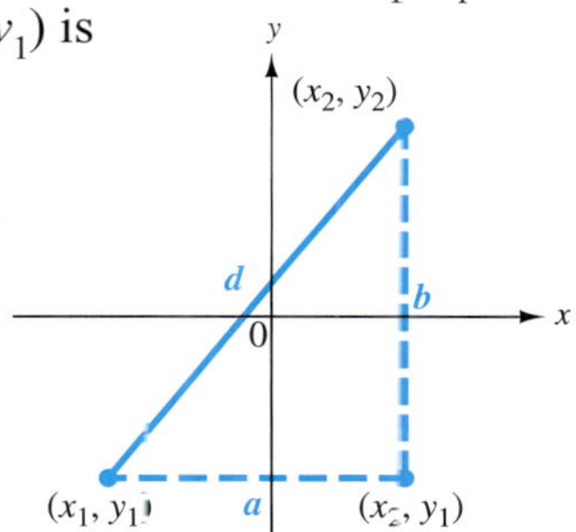

Figure 8

Choosing the principal square root gives the **distance formula.**

Distance Formula

The distance between the points (x_1, y_1) and (x_2, y_2) is

$$d = \sqrt{(x_2 - x_1)^2 + (y_2 - y_1)^2}.$$

EXAMPLE 9 Using the Distance Formula

Find the distance between $(-3, 5)$ and $(6, 4)$.

When using the distance formula to find the distance between two points, designating the points as (x_1, y_1) and (x_2, y_2) is arbitrary. We choose $(x_1, y_1) = (-3, 5)$ and $(x_2, y_2) = (6, 4)$.

$$d = \sqrt{(x_2 - x_1)^2 + (y_2 - y_1)^2} \quad \text{Distance formula}$$

$$d = \sqrt{(6 - (-3))^2 + (4 - 5)^2} \quad x_2 = 6, y_2 = 4, x_1 = -3, y_1 = 5$$

Substitute carefully.

$$d = \sqrt{9^2 + (-1)^2}$$

$$d = \sqrt{82} \quad \text{Leave in radical form.}$$

Work Problem 9 at the Side. ▶

9 Find the distance between each pair of points.

(a) $(2, -1)$ and $(5, 3)$

(b) $(-3, 2)$ and $(0, -4)$

Answers

9. (a) 5 **(b)** $\sqrt{45}$, or $3\sqrt{5}$

Math in the Media

SO, DID THE SCARECROW REALLY GET A BRAIN?

Probably the most famous mathematical statement in the history of motion pictures is heard in the 1939 classic *The Wizard of Oz*. Ray Bolger's character, the Scarecrow, wants a brain. When the Wizard grants him his "Th.D." (Doctor of Thinkology), the Scarecrow replies with a statement that has made mathematics teachers shudder for 70 years.

Scarecrow: *The sum of the square roots of any two sides of an isosceles triangle is equal to the square root of the remaining side.*

His statement is quite impressive and sounds like the formula for the *Pythagorean Theorem* (page 531). Let's see why it is incorrect.

1. To what kind of triangle does the Scarecrow refer in his statement? To what kind of triangle does the Pythagorean Theorem actually refer?

2. In the Scarecrow's statement, he refers to square roots. In applying the formula for the Pythagorean Theorem, do you find square roots of the sides? If not, what do you find?

3. An isosceles triangle has two sides of equal length. Draw an isosceles triangle with two sides of length 9 units and remaining side of length 4 units. Now show that this triangle does not satisfy the Scarecrow's statement.

(This is called a *counterexample* and is sufficient to show that his statement is false in general.)

4. Use wording similar to that of the Scarecrow, but state the Pythagorean Theorem correctly.

9.3 Exercises

FOR EXTRA HELP MyMathLab Math XL PRACTICE WATCH DOWNLOAD READ REVIEW

Decide whether each statement is true *or* false *by using the product rule explained in this section. Then support your answer by finding a calculator approximation for each expression.*

1. $2\sqrt{12} = \sqrt{48}$

2. $\sqrt{72} = 2\sqrt{18}$

3. $3\sqrt{8} = 2\sqrt{18}$

4. $5\sqrt{72} = 6\sqrt{50}$

5. Explain why $\sqrt[3]{x} \cdot \sqrt[3]{x}$ is not equal to x. What is it equal to?

6. Explain why $\sqrt[4]{x} \cdot \sqrt[4]{x}$ is not equal to x, but *is* equal to $\sqrt{x}$, for $x \geq 0$.

7. Which one of the following is *not* equal to $\sqrt{\frac{1}{2}}$? (Do not use calculator approximations.)

A. $\sqrt{0.5}$ **B.** $\sqrt{\frac{2}{4}}$ **C.** $\sqrt{\frac{3}{6}}$ **D.** $\frac{\sqrt{4}}{\sqrt{16}}$

8. Use the π key on your calculator to get a value for π. Now find an approximation for $\sqrt[4]{\frac{2143}{22}}$. Does the result mean that π is actually equal to $\sqrt[4]{\frac{2143}{22}}$? Why or why not?

Multiply using the product rule. Assume all variables represent positive real numbers. See Examples 1 and 2.

9. $\sqrt{5} \cdot \sqrt{6}$

10. $\sqrt{10} \cdot \sqrt{3}$

11. $\sqrt{14} \cdot \sqrt{x}$

12. $\sqrt{23} \cdot \sqrt{t}$

13. $\sqrt{14} \cdot \sqrt{3pqr}$

14. $\sqrt{7} \cdot \sqrt{5xt}$

15. $\sqrt[3]{7x} \cdot \sqrt[3]{2y}$

16. $\sqrt[3]{9x} \cdot \sqrt[3]{4y}$

17. $\sqrt[4]{11} \cdot \sqrt[4]{3}$

18. $\sqrt[4]{6} \cdot \sqrt[4]{9}$

19. $\sqrt[4]{2x} \cdot \sqrt[4]{3y^2}$

20. $\sqrt[4]{3y^2} \cdot \sqrt[4]{6yz}$

21. $\sqrt[3]{7} \cdot \sqrt[4]{3}$

22. $\sqrt[5]{8} \cdot \sqrt[6]{12}$

Simplify. Assume that all variables represent positive real numbers. See Example 3.

23. $\sqrt{\frac{64}{121}}$

24. $\sqrt{\frac{16}{49}}$

25. $\sqrt{\frac{3}{25}}$

26. $\sqrt{\frac{13}{49}}$

27. $\sqrt{\frac{x}{25}}$

28. $\sqrt{\frac{k}{100}}$

29. $\sqrt{\frac{p^6}{81}}$

30. $\sqrt{\frac{w^{10}}{36}}$

31. $\sqrt[3]{-\frac{27}{64}}$

32. $\sqrt[3]{-\frac{216}{125}}$

33. $\sqrt[3]{\frac{r^2}{8}}$

34. $\sqrt[3]{\frac{t}{125}}$

35. $-\sqrt[4]{\frac{81}{x^4}}$

36. $-\sqrt[4]{\frac{625}{y^4}}$

37. $\sqrt[5]{\frac{1}{x^{15}}}$

38. $\sqrt[5]{\frac{32}{y^{20}}}$

Express each radical in simplified form. See Example 4.

39. $\sqrt{12}$

40. $\sqrt{18}$

41. $\sqrt{288}$

42. $\sqrt{72}$

43. $-\sqrt{32}$

44. $-\sqrt{48}$

45. $-\sqrt{28}$

46. $-\sqrt{24}$

47. $\sqrt{30}$

48. $\sqrt{46}$

49. $\sqrt[3]{128}$

50. $\sqrt[3]{24}$

51. $\sqrt[3]{-16}$ **52.** $\sqrt[3]{-250}$ **53.** $\sqrt[3]{40}$ **54.** $\sqrt[3]{375}$

55. $-\sqrt[4]{512}$ **56.** $-\sqrt[4]{1250}$ **57.** $\sqrt[5]{64}$ **58.** $\sqrt[5]{128}$

59. A student claimed that $\sqrt[3]{14}$ is not in simplified form, since $14 = 8 + 6$, and 8 is a perfect cube. Was his reasoning correct? Why or why not?

60. Explain in your own words why $\sqrt[3]{k^4}$ is not a simplified radical.

Express each radical in simplified form. Assume that all variables represent positive real numbers. See Example 5.

61. $\sqrt{72k^2}$ **62.** $\sqrt{18m^2}$ **63.** $\sqrt{144x^3y^9}$ **64.** $\sqrt{169s^5t^{10}}$

65. $\sqrt{121x^6}$ **66.** $\sqrt{256z^{12}}$ **67.** $-\sqrt[3]{27t^{12}}$ **68.** $-\sqrt[3]{64y^{18}}$

69. $-\sqrt{100m^8z^4}$ **70.** $-\sqrt{25t^6s^{20}}$ **71.** $-\sqrt[3]{-125a^6b^9c^{12}}$ **72.** $-\sqrt[3]{-216y^{15}x^6z^3}$

73. $\sqrt[4]{\frac{1}{16}r^8t^{20}}$ **74.** $\sqrt[4]{\frac{81}{256}t^{12}u^8}$ **75.** $\sqrt{50x^3}$ **76.** $\sqrt{300z^3}$

77. $-\sqrt{500r^{11}}$ **78.** $-\sqrt{200p^{13}}$ **79.** $\sqrt{13x^7y^8}$ **80.** $\sqrt{23k^9p^{14}}$

81. $\sqrt[3]{8z^6w^9}$

82. $\sqrt[3]{64a^{15}b^{12}}$

83. $\sqrt[3]{-16z^5t^7}$

84. $\sqrt[3]{-81m^4n^{10}}$

85. $\sqrt[4]{81x^{12}y^{16}}$

86. $\sqrt[4]{81t^8u^{28}}$

87. $-\sqrt[4]{162r^{15}s^{10}}$

88. $-\sqrt[4]{32k^5m^{10}}$

89. $\sqrt{\dfrac{y^{11}}{36}}$

90. $\sqrt{\dfrac{v^{13}}{49}}$

91. $\sqrt[3]{\dfrac{x^{16}}{27}}$

92. $\sqrt[3]{\dfrac{y^{17}}{125}}$

Simplify. Assume that $x \geq 0$. *See Example 6.*

93. $\sqrt[4]{48^2}$

94. $\sqrt[4]{50^2}$

95. $\sqrt[4]{25}$

96. $\sqrt[6]{8}$

97. $\sqrt[10]{x^{25}}$

98. $\sqrt[12]{x^{44}}$

Simplify by first writing the radicals as radicals with the same index. Then multiply. Assume that $x \geq 0$. *See Example 7.*

99. $\sqrt[3]{4} \cdot \sqrt{3}$

100. $\sqrt[3]{5} \cdot \sqrt{6}$

101. $\sqrt[4]{3} \cdot \sqrt[3]{4}$

102. $\sqrt[3]{2} \cdot \sqrt[5]{3}$

103. $\sqrt{x} \cdot \sqrt[3]{x}$

104. $\sqrt[3]{x} \cdot \sqrt[4]{x}$

Find the unknown length in each right triangle. Simplify the answer if necessary. See Example 8.

105.

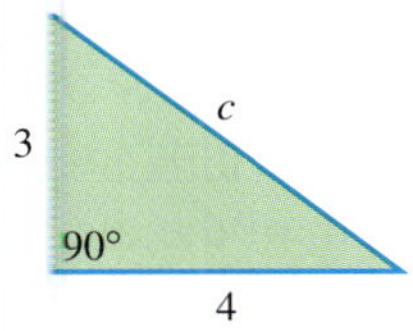

106.

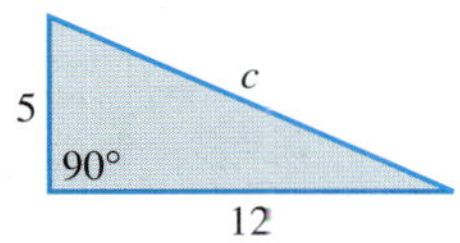

107.

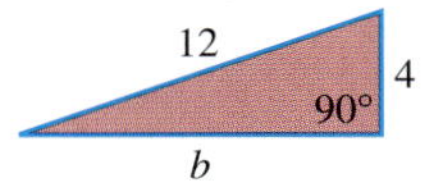

108.

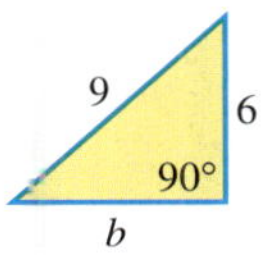

109.

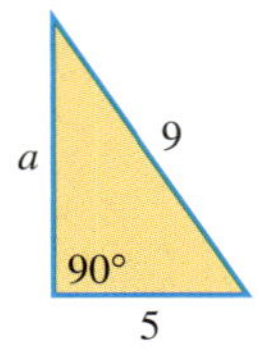

110.

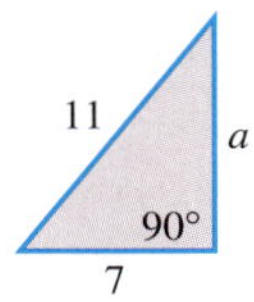

Find the distance between each pair of points. See Example 9.

111. $(6, 13)$ and $(1, 1)$

112. $(8, 13)$ and $(2, 5)$

113. $(-6, 5)$ and $(3, -4)$

114. $(-1, 5)$ and $(-7, 7)$

115. $(-8, 2)$ and $(-4, 1)$

116. $(-1, 2)$ and $(5, 3)$

117. $(4.7, 2.3)$ and $(1.7, -1.7)$

118. $(-2.9, 18.2)$ and $(2.1, 6.2)$

119. $\left(\sqrt{2}, \sqrt{6}\right)$ and $\left(-2\sqrt{2}, 4\sqrt{6}\right)$

120. $\left(\sqrt{7}, 9\sqrt{3}\right)$ and $\left(-\sqrt{7}, 4\sqrt{3}\right)$

121. $(x + y, y)$ and $(x - y, x)$

122. $(c, c - d)$ and $(d, c + d)$

Solve each problem.

123. A Sanyo color television, model AVM-2755, has a rectangular screen with a 21.7-in. width. Its height is 16 in. What is the diagonal of the screen to the nearest tenth of an inch? (*Source:* Actual measurements of the author's television.)

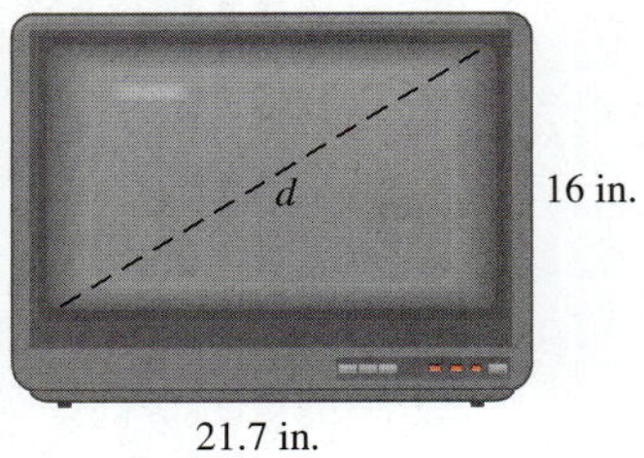

124. The length of the diagonal of a box is given by

$$D = \sqrt{L^2 + W^2 + H^2},$$

where L, W, and H are the length, width, and height of the box. Find the length of the diagonal, D, of a box that is 4 ft long, 3 ft high, and 2 ft wide. Give the exact value, and then round to the nearest tenth of a foot.

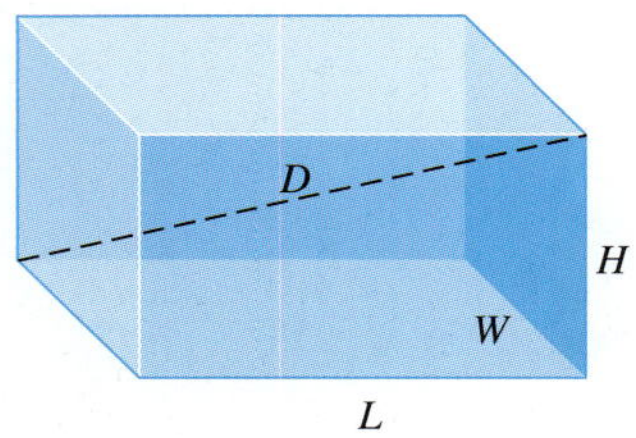

125. A formula from electronics dealing with impedance of parallel resonant circuits is

$$I = \frac{E}{\sqrt{R^2 + \omega^2 L^2}},$$

where the variables are in appropriate units. Find I if $E = 282$, $R = 100$, $L = 264$, and $\omega = 120\pi$. Give your answer to the nearest thousandth.

126. In the study of sound, one version of the law of tensions is

$$f_1 = f_2\sqrt{\frac{F_1}{F_2}}.$$

If $F_1 = 300$, $F_2 = 60$, and $f_2 = 260$, find f_1 to the nearest unit.

127. The following letter appeared in the column "Ask Tom Why," written by Tom Skilling of the *Chicago Tribune*.

Dear Tom,

I cannot remember the formula to calculate the distance to the horizon. I have a stunning view from my 14th floor condo, 150 feet above the ground. How far can I see?

Ted Fleischaker; Indianapolis, Ind.

Skilling's answer was as follows.

To find the distance to the horizon in miles, take the square root of the height of your view in feet and multiply that result by 1.224. Your answer will be the number of miles to the horizon. (*Source: Chicago Tribune,* August 17, 2002.)

Assuming Ted's eyes are 6 ft above the ground, the total height from the ground is $150 + 6 = 156$ ft. To the nearest tenth of a mile, how far can he see to the horizon?

9.4 Adding and Subtracting Radical Expressions

OBJECTIVE 1 Simplify radical expressions involving addition and subtraction. Expressions such as $4\sqrt{2} + 3\sqrt{2}$ and $2\sqrt{3} - 5\sqrt{3}$ can be simplified by using the distributive property.

$$4\sqrt{2} + 3\sqrt{2} = (4 + 3)\sqrt{2} = 7\sqrt{2}$$

This is similar to simplifying $4x + 3x$ as $7x$.

$$2\sqrt{3} - 5\sqrt{3} = (2 - 5)\sqrt{3} = -3\sqrt{3}$$

This is similar to simplifying $2x - 5x$ as $-3x$.

CAUTION

Only radical expressions with the same index and the same radicand may be combined. Expressions such as $5\sqrt{3} + 2\sqrt{2}$ or $3\sqrt{3} + 2\sqrt[3]{3}$ cannot be simplified by combining terms.

EXAMPLE 1 Adding and Subtracting Radicals

Add or subtract to simplify each radical expression.

(a) $3\sqrt{24} + \sqrt{54}$

Simplify each radical; then use the distributive property to combine terms.

$$\begin{aligned} &3\sqrt{24} + \sqrt{54} \\ &= 3\sqrt{4} \cdot \sqrt{6} + \sqrt{9} \cdot \sqrt{6} && \text{Product rule} \\ &= 3 \cdot 2\sqrt{6} + 3\sqrt{6} \\ &= 6\sqrt{6} + 3\sqrt{6} \\ &= 9\sqrt{6} && \text{Combine like terms.} \end{aligned}$$

(b) $2\sqrt{20x} - \sqrt{45x}, \quad x \geq 0$

$$\begin{aligned} &= 2\sqrt{4} \cdot \sqrt{5x} - \sqrt{9} \cdot \sqrt{5x} && \text{Product rule} \\ &= 2 \cdot 2\sqrt{5x} - 3\sqrt{5x} \\ &= 4\sqrt{5x} - 3\sqrt{5x} \\ &= \sqrt{5x} && \text{Combine like terms.} \end{aligned}$$

(c) $2\sqrt{3} - 4\sqrt{5}$

Here the radicals differ and are already simplified, so $2\sqrt{3} - 4\sqrt{5}$ cannot be simplified further.

Work Problem 1 at the Side. ▶

CAUTION

Do not confuse the product rule with combining like terms. ***The root of a sum does not equal the sum of the roots.*** For example,

$$\sqrt{9 + 16} \neq \sqrt{9} + \sqrt{16},$$

since $\sqrt{9 + 16} = \sqrt{25} = 5$, but $\sqrt{9} + \sqrt{16} = 3 + 4 = 7$.

OBJECTIVE

1 Simplify radical expressions involving addition and subtraction.

1 Add or subtract to simplify each radical expression.

(a) $3\sqrt{5} + 7\sqrt{5}$

(b) $2\sqrt{11} - \sqrt{11} + 3\sqrt{44}$

(c) $5\sqrt{12y} + 6\sqrt{75y}, \ y \geq 0$

(d) $3\sqrt{8} - 6\sqrt{50} + 2\sqrt{200}$

(e) $9\sqrt{5} - 4\sqrt{10}$

ANSWERS

1. (a) $10\sqrt{5}$ **(b)** $7\sqrt{11}$ **(c)** $40\sqrt{3y}$ **(d)** $-4\sqrt{2}$ **(e)** cannot be simplified further

2 Simplify. Assume that all variables represent positive real numbers.

(a) $7\sqrt[3]{81} + 3\sqrt[3]{24}$

(b) $-2\sqrt[4]{32} - 7\sqrt[4]{162}$

(c) $\sqrt[3]{p^4q^7} - \sqrt[3]{64pq}$

3 Simplify. Assume that all variables represent positive real numbers.

(a) $2\sqrt{\frac{8}{9}} - 2\frac{\sqrt{27}}{\sqrt{108}}$

(b) $\sqrt{\frac{80}{y^4}} + \sqrt{\frac{81}{y^{10}}}$

ANSWERS

2. (a) $27\sqrt[3]{3}$ (b) $-25\sqrt[4]{2}$ (c) $(pq^2 - 4)\sqrt[3]{pq}$

3. (a) $\frac{4\sqrt{2} - 3}{3}$ (b) $\frac{4y^3\sqrt{5} + 9}{y^5}$

EXAMPLE 2 Adding and Subtracting Radicals

Simplify. Assume that all variables represent positive real numbers.

(a) $2\sqrt[3]{16} - 5\sqrt[3]{54}$

Remember to write the index with each radical.

$$= 2\sqrt[3]{8 \cdot 2} - 5\sqrt[3]{27 \cdot 2} \quad \text{Factor.}$$
$$= 2\sqrt[3]{8} \cdot \sqrt[3]{2} - 5\sqrt[3]{27} \cdot \sqrt[3]{2} \quad \text{Product rule}$$
$$= 2 \cdot 2 \cdot \sqrt[3]{2} - 5 \cdot 3 \cdot \sqrt[3]{2}$$
$$= 4\sqrt[3]{2} - 15\sqrt[3]{2}$$
$$= (4 - 15)\sqrt[3]{2} \quad \text{Distributive property}$$
$$= -11\sqrt[3]{2} \quad \text{Combine like terms.}$$

(b) $2\sqrt[3]{x^2y} + \sqrt[3]{8x^5y^4}$

$$= 2\sqrt[3]{x^2y} + \sqrt[3]{(8x^3y^3)x^2y} \quad \text{Factor.}$$
$$= 2\sqrt[3]{x^2y} + \sqrt[3]{8x^3y^3} \cdot \sqrt[3]{x^2y} \quad \text{Product rule}$$
$$= 2\sqrt[3]{x^2y} + 2xy\sqrt[3]{x^2y}$$
$$= (2 + 2xy)\sqrt[3]{x^2y} \quad \text{Distributive property}$$

This result cannot be simplified further.

◀ Work Problem **2** at the Side.

EXAMPLE 3 Adding and Subtracting Radicals with Fractions

Simplify. Assume that all variables represent positive real numbers.

(a) $2\sqrt{\frac{75}{16}} + 4\frac{\sqrt{8}}{\sqrt{32}}$

$$= 2\frac{\sqrt{25 \cdot 3}}{\sqrt{16}} + 4\frac{\sqrt{4 \cdot 2}}{\sqrt{16 \cdot 2}} \quad \text{Quotient rule; factor.}$$
$$= 2\left(\frac{5\sqrt{3}}{4}\right) + 4\left(\frac{2\sqrt{2}}{4\sqrt{2}}\right) \quad \text{Product rule; take square roots.}$$
$$= \frac{5\sqrt{3}}{2} + 2 \quad \text{Multiply; } \frac{\sqrt{2}}{\sqrt{2}} = 1.$$
$$= \frac{5\sqrt{3}}{2} + \frac{4}{2} \quad \text{Write with a common denominator.}$$
$$= \frac{5\sqrt{3} + 4}{2} \quad \text{Add fractions.}$$

(b) $10\sqrt[3]{\frac{5}{x^6}} - 3\sqrt[3]{\frac{4}{x^9}}$

$$= 10\frac{\sqrt[3]{5}}{\sqrt[3]{x^6}} - 3\frac{\sqrt[3]{4}}{\sqrt[3]{x^9}} \quad \text{Quotient rule}$$
$$= \frac{10\sqrt[3]{5}}{x^2} - \frac{3\sqrt[3]{4}}{x^3} \quad \text{Simplify denominators.}$$
$$= \frac{10\sqrt[3]{5} \cdot x}{x^2 \cdot x} - \frac{3\sqrt[3]{4}}{x^3} \quad \text{Write with a common denominator.}$$
$$= \frac{10x\sqrt[3]{5} - 3\sqrt[3]{4}}{x^3} \quad \text{Subtract fractions.}$$

◀ Work Problem **3** at the Side.

9.4 Exercises

FOR EXTRA HELP MyMathLab Math XL PRACTICE WATCH DOWNLOAD READ REVIEW

1. Which one of the following sums could be simplified without first simplifying the individual radical expressions?
   A. $\sqrt{50}+\sqrt{32}$ **B.** $3\sqrt{6}+9\sqrt{6}$ **C.** $\sqrt[3]{32}-\sqrt[3]{108}$ **D.** $\sqrt[5]{6}-\sqrt[5]{192}$

2. Let $a=1$ and $b=64$.
   (a) Evaluate $\sqrt{a}+\sqrt{b}$. Then find $\sqrt{a+b}$. Are they equal?
   (b) Evaluate $\sqrt[3]{a}+\sqrt[3]{b}$. Then find $\sqrt[3]{a+b}$. Are they equal?
   (c) Complete the following: In general, $\sqrt[n]{a}+\sqrt[n]{b}\neq$ ______________, based on the observations in parts (a) and (b) of this exercise.

3. Even though the indexes of the terms are not equal, the sum $\sqrt{64}+\sqrt[3]{125}+\sqrt[4]{16}$ can be simplified quite easily. What is this sum? Why can these terms be combined so easily?

4. Explain why $28-4\sqrt{2}$ ***is not equal to*** $24\sqrt{2}$. (This is a common error among algebra students.)

Simplify. Assume that all variables represent positive real numbers. See Examples 1 and 2.

5. $\sqrt{36}-\sqrt{100}$
6. $\sqrt{25}-\sqrt{81}$
7. $-2\sqrt{48}+3\sqrt{75}$
8. $4\sqrt{32}-2\sqrt{8}$
9. $\sqrt[3]{16}+4\sqrt[3]{54}$
10. $3\sqrt[3]{24}-2\sqrt[3]{192}$
11. $\sqrt[4]{32}+3\sqrt[4]{2}$
12. $\sqrt[4]{405}-2\sqrt[4]{5}$
13. $6\sqrt{18}-\sqrt{32}+2\sqrt{50}$
14. $5\sqrt{8}+3\sqrt{72}-3\sqrt{50}$
15. $5\sqrt{6}+2\sqrt{10}$
16. $3\sqrt{11}-5\sqrt{13}$
17. $2\sqrt{5}+3\sqrt{20}+4\sqrt{45}$
18. $5\sqrt{54}-2\sqrt{24}-2\sqrt{96}$
19. $8\sqrt{2x}-\sqrt{8x}+\sqrt{72x}$
20. $4\sqrt{18k}-\sqrt{72k}+\sqrt{50k}$
21. $3\sqrt{72m^2}-5\sqrt{32m^2}$
22. $9\sqrt{27p^2}-14\sqrt{108p^2}$
23. $-\sqrt[3]{54}+2\sqrt[3]{16}$
24. $15\sqrt[3]{81}-4\sqrt[3]{24}$
25. $2\sqrt[3]{27x}-2\sqrt[3]{8x}$
26. $6\sqrt[3]{128m}+3\sqrt[3]{16m}$
27. $\sqrt[3]{x^2y}-\sqrt[3]{8x^2y}$
28. $3\sqrt[3]{x^2y^2}-2\sqrt[3]{64x^2y^2}$
29. $3x\sqrt[3]{xy^2}-2\sqrt[3]{8x^4y^2}$
30. $6q^2\sqrt[3]{5q}-2q\sqrt[3]{40q^4}$

31. $5\sqrt[4]{32} + 3\sqrt[4]{162}$

32. $2\sqrt[4]{512} + 4\sqrt[4]{32}$

33. $3\sqrt[4]{x^5y} - 2x\sqrt[4]{xy}$

34. $2\sqrt[4]{m^9p^6} - 3m^2p\sqrt[4]{mp^2}$

35. $2\sqrt[4]{32a^3} + 5\sqrt[4]{2a^3}$

36. $-\sqrt[4]{16r} + 5\sqrt[4]{r}$

Simplify. Assume that all variables represent positive real numbers. See Example 3.

37. $\frac{2\sqrt{5}}{3} + \frac{\sqrt{5}}{6}$

38. $\frac{4\sqrt{3}}{3} + \frac{2\sqrt{3}}{9}$

39. $\sqrt{\frac{8}{9}} + \sqrt{\frac{18}{36}}$

40. $\sqrt{\frac{12}{16}} + \sqrt{\frac{48}{64}}$

41. $\frac{\sqrt{32}}{3} + \frac{2\sqrt{2}}{3} - \frac{\sqrt{2}}{\sqrt{9}}$

42. $\frac{\sqrt{27}}{2} - \frac{3\sqrt{3}}{2} + \frac{\sqrt{3}}{\sqrt{4}}$

43. $3\sqrt{\frac{50}{9}} + 8\frac{\sqrt{2}}{\sqrt{8}}$

44. $9\sqrt{\frac{48}{25}} - 2\frac{\sqrt{2}}{\sqrt{98}}$

45. $\sqrt{\frac{25}{x^8}} - \sqrt{\frac{9}{x^6}}$

46. $\sqrt{\frac{100}{y^4}} + \sqrt{\frac{81}{y^{10}}}$

47. $3\sqrt[3]{\frac{m^5}{27}} - 2m\sqrt[3]{\frac{m^2}{64}}$

48. $2a\sqrt[4]{\frac{a}{16}} - 5a\sqrt[4]{\frac{a}{81}}$

49. $3\sqrt[3]{\frac{2}{x^6}} - 4\sqrt[3]{\frac{5}{x^9}}$

50. $-4\sqrt[3]{\frac{4}{t^9}} + 3\sqrt[3]{\frac{9}{t^{12}}}$

Solve each problem. Give answers as simplified radical expressions.

51. Find the perimeter of the triangle.

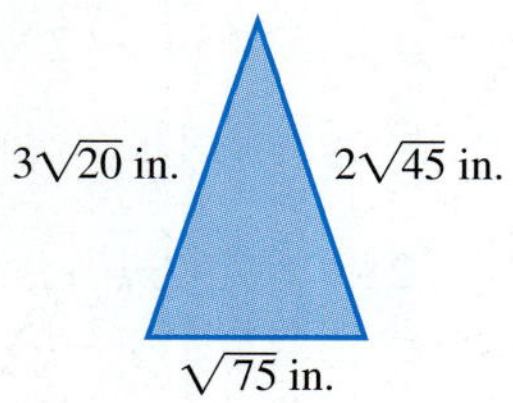

52. Find the perimeter of the rectangle.

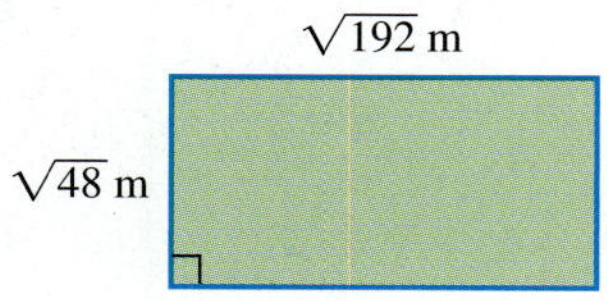

53. What is the perimeter of the computer graphic?

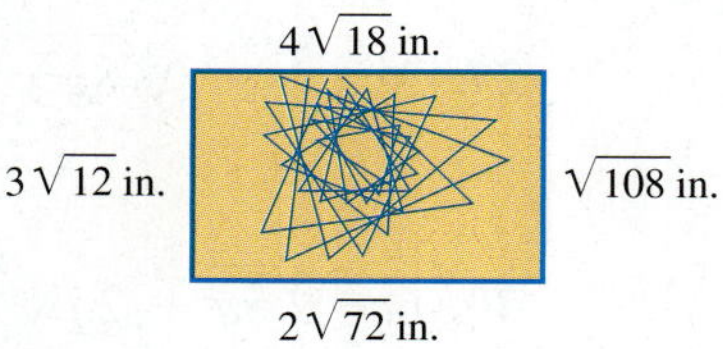

54. Find the area of the trapezoid.

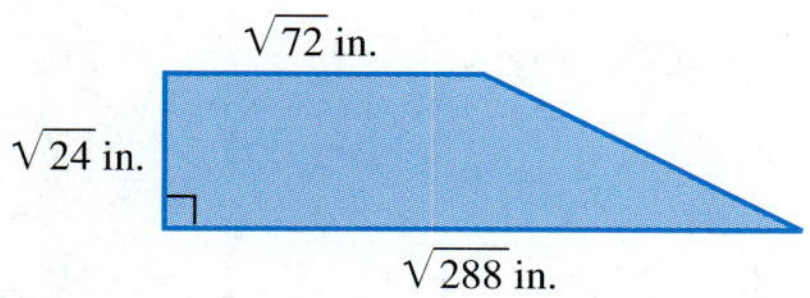

9.5 Multiplying and Dividing Radical Expressions

OBJECTIVES

1. Multiply radical expressions.
2. Rationalize denominators with one radical term.
3. Rationalize denominators with binomials involving radicals.
4. Write radical quotients in lowest terms.

OBJECTIVE 1 Multiply radical expressions. We multiply binomial expressions involving radicals by using the FOIL method from **Section 6.4.** For example, we find the product of $\sqrt{5}+3$ and $\sqrt{6}+1$ as follows:

$$\left(\sqrt{5}+3\right)\left(\sqrt{6}+1\right)$$

$$= \underbrace{\sqrt{5}\cdot\sqrt{6}}_{\text{First}} + \underbrace{\sqrt{5}\cdot 1}_{\text{Outer}} + \underbrace{3\cdot\sqrt{6}}_{\text{Inner}} + \underbrace{3\cdot 1}_{\text{Last}}$$

$$= \sqrt{30}+\sqrt{5}+3\sqrt{6}+3.$$

This result cannot be simplified further.

EXAMPLE 1 Multiplying Binomials Involving Radical Expressions

Multiply, using the FOIL method.

(a) $\left(7-\sqrt{3}\right)\left(\sqrt{5}+\sqrt{2}\right)$

$$= \overset{\text{F}}{7\sqrt{5}} + \overset{\text{O}}{7\sqrt{2}} - \overset{\text{I}}{\sqrt{3}\cdot\sqrt{5}} - \overset{\text{L}}{\sqrt{3}\cdot\sqrt{2}}$$

$$= 7\sqrt{5}+7\sqrt{2}-\sqrt{15}-\sqrt{6}$$

(b) $\left(\sqrt{10}+\sqrt{3}\right)\left(\sqrt{10}-\sqrt{3}\right)$

$$= \sqrt{10}\cdot\sqrt{10}-\sqrt{10}\cdot\sqrt{3}+\sqrt{3}\cdot\sqrt{10}-\sqrt{3}\cdot\sqrt{3}$$

$$= 10-3$$

$$= 7$$

The product $\left(\sqrt{10}+\sqrt{3}\right)\left(\sqrt{10}-\sqrt{3}\right)=\left(\sqrt{10}\right)^2-\left(\sqrt{3}\right)^2$ is the difference of squares:

$$(x+y)(x-y)=x^2-y^2. \quad \text{Here, } x=\sqrt{10} \text{ and } y=\sqrt{3}.$$

(c) $\left(\sqrt{7}-3\right)^2$

$$= \left(\sqrt{7}-3\right)\left(\sqrt{7}-3\right)$$

$$= \sqrt{7}\cdot\sqrt{7}-3\sqrt{7}-3\sqrt{7}+3\cdot 3$$

$$= 7-6\sqrt{7}+9$$

$$= 16-6\sqrt{7}$$

Be careful! These terms cannot be combined.

(d) $\left(5-\sqrt[3]{3}\right)\left(5+\sqrt[3]{3}\right)$

$$= 5\cdot 5+5\sqrt[3]{3}-5\sqrt[3]{3}-\sqrt[3]{3}\cdot\sqrt[3]{3}$$

$$= 25-\sqrt[3]{3^2}$$

$$= 25-\sqrt[3]{9}$$

Remember to write the index 3 in *each* radical.

(e) $\left(\sqrt{k}+\sqrt{y}\right)\left(\sqrt{k}-\sqrt{y}\right)$

$$= \left(\sqrt{k}\right)^2-\left(\sqrt{y}\right)^2 \qquad \text{Difference of squares}$$

$$= k-y, \quad k\ge 0 \text{ and } y\ge 0$$

Note

In Example 1(c) we could have used the formula for the square of a binomial,

$$(x - y)^2 = x^2 - 2xy + y^2,$$

to obtain the same result:

$$\begin{aligned}&\left(\sqrt{7} - 3\right)^2\\ &= \left(\sqrt{7}\right)^2 - 2\left(\sqrt{7}\right)(3) + 3^2\\ &= 7 - 6\sqrt{7} + 9\\ &= 16 - 6\sqrt{7}.\end{aligned}$$

◀ Work Problem 1 at the Side.

OBJECTIVE 2 Rationalize denominators with one radical term. As defined earlier, a simplified radical expression will have no radical in the denominator. The origin of this agreement no doubt occurred before the days of high-speed calculation, when computation was a tedious process performed by hand.

For example, consider the radical expression $\frac{1}{\sqrt{2}}$. To find a decimal approximation by hand, it would be necessary to divide 1 by a decimal approximation for $\sqrt{2}$, such as 1.414. It would be much easier if the divisor were a whole number. This can be accomplished by multiplying $\frac{1}{\sqrt{2}}$ by 1 in the form $\frac{\sqrt{2}}{\sqrt{2}}$. ***Multiplying by 1 in any form does not change the value of the original expression.***

$$\frac{1}{\sqrt{2}} \cdot \frac{\sqrt{2}}{\sqrt{2}} = \frac{\sqrt{2}}{2} \qquad \text{Multiply by 1; } \frac{\sqrt{2}}{\sqrt{2}} = 1.$$

Now the computation would require dividing 1.414 by 2 to obtain 0.707, a much easier task.

With current technology, either form of this fraction can be approximated with the same number of keystrokes. See Figure 9, which shows how a calculator gives the same approximation for both forms of the expression.

Figure 9

Rationalizing a Denominator

A common way of "standardizing" the form of a radical expression is to have the denominator contain no radicals. The process of removing radicals from a denominator so that the denominator contains only rational numbers is called **rationalizing the denominator.**

---

1 Multiply, using the FOIL method.

(a) $\left(2 + \sqrt{3}\right)\left(1 + \sqrt{5}\right)$

(b) $\left(4 + \sqrt{3}\right)\left(4 - \sqrt{3}\right)$

(c) $\left(\sqrt{13} - 2\right)^2$

(d) $\left(4 + \sqrt[3]{7}\right)\left(4 - \sqrt[3]{7}\right)$

(e) $\left(\sqrt{p} + \sqrt{s}\right)\left(\sqrt{p} - \sqrt{s}\right)$, $p \geq 0$ and $s \geq 0$

Answers

1. (a) $2 + 2\sqrt{5} + \sqrt{3} + \sqrt{15}$ **(b)** 13 **(c)** $17 - 4\sqrt{13}$ **(d)** $16 - \sqrt[3]{49}$ **(e)** $p - s$

EXAMPLE 2 Rationalizing Denominators with Square Roots

Rationalize each denominator.

(a) $\dfrac{3}{\sqrt{7}}$

Multiply by $\frac{\sqrt{7}}{\sqrt{7}}$. This is an application of the multiplicative identity property.

$$\frac{3}{\sqrt{7}} = \frac{3 \cdot \sqrt{7}}{\sqrt{7} \cdot \sqrt{7}} = \frac{3\sqrt{7}}{7}$$

In the denominator, $\sqrt{7} \cdot \sqrt{7} = \sqrt{7 \cdot 7} = \sqrt{49} = 7$. The final denominator is now a rational number.

(b) $\dfrac{5\sqrt{2}}{\sqrt{5}} = \dfrac{5\sqrt{2} \cdot \sqrt{5}}{\sqrt{5} \cdot \sqrt{5}} = \dfrac{5\sqrt{10}}{5} = \sqrt{10}$

(c) $\dfrac{-6}{\sqrt{12}}$

Less work is involved if the radical in the denominator is simplified first.

$$\frac{-6}{\sqrt{12}} = \frac{-6}{\sqrt{4 \cdot 3}} = \frac{-6}{2\sqrt{3}} = \frac{-3}{\sqrt{3}}$$

Now rationalize the denominator.

$$\frac{-3}{\sqrt{3}} = \frac{-3 \cdot \sqrt{3}}{\sqrt{3} \cdot \sqrt{3}} = \frac{-3\sqrt{3}}{3} = -\sqrt{3}$$

Work Problem **2** at the Side. ▶

2 Rationalize each denominator.

(a) $\dfrac{8}{\sqrt{3}}$

(b) $\dfrac{5\sqrt{6}}{\sqrt{5}}$

(c) $\dfrac{3}{\sqrt{48}}$

(d) $\dfrac{-16}{\sqrt{32}}$

EXAMPLE 3 Rationalizing Denominators in Roots of Fractions

Simplify each radical. Assume that all variables represent positive real numbers.

(a) $\sqrt{\dfrac{18}{125}}$

$= \dfrac{\sqrt{18}}{\sqrt{125}}$ Quotient rule

$= \dfrac{\sqrt{9 \cdot 2}}{\sqrt{25 \cdot 5}}$ Factor.

$= \dfrac{3\sqrt{2}}{5\sqrt{5}}$ Product rule

$= \dfrac{3\sqrt{2} \cdot \sqrt{5}}{5\sqrt{5} \cdot \sqrt{5}}$ Multiply by $\frac{\sqrt{5}}{\sqrt{5}}$.

$= \dfrac{3\sqrt{10}}{5 \cdot 5}$ Product rule

$= \dfrac{3\sqrt{10}}{25}$ Multiply.

Continued on Next Page

ANSWERS

2. (a) $\dfrac{8\sqrt{3}}{3}$ (b) $\sqrt{30}$ (c) $\dfrac{\sqrt{3}}{4}$ (d) $-2\sqrt{2}$

3 Simplify each radical. Assume that all variables represent positive real numbers.

(a) $\sqrt{\dfrac{8}{45}}$

(b) $\sqrt{\dfrac{72}{y}}$

(c) $\sqrt{\dfrac{200k^6}{y^7}}$

4 Simplify.

(a) $\sqrt[3]{\dfrac{15}{32}}$

(b) $\sqrt[3]{\dfrac{m^{12}}{n}}, \quad n \neq 0$

(c) $\sqrt[4]{\dfrac{6y}{w^2}}, \quad y \geq 0, w \neq 0$

(b) $\sqrt{\dfrac{50m^4}{p^5}}$

$= \dfrac{\sqrt{50m^4}}{\sqrt{p^5}}$ Quotient rule

$= \dfrac{\sqrt{25m^4 \cdot 2}}{\sqrt{p^4 \cdot p}}$ Factor.

$= \dfrac{5m^2\sqrt{2}}{p^2\sqrt{p}}$ Product rule

$= \dfrac{5m^2\sqrt{2} \cdot \sqrt{p}}{p^2\sqrt{p} \cdot \sqrt{p}}$ Multiply by $\frac{\sqrt{p}}{\sqrt{p}}$.

$= \dfrac{5m^2\sqrt{2p}}{p^2 \cdot p}$ Product rule

$= \dfrac{5m^2\sqrt{2p}}{p^3}$ Multiply.

◀ Work Problem **3** at the Side.

EXAMPLE 4 Rationalizing Denominators with Higher Roots

Simplify.

(a) $\sqrt[3]{\dfrac{27}{16}}$

Use the quotient rule and simplify the numerator and denominator.

$$\sqrt[3]{\frac{27}{16}} = \frac{\sqrt[3]{27}}{\sqrt[3]{16}} = \frac{3}{\sqrt[3]{8} \cdot \sqrt[3]{2}} = \frac{3}{2\sqrt[3]{2}}$$

To get a rational denominator, multiply the numerator and denominator by a number that will result in a perfect cube in the radicand in the denominator. Since $2 \cdot 4 = 8$, a perfect cube, apply the multiplicative identity property and multiply the numerator and denominator by $\sqrt[3]{4}$.

$$\sqrt[3]{\frac{27}{16}} = \frac{3}{2\sqrt[3]{2}} = \frac{3 \cdot \sqrt[3]{4}}{2\sqrt[3]{2} \cdot \sqrt[3]{4}} = \frac{3\sqrt[3]{4}}{2\sqrt[3]{8}} = \frac{3\sqrt[3]{4}}{2 \cdot 2} = \frac{3\sqrt[3]{4}}{4}$$

(b) $\sqrt[4]{\dfrac{5x}{z}} = \dfrac{\sqrt[4]{5x} \cdot \sqrt[4]{z^3}}{\sqrt[4]{z} \cdot \sqrt[4]{z^3}} = \dfrac{\sqrt[4]{5xz^3}}{\sqrt[4]{z^4}} = \dfrac{\sqrt[4]{5xz^3}}{z}, \quad x \geq 0, z > 0$

CAUTION

In Example 4(a), a typical error is to multiply the numerator and denominator by $\sqrt[3]{2}$, forgetting that $\sqrt[3]{2} \cdot \sqrt[3]{2} = \sqrt[3]{2^2}$, which does *not* equal 2. We need ***three*** factors of 2 to get 2^3 under the radical.

$$\sqrt[3]{2} \cdot \sqrt[3]{2} \cdot \sqrt[3]{2} = \sqrt[3]{2^3}, \quad \text{which does equal} \quad 2.$$

◀ Work Problem **4** at the Side.

ANSWERS

3. (a) $\dfrac{2\sqrt{10}}{15}$ (b) $\dfrac{6\sqrt{2y}}{y}$ (c) $\dfrac{10k^3\sqrt{2y}}{y^4}$

4. (a) $\dfrac{\sqrt[3]{30}}{4}$ (b) $\dfrac{m^4\sqrt[3]{n^2}}{n}$ (c) $\dfrac{\sqrt[4]{6yw^2}}{w}$

OBJECTIVE 3 Rationalize denominators with binomials involving radicals. Recall the special product

$$(x + y)(x - y) = x^2 - y^2.$$

To rationalize a denominator that contains a binomial expression (one that contains exactly two terms) involving radicals, such as

$$\frac{3}{1 + \sqrt{2}},$$

we must use *conjugates*. The conjugate of $1 + \sqrt{2}$ is $1 - \sqrt{2}$. In general, $x + y$ and $x - y$ are **conjugates.**

Rationalizing a Binomial Denominator

If a radical expression has a sum or difference with square root radicals in the denominator, rationalize the denominator by multiplying both the numerator and denominator by the conjugate of the denominator.

For $\frac{3}{1 + \sqrt{2}}$, we rationalize the denominator by multiplying both the numerator and denominator by $1 - \sqrt{2}$, the conjugate of the denominator.

$$\frac{3}{1 + \sqrt{2}}$$

$$= \frac{3(1 - \sqrt{2})}{(1 + \sqrt{2})(1 - \sqrt{2})}$$

$$= \frac{3(1 - \sqrt{2})}{-1}$$

The denominator is now a rational number.

$(1 + \sqrt{2})(1 - \sqrt{2})$
$= 1^2 - (\sqrt{2})^2$
$= 1 - 2 = -1$

$$= \frac{3}{-1}(1 - \sqrt{2})$$

$$= -3(1 - \sqrt{2}), \quad \text{or} \quad -3 + 3\sqrt{2}$$

EXAMPLE 5 Rationalizing Binomial Denominators

Rationalize each denominator.

(a) $\frac{5}{4 - \sqrt{3}}$

$= \frac{5(4 + \sqrt{3})}{(4 - \sqrt{3})(4 + \sqrt{3})}$ Multiply the numerator and denominator by $4 + \sqrt{3}$.

$= \frac{5(4 + \sqrt{3})}{16 - 3}$ Multiply in the denominator.

$= \frac{5(4 + \sqrt{3})}{13}$ Subtract.

Notice that the numerator is left in factored form. This makes it easier to determine whether the expression is written in lowest terms.

Continued on Next Page

5 Rationalize each denominator.

(a) $\dfrac{-4}{\sqrt{5}+2}$

(b) $\dfrac{15}{\sqrt{7}+\sqrt{2}}$

(c) $\dfrac{\sqrt{3}+\sqrt{5}}{\sqrt{2}-\sqrt{7}}$

(d) $\dfrac{2}{\sqrt{k}+\sqrt{z}}$, $k \neq z, k > 0, z > 0$

6 Write each quotient in lowest terms.

(a) $\dfrac{24-36\sqrt{7}}{16}$

(b) $\dfrac{2x+\sqrt{32x^2}}{6x}$, $x > 0$

Answers

5. **(a)** $-4(\sqrt{5}-2)$ **(b)** $3(\sqrt{7}-\sqrt{2})$
(c) $\dfrac{-(\sqrt{6}+\sqrt{21}+\sqrt{10}+\sqrt{35})}{5}$
(d) $\dfrac{2(\sqrt{k}-\sqrt{z})}{k-z}$

6. **(a)** $\dfrac{6-9\sqrt{7}}{4}$ **(b)** $\dfrac{1+2\sqrt{2}}{3}$

(b) $\dfrac{\sqrt{2}-\sqrt{3}}{\sqrt{5}+\sqrt{3}}$

$$= \frac{(\sqrt{2}-\sqrt{3})(\sqrt{5}-\sqrt{3})}{(\sqrt{5}+\sqrt{3})(\sqrt{5}-\sqrt{3})}$$ Multiply the numerator and denominator by $\sqrt{5}-\sqrt{3}$.

$$= \frac{\sqrt{10}-\sqrt{6}-\sqrt{15}+3}{5-3}$$ Multiply.

$$= \frac{\sqrt{10}-\sqrt{6}-\sqrt{15}+3}{2}$$ Subtract in the denominator.

(c) $\dfrac{3}{\sqrt{5m}-\sqrt{p}}$, $5m \neq p, m > 0, p > 0$

$$= \frac{3(\sqrt{5m}+\sqrt{p})}{(\sqrt{5m}-\sqrt{p})(\sqrt{5m}+\sqrt{p})}$$

$$= \frac{3(\sqrt{5m}+\sqrt{p})}{5m-p}$$

◀ Work Problem 5 at the Side.

OBJECTIVE 4 Write radical quotients in lowest terms.

EXAMPLE 6 Writing Radical Quotients in Lowest Terms

Write each quotient in lowest terms.

(a) $\dfrac{6+2\sqrt{5}}{4}$

$$= \frac{2(3+\sqrt{5})}{2 \cdot 2}$$ Factor the numerator and denominator. (This is a key step.)

$$= \frac{3+\sqrt{5}}{2}$$ Divide out the common factor.

Here is an alternative method for writing this expression in lowest terms.

$$\frac{6+2\sqrt{5}}{4} = \frac{6}{4} + \frac{2\sqrt{5}}{4} = \frac{3}{2} + \frac{\sqrt{5}}{2} = \frac{3+\sqrt{5}}{2}$$

(b) $\dfrac{5y-\sqrt{8y^2}}{6y}$, $y > 0$

$$= \frac{5y-2y\sqrt{2}}{6y}$$ $\sqrt{8y^2} = \sqrt{4y^2 \cdot 2} = 2y\sqrt{2}$

$$= \frac{y(5-2\sqrt{2})}{6y}$$ Factor the numerator.

$$= \frac{5-2\sqrt{2}}{6}$$ Divide out the common factor.

◀ Work Problem 6 at the Side.

9.5 Exercises

FOR EXTRA HELP MyMathLab MathXL PRACTICE WATCH DOWNLOAD READ REVIEW

Match each part of a rule for a special product in Column I with the part it equals in Column II.

I	II
1. $(x+\sqrt{y})(x-\sqrt{y})$	A. $x-y$
2. $(\sqrt{x}+y)(\sqrt{x}-y)$	B. $x+2y\sqrt{x}+y^2$
3. $(\sqrt{x}+\sqrt{y})(\sqrt{x}-\sqrt{y})$	C. $x-y^2$
4. $(\sqrt{x}+\sqrt{y})^2$	D. $x-2\sqrt{xy}+y$
5. $(\sqrt{x}-\sqrt{y})^2$	E. x^2-y
6. $(\sqrt{x}+y)^2$	F. $x+2\sqrt{xy}+y$

Multiply, and then simplify each product. Assume that all variables represent positive real numbers. See Example 1.

7. $\sqrt{3}(\sqrt{12}-4)$

8. $\sqrt{5}(\sqrt{125}-6)$

9. $\sqrt{2}(\sqrt{18}-\sqrt{3})$

10. $\sqrt{5}(\sqrt{15}+\sqrt{5})$

11. $(\sqrt{6}+2)(\sqrt{6}-2)$

12. $(\sqrt{7}+8)(\sqrt{7}-8)$

13. $(\sqrt{12}-\sqrt{3})(\sqrt{12}+\sqrt{3})$

14. $(\sqrt{18}+\sqrt{8})(\sqrt{18}-\sqrt{8})$

15. $(\sqrt{3}+2)(\sqrt{6}-5)$

16. $(\sqrt{7}+1)(\sqrt{2}-4)$

17. $(\sqrt{3x}+2)(\sqrt{3x}-2)$

18. $(\sqrt{6y}-4)(\sqrt{6y}+4)$

19. $(2\sqrt{x}+\sqrt{y})(2\sqrt{x}-\sqrt{y})$

20. $(\sqrt{p}+5\sqrt{s})(\sqrt{p}-5\sqrt{s})$

21. $(4\sqrt{x}+3)^2$

22. $(5\sqrt{p}-6)^2$

23. $(9-\sqrt[3]{2})(9+\sqrt[3]{2})$

24. $(7+\sqrt[3]{6})(7-\sqrt[3]{6})$

25. The correct answer to Exercise 7 is $6 - 4\sqrt{3}$. Explain why this is not equal to $2\sqrt{3}$.

26. When we rationalize the denominator in the radical expression $\frac{1}{\sqrt{2}}$, we multiply both the numerator and denominator by $\sqrt{2}$. What property of real numbers covered in **Section 1.4** justifies this procedure?

Rationalize the denominator in each expression. Assume that all variables represent positive real numbers. See Example 2.

27. $\frac{7}{\sqrt{7}}$ **28.** $\frac{11}{\sqrt{11}}$ **29.** $\frac{15}{\sqrt{3}}$ **30.** $\frac{12}{\sqrt{6}}$ **31.** $\frac{\sqrt{3}}{\sqrt{2}}$

32. $\frac{\sqrt{7}}{\sqrt{6}}$ **33.** $\frac{9\sqrt{3}}{\sqrt{5}}$ **34.** $\frac{3\sqrt{2}}{\sqrt{11}}$ **35.** $\frac{-6}{\sqrt{18}}$ **36.** $\frac{-5}{\sqrt{24}}$

37. $\frac{-8\sqrt{3}}{\sqrt{k}}$ **38.** $\frac{-4\sqrt{13}}{\sqrt{m}}$ **39.** $\frac{6\sqrt{3y}}{\sqrt{y^3}}$ **40.** $\frac{-8\sqrt{5y}}{\sqrt{y^5}}$

41. Explain why $\frac{1}{\sqrt[3]{2}}$ would not be written with the denominator rationalized if you begin by multiplying both the numerator and denominator by $\sqrt[3]{2}$. By what should you multiply them both to achieve the desired result?

42. Look again at the expression in Exercise 39. Start by multiplying both the numerator and the denominator by $\sqrt{y}$, to obtain the final answer. Then start over, multiplying both the numerator and denominator by $\sqrt{y^3}$, to obtain the same answer. Which method do you prefer? Why?

Simplify. Assume that all variables represent positive real numbers. See Examples 3 and 4.

43. $\sqrt{\frac{7}{2}}$ **44.** $\sqrt{\frac{10}{3}}$ **45.** $-\sqrt{\frac{7}{50}}$ **46.** $-\sqrt{\frac{13}{75}}$ **47.** $\sqrt{\frac{24}{x}}$

48. $\sqrt{\frac{52}{y}}$ **49.** $-\sqrt{\frac{98r^3}{s}}$ **50.** $-\sqrt{\frac{150m^5}{n}}$ **51.** $\sqrt{\frac{288x^7}{y^9}}$ **52.** $\sqrt{\frac{242t^9}{u^{11}}}$

53. $\sqrt[3]{\frac{2}{3}}$ **54.** $\sqrt[3]{\frac{4}{5}}$ **55.** $\sqrt[3]{\frac{4}{9}}$ **56.** $\sqrt[3]{\frac{5}{16}}$ **57.** $-\sqrt[3]{\frac{2p}{r^2}}$

58. $-\sqrt[3]{\frac{6x}{y^2}}$ **59.** $\sqrt[4]{\frac{16}{x}}$ **60.** $\sqrt[4]{\frac{81}{y}}$ **61.** $\sqrt[4]{\frac{2y}{z}}$ **62.** $\sqrt[4]{\frac{7t}{s^2}}$

Rationalize the denominator in each expression. Assume that all variables represent positive real numbers and that no denominators are 0. See Example 5.

63. $\frac{2}{4+\sqrt{3}}$ **64.** $\frac{6}{5+\sqrt{2}}$ **65.** $\frac{6}{\sqrt{5}+\sqrt{3}}$ **66.** $\frac{12}{\sqrt{6}+\sqrt{3}}$

67. $\frac{-4}{\sqrt{3}-\sqrt{7}}$ **68.** $\frac{-3}{\sqrt{2}+\sqrt{5}}$ **69.** $\frac{1-\sqrt{2}}{\sqrt{7}+\sqrt{6}}$ **70.** $\frac{-1-\sqrt{3}}{\sqrt{6}+\sqrt{5}}$

71. $\frac{\sqrt{2}-\sqrt{3}}{\sqrt{6}-\sqrt{5}}$ **72.** $\frac{\sqrt{5}+\sqrt{6}}{\sqrt{3}-\sqrt{2}}$ **73.** $\frac{4}{\sqrt{x}-2\sqrt{y}}$

74. $\frac{5}{3\sqrt{r}+\sqrt{s}}$ **75.** $\frac{\sqrt{x}-\sqrt{y}}{\sqrt{2x}+\sqrt{3y}}$ **76.** $\frac{\sqrt{a}+\sqrt{b}}{\sqrt{5a}-\sqrt{2b}}$

Write each quotient in lowest terms. Assume that all variables represent positive real numbers. See Example 6.

77. $\dfrac{25 + 10\sqrt{6}}{20}$

78. $\dfrac{12 - 6\sqrt{2}}{24}$

79. $\dfrac{16 + 4\sqrt{8}}{12}$

80. $\dfrac{12 + 9\sqrt{72}}{18}$

81. $\dfrac{6x + \sqrt{24x^3}}{3x}$

82. $\dfrac{11y + \sqrt{242y^5}}{22y}$

Relating Concepts (Exercises 83–86) For Individual or Group Work

Sometimes it is desirable to ***rationalize the numerator*** *in an expression. The procedure is similar to rationalizing the denominator. For example, to rationalize the numerator in the following expression, we multiply both the numerator and denominator by the conjugate of the numerator,* $6 + \sqrt{2}$.

$$\frac{6 - \sqrt{2}}{3}$$

$$= \frac{(6 - \sqrt{2})(6 + \sqrt{2})}{3(6 + \sqrt{2})}$$

$$= \frac{36 - 2}{3(6 + \sqrt{2})}$$

$$= \frac{34}{3(6 + \sqrt{2})}$$

In the final expression, the numerator is rationalized. ***Work Exercises 83–86 in order.***

83. Rationalize the numerator of $\dfrac{8\sqrt{5} - 1}{6}$.

84. Rationalize the numerator of $\dfrac{3\sqrt{a} + \sqrt{b}}{\sqrt{b} - \sqrt{a}}$. Assume a and b are positive and $a \neq b$.

85. Rationalize the denominator of the expression in Exercise 84.

86. Describe the difference in the procedures used in Exercises 84 and 85.

Summary Exercises on Operations with Radicals and Rational Exponents

Recall that a simplified radical satisfies the following conditions.

Conditions for a Simplified Radical

1. The radicand has no factor raised to a power greater than or equal to the index.
2. The radicand has no fractions.
3. No denominator has a radical.
4. Exponents in the radicand and the index of the radical have greatest common factor 1.

Perform all indicated operations, and express each answer in simplest form with positive exponents. Assume that all variables represent positive real numbers.

1. $6\sqrt{10} - 12\sqrt{10}$

2. $\sqrt{7}\left(\sqrt{7} - \sqrt{2}\right)$

3. $\left(1 - \sqrt{3}\right)\left(2 + \sqrt{6}\right)$

4. $\sqrt{50} - \sqrt{98} + \sqrt{72}$

5. $\left(3\sqrt{5} + 2\sqrt{7}\right)^2$

6. $\dfrac{-3}{\sqrt{6}}$

7. $\dfrac{8}{\sqrt{7} + \sqrt{5}}$

8. $\sqrt[3]{16x^2} - \sqrt[3]{54x^2} + \sqrt[3]{128x^2}$

9. $\dfrac{1 - \sqrt{2}}{1 + \sqrt{2}}$

10. $\left(1 - \sqrt[3]{3}\right)\left(1 + \sqrt[3]{3} + \sqrt[3]{9}\right)$

11. $\left(\sqrt{5} + 7\right)\left(\sqrt{5} - 7\right)$

12. $\dfrac{1}{\sqrt{x} - \sqrt{5}}, \quad x \neq 5$

13. $\sqrt[3]{8a^3b^5c^9}$

14. $\dfrac{15}{\sqrt[3]{9}}$

15. $\dfrac{3}{\sqrt{5} + 2}$

16. $\sqrt{\dfrac{3}{5x}}$

17. $\dfrac{16\sqrt{3}}{5\sqrt{12}}$

18. $\dfrac{2\sqrt{25}}{8\sqrt{50}}$

19. $\dfrac{-10}{\sqrt[3]{10}}$

20. $\dfrac{\sqrt{6}+\sqrt{5}}{\sqrt{6}-\sqrt{5}}$

21. $\sqrt{12x} - \sqrt{75x}$

22. $(5 - 3\sqrt{3})^2$

23. $(\sqrt{74} - \sqrt{73})(\sqrt{74} + \sqrt{73})$

24. $\sqrt[3]{\dfrac{13}{81}}$

25. $-t^2\sqrt[4]{t} + 3\sqrt[4]{t^9} - t\sqrt[4]{t^5}$

26. $\dfrac{\sqrt{3}+\sqrt{7}}{\sqrt{6}-\sqrt{5}}$

27. $\dfrac{6}{\sqrt[4]{3}}$

28. $\dfrac{1}{1-\sqrt[3]{3}}$

29. $\sqrt[3]{\dfrac{x^2y}{x^{-3}y^4}}$

30. $\sqrt{12} - \sqrt{108} - \sqrt[3]{27}$

31. $\dfrac{x^{-2/3}y^{4/5}}{x^{-5/3}y^{-2/5}}$

32. $\left(\dfrac{x^{3/4}y^{2/3}}{x^{1/3}y^{5/8}}\right)^{24}$

33. $(125x^3)^{-2/3}$

34. $(3x^{-2/3}y^{1/2})(-2x^{5/8}y^{-1/3})$

35. $\dfrac{4^{1/2}+3^{1/2}}{4^{1/2}-3^{1/2}}$

36. $(\sqrt{6} - \sqrt{5})^2(\sqrt{6} + \sqrt{5})^2$

9.6 Solving Equations with Radicals

An equation that includes one or more radical expressions with a variable is called a **radical equation.** Some examples of radical equations are

$$\sqrt{x-4}=8,\quad \sqrt{5x+12}=3\sqrt{2x-1},\quad \text{and}\quad \sqrt[3]{6+x}=27.$$

Radical equations

OBJECTIVES

1. **Solve radical equations using the power rule.**
2. **Solve radical equations that require additional steps.**
3. **Solve radical equations with indexes greater than 2.**

OBJECTIVE 1 Solve radical equations using the power rule. The equation $x=1$ has only one solution. Its solution set is $\{1\}$. If we square both sides of this equation, we get $x^2=1$. This new equation has two solutions: -1 and 1. Notice that the solution of the original equation is also a solution of the squared equation. However, the squared equation has another solution, -1, that is *not* a solution of the original equation. When solving equations with radicals, we use this idea of raising both sides to a power. This is an application of the **power rule.**

Power Rule for Solving Equations with Radicals

If both sides of an equation are raised to the same power, all solutions of the original equation are also solutions of the new equation.

Read the power rule carefully; it does not say that all solutions of the new equation are solutions of the original equation. They may or may not be. Solutions that do not satisfy the original equation are called **extraneous solutions;** they must be discarded.

CAUTION

When the power rule is used to solve an equation, ***every solution of the new equation* must *be checked in the original equation.***

EXAMPLE 1 Using the Power Rule

Solve $\sqrt{3x+4}=8$.

Use the power rule and square both sides to get

$$\left(\sqrt{3x+4}\right)^2=8^2$$
$$3x+4=64$$
$$3x=60 \quad \text{Subtract 4.}$$
$$x=20. \quad \text{Divide by 3.}$$

To check, substitute the proposed solution in the *original* equation.

Check

$$\sqrt{3x+4}=8 \quad \text{Original equation}$$
$$\sqrt{3\cdot 20+4}\stackrel{?}{=}8 \quad \text{Let } x=20.$$
$$\sqrt{64}\stackrel{?}{=}8$$
$$8=8 \quad \text{True}$$

Since 20 satisfies the *original* equation, the solution set is $\{20\}$.

Work Problem 1 at the Side. ▶

1 Solve each equation.

(a) $\sqrt{r}=3$

(b) $\sqrt{5x+1}=4$

ANSWERS

1. (a) $\{9\}$ (b) $\{3\}$

The solution of the equation in Example 1 can be generalized.

2 Solve each equation.

(a) $\sqrt{5x+3}+2=0$

(b) $\sqrt{x-9}-3=0$

Solving an Equation with Radicals

Step 1 **Isolate the radical.** Make sure that one radical term is alone on one side of the equation.

Step 2 **Apply the power rule.** Raise both sides of the equation to a power that is the same as the index of the radical.

Step 3 **Solve.** Solve the resulting equation; if it still contains a radical, repeat Steps 1 and 2.

Step 4 **Check** all proposed solutions in the original equation.

CAUTION
Remember to check (Step 4) or you may get an incorrect solution set.

EXAMPLE 2 **Using the Power Rule**

Solve $\sqrt{5x-1}+3=0$.

Step 1 To isolate the radical on one side, subtract 3 from each side.

$$\sqrt{5x-1}=-3$$

Step 2 Now square both sides.

$$\left(\sqrt{5x-1}\right)^2=(-3)^2$$

Step 3

$$5x-1=9$$

$$5x=10 \quad \text{Add 1.}$$

$$x=2 \quad \text{Divide by 5.}$$

Step 4 Check the proposed solution, 2, by substituting it in the original equation.

Check

$$\sqrt{5x-1}+3=0 \quad \text{Original equation}$$

$$\sqrt{5\cdot 2-1}+3\stackrel{?}{=}0 \quad \text{Let } x=2.$$

$$3+3=0 \quad \text{False}$$

This false result shows that 2 is *not* a solution of the original equation; it is extraneous. The solution set is $\emptyset$.

Note
We could have determined after Step 1 that the equation in Example 2 has no solution because the expression on the left cannot equal a negative number.

◀ *Work Problem* **2** *at the Side.*

OBJECTIVE 2 **Solve radical equations that require additional steps.** The next examples involve finding the square of a binomial. Recall that

$$(x+y)^2=x^2+2xy+y^2.$$

ANSWERS

2. (a) $\emptyset$ (b) $\{18\}$

EXAMPLE 3 Using the Power Rule; Squaring a Binomial

Solve $\sqrt{4-x} = x + 2$.

Step 1 The radical is alone on the left side of the equation.

Step 2 Square both sides. On the right, $(x+2)^2 = x^2 + 2(x)(2) + 2^2$.

$$\left(\sqrt{4-x}\right)^2 = (x+2)^2 \quad \text{Remember the middle term.}$$

$$4 - x = x^2 + 4x + 4$$

Pay careful attention here. — Twice the product of 2 and x

Step 3 The new equation is quadratic, so get 0 on one side.

$$0 = x^2 + 5x \qquad \text{Subtract 4; add } x.$$

$$0 = x(x+5) \qquad \text{Factor.}$$

Set each factor equal to 0.

$$x = 0 \quad \text{or} \quad x + 5 = 0 \qquad \text{Zero-factor property}$$

$$x = -5 \qquad \text{Solve.}$$

Step 4 Check each proposed solution in the original equation.

Check If $x = 0$, then

$$\sqrt{4-x} = x + 2$$

$$\sqrt{4-0} \stackrel{?}{=} 0 + 2$$

$$\sqrt{4} \stackrel{?}{=} 2$$

$$2 = 2. \qquad \text{True}$$

If $x = -5$, then

$$\sqrt{4-x} = x + 2$$

$$\sqrt{4-(-5)} \stackrel{?}{=} -5 + 2$$

$$\sqrt{9} \stackrel{?}{=} -3$$

$$3 = -3. \qquad \text{False}$$

The solution set is $\{0\}$. The other proposed solution, -5, is extraneous.

Work Problem 3 at the Side. ▶

EXAMPLE 4 Using the Power Rule; Squaring a Binomial

Solve $\sqrt{x^2 - 4x + 9} = x - 1$.

Square both sides. On the right, $(x-1)^2 = x^2 - 2(x)(1) + 1^2$.

$$\left(\sqrt{x^2 - 4x + 9}\right)^2 = (x-1)^2 \quad \text{Remember the middle term.}$$

$$x^2 - 4x + 9 = x^2 - 2x + 1$$

Twice the product of x and -1

$$-2x = -8 \qquad \text{Subtract } x^2 \text{ and 9; add } 2x.$$

$$x = 4 \qquad \text{Divide by } -2.$$

Check

$$\sqrt{x^2 - 4x + 9} = x - 1 \qquad \text{Original equation}$$

$$\sqrt{4^2 - 4 \cdot 4 + 9} \stackrel{?}{=} 4 - 1 \qquad \text{Let } x = 4.$$

$$3 = 3 \qquad \text{True}$$

The solution set of the original equation is $\{4\}$.

Work Problem 4 at the Side. ▶

CAUTION

When a radical equation requires squaring a binomial, as in Examples 3 and 4, ***remember to include the middle term.***

3 Solve.

(a) $\sqrt{3x-5} = x - 1$

(b) $x + 1 = \sqrt{-2x-2}$

4 Solve.

$$\sqrt{4x^2 + 2x - 3} = 2x + 7$$

ANSWERS

3. (a) $\{2, 3\}$ (b) $\{-1\}$

4. $\{-2\}$

EXAMPLE 5 **Using the Power Rule; Squaring Twice**

Solve $\sqrt{5x+6} + \sqrt{3x+4} = 2$.

Start by isolating one radical on one side of the equation. Do this by subtracting $\sqrt{3x+4}$ from each side.

$$\sqrt{5x+6} = 2 - \sqrt{3x+4} \quad \text{Subtract } \sqrt{3x+4}.$$

$$\left(\sqrt{5x+6}\right)^2 = \left(2 - \sqrt{3x+4}\right)^2 \quad \text{Square both sides.}$$

$$5x+6 = 4 - 4\sqrt{3x+4} + (3x+4)$$

Be careful here.

Remember the middle term.

Twice the product of 2 and $-\sqrt{3x+4}$

This equation still contains a radical, so isolate the radical term on the right and square both sides again.

$$5x+6 = 8 + 3x - 4\sqrt{3x+4} \quad \text{Combine like terms.}$$

$$2x - 2 = -4\sqrt{3x+4} \quad \text{Subtract 8 and } 3x.$$

Divide *each* term by 2.

$$x - 1 = -2\sqrt{3x+4} \quad \text{Divide by 2 to make the numbers smaller.}$$

$$(x-1)^2 = \left(-2\sqrt{3x+4}\right)^2 \quad \text{Square both sides again.}$$

$$x^2 - 2x + 1 = (-2)^2\left(\sqrt{3x+4}\right)^2 \quad \text{On the right, } (ab)^2 = a^2b^2.$$

$$x^2 - 2x + 1 = 4(3x+4) \quad \text{Apply the exponents.}$$

$$x^2 - 2x + 1 = 12x + 16 \quad \text{Distributive property}$$

$$x^2 - 14x - 15 = 0 \quad \text{Standard form}$$

$$(x+1)(x-15) = 0 \quad \text{Factor.}$$

$$x + 1 = 0 \quad \text{or} \quad x - 15 = 0 \quad \text{Zero-factor property}$$

$$x = -1 \quad \text{or} \quad x = 15 \quad \text{Solve each equation.}$$

Check each of these proposed solutions in the original equation. Only -1 satisfies the equation, so the solution set, $\{-1\}$, has only one element.

◀ Work Problem **5** at the Side.

5 **(a)** Verify that 15 is an extraneous solution of the equation in Example 5 and must be discarded.

(b) Solve.

$$\sqrt{2x+3} + \sqrt{x+1} = 1$$

OBJECTIVE 3 **Solve radical equations with indexes greater than 2.** The power rule also works for powers greater than 2.

EXAMPLE 6 **Using the Power Rule for a Power Greater than 2**

Solve $\sqrt[3]{x+5} = \sqrt[3]{2x-6}$.

Raise both sides to the third power.

$$\left(\sqrt[3]{x+5}\right)^3 = \left(\sqrt[3]{2x-6}\right)^3$$

$$x + 5 = 2x - 6$$

$$11 = x \quad \text{Subtract } x\text{; add 6.}$$

Check

$$\sqrt[3]{x+5} = \sqrt[3]{2x-6} \quad \text{Original equation}$$

$$\sqrt[3]{11+5} \stackrel{?}{=} \sqrt[3]{2\cdot 11 - 6} \quad \text{Let } x = 11.$$

$$\sqrt[3]{16} = \sqrt[3]{16} \quad \text{True}$$

The solution set is $\{11\}$.

◀ Work Problem **6** at the Side.

6 Solve each equation.

(a) $\sqrt[3]{2x+7} = \sqrt[3]{3x-2}$

(b) $\sqrt[4]{2x+5} + 1 = 0$

ANSWERS

5. **(a)** The final step in the check leads to $16 = 2$, which is false.
(b) $\{-1\}$

6. **(a)** $\{9\}$ **(b)** $\emptyset$

9.6 Exercises

FOR EXTRA HELP

MyMathLab | MathXL PRACTICE | WATCH | DOWNLOAD | READ | REVIEW

Check each equation to see if the given value for x is a solution.

1. $\sqrt{3x+18}=x$

(a) 6 **(b)** -3

2. $\sqrt{3x-3}=x-1$

(a) 1 **(b)** 4

3. $\sqrt{x+2}=\sqrt{9x-2}-2\sqrt{x-1}$

(a) 2 **(b)** 7

4. $\sqrt{8x-3}=2x$

(a) $\frac{3}{2}$ **(b)** $\frac{1}{2}$

5. Is 9 a solution of the equation $\sqrt{x}=-3$? If not, what is the solution of this equation? Explain.

6. Before even attempting to solve $\sqrt{3x+18}=x$, how can you be sure that the equation cannot have a negative solution?

Solve each equation. See Examples 1–4.

7. $\sqrt{x-2}=3$

8. $\sqrt{x+1}=7$

9. $\sqrt{6x-1}=1$

10. $\sqrt{7x-3}=5$

11. $\sqrt{4x+3}+1=0$

12. $\sqrt{5x-3}+2=0$

13. $\sqrt{3k+1}-4=0$

14. $\sqrt{5z+1}-11=0$

15. $4-\sqrt{x-2}=0$

16. $9 - \sqrt{4k + 1} = 0$

17. $\sqrt{9a - 4} = \sqrt{8a + 1}$

18. $\sqrt{4p - 2} = \sqrt{3p + 5}$

19. $2\sqrt{x} = \sqrt{3x + 4}$

20. $2\sqrt{m} = \sqrt{5m - 16}$

21. $3\sqrt{z - 1} = 2\sqrt{2z + 2}$

22. $5\sqrt{4x + 1} = 3\sqrt{10x + 25}$

23. $k = \sqrt{k^2 + 4k - 20}$

24. $p = \sqrt{p^2 - 3p + 18}$

25. $x = \sqrt{x^2 + 3x + 9}$

26. $z = \sqrt{z^2 - 4z - 8}$

27. $\sqrt{9 - x} = x + 3$

28. $\sqrt{5 - x} = x + 1$

29. $\sqrt{k^2 + 2k + 9} = k + 3$

30. $\sqrt{x^2 - 3x + 3} = x - 1$

31. $\sqrt{r^2 + 9r + 3} = -r$

32. $\sqrt{p^2 - 15p + 15} = p - 5$

33. $\sqrt{z^2 + 12z - 4} + 4 - z = 0$

34. $\sqrt{m^2 + 3m + 12} - m - 2 = 0$

35. A student wrote the following as his first step in solving $\sqrt{3x + 4} = 8 - x$.

$$3x + 4 = 64 + x^2$$

WHAT WENT WRONG? Solve the given equation correctly.

36. A student wrote the following as her first step in solving $\sqrt{5x + 6} = \sqrt{x + 3} + 3$.

$$5x + 6 = x + 3 + 9$$

WHAT WENT WRONG? Solve the given equation correctly.

Solve each equation. See Examples 5 and 6.

37. $\sqrt[3]{2x+5} = \sqrt[3]{6x+1}$

38. $\sqrt[3]{p-1} = 2$

39. $\sqrt[3]{a^2+5a+1} = \sqrt[3]{a^2+4a}$

40. $\sqrt[3]{r^2+2r+8} = \sqrt[3]{r^2}$

41. $\sqrt[3]{2m-1} = \sqrt[3]{m+13}$

42. $\sqrt[3]{2k-11} - \sqrt[3]{5k+1} = 0$

43. $\sqrt[4]{a+8} = \sqrt[4]{2a}$

44. $\sqrt[4]{z+11} = \sqrt[4]{2z+6}$

45. $\sqrt[3]{x-8} + 2 = 0$

46. $\sqrt[3]{r+1} + 1 = 0$

47. $\sqrt[4]{2k-5} + 4 = 0$

48. $\sqrt[4]{8z-3} + 2 = 0$

49. $\sqrt{k+2} - \sqrt{k-3} = 1$

50. $\sqrt{r+6} - \sqrt{r-2} = 2$

51. $\sqrt{2r+11} - \sqrt{5r+1} = -1$

52. $\sqrt{3x-2} - \sqrt{x+3} = 1$

53. $\sqrt{3p+4} - \sqrt{2p-4} = 2$

54. $\sqrt{4x+5} - \sqrt{2x+2} = 1$

55. $\sqrt{3-3p} - 3 = \sqrt{3p+2}$

56. $\sqrt{4x+7} - 4 = \sqrt{4x-1}$

57. $\sqrt{2\sqrt{x+11}} = \sqrt{4x+2}$

58. $\sqrt{1+\sqrt{24-10x}} = \sqrt{3x+5}$

For each equation, rewrite the expressions with rational exponents as radical expressions, and then solve using the procedures explained in this section.

59. $(2x - 9)^{1/2} = 2 + (x - 8)^{1/2}$

60. $(3w + 7)^{1/2} = 1 + (w + 2)^{1/2}$

61. $(2w - 1)^{2/3} - w^{1/3} = 0$

62. $(x^2 - 2x)^{1/3} - x^{1/3} = 0$

Solve each formula from electricity and radio for the indicated variable. (Source: Cooke, Nelson M., and Joseph B. Orleans, Mathematics Essential to Electricity and Radio, *McGraw-Hill, 1943.)*

63. $V = \sqrt{\frac{2K}{m}}$ for K

64. $V = \sqrt{\frac{2K}{m}}$ for m

65. $Z = \sqrt{\frac{L}{C}}$ for C

66. $Z = \sqrt{\frac{L}{C}}$ for L

67. $f = \frac{1}{2\pi\sqrt{LC}}$ for L

68. $r = \sqrt{\frac{Mm}{F}}$ for F

A number of useful formulas involve radicals or radical expressions. Many occur in the mathematics needed for working with objects in space. The formula

$$N = \frac{1}{2\pi}\sqrt{\frac{a}{r}}$$

is used to find the rotational rate N of a space station. Here a is the acceleration and r represents the radius of the space station in meters. To find the value of r that will make N simulate the effect of gravity on Earth, the equation must be solved for r, using the required value of N. (Source: Kastner, Bernice, Space Mathematics, *NASA, 1972.)*

69. Solve the equation for r.

70. If $a = 9.8$ m per sec^2, find the value of r (to the nearest tenth) using each value of N.

(a) $N = 0.063$ rotation per sec

(b) $N = 0.04$ rotation per sec

9.7 Complex Numbers

As we saw in **Section 1.1,** the set of real numbers includes many other number sets (the rational numbers, integers, and natural numbers, for example). In this section, a new set of numbers is introduced that includes the set of real numbers, as well as numbers that are even roots of negative numbers, like $\sqrt{-2}$.

OBJECTIVES

1. Simplify numbers of the form $\sqrt{-b}$, where $b > 0$.
2. Recognize subsets of the complex numbers.
3. Add and subtract complex numbers.
4. Multiply complex numbers.
5. Divide complex numbers.
6. Find powers of i.

OBJECTIVE 1 Simplify numbers of the form $\sqrt{-b}$, where $b > 0$. The equation $x^2 + 1 = 0$ has no real number solution since any solution must be a number whose square is -1. In the set of real numbers, all squares are nonnegative numbers because the product of two positive numbers or two negative numbers is positive and $0^2 = 0$. To provide a solution for the equation $x^2 + 1 = 0$, we introduce a new number i.

Imaginary Unit i

The **imaginary unit i** is defined as

$$i = \sqrt{-1}, \quad \text{where} \quad i^2 = -1.$$

In words, i is the principal square root of -1.

This definition of i makes it possible to define any square root of a negative number as follows.

$\sqrt{-b}$

For any positive number b, $\quad \sqrt{-b} = i\sqrt{b}.$

EXAMPLE 1 Simplifying Square Roots of Negative Numbers

Write each number as a product of a real number and i.

(a) $\sqrt{-100} = i\sqrt{100} = 10i$

(b) $-\sqrt{-36} = -i\sqrt{36} = -6i$

(c) $\sqrt{-2} = i\sqrt{2}$

(d) $\sqrt{-8} = i\sqrt{8} = i\sqrt{4 \cdot 2} = 2i\sqrt{2}$

CAUTION
It is easy to mistake $\sqrt{2}i$ for $\sqrt{2i}$, with the i under the radical. For this reason, we usually write $\sqrt{2}i$ as $i\sqrt{2}$, as in the definition of $\sqrt{-b}$.

Work Problem 1 at the Side. ▶

When finding a product such as $\sqrt{-4} \cdot \sqrt{-9}$, we cannot use the product rule for radicals because it applies only to nonnegative radicands. ***For this reason, we change $\sqrt{-b}$ to the form $i\sqrt{b}$ before performing any multiplications or divisions.***

1 Write each number as a product of a real number and i.

(a) $\sqrt{-16}$

(b) $-\sqrt{-81}$

(c) $\sqrt{-7}$

(d) $\sqrt{-32}$

ANSWERS

1. (a) $4i$ (b) $-9i$ (c) $i\sqrt{7}$ (d) $4i\sqrt{2}$

For example,

$$\begin{aligned} &\sqrt{-4} \cdot \sqrt{-9} \\ &= i\sqrt{4} \cdot i\sqrt{9} \quad \sqrt{-b} = i\sqrt{b} \\ &= i \cdot 2 \cdot i \cdot 3 \\ &= 6i^2 \\ &= 6(-1) \quad \text{Substitute: } i^2 = -1. \\ &= -6. \end{aligned}$$

First write all square roots in terms of i.

CAUTION

Using the product rule for radicals *before* using the definition of $\sqrt{-b}$ gives a *wrong* answer. The preceding example shows that

$$\sqrt{-4} \cdot \sqrt{-9} = -6, \quad \text{Correct}$$

but $\sqrt{-4(-9)} = \sqrt{36} = 6$, Incorrect

so $\sqrt{-4} \cdot \sqrt{-9} \neq \sqrt{-4(-9)}$.

EXAMPLE 2 Multiplying Square Roots of Negative Numbers

Multiply.

(a)

$$\begin{aligned} &\sqrt{-3} \cdot \sqrt{-7} \\ &= i\sqrt{3} \cdot i\sqrt{7} \quad \sqrt{-b} = i\sqrt{b} \\ &= i^2\sqrt{3 \cdot 7} \quad \text{Product rule} \\ &= (-1)\sqrt{21} \quad \text{Substitute: } i^2 = -1. \\ &= -\sqrt{21} \end{aligned}$$

First write all square roots in terms of i.

(b) $\sqrt{-2} \cdot \sqrt{-8}$

$$\begin{aligned} &= i\sqrt{2} \cdot i\sqrt{8} \\ &= i^2\sqrt{2 \cdot 8} \\ &= (-1)\sqrt{16} \\ &= (-1)4, \quad \text{or} \quad -4 \end{aligned}$$

(c) $\sqrt{-5} \cdot \sqrt{6}$

$$\begin{aligned} &= i\sqrt{5} \cdot \sqrt{6} \\ &= i\sqrt{30} \end{aligned}$$

◀ Work Problem **2** at the Side.

EXAMPLE 3 Dividing Square Roots of Negative Numbers

Divide.

(a)

$$\begin{aligned} &\frac{\sqrt{-75}}{\sqrt{-3}} \\ &= \frac{i\sqrt{75}}{i\sqrt{3}} \\ &= \sqrt{\frac{75}{3}} \quad \text{Quotient rule} \\ &= \sqrt{25} \quad \text{Divide.} \\ &= 5 \end{aligned}$$

First write all square roots in terms of i.

(b)

$$\begin{aligned} &\frac{\sqrt{-32}}{\sqrt{8}} \\ &= \frac{i\sqrt{32}}{\sqrt{8}} \\ &= i\sqrt{\frac{32}{8}} \\ &= i\sqrt{4} \\ &= 2i \end{aligned}$$

◀ Work Problem **3** at the Side.

2 Multiply.

(a) $\sqrt{-7} \cdot \sqrt{-5}$

(b) $\sqrt{-5} \cdot \sqrt{-10}$

(c) $\sqrt{-15} \cdot \sqrt{2}$

3 Divide.

(a) $\dfrac{\sqrt{-32}}{\sqrt{-2}}$

(b) $\dfrac{\sqrt{-27}}{\sqrt{-3}}$

(c) $\dfrac{\sqrt{-40}}{\sqrt{10}}$

ANSWERS

2. **(a)** $-\sqrt{35}$ **(b)** $-5\sqrt{2}$ **(c)** $i\sqrt{30}$

3. **(a)** 4 **(b)** 3 **(c)** $2i$

OBJECTIVE 2 Recognize subsets of the complex numbers. With the imaginary unit i and the real numbers, a new set of numbers can be formed that includes the real numbers as a subset. The *complex numbers* are defined as follows.

Complex Number

If a and b are real numbers, then any number of the form $\boldsymbol{a + bi}$ is called a **complex number.** In the complex number $a + bi$, the number a is called the **real part** and b is called the **imaginary part.***

For a complex number $a + bi$, if $b = 0$, then $a + bi = a$, which is a real number. ***Thus, the set of real numbers is a subset of the set of complex numbers.*** If $a = 0$ and $b \neq 0$, the complex number is said to be a **pure imaginary number.** For example, $3i$ is a pure imaginary number. A number such as $7 + 2i$ is a **nonreal complex number.** These numbers are very useful in applications, particularly in work with electricity.

The relationships among the sets of numbers are shown in Figure 10.

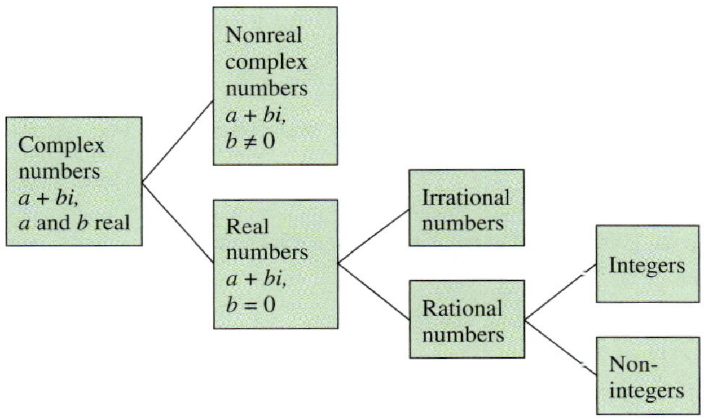

Figure 10

OBJECTIVE 3 Add and subtract complex numbers. The commutative, associative, and distributive properties for real numbers are also valid for complex numbers. ***Thus, to add complex numbers, we add their real parts and add their imaginary parts.***

EXAMPLE 4 Adding Complex Numbers

Add.

(a) $(2 + 3i) + (6 + 4i)$

$= (2 + 6) + (3 + 4)i$ Commutative, associative, and distributive properties

$= 8 + 7i$ Add real parts; add imaginary parts.

(b) $5 + (9 - 3i)$

$= (5 + 9) - 3i$ Associative property

$= 14 - 3i$ Add real parts.

Work Problem 4 at the Side. ▶

*Some texts define bi as the imaginary part of the complex number $a + bi$.

4 Add.

(a) $(4 + 6i) + (-3 + 5i)$

(b) $(-1 + 8i) + (9 - 3i)$

ANSWERS

4. (a) $1 + 11i$ **(b)** $8 + 5i$

To subtract complex numbers, we subtract their real parts and subtract their imaginary parts.

EXAMPLE 5 **Subtracting Complex Numbers**

Subtract.

(a) $(6 + 5i) - (3 + 2i)$

$= (6 - 3) + (5 - 2)i$ Properties of real numbers

$= 3 + 3i$ Subtract real parts; subtract imaginary parts.

(b) $(7 - 3i) - (8 - 6i)$

$= (7 - 8) + [-3 - (-6)]i$

$= -1 + 3i$

(c) $(-9 + 4i) - (-9 + 8i)$

$= (-9 + 9) + (4 - 8)i$

$= 0 - 4i$

$= -4i$

◀ Work Problem **5** at the Side.

In Example 5(c), the answer was written as $0 - 4i$ and then as just $-4i$. A complex number written in the form $a + bi$, like $0 - 4i$, is in **standard form.** In this section, most answers will be given in standard form, but if a or b is 0, we consider answers such as a or bi to be in standard form.

OBJECTIVE 4 **Multiply complex numbers.** We multiply complex numbers as we multiply polynomials.

EXAMPLE 6 **Multiplying Complex Numbers**

Multiply.

(a) $4i(2 + 3i)$

$= 4i(2) + 4i(3i)$ Distributive property

$= 8i + 12i^2$ Multiply.

$= 8i + 12(-1)$ Substitute: $i^2 = -1$.

$= -12 + 8i$ Standard form

(b) $(3 + 5i)(4 - 2i)$

$= \underbrace{3(4)}_{\text{First}} + \underbrace{3(-2i)}_{\text{Outer}} + \underbrace{5i(4)}_{\text{Inner}} + \underbrace{5i(-2i)}_{\text{Last}}$ Use the FOIL method for multiplying binomials. **(Section 6.4)**

$= 12 - 6i + 20i - 10i^2$ Multiply.

$= 12 + 14i - 10(-1)$ Combine imaginary terms; $i^2 = -1$.

$= 12 + 14i + 10$ Multiply.

$= 22 + 14i$ Combine real terms.

(c) $(2 + 3i)(1 - 5i)$

$= 2(1) + 2(-5i) + 3i(1) + 3i(-5i)$ FOIL

$= 2 - 10i + 3i - 15i^2$

$= 2 - 7i - 15(-1)$ Use parentheses around -1 to avoid errors.

$= 2 - 7i + 15$

$= 17 - 7i$

◀ Work Problem **6** at the Side.

5 Subtract.

(a) $(7 + 3i) - (4 + 2i)$

(b) $(-6 - i) - (-5 - 4i)$

(c) $8 - (3 - 2i)$

6 Multiply.

(a) $6i(4 + 3i)$

(b) $(6 - 4i)(2 + 4i)$

(c) $(3 - 2i)(3 + 2i)$

ANSWERS

5. **(a)** $3 + i$ **(b)** $-1 + 3i$ **(c)** $5 + 2i$

6. **(a)** $-18 + 24i$ **(b)** $28 + 16i$ **(c)** 13

The two complex numbers $\boldsymbol{a + bi}$ and $\boldsymbol{a - bi}$ are called *complex conjugates,* or simply *conjugates,* of each other. ***The product of a complex number and its conjugate is always a real number,*** as shown here.

$$\begin{aligned}(a + bi)(a - bi) &= a^2 - abi + abi - b^2i^2 \\ &= a^2 - b^2(-1) \\ \boldsymbol{(a + bi)(a - bi)} &\boldsymbol{= a^2 + b^2}\end{aligned}$$

The product eliminates i.

For example, $(3 + 7i)(3 - 7i) = 3^2 + 7^2 = 9 + 49 = 58.$

OBJECTIVE 5 Divide complex numbers. The quotient of two complex numbers should be a complex number. To write the quotient as a complex number, we need to eliminate i in the denominator. We use conjugates and a process like that for rationalizing a denominator to do this.

EXAMPLE 7 Dividing Complex Numbers

Find each quotient.

(a) $\dfrac{8 + 9i}{5 + 2i}$

Multiply both the numerator and denominator by the conjugate of the denominator. The conjugate of $5 + 2i$ is $5 - 2i$.

$$\frac{8 + 9i}{5 + 2i}$$

$= \dfrac{(8 + 9i)\boldsymbol{(5 - 2i)}}{(5 + 2i)\boldsymbol{(5 - 2i)}}$ $\quad \frac{5 - 2i}{5 - 2i} = 1$

$= \dfrac{40 - 16i + 45i - 18i^2}{5^2 + 2^2}$ $\quad$ In the denominator, $(a + bi)(a - bi) = a^2 + b^2$.

$= \dfrac{58 + 29i}{29}$ $\quad -18i^2 = -18(-1) = 18$; Combine like terms.

$= \dfrac{\boldsymbol{29}(2 + i)}{\boldsymbol{29}}$ $\quad$ Factor the numerator.

Factor first; then divide out the common factor.

$= 2 + i$ $\quad$ Lowest terms

(b) $\dfrac{1 + i}{i}$

$= \dfrac{(1 + i)\boldsymbol{(-i)}}{i\boldsymbol{(-i)}}$ $\quad$ Multiply numerator and denominator by $-i$, the conjugate of i.

$= \dfrac{-i - i^2}{-i^2}$ $\quad$ Distributive property; multiply.

$= \dfrac{-i - (-1)}{-(-1)}$ $\quad$ Substitute: $i^2 = -1$.

Use parentheses to avoid errors.

$= \dfrac{-i + 1}{1}$

$= 1 - i$

Work Problem 7 at the Side. ▶

7 Find each quotient.

(a) $\dfrac{2 + i}{3 - i}$

(b) $\dfrac{8 - 4i}{1 - i}$

(c) $\dfrac{5}{3 - 2i}$

(d) $\dfrac{5 - i}{i}$

Answers

7. (a) $\frac{1}{2} + \frac{1}{2}i$ **(b)** $6 + 2i$ **(c)** $\frac{15}{13} + \frac{10}{13}i$ **(d)** $-1 - 5i$

Calculator Tip In Examples 4–7, we showed how complex numbers can be added, subtracted, multiplied, and divided algebraically. Many current models of graphing calculators can perform these operations. Figure 11 shows how the computations in parts of Examples 4–7 are displayed on a TI-83/84 Plus calculator. Be sure to use parentheses as shown.

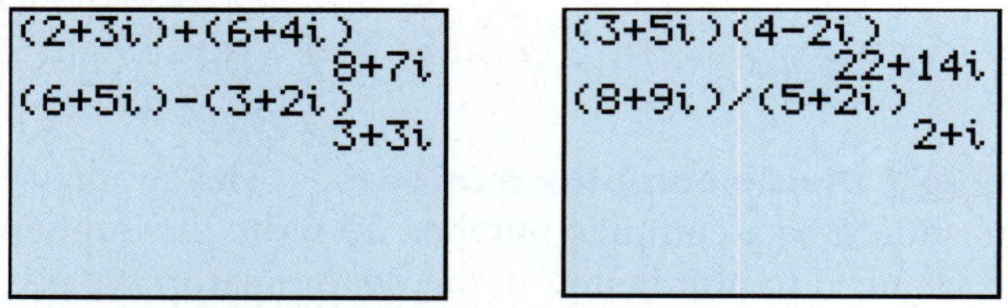

Figure 11

OBJECTIVE 6 Find powers of *i*. Because i^2 is defined to be -1, we can find higher powers of i as shown in the following examples.

$$i^3 = i \cdot i^2 = i(-1) = -i \qquad i^6 = i^2 \cdot i^4 = (-1) \cdot 1 = -1$$

$$i^4 = i^2 \cdot i^2 = (-1)(-1) = 1 \qquad i^7 = i^3 \cdot i^4 = (-i) \cdot 1 = -i$$

$$i^5 = i \cdot i^4 = i \cdot 1 = i \qquad i^8 = i^4 \cdot i^4 = 1 \cdot 1 = 1$$

As these examples suggest, the powers of i rotate through the four numbers i, -1, $-i$, and 1. Larger powers of i can be simplified by using the fact that $i^4 = 1$. For example,

$$i^{75} = (i^4)^{18} \cdot i^3 = 1^{18} \cdot i^3 = 1 \cdot i^3 = i^3 = -i.$$

EXAMPLE 8 Simplifying Powers of *i*

Find each power of i.

(a) $i^{12} = (i^4)^3 = 1^3 = 1$

(b) $i^{39} = i^{36} \cdot i^3 = (i^4)^9 \cdot i^3 = 1^9 \cdot (-i) = -i$

(c) $i^{-2} = \dfrac{1}{i^2} = \dfrac{1}{-1} = -1$

(d) $i^{-1} = \dfrac{1}{i} = \dfrac{1(-i)}{i(-i)} = \dfrac{-i}{-i^2} = \dfrac{-i}{-(-1)} = \dfrac{-i}{1} = -i$

◀ Work Problem **8** at the Side.

8 Find each power of i.

(a) i^{21}

(b) i^{36}

(c) i^{50}

(d) i^{-9}

Answers

8. (a) i **(b)** 1 **(c)** -1 **(d)** $-i$

9.7 Exercises

Decide whether each expression is equal to 1, -1, *i, or* $-i$.

1. $\sqrt{-1}$
2. $-i^2$
3. $\frac{1}{i}$
4. $(-i)^2$
5. Every real number is a complex number. Explain why this is so.
6. Not every complex number is a real number. Give an example of this, and explain why this statement is true.

Write each number as a product of a real number and i. Simplify all radical expressions. See Example 1.

7. $\sqrt{-169}$
8. $\sqrt{-225}$
9. $-\sqrt{-144}$
10. $-\sqrt{-196}$
11. $\sqrt{-5}$
12. $\sqrt{-21}$
13. $\sqrt{-48}$
14. $\sqrt{-96}$

Multiply or divide as indicated. See Examples 2 and 3.

15. $\sqrt{-15} \cdot \sqrt{-15}$
16. $\sqrt{-19} \cdot \sqrt{-19}$
17. $\sqrt{-3} \cdot \sqrt{-19}$
18. $\sqrt{-7} \cdot \sqrt{-15}$
19. $\sqrt{-4} \cdot \sqrt{-25}$
20. $\sqrt{-9} \cdot \sqrt{-81}$
21. $\sqrt{-3} \cdot \sqrt{11}$
22. $\sqrt{-5} \cdot \sqrt{13}$
23. $\frac{\sqrt{-300}}{\sqrt{-100}}$
24. $\frac{\sqrt{-40}}{\sqrt{-10}}$
25. $\frac{\sqrt{-75}}{\sqrt{3}}$
26. $\frac{\sqrt{-160}}{\sqrt{10}}$

Add or subtract as indicated. Write your answers in standard form. See Examples 4 and 5.

27. $(3 + 2i) + (-4 + 5i)$
28. $(7 + 15i) + (-11 + 14i)$
29. $(5 - i) + (-5 + i)$
30. $(-2 + 6i) + (2 - 6i)$
31. $(4 + i) - (-3 - 2i)$
32. $(9 + i) - (3 + 2i)$

33. $(-3 - 4i) - (-1 - 4i)$

34. $(-2 - 3i) - (-5 - 3i)$

35. $(-4 + 11i) + (-2 - 4i) + (7 + 6i)$

36. $(-1 + i) + (2 + 5i) + (3 + 2i)$

37. $[(7 + 3i) - (4 - 2i)] + (3 + i)$

38. $[(7 + 2i) + (-4 - i)] - (2 + 5i)$

39. Fill in the blank with the correct response: Because $(4 + 2i) - (3 + i) = 1 + i$, using the definition of subtraction we can check this to find that

$$(1 + i) + (3 + i) = \_\_\_\_\_\_.$$

40. Fill in the blank with the correct response: Because $\frac{-5}{2 - i} = -2 - i$, using the definition of division we can check this to find that

$$(-2 - i)(2 - i) = \_\_\_\_\_\_.$$

Multiply. See Example 6.

41. $(3i)(27i)$

42. $(5i)(125i)$

43. $(-8i)(-2i)$

44. $(-32i)(-2i)$

45. $5i(-6 + 2i)$

46. $3i(4 + 9i)$

47. $(4 + 3i)(1 - 2i)$

48. $(7 - 2i)(3 + i)$

49. $(4 + 5i)^2$

50. $(3 + 2i)^2$

51. $(12 + 3i)(12 - 3i)$

52. $(6 + 7i)(6 - 7i)$

53. (a) What is the conjugate of $a + bi$?

(b) If we multiply $a + bi$ by its conjugate, we get, ______ + ______, which is always a real number.

54. Explain the procedure you would use to find the quotient

$$\frac{-1 + 5i}{3 + 2i}.$$

Write each quotient in the form $a + bi$. See Example 7.

55. $\dfrac{2}{1 - i}$

56. $\dfrac{29}{5 + 2i}$

57. $\dfrac{-7 + 4i}{3 + 2i}$

58. $\dfrac{-38 - 8i}{7 + 3i}$

59. $\dfrac{3i}{2 + 2i}$

60. $\dfrac{-8i}{1 + i}$

61. $\dfrac{2 - 3i}{2 + 3i}$

62. $\dfrac{-1 + 5i}{3 + 2i}$

Relating Concepts (Exercises 63–68) For Individual or Group Work

Consider these expressions:

Binomials	**Complex Numbers**
$x + 2, \quad 3x - 1$	$1 + 2i, \quad 3 - i.$

When we add, subtract, or multiply complex numbers in standard form, the rules are the same as those for the corresponding operations on binomials. That is, we add or subtract like terms, and we use FOIL to multiply. Division, however, is comparable to division by the sum or difference of radicals, where we multiply by the conjugate of the denominator to get a rational denominator. To express the quotient of two complex numbers in standard form, we also multiply by the conjugate of the denominator. ***Work Exercises 63–68 in order,*** *to better understand these ideas.*

63. (a) Add the two binomials.

(b) Add the two complex numbers.

64. (a) Subtract the second binomial from the first.

(b) Subtract the second complex number from the first.

65. (a) Multiply the two binomials.

(b) Multiply the two complex numbers.

66. (a) Rationalize the denominator: $\dfrac{\sqrt{3} - 1}{1 + \sqrt{2}}$.

(b) Write in standard form: $\dfrac{3 - i}{1 + 2i}$.

67. Explain why the answers for parts (a) and (b) in Exercise 65 do not correspond as the answers in Exercises 63 and 64 do.

68. Explain why the answers for parts (a) and (b) in Exercise 66 do not correspond as the answers in Exercises 63 and 64 do.

69. Recall that if $a \neq 0$, then $\frac{1}{a}$ is called the reciprocal of a. Use this definition to express the reciprocal of $5 - 4i$ in the form $a + bi$.

70. Recall that if $a \neq 0$, then a^{-1} is defined to be $\frac{1}{a}$. Use this definition to express $(4 - 3i)^{-1}$ in the form $a + bi$.

Find each power of i. See Example 8.

71. i^{18}

72. i^{26}

73. i^{89}

74. i^{45}

75. i^{96}

76. i^{48}

77. i^{-5}

78. i^{-17}

79. A student simplified i^{-18} as follows:

$$i^{-18} = i^{-18} \cdot i^{20} = i^{-18+20} = i^2 = -1.$$

Explain the mathematical justification for this correct work.

80. Explain why

$$(46 + 25i)(3 - 6i) \quad \text{and} \quad (46 + 25i)(3 - 6i)i^{12}$$

must be equal. (Do not actually perform the computation.)

Ohm's law *for the current I in a circuit with voltage E, resistance R, capacitance reactance X_c, and inductive reactance X_L is*

$$I = \frac{E}{R + (X_L - X_c)i}.$$

Use this law to work Exercises 81 and 82.

81. Find I if $E = 2 + 3i$, $R = 5$, $X_L = 4$, and $X_c = 3$.

82. Find E if $I = 1 - i$, $R = 2$, $X_L = 3$, and $X_c = 1$.

83. Show that $1 + 5i$ is a solution of

$$x^2 - 2x + 26 = 0.$$

84. Show that $3 + 2i$ is a solution of

$$x^2 - 6x + 13 = 0.$$

Chapter 9 Summary

▸ Key Terms

9.1	**radicand, index**	In the expression $\sqrt[n]{a}$, a is the radicand and n is the index (**order**).
	radical	The expression $\sqrt[n]{a}$ is a radical.
	principal root	If a is positive and n is even, the principal nth root of a is the positive root.
	radical expression	A radical expression is an algebraic expression that contains radicals.
9.5	**rationalizing the denominator**	The process of removing radicals from the denominator so that the denominator contains only rational quantities is called rationalizing the denominator.
	conjugate	The conjugate of $a + b$ is $a - b$.
9.5	**radical equation**	A radical equation is an equation that includes one or more radical expressions with variables.
	extraneous solution	An extraneous solution of a radical equation is a solution found after applying the power rule that is not a solution of the original equation.
9.7	**complex number**	A complex number is a number that can be written in the form $a + bi$, where a and b are real numbers.
	real part	The real part of $a + bi$ is a.
	imaginary part	The imaginary part of $a + bi$ is b.
	pure imaginary number	A complex number $a + bi$ with $a = 0$ and $b \neq 0$ is called a pure imaginary number.
	nonreal complex number	A complex number $a + bi$ with $b \neq 0$ is called a nonreal complex number.
	standard form (of a complex number)	A complex number is in standard form if it is written in the form $a + bi$.
	complex conjugates	The complex conjugate of $a + bi$ is $a - bi$.

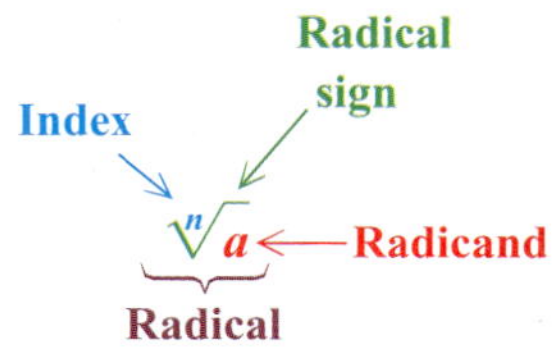

▸ New Symbols

$\sqrt{}$	radical sign
$\sqrt[n]{a}$	radical; principal nth root of a
$\pm$	"positive or negative," or "plus or minus"
$\approx$	is approximately equal to
$a^{1/n}$	a to the power $\frac{1}{n}$
$a^{m/n}$	a to the power $\frac{m}{n}$
i	imaginary unit

▶ Test Your Word Power

See how well you have learned the vocabulary in this chapter. Answers, with examples, follow the Quick Review.

1. A **radicand** is
   - **A.** the index of a radical
   - **B.** the number or expression under the radical sign
   - **C.** the positive root of a number
   - **D.** the radical sign.

2. A **hypotenuse** is
   - **A.** either of the two shorter sides of a triangle
   - **B.** the shortest side of a triangle
   - **C.** the side opposite the right angle in a right triangle
   - **D.** the longest side in any triangle.

3. **Rationalizing the denominator** is the process of
   - **A.** eliminating fractions from a radical expression
   - **B.** changing the denominator of a fraction from a radical expression to a rational number
   - **C.** clearing a radical expression of radicals
   - **D.** multiplying radical expressions.

4. An **extraneous solution** is a solution
   - **A.** that does not satisfy the original equation
   - **B.** that makes an equation true
   - **C.** that makes an expression equal 0
   - **D.** that checks in the original equation.

5. A **complex number** is
   - **A.** a real number that includes a complex fraction
   - **B.** a zero multiple of i
   - **C.** a number of the form $a + bi$, where a and b are real numbers
   - **D.** the square root of -1.

▶ Quick Review

Concepts	Examples

9.1 Radical Expressions and Graphs

Concepts

$\sqrt[n]{a} = b$ **means** $b^n = a$.

$\sqrt[n]{a}$ is the principal nth root of a.

$\sqrt[n]{a^n} = |a|$ if n is even; $\sqrt[n]{a^n} = a$ if n is odd.

Functions Defined by Radical Expressions

The square root function defined by $f(x) = \sqrt{x}$ and the cube root function defined by $f(x) = \sqrt[3]{x}$ are two important functions defined by radical expressions.

Examples

The two square roots of 64 are $\sqrt{64} = 8$, the principal square root, and $-\sqrt{64} = -8$.

$$\sqrt[4]{(-2)^4} = |-2| = 2 \qquad \sqrt[3]{-27} = -3$$

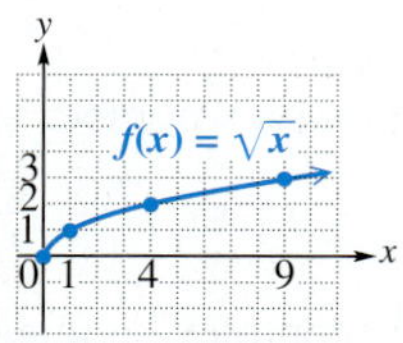

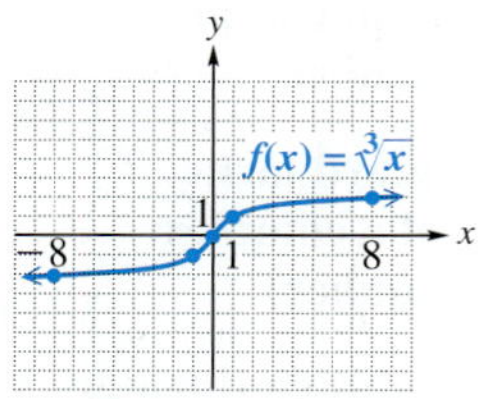

9.2 Rational Exponents

Concepts

$a^{1/n} = \sqrt[n]{a}$ whenever $\sqrt[n]{a}$ exists.

If m and n are positive integers with m/n in lowest terms, then $a^{m/n} = (a^{1/n})^m$, provided that $a^{1/n}$ is a real number.

All of the usual definitions and rules for exponents are valid for rational exponents.

Examples

$$81^{1/2} = \sqrt{81} = 9 \qquad -64^{1/3} = -\sqrt[3]{64} = -4$$

$$8^{5/3} = (8^{1/3})^5 = 2^5 = 32$$

$$5^{-1/2} \cdot 5^{1/4} = 5^{-1/2+1/4} = 5^{-1/4} = \frac{1}{5^{1/4}} \qquad (y^{2/5})^{10} = y^4$$

$$\frac{x^{-1/3}}{x^{-1/2}} = x^{-1/3-(-1/2)} = x^{-1/3+1/2} = x^{1/6}, \quad x > 0$$

9.3 Simplifying Radical Expressions

Concepts

Product and Quotient Rules for Radicals

If $\sqrt[n]{a}$ and $\sqrt[n]{b}$ are real numbers and n is a natural number,

$$\sqrt[n]{a} \cdot \sqrt[n]{b} = \sqrt[n]{ab} \quad \text{and} \quad \sqrt[n]{\frac{a}{b}} = \frac{\sqrt[n]{a}}{\sqrt[n]{b}}, \quad b \neq 0.$$

Examples

$$\sqrt{3} \cdot \sqrt{7} = \sqrt{21} \qquad \sqrt[5]{x^3y} \cdot \sqrt[5]{xy^2} = \sqrt[5]{x^4y^3}$$

$$\frac{\sqrt{x^5}}{\sqrt{x^4}} = \sqrt{\frac{x^5}{x^4}} = \sqrt{x}, \quad x > 0$$

Concepts | **Examples**

9.3 Simplifying Radical Expressions (*continued*)

Conditions for a Simplified Radical

1. The radicand has no factor raised to a power greater than or equal to the index.
2. The radicand has no fractions.
3. No denominator has a radical.
4. Exponents in the radicand and the index of the radical have greatest common factor 1.

$$\sqrt{18} = \sqrt{9 \cdot 2} = 3\sqrt{2}$$

$$\sqrt[3]{54x^5y^3} = \sqrt[3]{27x^3y^3 \cdot 2x^2} = 3xy\sqrt[3]{2x^2}$$

$$\sqrt{\frac{7}{4}} = \frac{\sqrt{7}}{\sqrt{4}} = \frac{\sqrt{7}}{2}$$

$$\sqrt[9]{x^3} = x^{3/9} = x^{1/3}, \quad \text{or} \quad \sqrt[3]{x}$$

Pythagorean Formula

If a and b are the lengths of the shorter sides of a right triangle and c is the length of the longest side, then

$$a^2 + b^2 = c^2.$$

The two shorter sides are the legs of the triangle, and the longest side is the hypotenuse. The hypotenuse is opposite the right angle.

Find b for the triangle in the figure.

$$10^2 + b^2 = \left(2\sqrt{61}\right)^2$$

$$b^2 = 4(61) - 100$$

$$b^2 = 144$$

$$b = 12$$

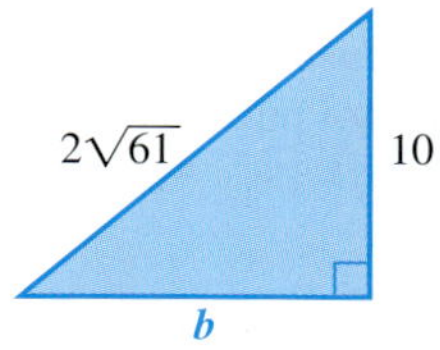

Distance Formula

The distance between (x_1, y_1) and (x_2, y_2) is

$$d = \sqrt{(x_2 - x_1)^2 + (y_2 - y_1)^2}.$$

The distance between $(3, -2)$ and $(-1, 1)$ is

$$\sqrt{(-1 - 3)^2 + [1 - (-2)]^2}$$

$$= \sqrt{(-4)^2 + 3^2}$$

$$= \sqrt{16 + 9}$$

$$= \sqrt{25}, \quad \text{or} \quad 5.$$

9.4 Adding and Subtracting Radical Expressions

Only radical expressions with the same index and the same radicand may be combined.

$$3\sqrt{17} + 2\sqrt{17} - 8\sqrt{17}$$

$$= (3 + 2 - 8)\sqrt{17}$$

$$= -3\sqrt{17}$$

$$\left.\begin{array}{l}\sqrt{15} + \sqrt{30} \\ \sqrt{3} + \sqrt[3]{9}\end{array}\right\} \text{cannot be simplified further}$$

9.5 Multiplying and Dividing Radical Expressions

Multiply binomial radical expressions by using the FOIL method. Special products from **Section 6.4** may apply.

$$\left(\sqrt{2} + \sqrt{7}\right)\left(\sqrt{3} - \sqrt{6}\right)$$

$$= \sqrt{6} - 2\sqrt{3} + \sqrt{21} - \sqrt{42} \qquad \sqrt{12} = 2\sqrt{3}$$

$$\left(\sqrt{5} - \sqrt{10}\right)\left(\sqrt{5} + \sqrt{10}\right)$$

$$= 5 - 10, \quad \text{or} \quad -5$$

$$\left(\sqrt{3} - \sqrt{2}\right)^2$$

$$= 3 - 2\sqrt{3} \cdot \sqrt{2} + 2$$

$$= 5 - 2\sqrt{6}$$

Rationalize the denominator by multiplying both the numerator and denominator by the same expression, one that will yield a rational number in the final denominator.

$$\frac{\sqrt{7}}{\sqrt{5}} = \frac{\sqrt{7} \cdot \sqrt{5}}{\sqrt{5} \cdot \sqrt{5}} = \frac{\sqrt{35}}{5}$$

$$\frac{4}{\sqrt{5} - \sqrt{2}} = \frac{4\left(\sqrt{5} + \sqrt{2}\right)}{\left(\sqrt{5} - \sqrt{2}\right)\left(\sqrt{5} + \sqrt{2}\right)}$$

$$= \frac{4\left(\sqrt{5} + \sqrt{2}\right)}{5 - 2} = \frac{4\left(\sqrt{5} + \sqrt{2}\right)}{3}$$

Concepts	Examples

9.6 Solving Equations with Radicals

Solving an Equation with Radicals

Step 1 Isolate one radical on one side of the equation.

Step 2 Raise each side of the equation to a power that is the same as the index of the radical.

Step 3 Solve the resulting equation; if it still contains a radical, repeat Steps 1 and 2.

Step 4 Check all proposed solutions in the *original* equation.

Proposed solutions that do not check are extraneous; they are not part of the solution set.

Solve $\sqrt{2x+3} - x = 0$.

$$\begin{aligned} \sqrt{2x+3} &= x && \text{Add } x. \\ \left(\sqrt{2x+3}\right)^2 &= x^2 && \text{Square both sides} \\ 2x+3 &= x^2 \\ x^2 - 2x - 3 &= 0 && \text{Standard form} \\ (x+1)(x-3) &= 0 && \text{Factor.} \\ x+1 = 0 \quad \text{or} \quad x-3 &= 0 && \text{Zero-factor property} \\ x = -1 \quad \text{or} \quad x &= 3 && \text{Solve each equation.} \end{aligned}$$

A check shows that 3 is a solution, but -1 is extraneous (as it leads to $2 = 0$, a false statement). The solution set is $\{3\}$.

9.7 Complex Numbers

$\boldsymbol{i = \sqrt{-1}}$, where $\boldsymbol{i^2 = -1}$.

For any positive number b, $\boldsymbol{\sqrt{-b} = i\sqrt{b}}$.

To multiply radicals with negative radicands, first change each factor to the form $i\sqrt{b}$, and then multiply. The same procedure applies to quotients.

$$\sqrt{-25} = i\sqrt{25} = 5i$$

$$\begin{aligned} &\sqrt{-3} \cdot \sqrt{-27} \\ &= i\sqrt{3} \cdot i\sqrt{27} \\ &= i^2\sqrt{81} \\ &= -1 \cdot 9 \\ &= -9 \end{aligned}$$

$$\frac{\sqrt{-18}}{\sqrt{-2}} = \frac{i\sqrt{18}}{i\sqrt{2}} = \sqrt{\frac{18}{2}} = \sqrt{9} = 3$$

Adding and Subtracting Complex Numbers
Add (or subtract) the real parts and add (or subtract) the imaginary parts.

$$(5+3i) + (8-7i) = 13 - 4i \qquad (5+3i) - (8-7i) = -3 + 10i$$

Multiplying Complex Numbers
Multiply complex numbers by using the FOIL method.

$$\begin{aligned} &(2+i)(5-3i) \\ &= 10 - 6i + 5i - 3i^2 && \text{FOIL} \\ &= 10 - i - 3(-1) && i^2 = -1 \\ &= 10 - i + 3 \\ &= 13 - i \end{aligned}$$

Dividing Complex Numbers
Divide complex numbers by multiplying the numerator and the denominator by the conjugate of the denominator.

$$\frac{2}{3+i} = \frac{2(3-i)}{(3+i)(3-i)} = \frac{2(3-i)}{9-i^2}$$
$$= \frac{2(3-i)}{10} = \frac{3-i}{5} = \frac{3}{5} - \frac{1}{5}i$$

ANSWERS TO TEST YOUR WORD POWER

1. B; *Example:* In $\sqrt{3xy}$, $3xy$ is the radicand.
2. C; *Example:* In a right triangle where the sides measure 9, 12, and 15 units, the hypotenuse is the side with measure 15 units.
3. B; *Example:* To rationalize the denominator of $\frac{5}{\sqrt{3}+1}$ multiply both the numerator and denominator by $\sqrt{3} - 1$ to get $\frac{5(\sqrt{3}-1)}{2}$.
4. A; *Example:* The proposed solution 2 is extraneous in $\sqrt{5x-1} + 3 = 0$, as it leads to $6 = 0$, a false statement.
5. C; *Examples:* -5 (or $-5 + 0i$), $7i$ (or $0 + 7i$), and $\sqrt{2} - 4i$.

Chapter 9 Review Exercises

[9.1] *Find each real number root. Use a calculator as necessary.*

1. $\sqrt{1764}$
2. $-\sqrt{289}$
3. $-\sqrt{-841}$
4. $\sqrt[3]{216}$
5. $\sqrt[5]{-32}$
6. $\sqrt{x^2}$
7. $\sqrt[3]{x^3}$
8. $\sqrt[4]{x^{20}}$

Graph each function. Give the domain and the range.

9. $f(x) = \sqrt{x - 1}$

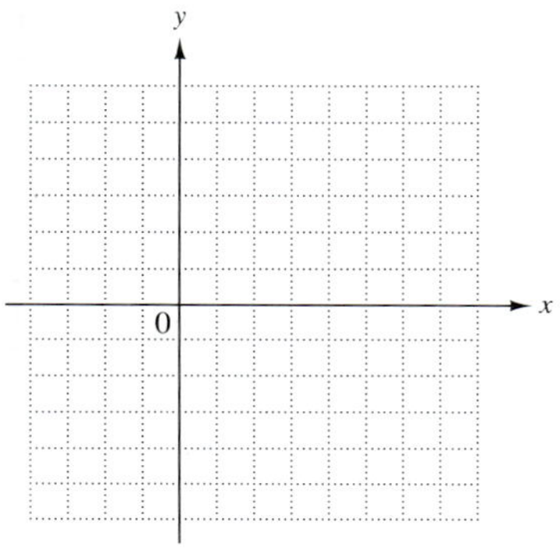

10. $f(x) = \sqrt[3]{x} + 4$

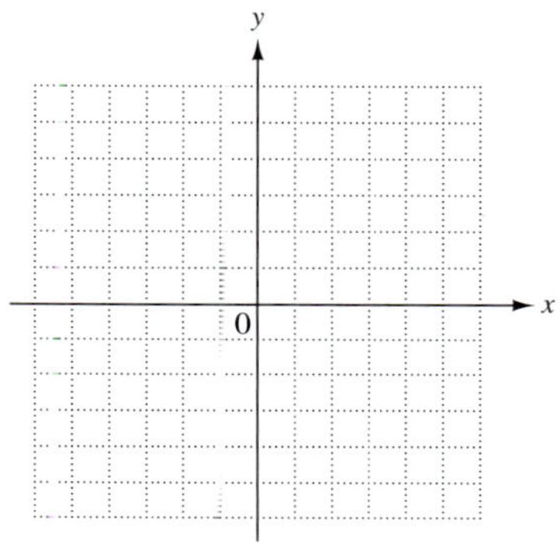

11. Under what conditions is $\sqrt[n]{a}$ not a real number?

12. If a is negative and n is even, what can be said about $a^{1/n}$?

Use a calculator to find a decimal approximation for each radical. Round to the nearest thousandth.

13. $\sqrt{40}$
14. $\sqrt{77}$
15. $\sqrt{310}$

16. Use the formula for the time for one complete swing of a pendulum from Exercise 82 in **Section 9.1,** $t = 2\pi\sqrt{\frac{L}{g}}$, to find the time to the nearest tenth of a second of a complete swing if the pendulum is 3 ft long and g is 32 ft per sec^2.

17. Use Heron's formula from Exercise 83 in **Section 9.1,**

$$A = \sqrt{s(s-a)(s-b)(s-c)},$$

where $s = \frac{1}{2}(a + b + c)$, to find the area of a triangle with sides of lengths 11, 13, and 20 in.

[9.2] *Find each real number root.*

18. $49^{1/2}$

19. $-8^{1/3}$

20. $(-16)^{1/4}$

21. Explain the relationship between the expressions $a^{m/n}$ and $\sqrt[n]{a^m}$.

Simplify each expression. Assume that all variables represent positive real numbers.

22. $16^{5/4}$

23. $-8^{2/3}$

24. $-\left(\frac{36}{25}\right)^{3/2}$

25. $\left(-\frac{1}{8}\right)^{-5/3}$

26. $\left(\frac{81}{10{,}000}\right)^{-3/4}$

27. $7^{1/3} \cdot 7^{5/3}$

28. $\frac{96^{2/3}}{96^{-1/3}}$

29. $\frac{k^{2/3}k^{-1/2}k^{3/4}}{2\,(k^2)^{-1/4}}$

30. Write $2^{4/5}$ as a radical.

Simplify each expression. Write answers in radical form. Assume that all variables represent positive real numbers.

31. $\sqrt{3^{18}}$

32. $\sqrt{7^9}$

33. $\sqrt[3]{m^5} \cdot \sqrt[3]{m^8}$

34. $\sqrt[4]{k^2} \cdot \sqrt[4]{k^7}$

35. $\sqrt[3]{\sqrt{m}}$

36. $\sqrt[4]{16y^5}$

37. $\sqrt[5]{y} \cdot \sqrt[3]{y}$

38. $\frac{\sqrt[3]{y^2}}{\sqrt[4]{y}}$

[9.3] *Simplify each expression. Assume that all variables represent positive real numbers.*

39. $\sqrt{6} \cdot \sqrt{11}$

40. $\sqrt{5} \cdot \sqrt{r}$

41. $\sqrt[3]{6} \cdot \sqrt[3]{5}$

42. $\sqrt[4]{7} \cdot \sqrt[4]{3}$

43. $\sqrt{20}$

44. $-\sqrt{125}$

45. $\sqrt[3]{-108x^4y}$

46. $\sqrt[3]{64p^4q^6}$

47. $\sqrt{\frac{49}{81}}$

48. $\sqrt{\frac{y^3}{144}}$

49. $\sqrt[3]{\frac{m^{15}}{27}}$

50. $\sqrt[3]{\frac{r^2}{8}}$

51. $\frac{\sqrt[3]{2^4}}{\sqrt[4]{32}}$

52. $\frac{\sqrt{x}}{\sqrt[5]{x}}$

Find the distance between each pair of points.

53. $(2, 7)$ and $(-1, -4)$

54. $(-3, -5)$ and $(4, -3)$

[9.4] *Perform the indicated operations. Assume that all variables represent positive real numbers.*

55. $2\sqrt{8} - 3\sqrt{50}$

56. $8\sqrt{80} - 3\sqrt{45}$

57. $-\sqrt{27y} + 2\sqrt{75y}$

58. $2\sqrt{54m^3} + 5\sqrt{96m^3}$

59. $3\sqrt[3]{54} + 5\sqrt[3]{16}$

60. $-6\sqrt[4]{32} + \sqrt[4]{512}$

[9.5] *Multiply, and then simplify the products.*

61. $(\sqrt{3} + 1)(\sqrt{3} - 2)$

62. $(\sqrt{7} + \sqrt{5})(\sqrt{7} - \sqrt{5})$

63. $(3\sqrt{2} + 1)(2\sqrt{2} - 3)$

64. $(\sqrt{11} + 3\sqrt{5})(\sqrt{11} + 5\sqrt{5})$

65. $(\sqrt{13} - \sqrt{2})^2$

66. $(\sqrt{5} - \sqrt{7})^2$

Rationalize each denominator. Assume that all variables represent positive real numbers.

67. $\dfrac{-6\sqrt{3}}{\sqrt{2}}$

68. $\dfrac{3\sqrt{7p}}{\sqrt{y}}$

69. $-\sqrt[3]{\dfrac{9}{25}}$

70. $\sqrt[3]{\dfrac{108m^3}{n^5}}$

71. $\dfrac{1}{\sqrt{2} + \sqrt{7}}$

72. $\dfrac{-5}{\sqrt{6} - \sqrt{3}}$

[9.6] *Solve each equation.*

73. $\sqrt{8x + 9} = 5$

74. $\sqrt{2z - 3} - 3 = 0$

75. $\sqrt{3m + 1} = -1$

76. $\sqrt{7z + 1} = z + 1$

77. $3\sqrt{m} = \sqrt{10m - 9}$

78. $\sqrt{p^2 + 3p + 7} = p + 2$

79. $\sqrt{x + 2} - \sqrt{x - 3} = 1$

80. $\sqrt[3]{5m - 1} = \sqrt[3]{3m - 2}$

81. $\sqrt[4]{x + 6} = \sqrt[4]{2x}$

[9.7] *Write as a product of a real number and i.*

82. $\sqrt{-25}$

83. $\sqrt{-200}$

84. $\sqrt{-160}$

Perform the indicated operations. Write answers in standard form.

85. $(-2 + 5i) + (-8 - 7i)$

86. $(5 + 4i) - (-9 - 3i)$

87. $\sqrt{-5} \cdot \sqrt{-7}$

88. $\sqrt{-25} \cdot \sqrt{-81}$

89. $\dfrac{\sqrt{-72}}{\sqrt{-8}}$

90. $(2 + 3i)(1 - i)$

91. $(6 - 2i)^2$

92. $\frac{3 - i}{2 + i}$

93. $\frac{5 + 14i}{2 + 3i}$

Find each power of i.

94. i^{11}

95. i^{52}

96. i^{-13}

Mixed Review Exercises

Simplify. Assume that all variables represent positive real numbers.

97. $-\sqrt{169a^2b^4}$

98. $1000^{-2/3}$

99. $\frac{y^{-1/3} \cdot y^{5/6}}{y}$

100. $\frac{z^{-1/4}x^{1/2}}{z^{1/2}x^{-1/4}}$

101. $\sqrt[4]{k^{24}}$

102. $\sqrt[3]{54z^9t^8}$

103. $-5\sqrt{18} + 12\sqrt{72}$

104. $8\sqrt[3]{x^3y^2} - 2x\sqrt[3]{y^2}$

105. $(\sqrt{5} - \sqrt{3})(\sqrt{7} + \sqrt{3})$

106. $\frac{-1}{\sqrt{12}}$

107. $\sqrt[3]{\frac{12}{25}}$

108. $\frac{2\sqrt{z}}{\sqrt{z} - 2}$

109. $\sqrt{-49}$

110. $(4 - 9i) + (-1 + 2i)$

111. $\frac{\sqrt{50}}{\sqrt{-2}}$

Solve each equation.

112. $\sqrt{x + 4} = x - 2$

113. $\sqrt{6 + 2x} - 1 = \sqrt{7 - 2x}$

Solve each problem.

114. Carpenters stabilize wall frames with a diagonal brace as shown in the figure. The length of the brace is given by $L = \sqrt{H^2 + W^2}$. If the bottom of the brace is attached 9 ft from the corner and the brace is 12 ft long, how far up the corner post should it be nailed (to the nearest tenth of a foot)?

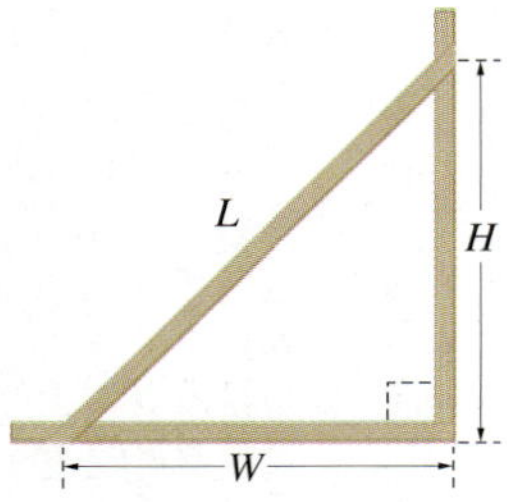

115. Find the perimeter of a triangular electronic highway road sign having the dimensions shown in the figure.

Chapter 9 Test

Chapter Test Prep Video CD — Use the Chapter Test Prep Video CD to see fully worked-out solutions to any of the exercises you want to review

Find each root. Use a calculator as necessary.

1. $-\sqrt{841}$ **2.** $\sqrt[3]{-512}$ **3.** $125^{1/3}$

4. For $\sqrt{146.25}$, which choice gives the best estimate?

A. 10 **B.** 11 **C.** 12 **D.** 13

Use a calculator to approximate each root to the nearest thousandth.

5. $\sqrt{478}$ **6.** $\sqrt[3]{-832}$

7. Graph the function defined by $f(x) = \sqrt{x+6}$, and give the domain and the range.

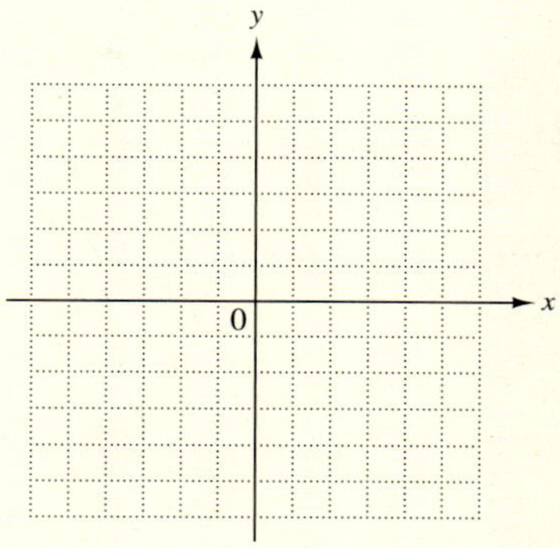

Simplify each expression. Assume that all variables represent positive real numbers.

8. $(-64)^{-4/3}$

9. $\dfrac{3^{2/5}x^{-1/4}y^{2/5}}{3^{-8/5}x^{7/4}y^{1/10}}$

10. $\sqrt{54x^5y^6}$

11. $\sqrt[4]{32a^7b^{13}}$

12. $\sqrt{2} \cdot \sqrt[3]{5}$

13. $3\sqrt{20} - 5\sqrt{80} + 4\sqrt{500}$

14. $(7\sqrt{5}+4)(2\sqrt{5}-1)$

1. ________
2. ________
3. ________
4. ________
5. ________
6. ________
7. ________
8. ________
9. ________
10. ________
11. ________
12. ________
13. ________
14. ________

15. ______________________

16. ______________________

17. ______________________

18. ______________________

19. ______________________

20. ______________________

21. ______________________

22. ______________________

23. ______________________

24. ______________________

25. ______________________

26. ______________________

27. ______________________

28. ______________________

15. $\left(\sqrt{3} - 2\sqrt{5}\right)^2$

16. $\dfrac{-5}{\sqrt{40}}$

17. $\dfrac{2}{\sqrt[3]{5}}$

18. $\dfrac{-4}{\sqrt{7} + \sqrt{5}}$

19. Write $\dfrac{6 + \sqrt{24}}{2}$ in lowest terms.

20. Find the distance between the points $(-3, 8)$ and $(2, 7)$.

21. Use the Pythagorean formula to find the exact length of side b in the figure.

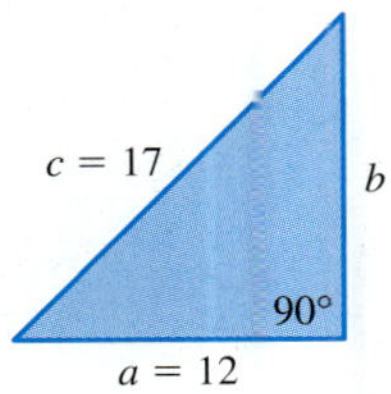

Solve each equation.

22. $\sqrt[3]{5x} = \sqrt[3]{2x - 3}$

23. $\sqrt{7 - x} + 5 = x$

24. $\sqrt{x + 4} - \sqrt{1 - x} = -1$

Perform the indicated operations. Express answers in the form $a + bi$.

25. $(-2 + 5i) - (3 + 6i) - 7i$

26. $(-4 + 2i)(3 - i)$

27. $\dfrac{7 + i}{1 - i}$

28. Simplify i^{35}.

Cumulative Review Exercises Chapters 1–9

Solve each equation or inequality.

1. $7 - (4 + 3t) + 2t = -6(t - 2) - 5$

2. $\frac{1}{3}x + \frac{1}{4}(x + 8) = x + 7$

3. $|6x - 9| = |-4x + 2|$

4. $-5 - 3(x - 2) < 11 - 2(x + 2)$

5. $1 + 4x > 5$ and $-2x > -6$

6. $-2 < 1 - 3x < 7$

7. Write an equation of the line through the points $(-4, 6)$ and $(7, -6)$.

8. Choose the correct response: The lines with equations $2x + 3y = 8$ and $6y = 4x + 16$ are

A. parallel **B.** perpendicular **C.** neither.

9. For the graph of $f(x) = -3x + 6$,

(a) what is the y-intercept? **(b)** what is the x-intercept?

10. Graph the inequality $-2x + y < -6$.

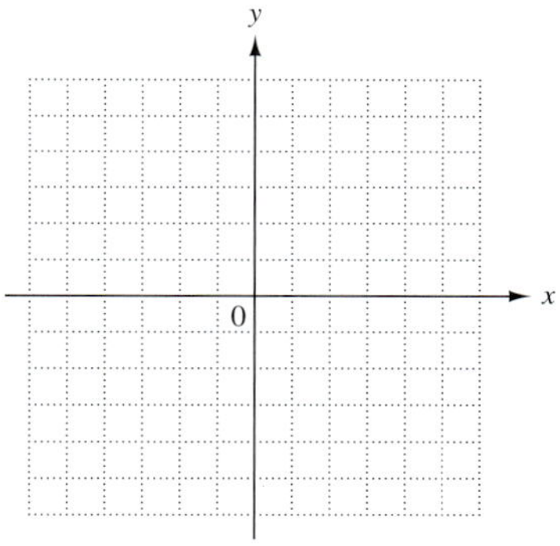

11. Find the measures of the marked angles.

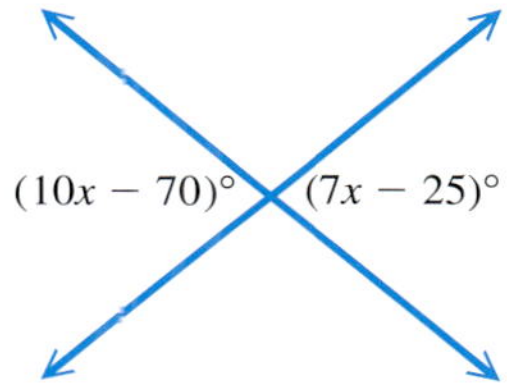

Solve each system.

12. $\begin{aligned} 3x - y &= 23 \\ 2x + 3y &= 8 \end{aligned}$

13. $\begin{aligned} 5x + 2y &= 7 \\ 10x + 4y &= 12 \end{aligned}$

14. $\begin{aligned} 2x + y - z &= 5 \\ 6x + 3y - 3z &= 15 \\ 4x + 2y - 2z &= 10 \end{aligned}$

15. In 2006, if you had sent five 2-oz letters and three 3-oz letters by first-class mail, it would have cost \$5.76. Sending three 2-oz letters and five 3-oz letters would have cost \$6.24. What was the 2006 postage rate for one 2-oz letter and for one 3-oz letter? (*Source:* U.S. Postal Service.)

Perform the indicated operations.

16. $(3k^3 - 5k^2 + 8k - 2) - (4k^3 + 11k + 7) + (2k^2 - 5k)$

17. $(8x - 7)(x + 3)$

18. $\frac{8z^3 - 16z^2 + 24z}{8z^2}$

19. $\frac{6y^4 - 3y^3 + 5y^2 + 6y - 9}{2y + 1}$

Factor each polynomial completely.

20. $2p^2 - 5pq + 3q^2$

21. $18k^4 + 9k^2 - 20$

22. $x^3 + 512$

Perform each operation and express answers in lowest terms.

23. $\dfrac{y^2 + y - 12}{y^3 + 9y^2 + 20y} \div \dfrac{y^2 - 9}{y^3 + 3y^2}$

24. $\dfrac{1}{x + y} + \dfrac{3}{x - y}$

Simplify each complex fraction.

25. $\dfrac{\dfrac{-6}{x - 2}}{\dfrac{8}{3x - 6}}$

26. $\dfrac{\dfrac{1}{a} - \dfrac{1}{b}}{\dfrac{a}{b} - \dfrac{b}{a}}$

Solve.

27. $2x^2 + 11x + 15 = 0$

28. $5t(t - 1) = 2(1 - t)$

Simplify.

29. $27^{-5/3}$

30. $\dfrac{x^{-2/3}}{x^{-3/4}}, \quad x \neq 0$

31. $8\sqrt{20} + 3\sqrt{80} - 2\sqrt{500}$

32. $\dfrac{-9}{\sqrt{80}}$

33. $\dfrac{4}{\sqrt{6} - \sqrt{5}}$

34. $\dfrac{12}{\sqrt[3]{2}}$

35. Find the distance between the points $(-4, 4)$ and $(-2, 9)$.

36. Solve $\sqrt{8x - 4} - \sqrt{7x + 2} = 0$.

Solve each problem.

37. The current of a river runs at 3 mph. Brent's boat can go 36 mi downstream in the same time that it takes to go 24 mi upstream. Find the speed of the boat in still water.

38. How many liters of pure alcohol must be mixed with 40 L of 18% alcohol to obtain a 22% alcohol solution?

39. A jar containing only dimes and quarters has 29 coins with a face value of \$4.70. How many of each denomination are there?

40. Brenda rides her bike 4 mph faster than her husband, Chuck. If Brenda can ride 48 mi in the same time that Chuck can ride 24 mi, what are their speeds?

Study Skills

PREPARING FOR YOUR MATH FINAL EXAM

Your math final exam is likely to be a **comprehensive exam.** This means that it will cover material from the entire term. The end of the term will be less stressful if you make a plan for how you will prepare for each of your exams.

First, figure out the grade you need to earn on the final exam to get the course grade you are aiming for. Check your course syllabus for grading policies, or ask your instructor if you are not sure of them. This allows you to set a goal for yourself.

How many points do you need to earn on your mathematics final exam to get the grade you want? ____________________

Second, create a final exam week plan for your work and personal life. If you need to make an adjustment in your work schedule, do it in advance. If you have family members to care for, perhaps enlist some help from others so you can spend extra time studying. Try to plan in advance so you don't create additional stress for yourself. You will have to set some priorities, and studying has to be at the top of the list. Although life doesn't stop for finals, some things can be ignored for a short time. **Get enough sleep and healthy food so you can perform your best.**

What adjustments in your personal life do you need to make for final exam week? ____________________

Third, use the following suggestions to guide your studying and reviewing.

- **Know exactly which chapters and sections will be on the final exam.**
- **Divide up the chapters,** and decide how much you will review each day.
- **Begin your reviewing several days before the exam.**
- **Use returned quizzes and tests to review earlier material** (if you have them).
- **Practice all types of problems,** but emphasize the types that are most difficult for you. Use the Cumulative Reviews that are at the end of each chapter in your textbook.
- **Rewrite your notes or make mind maps** to create summaries of important information.
- **Make study cards for all types of problems.** Carry the cards with you, and review them whenever you have a few spare minutes.

Of course, a week of final exams produces stress. **Students who develop skills for reducing and managing stress do better on their final exams.** You will feel better if you make a conscious effort to reduce your stress level. Even if it takes you away from studying for a little while each day, the time will be well spent.

OBJECTIVES

1. **Create a final exam week plan.**
2. **Break studying into chunks and study over several days.**
3. **Practice all types of problems.**

Managing Stress

Reducing Physical Stress

Examples of ways to reduce physical stress are listed below.

- *Laugh.* Watch your favorite funny movie, exchange a joke with a friend, or view a comedy bit on the Internet.
- *Exercise for 20 to 30 minutes.* If you normally exercise regularly, do NOT stop during final exam week.
- *Practice deep breathing.* Several minutes of deep breathing will calm you.
- *Visualize a relaxing scene.* Choose something that you find peaceful and picture it. Imagine what it feels like and sounds like. Try to put yourself in the picture.
- If you feel stress in your muscles, such as your shoulders or back, *slowly squeeze the muscles as much as you can, and then release them.* Sometimes we don't realize we are clenching our teeth or holding tension in our shoulders until we consciously work with them. Try to notice what it feels like when they are relaxed and loose.

Reducing Mental Stress

Mental stress reduction is also a powerful tool both before and during an exam.

- *Talk positively to yourself.* Tell yourself you will get through it.
- *Reward yourself.* Give yourself small breaks, a treat—something that makes you feel cared for—every day of final exam week.
- *Make a list of things to do* and feel the sense of accomplishment when you cross each item off.
- *Relax your mind* by using it for something *completely* different from the kind of thinking you do when you study. Play your favorite music, walk your dog, read a good book.
- *Visualize.* Picture yourself completing exams and projects successfully. Picture yourself taking tests calmly and confidently.

Finally, DON'T stay up all night the night before an exam—get a good night's sleep.

Now Try This ▶▶▶

Choose three techniques for reducing stress from the lists above to try during final exam week. Include other ideas as well.

1. ______________________________

2. ______________________________

3. ______________________________

Other ideas ______________________________

10

Quadratic Equations, Inequalities, and Functions

10.1 The Square Root Property and Completing the Square

10.2 The Quadratic Formula

10.3 Equations Quadratic in Form

Summary Exercises on Solving Quadratic Equations

10.4 Formulas and Further Applications

10.5 Graphs of Quadratic Functions

10.6 More about Parabolas and Their Applications

10.7 Polynomial and Rational Inequalities

In 2008 the prices of food, gasoline, and other products increased throughout the world. In particular, escalating oil prices caused increases in transportation and shipping costs, which trickled down to affect prices of a variety of goods and services.

Although prices tend to go up over time, the rate at which they increase (the inflation rate) varies considerably. The Consumer Price Index (CPI) used by the U.S. government measures changes in prices for goods purchased by typical American families over time. In Example 6 of Section 10.4, we use a *quadratic function* to model the CPI.

10.1 The Square Root Property and Completing the Square

OBJECTIVES

1. Review the zero-factor property.
2. Learn the square root property.
3. Solve quadratic equations of the form $(ax + b)^2 = c$ by using the square root property.
4. Solve quadratic equations by completing the square.
5. Solve quadratic equations with nonreal complex solutions.

We introduced quadratic equations in **Section 7.5.** Recall that a *quadratic equation* is defined as follows.

Quadratic Equation

An equation that can be written in the form

$$ax^2 + bx + c = 0,$$

where a, b, and c are real numbers, with $a \neq 0$, is a **quadratic equation.** The given form is called **standard form.**

A quadratic equation is a *second-degree equation,* that is, an equation with a squared term and no terms of higher degree. For example,

$$4m^2 + 4m - 5 = 0 \quad \text{and} \quad 3x^2 = 4x - 8 \qquad \text{Quadratic equations}$$

are quadratic equations, with the first equation in standard form.

◀ *Work Problem* **1** *at the Side.*

OBJECTIVE 1 Review the zero-factor property. In **Section 7.5** we used factoring and the zero-factor property to solve quadratic equations.

Zero-Factor Property

If two numbers have a product of 0, then at least one of the numbers must be 0. That is, if $ab = 0$, then $a = 0$ or $b = 0$.

We solved a quadratic equation such as $3x^2 - 5x - 28 = 0$ using the zero-factor property as follows.

$$3x^2 - 5x - 28 = 0$$

$$(3x + 7)(x - 4) = 0 \qquad \text{Factor.}$$

$$3x + 7 = 0 \quad \text{or} \quad x - 4 = 0 \qquad \text{Zero-factor property}$$

$$3x = -7 \quad \text{or} \quad x = 4 \qquad \text{Solve each equation.}$$

$$x = -\frac{7}{3}$$

The solution set is $\left\{-\frac{7}{3}, 4\right\}$.

◀ *Work Problem* **2** *at the Side.*

OBJECTIVE 2 Learn the square root property. Although factoring is the simplest way to solve quadratic equations, not every quadratic equation can be solved easily by factoring. We now develop other methods of solving quadratic equations based on the following property.

Square Root Property

If x and k are complex numbers and $x^2 = k$, then

$$x = \sqrt{k} \quad \text{or} \quad x = -\sqrt{k}.$$

1 (a) Which of the following are quadratic equations?

A. $x + 2y = 0$

B. $x^2 - 8x + 16 = 0$

C. $2t^2 - 5t = 3$

D. $x^3 + x^2 + 4 = 0$

(b) Which quadratic equation identified in part (a) is in standard form?

2 Solve each equation by factoring.

(a) $x^2 + 3x + 2 = 0$

(b) $3m^2 = 3 - 8m$

(*Hint:* Remember to write the equation in standard form first.)

ANSWERS

1. (a) B, C (b) B
2. (a) $\{-2, -1\}$ (b) $\left\{-3, \frac{1}{3}\right\}$

The following steps justify the square root property.

$$x^2 = k$$

$$x^2 - k = 0 \quad \text{Subtract } k.$$

$$\left(x - \sqrt{k}\right)\left(x + \sqrt{k}\right) = 0 \quad \text{Factor.}$$

$$x - \sqrt{k} = 0 \quad \text{or} \quad x + \sqrt{k} = 0 \quad \text{Zero-factor property}$$

$$\boldsymbol{x = \sqrt{k}} \quad \text{or} \quad \boldsymbol{x = -\sqrt{k}} \quad \text{Solve each equation.}$$

Thus, the solutions of the equation $x^2 = k$ are $\sqrt{k}$ and $-\sqrt{k}$.

CAUTION

If $k \neq 0$, then using the square root property always produces *two* square roots, one positive and one negative.

EXAMPLE 1 **Using the Square Root Property**

Solve each equation.

(a) $r^2 = 5$

By the square root property, if $r^2 = 5$, then

$$r = \sqrt{5} \quad \text{or} \quad r = -\sqrt{5},$$

Don't forget the negative solution

and the solution set is $\left\{\sqrt{5}, -\sqrt{5}\right\}$.

(b)

$$4x^2 - 48 = 0$$

$$4x^2 = 48 \quad \text{Add 48.}$$

$$x^2 = 12 \quad \text{Divide by 4.}$$

$$x = \sqrt{12} \quad \text{or} \quad x = -\sqrt{12} \quad \text{Square root property}$$

$$x = 2\sqrt{3} \quad \text{or} \quad x = -2\sqrt{3} \quad \sqrt{12} = \sqrt{4} \cdot \sqrt{3} = 2\sqrt{3}$$

The solutions are $2\sqrt{3}$ and $-2\sqrt{3}$. Check each in the original equation.

Check $4x^2 - 48 = 0$ Original equation

$4\left(2\sqrt{3}\right)^2 - 48 \stackrel{?}{=} 0$	$4\left(-2\sqrt{3}\right)^2 - 48 \stackrel{?}{=} 0$
$4(12) - 48 \stackrel{?}{=} 0$	$4(12) - 48 \stackrel{?}{=} 0$
$48 - 48 \stackrel{?}{=} 0$	$48 - 48 \stackrel{?}{=} 0$
$0 = 0$ True	$0 = 0$ True

The solution set is $\left\{2\sqrt{3}, -2\sqrt{3}\right\}$.

Work Problem **3** *at the Side.* ▶

Note

Recall that solutions such as those in Example 1 are sometimes abbreviated with the symbol $\pm$ (read "positive or negative" or "plus or minus.") With this symbol, the solutions in Example 1 would be written $\pm\sqrt{5}$ and $\pm 2\sqrt{3}$.

3 Solve each equation.

(a) $m^2 = 64$

(b) $p^2 = 7$

(c) $3x^2 - 54 = 0$

ANSWERS

3. **(a)** $\{8, -8\}$ **(b)** $\left\{\sqrt{7}, -\sqrt{7}\right\}$
(c) $\left\{3\sqrt{2}, -3\sqrt{2}\right\}$

4 Solve the problem.

An expert marksman can hold a silver dollar at forehead level, drop it, draw his gun, and shoot the coin as it passes waist level. If the coin falls about 4 ft, use the formula in Example 2 to find the time that elapses between the dropping of the coin and the shot.

Answer

4. 0.5 sec

EXAMPLE 2 Using the Square Root Property in an Application

Galileo Galilei (1564–1642) developed a formula for freely falling objects described by

$$d = 16t^2,$$

where d is the distance in feet that an object falls (disregarding air resistance) in t seconds, regardless of weight. Galileo dropped objects from the Leaning Tower of Pisa to develop this formula. If the Leaning Tower is about 180 ft tall, use Galileo's formula to determine how long it would take an object dropped from the tower to fall to the ground. (*Source: Microsoft Encarta Encyclopedia.*)

$$d = 16t^2$$

$$180 = 16t^2 \quad \text{Let } d = 180.$$

$$11.25 = t^2 \quad \text{Divide by 16.}$$

$$t = \sqrt{11.25} \quad \text{or} \quad t = -\sqrt{11.25} \quad \text{Square root property}$$

Time cannot be negative, so we discard $t = -\sqrt{11.25}$. Using a calculator, $\sqrt{11.25} \approx 3.4$ so $t \approx 3.4$. The object would fall to the ground in about 3.4 sec.

◀ Work Problem 4 at the Side.

OBJECTIVE 3 Solve quadratic equations of the form $(ax + b)^2 = c$ by using the square root property. To solve more complicated equations using the square root property, such as

$$(x - 5)^2 = 36,$$

substitute $(x - 5)^2$ for x^2 and 36 for k, to get

$$x - 5 = \sqrt{36} \quad \text{or} \quad x - 5 = -\sqrt{36}$$

$$x - 5 = 6 \quad \text{or} \quad x - 5 = -6$$

$$x = 11 \quad \text{or} \quad x = -1.$$

Check $(x - 5)^2 = 36$ Original equation

$$(11 - 5)^2 \stackrel{?}{=} 36 \qquad (-1 - 5)^2 \stackrel{?}{=} 36$$

$$6^2 \stackrel{?}{=} 36 \qquad (-6)^2 \stackrel{?}{=} 36$$

$$36 = 36 \quad \text{True} \qquad 36 = 36 \quad \text{True}$$

Both solutions satisfy the original equation. The solution set is $\{-1, 11\}$.

EXAMPLE 3 Using the Square Root Property

Solve $(2x - 3)^2 = 18$.

$$2x - 3 = \sqrt{18} \quad \text{or} \quad 2x - 3 = -\sqrt{18} \quad \text{Square root property}$$

$$2x = 3 + \sqrt{18} \quad \text{or} \quad 2x = 3 - \sqrt{18} \quad \text{Add 3.}$$

$$x = \frac{3 + \sqrt{18}}{2} \quad \text{or} \quad x = \frac{3 - \sqrt{18}}{2} \quad \text{Divide by 2.}$$

$$x = \frac{3 + 3\sqrt{2}}{2} \quad \text{or} \quad x = \frac{3 - 3\sqrt{2}}{2} \quad \sqrt{18} = \sqrt{9} \cdot \sqrt{2} = 3\sqrt{2}$$

Continued on Next Page

We show the check for the first solution. The check for the second solution is similar.

Check $(2x - 3)^2 = 18$ Original equation

$\left[2\left(\frac{3+3\sqrt{2}}{2}\right) - 3\right]^2 \stackrel{?}{=} 18$ Let $x = \frac{3+3\sqrt{2}}{2}$.

$\left(3 + 3\sqrt{2} - 3\right)^2 \stackrel{?}{=} 18$ Multiply.

$\left(3\sqrt{2}\right)^2 \stackrel{?}{=} 18$ Simplify.

$18 = 18$ True

The solution set is $\left\{\frac{3+3\sqrt{2}}{2}, \frac{3-3\sqrt{2}}{2}\right\}$.

Work Problem 5 at the Side. ▶

OBJECTIVE 4 Solve quadratic equations by completing the square. We can use the square root property to solve *any* quadratic equation by writing it in the form $(x + k)^2 = n$. That is, we must write the left side of the equation as a perfect square trinomial that can be factored as $(x + k)^2$, the square of a binomial, and the right side must be a constant. Rewriting a quadratic equation in this form is called **completing the square.**

Recall that the perfect square trinomial

$$x^2 + 10x + 25$$

can be factored as $(x + 5)^2$. In the trinomial, the coefficient of x (the first-degree term) is 10 and the constant term is 25. Notice that if we take half of 10 and square it, we get the constant term, 25.

Coefficient of x ↓ Constant ↓

$$\left[\frac{1}{2}(10)\right]^2 = 5^2 = 25$$

Similarly, in

$$x^2 + 12x + 36, \quad \left[\frac{1}{2}(12)\right]^2 = 6^2 = 36,$$

and in

$$m^2 - 6m + 9, \quad \left[\frac{1}{2}(-6)\right]^2 = (-3)^2 = 9.$$

This relationship is true in general and is the idea behind completing the square.

Work Problem 6 at the Side. ▶

EXAMPLE 4 Solving a Quadratic Equation by Completing the Square

Solve $x^2 + 8x + 10 = 0$.

This quadratic equation cannot be solved easily by factoring, and it is not in the correct form to solve using the square root property. To solve it by completing the square, we need a perfect square trinomial on the left side of the equation. To get this form, we first subtract 10 from each side.

Continued on Next Page

5 Solve each equation.

(a) $(x - 3)^2 = 25$

(b) $(3k + 1)^2 = 2$

(c) $(2r + 3)^2 = 8$

6 Find the constant to be added to get a perfect square trinomial. In each case, take half the coefficient of the first-degree term and square the result.

(a) $x^2 + 4x +$ ____

(b) $t^2 - 2t +$ ____

(c) $m^2 + 5m +$ ____

(d) $x^2 - \frac{2}{3}x +$ ____

ANSWERS

5. (a) $\{-2, 8\}$

(b) $\left\{\frac{-1+\sqrt{2}}{3}, \frac{-1-\sqrt{2}}{3}\right\}$

(c) $\left\{\frac{-3+2\sqrt{2}}{2}, \frac{-3-2\sqrt{2}}{2}\right\}$

6. (a) 4 **(b)** 1 **(c)** $\frac{25}{4}$ **(d)** $\frac{1}{9}$

7 Solve $n^2 + 6n + 4 = 0$ by completing the square.

$$x^2 + 8x + 10 = 0 \quad \text{Original equation}$$
$$x^2 + 8x = -10 \quad \text{Subtract 10.}$$

We must add a constant to get a perfect square trinomial on the left.

$$\underbrace{x^2 + 8x + \underline{\ ?\ }}_{\text{Needs to be a perfect square trinomial}}$$

Take half the coefficient of the first-degree term and square the result.

$$\left[\frac{1}{2}(8)\right]^2 = 4^2 = 16 \leftarrow \text{Desired constant}$$

We add this constant, 16, to *each* side of the equation.

This is a key step.

$$x^2 + 8x + 16 = -10 + 16$$

Next we factor on the left side and add on the right.

$$(x + 4)^2 = 6$$

We can now use the square root property.

$$x + 4 = \sqrt{6} \quad \text{or} \quad x + 4 = -\sqrt{6}$$
$$x = -4 + \sqrt{6} \quad \text{or} \quad x = -4 - \sqrt{6}$$

Check

$$x^2 + 8x + 10 = 0 \quad \text{Original equation}$$
$$\left(-4 + \sqrt{6}\right)^2 + 8\left(-4 + \sqrt{6}\right) + 10 \stackrel{?}{=} 0 \quad \text{Let } x = -4 + \sqrt{6}.$$
$$16 - 8\sqrt{6} + 6 - 32 + 8\sqrt{6} + 10 \stackrel{?}{=} 0$$
$$0 = 0 \quad \text{True}$$

Remember the middle term when squaring $-4 + \sqrt{6}$.

The check of the other solution is similar. Thus, $\left\{-4 + \sqrt{6}, -4 - \sqrt{6}\right\}$ is the solution set.

◀ *Work Problem* **7** *at the Side.*

Completing the Square

Solve $ax^2 + bx + c = 0$ $(a \neq 0)$ by completing the square as follows:

Step 1 **Be sure the second-degree (squared) term has coefficient 1.** If the coefficient of the squared term is 1, proceed to Step 2. If the coefficient of the squared term is not 1 but some other nonzero number a, divide each side of the equation by a.

Step 2 **Write the equation in correct form** so that terms with variables are on one side of the equals sign and the constant is on the other side.

Step 3 **Square half the coefficient of the first-degree (linear) term.**

Step 4 **Add the square to each side.**

Step 5 **Factor the perfect square trinomial.** One side should now be a perfect square trinomial. Factor it as the square of a binomial. Simplify the other side.

Step 6 **Solve the equation.** Apply the square root property to complete the solution.

Answer

7. $\left\{-3 + \sqrt{5}, -3 - \sqrt{5}\right\}$

EXAMPLE 5 **Solving a Quadratic Equation by Completing the Square ($a = 1$)**

Solve $k^2 + 5k - 1 = 0$.

Since the coefficient of the squared term is 1, begin with Step 2.

Step 2 $k^2 + 5k = 1$ Add 1 to each side.

Step 3 Take half the coefficient of the first-degree term and square the result.

$$\left[\frac{1}{2}(5)\right]^2 = \left(\frac{5}{2}\right)^2 = \frac{25}{4}$$

Step 4 $k^2 + 5k + \frac{25}{4} = 1 + \frac{25}{4}$ Add the square to each side of the equation.

Step 5 $\left(k + \frac{5}{2}\right)^2 = \frac{29}{4}$ Factor on the left; add on the right.

Step 6 $k + \frac{5}{2} = \sqrt{\frac{29}{4}}$ or $k + \frac{5}{2} = -\sqrt{\frac{29}{4}}$ Square root property

$k + \frac{5}{2} = \frac{\sqrt{29}}{2}$ or $k + \frac{5}{2} = -\frac{\sqrt{29}}{2}$ $\sqrt{\frac{a}{b}} = \frac{\sqrt{a}}{\sqrt{b}}$

$k = -\frac{5}{2} + \frac{\sqrt{29}}{2}$ or $k = -\frac{5}{2} - \frac{\sqrt{29}}{2}$ Add $-\frac{5}{2}$.

$k = \frac{-5 + \sqrt{29}}{2}$ or $k = \frac{-5 - \sqrt{29}}{2}$ $\frac{a}{c} + \frac{b}{c} = \frac{a+b}{c}$

Check that the solution set is $\left\{\frac{-5 + \sqrt{29}}{2}, \frac{-5 - \sqrt{29}}{2}\right\}$.

Work Problem **8** *at the Side.* ▶

EXAMPLE 6 **Solving a Quadratic Equation by Completing the Square ($a \neq 1$)**

Solve $2x^2 - 4x - 5 = 0$.

First divide each side by 2 to get 1 as the coefficient of the squared term.

$x^2 - 2x - \frac{5}{2} = 0$ Step 1

$x^2 - 2x = \frac{5}{2}$ Step 2

$\left[\frac{1}{2}(-2)\right]^2 = (-1)^2 = 1$ Step 3

$x^2 - 2x + 1 = \frac{5}{2} + 1$ Step 4

$(x - 1)^2 = \frac{7}{2}$ Step 5

$x - 1 = \sqrt{\frac{7}{2}}$ or $x - 1 = -\sqrt{\frac{7}{2}}$ Step 6

Continued on Next Page

8 Solve each equation by completing the square.

(a) $x^2 + 2x - 10 = 0$

(b) $r^2 + 3r - 1 = 0$

ANSWERS

8. (a) $\{-1 + \sqrt{11}, -1 - \sqrt{11}\}$

(b) $\left\{\frac{-3 + \sqrt{13}}{2}, \frac{-3 - \sqrt{13}}{2}\right\}$

$$x = 1 + \sqrt{\frac{7}{2}} \quad \text{or} \quad x = 1 - \sqrt{\frac{7}{2}} \qquad \text{Add 1.}$$

$$x = 1 + \frac{\sqrt{14}}{2} \quad \text{or} \quad x = 1 - \frac{\sqrt{14}}{2} \qquad \sqrt{\frac{7}{2}} = \frac{\sqrt{7}}{\sqrt{2}} = \frac{\sqrt{7}}{\sqrt{2}} \cdot \frac{\sqrt{2}}{\sqrt{2}} = \frac{\sqrt{14}}{2}$$

Add the two terms in each solution as follows:

$$1 + \frac{\sqrt{14}}{2} = \frac{2}{2} + \frac{\sqrt{14}}{2} = \frac{2 + \sqrt{14}}{2} \qquad 1 = \frac{2}{2}$$

$$1 - \frac{\sqrt{14}}{2} = \frac{2}{2} - \frac{\sqrt{14}}{2} = \frac{2 - \sqrt{14}}{2}.$$

Check that the solution set is $\left\{\frac{2 + \sqrt{14}}{2}, \frac{2 - \sqrt{14}}{2}\right\}$.

◀ Work Problem 9 at the Side.

OBJECTIVE 5 Solve quadratic equations with nonreal complex solutions. In the equation $x^2 = k$, if $k < 0$, there will be two nonreal complex solutions.

EXAMPLE 7 Solving for Nonreal Complex Solutions

Solve each equation.

(a)
$$x^2 = -15$$

$$x = \sqrt{-15} \quad \text{or} \quad x = -\sqrt{-15} \qquad \text{Square root property}$$

$$x = i\sqrt{15} \quad \text{or} \quad x = -i\sqrt{15} \qquad \sqrt{-1} = i$$

The solution set is $\{i\sqrt{15}, -i\sqrt{15}\}$.

(b)
$$(t + 2)^2 = -16$$

$$t + 2 = \sqrt{-16} \quad \text{or} \quad t + 2 = -\sqrt{-16} \qquad \text{Square root property}$$

$$t + 2 = 4i \quad \text{or} \quad t + 2 = -4i \qquad \sqrt{-16} = 4i$$

$$t = -2 + 4i \quad \text{or} \quad t = -2 - 4i \qquad \text{Subtract 2.}$$

The solution set is $\{-2 + 4i, -2 - 4i\}$.

(c) $x^2 + 2x + 7 = 0$

$$x^2 + 2x = -7 \qquad \text{Subtract 7.}$$

$$x^2 + 2x + 1 = -7 + 1 \qquad \left[\tfrac{1}{2}(2)\right]^2 = 1\text{; add 1 to each side.}$$

$$(x + 1)^2 = -6 \qquad \text{Factor on the left; add on the right.}$$

$$x + 1 = \pm i\sqrt{6} \qquad \text{Square root property}$$

$$x = -1 \pm i\sqrt{6} \qquad \text{Subtract 1.}$$

The solution set is $\{-1 + i\sqrt{6}, -1 - i\sqrt{6}\}$.

◀ Work Problem 10 at the Side.

Note
We will use completing the square in **Section 10.6** when we graph quadratic equations and in **Section 12.2** when we work with circles.

9 Solve each equation by completing the square.

(a) $2r^2 - 4r + 1 = 0$

(b) $3z^2 - 6z - 2 = 0$

(c) $8x^2 - 4x - 2 = 0$

10 Solve each equation.

(a) $x^2 = -17$

(b) $(k + 5)^2 = -100$

(c) $5t^2 - 15t + 12 = 0$

Answers

9. (a) $\left\{\frac{2 + \sqrt{2}}{2}, \frac{2 - \sqrt{2}}{2}\right\}$
(b) $\left\{\frac{3 + \sqrt{15}}{3}, \frac{3 - \sqrt{15}}{3}\right\}$
(c) $\left\{\frac{1 + \sqrt{5}}{4}, \frac{1 - \sqrt{5}}{4}\right\}$

10. (a) $\{i\sqrt{17}, -i\sqrt{17}\}$
(b) $\{-5 + 10i, -5 - 10i\}$
(c) $\left\{\frac{3}{2} + \frac{\sqrt{15}}{10}i, \frac{3}{2} - \frac{\sqrt{15}}{10}i\right\}$

10.1 Exercises

FOR EXTRA HELP MyMathLab Math XL PRACTICE WATCH DOWNLOAD READ REVIEW

1. A student was asked to solve the quadratic equation $x^2 = 16$ and did not get full credit for the solution set $\{4\}$. Why?

2. Why can't the zero-factor property be used to solve every quadratic equation?

3. Give a one-sentence description or explanation of each of the following.
   (a) Quadratic equation in standard form
   (b) Zero-factor property
   (c) Square root property

4. A student tried to solve $x^2 - x - 2 = 5$ as follows.

$$x^2 - x - 2 = 5$$

$$(x - 2)(x + 1) = 5 \quad \text{Factor.}$$

$$x - 2 = 5 \quad \text{or} \quad x + 1 = 5 \quad \text{Zero-factor property}$$

$$x = 7 \quad \text{or} \quad x = 4 \quad \text{Solve each equation.}$$

This method is incorrect. ***WHAT WENT WRONG?***

Use the square root property to solve each equation. See Examples 1 and 3.

5. $x^2 = 81$

6. $z^2 = 225$

7. $t^2 = 17$

8. $k^2 = 19$

9. $m^2 = 32$

10. $x^2 = 54$

11. $t^2 - 20 = 0$

12. $p^2 - 50 = 0$

13. $3n^2 - 72 = 0$

14. $5z^2 - 200 = 0$

15. $(x + 2)^2 = 25$

16. $(t + 8)^2 = 9$

17. $(x - 4)^2 = 3$

18. $(x + 3)^2 = 11$

19. $(t + 5)^2 = 48$

20. $(m-6)^2 = 27$

21. $(3k-1)^2 = 7$

22. $(2x+4)^2 = 10$

23. $(4p+1)^2 = 24$

24. $(5k-2)^2 = 12$

Solve Exercises 25 and 26 using Galileo's formula, $d = 16t^2$. Round answers to the nearest tenth. See Example 2.

25. Mount Rushmore National Memorial in South Dakota features a sculpture of four of America's favorite presidents carved into the rim of the mountain, 500 ft above the valley floor. How long would it take a rock dropped from the top of the sculpture to fall to the ground? (*Source: Microsoft Encarta Encyclopedia.*)

26. The Gateway Arch in St. Louis, Missouri, the tallest national monument in the United States, is 630 ft tall. How long would it take an object dropped from the top of it to fall to the ground? (*Source:* www.gatewayarch.com)

27. Of the two equations

$$(2x+1)^2 = 5 \quad \text{and} \quad x^2 + 4x = 12,$$

one is more suitable for solving by the square root property, and the other is more suitable for solving by completing the square. Which method do you think most students would use for each equation?

28. Why would most students find the equation $x^2 + 4x = 20$ easier to solve by completing the square than the equation $5x^2 + 2x = 3$?

29. Decide what number must be added to make each expression a perfect square trinomial.

(a) $x^2 + 6x +$ ___ **(b)** $x^2 + 14x +$ ___

(c) $p^2 - 12p +$ ___ **(d)** $x^2 + 3x +$ ___

(e) $q^2 - 9q +$ ___ **(f)** $t^2 - \frac{1}{2}t +$ ___

30. What would be the first step in solving

$$2x^2 + 8x = 9$$

by completing the square?

Determine the number that will complete the square to solve each equation after the constant term has been written on the right side. Do not actually solve. See Examples 4–6.

31. $x^2 + 4x - 2 = 0$

32. $t^2 + 2t - 1 = 0$

33. $x^2 + 10x + 18 = 0$

34. $x^2 + 8x + 11 = 0$

35. $3w^2 - w - 24 = 0$

36. $4z^2 - z - 39 = 0$

Solve each equation by completing the square. Use the results of Exercises 31–36 to solve Exercises 39–44. See Examples 4–6.

37. $x^2 - 2x - 24 = 0$

38. $m^2 - 4m - 32 = 0$

39. $x^2 + 4x - 2 = 0$

40. $t^2 + 2t - 1 = 0$

41. $x^2 + 10x + 18 = 0$

42. $x^2 + 8x + 11 = 0$

43. $3w^2 - w = 24$

44. $4z^2 - z = 39$

45. $2k^2 + 5k - 2 = 0$

46. $3r^2 + 2r - 2 = 0$

47. $5x^2 - 10x + 2 = 0$

48. $2x^2 - 16x + 25 = 0$

49. $9x^2 - 24x = -13$

50. $25n^2 - 20n = 1$

51. $z^2 - \frac{4}{3}z = -\frac{1}{9}$

52. $p^2 - \frac{8}{3}p = -1$

53. $0.1x^2 - 0.2x - 0.1 = 0$ (*Hint:* First clear the decimals.)

54. $0.1p^2 - 0.4p + 0.1 = 0$ (*Hint:* First clear the decimals.)

Find all complex solutions of each equation. See Example 7.

55. $x^2 = -12$

56. $x^2 = -18$

57. $(r - 5)^2 = -3$

58. $(t + 6)^2 = -5$

59. $(6k - 1)^2 = -8$

60. $(4m - 7)^2 = -27$

61. $m^2 + 4m + 13 = 0$

62. $t^2 + 6t + 10 = 0$

63. $3r^2 + 4r + 4 = 0$

64. $4x^2 + 5x + 5 = 0$

65. $-m^2 - 6m - 12 = 0$

66. $-k^2 - 5k - 10 = 0$

Relating Concepts (Exercises 67–72) For Individual or Group Work

The Greeks had a method of completing the square geometrically in which they literally changed a figure into a square. For example, to complete the square for $x^2 + 6x$, we begin with a square of side x, as in the figure. We add three rectangles of width 1 to the right side and the bottom to get a region with area $x^2 + 6x$. To fill in the corner (complete the square), we must add 9 1-by-1 squares as shown.

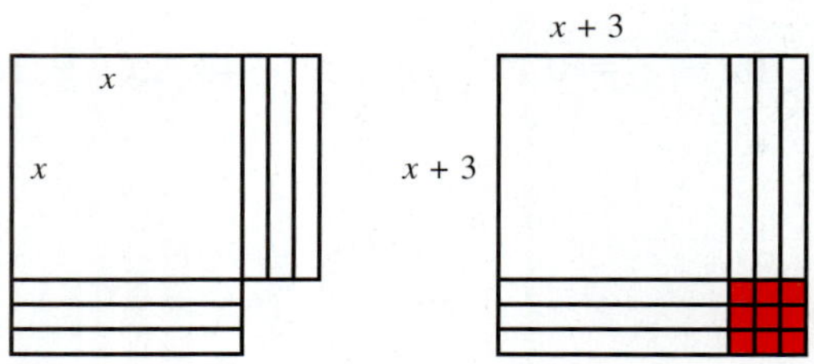

Work Exercises 67–72 in order.

67. What is the area of the original square?

68. What is the area of each strip?

69. What is the total area of the six strips?

70. What is the area of each small square in the corner of the second figure?

71. What is the total area of the small squares?

72. What is the area of the new, larger square?

10.2 The Quadratic Formula

OBJECTIVES

1. Derive the quadratic formula.
2. Solve quadratic equations by using the quadratic formula.
3. Use the discriminant to determine the number and type of solutions.

In this section, we complete the square to solve the general quadratic equation

$$ax^2 + bx + c = 0,$$

where a, b, and c are complex numbers and $a \neq 0$. The solution of this general equation gives a formula for finding the solution of *any* specific quadratic equation.

OBJECTIVE 1 Derive the quadratic formula. To solve the general quadratic equation $ax^2 + bx + c = 0$ by completing the square (assuming $a > 0$), we follow the steps given in **Section 10.1.**

$$ax^2 + bx + c = 0$$

$$x^2 + \frac{b}{a}x + \frac{c}{a} = 0 \qquad \text{Divide by } a. \text{ (Step 1)}$$

$$x^2 + \frac{b}{a}x = -\frac{c}{a} \qquad \text{Subtract } \tfrac{c}{a}. \text{ (Step 2)}$$

$$\left[\frac{1}{2}\left(\frac{b}{a}\right)\right]^2 = \left(\frac{b}{2a}\right)^2 = \frac{b^2}{4a^2} \qquad \text{(Step 3)}$$

$$x^2 + \frac{b}{a}x + \frac{b^2}{4a^2} = -\frac{c}{a} + \frac{b^2}{4a^2} \qquad \text{Add } \frac{b^2}{4a^2} \text{ to each side. (Step 4)}$$

Write the left side as a perfect square, and rearrange the right side.

$$\left(x + \frac{b}{2a}\right)^2 = \frac{b^2}{4a^2} + \frac{-c}{a} \qquad \text{(Step 5)}$$

$$\left(x + \frac{b}{2a}\right)^2 = \frac{b^2}{4a^2} + \frac{-4ac}{4a^2} \qquad \text{Write with a common denominator.}$$

$$\left(x + \frac{b}{2a}\right)^2 = \frac{b^2 - 4ac}{4a^2} \qquad \text{Add fractions.}$$

$$x + \frac{b}{2a} = \sqrt{\frac{b^2 - 4ac}{4a^2}} \quad \text{or} \quad x + \frac{b}{2a} = -\sqrt{\frac{b^2 - 4ac}{4a^2}} \qquad \text{Square root property (Step 6)}$$

Since

$$\sqrt{\frac{b^2 - 4ac}{4a^2}} = \frac{\sqrt{b^2 - 4ac}}{\sqrt{4a^2}} = \frac{\sqrt{b^2 - 4ac}}{2a},$$

the right sides of these equations can be expressed as

$$x + \frac{b}{2a} = \frac{\sqrt{b^2 - 4ac}}{2a} \quad \text{or} \quad x + \frac{b}{2a} = \frac{-\sqrt{b^2 - 4ac}}{2a}$$

$$x = \frac{-b}{2a} + \frac{\sqrt{b^2 - 4ac}}{2a} \quad \text{or} \quad x = \frac{-b}{2a} - \frac{\sqrt{b^2 - 4ac}}{2a}$$

$$\mathbf{x = \frac{-b + \sqrt{b^2 - 4ac}}{2a}} \quad \text{or} \quad \mathbf{x = \frac{-b - \sqrt{b^2 - 4ac}}{2a}}.$$

If $a < 0$, the same two solutions are obtained. The result is the **quadratic formula,** which is abbreviated as shown on the next page.

Quadratic Formula

The solutions of $ax^2 + bx + c = 0$ $(a \neq 0)$ are given by

$$x = \frac{-b \pm \sqrt{b^2 - 4ac}}{2a}.$$

CAUTION

In the quadratic formula, $x = \frac{-b \pm \sqrt{b^2 - 4ac}}{2a}$, ***the square root is added to or subtracted from the value of $-b$ before dividing by $2a$.*** A common student error is to fail to divide ***all*** of the expression $-b \pm \sqrt{b^2 - 4ac}$ by $2a$.

OBJECTIVE 2 Solve quadratic equations by using the quadratic formula. To use the quadratic formula, first write the equation in standard form

$$ax^2 + bx + c = 0.$$

Then identify the values of a, b, and c and substitute them into the quadratic formula.

◀ *Work Problem* **1** *at the Side.*

1 Identify the values of a, b, and c. (*Hint:* If necessary, first write the equation in standard form with 0 on the right side.) *Do not actually solve.*

(a) $-3x^2 + 9x - 4 = 0$

(b) $3x^2 = 6x + 2$

EXAMPLE 1 Using the Quadratic Formula (Rational Solutions)

Solve $6x^2 - 5x - 4 = 0$.

Here a, the coefficient of the second-degree term, is 6, while b, the coefficient of the first-degree term, is -5, and the constant c is -4. Substitute these values into the quadratic formula.

$$x = \frac{-b \pm \sqrt{b^2 - 4ac}}{2a} \quad \text{Quadratic formula}$$

$$x = \frac{-(-5) \pm \sqrt{(-5)^2 - 4(6)(-4)}}{2(6)} \quad a = 6,\ b = -5,\ c = -4$$

Use parentheses and substitute carefully to avoid errors.

$$x = \frac{5 \pm \sqrt{25 + 96}}{12}$$

$$x = \frac{5 \pm \sqrt{121}}{12}$$

$$x = \frac{5 \pm 11}{12}$$

This last statement leads to two solutions, one from the + sign and one from the − sign.

$$x = \frac{5 + 11}{12} = \frac{16}{12} = \frac{4}{3} \quad \text{or} \quad x = \frac{5 - 11}{12} = \frac{-6}{12} = -\frac{1}{2}$$

Check each solution in the original equation. The solution set is $\left\{-\frac{1}{2}, \frac{4}{3}\right\}$.

◀ *Work Problem* **2** *at the Side.*

2 Solve $4x^2 - 11x - 3 = 0$ using the quadratic formula.

ANSWERS

1. (a) -3; 9; -4 **(b)** 3; -6; -2

2. $\left\{-\frac{1}{4}, 3\right\}$

We could have used factoring to solve the equation in Example 1.

$$6x^2 - 5x - 4 = 0$$

$$(3x - 4)(2x + 1) = 0 \quad \text{Factor.}$$

$$3x - 4 = 0 \quad \text{or} \quad 2x + 1 = 0 \quad \text{Zero-factor property}$$

$$3x = 4 \quad \text{or} \quad 2x = -1 \quad \text{Solve each equation.}$$

$$x = \frac{4}{3} \quad \text{or} \quad x = -\frac{1}{2} \quad \text{Same solutions as in Example 1}$$

When solving quadratic equations, it is a good idea to try factoring first. If the equation cannot be factored or if factoring is difficult, then use the quadratic formula. Later in this section, we will show a way to determine whether factoring can be used to solve a quadratic equation.

EXAMPLE 2 Using the Quadratic Formula (Irrational Solutions)

Solve $4x^2 = 8x - 1$.

Write the equation in standard form as $4x^2 - 8x + 1 = 0$.

$$x = \frac{-b \pm \sqrt{b^2 - 4ac}}{2a} \quad \text{Quadratic formula}$$

$$x = \frac{-(-8) \pm \sqrt{(-8)^2 - 4(4)(1)}}{2(4)} \quad a = 4,\ b = -8,\ c = 1$$

$$x = \frac{8 \pm \sqrt{64 - 16}}{8}$$

$$x = \frac{8 \pm \sqrt{48}}{8}$$

$$x = \frac{8 \pm 4\sqrt{3}}{8} \quad \sqrt{48} = \sqrt{16} \cdot \sqrt{3} = 4\sqrt{3}$$

$$x = \frac{4\left(2 \pm \sqrt{3}\right)}{4(2)} \quad \text{Factor.}$$

Factor first; then divide out the common factor.

$$x = \frac{2 \pm \sqrt{3}}{2} \quad \text{Lowest terms}$$

The solution set is $\left\{\frac{2 + \sqrt{3}}{2}, \frac{2 - \sqrt{3}}{2}\right\}$.

CAUTION

1. ***Every quadratic equation must be written in standard form $ax^2 + bx + c = 0$ before we begin to solve it,*** whether we use factoring or the quadratic formula.
2. ***When writing solutions in lowest terms, be sure to factor first; then divide out the common factor,*** as shown in the last two steps in Example 2.

Work Problem 3 at the Side. ▶

3 Solve each equation using the quadratic formula.

(a) $6x^2 + 4x - 1 = 0$

(b) $2x^2 + 19 = 14x$

ANSWERS

3. (a) $\left\{\frac{-2 + \sqrt{10}}{6}, \frac{-2 - \sqrt{10}}{6}\right\}$

(b) $\left\{\frac{7 + \sqrt{11}}{2}, \frac{7 - \sqrt{11}}{2}\right\}$

4 Solve each equation using the quadratic formula.

(a) $x^2 + x + 1 = 0$

(b) $(x + 2)(x - 6) = -17$

EXAMPLE 3 Using the Quadratic Formula (Nonreal Complex Solutions)

Solve $(9x + 3)(x - 1) = -8$.

To write this equation in standard form, we first multiply and collect all nonzero terms on the left.

$$(9x + 3)(x - 1) = -8$$

$$9x^2 - 6x - 3 = -8 \qquad \text{Multiply.}$$

$$9x^2 - 6x + 5 = 0 \qquad \text{Add 8.}$$

From the equation $9x^2 - 6x + 5 = 0$, we identify $a = 9$, $b = -6$, and $c = 5$.

$$x = \frac{-b \pm \sqrt{b^2 - 4ac}}{2a} \qquad \text{Quadratic formula}$$

$$x = \frac{-(-6) \pm \sqrt{(-6)^2 - 4(9)(5)}}{2(9)} \qquad \text{Substitute.}$$

$$x = \frac{6 \pm \sqrt{-144}}{18}$$

$$x = \frac{6 \pm 12i}{18} \qquad \sqrt{-144} = 12i$$

$$x = \frac{6(1 \pm 2i)}{6(3)} \qquad \text{Factor.}$$

$$x = \frac{1 \pm 2i}{3} \qquad \text{Lowest terms}$$

$$x = \frac{1}{3} \pm \frac{2}{3}i \qquad \text{Standard form } a + bi \text{ for a complex number}$$

The solution set is $\left\{\frac{1}{3} + \frac{2}{3}i, \frac{1}{3} - \frac{2}{3}i\right\}$.

◀ Work Problem 4 at the Side.

OBJECTIVE 3 Use the discriminant to determine the number and type of solutions. The solutions of the quadratic equation $ax^2 + bx + c = 0$ are given by

$$x = \frac{-b \pm \sqrt{b^2 - 4ac}}{2a}. \quad \leftarrow \text{Discriminant}$$

If a, b, and c are integers, the type of solutions of a quadratic equation—that is, rational, irrational, or nonreal complex—is determined by the expression under the radical sign, $b^2 - 4ac$. Because it distinguishes among the three types of solutions, $b^2 - 4ac$ is called the *discriminant*. By calculating the discriminant before solving a quadratic equation, we can predict whether the solutions will be rational numbers, irrational numbers, or nonreal complex numbers.

ANSWERS

4. (a) $\left\{-\frac{1}{2} + \frac{\sqrt{3}}{2}i, -\frac{1}{2} - \frac{\sqrt{3}}{2}i\right\}$

(b) $\{2 + i, 2 - i\}$

Discriminant

The **discriminant** of $ax^2 + bx + c = 0$ is $\mathbf{b^2 - 4ac}$. If a, b, and c are integers, then the number and type of solutions are determined as follows.

Discriminant	Number and Type of Solutions
Positive, and the square of an integer	Two rational solutions
Positive, but not the square of an integer	Two irrational solutions
Zero	One rational solution
Negative	Two nonreal complex solutions

Calculating the discriminant can also help you decide whether to solve a quadratic equation by factoring or by using the quadratic formula. ***If the discriminant is a perfect square (including 0), then the equation can be solved by factoring. Otherwise, the quadratic formula (or completing the square) should be used.***

EXAMPLE 4 **Using the Discriminant**

Find the discriminant. Use it to predict the number and type of solutions for each equation. Tell whether the equation can be solved by factoring or whether the quadratic formula should be used.

(a) $6x^2 - x - 15 = 0$

First identify the values of a, b, and c. Because $-x = -1x$, the value of b is -1. We find the discriminant by evaluating $b^2 - 4ac$.

$$\begin{aligned} & b^2 - 4ac \\ &= (-1)^2 - 4(6)(-15) \qquad a = 6, b = -1, c = -15 \\ &= 1 + 360 \\ &= 361 \end{aligned}$$

Use parentheses and substitute carefully.

A calculator shows that $361 = 19^2$, a perfect square. Since a, b, and c are integers and the discriminant is a perfect square, there will be two rational solutions and the equation can be solved by factoring.

(b) $3x^2 - 4x = 5$

Write the equation in standard form as

$$3x^2 - 4x - 5 = 0 \qquad \text{Subtract 5.}$$

to find $a = 3$, $b = -4$, and $c = -5$. The discriminant is

$$\begin{aligned} & b^2 - 4ac \\ &= (-4)^2 - 4(3)(-5) \\ &= 16 + 60 \\ &= 76. \end{aligned}$$

Because 76 is positive but *not* the square of an integer and a, b, and c are integers, the equation will have two irrational solutions and is best solved using the quadratic formula.

Continued on Next Page

(c) $4x^2 + x + 1 = 0$

Since $a = 4$, $b = 1$, and $c = 1$, the discriminant is

$$\begin{aligned} & b^2 - 4ac \\ &= 1^2 - 4(4)(1) \\ &= 1 - 16 \\ &= -15. \end{aligned}$$

Since the discriminant is negative and a, b, and c are integers, this quadratic equation will have two nonreal complex solutions. The quadratic formula should be used to solve it.

(d) $4x^2 + 9 = 12x$

Write the equation in standard form as

$$4x^2 - 12x + 9 = 0 \quad \text{Subtract } 12x.$$

to find $a = 4$, $b = -12$, and $c = 9$. The discriminant is

$$\begin{aligned} & b^2 - 4ac \\ &= (-12)^2 - 4(4)(9) \\ &= 144 - 144 \\ &= 0. \end{aligned}$$

Because the discriminant is 0, the quantity under the radical in the quadratic formula is 0, and there is only one rational solution. Again, the equation can be solved by factoring.

◀ *Work Problem* **5** *at the Side.*

Note

In **Section 10.6** we will see how the discriminant can be used to determine the number of x-intercepts of the graph of a quadratic function.

5 Find the discriminant. Use it to predict the number and type of solutions for each equation.

(a) $2x^2 + 3x = 4$

(b) $2x^2 + 3x + 4 = 0$

(c) $x^2 + 20x + 100 = 0$

(d) $15x^2 + 11x = 14$

(e) Which of the equations in parts (a)–(d) can be solved by factoring?

Answers

5. (a) 41; two; irrational
(b) -23; two; nonreal complex
(c) 0; one; rational
(d) 961; two; rational **(e)** (c) and (d)

10.2 Exercises

1. A student wrote the following as the quadratic formula for solving $ax^2 + bx + c = 0, a \neq 0$:

$$x = -b \pm \frac{\sqrt{b^2 - 4ac}}{2a}.$$

This is incorrect. ***WHAT WENT WRONG?***

2. A student attempted to solve the equation $5x^2 - 5x + 1 = 0$ as follows.

$$x = \frac{5 \pm \sqrt{25 - 4(5)(1)}}{2(5)} \qquad a = 5, b = -5, c = 1$$

$$x = \frac{5 \pm \sqrt{5}}{10}$$

$$x = \frac{1}{2} \pm \sqrt{5}$$

This is incorrect. ***WHAT WENT WRONG?***

Use the quadratic formula to solve each equation. (All solutions for these equations are real numbers.) See Examples 1 and 2.

3. $x^2 - 8x + 15 = 0$
4. $x^2 + 3x - 28 = 0$
5. $2x^2 + 4x + 1 = 0$
6. $2x^2 + 3x - 1 = 0$
7. $2x^2 - 2x = 1$
8. $9x^2 + 6x = 1$
9. $x^2 + 18 = 10x$
10. $x^2 - 4 = 2x$
11. $4k^2 + 4k - 1 = 0$
12. $4r^2 - 4r - 19 = 0$
13. $2 - 2x = 3x^2$
14. $26r - 2 = 3r^2$

15. $\frac{x^2}{4} - \frac{x}{2} = 1$
(*Hint:* First clear the fractions.)

16. $p^2 + \frac{p}{3} = \frac{1}{6}$
(*Hint:* First clear the fractions.)

17. $-2t(t + 2) = -3$

18. $-3x(x + 2) = -4$

19. $(r - 3)(r + 5) = 2$

20. $(k + 1)(k - 7) = 1$

Use the quadratic formula to solve each equation. (All solutions for these equations are nonreal complex numbers.) See Example 3.

21. $x^2 - 3x + 17 = 0$

22. $x^2 - 5x + 20 = 0$

23. $r^2 - 6r + 14 = 0$

24. $t^2 + 4t + 11 = 0$

25. $4x^2 - 4x = -7$

26. $9x^2 - 6x = -7$

27. $x(3x + 4) = -2$

28. $p(2p + 3) = -2$

Use the discriminant to determine whether the solutions for each equation are

A. *two rational numbers,* **B.** *one rational number,*
C. *two irrational numbers,* **D.** *two nonreal complex numbers.*

Do not actually solve. See Example 4.

29. $25x^2 + 70x + 49 = 0$

30. $4k^2 - 28k + 49 = 0$

31. $x^2 + 4x + 2 = 0$

32. $9x^2 - 12x - 1 = 0$

33. $3x^2 = 5x + 2$

34. $4x^2 = 4x + 3$

35. $3m^2 - 10m + 15 = 0$

36. $18x^2 + 60x + 82 = 0$

37. Using the discriminant, which equations in Exercises 29–36 can be solved by factoring?

38. Based on your answer in Exercise 37, solve the equation given in each exercise.
(a) Exercise 29 **(b)** Exercise 33

10.3 Equations Quadratic in Form

OBJECTIVES

1. **Solve an equation with fractions by writing it in quadratic form.**
2. **Use quadratic equations to solve applied problems.**
3. **Solve an equation with radicals by writing it in quadratic form.**
4. **Solve an equation that is quadratic in form by substitution.**

OBJECTIVE 1 Solve an equation with fractions by writing it in quadratic form. A variety of nonquadratic equations can be written in the form of a quadratic equation and solved by using one of the methods from **Sections 10.1 and 10.2.**

EXAMPLE 1 Solving an Equation with Fractions That Leads to a Quadratic Equation

Solve $\frac{1}{x} + \frac{1}{x-1} = \frac{7}{12}$.

Clear fractions by multiplying each side by the least common denominator, $12x(x-1)$. (Note that the domain must be restricted to $x \neq 0$ and $x \neq 1$.)

$$12x(x-1)\left(\frac{1}{x} + \frac{1}{x+1}\right) = 12x(x-1)\left(\frac{7}{12}\right)$$

$$12x(x-1)\frac{1}{x} + 12x(x-1)\frac{1}{x-1} = 12x(x-1)\left(\frac{7}{12}\right) \quad \text{Distributive property}$$

$$12(x-1) + 12x = 7x(x-1)$$

$$12x - 12 + 12x = 7x^2 - 7x \quad \text{Distributive property}$$

$$24x - 12 = 7x^2 - 7x \quad \text{Combine like terms.}$$

$$7x^2 - 31x + 12 = 0 \quad \text{Standard form}$$

$$(7x-3)(x-4) = 0 \quad \text{Factor.}$$

$$7x - 3 = 0 \quad \text{or} \quad x - 4 = 0 \quad \text{Zero-factor property}$$

$$x = \frac{3}{7} \quad \text{or} \quad x = 4 \quad \text{Solve each equation.}$$

Check by substituting these solutions in the original equation. The solution set is $\left\{\frac{3}{7}, 4\right\}$.

Work Problem 1 at the Side. ▶

1 Solve each equation. Check your solutions.

(a) $\frac{5}{m} + \frac{12}{m^2} = 2$

(b) $\frac{2}{x} + \frac{1}{x-2} = \frac{5}{3}$

(c) $\frac{4}{m-1} + 9 = -\frac{7}{m}$

OBJECTIVE 2 Use quadratic equations to solve applied problems. In **Sections 2.4 and 8.5** we solved distance-rate-time (or motion) problems that led to linear equations or rational equations. Now we can extend that work to motion problems that lead to quadratic equations. We continue to use the six-step problem-solving method from **Section 2.3.**

EXAMPLE 2 Solving a Motion Problem

A riverboat for tourists averages 12 mph in still water. It takes the boat 1 hr, 4 min to go 6 mi upstream and return. Find the speed of the current.

Step 1 **Read** the problem carefully.

Step 2 **Assign a variable.** Let x = the speed of the current.

The current slows down the boat when it is going upstream, so the rate (or speed) upstream is the speed of the boat in still water less the speed of the current, or $(12 - x)$ mph. See Figure 1 on the next page.

Continued on Next Page

ANSWERS

1. (a) $\left\{-\frac{3}{2}, 4\right\}$ (b) $\left\{\frac{4}{5}, 3\right\}$ (c) $\left\{\frac{7}{9}, -1\right\}$

Riverboat traveling *upstream*—the current slows it down.

Figure 1

Similarly, the current speeds up the boat as it travels downstream, so its speed downstream is $(12 + x)$ mph. Thus,

$$12 - x = \text{the rate upstream in miles per hour;}$$
$$12 + x = \text{the rate downstream in miles per hour.}$$

This information can be used to complete a table. We use the distance formula, $d = rt$, solved for time t, $t = \frac{d}{r}$, to write expressions for t.

	d	r	t
Upstream	6	$12 - x$	$\frac{6}{12 - x}$
Downstream	6	$12 + x$	$\frac{6}{12 + x}$

Times in hours

Step 3 **Write an equation.** The total time, 1 hr and 4 min, can be written as

$$1 + \frac{4}{60} = 1 + \frac{1}{15} = \frac{16}{15} \text{ hr.}$$

Because the time upstream plus the time downstream equals $\frac{16}{15}$ hr,

Time upstream + time downstream = total time.

$$\frac{6}{12 - x} + \frac{6}{12 + x} = \frac{16}{15}.$$

Step 4 **Solve** the equation. Multiply each side by $15(12 - x)(12 + x)$, the LCD, and solve the resulting quadratic equation.

$$15(12 + x)6 + 15(12 - x)6 = 16(12 - x)(12 + x)$$
$$90(12 + x) + 90(12 - x) = 16(144 - x^2)$$
$$1080 + 90x + 1080 - 90x = 2304 - 16x^2 \quad \text{Distributive property}$$
$$2160 = 2304 - 16x^2 \quad \text{Combine like terms.}$$
$$16x^2 = 144$$
$$x^2 = 9 \quad \text{Divide by 16.}$$
$$x = 3 \quad \text{or} \quad x = -3 \quad \text{Square root property}$$

Step 5 **State the answer.** The speed of the current cannot be -3, so the answer is 3 mph.

Step 6 **Check** that this value satisfies the original problem.

CAUTION

As shown in Example 2, when a quadratic equation is used to solve an applied problem, sometimes only *one* answer satisfies the application. ***Always check each answer in the words of the original problem.***

Work Problem **2** *at the Side.* ▶

Recall from **Section 8.5** that a person's work rate is $\frac{1}{t}$ part of the job per hour, where t is the time in hours required to do the complete job. Thus, the part of the job the person will do in x hours is $\frac{1}{t}x$.

EXAMPLE 3 Solving a Work Problem

It takes two carpet layers 4 hr to carpet a room. If each worked alone, one of them could do the job in 1 hr less time than the other. How long would it take each carpet layer to complete the job alone?

Step 1 **Read** the problem again. There will be two answers.

Step 2 **Assign a variable.** Let $x =$ the number of hours for the slower carpet layer to complete the job alone. Then the faster carpet layer could do the entire job in $(x - 1)$ hours. The slower person's rate is $\frac{1}{x}$, and the faster person's rate is $\frac{1}{x-1}$. Together, they can do the job in 4 hr. Complete a table as shown.

	Rate	Time Working Together	Fractional Part of the Job Done
Slower Worker	$\frac{1}{x}$	4	$\frac{1}{x}(4)$
Faster Worker	$\frac{1}{x-1}$	4	$\frac{1}{x-1}(4)$

Sum is 1 whole job.

Step 3 **Write an equation.** The sum of the fractional parts done by the workers should equal 1 (the whole job).

Part done by slower worker + part done by faster worker = 1 whole job.

$$\frac{4}{x} + \frac{4}{x-1} = 1$$

Step 4 **Solve** the equation. Multiply by $x(x - 1)$, the LCD.

$$x(x-1)\left(\frac{4}{x} + \frac{4}{x-1}\right) = x(x-1)(1) \quad \text{Multiply by the LCD.}$$

$$4(x-1) + 4x = x(x-1) \quad \text{Distributive property}$$

$$4x - 4 + 4x = x^2 - x \quad \text{Distributive property}$$

$$x^2 - 9x + 4 = 0 \quad \text{Standard form}$$

Continued on Next Page

2 Solve each problem.

(a) In 4 hr, Kerrie can go 15 mi upriver and come back. The speed of the current is 5 mph. Complete this table.

	d	r	t
Up			
Down			

(b) Find the speed of the boat from part (a) in still water.

(c) In $1\frac{3}{4}$ hr, Ken rows his boat 5 mi upriver and comes back. The speed of the current is 3 mph. How fast does Ken row?

ANSWERS

2. (a) row 1: 15; $x - 5$; $\frac{15}{x-5}$
row 2: 15; $x + 5$; $\frac{15}{x+5}$
(b) 10 mph (c) 7 mph

3 Solve each problem. Round answers to the nearest tenth.

(a) Carlos can complete a certain lab test in 2 hr less time than Jaime can. If they can finish the job together in 2 hr, how long would it take each of them working alone?

	Rate	Time Working Together	Fractional Part of the Job Done
Carlos			
Jaime			

(b) Two chefs are preparing a banquet. One chef could prepare the banquet in 2 hr less time than the other. Together, they complete the job in 5 hr. How long would it take the faster chef working alone?

Answers

3. (a) Jaime: 5.2 hr; Carlos: 3.2 hr **(b)** 9.1 hr

The resulting equation $x^2 - 9x + 4 = 0$ cannot be solved by factoring, so use the quadratic formula.

$$x = \frac{-b \pm \sqrt{b^2 - 4ac}}{2a} \qquad \text{Quadratic formula}$$

$$x = \frac{-(-9) \pm \sqrt{(-9)^2 - 4(1)(4)}}{2(1)} \qquad a = 1, b = -9, c = 4$$

$$x = \frac{9 \pm \sqrt{65}}{2} \qquad \text{Simplify.}$$

$$x = \frac{9 + \sqrt{65}}{2} \approx 8.5 \quad \text{or} \quad x = \frac{9 - \sqrt{65}}{2} \approx 0.5 \qquad \text{Use a calculator.}$$

Step 5 **State the answer.** Only the solution 8.5 makes sense in the original problem. (Why?) Thus, the slower worker can do the job in about 8.5 hr and the faster in about $8.5 - 1 = 7.5$ hr.

Step 6 **Check** that these results satisfy the original problem.

◀ Work Problem **3** at the Side.

OBJECTIVE 3 Solve an equation with radicals by writing it in quadratic form.

EXAMPLE 4 Solving Radical Equations That Lead to Quadratic Equations

Solve each equation.

(a) $k = \sqrt{6k - 8}$

This equation is not quadratic. However, squaring both sides of the equation gives a quadratic equation that can be solved by factoring.

$$k^2 = 6k - 8 \qquad \text{Square each side.}$$

$$k^2 - 6k + 8 = 0 \qquad \text{Standard form}$$

$$(k - 4)(k - 2) = 0 \qquad \text{Factor.}$$

$$k - 4 = 0 \quad \text{or} \quad k - 2 = 0 \qquad \text{Zero-factor property}$$

$$k = 4 \quad \text{or} \quad k = 2 \qquad \text{Potential solutions}$$

Recall from **Section 9.6** that squaring both sides of a radical equation can introduce extraneous solutions that do not satisfy the original equation. ***All proposed solutions must be checked in the original (not the squared) equation.***

If $k = 4$, then

$$k = \sqrt{6k - 8}$$

$$4 \stackrel{?}{=} \sqrt{6(4) - 8}$$

$$4 \stackrel{?}{=} \sqrt{16}$$

$$4 = 4. \qquad \text{True}$$

If $k = 2$, then

$$k = \sqrt{6k - 8}$$

$$2 \stackrel{?}{=} \sqrt{6(2) - 8}$$

$$2 \stackrel{?}{=} \sqrt{4}$$

$$2 = 2. \qquad \text{True}$$

Both solutions check, so the solution set is $\{2, 4\}$.

Continued on Next Page

(b) $x + \sqrt{x} = 6$

$(a - b)^2 = a^2 - 2ab + b^2$

$\sqrt{x} = 6 - x$ Isolate the radical on one side.

$x = 36 - 12x + x^2$ Square each side.

$0 = x^2 - 13x + 36$ Standard form

$0 = (x - 4)(x - 9)$ Factor.

$x - 4 = 0$ or $x - 9 = 0$ Zero-factor property

$x = 4$ or $x = 9$ Proposed solutions

Check both proposed solutions in the *original* equation.

If $x = 4$, then	If $x = 9$, then
$x + \sqrt{x} = 6$	$x + \sqrt{x} = 6$
$4 + \sqrt{4} \stackrel{?}{=} 6$	$9 + \sqrt{9} \stackrel{?}{=} 6$
$6 = 6.$ True	$12 = 6.$ False

Only the solution 4 checks, so the solution set is $\{4\}$.

Work Problem **4** at the Side. ▶

OBJECTIVE 4 Solve an equation that is quadratic in form by substitution. A nonquadratic equation that can be written in the form

$$au^2 + bu + c = 0,$$

for $a \neq 0$ and an algebraic expression u, is called **quadratic in form.**

Many equations that are quadratic in form can be solved more easily by defining and substituting a "temporary" variable u for an expression involving the variable in the original equation. The first step is to define this temporary variable u.

EXAMPLE 5 Defining Substitution Variables

Define a variable u, and write each equation in the form $au^2 + bu + c = 0$.

(a) $x^4 - 13x^2 + 36 = 0$

Look at the two terms involving the variable x, ignoring their coefficients. Try to find one variable expression that is the square of the other. We see that $x^4 = (x^2)^2$, so we define $u = x^2$, and rewrite the original equation as the quadratic equation

$$u^2 - 13u + 36 = 0. \quad \text{Here, } u = x^2.$$

(b) $2(4m - 3)^2 + 7(4m - 3) + 5 = 0$

Because this equation involves both $(4m - 3)^2$ and $(4m - 3)$, choose $u = 4m - 3$. Substituting u for $4m - 3$ gives the quadratic equation

$$2u^2 + 7u + 5 = 0. \quad \text{Here, } u = 4m - 3.$$

(c) $2x^{2/3} - 11x^{1/3} + 12 = 0$

Here we apply one of the power rules for exponents from **Section 6.1:** $(a^m)^n = a^{mn}$. Because $(x^{1/3})^2 = x^{2/3}$, we define $u = x^{1/3}$. With this substitution, the original equation becomes

$$2u^2 - 11u + 12 = 0. \quad \text{Here, } u = x^{1/3}.$$

Work Problem **5** at the Side. ▶

4 Solve each equation. Check your solutions.

(a) $x = \sqrt{7x - 10}$

(b) $2x = \sqrt{x} + 1$

5 Define a variable u, and write each equation in the form $au^2 + bu + c = 0$.

(a) $2x^4 + 5x^2 - 12 = 0$

(b) $2(x + 5)^2 - 7(x + 5) + 6 = 0$

(c) $x^{4/3} - 8x^{2/3} + 16 = 0$

Answers

4. (a) $\{2, 5\}$ **(b)** $\{1\}$

5. (a) $u = x^2$; $2u^2 + 5u - 12 = 0$
(b) $u = x + 5$; $2u^2 - 7u + 6 = 0$
(c) $u = x^{2/3}$; $u^2 - 8u + 16 = 0$

EXAMPLE 6 Solving Equations That Are Quadratic in Form

Solve each equation.

(a) $x^4 - 13x^2 + 36 = 0$

From Example 5(a), we write this equation in quadratic form by substituting u for x^2.

$$x^4 - 13x^2 + 36 = 0$$

$$(x^2)^2 - 13x^2 + 36 = 0 \qquad x^4 = (x^2)^2$$

$$u^2 - 13u + 36 = 0 \qquad \text{Let } u = x^2.$$

$$(u - 4)(u - 9) = 0 \qquad \text{Factor.}$$

$$u - 4 = 0 \quad \text{or} \quad u - 9 = 0 \qquad \text{Zero-factor property}$$

Don't stop here.

$$u = 4 \quad \text{or} \quad u = 9 \qquad \text{Solve.}$$

$$x^2 = 4 \quad \text{or} \quad x^2 = 9 \qquad \text{Substitute } x^2 \text{ for } u.$$

$$x = \pm 2 \quad \text{or} \quad x = \pm 3 \qquad \text{Square root property}$$

The equation $x^4 - 13x^2 + 36 = 0$, a fourth-degree equation, has four solutions.* The solution set is $\{-3, -2, 2, 3\}$. Check each of the four solutions by substitution.

(b)

$$4x^4 + 1 = 5x^2$$

$$4(x^2)^2 + 1 = 5x^2 \qquad x^4 = (x^2)^2$$

$$4u^2 + 1 = 5u \qquad \text{Let } u = x^2.$$

$$4u^2 - 5u + 1 = 0 \qquad \text{Standard form}$$

$$(4u - 1)(u - 1) = 0 \qquad \text{Factor.}$$

$$4u - 1 = 0 \quad \text{or} \quad u - 1 = 0 \qquad \text{Zero-factor property}$$

$$u = \frac{1}{4} \quad \text{or} \quad u = 1 \qquad \text{Solve.}$$

This is a key step.

$$x^2 = \frac{1}{4} \quad \text{or} \quad x^2 = 1 \qquad \text{Substitute } x^2 \text{ for } u.$$

$$x = \pm\frac{1}{2} \quad \text{or} \quad x = \pm 1 \qquad \text{Square root property}$$

Check that the solution set is $\left\{-1, -\frac{1}{2}, \frac{1}{2}, 1\right\}$.

(c) $x^4 = 6x^2 - 3$

First write the equation as

$$x^4 - 6x^2 + 3 = 0 \quad \text{or} \quad (x^2)^2 - 6x^2 + 3 = 0,$$

which is quadratic in form with $u = x^2$. Substitute u for x^2 and u^2 for x^4 to get

$$u^2 - 6u + 3 = 0.$$

Since this equation cannot be solved by factoring, use the quadratic formula.

Continued on Next Page

*In general, an equation in which an nth-degree polynomial equals 0 has n solutions, although some of them may be repeated.

$$u = \frac{-(-6) \pm \sqrt{(-6)^2 - 4(1)(3)}}{2(1)} \qquad a = 1, b = -6, c = 3$$

$$u = \frac{6 \pm \sqrt{24}}{2} \qquad \text{Simplify.}$$

$$u = \frac{6 \pm 2\sqrt{6}}{2} \qquad \sqrt{24} = \sqrt{4} \cdot \sqrt{6} = 2\sqrt{6}$$

$$u = \frac{2(3 \pm \sqrt{6})}{2} \qquad \text{Factor.}$$

$$u = 3 \pm \sqrt{6} \qquad \text{Lowest terms}$$

$$x^2 = 3 + \sqrt{6} \quad \text{or} \quad x^2 = 3 - \sqrt{6} \qquad \text{Substitute } x^2 \text{ for } u.$$

Find *both* square roots in each case.

$$x = \pm\sqrt{3 + \sqrt{6}} \quad \text{or} \quad x = \pm\sqrt{3 - \sqrt{6}} \qquad \text{Square root property}$$

The solution set contains four numbers:

$$\left\{\sqrt{3 + \sqrt{6}}, -\sqrt{3 + \sqrt{6}}, \sqrt{3 - \sqrt{6}}, -\sqrt{3 - \sqrt{6}}\right\}.$$

Note

Some students prefer to solve equations like those in Examples 6(a) and (b) by factoring directly. For example,

$$x^4 - 13x^2 + 36 = 0 \qquad \text{Example 6(a) equation}$$

$$(x^2 - 9)(x^2 - 4) = 0 \qquad \text{Factor.}$$

$$(x + 3)(x - 3)(x + 2)(x - 2) = 0. \qquad \text{Factor again.}$$

Using the zero-factor property gives the same solutions obtained in Example 6(a). Equations that cannot be solved by factoring (as in Example 6(c)) must be solved by substitution and the quadratic formula.

Work Problem 6 *at the Side.* ▶

The method used in Example 6 can be generalized.

Solving an Equation That Is Quadratic in Form by Substitution

Step 1 Define a temporary variable u, based on the relationship between the variable expressions in the given equation. Substitute u in the original equation and rewrite the equation in the form $au^2 + bu + c = 0$.

Step 2 **Solve the quadratic equation obtained in Step 1** by factoring or the quadratic formula.

Step 3 **Replace u with the expression it defined in Step 1.**

Step 4 **Solve the resulting equations for the original variable.**

Step 5 **Check** all solutions by substituting them in the original equation.

6 Solve each equation. Check your solutions.

(a) $m^4 - 10m^2 + 9 = 0$

(b) $9k^4 - 37k^2 + 4 = 0$

(c) $x^4 - 4x^2 = -2$

Answers

6. (a) $\{-3, -1, 1, 3\}$ (b) $\left\{-2, -\frac{1}{3}, \frac{1}{3}, 2\right\}$
(c) $\left\{\sqrt{2 + \sqrt{2}}, -\sqrt{2 + \sqrt{2}}, \sqrt{2 - \sqrt{2}}, -\sqrt{2 - \sqrt{2}}\right\}$

7 Solve each equation. Check your solutions.

(a) $5(r+3)^2 + 9(r+3) = 2$

(b) $4m^{2/3} = 3m^{1/3} + 1$

EXAMPLE 7 Solving Equations That Are Quadratic in Form

Solve each equation.

(a) $2(4m-3)^2 + 7(4m-3) + 5 = 0$

Step 1 Because of the repeated quantity $4m - 3$, substitute u for $4m - 3$ as in Example 5(b).

$$2(4m-3)^2 + 7(4m-3) + 5 = 0$$

$$2u^2 + 7u + 5 = 0 \quad \text{Let } u = 4m - 3.$$

Step 2

$$(2u+5)(u+1) = 0 \quad \text{Factor.}$$

$$2u + 5 = 0 \quad \text{or} \quad u + 1 = 0 \quad \text{Zero-factor property}$$

Don't stop here.

$$u = -\frac{5}{2} \quad \text{or} \quad u = -1 \quad \text{Solve for } u.$$

Step 3

$$4m - 3 = -\frac{5}{2} \quad \text{or} \quad 4m - 3 = -1 \quad \text{Substitute } 4m - 3 \text{ for } u.$$

Step 4

$$4m = \frac{1}{2} \quad \text{or} \quad 4m = 2 \quad \text{Solve for } m.$$

$$m = \frac{1}{8} \quad \text{or} \quad m = \frac{1}{2}$$

Step 5 Check that the solution set of the original equation is $\left\{\frac{1}{8}, \frac{1}{2}\right\}$.

(b) $2x^{2/3} - 11x^{1/3} + 12 = 0$

From Example 5(c), substitute u for $x^{1/3}$.

$$2u^2 - 11u + 12 = 0 \quad \text{Let } x^{1/3} = u;\ x^{2/3} = u^2.$$

$$(2u-3)(u-4) = 0 \quad \text{Factor.}$$

$$2u - 3 = 0 \quad \text{or} \quad u - 4 = 0 \quad \text{Zero-factor property}$$

$$u = \frac{3}{2} \quad \text{or} \quad u = 4 \quad \text{Solve for } u.$$

$$x^{1/3} = \frac{3}{2} \quad \text{or} \quad x^{1/3} = 4 \quad u = x^{1/3}$$

$$(x^{1/3})^3 = \left(\frac{3}{2}\right)^3 \quad \text{or} \quad (x^{1/3})^3 = 4^3 \quad \text{Cube each side.}$$

$$x = \frac{27}{8} \quad \text{or} \quad x = 64$$

Check that the solution set is $\left\{\frac{27}{8}, 64\right\}$.

CAUTION

A common error when solving problems like those in Examples 6 and 7 is to stop too soon. ***Once you have solved for u, remember to substitute and solve for the values of the original variable.*** Keep in mind that u is just a temporary variable that helps you solve the given equation. As in any equation, you must solve for the variable in the *original* equation.

◀ Work Problem 7 at the Side.

ANSWERS

7. (a) $\left\{-5, -\frac{14}{5}\right\}$ (b) $\left\{-\frac{1}{64}, 1\right\}$

10.3 Exercises

FOR EXTRA HELP

MyMathLab

Math XL PRACTICE   WATCH   DOWNLOAD   READ   REVIEW

Based on the discussion and examples of this section, write a sentence describing the first step you would take to solve each equation. Do not actually solve.

1. $\frac{14}{x} = x - 5$

2. $\sqrt{1 + x} + x = 5$

3. $(r^2 + r)^2 - 8(r^2 + r) + 12 = 0$

4. $3t = \sqrt{16 - 10t}$

5. Read this incorrect "solution" carefully. ***WHAT WENT WRONG?***

$$x = \sqrt{3x + 4}$$
$$x^2 = 3x + 4 \quad \text{Square both sides.}$$
$$x^2 - 3x - 4 = 0$$
$$(x - 4)(x + 1) = 0$$
$$x - 4 = 0 \quad \text{or} \quad x + 1 = 0$$
$$x = 4 \quad \text{or} \quad x = -1$$

Solution set: $\{4, -1\}$

6. Read this incorrect "solution" carefully. ***WHAT WENT WRONG?***

$$2(m - 1)^2 - 3(m - 1) + 1 = 0$$
$$2u^2 - 3u + 1 = 0 \quad \text{Let } u = m - 1.$$
$$(2u - 1)(u - 1) = 0$$
$$2u - 1 = 0 \quad \text{or} \quad u - 1 = 0$$
$$u = \frac{1}{2} \quad \text{or} \quad u = 1$$

Solution set: $\left\{\frac{1}{2}, 1\right\}$

Solve each equation. Check your solutions. See Example 1.

7. $1 - \frac{3}{x} - \frac{28}{x^2} = 0$

8. $4 - \frac{7}{r} - \frac{2}{r^2} = 0$

9. $3 - \frac{1}{t} = \frac{2}{t^2}$

10. $1 + \frac{2}{k} = \frac{3}{k^2}$

11. $\frac{1}{x} + \frac{2}{x + 2} = \frac{17}{35}$

12. $\frac{2}{m} + \frac{3}{m + 9} = \frac{11}{4}$

13. $\frac{2}{x + 1} + \frac{3}{x + 2} = \frac{7}{2}$

14. $\frac{4}{3 - p} + \frac{2}{5 - p} = \frac{26}{15}$

15. $\frac{3}{2x} - \frac{1}{2(x + 2)} = 1$

16. $\frac{4}{3x} - \frac{1}{2(x + 1)} = 1$

17. $\frac{6}{p} = 2 + \frac{p}{p + 1}$

18. $\frac{k}{2 - k} + \frac{2}{k} = 5$

19. A boat goes 20 mph in still water, and the rate of the current is t mph.

(a) What is the rate of the boat when it travels upstream?

(b) What is the rate of the boat when it travels downstream?

20. It takes m hours to grade a set of papers.

(a) What is the grader's rate (in job per hour)?

(b) How much of the job will the grader do in 2 hr?

Solve each problem. See Examples 2 and 3.

21. On a windy day Yoshiaki found that he could go 16 mi downstream and then 4 mi back upstream at top speed in a total of 48 min. What was the top speed of Yoshiaki's boat if the current was 15 mph?

	d	r	t
Upstream	4	$x - 15$	
Downstream	16		

22. Lekesha flew her plane for 6 hr at a constant speed. She traveled 810 mi with the wind, then turned around and traveled 720 mi against the wind. The wind speed was a constant 15 mph. Find the speed of the plane.

	d	r	t
With Wind	810		
Against Wind	720		

23. In Canada, Medicine Hat and Cranbrook are 300 km apart. Harry rides his Honda 20 km per hr faster than Yoshi rides his Yamaha. Find Harry's average speed if he travels from Cranbrook to Medicine Hat in $1\frac{1}{4}$ hr less time than Yoshi. (*Source: State Farm Road Atlas.*)

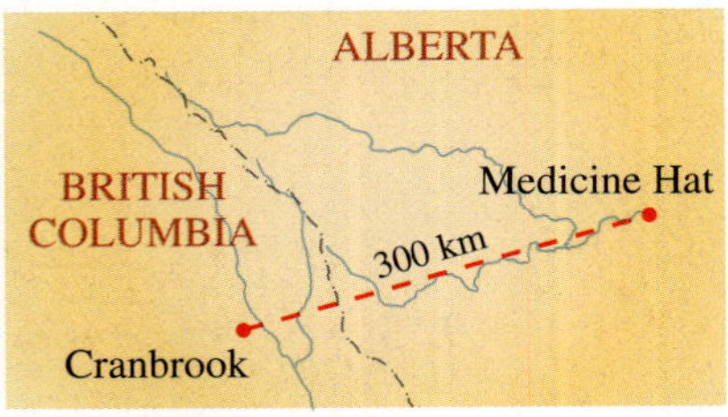

24. In California, the distance from Jackson to Lodi is about 40 mi, as is the distance from Lodi to Manteca. Rico drove from Jackson to Lodi during the rush hour, stopped in Lodi for a root beer, and then drove on to Manteca at 10 mph faster. Driving time for the entire trip was 88 min. Find his speed from Jackson to Lodi. (*Source: State Farm Road Atlas.*)

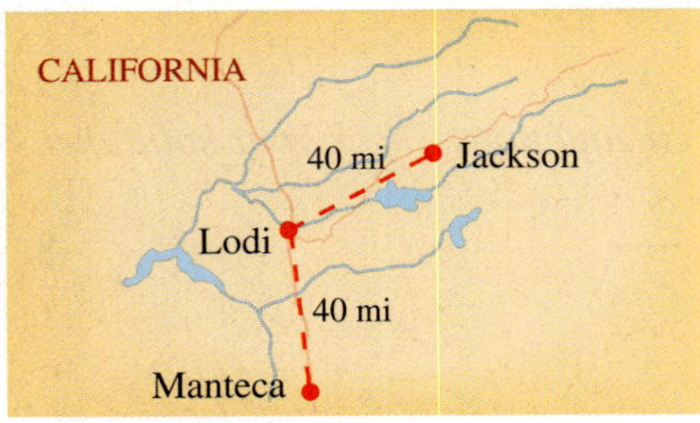

25. Working together, two people can cut a large lawn in 2 hr. One person can do the job alone in 1 hr less time than the other. How long (to the nearest tenth) would it take the faster person to do the job? (*Hint:* x is the time of the faster person.)

	Rate	Time Working Together	Fractional Part of the Job Done
Faster Worker	$\frac{1}{x}$	2	
Slower Worker		2	

26. A janitorial service provides two people to clean an office building. Working together, the two can clean the building in 5 hr. One person is new to the job and would take 2 hr longer than the other person to clean the building alone. How long (to the nearest tenth) would it take the new worker to clean the building alone?

	Rate	Time Working Together	Fractional Part of the Job Done
Faster Worker			
Slower Worker			

27. A washing machine can be filled in 6 min if both the hot and cold water taps are fully opened. Filling the washer with hot water alone takes 9 min longer than filling it with cold water alone. How long does it take to fill the washer with cold water?

28. Two pipes together can fill a large tank in 2 hr. One of the pipes, used alone, takes 3 hr longer than the other to fill the tank. How long would each pipe take to fill the tank alone?

Solve each equation. Check your solutions. See Example 4.

29. $z = \sqrt{5z - 4}$

30. $x = \sqrt{9x - 14}$

31. $2x = \sqrt{11x + 3}$

32. $4x = \sqrt{6x + 1}$

33. $3x = \sqrt{16 - 10x}$

34. $4t = \sqrt{8t + 3}$

35. $p - 2\sqrt{p} = 8$

36. $k + \sqrt{k} = 12$

37. $m = \sqrt{\dfrac{6 - 13m}{5}}$

38. $r = \sqrt{\dfrac{20 - 19r}{6}}$

Solve each equation. Check your solutions. See Examples 5–7.

39. $t^4 - 18t^2 + 81 = 0$

40. $x^4 - 8x^2 - 16 = 0$

41. $4k^4 - 13k^2 + 9 = 0$

42. $9x^4 - 25x^2 + 16 = 0$

43. $x^4 + 48 = 16x^2$

44. $z^4 = 17z^2 - 72$

45. $2x^4 - 9x^2 = -2$

46. $8x^4 + 1 = 11x^2$

47. $(x + 3)^2 + 5(x + 3) + 6 = 0$

48. $(k - 4)^2 + (k - 4) - 20 = 0$

49. $(t + 5)^2 + 6 = 7(t + 5)$

50. $3(m + 4)^2 - 8 = 2(m + 4)$

51. $2 + \dfrac{5}{3k - 1} = \dfrac{-2}{(3k - 1)^2}$

52. $3 - \dfrac{7}{2p + 2} = \dfrac{6}{(2p + 2)^2}$

53. $x^{2/3} + x^{1/3} - 2 = 0$

54. $x^{2/3} - 2x^{1/3} - 3 = 0$

55. $r^{2/3} + r^{1/3} - 12 = 0$

56. $3x^{2/3} - x^{1/3} - 24 = 0$

57. $2\left(1 + \sqrt{r}\right)^2 = 13\left(1 + \sqrt{r}\right) - 6$

58. $(k^2 + k)^2 + 12 = 8(k^2 + k)$

Relating Concepts (Exercises 59–64) For Individual or Group Work

Consider the following equation, which contains variable expressions in the denominators. ***Work Exercises 59–64 in order.***

$$\frac{x^2}{(x-3)^2} + \frac{3x}{x-3} - 4 = 0$$

59. Why must 3 be excluded from the domain of this equation?

60. Multiply each side of the equation by the LCD, $(x - 3)^2$, and solve. There is only one solution—what is it?

61. Write the equation in a different manner so that it is quadratic in form using the expression $\frac{x}{x-3}$.

62. In your own words, explain why the expression $\frac{x}{x-3}$ cannot equal 1.

63. Solve the equation from Exercise 61 by making the substitution $t = \frac{x}{x-3}$. You should get two values for t. Why is one of them impossible for this equation?

64. Solve the equation $x^2(x - 3)^{-2} + 3x(x - 3)^{-1} - 4 = 0$ by letting $s = (x - 3)^{-1}$. You should get two values for s. Why is this impossible for this equation?

Summary Exercises on Solving Quadratic Equations

We have introduced four methods for solving quadratic equations written in standard form $ax^2 + bx + c = 0$. *The following table lists some advantages and disadvantages of each method.*

METHODS FOR SOLVING QUADRATIC EQUATIONS

Method	Advantages	Disadvantages
Factoring	This is usually the fastest method.	Not all polynomials are factorable; some factorable polynomials are hard to factor.
Square root property	This is the simplest method for solving equations of the form $(ax + b)^2 = c$.	Few equations are given in this form.
Completing the square	This method can always be used, although many people prefer the quadratic formula.	It requires more steps than other methods.
Quadratic formula	This method can always be used.	It is more difficult than factoring because of the square root, although calculators can simplify its use.

Refer to the preceding box. Decide whether factoring, the square root property, *or* the quadratic formula *is most appropriate for solving each quadratic equation. Do not actually solve the equations.*

1. $(2x + 3)^2 = 4$

2. $4x^2 - 3x = 1$

3. $z^2 + 5z - 8 = 0$

4. $2k^2 + 3k = 1$

5. $3m^2 = 2 - 5m$

6. $p^2 = 5$

Solve each quadratic equation by the method of your choice. Check your solutions.

7. $p^2 = 47$

8. $6x^2 - x - 15 = 0$

9. $n^2 + 8n + 6 = 0$

10. $(x - 4)^2 = 49$

11. $\frac{9}{m} + \frac{5}{m^2} = 2$

12. $3m^2 = 3 - 8m$

13. $3x^2 - 9x + 4 = 0$

***14.** $x^2 = -12$

15. $x\sqrt{2} = \sqrt{5x - 2}$

16. $12x^4 - 11x^2 + 2 = 0$

17. $(2k + 5)^2 = 12$

18. $\frac{2}{x} + \frac{1}{x - 2} - \frac{5}{3} = 0$

19. $t^4 + 14 = 9t^2$

20. $2x^2 + 4x = 5$

***21.** $z^2 + z + 2 = 0$

22. $x^4 - 8x^2 = -1$

23. $4t^2 - 12t + 9 = 0$

24. $x\sqrt{3} = \sqrt{2 - x}$

25. $r^2 - 72 = 0$

26. $-3x^2 + 4x = -4$

27. $x^2 - 5x - 36 = 0$

28. $w^2 = 169$

***29.** $3p^2 = 6p - 4$

30. $z = \sqrt{\frac{5z + 3}{2}}$

31. $2(3k - 1)^2 + 5(3k - 1) = -2$

***32.** $\frac{4}{r^2} + 3 = \frac{1}{r}$

33. $x - \sqrt{15 - 2x} = 0$

34. $3 = \frac{1}{t + 2} + \frac{2}{(t + 2)^2}$

***35.** $4k^4 + 5k^2 + 1 = 0$

36. $(x + 1)^{2/3} - (x + 1)^{1/3} = 2$

*This exercise requires knowledge of complex numbers.

10.4 Formulas and Further Applications

OBJECTIVES

1. Solve formulas for variables involving squares and square roots.
2. Solve applied problems using the Pythagorean formula.
3. Solve applied problems using area formulas.
4. Solve applied problems using quadratic functions as models.

OBJECTIVE 1 Solve formulas for variables involving squares and square roots. The methods presented earlier in this chapter and the previous one can be used to solve formulas with squares and square roots.

EXAMPLE 1 Solving for Variables Involving Squares or Square Roots

Solve each formula for the given variable.

(a) $w = \frac{kFr}{v^2}$ for v

$$w = \frac{kFr}{v^2}$$ The goal is to isolate v on one side.

$$v^2 w = kFr \qquad \text{Multiply by } v^2.$$

$$v^2 = \frac{kFr}{w} \qquad \text{Divide by } w.$$

$$v = \pm\sqrt{\frac{kFr}{w}} \qquad \text{Square root property}$$

$$v = \frac{\pm\sqrt{kFr}}{\sqrt{w}} \cdot \frac{\sqrt{w}}{\sqrt{w}} = \frac{\pm\sqrt{kFrw}}{w} \qquad \text{Rationalize the denominator.}$$

(b) $d = \sqrt{\frac{4A}{\pi}}$ for A

$$d = \sqrt{\frac{4A}{\pi}}$$ The goal is to isolate A on one side.

$$d^2 = \frac{4A}{\pi} \qquad \text{Square both sides.}$$

$$\pi d^2 = 4A \qquad \text{Multiply by } \pi.$$

$$\frac{\pi d^2}{4} = A \qquad \text{Divide by 4.}$$

Work Problem **1** at the Side. ▶

1 Solve each formula for the given variable.

(a) $A = \pi r^2$ for r

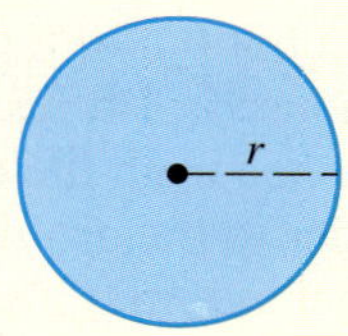

(b) $s = 30\sqrt{\frac{a}{p}}$ for a

Note

In formulas like $v = \frac{\pm\sqrt{kFrw}}{w}$ in Example 1(a), we will include both positive and negative values.

EXAMPLE 2 Solving for a Second-Degree Variable

Solve $s = 2t^2 + kt$ for t.

Since the equation has terms with t^2 and t, write it in standard form $ax^2 + bx + c = 0$, with t as the variable instead of x.

$$s = 2t^2 + kt$$

$$0 = 2t^2 + kt - s \qquad \text{Subtract } s.$$

$$2t^2 + kt - s = 0 \qquad \text{Standard form}$$

Continued on Next Page

ANSWERS

1. (a) $r = \frac{\pm\sqrt{A\pi}}{\pi}$ (b) $a = \frac{ps^2}{900}$

Now solve $2t^2 + kt - s = 0$ for t using the quadratic formula.

$$t = \frac{-k \pm \sqrt{k^2 - 4(2)(-s)}}{2(2)} \quad \text{Let } a = 2, b = k, \text{ and } c = -s.$$

$$t = \frac{-k \pm \sqrt{k^2 + 8s}}{4} \quad \text{Simplify.}$$

The solutions are $t = \dfrac{-k + \sqrt{k^2 + 8s}}{4}$ and $t = \dfrac{-k - \sqrt{k^2 + 8s}}{4}$.

◀ Work Problem **2** at the Side.

2 Solve $2t^2 - 5t + k = 0$ for t.

OBJECTIVE 2 Solve applied problems using the Pythagorean formula. The Pythagorean formula

$$a^2 + b^2 = c^2,$$

illustrated by the figure in the margin, was introduced in **Section 9.3** and is used to solve applications involving right triangles. Such problems often require solving quadratic equations.

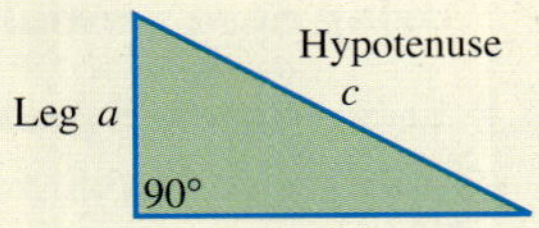

$a^2 + b^2 = c^2$
Pythagorean Formula

EXAMPLE 3 Using the Pythagorean Formula

Two cars left an intersection at the same time, one heading due north, the other due west. Some time later, they were exactly 100 mi apart. The car headed north had gone 20 mi farther than the car headed west. How far had each car traveled?

Step 1 **Read** the problem carefully.

Step 2 **Assign a variable.**

Let x = the distance traveled by the car headed west.

Then $x + 20$ = the distance traveled by the car headed north.

See Figure 2. The cars are 100 mi apart, so the hypotenuse of the right triangle equals 100.

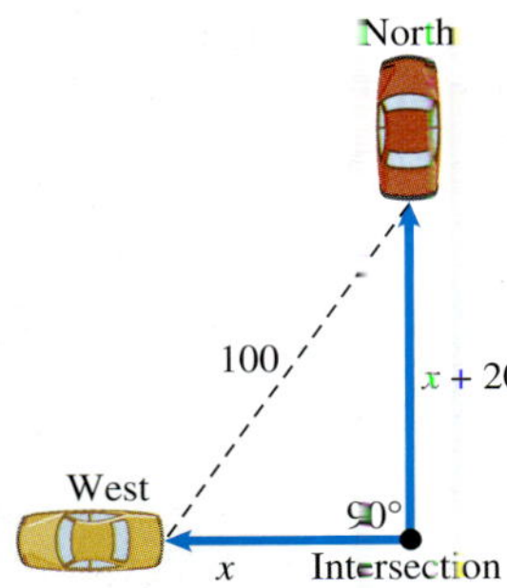

Figure 2

Step 3 **Write an equation.** Use the Pythagorean formula.

$$a^2 + b^2 = c^2$$

$$x^2 + (x + 20)^2 = 100^2$$

$(x + y)^2 = x^2 + 2xy + y^2$

Step 4 **Solve.**

$$x^2 + x^2 + 40x + 400 = 10{,}000 \quad \text{Square the binomial.}$$

$$2x^2 + 40x - 9600 = 0 \quad \text{Standard form}$$

$$x^2 + 20x - 4800 = 0 \quad \text{Divide by 2.}$$

$$(x + 80)(x - 60) = 0 \quad \text{Factor.}$$

$$x + 80 = 0 \quad \text{or} \quad x - 60 = 0 \quad \text{Zero-factor property}$$

$$x = -80 \quad \text{or} \quad x = 60 \quad \text{Solve for } x.$$

Step 5 **State the answer.** Distance cannot be negative, so discard the negative solution. The distances are 60 mi and $60 + 20 = 80$ mi.

Step 6 **Check.** Since $60^2 + 80^2 = 100^2$, the answers are correct.

◀ Work Problem **3** at the Side.

3 Solve the problem.

A 13-ft ladder is leaning against a house. The distance from the bottom of the ladder to the house is 7 ft less than the distance from the top of the ladder to the ground. How far is the bottom of the ladder from the house?

ANSWERS

2. $t = \dfrac{5 + \sqrt{25 - 8k}}{4}, t = \dfrac{5 - \sqrt{25 - 8k}}{4}$

3. 5 ft

OBJECTIVE 3 **Solve applied problems using area formulas.**

EXAMPLE 4 **Solving an Area Problem**

A rectangular reflecting pool in a park is 20 ft wide and 30 ft long. The park gardener wants to plant a strip of grass of uniform width around the edge of the pool. She has enough seed to cover 336 ft^2. How wide will the strip be?

Step 1 **Read** the problem carefully.

Step 2 **Assign a variable.** The pool is shown in Figure 3. If x represents the unknown width of the grass strip, the width of the large rectangle is given by $20 + 2x$ (the width of the pool plus two grass strips), and the length is given by $30 + 2x$.

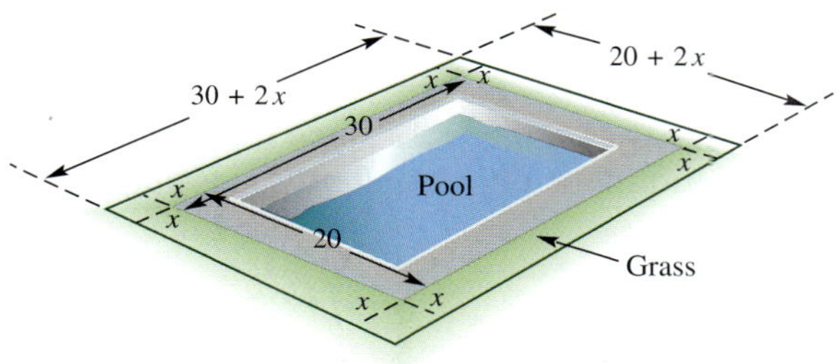

Figure 3

Step 3 **Write an equation.** The area of the large rectangle is given by the product of its length and width, $(30 + 2x)(20 + 2x)$. The area of the pool is $30 \cdot 20 = 600$ ft^2. The area of the large rectangle, minus the area of the pool, should equal the area of the grass strip. Since the area of the grass strip is to be 336 ft^2, the equation is

$$\underbrace{(30 + 2x)(20 + 2x)}_{\text{Area of rectangle}} - \underbrace{600}_{\text{area of pool}} = \underbrace{336.}_{\text{area of grass.}}$$

Step 4 **Solve.**

$$600 + 100x + 4x^2 - 600 = 336 \quad \text{Multiply.}$$
$$4x^2 + 100x - 336 = 0 \quad \text{Standard form}$$
$$x^2 + 25x - 84 = 0 \quad \text{Divide by 4.}$$
$$(x + 28)(x - 3) = 0 \quad \text{Factor.}$$
$$x = -28 \quad \text{or} \quad x = 3 \quad \text{Zero-factor property}$$

Step 5 **State the answer.** The width cannot be -28 ft, so the grass strip should be 3 ft wide.

Step 6 **Check.** If $x = 3$, then the area of the large rectangle is

$$(30 + 2 \cdot 3)(20 + 2 \cdot 3) = 36 \cdot 26 = 936 \text{ ft}^2. \quad \text{Area of pool and strip}$$

The area of the pool is $30 \cdot 20 = 600$ ft^2. So, the area of the grass strip is $936 - 600 = 336$ ft^2, as required. The answer is correct.

Work Problem 4 at the Side. ▶

OBJECTIVE 4 **Solve applied problems using quadratic functions as models.** Some applied problems can be modeled by *quadratic functions,* which can be written in the form

$$f(x) = ax^2 + bx + c,$$

for real numbers a, b, and c, with $a \neq 0$.

4 Solve the problem.

Suppose the pool in Example 4 is 20 ft by 40 ft and there is enough seed to cover 700 ft^2. How wide should the grass strip be?

ANSWER

4. 5 ft

5 Solve the problem.

A ball is projected vertically upward from the ground. Its distance in feet from the ground at t seconds is

$$s(t) = -16t^2 + 64t.$$

At what times will the ball be 32 ft from the ground? Use a calculator and round answers to the nearest tenth. (*Hint:* There are two answers.)

EXAMPLE 5 **Solving an Applied Problem Using a Quadratic Function**

If an object is projected upward from the top of a 144-ft building at 112 ft per sec, its position (in feet above the ground) is given by

$$s(t) = -16t^2 + 112t + 144,$$

where t is time in seconds after it was propelled. When does it hit the ground?

When the object hits the ground, its distance above the ground is 0. We must find the value of t that makes $s(t) = 0$.

$0 = -16t^2 + 112t + 144$ Let $s(t) = 0$.

$0 = t^2 - 7t - 9$ Divide by -16.

$t = \frac{-(-7) \pm \sqrt{(-7)^2 - 4(1)(-9)}}{2(1)} = \frac{7 \pm \sqrt{85}}{2} \approx \frac{7 \pm 9.2}{2}$ Use the quadratic formula and a calculator.

The solutions are $t \approx 8.1$ or $t \approx -1.1$. Since time cannot be negative, discard the negative solution. The object will hit the ground about 8.1 sec after it is projected.

◀ Work Problem **5** at the Side.

EXAMPLE 6 **Using a Quadratic Function to Model the CPI**

The Consumer Price Index (CPI) is used to measure trends in prices for a "basket" of goods purchased by typical American families. This index uses a base year of 1967, which means that the index number for 1967 is 100. The quadratic function defined by

$$f(x) = -0.065x^2 + 14.8x + 249$$

approximates the CPI for the years 1980–2005, where x is the number of years that have elapsed since 1980. (*Source:* Bureau of Labor Statistics.)

(a) Use the model to approximate the CPI for 1995.
For 1995, $x = 1995 - 1980 = 15$, so find $f(15)$.

$f(x) = -0.065x^2 + 14.8x + 249$ Given model

$f(15) = -0.065(15)^2 + 14.8(15) + 249$ Let $x = 15$.

$f(15) \approx 456$ Nearest whole number

The CPI for 1995 was about 456.

(b) In what year did the CPI reach 550?
Find the value of x that makes $f(x) = 550$.

$f(x) = -0.065x^2 + 14.8x + 249$ Given model

$550 = -0.065x^2 + 14.8x + 249$ Let $f(x) = 550$.

$0 = -0.065x^2 + 14.8x - 301$ Standard form

Now use $a = -0.065$, $b = 14.8$, and $c = -301$ in the quadratic formula.

◀ Work Problem **6** at the Side.

6 Use a calculator to evaluate

$$\frac{-14.8 \pm \sqrt{14.8^2 - 4(-0.065)(-301)}}{2(-0.065)}$$

for both solutions. Round to the nearest whole number. Which solution is valid for this problem?

The first solution is $x \approx 23$. Rounding up to the next whole number, the CPI first reached 550 in $1980 + 23 = 2003$. (Reject the solution $x \approx 205$, as this corresponds to a year far beyond the period covered by the model.)

ANSWERS

5. at 0.6 sec and at 3.4 sec
6. 23; 205; 23

10.4 Exercises

FOR EXTRA HELP

MyMathLab

Math XL PRACTICE | WATCH | DOWNLOAD | READ | REVIEW

1. What is the first step in solving a formula like $gw^2 = 2r$ for w?

2. What is the first step in solving a formula like $gw^2 = kw + 24$ for w?

In Exercises 3 and 4, solve for m in terms of the other variables ($m > 0$).

3.

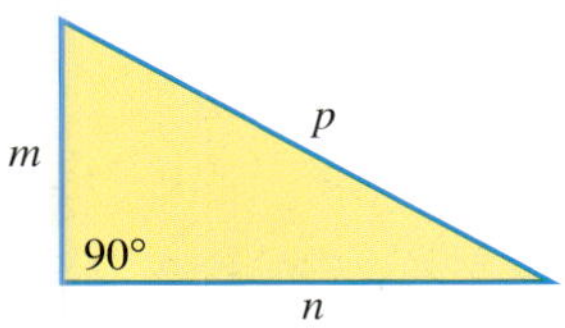

4.

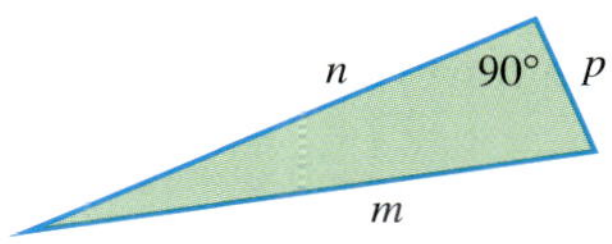

Solve each equation for the indicated variable. (Leave ± in your answers.) See Examples 1 and 2.

5. $d = kt^2$ for t

6. $s = kwd^2$ for d

7. $I = \dfrac{ks}{d^2}$ for d

8. $R = \dfrac{k}{d^2}$ for d

9. $F = \dfrac{kA}{v^2}$ for v

10. $L = \dfrac{kd^4}{h^2}$ for h

11. $V = \dfrac{1}{3}\pi r^2 h$ for r

12. $V = \pi(r^2 + R^2)h$ for r

13. $At^2 + Bt = -C$ for t

14. $S = 2\pi rh + \pi r^2$ for r

15. $D = \sqrt{kh}$ for h

16. $F = \dfrac{k}{\sqrt{d}}$ for d

17. $p = \sqrt{\dfrac{k\ell}{g}}$ for ℓ

18. $p = \sqrt{\dfrac{k\ell}{g}}$ for g

Solve each problem. When appropriate, round answers to the nearest tenth. See Example 3.

19. Find the lengths of the sides of the triangle.

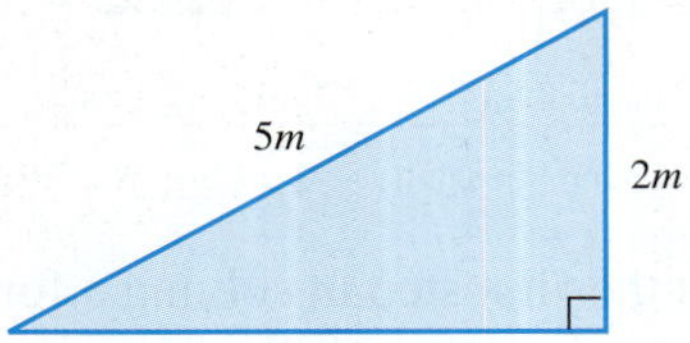

20. Find the lengths of the sides of the triangle.

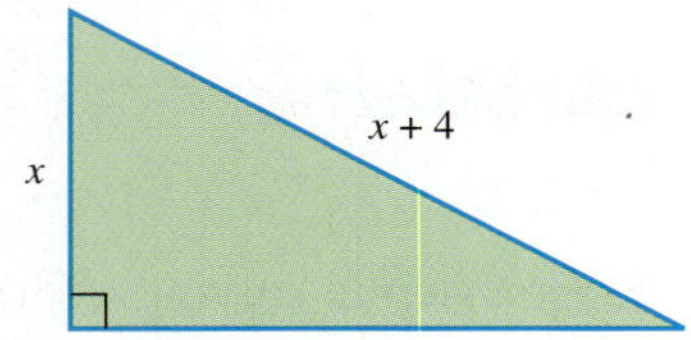

21. Two ships leave port at the same time, one heading due south and the other heading due east. Several hours later, they are 170 mi apart. If the ship traveling south traveled 70 mi farther than the other, how many miles did they each travel?

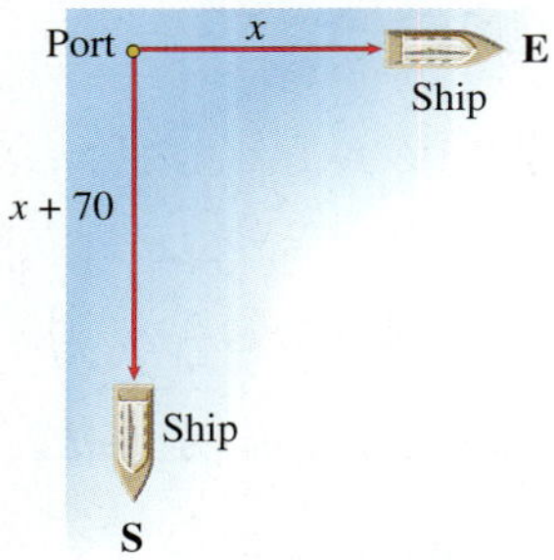

22. Faith Varnado is flying a kite that is 30 ft farther above her hand than its horizontal distance from her. The string from her hand to the kite is 150 ft long. How high is the kite?

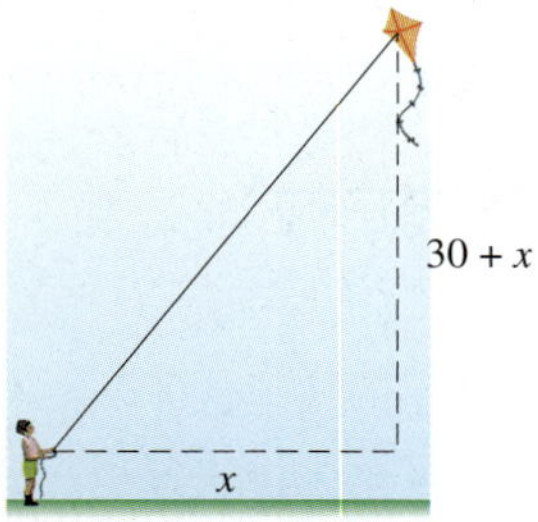

Solve each problem. See Example 4.

23. A couple wants to buy a rug for a room that is 20 ft long and 15 ft wide. They want to leave an even strip of flooring uncovered around the edges of the room. How wide a strip will they have if they buy a rug with an area of 234 ft^2?

24. A club swimming pool is 30 ft wide and 40 ft long. The club members want an exposed aggregate border in a strip of uniform width around the pool. They have enough material for 296 ft^2. How wide can the strip be?

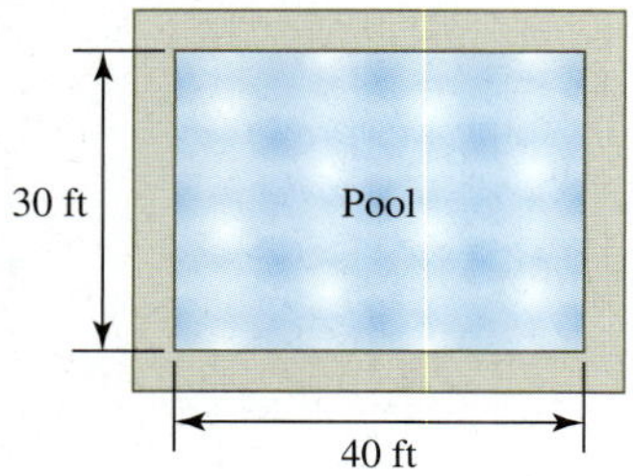

25. A rectangular piece of sheet metal has a length that is 4 in. less than twice the width. A square piece 2 in. on a side is cut from each corner. The sides are then turned up to form an uncovered box of volume 256 in.^3. Find the length and width of the original piece of metal.

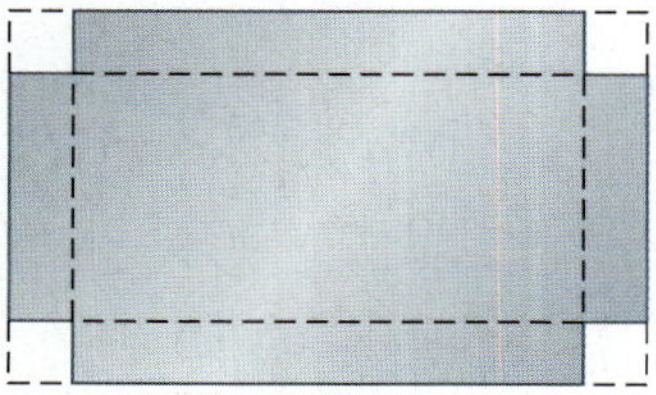

26. Another rectangular piece of sheet metal is 2 in. longer than it is wide. A square piece 3 in. on a side is cut from each corner. The sides are then turned up to form an uncovered box of volume 765 in.^3. Find the dimensions of the original piece of metal.

Solve each problem. Round answers to the nearest tenth. See Example 5.

27. A ball is projected upward from the ground. Its distance in feet from the ground in t seconds is given by

$$s(t) = -16t^2 + 128t.$$

At what times will the ball be 213 ft from the ground?

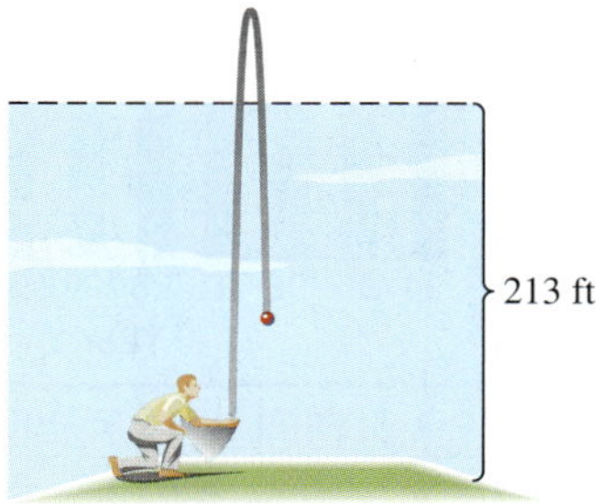

28. A toy rocket is launched from ground level. Its distance in feet from the ground in t seconds is given by

$$s(t) = -16t^2 + 208t.$$

At what times will the rocket be 550 ft from the ground?

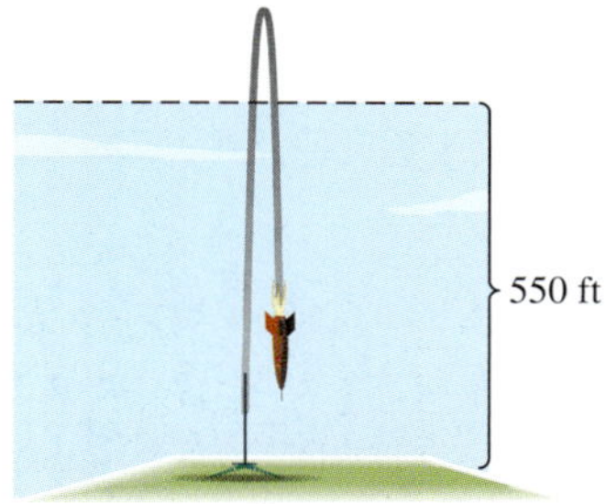

29. The function defined by

$$D(t) = 13t^2 - 100t$$

gives the distance in feet a car going approximately 68 mph will skid in t seconds. Find the time it would take for the car to skid 180 ft.

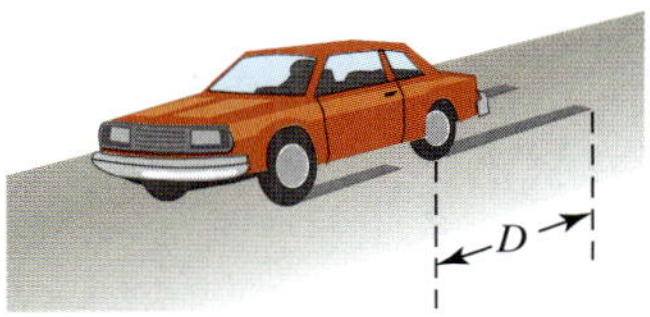

30. The function given in Exercise 29 becomes

$$D(t) = 13t^2 - 73t$$

for a car going 50 mph. Find the time for this car to skid 218 ft.

A ball is projected upward from ground level, and its distance in feet from the ground in t seconds is given by $s(t) = -16t^2 + 160t$. Use algebra and a short explanation to answer Exercises 31 and 32.

31. After how many seconds does it reach a height of 400 ft? How would you describe in words its position at this height?

32. After how many seconds does it reach a height of 425 ft? How would you interpret the mathematical result here?

Solve each problem using a quadratic equation.

33. A certain bakery has found that the daily demand for blueberry muffins is $\frac{3200}{p}$, where p is the price of a muffin in cents. The daily supply is $3p - 200$. Find the price at which supply and demand are equal.

34. In one area the demand for compact discs is $\frac{700}{P}$ per day, where P is the price in dollars per disc. The supply is $5P - 1$ per day. At what price, to the nearest cent, does supply equal demand?

The total number of miles traveled by all motor vehicles in the United States for the years 1994–2003 are shown in the bar graph and can be modeled by the quadratic function defined by

$$f(x) = -1.705x^2 + 75.93x + 2351.$$

Here, $x = 0$ represents 1994, $x = 1$ represents 1995, and so on. Use the graph and the model to work Exercises 35–38. See Example 6.

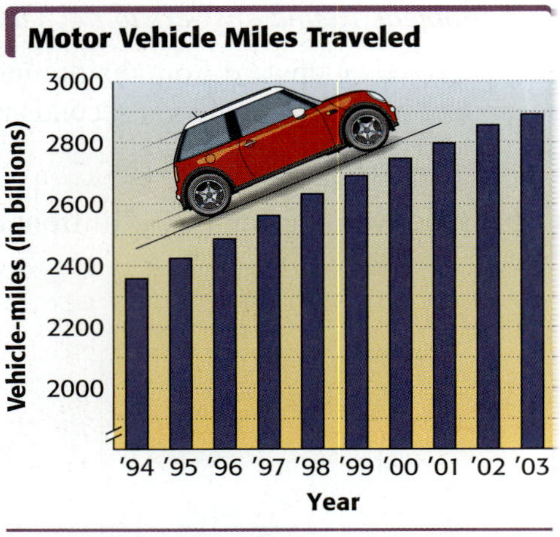

Source: U.S. Federal Highway Administration.

35. **(a)** Use the graph to estimate miles traveled in 2000 to the nearest ten billion.

(b) Use the model to approximate miles traveled in 2000 to the nearest ten billion. How does this result compare to your estimate from part (a)?

36. Based on the model, in what year did miles traveled reach 2600 billion? (Round down to the nearest year.) How does this result compare to the vehicle-miles shown in the graph?

37. Based on the model, in what year did miles traveled reach 2800 billion? (Round down to the nearest year.) How does this result compare to the vehicle-miles shown in the graph?

38. If these data were modeled by a *linear* function defined by $f(x) = ax + b$, would the value of a be positive or negative? Explain.

William Froude was a 19th-century naval architect who used the expression

$$\frac{v^2}{g\ell}$$

*in shipbuilding. This expression, known as the **Froude number,** was also used by R. McNeill Alexander in his research on dinosaurs.* (Source: *"How Dinosaurs Ran,"* Scientific American, *April 1991.) In Exercises 39 and 40, find to the nearest tenth the value of v (in meters per second), given that g* = 9.8 m per sec^2.

39. Rhinoceros: $\ell = 1.2$; Froude number $= 2.57$

40. Triceratops: $\ell = 2.8$; Froude number $= 0.16$

*Recall from the **Section 8.5** exercises that corresponding sides of similar triangles are proportional. Use this fact to find the lengths of the indicated sides of each pair of similar triangles. Check all possible solutions in both triangles. Sides of a triangle cannot be negative (and are not drawn to scale here).*

41. Side AC

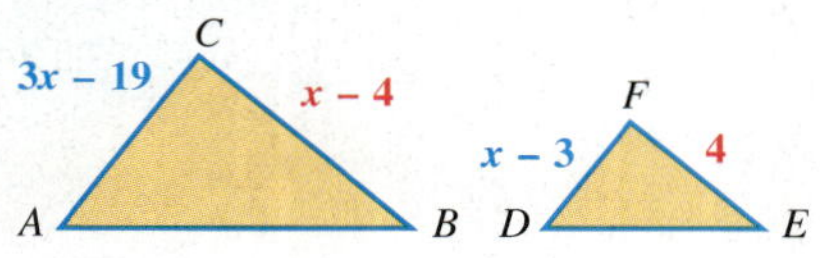

42. Side RQ

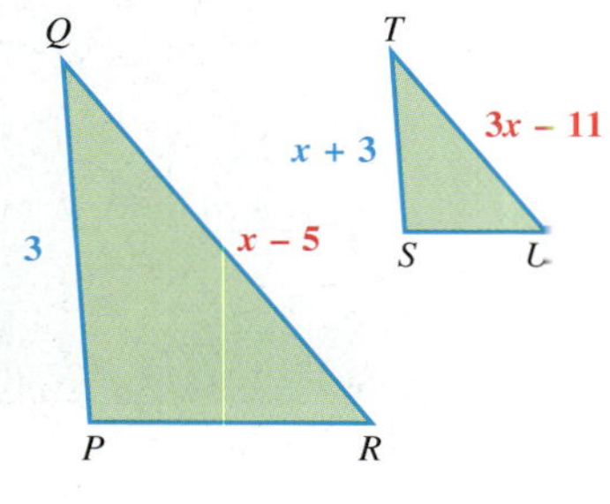

10.5 Graphs of Quadratic Functions

OBJECTIVES

1. Graph a quadratic function.
2. Graph parabolas with horizontal and vertical shifts.
3. Use the coefficient of x^2 to predict the shape and direction in which a parabola opens.
4. Find a quadratic function to model data.

OBJECTIVE 1 Graph a quadratic function. Figure 4 gives a graph of the simplest *quadratic function,* defined by $y = x^2$.

x	y
−2	4
−1	1
0	0
1	1
2	4

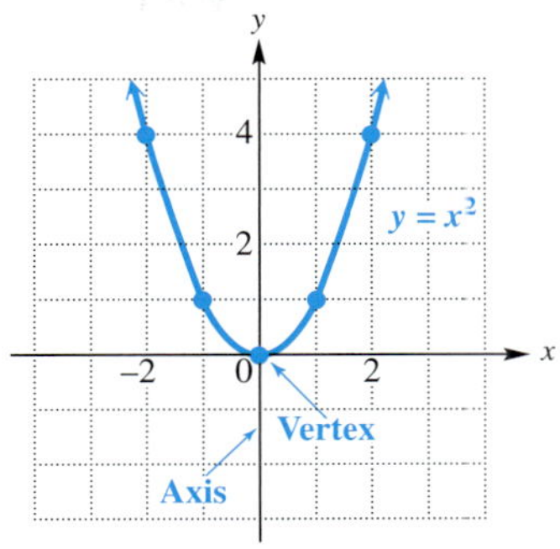

Figure 4

As mentioned in **Section 6.3,** this graph is called a **parabola.** The point (0, 0), the lowest point on the curve, is the **vertex** of this parabola. The vertical line through the vertex is the **axis** of the parabola, here $x = 0$. A parabola is **symmetric about its axis;** that is, if the graph were folded along the axis, the two portions of the curve would coincide. As Figure 4 suggests, x can be any real number, so the domain of the function defined by $y = x^2$ is $(-\infty, \infty)$. Since y is always nonnegative, the range is $[0, \infty)$.

In **Section 10.4,** we solved applications modeled by quadratic functions.

Quadratic Function

A function that can be written in the form

$$f(x) = ax^2 + bx + c$$

for real numbers a, b, and c, with $a \neq 0$, is a **quadratic function.**

The graph of any quadratic function is a parabola with a vertical axis.

Note

We use the variable y and function notation $f(x)$ interchangeably. Although we use the letter f most often to name quadratic functions, other letters can be used. We use the capital letter F to distinguish between different parabolas graphed on the same coordinate axes.

Parabolas, which are a type of *conic section* **(Chapter 12),** have many applications. Cross sections of telescopes, satellite dishes and automobile headlights form parabolas, as do the cables that support suspension bridges.

OBJECTIVE 2 Graph parabolas with horizontal and vertical shifts. Parabolas need not have their vertices at the origin, as does the graph of $f(x) = x^2$. To graph a parabola of the form

$$F(x) = x^2 + k, \qquad \text{Vertical shift } k$$

select sample values of x like those that were used to graph $f(x) = x^2$. The corresponding values of $F(x)$ in $F(x) = x^2 + k$ differ by k from those of $f(x) = x^2$. For this reason, the graph of $F(x) = x^2 + k$ is *shifted,* or *translated,* k units vertically compared with that of $f(x) = x^2$.

1 Graph each parabola. Give the vertex, domain, and range.

(a) $f(x) = x^2 + 3$

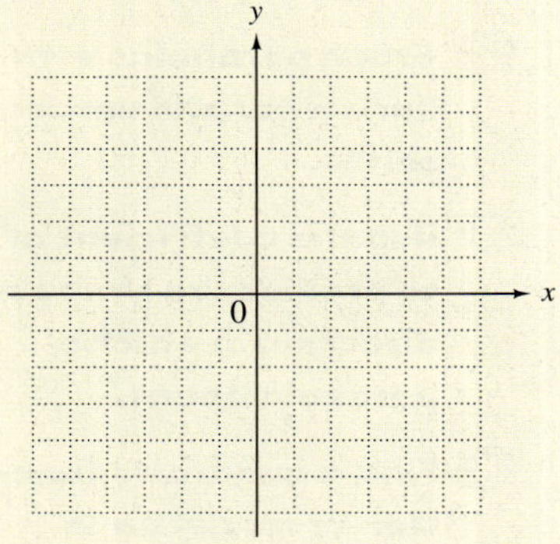

(b) $f(x) = x^2 - 1$

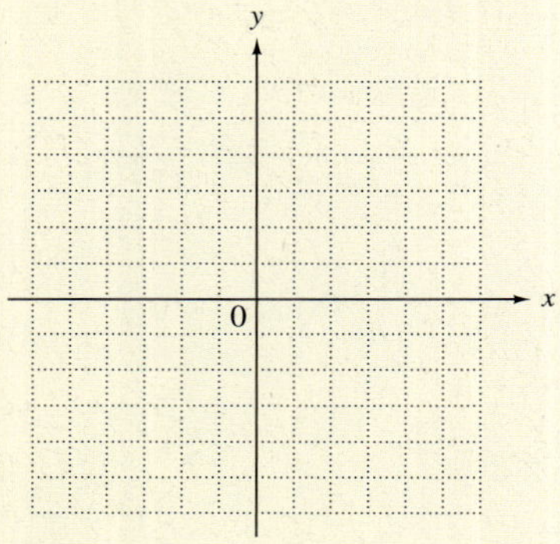

EXAMPLE 1 Graphing a Parabola with a Vertical Shift

Graph $F(x) = x^2 - 2$.

This graph has the same shape as that of $f(x) = x^2$, but since k here is -2, the graph is shifted 2 units down, with vertex $(0, -2)$. Every function value is 2 less than the corresponding function value of $f(x) = x^2$. Plotting points on both sides of the vertex gives the graph in Figure 5.

x	$f(x) = x^2$	$F(x) = x^2 - 2$
-2	4	2
-1	1	-1
0	0	-2
1	1	-1
2	4	2

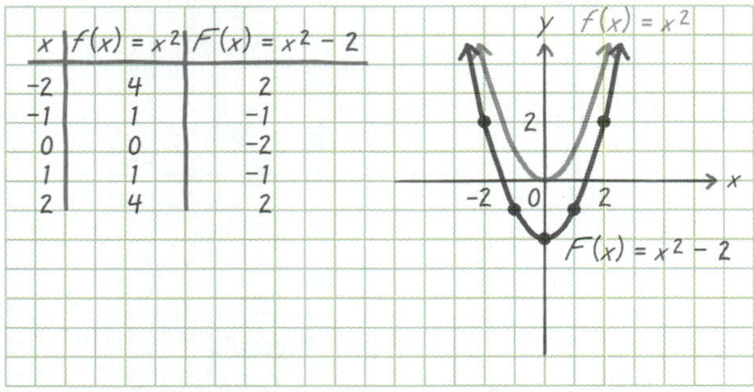

Figure 5

Notice that since the parabola is symmetric about its axis $x = 0$, the plotted points are "mirror images" of each other. Since x can be any real number, the domain is still $(-\infty, \infty)$. The value of y (or $F(x)$) is always greater than or equal to -2, so the range is $[-2, \infty)$. The graph of $f(x) = x^2$ is shown for comparison.

Vertical Shift

The graph of $\boldsymbol{F(x) = x^2 + k}$ is a parabola with the same shape as the graph of $f(x) = x^2$. The parabola is shifted vertically: k units up if $k > 0$, and $|k|$ units down if $k < 0$. The vertex is $(0, k)$.

◀ *Work Problem* **1** *at the Side.*

The graph of the function defined by

$$\boldsymbol{F(x) = (x - h)^2} \qquad \text{Horizontal shift } h$$

is also a parabola with the same shape as that of $f(x) = x^2$. Because $(x - h)^2 \geq 0$ for all x, the vertex of $F(x) = (x - h)^2$ is the lowest point on the parabola. The lowest point occurs here when $F(x)$ is 0. To get $F(x)$ equal to 0, let $x = h$ so the vertex of $F(x) = (x - h)^2$ is $(h, 0)$. Based on this, the graph of $F(x) = (x - h)^2$ is shifted h units horizontally compared with that of $f(x) = x^2$.

EXAMPLE 2 Graphing a Parabola with a Horizontal Shift

Graph $F(x) = (x - 2)^2$.

If $x = 2$, then $F(x) = 0$, which gives the vertex $(2, 0)$. The graph of $F(x) = (x - 2)^2$ has the same shape as that of $f(x) = x^2$ but is shifted 2 units to the right. Plotting several points on one side of the vertex and using symmetry about the axis $x = 2$ to find corresponding points on the other side of the vertex gives the graph in Figure 6 on the next page. Again, the domain is $(-\infty, \infty)$; the range is $[0, \infty)$.

Continued on Next Page

ANSWERS

1. (a)

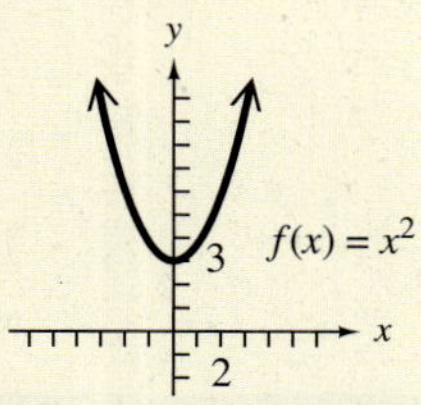

vertex: $(0, 3)$; domain: $(-\infty, \infty)$; range: $[3, \infty)$

(b)

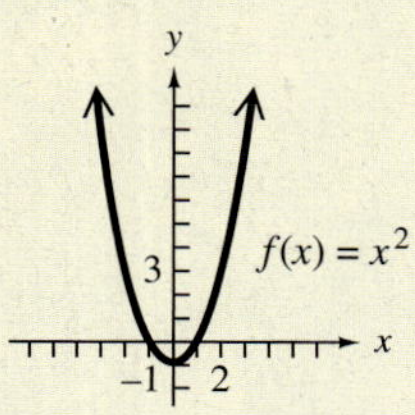

vertex: $(0, -1)$; domain: $(-\infty, \infty)$; range: $[-1, \infty)$

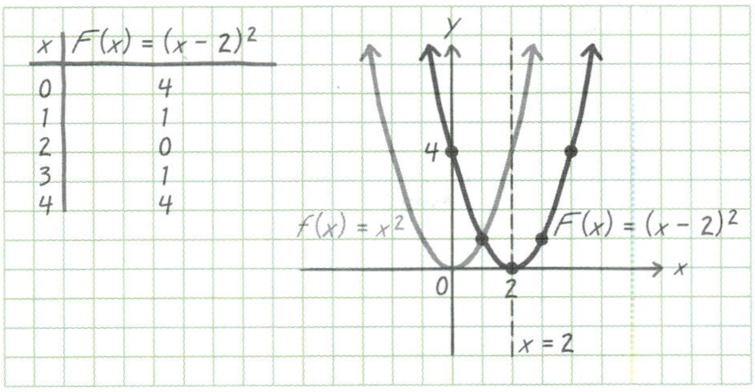

Figure 6

Horizontal Shift

The graph of $F(x) = (x - h)^2$ is a parabola with the same shape as the graph of $f(x) = x^2$. The parabola is shifted h units horizontally: h units to the right if $h > 0$, and $|h|$ units to the left if $h < 0$. The vertex is $(h, 0)$.

CAUTION

Errors frequently occur when horizontal shifts are involved. To determine the direction and magnitude of a horizontal shift, find the value that would cause the expression $x - h$ to equal 0. For example, the graph of $F(x) = (x - 5)^2$ would be shifted 5 units to the *right,* because $+5$ would cause $x - 5$ to equal 0. On the other hand, the graph of $F(x) = (x + 5)^2$ would be shifted 5 units to the *left,* because -5 would cause $x + 5$ to equal 0.

Work Problem **2** at the Side. ▶

A parabola can have both horizontal and vertical shifts.

EXAMPLE 3 Graphing a Parabola with Horizontal and Vertical Shifts

Graph $F(x) = (x + 3)^2 - 2$.

This graph has the same shape as that of $f(x) = x^2$, but is shifted 3 units to the left (since $x + 3 = 0$ if $x = -3$) and 2 units down (because of the -2). See Figure 7. The vertex is $(-3, -2)$, with axis $x = -3$. This function has domain $(-\infty, \infty)$ and range $[-2, \infty)$.

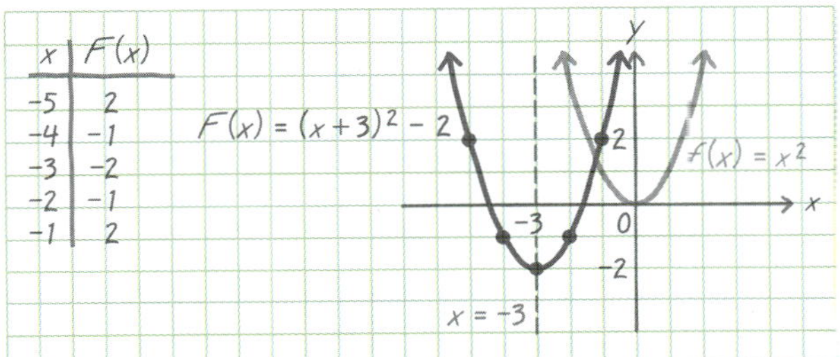

Figure 7

2 Graph each parabola. Give the vertex, axis, domain, and range.

(a) $f(x) = (x - 3)^2$

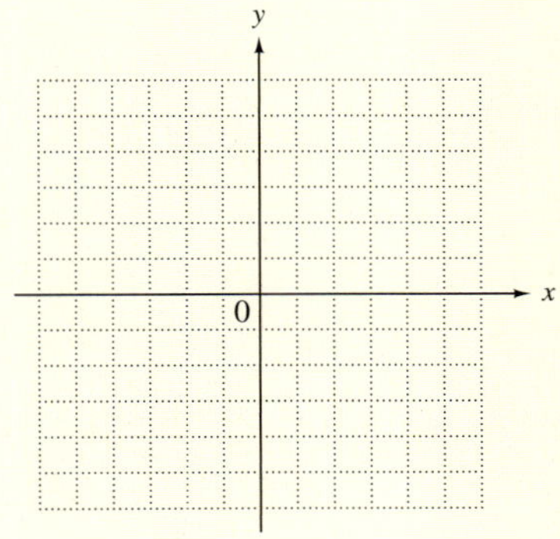

(b) $f(x) = (x + 2)^2$

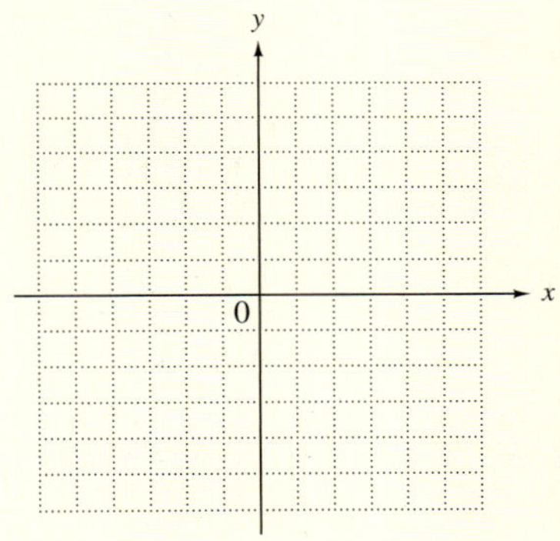

ANSWERS

2. (a)

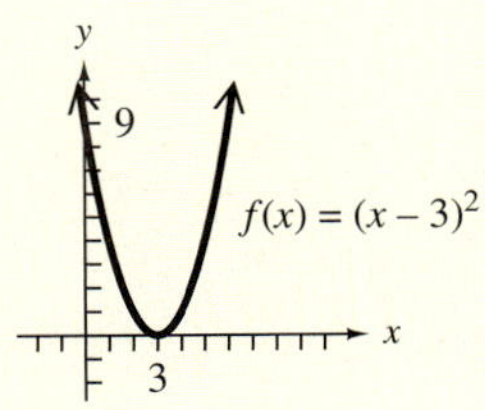

vertex: $(3, 0)$; axis: $x = 3$; domain: $(-\infty, \infty)$; range: $[0, \infty)$

(b)

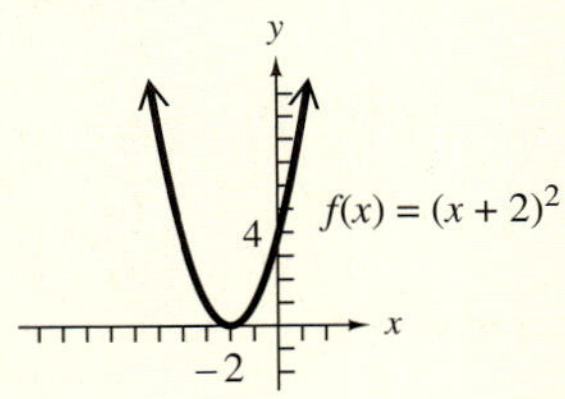

vertex: $(-2, 0)$; axis: $x = -2$; domain: $(-\infty, \infty)$; range: $[0, \infty)$

3 Graph each parabola. Give the vertex, axis, domain, and range.

(a) $f(x) = (x + 2)^2 - 1$

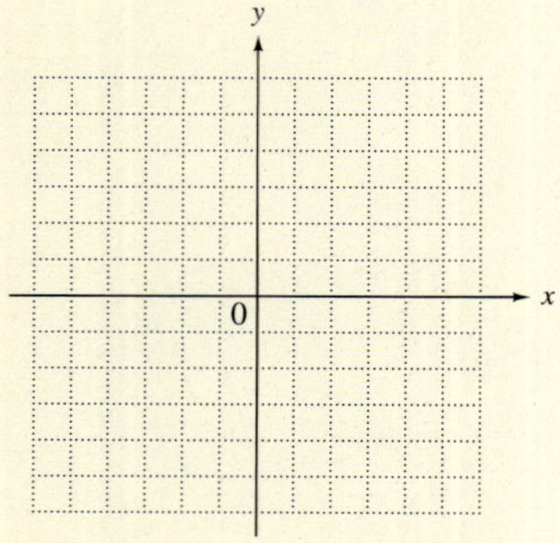

(b) $f(x) = (x - 2)^2 + 5$

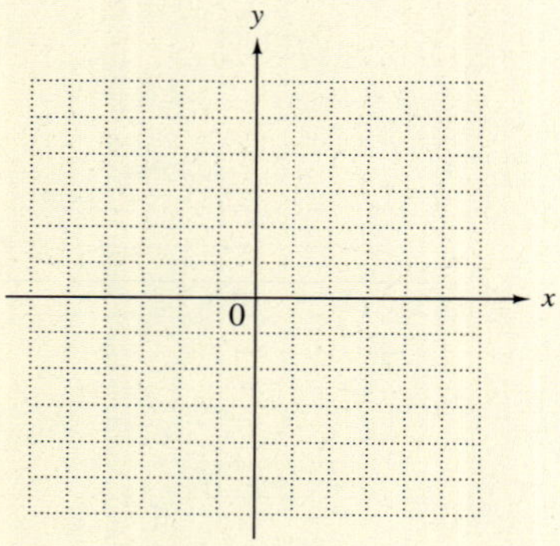

The characteristics of the graph of a parabola with equation of the form $F(x) = (x - h)^2 + k$ are summarized as follows.

Vertex and Axis of a Parabola

The graph of $\mathbf{F(x) = (x - h)^2 + k}$ is a parabola with the same shape as the graph of $f(x) = x^2$ with vertex (h, k). The axis is the vertical line $x = h$.

◀ *Work Problem* **3** *at the Side.*

OBJECTIVE 3 Use the coefficient of x^2 to predict the shape and direction in which a parabola opens. Not all parabolas open up, and not all parabolas have the same shape as the graph of $f(x) = x^2$.

EXAMPLE 4 Graphing a Parabola That Opens Down

Graph $f(x) = -\frac{1}{2}x^2$.

This parabola is shown in Figure 8. The coefficient $-\frac{1}{2}$ affects the shape of the graph; the $\frac{1}{2}$ makes the parabola wider (since the values of $\frac{1}{2}x^2$ increase more slowly than those of x^2), and the negative sign makes the parabola open down. The graph is not shifted in any direction; the vertex is still (0, 0) and the axis has equation $x = 0$. Unlike the parabolas graphed in Examples 1–3, the vertex here has the *greatest* function value of any point on the graph. The domain is $(-\infty, \infty)$; the range is $(-\infty, 0]$.

x	$f(x)$
-2	-2
-1	$-\frac{1}{2}$
0	0
1	$-\frac{1}{2}$
2	-2

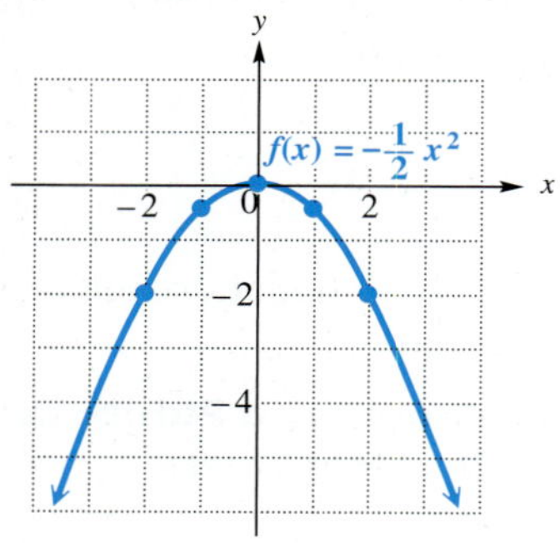

Figure 8

Some general principles concerning the graph of $F(x) = a(x - h)^2 + k$ are summarized as follows.

General Principles

1. The graph of the quadratic function defined by
$$\mathbf{F(x) = a(x - h)^2 + k, \quad a \neq 0,}$$
is a parabola with vertex (h, k) and the vertical line $x = h$ as axis.
2. The graph opens up if a is positive and down if a is negative.
3. The graph is wider than that of $f(x) = x^2$ if $0 < |a| < 1$. The graph is narrower than that of $f(x) = x^2$ if $|a| > 1$.

ANSWERS

3. (a)

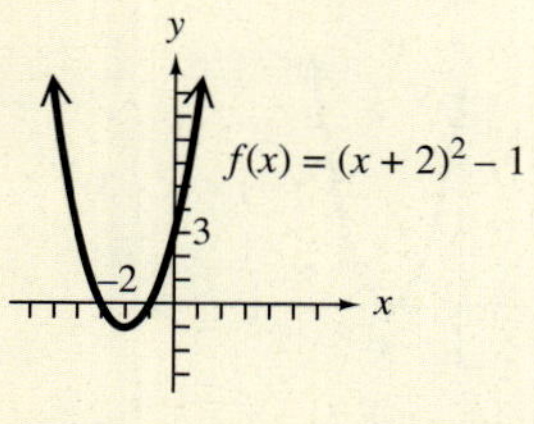

vertex: $(-2, -1)$; axis: $x = -2$; domain: $(-\infty, \infty)$; range: $[-1, \infty)$

(b)

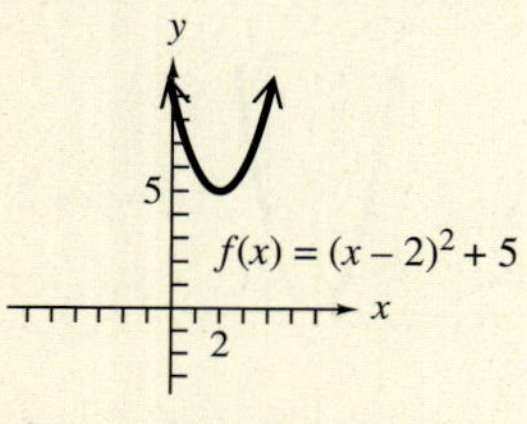

vertex: (2, 5); axis: $x = 2$; domain: $(-\infty, \infty)$; range: $[5, \infty)$

Work Problems **4** *and* **5** *at the Side.* ▶

EXAMPLE 5 Using the General Principles to Graph a Parabola

Graph $F(x) = -2(x + 3)^2 + 4$. Give the domain and the range.

The parabola opens down (because $a < 0$), and is narrower than the graph of $f(x) = x^2$, since $|-2| = 2 > 1$, causing values of $F(x)$ to decrease more quickly than those of $f(x) = -x^2$. This parabola has vertex $(-3, 4)$ as shown in Figure 9. To complete the graph, we plotted the ordered pairs $(-4, 2)$ and, by symmetry, $(-2, 2)$. Symmetry can be used to find additional ordered pairs that satisfy the equation, if desired. The domain is $(-\infty, \infty)$ and the range is $(-\infty, 4]$.

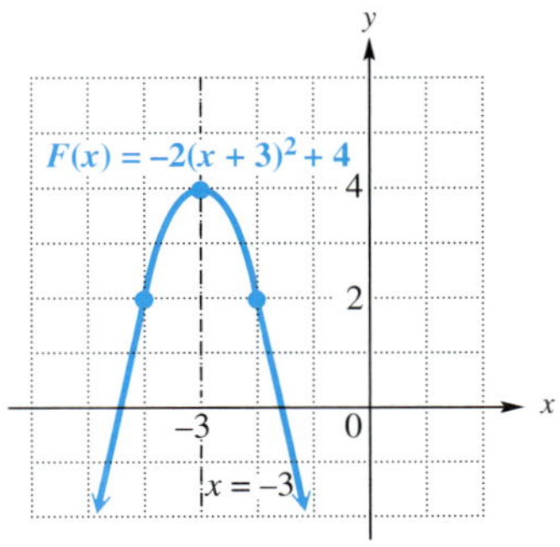

Figure 9

Work Problem **6** *at the Side.* ▶

OBJECTIVE 4 Find a quadratic function to model data.

EXAMPLE 6 Modeling the Number of Multiple Births

After rising steadily over several decades, the number of higher-order multiple births (triplets or more) in the United States started to decline during the first decade of the 21st century. Let x represent the number of years since 1995 and y represent the number of higher-order multiple births. Data for selected years are shown in the table.

Year	x	y
1995	0	4973
1996	1	5939
1997	2	6737
1999	4	7321
2001	6	7471
2003	8	7663
2004	9	7275
2005	10	6694

Source: National Center for Health Statistics.

Find a quadratic function that models the data.

A scatter diagram of the ordered pairs (x, y) is shown in Figure 10 on the next page. The general shape suggested by the scatter diagram indicates that a parabola should approximate these points, as shown by the dashed curve in Figure 11. The equation for such a parabola would have a negative coefficient for x^2 since the graph opens down.

Continued on Next Page

4 Decide whether each parabola opens up or down.

(a) $f(x) = -\frac{2}{3}x^2$

(b) $f(x) = \frac{3}{4}x^2 + 1$

(c) $f(x) = -2x^2 - 3$

(d) $f(x) = 3x^2 + 2$

5 Decide whether each parabola in Problem 4 is wider or narrower than the graph of $f(x) = x^2$.

6 Graph

$$f(x) = \frac{1}{2}(x - 2)^2 + 1.$$

Give the domain and the range.

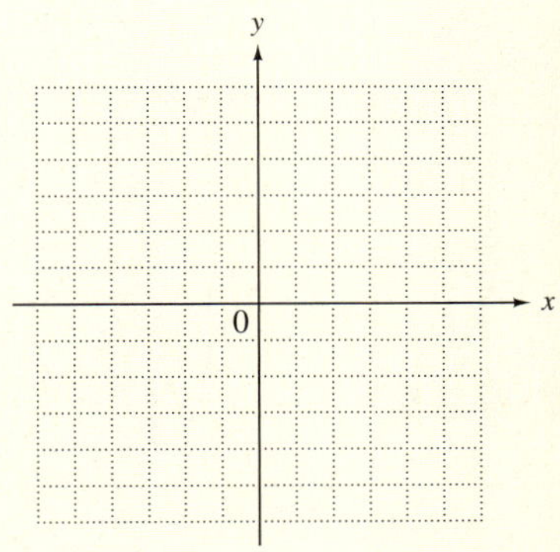

ANSWERS

4. (a) down **(b)** up **(c)** down **(d)** up

5. (a) wider **(b)** wider **(c)** narrower **(d)** narrower

6.

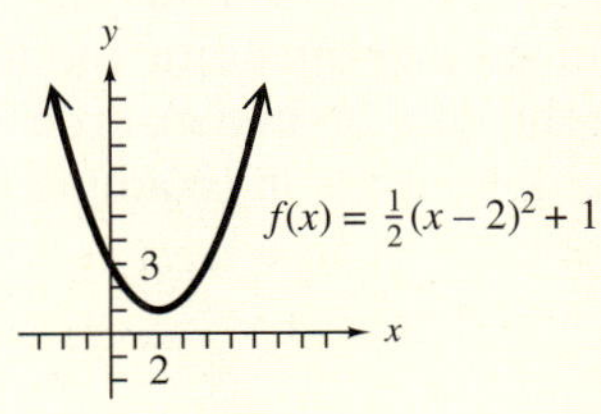

domain: $(-\infty, \infty)$
range: $[1, \infty)$

7 Tell whether a linear or quadratic function would be a more appropriate model for each set of graphed data. If linear, tell whether the slope should be positive or negative. If quadratic, tell whether the coefficient a of x^2 should be positive or negative.

(a)

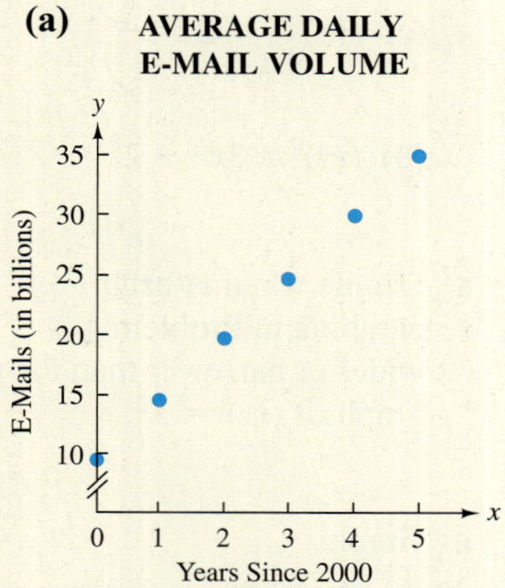

Source: General Accounting Office.

(b)

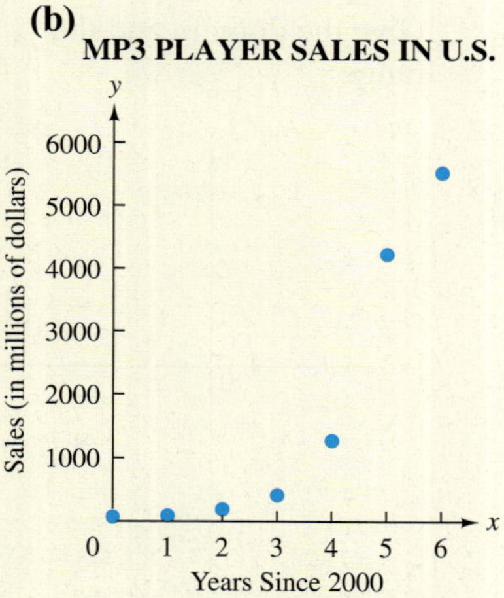

Source: Consumer Electronics Association.

8 Using the points (1, 5939), (6, 7471), and (10, 6694), find another quadratic model for the data on higher-order multiple births in Example 6.

Answers

7. **(a)** linear; positive **(b)** quadratic; positive
8. $y = -55.63x^2 + 695.80x + 5299$

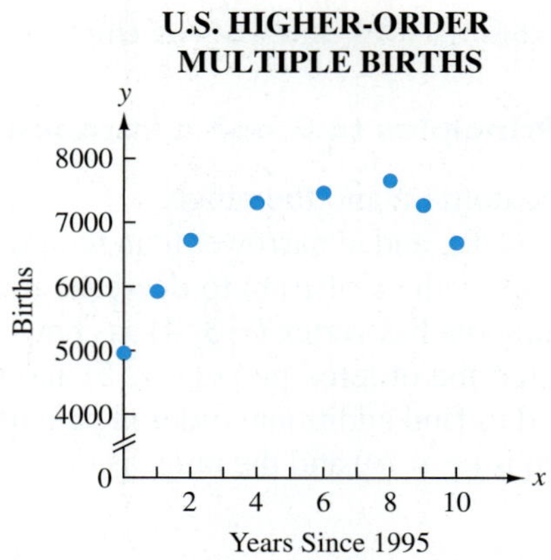

Figure 10

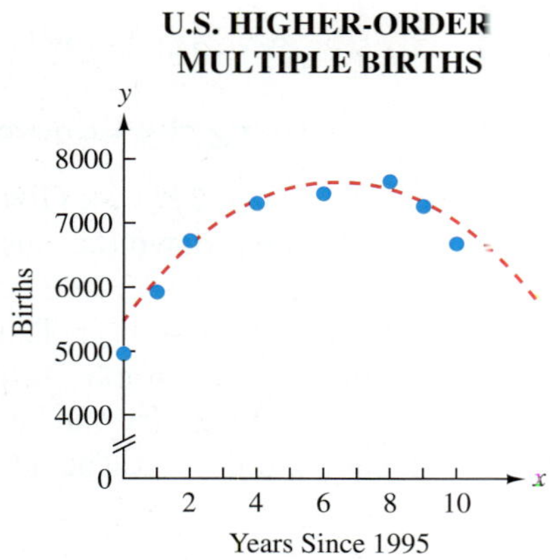

Figure 11

To find a quadratic function of the form

$$y = ax^2 + bx + c$$

that models, or *fits,* these data, we choose three representative ordered pairs and use them to write a system of three equations. Using (0, 4973), (4, 7321), and (10, 6694), we substitute the x- and y-values from the ordered pairs into the quadratic form $y = ax^2 + bx + c$ to get the three equations

$$a(0)^2 + b(0) + c = 4973 \quad \text{or} \quad c = 4973 \quad (1)$$
$$a(4)^2 + b(4) + c = 7321 \quad \text{or} \quad 16a + 4b + c = 7321 \quad (2)$$
$$a(10)^2 + b(10) + c = 6694 \quad \text{or} \quad 100a + 10b + c = 6694. \quad (3)$$

We can find the values of a, b, and c by solving this system of three equations in three variables using the methods of **Section 5.2.** From equation (1), $c = 4973$. Substitute 4973 for c in equations (2) and (3) to obtain the equations

$$16a + 4b + 4973 = 7321, \quad \text{or} \quad 16a + 4b = 2348 \quad (4)$$
$$100a + 10b + 4973 = 6694, \quad \text{or} \quad 100a + 10b = 1721. \quad (5)$$

We can eliminate b from this system of equations in two variables by multiplying equation (4) by -5 and equation (5) by 2, and adding the results to get

$$120a = -8298$$
$$a = -69.15. \quad \text{Divide by 120; use a calculator.}$$

We substitute -69.15 for a in equation (4) or (5) to find that $b = 863.6$. Using the values we have found for a, b, and c, our model is defined by

$$y = -69.15x^2 + 863.6b + 4973.$$

◀ *Work Problems* **7** *and* **8** *at the Side.*

Note
If we had chosen three different ordered pairs of data in Example 6, a slightly different model would have resulted, as in Problem 8 at the side.

Calculator Tip The *quadratic regression* feature on a graphing calculator can be used to generate a quadratic model that fits given data. See your owner's manual for details on how to do this.

10.5 Exercises

1. Match each quadratic function with its graph from choices A–D.

(a) $f(x) = (x + 2)^2 - 1$ (b) $f(x) = (x + 2)^2 + 1$ (c) $f(x) = (x - 2)^2 - 1$ (d) $f(x) = (x - 2)^2 + 1$

A.
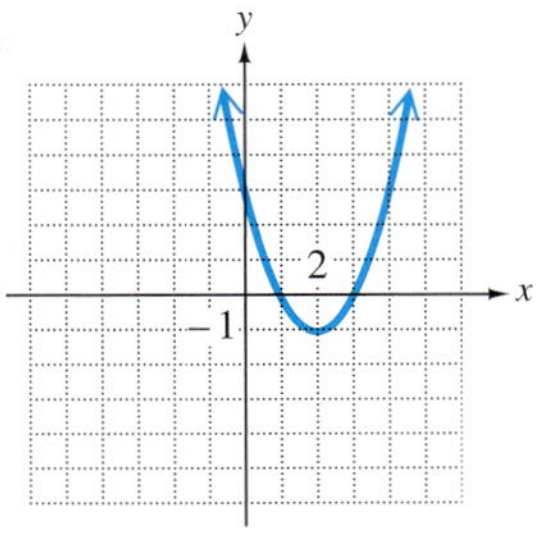

B.
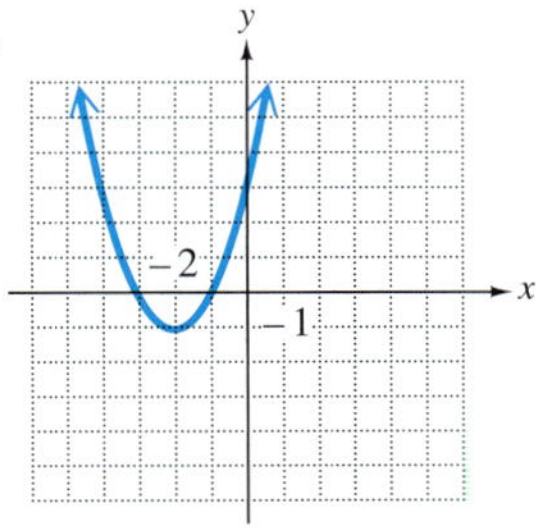

C.
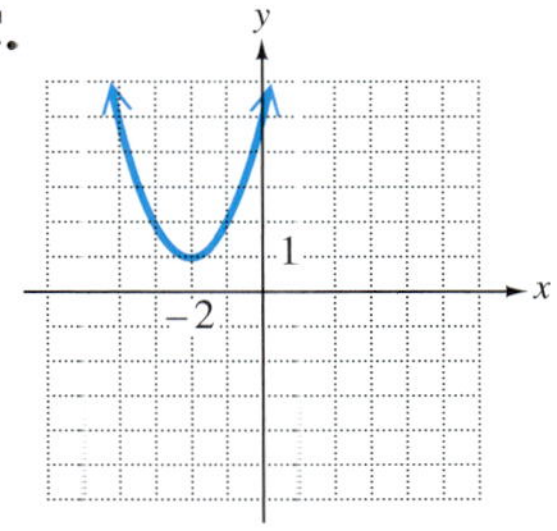

D.
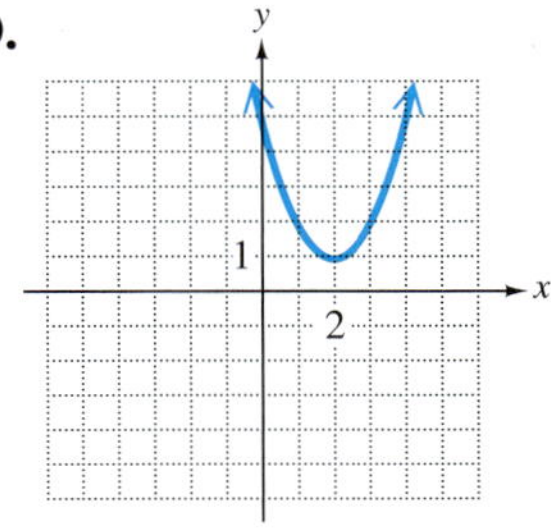

2. Match each quadratic function with its graph from choices A–D.

(a) $f(x) = -x^2 + 2$ (b) $f(x) = -x^2 - 2$ (c) $f(x) = -(x + 2)^2$ (d) $f(x) = -(x - 2)^2$

A.
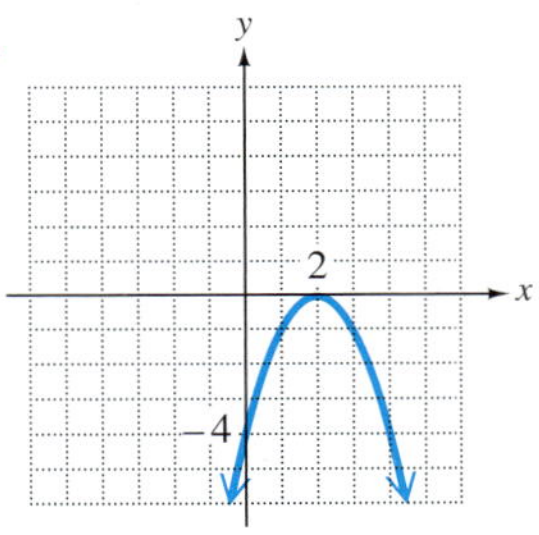

B.
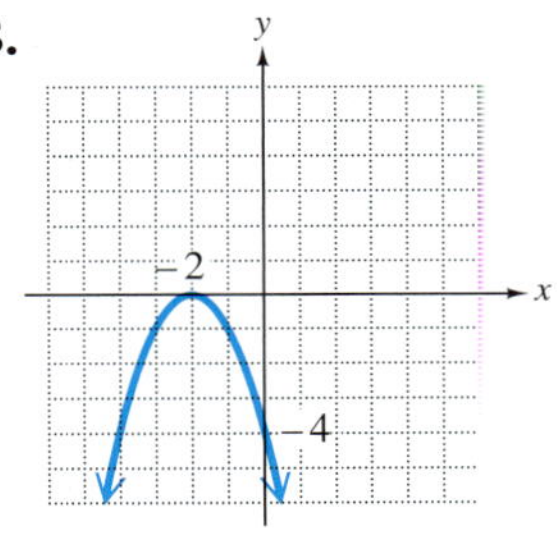

C.
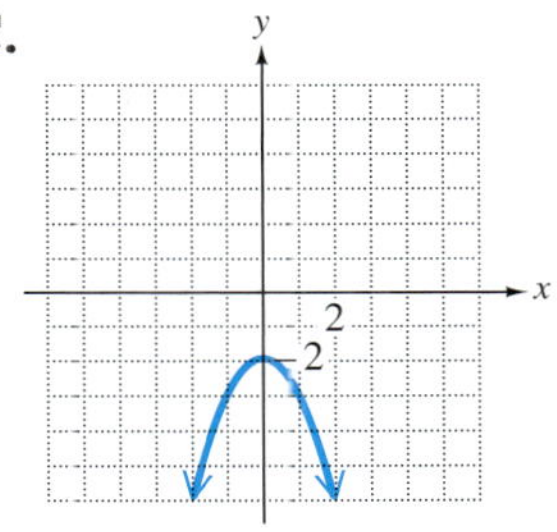

D.
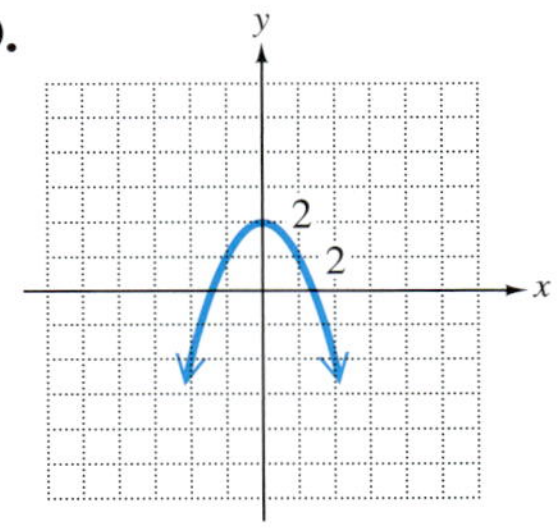

Identify the vertex of each parabola. See Examples 1–4.

3. $f(x) = -3x^2$

4. $f(x) = \frac{1}{2}x^2$

5. $f(x) = x^2 + 4$

6. $f(x) = x^2 - 4$

7. $f(x) = (x - 1)^2$

8. $f(x) = (x + 3)^2$

9. $f(x) = (x + 3)^2 - 4$

10. $f(x) = (x - 5)^2 - 8$

11. Describe how each of the parabolas in Exercises 9 and 10 is shifted compared to the graph of $f(x) = x^2$.

12. What does the value of a in $F(x) = a(x - h)^2 + k$ tell you about the graph of the equation compared to the graph of $f(x) = x^2$?

For each quadratic function, tell whether the graph opens up or down and whether the graph is wider, narrower, or the same shape as the graph of $f(x) = x^2$. See Examples 4 and 5.

13. $f(x) = -\frac{2}{5}x^2$

14. $f(x) = -2x^2$

15. $f(x) = 3x^2 + 1$

16. $f(x) = \frac{2}{3}x^2 - 4$

17. For $f(x) = a(x - h)^2 + k$, in what quadrant is the vertex if

(a) $h > 0, k > 0$;
(b) $h > 0, k < 0$;
(c) $h < 0, k > 0$;
(d) $h < 0, k < 0$?

18. Match each quadratic function with the description of the parabola that is its graph.

(a) $f(x) = (x - 4)^2 - 2$
(b) $f(x) = (x - 2)^2 - 4$
(c) $f(x) = -(x - 4)^2 - 2$
(d) $f(x) = -(x - 2)^2 - 4$

A. Vertex $(2, -4)$, opens down
B. Vertex $(2, -4)$, opens up
C. Vertex $(4, -2)$, opens down
D. Vertex $(4, -2)$, opens up

Sketch the graph of each parabola. Plot at least two points in addition to the vertex. In Exercises 25–32, give the vertex, axis, domain, and range of the parabola. See Examples 1–5.

19. $f(x) = -2x^2$

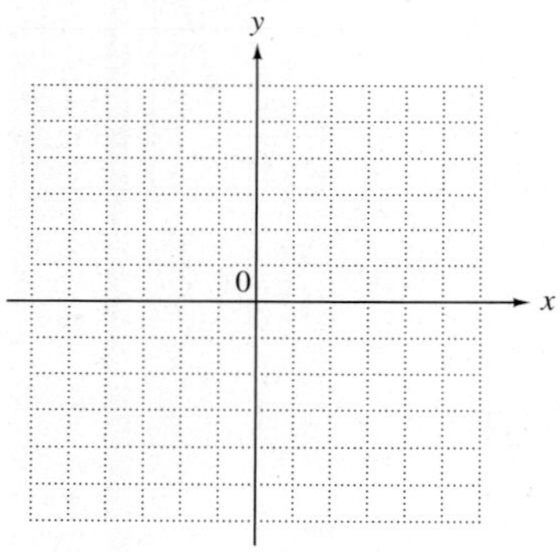

20. $f(x) = \frac{1}{3}x^2$

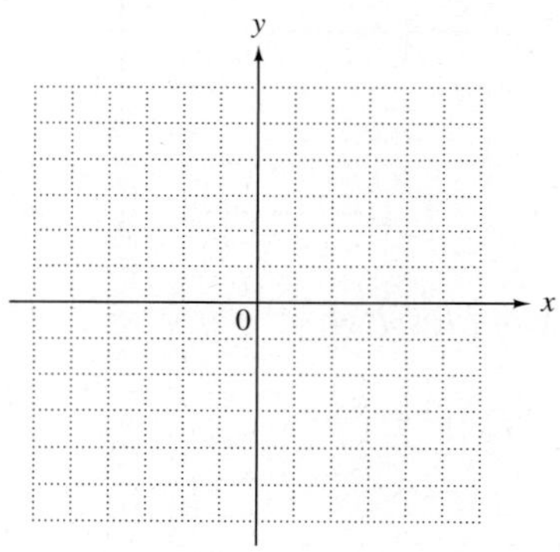

21. $f(x) = x^2 - 1$

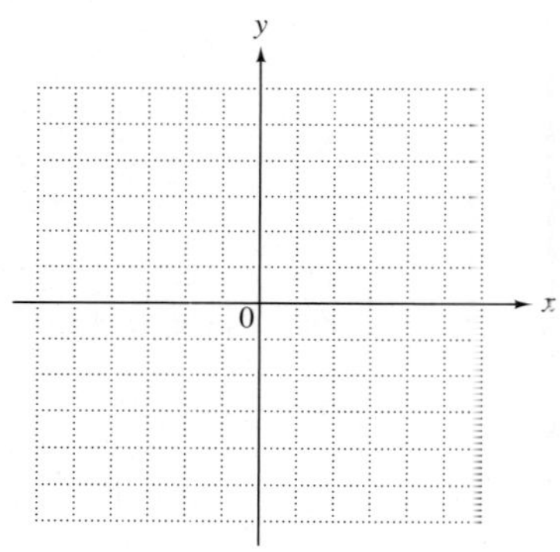

22. $f(x) = x^2 + 3$

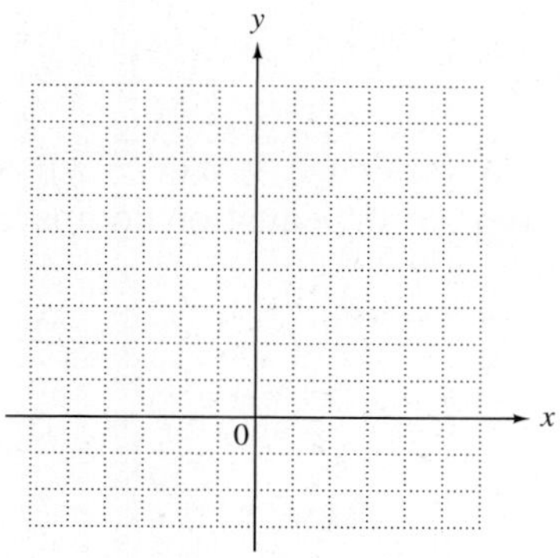

23. $f(x) = -x^2 + 2$

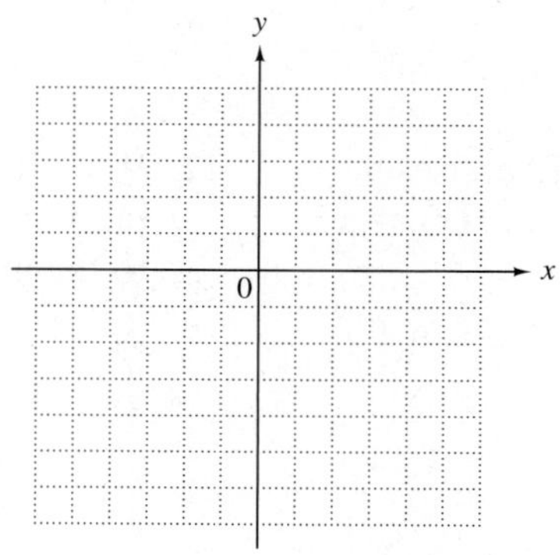

24. $f(x) = 2x^2 - 2$

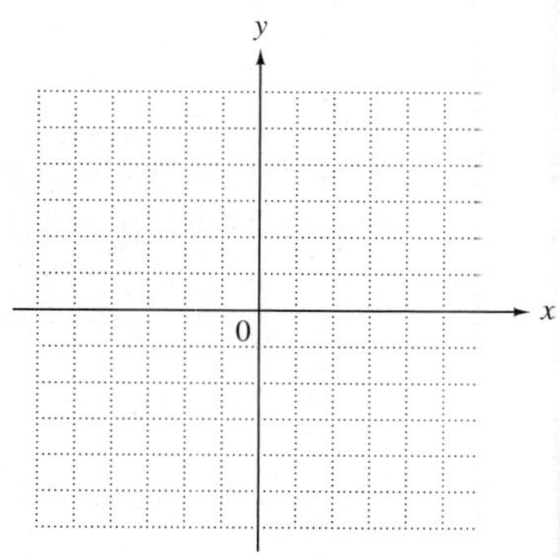

25. $f(x) = \frac{1}{2}(x - 4)^2$

vertex:
axis:
domain:
range:

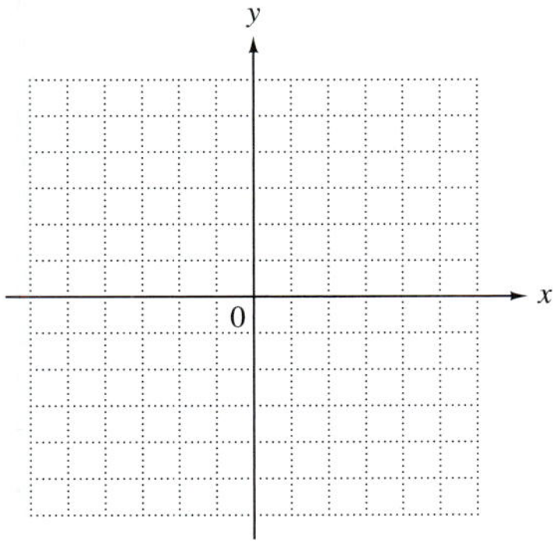

26. $f(x) = -2(x + 1)^2$

vertex:
axis:
domain:
range:

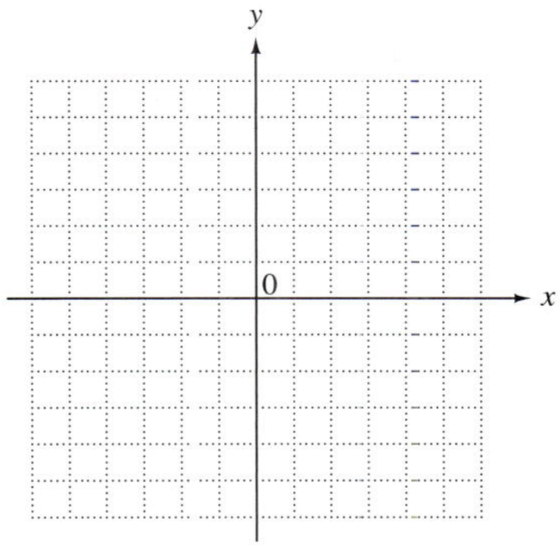

27. $f(x) = (x + 2)^2 - 1$

vertex:
axis:
domain:
range:

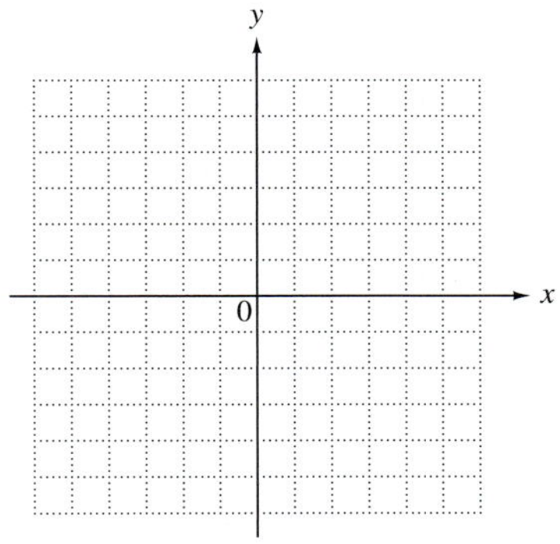

28. $f(x) = (x - 1)^2 + 2$

vertex:
axis:
domain:
range:

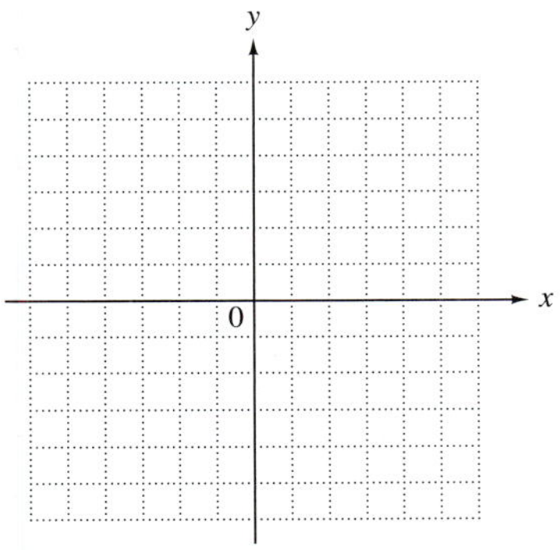

29. $f(x) = -2(x + 3)^2 + 4$

vertex:
axis:
domain:
range:

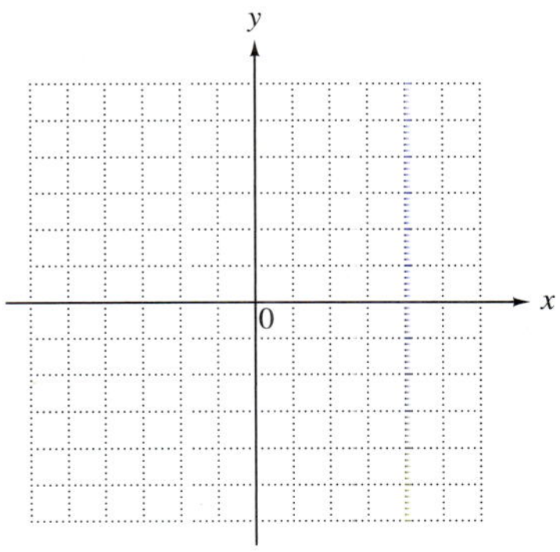

30. $f(x) = 2(x - 2)^2 - 3$

vertex:
axis:
domain:
range:

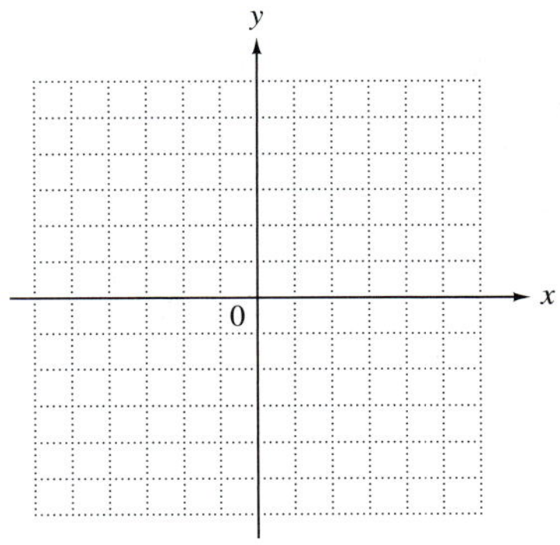

31. $f(x) = -\frac{2}{3}(x + 2)^2 + 1$

vertex:
axis:
domain:
range:

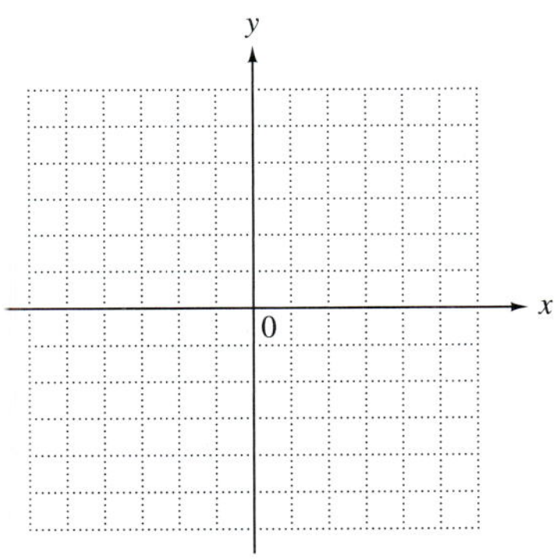

32. $f(x) = -\frac{1}{2}(x + 1)^2 + 2$

vertex:
axis:
domain:
range:

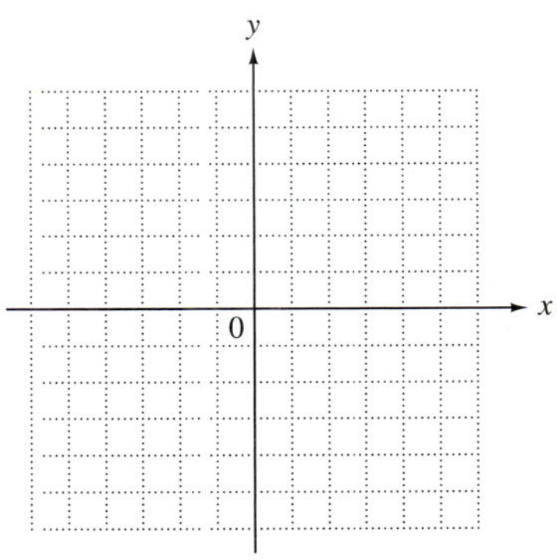

Relating Concepts (Exercises 33–38) For Individual or Group Work

The procedures described in this section that allow the graph of $f(x) = x^2$ to be shifted vertically and horizontally are applicable to other types of functions. In **Section 4.5** *we introduced linear functions of the form $g(x) = ax + b$. Consider the graph of the simplest linear function defined by $g(x) = x$, shown here, and then* **work Exercises 33–38 in order.**

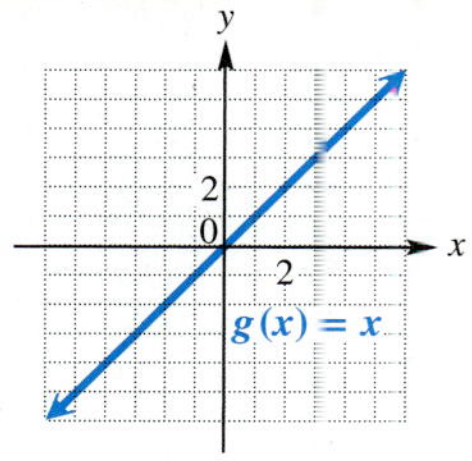

33. Based on the concepts of this section, how does the graph of $F(x) = x^2 + 6$ compare to the graph of $f(x) = x^2$ if a *vertical* shift is considered?

34. Graph the linear function defined by $G(x) = x + 6$.

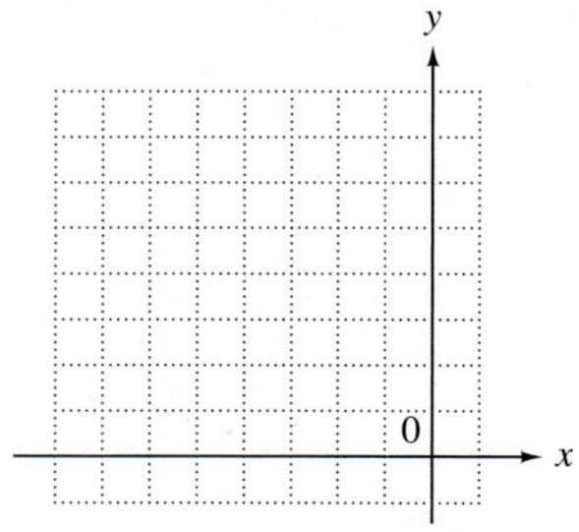

35. Based on the concepts of **Chapter 4,** how does the graph of $G(x) = x + 6$ compare to the graph of $g(x) = x$ if a *vertical* shift is considered? (*Hint:* Look at the y-intercept.)

36. Based on the concepts of this section, how does the graph of $F(x) = (x - 6)^2$ compare to the graph of $f(x) = x^2$ if a *horizontal* shift is considered?

37. Graph the linear function defined by $G(x) = x - 6$.

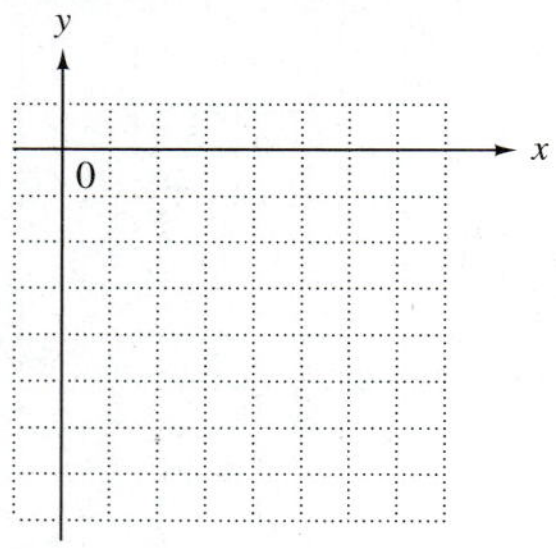

38. Based on the concepts of **Chapter 4,** how does the graph of $G(x) = x - 6$ compare to the graph of $g(x) = x$ if a *horizontal* shift is considered? (*Hint:* Look at the x-intercept.)

In Exercises 39–44, tell whether a linear or quadratic function would be a more appropriate model for each set of graphed data. If linear, tell whether the slope should be positive or negative. If quadratic, tell whether the coefficient of x^2 should be positive or negative. See Example 6.

39. **PLASMA TV SALES IN U.S.**

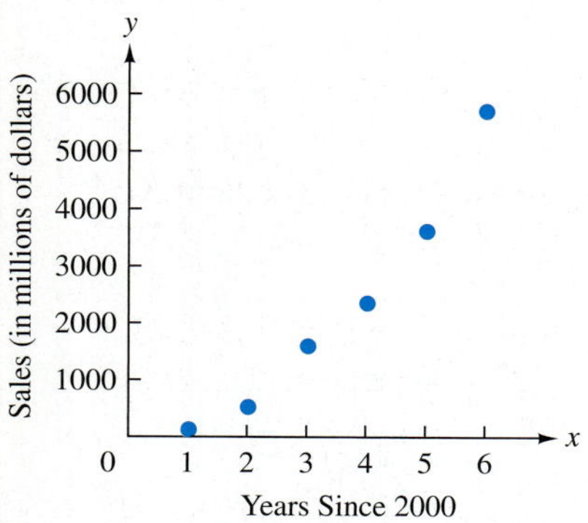

Source: Consumer Electronics Association.

40. **AVERAGE DAILY VOLUME OF FIRST-CLASS MAIL**

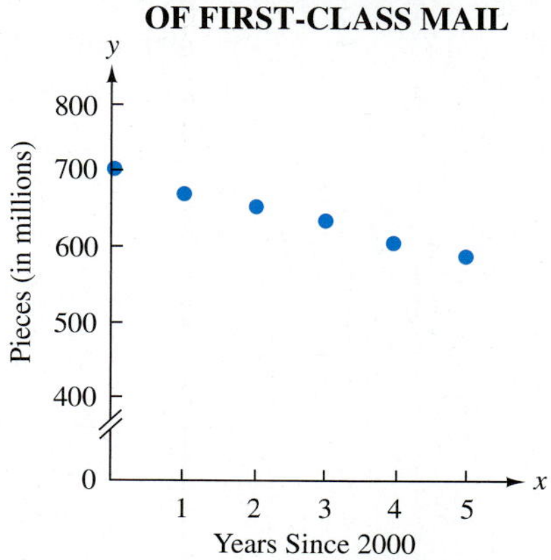

Source: General Accounting Office.

41. **SOCIAL SECURITY ASSETS***

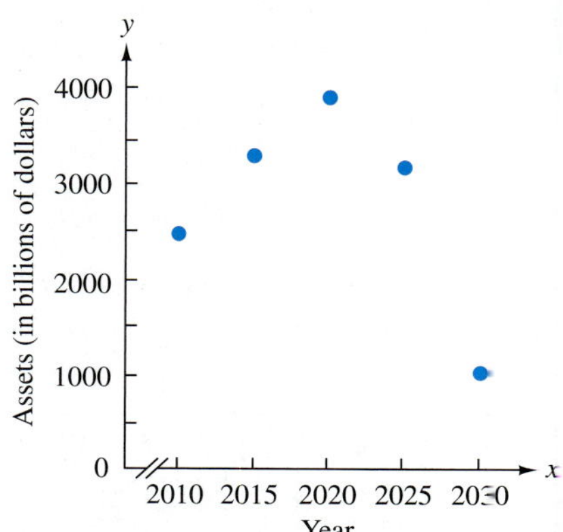

*Projected

Source: Social Security Administration.

42. FOOD ASSISTANCE SPENDING IN IOWA

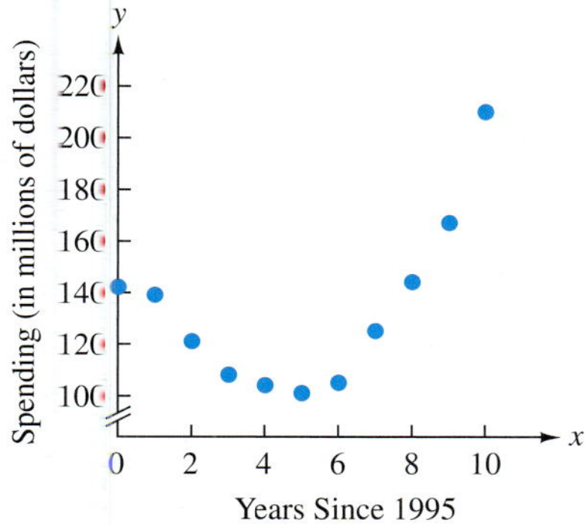

Source: Iowa Department of Human Services.

43. TIME SPENT PLAYING VIDEO GAMES*

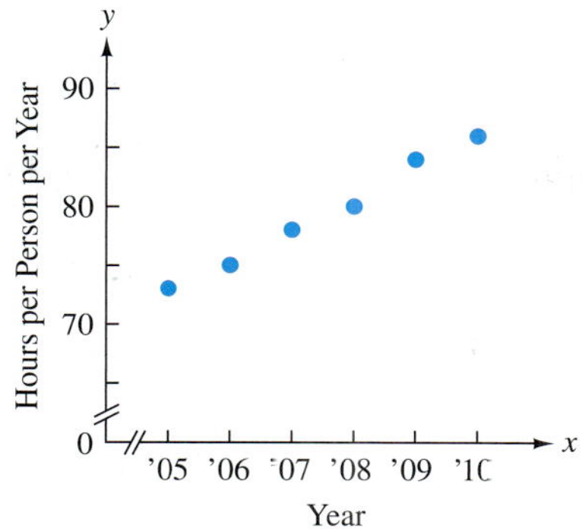

*Later years projected

Source: Veronis Suhler Stevenson.

44. SALES OF MUSIC CASSETTE TAPES

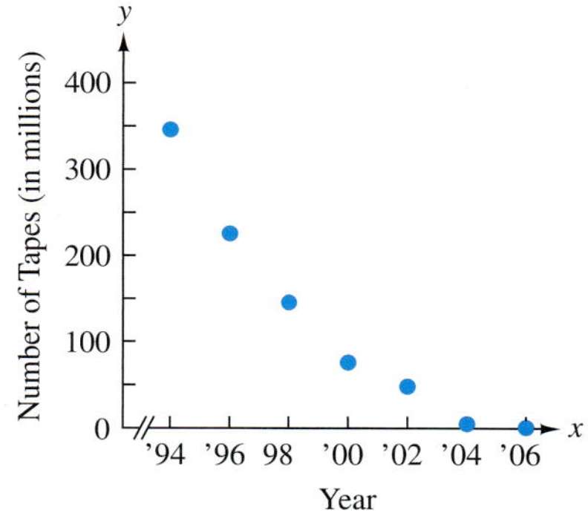

Source: Recording Industry Association of America.

Solve each problem. See Example 6.

45. Sales of digital cameras in the United States (in millions of dollars) between 2000 and 2006 are shown in the table. In the year column, 0 represents 2000, 1 represents 2001, and so on.

SALES OF DIGITAL CAMERAS

Year	Sales
0	1825
1	1972
2	2794
3	3921
4	4739
5	5611
6	7805

Source: Consumer Electronics Association.

(a) Use the ordered pairs (year, sales) to make a scatter diagram of the data.

DIGITAL CAMERA SALES IN U.S.

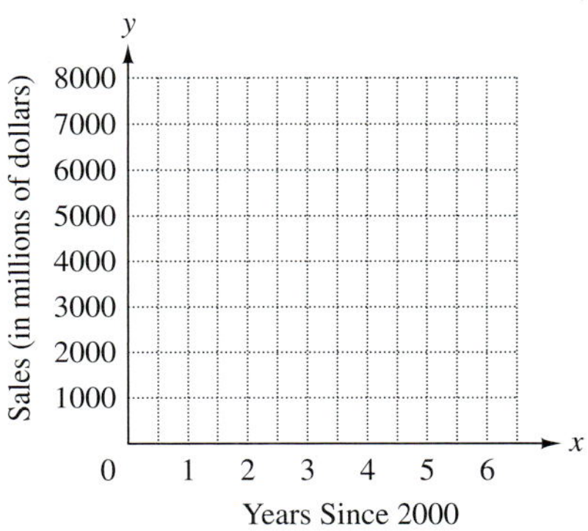

(b) Use the scatter diagram to decide whether a linear or quadratic function would better model the data. If quadratic, should the coefficient a of x^2 be positive or negative?

(c) Use the ordered pairs (0, 1825), (3, 3921), and (6, 7805) to find a quadratic function that models the data. Round the values of a, b, and c in your model to the nearest tenth, as necessary.

(d) Use your model from part (c) to approximate the sales of digital cameras in the United States in 2007. Round your answer to the nearest whole number (of millions).

(e) Sales of digital cameras were projected to be \$6945 million in 2007. Based on this, is the model valid for 2007? Explain.

46. The percent of U.S. high school students in grades 9–12 who smoke is shown in the table for selected years. In the year column, 1 represents 1991, 3 represents 1993, and so on.

HIGH SCHOOL STUDENTS WHO SMOKE

Year	Percent of Students
1	28
3	31
5	35
7	36
9	35
11	29
13	22

Source: Centers for Disease Control and Prevention.

(a) Use the ordered pairs (year, percent of students) to make a scatter diagram of the data.

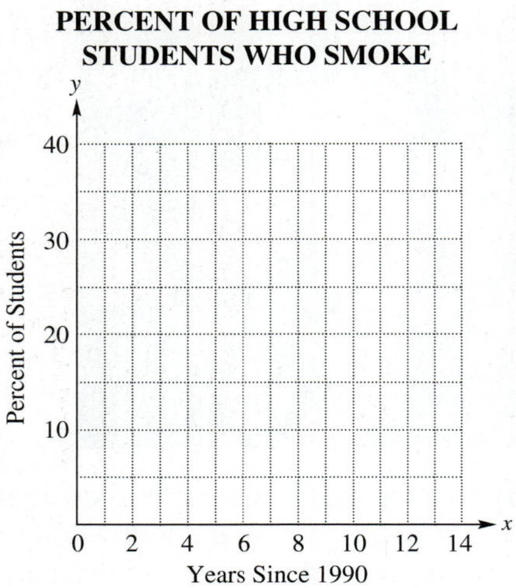

(b) Would a linear or quadratic function better model the data?

(c) Should the coefficient a of x^2 in a quadratic model be positive or negative?

(d) Use the ordered pairs (1, 28), (7, 36), and (11, 29) to find a quadratic function that models the data. Round the values of a, b, and c in your model to the nearest tenth, as necessary.

(e) Use your model from part (d) to approximate the percent of high school students who smoked during 1995 and 2003 to the nearest percent. How well does the model approximate the actual data from the table?

47. In Exercise 46(d), we determined that the quadratic function defined by

$$y = -0.3x^2 + 3.8x + 24.5$$

modeled the percent of U.S. high school students who smoked in the years 1991–2003.

(a) Use this model to approximate the number of high school students who smoked in 2005 and 2007.

(b) The actual smoking rates for high school students were 23% in 2005 and 20% in 2007. (*Source:* Centers for Disease Control and Prevention.) How do the approximations using the model compare to the actual rates for these two years?

48. Should the model from Exercise 46(d) be used to approximate the smoking rate for high school students in years after 2003? Explain.

10.6 More about Parabolas and Their Applications

OBJECTIVE 1 Find the vertex of a vertical parabola. When the equation of a parabola is given in the form $f(x) = ax^2 + bx + c$, we need to locate the vertex in order to sketch an accurate graph. There are two ways to do this:

1. Complete the square, as shown in Examples 1 and 2, or
2. Use a formula derived by completing the square, as shown in Example 3.

EXAMPLE 1 Completing the Square to Find the Vertex ($a = 1$)

Find the vertex of the graph of $f(x) = x^2 - 4x + 5$.

To find the vertex, we need to write the expression $x^2 - 4x + 5$ in the form $(x - h)^2 + k$. We do this by completing the square on $x^2 - 4x$, as in **Section 10.1.** The process is slightly different here because we want to keep $f(x)$ alone on one side of the equation. Instead of adding the appropriate number to each side, we *add and subtract* it on the right. This is equivalent to adding 0.

$f(x) = x^2 - 4x + 5$

$f(x) = (x^2 - 4x \quad) + 5$ Group the variable terms.

$\left[\frac{1}{2}(-4)\right]^2 = (-2)^2 = 4$

$f(x) = (x^2 - 4x + 4 - 4) + 5$ Add and subtract 4.

$f(x) = (x^2 - 4x + 4) - 4 + 5$ Bring -4 outside the parentheses.

$f(x) = (x - 2)^2 + 1$ Factor; combine like terms.

The vertex of this parabola is $(2, 1)$.

Work Problem 1 at the Side. ▶

EXAMPLE 2 Completing the Square to Find the Vertex ($a \neq 1$)

Find the vertex of the graph of $f(x) = -3x^2 + 6x - 1$.

We must complete the square on $-3x^2 + 6x$. Because the x^2-term has a coefficient other than 1, we factor that coefficient out of the first two terms and then proceed as in Example 1.

$f(x) = -3x^2 + 6x - 1$

$f(x) = -3(x^2 - 2x) - 1$ Factor out -3.

$\left[\frac{1}{2}(-2)\right]^2 = (-1)^2 = 1$

$f(x) = -3(x^2 - 2x + 1 - 1) - 1$ Add and subtract 1.

Now bring −1 outside the parentheses; be sure to multiply it by −3.

$f(x) = -3(x^2 - 2x + 1) + (-3)(-1) - 1$ Distributive property

This is a key step.

$f(x) = -3(x^2 - 2x + 1) + 3 - 1$

$f(x) = -3(x - 1)^2 + 2$ Factor; combine like terms.

The vertex is $(1, 2)$.

Work Problem 2 at the Side. ▶

OBJECTIVES

1. Find the vertex of a vertical parabola.
2. Graph a quadratic function.
3. Use the discriminant to find the number of x-intercepts of a parabola with a vertical axis.
4. Use quadratic functions to solve problems involving maximum or minimum value.
5. Graph parabolas with horizontal axes.

1 Find the vertex of the graph of each quadratic function.

(a) $f(x) = x^2 - 6x + 7$

(b) $f(x) = x^2 + 4x - 9$

2 Find the vertex of the graph of each quadratic function.

(a) $f(x) = 2x^2 - 4x + 1$

(b) $f(x) = -\frac{1}{2}x^2 + 2x - 3$

ANSWERS

1. (a) $(3, -2)$ **(b)** $(-2, -13)$
2. (a) $(1, -1)$ **(b)** $(2, -1)$

3 Use the formula to find the vertex of the graph of each quadratic function.

(a) $f(x) = -2x^2 + 3x - 1$

(b) $f(x) = 4x^2 - x + 5$

To derive a formula for the vertex of the graph of the quadratic function defined by $f(x) = ax^2 + bx + c$, complete the square.

$$f(x) = ax^2 + bx + c \quad (a \neq 0) \qquad \text{Standard form}$$

$$f(x) = a\left(x^2 + \frac{b}{a}x\right) + c \qquad \text{Factor } a \text{ from the first two terms.}$$

$$\left[\frac{1}{2}\left(\frac{b}{a}\right)\right]^2 = \left(\frac{b}{2a}\right)^2 = \frac{b^2}{4a^2}$$

$$f(x) = a\left(x^2 + \frac{b}{a}x + \frac{b^2}{4a^2} - \frac{b^2}{4a^2}\right) + c \qquad \text{Add and subtract } \frac{b^2}{4a^2}.$$

$$f(x) = a\left(x^2 + \frac{b}{a}x + \frac{b^2}{4a^2}\right) + a\left(-\frac{b^2}{4a^2}\right) + c \qquad \text{Distributive property}$$

$$f(x) = a\left(x^2 + \frac{b}{a}x + \frac{b^2}{4a^2}\right) - \frac{b^2}{4a} + c \qquad -\frac{ab^2}{4a^2} = -\frac{b^2}{4a}$$

$$f(x) = a\left(x + \frac{b}{2a}\right)^2 + \frac{4ac - b^2}{4a} \qquad \text{Factor; rewrite terms with a common denominator.}$$

$$f(x) = a\left[x - \underbrace{\left(\frac{-b}{2a}\right)}_{h}\right]^2 + \underbrace{\frac{4ac - b^2}{4a}}_{k} \qquad f(x) = (x - h)^2 + k$$

Thus, the vertex (h, k) can be expressed in terms of a, b, and c. It is not necessary to remember the expression for k, since it can be found by replacing x with $\frac{-b}{2a}$. Using function notation, if $y = f(x)$, then the y-value of the vertex is $f\left(\frac{-b}{2a}\right)$.

Vertex Formula

The graph of the quadratic function defined by $f(x) = ax^2 + bx + c$ has vertex

$$\left(\frac{-b}{2a}, f\left(\frac{-b}{2a}\right)\right).$$

The axis of the parabola is the line

$$x = \frac{-b}{2a}.$$

EXAMPLE 3 Using the Formula to Find the Vertex

Use the vertex formula to find the vertex of the graph of $f(x) = x^2 - x - 6$.

For this function, $a = 1$, $b = -1$, and $c = -6$. The x-coordinate of the vertex of the parabola is given by

$$\frac{-b}{2a} = \frac{-(-1)}{2(1)} = \frac{1}{2}.$$

The y-coordinate is $f\left(\frac{-b}{2a}\right) = f\left(\frac{1}{2}\right)$.

$$f\left(\frac{1}{2}\right) = \left(\frac{1}{2}\right)^2 - \frac{1}{2} - 6 = \frac{1}{4} - \frac{1}{2} - 6 = -\frac{25}{4}$$

The vertex is $\left(\frac{1}{2}, -\frac{25}{4}\right)$.

◀ Work Problem **3** at the Side.

ANSWERS

3. (a) $\left(\frac{3}{4}, \frac{1}{8}\right)$ (b) $\left(\frac{1}{8}, \frac{79}{16}\right)$

OBJECTIVE 2 Graph a quadratic function.

Graphing a Quadratic Function f

Step 1 **Determine whether the graph opens up or down.** If $a > 0$, the parabola opens up; if $a < 0$, it opens down.

Step 2 **Find the vertex.** Use either the vertex formula or completing the square.

Step 3 **Find any intercepts.** To find the x-intercepts (if any), solve $f(x) = 0$. To find the y-intercept, evaluate $f(0)$.

Step 4 **Complete the graph.** Plot the points found so far. Find and plot additional points as needed, using symmetry about the axis.

EXAMPLE 4 Graphing a Quadratic Function

Graph the quadratic function defined by $f(x) = x^2 - x - 6$.

Step 1 From the equation, $a = 1$, so the graph of the function opens up.

Step 2 The vertex, $\left(\frac{1}{2}, -\frac{25}{4}\right)$, was found in Example 3 by substituting the values $a = 1$, $b = -1$, and $c = -6$ in the vertex formula.

Step 3 Now find any intercepts. Since the vertex, $\left(\frac{1}{2}, -\frac{25}{4}\right)$, is in quadrant IV and the graph opens up, there will be two x-intercepts. To find them, let $f(x) = 0$ and solve the equation.

$$f(x) = x^2 - x - 6$$

$$0 = x^2 - x - 6 \qquad \text{Let } f(x) = 0.$$

$$0 = (x - 3)(x + 2) \qquad \text{Factor.}$$

$$x - 3 = 0 \quad \text{or} \quad x + 2 = 0 \qquad \text{Zero-factor property}$$

$$x = 3 \quad \text{or} \quad x = -2 \qquad \text{Solve each equation.}$$

The x-intercepts are $(3, 0)$ and $(-2, 0)$. Find the y-intercept.

$$f(x) = x^2 - x - 6$$

$$f(0) = 0^2 - 0 - 6 \qquad \text{Let } x = 0.$$

$$f(0) = -6$$

The y-intercept is $(0, -6)$.

Step 4 Plot the points found so far and additional points as needed using symmetry about the axis $x = \frac{1}{2}$. The graph is shown in Figure 12. The domain is $(-\infty, \infty)$, and the range is $\left[-\frac{25}{4}, \infty\right)$.

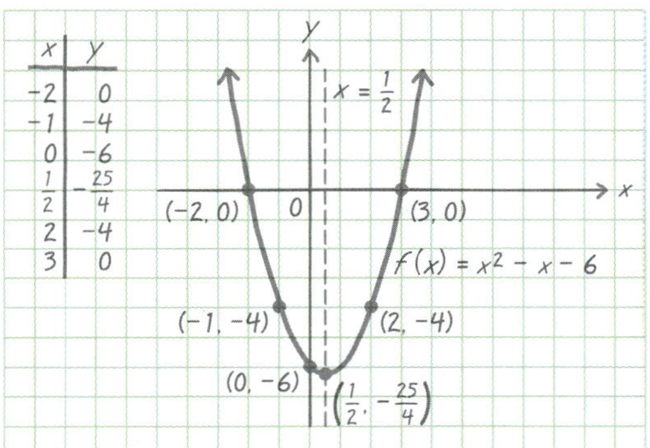

Figure 12

Work Problem 4 at the Side. ▶

4 Graph the quadratic function defined by

$$f(x) = x^2 - 6x + 5.$$

Give the vertex, axis, domain, and range.

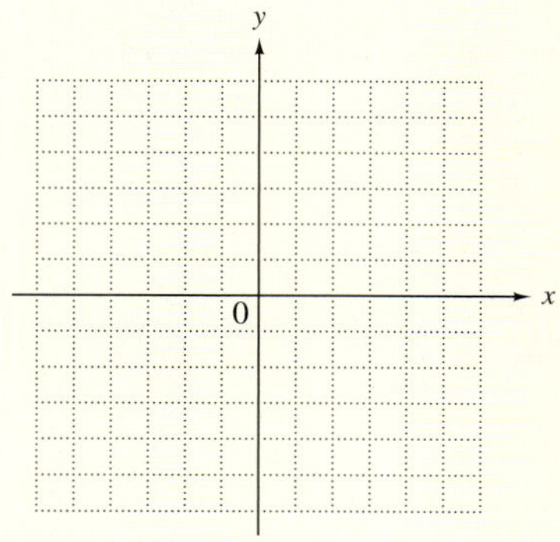

ANSWER

4.

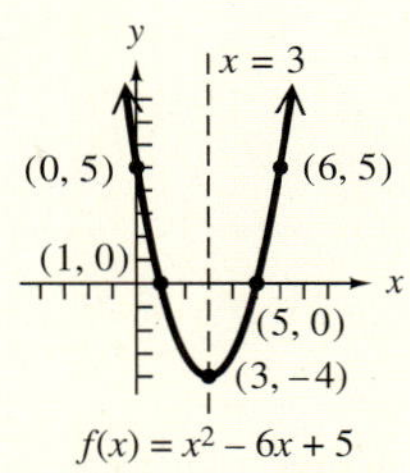

vertex: $(3, -4)$; axis: $x = 3$; domain: $(-\infty, \infty)$; range: $[-4, \infty)$

5 Use the discriminant to determine the number of x-intercepts of the graph of each quadratic function.

(a) $f(x) = 4x^2 - 20x + 25$

(b) $f(x) = 2x^2 + 3x + 5$

(c) $f(x) = -3x^2 - x + 2$

OBJECTIVE 3 Use the discriminant to find the number of x-intercepts of a parabola with a vertical axis. Recall from **Section 10.2** that the expression $b^2 - 4ac$ is called the **discriminant** of the quadratic *equation* $ax^2 + bx + c = 0$ and that we can use it to determine the number of real solutions of a quadratic equation.

In a similar way, we can use the discriminant of a quadratic *function* to determine the number of x-intercepts of its graph. See Figure 13. If the discriminant is positive, the parabola will have two x-intercepts. If the discriminant is 0, there will be only one x-intercept, and it will be the vertex of the parabola. If the discriminant is negative, the graph will have no x-intercepts.

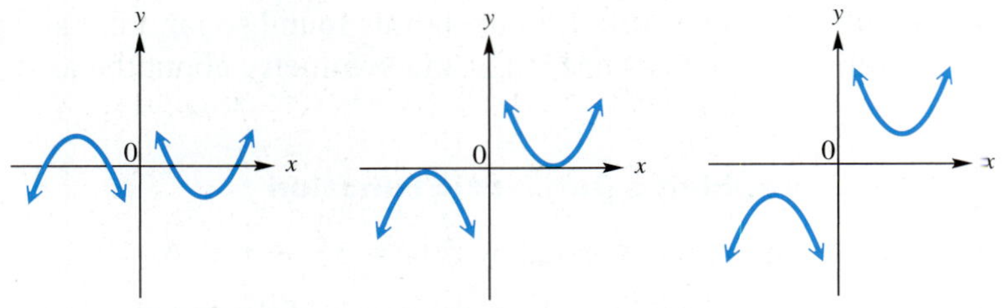

$b^2 - 4ac > 0$ Two x-intercepts | $b^2 - 4ac = 0$ One x-intercept | $b^2 - 4ac < 0$ No x-intercepts

Figure 13

EXAMPLE 5 Using the Discriminant to Determine the Number of x-Intercepts

Use the discriminant to determine the number of x-intercepts of the graph of each quadratic function.

(a) $f(x) = 2x^2 + 3x - 5$

The discriminant is $b^2 - 4ac$. Here $a = 2$, $b = 3$, and $c = -5$, so

$$\begin{aligned} & b^2 - 4ac \\ &= 3^2 - 4(2)(-5) && \text{Substitute.} \\ &= 9 - (-40) && \text{Apply the exponent; multiply.} \\ &= 49. && \text{Subtract.} \end{aligned}$$

Since the discriminant is positive, the parabola has two x-intercepts.

(b) $f(x) = -3x^2 - 1$

In this equation, $a = -3$, $b = 0$, and $c = -1$. The discriminant is

$$\begin{aligned} & b^2 - 4ac \\ &= 0^2 - 4(-3)(-1) \\ &= -12. \end{aligned}$$

The discriminant is negative, so the graph has no x-intercepts.

(c) $f(x) = 9x^2 + 6x + 1$

Here, $a = 9$, $b = 6$, and $c = 1$. The discriminant is

$$\begin{aligned} & b^2 - 4ac \\ &= 6^2 - 4(9)(1) \\ &= 0. \end{aligned}$$

The parabola has only one x-intercept (its vertex) because the value of the discriminant is 0.

◀ *Work Problem* **5** *at the Side.*

ANSWERS

5. (a) discriminant is 0; one x-intercept
(b) discriminant is -31; no x-intercepts
(c) discriminant is 25; two x-intercepts

OBJECTIVE 4 Use quadratic functions to solve problems involving maximum or minimum value. The vertex of a parabola is either the highest or the lowest point on the parabola. The y-value of the vertex gives the maximum or minimum value of y, while the x-value tells where that maximum or minimum occurs.

Problem-Solving Hint

In many applied problems we must find the least or greatest value of some quantity. When we can express that quantity as a quadratic function, the value of k in the vertex (h, k) gives that optimum value.

EXAMPLE 6 Finding the Maximum Area of a Rectangular Region

A farmer has 120 ft of fencing to enclose a rectangular area next to a building. See Figure 14. Find the maximum area he can enclose and the width required to produce this maximum area.

Figure 14

Let x represent the width of the rectangle. Since he has 120 ft of fencing,

$$x + x + \text{length} = 120 \quad \text{Sum of the sides is 120 ft.}$$
$$2x + \text{length} = 120 \quad \text{Combine like terms.}$$
$$\text{length} = 120 - 2x. \quad \text{Subtract } 2x.$$

The area $A(x)$ is given by the product of the width and length, so

$$A(x) = x(120 - 2x)$$
$$A(x) = 120x - 2x^2.$$

To determine the maximum area, find the vertex of the parabola given by $A(x) = 120x - 2x^2$ using the vertex formula. Writing the equation in standard form as

$$A(x) = -2x^2 + 120x$$

gives $a = -2$, $b = 120$, and $c = 0$, so

$$h = \frac{-b}{2a} = \frac{-120}{2(-2)} = \frac{-120}{-4} = 30;$$

$$A(30) = -2(30)^2 + 120(30) = -2(900) + 3600 = 1800.$$

The graph is a parabola that opens down, and its vertex is $(30, 1800)$. Thus, the maximum area will be 1800 ft^2. This area will occur if x, the width of the rectangle, is 30 ft.

Work Problem 6 at the Side. ▶

6 Solve Example 6 if the farmer has only 100 ft of fencing.

ANSWER

6. The rectangle should be 25 ft by 50 ft with a maximum area of 1250 ft^2.

7 Solve the problem.

A toy rocket is launched from the ground so that its distance in feet above the ground after t seconds is

$$s(t) = -16t^2 + 208t.$$

Find the maximum height it reaches and the number of seconds it takes to reach that height.

CAUTION

Be careful when interpreting the meanings of the coordinates of the vertex. The first coordinate, x, gives the value for which the *function value* is a maximum or a minimum. Be sure to read the problem carefully to determine whether you are asked to find the value of the independent variable, the function value, or both.

EXAMPLE 7 Finding the Maximum Height Attained by a Projectile

If air resistance is neglected, a projectile on Earth shot straight upward with an initial velocity of 40 m per sec will be at a height s in meters given by

$$s(t) = -4.9t^2 + 40t,$$

where t is the number of seconds elapsed after projection. After how many seconds will it reach its maximum height, and what is this maximum height?

For this function, $a = -4.9$, $b = 40$, and $c = 0$. Use the vertex formula.

$$h = \frac{-b}{2a} = \frac{-40}{2(-4.9)} \approx 4.1 \quad \text{Use a calculator.}$$

Thus, the maximum height is attained at 4.1 sec. To find this maximum height, calculate $s(4.1)$.

$$s(4.1) = -4.9(4.1)^2 + 40(4.1) \approx 81.6 \quad \text{Use a calculator.}$$

The projectile will attain a maximum height of approximately 81.6 m.

◀ Work Problem 7 at the Side.

OBJECTIVE 5 Graph parabolas with horizontal axes. If x and y are interchanged in the equation $y = ax^2 + bx + c$, the equation becomes

$$x = ay^2 + by + c.$$

Because of the interchange of the roles of x and y, these parabolas are horizontal (with horizontal lines as axes).

Graph of a Horizontal Parabola

The graph of

$$\mathbf{x = ay^2 + by + c} \quad \text{or} \quad \mathbf{x = a(y - k)^2 + h}$$

is a parabola with vertex (h, k) and the horizontal line $y = k$ as axis. The graph opens to the right if $a > 0$ and to the left if $a < 0$.

EXAMPLE 8 Graphing a Horizontal Parabola

Graph $x = (y - 2)^2 - 3$. Give the vertex, axis, domain, and range.

This graph has its vertex at $(-3, 2)$, since the roles of x and y are reversed. It opens to the right, the positive x-direction, and has the same shape as $y = x^2$. Plotting a few additional points gives the graph shown in Figure 15 on the next page. Note that the graph is symmetric about its axis, $y = 2$. The domain is $[-3, \infty)$, and the range is $(-\infty, \infty)$.

Continued on Next Page

ANSWER

7. 676 ft; 6.5 sec

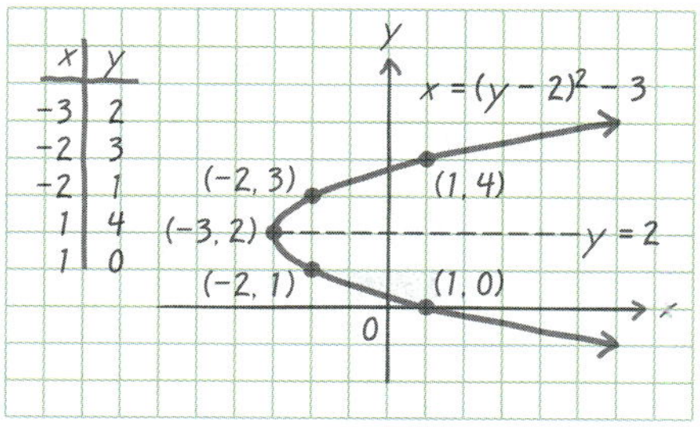

Figure 15

Work Problem **8** at the Side. ▶

When a quadratic equation is given in the form $x = ay^2 + by + c$, completing the square on y will allow us to find the vertex.

EXAMPLE 9 **Completing the Square to Graph a Horizontal Parabola**

Graph $x = -2y^2 + 4y - 3$. Give the vertex, axis, domain, and range.

$x = -2y^2 + 4y - 3$	
$x = -2(y^2 - 2y) - 3$	Factor out -2.
$x = -2(y^2 - 2y + 1 - 1) - 3$	Complete the square within the parentheses; add and subtract 1.
$x = -2(y^2 - 2y + 1) + (-2)(-1) - 3$	Distributive property
$x = -2(y - 1)^2 - 1$	Factor; simplify.

Be careful here.

Because of the negative coefficient (-2) in $x = -2(y - 1)^2 - 1$, the graph opens to the left (the negative x-direction). The graph is narrower than the graph of $y = x^2$ because $|-2| > 1$. As shown in Figure 16, the vertex is $(-1, 1)$ and the axis is $y = 1$. The domain is $(-\infty, -1]$, and the range is $(-\infty, \infty)$.

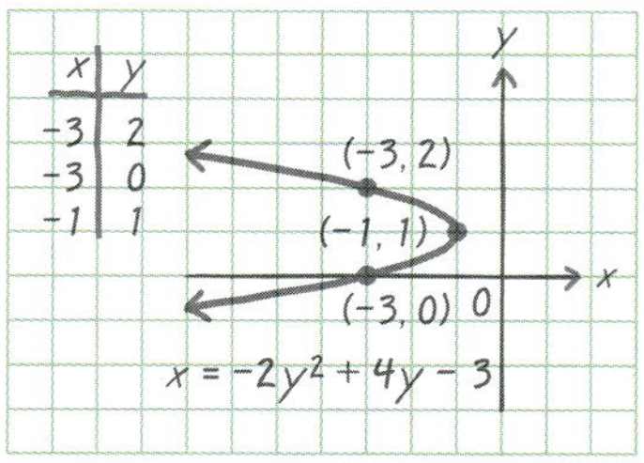

Figure 16

Work Problem **9** at the Side. ▶

CAUTION

Only quadratic equations solved for y (whose graphs are vertical parabolas) are examples of functions. The horizontal parabolas in Examples 8 and 9 are ***not*** graphs of functions, because they do not satisfy the vertical line test.

8 Graph $x = (y + 1)^2 - 4$. Give the vertex, axis, domain, and range.

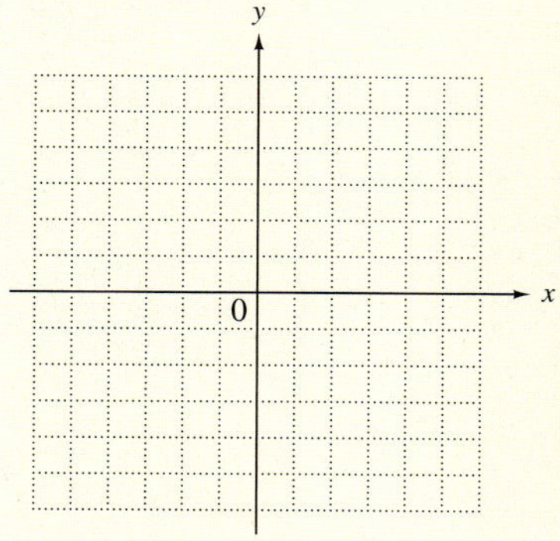

9 Graph $x = -y^2 + 2y + 5$. Give the vertex, axis, domain, and range.

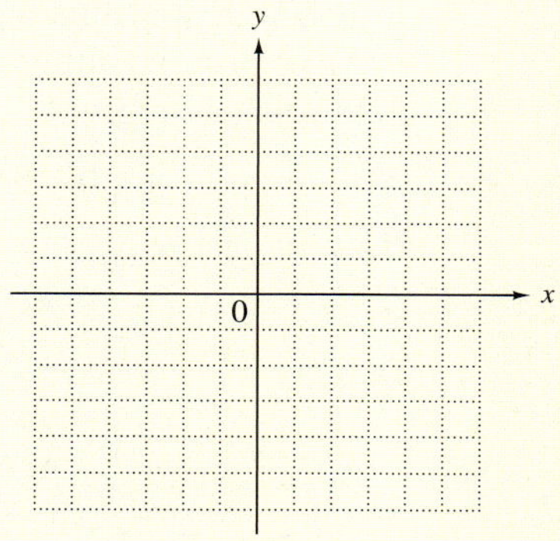

ANSWERS

8.

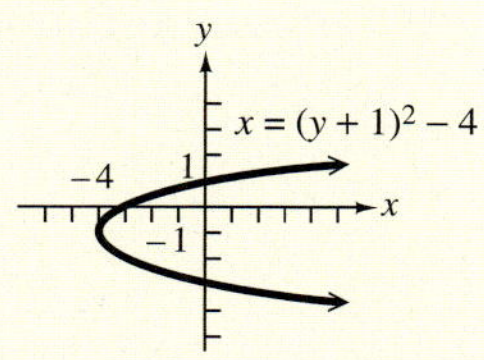

vertex: $(-4, -1)$; axis: $y = -1$; domain: $[-4, \infty)$; range: $(-\infty, \infty)$

9.

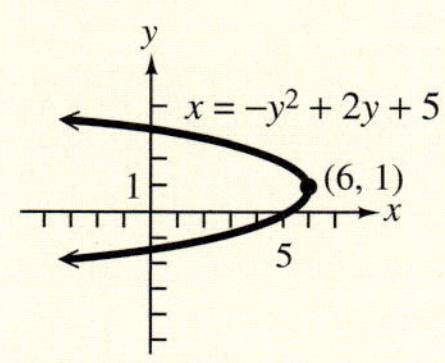

vertex: $(6, 1)$; axis: $y = 1$; domain: $(-\infty, 6]$; range: $(-\infty, \infty)$

10 Find the vertex of each parabola. Tell whether the graph opens to the right or to the left. Give the domain and range.

(a) $x = 2y^2 - 6y + 5$

(b) $x = -y^2 + 2y + 5$

11 **(a)** Tell whether each of the following equations has a vertical or horizontal parabola as its graph.

A. $y = -x^2 + 20x + 80$

B. $x = 2y^2 + 6y + 5$

C. $x + 1 = (y + 2)^2$

D. $f(x) = (x - 4)^2$

(b) Which of the equations in part (a) represent functions?

In summary, the graphs of parabolas studied in **Sections 10.5 and 10.6** fall into the following categories.

GRAPHS OF PARABOLAS

Equation	Graph
$y = ax^2 + bx + c$ $y = a(x - h)^2 + k$	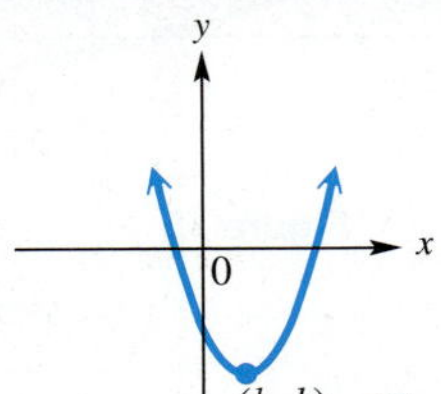 These graphs represent functions. 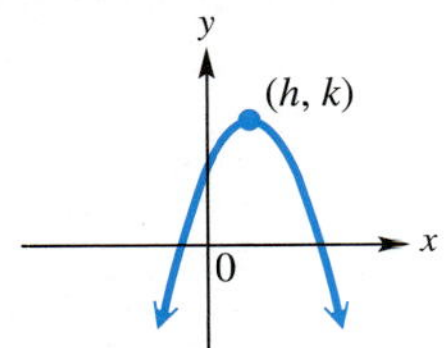
$x = ay^2 + by + c$ $x = a(y - k)^2 + h$	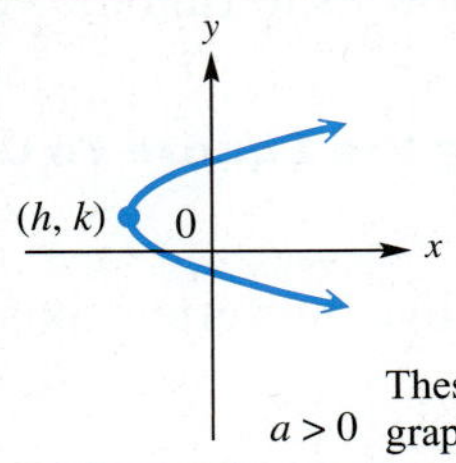 These graphs are not graphs of functions. 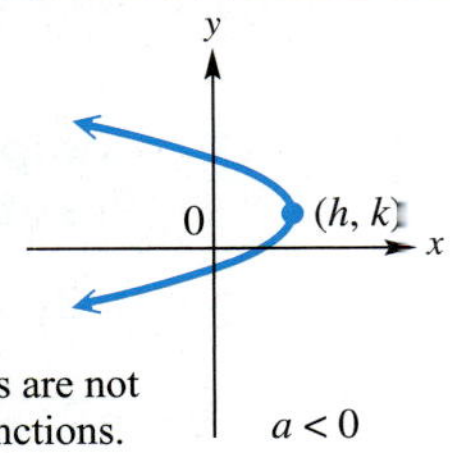

◀ *Work Problems* **10** *and* **11** *at the Side.*

ANSWERS

10. (a) $\left(\frac{1}{2}, \frac{3}{2}\right)$; right; domain: $\left[\frac{1}{2}, \infty\right)$; range: $(-\infty, \infty)$

(b) $(6, 1)$; left; domain; $(-\infty, 6]$; range: $(-\infty, \infty)$

11. (a) A, D are vertical parabolas; B, C are horizontal parabolas.

(b) A, D

10.6 Exercises

FOR EXTRA HELP — MyMathLab — MathXL PRACTICE — WATCH — DOWNLOAD — READ — REVIEW

1. How can you determine just by looking at the equation of a parabola whether it has a vertical or a horizontal axis?

2. Why can't the graph of a quadratic function be a horizontal parabola?

3. How can you determine the number of x-intercepts of the graph of a quadratic function without graphing the function?

4. If the vertex of the graph of a quadratic function is $(1, -3)$ and the graph opens down, how many x-intercepts does the graph have?

Find the vertex of each parabola. For each equation, decide whether the graph opens up, down, to the left, or to the right, and whether it is wider, narrower, or the same shape as the graph of $y = x^2$. If it is a vertical parabola, use the discriminant to determine the number of x-intercepts. See Examples 1–3, 5, 8, and 9.

5. $y = 2x^2 + 4x + 5$

6. $y = 3x^2 - 6x + 4$

7. $y = -x^2 + 5x + 3$

8. $x = -y^2 + 7y - 2$

9. $x = \frac{1}{3}y^2 + 6y + 24$

10. $x = \frac{1}{2}y^2 + 10y - 5$

Graph each parabola. Give the vertex, axis, domain, and range. See Examples 4, 8, and 9.

11. $f(x) = x^2 + 4x + 3$
vertex:
axis:
domain:
range:

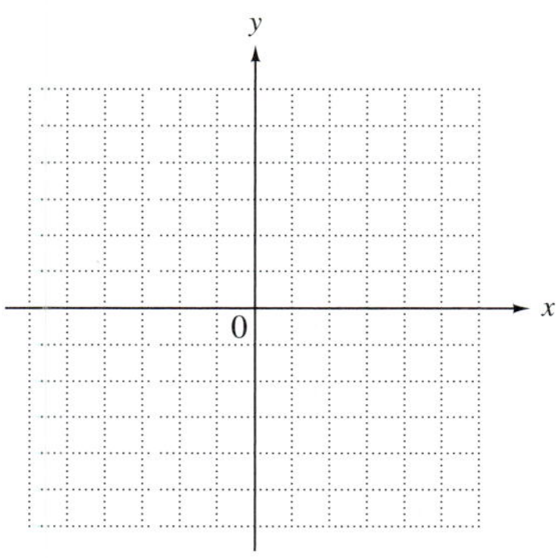

12. $f(x) = x^2 + 2x - 2$
vertex:
axis:
domain:
range:

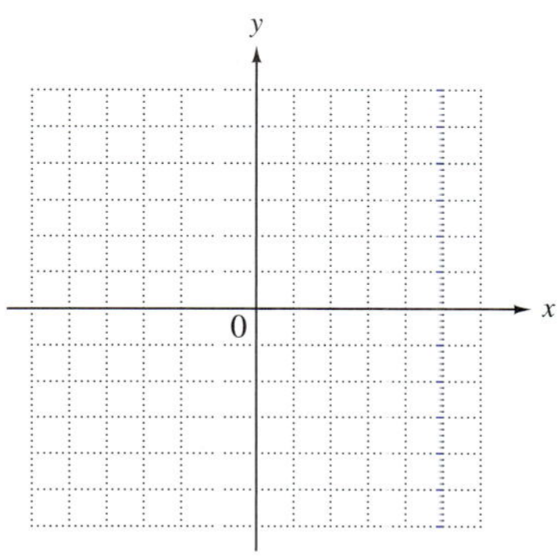

13. $f(x) = -2x^2 + 4x - 5$
vertex:
axis:
domain:
range:

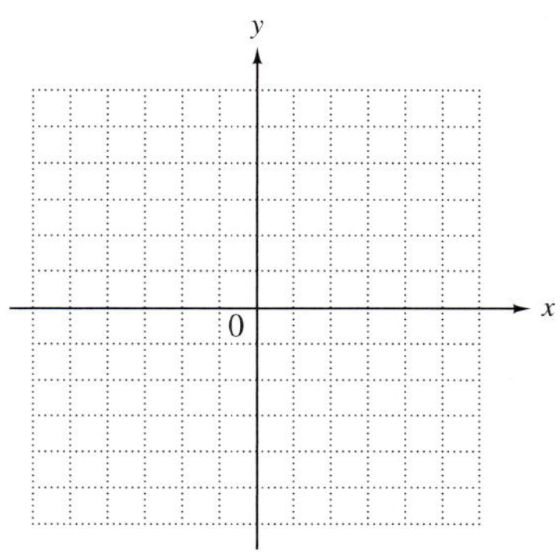

14. $f(x) = -3x^2 + 12x - 8$
vertex:
axis:
domain:
range:

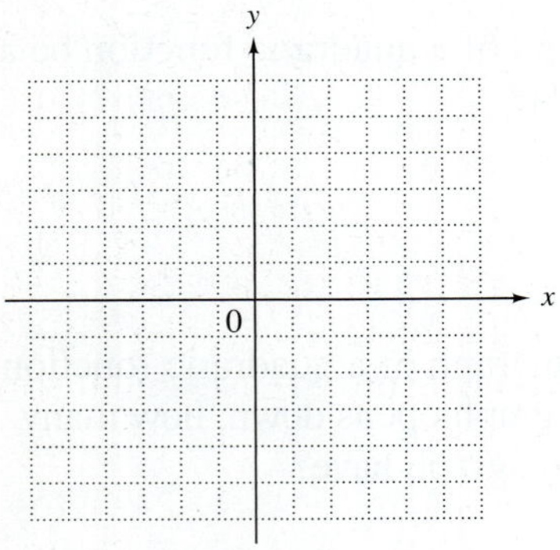

15. $x = -\frac{1}{5}y^2 + 2y - 4$
vertex:
axis:
domain:
range:

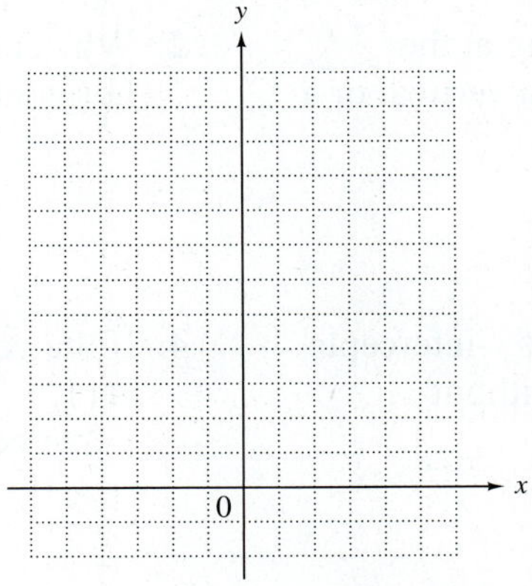

16. $x = -\frac{1}{2}y^2 - 4y - 6$
vertex:
axis:
domain:
range:

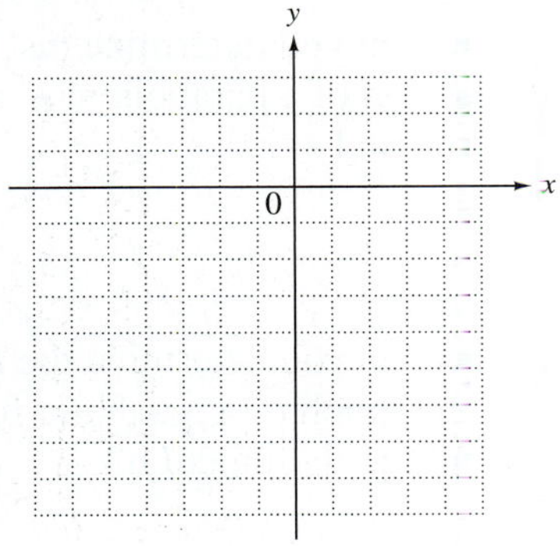

17. $x = 3y^2 + 12y + 5$
vertex:
axis:
domain:
range:

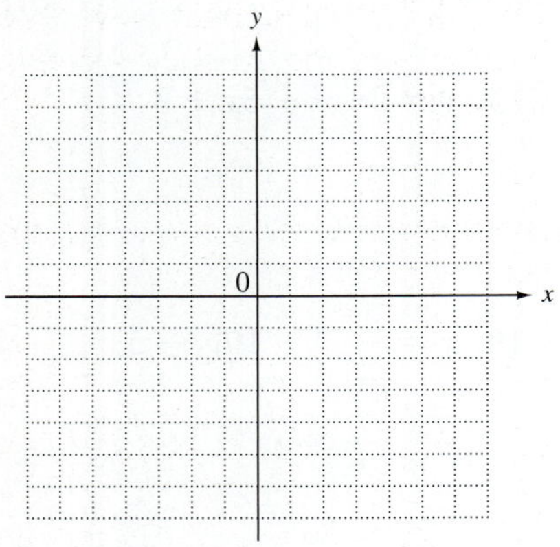

18. $x = 4y^2 + 16y + 11$
vertex:
axis:
domain:
range:

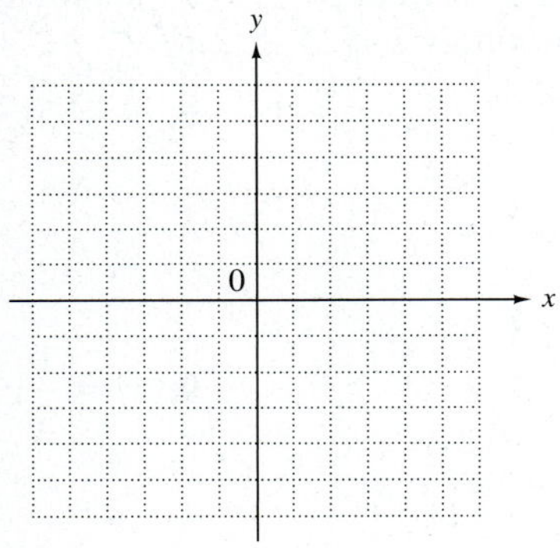

Use the concepts of this section to match each equation with its graph.

19. $y = 2x^2 + 4x - 3$

20. $y = -x^2 + 3x + 5$

21. $y = -\frac{1}{2}x^2 - x + 1$

22. $x = y^2 + 6y + 3$

23. $x = -y^2 - 2y + 4$

24. $x = 3y^2 + 6y + 5$

A.

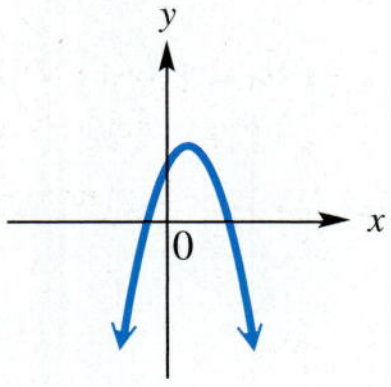

B.

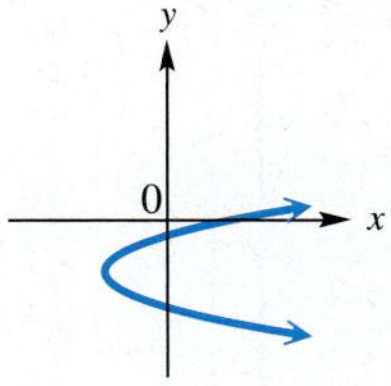

C.

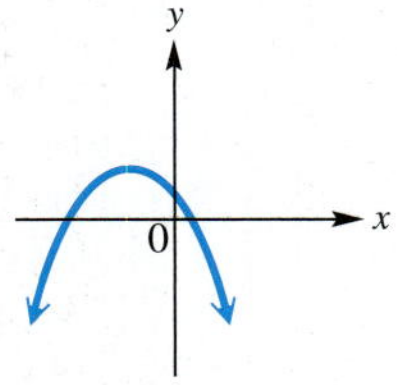

D.

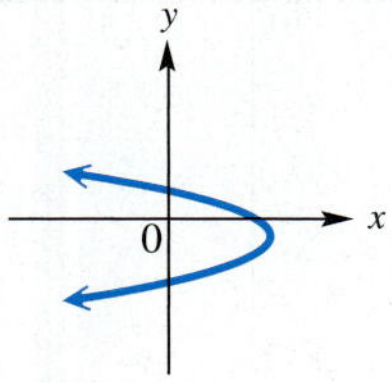

E.

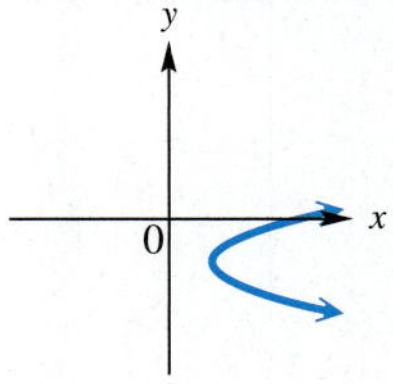

F.

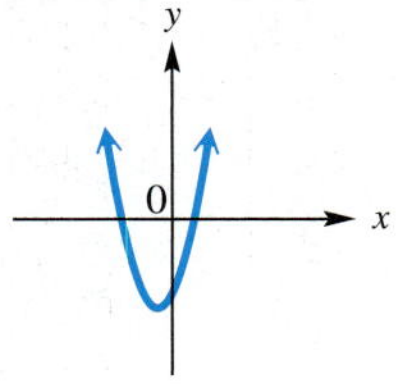

Solve each problem. See Examples 6 and 7.

25. Find the pair of numbers whose sum is 60 and whose product is a maximum. (*Hint:* Let x and $60 - x$ represent the two numbers.)

26. Find the pair of numbers whose sum is 10 and whose product is a maximum.

27. Palo Alto College is planning to construct a rectangular parking lot on land bordered on one side by a highway. The plan is to use 640 ft of fencing to fence off the other three sides. What should the dimensions of the lot be if the enclosed area is to be a maximum?

28. Keisha Hughes has 100 m of fencing material to enclose a rectangular exercise run for her dog. What width will give the enclosure the maximum area?

29. If an object on Earth is projected upward with an initial velocity of 32 ft per sec, then its height (in feet) after t seconds is given by

$$h(t) = 32t - 16t^2.$$

Find the maximum height attained by the object and the number of seconds it takes to hit the ground.

30. A projectile on Earth is fired straight upward so that its distance (in feet) above the ground t seconds after firing is given by

$$s(t) = -16t^2 + 400t.$$

Find the maximum height it reaches and the number of seconds it takes to reach that height.

31. A charter flight charges a fare of $200 per person, plus $4 per person for each unsold seat on the plane. If the plane holds 100 passengers and if x represents the number of unsold seats, find the following.

(a) A function defined by $R(x)$ that describes the total revenue received for the flight (*Hint:* Multiply the number of people flying, $100 - x$, by the price per ticket, $200 + 4x$.)

(b) The number of unsold seats that will produce the maximum revenue

(c) The maximum revenue

32. For a trip, a charter bus company charges a fare of $48 per person, plus $2 per person for each unsold seat on the bus. If the bus has 42 seats and x represents the number of unsold seats, find the following.

(a) A function defined by $R(x)$ that describes the total revenue from the trip (*Hint:* Multiply the total number riding, $42 - x$, by the price per ticket, $48 + 2x$.)

(b) The number of unsold seats that produces the maximum revenue

(c) The maximum revenue

33. The percent of births in the United States to teenage mothers in the years 1990–2002 can be modeled by the quadratic function defined by

$$f(x) = -0.0334x^2 + 0.2351x + 12.79,$$

where $x = 0$ represents 1990, $x = 1$ represents 1991, and so on. (*Source:* U.S. National Center for Health Statistics.)

(a) Since the coefficient of x^2 in the model is negative, the graph of this quadratic function is a parabola that opens down. Will the y-value of the vertex of this graph be a maximum or a minimum?

(b) In what year during this period was the percent of births in the United States to teenage mothers a maximum? (Round down to the nearest year.) Use the actual y-value of the vertex, to the nearest tenth, to find this percent.

34. The total receipts from individual income taxes by the U.S. Treasury in the years 2002–2006 can be modeled by the quadratic function defined by

$$f(x) = 33.79x^2 - 83.44x + 1036,$$

where $x = 0$ represents 2002, $x = 1$ represents 2003, and so on, and $f(x)$ is in billions of dollars. (*Source:* Internal Revenue Service.)

(a) Since the coefficient of x^2 given in the model is positive, the graph of this quadratic function is a parabola that opens up. Will the y-value of the vertex of this graph be a maximum or minimum?

(b) In what year during this period were total receipts from individual taxes a minimum? Use the actual x-value of the vertex, to the nearest tenth, to find this amount.

35. The graph shows how Social Security assets are expected to change as the number of retirees receiving benefits increases.

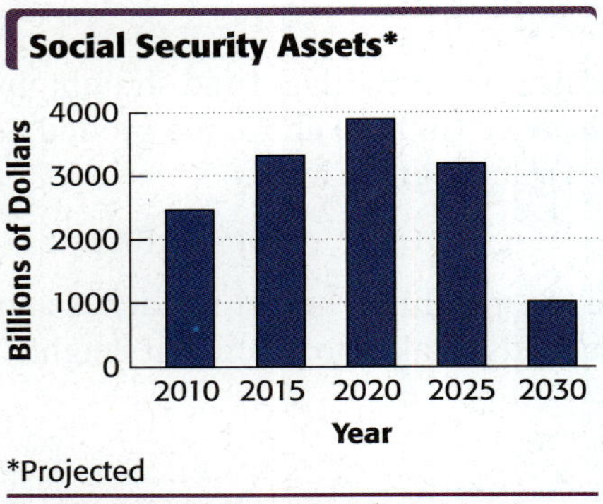

Source: Social Security Administration.

The graph suggests that a quadratic function would be a good fit to the data. The data are approximated by the function defined by

$$f(x) = -20.57x^2 + 758.9x - 3140.$$

In the model, $x = 10$ represents 2010, $x = 15$ represents 2015, and so on, and $f(x)$ is in billions of dollars.

(a) Explain why the coefficient of x^2 in the model is negative, based on the graph.

(b) Algebraically determine the vertex of the graph, with coordinates to four significant digits.

(c) Interpret the answer to part (b) as it applies to the application.

36. The graph shows the performance of investment portfolios with different mixtures of U.S. and foreign investments over a 25-yr period.

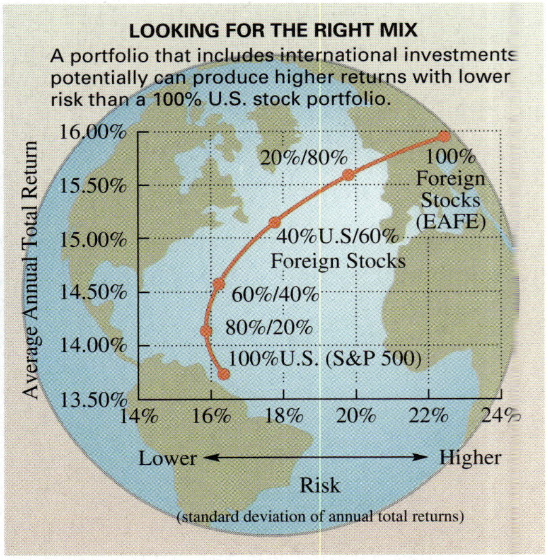

Source: *Financial Ink* Newsletter, Investment Management and Research, Inc., Feb. 1998. Thanks to David Van Geffen for this information.

(a) Is this the graph of a function? Explain.

(b) What investment mixture shown on the graph appears to represent the vertex? What relative amount of risk does this point represent? What return on investment does it provide?

(c) Which point on the graph represents the riskiest investment mixture? What return on investment does it provide?

10.7 Polynomial and Rational Inequalities

OBJECTIVES

1. Solve quadratic inequalities.
2. Solve polynomial inequalities of degree 3 or more.
3. Solve rational inequalities.

Now we combine methods of solving linear inequalities and methods of solving quadratic equations to solve *quadratic inequalities*.

Quadratic Inequality

A **quadratic inequality** can be written in the form

$$ax^2 + bx + c < 0 \quad \text{or} \quad ax^2 + bx + c > 0,$$

where a, b, and c are real numbers, with $a \neq 0$.

As before, $<$ and $>$ may be replaced with $\leq$ and $\geq$.

OBJECTIVE 1 Solve quadratic inequalities. One method for solving a quadratic inequality is by graphing the related quadratic function.

EXAMPLE 1 Solving Quadratic Inequalities by Graphing

Solve each inequality.

(a) $x^2 - x - 12 > 0$

To solve the inequality, we graph the related quadratic function defined by $f(x) = x^2 - x - 12$. We are particularly interested in the x-intercepts, which are found as in **Section 10.6** by letting $f(x) = 0$ and solving the quadratic equation

$$\begin{aligned} x^2 - x - 12 &= 0. \\ (x - 4)(x + 3) &= 0 && \text{Factor.} \\ x - 4 = 0 \quad &\text{or} \quad x + 3 = 0 && \text{Zero-factor property} \\ x = 4 \quad &\text{or} \quad x = -3 && \text{Solve each equation.} \end{aligned}$$

Thus, the x-intercepts are (4, 0) and $(-3, 0)$. The graph, which opens up since the coefficient of x^2 is positive, is shown in Figure 17(a). Notice from this graph that x-values less than -3 or greater than 4 result in y-values *greater than* 0. Therefore, the solution set of $x^2 - x - 12 > 0$, written in interval notation, is $(-\infty, -3) \cup (4, \infty)$.

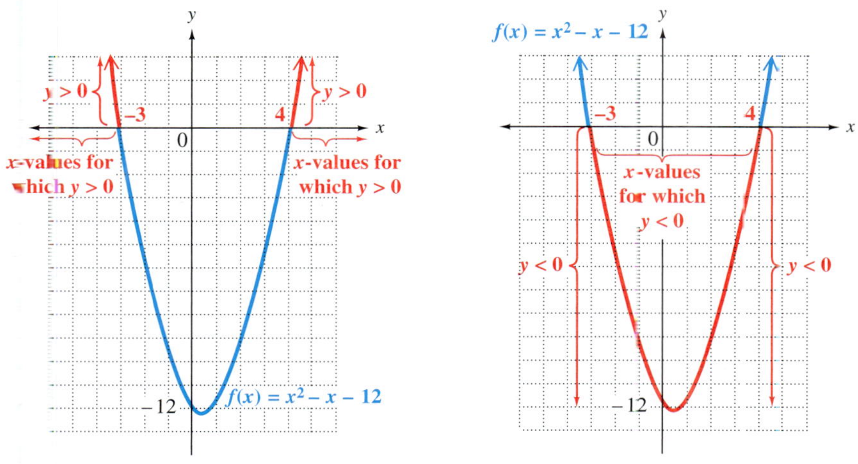

The graph is *above* the x-axis for $(-\infty, -3) \cup (4, \infty)$.

(a)

The graph is *below* the x-axis for $(-3, 4)$.

(b)

Figure 17

Continued on Next Page

1 Use the graph to solve each quadratic inequality.

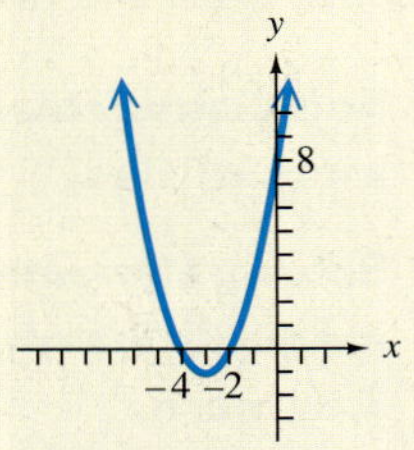

$f(x) = x^2 + 6x + 8$

(a) $x^2 + 6x + 8 > 0$

(b) $x^2 + 6x + 8 < 0$

2 Graph $f(x) = x^2 + 3x - 4$ and use the graph to solve each quadratic inequality.

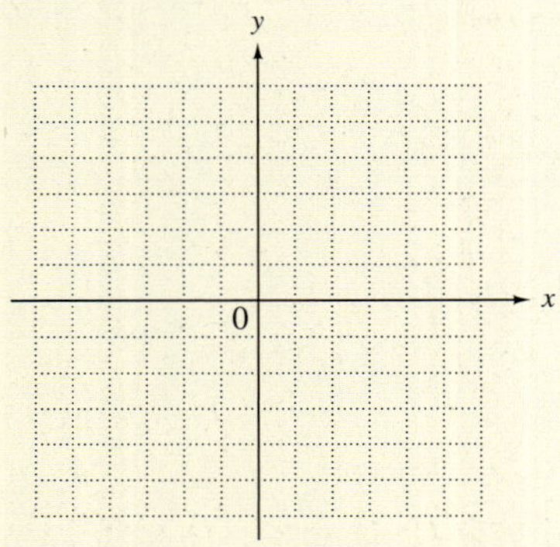

(a) $x^2 + 3x - 4 \geq 0$

(b) $x^2 + 3x - 4 \leq 0$

Answers

1. **(a)** $(-\infty, -4) \cup (-2, \infty)$ **(b)** $(-4, -2)$
2. **(a)** $(-\infty, -4] \cup [1, \infty)$ **(b)** $[-4, 1]$

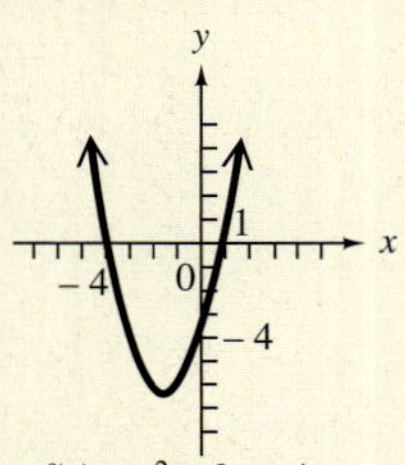

$f(x) = x^2 + 3x - 4$

(b) $x^2 - x - 12 < 0$

Here we want values of y that are *less than* 0. Referring to Figure 17(b) on the previous page, we notice from the graph that x-values between -3 and 4 result in y-values less than 0. Therefore, the solution set of the inequality $x^2 - x - 12 < 0$, written in interval notation, is $(-3, 4)$.

Note

If the inequalities in Example 1 had used $\geq$ and $\leq$, the solution sets would have included the x-values of the intercepts, which make the quadratic expression equal to 0, and been written in interval notation as $(-\infty, -3] \cup [4, \infty)$ for Example 1(a) and $[-3, 4]$ for Example 1(b). Square brackets would indicate that the endpoints -3 and 4 are *included* in the solution sets.

◀ *Work Problems* **1** *and* **2** *at the Side.*

In Example 1, we used graphing to divide the x-axis into intervals. Then using the graphs in Figure 17, we determined which x-values resulted in y-values that were either greater than or less than 0. Another method for solving a quadratic inequality uses these basic ideas without actually graphing the related quadratic function.

EXAMPLE 2 **Solving a Quadratic Inequality Using Test Numbers**

Solve $x^2 - x - 12 > 0$.

First solve the quadratic equation $x^2 - x - 12 = 0$ by factoring, as in Example 1(a).

$$x^2 - x - 12 = 0$$

$$(x - 4)(x + 3) = 0 \qquad \text{Factor.}$$

$$x - 4 = 0 \quad \text{or} \quad x + 3 = 0 \qquad \text{Zero-factor property}$$

$$x = 4 \quad \text{or} \quad x = -3 \qquad \text{Solve each equation.}$$

The numbers 4 and -3 divide the number line into the three intervals shown in Figure 18. ***Be careful to put the lesser number on the left.*** (Notice the similarity between Figure 18 and the x-axis with intercepts $(-3, 0)$ and $(4, 0)$ in Figure 17(a).)

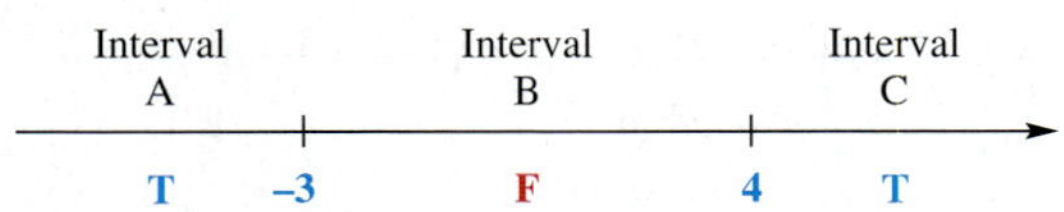

Figure 18

The numbers 4 and -3 are the only numbers that make the expression $x^2 - x - 12$ equal to 0. All other numbers make the expression either positive or negative. The sign of the expression can change from positive to negative or from negative to positive only at a number that makes it 0. Therefore, if one number in an interval satisfies the inequality, then all the numbers in that interval will satisfy the inequality.

To see if the numbers in Interval A satisfy the inequality, choose any number from Interval A in Figure 18 (that is, any number less than -3). Substitute this test number for x in the original inequality $x^2 - x - 12 > 0$. If the result is *true,* then all numbers in Interval A satisfy the inequality.

Continued on Next Page

We try -5 from Interval A. Substitute -5 for x.

$$x^2 - x - 12 > 0 \quad \text{Original inequality}$$

Use parentheses to avoid sign errors.

$$(-5)^2 - (-5) - 12 \overset{?}{>} 0 \quad \text{Let } x = -5.$$

$$25 + 5 - 12 \overset{?}{>} 0 \quad \text{Simplify.}$$

$$18 > 0 \quad \text{True}$$

Because -5 from Interval A satisfies the inequality, all numbers from Interval A are solutions.

Try 0 from Interval B. If $x = 0$, then

$$0^2 - 0 - 12 \overset{?}{>} 0 \quad \text{Let } x = 0.$$

$$-12 > 0 \quad \text{False}$$

The numbers in Interval B are *not* solutions.

Work Problem 3 at the Side. ▶

In Problem 3 at the side, the test number 5 satisfies the inequality, so the numbers in Interval C are also solutions.

Based on these results (shown by the colored letters in Figure 18), the solution set includes the numbers in Intervals A and C, as shown on the graph in Figure 19. The solution set is written in interval notation as

$$(-\infty, -3) \cup (4, \infty).$$

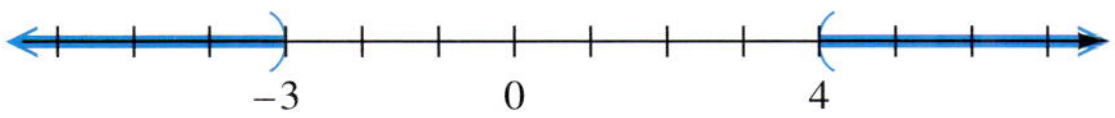

Figure 19

This agrees with the solution set we found by graphing the related quadratic function in Example 1(a).

In summary, a quadratic inequality is solved by following these steps.

Solving a Quadratic Inequality

Step 1 **Write the inequality as an equation and solve it.**

Step 2 **Use the solutions from Step 1 to determine intervals.** Graph the numbers found in Step 1 on a number line. These numbers divide the number line into intervals.

Step 3 **Find the intervals that satisfy the inequality.** Substitute a test number from each interval into the original inequality to determine the intervals that satisfy the inequality. All numbers in those intervals are in the solution set. A graph of the solution set will usually look like one of these. (Square brackets might be used instead of parentheses.)

 or 

Step 4 **Consider the endpoints separately.** The numbers from Step 1 are included in the solution set if the inequality is $\leq$ or $\geq$; they are not included if it is $<$ or $>$.

Work Problem 4 at the Side. ▶

3 Does the number 5 from Interval C satisfy $x^2 - x - 12 > 0$?

4 Solve each inequality, and graph the solution set.

(a) $x^2 + x - 6 > 0$

(b) $3m^2 - 13m - 10 \leq 0$

ANSWERS

3. yes

4. (a) $(-\infty, -3) \cup (2, \infty)$

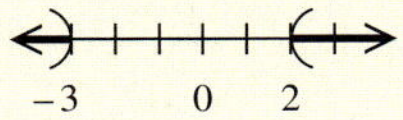

(b) $\left[-\frac{2}{3}, 5\right]$

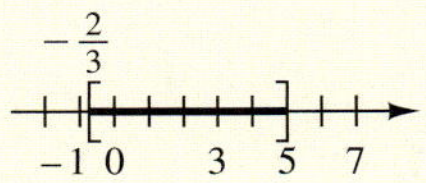

5 Solve each inequality.

(a) $(3x - 2)^2 > -2$

(b) $(3x - 2)^2 < -2$

Special cases of quadratic inequalities may occur, as in the next example.

EXAMPLE 3 **Solving Special Cases**

Solve each inequality.

(a) $(2x - 3)^2 > -1$

Because $(2x - 3)^2$ is never negative, it is greater than -1 for all replacements for x. Thus, the solution set is all real numbers, $(-\infty, \infty)$.

(b) $(2x - 3)^2 < -1$

Using the same reasoning as in part (a), there is no solution for this inequality. The solution set is $\emptyset$.

◀ *Work Problem* **5** *at the Side.*

OBJECTIVE 2 **Solve polynomial inequalities of degree 3 or more.** Higher-degree inequalities that have factorable polynomials are solved using a method similar to that of solving quadratic inequalities.

EXAMPLE 4 **Solving a Third-Degree Polynomial Inequality**

Solve $(x - 1)(x + 2)(x - 4) \leq 0$.

This is a *cubic* (third-degree) inequality rather than a quadratic inequality, but it can be solved using the method shown in the box by extending the zero-factor property to more than two factors. Begin by setting the factored polynomial *equal* to 0 and solving the equation (Step 1).

$$(x - 1)(x + 2)(x - 4) = 0$$

$$x - 1 = 0 \quad \text{or} \quad x + 2 = 0 \quad \text{or} \quad x - 4 = 0$$

$$x = 1 \quad \text{or} \quad x = -2 \quad \text{or} \quad x = 4$$

Locate the numbers -2, 1, and 4 on a number line, as in Figure 20, to determine the Intervals A, B, C, and D (Step 2).

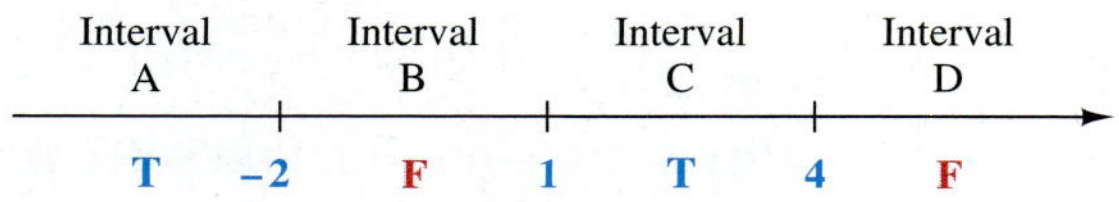

Figure 20

Substitute a test number from each interval in the *original* inequality to determine which intervals satisfy the inequality (Step 3). It is helpful to organize this information in a table.

Interval	Test Number	Test of Inequality	True or False?
A	-3	$-28 \leq 0$	T
B	0	$8 \leq 0$	F
C	2	$-8 \leq 0$	T
D	5	$28 \leq 0$	F

Verify the information given in the table. The numbers in Intervals A and C are in the solution set, which is written as

$$(-\infty, -2] \cup [1, 4].$$

The three endpoints are included since the inequality symbol is $\leq$ (Step 4).

Continued on Next Page

ANSWERS

5. **(a)** $(-\infty, \infty)$ **(b)** $\emptyset$

The solution set is graphed in Figure 21.

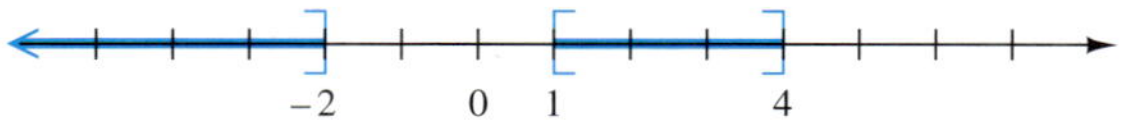

Figure 21

Work Problem 6 at the Side. ▶

OBJECTIVE 3 Solve rational inequalities. Inequalities that involve rational expressions, called **rational inequalities,** are solved similarly using the following steps.

Solving a Rational Inequality

Step 1 **Write the inequality** so that 0 is on one side and there is a single fraction on the other side.

Step 2 **Determine the numbers that make the numerator and denominator equal to 0.**

Step 3 **Divide a number line into intervals.** Use the numbers from Step 2.

Step 4 **Find the intervals that satisfy the inequality.** Test a number from each interval by substituting it into the *original* inequality.

Step 5 **Consider the endpoints separately.** Exclude any values that make the denominator 0.

EXAMPLE 5 Solving a Rational Inequality

Solve $\dfrac{-1}{x-3} > 1$.

Write the inequality so that 0 is on one side (Step 1).

$$\frac{-1}{x-3} - 1 > 0 \qquad \text{Subtract 1.}$$

$$\frac{-1}{x-3} - \frac{x-3}{x-3} > 0 \qquad \text{Use } x - 3 \text{ as the common denominator.}$$

Be careful with signs.

$$\frac{-1 - x + 3}{x-3} > 0 \qquad \text{Write the left side as a single fraction.}$$

$$\frac{-x+2}{x-3} > 0 \qquad \text{Combine like terms in the numerator.}$$

The sign of the rational expression $\frac{-x+2}{x-3}$ will change from positive to negative or negative to positive only at those numbers that make the numerator or denominator 0. The number 2 makes the numerator 0, and 3 makes the denominator 0 (Step 2). These two numbers, 2 and 3, divide a number line into three intervals. See Figure 22 (Step 3).

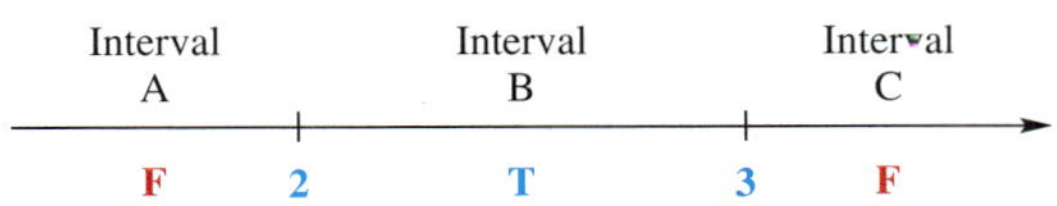

Figure 22

Continued on Next Page

6 Solve each inequality, and graph the solution set.

(a) $(x-3)(x+2)(x+1) > 0$

(b) $(x-5)(x+1)(x-3) \le 0$

Answers

6. (a) $(-2, -1) \cup (3, \infty)$

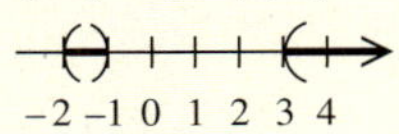

(b) $(-\infty, -1] \cup [3, 5]$

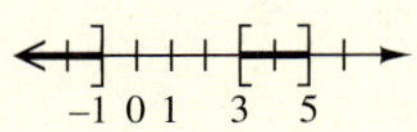

7 Solve each inequality, and graph the solution set.

(a) $\dfrac{2}{x-4} < 3$

(b) $\dfrac{5}{x+1} > 4$

8 Solve $\dfrac{x+2}{x-1} \leq 5$, and graph the solution set.

Testing a number from each interval in the *original* inequality, $\frac{-1}{x-3} > 1$, gives the results shown in the table (Step 4).

Interval	Test Number	Test of Inequality	True or False?
A	0	$\frac{1}{3} > 1$	F
B	2.5	$2 > 1$	T
C	4	$-1 > 1$	F

The solution set of $\frac{-1}{x-3} > 1$ is the interval $(2, 3)$. This interval does not include 3 since it would make the denominator of the original inequality 0; 2 is not included either since the inequality symbol is $>$, which does not involve equality (Step 5). A graph of the solution set is given in Figure 23.

Figure 23

◀ *Work Problem* **7** *at the Side.*

CAUTION
When solving a rational inequality, any number that makes the denominator 0 must be excluded from the solution set.

EXAMPLE 6 **Solving a Rational Inequality**

Solve $\dfrac{x-2}{x+2} \leq 2$.

Write the inequality so that 0 is on one side (Step 1).

$$\frac{x-2}{x+2} - 2 \leq 0 \qquad \text{Subtract 2.}$$

$$\frac{x-2}{x+2} - \frac{2(x+2)}{x+2} \leq 0 \qquad \text{Use } x+2 \text{ as the common denominator.}$$

$$\frac{x-2}{x+2} - \frac{2x+4}{x+2} \leq 0 \qquad \text{Distributive property}$$

$$\frac{x-2-2x-4}{x+2} \leq 0 \qquad \text{Write as a single fraction.}$$

$$\frac{-x-6}{x+2} \leq 0 \qquad \text{Combine like terms in the numerator.}$$

The number -6 makes the numerator 0, and -2 makes the denominator 0 (Step 2). These two numbers determine three intervals (Step 3). Test one number from each interval (Step 4) to see that the solution set is the interval

$$(-\infty, -6] \cup (-2, \infty).$$

The number -6 satisfies the original inequality, but -2 cannot be used as a solution since it makes the denominator 0 (Step 5). A graph of the solution set is shown in Figure 24.

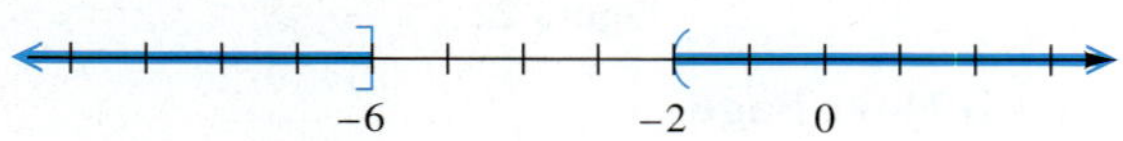

Figure 24

◀ *Work Problem* **8** *at the Side.*

ANSWERS

7. (a) $(-\infty, 4) \cup \left(\frac{14}{3}, \infty\right)$

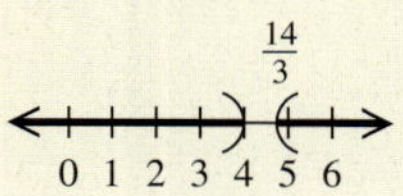

(b) $\left(-1, \frac{1}{4}\right)$

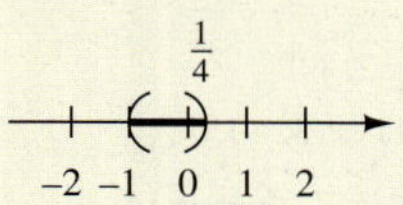

8. $(-\infty, 1) \cup \left[\frac{7}{4}, \infty\right)$

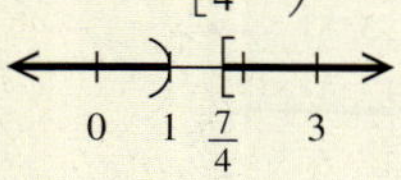

10.7 Exercises

FOR EXTRA HELP MyMathLab Math XL PRACTICE WATCH DOWNLOAD READ REVIEW

In Example 1, we determined the solution sets of the quadratic inequalities $x^2 - x - 12 > 0$ and $x^2 - x - 12 < 0$ by graphing $f(x) = x^2 - x - 12$. The x-intercepts of this graph indicated the solutions of the equation $x^2 - x - 12 = 0$. The x-values of the points on the graph that were **above** *the x-axis formed the solution set of $x^2 - x - 12 > 0$, and the x-values of the points on the graph that were* **below** *the x-axis formed the solution set of $x^2 - x - 12 < 0$.*

In Exercises 1–4, the graph of a quadratic function f is given. Use the graph to find the solution set of each equation or inequality. See Example 1.

1. **(a)** $x^2 - 4x + 3 = 0$
(b) $x^2 - 4x + 3 > 0$
(c) $x^2 - 4x + 3 < 0$

$f(x) = x^2 - 4x + 3$

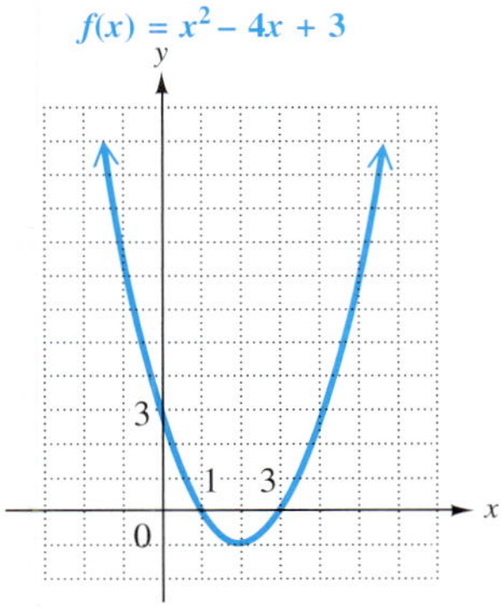

2. **(a)** $3x^2 + 10x - 8 = 0$
(b) $3x^2 + 10x - 8 \geq 0$
(c) $3x^2 + 10x - 8 < 0$

$f(x) = 3x^2 + 10x - 8$

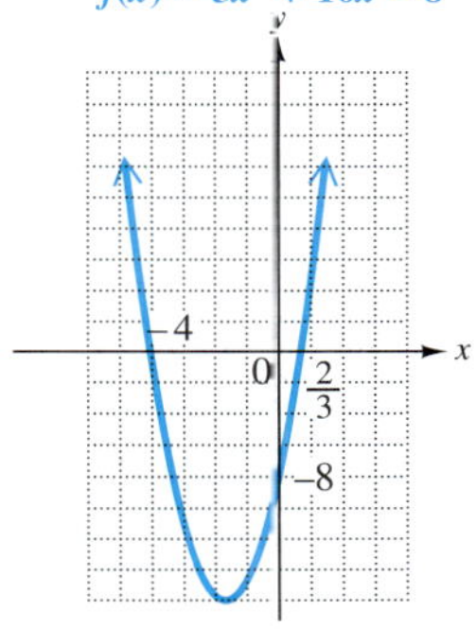

3. **(a)** $-2x^2 - x + 15 = 0$
(b) $-2x^2 - x + 15 \geq 0$
(c) $-2x^2 - x + 15 \leq 0$

$f(x) = -2x^2 - x + 15$

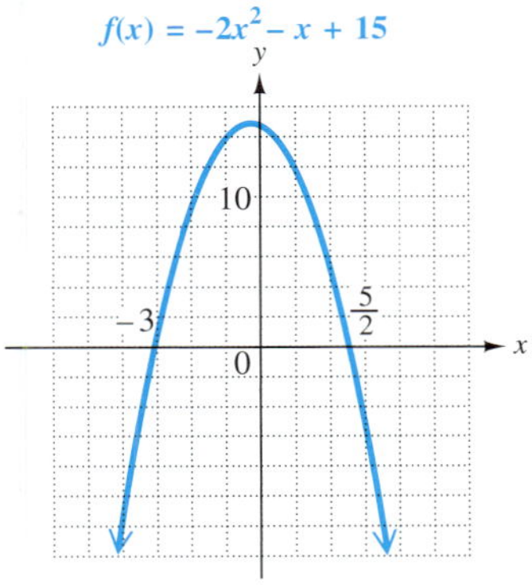

4. **(a)** $-x^2 + 3x + 10 = 0$
(b) $-x^2 + 3x + 10 \geq 0$
(c) $-x^2 + 3x + 10 \leq 0$

$f(x) = -x^2 - 3x + 10$

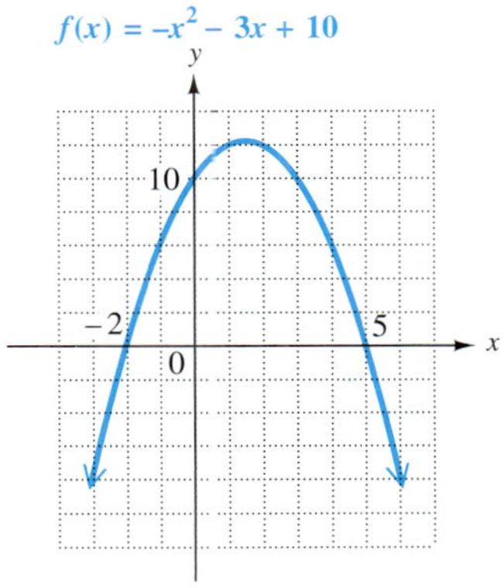

5. Explain how to determine whether to include or exclude endpoints when solving a quadratic or higher-degree inequality.

6. The solution set of the inequality $x^2 + x - 12 < 0$ is the interval $(-4, 3)$. Without actually performing any work, give the solution set of the inequality $x^2 + x - 12 \geq 0$.

Solve each inequality, and graph the solution set. See Example 2.

7. $(x + 1)(x - 5) > 0$

8. $(m + 6)(m - 2) > 0$

9. $(r + 4)(r - 6) < 0$

10. $(x + 4)(x - 8) < 0$

11. $x^2 - 4x + 3 \geq 0$

12. $m^2 - 3m - 10 \geq 0$

13. $10t^2 + 9t \geq 9$

14. $3r^2 + 10r \geq 8$

15. $9p^2 + 3p < 2$

16. $2x^2 + x < 15$

17. $6x^2 + x \geq 1$

18. $4m^2 + 7m \geq -3$

19. $x^2 - 6x + 6 \geq 0$
(*Hint:* Use the quadratic formula.)

20. $3k^2 - 6k + 2 \leq 0$
(*Hint:* Use the quadratic formula.)

Solve each inequality. See Example 3.

21. $(4 - 3x)^2 \geq -2$ **22.** $(6p + 7)^2 \geq -1$ **23.** $(3x + 5)^2 \leq -4$ **24.** $(8t + 5)^2 \leq -5$

Solve each inequality, and graph the solution set. See Example 4.

25. $(p - 1)(p - 2)(p - 4) < 0$

26. $(2r + 1)(3r - 2)(4r + 7) < 0$

27. $(x - 4)(2x + 3)(3x - 1) \geq 0$

28. $(z + 2)(4z - 3)(2z + 7) \geq 0$

Solve each inequality, and graph the solution set. See Examples 5 and 6.

29. $\dfrac{x - 1}{x - 4} > 0$ **30.** $\dfrac{x + 1}{x - 5} > 0$ **31.** $\dfrac{2n + 3}{n - 5} \leq 0$

32. $\dfrac{3t + 7}{t - 3} \leq 0$ **33.** $\dfrac{8}{x - 2} \geq 2$ **34.** $\dfrac{20}{x - 1} \geq 1$

35. $\dfrac{3}{2t - 1} < 2$ **36.** $\dfrac{6}{m - 1} < 1$ **37.** $\dfrac{w}{w + 2} \geq 2$

38. $\frac{m}{m+5} \geq 2$

39. $\frac{4k}{2k-1} < k$

40. $\frac{r}{r+2} < 2r$

41. $\frac{x-8}{x-4} \leq 3$

42. $\frac{2t-3}{t+1} \geq 4$

Relating Concepts (Exercises 43–46) For Individual or Group Work

A rock is projected vertically upward from the ground. Its distance s in feet above the ground after t seconds is given by the quadratic function defined by

$$s(t) = -16t^2 + 256t.$$

Work Exercises 43–46 in order, *to see how quadratic equations and inequalities are related.*

43. At what times will the rock be 624 ft above the ground? (*Hint:* Let $s(t) = 624$ and solve the quadratic *equation.*)

44. At what times will the rock be more than 624 ft above the ground? (*Hint:* Set $s(t) > 624$ and solve the quadratic *inequality.*)

45. At what times will the rock be at ground level? (*Hint:* Let $s(t) = 0$ and solve the quadratic *equation.*)

46. At what times will the rock be less than 624 ft above the ground? (*Hint:* Set $s(t) < 624$, solve the quadratic *inequality,* and observe the solutions in Exercises 44 and 45 to determine the least and greatest possible values of t.)

Chapter 10 Summary

Key Terms

10.1 **quadratic equation** A quadratic equation is an equation that can be written in the form $ax^2 + bx + c = 0$, where a, b, and c are real numbers, with $a \neq 0$. This form is called standard form.

10.2 **quadratic formula** The quadratic formula is a formula for solving quadratic equations.

discriminant The discriminant is the expression under the radical in the quadratic formula.

10.3 **quadratic in form** A nonquadratic equation that can be written as a quadratic equation is called quadratic in form.

10.5 **quadratic function** A function defined by $f(x) = ax^2 + bx + c$, for real numbers a, b, and c, with $a \neq 0$, is a quadratic function.

parabola The graph of a quadratic function is a parabola.

vertex The point on a parabola that has the least y-value (if the parabola opens up) or the greatest y-value (if the parabola opens down) is called the vertex of the parabola.

axis The vertical (or horizontal) line through the vertex of a vertical (or horizontal) parabola is its axis.

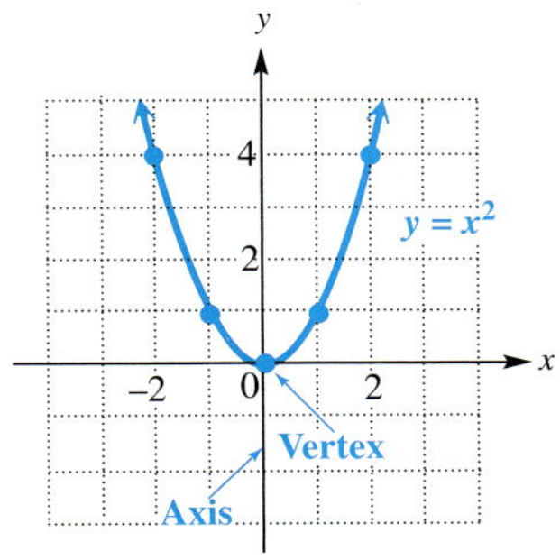

10.7 **quadratic inequality** A quadratic inequality is an inequality that can be written in the form $ax^2 + bx + c < 0$ or $ax^2 + bx + c > 0$ (or with $\leq$ or $\geq$) where a, b, and c are real numbers, with $a \neq 0$.

rational inequality An inequality that involves a rational expression is a rational inequality.

Test Your Word Power

See how well you have learned the vocabulary in this chapter. Answers, with examples, follow the Quick Review.

1. The **quadratic formula** is
- **A.** a formula to find the number of solutions of a quadratic equation
- **B.** a formula to find the type of solutions of a quadratic equation
- **C.** the standard form of a quadratic equation
- **D.** a general formula for solving any quadratic equation.

2. A **quadratic function** is a function that can be written in the form
- **A.** $f(x) = mx + b$ for real numbers m and b
- **B.** $f(x) = \frac{P(x)}{Q(x)}$, where $Q(x) \neq 0$
- **C.** $f(x) = ax^2 + bx + c$ for real numbers a, b, and c ($a \neq 0$)
- **D.** $f(x) = \sqrt{x}$ for $x \geq 0$.

3. A **parabola** is the graph of
- **A.** any equation in two variables
- **B.** a linear equation
- **C.** an equation of degree 3
- **D.** a quadratic equation in two variables.

4. The **vertex** of a parabola is
- **A.** the point where the graph intersects the y-axis
- **B.** the point where the graph intersects the x-axis
- **C.** the lowest point on a parabola that opens up or the highest point on a parabola that opens down
- **D.** the origin.

5. The **axis** of a parabola is
- **A.** either the x-axis or the y-axis
- **B.** the vertical line (of a vertical parabola) or the horizontal line (of a horizontal parabola) through the vertex
- **C.** the lowest or highest point on the graph of a parabola
- **D.** a line through the origin.

6. A parabola is **symmetric about its axis** since
- **A.** its graph is near the axis
- **B.** its graph is a mirror image on each side of the axis
- **C.** its graph looks different on each side of the axis
- **D.** its graph intersects the axis.

▶ Quick Review

Concepts	Examples

10.1 The Square Root Property and Completing the Square

Square Root Property

If x and k are complex numbers and $x^2 = k$, then

$$x = \sqrt{k} \quad \text{or} \quad x = -\sqrt{k}.$$

Solve $(x-1)^2 = 8$.

$$x - 1 = \sqrt{8} \quad \text{or} \quad x - 1 = -\sqrt{8}$$

$$x = 1 + 2\sqrt{2} \quad \text{or} \quad x = 1 - 2\sqrt{2}$$

Solution set: $\{1 + 2\sqrt{2}, 1 - 2\sqrt{2}\}$

Completing the Square

To solve $ax^2 + bx + c = 0$ $(a \neq 0)$:

Step 1 If $a \neq 1$, divide each side by a.

Step 2 Write the equation with the variable terms on one side and the constant on the other.

Step 3 Take half the coefficient of x and square it.

Step 4 Add the square to each side.

Step 5 Factor the perfect square trinomial, and write it as the square of a binomial. Simplify the other side.

Step 6 Use the square root property to complete the solution.

Solve $2x^2 - 4x - 18 = 0$.

$$x^2 - 2x - 9 = 0 \quad \text{Divide by 2.}$$

$$x^2 - 2x = 9 \quad \text{Add 9.}$$

$$\left[\frac{1}{2}(-2)\right]^2 = (-1)^2 = 1$$

$$x^2 - 2x + 1 = 9 + 1 \quad \text{Add 1.}$$

$$(x-1)^2 = 10 \quad \text{Factor; add.}$$

$$x - 1 = \sqrt{10} \quad \text{or} \quad x - 1 = -\sqrt{10}$$

$$x = 1 + \sqrt{10} \quad \text{or} \quad x = 1 - \sqrt{10}$$

Solution set: $\{1 + \sqrt{10}, 1 - \sqrt{10}\}$

10.2 The Quadratic Formula

Quadratic Formula

The solutions of $ax^2 + bx + c = 0$ $(a \neq 0)$ are given by

$$x = \frac{-b \pm \sqrt{b^2 - 4ac}}{2a}.$$

Solve $3x^2 + 5x + 2 = 0$.

$$x = \frac{-5 \pm \sqrt{5^2 - 4(3)(2)}}{2(3)}$$

$$x = \frac{-5 \pm 1}{6}$$

$$x = -1 \quad \text{or} \quad x = -\frac{2}{3}$$

Solution set: $\{-1, -\frac{2}{3}\}$

The Discriminant

If a, b, and c are integers, then the discriminant, $b^2 - 4ac$, of $ax^2 + bx + c = 0$ determines the number and type of solutions as follows.

Discriminant	Number and Type of Solutions
Positive, the square of an integer	Two rational solutions
Positive, not the square of an integer	Two irrational solutions
Zero	One rational solution
Negative	Two nonreal complex solutions

For $x^2 + 3x - 10 = 0$, the discriminant is

$$3^2 - 4(1)(-10) = 49. \quad \text{Two rational solutions}$$

For $4x^2 + x + 1 = 0$, the discriminant is

$$1^2 - 4(4)(1) = -15. \quad \text{Two nonreal complex solutions}$$

Concepts	Examples

10.3 Equations Quadratic in Form

A nonquadratic equation that can be written in the form

$$au^2 + bu + c = 0,$$

for $a \neq 0$ and an algebraic expression u, is called quadratic in form. Substitute u for the expression, solve for u, and then solve for the variable in the expression.

Solve $3(x + 5)^2 + 7(x + 5) + 2 = 0$.

$3u^2 + 7u + 2 = 0$ Let $u = x + 5$.

$(3u + 1)(u + 2) = 0$ Factor.

$u = -\frac{1}{3}$ or $u = -2$

$x + 5 = -\frac{1}{3}$ or $x + 5 = -2$ $x + 5 = u$

$x = -\frac{16}{3}$ or $x = -7$

Solution set: $\left\{-7, -\frac{16}{3}\right\}$

10.4 Formulas and Further Applications

To solve a formula for a second-degree variable, proceed as follows.

(a) If the variable appears only to the second power: Isolate the second-degree variable on one side of the equation, and then use the square root property.

Solve $A = \frac{2mp}{r^2}$ for r.

$r^2A = 2mp$ Multiply by r^2.

$r^2 = \frac{2mp}{A}$ Divide by A.

$r = \pm\sqrt{\frac{2mp}{A}}$ Square root property

$r = \frac{\pm\sqrt{2mpA}}{A}$ Rationalize the denominator.

(b) If the variable appears to the first and second powers: Write the equation in standard form, and then use the quadratic formula.

Solve $m^2 + rm = t$ for m.

$m^2 + rm - t = 0$ Standard form

$m = \frac{-r \pm \sqrt{r^2 - 4(1)(-t)}}{2(1)} = \frac{-r \pm \sqrt{r^2 + 4t}}{2}$

$a = 1, b = r, c = -t$

10.5 Graphs of Quadratic Functions

1. The graph of the quadratic function defined by $F(x) = a(x - h)^2 + k$, $a \neq 0$, is a parabola with vertex at (h, k) and the vertical line $x = h$ as axis.
2. The graph opens up if a is positive and down if a is negative.
3. The graph is wider than the graph of $f(x) = x^2$ if $0 < |a| < 1$ and narrower if $|a| > 1$.

Graph $f(x) = -(x + 3)^2 + 1$.

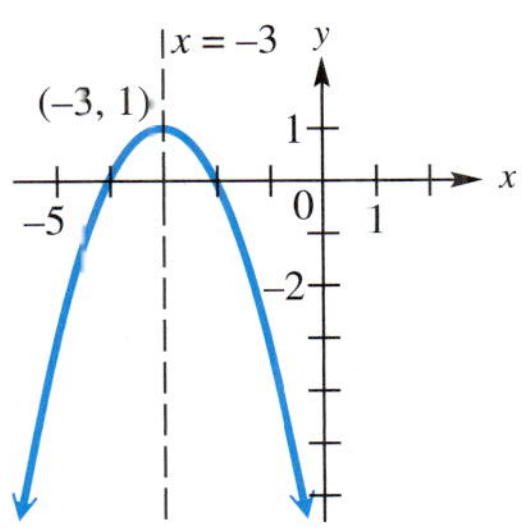

The graph opens down since $a < 0$. It is shifted 3 units left and 1 unit up, so the vertex is $(-3, 1)$, with axis $x = -3$. The domain is $(-\infty, \infty)$; the range is $(-\infty, 1]$.

10.6 More about Parabolas and Their Applications

The vertex of the graph of $f(x) = ax^2 + bx + c$, $a \neq 0$, has coordinates

$$\left(\frac{-b}{2a}, f\left(\frac{-b}{2a}\right)\right).$$

Graphing a Quadratic Function

Step 1 Determine whether the graph opens up or down.

Step 2 Find the vertex.

Step 3 Find the x-intercepts (if any). Find the y-intercept.

Step 4 Find and plot additional points as needed.

Graph $f(x) = x^2 + 4x + 3$.

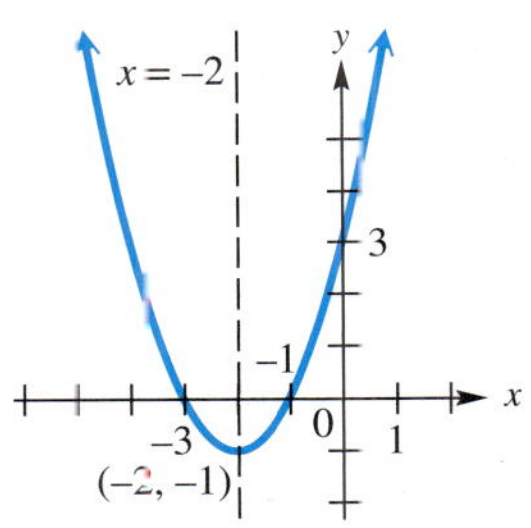

The graph opens up since $a > 0$. The vertex is $(-2, -1)$. The solutions of $x^2 + 4x + 3 = 0$ are -1 and -3, so the x-intercepts are $(-1, 0)$ and $(-3, 0)$. Since $f(0) = 3$, the y-intercept is $(0, 3)$. The domain is $(-\infty, \infty)$; the range is $[-1, \infty)$.

Concepts	Examples

10.6 More about Parabolas and Their Applications (continued)

Horizontal Parabolas

The graph of

$$x = ay^2 + by + c \quad \text{or} \quad x = a(y - k)^2 + h$$

is a horizontal parabola with vertex (h, k) and the horizontal line $y = k$ as axis. The graph opens to the right if $a > 0$ and to the left if $a < 0$.

Horizontal parabolas do not represent functions.

Graph $x = 2y^2 + 6y + 5$.

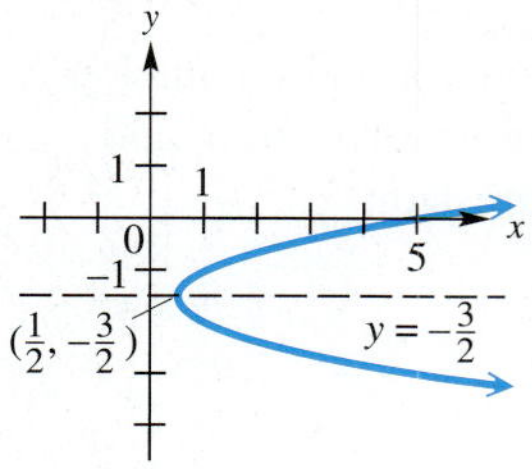

The graph opens to the right since $a > 0$. The vertex is $\left(\frac{1}{2}, -\frac{3}{2}\right)$. The axis is $y = -\frac{3}{2}$. The domain is $\left[\frac{1}{2}, \infty\right)$; the range is $(-\infty, \infty)$.

10.7 Polynomial and Rational Inequalities

Solving a Quadratic Inequality

Step 1 Write the inequality as an equation and solve.

Step 2 Use the numbers found in Step 1 to divide a number line into intervals.

Step 3 Substitute a number from each interval into the original inequality to determine the intervals that belong in the solution set.

Step 4 Consider the endpoints separately.

Solve $2x^2 + 5x + 2 < 0$.

$$2x^2 + 5x + 2 = 0$$

$$x = -\frac{1}{2} \quad \text{or} \quad x = -2$$

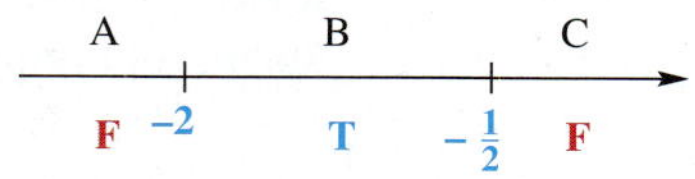

Use the method of Example 2 on pages 656–657 to find that $x = -3$ makes the original inequality false, $x = -1$ makes it true, and $x = 0$ makes it false.

The solution set is the interval $\left(-2, -\frac{1}{2}\right)$.

Solving a Rational Inequality

Step 1 Write the inequality so that 0 is on one side and there is a single fraction on the other side.

Step 2 Determine the numbers that make the numerator and denominator 0.

Step 3 Use the numbers from Step 2 to divide a number line into intervals.

Step 4 Substitute a number from each interval into the original inequality to determine the intervals that belong in the solution set.

Step 5 Consider the endpoints separately.

Solve $\frac{x}{x+2} \geq 4$.

$$\frac{x}{x+2} - 4 \geq 0 \qquad \text{Subtract 4.}$$

$$\frac{x}{x+2} - \frac{4(x+2)}{x+2} \geq 0 \qquad \text{Write with a common denominator.}$$

$$\frac{-3x - 8}{x+2} \geq 0 \qquad \text{Subtract fractions.}$$

$-\frac{8}{3}$ makes the numerator 0 and -2 makes the denominator 0.

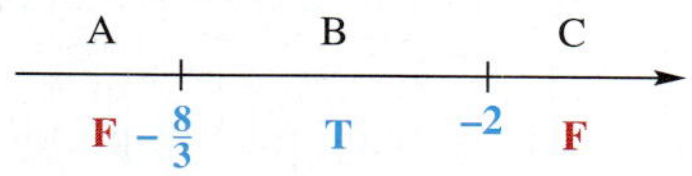

Use the method of Example 5 on pages 659–660 to find that $x = -4$ makes the original inequality false, $x = -\frac{7}{3}$ makes it true, and $x = 0$ makes it false.

The solution set is the interval $\left[-\frac{8}{3}, -2\right)$. Note that -2 is excluded since it makes the denominator 0.

ANSWERS TO TEST YOUR WORD POWER

1. D; *Example:* The solutions of $ax^2 + bx + c = 0$ $(a \neq 0)$ are given by $x = \frac{-b \pm \sqrt{b^2 - 4ac}}{2a}$.
2. C; *Examples:* $f(x) = x^2 - 2$, $f(x) = (x + 4)^2 + 1$, $f(x) = x^2 - 4x + 5$ **3.** D; *Examples:* See the figures in the Quick Review for **Sections 10.5 and 10.6.** **4.** C; *Example:* The graph of $y = (x + 3)^2$ has vertex $(-3, 0)$, which is the lowest point on the graph. **5.** B; *Example:* The axis of $y = (x + 3)^2$ is the vertical line $x = -3$. **6.** B; *Example:* Since the graph of $y = (x + 3)^2$ is symmetric about its axis $x = -3$, the points $(-2, 1)$ and $(-4, 1)$ are on this graph.

Chapter 10 Review Exercises

[10.1] *Solve each equation by using the square root property or completing the square.*

1. $t^2 = 121$

2. $p^2 = 3$

3. $(2x + 5)^2 = 100$

***4.** $(3k - 2)^2 = -25$

5. $x^2 + 4x = 15$

6. $2m^2 - 3m = -1$

7. A student gave the following "solution" to the equation $x^2 = 12$.

$$x^2 = 12$$
$$x = \sqrt{12} \quad \text{Square root property}$$
$$x = 2\sqrt{3}$$

Solution set: $\{2\sqrt{3}\}$

The answer is not correct. ***WHAT WENT WRONG?*** Give the correct solution set.

8. When it opened in 2008, the Singapore Flyer became the world's largest Ferris wheel, with a height of 165 m. Use the metric version of Galileo's formula, $d = 4.9t^2$ (where d is in meters), to find how long it would take a wallet dropped from the top of the Singapore Flyer to reach the ground. Round your answer to the nearest tenth of a second. (*Source:* www.singaporeflyer.com)

[10.2] *Solve each equation using the quadratic formula.*

9. $2x^2 + x - 21 = 0$

10. $k^2 + 5k = 7$

11. $(t + 3)(t - 4) = -2$

***12.** $2x^2 + 3x + 4 = 0$

***13.** $3p^2 = 2(2p - 1)$

14. $m(2m - 7) = 3m^2 + 3$

**Use the discriminant to predict whether the solutions to each equation are*

A. *two rational numbers;* **B.** *one rational number;*
C. *two irrational numbers;* **D.** *two nonreal complex numbers.*

15. $x^2 + 5x + 2 = 0$

16. $4t^2 = 3 - 4t$

17. $4x^2 = 6x - 8$

18. $9z^2 + 30z + 25 = 0$

*This exercise requires knowledge of complex numbers.

[10.3] *Solve each equation.*

19. $\frac{15}{x} = 2x - 1$

20. $\frac{1}{n} + \frac{2}{n+1} = 2$

21. $-2r = \sqrt{\frac{48 - 20r}{2}}$

22. $8(3x + 5)^2 + 2(3x + 5) - 1 = 0$

23. $2x^{2/3} - x^{1/3} - 28 = 0$

24. $p^4 - 5p^2 + 4 = 0$

Solve each problem. Round answers to the nearest tenth, as necessary.

25. Matthew Sudak drove 8 mi to pick up his cousin Jack, and then drove 11 mi to a mall at a speed 15 mph faster. If Matthew's total travel time was 24 min, what was his speed on the trip to pick up Jack?

26. An old machine processes a batch of checks in 1 hr more time than a new one. How long would it take the old machine to process a batch of checks that the two machines together process in 2 hr?

[10.4] *Solve each formula for the indicated variable. (Give answers with ± when applicable.)*

27. $k = \frac{rF}{wv^2}$ for v

28. $mt^2 = 3mt + 6$ for t

Solve each problem. Round answers to the nearest tenth, as necessary.

29. A large machine requires a part in the shape of a right triangle with a hypotenuse 9 ft less than twice the length of the longer leg. The shorter leg must be $\frac{3}{4}$ the length of the longer leg. Find the lengths of the three sides of the part.

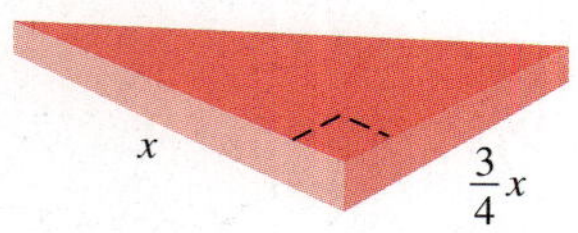

30. A square has an area of 256 cm^2. If the same amount is removed from one dimension and added to the other, the resulting rectangle has an area 16 cm^2 less. Find the dimensions of the rectangle.

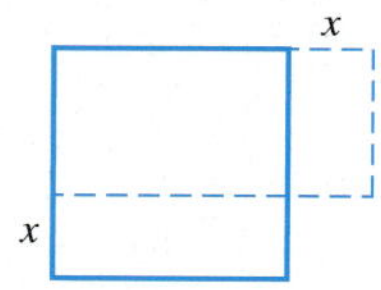

31. Nancy wants to buy a mat for a photograph that measures 14 in. by 20 in. She wants to have an even border around the picture when it is mounted on the mat. If the area of the mat she chooses is 352 in.^{2}, how wide will the border be?

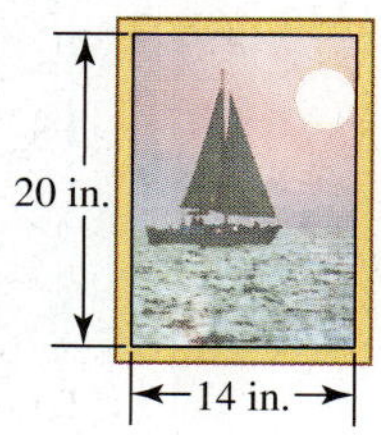

32. Lewis Tower, built in Philadelphia, Pennsylvania, in 1929, is 400 ft high. Suppose that a ball is projected upward from the top of the Tower, and its position in feet above the ground is given by the quadratic function defined by

$$f(t) = -16t^2 + 45t + 400,$$

where t is the number of seconds elapsed. How long will it take for the ball to reach a height of 200 ft above the ground? (*Source: World Almanac and Book of Facts.*)

[10.5–10.6] *Identify the vertex of the graph of each parabola.*

33. $f(x) = -(x - 1)^2$ **34.** $f(x) = (x - 3)^2 + 7$ **35.** $y = -3x^2 + 4x - 2$ **36.** $x = (y - 3)^2 - 4$

Graph each parabola. Give the vertex, axis, domain, and range.

37. $y = 2(x - 2)^2 - 3$

vertex: domain:
axis: range:

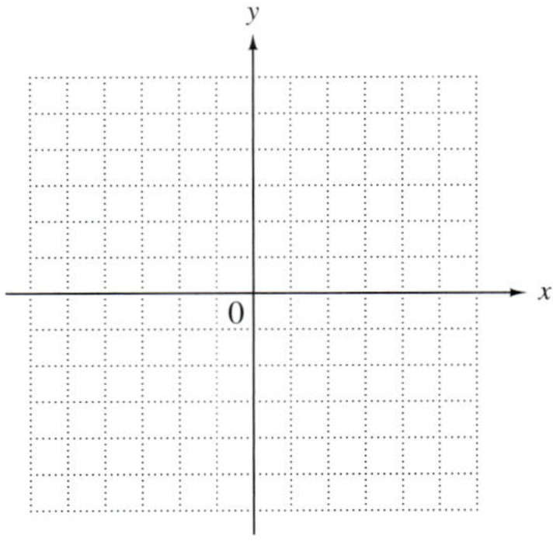

38. $f(x) = -2x^2 + 8x - 5$

vertex: domain:
axis: range:

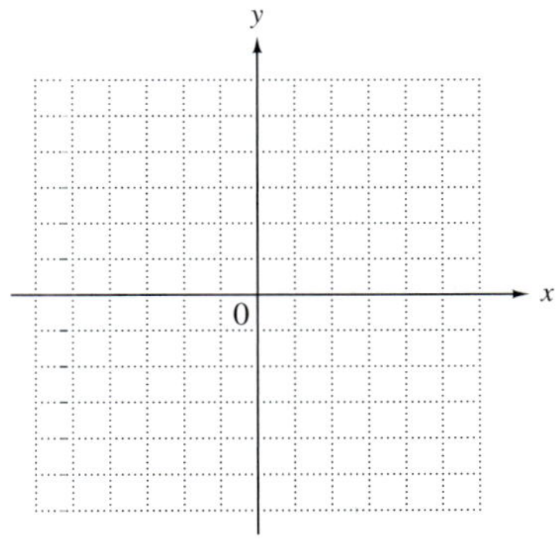

39. $x = 2(y + 3)^2 - 4$

vertex: domain:
axis: range:

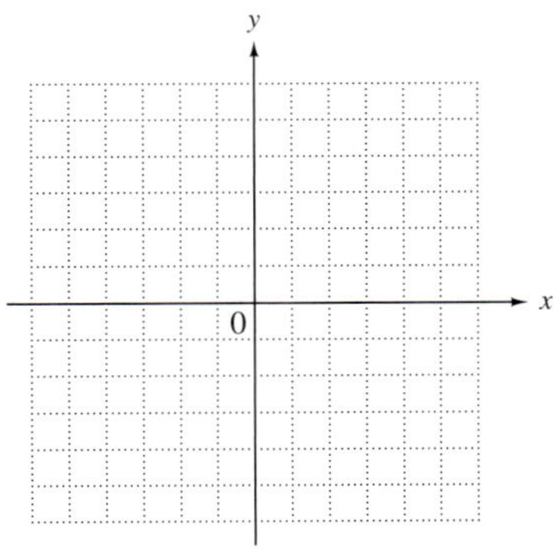

40. $x = -\frac{1}{2}y^2 + 6y - 14$

vertex: domain:
axis: range:

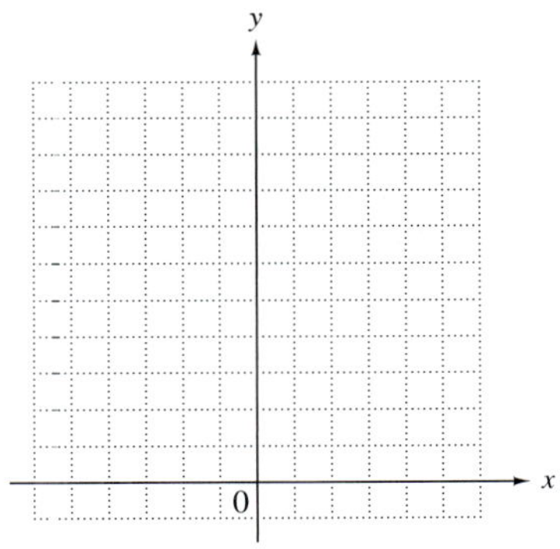

Solve each problem.

41. Total consumer spending on computers, peripherals, and software in the United States for selected years is given in the table. Let $x = 0$ represent 1985, $x = 5$ represent 1990, and so on.

(a) Use the data for 1985, 1995, and 2005 in the quadratic form $ax^2 + bx + c = y$ to write a system of three equations.

(b) Solve the system from part (a) to get a quadratic function f that models the data.

(c) Use the model found in part (b) to approximate consumer spending for computers, peripherals, and software games in 2006 to the nearest tenth. How does your answer compare to the actual data from the table?

CONSUMER SPENDING ON COMPUTERS, PERIPHERALS, AND SOFTWARE

Year	Spending (billions of dollars)
1985	2.9
1990	8.9
1995	24.3
2000	43.8
2004	51.6
2005	56.5
2006	61.4

Source: Bureau of Economic Analysis.

42. The height (in feet) of a projectile t seconds after being fired from Earth into the air is given by

$$f(t) = -16t^2 + 160t.$$

Find the number of seconds required for the projectile to reach maximum height. What is the maximum height?

43. Find the length and width of a rectangle having a perimeter of 200 m if the area is to be a maximum.

[10.7] *Solve each inequality, and graph the solution set.*

44. $(x - 4)(2x + 3) > 0$

45. $x^2 + x \leq 12$

46. $(x + 2)(x - 3)(x + 5) \leq 0$

47. $(4m + 3)^2 \leq -4$

48. $\dfrac{6}{2z - 1} < 2$

49. $\dfrac{3t + 4}{t - 2} \leq 1$

Mixed Review Exercises

Solve each equation or inequality.

50. $V = r^2 + R^2h$ for R

***51.** $3t^2 - 6t = -4$

52. $(x^2 - 2x)^2 = 11(x^2 - 2x) - 24$

53. $(r - 1)(2r + 3)(r + 6) < 0$

54. $(3k + 11)^2 = 7$

55. $S = \dfrac{Id^2}{k}$ for d

56. $2x - \sqrt{x} = 6$

57. $6 + \dfrac{15}{s^2} = -\dfrac{19}{s}$

58. $\dfrac{-2}{x + 5} \leq -5$

*This exercise requires knowledge of complex numbers.

Chapter 10 Test

Chapter Test Prep Video CD — Use the Chapter Test Prep Video CD to see fully worked-out solutions to any of the exercises you want to review

Solve by using either the square root property or completing the square.

1. $t^2 = 54$ 1. ______

2. $(7x + 3)^2 = 25$ 2. ______

3. $x^2 + 2x = 1$ 3. ______

Solve using the quadratic formula.

4. $2x^2 - 3x - 1 = 0$ 4. ______

*5. $3t^2 - 4t = -5$ 5. ______

6. $3x = \sqrt{\dfrac{9x + 2}{2}}$ 6. ______

*7. If k is a negative number, then which one of the following equations will have two nonreal complex solutions? 7. ______

A. $x^2 = 4k$ **B.** $x^2 = -4k$ **C.** $(x + 2)^2 = -k$

8. What is the discriminant for $2x^2 - 8x - 3 = 0$? How many and what type of solutions does this equation have? (Do not actually solve.) 8. ______

Solve by any method.

9. $3 - \dfrac{16}{x} - \dfrac{12}{x^2} = 0$ 9. ______

10. $4x^2 + 7x - 3 = 0$ 10. ______

*This exercise requires knowledge of complex numbers.

11. ______________________

11. $9x^4 + 4 = 37x^2$

12. ______________________

12. $12 = (2n + 1)^2 + (2n + 1)$

13. ______________________

13. Solve for r: $S = 4\pi r^2$. (Leave $\pm$ in your answer.)

Solve each problem.

14. ______________________

14. Andrew and Kent do desktop publishing. Kent can prepare a certain prospectus 2 hr faster than Andrew. If they work together, they can do the entire prospectus in 5 hr. How long will it take each of them working alone to prepare the prospectus? Round your answers to the nearest tenth of an hour.

15. ______________________

15. Bryn Ruhberg paddled her canoe 10 mi upstream, and then paddled back to her starting point. If the rate of the current was 3 mph and the entire trip took $3\frac{1}{2}$ hr, what was Bryn's rate?

16. ______________________

16. Tyler McGinnis has a pool 24 ft long and 10 ft wide. He wants to construct a concrete walk around the pool. If he plans for the walk to be of uniform width and cover 152 ft^2, what will the width of the walk be?

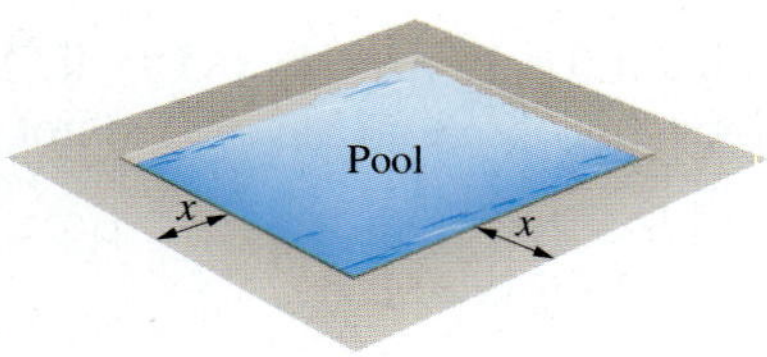

17. ______________________

17. At a point 30 m from the base of a tower, the distance to the top of the tower is 2 m more than twice the height of the tower. Find the height of the tower.

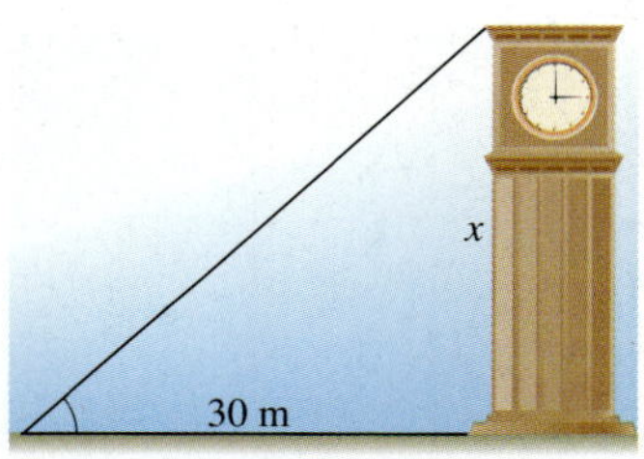

18. Professor Bernstein has found that the number of students attending her intermediate algebra class is approximated by

$$S(x) = -x^2 + 20x + 80,$$

where x is the number of hours that the Campus Center is open daily.

(a) Find the number of hours that the center should be open so that the number of students attending class is a maximum.

(b) What is this maximum number of students?

18. (a) ______________

(b) ______________

19. Which one of the following most closely resembles the graph of $f(x) = a(x - h)^2 + k$ if $a < 0$, $h > 0$, and $k < 0$?

A.

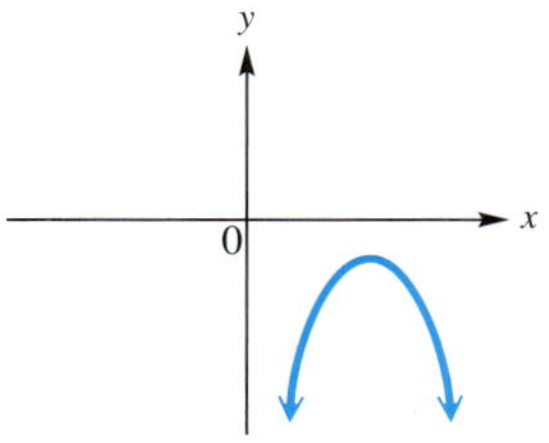

B.

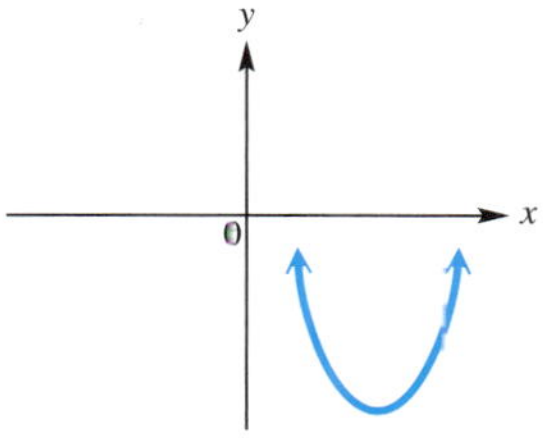

C.

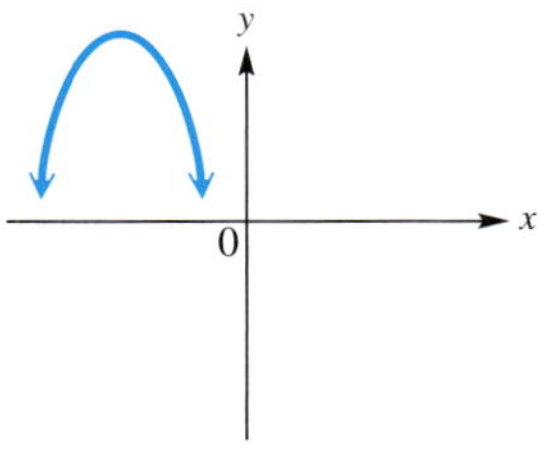

D.

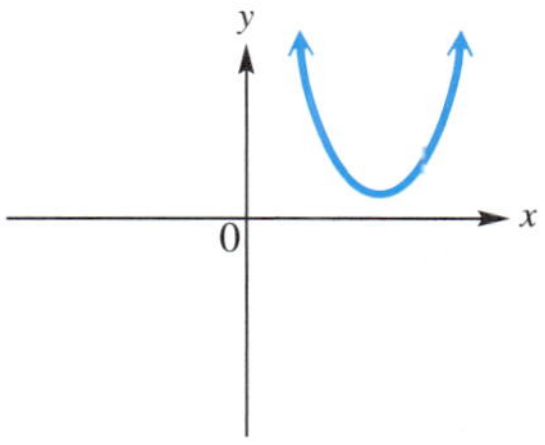

19. ______________

Graph each parabola. Give the vertex, axis, domain, and range.

20. $f(x) = \frac{1}{2}x^2 - 2$

20. ______________

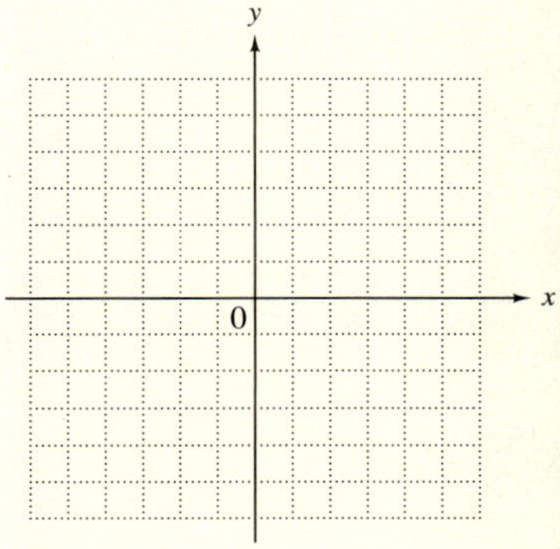

21. ______________

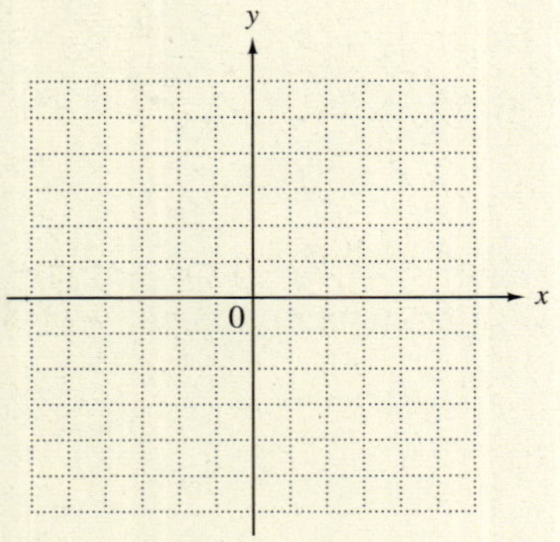

21. $f(x) = -x^2 + 4x - 1$

22. ______________

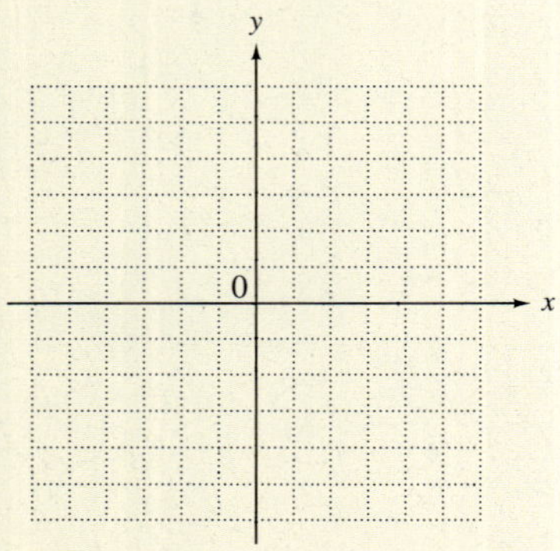

22. $x = 2y^2 + 8y + 3$

23. ______________

23. The manager of Morgan's Department Store wants to construct a rectangular parking lot on land bordered on one side by a highway. The store has 280 ft of fencing that is to be used to fence off the other three sides. What should be the dimensions of the lot if the enclosed area is to be a maximum? What is the maximum area?

Solve. Graph each solution set.

24. ______________

24. $2x^2 + 7x > 15$

25. ______________

25. $\dfrac{5}{t - 4} \leq 1$

Cumulative Review Exercises Chapters 1–10

Solve each equation or inequality.

1. $-2x + 4 = 5(x - 4) + 17$

2. $-2x + 4 \leq -x + 3$

3. $|3x - 7| \leq 1$

4. Find the slope and y-intercept of the line with equation $2x - 4y = 7$.

5. Write the equation in standard form of the line through $(2, -1)$ and perpendicular to $-3x + y = 5$.

Graph each relation. Tell whether or not each is a function, and if it is, give its domain and range.

6. $4x - 5y = 15$

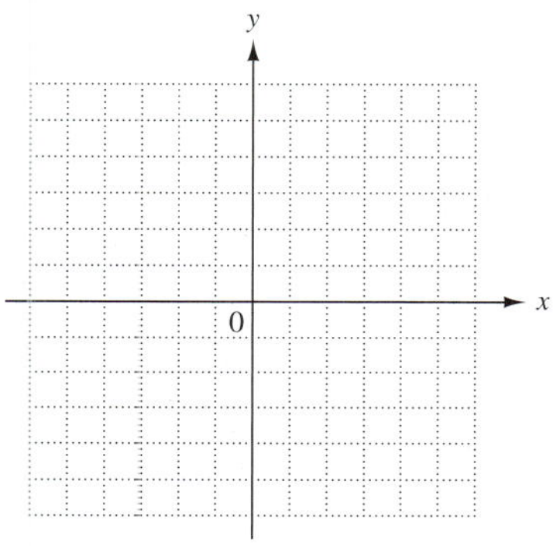

7. $4x - 5y < 15$

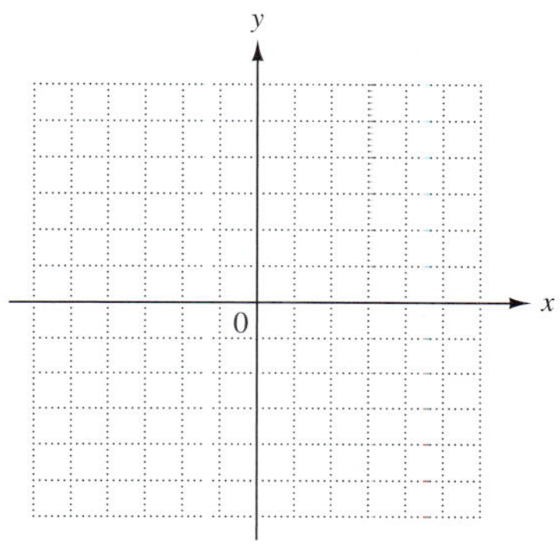

8. $y = -2(x - 1)^2 + 3$

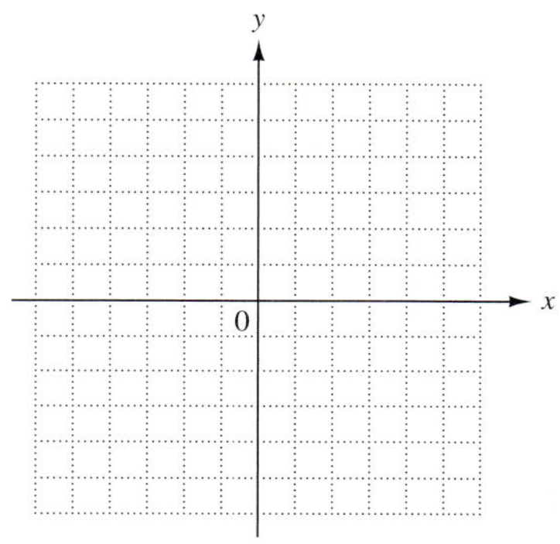

Solve each system of equations.

9. $\begin{aligned} 2x - 4y &= 10 \\ 9x + 3y &= 3 \end{aligned}$

10. $\begin{aligned} x + y + 2z &= 3 \\ -x + y + z &= -5 \\ 2x + 3y - z &= -8 \end{aligned}$

Write with positive exponents only. Assume that variables represent positive real numbers.

11. $\left(\dfrac{x^{-3}y^2}{x^5y^{-2}}\right)^{-1}$

12. $\dfrac{(4x^{-2})^2(2y^3)}{3x^{-3}y^5}$

13. Multiply: $\left(\dfrac{2}{3}t + 9\right)^2$.

14. Divide $4x^3 + 2x^2 - x + 26$ by $x + 2$.

Factor completely.

15. $16x - x^3$

16. $24m^2 + 2m - 15$

17. $9x^2 - 30xy + 25y^2$

Perform the operations, and express answers in lowest terms. Assume that denominators represent nonzero real numbers.

18. $\frac{5t + 2}{-6} \div \frac{15t + 6}{5}$

19. $\frac{3}{2 - k} - \frac{5}{k} + \frac{6}{k^2 - 2k}$

20. $\frac{\frac{r}{s} - \frac{s}{r}}{\frac{r}{s} + 1}$

Simplify each radical expression.

21. $\sqrt[3]{\frac{27}{16}}$

22. $\frac{2}{\sqrt{7} - \sqrt{5}}$

Solve each equation.

23. $2x = \sqrt{\frac{5x + 2}{3}}$

24. $2x^2 - 4x - 3 = 0$

25. $z^2 - 2z = 15$

26. $\frac{3}{x - 3} - \frac{2}{x - 2} = \frac{3}{x^2 - 5x + 6}$

27. $p^4 - 10p^2 + 9 = 0$

28. Two cars left an intersection at the same time, one heading due south and the other due east. Later they were exactly 95 mi apart. The car heading east had gone 38 mi less than twice as far as the car heading south. How far had each car traveled?

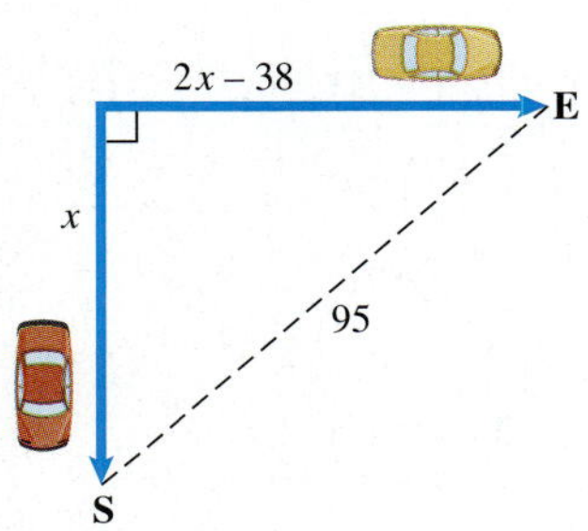

11

Inverse, Exponential, and Logarithmic Functions

11.1 Inverse Functions
11.2 Exponential Functions
11.3 Logarithmic Functions
11.4 Properties of Logarithms
11.5 Common and Natural Logarithms
11.6 Exponential and Logarithmic Equations; Further Applications

In 2001, Apple Computer Inc., introduced the iPod. Since then, the company has sold over 40 million of the popular music players, in spite of warnings by experts that listening to the devices at high volumes may put people at increased risk of hearing loss. In 2006, a federal class-action lawsuit was filed against the company, accusing it of not taking adequate steps to protect the hearing of iPod users. As a result, Apple issued a software update that allows listeners to set maximum volume limits on some of the newer iPod models. (*Source: Sacramento Bee, USA Today.*)

In Example 4 of Section 11.5, we use a *logarithmic function* to calculate the volume level, in *decibels,* of an iPod.

11.1 Inverse Functions

OBJECTIVES

1. Decide whether a function is one-to-one and, if it is, find its inverse.
2. Use the horizontal line test to determine whether a function is one-to-one.
3. Find the equation of the inverse of a function.
4. Graph f^{-1} from the graph of f.

In this chapter we will study two important types of functions, *exponential* and *logarithmic*. These functions are related in a special way: They are *inverses* of one another. We begin by discussing inverse functions in general.

Calculator Tip A calculator with the following keys will be essential in this chapter.

y^x, 10^x or LOG, e^x or LN

We will explain how these keys are used at appropriate places in the chapter.

OBJECTIVE 1 Decide whether a function is one-to-one and, if it is, find its inverse. Suppose we define the function

$$G = \{(-2, 2), (-1, 1), (0, 0), (1, 3), (2, 5)\}.$$

We can form another set of ordered pairs from G by interchanging the x- and y-values of each pair in G. Call this set F, with

$$F = \{(2, -2), (1, -1), (0, 0), (3, 1), (5, 2)\}.$$

To show that these two sets are related, F is called the *inverse* of G. For a function f to have an inverse function, f must be *one-to-one.*

One-to-One Function

In a **one-to-one function,** each x-value corresponds to just one y-value, and each y-value corresponds to just one x-value.

The function shown in Figure 1(a) is not one-to-one because the y-value 7 corresponds to *two* x-values, 2 and 3. That is, the ordered pairs (2, 7) and (3, 7) both appear in the function. The function in Figure 1(b) is one-to-one.

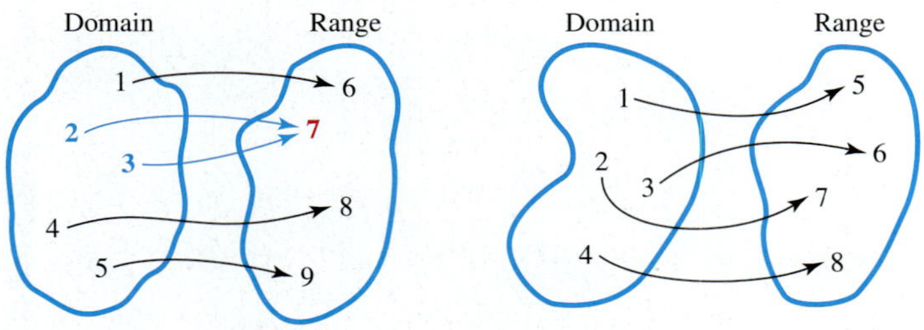

Not one-to-one
(a)

One-to-one
(b)

Figure 1

The *inverse* of any one-to-one function f is found by interchanging the components of the ordered pairs of f. The inverse of f is written f^{-1}**.** Read f^{-1} as **"the inverse of f"** or **"f-inverse."**

CAUTION

The symbol $f^{-1}(x)$ does not represent $\dfrac{1}{f(x)}$.

The definition of the inverse of a function follows.

Inverse of a Function

The **inverse** of a one-to-one function f, written f^{-1}, is the set of all ordered pairs of the form (y, x), where (x, y) belongs to f. ***Since the inverse is formed by interchanging x and y, the domain of f becomes the range of f^{-1} and the range of f becomes the domain of f^{-1}.***

For inverses f and f^{-1}, it follows that

$$f(f^{-1}(x)) = x \quad \text{and} \quad f^{-1}(f(x)) = x.$$

EXAMPLE 1 Finding the Inverses of One-to-One Functions

Find the inverse of each function that is one-to-one.

(a) $F = \{(-2, 1), (-1, 0), (0, 1), (1, 2), (2, 2)\}$

Each x-value in F corresponds to just one y-value. However, the y-value 2 corresponds to two x-values, 1 and 2. Also, the y-value 1 corresponds to both -2 and 0. Because some y-values correspond to more than one x-value, F is not one-to-one and does not have an inverse function.

(b) $G = \{(3, 1), (0, 2), (2, 3), (4, 0)\}$

Every x-value in G corresponds to only one y-value, and every y-value corresponds to only one x-value, so G is a one-to-one function. The inverse function is found by interchanging the x- and y-values in each ordered pair.

$$G^{-1} = \{(1, 3), (2, 0), (3, 2), (0, 4)\}$$

Notice how the domain and range of G become the range and domain, respectively, of G^{-1}.

(c) Ozone is the major component in smog. Breathing ozone may cause serious risk of respiratory problems. In 1997, the U.S. Environmental Protection Agency set an 8-hour average ground-level ozone standard of 0.08 parts per million (ppm). The table shows the number of days in which the air in Connecticut exceeded this standard for the years 1995–2006.

Year	Number of Days Exceeding Standard	Year	Number of Days Exceeding Standard
1995	24	2001	26
1996	16	2002	36
1997	27	2003	14
1998	25	2004	6
1999	33	2005	20
2000	13	2006	13

Source: U.S. Environmental Protection Agency.

Let f be the function defined in the table, with the years forming the domain and the numbers of days exceeding the ozone standard forming the range. Then f is not one-to-one, because in two different years (2000 and 2006), the number of days with unacceptable ozone levels was the same, 13.

Work Problem 1 at the Side. ▶

1 Find the inverse of each function that is one-to-one.

(a) $\{(1, 2), (2, 4), (3, 3), (4, 5)\}$

(b) $\{(0, 3), (-1, 2), (1, 3)\}$

(c) A Norwegian physiologist has developed a rule for predicting running times based on the time to run 5 km (5K). An example for one runner is shown here. (*Source:* Stephen Seiler, Agder College, Kristiansand, Norway.)

Distance	Time
1.5K	4:22
3K	9:18
5K	16:00
10K	33:40

ANSWERS

1. (a) $\{(2, 1), (4, 2), (3, 3), (5, 4)\}$
(b) not a one-to-one function
(c)

Time	Distance
4:22	1.5K
9:18	3K
16:00	5K
33:40	10K

2 Use the horizontal line test to determine whether each graph is the graph of a one-to-one function.

(a)

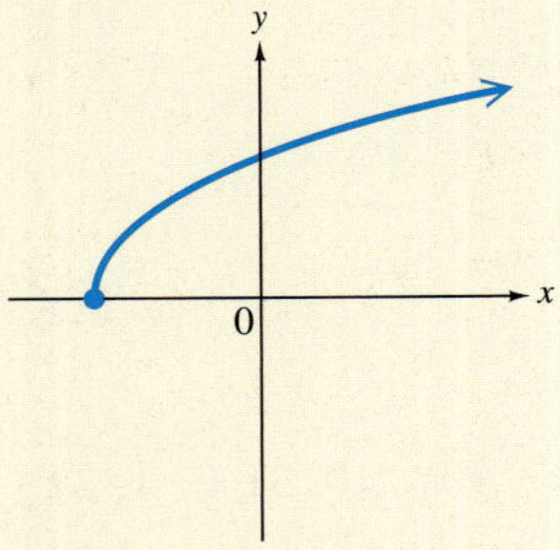

(b)

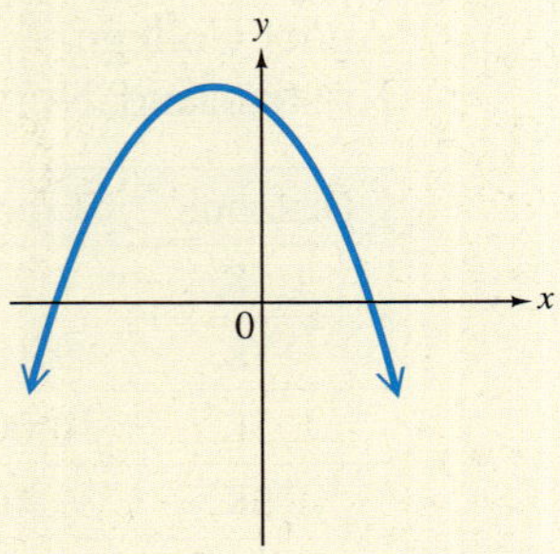

OBJECTIVE 2 Use the horizontal line test to determine whether a function is one-to-one. It may be difficult to decide whether a function is one-to-one just by looking at the equation that defines the function. However, by graphing the function and observing the graph, we can use the *horizontal line test* to tell whether the function is one-to-one.

> **Horizontal Line Test**
>
> A function is one-to-one if every horizontal line intersects the graph of the function at most once.

The horizontal line test follows from the definition of a one-to-one function. Any two points that lie on the same horizontal line have the same y-coordinate. No two ordered pairs that belong to a one-to-one function may have the same y-coordinate, and, therefore, no horizontal line will intersect the graph of a one-to-one function more than once.

EXAMPLE 2 Using the Horizontal Line Test

Use the horizontal line test to determine whether the graphs in Figures 2 and 3 are graphs of one-to-one functions.

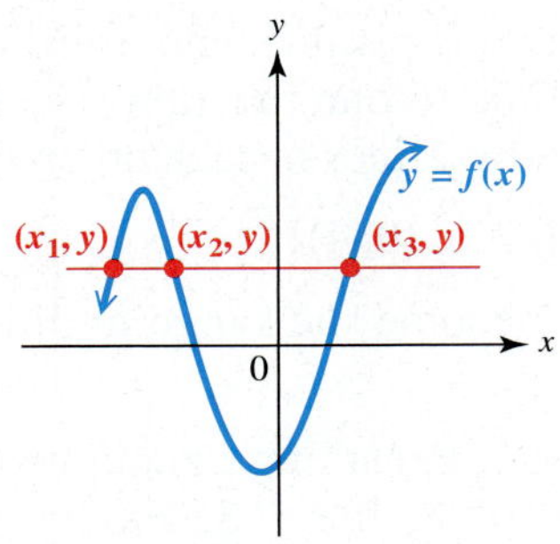

Figure 2

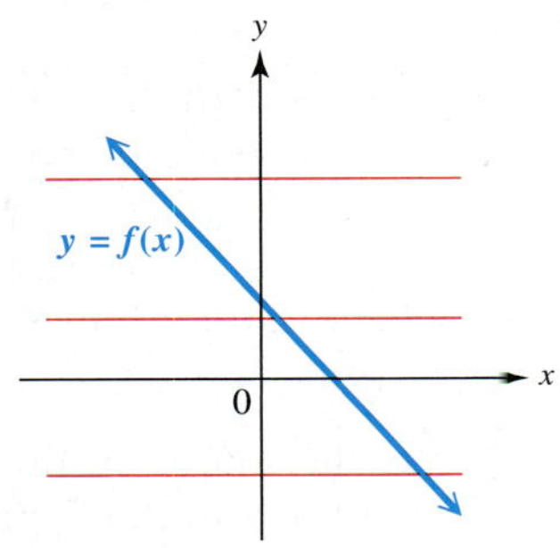

Figure 3

Because the red horizontal line shown in Figure 2 intersects the graph in more than one point (actually three points), the function is not one-to-one.

Every horizontal line will intersect the graph in Figure 3 in exactly one point. This function is one-to-one.

◀ *Work Problem* **2** *at the Side.*

OBJECTIVE 3 Find the equation of the inverse of a function. By definition, the inverse of a one-to-one function is found by interchanging the x- and y-values of each of its ordered pairs. The equation of the inverse of a function defined by $y = f(x)$ is found in the same way.

> **Finding the Equation of the Inverse of $y = f(x)$**
>
> For a one-to-one function f defined by an equation $y = f(x)$, find the defining equation of the inverse as follows.
>
> ***Step 1*** Interchange x and y.
>
> ***Step 2*** Solve for y.
>
> ***Step 3*** Replace y with $f^{-1}(x)$.

ANSWERS

2. (a) one-to-one **(b)** not one-to-one

EXAMPLE 3 Finding Equations of Inverses

Decide whether each equation defines a one-to-one function. If so, find the equation that defines the inverse.

(a) $f(x) = 2x + 5$

The graph of $y = 2x + 5$ is a nonvertical line, so by the horizontal line test, f is a one-to-one function. To find the inverse, let $y = f(x)$ so that

$$\begin{aligned} y &= 2x + 5 \\ x &= 2y + 5 && \text{Interchange } x \text{ and } y. \text{ (Step 1)} \\ 2y &= x - 5 && \text{Solve for } y. \text{ (Step 2)} \\ y &= \frac{x-5}{2} \\ f^{-1}(x) &= \frac{x-5}{2}, && \text{Replace } y \text{ with } f^{-1}(x). \text{ (Step 3)} \end{aligned}$$

which can also be written

$$f^{-1}(x) = \frac{x}{2} - \frac{5}{2}, \quad \text{or} \quad f^{-1}(x) = \frac{1}{2}x - \frac{5}{2}. \qquad \frac{a-b}{c} = \frac{a}{c} - \frac{b}{c}$$

Thus, f^{-1} is a linear function. In the function with $y = 2x + 5$, the value of y is found by starting with a value of x, multiplying by 2, and adding 5. The equation $f^{-1}(x) = \frac{x-5}{2}$ for the inverse has us *subtract* 5, and then *divide* by 2. This shows how an inverse is used to "undo" what a function does to the variable x.

(b) $y = x^2 + 2$

This equation has a vertical parabola as its graph, so some horizontal lines will intersect the graph at two points. For example, both $x = 3$ and $x = -3$ correspond to $y = 11$. Because of the x^2-term, there are many pairs of x-values that correspond to the same y-value. This means that the function defined by $y = x^2 + 2$ is not one-to-one and does not have an inverse function.

Following the steps for finding the equation of an inverse leads to

$$\begin{aligned} y &= x^2 + 2 \\ x &= y^2 + 2 && \text{Interchange } x \text{ and } y. \\ y^2 &= x - 2 && \text{Solve for } y. \\ y &= \pm\sqrt{x-2}. && \text{Square root property} \end{aligned}$$

The last step shows that there are two y-values for each choice of $x > 2$, so the given function is not one-to-one and cannot have an inverse.

(c) $f(x) = (x - 2)^3$

Refer to **Section 6.3** to see from its graph that a cubing function like this is a one-to-one function.

$$\begin{aligned} y &= (x-2)^3 && \text{Replace } f(x) \text{ with } y. \\ x &= (y-2)^3 && \text{Interchange } x \text{ and } y. \\ \sqrt[3]{x} &= \sqrt[3]{(y-2)^3} && \text{Take the cube root on each side.} \\ \sqrt[3]{x} &= y - 2 \\ y &= \sqrt[3]{x} + 2 && \text{Solve for } y. \\ f^{-1}(x) &= \sqrt[3]{x} + 2 && \text{Replace } y \text{ with } f^{-1}(x). \end{aligned}$$

Work Problem 3 at the Side. ▶

3 Decide whether each equation defines a one-to-one function. If so, find the equation that defines the inverse.

(a) $f(x) = 3x - 4$

(b) $f(x) = x^3 + 1$

(c) $f(x) = (x - 3)^2$

ANSWERS

3. (a) one-to-one function; $f^{-1}(x) = \frac{x+4}{3}$, or $f^{-1}(x) = \frac{1}{3}x + \frac{4}{3}$
(b) one-to-one function; $f^{-1}(x) = \sqrt[3]{x-1}$
(c) not a one-to-one function

4 Use the given graphs to graph each inverse.

(a)

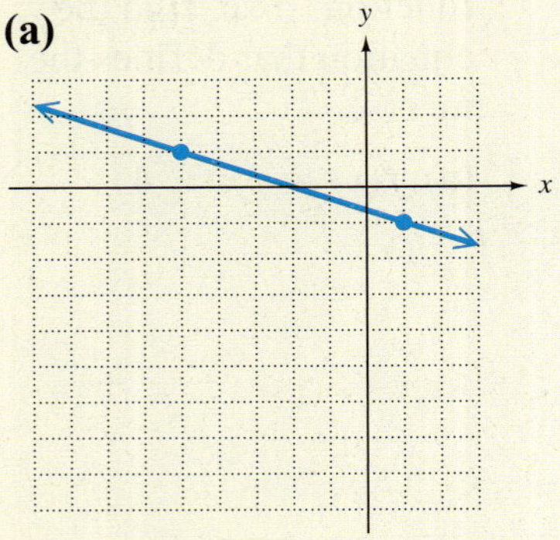

(b)

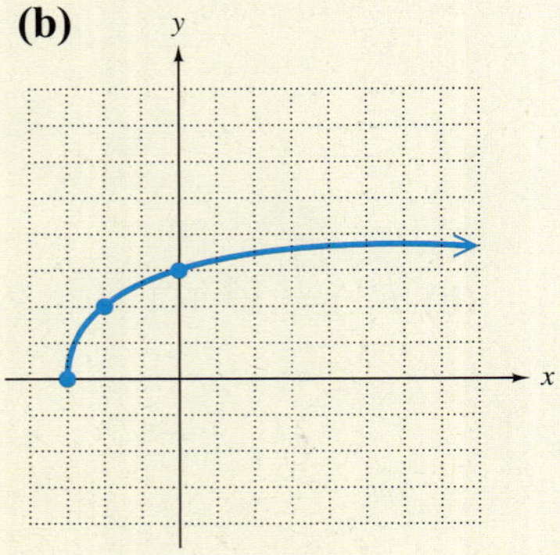

(c)

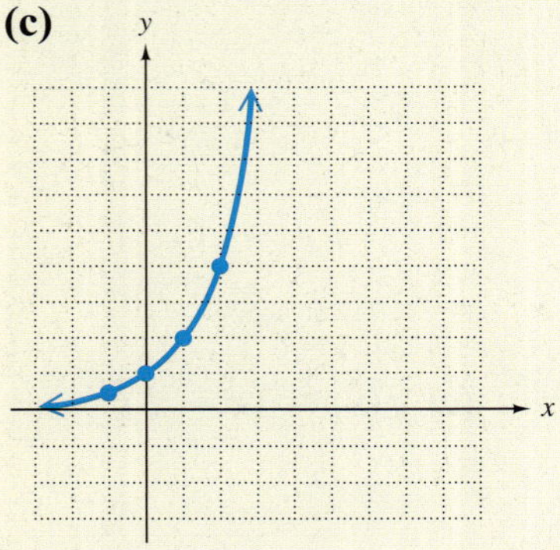

OBJECTIVE 4 Graph f^{-1} from the graph of f. One way to graph the inverse of a function f whose equation is known is to find some ordered pairs that belong to f, interchange x and y to get ordered pairs that belong to f^{-1}, plot those points, and sketch the graph of f^{-1} through the points. A simpler way is to select points on the graph of f and use symmetry to find corresponding points on the graph of f^{-1}.

For example, suppose the point (a, b) shown in Figure 4 belongs to a one-to-one function f. Then the point (b, a) belongs to f^{-1}. The line segment connecting (a, b) and (b, a) is perpendicular to, and cut in half by, the line $y = x$. The points (a, b) and (b, a) are "mirror images" of each other with respect to $y = x$. ***Thus, we can find the graph of f^{-1} from the graph of f by locating the mirror image of each point in f with respect to the line $y = x$.***

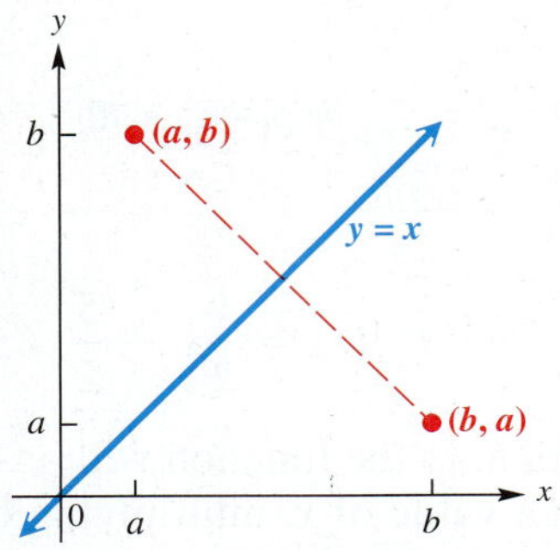

Figure 4

EXAMPLE 4 Graphing the Inverse

Graph the inverses of the functions shown in Figure 5.

In Figure 5 the graphs of two functions are shown in blue. Their inverses are shown in red. In each case, the graph of f^{-1} is symmetric to the graph of f with respect to the line $y = x$.

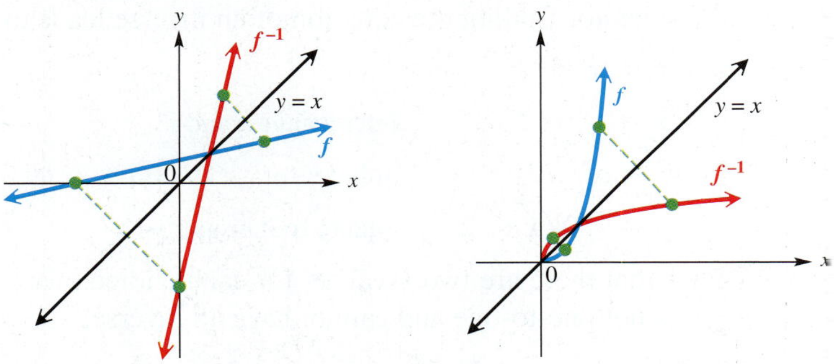

Figure 5

◀ Work Problem **4** at the Side.

ANSWERS

4. (a)

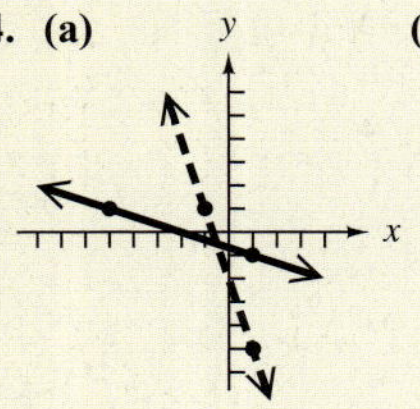

(b)

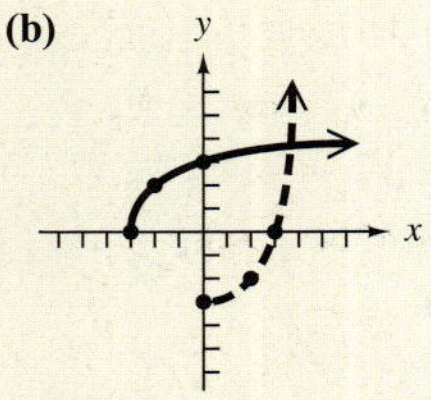

(c)

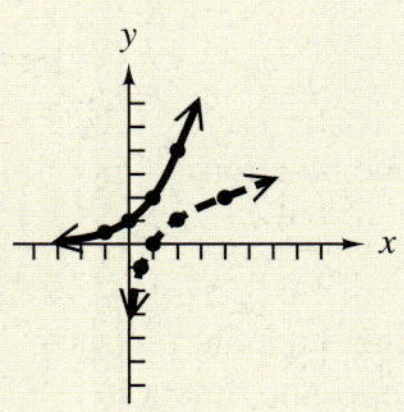

11.1 Exercises

FOR EXTRA HELP MyMathLab Math XL PRACTICE WATCH DOWNLOAD READ REVIEW

Answer each question. See Example 1.

1. A new study found that the trans fat content in fast-food products varied widely around the world, based on the type of frying oil used, as shown in the table.

If the set of countries is the domain and the set of trans fat percentages is the range of a function, is it one-to-one? Why or why not?

Country	Percentage of Trans Fat in McDonald's Chicken
Scotland	14
France	11
United States	11
Peru	9
Russia	5
Denmark	1

Source: New England Journal of Medicine.

2. The table shows concentrations of a major air pollutant, carbon monoxide, in the United States for the years 2000–2005.

If this correspondence is considered to be a function that pairs each year with its concentration, is it one-to-one? If not, explain why.

Year	Concentration (in parts per million)
2000	3.5
2001	3.2
2002	3.0
2003	2.8
2004	2.6
2005	2.4

Source: E.P.A.

3. Suppose you consider the set of ordered pairs (x, y) such that x represents a person in your mathematics class and y represents that person's mother. How might this function not be a one-to-one function?

4. The road mileage between Denver, Colorado, and several selected U.S. cities is shown in the table.

If we consider this as a function that pairs each city with a distance, is it one-to-one? How could we change the answer to this question by adding 1 mile to one of the distances shown?

City	Distance to Denver (in miles)
Atlanta	1398
Dallas	781
Indianapolis	1058
Kansas City, MO	600
Los Angeles	1059

Choose the correct response from the given list.

5. If a function is made up of ordered pairs in such a way that the same y-value appears in a correspondence with two different x-values, then

A. the function is one-to-one

B. the function is not one-to-one

C. its graph does not pass the vertical line test

D. it has an inverse function associated with it.

6. Which equation defines a one-to-one function? Explain why the others are not, using specific examples.

A. $f(x) = x$ **B.** $f(x) = x^2$

C. $f(x) = |x|$ **D.** $f(x) = -x^2 + 2x - 1$

7. Only one of the graphs illustrates a one-to-one function. Which one is it?

A.

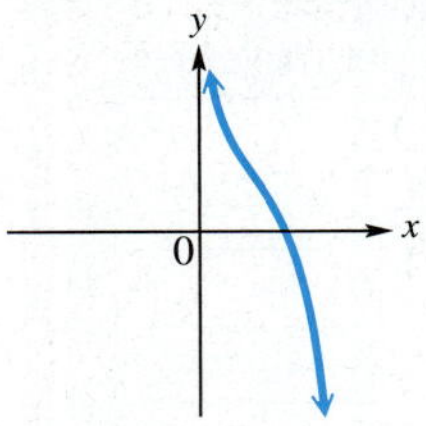

B.

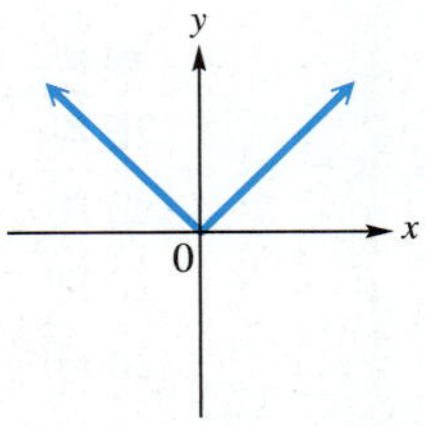

C.

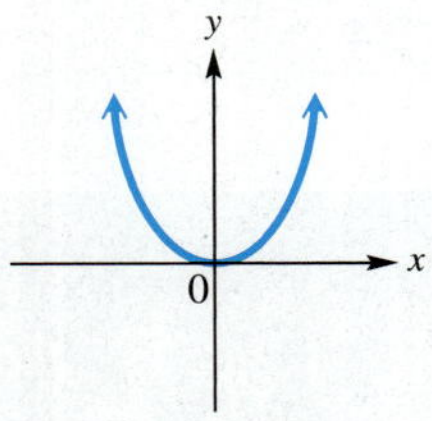

D.

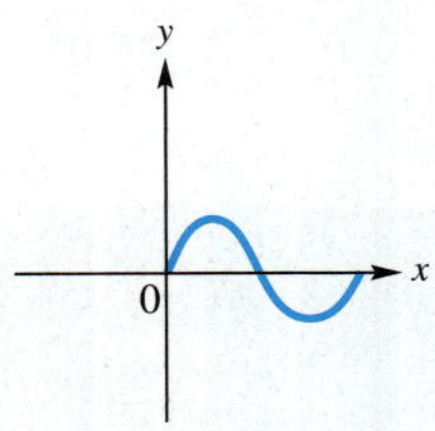

8. If a function f is one-to-one and the point (p, q) lies on the graph of f, then which point *must* lie on the graph of f^{-1}?

A. $(-p, q)$ **B.** $(-q, -p)$

C. $(p, -q)$ **D.** (q, p)

If the function is one-to-one, find its inverse. See Examples 1–3.

9. $\{(3, 6), (2, 10), (5, 12)\}$

10. $\left\{(-1, 3), (0, 5), (5, 0), \left(7, -\frac{1}{2}\right)\right\}$

11. $\{(-1, 3), (2, 7), (4, 3), (5, 8)\}$

12. $\{(-8, 6), (-4, 3), (0, 6), (5, 10)\}$

13. $f(x) = 2x + 4$

14. $f(x) = 3x + 1$

15. $g(x) = \sqrt{x - 3}, \quad x \geq 3$

16. $g(x) = \sqrt{x + 2}, \quad x \geq -2$

17. $f(x) = 3x^2 + 2$

18. $f(x) = -4x^2 - 1$

19. $f(x) = x^3 - 4$

20. $f(x) = x^3 - 3$

Let $f(x) = 2^x$. We will see in the next section that the function f is one-to-one. Find each value, always working part (a) before part (b).

21. **(a)** $f(3)$ **(b)** $f^{-1}(8)$

22. **(a)** $f(4)$ **(b)** $f^{-1}(16)$

23. **(a)** $f(0)$ **(b)** $f^{-1}(1)$

24. **(a)** $f(-2)$ **(b)** $f^{-1}\left(\frac{1}{4}\right)$

*The graphs of some functions are given in Exercises 25–30. **(a)** Use the horizontal line test to determine whether each function is one-to-one. **(b)** If the function is one-to-one, graph the inverse of the function with a dashed line (or curve) on the same set of axes. (Remember that if f is one-to-one and $f(a) = b$, then $f^{-1}(b) = a$.) See Example 4.*

25.

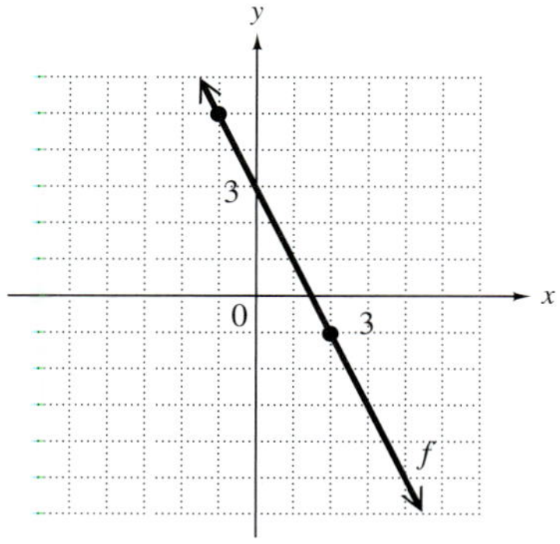

26.

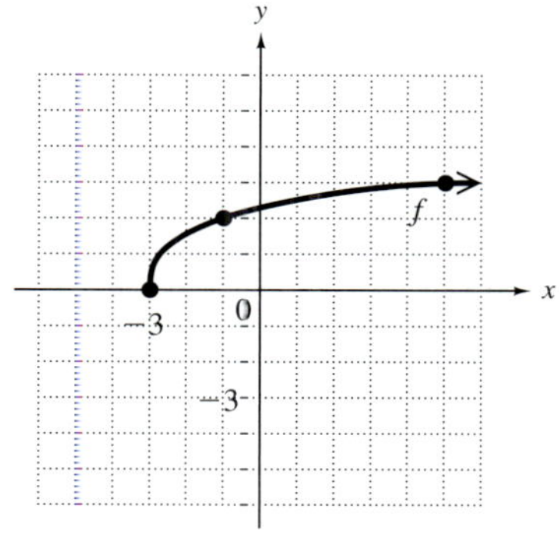

27.

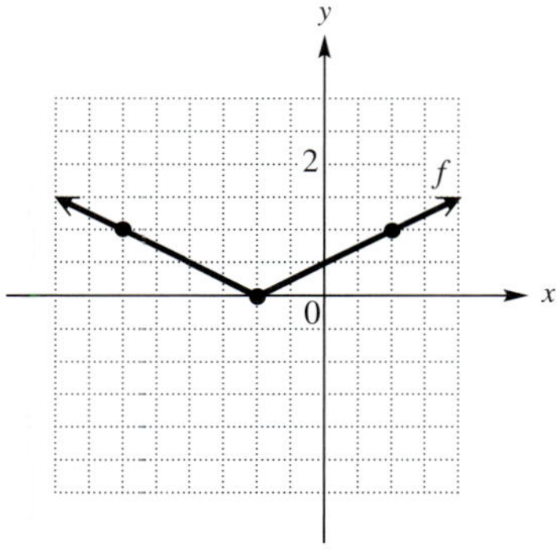

28.

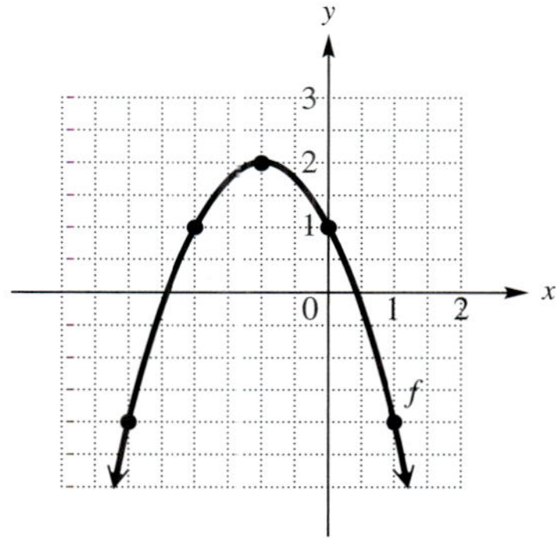

29.

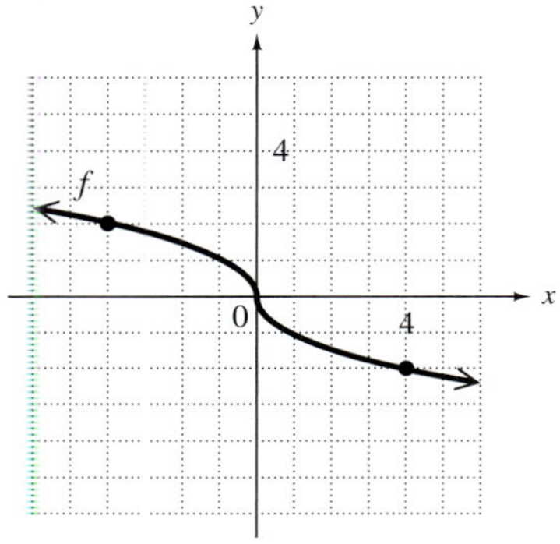

30.

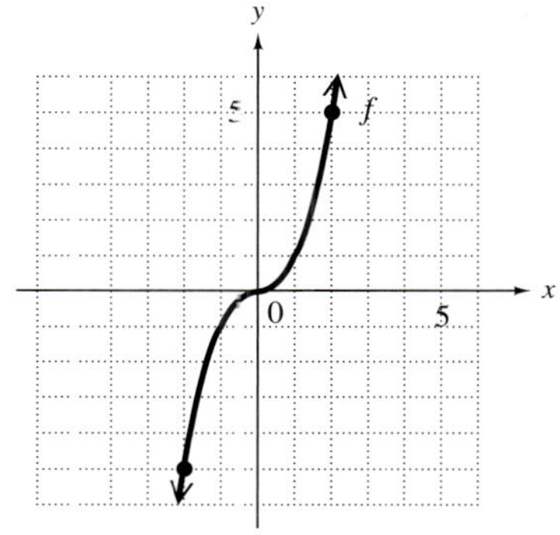

Each function defined in Exercises 31–38 is a one-to-one function. Graph the function as a solid line (or curve), and then graph its inverse on the same set of axes as a dashed line (or curve). In Exercises 35–38 you are given a table to complete so that graphing the function will be easier. See Example 4.

31. $f(x) = 2x - 1$

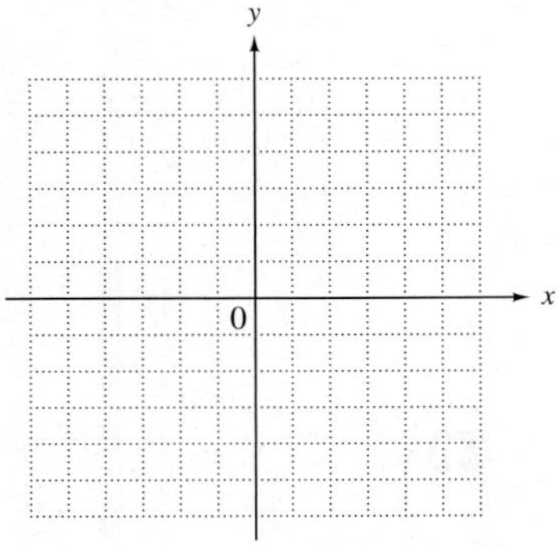

32. $f(x) = 2x + 3$

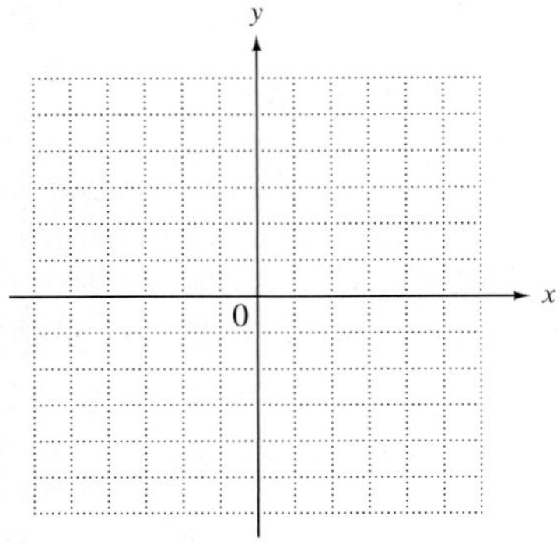

33. $g(x) = -4x$

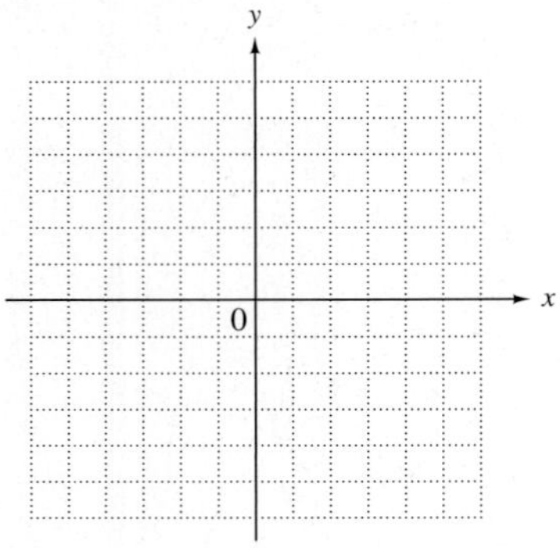

34. $g(x) = -2x$

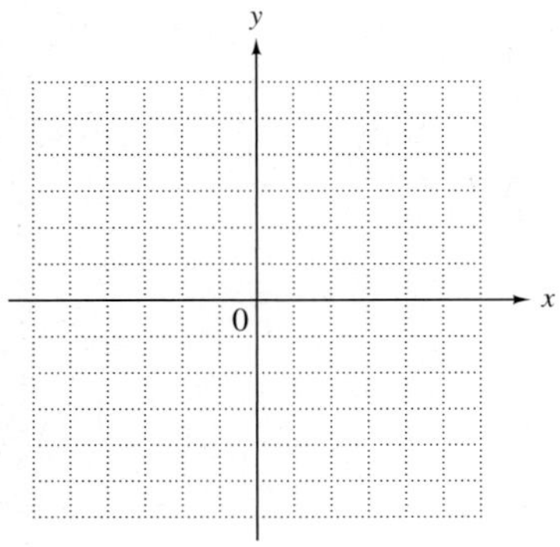

35. $f(x) = y = \sqrt{x}, x \geq 0$

x	f(x) = y
0	
1	
4	

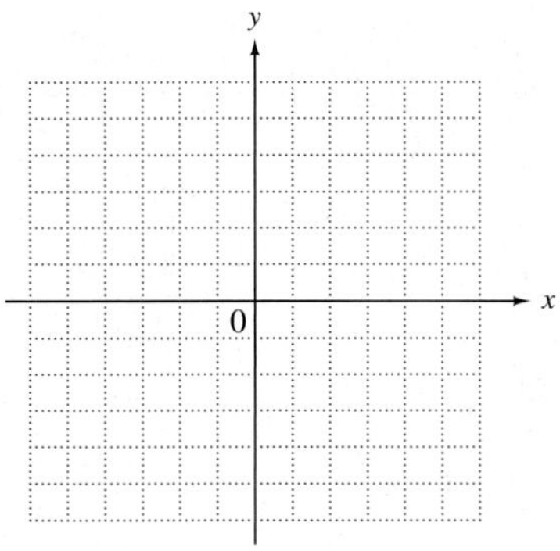

36. $f(x) = y = -\sqrt{x}, x \geq 0$

x	f(x) = y
0	
1	
4	

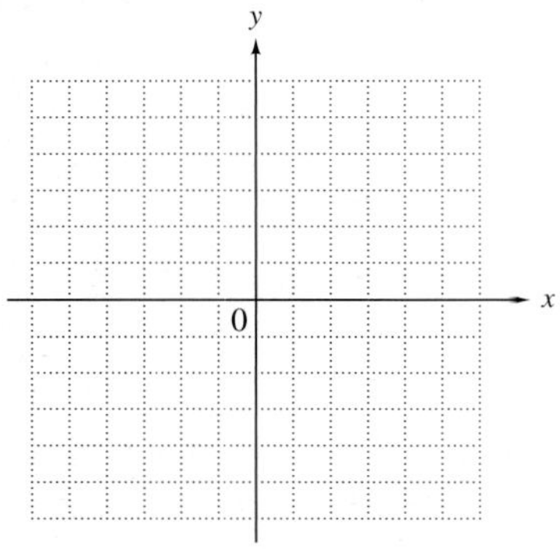

37. $f(x) = y = x^3 - 2$

x	f(x) = y
−1	
0	
1	
2	

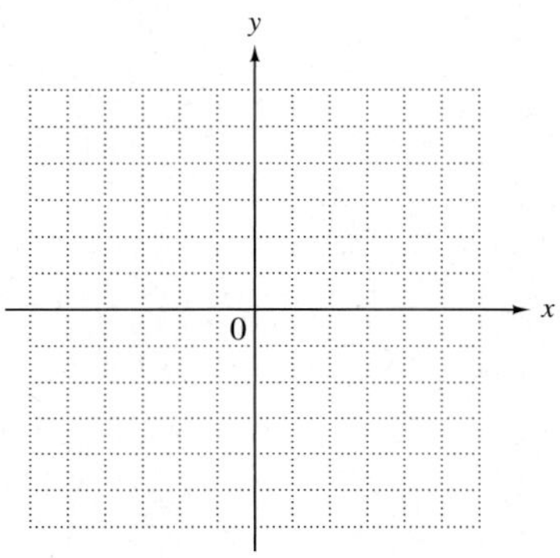

38. $f(x) = y = x^3 + 3$

x	f(x) = y
−2	
−1	
0	
1	

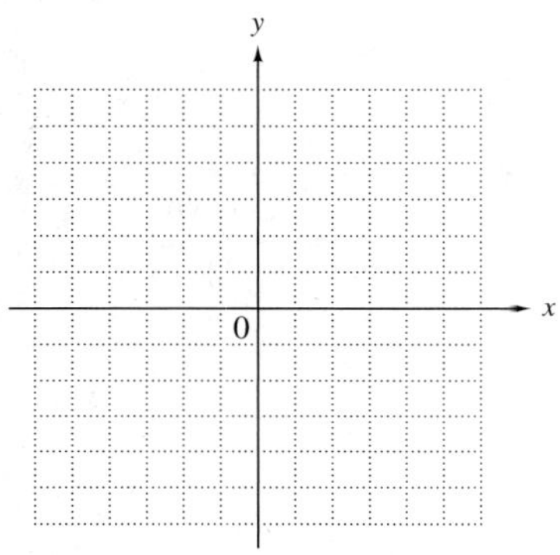

11.2 Exponential Functions

OBJECTIVES

1. Define exponential functions.
2. Graph exponential functions.
3. Solve exponential equations of the form $a^x = a^k$ for x.
4. Use exponential functions in applications involving growth or decay.

OBJECTIVE 1 Define exponential functions. In **Section 9.2,** we showed how to evaluate 2^x for rational values of x. For example,

$$2^3 = 8, \qquad 2^{-1} = \frac{1}{2}, \qquad 2^{1/2} = \sqrt{2}, \qquad 2^{3/4} = \sqrt[4]{2^3} = \sqrt[4]{8}.$$

In more advanced courses it is shown that 2^x exists for all real number values of x, both rational and irrational. The following definition of an exponential function assumes that a^x exists for all real numbers x.

Exponential Function

For $a > 0$, $a \neq 1$, and all real numbers x,

$$F(x) = a^x$$

defines the **exponential function with base a.**

Note

The two restrictions on the value of a in the definition of an exponential function $F(x) = a^x$ are important.

1. The restriction $a > 0$ is necessary so that the function can be defined for all real numbers x. Letting a be negative ($a = -2$, for instance) and letting $x = \frac{1}{2}$ would give $(-2)^{1/2}$, which is not real.
2. The restriction $a \neq 1$ is necessary because 1 raised to any power is equal to 1, and the function would then be the linear function defined by $F(x) = 1$.

OBJECTIVE 2 Graph exponential functions. When graphing exponential functions of the form $F(x) = a^x$, pay particular attention to whether $a > 1$ or $a < 1$.

EXAMPLE 1 Graphing an Exponential Function ($a > 1$)

Graph $f(x) = 2^x$.

Choose some values of x, and find the corresponding values of $f(x)$. Plotting these points and drawing a smooth curve through them gives the graph of $f(x) = 2^x$ shown in Figure 6.

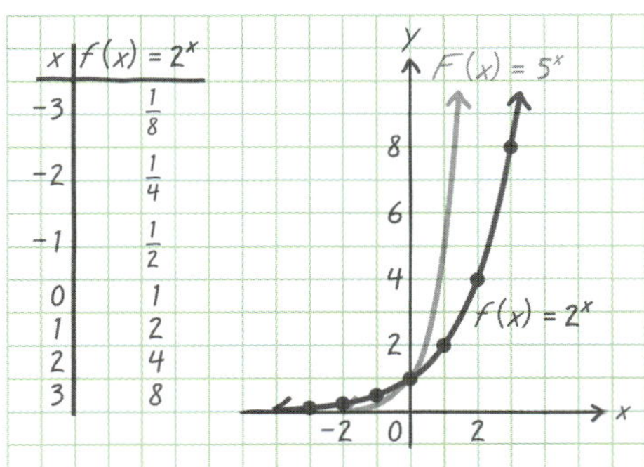

Figure 6

Continued on Next Page

1 Graph.

(a) $f(x) = 10^x$

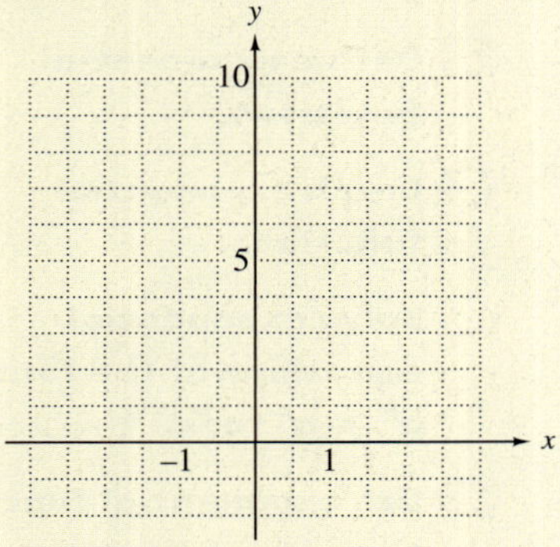

(b) $g(x) = \left(\frac{1}{10}\right)^x$

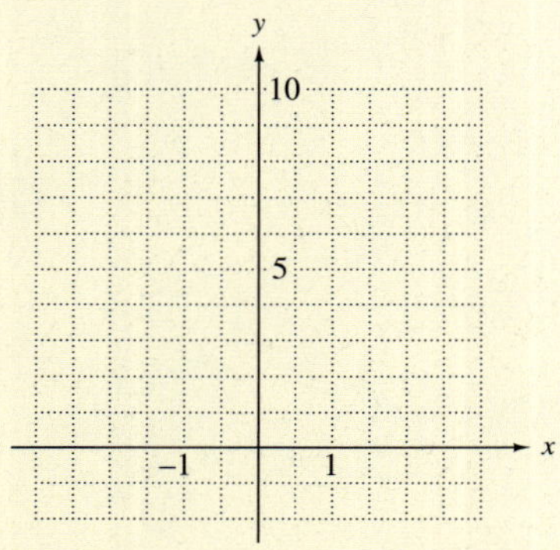

ANSWERS

1. (a)

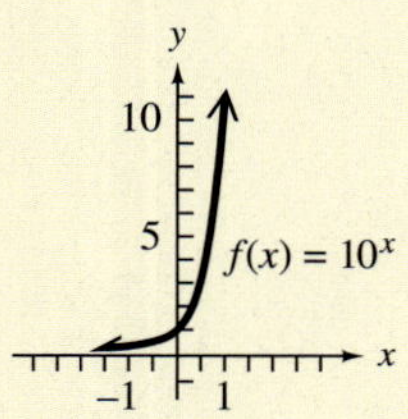

(b)

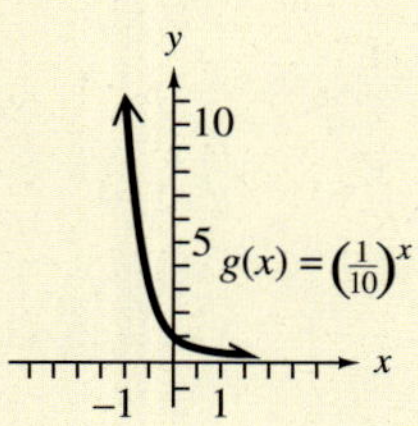

This graph is typical of the graphs of exponential functions of the form $F(x) = a^x$, where $a > 1$. ***The larger the value of a, the faster the graph rises.*** To see this, compare the graph of $F(x) = 5^x$ with the graph of $f(x) = 2^x$ in Figure 6. When graphing such functions, be sure to plot a sufficient number of points to see how rapidly the graph rises.

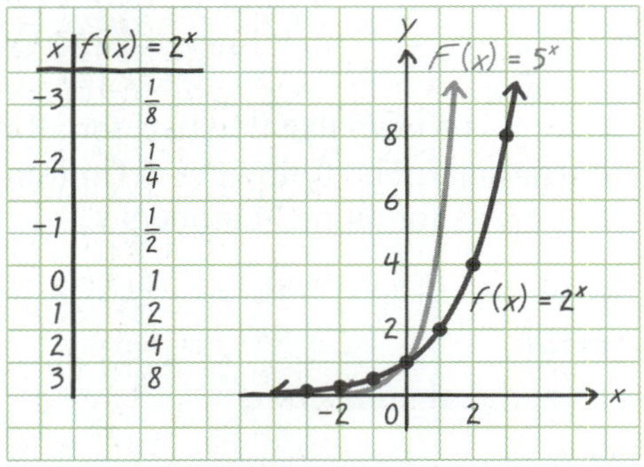

Figure 6 (repeated)

By the vertical line test, the graphs in Figure 6 represent functions. As these graphs suggest, the domain of an exponential function includes all real numbers. Because y is always positive, the range is $(0, \infty)$. Figure 6 also shows an important characteristic of exponential functions where $a > 1$: ***As x gets larger, y increases at a faster and faster rate.***

EXAMPLE 2 Graphing an Exponential Function ($a < 1$)

Graph $g(x) = \left(\frac{1}{2}\right)^x$.

Again, find some points on the graph. The graph, shown in Figure 7, is very similar to that of $f(x) = 2^x$ (Figure 6) with the same domain and range, except that here ***as x gets larger, y decreases.*** This graph is typical of the graphs of exponential functions of the form $F(x) = a^x$, where $0 < a < 1$.

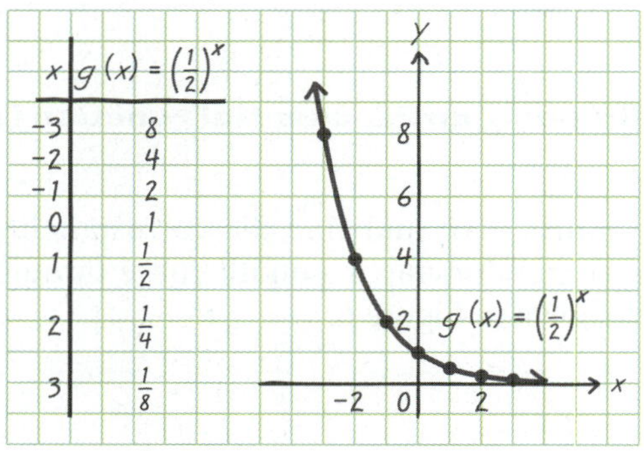

Figure 7

◀ *Work Problem* **1** *at the Side.*

CAUTION
The graph of an exponential function *approaches* the x-axis, but does ***not*** touch it.

Based on Examples 1 and 2, we make the following generalizations about the graphs of exponential functions of the form $F(x) = a^x$.

Characteristics of the Graph of $F(x) = a^x$

1. The graph contains the point (0, 1).
2. When $a > 1$, the graph *rises* from left to right. (See Figure 6.) When $0 < a < 1$, the graph *falls* from left to right. (See Figure 7.) In both cases, the graph goes from the second quadrant to the first.
3. The graph approaches the x-axis, but never touches it. (Recall from **Section 8.4** that such a line is called an **asymptote.**)
4. The domain is $(-\infty, \infty)$, and the range is $(0, \infty)$.

EXAMPLE 3 Graphing a More Complicated Exponential Function

Graph $f(x) = 3^{2x-4}$.

Find some ordered pairs.

$$\text{If } x = 0, \text{ then } y = 3^{2(0)-4} = 3^{-4} = \frac{1}{81}.$$

$$\text{If } x = 2, \text{ then } y = 3^{2(2)-4} = 3^{0} = 1.$$

These ordered pairs, $(0, \frac{1}{81})$ and (2, 1), along with the other ordered pairs shown in the table, lead to the graph in Figure 8. The graph is similar to the graph of $f(x) = 3^x$ except that it is shifted to the right and rises more rapidly.

x	y
0	$\frac{1}{81}$
1	$\frac{1}{9}$
2	1
3	9

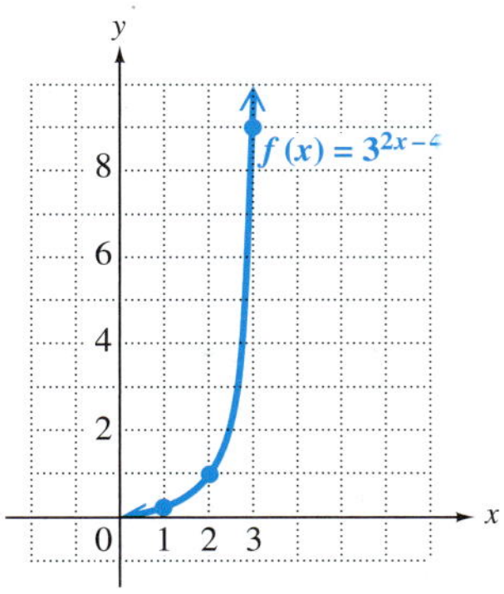

Figure 8

Work Problem **2** *at the Side.* ▶

OBJECTIVE 3 Solve exponential equations of the form $a^x = a^k$ for x. Until this chapter, we have solved only equations that had the variable as a base, like $x^2 = 8$; all exponents have been constants. An **exponential equation** is an equation that has a variable in an exponent, such as

$$9^x = 27.$$

By the horizontal line test, the exponential function defined by $F(x) = a^x$ is a one-to-one function, so we can use the following property to solve many exponential equations.

Property for Solving an Exponential Equation

For $a > 0$ and $a \neq 1$, if $\quad a^x = a^y \quad$ then $\quad x = y.$

This property would not necessarily be true if $a = 1$.

2 Graph $y = 2^{4x-3}$.

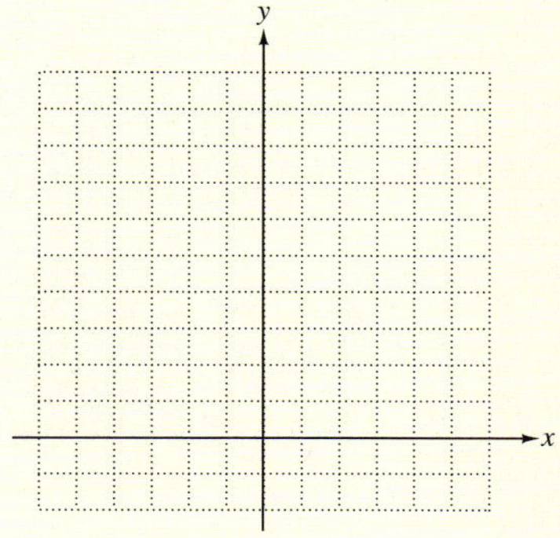

ANSWER

2.

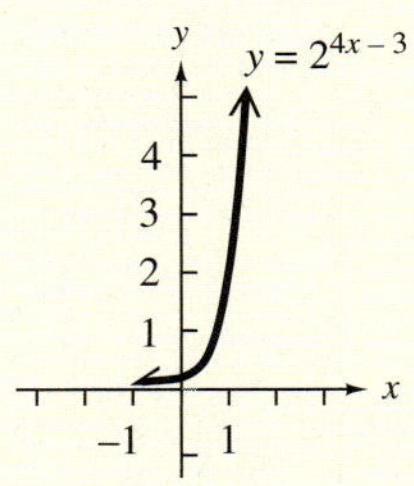

3 Solve each equation and check the solution.

(a) $25^x = 125$

(b) $4^x = 32$

(c) $81^p = 27$

To solve an exponential equation using this property, follow these steps.

Solving an Exponential Equation

Step 1 **Each side must have the same base.** If the two sides of the equation do not have the same base, express each as a power of the same base.

Step 2 **Simplify exponents,** if necessary, using the rules of exponents.

Step 3 **Set exponents equal** using the property given in this section.

Step 4 **Solve** the equation obtained in Step 3.

Note

These steps cannot be applied to an exponential equation like

$$3^x = 12$$

because Step 1 cannot easily be done. A method for solving such equations is given in **Section 11.6.**

EXAMPLE 4 Solving an Exponential Equation

Solve the equation $9^x = 27$.

We can use the property given in the box if both sides are written with the same base. Since $9 = 3^2$ and $27 = 3^3$,

$$9^x = 27$$

$$(3^2)^x = 3^3 \quad \text{Write with the same base. (Step 1)}$$

$$3^{2x} = 3^3 \quad \text{Power rule for exponents (Step 2)}$$

$$2x = 3 \quad \text{If } a^x = a^y, \text{ then } x = y. \text{ (Step 3)}$$

$$x = \frac{3}{2}. \quad \text{Solve for } x. \text{ (Step 4)}$$

Check Substitute $\frac{3}{2}$ for x:

$$9^x = 9^{3/2} = (9^{1/2})^3 = 3^3 = 27, \quad \text{as required.}$$

The solution set is $\left\{\frac{3}{2}\right\}$.

◀ Work Problem 3 at the Side.

EXAMPLE 5 Solving Exponential Equations

Solve each equation.

(a)

$$4^{3x-1} = 16^{x+2}$$

Be careful multiplying the exponents.

$$4^{3x-1} = (4^2)^{x+2} \quad \text{Write with the same base; } 16 = 4^2.$$

$$4^{3x-1} = 4^{2x+4} \quad \text{Power rule for exponents}$$

$$3x - 1 = 2x + 4 \quad \text{Set exponents equal.}$$

$$x = 5 \quad \text{Subtract } 2x\text{; add 1.}$$

Verify that the solution set is $\{5\}$.

Continued on Next Page

ANSWERS

3. (a) $\left\{\frac{3}{2}\right\}$ (b) $\left\{\frac{5}{2}\right\}$ (c) $\left\{\frac{3}{4}\right\}$

(b) $6^x = \frac{1}{216}$

$6^x = \frac{1}{6^3}$ $\quad 216 = 6^3$

$6^x = 6^{-3}$ $\quad$ Write with the same base; $\frac{1}{6^3} = 6^{-3}$.

$x = -3$ $\quad$ Set exponents equal.

Check Substitute -3 for x:

$$6^x = 6^{-3} = \frac{1}{6^3} = \frac{1}{216}, \text{ as required.}$$

The solution set is $\{-3\}$.

(c) $\left(\frac{2}{3}\right)^x = \frac{9}{4}$

$\left(\frac{2}{3}\right)^x = \left(\frac{4}{9}\right)^{-1}$ $\quad \frac{9}{4} = \left(\frac{4}{9}\right)^{-1}$

$\left(\frac{2}{3}\right)^x = \left[\left(\frac{2}{3}\right)^2\right]^{-1}$ $\quad$ Write with the same base.

$\left(\frac{2}{3}\right)^x = \left(\frac{2}{3}\right)^{-2}$ $\quad$ Power rule for exponents

$x = -2$ $\quad$ Set exponents equal.

Check that the solution set is $\{-2\}$.

Work Problem **4** *at the Side.* ▶

4 Solve each equation and check the solution.

(a) $25^{x-2} = 125^x$

(b) $4^x = \frac{1}{32}$

(c) $\left(\frac{3}{4}\right)^x = \frac{16}{9}$

OBJECTIVE 4 Use exponential functions in applications involving growth or decay.

EXAMPLE 6 Solving an Application of Exponential Growth

The graph in Figure 9 shows the concentration of carbon dioxide (in parts per million) in the air. This concentration is increasing exponentially.

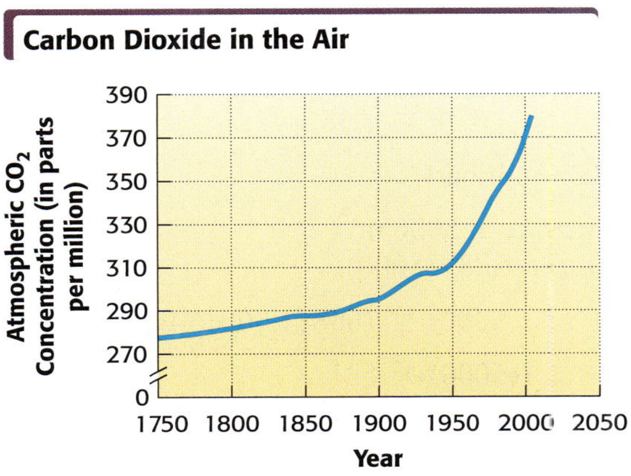

Source: *Sacramento Bee*; National Oceanic and Atmospheric Administration.

Figure 9

Continued on Next Page

ANSWERS

4. (a) $\{-4\}$ (b) $\left\{-\frac{5}{2}\right\}$ (c) $\{-2\}$

5 Use the exponential function in Example 6 to approximate the carbon dioxide concentration in 1925.

6 Use the exponential function in Example 7 to find the pressure at 8000 m.

The data are approximated by the exponential function defined by

$$f(x) = 266(1.001)^x,$$

where x is the number of years since 1750. Use this function and a calculator to approximate the concentration of carbon dioxide in parts per million for each year.

(a) 1900

Since x represents the number of years since 1750, in this case we have $x = 1900 - 1750 = 150$. Thus, evaluate $f(150)$.

$$f(x) = 266(1.001)^x$$

$$f(150) = 266(1.001)^{150} \quad \text{Let } x = 150.$$

$$f(150) \approx 309 \quad \text{Use a calculator.}$$

The concentration in 1900 was about 309 parts per million.

(b) 1950

Use $x = 1950 - 1750 = 200$.

$$f(200) = 266(1.001)^{200} \quad \text{Let } x = 200.$$

$$f(200) \approx 325 \quad \text{Use a calculator.}$$

The concentration in 1950 was about 325 parts per million.

◀ *Work Problem* **5** *at the Side.*

EXAMPLE 7 Applying an Exponential Decay Function

The atmospheric pressure (in millibars) at a given altitude x, in meters, can be approximated by the exponential function defined by

$$f(x) = 1038(1.000134)^{-x},$$

for values of x between 0 and 10,000. Because the base is greater than 1 and the coefficient of x in the exponent is negative, the function values decrease as x increases. This means that as the altitude increases, the atmospheric pressure decreases. (*Source:* Miller, A. and J. Thompson, *Elements of Meteorology,* Fourth Edition, Charles E. Merrill Publishing Company, 1993.)

(a) According to this function, what is the pressure at ground level?

At ground level, $x = 0$, so

$$f(0) = 1038(1.000134)^{-0}$$

$$f(0) = 1038(1)$$

$$f(0) = 1038.$$

The pressure is 1038 millibars.

(b) What is the pressure at 5000 m?

Use a calculator to find $f(5000)$.

$$f(5000) = 1038(1.000134)^{-5000}$$

$$f(5000) \approx 531$$

The pressure is approximately 531 millibars.

◀ *Work Problem* **6** *at the Side.*

Answers

5. 317 parts per million
6. approximately 355 millibars

11.2 Exercises

FOR EXTRA HELP MyMathLab MathXL PRACTICE WATCH DOWNLOAD READ REVIEW

Choose the correct response in Exercises 1–4.

1. Which point lies on the graph of $f(x) = 2^x$?

A. $(1, 0)$ **B.** $(2, 1)$

C. $(0, 1)$ **D.** $\left(\sqrt{2}, \frac{1}{2}\right)$

2. The asymptote of the graph of $F(x) = a^x$

A. is the x-axis. **B.** is the y-axis.

C. has equation $x = 1$. **D.** has equation $y = 1$.

3. Which statement is true?

A. The y-intercept of the graph of $f(x) = 10^x$ is $(0, 10)$.

B. For any $a > 1$, the graph of $f(x) = a^x$ falls from left to right.

C. The point $\left(\frac{1}{2}, \sqrt{5}\right)$ lies on the graph of $f(x) = 5^x$.

D. The graph of $y = 4^x$ rises at a faster rate than the graph of $y = 10^x$.

4. Which equation is graphed here?

A. $y = 1000\left(\frac{1}{2}\right)^{0.3x}$

B. $y = 1000\left(\frac{1}{2}\right)^{x}$

C. $y = 1000(2)^{0.3x}$

D. $y = 1000^x$

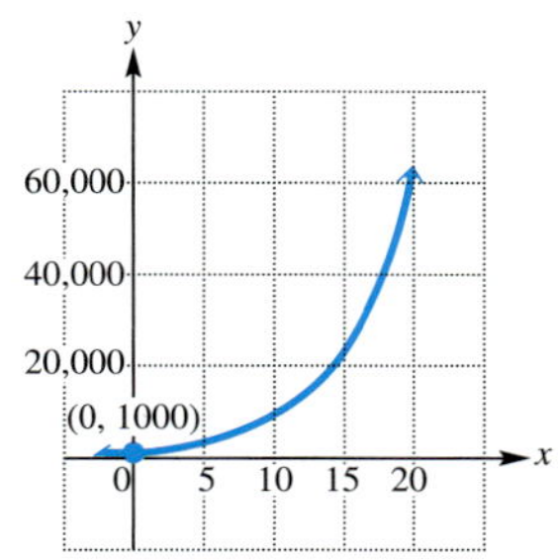

Graph each exponential function. See Examples 1–3.

5. $f(x) = 3^x$

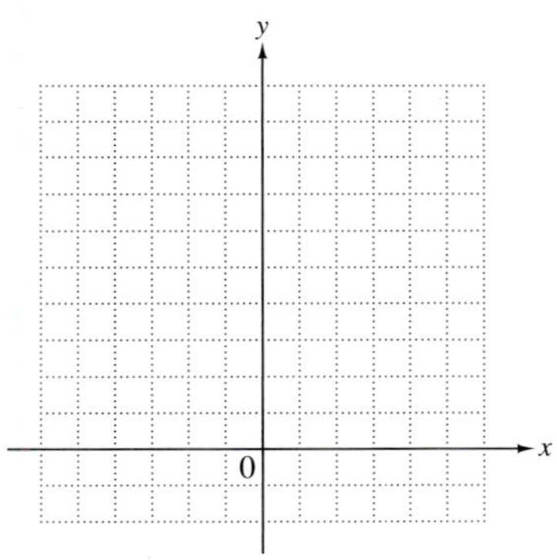

6. $f(x) = 5^x$

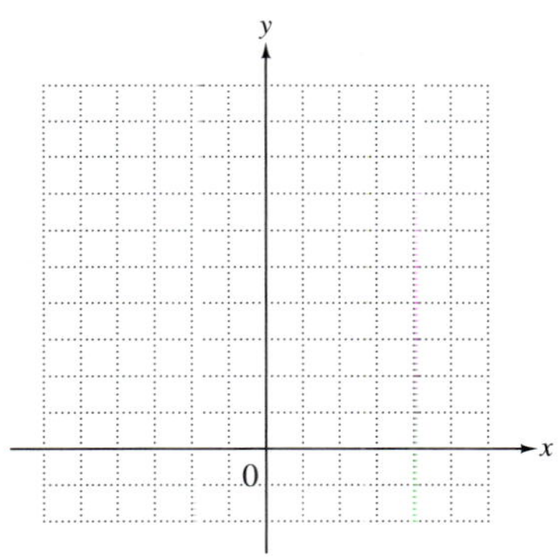

7. $g(x) = \left(\frac{1}{3}\right)^x$

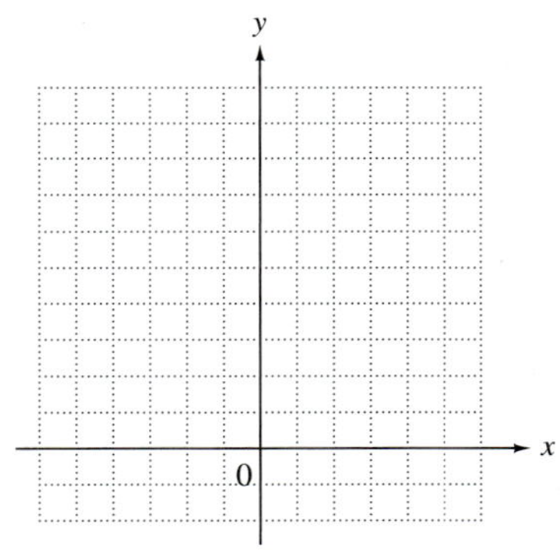

8. $g(x) = \left(\frac{1}{5}\right)^x$

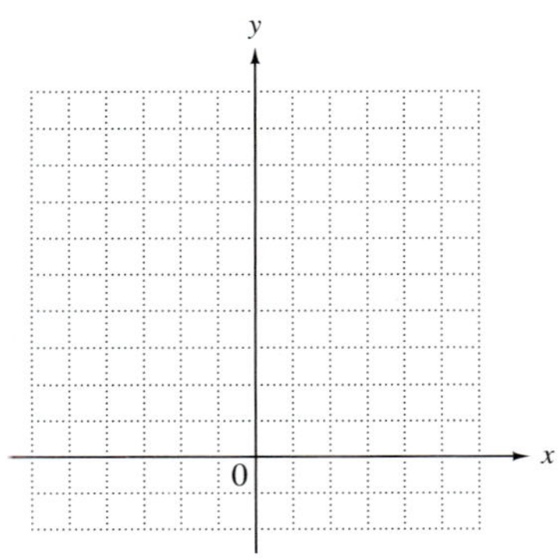

9. $y = 2^{2x-2}$

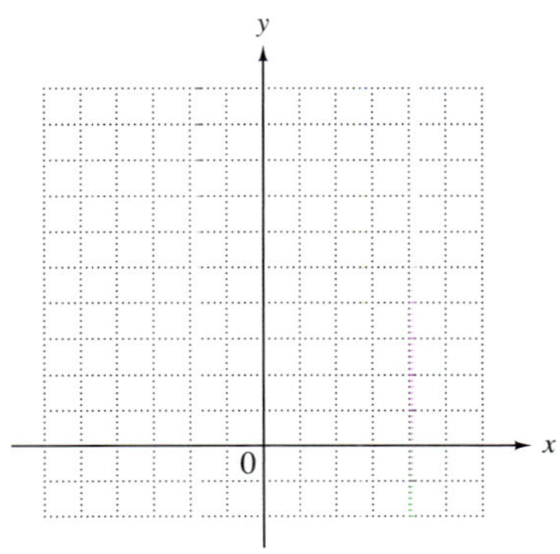

10. $y = 2^{2x+1}$

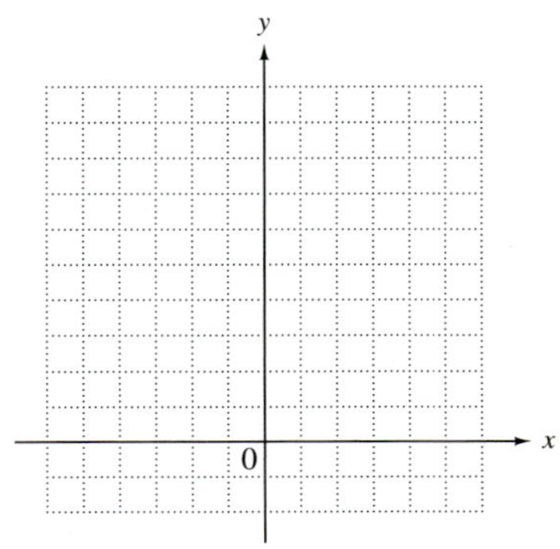

Solve each equation. See Examples 4 and 5.

11. $6^x = 36$

12. $8^x = 64$

13. $100^x = 1000$

14. $8^x = 4$

15. $16^{2x+1} = 64^{x+3}$

16. $9^{2x-8} = 27^{x-4}$

17. $5^x = \frac{1}{125}$

18. $3^x = \frac{1}{81}$

19. $5^x = 0.2$

20. $10^x = 0.1$

21. $\left(\frac{3}{2}\right)^x = \frac{8}{27}$

22. $\left(\frac{4}{3}\right)^x = \frac{27}{64}$

23. **(a)** For an exponential function defined by $f(x) = a^x$, if $a > 1$, then the graph ________ (rises/falls) from left to right. If $0 < a < 1$, then the graph ________ (rises/falls) from left to right.

(b) Based on your answers in part (a), make a conjecture (an educated guess) concerning whether an exponential function defined by $f(x) = a^x$ is one-to-one. Then decide whether it has an inverse based on the concepts of **Section 11.1.**

Solve each problem. See Examples 6 and 7.

The figure shown here accompanied the article "Is Our World Warming?", which appeared in the October 1990 issue of *National Geographic*. It shows projected temperature increases using two graphs: one an exponential-type curve and the other linear. From the figure, approximate the increase **(a)** for the exponential curve, and **(b)** for the linear graph for each year.

24. 2000

25. 2010

26. 2020

27. 2040

IS OUR WORLD WARMING?

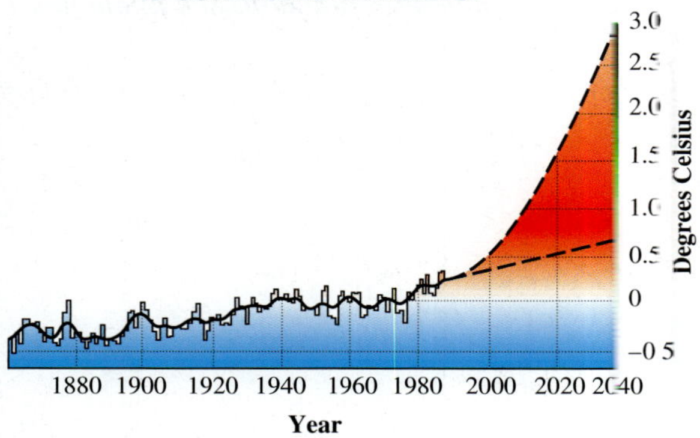

Graph, "Zero Equals Average Global Temperature for the Period 1950–1979." Dale D. Glasgow, © National Geographic Society. Reprinted by permission.

28. A small business estimates that the value $V(t)$ of a copy machine is decreasing according to the function defined by

$$V(t) = 5000(2)^{-0.15t},$$

where t is the number of years that have elapsed since the machine was purchased and $V(t)$ is in dollars.

(a) What was the original value of the machine?

(b) What is the value of the machine 5 yr after purchase? Give your answer to the nearest dollar.

(c) What is the value of the machine 10 yr after purchase? Give your answer to the nearest dollar.

(d) Graph the function.

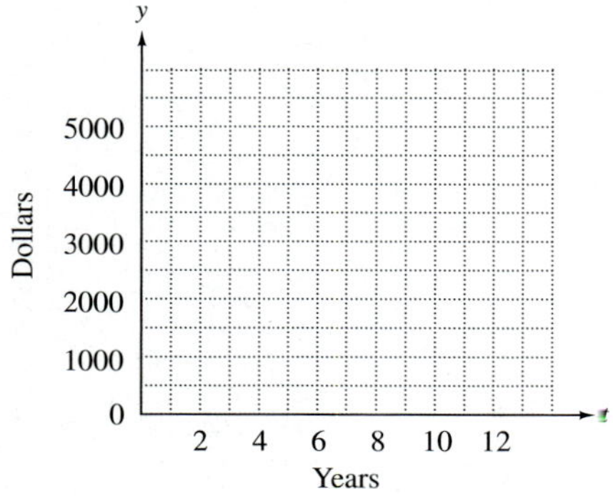

11.3 Logarithmic Functions

OBJECTIVES

1. Define a logarithm.
2. Convert between exponential and logarithmic forms.
3. Solve logarithmic equations of the form $\log_a b = k$ for a, b, or k.
4. Define and graph logarithmic functions.
5. Use logarithmic functions in applications involving growth or decay.

The graph of $y = 2^x$ is the curve shown in blue in Figure 10. Because $y = 2^x$ defines a one-to-one function, it has an inverse function. Interchanging x and y gives

$$x = 2^y, \quad \text{the inverse of} \quad y = 2^x.$$

As we saw in **Section 11.1,** the graph of the inverse is found by reflecting the graph of $y = 2^x$ about the line $y = x$. The graph of $x = 2^y$ is shown as a red curve in Figure 10.

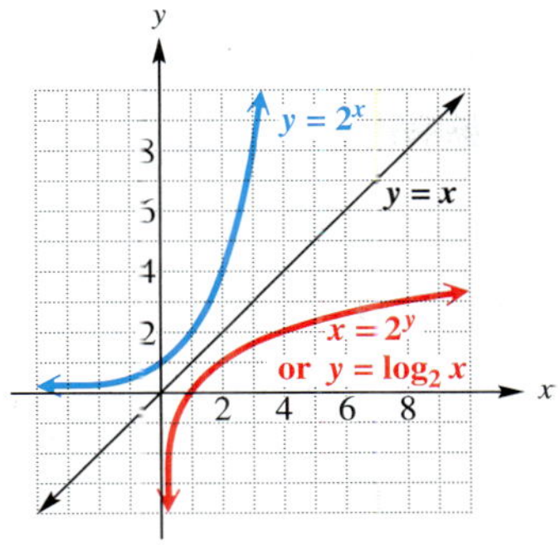

Figure 10

OBJECTIVE 1 Define a logarithm. We cannot solve the equation $x = 2^y$ for the dependent variable y with the methods presented up to now. The following definition is used to solve $x = 2^y$ for y.

> **Logarithm**
>
> For all positive numbers a, with $a \neq 1$, and all positive numbers x,
>
> $$y = \log_a x \quad \text{means the same as} \quad x = a^y.$$

This key statement should be memorized. The abbreviation **log** is used for **logarithm.** Read $\log_a x$ as **"the logarithm of x with base a"** or **"the base a logarithm of x."** To remember the location of the base and the exponent in each form, refer to the following diagrams.

Logarithmic form: $y = \log_a x$ (Exponent: y; Base: a) | Exponential form: $x = a^y$ (Exponent: y; Base: a)

In work with logarithmic and exponential forms, remember the following.

> **Meaning of $\log_a x$**
>
> A logarithm is an exponent. ***The expression $\log_a x$ represents the exponent to which the base a must be raised to obtain x.***

OBJECTIVE 2 Convert between exponential and logarithmic forms. We use the definition of a logarithm to convert between exponential and logarithmic forms. The table shows pairs of equivalent statements.

Exponential Form	Logarithmic Form
$3^2 = 9$	$\log_3 9 = 2$
$\left(\frac{1}{5}\right)^{-2} = 25$	$\log_{1/5} 25 = -2$
$10^5 = 100{,}000$	$\log_{10} 100{,}000 = 5$
$4^{-3} = \frac{1}{64}$	$\log_4 \frac{1}{64} = -3$

Work Problem 1 at the Side. ▶

1 Complete the table.

Exponential Form	Logarithmic Form
$2^5 = 32$	
$100^{1/2} = 10$	
	$\log_8 4 = \frac{2}{3}$
	$\log_6 \frac{1}{1296} = -4$

ANSWERS

1. $\log_2 32 = 5$; $\log_{100} 10 = \dfrac{1}{2}$; $8^{2/3} = 4$; $6^{-4} = \dfrac{1}{1296}$

OBJECTIVE 3 Solve logarithmic equations of the form $\log_a b = k$ for a, b, or k. A **logarithmic equation** is an equation with a logarithm in at least one term. To solve, first write the equation in exponential form.

EXAMPLE 1 Solving Logarithmic Equations

Solve each equation.

(a) $\log_4 x = -2$

By definition, $\log_4 x = -2$ is equivalent to $x = 4^{-2}$.

$$x = 4^{-2} = \frac{1}{16}$$

The solution set is $\left\{\frac{1}{16}\right\}$.

(b) $\log_{1/2}(3x + 1) = 2$

$$3x + 1 = \left(\frac{1}{2}\right)^2 \quad \text{Write in exponential form.}$$

This is a key step.

$$3x + 1 = \frac{1}{4} \quad \text{Apply the exponent.}$$

$$12x + 4 = 1 \quad \text{Multiply each term by 4.}$$

$$12x = -3 \quad \text{Subtract 4.}$$

$$x = -\frac{1}{4} \quad \text{Divide by 12; lowest terms}$$

Check $\log_{1/2}\left(3\left(-\frac{1}{4}\right) + 1\right) \stackrel{?}{=} 2$ Let $x = -\frac{1}{4}$.

$$\log_{1/2}\frac{1}{4} \stackrel{?}{=} 2 \quad \text{Simplify within parentheses}$$

$$\left(\frac{1}{2}\right)^2 = \frac{1}{4} \quad \text{Exponential form; true}$$

The solution set is $\left\{-\frac{1}{4}\right\}$.

(c) $\log_x 3 = 2$

$$x^2 = 3 \quad \text{Write in exponential form.}$$

$$x = \pm\sqrt{3} \quad \text{Take square roots.}$$

Notice that only the *principal* square root satisfies the equation, however, since the base x must be a positive number. The solution set is $\{\sqrt{3}\}$.

(d) $\log_{49} \sqrt[3]{7} = x$

$$49^x = \sqrt[3]{7} \quad \text{Write in exponential form.}$$

$$(7^2)^x = 7^{1/3} \quad \text{Write with the same base.}$$

$$7^{2x} = 7^{1/3} \quad \text{Power rule for exponents}$$

$$2x = \frac{1}{3} \quad \text{Set exponents equal.}$$

$$x = \frac{1}{6} \quad \text{Divide by 2.}$$

The solution set is $\left\{\frac{1}{6}\right\}$.

◀ Work Problem 2 at the Side.

2 Solve each equation.

(a) $\log_3 27 = x$

(b) $\log_5 p = 2$

(c) $\log_m \frac{1}{16} = -4$

(d) $\log_x 12 = 3$

Answers

2. (a) $\{3\}$ (b) $\{25\}$ (c) $\{2\}$ (d) $\{\sqrt[3]{12}\}$

For any real positive number b, we know that $b^1 = b$ and $b^0 = 1$. Writing these two statements in logarithmic form gives the following two properties of logarithms.

> **Properties of Logarithms**
>
> For any positive real number b, with $b \neq 1$,
>
> $$\log_b b = 1 \quad \text{and} \quad \log_b 1 = 0.$$

EXAMPLE 2 Using Properties of Logarithms

Use the preceding two properties of logarithms to evaluate each logarithm.

(a) $\log_7 7 = 1$ **(b)** $\log_{\sqrt{2}} \sqrt{2} = 1$

(c) $\log_9 1 = 0$ **(d)** $\log_{0.2} 1 = 0$

Work Problem 3 at the Side. ▶

OBJECTIVE 4 Define and graph logarithmic functions. Now we define the logarithmic function with base a.

> **Logarithmic Function**
>
> If a and x are positive numbers, with $a \neq 1$, then
>
> $$G(x) = \log_a x$$
>
> defines the **logarithmic function with base a.**

EXAMPLE 3 Graphing a Logarithmic Function ($a > 1$)

Graph $f(x) = \log_2 x$.

By writing $y = f(x) = \log_2 x$ in exponential form as $x = 2^y$, we can identify ordered pairs that satisfy the equation. It is easier to choose values for y and find the corresponding values of x. Plotting the points in the table of ordered pairs and connecting them with a smooth curve gives the graph in Figure 11. This graph is typical of logarithmic functions with base $a > 1$.

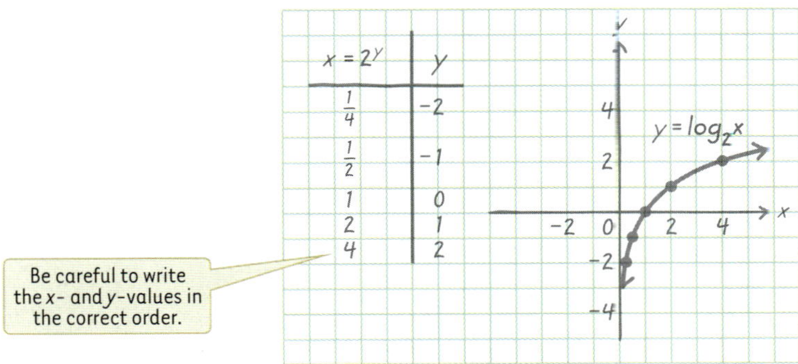

Figure 11

As the table and graph suggest, x is always positive, so the domain of a logarithmic function is $(0, \infty)$. The range includes all real numbers, $(-\infty, \infty)$.

Work Problem 4 at the Side. ▶

3 Evaluate each logarithm.

(a) $\log_{2/5} \frac{2}{5}$

(b) $\log_\pi \pi$

(c) $\log_{0.4} 1$

(d) $\log_6 1$

4 Graph $y = \log_{10} x$.

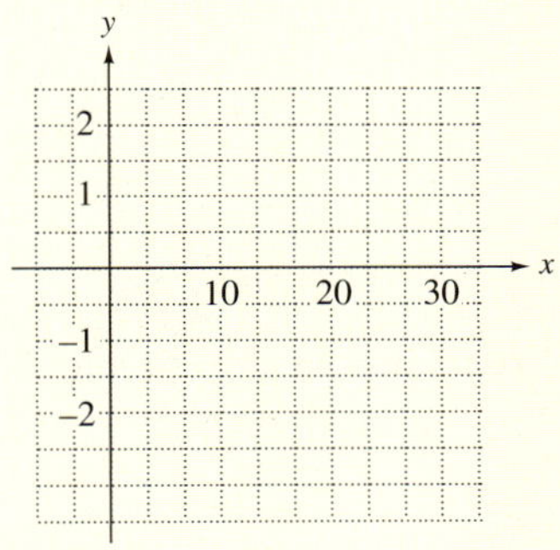

ANSWERS

3. (a) 1 **(b)** 1 **(c)** 0 **(d)** 0

4.

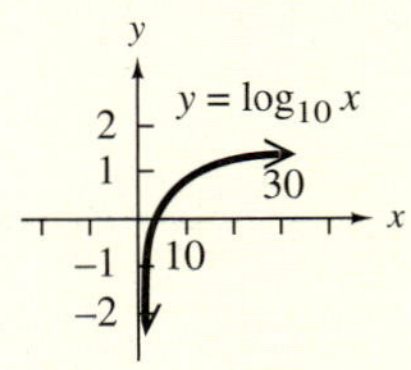

5 Graph $y = \log_{1/10} x$.

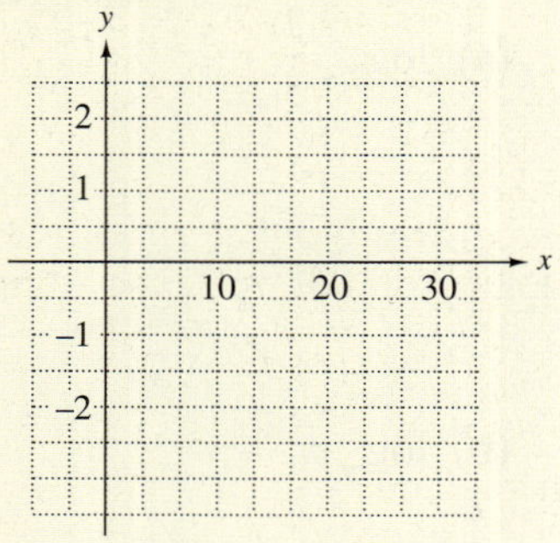

EXAMPLE 4 **Graphing a Logarithmic Function ($0 < a < 1$)**

Graph $g(x) = \log_{1/2} x$.

We write $y = g(x) = \log_{1/2} x$ in exponential form as $x = \left(\frac{1}{2}\right)^y$; then choose values for y and find the corresponding values of x. Plotting these points and connecting them with a smooth curve gives the graph in Figure 12. This graph is typical of logarithmic functions with $0 < a < 1$.

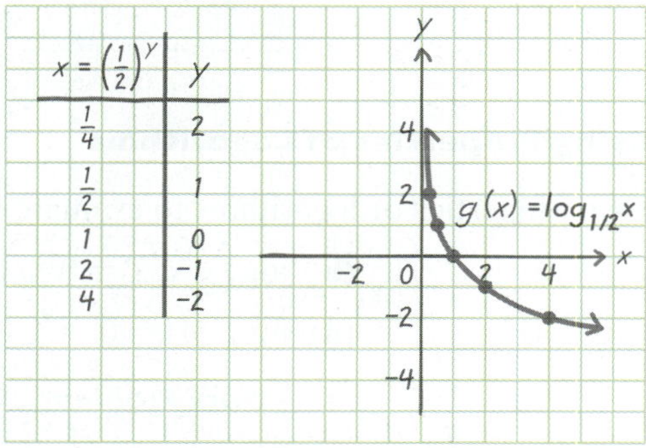

Figure 12

◀ *Work Problem* **5** *at the Side.*

6 Solve the problem.

A population of mites in a laboratory is growing according to the logarithmic function defined by

$$P(t) = 80 \log_{10}(t + 10),$$

where t is the number of days after a study is begun.

(a) Find the number of mites at the beginning of the study.

(b) Find the number present after 90 days.

Characteristics of the Graph of $G(x) = \log_a x$

1. The graph contains the point (1, 0).
2. When $a > 1$, the graph *rises* from left to right, from the fourth quadrant to the first. (See Figure 11.) When $0 < a < 1$, the graph *falls* from left to right, from the first quadrant to the fourth. (See Figure 12.)
3. The graph approaches the y-axis, but never touches it. (The y-axis is an asymptote.)
4. The domain is $(0, \infty)$, and the range is $(-\infty, \infty)$.

OBJECTIVE 5 **Use logarithmic functions in applications involving growth or decay.** Logarithmic functions, like exponential functions, can be applied to growth or decay of real-world phenomena.

EXAMPLE 5 **Solving an Application of a Logarithmic Function**

The logarithmic function defined by

$$f(x) = 27 + 1.105 \log_{10}(x + 1)$$

approximates the barometric pressure in inches of mercury at a distance of x miles from the eye of a typical hurricane. (*Source:* Miller, A. and R. Anthes, *Meteorology,* Fifth Edition, Charles E. Merrill Publishing Company, 1985.) Approximate the pressure 9 mi from the eye of the hurricane.

$$f(9) = 27 + 1.105 \log_{10}(9 + 1) \quad \text{Let } x = 9.$$

$$f(9) = 27 + 1.105 \log_{10} 10 \quad \text{Add inside parentheses.}$$

$$f(9) = 27 + 1.105(1) \quad \log_{10} 10 = 1$$

$$f(9) = 28.105 \quad \text{Add.}$$

The pressure 9 mi from the eye of the hurricane is 28.105 in.

◀ *Work Problem* **6** *at the Side.*

ANSWERS

5.

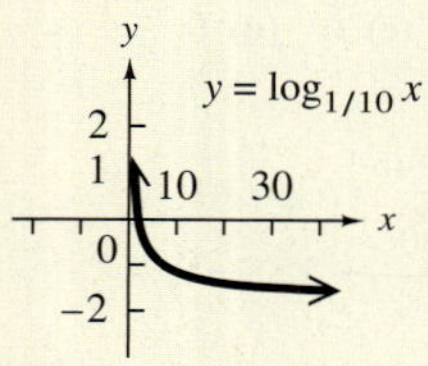

6. (a) 80 **(b)** 160

11.3 Exercises

1. By definition, $\log_a x$ is the exponent to which the base a must be raised in order to obtain x. Use this definition to match the logarithm in Column I with its value in Column II. (*Example:* $\log_3 9$ is equal to 2 because 2 is the exponent to which 3 must be raised in order to obtain 9.)

I	II
(a) $\log_4 16$	**A.** -2
(b) $\log_3 81$	**B.** -1
(c) $\log_3 \left(\frac{1}{3}\right)$	**C.** 2
(d) $\log_{10} 0.01$	**D.** 0
(e) $\log_5 \sqrt{5}$	**E.** $\frac{1}{2}$
(f) $\log_{13} 1$	**F.** 4

2. Match the logarithmic equation in Column I with the corresponding exponential equation from Column II.

I	II
(a) $\log_{1/3} 3 = -1$	**A.** $8^{1/3} = \sqrt[3]{8}$
(b) $\log_5 1 = 0$	**B.** $\left(\frac{1}{3}\right)^{-1} = 3$
(c) $\log_2 \sqrt{2} = \frac{1}{2}$	**C.** $4^1 = 4$
(d) $\log_{10} 1000 = 3$	**D.** $2^{1/2} = \sqrt{2}$
(e) $\log_8 \sqrt[3]{8} = \frac{1}{3}$	**E.** $5^0 = 1$
(f) $\log_4 4 = 1$	**F.** $10^3 = 1000$

Write in logarithmic form. See the table in Objective 2.

3. $4^5 = 1024$

4. $3^6 = 729$

5. $\left(\frac{1}{2}\right)^{-3} = 8$

6. $\left(\frac{1}{6}\right)^{-3} = 216$

7. $10^{-3} = 0.001$

8. $36^{1/2} = 6$

9. $\sqrt[4]{625} = 5$

10. $\sqrt[3]{343} = 7$

Write in exponential form. See the table in Objective 2

11. $\log_4 64 = 3$

12. $\log_2 512 = 9$

13. $\log_{10} \frac{1}{10{,}000} = -4$

14. $\log_{100} 100 = 1$

15. $\log_6 1 = 0$

16. $\log_\pi 1 = 0$

17. $\log_9 3 = \frac{1}{2}$

18. $\log_{64} 2 = \frac{1}{6}$

19. When a student asked his teacher to explain to him how to evaluate $\log_9 3$ without showing any work, his teacher told him, "Think radically." Explain what the teacher meant by this hint.

20. A student told her teacher, "I know that $\log_2 1$ is the exponent to which 2 must be raised in order to obtain 1, but I can't think of any such number." How would you explain to the student that the value of $\log_2 1$ is 0?

Solve each equation for x. See Examples 1 and 2.

21. $x = \log_{27} 3$

22. $x = \log_{125} 5$

23. $\log_x 9 = \frac{1}{2}$

24. $\log_x 5 = \frac{1}{2}$

25. $\log_x 125 = -3$

26. $\log_x 64 = -6$

27. $\log_{12} x = 0$

28. $\log_4 x = 0$

29. $\log_x x = 1$

30. $\log_x 1 = 0$

31. $\log_x \frac{1}{25} = -2$

32. $\log_x \frac{1}{10} = -1$

33. $\log_8 32 = x$

34. $\log_{81} 27 = x$

35. $\log_\pi \pi^4 = x$

36. $\log_{\sqrt{2}} (\sqrt{2})^9 = x$

37. $\log_6 \sqrt{216} = x$

38. $\log_4 \sqrt{64} = x$

If the point (p, q) is on the graph of $f(x) = a^x$ (for $a > 0$ and $a \neq 1$), then the point (q, p) is on the graph of $f^{-1}(x) = \log_a x$. Use this fact and refer to the graphs required in Exercises 5–8 in **Section 11.2** *to graph each logarithmic function. See Examples 3 and 4.*

39. $y = \log_3 x$

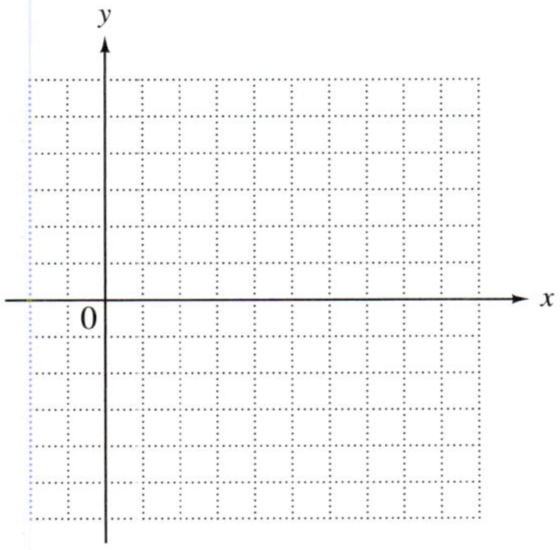

40. $y = \log_5 x$

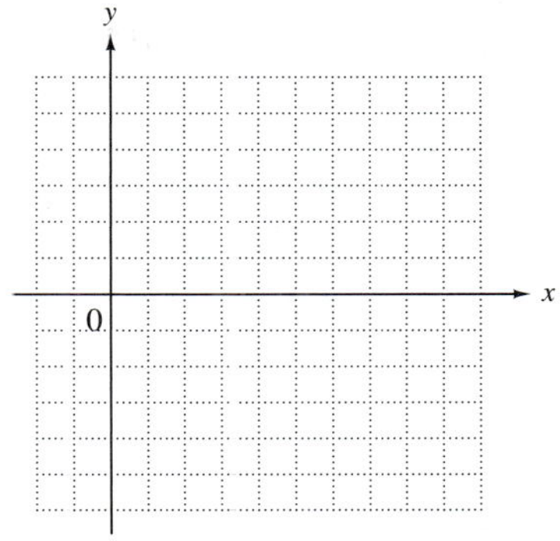

41. $y = \log_{1/3} x$

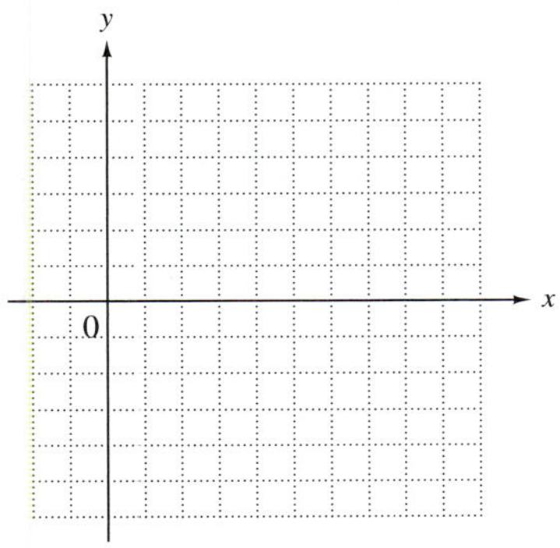

42. $y = \log_{1/5} x$

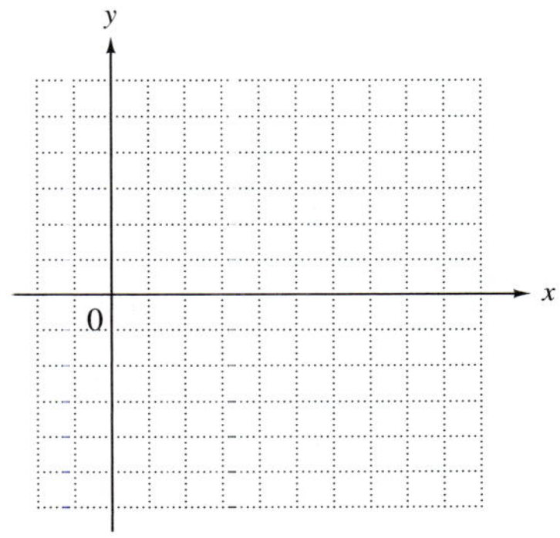

43. Compare the summary of characteristics of the graph of $F(x) = a^x$ in **Section 11.2** with the similar summary of characteristics of the graph of $G(x) = \log_a x$ in this section. Make a list of the characteristics that reinforce the concept that F and G are inverse functions.

44. The domain of $F(x) = a^x$ is $(-\infty, \infty)$, while the range is $(0, \infty)$. Therefore, since $G(x) = \log_a x$ defines the inverse of F, the domain of G is ________, while the range of G is ________.

Use the graph to predict the value of $f(t)$ for each value of t.

45. $t = 0$

46. $t = 10$

47. $t = 60$

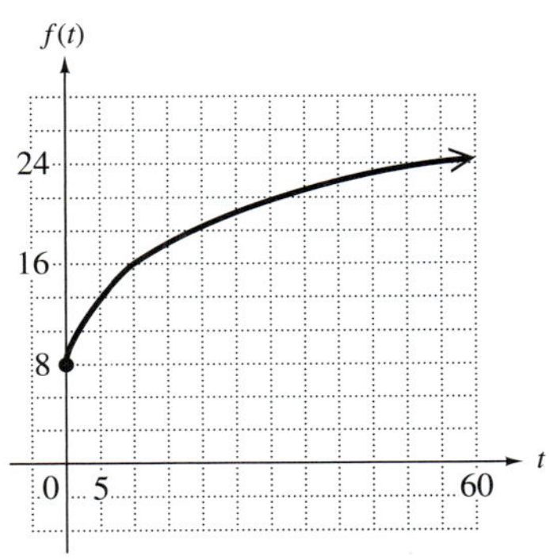

48. Show that the points determined in Exercises 45–47 lie on the graph of $f(t) = 8 \log_5 (2t + 5)$.

49. Explain why 1 is not allowed as a base for a logarithmic function.

50. Explain why $\log_a 1$ is 0 for any value of a that is allowed as the base of a logarithm. Use a rule of exponents introduced earlier in your explanation.

51. The graphs of both $f(x) = 3^x$ and $g(x) = \log_3 x$ rise from left to right. Which one rises at a faster rate?

52. Use the exponential key of your calculator to find approximations for the expression $(1 + \frac{1}{x})^x$, using x values of 1, 10, 100, 1000, and 10,000. Explain what seems to be happening as x gets larger and larger.

Solve each application of a logarithmic function. See Example 5.

53. According to selected figures from 1981 through 2003, the number of billion cubic feet of natural gas gross withdrawals from crude oil wells in the United States can be approximated by the function defined by

$$f(x) = 3800 + 585 \log_2 x,$$

where $x = 1$ represents 1981, $x = 2$ represents 1982, and so on. (*Source:* Energy Information Administration, Annual Energy Review 2003.) Use this function to approximate the number of cubic feet withdrawn in each of the following years.

(a) 1982

(b) 1988

(c) 1996

54. According to selected figures from the last two decades of the twentieth century, the number of trillion cubic feet of dry natural gas consumed worldwide was approximated by the function defined by

$$f(x) = 51.47 + 6.044 \log_2 (x + 1),$$

where $x = 0$ corresponds to 1980, $x = 1$ to 1981, and so on. (*Source:* Energy Information Administration.) Use the function to approximate consumption in each of the following years.

(a) 1980

(b) 1987

(c) 1995

In the United States, the intensity of an earthquake is rated using the ***Richter scale.*** *The Richter scale rating of an earthquake of intensity x is given by*

$$R = \log_{10} \frac{x}{x_0},$$

where x_0 is the intensity of an earthquake of a certain (small) size. The figure shows Richter scale ratings for major Southern California earthquakes since 1920. As the figure indicates, earthquakes "come in bunches" and the 1990s were an especially busy time.

Major Southern California Earthquakes
(with magnitudes greater than 4.7)

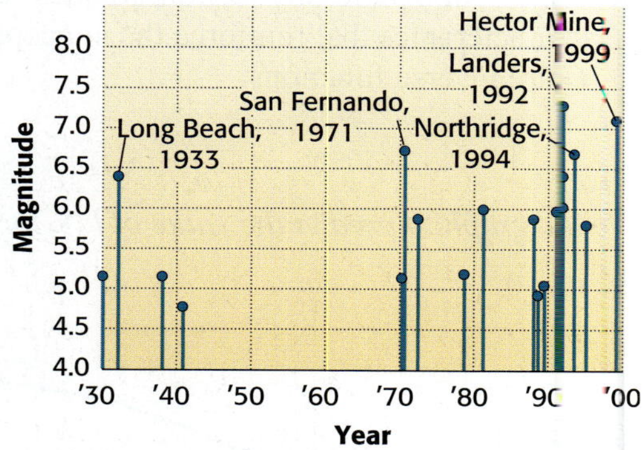

Source: Caltech; U.S. Geological Survey.

55. The 1994 Northridge earthquake had a Richter scale rating of 6.7; the 1992 Landers earthquake had a rating of 7.3. How much more powerful was the Landers earthquake than the Northridge earthquake?

56. Compare the smallest rated earthquake in the figure (at 4.8) with the Landers quake. How much more powerful was the Landers quake?

11.4 Properties of Logarithms

OBJECTIVES

1. Use the product rule for logarithms.
2. Use the quotient rule for logarithms.
3. Use the power rule for logarithms.
4. Use properties to write alternative forms of logarithmic expressions.

Logarithms have been used as an aid to numerical calculation for several hundred years. Today the widespread use of calculators has made the use of logarithms for calculation obsolete. However, logarithms are still very important in applications and in further work in mathematics.

OBJECTIVE 1 Use the product rule for logarithms. One way in which logarithms simplify problems is by changing a problem of multiplication into one of addition. We know that $\log_2 4 = 2$, $\log_2 8 = 3$, and $\log_2 32 = 5$. Since $2 + 3 = 5$,

$$\log_2 32 = \log_2 4 + \log_2 8$$
$$\log_2 (4 \cdot 8) = \log_2 4 + \log_2 8.$$

This is true in general.

Product Rule for Logarithms

If x, y, and b are positive real numbers, where $b \neq 1$, then

$$\log_b xy = \log_b x + \log_b y.$$

In words, the logarithm of a product is the sum of the logarithms of the factors.

Note

The word statement of the product rule can be restated by replacing "logarithm" with "exponent." The rule then becomes the familiar rule for multiplying exponential expressions: The *exponent* of a product is equal to the sum of the *exponents* of the factors.

To prove this rule, let $m = \log_b x$ and $n = \log_b y$, and recall that

$$\log_b x = m \quad \text{means} \quad b^m = x.$$
$$\log_b y = n \quad \text{means} \quad b^n = y.$$

Now consider the product xy.

$xy = b^m \cdot b^n$	Substitute.
$xy = b^{m+n}$	Product rule for exponents
$\log_b xy = m + n$	Convert to logarithmic form.
$\log_b xy = \log_b x + \log_b y$	Substitute.

The last statement is the result we wished to prove.

EXAMPLE 1 Using the Product Rule

Use the product rule to rewrite each expression. Assume $x > 0$.

(a) $\log_5 (6 \cdot 9)$
$= \log_5 6 + \log_5 9$

(b) $\log_7 8 + \log_7 12$
$= \log_7 (8 \cdot 12)$
$= \log_7 96$

Continued on Next Page

(c) $\log_3 (3x)$

$= \log_3 3 + \log_3 x$

$= 1 + \log_3 x$ $\quad \log_3 3 = 1$

(d) $\log_4 x^3$

$= \log_4 (x \cdot x \cdot x)$ $\quad x^3 = x \cdot x \cdot x$

$= \log_4 x + \log_4 x + \log_4 x$ $\quad$ Product rule

$= 3 \log_4 x$

◀ Work Problem **1** at the Side.

OBJECTIVE 2 Use the quotient rule for logarithms. The rule for division is similar to the rule for multiplication.

Quotient Rule for Logarithms

If x, y, and b are positive real numbers, where $b \neq 1$, then

$$\log_b \frac{x}{y} = \log_b x - \log_b y.$$

In words, the logarithm of a quotient is the difference between the logarithm of the numerator and the logarithm of the denominator.

The proof of this rule is very similar to the proof of the product rule.

EXAMPLE 2 Using the Quotient Rule

Use the quotient rule to rewrite each expression.

(a) $\log_4 \frac{7}{9}$

$= \log_4 7 - \log_4 9$

(b) $\log_5 6 - \log_5 x$

$= \log_5 \frac{6}{x}, \quad x > 0$

(c) $\log_3 \frac{27}{5}$

$= \log_3 27 - \log_3 5$

$= 3 - \log_3 5$ $\quad \log_3 27 = 3$

CAUTION

There is no property of logarithms to rewrite the logarithm of a sum or difference. For example, we ***cannot*** write $\log_b (x + y)$ in terms of $\log_b x$ and $\log_b y$. Also,

$$\log_b \frac{x}{y} \neq \frac{\log_b x}{\log_b y}.$$

◀ Work Problem **2** at the Side.

1 Use the product rule to rewrite each expression.

(a) $\log_6 (5 \cdot 8)$

(b) $\log_4 3 + \log_4 7$

(c) $\log_8 8k, \quad k > 0$

(d) $\log_5 m^2, \quad m > 0$

2 Use the quotient rule to rewrite each expression.

(a) $\log_7 \frac{9}{4}$

(b) $\log_3 p - \log_3 q$,
$p > 0, \quad q > 0$

(c) $\log_4 \frac{3}{16}$

ANSWERS

1. **(a)** $\log_6 5 + \log_6 8$ **(b)** $\log_4 21$
(c) $1 + \log_8 k$ **(d)** $2 \log_5 m$

2. **(a)** $\log_7 9 - \log_7 4$ **(b)** $\log_3 \frac{p}{q}$
(c) $\log_4 3 - 2$

OBJECTIVE 3 Use the power rule for logarithms. An exponential expression such as 2^3 means $2 \cdot 2 \cdot 2$; the base is used as a factor 3 times. Similarly, the product rule can be extended to rewrite the logarithm of a power as the product of the exponent and the logarithm of the base. For example, by the product rule for logarithms,

$\log_5 2^3$	$\log_2 7^4$
$= \log_5 (2 \cdot 2 \cdot 2)$	$= \log_2 (7 \cdot 7 \cdot 7 \cdot 7)$
$= \log_5 2 + \log_5 2 + \log_5 2$	$= \log_2 7 + \log_2 7 + \log_2 7 + \log_2 7$
$= 3 \log_5 2.$	$= 4 \log_2 7.$

Furthermore, we saw in Example 1(d) that $\log_4 x^3 = 3 \log_4 x$. These examples suggest the following rule.

Power Rule for Logarithms

If x and b are positive real numbers, where $b \neq 1$, and if r is any real number, then

$$\log_b x^r = r \log_b x.$$

In words, the logarithm of a number to a power equals the exponent times the logarithm of the number.

As further examples of this rule, we have

$$\log_b m^5 = 5 \log_b m \quad \text{and} \quad \log_3 5^4 = 4 \log_3 5.$$

To prove the power rule, let $\log_b x = m$.

$$\log_b x = m$$

$$b^m = x \qquad \text{Convert to exponential form.}$$

$$(b^m)^r = x^r \qquad \text{Raise to the power } r.$$

$$b^{mr} = x^r \qquad \text{Power rule for exponents}$$

$$\log_b x^r = mr \qquad \text{Convert to logarithmic form.}$$

$$\log_b x^r = rm \qquad \text{Commutative property}$$

$$\log_b x^r = r \log_b x \qquad m = \log_b x$$

This is the statement to be proved.

As a special case of the power rule, let $r = \frac{1}{p}$, so

$$\log_b \sqrt[p]{x} = \log_b x^{1/p} = \frac{1}{p} \log_b x.$$

For example, using this result, with $x > 0$,

$$\log_b \sqrt[5]{x} = \log_b x^{1/5} = \frac{1}{5} \log_b x \quad \text{and} \quad \log_b \sqrt[3]{x^4} = \log_b x^{4/3} = \frac{4}{3} \log_b x.$$

Another special case is

$$\log_b \frac{1}{x} = \log_b x^{-1} = -\log_b x.$$

Note

For a review of rational exponents, refer to **Section 9.2.**

EXAMPLE 3 **Using the Power Rule**

Use the power rule to rewrite each logarithm. Assume that $b > 0$, $x > 0$, and $b \neq 1$.

(a) $\log_5 4^2$
$= 2 \log_5 4$

(b) $\log_b x^5$
$= 5 \log_b x$

(c) $\log_b \sqrt{7}$

When using the power rule with logarithms of expressions involving radicals, begin by rewriting the radical expression with a rational exponent.

$$\log_b \sqrt{7}$$
$$= \log_b 7^{1/2} \qquad \sqrt{x} = x^{1/2}$$
$$= \frac{1}{2} \log_b 7 \qquad \text{Power rule}$$

(d) $\log_2 \sqrt[5]{x^2}$

$$= \log_2 x^{2/5} \qquad \sqrt[5]{x^2} = x^{2/5}$$
$$= \frac{2}{5} \log_2 x \qquad \text{Power rule}$$

◀ Work Problem **3** at the Side.

Two special properties involving both exponential and logarithmic expressions come directly from the fact that logarithmic and exponential functions are inverses of each other.

Special Properties

If $b > 0$ and $b \neq 1$, then

$$b^{\log_b x} = x, \; x > 0 \quad \text{and} \quad \log_b b^x = x.$$

To prove the first statement, let $y = \log_b x$.

$$y = \log_b x$$
$$b^y = x \qquad \text{Convert to exponential form.}$$
$$b^{\log_b x} = x \qquad \text{Replace } y \text{ with } \log_b x.$$

The proof of the second statement is similar.

EXAMPLE 4 **Using the Special Properties**

Find the value of each logarithmic expression.

(a) $\log_5 5^4 = 4$, since $\log_b b^x = x$.

(b) $\log_3 9$
$= \log_3 3^2 \qquad 9 = 3^2$
$= 2$

(c) $4^{\log_4 10} = 10$, since $b^{\log_b x} = x$.

◀ Work Problem **4** at the Side.

3 Use the power rule to rewrite each logarithm. Assume that $a > 0$, $b > 0$, $x > 0$, $a \neq 1$, and $b \neq 1$.

(a) $\log_3 5^2$

(b) $\log_a x^4$

(c) $\log_b \sqrt{8}$

(d) $\log_2 \sqrt[3]{2}$

4 Find the value of each logarithmic expression.

(a) $\log_{10} 10^3$

(b) $\log_2 8$

(c) $5^{\log_5 3}$

ANSWERS

3. **(a)** $2 \log_3 5$ **(b)** $4 \log_a x$ **(c)** $\frac{1}{2} \log_b 8$ **(d)** $\frac{1}{3}$

4. **(a)** 3 **(b)** 3 **(c)** 3

Here is a summary of the properties of logarithms.

Properties of Logarithms

If x, y, and b are positive real numbers, where $b \neq 1$, and r is any real number, then

Product Rule $\log_b xy = \log_b x + \log_b y$

Quotient Rule $\log_b \frac{x}{y} = \log_b x - \log_b y$

Power Rule $\log_b x^r = r \log_b x$

Special Properties $b^{\log_b x} = x$ and $\log_b b^x = x$.

OBJECTIVE 4 Use properties to write alternative forms of logarithmic expressions. Applying the properties of logarithms is important for solving equations with logarithms.

EXAMPLE 5 Writing Logarithms in Alternative Forms

Use the properties of logarithms to rewrite each expression. Assume that all variables represent positive real numbers.

(a) $\log_4 4x^3$

$= \log_4 4 + \log_4 x^3$ Product rule

$= 1 + 3 \log_4 x$ $\log_4 4 = 1$; power rule

(b) $\log_7 \sqrt{\frac{m}{n}}$

$= \log_7 \left(\frac{m}{n}\right)^{1/2}$ Write the radical expression with a rational exponent.

$= \frac{1}{2} \log_7 \frac{m}{n}$ Power rule

$= \frac{1}{2}(\log_7 m - \log_7 n)$ Quotient rule

(c) $\log_5 \frac{a^2}{bc}$

$= \log_5 a^2 - \log_5 bc$ Quotient rule

$= 2 \log_5 a - \log_5 bc$ Power rule

$= 2 \log_5 a - (\log_5 b + \log_5 c)$ Product rule

$= 2 \log_5 a - \log_5 b - \log_5 c$ Use parentheses to avoid errors.

(d) $4 \log_b m - \log_b n$

$= \log_b m^4 - \log_b n$ Power rule

$= \log_b \frac{m^4}{n}$ Quotient rule

Continued on Next Page

5 Use the properties of logarithms to rewrite each expression. Assume that all variables represent positive real numbers.

(a) $\log_6 36m^5$

(b) $\log_2 \sqrt{9z}$

(c) $\log_q \frac{8r^2}{m-1}, m > 1, q \neq 1$

(d) $2 \log_a x + 3 \log_a y, a \neq 1$

(e) $\log_4 (3x + y)$

6 Decide whether each statement is *true* or *false*.

(a) $\log_6 36 - \log_6 6 = \log_6 30$

(b) $\log_4 (\log_2 16) = \frac{\log_6 6}{\log_6 36}$

Answers

5. (a) $2 + 5 \log_6 m$ **(b)** $\log_2 3 + \frac{1}{2} \log_2 z$
(c) $\log_q 8 + 2 \log_q r - \log_q (m - 1)$
(d) $\log_a x^2y^3$ **(e)** cannot be rewritten

6. (a) false **(b)** false

(e) $\log_b (x+1) + \log_b (2x+1) - \frac{2}{3} \log_b x$

$= \log_b (x+1) + \log_b (2x+1) - \log_b x^{2/3}$ Power rule

$= \log_b \frac{(x+1)(2x+1)}{x^{2/3}}$ Product and quotient rules

$= \log_b \frac{2x^2 + 3x + 1}{x^{2/3}}$ Multiply in the numerator.

(f) $\log_8 (2p + 3r)$ cannot be rewritten using the properties of logarithms. There is no property of logarithms that allows us to rewrite the logarithm of a sum.

◀ Work Problem **5** at the Side.

EXAMPLE 6 Deciding Whether Statements about Logarithms Are True

Decide whether each statement is *true* or *false.*

(a) $\log_2 8 - \log_2 4 = \log_2 4$

Evaluate both sides.

Left side: $\log_2 8 - \log_2 4 = \log_2 2^3 - \log_2 2^2 = 3 - 2 = 1$

Right side: $\log_2 4 = \log_2 2^2 = 2$

The statement is false because $1 \neq 2$.

(b) $\log_3 (\log_2 8) = \frac{\log_7 49}{\log_8 64}$

Evaluate both sides.

Left side: $\log_3 (\log_2 8) = \log_3 3 = 1$

Right side: $\frac{\log_7 49}{\log_8 64} = \frac{\log_7 7^2}{\log_8 8^2} = \frac{2}{2} = 1$

The statement is true because $1 = 1$.

◀ Work Problem **6** at the Side.

11.4 Exercises

FOR EXTRA HELP MyMathLab Math XL PRACTICE WATCH DOWNLOAD READ REVIEW

Decide whether each statement of a logarithmic property is true *or* false. *If it is* false, *correct it by changing the right side of the equation.*

1. $\log_b x + \log_b y = \log_b (x + y)$

2. $\log_b \frac{x}{y} = \log_b x - \log_b y$

3. $\log_b b^x = x$

4. $\log_b x^r = \log_b rx$

Use the properties of logarithms introduced in this section to express each logarithm as a sum or difference of logarithms, or as a single number if possible. Assume that all variables represent positive real numbers. See Examples 1–5.

5. $\log_7 \frac{4}{5}$

6. $\log_8 \frac{9}{11}$

7. $\log_2 8^{1/4}$

8. $\log_3 9^{3/4}$

9. $\log_4 \frac{3\sqrt{x}}{y}$

10. $\log_5 \frac{6\sqrt{z}}{w}$

11. $\log_3 \frac{\sqrt[3]{4}}{x^2 y}$

12. $\log_7 \frac{\sqrt[3]{13}}{pq^2}$

13. $\log_3 \sqrt{\frac{xy}{5}}$

14. $\log_6 \sqrt{\frac{pq}{7}}$

15. $\log_2 \frac{\sqrt[3]{x} \cdot \sqrt[5]{y}}{r^2}$

16. $\log_4 \frac{\sqrt[4]{z} \cdot \sqrt[5]{w}}{s^2}$

17. A student erroneously wrote

$$\log_a (x + y) = \log_a x + \log_a y.$$

When his teacher explained that this was wrong, the student claimed he had used the distributive property. Write a few sentences explaining why the distributive property does not apply in this case.

18. Write a few sentences explaining how the rules for multiplying and dividing powers of the same base are similar to the rules for finding logarithms of products and quotients.

Use the properties of logarithms introduced in this section to rewrite each expression as a single logarithm. Assume that all variables are defined in such a way that the variable expressions are positive, and bases are positive numbers not equal to 1. *See Examples 1–5.*

19. $\log_b x + \log_b y$

20. $\log_b 2 + \log_b z$

21. $3 \log_a m - \log_a n$

22. $5 \log_b x - \log_b y$

23. $(\log_a r - \log_a s) + 3 \log_a t$

24. $(\log_a p - \log_a q) + 2 \log_a r$

25. $3 \log_a 5 - 4 \log_a 3$

26. $3 \log_a 5 + \frac{1}{2} \log_a 9$

27. $\log_{10}(x + 3) + \log_{10}(x - 3)$

28. $\log_{10}(y + 4) + \log_{10}(y - 4)$

29. $3 \log_p x + \frac{1}{2} \log_p y - \frac{3}{2} \log_p z - 3 \log_p a$

30. $\frac{1}{3} \log_b x + \frac{2}{3} \log_b y - \frac{3}{4} \log_b s - \frac{2}{3} \log_b t$

Decide whether each statement is true *or* false. *See Example 6.*

31. $\log_2(8 + 32) = \log_2 8 + \log_2 32$

32. $\log_2(64 - 16) = \log_2 64 - \log_2 16$

33. $\log_3 7 + \log_3 7^{-1} = 0$

34. $\log_9 14 - \log_{14} 9 = 0$

35. $\log_6 60 - \log_6 10 = 1$

36. $\log_3 8 + \log_3 \frac{1}{8} = 0$

37. $\frac{\log_{10} 7}{\log_{10} 14} = \frac{1}{2}$

38. $\frac{\log_{10} 10}{\log_{10} 100} = \frac{1}{10}$

Relating Concepts (Exercises 39–44) For Individual or Group Work

Work Exercises 39–44 in order.

39. Evaluate $\log_3 81$.

40. Write the *meaning* of the expression $\log_3 81$.

41. Evaluate $3^{\log_3 81}$.

42. Write the *meaning* of the expression $\log_2 19$.

43. Evaluate $2^{\log_2 19}$.

44. Keeping in mind that a logarithm is an exponent, and using the results from Exercises 39–43, what is the simplest form of the expression $k^{\log_k m}$?

11.5 Common and Natural Logarithms

Logarithms are important in many applications in biology, engineering, economics, and social science. In this section we find numerical approximations for logarithms. Traditionally, base 10 logarithms were used most often because our number system is base 10. Logarithms to base 10 are called **common logarithms,** and $\log_{10} x$ is abbreviated as simply $\log x$, where the base is understood to be 10.

OBJECTIVES

1. Evaluate common logarithms using a calculator.
2. Use common logarithms in applications.
3. Evaluate natural logarithms using a calculator.
4. Use natural logarithms in applications.

OBJECTIVE 1 Evaluate common logarithms using a calculator.

Calculator Tip In Example 1, we give the results of evaluating some common logarithms using a calculator with a LOG key. (This may be a second function key on some calculators.) For simple scientific calculators, just enter the number, then press the LOG key. For graphing calculators, these steps are reversed. In this section, we give calculator approximations for logarithms to four decimal places.

EXAMPLE 1 Evaluating Common Logarithms

Evaluate each logarithm using a calculator.

(a) $\log 327.1 \approx 2.5147$ **(b)** $\log 437{,}000 \approx 5.6405$

(c) $\log 0.0615 \approx -1.2111$

Notice in part (c) that $\log 0.0615 \approx -1.2111$, a negative result. ***The common logarithm of a number between 0 and 1 is always negative*** because the logarithm is the exponent on 10 that produces the number. Thus,

$$10^{-1.2111} \approx 0.0615.$$

If the exponent (the logarithm) were positive, the result would be greater than 1. See Figure 13.

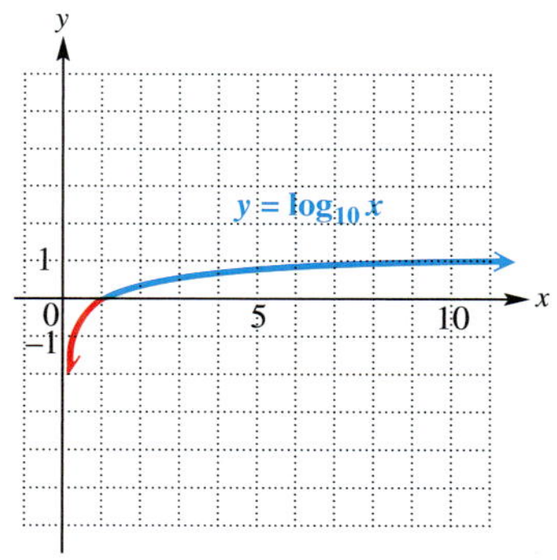

Figure 13

Work Problem 1 at the Side. ▶

1 Evaluate each logarithm to four decimal places using a calculator.

(a) $\log 41{,}600$

(b) $\log 43.5$

(c) $\log 0.442$

OBJECTIVE 2 Use common logarithms in applications. In chemistry, pH is a measure of the acidity or alkalinity of a solution; water, for example, has pH 7. In general, acids have pH numbers less than 7, and alkaline solutions have pH values greater than 7. The **pH** of a solution is defined as

$$\mathbf{pH = -\log[H_3O^+]},$$

where $[H_3O^+]$ is the hydronium ion concentration in moles per liter. It is customary to round pH values to the nearest tenth.

ANSWERS

1. (a) 4.6191 (b) 1.6385 (c) −0.3546

Figure 14 illustrates the pH scale.

Figure 14 **pH Scale**

EXAMPLE 2 Using pH in an Application

Wetlands are classified as *bogs, fens, marshes,* and *swamps*. These classifications are based on pH values. A pH value between 6.0 and 7.5, such as that of Summerby Swamp in Michigan's Hiawatha National Forest, indicates that the wetland is a "rich fen." When the pH is between 4.0 and 6.0, the wetland is a "poor fen," and if the pH falls to 3.0 or less, it is a "bog." (*Source:* Mohlenbrock, R., "Summerby Swamp, Michigan," *Natural History,* March 1994.)

Suppose that the hydronium ion concentration of a sample of water from a wetland is 6.3×10^{-3}. How would this wetland be classified?

Use the definition of pH.

$$\begin{aligned} pH &= -\log(6.3 \times 10^{-3}) && \text{Definition of pH} \\ pH &= -(\log 6.3 + \log 10^{-3}) && \text{Product rule} \\ pH &= -[0.7993 - 3(1)] && \text{Use a calculator to find log 6.3.} \\ pH &= -0.7993 + 3 && \text{Distributive property} \\ pH &\approx 2.2 \end{aligned}$$

Since the pH is less than 3.0, the wetland is a bog.

◀ Work Problem **2** at the Side.

EXAMPLE 3 Finding Hydronium Ion Concentration

Find the hydronium ion concentration of drinking water with pH 6.5.

$$\begin{aligned} \mathbf{pH} &= -\log[H_3O^+] \\ \mathbf{6.5} &= -\log[H_3O^+] && \text{Let pH} = 6.5. \\ \log[H_3O^+] &= -6.5 && \text{Multiply by } -1. \end{aligned}$$

Solve for $[H_3O^+]$ by writing the equation in exponential form, remembering that the base is 10.

$$\begin{aligned} [H_3O^+] &= 10^{-6.5} \\ [H_3O^+] &\approx 3.2 \times 10^{-7} && \text{Use a calculator.} \end{aligned}$$

◀ Work Problem **3** at the Side.

2 Solve the problem.

Find the pH of water with a hydronium ion concentration of 1.2×10^{-3}. If this water had been taken from a wetland, is the wetland a rich fen, a poor fen, or a bog?

3 Find the hydronium ion concentrations of solutions with the following pH values.

(a) 4.6

(b) 7.5

ANSWERS

2. 2.9; bog
3. (a) 2.5×10^{-5} (b) 3.2×10^{-8}

The loudness of sound is measured in a unit called a decibel, abbreviated **dB.** To measure with this unit, we first assign an intensity of I_0 to a very faint sound, called the **threshold sound.** If a particular sound has intensity I, then the decibel level of this louder sound is

$$D = 10 \log \left(\frac{I}{I_0} \right).$$

The table in the margin gives average decibel levels for some common sounds. Any sound over 85 dB exceeds what hearing experts consider safe. Permanent hearing damage can be suffered at levels above 150 dB.

Decibel Level	Example
60	Normal conversation
90	Rush hour traffic, lawn mower
100	Garbage truck, chain saw, pneumatic drill
120	Rock concert, thunderclap
140	Gunshot blast, jet engine
180	Rocket launching pad

Source: Deafness Research Foundation.

EXAMPLE 4 Measuring the Loudness of Sound

If music downloaded from a computer and delivered through the earphones of an iPod has intensity I of $(3.162 \times 10^{11}) I_0$, find the average decibel level. (*Source: Sacramento Bee.*)

$$D = 10 \log \left(\frac{I}{I_0} \right)$$

Substitute the given value for I.

$$D = 10 \log \left(\frac{(3.162 \times 10^{11}) I_0}{I_0} \right)$$

$$D = 10 \log (3.162 \times 10^{11})$$

$$D \approx 115 \text{ dB}$$

Work Problem 4 at the Side. ▶

4 Find the decibel level to the nearest whole number of a whisper with intensity I of $115\ I_0$.

OBJECTIVE 3 Evaluate natural logarithms using a calculator. The most important logarithms used in applications are **natural logarithms,** which have as base the number e. The number e is a fundamental number in our universe. For this reason e, like π, is called a **universal constant.** The letter e is used to honor Leonhard Euler, who published extensive results on the number in 1748. Since it is an irrational number, its decimal expansion never terminates and never repeats. The first few digits of the decimal value of e are **2.718281828.**

Calculator Tip A calculator key e^x or the two keys INV and LN are used to approximate powers of e. For example, a calculator gives

$$e^2 \approx 7.389056099, \quad e^3 \approx 20.08553692, \quad \text{and} \quad e^{0.6} \approx 1.8221188.$$

Logarithms to base e are called natural logarithms because they occur in biology and the social sciences in natural situations that involve growth or decay. The base e logarithm of x is written $\ln x$ (read **"el en x"**). A graph of $y = \ln x$, the equation that defines the natural logarithmic function, is given in Figure 15.

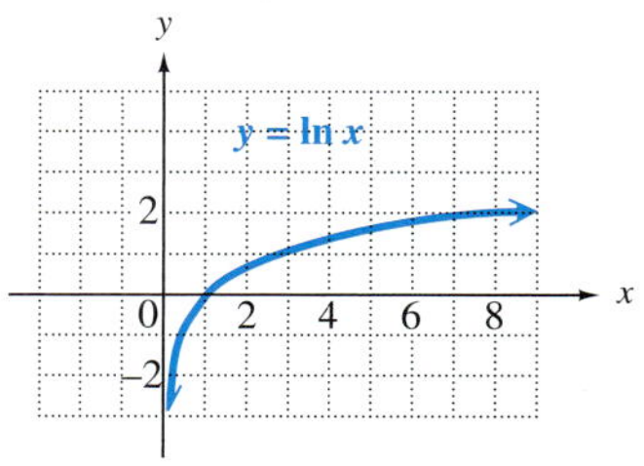

Figure 15

ANSWER

4. 21 dB

5 Find each logarithm to four decimal places.

(a) ln 0.01

(b) ln 27

(c) ln 529

6 Use the natural logarithmic function in Example 6 to approximate the altitude at 700 millibars of pressure.

Answers

5. (a) −4.6052 **(b)** 3.2958 **(c)** 6.2710
6. approximately 2700 m

Calculator Tip A calculator key labeled LN is used to evaluate natural logarithms. If your scientific calculator has an e^x key, but not a key labeled LN, find natural logarithms by entering the number, pressing the INV key, and then pressing the e^x key. This works because $y = e^x$ defines the inverse function of $y = \ln x$ (or $y = \log_e x$).

EXAMPLE 5 Finding Natural Logarithms

Find each logarithm to four decimal places.

(a) $\ln 0.5841 \approx -0.5377$

As with common logarithms, ***a number between 0 and 1 has a negative natural logarithm.***

(b) $\ln 192.7 \approx 5.2611$ **(c)** $\ln 10.84 \approx 2.3832$

◀ Work Problem 5 at the Side.

OBJECTIVE 4 Use natural logarithms in applications. A common application of natural logarithmic functions is to express growth or decay of a quantity, as in the next example.

EXAMPLE 6 Applying Natural Logarithms

The altitude in meters that corresponds to an atmospheric pressure of x millibars is given by the natural logarithmic function defined by

$$f(x) = 51{,}600 - 7457 \ln x.$$

(*Source:* Miller, A. and J. Thompson, *Elements of Meteorology,* Fourth Edition, Charles E. Merrill Publishing Company, 1993.) Use this function to find the altitude when atmospheric pressure is 400 millibars.

Let $x = 400$ and substitute in the expression for $f(x)$.

$$f(400) = 51{,}600 - 7457 \ln 400$$

$$f(400) \approx 6900 \text{ (to the nearest hundred)}$$

Atmospheric pressure is 400 millibars at approximately 6900 m.

Calculator Tip In Example 6, the final answer was obtained using a calculator *without* rounding the intermediate values. In general, it is best to wait until the final step to round the answer; otherwise, a buildup of round-off error may cause the final answer to have an incorrect final decimal place digit.

◀ Work Problem 6 at the Side.

11.5 Exercises

Choose the correct response in Exercises 1–4.

1. What is the base in the expression $\log x$?

A. e **B.** 1 **C.** 10 **D.** x

2. What is the base in the expression $\ln x$?

A. e **B.** 1 **C.** 10 **D.** x

3. Since $10^0 = 1$ and $10^1 = 10$, between what two consecutive integers is the value of $\log 5.6$?

A. 5 and 6 **B.** 10 and 11 **C.** 0 and 1 **D.** -1 and 0

4. Since $e^1 \approx 2.718$ and $e^2 \approx 7.389$, between what two consecutive integers is the value of $\ln 5.6$?

A. 5 and 6 **B.** 2 and 3 **C.** 1 and 2 **D.** 0 and 1

5. Without using a calculator, give the value of $\log 10^{19.2}$.

6. Without using a calculator, give the value of $\ln e^{\sqrt{2}}$.

You will need a calculator for the remaining exercises in this set.

Find each logarithm. Give an approximation to four decimal places. See Examples 1 and 5.

7. $\log 328.4$

8. $\log 457.2$

9. $\log 0.0326$

10. $\log 0.1741$

11. $\log (4.76 \times 10^9)$

12. $\log (2.13 \times 10^4)$

13. $\ln 7.84$

14. $\ln 8.32$

15. $\ln 0.0556$

16. $\ln 0.0217$

17. $\ln 10$

18. $\log e$

Refer to Example 2. In Exercises 19–21, suppose that water from a wetland area is sampled and found to have the given hydronium ion concentration. Determine whether the wetland is a rich fen, *a* poor fen, *or a* bog.

19. 2.5×10^{-5}

20. 2.5×10^{-2}

21. 2.5×10^{-7}

Find the pH *of the substance with the given hydronium ion concentration. See Example 2.*

22. Ammonia, 2.5×10^{-12}

23. Tuna, 1.3×10^{-6}

24. Grapes, 5.0×10^{-5}

Use the formula for pH *to find the hydronium ion concentration of the substance with the given* pH. *See Example 3.*

25. Human gastric contents, 2.0

26. Human blood plasma, 7.4

27. Bananas, 4.6

28. Spinach, 5.4

Solve each problem. See Examples 4 and 6.

29. The time t in years for an amount increasing at a rate of r (in decimal form) to double is given by

$$t = \frac{\ln 2}{\ln(1 + r)}.$$

This is called **doubling time.** Find the doubling time to the nearest tenth for an investment at each interest rate.

(a) $2\% = 0.02$ **(b)** $5\% = 0.05$

30. The number of years, $N(r)$, since two independently evolving languages split off from a common ancestral language is approximated by

$$N(r) = -5000 \ln r,$$

where r is the percent of words (as a decimal) from the ancestral language common to both languages now. Find the number of years (to the nearest hundred) since the split for each percent of common words.

(a) 85% (or 0.85) **(b)** 35% (or 0.35)

31. In the central Sierra Nevada of California, the percent of moisture p that falls as snow rather than rain is approximated reasonably well by

$$p(h) = 86.3 \ln h - 680,$$

where h is the altitude in feet.

(a) What percent of the moisture at 5000 ft falls as snow?

(b) What percent at 7500 ft falls as snow?

32. Use the formula from Example 4 to find the decibel level of each sound. (*Source:* The Canadian Society of Otolaryngology.)

(a) noisy restaurant: $I = 10^8 I_0$

(b) farm tractor: $I = (6.310 \times 10^9) I_0$

(c) snowmobile: $I = 31{,}622{,}776{,}600\, I_0$

33. The age in years of a female blue whale is approximated by

$$t = -2.57 \ln\left(\frac{87 - L}{63}\right),$$

where L is its length in feet.

(a) How old is a female blue whale that measures 80 ft?

(b) The equation that defines t has domain $24 < L < 87$. Explain why.

34. The **cost-benefit equation**

$$T = -0.642 - 189 \ln(1 - p)$$

describes the approximate tax T, in dollars per ton, that would result in a $p\%$ (in decimal form) reduction in carbon dioxide emissions.

(a) What tax will reduce emissions 25%?

(b) Explain why the equation is not valid for $p = 0$ or $p = 1$.

11.6 Exponential and Logarithmic Equations; Further Applications

We solved exponential and logarithmic equations in **Sections 11.2 and 11.3.** General methods for solving these equations depend on the following properties.

Properties for Solving Exponential and Logarithmic Equations

For all real numbers $b > 0$, $b \neq 1$, and any real numbers x and y:

1. If $x = y$, then $b^x = b^y$.
2. If $b^x = b^y$, then $x = y$.
3. If $x = y$, and $x > 0$, $y > 0$, then $\log_b x = \log_b y$.
4. If $x > 0$, $y > 0$, and $\log_b x = \log_b y$, then $x = y$.

We used Property 2 to solve exponential equations in **Section 11.2.**

OBJECTIVES

1. Solve equations involving variables in the exponents.
2. Solve equations involving logarithms.
3. Solve applications of compound interest.
4. Solve applications involving base *e* exponential growth and decay.
5. Use the change-of-base rule.

OBJECTIVE 1 Solve equations involving variables in the exponents. In Examples 1 and 2, we use Property 3.

EXAMPLE 1 Solving an Exponential Equation

Solve $3^x = 12$.

$$3^x = 12$$

$$\log 3^x = \log 12 \quad \text{Property 3 (common logs)}$$

$$x \log 3 = \log 12 \quad \text{Power rule}$$

$$\text{Exact solution} \longrightarrow x = \frac{\log 12}{\log 3} \quad \text{Divide by log 3.}$$

$$\text{Decimal approximation} \longrightarrow x \approx 2.262 \quad \text{Use a calculator.}$$

The solution set is $\{2.262\}$. Check with a calculator that $3^{2.262} \approx 12$.

Work Problem 1 at the Side. ▶

CAUTION

Be careful: $\frac{\log 12}{\log 3}$ is ***not*** equal to log 4. Note that $\log 4 \approx 0.6021$, but $\frac{\log 12}{\log 3} \approx 2.262$.

EXAMPLE 2 Solving an Exponential Equation (Base *e*)

Solve $e^{0.003x} = 40$.

$$\ln e^{0.003x} = \ln 40 \quad \text{Property 3 (natural logs)}$$

$$0.003x \ln e = \ln 40 \quad \text{Power rule}$$

$$0.003x = \ln 40 \quad \ln e = \ln e^1 = 1$$

$$x = \frac{\ln 40}{0.003} \quad \text{Divide by 0.003.}$$

$$x \approx 1230 \quad \text{Use a calculator.}$$

The solution set is $\{1230\}$. Check that $e^{0.003(1230)} \approx 40$.

Work Problem 2 at the Side. ▶

1 Solve each equation and give the decimal approximation to three places.

(a) $2^x = 9$

(b) $10^x = 4$

2 Solve $e^{-0.01t} = 0.38$.

ANSWERS

1. **(a)** $\{3.170\}$ **(b)** $\{0.602\}$
2. $\{96.8\}$

3 Solve $\log_3 (x + 1)^5 = 3$. Give the exact solution.

General Method for Solving an Exponential Equation

Take logarithms of the same base on both sides of the equation and then use the power rule of logarithms or the special property $\log_b b^x = x$. (See Examples 1 and 2.) As a special case, if both sides can be written as exponentials with the same base, do so, and then set the exponents equal. (See **Section 11.2.**)

OBJECTIVE 2 Solve equations involving logarithms. The properties of logarithms from **Section 11.4** are useful here, as is using the definition of a logarithm to change the equation to exponential form.

EXAMPLE 3 Solving a Logarithmic Equation

Solve $\log_2 (x + 5)^3 = 4$. Give the exact solution.

$$(x + 5)^3 = 2^4 \quad \text{Convert to exponential form.}$$

$$(x + 5)^3 = 16$$

$$x + 5 = \sqrt[3]{16} \quad \text{Take the cube root on each side.}$$

$$x = -5 + \sqrt[3]{16} \quad \text{Subtract 5.}$$

$$x = -5 + 2\sqrt[3]{2} \quad \sqrt[3]{16} = \sqrt[3]{8 \cdot 2} = \sqrt[3]{8} \cdot \sqrt[3]{2} = 2\sqrt[3]{2}$$

Verify that the solution satisfies the equation, so the solution set is $\{-5 + 2\sqrt[3]{2}\}$.

◀ Work Problem **3** at the Side.

4 Solve

$\log_8 (2x + 5) + \log_8 3 = \log_8 33$.

CAUTION

Recall that the domain of $y = \log_b x$ is $(0, \infty)$. For this reason, ***it is always necessary to check that the solution of an equation with logarithms yields only logarithms of positive numbers in the original equation.***

EXAMPLE 4 Solving a Logarithmic Equation

Solve $\log_2 (x + 1) - \log_2 x = \log_2 7$.

$$\log_2 (x + 1) - \log_2 x = \log_2 7$$

$$\log_2 \frac{x + 1}{x} = \log_2 7 \quad \text{Quotient rule}$$

Transform the left side to an expression with only *one* logarithm.

$$\frac{x + 1}{x} = 7 \quad \text{Property 4}$$

$$x + 1 = 7x \quad \text{Multiply by } x.$$

$$1 = 6x \quad \text{Subtract } x.$$

$$\frac{1}{6} = x \quad \text{Divide by 6.}$$

Check this solution by substituting in the original equation. Since we cannot take the logarithm of a *nonpositive* number, both $x + 1$ and x must be positive. If $x = \frac{1}{6}$, this condition is satisfied, so the solution set is $\{\frac{1}{6}\}$.

◀ Work Problem **4** at the Side.

ANSWERS

3. $\{-1 + \sqrt[5]{27}\}$
4. $\{3\}$

EXAMPLE 5 Solving a Logarithmic Equation

Solve $\log x + \log(x - 21) = 2$.

$$\log x + \log(x - 21) = 2$$

$$\log x(x - 21) = 2 \quad \text{Product rule}$$

$$x(x - 21) = 10^2 \quad \text{Write in exponential form.}$$

The base is 10.

$$x^2 - 21x = 100 \quad \text{Distributive property; multiply.}$$

$$x^2 - 21x - 100 = 0 \quad \text{Subtract 100; standard form.}$$

$$(x - 25)(x + 4) = 0 \quad \text{Factor.}$$

$$x - 25 = 0 \quad \text{or} \quad x + 4 = 0 \quad \text{Zero-factor property}$$

$$x = 25 \quad \text{or} \quad x = -4 \quad \text{Solve each equation.}$$

The value -4 must be rejected as a solution since it leads to the logarithm of at least one negative number in the original equation.

$$\log(-4) + \log(-4 - 21) = 2 \quad \text{The left side is undefined.}$$

Check that the only solution is 25, so the solution set is $\{25\}$.

CAUTION

Do not reject a proposed solution just because it is nonpositive. Reject any value that leads to the logarithm of a nonpositive number.

Work Problem 5 at the Side. ▶

5 Solve

$$\log_3 2x - \log_3(3x + 15) = -2.$$

Solving a Logarithmic Equation

Step 1 **Transform the equation so that a single logarithm appears on one side.** Use the product rule or quotient rule of logarithms to do this.

Step 2 (a) **Use Property 4.** If $\log_b x = \log_b y$, then $x = y$. (See Example 4.)

(b) **Write the equation in exponential form.** If $\log_b x = k$, then $x = b^k$. (See Examples 3 and 5.)

OBJECTIVE 3 Solve applications of compound interest. So far in this book, applications involving interest have been limited to simple interest using the formula $I = prt$. In most cases, interest paid or charged is **compound interest** (interest paid on both principal and interest). The formula for compound interest is an application of exponential functions.

Compound Interest Formula (for a Finite Number of Periods)

If a principal of P dollars is deposited at an annual rate of interest r compounded (paid) n times per year, the account will contain

$$A = P\left(1 + \frac{r}{n}\right)^{nt}$$

dollars after t years. (In this formula, r is expressed as a decimal.)

ANSWER

5. $\{1\}$

6 Find the value of \$2000 deposited at 5% compounded annually for 10 yr.

7 Find the number of years it will take for \$500 deposited in an account paying 4% interest compounded semiannually to double.

EXAMPLE 6 Solving a Compound Interest Problem for *A*

How much money will there be in an account at the end of 5 yr if \$1000 is deposited at 6% compounded quarterly? (Assume no withdrawals are made.)

Because interest is compounded quarterly, $n = 4$. The other values given in the problem are $P = 1000$, $r = 0.06$ (because $6\% = 0.06$), and $t = 5$.

$$A = P\left(1 + \frac{r}{n}\right)^{nt} \quad \text{Compound interest formula}$$

$$A = 1000\left(1 + \frac{0.06}{4}\right)^{4 \cdot 5} \quad \text{Substitute the given values.}$$

$$A = 1000(1.015)^{20}$$

$$A \approx 1346.86 \quad \text{Use a calculator.}$$

To the nearest cent, the account will contain \$1346.86. (The actual amount of interest earned is $\$1346.86 - \$1000 = \$346.86$. Why?)

◀ Work Problem **6** at the Side.

EXAMPLE 7 Solving a Compound Interest Problem for *t*

Suppose inflation is averaging 3% per year. How many years will it take for prices to double?

We want to find the number of years t for \$1 to grow to \$2 at a rate of 3% per year. In the compound interest formula, we let $A = 2$, $P = 1$, $r = 0.03$, and $n = 1$.

$$2 = 1\left(1 + \frac{0.03}{1}\right)^{1t} \quad \text{Substitute in the compound interest formula.}$$

$$2 = (1.03)^t \quad \text{Simplify.}$$

$$\log 2 = \log (1.03)^t \quad \text{Property 3}$$

$$\log 2 = t \log 1.03 \quad \text{Power rule}$$

$$t = \frac{\log 2}{\log 1.03} \quad \text{Divide by log 1.03; rewrite.}$$

$$t \approx 23.45 \quad \text{Use a calculator.}$$

Prices will double in about 23 yr. (This is called the **doubling time** of the money.) To check, verify that $1.03^{23.45} \approx 2$.

◀ Work Problem **7** at the Side.

Interest can be compounded annually, semiannually, quarterly, daily, and so on. If the number of compounding periods n is allowed to approach infinity, we have an example of **continuous compounding.** However, the compound interest formula above cannot be used for continuous compounding since there is no finite value for n. The formula for continuous compounding is an example of exponential growth involving the number e.

Continuous Compound Interest Formula

If a principal of P dollars is deposited at an annual rate of interest r compounded continuously for t years, the final amount on deposit is

$$A = Pe^{rt}.$$

Answers

6. about \$3257.79
7. about 17.50 yr

EXAMPLE 8 Solving a Continuous Interest Problem

(a) In Example 6, we found that \$1000 invested for 5 yr at 6% interest compounded quarterly would grow to \$1346.86. How much would this same investment grow to if compounded continuously?

$A = Pe^{rt}$	Continuous compound interest formula
$A = 1000e^{(0.06)5}$	Let $P = 1000$, $r = 0.06$, and $t = 5$.
$A \approx 1349.86$	Use a calculator; round to two decimal places.

The account will grow to \$1349.86.

(b) How long would it take for the initial investment amount to double?
We must find the value of t that will cause A to be $2(\$1000) = \$2000$.

$A = Pe^{rt}$	
$2000 = 1000e^{0.06t}$	Let $A = 2P = 2000$.
$2 = e^{0.06t}$	Divide by 1000.
$\ln 2 = 0.06t$	Take natural logarithms; $\ln e^k = k$.
$t = \dfrac{\ln 2}{0.06}$	Divide by 0.06; rewrite.
$t \approx 11.55$	Use a calculator.

It would take about 11.55 yr for the original investment to double.

Work Problem **8** *at the Side.* ▶

8 (a) How much will \$2500 grow to at 4% interest compounded continuously for 3 yr?

(b) How long would it take for the initial investment in part (a) to double?

OBJECTIVE 4 Solve applications involving base *e* exponential growth and decay.

EXAMPLE 9 Solving an Exponential Decay Application

Carbon-14 is a radioactive form of carbon that is found in all living plants and animals. After a plant or animal dies, the amount of radioactive carbon-14 disintegrates according to the natural logarithmic function defined by

$$y = y_0e^{-0.000121t},$$

where t is time in years, y is the amount of the sample at time t, and y_0 is the initial amount present at $t = 0$.

(a) If an initial sample contains $y_0 = 10$ g of carbon-14, how many grams will be present after 3000 yr?
Let $y_0 = 10$ and $t = 3000$ in the formula, and use a calculator.

$$y = 10e^{-0.000121(3000)} \approx 6.96 \text{ g}$$

(b) How long would it take for the initial sample to decay to half of its original amount? (This is called the **half-life.**)
Let $y = \frac{1}{2}(10) = 5$, and solve for t.

$5 = 10e^{-0.000121t}$	Substitute.
$\dfrac{1}{2} = e^{-0.000121t}$	Divide by 10.
$\ln \dfrac{1}{2} = -0.000121t$	Take natural logarithms; $\ln e^k = k$.
$t = \dfrac{\ln \frac{1}{2}}{-0.000121}$	Divide by -0.000121; rewrite.
$t \approx 5728$	Use a calculator.

The half-life is just over 5700 yr.

Work Problem **9** *at the Side.* ▶

9 Radioactive strontium decays according to the natural logarithmic function defined by

$$y = y_0e^{-0.0239t},$$

where t is time in years.

(a) If an initial sample contains $y_0 = 12$ g of radioactive strontium, how many grams will be present after 35 yr?

(b) What is the half-life of radioactive strontium?

ANSWERS

8. (a) \$2818.74 **(b)** about 17.33 yr
9. (a) 5.20 g **(b)** 29 yr

10 **(a)** Find $\log_3 17$ using common logarithms.

(b) Find $\log_3 17$ using natural logarithms.

OBJECTIVE 5 Use the change-of-base rule. In **Section 11.5**, we used a calculator to approximate the values of common logarithms (base 10) or natural logarithms (base e). The following rule is used to convert logarithms from one base to another.

Change-of-Base Rule

If $a > 0$, $a \neq 1$, $b > 0$, $b \neq 1$, and $x > 0$, then

$$\log_a x = \frac{\log_b x}{\log_b a}.$$

Note

Any positive number other than 1 can be used for base b in the change-of-base rule, but usually the only practical bases are e and 10 because calculators give logarithms only for these two bases.

To derive the change-of-base rule, let $\log_a x = m$.

$$\begin{aligned} \log_a x &= m \\ a^m &= x && \text{Change to exponential form.} \\ \log_b (a^m) &= \log_b x && \text{Property 3} \\ m \log_b a &= \log_b x && \text{Power rule} \\ (\log_a x)(\log_b a) &= \log_b x && \text{Substitute for } m. \\ \log_a x &= \frac{\log_b x}{\log_b a} && \text{Divide by } \log_b a. \end{aligned}$$

The last step gives the change-of-base rule.

EXAMPLE 10 Using the Change-of-Base Rule

Find $\log_5 12$.

Use common logarithms and the change-of-base rule.

$$\log_5 12 = \frac{\log 12}{\log 5}$$

$$\log_5 12 \approx 1.5440 \quad \text{Use a calculator.}$$

Note

Either common or natural logarithms can be used when applying the change-of-base rule. Verify that the same value is found in Example 10 if natural logarithms are used.

◀ Work Problem 10 at the Side.

Answers

10. (a) 2.5789 **(b)** 2.5789

11.6 Exercises

FOR EXTRA HELP MyMathLab Math XL PRACTICE WATCH DOWNLOAD READ REVIEW

Relating Concepts (Exercises 1–4) For Individual or Group Work

*In **Section 11.2,** we solved an equation such as $5^x = 125$ by writing each side as a power of the same base, setting exponents equal, and then solving the resulting equation. The equation is solved as follows.*

$$5^x = 125 \quad \text{Original equation}$$
$$5^x = 5^3 \quad 125 = 5^3$$
$$x = 3 \quad \text{Set exponents equal.}$$

Solution set: $\{3\}$

The method described in this section can also be used to solve this equation. ***Work Exercises 1–4 in order,*** *to see how this is done.*

1. Take common logarithms on both sides, and write this equation.

2. Apply the power rule for logarithms on the left.

3. Write the equation so that x is alone on the left.

4. Use a calculator to find the decimal form of the solution. What is the solution set?

Many of the problems in the remaining exercises require a scientific calculator.

Solve each equation. Give solutions to three decimal places. See Example 1.

5. $7^x = 5$

6. $4^x = 3$

7. $9^{-x+2} = 13$

8. $6^{-t+1} = 22$

9. $3^{2x} = 14$

10. $5^{0.3x} = 11$

11. $2^{y+3} = 5^y$

12. $6^{m+3} = 4^m$

13. $2^{x+3} = 3^{x-4}$

Solve each equation. Use natural logarithms. Give solutions to three decimal places. See Example 2.

14. $e^{0.006x} = 30$

15. $e^{0.012x} = 23$

16. $e^{-0.103x} = 7$

17. $e^{-0.205x} = 9$

18. $\ln e^{x} = 4$

19. $\ln e^{3x} = 9$

20. $\ln e^{0.04x} = \sqrt{3}$

21. $\ln e^{0.45x} = \sqrt{7}$

22. $\ln e^{2x} = \pi$

23. Try solving one of the equations in Exercises 14–17 using common logarithms rather than natural logarithms. (You should get the same solution.) Explain why using natural logarithms is a better choice.

24. If you were asked to solve

$$10^{0.0025x} = 75,$$

would natural or common logarithms be a better choice? Explain.

Solve each equation. Give exact solutions. See Example 3.

25. $\log_3 (6x + 5) = 2$

26. $\log_5 (12x - 8) = 3$

27. $\log_2 (2x - 1) = 5$

28. $\log_6 (4x + 2) = 2$

29. $\log_7 (x + 1)^3 = 2$

30. $\log_4 (x - 3)^3 = 4$

31. Suppose that in solving a logarithmic equation having the term $\log (x - 3)$ you obtain a proposed solution of 2. All algebraic work is correct. Explain why you must reject 2 as a solution of the equation.

32. Suppose that in solving a logarithmic equation having the term $\log (3 - x)$ you obtain a proposed solution of -4. All algebraic work is correct. Should you reject -4 as a solution of the equation? Explain why or why not.

Solve each equation. Give exact solutions. See Examples 4 and 5.

33. $\log (6x + 1) = \log 3$

34. $\log (7 - x) = \log 12$

35. $\log_5 (3t + 2) - \log_5 t = \log_5 4$

36. $\log_2 (x + 5) - \log_2 (x - 1) = \log_2 3$

37. $\log 4x - \log (x - 3) = \log 2$

38. $\log (-x) + \log 3 = \log (2x - 15)$

39. $\log_2 x + \log_2 (x - 7) = 3$

40. $\log (2x - 1) + \log 10x = \log 10$

41. $\log 5x - \log (2x - 1) = \log 4$

42. $\log_3 x + \log_3 (2x + 5) = 1$

43. $\log_2 x + \log_2 (x - 6) = 4$

44. $\log_2 x + \log_2 (x + 4) = 5$

Solve each problem. See Examples 6–8.

45. (a) How much money will there be in an account at the end of 6 yr if \$2000 is deposited at 4% compounded quarterly? (Assume no withdrawals are made.)

(b) To one decimal place, how long will it take for the account to grow to \$3000?

46. (a) How much money will there be in an account at the end of 7 yr if \$3000 is deposited at 3.5% compounded quarterly? (Assume no withdrawals are made.)

(b) To one decimal place, how long will it take for the account to grow to \$5000?

47. What will be the amount A in an account with initial principal \$4000 if interest is compounded continuously at an annual rate of 3.5% for 6 yr?

48. Refer to Exercise 46. Does the money grow to a larger value under those conditions, or when invested for 7 yr at 3% compounded continuously?

49. How long would it take an initial principal P to double if it is invested at 4.5% compounded continuously?

50. How long would it take \$4000 to double at 3.25% compounded continuously?

Solve each problem. See Example 9.

51. A sample of 400 g of lead-210 decays to polonium-210 according to the function defined by

$$A(t) = 400e^{-0.032t},$$

where t is time in years. How much lead will be left in the sample after 25 yr?

52. How long will it take the initial sample of lead in Exercise 51 to decay to half of its original amount?

Use the change-of-base rule (with either common or natural logarithms) to find each logarithm. Give approximations to four decimal places. See Example 10.

53. $\log_6 13$

54. $\log_7 19$

55. $\log_{\sqrt{2}} \pi$

56. $\log_\pi \sqrt{2}$

57. $\log_{21} 0.7496$

58. $\log_{19} 0.8325$

59. $\log_{1/2} 5$

60. $\log_{1/3} 7$

61. $\log_{0.3} 12$

One measure of the diversity of the species in an ecological community is the **index of diversity,** *the logarithmic expression*

$$-(p_1 \ln p_1 + p_2 \ln p_2 + \cdots + p_n \ln p_n),$$

where $p_1, p_2, \ldots, p_n$ are the proportions of a sample belonging to each of n species in the sample. (Source: Ludwig, John and James Reynolds, Statistical Ecology: A Primer on Methods and Computing, *New York, Wiley, 1988.) Find the index of diversity to three decimal places if a sample of 100 from a community produces the following numbers.*

62. 90 of one species, 10 of another

63. 60 of one species, 40 of another

Chapter 11 Summary

Key Terms

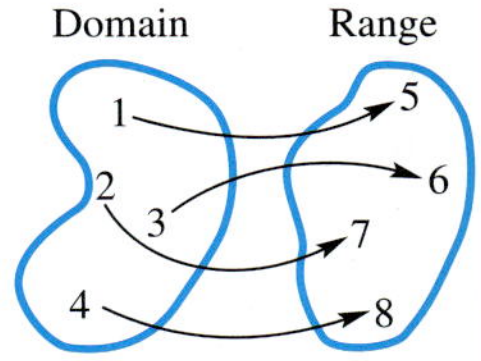

11.1 **one-to-one function** A one-to-one function is a function in which each x-value corresponds to just one y-value and each y-value corresponds to just one x-value.

inverse of a function f If f is a one-to-one function, the inverse of f is the set of all ordered pairs of the form (y, x), where (x, y) belongs to f.

11.2 **exponential equation** An equation involving an exponential, where the variable is in the exponent, is an exponential equation.

11.3 **logarithm** A logarithm is an exponent. The expression $\log_a x$ represents the exponent on the base a that gives the number x.

logarithmic equation A logarithmic equation is an equation with a logarithm in at least one term.

11.5 **common logarithm** A common logarithm is a logarithm with base 10.

natural logarithm A natural logarithm is a logarithm with base e.

New Symbols

f^{-1}	inverse of f
$\log_a x$	logarithm of x with base a
$\log x$	common (base 10) logarithm of x
$\ln x$	natural (base e) logarithm of x
e	a constant, approximately 2.718281828

Test Your Word Power

See how well you have learned the vocabulary in this chapter. Answers, with examples, follow the Quick Review.

1. In a **one-to-one function**
   - **A.** each x-value corresponds to only one y-value
   - **B.** each x-value corresponds to one or more y-values
   - **C.** each x-value is the same as each y-value
   - **D.** each x-value corresponds to only one y-value and each y-value corresponds to only one x-value.

2. If f is a one-to-one function, then the **inverse** of f is
   - **A.** the set of all solutions of f
   - **B.** the set of all ordered pairs formed by interchanging the coordinates of the ordered pairs of f
   - **C.** an equation involving an exponential expression
   - **D.** the set of all ordered pairs that are the opposite (negative) of the coordinates of the ordered pairs of f.

3. An **exponential function** is a function defined by an expression of the form
   - **A.** $f(x) = ax^2 + bx + c$ for real numbers a, b, c $(a \neq 0)$
   - **B.** $f(x) = \log_a x$, for a and x positive numbers $(a \neq 1)$
   - **C.** $f(x) = a^x$ for all real numbers x $(a > 0, a \neq 1)$
   - **D.** $f(x) = \sqrt{x}$ for $x \geq 0$.

4. A **logarithm** is
   - **A.** an exponent
   - **B.** a base
   - **C.** an equation
   - **D.** a radical expression.

5. A **logarithmic function** is a function defined by an expression of the form
   - **A.** $f(x) = ax^2 + bx + c$ for real numbers a, b, c $(a \neq 0)$
   - **B.** $f(x) = \log_a x$, for a and x positive numbers $(a \neq 1)$
   - **C.** $f(x) = a^x$ for all real numbers x $(a > 0, a \neq 1)$
   - **D.** $f(x) = \sqrt{x}$ for $x \geq 0$.

Quick Review

Concepts	Examples

11.1 Inverse Functions

Horizontal Line Test
If a horizontal line intersects the graph of a function in no more than one point, then the function is one-to-one.

Find f^{-1} if $f(x) = 2x - 3$. The graph of f is a straight line, so f is one-to-one by the horizontal line test.

Inverse Functions
For a one-to-one function f defined by an equation $y = f(x)$, the equation that defines the inverse function f^{-1} is found by interchanging x and y, solving for y, and replacing y with $f^{-1}(x)$.

Interchange x and y in the equation $y = 2x - 3$.

$$x = 2y - 3$$

Solve for y to get $y = \frac{x+3}{2}$.

Therefore, $f^{-1}(x) = \frac{x+3}{2}$, or $f^{-1}(x) = \frac{1}{2}x + \frac{3}{2}$.

In general, the graph of f^{-1} is the mirror image of the graph of f with respect to the line $y = x$.

The graphs of a nonlinear function f and its inverse f^{-1} are shown here.

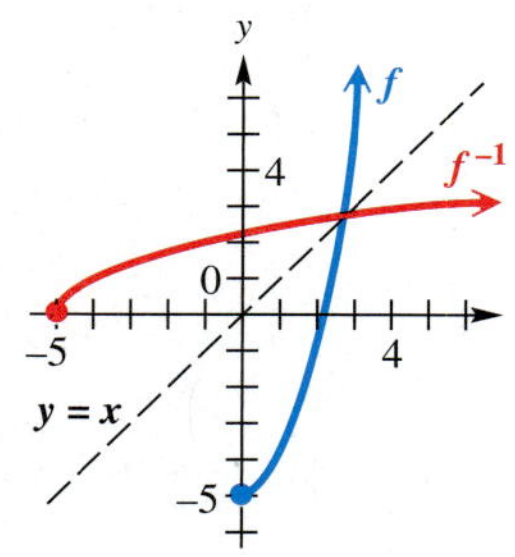

11.2 Exponential Functions

For $a > 0$, $a \neq 1$, $F(x) = a^x$ defines the exponential function with base a.

$F(x) = 3^x$ defines the exponential function with base 3.

Characteristics of the Graph of $F(x) = a^x$

1. The graph contains the point $(0, 1)$.
2. When $a > 1$, the graph rises from left to right. When $0 < a < 1$, the graph falls from left to right.
3. The x-axis is an asymptote.
4. The domain is $(-\infty, \infty)$; the range is $(0, \infty)$.

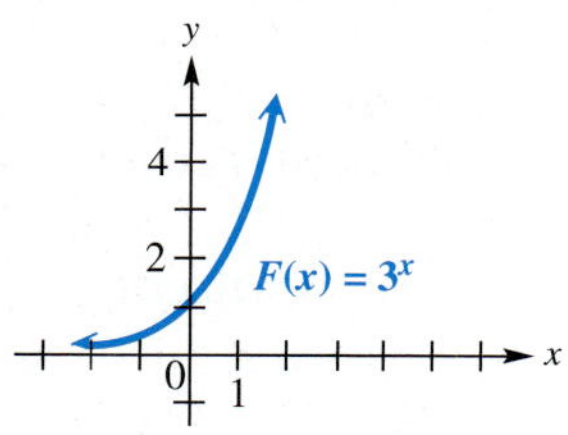

11.3 Logarithmic Functions

$y = \log_a x$ means $x = a^y$.

$y = \log_2 x$ means $x = 2^y$.

For $b > 0$, $b \neq 1$, $\log_b b = 1$ and $\log_b 1 = 0$.

$$\log_3 3 = 1 \qquad \log_5 1 = 0$$

For $a > 0$, $a \neq 1$, $x > 0$, $G(x) = \log_a x$ defines the logarithmic function with base a.

$G(x) = \log_3 x$ defines the logarithmic function with base 3.

Characteristics of the Graph of $G(x) = \log_a x$

1. The graph contains the point $(1, 0)$.
2. When $a > 1$, the graph rises from left to right. When $0 < a < 1$, the graph falls from left to right.
3. The y-axis is an asymptote.
4. The domain is $(0, \infty)$; the range is $(-\infty, \infty)$.

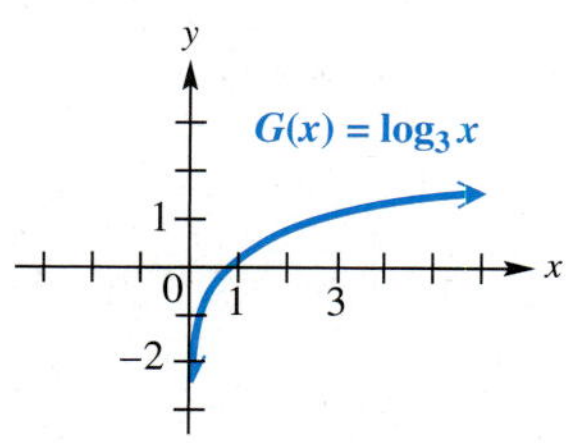

Concepts	Examples

11.4 Properties of Logarithms

Product Rule

$$\log_a xy = \log_a x + \log_a y$$

$$\log_2 (3m) = \log_2 3 + \log_2 m$$

Quotient Rule

$$\log_a \frac{x}{y} = \log_a x - \log_a y$$

$$\log_5 \frac{9}{4} = \log_5 9 - \log_5 4$$

Power Rule

$$\log_a x^r = r \log_a x$$

$$\log_{10} 2^3 = 3 \log_{10} 2$$

Special Properties

$$b^{\log_b x} = x \quad \text{and} \quad \log_b b^x = x$$

$$6^{\log_6 10} = 10 \qquad \log_3 3^4 = 4$$

11.5 Common and Natural Logarithms

Common logarithms (base 10) are used in applications such as pH, sound level, and intensity of an earthquake. Use the LOG key of a calculator to evaluate common logarithms.

Use the formula $\text{pH} = -\log [H_3O^+]$ to find the pH (to one decimal place) of grapes with hydronium ion concentration 5.0×10^{-5}.

$\text{pH} = -\log (5.0 \times 10^{-5})$ Substitute.

$\text{pH} = -(\log 5.0 + \log 10^{-5})$ Property of logarithms

$\text{pH} \approx 4.3$ Evaluate.

Natural logarithms (base e) are most often used in applications of growth and decay, such as time for money invested to double, decay of chemical compounds, and biological growth. Use the LN key or both the INV and e^x keys to evaluate natural logarithms.

Use the formula for doubling time (in years) $t = \frac{\ln 2}{\ln(1 + r)}$ to find the doubling time to the nearest tenth at an interest rate of 4% compounded annually.

$t = \frac{\ln 2}{\ln(1 + 0.04)}$ Substitute.

$t \approx 17.7$ Evaluate.

The doubling time is about 17.7 yr.

11.6 Exponential and Logarithmic Equations; Further Applications

To solve exponential equations, use these properties ($b > 0, b \neq 1$).

1. If $b^x = b^y$, then $x = y$.

Solve.

$$2^{3x} = 2^5$$
$$3x = 5$$
$$x = \frac{5}{3}$$

The solution set is $\left\{\frac{5}{3}\right\}$.

2. If $x = y$ ($x > 0, y > 0$), then $\log_b x = \log_b y$.

Solve.

$$5^x = 8$$
$$\log 5^x = \log 8$$
$$x \log 5 = \log 8$$
$$x = \frac{\log 8}{\log 5}$$
$$x \approx 1.2920$$

The solution set is $\{1.2920\}$.

(continued)

Concepts	Examples
11.6 Exponential and Logarithmic Equations; Further Applications *(continued)*	
To solve logarithmic equations, use these properties, where $b > 0, b \neq 1, x > 0, y > 0$. First use the properties of **Section 11.4,** if necessary, to write the equation in the proper form.	
1. If $\log_b x = \log_b y$, then $x = y$.	Solve. $\log_3 2x = \log_3 (x + 1)$ $2x = x + 1$ $x = 1$ The solution set is $\{1\}$.
2. If $\log_b x = y$, then $b^y = x$.	Solve. $\log_2 (3x - 1) = 4$ $3x - 1 = 2^4$ $3x - 1 = 16$ $3x = 17$ $x = \frac{17}{3}$ The solution set is $\left\{\frac{17}{3}\right\}$.
Change-of-Base Rule If $a > 0, a \neq 1, b > 0, b \neq 1, x > 0$, then $\log_a x = \frac{\log_b x}{\log_b a}$.	Approximate $\log_3 37$. $\log_3 37 = \frac{\ln 37}{\ln 3} = \frac{\log 37}{\log 3} \approx 3.2868$

ANSWERS TO TEST YOUR WORD POWER

1. D; *Example:* The function $f = \{(0, 2), (1, -1), (3, 5), (-2, 3)\}$ is one-to-one.
2. B; *Example:* The inverse of the one-to-one function f defined in Answer 1 is $f^{-1} = \{(2, 0), (-1, 1), (5, 3), (3, -2)\}$.
3. C; *Examples:* $f(x) = 4^x, g(x) = \left(\frac{1}{2}\right)^x$
4. A; *Example:* $\log_a x$ is the exponent to which a must be raised to obtain x; $\log_3 9 = 2$ since $3^2 = 9$.
5. B; *Examples:* $y = \log_3 x, y = \log_{1/3} x$

Chapter 11 Review Exercises

[11.1] *Determine whether each graph is the graph of a one-to-one function.*

1.

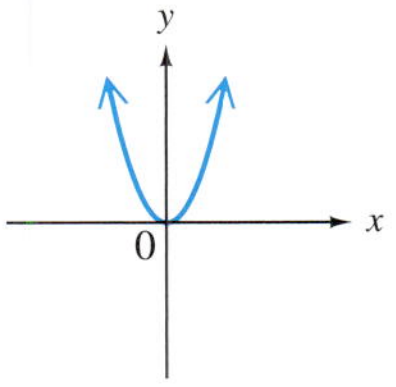

2.

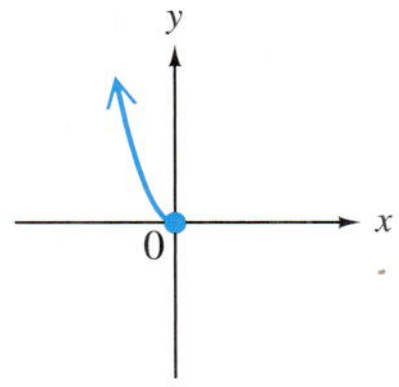

3. The table lists caffeine amounts in several popular 12-oz soft drinks. If the set of sodas is the domain and the set of caffeine amounts is the range of the function consisting of the six pairs listed, is it a one-to-one function? Why or why not?

Soda	Caffeine (mg)
Mountain Dew	55
Diet Coke	45
Dr. Pepper	41
Sunkist Orange Soda	41
Diet Pepsi-Cola	36
Coca-Cola Classic	34

Source: National Soft Drink Association.

Determine whether each function is one-to-one. If it is, find its inverse.

4. $f(x) = -3x + 7$

5. $f(x) = \sqrt[3]{5x - 4}$

6. $f(x) = -x^2 + 3$

Each function graphed is one-to-one. Graph its inverse.

7.

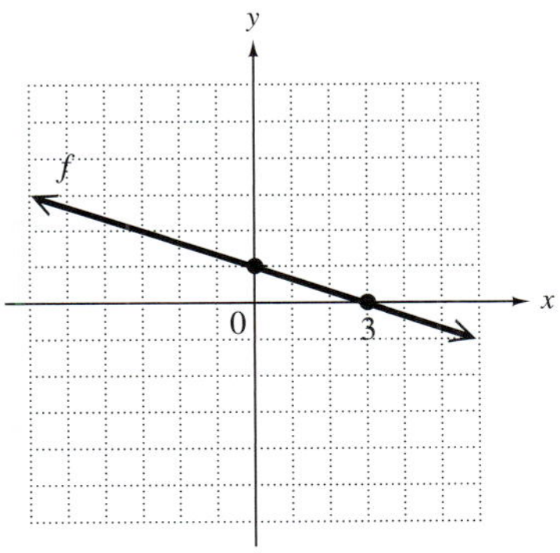

8.

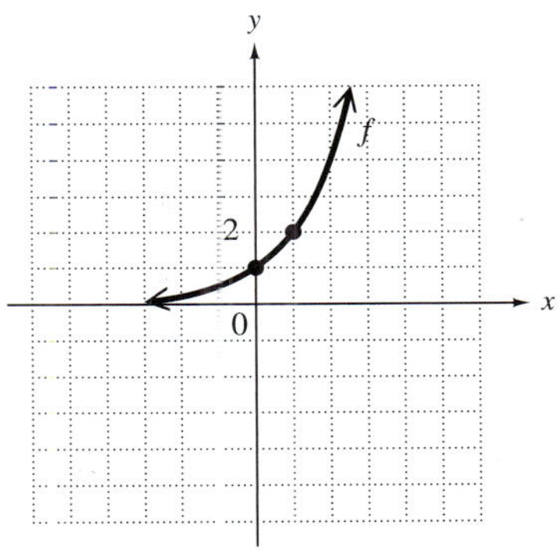

[11.2] *Graph each function.*

9. $f(x) = 4^x$

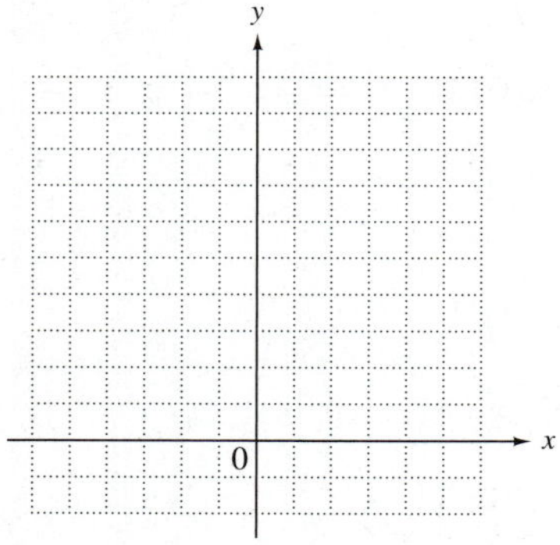

10. $f(x) = \left(\frac{1}{4}\right)^x$

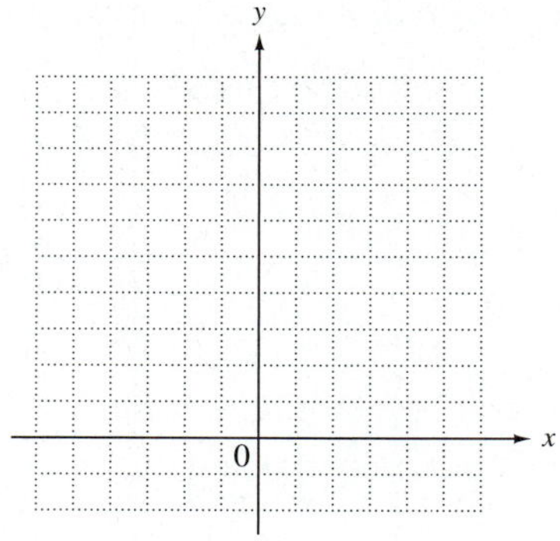

11. $f(x) = 4^{x+1}$

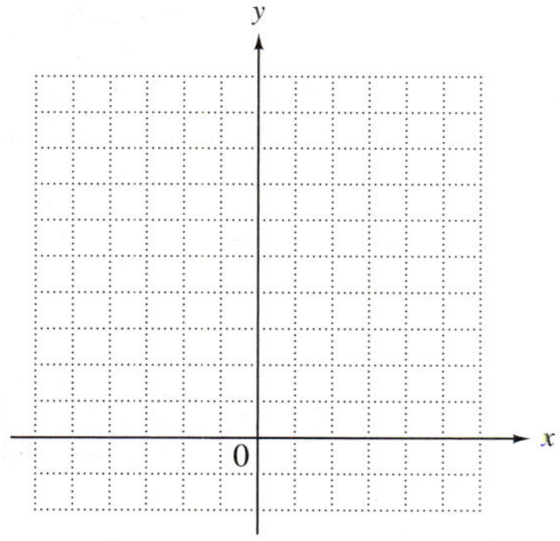

Solve each equation.

12. $4^{3x} = 8^{x+4}$

13. $\left(\frac{1}{27}\right)^{x-1} = 9^{2x}$

14. $5^x = 1$

In the remainder of the Chapter Review, many exercises will require a scientific calculator. We do not mark each such exercise.

15. A recent report predicts that the U.S. Hispanic population will increase from 35.6 million in 2000 to 102.8 million in 2050. (*Source:* U.S. Census Bureau.) Assuming an exponential growth pattern, the population is approximated by

$$f(x) = 35.6(2)^{0.0306x},$$

where x represents the number of years since 2000. Use this function to estimate the Hispanic population in each year.

(a) 2015 **(b)** 2030

[11.3]

16. (a) Write in exponential form: $\log_5 625 = 4$.

(b) Write in logarithmic form: $5^{-2} = 0.04$.

17. (a) In your own words, explain the meaning of $\log_b a$.

(b) Based on the meaning of $\log_b a$, what is the simplest form of $b^{\log_b a}$?

Graph each function.

18. $g(x) = \log_4 x$ (*Hint:* See Exercise 9.)

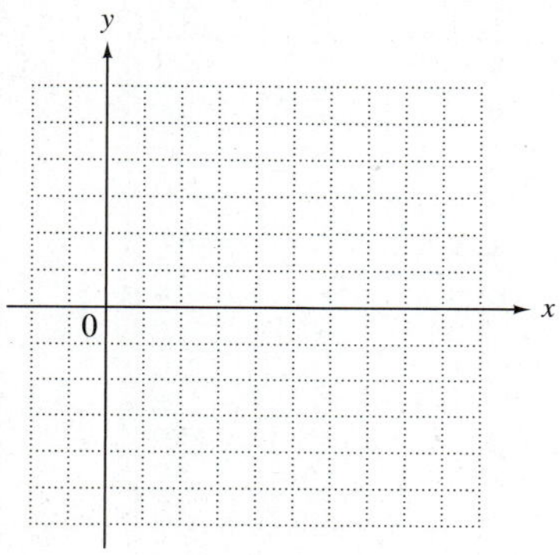

19. $g(x) = \log_{1/4} x$ (*Hint:* See Exercise 10.)

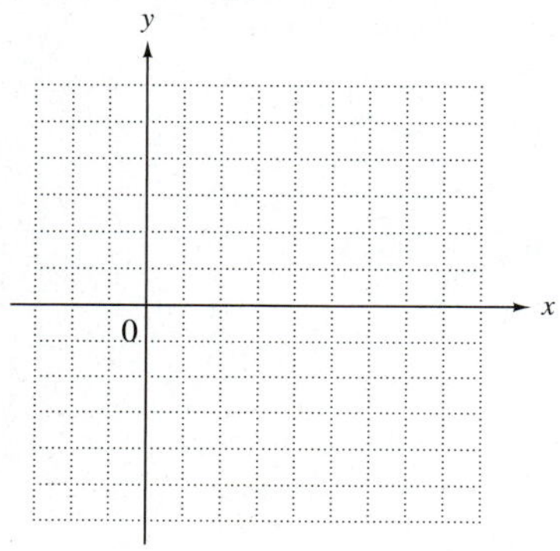

Solve each equation.

20. $\log_8 64 = x$ **21.** $\log_7 \frac{1}{49} = x$ **22.** $\log_4 x = \frac{3}{2}$ **23.** $\log_b b^2 = 2$

[11.4] *Apply the properties of logarithms to express each logarithm as a sum or difference of logarithms. Assume that all variables represent positive real numbers.*

24. $\log_4 3x^2$ **25.** $\log_2 \frac{p^2 r}{\sqrt{z}}$

Use the properties of logarithms to write each expression as a single logarithm. Assume that all variables represent positive real numbers, b ≠ 1.

26. $\log_b 3 + \log_b x - 2\log_b y$ **27.** $\log_3 (x + 7) - \log_3 (4x + 6)$

[11.5] *Evaluate each logarithm. Give approximations to four decimal places.*

28. log 28.9 **29.** log 0.257 **30.** ln 28.9 **31.** ln 0.257

Find the pH *of each substance with the given hydronium ion concentration.*

32. Milk, 4.0×10^{-7} **33.** Crackers, 3.8×10^{-9}

34. If orange juice has pH 4.6, what is its hydronium ion concentration?

Solve each problem.

35. Section 11.5 Exercise 29 introduced the formula for doubling time,

$$t = \frac{\ln 2}{\ln(1 + r)},$$

which gives the number of years required to double your money when it is invested at interest rate r (in decimal form) compounded annually. How long does it take to double your money at each rate? Round answers to the nearest year.

(a) 4% **(b)** 6%

(c) 10% **(d)** 12%

(e) Compare each answer in parts (a) – (d) with the following numbers. What do you find?

$$\frac{72}{4}, \frac{72}{6}, \frac{72}{10}, \frac{72}{12}$$

36. The concentration of a drug injected into the bloodstream decreases with time. The intervals of time T when the drug should be administered are given by

$$T = \frac{1}{k} \ln \frac{C_2}{C_1},$$

where k is a constant determined by the drug in use, C_2 is the concentration at which the drug is harmful, and C_1 is the concentration below which the drug is ineffective. (*Source:* Horelick, Brindell and Sinan Koont, "Applications of Calculus to Medicine: Prescribing Safe and Effective Dosage," *UMAP Module 202,* 1977.) Thus, if $T = 4$, the drug should be administered every 4 hr. For a certain drug, $k = \frac{1}{3}$, $C_2 = 5$, and $C_1 = 2$. How often should the drug be administered? (*Hint:* Round down.)

[11.6] *Solve each equation. Give solutions to three decimal places.*

37. $3^x = 9.42$

38. $2^{x-1} = 15$

39. $e^{0.06x} = 3$

Solve each equation. Give exact solutions.

40. $\log_3 (9x + 8) = 2$

41. $\log_5 (x + 6)^3 = 2$

42. $\log_3 (p + 2) - \log_3 p = \log_3 2$

43. $\log (2x + 3) - \log x = 1$

44. $\log_4 x + \log_4 (8 - x) = 2$

45. $\log_2 x + \log_2 (x + 15) = 4$

Solve each problem.

46. How much would be in an account after 3 yr if \$6500.00 was invested at 3% annual interest, compounded daily (use $n = 365$)?

47. Which is a better plan?

Plan A: Invest \$1000.00 at 4% compounded quarterly for 3 yr

Plan B: Invest \$1000.00 at 3.9% compounded monthly for 3 yr

A machine purchased for business use ***depreciates,*** *or loses value, over a period of years. The value of the machine at the end of its useful life is called its* ***scrap value.*** *By one method of depreciation (where it is assumed a constant percentage of the value depreciates annually), the scrap value, S, is given by*

$$S = C(1 - r)^n,$$

where C is the original cost, n is the useful life in years, and r is the constant percent of depreciation.

48. Find the scrap value of a machine costing \$30,000, having a useful life of 12 yr and a constant annual rate of depreciation of 15%.

49. A machine has a "half-life" of 6 yr. Find the constant annual rate of depreciation.

Use the change-of-base rule (with either common or natural logarithms) to find each logarithm. Give approximations to four decimal places.

50. $\log_{16} 13$

51. $\log_4 12$

52. $\log_{\sqrt{6}} \sqrt{13}$

Mixed Review Exercises

Solve.

53. $\log_3 (x + 9) = 4$

54. $\log_2 32 = x$

55. $\log_x \frac{1}{81} = 2$

56. $27^x = 81$

57. $2^{2x-3} = 8$

58. $\log_3 (x + 1) - \log_3 x = 2$

59. $\log (3x - 1) = \log 10$

60. Find the value of n in the equation for Exercise 48 if the scrap value is \$10,000, the cost is \$30,000, and the depreciation rate is 15%.

Chapter 11 Test

Use the Chapter Test Prep Video CD to see fully worked-out solutions to any of the exercises you want to review

1. Decide whether each function is one-to-one.

(a) $f(x) = x^2 + 9$

(b)

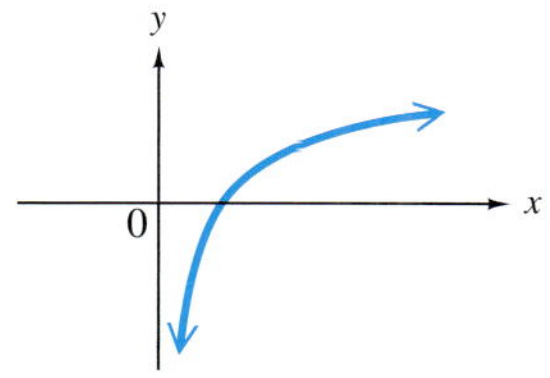

1. (a) ______

(b) ______

2. Find $f^{-1}(x)$ for the one-to-one function defined by $f(x) = \sqrt[3]{x + 7}$.

2. ______

3. Graph the inverse of f, given the graph of f here.

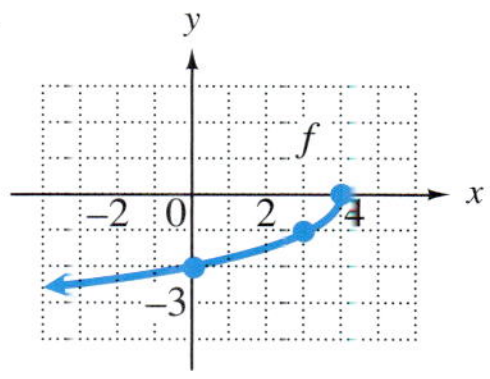

3.

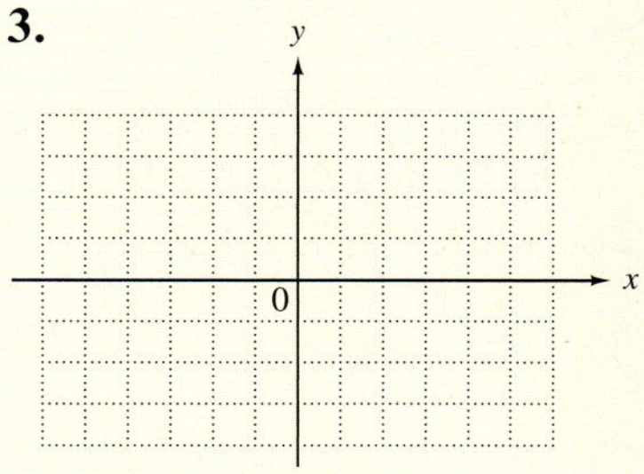

Graph each function.

4. $y = 6^x$

4.

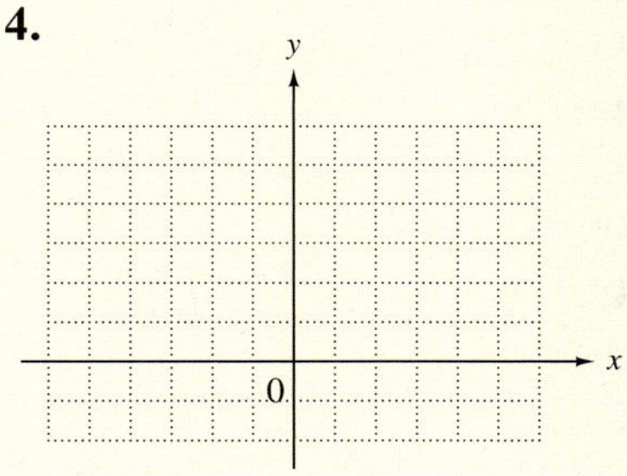

5. $y = \log_6 x$

5.

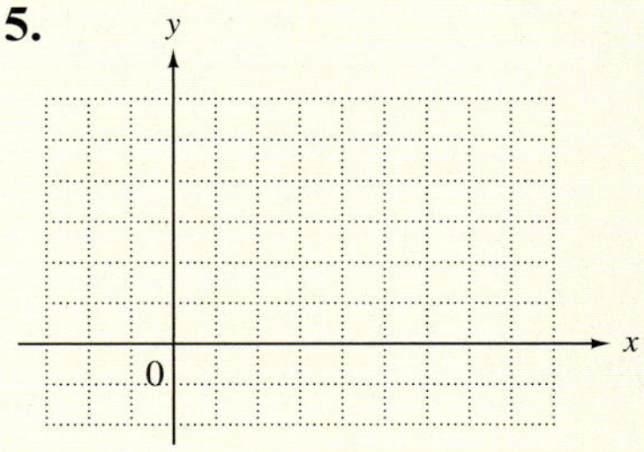

6. Explain how the graph of the function in Exercise 5 can be obtained from the graph of the function in Exercise 4.

6. ______

Solve each equation. Give exact solutions.

7. $5^x = \frac{1}{625}$

8. $2^{3x-7} = 8^{2x+2}$

7. ______

8. ______

9. (a) ____________________

(b) ____________________

10. ____________________

11. ____________________

12. ____________________

13. ____________________

14. ____________________

15. ____________________

16. ____________________

17. (a) ____________________

(b) ____________________

(c) ____________________

18. ____________________

19. ____________________

20. (a) ____________________

(b) ____________________

9. The atmospheric pressure (in millibars) at a given altitude x (in meters) is approximated by

$$f(x) = 1013e^{-0.0001341x}.$$

Use this function to approximate the atmospheric pressure at

(a) 2000 m **(b)** 10,000 m.

10. Write in logarithmic form: $4^{-2} = 0.0625$.

11. Write in exponential form: $\log_7 49 = 2$.

Solve each equation.

12. $\log_{1/2} x = -5$

13. $x = \log_9 3$

14. $\log_x 16 = 4$

15. Use properties of logarithms to write $\log_3 x^2y$ as a sum or difference of logarithms. Assume that the variables represent positive real numbers.

16. Use properties of logarithms to write $\frac{1}{4}\log_b r + 2\log_b s - \frac{2}{3}\log_b t$ as a single logarithm. Assume that the variables represent positive real numbers, $b \neq 1$.

17. Use a calculator to find an approximation to four decimal places for each logarithm.

(a) $\log 21.3$ **(b)** $\ln 0.43$ **(c)** $\log_6 45$

18. Solve $3^x = 78$, giving the solution to four decimal places.

19. Solve $\log_8(x + 5) + \log_8(x - 2) = \log_8 8$.

20. Suppose that $10,000 is invested at 4.5% annual interest, compounded quarterly.

(a) How much will be in the account in 5 yr if no money is withdrawn?

(b) How long will it take for the initial principal to double?

Cumulative Review Exercises — Chapters 1–11

Let $S = \left\{-\frac{9}{4}, -2, -\sqrt{2}, 0, 0.6, \sqrt{11}, \sqrt{-8}, 6, \frac{30}{3}\right\}$. *List the elements of S that are elements of each set.*

1. Integers

2. Rational numbers

3. Irrational numbers

Solve each equation or inequality.

4. $7 - (3 + 4x) + 2x = -5(x - 1) - 3$

5. $2x + 2 \leq 5x - 1$

6. $|2x - 5| = 9$

7. $|4x + 2| > 10$

8. The graph indicates that the number of international travelers to the United States increased from 41,218 thousand in 2003 to 50,980 thousand in 2006.

(a) Is this the graph of a function?

(b) What is the slope of the line in the graph? Interpret the slope in the context of U.S. travelers to foreign countries.

International Travelers to the U.S.

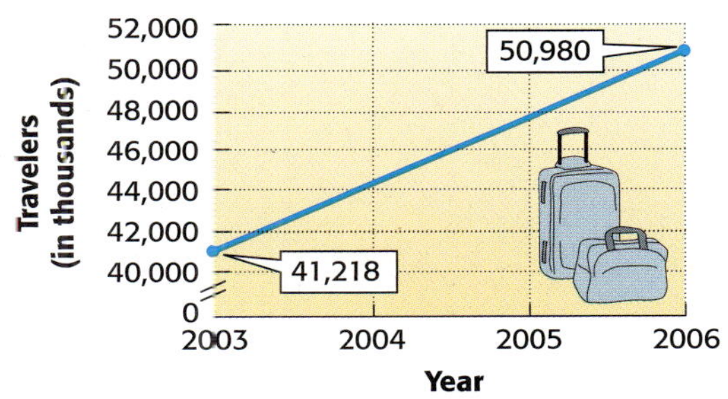

Source: U.S. Department of Commerce.

Solve each system of equations.

9. $\begin{aligned} 5x - 3y &= 14 \\ 2x + 5y &= 18 \end{aligned}$

10. $\begin{aligned} x + 2y + 3z &= 11 \\ 3x - y + z &= 8 \\ 2x + 2y - 3z &= -12 \end{aligned}$

Perform the indicated operations.

11. $(2p + 3)(3p - 1)$

12. $(4k - 3)^2$

13. $(3m^3 + 2m^2 - 5m) - (8m^3 + 2m - 4)$

14. Divide $6t^4 + 17t^3 - 4t^2 + 9t + 4$ by $3t + 1$.

Factor completely.

15. $5z^3 - 19z^2 - 4z$

16. $16a^2 - 25b^4$

17. $8c^3 + d^3$

Perform the indicated operations.

18. $\dfrac{(5p^3)^4(-3p^7)}{2p^2(4p^4)}$

19. $\dfrac{x^2 - 9}{x^2 + 7x + 12} \div \dfrac{x - 3}{x + 5}$

20. $\dfrac{2}{k + 3} - \dfrac{5}{k - 2}$

Simplify.

21. $\sqrt{288}$

22. $\dfrac{-8^{4/3}}{8^2}$

23. $2\sqrt{32} - 5\sqrt{98}$

24. Solve $\sqrt{2x + 1} - \sqrt{x} = 1$.

25. Multiply $(5 + 4i)(5 - 4i)$.

Solve each equation or inequality.

26. $3x^2 = x + 1$

27. $x^2 + 2x - 8 > 0$

28. $x^4 - 5x^2 + 4 = 0$

Solve.

29. $5^{x+3} = \left(\dfrac{1}{25}\right)^{3x+2}$

30. $\log_5 x + \log_5 (x + 4) = 1$

31. Write $\log_5 125 = 3$ in exponential form.

32. Rewrite the following using the product, quotient, and power rules for logarithms:

$$\log \frac{x^3\sqrt{y}}{z}.$$

Graph.

33. $y = \dfrac{1}{3}(x - 1)^2 + 2$

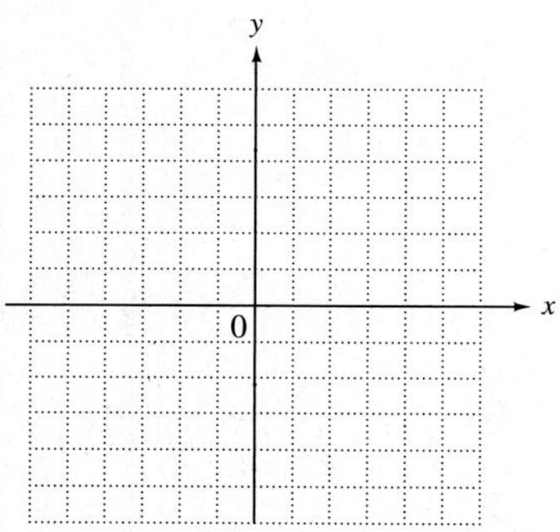

34. $f(x) = 2^x$

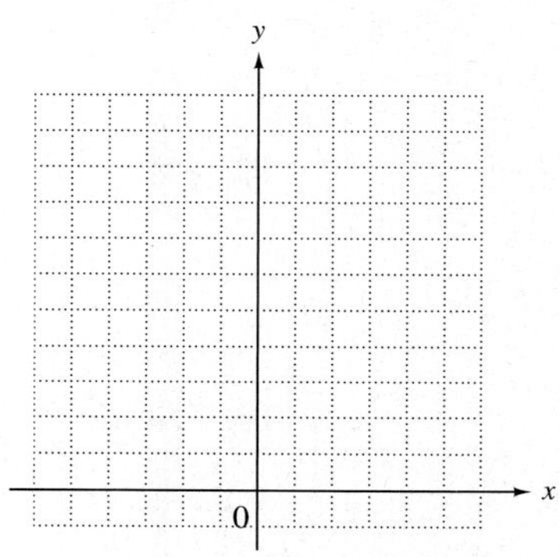

35. $f(x) = \log_3 x$

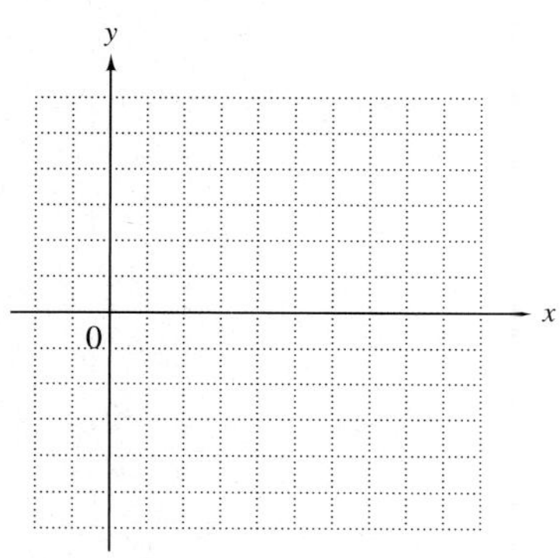

12

Nonlinear Functions, Conic Sections, and Nonlinear Systems

12.1 Additional Graphs of Functions; Composition

12.2 The Circle and the Ellipse

12.3 The Hyperbola and Other Functions Defined by Radicals

12.4 Nonlinear Systems of Equations

12.5 Second-Degree Inequalities and Systems of Inequalities

In this chapter, we study a group of curves known as *conic sections.* One conic section, the *ellipse,* has a special reflecting property responsible for "whispering galleries." In a whispering gallery, a person whispering at a certain point in the room can be heard clearly at another point across the room.

The Old House Chamber of the U.S. Capitol, now called Statuary Hall, is a whispering gallery. History has it that John Quincy Adams, whose desk was positioned at exactly the right point beneath the ellipsoidal ceiling, often pretended to sleep there as he listened to political opponents whispering strategies across the room. (*Source: We, the People, The Story of the United States Capitol,* 1991.)

In Section 12.2, we investigate ellipses.

12.1 Additional Graphs of Functions; Composition

OBJECTIVES

1. Recognize the graphs of the elementary functions defined by $|x|$, $\frac{1}{x}$, and $\sqrt{x}$, and graph their translations.
2. Recognize and graph step functions.
3. Find the composition of functions.

In earlier chapters we introduced the function defined by $f(x) = x^2$, sometimes called the **squaring function.** This is one of the most important elementary functions in algebra.

OBJECTIVE 1 Recognize the graphs of the elementary functions defined by $|x|$, $\frac{1}{x}$, and $\sqrt{x}$, and graph their translations. Another one of the elementary functions, defined by $f(x) = |x|$, is called the **absolute value function.** Its graph, along with a table of selected ordered pairs, is shown in Figure 1. Its domain is $(-\infty, \infty)$, and its range is $[0, \infty)$.

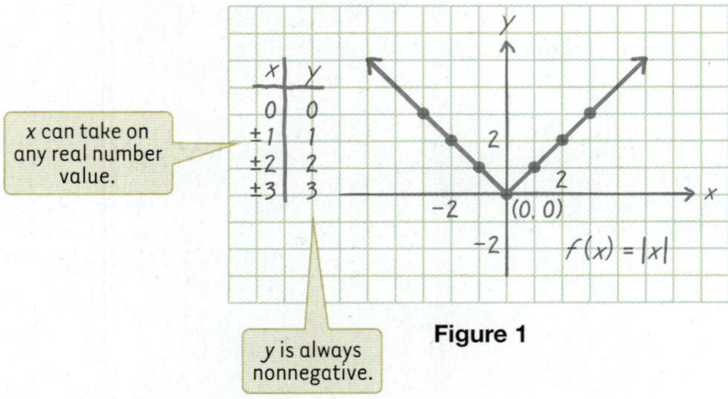

Figure 1

The **reciprocal function,** defined by $f(x) = \frac{1}{x}$, was introduced in **Section 8.4.** Its graph is shown in Figure 2, along with a table of selected ordered pairs. Notice that x can never equal 0 for this function, and as a result, as x gets closer and closer to 0, the graph approaches either ∞ or $-\infty$. Also, $\frac{1}{x}$ can never equal 0, and as x approaches ∞ or $-\infty$, $\frac{1}{x}$ approaches 0. The axes are called **asymptotes** for the function. For the reciprocal function, the domain and the range are both $(-\infty, 0) \cup (0, \infty)$.

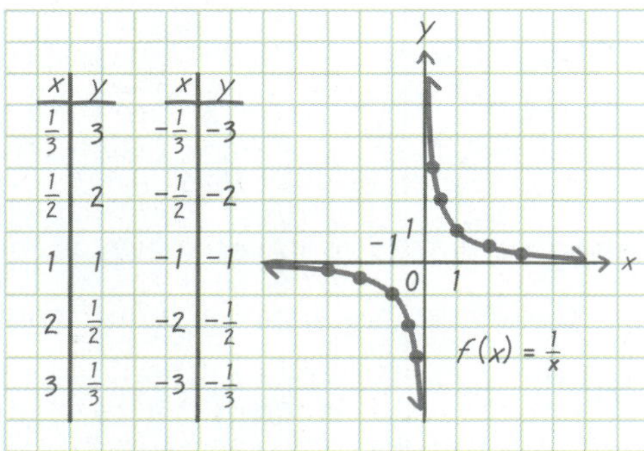

Figure 2

The **square root function**, defined by $f(x) = \sqrt{x}$, was introduced in **Section 9.1.** Its graph is shown in Figure 3 on the next page. Since we restrict function values to be real numbers, x cannot take on negative values. Thus, the domain of the square root function is $[0, \infty)$. Because the principal square root is always nonnegative, the range is also $[0, \infty)$. A table of values is shown along with the graph.

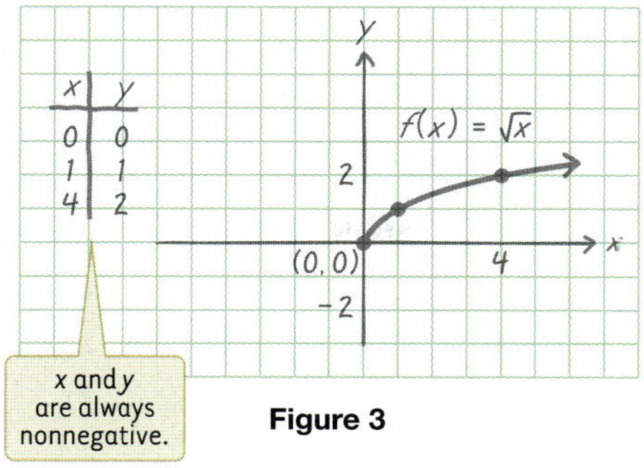

Figure 3

Just as the graph of $f(x) = x^2$ can be shifted, or translated, as we saw in **Section 10.5,** so can the graphs of these other elementary functions.

EXAMPLE 1 Applying a Horizontal Shift

Graph $f(x) = |x - 2|$.

The graph of $y = (x - 2)^2$ is obtained by shifting the graph of $y = x^2$ two units to the right. In a similar manner, the graph of $f(x) = |x - 2|$ is found by shifting the graph of $y = |x|$ two units to the right, as shown in Figure 4. The table of ordered pairs accompanying the graph supports this, as can be seen by comparing it to the table with Figure 1. The domain of this function is $(-\infty, \infty)$, and its range is $[0, \infty)$.

x	y
0	2
1	1
2	0
3	1
4	2

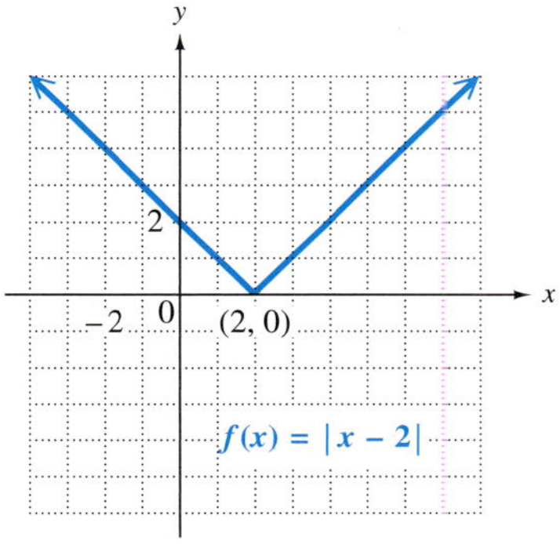

Figure 4

Work Problem **1** at the Side. ▶

EXAMPLE 2 Applying a Vertical Shift

Graph $f(x) = \frac{1}{x} + 3$.

The graph of this function is found by shifting the graph of $y = \frac{1}{x}$ three units up. See Figure 5 on the next page. The domain is

$$(-\infty, 0) \cup (0, \infty),$$

and the range is

$$(-\infty, 3) \cup (3, \infty).$$

Continued on Next Page

1 Graph $f(x) = \sqrt{x + 4}$. Give the domain and range.

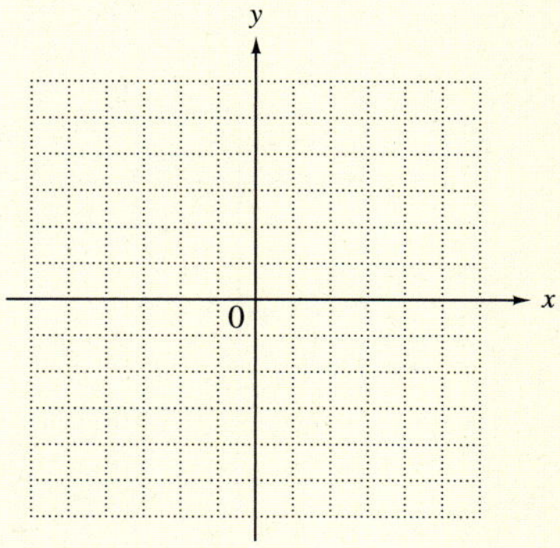

ANSWER

1.

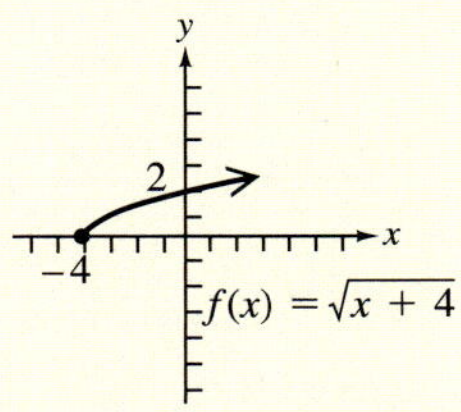

domain: $[-4, \infty)$; range: $[0, \infty)$

2 Graph $f(x) = \frac{1}{x} - 2$.

Give the domain and range.

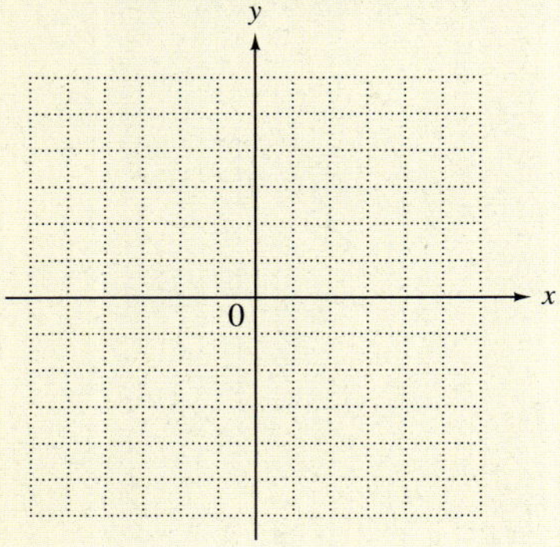

3 Graph $f(x) = |x + 2| + 1$.
Give the domain and range.

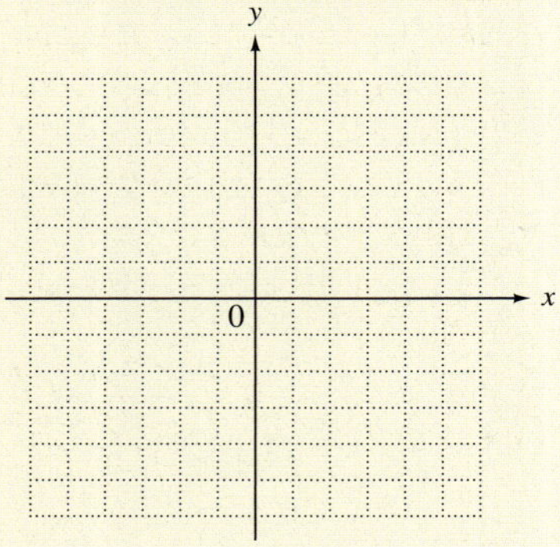

4 Find each of the following.

(a) $[\![18]\!]$ **(b)** $[\![8.7]\!]$

(c) $[\![-5]\!]$ **(d)** $[\![-6.9]\!]$

(e) $\left[\!\!\left[\frac{1}{2}\right]\!\!\right]$

x	y
$\frac{1}{3}$	6
$\frac{1}{2}$	5
1	4
2	3.5

x	y
$-\frac{1}{3}$	0
$-\frac{1}{2}$	1
-1	2
-2	2.5

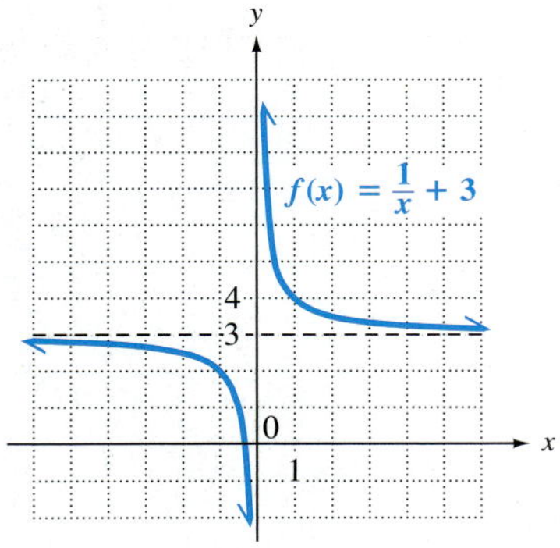

Figure 5

◀ *Work Problem* **2** *at the Side.*

EXAMPLE 3 Applying Both Horizontal and Vertical Shifts

Graph $f(x) = \sqrt{x + 1} - 4$.

The graph of $y = (x + 1)^2 - 4$ is obtained by shifting the graph of $y = x^2$ one unit to the left and four units down. Following this pattern here, we shift the graph of $y = \sqrt{x}$ one unit to the left and four units down to get the graph of $f(x) = \sqrt{x + 1} - 4$. See Figure 6. The domain is $[-1, \infty)$, and the range is $[-4, \infty)$.

x	y
-1	-4
0	-3
3	-2

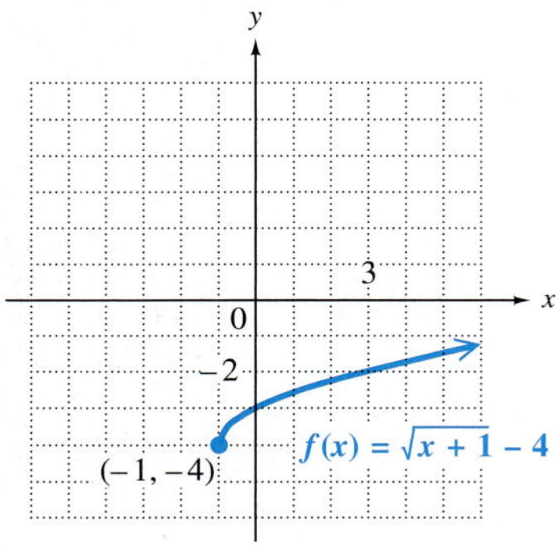

Figure 6

◀ *Work Problem* **3** *at the Side.*

OBJECTIVE 2 Recognize and graph step functions. The **greatest integer function,** usually written $f(x) = [\![x]\!]$, is defined as follows:

$[\![x]\!]$ denotes the greatest integer that is less than or equal to x.

For example,

$$[\![8]\!] = 8, \quad [\![7.45]\!] = 7, \quad [\![\pi]\!] = 3, \quad [\![-1]\!] = -1, \quad \text{and} \quad [\![-2.6]\!] = -3.$$

In general, if $f(x) = [\![x]\!]$, then

$$\begin{aligned} &\text{for} \quad -2 \le x < -1, \quad f(x) = -2, \\ &\text{for} \quad -1 \le x < 0, \quad f(x) = -1, \\ &\text{for} \quad 0 \le x < 1, \quad f(x) = 0, \\ &\text{for} \quad 1 \le x < 2, \quad f(x) = 1, \\ &\text{for} \quad 2 \le x < 3, \quad f(x) = 2, \end{aligned}$$

and so on.

◀ *Work Problem* **4** *at the Side.*

ANSWERS

2.

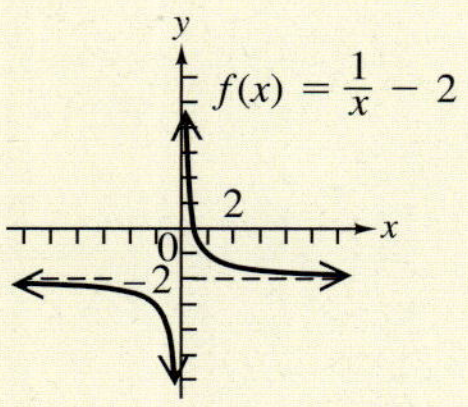

domain: $(-\infty, 0) \cup (0, \infty)$;
range: $(-\infty, -2) \cup (-2, \infty)$

3.

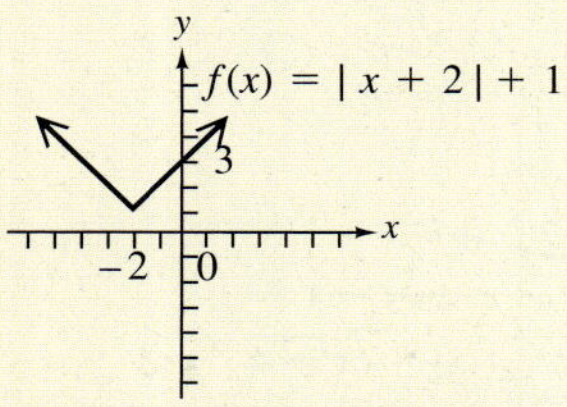

domain: $(-\infty, \infty)$; range: $[1, \infty)$

4. (a) 18 **(b)** 8 **(c)** -5 **(d)** -7 **(e)** 0

EXAMPLE 4 Graphing the Greatest Integer Function

Graph $f(x) = [\![x]\!]$.

For $[\![x]\!]$,

$$\begin{aligned} &\text{if} \quad -1 \le x < 0, \quad \text{then} \quad [\![x]\!] = -1; \\ &\text{if} \quad 0 \le x < 1, \quad \text{then} \quad [\![x]\!] = 0; \\ &\text{if} \quad 1 \le x < 2, \quad \text{then} \quad [\![x]\!] = 1, \end{aligned}$$

and so on. Thus, the graph, as shown in Figure 7, consists of a series of horizontal line segments. In each one, the left endpoint is included and the right endpoint is excluded. These segments continue infinitely following this pattern to the left and right. Since x can take any real number value, the domain is $(-\infty, \infty)$. The range is the set of integers $\{\ldots, -4, -3, -2, -1, 0, 1, 2, 3, 4, \ldots\}$. The appearance of the graph is the reason that this function is called a **step function.**

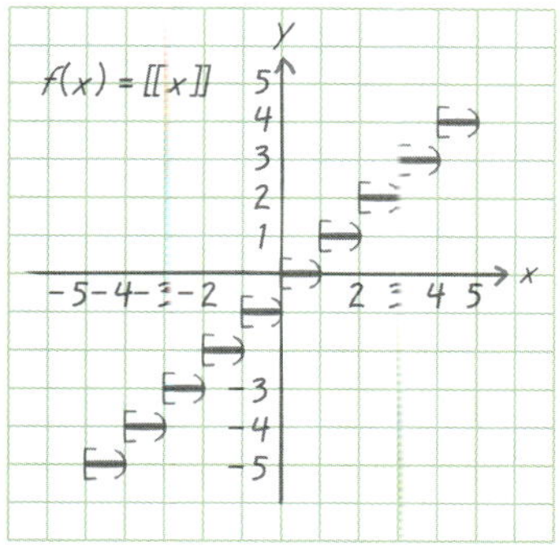

Figure 7

The graph of a step function also may be shifted. For example, the graph of $h(x) = [\![x - 2]\!]$ is the same as the graph of $f(x) = [\![x]\!]$ shifted two units to the right. Similarly, the graph of $g(x) = [\![x]\!] + 2$ is the graph of $f(x)$ shifted two units up.

Work Problem **5** *at the Side.* ▶

EXAMPLE 5 Applying a Greatest Integer Function

An overnight delivery service charges \$25 for a package weighing up to 2 lb. For each additional pound or fraction of a pound there is an additional charge of \$3. Let $D(x)$ represent the cost to send a package weighing x pounds. Graph $D(x)$ for x in the interval $(0, 6]$.

For x in the interval $(0, 2]$, $\quad y = 25.$

For x in the interval $(2, 3]$, $\quad y = 25 + 3 = 28$

For x in the interval $(3, 4]$, $\quad y = 28 + 3 = 31$, and so on.

The graph, which is that of a step function, is shown in Figure 8.

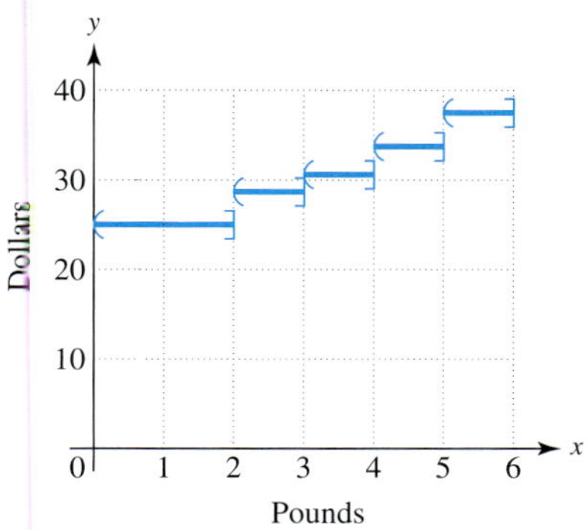

Figure 8

Work Problem **6** *at the Side.* ▶

5 Graph $f(x) = [\![x + 1]\!]$. Give the domain and range.

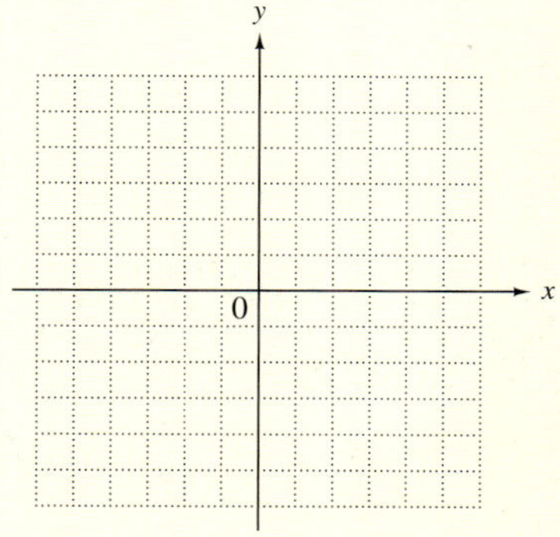

6 Assume that the post office charges 80¢ per oz (or fraction of an ounce) to mail a letter to Europe. Graph the ordered pairs (ounces, cost) for x in the interval $(0, 4]$.

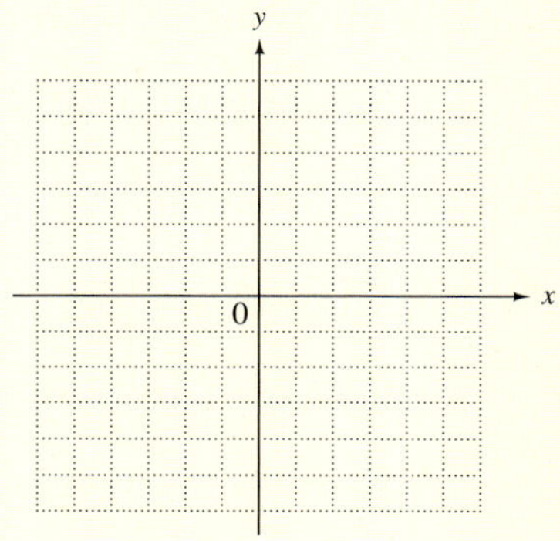

ANSWERS

5.

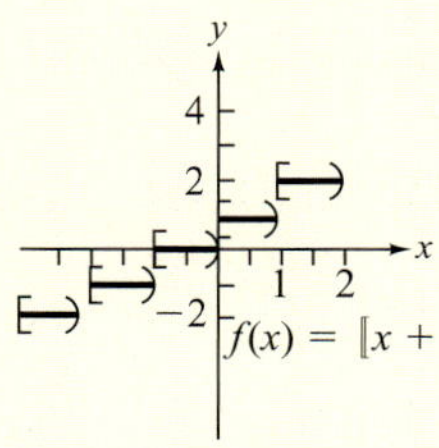

domain: $(-\infty, \infty)$;
range: $\{\ldots, -2, -1, 0, 1, 2, \ldots\}$

6.

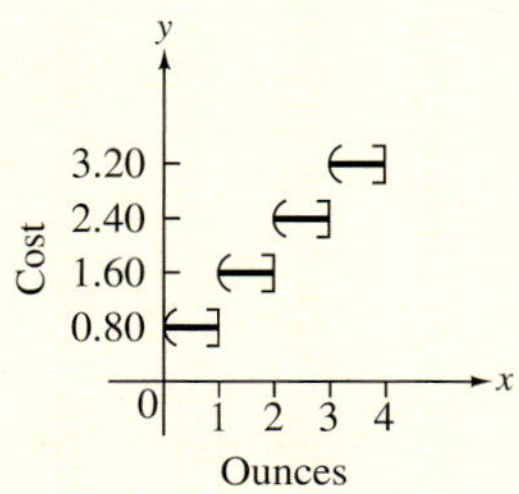

OBJECTIVE 3 Find the composition of functions. The diagram in Figure 9 shows a function f that assigns, to each element x of set X, some element y of set Y. Suppose that a function g takes each element of set Y and assigns a value z of set Z. Then f and g together assign an element x in X to an element z in Z. The result of this process is a new function h, which takes an element x in X and assigns it an element z in Z.

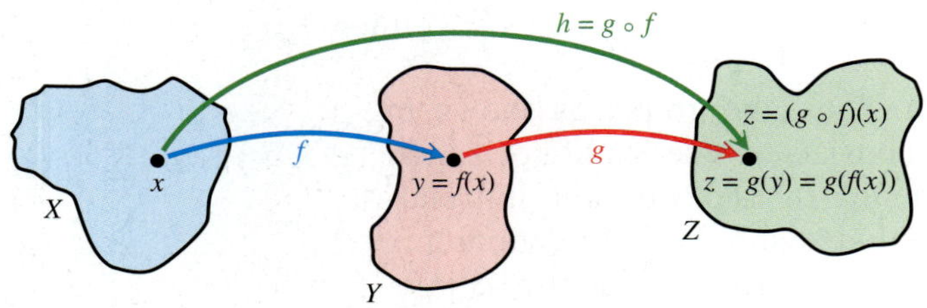

Figure 9

This function h is called the *composition* of functions g and f, written $g \circ f$, and is defined as follows.

Composition of Functions

If f and g are functions, then the **composite function,** or **composition,** of g and f is defined by

$$(g \circ f)(x) = g(f(x))$$

for all x in the domain of f such that $f(x)$ is in the domain of g.

Read $g \circ f$ as "g of f."

As a real-life example of how composite functions occur, consider the following retail situation.

A \$40 pair of blue jeans is on sale for 25% off. If you purchase the jeans before noon, the retailer offers an additional 10% off. What is the final sale price of the blue jeans?

You might be tempted to say that the jeans are 35% off and calculate \$40(0.35) = \$14, giving a final sale price of \$40 − \$14 = \$26 for the jeans. ***This is not correct.*** To find the final sale price, we must first find the price after taking 25% off, and then take an additional 10% off that price.

\$40(0.25) = \$10, giving a sale price of \$40 − \$10 = \$30. Take 25% off original price.

\$30(0.10) = \$3, giving a ***final sale price*** of \$30 − \$3 = \$27. Take additional 10% off.

This is the idea behind composition of functions.

As another example of composition, suppose an oil well off the Louisiana coast is leaking, with the leak spreading oil in a circular layer over the surface. See Figure 10.

Figure 10

At any time t, in minutes, after the beginning of the leak, the radius of the circular oil slick is given by $r(t) = 5t$ feet. Since $A(r) = \pi r^2$ gives the area of a circle of radius r, the area can be expressed as a function of time by substituting $5t$ for r in $A(r) = \pi r^2$ to get

$$A(r) = \pi r^2$$
$$A(r(t)) = \pi(5t)^2$$
$$A(r(t)) = 25\pi t^2.$$

The function $A(r(t))$ is a composite function of the functions A and r.

EXAMPLE 6 **Finding a Composite Function**

Let $f(x) = x^2$ and $g(x) = x + 3$. Find $(f \circ g)(4)$.

$(f \circ g)(4)$	$= f(g(4))$	Definition
Evaluate the "inside" function value first.	$= f(4 + 3)$	Use the rule for $g(x)$; $g(4) = 4 + 3$.
	$= f(7)$	Add.
Now evaluate the "outside" function.	$= 7^2$	Use the rule for $f(x)$; $f(7) = 7^2$.
	$= 49$	Square 7.

Work Problem 7 at the Side. ▶

Notice in Example 6 that if we interchange the order of the functions, the composition of g and f is defined by $g(f(x))$. Once again, letting $x = 4$, we have

$(g \circ f)(4)$	$= g(f(4))$	Definition
	$= g(4^2)$	Use the rule for $f(x)$; $f(4) = 4^2$.
	$= g(16)$	Square 4.
	$= 16 + 3$	Use the rule for $g(x)$; $g(16) = 16 + 3$.
	$= 19.$	Add.

Here we see that $(f \circ g)(4) \neq (g \circ f)(4)$ because $49 \neq 19$. In general,

$$(f \circ g)(x) \neq (g \circ f)(x).$$

7 Let $f(x) = x - 4$ and $g(x) = x^2$. Find $(f \circ g)(3)$.

ANSWER

7. 5

8 Let $f(x) = 3x + 6$ and $g(x) = x^3$. Find each of the following.

(a) $(f \circ g)(2)$

(b) $(g \circ f)(2)$

(c) $(f \circ g)(x)$

(d) $(g \circ f)(x)$

EXAMPLE 7 Finding Composite Functions

Let $f(x) = 4x - 1$ and $g(x) = x^2 + 5$. Find each of the following.

(a) $(f \circ g)(2)$

$$
\begin{aligned}
(f \circ g)(2) &= f(g(2)) \\
&= f(2^2 + 5) && g(x) = x^2 + 5 \\
&= f(9) && \text{Work inside the parentheses.} \\
&= 4(9) - 1 && f(x) = 4x - 1 \\
&= 35 && \text{Multiply; subtract.}
\end{aligned}
$$

(b) $(f \circ g)(x)$
Here, use $g(x)$ as the input for the function f.

$$
\begin{aligned}
(f \circ g)(x) &= f(g(x)) \\
&= 4(g(x)) - 1 && \text{Use the rule for } f(x);\ f(x) = 4x - 1. \\
&= 4(x^2 + 5) - 1 && g(x) = x^2 + 5 \\
&= 4x^2 + 20 - 1 && \text{Distributive property} \\
&= 4x^2 + 19 && \text{Combine like terms.}
\end{aligned}
$$

(c) Find $(f \circ g)(2)$ again, this time using the rule obtained in part (b).

$$
\begin{aligned}
(f \circ g)(x) &= 4x^2 + 19 && \text{From part (b)} \\
(f \circ g)(2) &= 4(2)^2 + 19 && \text{Let } x = 2. \\
&= 4(4) + 19 && \text{Square 2.} \\
&= 16 + 19 && \text{Multiply.} \\
&= 35 && \text{Add.}
\end{aligned}
$$

The result, 35, is the same as the result in part (a).

(d) $(g \circ f)(x)$
Here, use $f(x)$ as the input for the function g.

$$
\begin{aligned}
(g \circ f)(x) &= g(f(x)) \\
&= (f(x))^2 + 5 && \text{Use the rule for } g(x);\ g(x) = x^2 + 5. \\
&= (4x - 1)^2 + 5 && f(x) = 4x - 1 \\
&= 16x^2 - 8x + 1 + 5 && (x - y)^2 = x^2 - 2xy + y^2 \\
&= 16x^2 - 8x + 6 && \text{Combine like terms.}
\end{aligned}
$$

Compare this result to that in part (b). Again, $(f \circ g)(x) \neq (g \circ f)(x)$.

◀ Work Problem **8** at the Side.

ANSWERS

8. **(a)** 30 **(b)** 1728 **(c)** $3x^3 + 6$ **(d)** $(3x + 6)^3$

12.1 Exercises

FOR EXTRA HELP

MyMathLab

MathXL PRACTICE | WATCH | DOWNLOAD | READ | REVIEW

Fill in each blank with the correct response.

1. For the reciprocal function defined by $f(x) = \frac{1}{x}$, ______ is the only real number not in the domain.

2. The range of the square root function, given by $f(x) = \sqrt{x}$, is ______________.

3. The lowest point on the graph of $f(x) = |x|$ has coordinates (______, ______).

4. The range of $f(x) = [\![x]\!]$, the greatest integer function, is ______________.

Without actually plotting points, match each function defined by the absolute value expression with its graph. See Examples 1–3.

5. $f(x) = |x - 2| + 2$

6. $f(x) = |x + 2| + 2$

7. $f(x) = |x - 2| - 2$

8. $f(x) = |x + 2| - 2$

A.

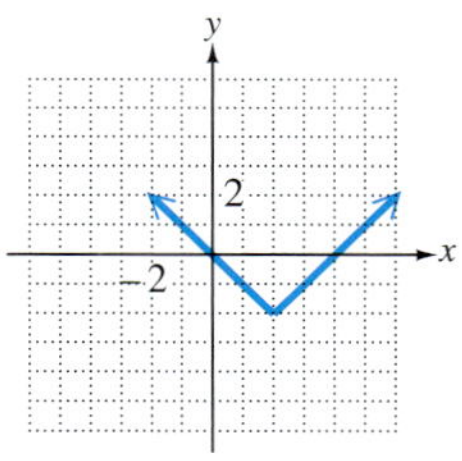

B.

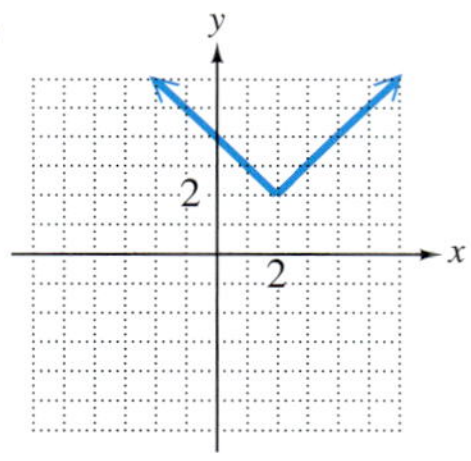

C.

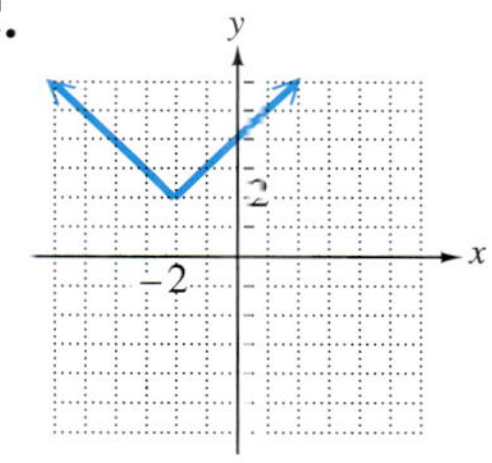

D.

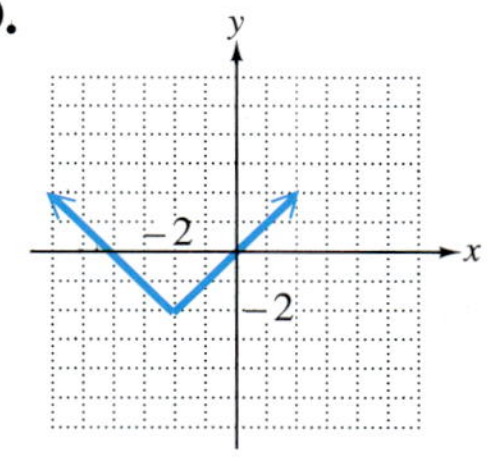

Graph each function. Give the domain and range. See Examples 1–3.

9. $f(x) = |x + 1|$

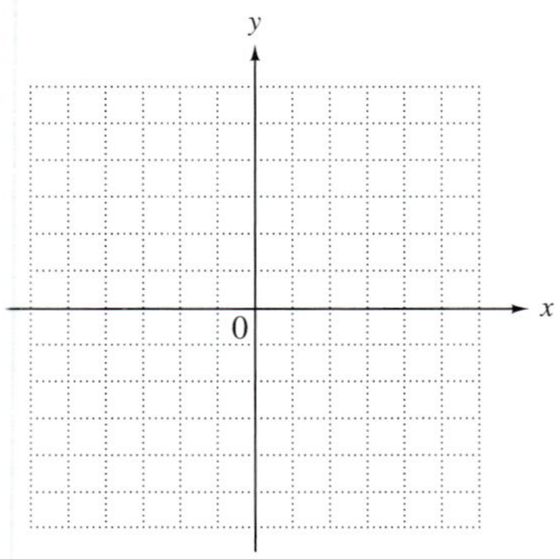

10. $f(x) = |x - 1|$

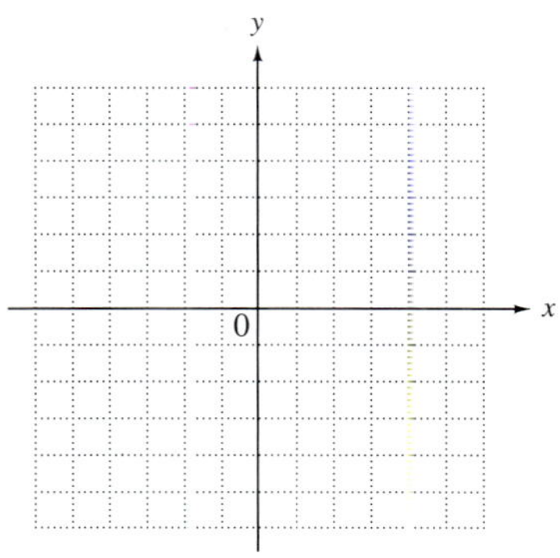

11. $f(x) = \frac{1}{x} + 1$

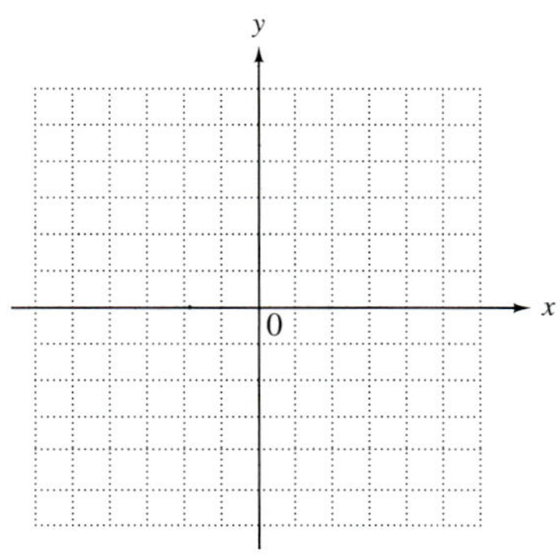

12. $f(x) = \frac{1}{x} - 1$

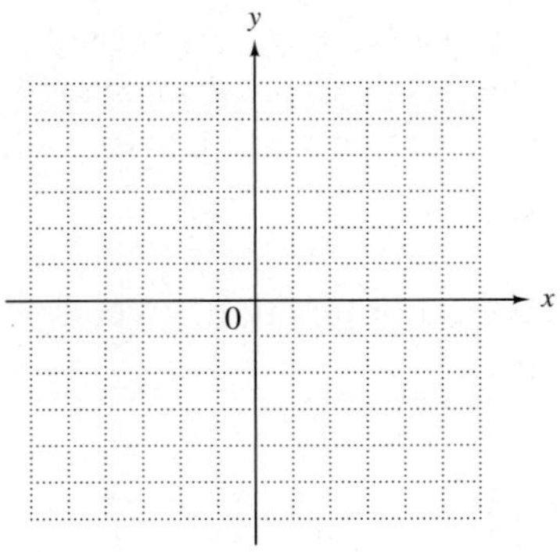

13. $f(x) = \sqrt{x - 2}$

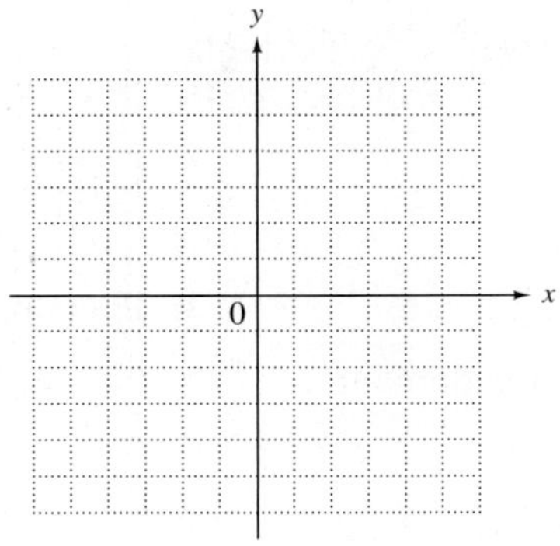

14. $f(x) = \sqrt{x + 5}$

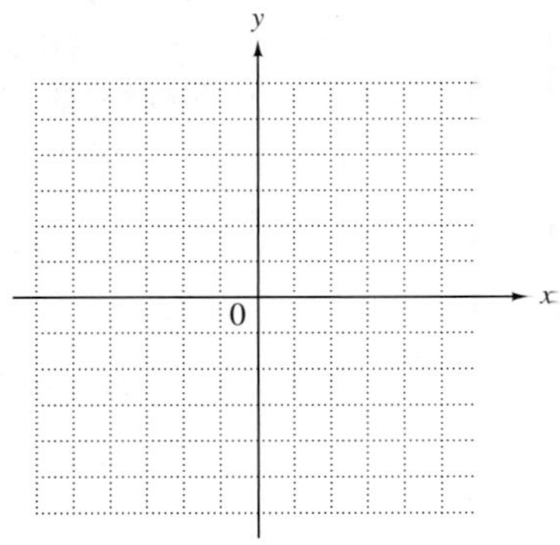

15. $f(x) = \frac{1}{x - 2}$

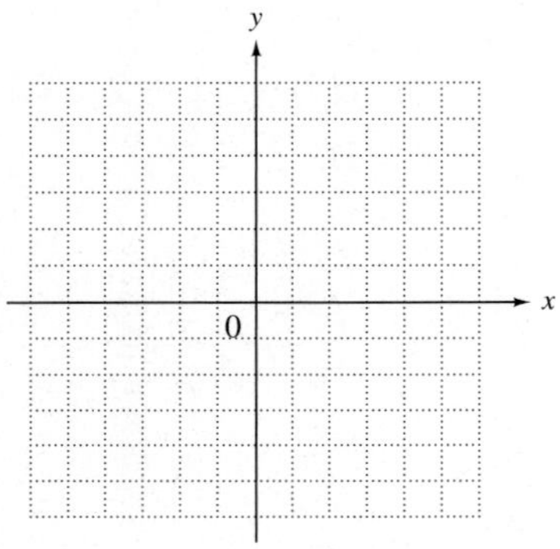

16. $f(x) = \frac{1}{x + 2}$

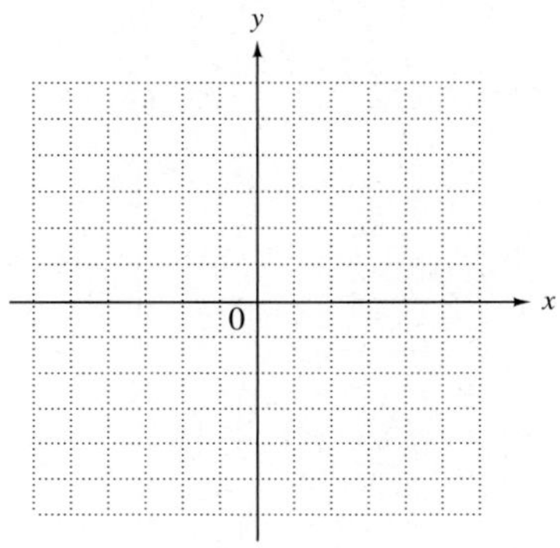

17. $f(x) = \sqrt{x + 3} - 3$

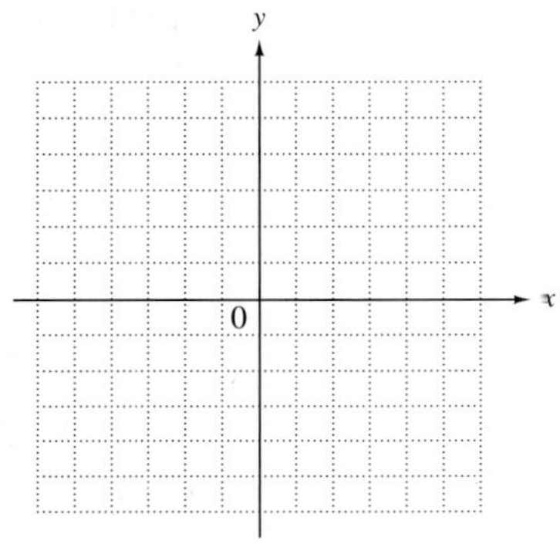

18. Explain how the graph of $f(x) = \frac{1}{x - 3} + 2$ is obtained from the graph of $g(x) = \frac{1}{x}$.

Graph each step function. See Examples 4 and 5.

19. $f(x) = [\![-x]\!]$

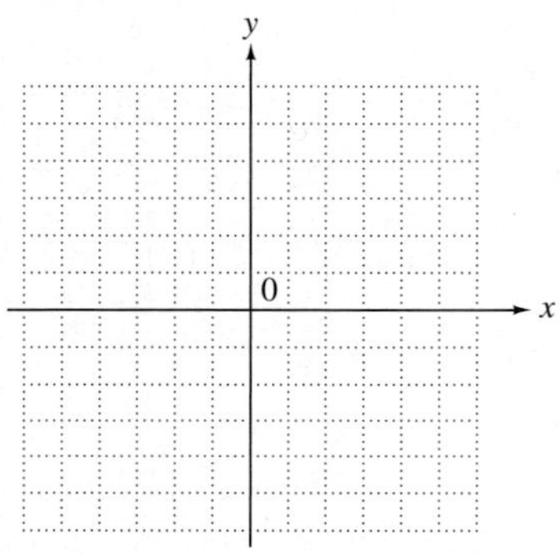

20. $g(x) = [\![x + 2]\!]$

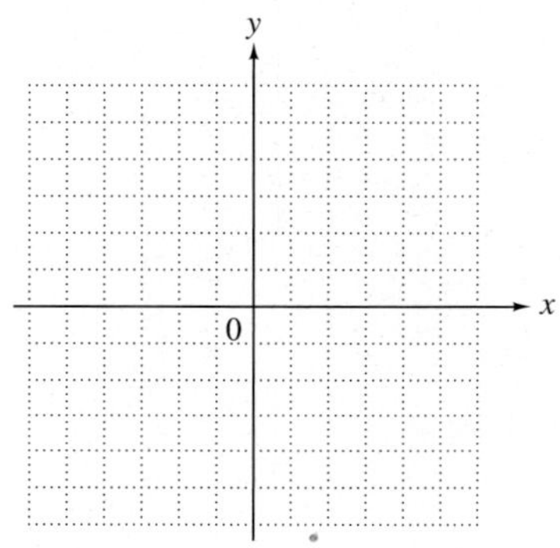

21. $f(x) = [\![x - 3]\!]$

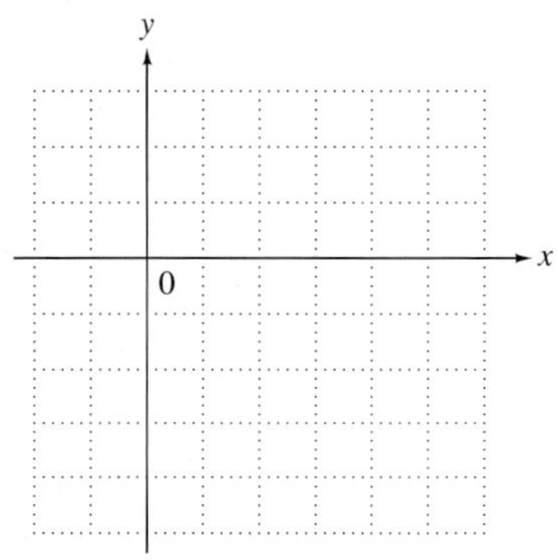

22. The cost of parking a car at an airport hourly parking lot is \$3 for the first half-hour and \$2 for each additional half-hour or fraction thereof. Graph the function defined by $f(x) =$ the cost of parking a car for x hours. Use the interval $(0, 2]$.

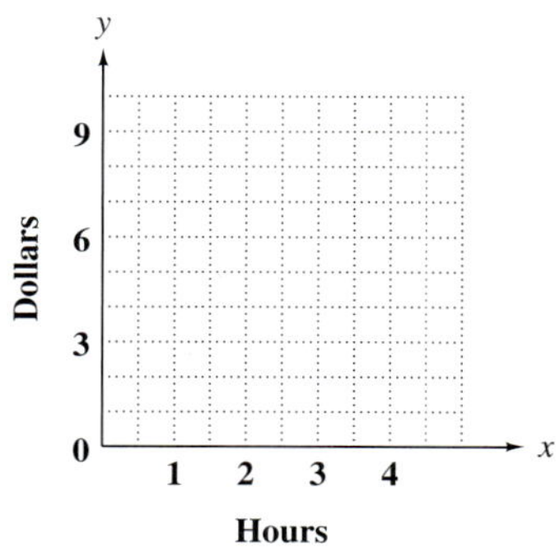

23. Assume that postage rates are 42¢ for the first ounce, plus 17¢ for each additional ounce, and that each letter carries one 42¢ stamp and as many 17¢ stamps as necessary. Graph the function defined by $p(x) =$ the number of stamps on a letter weighing x ounces. Use the interval $(0, 5]$.

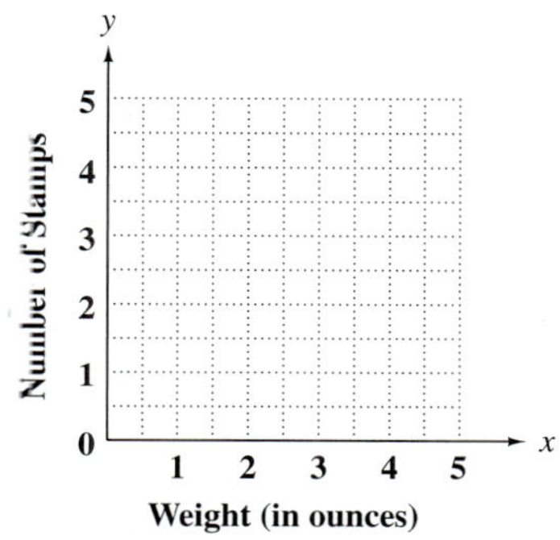

24. A certain long-distance carrier provides service between Podunk and Nowhereville. If x represents the number of minutes for the call, where $x > 0$, then the function f defined by

$$f(x) = 0.40[\![x]\!] + 0.75$$

gives the total cost of the call in dollars. Find the cost of a 5.5-min call.

Let $f(x) = x^2 + 4$, $g(x) = 2x + 3$, *and* $h(x) = x + 5$. *Find each value or expression. See Examples 6 and 7.*

25. $(h \circ g)(4)$

26. $(f \circ g)(4)$

27. $(g \circ f)(6)$

28. $(h \circ f)(6)$

29. $(f \circ h)(-2)$

30. $(h \circ g)(-2)$

31. $(f \circ g)(x)$

32. $(g \circ h)(x)$

33. $(f \circ h)(x)$

34. $(g \circ f)(x)$

35. $(h \circ g)(x)$

36. $(h \circ f)(x)$

37. $(f \circ h)\left(\frac{1}{2}\right)$

38. $(h \circ f)\left(\frac{1}{2}\right)$

39. $(f \circ g)\left(-\frac{1}{2}\right)$

40. $(g \circ f)\left(-\frac{1}{2}\right)$

Solve each problem.

41. The function defined by

$$f(x) = 12x$$

computes the number of inches in x feet and the function defined by

$$g(x) = 5280x$$

computes the number of feet in x miles. What is $(f \circ g)(x)$ and what does it compute?

42. The perimeter x of a square with sides of length s is given by the formula $x = 4s$.

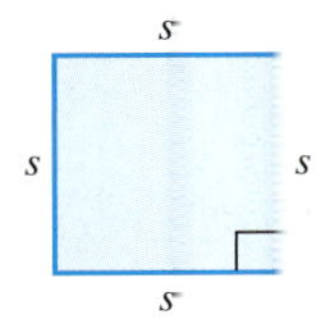

(a) Solve for s in terms of x.

(b) If y represents the area of this square, write y as a function of the perimeter x.

(c) Use the composite function of part (b) to find the area of a square with perimeter 6.

43. When a thermal inversion layer is over a city (as happens often in Los Angeles), pollutants cannot rise vertically but are trapped below the layer and must disperse horizontally. Assume that a factory smokestack begins emitting a pollutant at 8 A.M. Assume that the pollutant disperses horizontally over a circular area. Suppose that t represents the time, in hours, since the factory began emitting pollutants ($t = 0$ represents 8 A.M.), and assume that the radius of the circle of pollution is $r(t) = 2t$ miles. Let $A(r) = \pi r^2$ represent the area of a circle of radius r. Find and interpret $(A \circ r)(t)$.

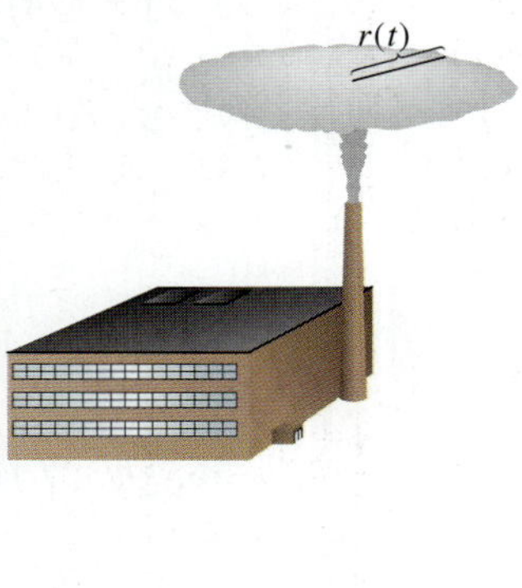

44. An oil well off the Gulf Coast is leaking, with the leak spreading oil over the surface as a circle. At any time t, in minutes, after the beginning of the leak, the radius of the circular oil slick on the surface is $r(t) = 4t$ feet. Let $A(r) = \pi r^2$ represent the area of a circle of radius r. Find and interpret $(A \circ r)(t)$.

12.2 The Circle and the Ellipse

When an infinite cone is intersected by a plane, the resulting figure is called a **conic section.** The parabola is one example of a conic section; *circles, ellipses,* and *hyperbolas* may also result. See Figure 11.

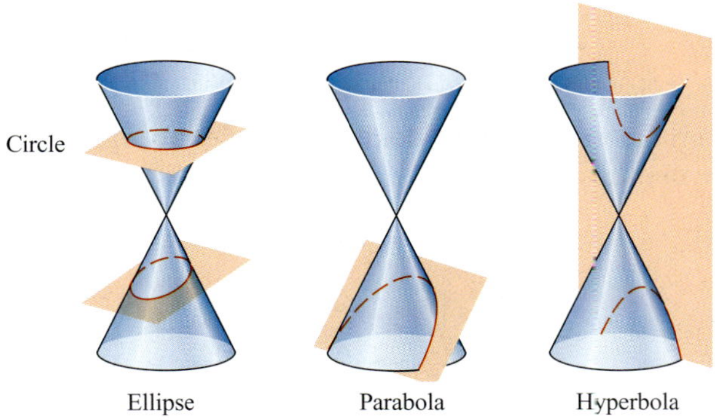

Figure 11

OBJECTIVES

1. Find an equation of a circle given the center and radius.
2. Determine the center and radius of a circle given its equation.
3. Recognize the equation of an ellipse.
4. Graph ellipses.

OBJECTIVE 1 Find an equation of a circle given the center and radius. A **circle** is the set of all points in a plane that lie a fixed distance from a fixed point. The fixed point is called the **center,** and the fixed distance is called the **radius.** We use the distance formula from **Section 9.3** to find an equation of a circle.

EXAMPLE 1 Finding an Equation of a Circle and Graphing It

Find an equation of the circle with radius 3 and center at (0, 0), and graph it.

If the point (x, y) is on the circle, the distance from (x, y) to the center (0, 0) is 3. By the distance formula,

$$\sqrt{(x_2 - x_1)^2 + (y_2 - y_1)^2} = d \quad \text{Distance formula}$$

$$\sqrt{(x - 0)^2 + (y - 0)^2} = 3 \quad \text{Let } x_1 = 0, y_1 = 0, \text{ and } d = 3.$$

$$x^2 + y^2 = 9. \quad \text{Square both sides.}$$

An equation of this circle is $x^2 + y^2 = 9$. The graph is shown in Figure 12.

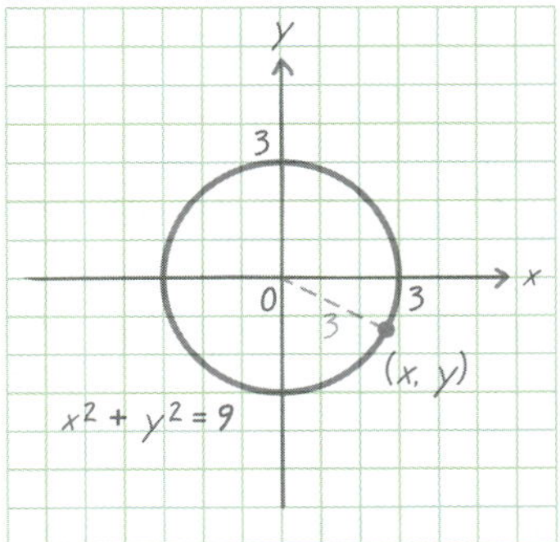

Figure 12

Work Problem 1 at the Side. ▶

1 Find an equation of the circle with radius 4 and center (0, 0). Sketch its graph.

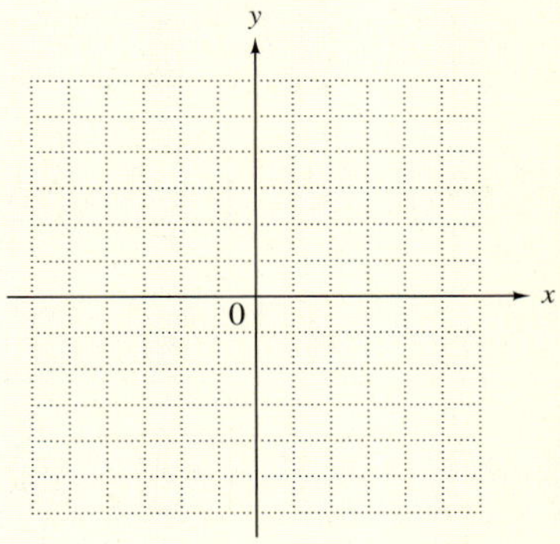

ANSWER

1. $x^2 + y^2 = 16$

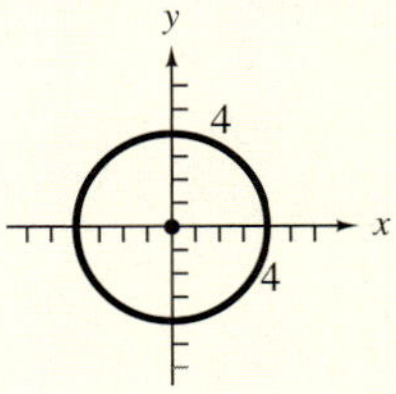

2 **(a)** Find an equation of the circle with center at $(3, -2)$ and radius 4. Graph the circle.

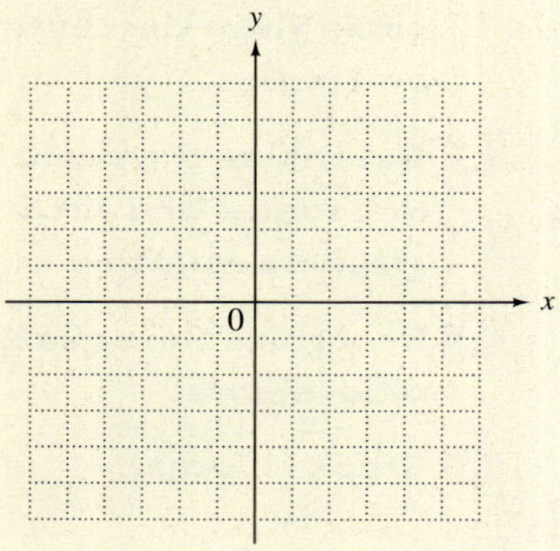

(b) Use the center-radius form to determine the center and radius of $(x-5)^2+(y+2)^2=9$, and then graph the circle.

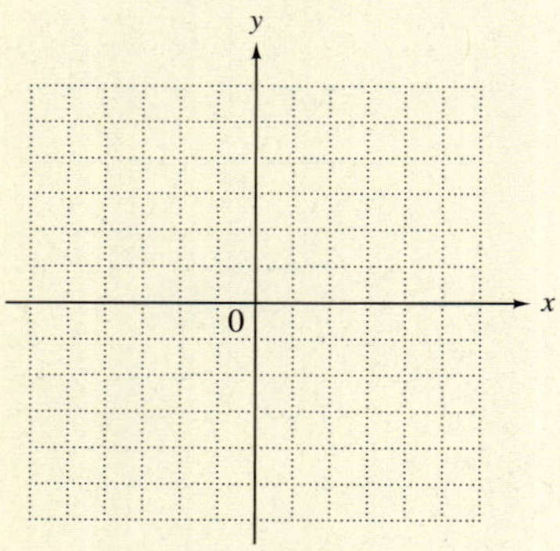

A circle may not be centered at the origin, as seen in the next example.

EXAMPLE 2 Finding an Equation of a Circle and Graphing It

Find an equation of the circle with center at $(4, -3)$ and radius 5, and graph it.

$$\sqrt{(x-4)^2+[y-(-3)]^2}=5 \quad \text{Distance formula}$$

$$(x-4)^2+(y+3)^2=25 \quad \text{Square both sides.}$$

To graph the circle, plot the center $(4, -3)$, then move 5 units right, left, up, and down from the center. Draw a smooth curve through these four points, sketching one quarter of the circle at a time. The graph of this circle is shown in Figure 13.

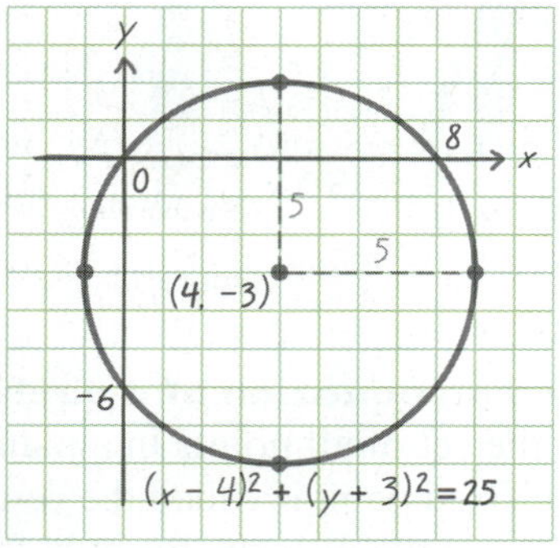

Figure 13

Examples 1 and 2 suggest the form of an equation of a circle with radius r and center at (h, k). If (x, y) is a point on the circle, then the distance from the center (h, k) to the point (x, y) is r. By the distance formula,

$$\sqrt{(x-h)^2+(y-k)^2}=r.$$

Squaring both sides gives us the **center-radius form** of the equation of a circle.

Equation of a Circle (Center-Radius Form)

An equation of a circle of radius r with center at (h, k) is

$$(x-h)^2+(y-k)^2=r^2.$$

EXAMPLE 3 Using the Center-Radius Form of the Equation of a Circle

Find an equation of the circle with center at $(-1, 2)$ and radius 4.

Use the center-radius form, with $h=-1$, $k=2$, and $r=4$.

$$(x-h)^2+(y-k)^2=r^2$$

$$[x-(-1)]^2+(y-2)^2=4^2$$

Pay attention to signs here.

$$(x+1)^2+(y-2)^2=16$$

◀ Work Problem **2** at the Side.

ANSWERS

2. (a) $(x-3)^2+(y+2)^2=16$

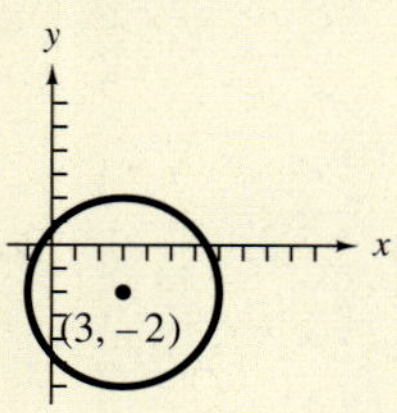

(b) center at $(5, -2)$; radius 3

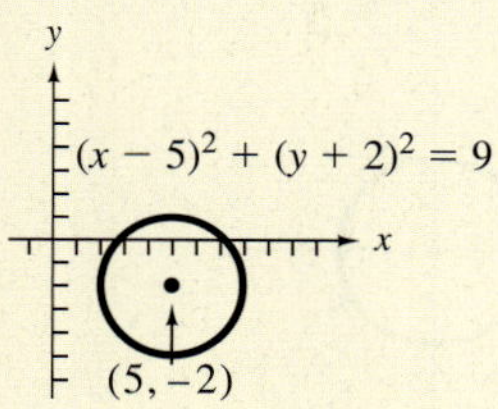

OBJECTIVE 2 Determine the center and radius of a circle given its equation. In the equation found in Example 2, multiplying out $(x-4)^2$ and $(y+3)^2$ and then combining like terms gives

$$(x-4)^2+(y+3)^2=25$$
$$x^2-8x+16+y^2+6y+9=25$$
$$x^2+y^2-8x+6y=0.$$

This general form suggests that an equation with both x^2- and y^2-terms with equal coefficients may represent a circle. The next example shows how to tell, by completing the square. This procedure was introduced in **Section 10.1.**

EXAMPLE 4 Completing the Square to Find the Center and Radius

Graph $x^2+y^2+2x+6y-15=0$.

Since the equation has x^2- and y^2-terms with equal coefficients, its graph might be that of a circle. To find the center and radius, complete the squares on x and y.

$$x^2+y^2+2x+6y=15 \quad \text{Transform so that the constant is on the right.}$$
$$(x^2+2x\quad)+(y^2+6y\quad)=15 \quad \text{Rewrite in anticipation of completing the square.}$$
$$\left[\frac{1}{2}(2)\right]^2=1 \qquad \left[\frac{1}{2}(6)\right]^2=9 \quad \text{Square half the coefficient of each middle term.}$$
$$(x^2+2x+1)+(y^2+6y+9)=15+1+9 \quad \text{Complete the squares on both } x \text{ and } y.$$
$$(x+1)^2+(y+3)^2=25 \quad \text{Factor on the left; add on the right.}$$
$$[x-(-1)]^2+[y-(-3)]^2=5^2 \quad \text{Center-radius form}$$

The final equation shows that the graph is a circle with center at $(-1,-3)$ and radius 5. The graph is shown in Figure 14.

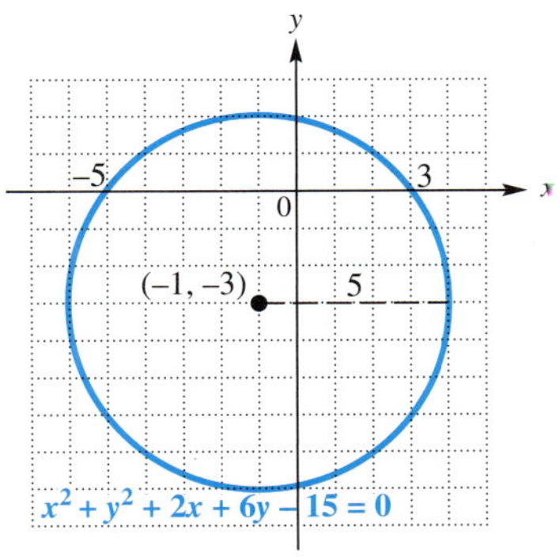

Figure 14

Note

If the procedure of Example 4 leads to an equation of the form $(x-h)^2+(y-k)^2=0$, then the graph is the single point (h,k). If the constant on the right side is negative, then the equation has no graph.

Work Problem 3 at the Side. ▶

3 Find the center and radius of the circle with equation

$$x^2+y^2-6x+8y-11=0.$$

ANSWER

3. center at $(3,-4)$; radius 6

OBJECTIVE 3 Recognize the equation of an ellipse. An ellipse is the set of all points in a plane the *sum* of whose distances from two fixed points is constant. These fixed points are called **foci** (singular: *focus*). Figure 15 shows an ellipse whose foci are $(c, 0)$ and $(-c, 0)$, with x-intercepts $(a, 0)$ and $(-a, 0)$ and y-intercepts $(0, b)$ and $(0, -b)$. It can be shown in more advanced courses that $c^2 = a^2 - b^2$ for an ellipse of this type. The origin is the **center** of the ellipse.

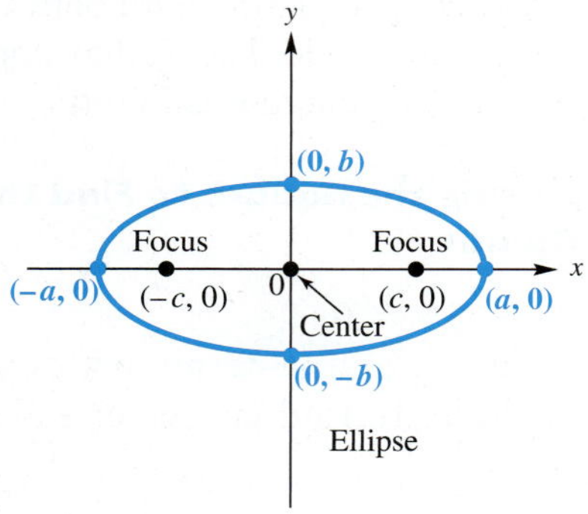

Figure 15

An ellipse has the following equation.

Equation of an Ellipse

The ellipse whose x-intercepts are $(a, 0)$ and $(-a, 0)$ and whose y-intercepts are $(0, b)$ and $(0, -b)$ has an equation of the form

$$\frac{x^2}{a^2} + \frac{y^2}{b^2} = 1.$$

Note

A circle is a special case of an ellipse, where $a^2 = b^2$.

When a ray of light or sound emanating from one focus of an ellipse bounces off the ellipse, it passes through the other focus. See the figure. As mentioned in the chapter introduction, this reflecting property is responsible for whispering galleries. John Quincy Adams was able to listen in on his opponents' conversations because his desk was positioned at one of the foci beneath the ellipsoidal ceiling and his opponents were located across the room at the other focus.

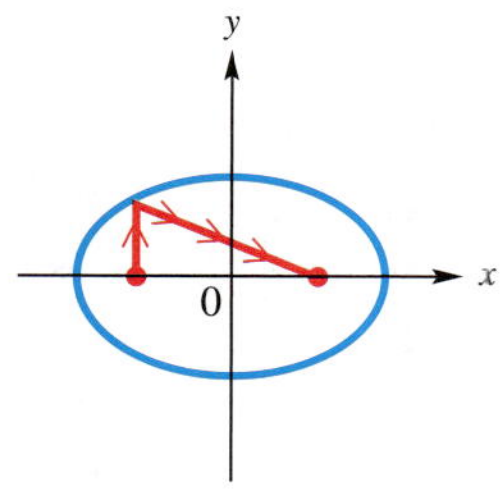

Reflecting property of an ellipse

The paths of Earth and other planets around the sun are approximately ellipses; the sun is at one focus and a point in space is at the other. The orbits of communication satellites and other space vehicles are also elliptical.

Elliptical bicycle gears are designed to respond to the legs' natural strengths and weaknesses. At the top and bottom of the powerstroke, where the legs have the least leverage, the gear offers little resistance, but as the gear rotates, the resistance increases. This allows the legs to apply more power where it is most naturally available. See Figure 16 on the next page.

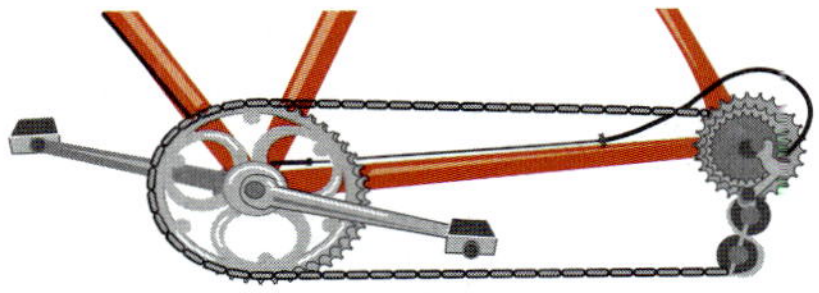

Figure 16

OBJECTIVE 4 Graph ellipses. To graph an ellipse centered at the origin, we plot the four intercepts and then sketch the ellipse through those points.

EXAMPLE 5 Graphing Ellipses

Graph each ellipse.

(a) $\frac{x^2}{49} + \frac{y^2}{36} = 1$

Here, $a^2 = 49$, so $a = 7$, and the x-intercepts for this ellipse are $(7, 0)$ and $(-7, 0)$. Similarly, $b^2 = 36$, so $b = 6$, and the y-intercepts are $(0, 6)$ and $(0, -6)$. Plotting the intercepts and sketching the ellipse through them gives the graph in Figure 17.

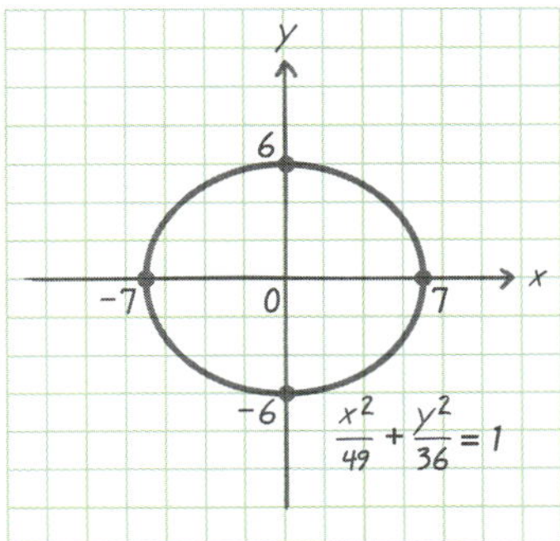

Figure 17

(b) $\frac{x^2}{36} + \frac{y^2}{121} = 1$

The x-intercepts for this ellipse are $(6, 0)$ and $(-6, 0)$, and the y-intercepts are $(0, 11)$ and $(0, -11)$. Join these intercepts with the smooth curve of an ellipse. The graph has been sketched in Figure 18.

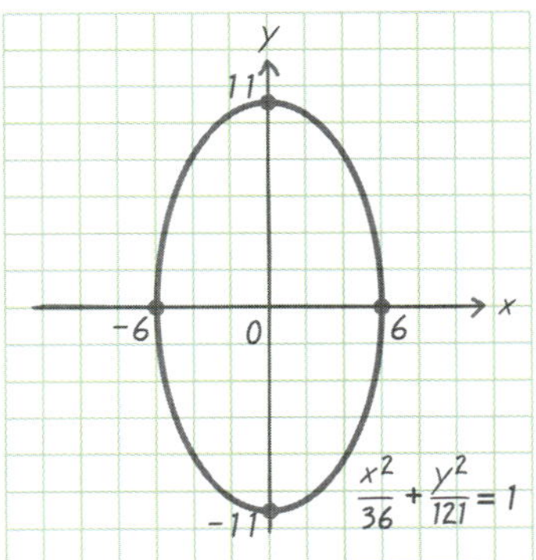

Figure 18

Work Problem 4 at the Side. ▶

4 Graph each ellipse.

(a) $\frac{x^2}{4} + \frac{y^2}{25} = 1$

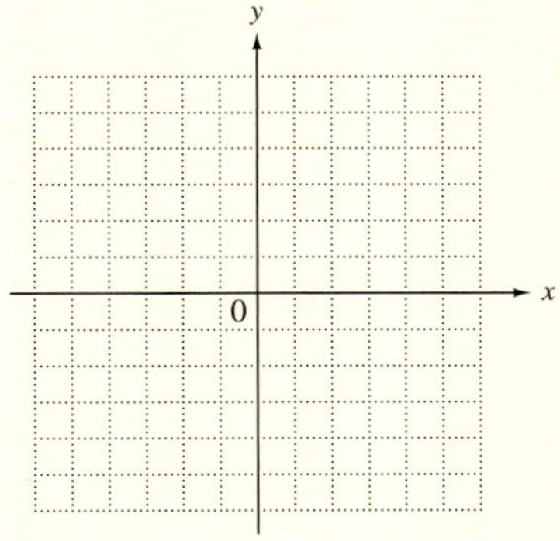

(b) $\frac{x^2}{64} + \frac{y^2}{49} = 1$

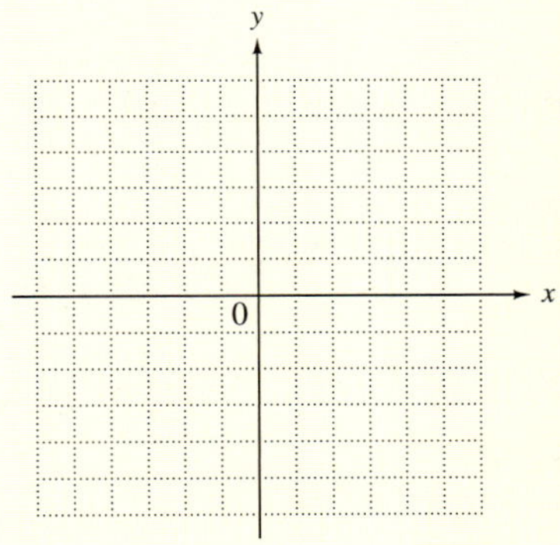

ANSWERS

4. (a)

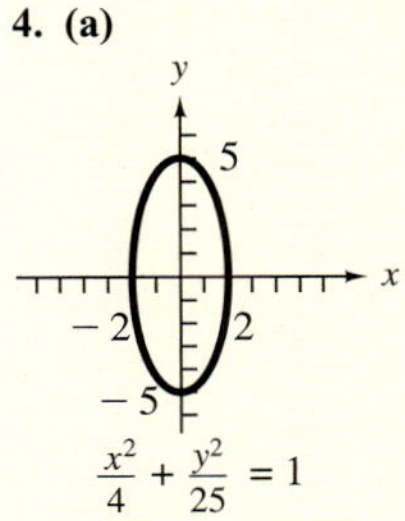

(b)

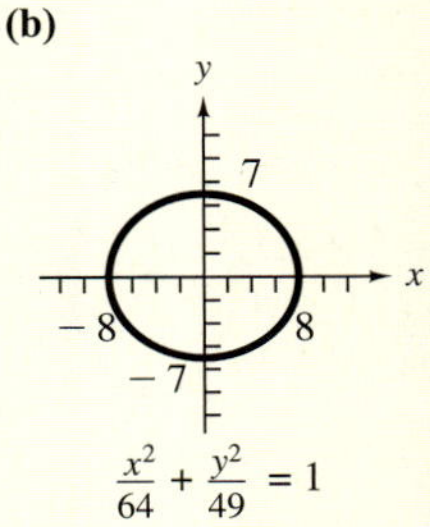

5 Graph

$$\frac{(x+4)^2}{16} + \frac{(y-1)^2}{36} = 1.$$

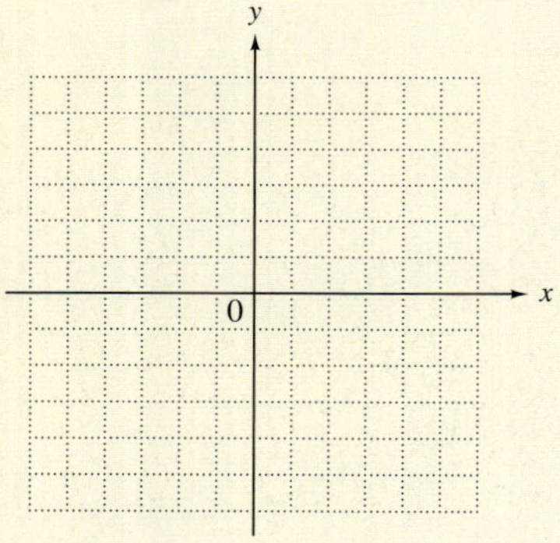

As with the graphs of parabolas and circles, the graph of an ellipse may be shifted horizontally and vertically, as in the next example.

EXAMPLE 6 Graphing an Ellipse Shifted Horizontally and Vertically

Graph $\frac{(x-2)^2}{25} + \frac{(y+3)^2}{49} = 1.$

Just as $(x - 2)^2$ and $(y + 3)^2$ would indicate that the center of a circle would be $(2, -3)$, so it is with this ellipse. Figure 19 shows that the graph goes through the four points $(2, 4)$, $(7, -3)$, $(2, -10)$, and $(-3, -3)$. The x-values of these points are found by adding $\pm a = \pm 5$ to 2, and the y-values come from adding $\pm b = \pm 7$ to -3.

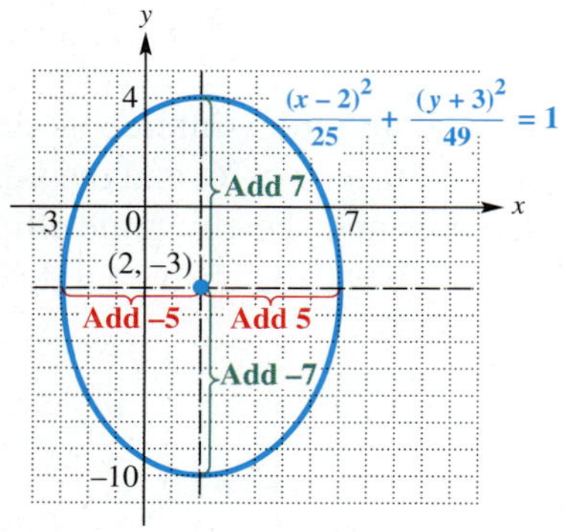

Figure 19

◀ Work Problem **5** at the Side.

Note

The graphs in this section are not graphs of functions. The only conic section whose graph is a function is the vertical parabola with equation $f(x) = ax^2 + bx + c$.

ANSWER

5.

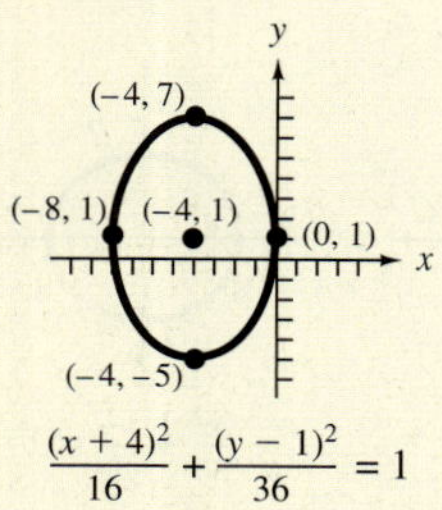

12.2 Exercises

FOR EXTRA HELP MyMathLab MathXL PRACTICE WATCH DOWNLOAD READ REVIEW

1. See Example 1. Consider the circle whose equation is $x^2 + y^2 = 25$.

(a) What are the coordinates of its center?

(b) What is its radius?

(c) Sketch its graph.

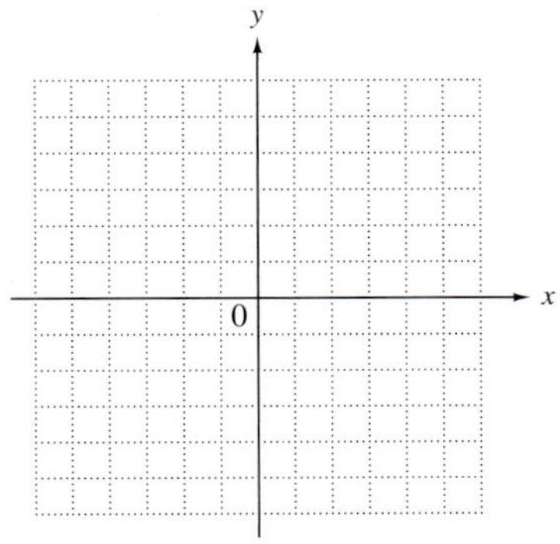

2. Explain why a set of points defined by a circle does not satisfy the definition of a function.

Match each equation with the correct graph. See Examples 1–3.

3. $(x - 3)^2 + (y - 2)^2 = 25$

4. $(x - 3)^2 + (y + 2)^2 = 25$

5. $(x + 3)^2 + (y - 2)^2 = 25$

6. $(x + 3)^2 + (y + 2)^2 = 25$

A.

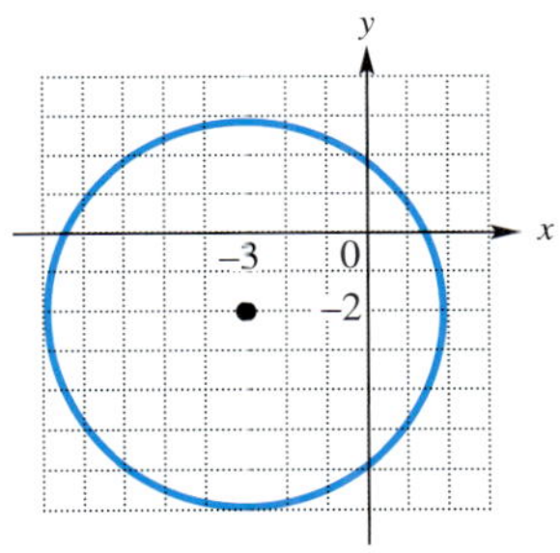

B.

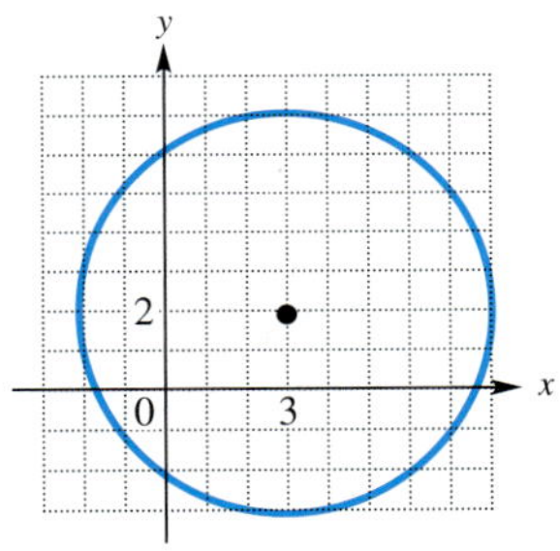

C.

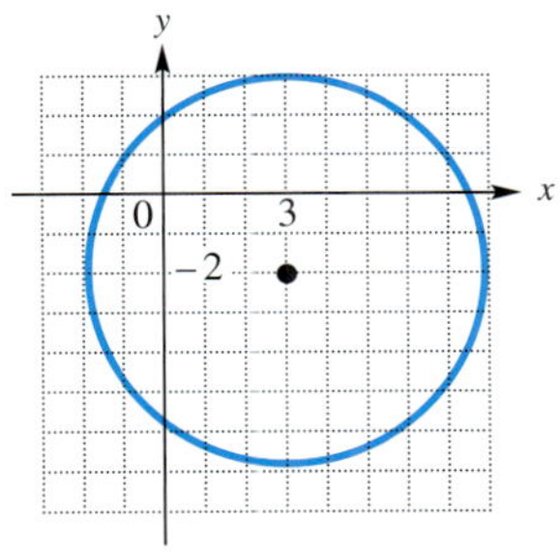

D.

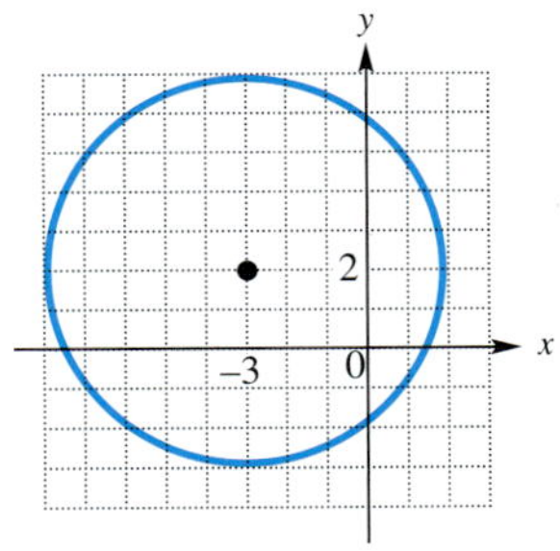

Find the equation of a circle satisfying the given conditions. See Examples 2 and 3.

7. Center: $(-4, 3)$; radius: 2

8. Center: $(5, -2)$; radius: 4

9. Center: $(-8, -5)$; radius: $\sqrt{5}$

10. Center: $(-12, 13)$; radius: $\sqrt{7}$

Find the center and radius of each circle. (Hint: In Exercises 15 and 16, divide each side by a common factor.) See Example 4.

11. $x^2 + y^2 + 4x + 6y + 9 = 0$

12. $x^2 + y^2 - 8x - 12y + 3 = 0$

13. $x^2 + y^2 + 10x - 14y - 7 = 0$

14. $x^2 + y^2 - 2x + 4y - 4 = 0$

15. $3x^2 + 3y^2 - 12x - 24y + 12 = 0$

16. $2x^2 + 2y^2 + 20x + 16y + 10 = 0$

17. A circle can be drawn on a piece of posterboard by fastening one end of a string with a thumbtack, pulling the string taut with a pencil, and tracing a curve, as shown in the figure. Explain why this method works.

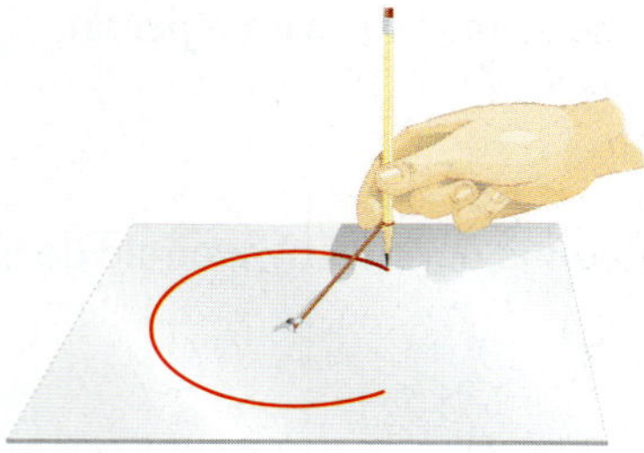

Graph each circle. See Examples 1–4.

18. $x^2 + y^2 = 9$

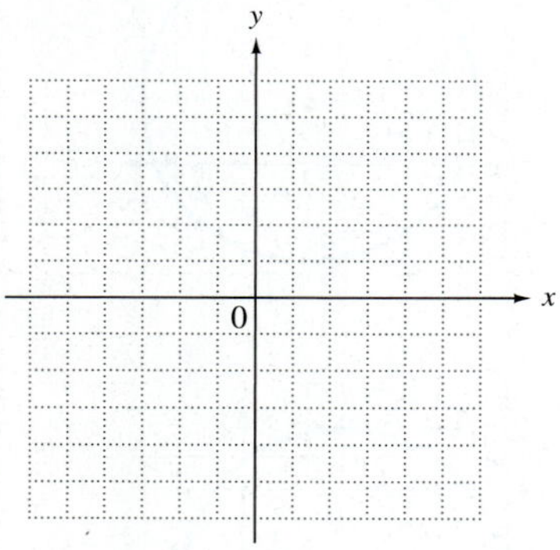

19. $x^2 + y^2 = 4$

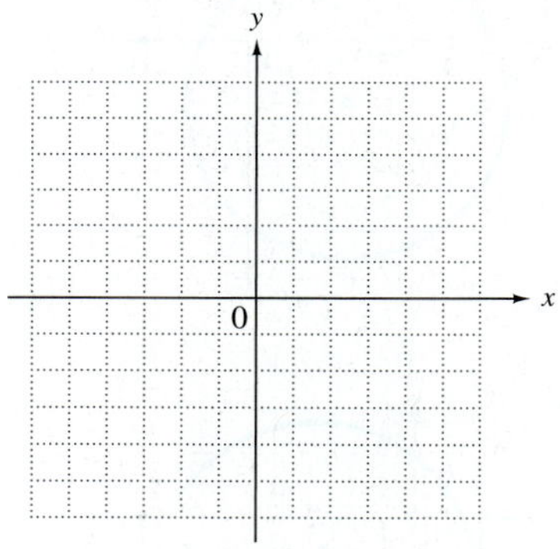

20. $2y^2 = 10 - 2x^2$

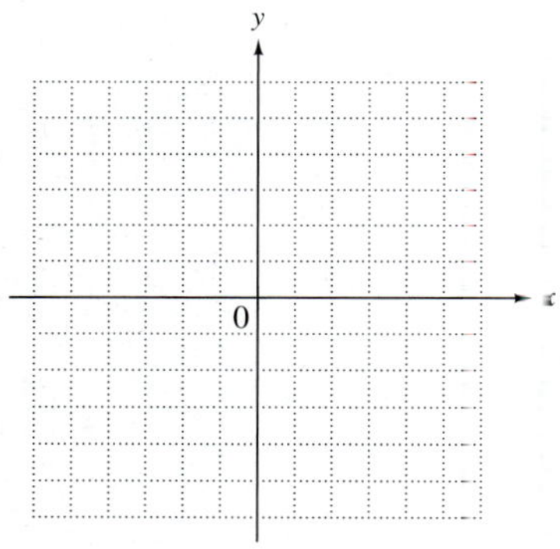

21. $3x^2 = 48 - 3y^2$

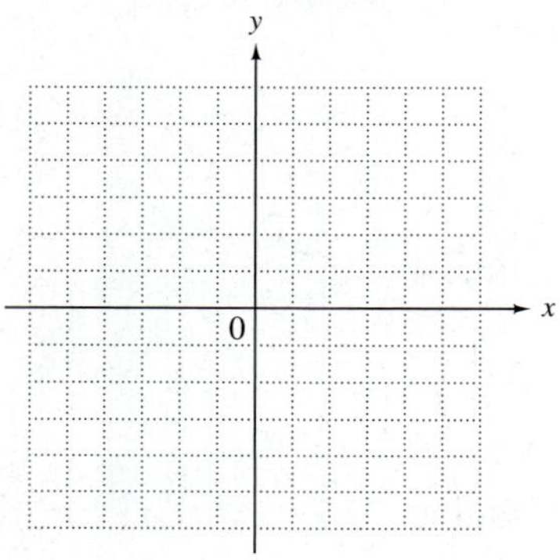

22. $(x + 3)^2 + (y - 2)^2 = 9$

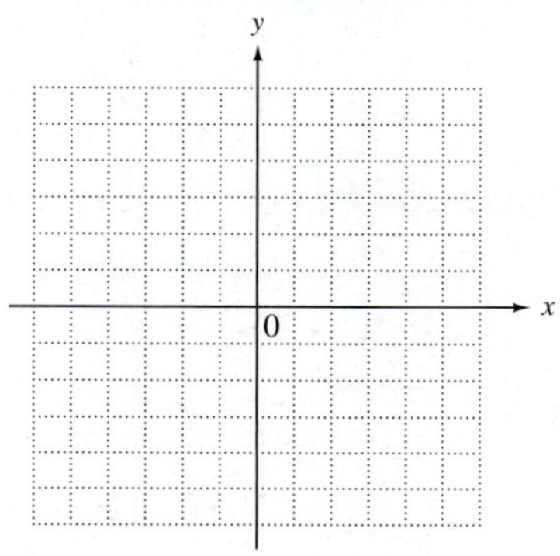

23. $(x - 1)^2 + (y + 3)^2 = 16$

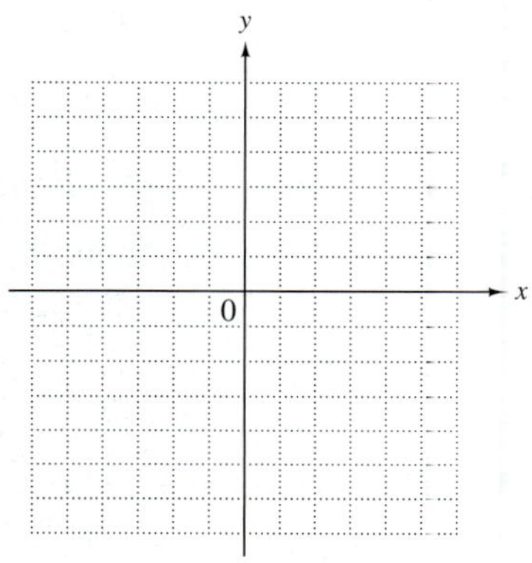

24. $x^2 + y^2 - 4x - 6y + 9 = 0$

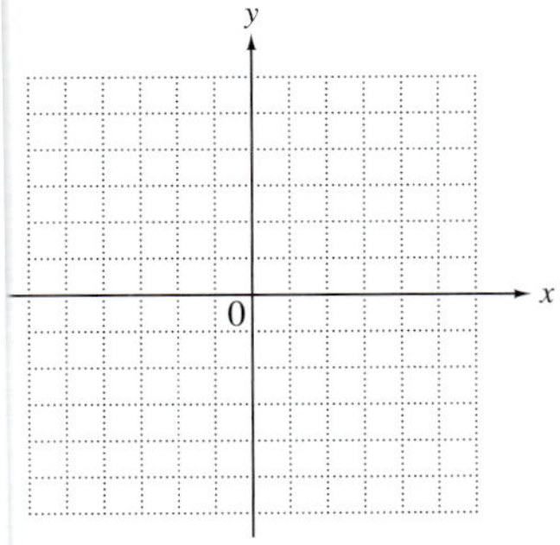

25. $x^2 + y^2 + 8x + 2y - 8 = 0$

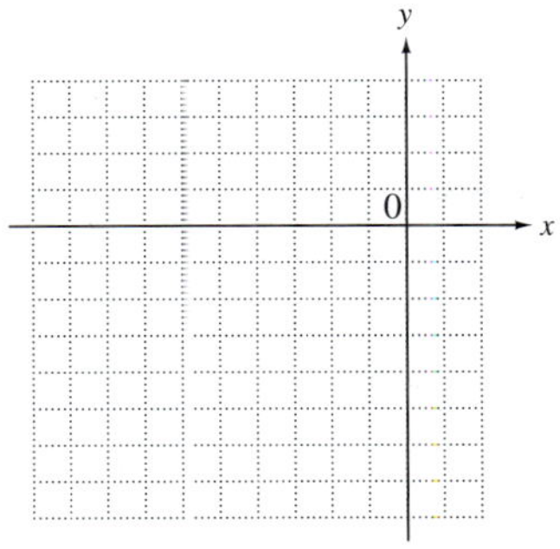

26. $x^2 + y^2 - 4x + 10y + 20 = 0$

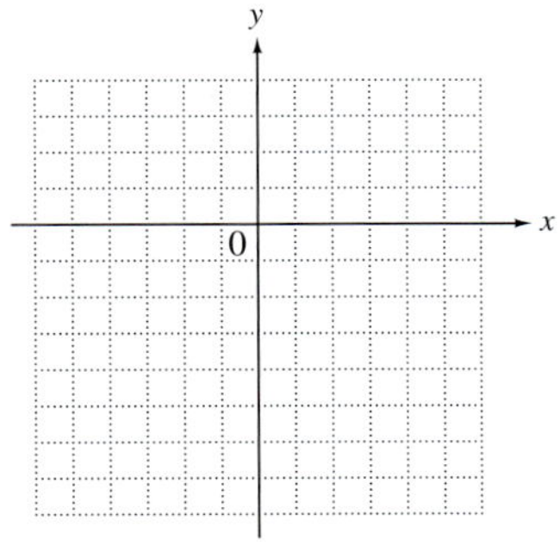

27. An ellipse can be drawn on a piece of posterboard by fastening two ends of a length of string with thumbtacks, pulling the string taut with a pencil, and tracing a curve, as shown in the figure. Explain why this method works.

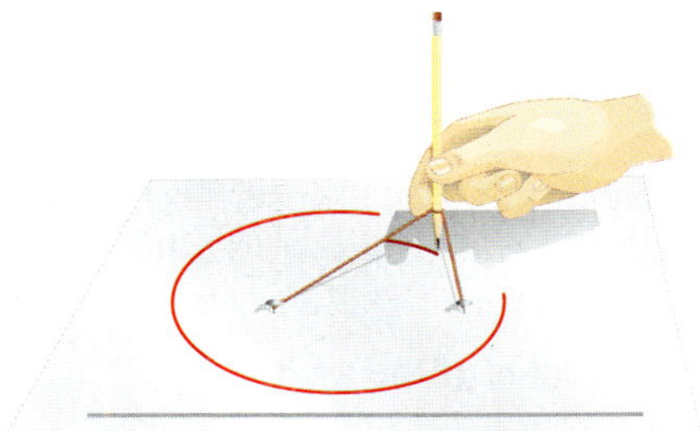

Graph each ellipse. See Examples 5 and 6.

28. $\frac{x^2}{9} + \frac{y^2}{16} = 1$

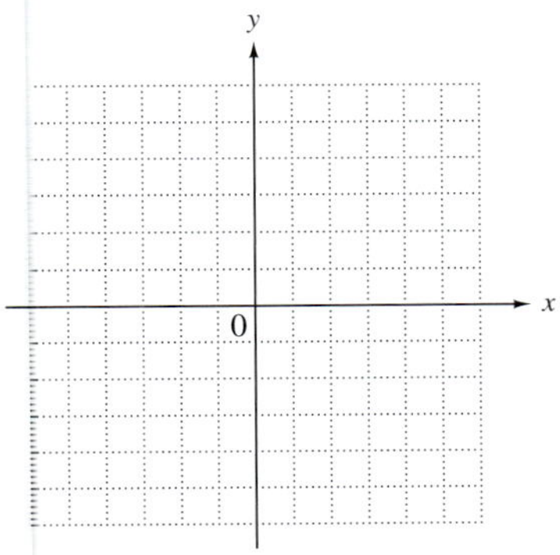

29. $\frac{x^2}{9} + \frac{y^2}{25} = 1$

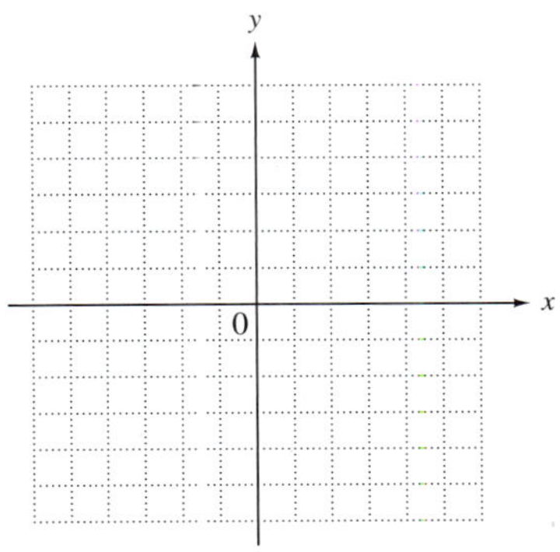

30. $\frac{x^2}{9} + \frac{y^2}{4} = 1$

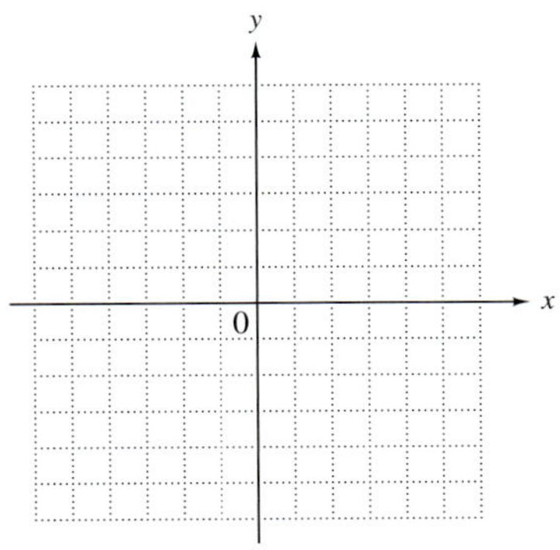

31. $\frac{x^2}{36} + \frac{y^2}{16} = 1$

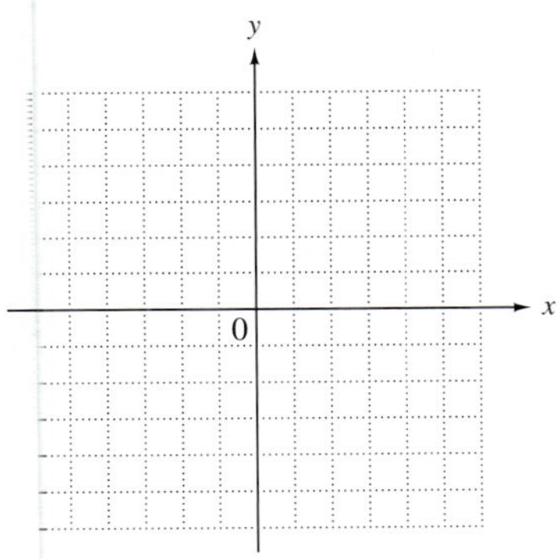

32. $\frac{x^2}{16} + \frac{y^2}{9} = 1$

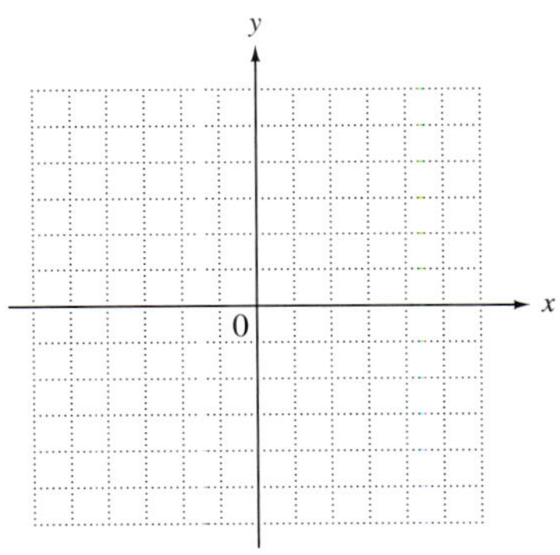

33. $\frac{x^2}{49} + \frac{y^2}{25} = 1$

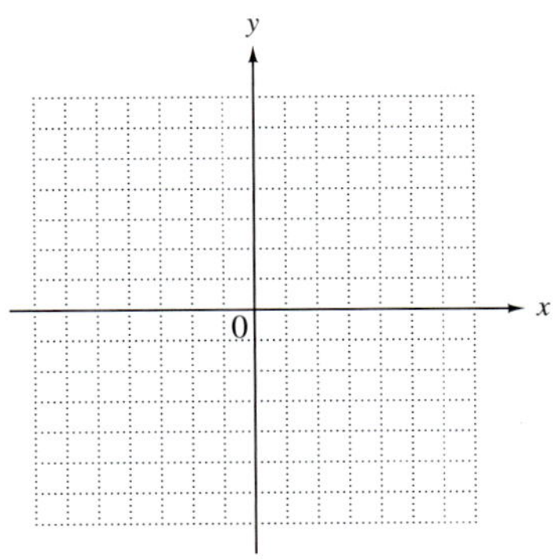

34. $\dfrac{(x-4)^2}{9} + \dfrac{(y+2)^2}{4} = 1$

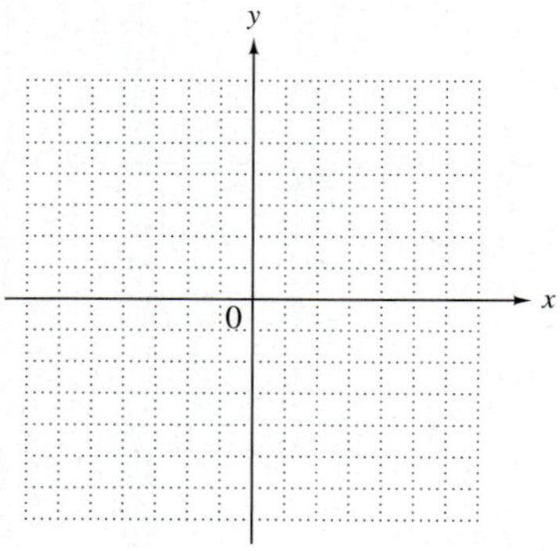

35. $\dfrac{(x-2)^2}{16} + \dfrac{(y-1)^2}{9} = 1$

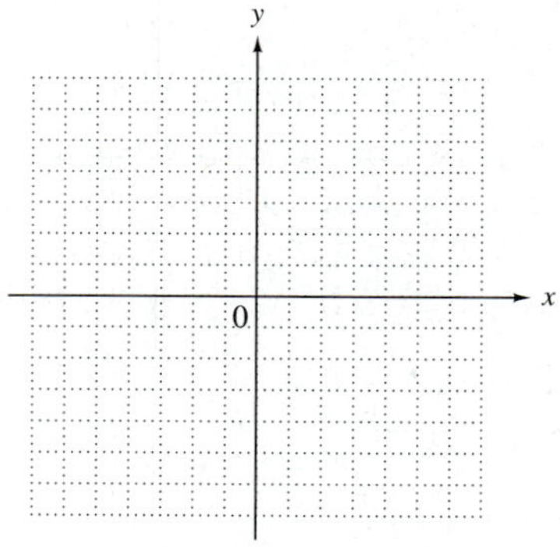

36. $\dfrac{(x+3)^2}{25} + \dfrac{(y+2)^2}{36} = 1$

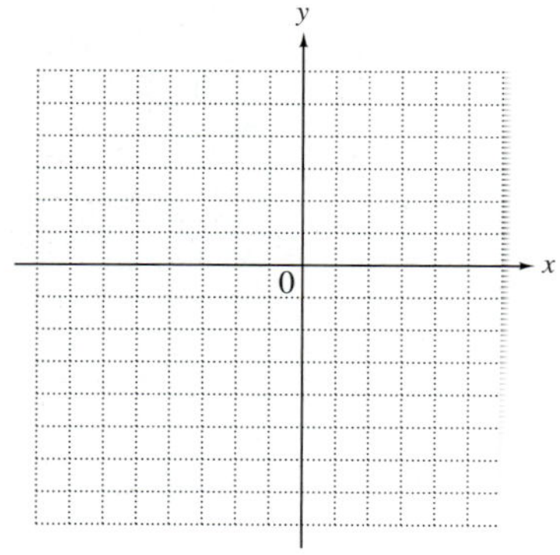

Solve each problem.

37. An arch has the shape of half an ellipse. The equation of the ellipse is $100x^2 + 324y^2 = 32{,}400$, where x and y are in meters.

(a) How high is the center of the arch?

(b) How wide is the arch across the bottom?

NOT TO SCALE

38. A one-way street passes under an overpass, which is in the form of the top half of an ellipse, as shown in the figure. Suppose that a truck 12 ft wide passes directly under the overpass. What is the maximum possible height of this truck?

NOT TO SCALE

In Exercises 39 and 40, see Figure 15 and use the fact that $c^2 = a^2 - b^2$, where $a^2 > b^2$.

39. The orbit of Mars is an ellipse with the sun at one focus. For x and y in millions of miles, the equation of the orbit is

$$\frac{x^2}{141.7^2} + \frac{y^2}{141.1^2} = 1.$$

(*Source:* Kaler, James B., *Astronomy!*, Addison-Wesley, 1997.)

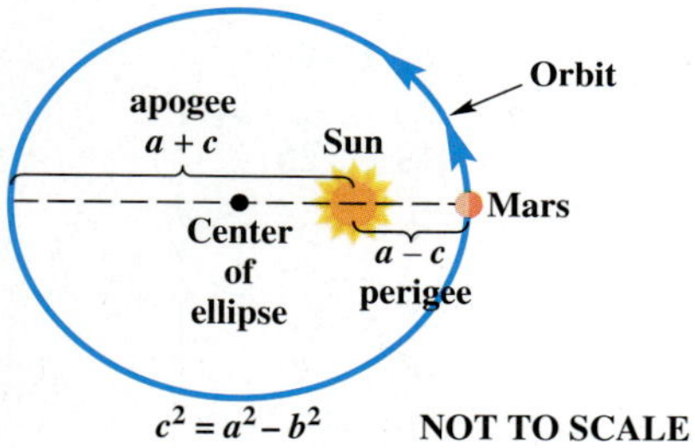

NOT TO SCALE

(a) Find the greatest distance (the **apogee**) from Mars to the sun.

(b) Find the least distance (the **perigee**) from Mars to the sun.

40. The orbit of Venus around the sun (one of the foci) is an ellipse with equation

$$\frac{x^2}{5013} + \frac{y^2}{4970} = 1,$$

where x and y are measured in millions of miles. (*Source:* Kaler, James B., *Astronomy!*, Addison-Wesley, 1997.)

(a) Find the greatest distance between Venus and the sun.

(b) Find the least distance between Venus and the sun.

12.3 The Hyperbola and Other Functions Defined by Radicals

OBJECTIVES

1. Recognize the equation of a hyperbola.
2. Graph hyperbolas by using asymptotes.
3. Identify conic sections by their equations.
4. Graph certain square root functions.

OBJECTIVE 1 Recognize the equation of a hyperbola. A **hyperbola** is the set of all points in a plane such that the absolute value of the *difference* of the distances from two fixed points (called *foci*) is constant. Figure 20 shows a hyperbola. Using the distance formula and the definition above, we can show that this hyperbola has equation $\frac{x^2}{16} - \frac{y^2}{12} = 1$.

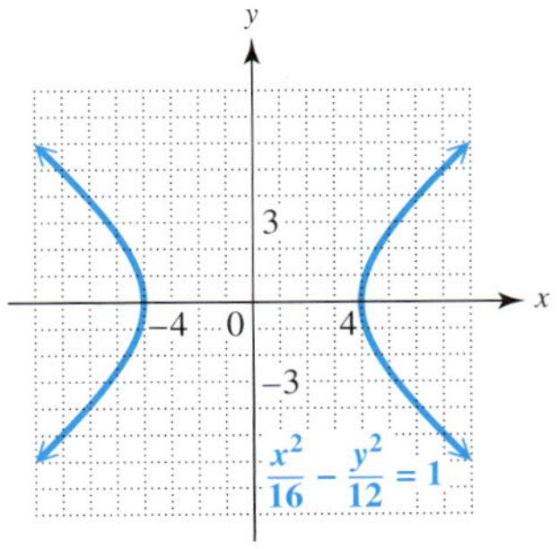

Figure 20

To graph hyperbolas centered at the origin, we need to find their intercepts. For the hyperbola in Figure 20, we proceed as follows.

x-Intercepts

Let $y = 0$.

$$\frac{x^2}{16} - \frac{0^2}{12} = 1 \quad \text{Let } y = 0.$$

$$\frac{x^2}{16} = 1$$

$$x^2 = 16 \quad \text{Multiply by 16.}$$

$$x = \pm 4$$

The x-intercepts of the graph are $(4, 0)$ and $(-4, 0)$.

y-Intercepts

Let $x = 0$.

$$\frac{0^2}{16} - \frac{y^2}{12} = 1 \quad \text{Let } x = 0.$$

$$-\frac{y^2}{12} = 1$$

$$y^2 = -12 \quad \text{Multiply by } -12.$$

Because there are no *real* solutions to $y^2 = -12$, the graph has no y-intercepts.

The graph of $\frac{x^2}{16} - \frac{y^2}{12} = 1$ in Figure 20 has no y-intercepts, while the hyperbola in Figure 21 has no x-intercepts. Its equation is $\frac{y^2}{25} - \frac{x^2}{9} = 1$, with y-intercepts $(0, 5)$ and $(0, -5)$.

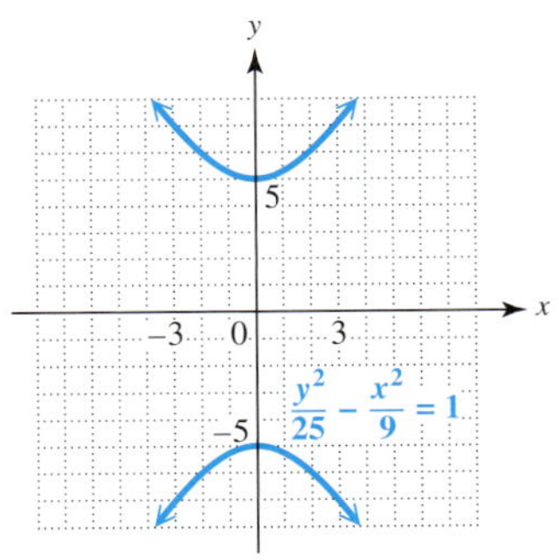

Figure 21

Equations of Hyperbolas

A hyperbola with x-intercepts $(a, 0)$ and $(-a, 0)$ has an equation of the form

$$\frac{x^2}{a^2} - \frac{y^2}{b^2} = 1,$$

and a hyperbola with y-intercepts $(0, b)$ and $(0, -b)$ has an equation of the form

$$\frac{y^2}{b^2} - \frac{x^2}{a^2} = 1.$$

OBJECTIVE 2 Graph hyperbolas by using asymptotes. The two branches of the graph of a hyperbola approach a pair of intersecting straight lines, which are its *asymptotes*. (See Figure 22 on the next page.) The asymptotes are useful for sketching the graph of the hyperbola.

Asymptotes of Hyperbolas

The extended diagonals of the rectangle with vertices (corners) at the points (a, b), $(-a, b)$, $(-a, -b)$, and $(a, -b)$ are the **asymptotes** of the hyperbolas

$$\frac{x^2}{a^2} - \frac{y^2}{b^2} = 1 \quad \text{and} \quad \frac{y^2}{b^2} - \frac{x^2}{a^2} = 1.$$

This rectangle is called the **fundamental rectangle.** Using the methods of **Chapter 4,** we could show that the equations of these asymptotes are

$$y = \frac{b}{a}x \quad \text{and} \quad y = -\frac{b}{a}x.$$

To graph hyperbolas, follow these steps.

Graphing a Hyperbola

Step 1 **Find the intercepts.** Locate the intercepts of the graph at $(a, 0)$ and $(-a, 0)$ if the x^2-term has a positive coefficient, or at $(0, b)$ and $(0, -b)$ if the y^2-term has a positive coefficient.

Step 2 **Find the fundamental rectangle.** Locate the vertices of the fundamental rectangle at (a, b), $(-a, b)$, $(-a, -b)$, and $(a, -b)$.

Step 3 **Sketch the asymptotes.** The extended diagonals of the rectangle are the asymptotes of the hyperbola. They have equations $y = \pm\frac{b}{a}x$.

Step 4 **Draw the graph.** Sketch each branch of the hyperbola through an intercept and approaching (but not touching) the asymptotes.

EXAMPLE 1 Graphing a Horizontal Hyperbola

Graph $\frac{x^2}{16} - \frac{y^2}{25} = 1$.

Step 1 Here $a = 4$ and $b = 5$. The x-intercepts are $(4, 0)$ and $(-4, 0)$.

Step 2 The four points $(4, 5)$, $(-4, 5)$, $(-4, -5)$, and $(4, -5)$ are the vertices of the fundamental rectangle, as shown in Figure 22.

Steps 3 and 4 The equations of the asymptotes are $y = \pm\frac{5}{4}x$. The hyperbola approaches these lines as x and y get larger in absolute value.

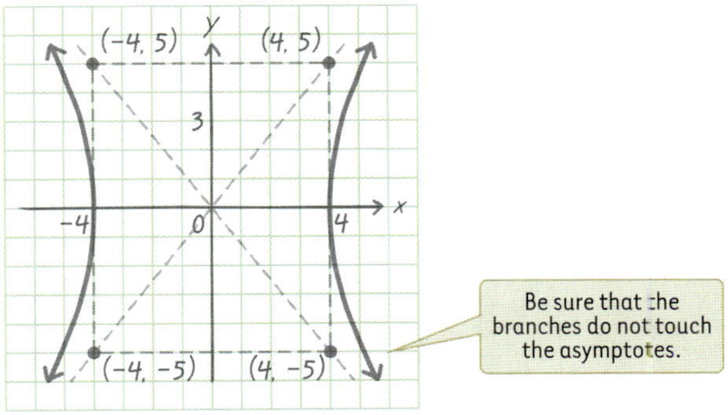

$\frac{x^2}{16} - \frac{y^2}{25} = 1$

Figure 22

Work Problem **1** at the Side. ▶

EXAMPLE 2 Graphing a Vertical Hyperbola

Graph $\frac{y^2}{49} - \frac{x^2}{16} = 1$.

This hyperbola has y-intercepts $(0, 7)$ and $(0, -7)$. The asymptotes are the extended diagonals of the rectangle with vertices at $(4, 7)$, $(-4, 7)$, $(-4, -7)$, and $(4, -7)$. Their equations are $y = \pm\frac{7}{4}x$. See Figure 23.

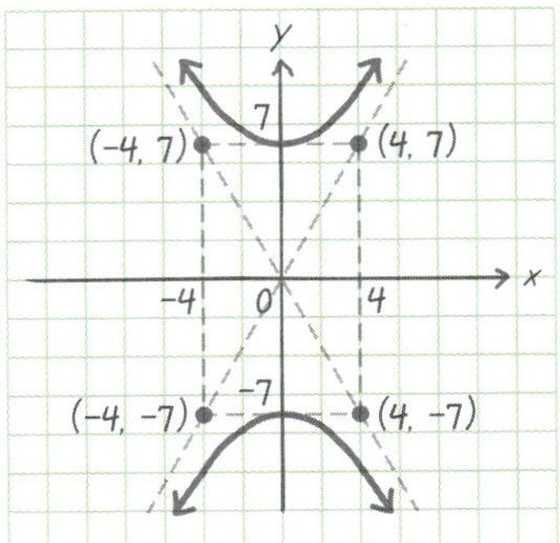

$\frac{y^2}{49} - \frac{x^2}{16} = 1$

Figure 23

Work Problem **2** at the Side. ▶

1 Graph $\frac{x^2}{4} - \frac{y^2}{25} = 1$.

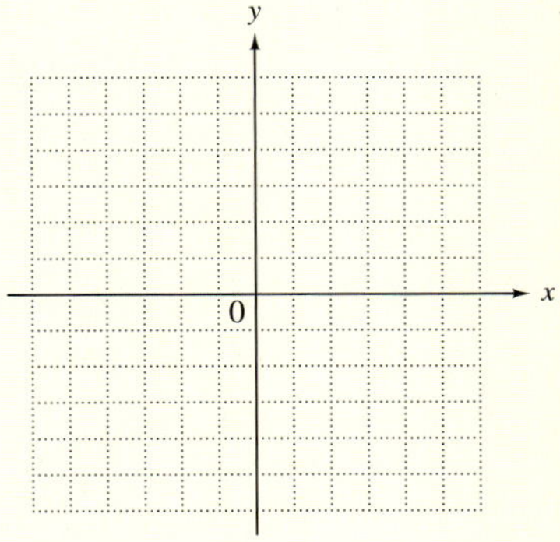

2 Graph $\frac{y^2}{81} - \frac{x^2}{64} = 1$.

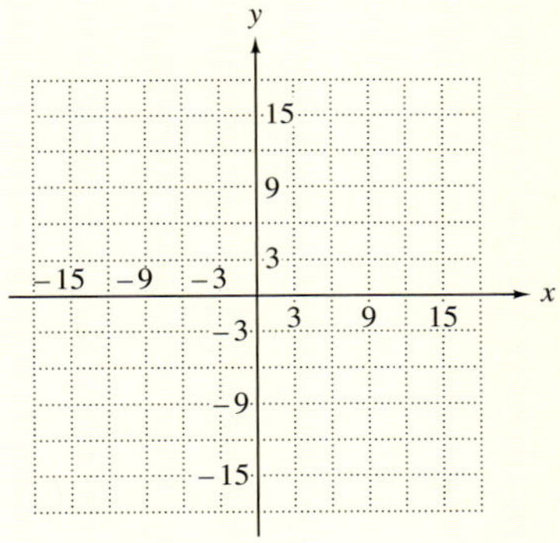

ANSWERS

1.

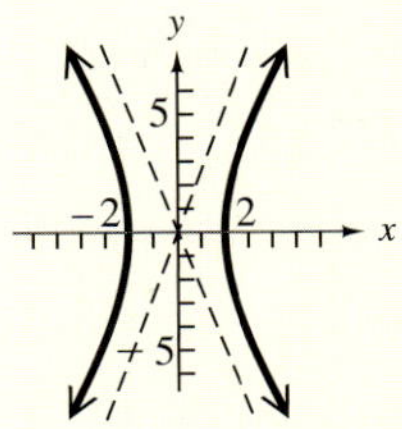

$\frac{x^2}{4} - \frac{y^2}{25} = 1$

2.

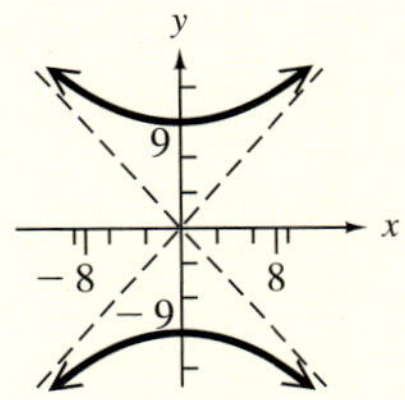

$\frac{y^2}{81} - \frac{x^2}{64} = 1$

OBJECTIVE 3 Identify conic sections by their equations Rewriting a second-degree equation in one of the forms given for ellipses, hyperbolas, circles, or parabolas makes it possible to identify the graph of the equation.

SUMMARY OF CONIC SECTIONS

Equation	Graph	Description	Identification
$y = ax^2 + bx + c$ or $y = a(x - h)^2 + k$	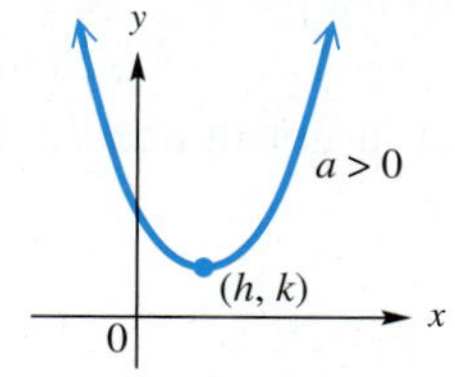 Parabola	It opens up if $a > 0$, down if $a < 0$. The vertex is (h, k).	It has an x^2-term. y is not squared.
$x = ay^2 + by + c$ or $x = a(y - k)^2 + h$	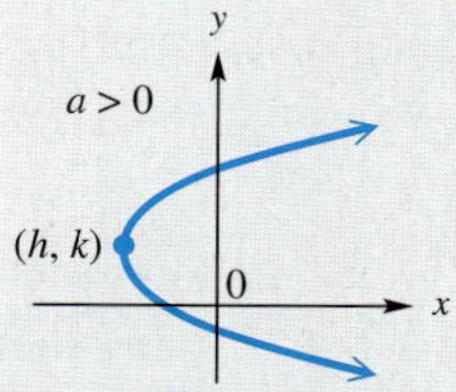 Parabola	It opens to the right if $a > 0$, to the left if $a < 0$. The vertex is (h, k).	It has a y^2-term. x is not squared.
$(x - h)^2 + (y - k)^2 = r^2$	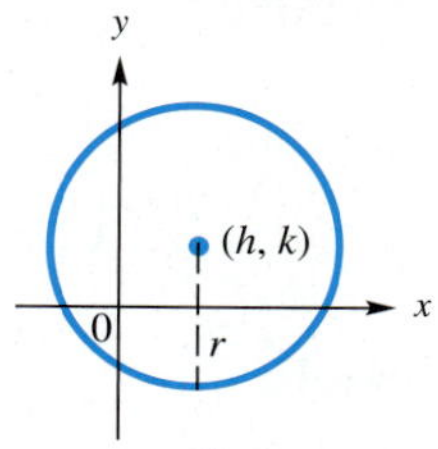 Circle	The center is (h, k), and the radius is r.	x^2- and y^2-terms have the same positive coefficient.
$\frac{x^2}{a^2} + \frac{y^2}{b^2} = 1$	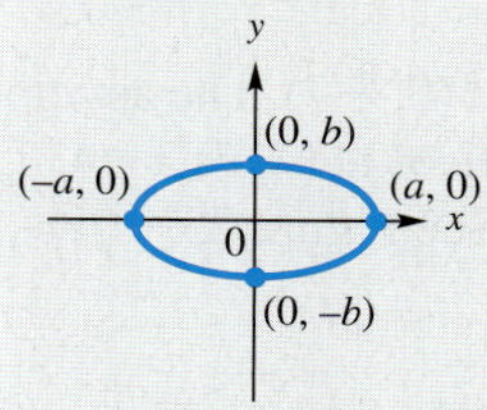 Ellipse	The x-intercepts are $(a, 0)$ and $(-a, 0)$. The y-intercepts are $(0, b)$ and $(0, -b)$.	x^2- and y^2-terms have different positive coefficients.
$\frac{x^2}{a^2} - \frac{y^2}{b^2} = 1$	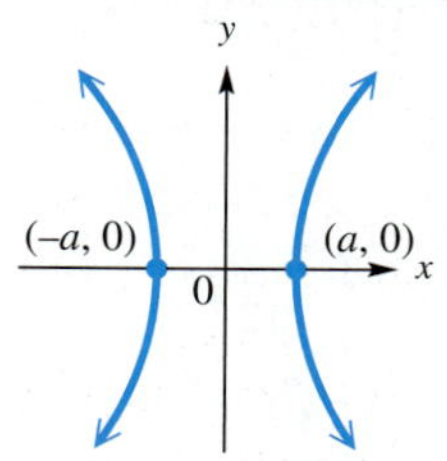 Hyperbola	The x-intercepts are $(a, 0)$ and $(-a, 0)$. The asymptotes are found from (a, b), $(a, -b)$, $(-a, -b)$, and $(-a, b)$.	x^2 has a positive coefficient. y^2 has a negative coefficient.
$\frac{y^2}{b^2} - \frac{x^2}{a^2} = 1$	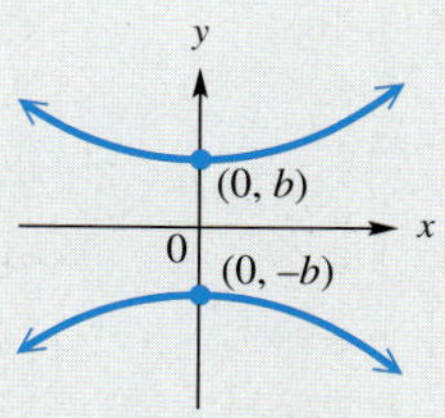 Hyperbola	The y-intercepts are $(0, b)$ and $(0, -b)$. The asymptotes are found from (a, b), $(a, -b)$, $(-a, -b)$, and $(-a, b)$.	y^2 has a positive coefficient. x^2 has a negative coefficient.

EXAMPLE 3 **Identifying the Graphs of Equations**

Identify the graph of each equation.

(a) $9x^2 = 108 + 12y^2$

Both variables are squared, so the graph is either an ellipse or a hyperbola. (This situation also occurs for a circle, which is a special case of the ellipse.) To see which one it is, rewrite the equation so that the x^2- and y^2-terms are on one side of the equation and 1 is on the other.

$$9x^2 - 12y^2 = 108 \qquad \text{Subtract } 12y^2.$$

$$\frac{x^2}{12} - \frac{y^2}{9} = 1 \qquad \text{Divide by 108.}$$

Because of the minus sign, the graph of this equation is a hyperbola.

(b) $x^2 = y - 3$

Only one of the two variables, x, is squared, so this is the vertical parabola $y = x^2 + 3$.

(c) $x^2 = 9 - y^2$

Write the variable terms on the same side of the equation.

$$x^2 + y^2 = 9 \qquad \text{Add } y^2.$$

The graph of this equation is a circle with center at the origin and radius 3.

Work Problem **3** at the Side. ▶

3 Identify the graph of each equation.

(a) $3x^2 = 27 - 4y^2$

(b) $6x^2 = 100 + 2y^2$

(c) $3x^2 = 27 - 4y$

(d) $3x^2 = 27 - 3y^2$

OBJECTIVE 4 **Graph certain square root functions.** Recall from **Section 4.5** that no vertical line will intersect the graph of a function in more than one point. Thus, horizontal parabolas and all circles, ellipses, and hyperbolas are examples of graphs that do not satisfy the conditions of a function. However, by considering only a part of the graph of each of these we have the graph of a function, as seen in Figure 24.

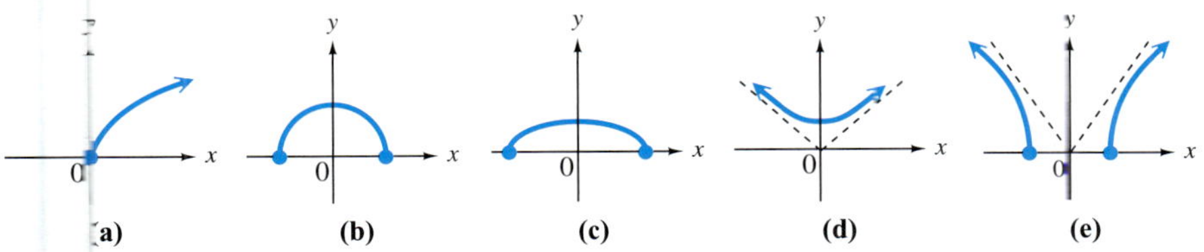

Figure 24

In parts (a)–(d) of Figure 24, the top portion of a conic section is shown (parabola, circle, ellipse, and hyperbola, respectively). In part (e), the top two portions of a hyperbola are shown. In each case, the graph is that of a function since the graph satisfies the conditions of the vertical line test.

In **Sections 9.1** and **12.1** we observed the square root function defined by $f(x) = \sqrt{x}$. To find equations for the types of graphs shown in Figure 24, we extend its definition.

Square Root Function

For an algebraic expression in x defined by u, with $u \geq 0$, a function of the form

$$f(x) = \sqrt{u}$$

is called a **square root function.**

ANSWERS

3. **(a)** ellipse **(b)** hyperbola **(c)** parabola **(d)** circle

EXAMPLE 4 **Graphing a Semicircle**

Graph $f(x) = \sqrt{25 - x^2}$. Give the domain and range.

Replace $f(x)$ with y and square both sides to get the equation

$$y^2 = 25 - x^2, \quad \text{or} \quad x^2 + y^2 = 25.$$

This is the graph of a circle with center at (0, 0) and radius 5. Since $f(x)$, or y, represents a principal square root in the original equation, $f(x)$ must be nonnegative. This restricts the graph to the upper half of the circle, as shown in Figure 25. Use the graph and the vertical line test to verify that it is indeed a function. The domain is $[-5, 5]$, and the range is $[0, 5]$.

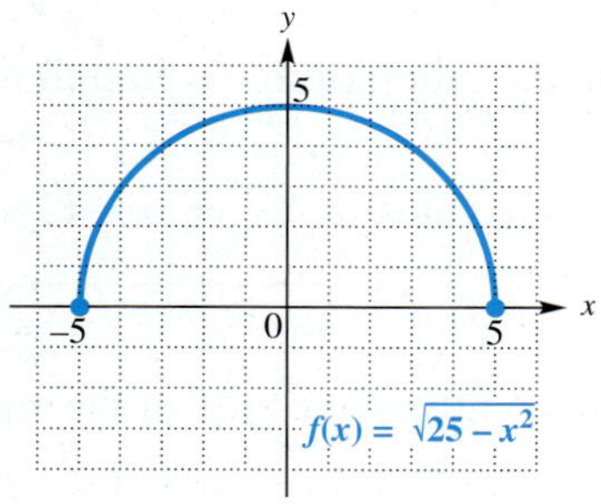

Figure 25

◀ Work Problem **4** at the Side.

4 Graph $f(x) = \sqrt{36 - x^2}$. Give the domain and range.

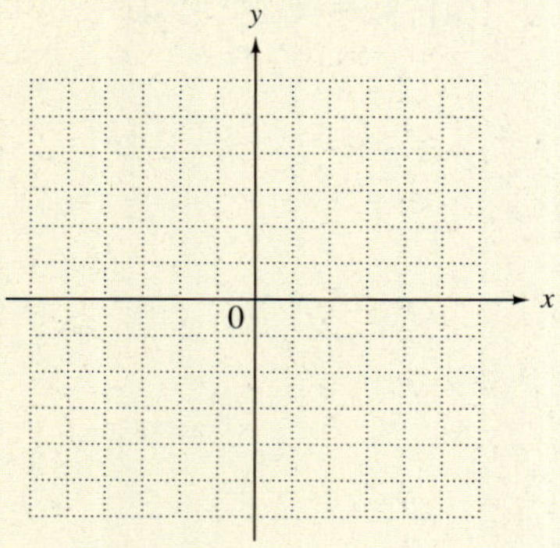

EXAMPLE 5 **Graphing a Portion of an Ellipse**

Graph $\frac{y}{6} = -\sqrt{1 - \frac{x^2}{16}}$. Give the domain and range.

Square both sides to get an equation whose form is known.

$$\frac{y^2}{36} = 1 - \frac{x^2}{16}$$

$$\frac{x^2}{16} + \frac{y^2}{36} = 1 \qquad \text{Add } \frac{x^2}{16}.$$

This is the equation of an ellipse with x-intercepts (4, 0) and (−4, 0) and y-intercepts (0, 6) and (0, −6). Since $\frac{y}{6}$ equals a negative square root in the original equation, y must be nonpositive, restricting the graph to the lower half of the ellipse, as shown in Figure 26. Verify that this is the graph of a function, using the vertical line test. The domain is $[-4, 4]$, and the range is $[-6, 0]$.

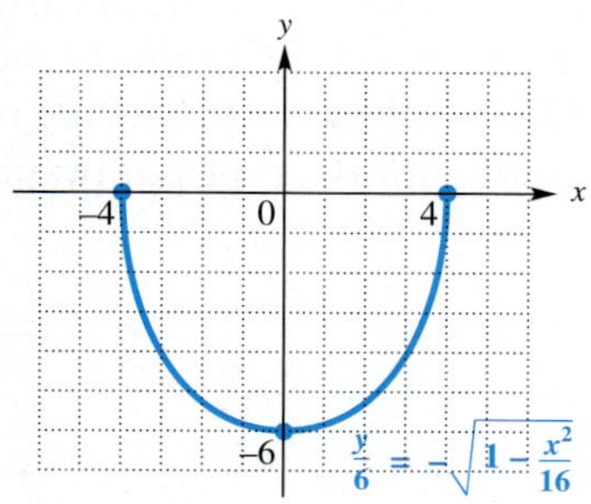

Figure 26

◀ Work Problem **5** at the Side.

5 Graph

$$\frac{y}{3} = -\sqrt{1 - \frac{x^2}{4}}.$$

Give the domain and range.

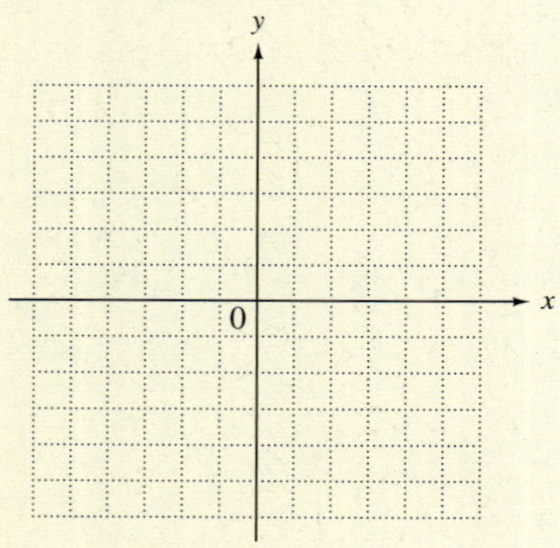

ANSWERS

4.

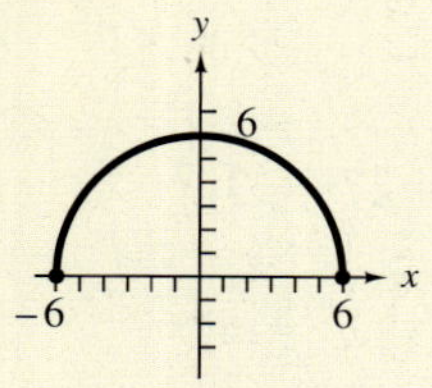

$f(x) = \sqrt{36 - x^2}$

domain: $[-6, 6]$; range: $[0, 6]$

5.

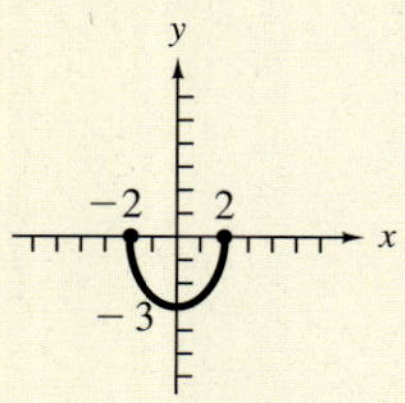

$\frac{y}{3} = -\sqrt{1 - \frac{x^2}{4}}$

domain: $[-2, 2]$; range: $[-3, 0]$

12.3 Exercises

FOR EXTRA HELP

MyMathLab | Math XL PRACTICE | WATCH | DOWNLOAD | READ | REVIEW

Based on the discussions of ellipses in **Section 12.2** *and of hyperbolas in this section, match each equation with its graph.*

1. $\frac{x^2}{25} + \frac{y^2}{9} = 1$

2. $\frac{x^2}{9} + \frac{y^2}{25} = 1$

3. $\frac{x^2}{9} - \frac{y^2}{25} = 1$

4. $\frac{x^2}{25} - \frac{y^2}{9} = 1$

A.

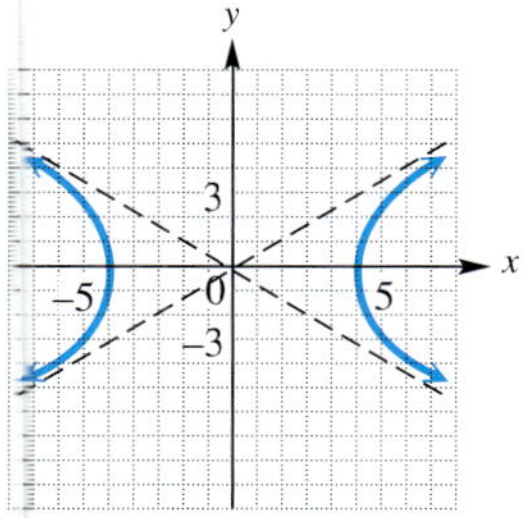

B.

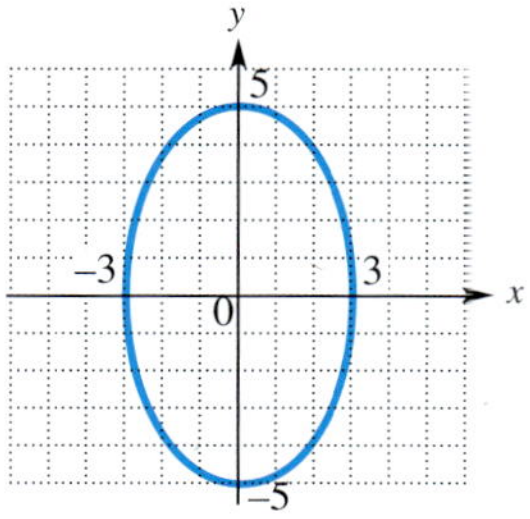

C.

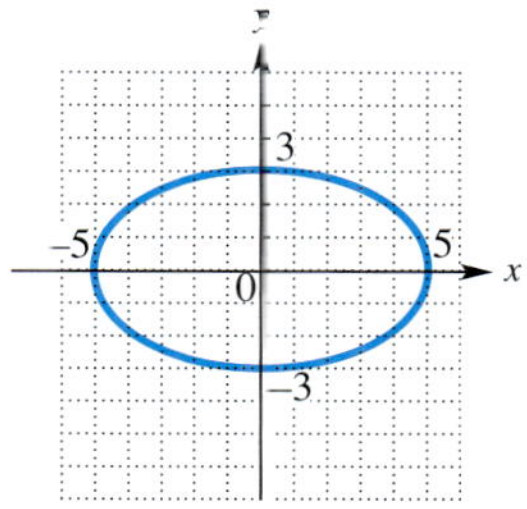

D.

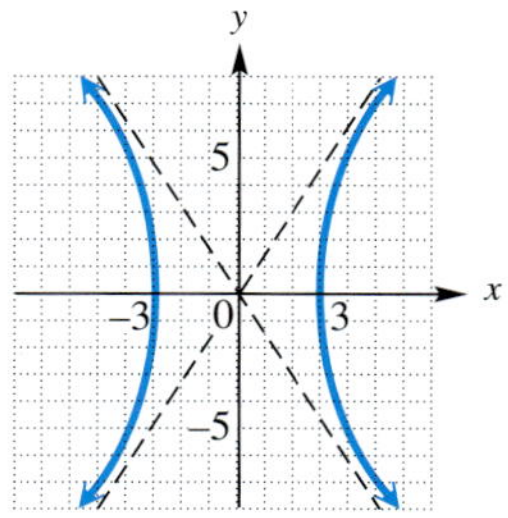

5. Write an explanation of how you can tell from the equation whether the branches of a hyperbola open up and down or left and right.

6. Describe how the fundamental rectangle is used to sketch a hyperbola.

Graph each hyperbola. See Examples 1 and 2.

7. $\frac{x^2}{16} - \frac{y^2}{9} = 1$

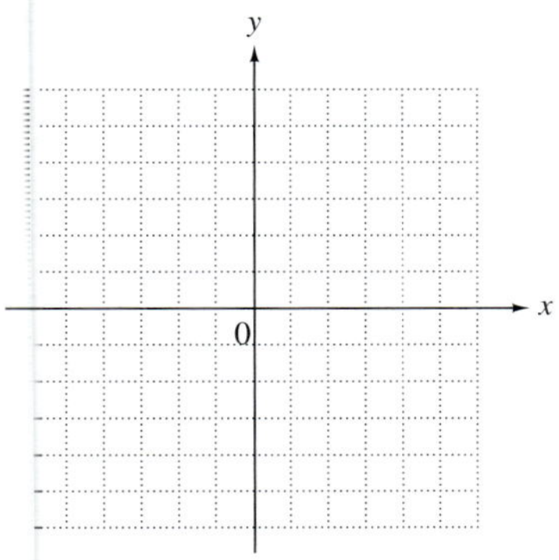

8. $\frac{y^2}{4} - \frac{x^2}{25} = 1$

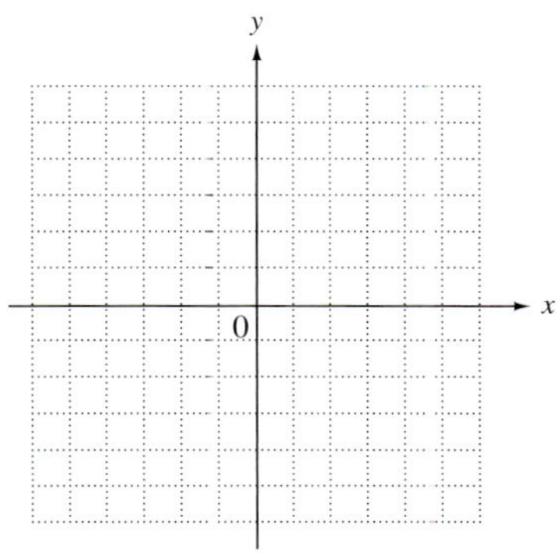

9. $\frac{y^2}{9} - \frac{x^2}{9} = 1$

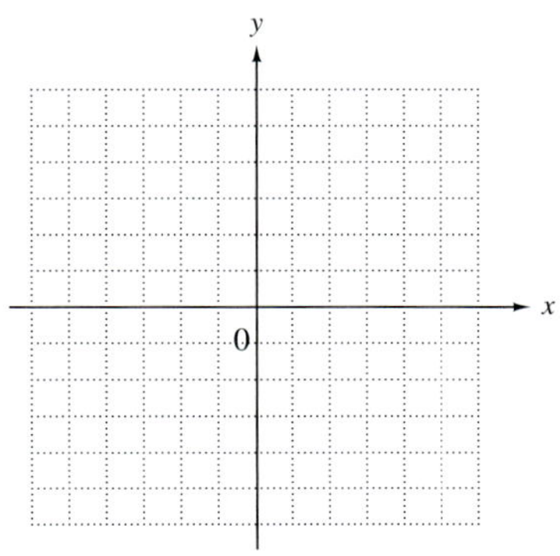

10. $\dfrac{x^2}{49} - \dfrac{y^2}{16} = 1$

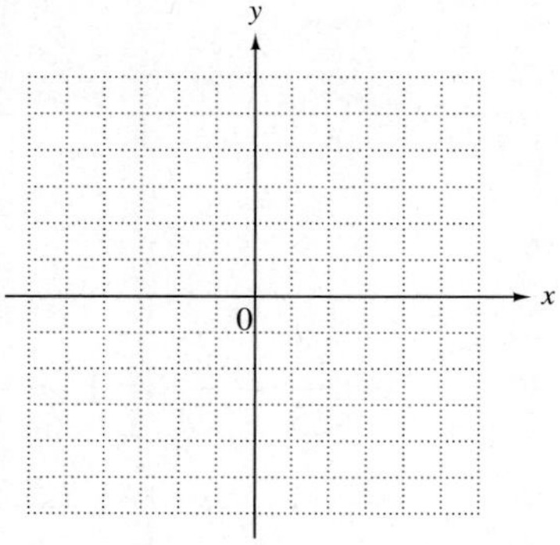

11. $\dfrac{x^2}{25} - \dfrac{y^2}{36} = 1$

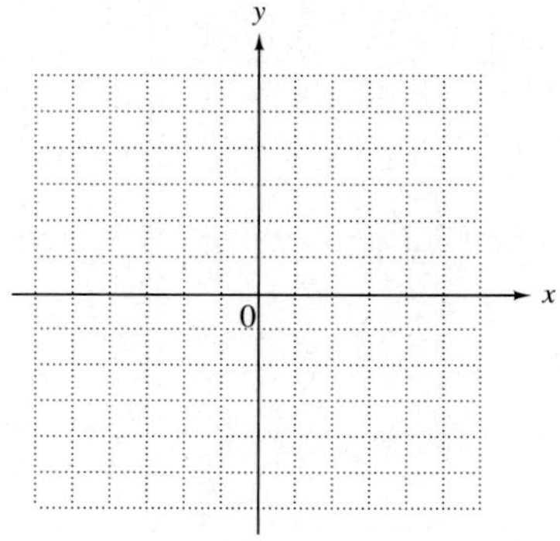

12. $\dfrac{y^2}{9} - \dfrac{x^2}{4} = 1$

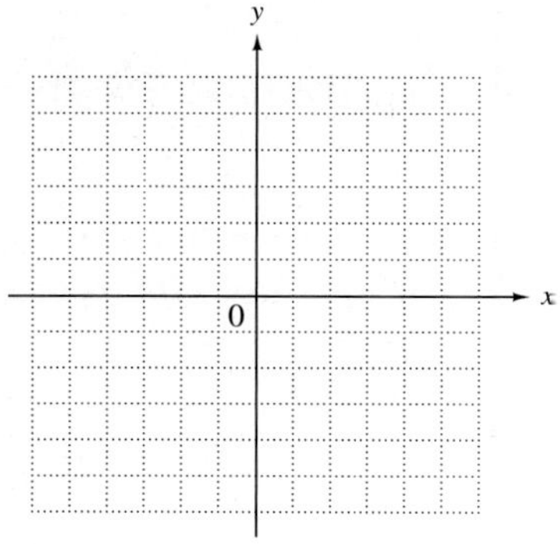

Identify the graph of each equation as a parabola, circle, ellipse, *or* hyperbola, *and sketch it. See Example 3.*

13. $x^2 - y^2 = 16$

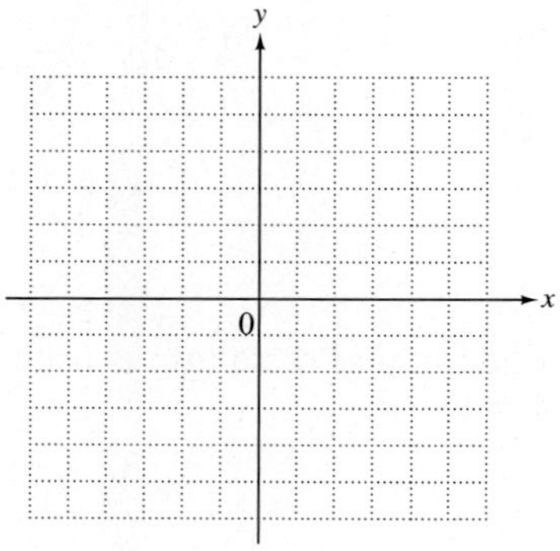

14. $x^2 + y^2 = 16$

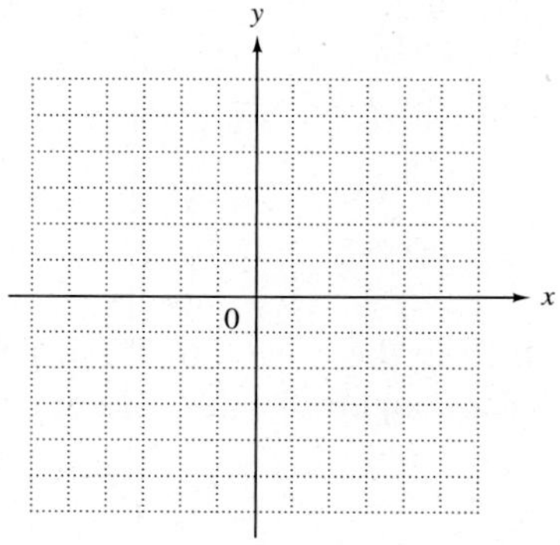

15. $4x^2 + y^2 = 16$

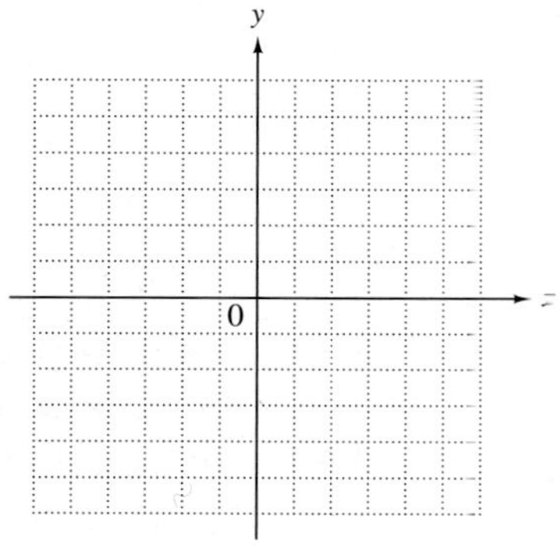

16. $x^2 - 2y = 0$

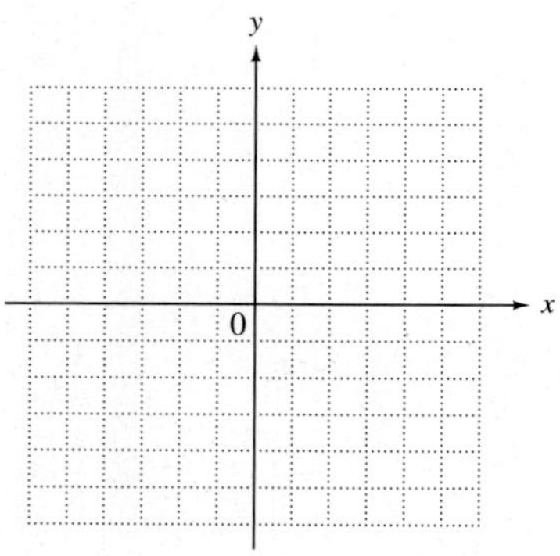

17. $y^2 = 36 - x^2$

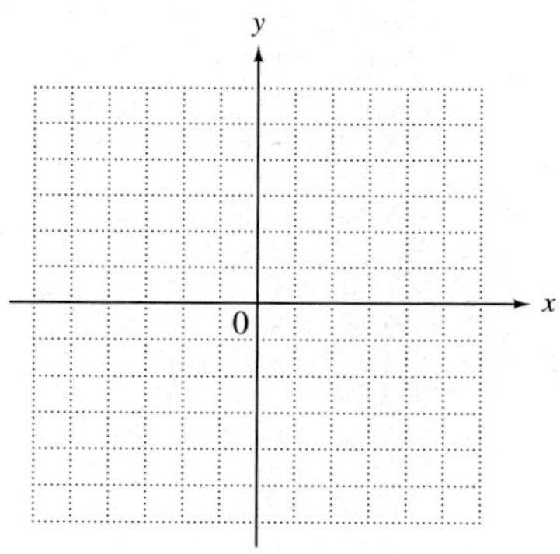

18. $9x^2 + 25y^2 = 225$

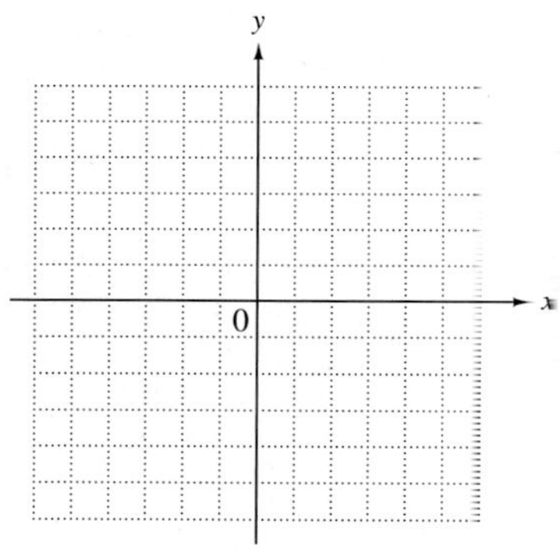

19. $9x^2 = 144 + 16y^2$

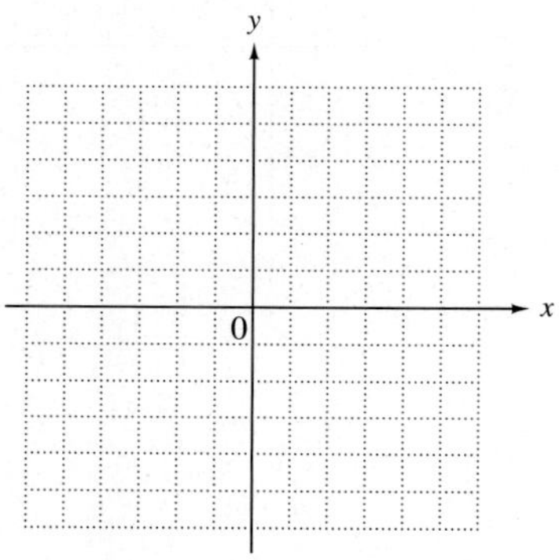

20. $y^2 = 4 + x^2$

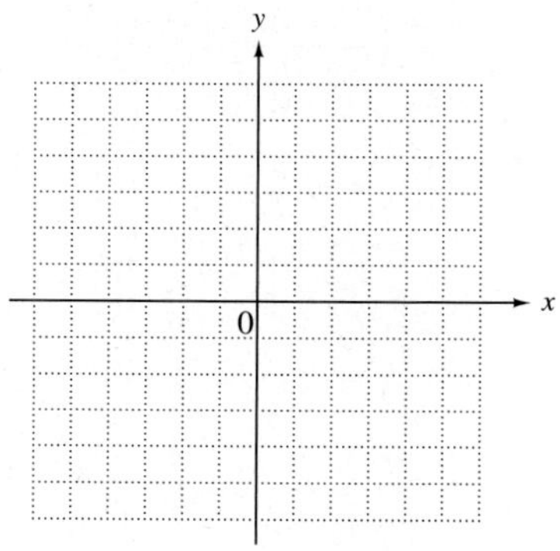

21. $x^2 + 9y^2 = 9$

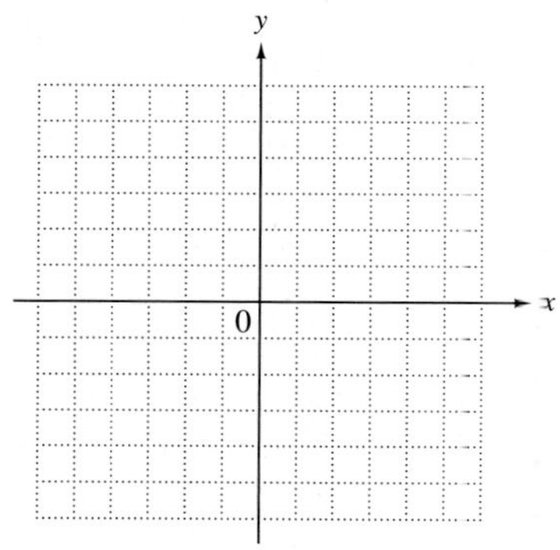

Graph each function defined by a radical expression. Give the domain and range. See Examples 4 and 5.

22. $f(x) = \sqrt{16 - x^2}$

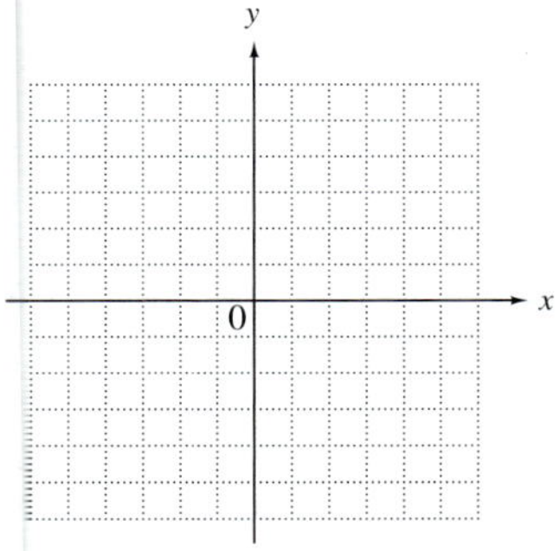

23. $f(x) = \sqrt{9 - x^2}$

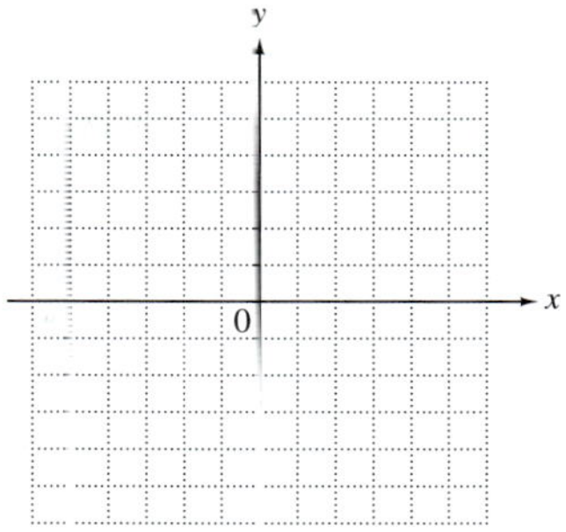

24. $f(x) = -\sqrt{36 - x^2}$

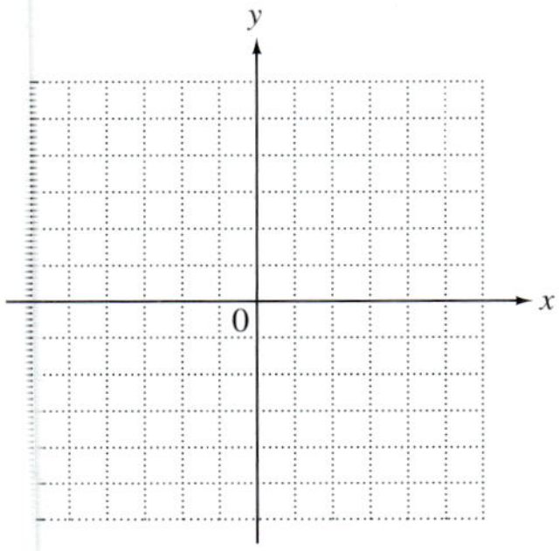

25. $f(x) = -\sqrt{25 - x^2}$

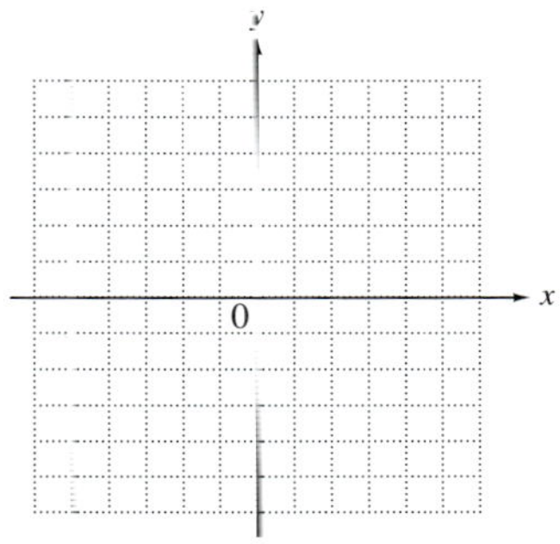

26. $\dfrac{y}{3} = \sqrt{1 + \dfrac{x^2}{9}}$

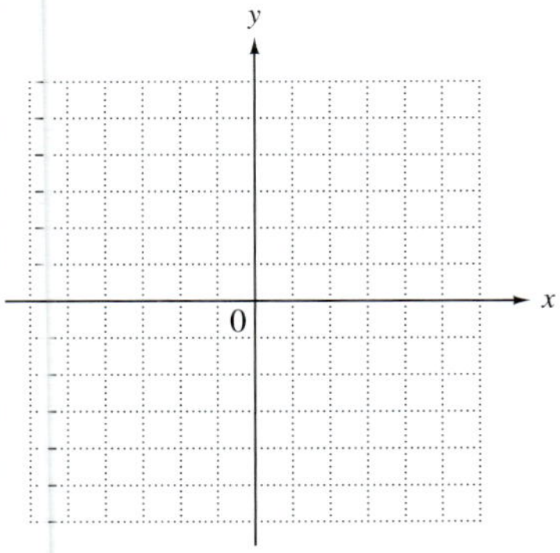

27. $y = \sqrt{\dfrac{x + 4}{2}}$

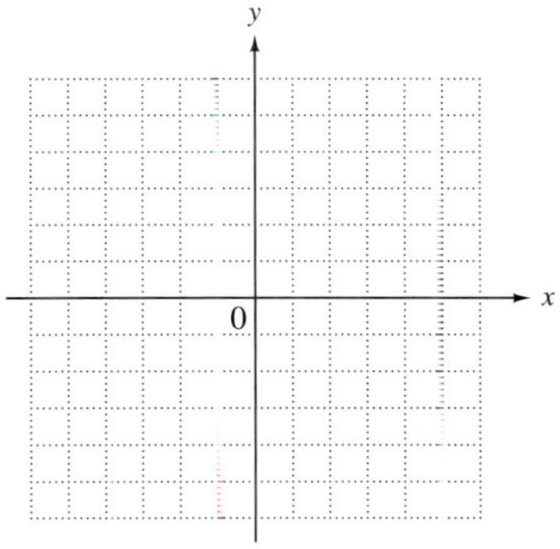

28. $y = -2\sqrt{\dfrac{9 - x^2}{9}}$

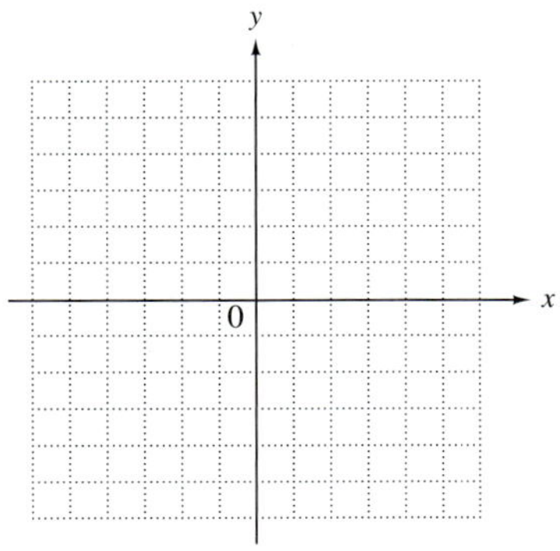

In Section 12.2, Example 6, we saw that the center of an ellipse may be shifted away from the origin. The same process can be applied to hyperbolas. For example, the hyperbola shown at the right,

$$\frac{(x+5)^2}{4} - \frac{(y-2)^2}{9} = 1,$$

has the same graph as $\frac{x^2}{4} - \frac{y^2}{9} = 1$, *but it is centered at* $(-5, 2)$. *Graph each hyperbola with center shifted away from the origin.*

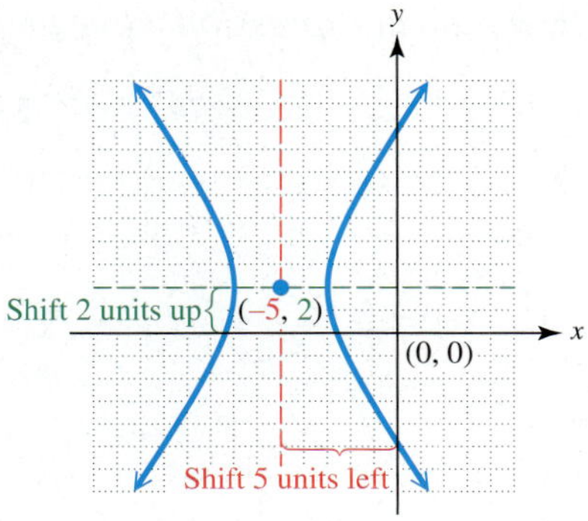

29. $\frac{(x-2)^2}{4} - \frac{(y+1)^2}{9} = 1$

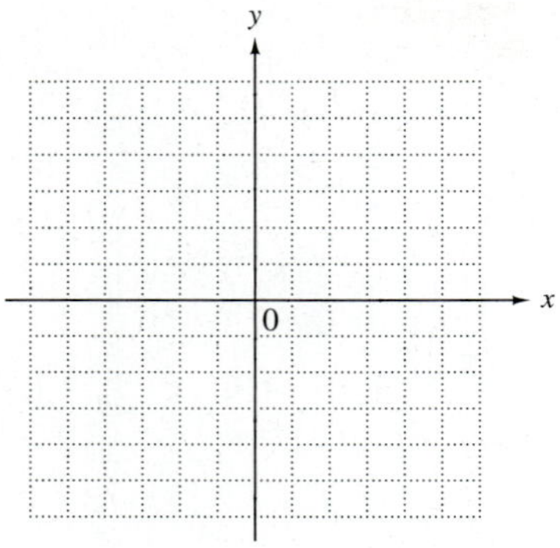

30. $\frac{(x+3)^2}{16} - \frac{(y-2)^2}{4} = 1$

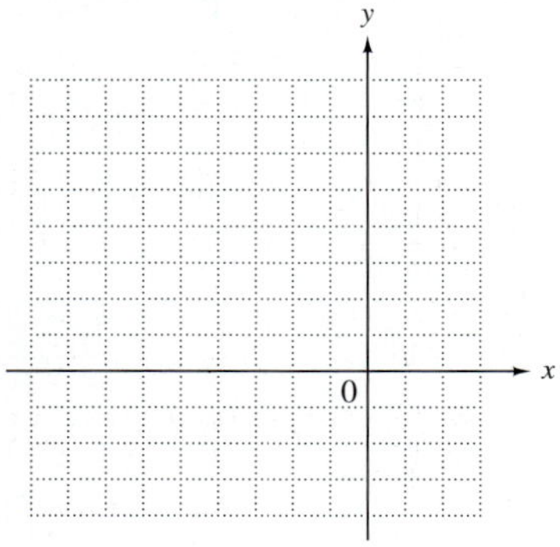

Solve each problem.

31. Two buildings in a sports complex are shaped and positioned like a portion of the branches of the hyperbola with equation

$$400x^2 - 625y^2 = 250{,}000,$$

where x and y are in meters.

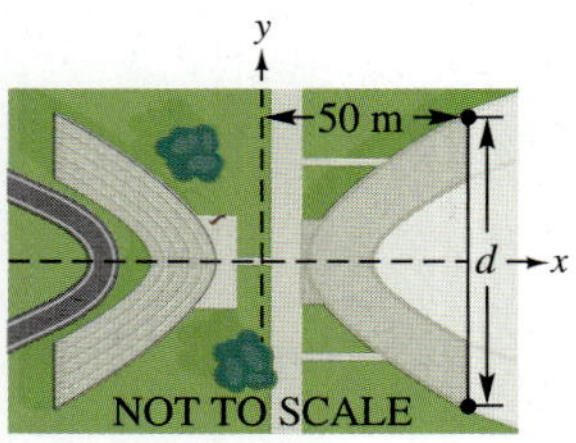

(a) How far apart are the buildings at their closest point?

(b) Find the distance d in the figure.

32. In rugby, after a *try* (similar to a touchdown in American football) the scoring team attempts a kick for extra points. The ball must be kicked from directly behind the point where the try was scored. The kicker can choose the distance but cannot move the ball sideways. It can be shown that the kicker's best choice is on the hyperbola with equation

$$\frac{x^2}{g^2} - \frac{y^2}{g^2} = 1,$$

where $2g$ is the distance between the goal posts. Since the hyperbola approaches its asymptotes, it is easier for the kicker to estimate points on the asymptotes instead of on the hyperbola. What are the equations of the asymptotes of this hyperbola? Why is it relatively easy to estimate them? (*Source:* Isaksen, Daniel C., "How to Kick a Field Goal," *The College Mathematics Journal,* September 1996.)

12.4 Nonlinear Systems of Equations

OBJECTIVES

1. Solve a nonlinear system by substitution.
2. Use the elimination method to solve a system with two second-degree equations.
3. Solve a system that requires a combination of methods.

An equation in which some terms have more than one variable or a variable of degree 2 or greater is called a **nonlinear equation.** A **nonlinear system of equations** includes at least one nonlinear equation.

When solving a nonlinear system, it helps to visualize the types of graphs of the equations of the system to determine the possible number of points of intersection. For example, if a system includes two equations where the graph of one is a circle and the graph of the other is a line, then there may be 0, 1, or 2 points of intersection, as illustrated in Figure 27.

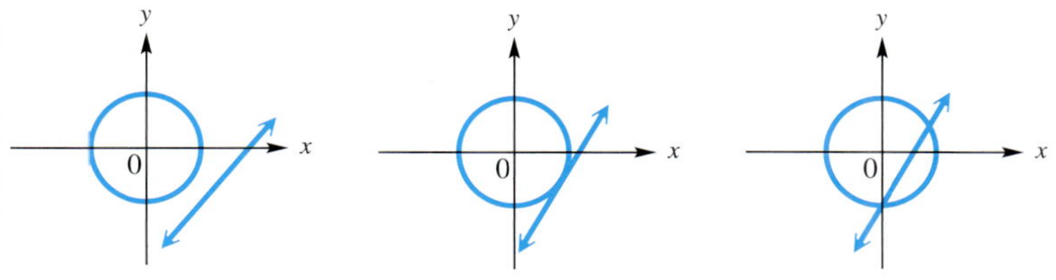

Figure 27

OBJECTIVE 1 Solve a nonlinear system by substitution. We solve nonlinear systems by the substitution method, the elimination method, or a combination of the two. The substitution method **(Section 5.1)** is usually appropriate when one of the equations is linear.

EXAMPLE 1 Solving a Nonlinear System by Substitution

Solve the system.

$$x^2 + y^2 = 9 \quad (1)$$
$$2x - y = 3 \quad (2)$$

The graph of (1) is a circle and the graph of (2) is a line. There may be 0, 1, or 2 points of intersection, as shown in Figure 27. First solve the linear equation for one of the two variables, and then substitute the resulting expression into the nonlinear equation to obtain an equation in one variable.

$$2x - y = 3 \quad (2)$$
$$y = 2x - 3 \quad (3)$$

Substitute $2x - 3$ for y in equation (1).

$$x^2 + y^2 = 9 \quad (1)$$
$$x^2 + (2x - 3)^2 = 9 \quad \text{Let } y = 2x - 3.$$
$$x^2 + 4x^2 - 12x + 9 = 9 \quad \text{Square } 2x - 3.$$
$$5x^2 - 12x = 0 \quad \text{Subtract 9; combine like terms.}$$
$$x(5x - 12) = 0 \quad \text{Factor; GCF is } x.$$

Set *both* factors equal to 0.

$$x = 0 \quad \text{or} \quad x = \frac{12}{5} \quad \text{Zero-factor property}$$

Let $x = 0$ in equation (3) to get $y = -3$. If $x = \frac{12}{5}$ in equation (3), then $y = \frac{9}{5}$. The solution set of the system is $\left\{(0, -3), \left(\frac{12}{5}, \frac{9}{5}\right)\right\}$. The graph in Figure 28 confirms the two points of intersection.

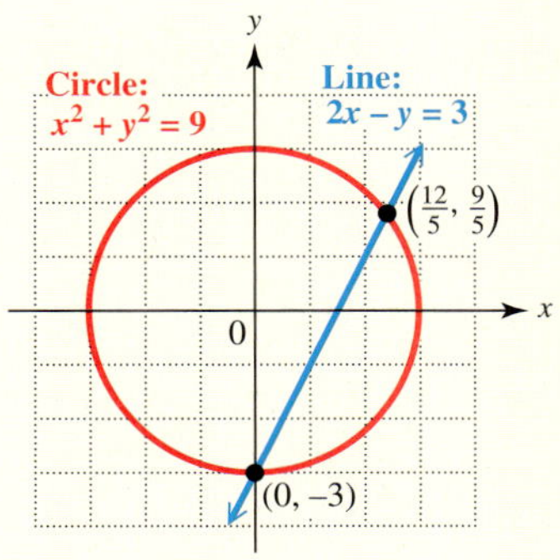

Figure 28

Work Problem 1 at the Side. ▶

1 Solve each system.

(a) $x^2 + y^2 = 10$
$x = y + 2$

(b) $x^2 - 2y^2 = 8$
$y + x = 6$

ANSWERS

1. **(a)** $\{(3, 1), (-1, -3)\}$
(b) $\{(4, 2), (20, -14)\}$

2 Solve each system.

(a) $xy = 8$
$x + y = 6$

(b) $xy + 10 = 0$
$4x + 9y = -2$

Answers

2. (a) $\{(4, 2), (2, 4)\}$

(b) $\left\{(-5, 2), \left(\frac{9}{2}, -\frac{20}{9}\right)\right\}$

EXAMPLE 2 Solving a Nonlinear System by Substitution

Solve the system.

$$6x - y = 5 \quad (1)$$
$$xy = 4 \quad (2)$$

The graph of (1) is a line. It can be shown by plotting points that the graph of (2) is a hyperbola. Visualizing a line and a hyperbola indicates that there may be 0, 1, or 2 points of intersection. We can solve either equation for one of the variables and then substitute the result into the other equation. Solving $xy = 4$ for x gives $x = \frac{4}{y}$. Substitute $\frac{4}{y}$ for x in equation (1).

$$6\left(\frac{4}{y}\right) - y = 5 \qquad \text{Let } x = \tfrac{4}{y}.$$
$$\frac{24}{y} - y = 5 \qquad \text{Multiply.}$$
$$24 - y^2 = 5y \qquad \text{Multiply by } y, y \neq 0.$$
$$0 = y^2 + 5y - 24 \qquad \text{Standard form}$$
$$0 = (y - 3)(y + 8) \qquad \text{Factor.}$$
$$y = 3 \quad \text{or} \quad y = -8 \qquad \text{Zero-factor property}$$

We substitute these results into $x = \frac{4}{y}$ to obtain the corresponding values of x.

$$\text{If } y = 3, \text{ then } x = \frac{4}{3}. \qquad \text{If } y = -8, \text{ then } x = -\frac{1}{2}.$$

The solution set of the system is $\left\{\left(\frac{4}{3}, 3\right), \left(-\frac{1}{2}, -8\right)\right\}$. See Figure 29.

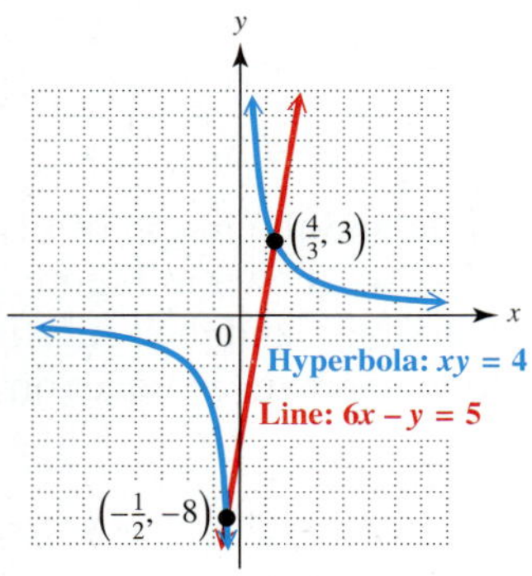

Figure 29

◀ Work Problem **2** at the Side.

OBJECTIVE 2 Use the elimination method to solve a system with two second-degree equations. The elimination method **(Section 5.1)** is often used when both equations are second degree.

EXAMPLE 3 Solving a Nonlinear System by Elimination

Solve the system.

$$x^2 + y^2 = 9 \quad (1)$$
$$2x^2 - y^2 = -6 \quad (2)$$

The graph of (1) is a circle, while the graph of (2) is a hyperbola. By analyzing the possibilities, we conclude that there may be 0, 1, 2, 3, or 4 points of intersection. Adding the two equations will eliminate y.

Continued on Next Page

$$\begin{aligned} x^2 + y^2 &= 9 && (1) \\ 2x^2 - y^2 &= -6 && (2) \\ \hline 3x^2 &= 3 && \text{Add.} \\ x^2 &= 1 && \text{Divide by 3.} \\ x = 1 \quad \text{or} \quad x &= -1 && \text{Square root property} \end{aligned}$$

Each value of x gives corresponding values for y when substituted into one of the original equations. Using equation (1) is easier since the coefficients of the x^2- and y^2-terms are 1.

If $x = 1$, then	**If $x = -1$, then**
$1^2 + y^2 = 9$	$(-1)^2 + y^2 = 9$
$y^2 = 8$	$y^2 = 8$
$y = \sqrt{8}$ or $y = -\sqrt{8}$	$y = \sqrt{8}$ or $y = -\sqrt{8}$
$y = 2\sqrt{2}$ or $y = -2\sqrt{2}$.	$y = 2\sqrt{2}$ or $y = -2\sqrt{2}$.

The solution set is

$$\{(1, 2\sqrt{2}), (1, -2\sqrt{2}), (-1, 2\sqrt{2}), (-1, -2\sqrt{2})\}.$$

Figure 30 shows the four points of intersection.

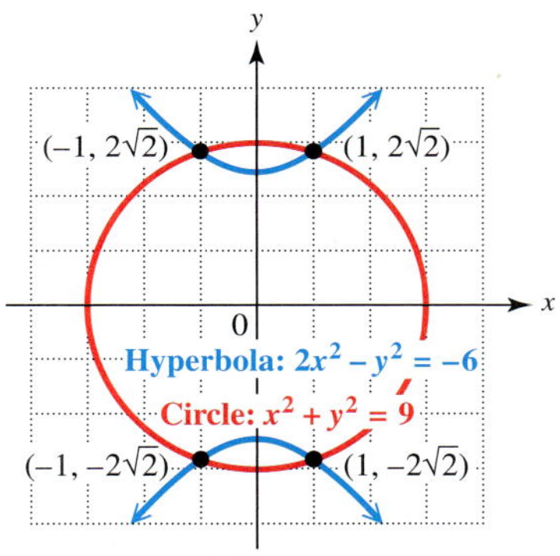

Figure 30

Work Problem **3** at the Side. ▶

OBJECTIVE 3 Solve a system that requires a combination of methods. Solving a system of second-degree equations may require a combination of methods.

EXAMPLE 4 Solving a Nonlinear System by a Combination of Methods

Solve the system.

$$\begin{aligned} x^2 + 2xy - y^2 &= 7 && (1) \\ x^2 - y^2 &= 3 && (2) \end{aligned}$$

While we have not graphed equations like (1), its graph is a hyperbola. The graph of (2) is also a hyperbola. Two hyperbolas may have 0, 1, 2, 3, or 4 points of intersection. We use the elimination method here in combination with the substitution method. We begin by eliminating the squared terms by multiplying each side of equation (2) by -1 and then adding the result to equation (1).

Continued on Next Page

3 Solve each system.

(a) $x^2 + y^2 = 41$
$x^2 - y^2 = 9$

(b) $x^2 + 3y^2 = 40$
$4x^2 - y^2 = 4$

ANSWERS

3. (a) $\{(5, 4), (5, -4), (-5, 4), (-5, -4)\}$
(b) $\{(2, 2\sqrt{3}), (2, -2\sqrt{3}), (-2, 2\sqrt{3}), (-2, -2\sqrt{3})\}$

4 Solve each system.

(a) $x^2 + xy + y^2 = 3$
$x^2 + y^2 = 5$

(b) $x^2 + 7xy - 2y^2 = -8$
$-2x^2 + 4y^2 = 16$

$$\begin{aligned} x^2 + 2xy - y^2 &= 7 && (1) \\ -x^2 \qquad + y^2 &= -3 && \text{Multiply (2) by } -1. \\ \hline 2xy \qquad &= 4 && \text{Add.} \end{aligned}$$

Next, we solve $2xy = 4$ for one of the variables. We choose y.

$$2xy = 4$$
$$y = \frac{2}{x} \qquad (3)$$

Now, we substitute $y = \frac{2}{x}$ into one of the original equations. It is easier to do this with equation (2).

$$\begin{aligned} x^2 - y^2 &= 3 && (2) \\ x^2 - \left(\frac{2}{x}\right)^2 &= 3 && \text{Let } y = \tfrac{2}{x}. \\ x^2 - \frac{4}{x^2} &= 3 && \text{Square } \tfrac{2}{x}. \\ x^4 - 4 &= 3x^2 && \text{Multiply by } x^2,\ x \neq 0. \\ x^4 - 3x^2 - 4 &= 0 && \text{Subtract } 3x^2. \\ (x^2 - 4)(x^2 + 1) &= 0 && \text{Factor.} \end{aligned}$$

$$x^2 - 4 = 0 \quad \text{or} \quad x^2 + 1 = 0$$
$$x^2 = 4 \quad \text{or} \quad x^2 = -1$$
$$x = 2 \quad \text{or} \quad x = -2 \qquad x = i \quad \text{or} \quad x = -i$$

Substituting these four values of x into equation (3) gives the corresponding values for y.

If $x = 2$, then $y = 1$. If $x = i$, then $y = -2i$.
If $x = -2$, then $y = -1$. If $x = -i$, then $y = 2i$.

Note that if we substitute the x-values we found into equation (1) or (2) instead of into equation (3), we get extraneous solutions. ***It is always wise to check all solutions in both of the given equations.*** There are four ordered pairs in the solution set, two with real values and two with nonreal complex values. The solution set is

$$\{(2, 1), (-2, -1), (i, -2i), (-i, 2i)\}.$$

The graph of the system, shown in Figure 31, shows only the two real intersection points because the graph is in the real number plane. The two ordered pairs with nonreal complex components are solutions of the system, but do not appear on the graph.

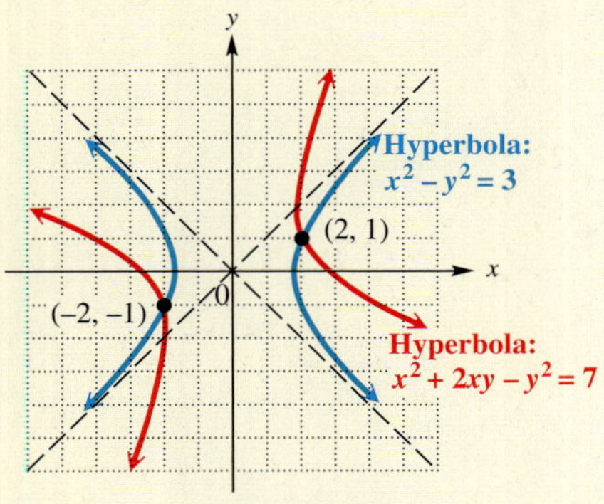

Figure 31

◀ Work Problem **4** at the Side.

Note

In the examples of this section, we analyzed the possible number of points of intersection of the graphs in each system. However, in Examples 2 and 4, we worked with equations whose graphs had not been studied. Keep in mind that it is not absolutely essential to visualize the number of points of intersection in order to solve the system. Furthermore, as in Example 4, there are sometimes nonreal complex solutions to nonlinear systems that do not appear as points of intersection in the real plane. Visualizing the geometry of the graphs is only an aid to solving these systems.

Answers

4. (a) $\{(1, -2), (-1, 2), (2, -1), (-2, 1)\}$
(b) $\{(0, 2), (0, -2), (2i\sqrt{2}, 0), (-2i\sqrt{2}, 0)\}$

12.4 Exercises

Each sketch represents the graphs of a pair of equations in a system. How many ordered pairs of real numbers are in each solution set?

1.

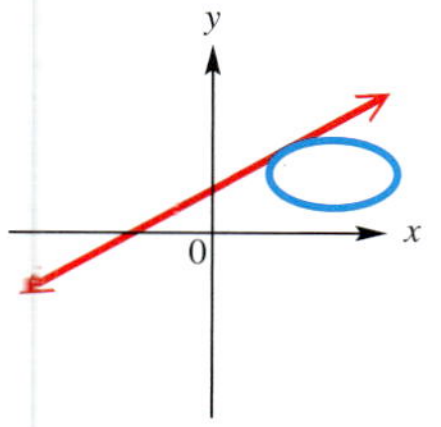

2.

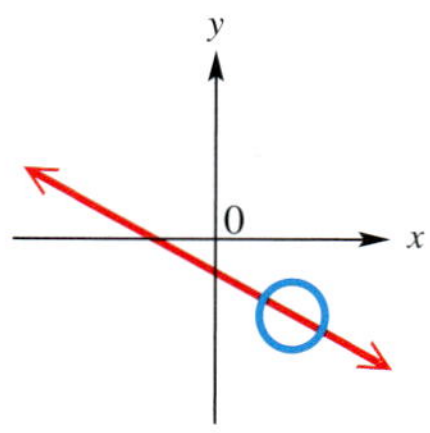

3.

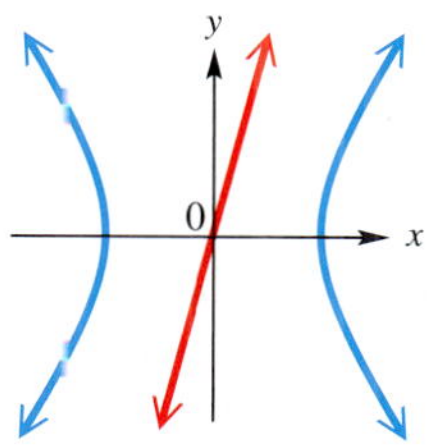

4.

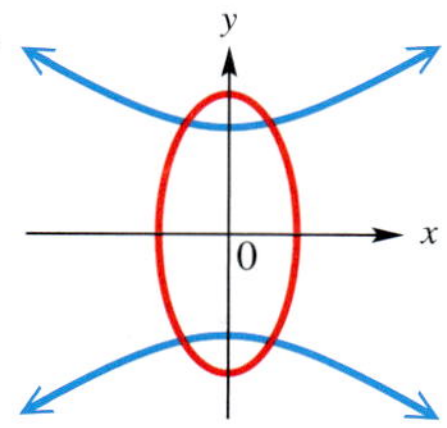

Suppose that a nonlinear system is composed of equations whose graphs are those described, and the number of points of intersection of the two graphs is as given. Make a sketch satisfying these conditions. (There may be more than one way to do this.)

5. A line and a circle; no points

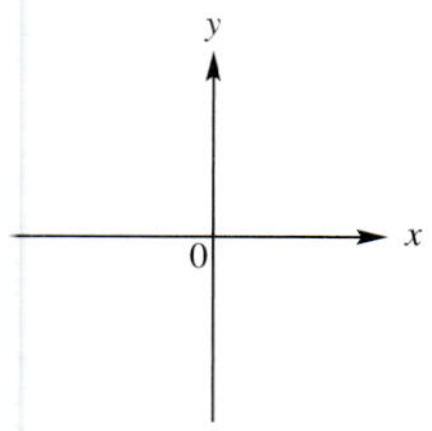

6. A line and a circle; one point

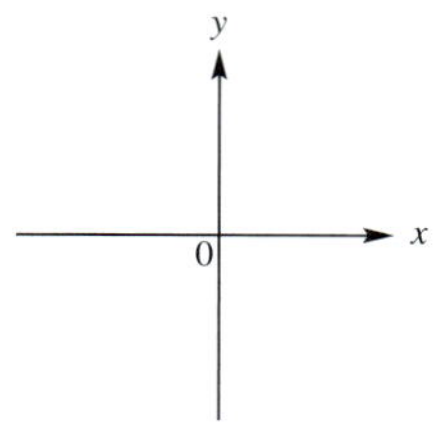

7. A line and an ellipse; two points

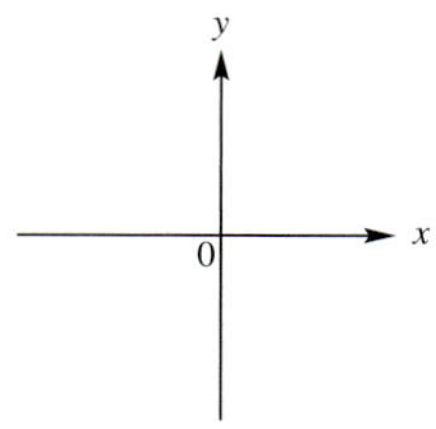

8. A line and a hyperbola; no points

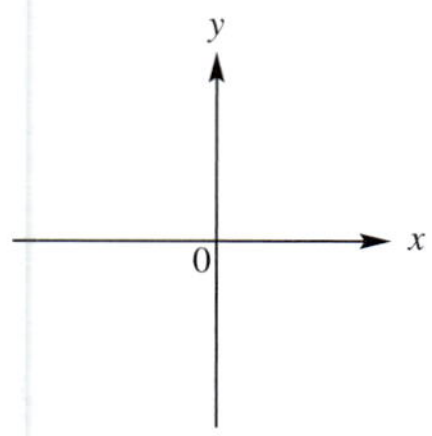

9. A circle and an ellipse; four points

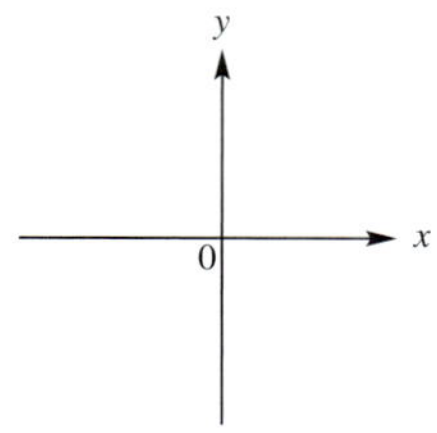

10. A parabola and an ellipse; one point

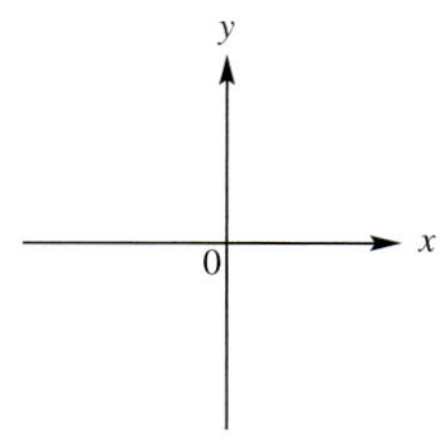

Solve each system by the substitution method. See Examples 1 and 2.

11. $y = 4x^2 - x$
$y = x$

12. $y = x^2 + 6x$
$3y = 12x$

13. $y = x^2 + 6x + 9$
$x + y = 3$

14. $y = x^2 + 8x + 16$
$x - y = -4$

15. $x^2 + y^2 = 2$
$2x + y = 1$

16. $2x^2 + 4y^2 = 4$
$x = 4y$

17. $xy = 4$
$3x + 2y = -10$

18. $xy = -5$
$2x + y = 3$

19. $xy = -3$
$x + y = -2$

20. $xy = 12$
$x + y = 8$

21. $y = 3x^2 + 6x$
$y = x^2 - x - 6$

22. $y = 2x^2 + 1$
$y = 5x^2 + 2x - 7$

23. $2x^2 - y^2 = 6$
$y = x^2 - 3$

24. $x^2 + y^2 = 4$
$y = x^2 - 2$

Solve each system using the elimination method or a combination of the elimination and substitution methods. See Examples 3 and 4.

25. $3x^2 + 2y^2 = 12$
$x^2 + 2y^2 = 4$

26. $2x^2 + y^2 = 28$
$4x^2 - 5y^2 = 28$

27. $xy = 6$
$3x^2 - y^2 = 12$

28. $xy = 5$
$2y^2 - x^2 = 5$

29. $2x^2 + 2y^2 = 8$
$3x^2 + 4y^2 = 24$

30. $5x^2 + 5y^2 = 20$
$x^2 + 2y^2 = 2$

31. $x^2 + xy + y^2 = 15$
$x^2 + y^2 = 10$

32. $2x^2 + 3xy + 2y^2 = 21$
$x^2 + y^2 = 6$

Solve each problem by using a nonlinear system.

33. The area of a rectangular rug is 84 ft^2 and its perimeter is 38 ft. Find the length and width of the rug.

34. Find the length and width of a rectangular room whose perimeter is 50 m and whose area is 100 m^2.

12.5 Second-Degree Inequalities and Systems of Inequalities

OBJECTIVES

1. **Graph second-degree inequalities.**
2. **Graph the solution set of a system of inequalities.**

OBJECTIVE 1 Graph second-degree inequalities. A **second-degree inequality** is an inequality with at least one variable of degree 2 and no variable with degree greater than 2. An example is $x^2 + y^2 \leq 36$. To graph this inequality, we first graph the boundary, as we did with linear inequalities in **Section 4.4.** The boundary of the inequality $x^2 + y^2 \leq 36$ is the graph of the equation $x^2 + y^2 = 36$, a circle with radius 6 and center at the origin, as shown in Figure 32.

The graph of the inequality $x^2 + y^2 \leq 36$ will include either the points outside the circle or the points inside the circle, as well as the boundary. We decide which region to shade by substituting any test point not on the circle, such as (0, 0), into the original inequality. Since $0^2 + 0^2 \leq 36$ is a true statement, the original inequality includes the points inside the circle, the shaded region in Figure 32, and the boundary.

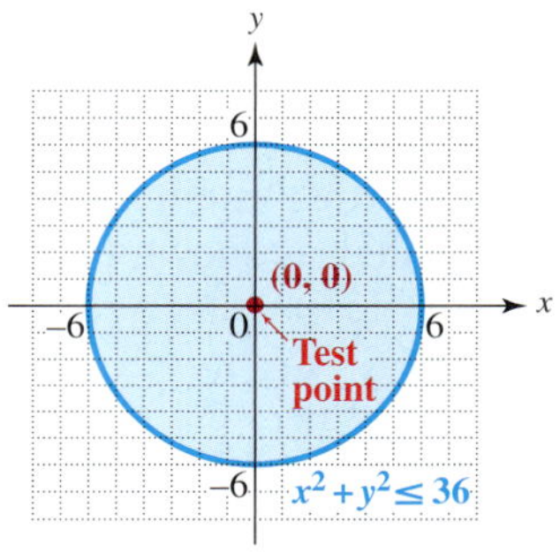

Figure 32

EXAMPLE 1 Graphing a Second-Degree Inequality

Graph $y < -2(x - 4)^2 - 3$.

The boundary, $y = -2(x - 4)^2 - 3$, is a parabola that opens down with vertex at $(4, -3)$. Using (0, 0) as a test point gives

$$0 \stackrel{?}{<} -2(0 - 4)^2 - 3$$

$$0 \stackrel{?}{<} -32 - 3$$

$$0 < -35. \quad \text{False}$$

Because the final inequality is a false statement, the points in the region containing (0, 0) do not satisfy the inequality. Figure 33 shows the final graph. The parabola is drawn as a dashed curve since the points of the parabola itself do not satisfy the inequality, and the region inside (or below) the parabola is shaded.

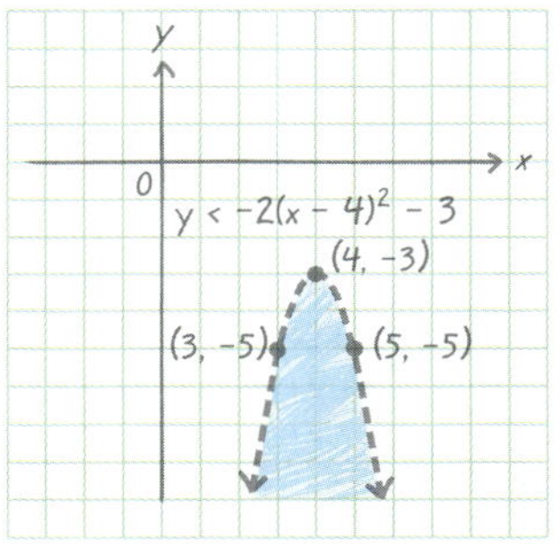

Figure 33

Work Problem 1 at the Side. ▶

1 Graph $y \geq (x + 1)^2 - 5$.

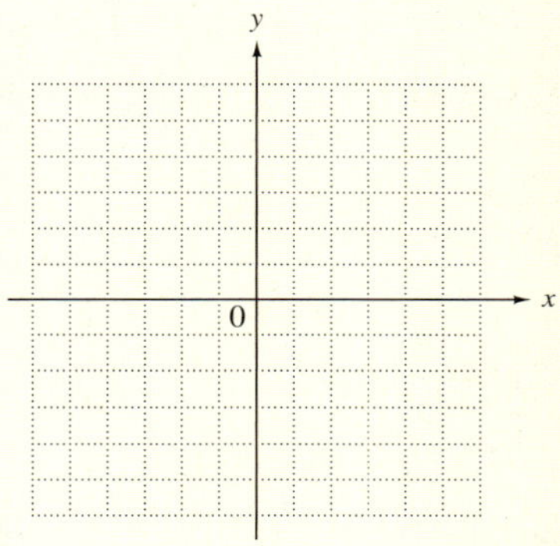

ANSWER

1\.

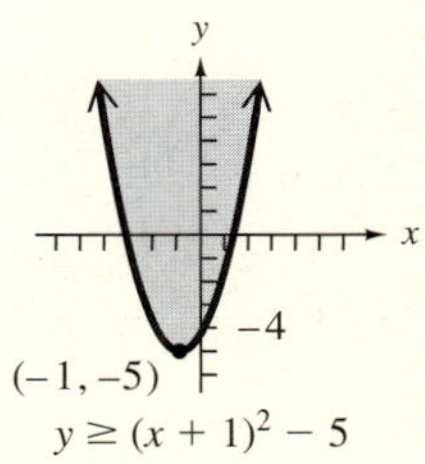

2 Graph $x^2 + 4y^2 > 36$.

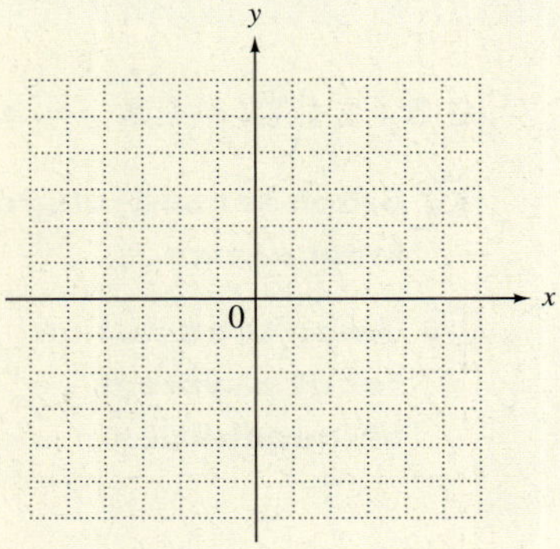

> **Note**
> Since the substitution is easy, the origin is the test point of choice when graphing second-degree inequalities unless the graph actually passes through (0, 0).

EXAMPLE 2 Graphing a Second-Degree Inequality

Graph $16y^2 \leq 144 + 9x^2$.

First rewrite the inequality as follows.

$$16y^2 - 9x^2 \leq 144 \qquad \text{Subtract } 9x^2.$$

$$\frac{y^2}{9} - \frac{x^2}{16} \leq 1 \qquad \text{Divide by 144.}$$

This form shows that the boundary is the hyperbola given by

$$\frac{y^2}{9} - \frac{x^2}{16} = 1.$$

Since the graph is a vertical hyperbola, the desired region will be either the region between the branches or the regions above the top branch and below the bottom branch. Choose (0, 0) as a test point. Substituting into the original inequality leads to $0 \leq 144$, a true statement, so the region between the branches containing (0, 0) is shaded, as shown in Figure 34.

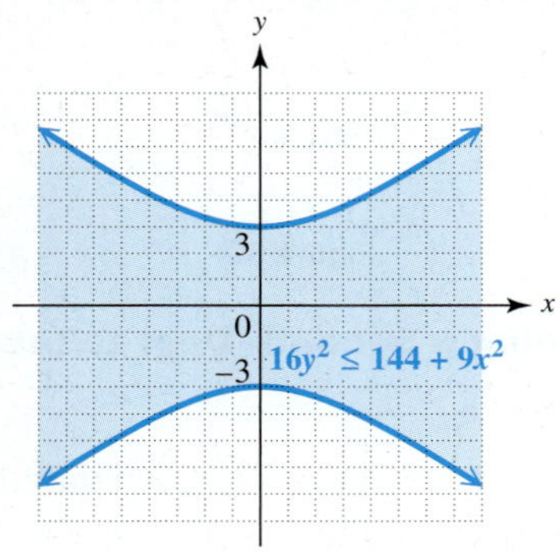

Figure 34

◀ *Work Problem* **2** *at the Side.*

OBJECTIVE 2 Graph the solution set of a system of inequalities.
If two or more inequalities are considered at the same time, we have a **system of inequalities.** To find the solution set of the system, we find the intersection of the graphs (solution sets) of the inequalities in the system.

EXAMPLE 3 Graphing a System of Two Inequalities

Graph the solution set of the system.

$$2x + 3y > 6$$
$$x^2 + y^2 < 16$$

Begin by graphing the solution set of $2x + 3y > 6$. The boundary line is the graph of $2x + 3y = 6$ and is a dashed line because equality is not included. The test point (0, 0) leads to a false statement in $2x + 3y > 6$, so shade the region above the line, as shown in Figure 35 on the next page.

Continued on Next Page

ANSWER

2.

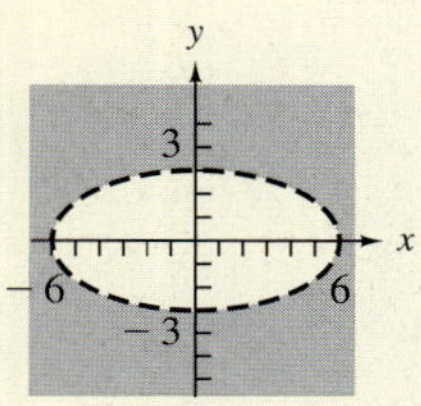

$x^2 + 4y^2 > 36$

The graph of $x^2 + y^2 < 16$ is the interior of a dashed circle centered at the origin with radius 4. This is shown in Figure 36.

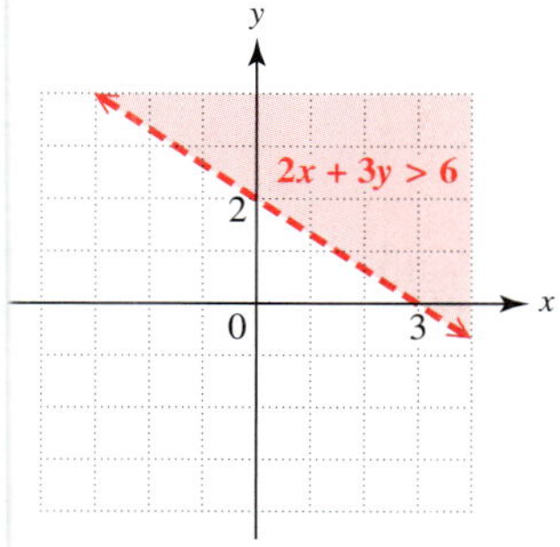

Figure 35

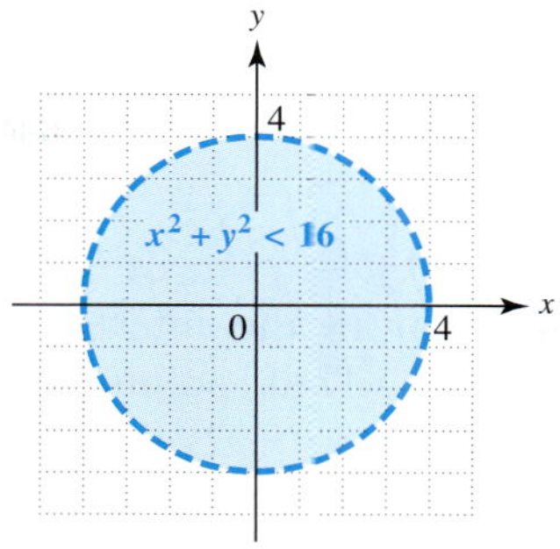

Figure 36

Finally, to show the graph of the solution set of the system, determine the intersection of the graphs of the two inequalities. The overlapping region in Figure 37 is the solution set.

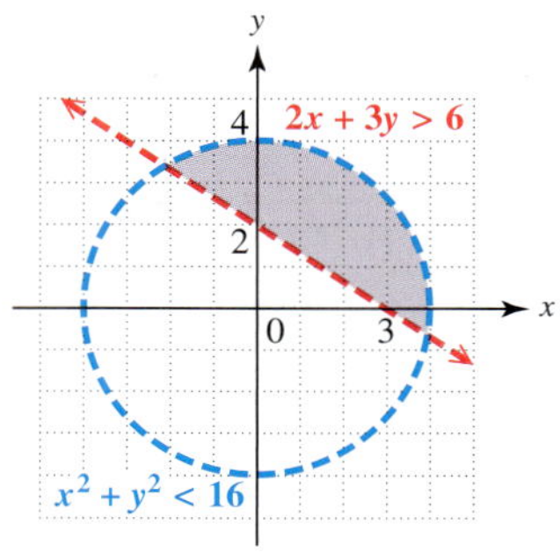

Figure 37

Work Problem **3** at the Side. ▶

3 Graph the solution set of the system.

$$\begin{aligned} x^2 + y^2 &\leq 25 \\ x + y &\leq 3 \end{aligned}$$

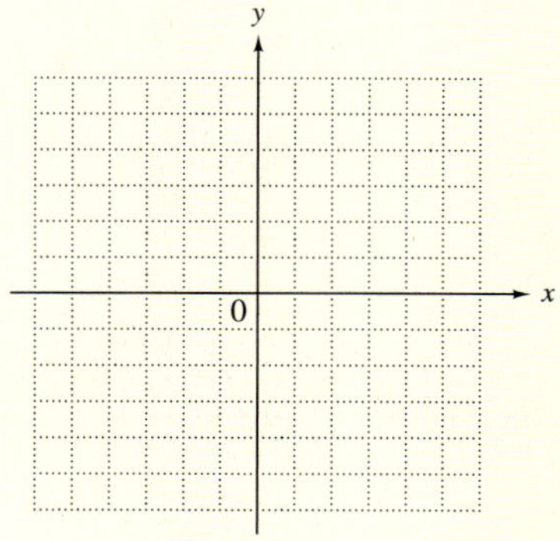

EXAMPLE 4 Graphing a System of Three Inequalities

Graph the solution set of the system.

$$\begin{aligned} x + y &< 1 \\ y &\leq 2x + 3 \\ y &\geq -2 \end{aligned}$$

Graph each inequality separately, on the same axes. The graph of $x + y < 1$ consists of all points below the dashed line $x + y = 1$. The graph of $y \leq 2x + 3$ is the region that lies below the solid line $y = 2x + 3$. Finally, the graph of $y \geq -2$ is the region above the solid horizontal line $y = -2$.

The graph of the system, the intersection of these three graphs, is the triangular region enclosed by the three boundary lines in Figure 38, including two of its boundaries.

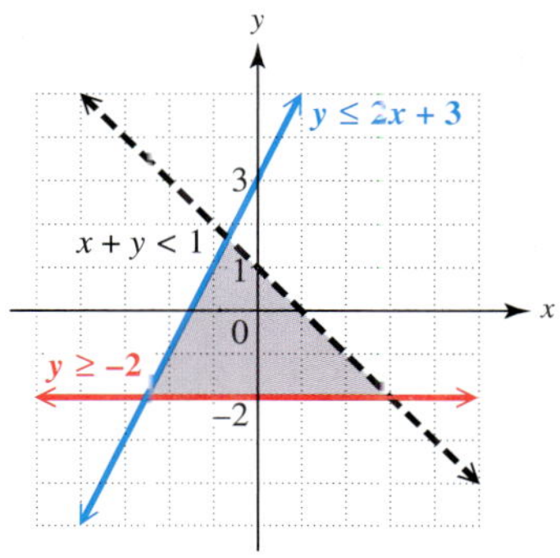

Figure 38

Work Problem **4** at the Side. ▶

4 Graph the solution set of the system.

$$\begin{aligned} 3x - 4y &\geq 12 \\ x + 3y &\geq 6 \\ y &\leq 2 \end{aligned}$$

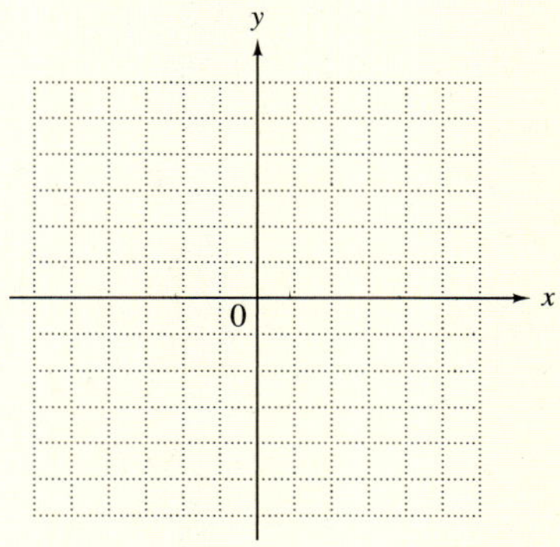

ANSWERS

3.

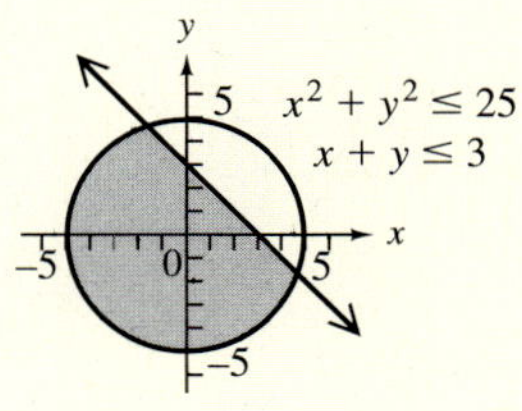

4.

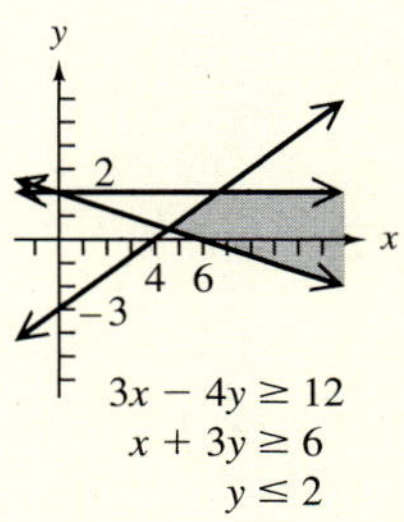

5 Graph the solution set of the system.

$$y \geq x^2 + 1$$
$$\frac{x^2}{9} + \frac{y^2}{4} \geq 1$$
$$y \leq 5$$

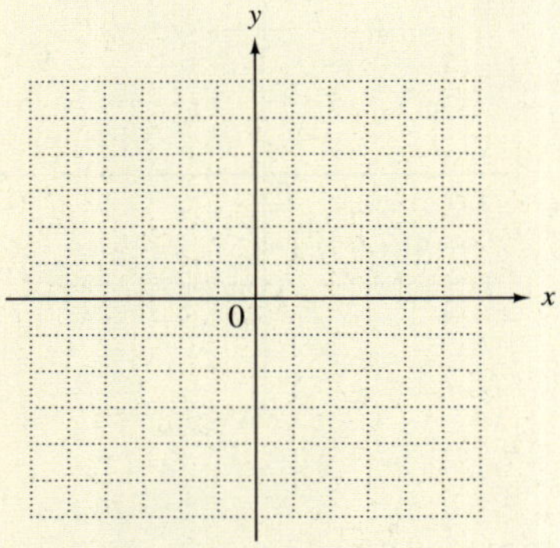

EXAMPLE 5 Graphing a System of Three Inequalities

Graph the solution set of the system.

$$y \geq x^2 - 2x + 1$$
$$2x^2 + y^2 > 4$$
$$y < 4$$

The graph of $y = x^2 - 2x + 1$ is a parabola with vertex at $(1, 0)$. Those points above (or in the interior of) the parabola satisfy the condition $y > x^2 - 2x + 1$. Thus, points on the parabola or in the interior are in the solution set of $y \geq x^2 - 2x + 1$.

The graph of the equation $2x^2 + y^2 = 4$ is an ellipse. We draw it as a dashed curve. To satisfy the inequality $2x^2 + y^2 > 4$, a point must lie outside the ellipse.

The graph of $y < 4$ includes all points below the dashed line $y = 4$. Finally, the graph of the system is the shaded region in Figure 39 that lies outside the ellipse, inside or on the boundary of the parabola, and below the line $y = 4$.

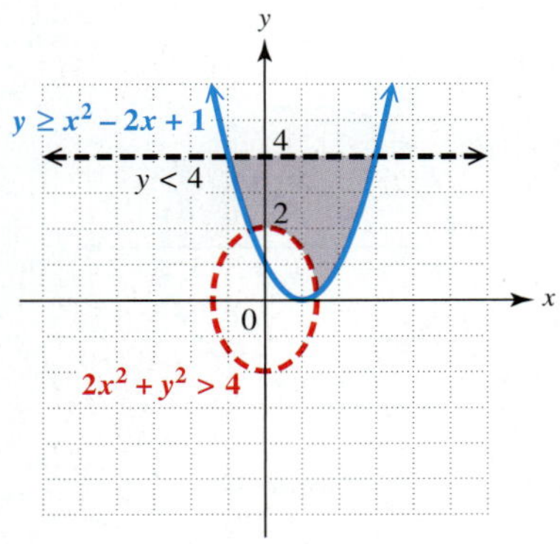

Figure 39

◀ *Work Problem* **5** *at the Side.*

ANSWER

5.

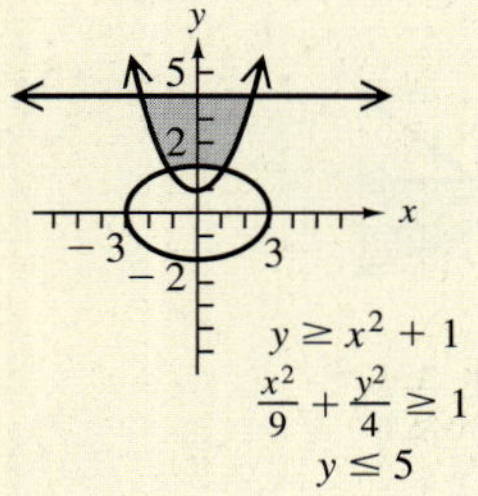

12.5 Exercises

Graph each inequality. See Examples 1 and 2.

1. $y > x^2 - 1$

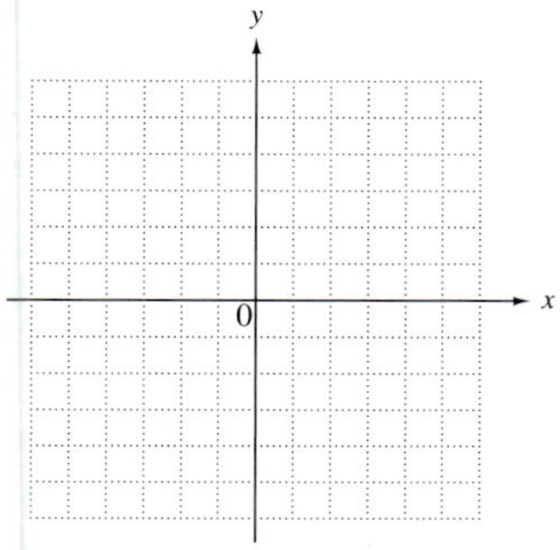

2. $y^2 > 4 + x^2$

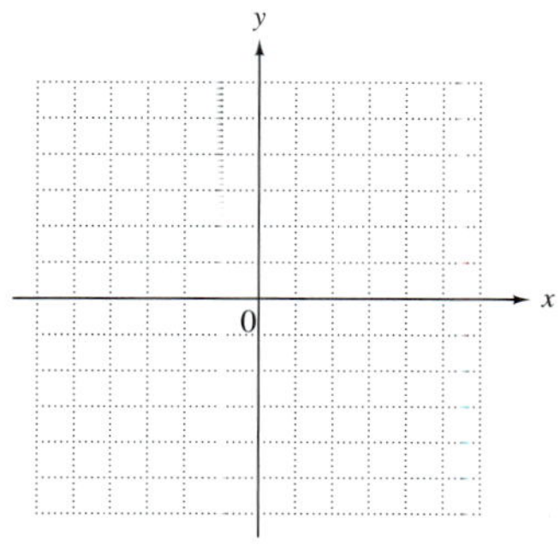

3. $y^2 \leq 4 - 2x^2$

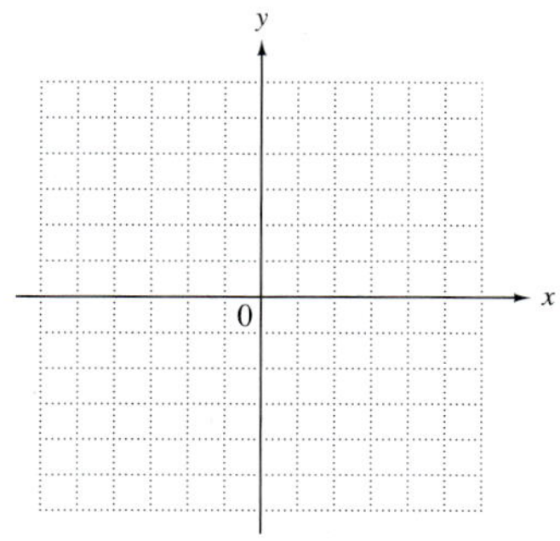

4. $y + 2 \geq x^2$

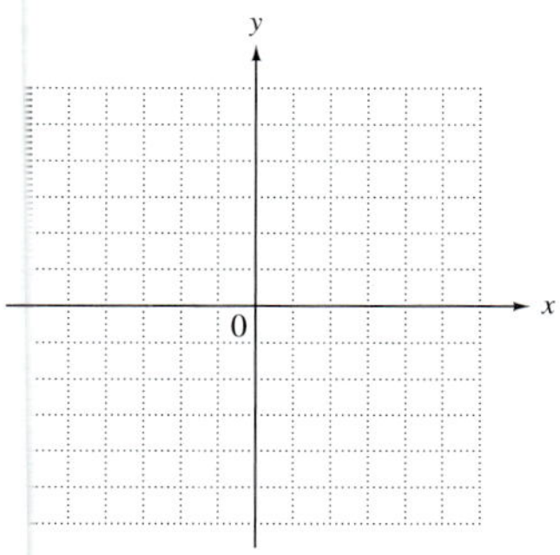

5. $x^2 \leq 16 - y^2$

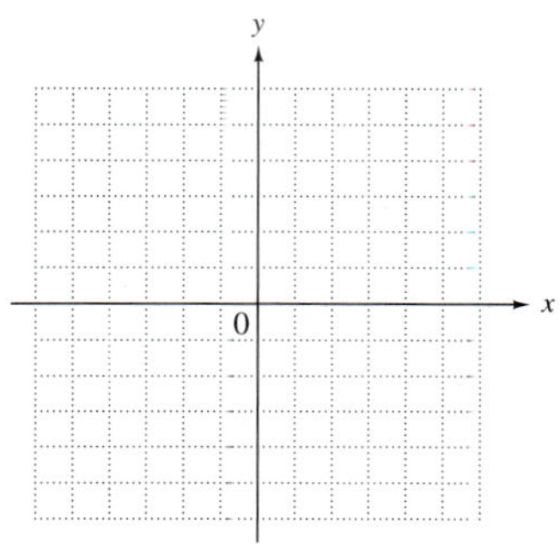

6. $2y^2 \geq 8 - x^2$

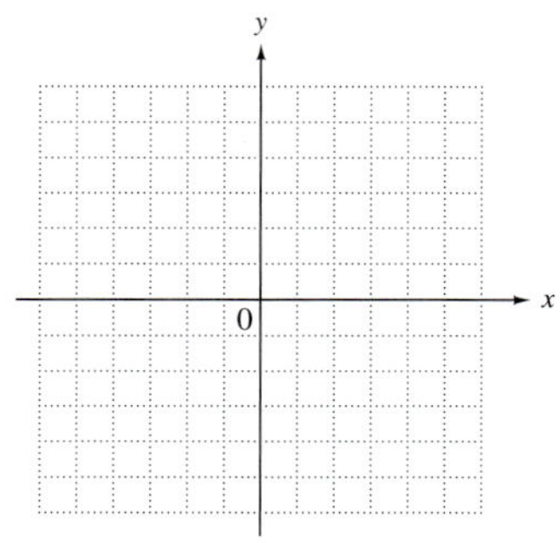

7. $x^2 \leq 16 + 4y^2$

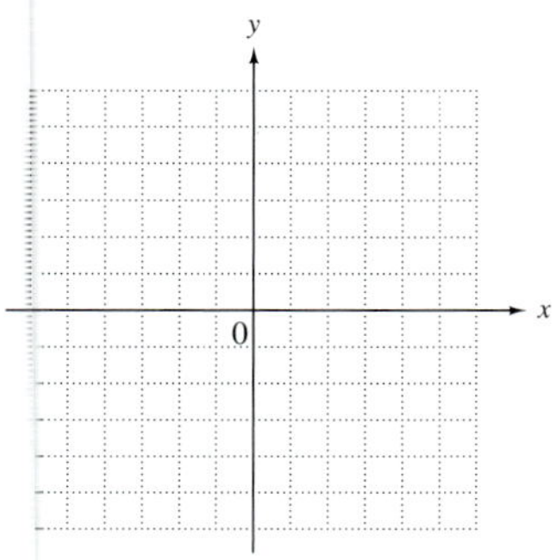

8. $y \leq x^2 + 4x + 2$

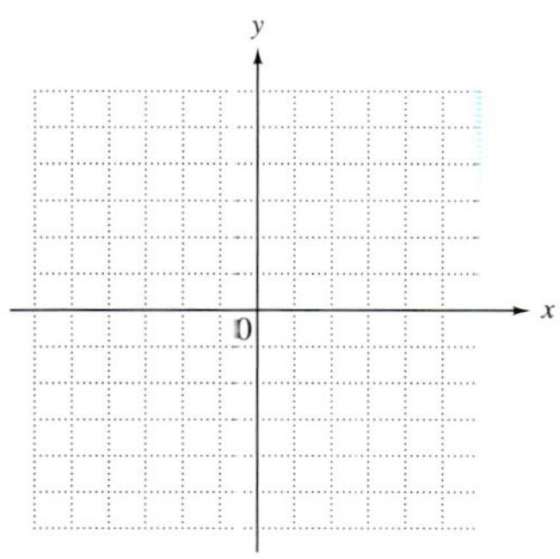

9. $9x^2 < 16y^2 - 144$

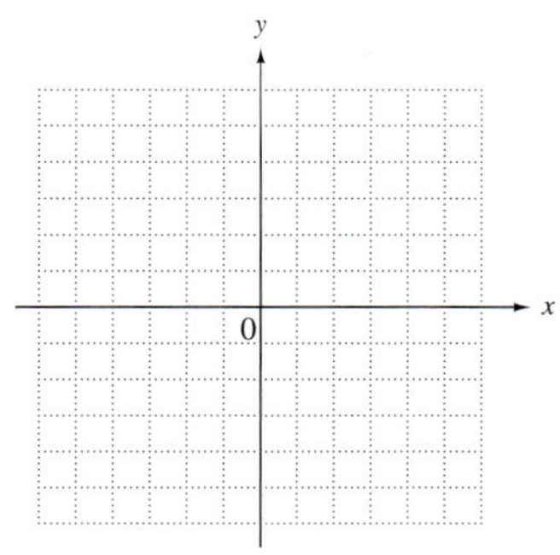

10. $9x^2 > 16y^2 + 144$

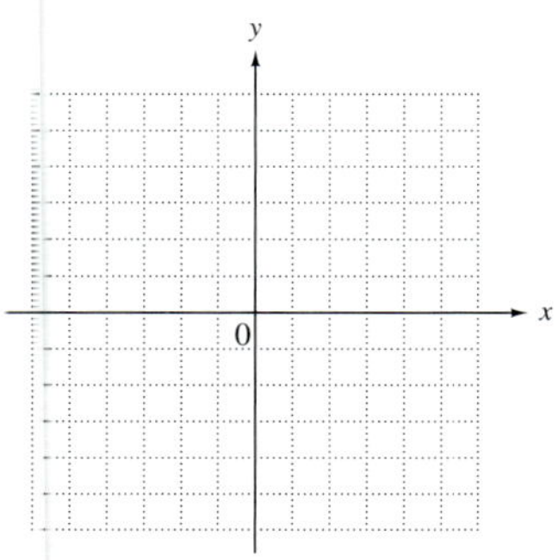

11. $4y^2 \leq 36 - 9x^2$

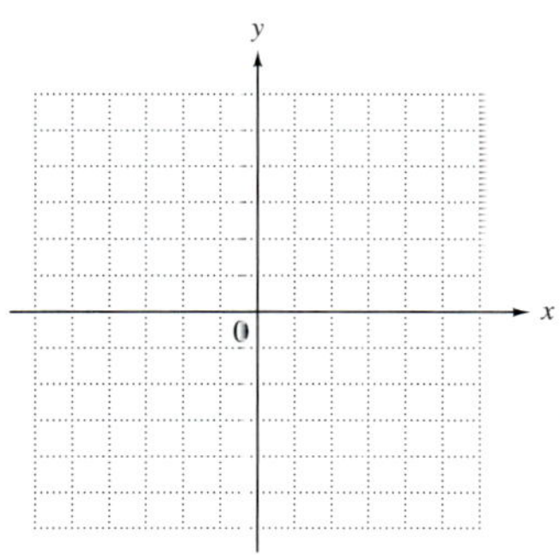

12. $x^2 - 4 \geq -4y^2$

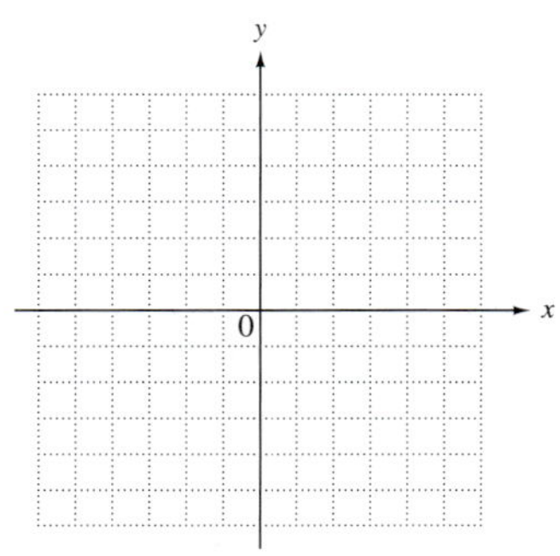

13. $x \geq y^2 - 8y + 14$

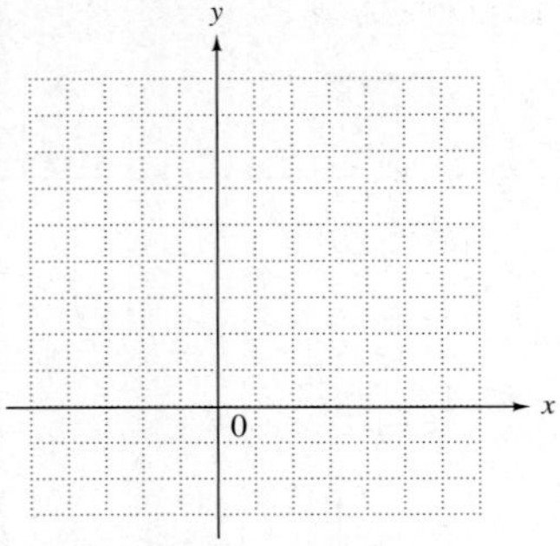

14. $x \leq -y^2 + 6y - 7$

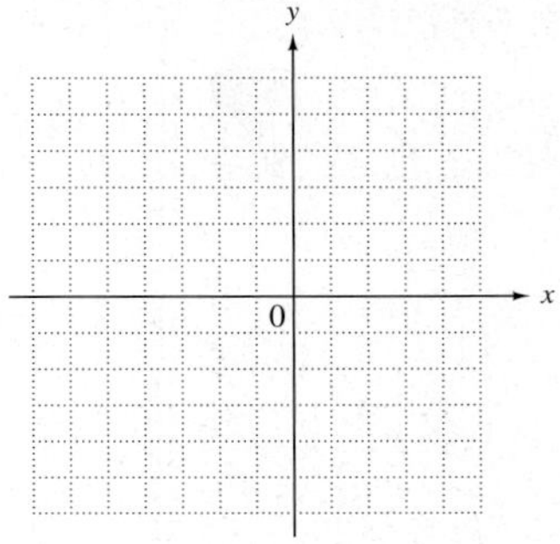

15. $25x^2 \leq 9y^2 + 225$

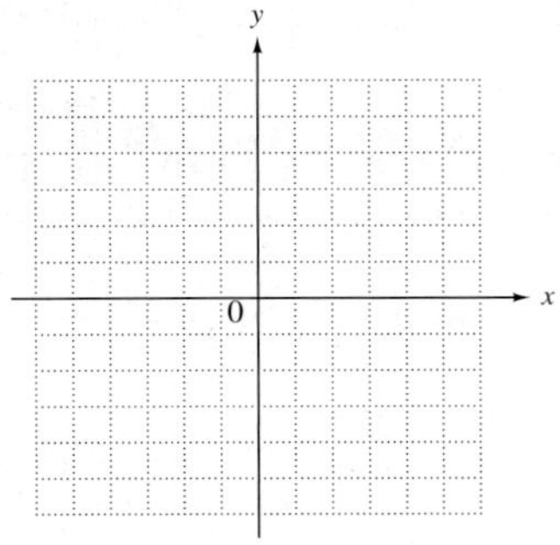

Graph each system of inequalities. See Examples 3–5.

16. $2x + 5y < 10$
$x - 2y < 4$

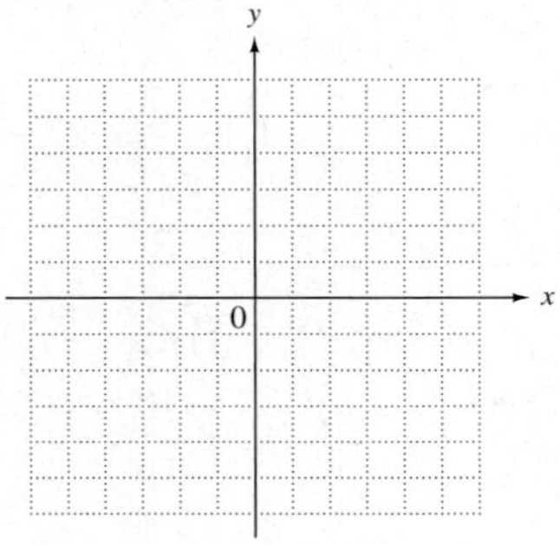

17. $3x - y > -6$
$4x + 3y > 12$

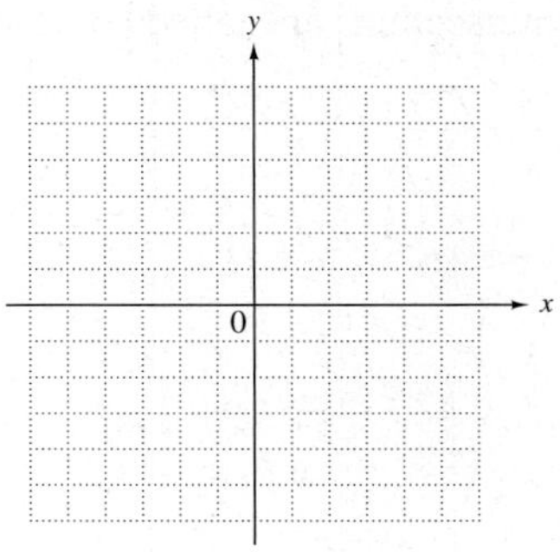

18. $5x - 3y \leq 15$
$4x + y \geq 4$

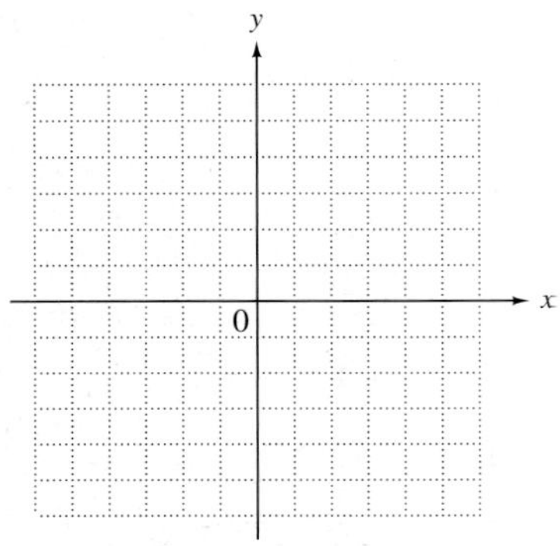

19. $4x - 3y \leq 0$
$x + y \leq 5$

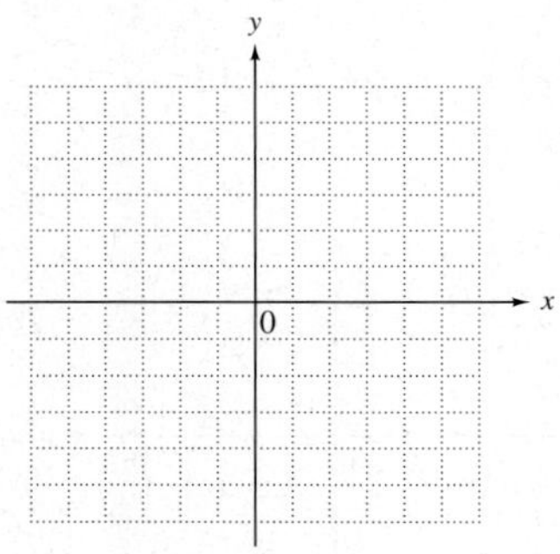

20. $y > x^2 - 4$
$y < -x^2 + 3$

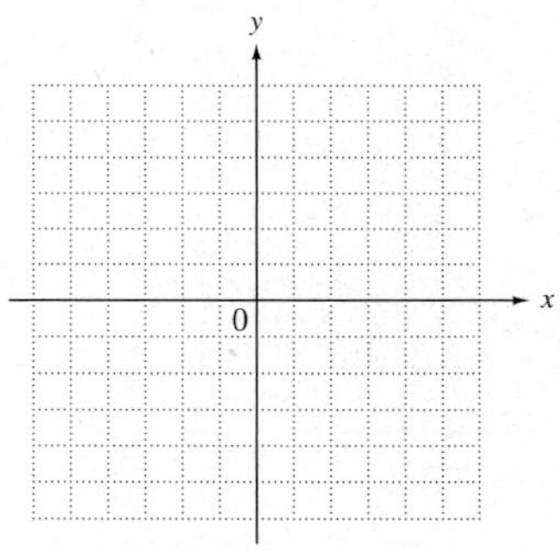

21. $x^2 - y^2 \geq 9$
$\frac{x^2}{16} + \frac{y^2}{9} \leq 1$

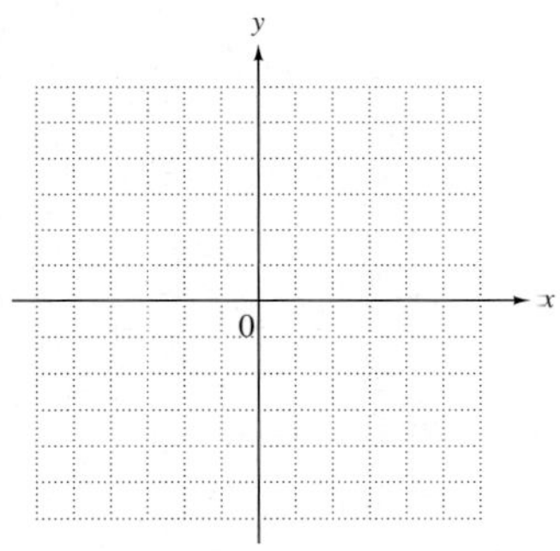

22. $y^2 - x^2 \geq 4$
$-5 \leq y \leq 5$

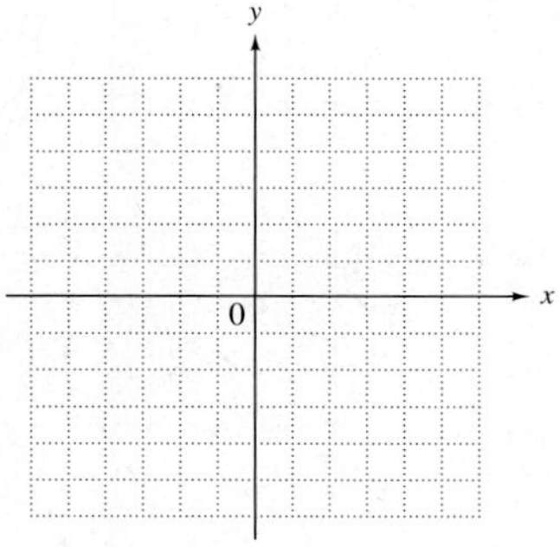

23. $y < x^2$
$y > -2$
$x + y < 3$
$3x - 2y > -6$

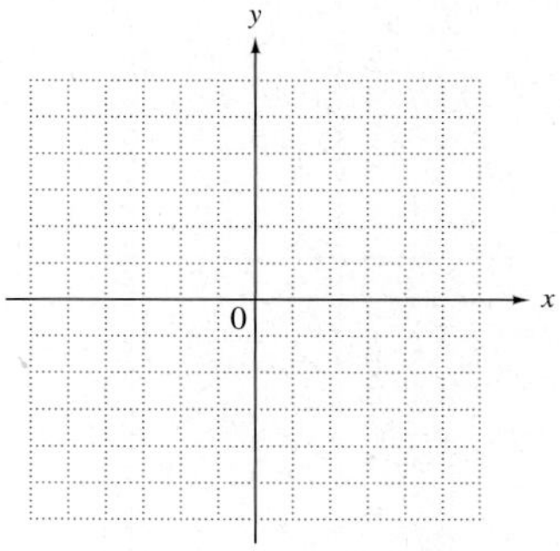

24. $y \leq -x^2$
$y \geq x - 3$
$y \leq -1$
$x < 1$

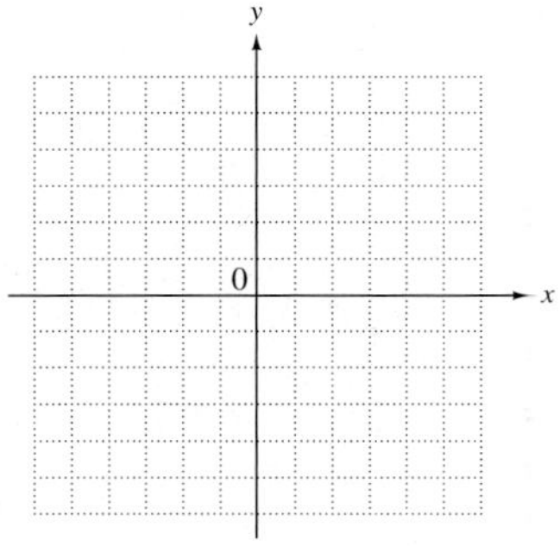

Chapter 12 Summary

Key Terms

12.1	**asymptotes**	Lines that a graph approaches, such as the x- and y-axes for the graph of the reciprocal function, are called asymptotes of the graph.
	greatest integer function	The function defined by $f(x) = [\![x]\!]$, where the symbol $[\![x]\!]$ represents the greatest integer less than or equal to x, is called the greatest integer function.
	step function	A step function is a function with a graph that looks like a series of steps.
	composition (composite function)	If f and g are functions, then the composition of g and f is defined by $(g \circ f)(x) = g(f(x))$ for all x in the domain of f such that $f(x)$ is in the domain of g.
12.2	**conic section**	When a plane intersects an infinite cone at different angles, the figures formed by the intersections are called conic sections.
	circle	A circle is the set of all points in a plane that lie a fixed distance from a fixed point.
	center	The fixed point discussed in the definition of a circle is the center of the circle.
	radius	The radius of a circle is the fixed distance between the center and any point on the circle.
	ellipse	An ellipse is the set of all points in a plane the sum of whose distances from two fixed points **(foci)** is constant.
12.3	**hyperbola**	A hyperbola is the set of all points in a plane such that the absolute value of the difference of the distances from two fixed points (foci) is constant.
	asymptotes of a hyperbola	The two intersecting lines that the branches of a hyperbola approach are called asymptotes of the hyperbola.
	fundamental rectangle	The asymptotes of a hyperbola are the extended diagonals of its fundamental rectangle.
12.4	**nonlinear equation**	An equation in which some terms have more than one variable or a variable of degree 2 or greater is called a nonlinear equation.
	nonlinear system of equations	A nonlinear system of equations is a system with at least one nonlinear equation.
12.5	**second-degree inequality**	A second-degree inequality is an inequality with at least one variable of degree 2 and no variable with degree greater than 2.
	system of inequalities	A system of inequalities consists of two or more inequalities to be solved at the same time.

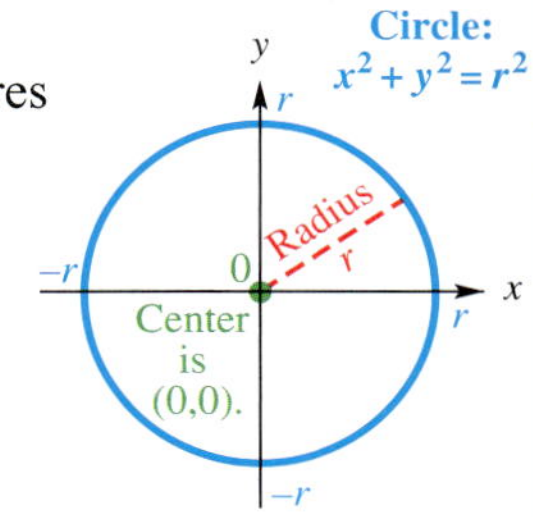

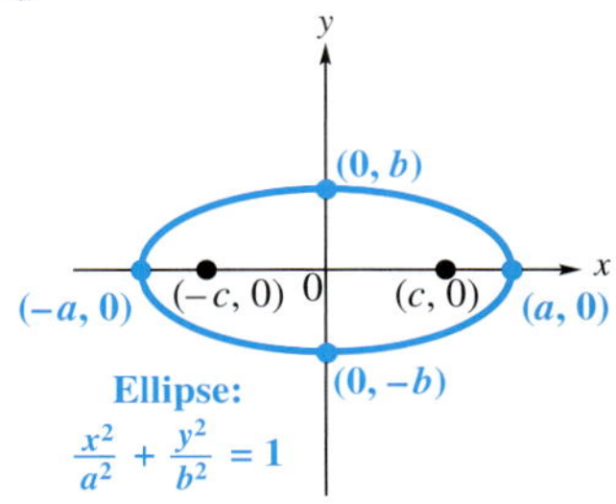

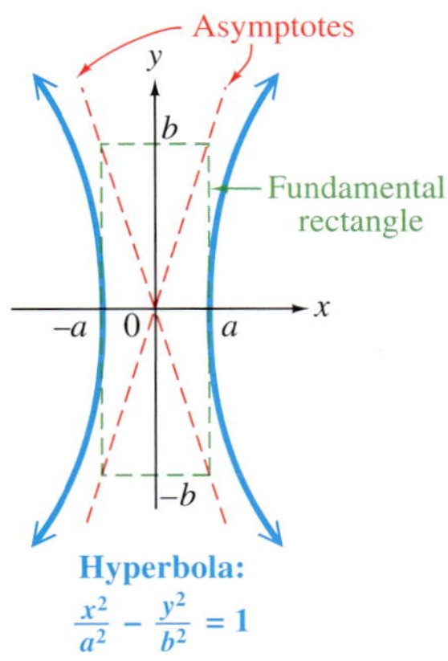

New Symbols

$[\![x]\!]$ greatest integer less than or equal to x

$(f \circ g)(x) = f(g(x))$ composite function

▸ Test Your Word Power

See how well you have learned the vocabulary in this chapter. Answers, with examples, follow the Quick Review.

1. Conic sections are
- **A.** graphs of first-degree equations
- **B.** the result of two or more intersecting planes
- **C.** graphs of first-degree inequalities
- **D.** figures that result from the intersection of an infinite cone with a plane.

2. A **circle** is the set of all points in a plane
- **A.** such that the absolute value of the difference of the distances from two fixed points is constant
- **B.** that lie a fixed distance from a fixed point
- **C.** the sum of whose distances from two fixed points is constant
- **D.** that make up the graph of any second-degree equation.

3. An **ellipse** is the set of all points in a plane
- **A.** such that the absolute value of the difference of the distances from two fixed points is constant
- **B.** that lie a fixed distance from a fixed point
- **C.** the sum of whose distances from two fixed points is constant
- **D.** that make up the graph of any second-degree equation.

4. A **hyperbola** is the set of all points in a plane
- **A.** such that the absolute value of the difference of the distances from two fixed points is constant
- **B.** that lie a fixed distance from a fixed point
- **C.** the sum of whose distances from two fixed points is constant
- **D.** that make up the graph of any second-degree equation.

5. A **nonlinear equation** is an equation
- **A.** in which some terms have more than one variable or a variable of degree 2 or greater
- **B.** in which the terms have only one variable
- **C.** of degree 1
- **D.** of a linear function.

6. A **nonlinear system of equations** is a system
- **A.** with at least one linear equation
- **B.** with two or more inequalities
- **C.** with at least one nonlinear equation
- **D.** with at least two linear equations.

▸ Quick Review

Concepts	Examples

12.1 Additional Graphs of Functions; Composition

Other Functions

In addition to the squaring function, some other important elementary functions in algebra are the absolute value function, defined by $f(x) = |x|$; the reciprocal function, defined by $f(x) = \frac{1}{x}$; the square root function, defined by $f(x) = \sqrt{x}$; and step functions, such as the greatest integer function defined by $f(x) = [\![x]\!]$. Their graphs can be translated.

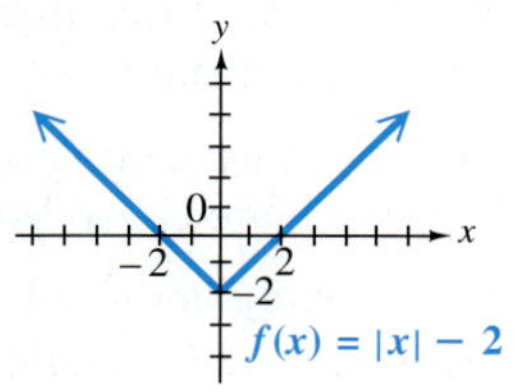

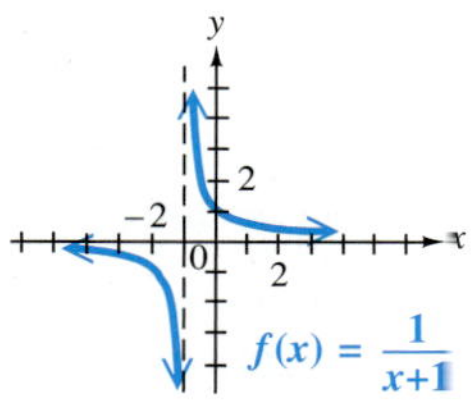

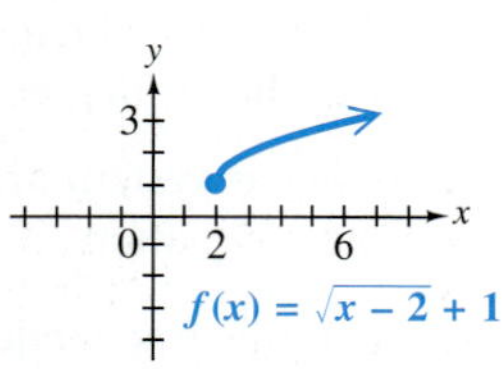

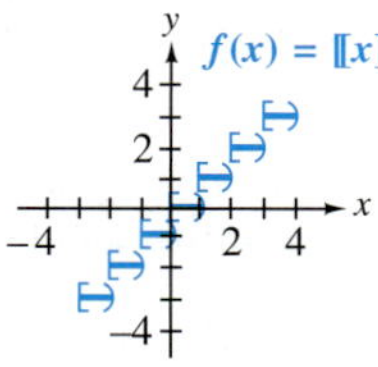

Composition of f and g

$$(f \circ g)(x) = f(g(x))$$

If $f(x) = x^2$ and $g(x) = 2x + 1$, then

$$\begin{aligned}(f \circ g)(x) &= f(g(x))\\ &= (2x + 1)^2\\ &= 4x^2 + 4x + 1\end{aligned}$$

and

$$\begin{aligned}(g \circ f)(x) &= g(f(x))\\ &= 2x^2 + 1.\end{aligned}$$

Concepts	Examples

12.2 The Circle and the Ellipse

Circle

The circle with radius r and center at (h, k) has an equation of the form

$$(x - h)^2 + (y - k)^2 = r^2.$$

The circle with equation $(x + 2)^2 + (y - 3)^2 = 25$, which can be written $[x - (-2)]^2 + (y - 3)^2 = 5^2$, has center $(-2, 3)$ and radius 5.

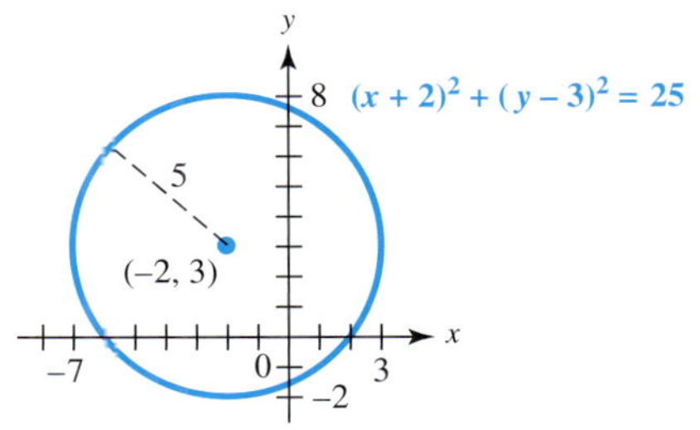

Ellipse

The ellipse whose x-intercepts are $(a, 0)$ and $(-a, 0)$ and whose y-intercepts are $(0, b)$ and $(0, -b)$ has an equation of the form

$$\frac{x^2}{a^2} + \frac{y^2}{b^2} = 1.$$

Graph $\frac{x^2}{9} + \frac{y^2}{4} = 1$.

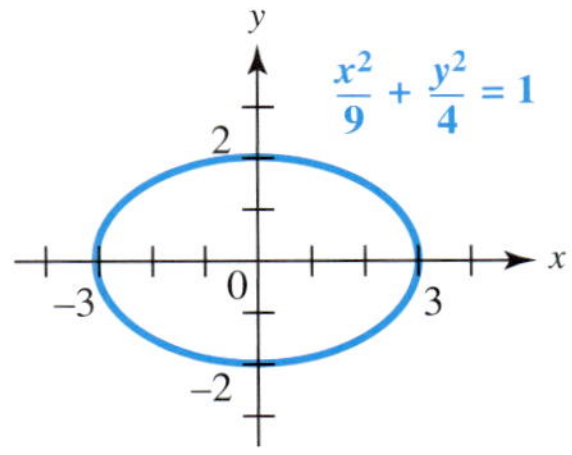

12.3 The Hyperbola and Other Functions Defined by Radicals

Hyperbola

A hyperbola with x-intercepts $(a, 0)$ and $(-a, 0)$ has an equation of the form

$$\frac{x^2}{a^2} - \frac{y^2}{b^2} = 1.$$

and a hyperbola with y-intercepts $(0, b)$ and $(0, -b)$ has an equation of the form

$$\frac{y^2}{b^2} - \frac{x^2}{a^2} = 1.$$

The extended diagonals of the fundamental rectangle with vertices at the points (a, b), $(-a, b)$, $(-a, -b)$, and $(a, -b)$ are the asymptotes of these hyperbolas.

Graph $\frac{x^2}{4} - \frac{y^2}{4} = 1$.

The graph has x-intercepts $(2, 0)$ and $(-2, 0)$.

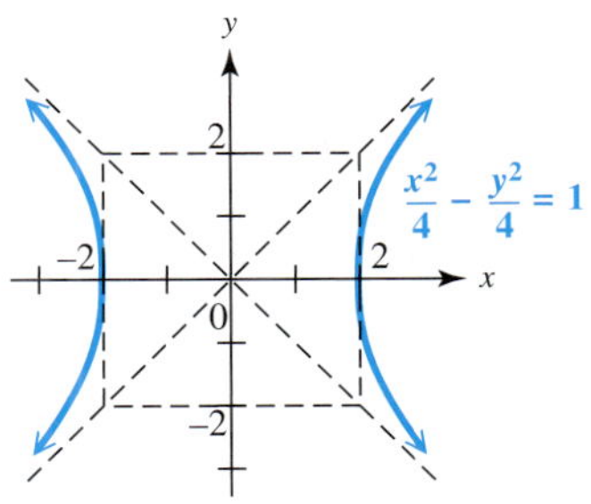

The fundamental rectangle has vertices at $(2, 2)$, $(-2, 2)$, $(-2, -2)$, and $(2, -2)$.

Square Root Function

For an algebraic expression in x defined by u, with $u \geq 0$, a function of the form

$$f(x) = \sqrt{u}$$

is called a square root function.

To graph a square root function, square both sides so that the equation can be easily recognized. Then graph only the part indicated by the original equation.

Graph $y = -\sqrt{4 - x^2}$.

Square both sides and rearrange terms to get

$$x^2 + y^2 = 4.$$

This equation has a circle as its graph. However, graph only the lower half of the circle, since the original equation indicates that y cannot be positive.

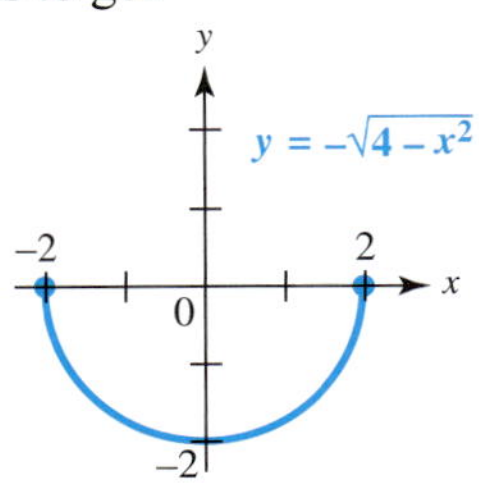

Concepts	Examples

12.4 Nonlinear Systems of Equations

Solving a Nonlinear System

A nonlinear system can be solved by the substitution method, the elimination method, or a combination of the two.

Solve the system.

$$x^2 + 2xy - y^2 = 14 \quad (1)$$
$$x^2 - y^2 = -16 \quad (2)$$

Multiply equation (2) by -1 and use elimination.

$$\begin{aligned} x^2 + 2xy - y^2 &= 14 \\ -x^2 \qquad + y^2 &= 16 \\ \hline 2xy \qquad &= 30 \\ xy &= 15 \end{aligned}$$

Solve for y to obtain $y = \frac{15}{x}$, and substitute into equation (2).

$$x^2 - \left(\frac{15}{x}\right)^2 = -16 \quad (2)$$
$$x^2 - \frac{225}{x^2} = -16$$
$$x^4 + 16x^2 - 225 = 0 \quad \text{Multiply by } x^2\text{; add } 16x^2.$$
$$(x^2 - 9)(x^2 + 25) = 0 \quad \text{Factor.}$$
$$x = \pm 3 \quad \text{or} \quad x = \pm 5i \quad \text{Zero-factor property}$$

Find corresponding y-values to obtain the solution set

$$\{(3, 5), (-3, -5), (5i, -3i), (-5i, 3i)\}.$$

12.5 Second-Degree Inequalities and Systems of Inequalities

Graphing a Second-Degree Inequality

To graph a second-degree inequality, graph the corresponding equation as a boundary and use test points to determine which region(s) form the solution set. Shade the appropriate region(s).

Graphing a System of Inequalities

The solution set of a system of inequalities is the intersection of the solution sets of the individual inequalities.

Graph $y \geq x^2 - 2x + 3$.

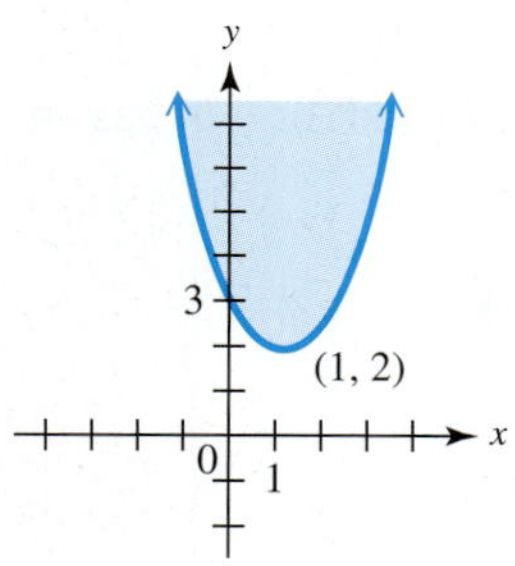

Graph the solution set of the system.

$$3x - 5y > -15$$
$$x^2 + y^2 \leq 25$$

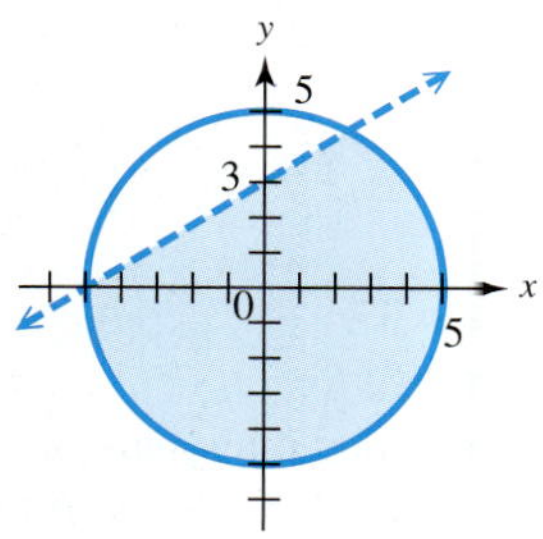

ANSWERS TO TEST YOUR WORD POWER

1. D; *Example*: Parabolas, circles, ellipses, and hyperbolas are conic sections.
2. B; *Example*: See the graph of $x^2 + y^2 = 9$ in Figure 12 of **Section 12.2.**
3. C; *Example*: See the graph of $\frac{x^2}{49} + \frac{y^2}{36} = 1$ in Figure 17 of **Section 12.2.**
4. A; *Example*: See the graph of $\frac{x^2}{16} - \frac{y^2}{12} = 1$ in Figure 20 of **Section 12.3.**
5. A; *Examples*: $y = x^2 + 8x + 16$, $xy = 5$, $2x^2 - y^2 = 6$
6. C; *Example*: $x^2 + y^2 = 2$
   $2x + y = 1$

Chapter 12 Review Exercises

[12.1] *Graph each function. Give the domain and range.*

1. $f(x) = |x + 4|$

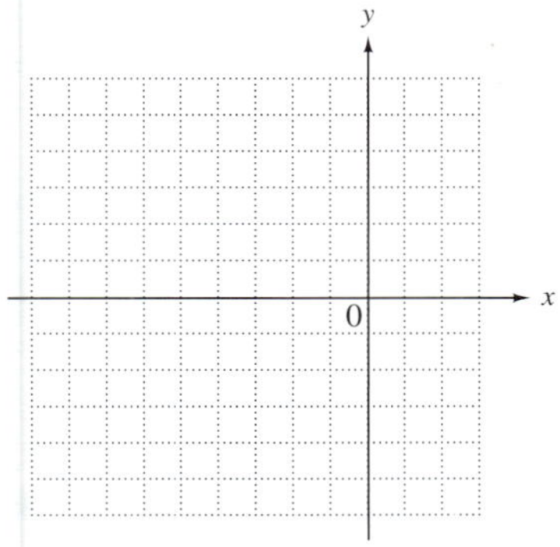

2. $f(x) = \dfrac{1}{x - 4}$

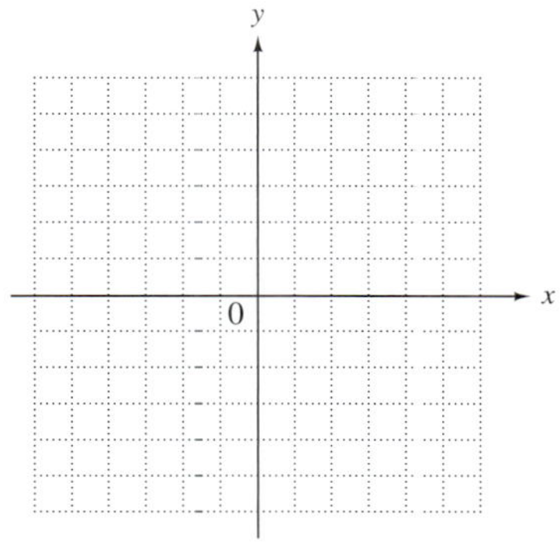

3. $f(x) = \sqrt{x} + 3$

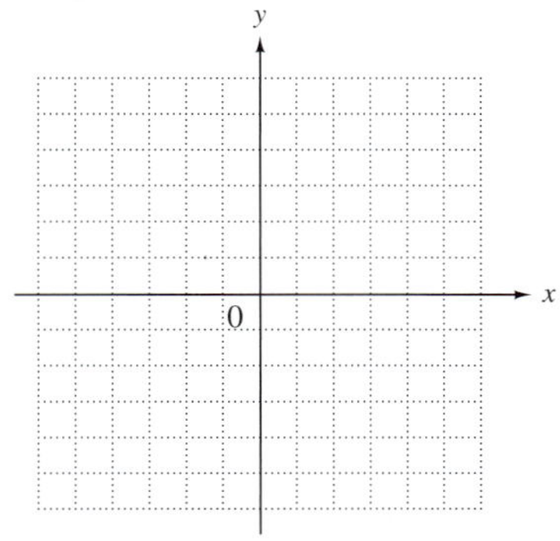

Find each of the following.

4. $[\![12]\!]$

5. $\left[\!\!\left[2\frac{1}{4}\right]\!\!\right]$

6. $[\![-4.75]\!]$

Let $f(x) = 3x^2 + 2x - 1$ *and* $g(x) = 5x + 7$. *Find each of the following.*

7. **(a)** $(g \circ f)(3)$ **(b)** $(f \circ g)(3)$

8. **(a)** $(f \circ g)(-2)$ **(b)** $(g \circ f)(-2)$

9. **(a)** $(f \circ g)(x)$ **(b)** $(g \circ f)(x)$

10. Based on your answers to Exercises 7–9, discuss whether composition of functions is a commutative operation.

[12.2] *Write an equation for each circle.*

11. Center $(-2, 4)$, $r = 3$

12. Center $(-1, -3)$, $r = 5$

13. Center $(4, 2)$, $r = 6$

Find the center and radius of each circle.

14. $x^2 + y^2 + 6x - 4y - 3 = 0$

15. $x^2 + y^2 - 8x - 2y + 13 = 0$

16. $2x^2 + 2y^2 + 4x + 20y = -34$

17. $4x^2 + 4y^2 - 24x + 16y = 48$

Graph each equation.

18. $x^2 + y^2 = 16$

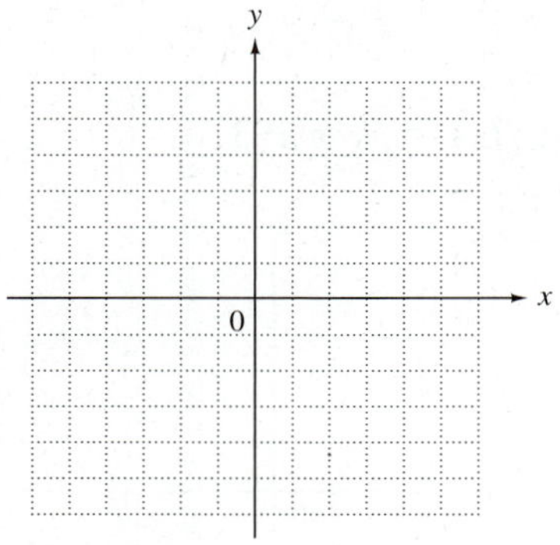

19. $\dfrac{x^2}{16} + \dfrac{y^2}{9} = 1$

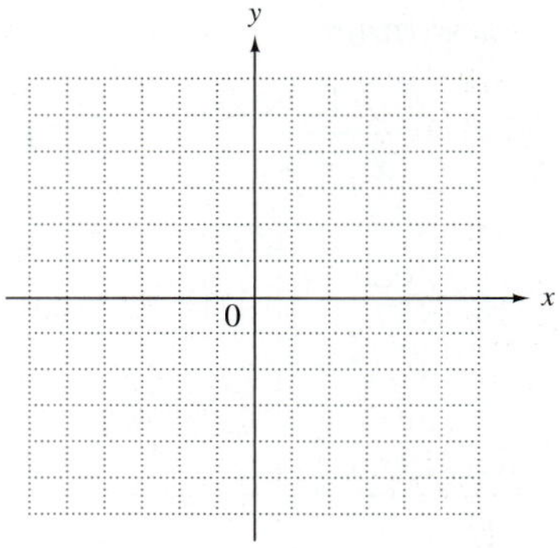

20. $\dfrac{x^2}{49} + \dfrac{y^2}{25} = 1$

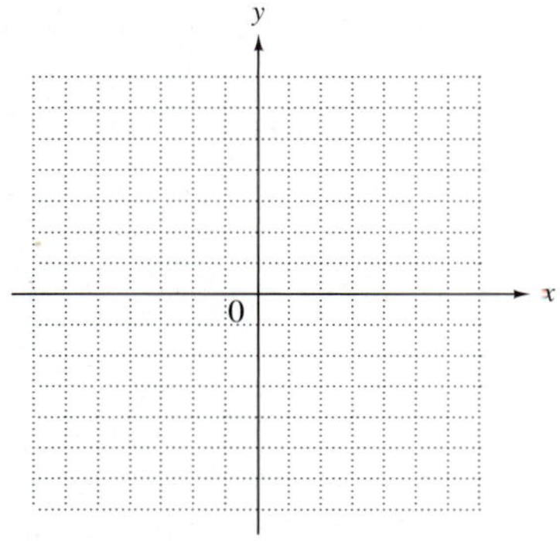

21. A satellite is in an elliptical orbit around Earth with perigee altitude of 160 km and apogee altitude of 16,000 km. See the figure. (*Source*: Kastner, Bernice, *Space Mathematics*, NASA, 1985.) Find the equation of the ellipse.

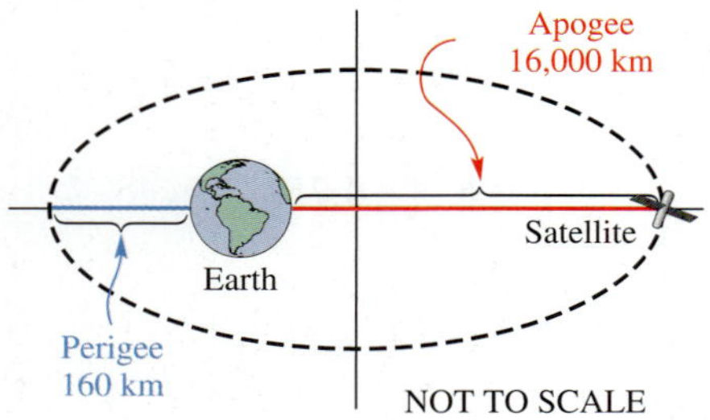

22. (a) The Roman Colosseum is an ellipse with $a = 310$ ft and $b = \frac{513}{2}$ ft. Find the distance between the foci of this ellipse to the nearest tenth of a foot.

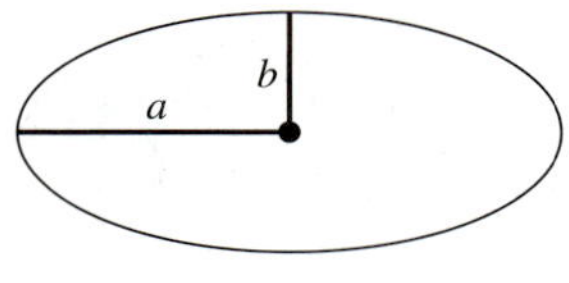

(b) A formula for the approximate circumference of an ellipse is

$$C \approx 2\pi\sqrt{\frac{a^2 + b^2}{2}},$$

where a and b are the lengths given in part (a). Use this formula to find the approximate circumference of the Roman Colosseum.

[12.3] *Graph each equation.*

23. $\dfrac{x^2}{16} - \dfrac{y^2}{25} = 1$

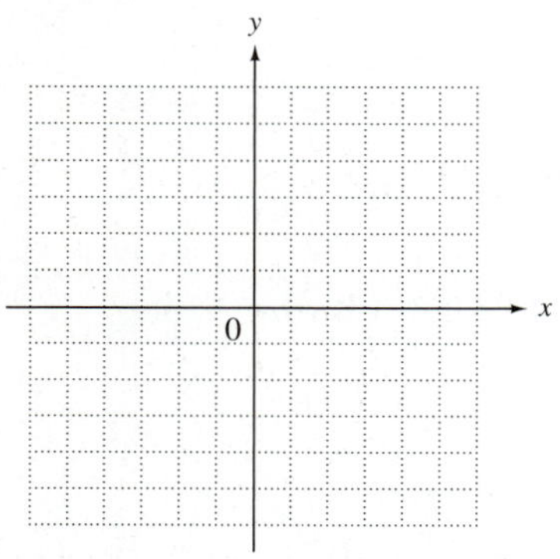

24. $\dfrac{y^2}{25} - \dfrac{x^2}{4} = 1$

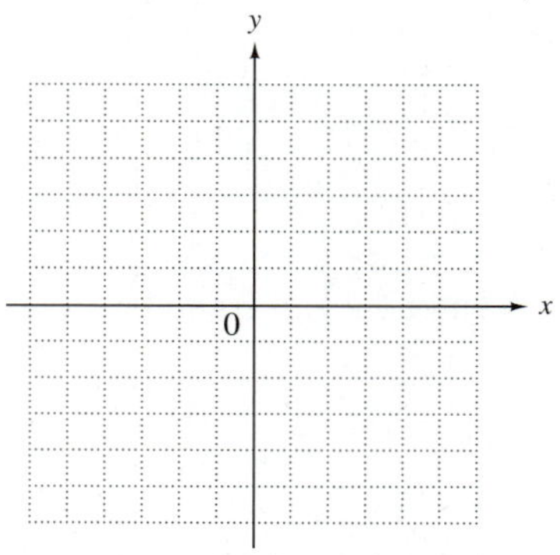

25. $f(x) = -\sqrt{16 - x^2}$

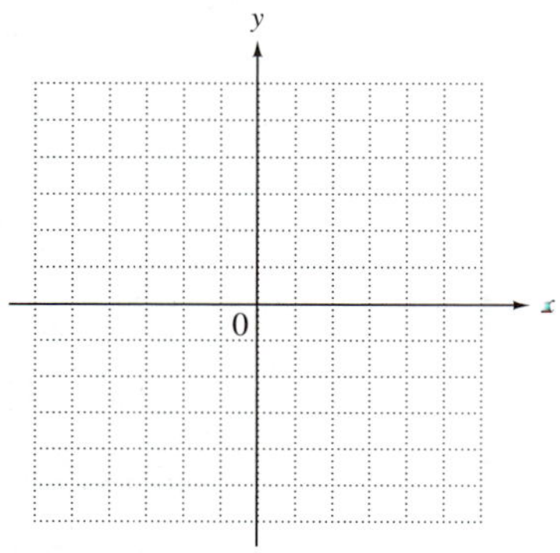

Identify the graph of each equation as a parabola, circle, ellipse, *or* hyperbola.

26. $x^2 + y^2 = 64$

27. $y = 2x^2 - 3$

28. $y^2 = 2x^2 - 8$

29. $y^2 = 8 - 2x^2$

30. $x = y^2 + 4$

31. $x^2 - y^2 = 64$

32. Ships and planes often use a location-finding system called LORAN. With this system, a radio transmitter at M sends out a series of pulses. (See the figure.) When each pulse is received at transmitter S, it then sends out a pulse. A ship at P receives pulses from both M and S. A receiver on the ship measures the difference in the arrival times of the pulses. A special map gives hyperbolas that correspond to the differences in arrival times (which give the distances d_1 and d_2 in the figure). The ship can then be located as lying on a branch of a particular hyperbola. Suppose $d_1 = 80$ mi and $d_2 = 30$ mi, and the distance between transmitters M and S is 100 mi. Use the definition to find an equation of the hyperbola on which the ship is located.

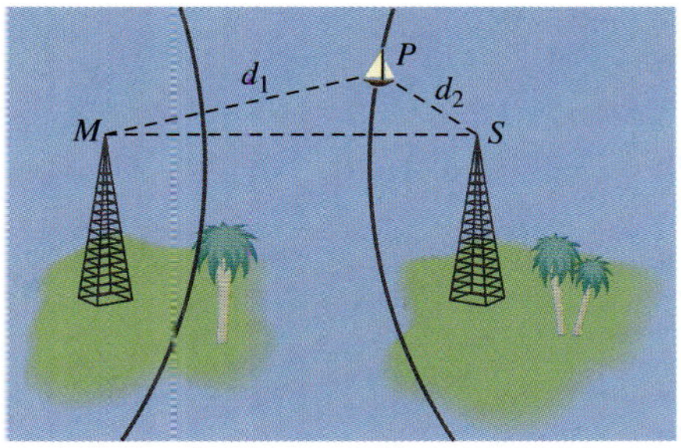

[12.4] *Solve each system.*

33. $2y = 3x - x^2$
$x + 2y = -12$

34. $y + 1 = x^2 + 2x$
$y + 2x = 4$

35. $x^2 + 3y^2 = 28$
$y - x = -2$

36. $xy = 8$
$x - 2y = 6$

37. $x^2 + y^2 = 6$
$x^2 - 2y^2 = -6$

38. $3x^2 - 2y^2 = 12$
$x^2 + 4y^2 = 18$

39. How many solutions are possible for a system of two equations whose graphs are a circle and a line?

40. How many solutions are possible for a system of two equations whose graphs are a parabola and a hyperbola?

[12.5] *Graph each inequality.*

41. $9x^2 \geq 16y^2 + 144$

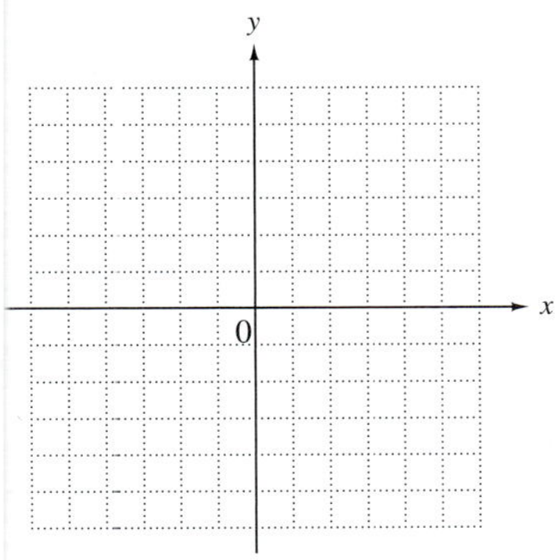

42. $4x^2 + y^2 \geq 16$

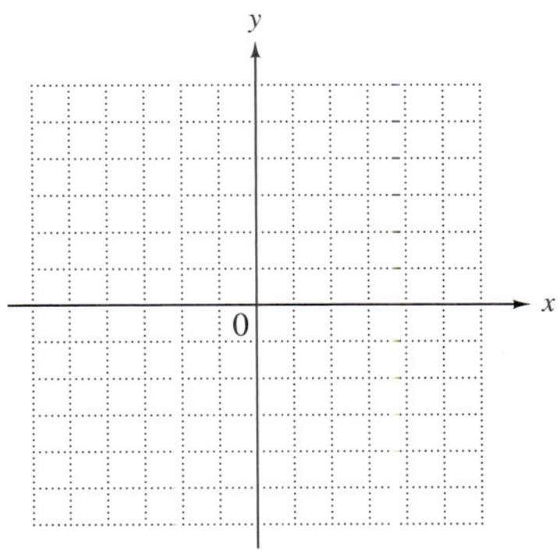

43. $y < -(x + 2)^2 + 1$

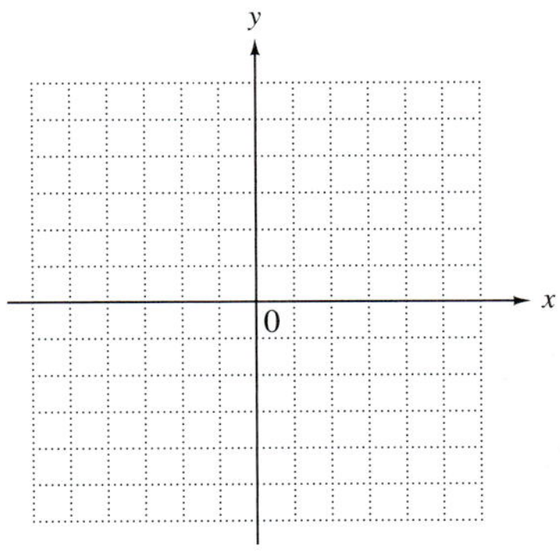

Graph each system of inequalities.

44. $2x + 5y \leq 10$
$3x - y \leq 6$

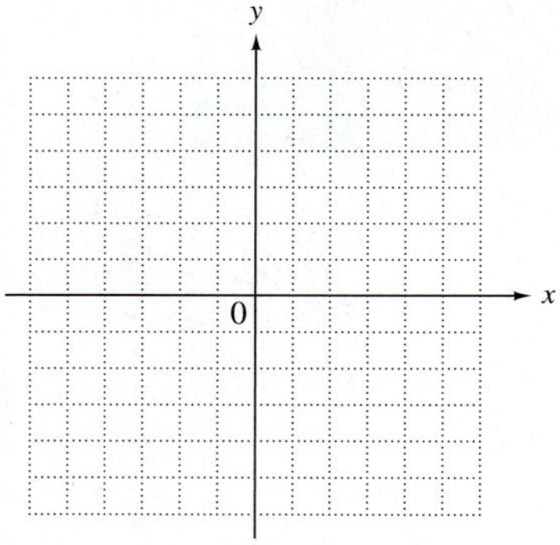

45. $|x| \leq 2$
$|y| > 1$
$4x^2 + 9y^2 \leq 36$

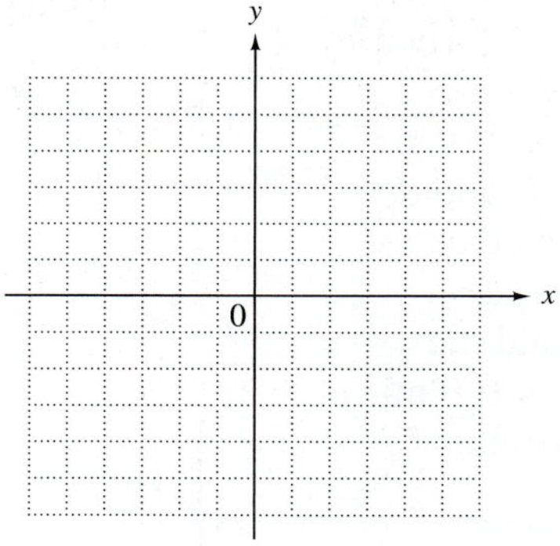

46. $9x^2 \leq 4y^2 + 36$
$x^2 + y^2 \leq 16$

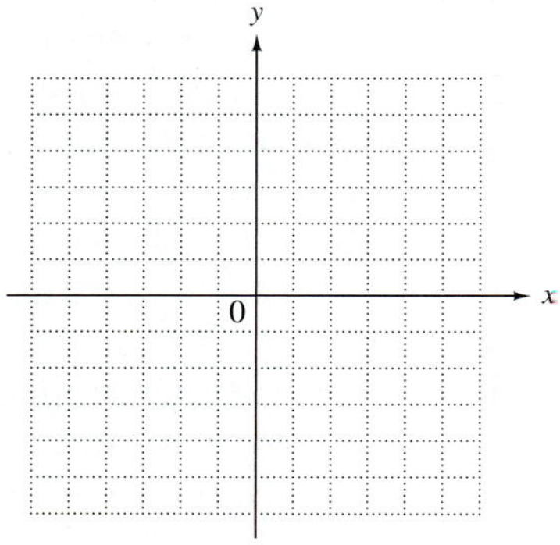

Mixed Review Exercises

Graph.

47. $x^2 + y^2 = 25$

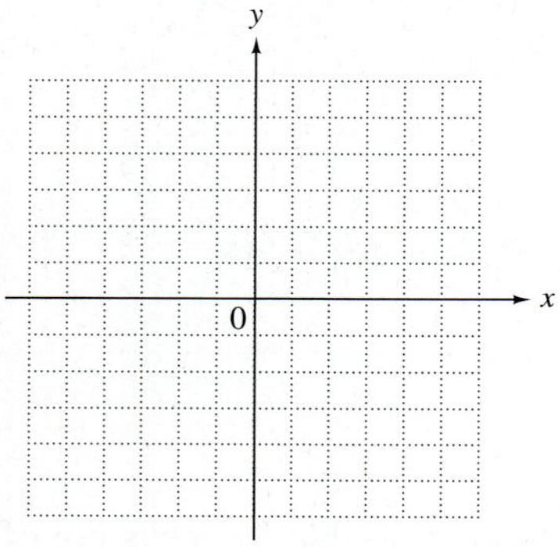

48. $x^2 + 9y^2 = 9$

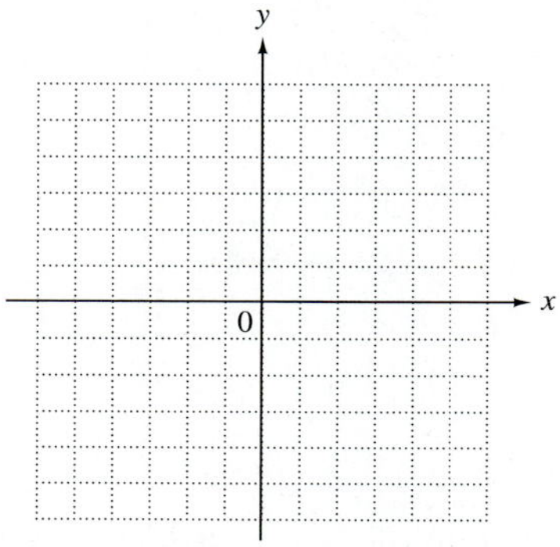

49. $x^2 - 9y^2 = 9$

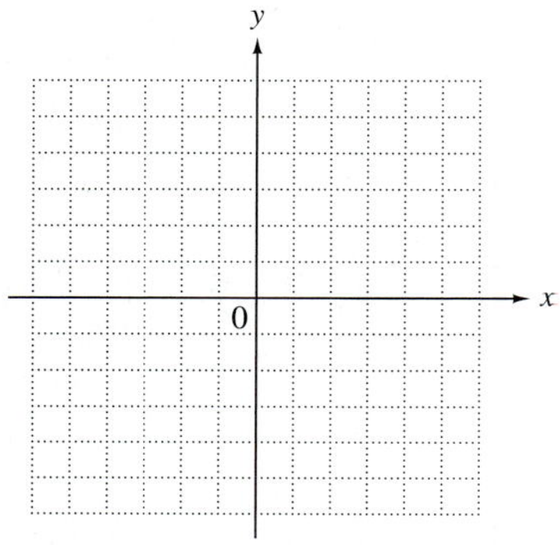

50. $f(x) = \sqrt{4 - x}$

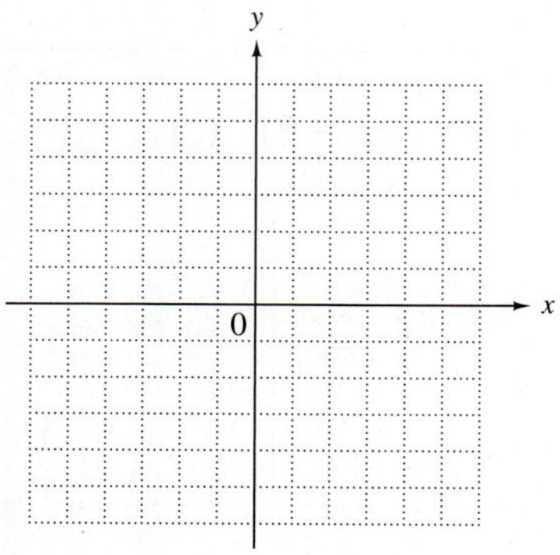

51. $f(x) = [\![x]\!] - 1$

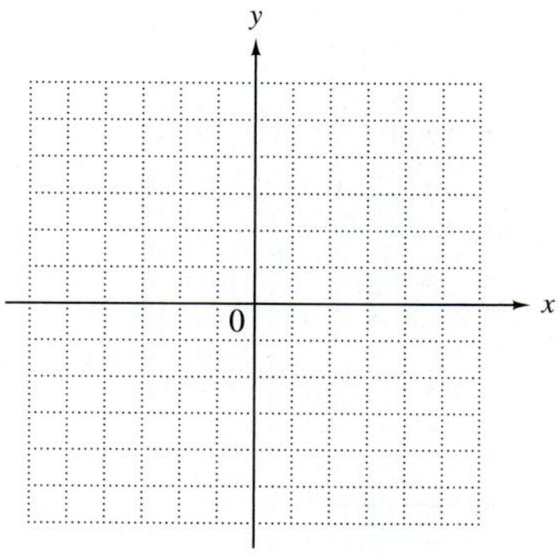

52. $4y > 3x - 12$
$x^2 < 16 - y^2$

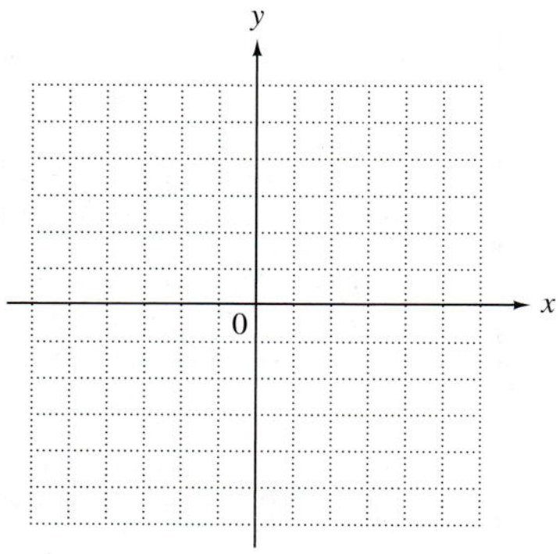

53. Explain why a set of points that form an ellipse does not satisfy the definition of a function.

Chapter 12 Test

Chapter Test Prep Video CD — Use the Chapter Test Prep Video CD to see fully worked-out solutions to any of the exercises you want to review

Match each function with its graph from choices A, B, C, and D.

1. $f(x) = \sqrt{x - 2}$

2. $f(x) = \sqrt{x + 2}$

3. $f(x) = \sqrt{x} + 2$

4. $f(x) = \sqrt{x} - 2$

A.

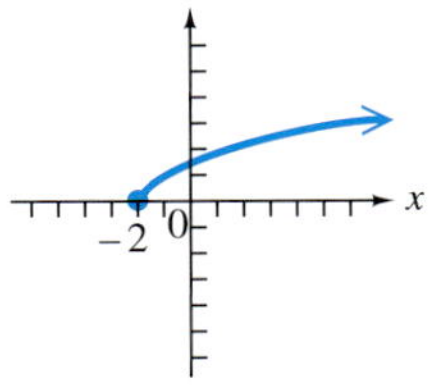

B.

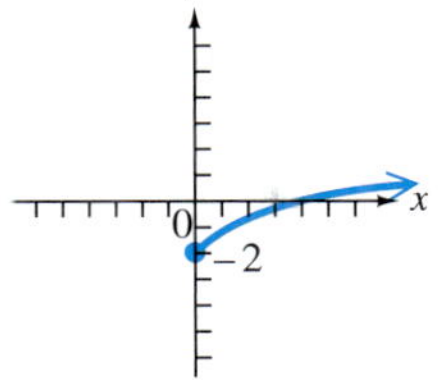

C.

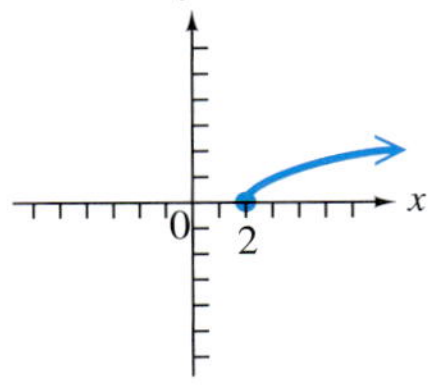

D.

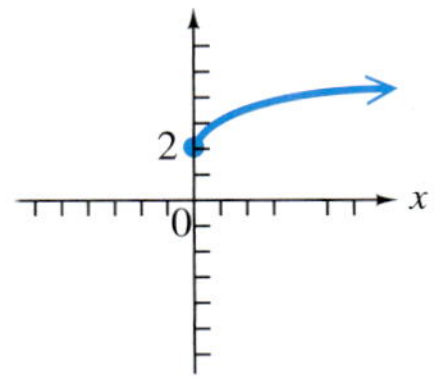

1. ______________

2. ______________

3. ______________

4. ______________

5. Sketch the graph of $f(x) = |x - 3| + 4$. Give the domain and range.

5. ______________

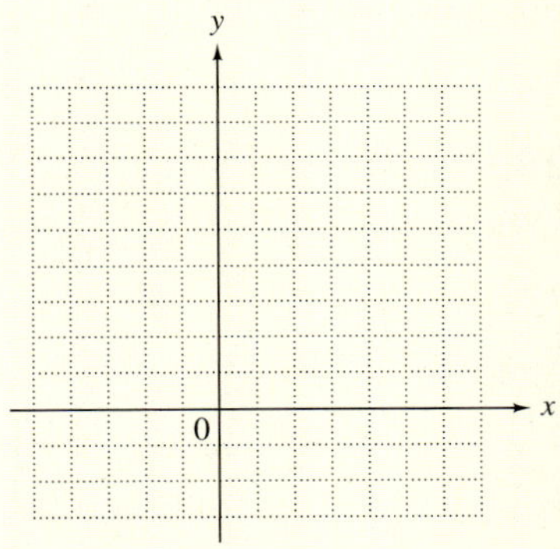

6. For $f(x) = 3x + 5$ and $g(x) = x^2 + 2$, find each of the following.

(a) $(f \circ g)(-2)$

(b) $(f \circ g)(x)$

(c) $(g \circ f)(x)$

6. (a) ______________

(b) ______________

(c) ______________

7. Find the center and radius of the circle whose equation is $(x - 2)^2 + (y + 3)^2 = 16$. Sketch the graph.

7. ______________

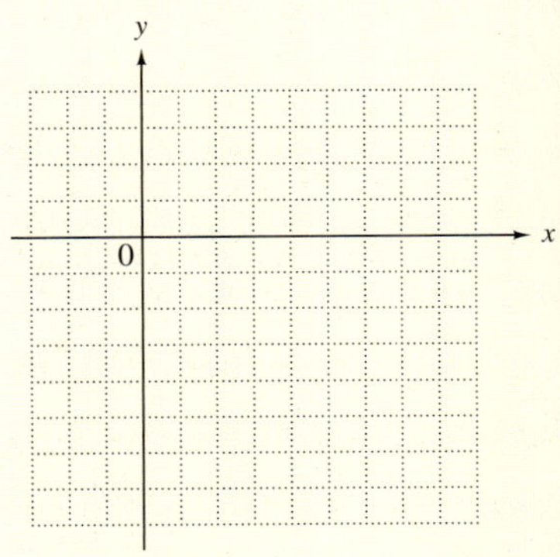

8. ______________________

8. Find the center and radius of the circle whose equation is $x^2 + y^2 + 8x - 2y = 8$.

Graph.

9.

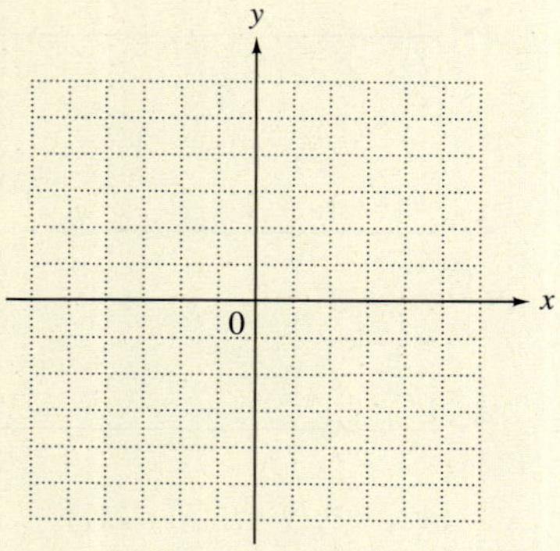

9. $f(x) = \sqrt{9 - x^2}$

10.

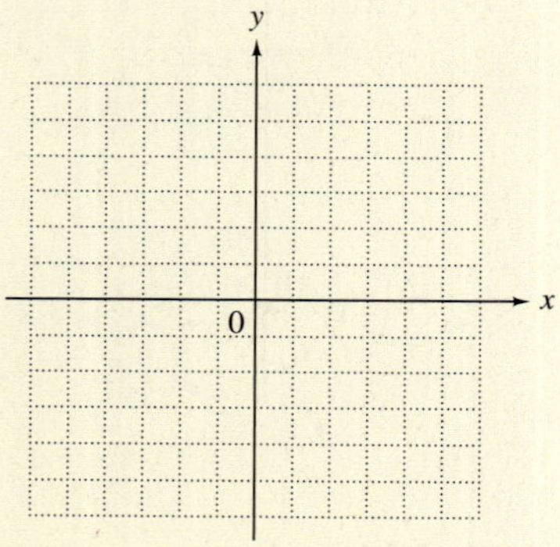

10. $4x^2 + 9y^2 = 36$

11.

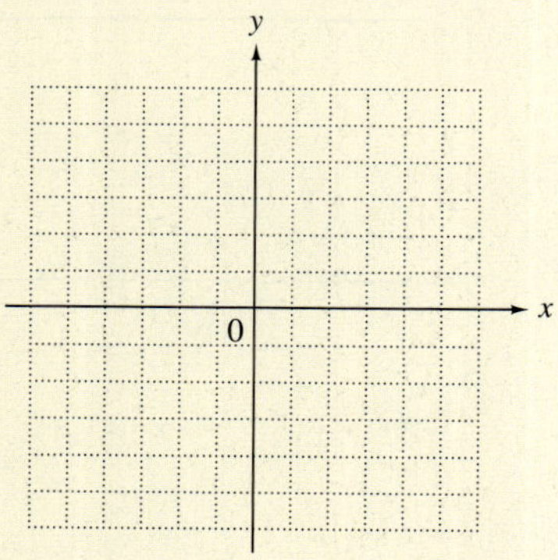

11. $16y^2 - 4x^2 = 64$

12. $\frac{y}{2} = -\sqrt{1 - \frac{x^2}{9}}$

12.

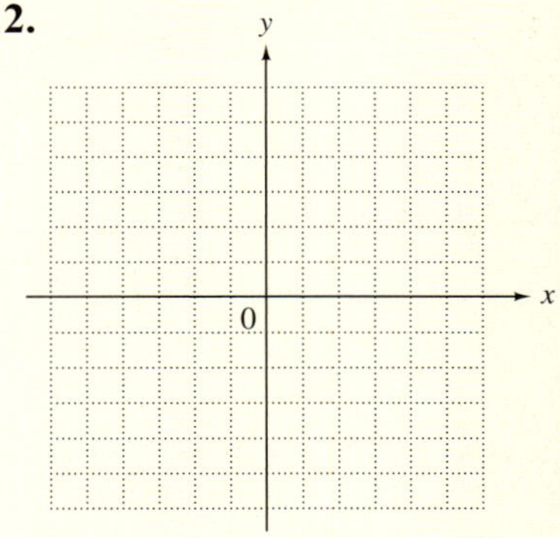

Identify the graph of each equation as a parabola, hyperbola, ellipse, *or* circle.

13. $6x^2 + 4y^2 = 12$

13. ______________________

14. $16x^2 = 144 + 9y^2$

14. ______________________

15. $4y^2 + 4x = 9$

15. ______________________

Solve each nonlinear system.

16. $2x - y = 9$

$xy = 5$

16. ______________________

17. ______________________

17. $x - 4 = 3y$
$x^2 + y^2 = 8$

18. ______________________

18. $x^2 + y^2 = 25$
$x^2 - 2y^2 = 16$

19.

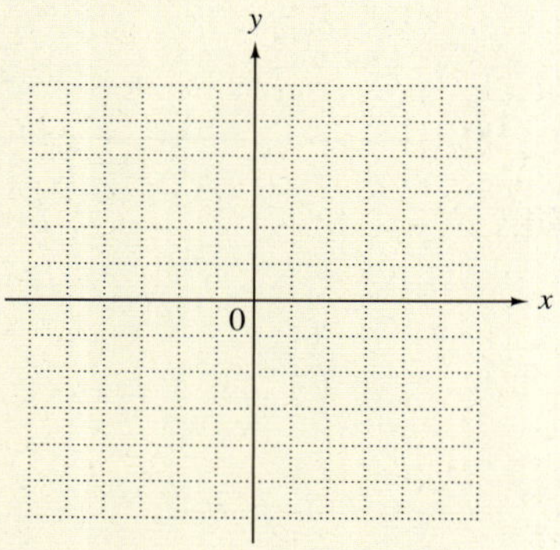

19. Graph the inequality $y < x^2 - 2$.

20.

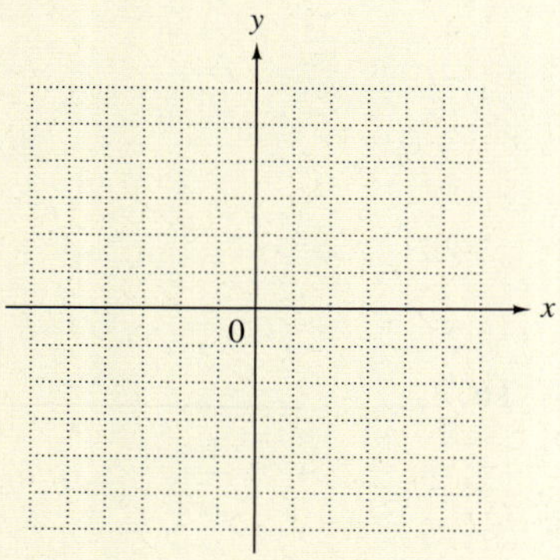

20. Graph the system $\begin{aligned} x^2 + 25y^2 &\le 25 \\ x^2 + y^2 &\le 9 \end{aligned}$.

Cumulative Review Exercises — Chapters 1–12

Solve.

1. $[illegible] - (2x + 3) + x = 5x - 3$

2. $-4k + 7 \geq 6k + 1$

3. $|5m| - 6 = 14$

4. Find the slope of the line through $(2, 5)$ and $(-4, 1)$.

5. Find the equation of the line through $(-3, -2)$ and perpendicular to the graph of $2x - 3y = 7$.

Solve each system.

6. $\begin{aligned} 3x - y &= 12 \\ 2x + 3y &= -3 \end{aligned}$

7. $\begin{aligned} x + y - 2z &= 9 \\ 2x + y + z &= 7 \\ 3x - y - z &= 13 \end{aligned}$

8. $\begin{aligned} xy &= -5 \\ 2x + y &= 3 \end{aligned}$

Perform the indicated operations.

9. $(5y - 3)^2$

10. $(2r + 7)(6r - 1)$

11. $\dfrac{8x^4 - 4x^3 + 2x^2 + 13x + 8}{2x + 1}$

Factor.

12. $12x^2 - 7x - 10$

13. $z^4 - 1$

14. $a^3 - 27b^3$

Perform each operation.

15. $\dfrac{y^2 - 4}{y^2 - y - 6} \div \dfrac{y^2 - 2y}{y - 1}$

16. $\dfrac{5}{c + 5} - \dfrac{2}{c + 3}$

17. $\dfrac{p}{p^2 + p} + \dfrac{1}{p^2 + p}$

Solve.

18. Kareem and Jamal want to clean their office. Kareem can do the job alone in 3 hr, while Jamal can do it alone in 2 hr. How long will it take them if they work together?

Simplify. Assume that all variables represent positive real numbers.

19. $\dfrac{(2a)^{-2}a^4}{a^{-3}}$

20. $4\sqrt[3]{16} - 2\sqrt[3]{54}$

21. $\dfrac{3\sqrt{5x}}{\sqrt{2x}}$

22. $\dfrac{5 + 3i}{2 - i}$

Solve.

23. $2\sqrt{k} = \sqrt{5k + 3}$

24. $10q^2 + 13q = 3$

25. $3k^2 - 3k - 2 = 0$

26. $2(x^2 - 3)^2 - 5(x^2 - 3) = 12$

27. $\log(x + 2) + \log(x - 1) = 1$

28. $F = \dfrac{kwv^2}{r}$ for v

29. If $f(x) = x^3 + 4$, find $f^{-1}(x)$.

30. Evaluate.

(a) $3^{\log_3 4}$

(b) $e^{\ln 7}$

31. Use properties of logarithms to write the following as a single logarithm.

$$2\log(3x + 7) - \log 4$$

32. If $f(x) = x^2 + 2x - 4$ and $g(x) = 3x + 2$, find **(a)** $(g \circ f)(1)$ **(b)** $(f \circ g)(x)$.

Graph.

33. $f(x) = -3x + 5$

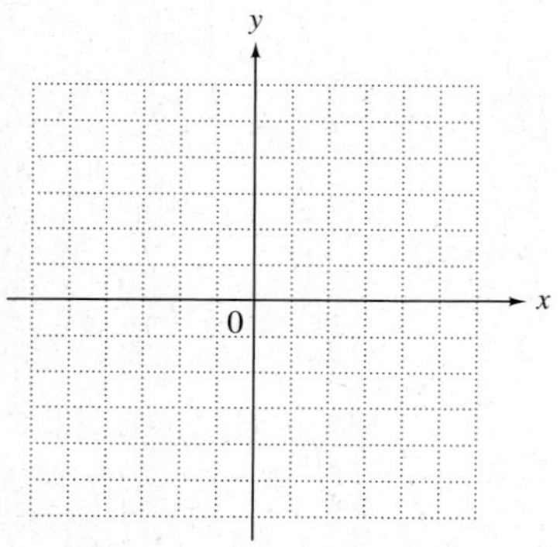

34. $f(x) = -2(x - 1)^2 + 3$

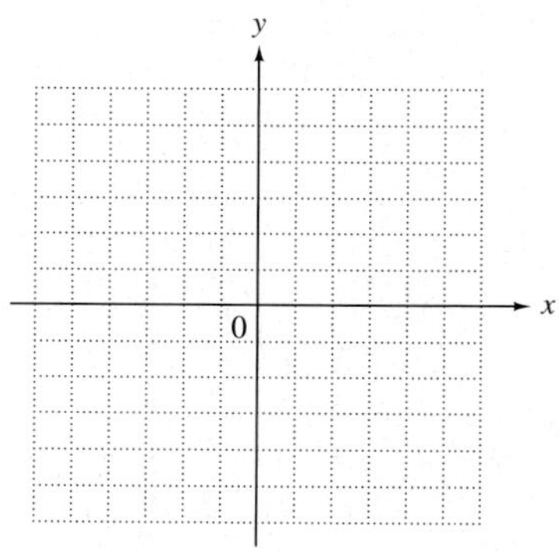

35. $\dfrac{x^2}{25} + \dfrac{y^2}{16} \le 1$

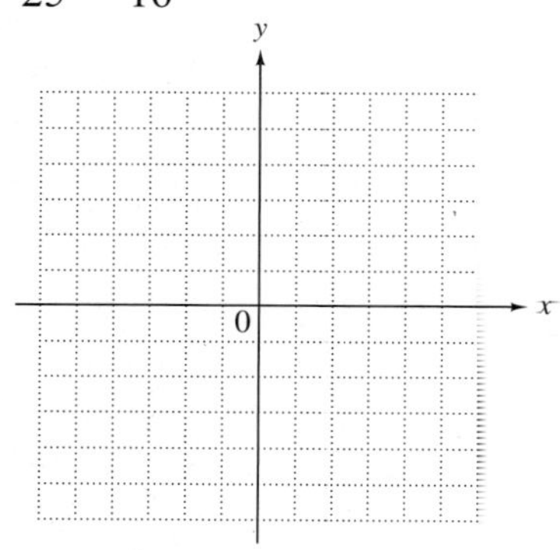

36. $f(x) = \sqrt{x - 2}$

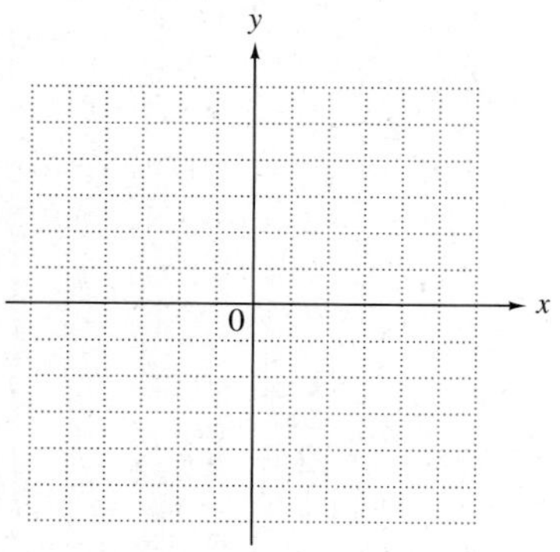

37. $\dfrac{x^2}{4} - \dfrac{y^2}{16} = 1$

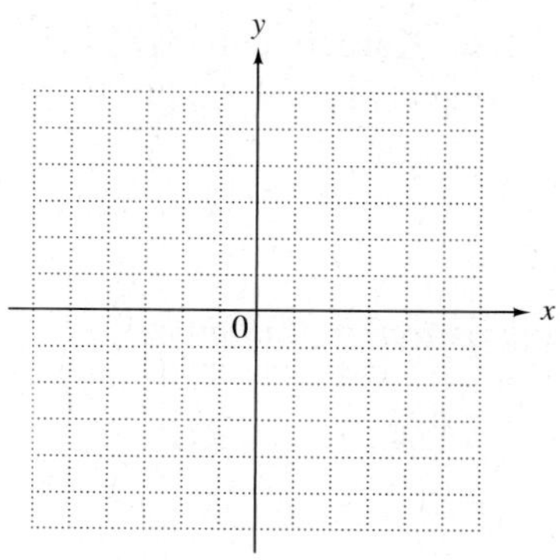

38. $f(x) = 3^x$

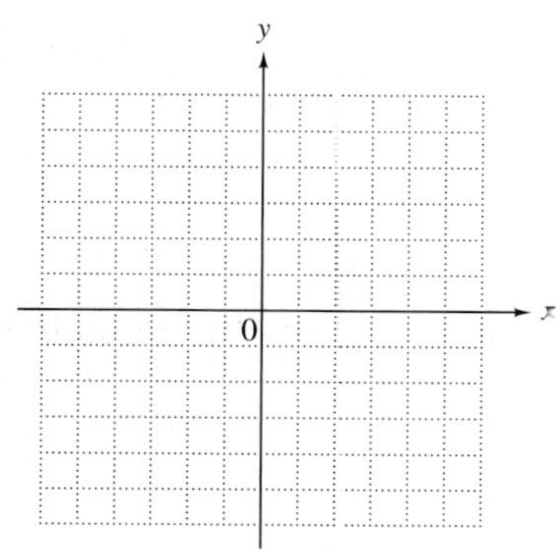

Appendix: Strategies for Problem Solving

A

OBJECTIVE 1 **Learn additional problem-solving strategies.** In **Section 2.3,** we introduce a six-step method for problem solving that we use with applications throughout this text. This method is based on a problem-solving process developed by Hungarian native George Polya, among whose many publications is the modern classic *How to Solve It.*

OBJECTIVE

1 **Learn additional problem-solving strategies.**

Polya's Four-Step Process for Problem Solving

Step 1 **Understand the problem.** You cannot solve a problem if you do not understand what you are asked to find. The problem must be read and analyzed carefully. You may need to read it several times. After you have done so, ask yourself, "What must I find?"

Step 2 **Devise a plan.** There are many ways to attack a problem. Decide what plan is appropriate for the particular problem you are solving.

Step 3 **Carry out the plan.** Once you know how to approach the problem, carry out your plan. You may run into "dead ends" and unforeseen roadblocks, but be persistent. If you are able to solve a problem without a struggle, it isn't much of a problem, is it?

Step 4 **Look back and check.** Check your answer to see that it is reasonable. Does it satisfy the conditions of the problem? Have you answered all the questions the problem asks? Can you solve the problem a different way and come up with the same answer?

George Polya (1887–1985)

Work Problem 1 *at the Side* ▶

1 Compare the six-step problem-solving method given in **Section 2.3** with Polya's four steps.

In Step 2 of Polya's problem-solving process, we are told to devise a plan. The box on the next page lists some strategies that may prove useful.

ANSWER

1. Step 1 compares to Polya's first step, Steps 2 and 3 compare to his second step, Step 4 compares to his third step, and Step 6 compares to his fourth step.

Problem-Solving Strategies

Make a table or a chart.
Look for a pattern.
Solve a similar simpler problem.
Draw a sketch.
Write an equation and solve it.
If a formula applies, use it.
Work backward.
Guess and check.
Use trial and error.
Use common sense.
Look for a "catch" if an answer seems too obvious or impossible.

The problem in Example 1 first appeared in *Liber Abaci,* a book written by the Italian mathematician Leonardo Pisano (also known as Fibonacci) in the year 1202. We apply Polya's four-step process to solve it.

EXAMPLE 1 Using a Table or a Chart

A man put a pair of rabbits in a cage. During the first month the rabbits produced no offspring, but each month thereafter produced one new pair of rabbits. If each new pair thus produced reproduces in the same manner, how many pairs of rabbits will there be at the end of one year?

Step 1 **Understand the problem.** We can reword it as follows:

How many pairs of rabbits will the man have at the end of one year if he starts with one pair, and they reproduce this way: During the first month of life, each pair produces no new rabbits, but each month thereafter each pair produces one new pair?

Step 2 **Devise a plan.** We can construct a table as shown below.

Month	Number of Pairs at Start	Number of New Pairs Produced	Number of Pairs at End of Month
1st			
2nd			
3rd			
4th			
5th			
6th			
7th			
8th			
9th			
10th			
11th			
12th			

Step 3 **Carry out the plan.** At the start of the first month there is only one pair of rabbits. No new pairs are produced during the first month, so there is $1 + 0 = 1$ pair present at the end of the first month. This pattern continues throughout the table on the next page. We add the number in the first column of numbers to the number in the second column to get the number in the third.

Continued on Next Page

Month	Number of Pairs at Start	+ Number of New Pairs Produced	= Number of Pairs at End of Month	
1st	1	0	1	$1 + 0 = 1$
2nd	1	1	2	$1 + 1 = 2$
3rd	2	1	3	$2 + 1 = 3$
4th	3	2	5	.
5th	5	3	8	.
6th	8	5	13	.
7th	13	8	21	.
8th	21	13	34	.
9th	34	21	55	.
10th	55	34	89	.
11th	89	55	144	.
12th	144	89	233	$144 + 89 = 233$

There will be 233 pairs of rabbits at the end of one year.

Step 4 **Look back and check.** This problem can be checked by going back and making sure that we have interpreted it correctly, which we have. Double-check the arithmetic. We have answered the question posed by the problem, so the problem is solved.

Work Problem 2 at the Side. ▶

EXAMPLE 2 Working Backward

Rob Zwettler goes to the racetrack with his buddies on a weekly basis. One week he tripled his money, but then lost \$12. He took his money back the next week, doubled it, but then lost \$40. The following week he tried again, taking his money back with him. He quadrupled it, and then played well enough to take that much home with him, a total of \$224. How much did he start with the first week?

This problem asks us to find Rob's starting amount. Since we know his final amount, the method of working backward can be applied. Since his final amount was \$224 and this represents four times the amount he started with on the third week, we *divide* \$224 by 4 to find that he started the third week with \$56. Before he lost \$40 the second week, he had this \$56 plus the \$40 he lost, giving him \$96. This represented double what he started with, so he started with \$96 *divided by* 2, or \$48, the second week. Repeating this process once more for the first week, before his \$12 loss he had

$$\$48 + \$12 = \$60,$$

which represents triple what he started with. Therefore, he started with

$$\$60 \div 3, \quad \text{or} \quad \$20. \longleftarrow \text{Answer}$$

To check our answer, \$20, observe the following equations:

First week: $(3 \times \$20) - \$12 = \$60 - \$12 = \$48$

Second week: $(2 \times \$48) - \$40 = \$96 - \$40 = \$56$

Third week: $(4 \times \$56) = \$224. \longleftarrow$ His final amount

Work Problem 3 at the Side. ▶

2 Refer to the completed table in Example 1, and observe the sequence of numbers in color, the *Fibonacci sequence*. Choose any four consecutive terms. Multiply the first one chosen by the fourth, and then write the product. Now multiply the two middle terms and write the product. Repeat this process a few more times. What do you notice when the two products are compared?

3 Solve each problem.

(a) Bonnie Boehme bought a book for \$10 and then spent half her remaining money on a train ticket. She then bought lunch for \$4 and spent half her remaining money at a bazaar. She left the bazaar with \$20. How much money did she start with?

(b) If a, b, and c are digits for which

$$\begin{array}{r} 7\ a\ 2 \\ -\ 4\ 8\ b \\ \hline c\ 7\ 3, \end{array}$$

then $a + b + c =$ ____.

A. 14 **B.** 15 **C.** 16
D. 17 **E.** 18

(*Source: Mathematics Teacher* calendar, September 22, 1999.)

ANSWERS

2. The products will always differ by 1.
3. **(a)** \$98 **(b)** D

Recall that $5^2 = 5 \cdot 5 = 25$, that is, 5 squared is 25. Thus, 25 is called a **perfect square.** Other perfect squares include

$$4, \quad 9, \quad 16, \quad 36, \quad \text{and so on.} \qquad \text{Perfect squares}$$

EXAMPLE 3 **Using Trial and Error**

The mathematician Augustus De Morgan lived in the nineteenth century. He once made the following statement: "I was x years old in the year x^2." In what year was De Morgan born?

We must find the year of De Morgan's birth. The problem tells us that he lived in the nineteenth century, which is another way of saying that he lived during the 1800s. One year of his life was a perfect square, so we must find a number between 1800 and 1900 that is a perfect square. Use trial and error.

$$42^2 = 42 \cdot 42 = 1764$$

$$43^2 = 43 \cdot 43 = \mathbf{1849} \qquad \text{1849 is between 1800 and 1900.}$$

$$44^2 = 44 \cdot 44 = 1936$$

The only natural number whose square is between 1800 and 1900 is 43, since $43^2 = 1849$. Therefore, De Morgan was 43 yr old in 1849. The final step in solving the problem is to subtract 43 from 1849 to find the year of his birth:

$$1849 - 43 = \mathbf{1806}. \qquad \text{He was born in 1806.}$$

While the following check may seem unorthodox, it works: Look up De Morgan's birth date in a book on mathematics history, such as *An Introduction to the History of Mathematics,* Sixth Edition, by Howard W. Eves.

◀ Work Problem 4 at the Side.

4 Solve each problem.

(a) Assuming that he lives that long, one of the authors of this book will be 76 yr old in the year x^2, where x is a counting number. In what year was he born?

(b) Place each of the digits 1, 2, 3, 4, 5, 6, 7, and 8 in separate boxes so that boxes that share common corners do not contain successive digits.
(*Source: Mathematics Teacher* calendar, November 29, 1997.)

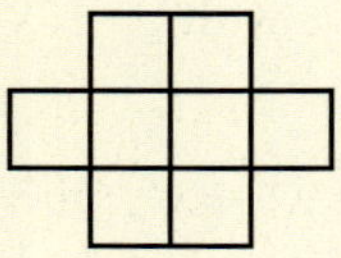

As mentioned above, $5^2 = 25$. The inverse (opposite) of squaring a number is called taking the **square root.** (See **Section 1.3.**) We indicate the positive square root using a **radical sign** $\sqrt{}$. Thus, $\sqrt{25} = 5$. Also,

$$\sqrt{4} = 2, \quad \sqrt{9} = 3, \quad \sqrt{16} = 4, \quad \text{and so on.} \qquad \text{Square roots}$$

The next problem dates back to Hindu mathematics, circa 850.

EXAMPLE 4 **Guessing and Checking**

One-fourth of a herd of camels was seen in the forest. Twice the square root of that herd had gone to the mountain slopes, and 3 times 5 camels remained on the riverbank. What is the numerical measure of that herd of camels?

The numerical measure of the herd of camels must be a natural number. Since the problem mentions "one-fourth of a herd" and "the square root of that herd," the number of camels must be both a multiple of 4 and a perfect square, so no fractions will be encountered. The smallest natural number that satisfies both conditions is 4. We write an equation where x represents the numerical measure of the herd.

"one-fourth of the herd" + "twice the square root of that herd" + "3 times 5 camels" = "the numerical measure of the herd"

$$\frac{1}{4}x \quad + \quad 2\sqrt{x} \quad + \quad 3 \cdot 5 \quad = \quad x$$

Continued on Next Page

Answers

4. (a) 1949 **(b)** Here is one possible solution.

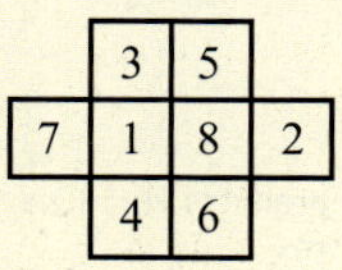

Now substitute 4 for x to see if it is a solution of the equation.

$$\frac{1}{4}(4) + 2\sqrt{4} + 3 \cdot 5 = 4 \qquad \text{Let } x = 4.$$

$$1 + 4 + 15 \stackrel{?}{=} 4 \qquad \sqrt{4} = 2$$

$$20 \neq 4 \qquad \text{Add.}$$

Since 4 is not a solution, try 16, the next perfect square that is a multiple of 4.

$$\frac{1}{4}(16) + 2\sqrt{16} + 3 \cdot 5 = 16 \qquad \text{Let } x = 16.$$

$$4 + 8 + 15 \stackrel{?}{=} 16 \qquad \sqrt{16} = 4$$

$$27 \neq 16 \qquad \text{Add.}$$

Since 16 is not a solution, try 36.

$$\frac{1}{4}(36) + 2\sqrt{36} + 3 \cdot 5 = 36 \qquad \text{Let } x = 36.$$

$$9 + 12 + 15 \stackrel{?}{=} 36 \qquad \sqrt{36} = 6$$

$$36 = 36 \qquad \text{Add.}$$

We see that 36 is the numerical measure of the herd. Check in the words of the problem: "One-fourth of 36, plus twice the square root of 36, plus 3 times 5" gives 9 plus 12 plus 15, which equals 36.

Work Problem **5** *at the Side.* ▶

EXAMPLE 5 **Considering a Similar Simpler Problem and Looking for a Pattern**

The digit farthest to the right in a natural number is called the *ones* or *units digit*, since it tells how many ones are contained in the number when grouping by tens is considered. What is the ones (or units) digit in 2^{4000}?

Recall that 2^{4000} means that 2 is used as a factor 4000 times:

$$2^{4000} = \underbrace{2 \times 2 \times 2 \times \cdots \times 2}_{4000 \text{ factors}}.$$

Certainly, we are not expected to evaluate this number. To answer the question, we examine some smaller powers of 2 and then look for a pattern. We start with the exponent 1 and look at the first twelve powers of 2.

$2^1 = 2$	$2^5 = 32$	$2^9 = 512$
$2^2 = 4$	$2^6 = 64$	$2^{10} = 1024$
$2^3 = 8$	$2^7 = 128$	$2^{11} = 2048$
$2^4 = 16$	$2^8 = 256$	$2^{12} = 4096$

Notice that in each of the four rows above, the ones digit is the same. The final row, which contains the exponents 4, 8, and 12, has the ones digit 6. Each of these exponents is divisible by 4, and since 4000 is divisible by 4, we observe the pattern to predict that the units digit in 2^{4000} is 6.

The units digit for any other power of 2 can be found if we divide the exponent by 4 and compare the remainder to the preceding list of powers. For example, to find the units digit of 2^{543}, we divide 543 by 4 to get a quotient of 135 and a remainder of 3. The units digit is the same as that of 2^3, which is 8.

Work Problem **6** *at the Side.* ▶

5 Solve each problem.

(a) I am thinking of a positive number. If I square it, double the result, take half of that result, and then add 12, I get 21. What is my number?

(b) The same author mentioned in Margin Problem 4(a) graduated from high school in the year that satisfies these conditions:

(1) The sum of the digits is 23;

(2) The hundreds digit is 3 more than the tens digit;

(3) No digit is an 8.

In what year did he graduate?

6 Solve each problem.

(a) What is the units digit in 7^{491}?

(b) What is the 103rd digit in the decimal representation for $\frac{1}{11}$?

Answers

5. (a) 3 **(b)** 1967
6. (a) 3 **(b)** 0

7 Solve each problem.

(a) What is the maximum number of small squares in which we may place a cross (×) and not have any row, column, or diagonal completely filled with crosses? Illustrate your answer.

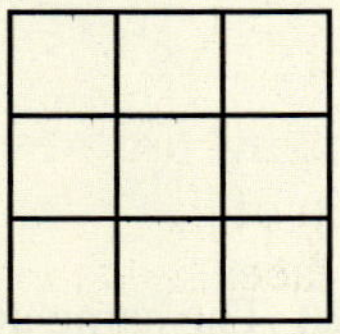

(b) By drawing two straight lines, divide the face of a clock into three regions such that the numbers in the regions have the same total. (*Source: Mathematics Teacher* calendar, October 28, 1998.)

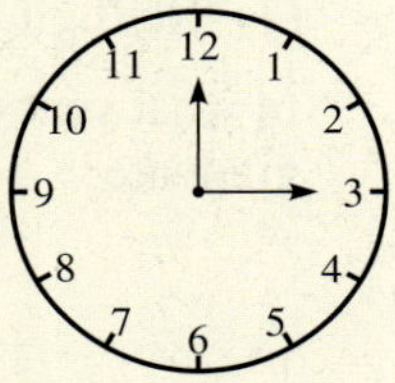

8 Solve each problem.

(a) Which is correct? Three cubed *is* nine or three cubed *are* nine?

(b) If you take 7 bowling pins from 10 bowling pins, what do you have?

ANSWERS

7. (a) 6

	X	X
X		X
X	X	

One of several possibilities

(b) Each region has a sum of 26.

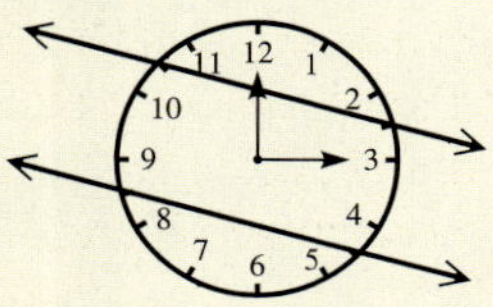

8. (a) Neither is correct, since $3^3 = 27$.
(b) 7 bowling pins

EXAMPLE 6 **Drawing a Sketch**

An array of nine dots is arranged in a 3×3 square, as shown in Figure 1. Is it possible to join the dots with exactly four straight lines if you are not allowed to pick up your pencil from the paper and may not trace over a line that has already been drawn? If so, show how.

Figure 1

Figure 2 shows three attempts. In each case, something is wrong. In the first sketch, one dot is not joined. In the second, the figure cannot be drawn without picking up your pencil from the paper or tracing over a line that has already been drawn. In the third figure, all dots have been joined, but you have used five lines as well as retraced over the figure.

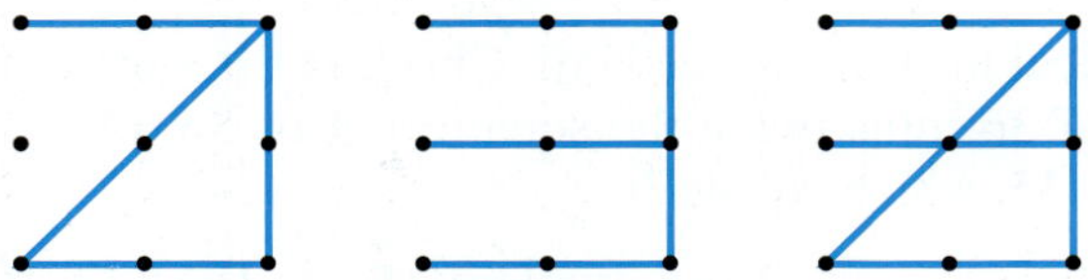

Figure 2

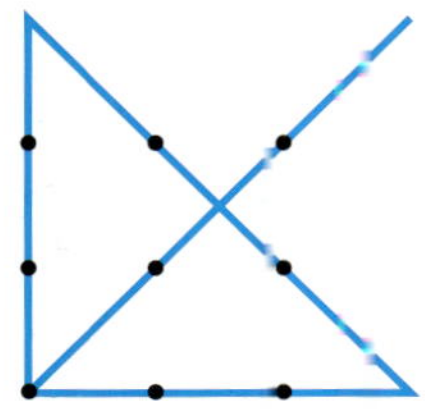

Figure 3

However, the conditions of the problem can be satisfied, as shown in Figure 3. We "went outside of the box," which was not prohibited by the conditions of the problem. This is an example of creative thinking—we used a strategy that is usually not considered at first, since our initial attempts involved "staying within the confines" of the figure.

◀ *Work Problem* **7** *at the Side.*

The final example falls into a category of problems that involve a "catch." Some of these problems seem too easy or perhaps impossible at first, because we tend to overlook an obvious situation. We must look carefully at the use of language in such problems. And, of course, we should never forget to use common sense.

EXAMPLE 7 **Using Common Sense**

Two currently minted U.S. coins together have a total value of $1.05. One is not a dollar. What are the two coins?

Our initial reaction might be, "The only way to have two such coins with a total of $1.05 is to have a nickel and a dollar, but the problem says that one of them is not a dollar." This statement is indeed true. What we must realize here is that the one that is not a dollar is the nickel, and the *other* coin is a dollar! So the two coins are a dollar and a nickel.

◀ *Work Problem* **8** *at the Side.*

Appendix A Exercises

Exercises 1–9 are from the popular monthly calendar feature in the journal Mathematics Teacher. *The authors wish to thank the many journal contributors for permission to use these problems. Original calendar dates are included.*

Use the various problem-solving strategies to solve each problem. In many cases, there is more than one possible approach, so be creative.

1. You are working in a store that has been very careless with the stock. Three boxes of socks are each incorrectly labeled. The labels say *red socks, green socks,* and *red and green socks*. How can you relabel the boxes correctly by taking only one sock out of one box, without looking inside the boxes? (October 22, 2001)

2. Three dice with faces numbered 1 through 6 are stacked as shown. Seven of the eighteen faces are visible, leaving eleven faces hidden on the back, on the bottom, and between faces. The total number of dots not visible in this view is _____.

A. 21
B. 22
C. 31
D. 41
E. 53

(September 17, 2001)

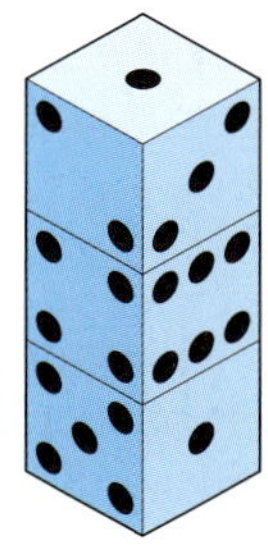

3. You and a friend are playing tick-tack-toe, where three in a row *loses*. You are O. If you want to win, what must your next move be? (October 21, 2001)

4. How can you connect each square with the triangle that has the same number? Lines cannot cross, enter a square or triangle, or go outside the diagram. (October 15, 1999)

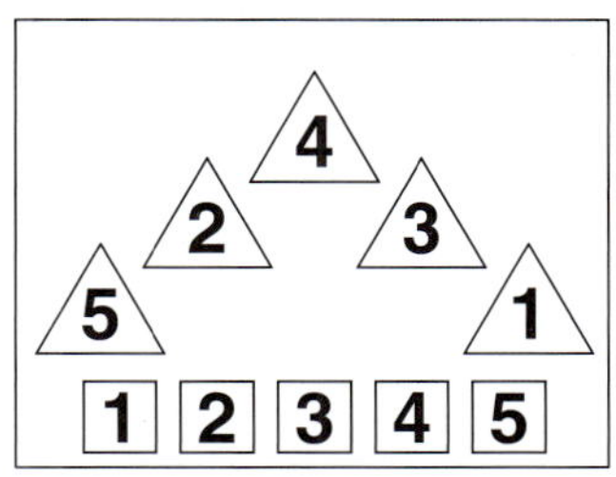

5. You have brought two unmarked buckets to a stream. The buckets hold 7 gal and 3 gal of water, respectively. How can you obtain exactly 5 gal of water to take home? (October 19, 1997)

6. Chip and Dale collected 32 acorns on Monday and stored them with their acorn supply. After Chip fell asleep, Dale ate half the acorns. This pattern continued through Friday night, with 32 acorns being added and half the supply being eaten. On Saturday morning, Chip counted the acorns and found that they had only 35. How many acorns had they started with on Monday morning? (March 12, 1997)

7. Pat and Chris have the same birthday. Pat is twice as old as Chris was when Pat was as old as Chris is now. If Pat is now 24 yr old, how old is Chris? (December 3, 2001)

8. Balls numbered 1 through 6 are arranged in a **difference triangle.** Note that in any row, the difference between the larger and the smaller of two successive balls is the number of the ball that appears below them. Arrange balls numbered 1 through 10 in a difference triangle. (May 6, 1998)

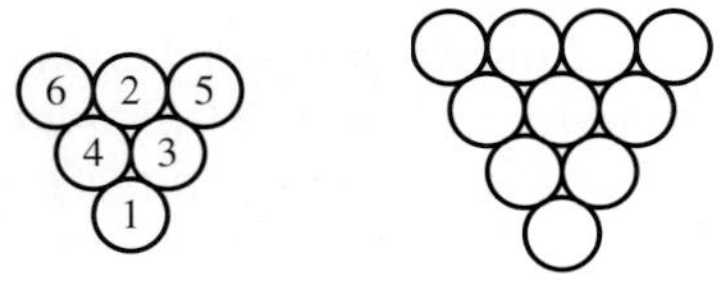

9. What are the final two digits of 7^{1997}? (November 29, 1997)

10. If you raise 3 to the 324th power, what is the units digit of the result?

11. A frog is at the bottom of a 20-ft well. Each day it crawls up 4 ft but each night it slips back 3 ft. After how many days will the frog reach the top of the well?

12. A lily pad grows so that each day it doubles its size. On the twentieth day of its life, it completely covers a pond. On what day was the pond half covered?

13. Some children are standing in a circular arrangement. They are evenly spaced and arranged in numerical order. The fourth child is standing directly opposite the twelfth child. How many children are there in the circle?

14. A **perfect number** is a natural number that is equal to the sum of all its counting number divisors except itself. For example, 28 is a perfect number, since its divisors other than itself are 1, 2, 4, 7, and 14, and $1 + 2 + 4 + 7 + 14 = 28$. What is the least perfect number?

15. A **magic square** is a square array of numbers that has the property that the sum of the numbers in any row, column, or diagonal is the same. Fill in the square below so that it becomes a magic square, and all digits 1, 2, 3, . . . , 9 are used exactly once.

6		8
	5	
		4

16. Refer to Exercise 15. Complete the magic square below so that all counting numbers 1, 2, 3, . . . , 15 are used exactly once, and the sum in each row, column, or diagonal is 34.

6			9
	15		14
11		10	
16		13	

17. What is the minimum number of pitches that a baseball player who pitches a complete game can make in a regulation 9-inning baseball game?

18. Draw a square in the following figure so that no two cats share the same region.

19. You have eight coins. Seven are genuine and one is a fake, which weighs a little less than the other seven. You have a balance scale, which you may use only three times. Tell how to locate the bad coin in three weighings. Then show how to detect the bad coin in only *two* weighings.

20. A person must take a wolf, a goat, and some cabbage across a river. The rowboat to be used has room for the person plus either the wolf, the goat, or the cabbage. If the person takes the cabbage in the boat, the wolf will eat the goat. While the wolf crosses in the boat, the cabbage will be eaten by the goat. The goat and cabbage are safe only when the person is present. Even so, the person gets everything across the river. Explain how. (This problem dates back to around the year 750.)

21. (This is an ancient Hindu problem.) Beautiful maiden with beaming eyes, tell me . . . which is the number that when multiplied by 3, then increased by $\frac{3}{4}$ the product, then divided by 7, diminished by $\frac{1}{3}$ of the quotient, multiplied by itself, diminished by 52, by the extraction of the square root, addition of 8, and division by 10 gives the number 2?

22. A teenager's age increased by 2 gives a perfect square. Her age decreased by 10 gives the square root of that perfect square. She is 5 yr older than her brother. How old is her brother?

23. Draw the following figure without picking up your pencil from the paper and without tracing over a line you have already drawn.

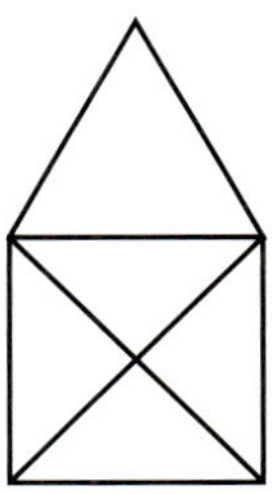

24. Repeat Exercise 23 for the figure shown here.

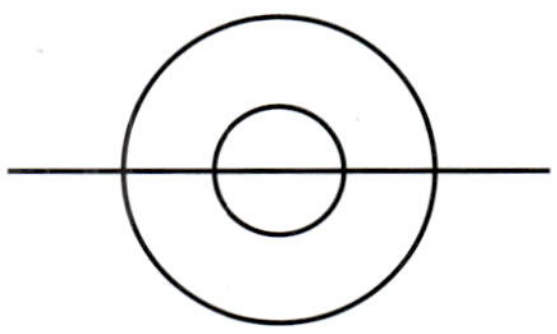

25. James, Dan, Jessica, and Cathy form a pair of married couples. Their ages are 36, 31, 30, and 29. Jessica is married to the oldest person in the group. James is older than Jessica but younger than Cathy. Who is married to whom, and what are their ages?

26. If a year has two consecutive months with Friday the thirteenth, what months must they be?

27. How much dirt is there in a cubical hole, 6 ft on each side?

28. Some months have 30 days and some have 31 days. How many months have 28 days?

29. Place one of the arithmetic operations $+$, $-$, $\times$, or $\div$ between each pair of successive numbers on the left side of this equation to make it true. Any operation may be used more than once or not at all.

$$1 \quad 2 \quad 3 \quad 4 \quad 5 \quad 6 \quad 7 \quad 8 \quad 9 = 100$$

30. In the addition problem below, some digits are missing as indicated by the blanks. If the problem is done correctly, what is the sum of the missing digits?

$$\begin{array}{rccc} & \_ & 3 & 5 \\ & 8 & \_ & 6 \\ + & 1 & 4 & \_ \\ \hline \_ & 4 & 0 & 8 \end{array}$$

31. Fill in the blanks so that the multiplication problem below uses all digits 0, 1, 2, 3, . . . , 9 exactly once, and is correctly worked.

$$\begin{array}{rrrrr} & & \_ & 0 & 2 \\ \times & & & 3 & \_ \\ \hline \_ & 5, & \_ & \_ & \_ \end{array}$$

32. Based on your knowledge of elementary arithmetic, describe the pattern that can be observed when the following operations are performed:

$$9 \times 1, \quad 9 \times 2, \quad 9 \times 3, \quad \ldots, \quad 9 \times 9.$$

(*Hint:* Add the digits in the answers. What do you notice?)

33. How many triangles are in the following figure?

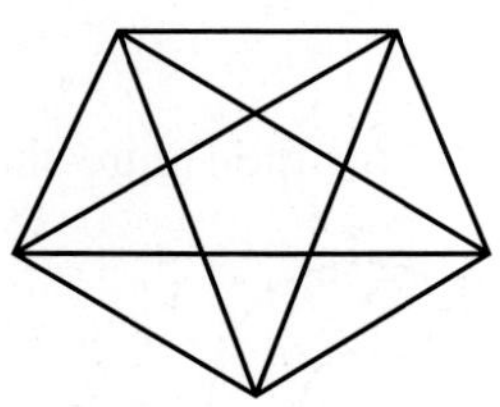

34. How many squares are in the following figure?

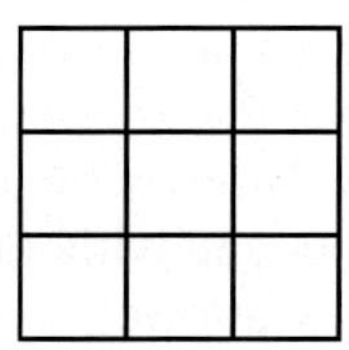

35. If it takes $7\frac{1}{2}$ min to boil an egg, how long does it take to boil 5 eggs?

36. The brother of the chief executive officer (CEO) of a major company died. The man who died had no brother. How is this possible?

Appendix: Review of Fractions

B

OBJECTIVES

1. Identify prime numbers.
2. Write numbers in prime factored form.
3. Write fractions in lowest terms.
4. Convert between improper fractions and mixed numbers.
5. Multiply and divide fractions.
6. Add and subtract fractions.

The numbers used most often in everyday life are the **whole numbers,**

$$0, 1, 2, 3, 4, 5, \ldots$$

and **fractions,** such as

$$\frac{1}{3}, \quad \frac{5}{4}, \quad \text{and} \quad \frac{11}{12}.$$

The parts of a fraction are named as follows.

$$\text{Fraction bar} \longrightarrow \frac{4}{7} \begin{array}{l} \longleftarrow \text{Numerator} \\ \longleftarrow \text{Denominator} \end{array}$$

If the numerator of a fraction is less than the denominator, we call it a **proper fraction.** A proper fraction has a value less than 1. If the numerator is greater than or equal to the denominator, the fraction is an **improper fraction.** Some examples follow.

$$\frac{1}{5}, \quad \frac{2}{7}, \quad \frac{9}{10}, \quad \frac{23}{25} \qquad \text{Proper fractions}$$

$$\frac{3}{2}, \quad \frac{5}{5}, \quad \frac{11}{7}, \quad \frac{28}{4} \qquad \text{Improper fractions}$$

OBJECTIVE 1 Identify prime numbers. In work with fractions, we will need to write the numerators and denominators as products. A **product** is the answer to a multiplication problem. When 12 is written as the product 2×6, for example, 2 and 6 are called **factors** of 12. Other factors of 12 are 1, 3, 4, and 12. A whole number is **prime** if it has exactly two different factors (itself and 1). The first dozen primes are listed here.

$$2, 3, 5, 7, 11, 13, 17, 19, 23, 29, 31, 37 \qquad \text{Prime numbers}$$

A whole number greater than 1 that is not prime is called a **composite number.** Some examples follow.

$$4, 6, 8, 9, 10, 12 \qquad \text{Composite numbers}$$

The number 1 is neither prime nor composite.

1 Tell whether each number is *prime* or *composite*.

(a) 12

(b) 13

(c) 27

(d) 59

(e) 1806

EXAMPLE 1 Distinguishing between Prime and Composite Numbers

Decide whether each number is *prime* or *composite*.

(a) 33

33 has factors of 3 and 11, as well as 1 and 33, so it is composite.

(b) 43

Since there are no numbers other than 1 and 43 itself that divide *evenly* into 43, the number 43 is prime.

(c) 9832

Since 9832 can be divided by 2, giving 2×4916, it is composite.

◀ Work Problem 1 at the Side.

OBJECTIVE 2 **Write numbers in prime factored form.** To factor a number means to write it as the product of two or more numbers. Factoring is the reverse of multiplying two numbers to get the product.

Multiplication	Factoring
$6 \cdot 3 = 18$	$18 = 6 \cdot 3$
Factors ↑ ↑ Product ↑	Product ↑ Factors ↑ ↑

In algebra, a dot $\cdot$ is used instead of the $\times$ symbol to indicate multiplication because $\times$ may be confused with the letter x. A composite number written using factors that are all prime numbers is in **prime factored form.**

2 Write each number in prime factored form.

(a) 70

(b) 72

(c) 693

(d) 97

EXAMPLE 2 Writing Numbers in Prime Factored Form

Write each number in prime factored form.

(a) 35

Factor 35 as the product of the prime factors 5 and 7, or as $35 = 5 \cdot 7$.

(b) 24

We use a factor tree, as shown below. The prime factors are circled.

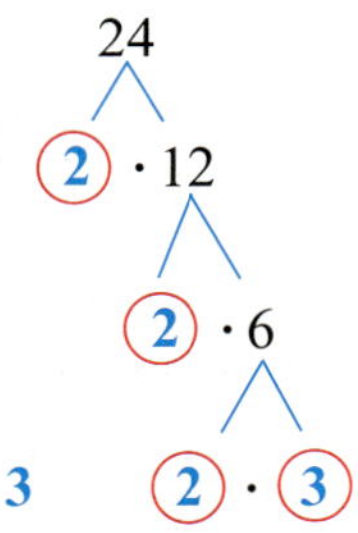

Divide by the smallest prime, 2, to get $24 = 2 \cdot 12$.

Now divide 12 by 2 to find factors of 12. $24 = 2 \cdot 2 \cdot 6$

Since 6 can be factored as $2 \cdot 3$, $24 = 2 \cdot 2 \cdot 2 \cdot 3$, where all factors are prime. $24 = 2 \cdot 2 \cdot 2 \cdot 3$

◀ Work Problem 2 at the Side.

OBJECTIVE 3 **Write fractions in lowest terms.** A fraction is in **lowest terms** when the numerator and denominator have no factors in common (other than 1). The following properties are useful.

Properties of 1

Any nonzero number divided by itself is equal to 1; for example, $\frac{3}{3} = 1$.

Any number multiplied by 1 remains the same; for example, $7 \cdot 1 = 7$.

Answers

1. (a) composite (b) prime (c) composite (d) prime (e) composite
2. (a) $2 \cdot 5 \cdot 7$ (b) $2 \cdot 2 \cdot 2 \cdot 3 \cdot 3$ (c) $3 \cdot 3 \cdot 7 \cdot 11$ (d) 97 is prime.

Writing a Fraction in Lowest Terms

Step 1 Write the numerator and denominator in prime factored form.

Step 2 Replace each pair of factors common to the numerator and denominator with 1.

Step 3 Multiply the remaining factors in the numerator and in the denominator.

(This procedure is sometimes called "simplifying the fraction.")

EXAMPLE 3 **Writing Fractions in Lowest Terms**

Write each fraction in lowest terms.

(a) $\frac{10}{15} = \frac{2 \cdot 5}{3 \cdot 5} = \frac{2}{3} \cdot \frac{5}{5} = \frac{2}{3} \cdot 1 = \frac{2}{3}$

Since 5 is a common factor of 10 and 15, we use the first property of 1 to replace $\frac{5}{5}$ with 1.

(b) $\frac{15}{45} = \frac{3 \cdot 5}{3 \cdot 3 \cdot 5} = \frac{1 \cdot 3 \cdot 5}{3 \cdot 3 \cdot 5} = \frac{1}{3} \cdot \frac{3}{3} \cdot \frac{5}{5} = \frac{1}{3} \cdot 1 \cdot 1 = \frac{1}{3}$

Multiplying by 1 in the numerator does not change the value of the numerator and makes it possible to rewrite the expression as the product of three fractions in the next step.

(c) $\frac{150}{200}$

It is not always necessary to factor into *prime* factors in Step 1. Here, if you see that 50 is a common factor of the numerator and the denominator, factor as follows:

$$\frac{150}{200} = \frac{3 \cdot 50}{4 \cdot 50} = \frac{3}{4} \cdot 1 = \frac{3}{4}.$$

Note

When writing a fraction in lowest terms, look for the largest common factor in the numerator and the denominator. If none is obvious, factor the numerator and the denominator into prime factors. *Any* common factor can be used and the fraction can be simplified in stages.

For example, $\frac{150}{200} = \frac{15 \cdot 10}{20 \cdot 10} = \frac{3 \cdot 5 \cdot 10}{4 \cdot 5 \cdot 10} = \frac{3}{4}.$

Work Problem 3 at the Side. ▶

OBJECTIVE 4 **Convert between improper fractions and mixed numbers.** A **mixed number** is a single number that represents the sum of a whole number and a fraction. For example, $5\frac{3}{4} = 5 + \frac{3}{4}$. Mixed numbers are commonly used in everyday life. For example, a child's age might be given as $2\frac{1}{2}$ years old, or a cookie recipe might call for $3\frac{1}{4}$ cups of flour.

3 Write each fraction in lowest terms.

(a) $\frac{8}{14}$

(b) $\frac{35}{42}$

(c) $\frac{72}{120}$

ANSWERS

3. (a) $\frac{4}{7}$ **(b)** $\frac{5}{6}$ **(c)** $\frac{3}{5}$

4 Write $\frac{92}{5}$ as a mixed number.

Any improper fraction whose value is not a whole number can be rewritten as a mixed number, and any mixed number can be rewritten as an improper fraction. It is convenient to be able to convert between these two forms of the same number.

EXAMPLE 4 Converting an Improper Fraction to a Mixed Number

Write $\frac{59}{8}$ as a mixed number.

To convert an improper fraction to a mixed number, divide the numerator by the denominator. Here, divide 59 by 8. Use the quotient and remainder to form the mixed number.

$$\begin{array}{r} 7 \\ 8\overline{)59} \\ \underline{56} \\ 3 \end{array}$$

Denominator of fraction (divisor) → 8; 7 ← Quotient; 59 ← Numerator of fraction (dividend); 3 ← Remainder

Thus, $\frac{59}{8} = 7\frac{3}{8}$.

◀ Work Problem 4 at the Side.

EXAMPLE 5 Converting a Mixed Number to an Improper Fraction

Write $6\frac{4}{7}$ as an improper fraction.

To convert a mixed number to an improper fraction, multiply the denominator of the fraction by the whole number and add the numerator of the fraction to get the numerator of the improper fraction. To write $6\frac{4}{7}$ as an improper fraction, the numerator is

$$7 \cdot 6 + 4 = 42 + 4 = 46.$$

The denominator of the improper fraction is the same as the denominator in the mixed number. The denominator is 7. Thus, $6\frac{4}{7} = \frac{46}{7}$.

◀ Work Problem 5 at the Side.

5 Write $11\frac{2}{3}$ as an improper fraction.

OBJECTIVE 5 Multiply and divide fractions.

Multiplying Fractions

To multiply two fractions, multiply the numerators to get the numerator of the product, and multiply the denominators to get the denominator of the product. ***The product should be written in lowest terms.***

EXAMPLE 6 Multiplying Fractions

Find each product, and write it in lowest terms.

(a) $\frac{3}{8} \cdot \frac{4}{9} = \frac{3 \cdot 4}{8 \cdot 9}$ Multiply numerators. Multiply denominators.

$= \frac{3 \cdot 4}{2 \cdot 4 \cdot 3 \cdot 3}$ Factor.

$= \frac{1}{2 \cdot 3} = \frac{1}{6}$ Write in lowest terms.

Continued on Next Page

Answers

4. $18\frac{2}{5}$

5. $\frac{35}{3}$

(b) $2\frac{1}{3} \cdot 5\frac{1}{2} = \frac{7}{3} \cdot \frac{11}{2}$ Write as improper fractions.

$= \frac{77}{6}$, or $12\frac{5}{6}$ Multiply numerators and denominators; write as a mixed number.

Work Problem 6 at the Side. ▶

6 Find each product, and write it in lowest terms.

(a) $\frac{5}{8} \cdot \frac{2}{10}$

(b) $\frac{1}{10} \cdot \frac{12}{5}$

(c) $\frac{7}{9} \cdot \frac{12}{14}$

(d) $3\frac{1}{3} \cdot 1\frac{3}{4}$

Two fractions are **reciprocals** of each other if their product is 1. For example, $\frac{3}{4}$ and $\frac{4}{3}$ are reciprocals because

$$\frac{3}{4} \cdot \frac{4}{3} = 1.$$

The numbers $\frac{7}{11}$ and $\frac{11}{7}$ are reciprocals also. Other examples are $\frac{1}{5}$ and 5 (which can be written as $\frac{5}{1}$), $\frac{4}{9}$ and $\frac{9}{4}$, and 16 and $\frac{1}{16}$.

Because division is the opposite, or inverse, of multiplication, we use reciprocals to divide fractions.

Dividing Fractions

To divide two fractions, multiply the first fraction by the reciprocal of the second. The result, called the **quotient,** should be written in lowest terms.

For example, we know that $20 \div 10 = 2$, and $20 \cdot \frac{1}{10} = 2$.

EXAMPLE 7 Dividing Fractions

Find each quotient, and write it in lowest terms.

(a) $\frac{3}{4} \div \frac{8}{5} = \frac{3}{4} \cdot \frac{5}{8} = \frac{3 \cdot 5}{4 \cdot 8} = \frac{15}{32}$

Multiply by the reciprocal of the second fraction.

(b) $\frac{3}{4} \div \frac{5}{8} = \frac{3}{4} \cdot \frac{8}{5} = \frac{3 \cdot 8}{4 \cdot 5} = \frac{3 \cdot 4 \cdot 2}{4 \cdot 5} = \frac{6}{5}$, or $1\frac{1}{5}$

The answer can be written as an improper fraction or as a mixed number.

(c) $\frac{5}{8} \div 10 = \frac{5}{8} \div \frac{10}{1} = \frac{5}{8} \cdot \frac{1}{10} = \frac{5 \cdot 1}{8 \cdot 10} = \frac{5 \cdot 1}{8 \cdot 2 \cdot 5} = \frac{1}{16}$

Write 10 as $\frac{10}{1}$.

Remember to write 1 in the numerator.

(d) $1\frac{2}{3} \div 4\frac{1}{2} = \frac{5}{3} \div \frac{9}{2}$ Write as improper fractions.

$= \frac{5}{3} \cdot \frac{2}{9}$ Multiply by the reciprocal of the second fraction.

$= \frac{10}{27}$ Multiply numerators and denominators.

Always check to make sure your final answer is in lowest terms.

Answers

6. (a) $\frac{1}{8}$ (b) $\frac{6}{25}$ (c) $\frac{2}{3}$ (d) $\frac{35}{6}$, or $5\frac{5}{6}$

7 Find each quotient, and write it in lowest terms.

(a) $\frac{3}{10} \div \frac{2}{7}$

(b) $\frac{3}{4} \div \frac{7}{16}$

(c) $\frac{4}{3} \div 6$

(d) $3\frac{1}{4} \div 1\frac{2}{5}$

8 Add. Write sums in lowest terms.

(a) $\frac{3}{5} + \frac{4}{5}$

(b) $\frac{5}{14} + \frac{3}{14}$

ANSWERS

7. (a) $\frac{21}{20}$, or $1\frac{1}{20}$ **(b)** $\frac{12}{7}$, or $1\frac{5}{7}$ **(c)** $\frac{2}{9}$ **(d)** $\frac{65}{28}$, or $2\frac{9}{28}$

8. (a) $\frac{7}{5}$, or $1\frac{2}{5}$ **(b)** $\frac{4}{7}$

CAUTION

Notice in Example 7 that *only* the second fraction (the divisor) is replaced by its reciprocal in the multiplication.

◀ Work Problem 7 at the Side.

OBJECTIVE 6 Add and subtract fractions. The result of adding two numbers is called the **sum** of the numbers. For example, since $2 + 3 = 5$, the sum of 2 and 3 is 5.

Adding Fractions

To find the sum of two fractions with the *same* denominator, add their numerators and ***keep the same denominator.***

EXAMPLE 8 Adding Fractions with the Same Denominator

Add. Write sums in lowest terms.

(a) $\frac{3}{7} + \frac{2}{7} = \frac{3+2}{7} = \frac{5}{7}$ Add numerators; keep the same denominator.

(b) $\frac{2}{10} + \frac{3}{10} = \frac{2+3}{10} = \frac{5}{10} = \frac{1}{2}$ Write in lowest terms.

◀ Work Problem 8 at the Side.

If the fractions to be added do not have the same denominator, the procedure above can still be used, but only *after* the fractions are rewritten with a common denominator. For example, to rewrite $\frac{3}{4}$ as a fraction with a denominator of 32,

$$\frac{3}{4} = \frac{?}{32},$$

we must find the number that can be multiplied by 4 to give 32. Since $4 \cdot 8 = 32$, we use the number 8. By the second property of 1, we can multiply the numerator and the denominator by 8.

$$\frac{3}{4} = \frac{3}{4} \cdot 1 = \frac{3}{4} \cdot \frac{8}{8} = \frac{3 \cdot 8}{4 \cdot 8} = \frac{24}{32}$$

This process is the reverse of writing a fraction in lowest terms.

Finding the Least Common Denominator (LCD)

Step 1 Factor all denominators to prime factored form.

Step 2 The LCD is the product of every (different) factor that appears in any of the factored denominators. If a factor is repeated, use the greatest number of repeats as factors of the LCD.

Step 3 Write each fraction with the LCD as the denominator, using the second property of 1.

EXAMPLE 9 **Adding Fractions with Different Denominators**

Add. Write sums in lowest terms.

(a) $\frac{4}{15} + \frac{5}{9}$

Step 1 To find the LCD, factor the denominators to prime factored form.

$$15 = 5 \cdot 3 \quad \text{and} \quad 9 = 3 \cdot 3$$

3 is a factor of both denominators.

Step 2

$$\text{LCD} = 5 \cdot 3 \cdot 3 = 45$$

In this example, the LCD needs one factor of 5 and two factors of 3 because the second denominator has two factors of 3.

Step 3 Now we can use the second property of 1 to write each fraction with 45 as the denominator.

$$\frac{4}{15} = \frac{4}{15} \cdot \frac{3}{3} = \frac{12}{45} \quad \text{and} \quad \frac{5}{9} = \frac{5}{9} \cdot \frac{5}{5} = \frac{25}{45}$$

At this stage, the fractions are *not* in lowest terms.

Now add the two equivalent fractions to get the required sum.

$$\frac{4}{15} + \frac{5}{9} = \frac{12}{45} + \frac{25}{45} = \frac{37}{45}$$

(b) $3\frac{1}{2} + 2\frac{3}{4} = \frac{7}{2} + \frac{11}{4}$ Change to improper fractions.

$= \frac{14}{4} + \frac{11}{4}$ Get a common denominator.

$= \frac{25}{4}$, or $6\frac{1}{4}$ Add; write as a mixed number.

(c) $45\frac{2}{3} + 73\frac{1}{2}$

We use a vertical method here.

$$\begin{array}{r} 45\frac{2}{3} = 45\frac{4}{6} \\ +\ 73\frac{1}{2} = 73\frac{3}{6} \\ \hline \end{array}$$

Add the whole numbers and the fractions separately.

$$118\frac{7}{6} = 118 + \left(1 + \frac{1}{6}\right) = 119\frac{1}{6}$$

Work Problem **9** at the Side. ▶

The **difference** between two numbers is found by subtracting the numbers. For example, $9 - 5 = 4$, so the difference between 9 and 5 is 4. We find the difference between two fractions as given in the box on the next page.

9 Add. Write sums in lowest terms.

(a) $\frac{7}{30} + \frac{2}{45}$

(b) $\frac{17}{10} + \frac{8}{27}$

(c) $2\frac{1}{8} + 1\frac{2}{3}$

(d) $132\frac{4}{5} + 28\frac{3}{4}$

ANSWERS

9. (a) $\frac{5}{18}$ **(b)** $\frac{539}{270}$, or $1\frac{269}{270}$ **(c)** $\frac{91}{24}$, or $3\frac{19}{24}$ **(d)** $161\frac{11}{20}$

10 Subtract.

(a) $\frac{9}{11} - \frac{3}{11}$

(b) $\frac{13}{15} - \frac{5}{6}$

(c) $2\frac{3}{8} - 1\frac{1}{2}$

(d) $50\frac{1}{4} - 32\frac{2}{3}$

Answers

10. (a) $\frac{6}{11}$ (b) $\frac{1}{30}$ (c) $\frac{7}{8}$ (d) $17\frac{7}{12}$

Subtracting Fractions

To find the difference between two fractions with the *same* denominator, subtract their numerators and ***keep the same denominator.***

If the fractions have *different* denominators, write them with a common denominator first.

EXAMPLE 10 Subtracting Fractions

Subtract. Write differences in lowest terms.

(a) $\frac{15}{8} - \frac{3}{8} = \frac{15-3}{8}$ Subtract numerators; keep the same denominator.

$= \frac{12}{8}$

$= \frac{3}{2}$, or $1\frac{1}{2}$ Write in lowest terms or as a mixed number.

(b) $\frac{15}{16} - \frac{4}{9}$

Since $16 = 2 \cdot 2 \cdot 2 \cdot 2$ and $9 = 3 \cdot 3$ have no common factors, the LCD is $16 \cdot 9 = 144$.

$\frac{15}{16} - \frac{4}{9} = \frac{15 \cdot 9}{16 \cdot 9} - \frac{4 \cdot 16}{9 \cdot 16}$ Get a common denominator.

$= \frac{135}{144} - \frac{64}{144}$

$= \frac{71}{144}$ Subtract numerators; keep the same denominator.

(c) $2\frac{1}{2} - 1\frac{3}{4} = \frac{5}{2} - \frac{7}{4}$ Write as improper fractions.

$= \frac{10}{4} - \frac{7}{4}$ Get a common denominator.

$= \frac{3}{4}$ Subtract.

Alternatively, we could use a vertical method.

$$\begin{array}{rcrcr} 2\frac{1}{2} & = & 2\frac{2}{4} & = & 1\frac{6}{4} \\ -\,1\frac{3}{4} & = & 1\frac{3}{4} & = & 1\frac{3}{4} \\ \hline & & & & \frac{3}{4} \end{array}$$

◀ Work Problem **10** at the Side.

Appendix B Exercises

FOR EXTRA HELP MyMathLab Math XL PRACTICE WATCH DOWNLOAD READ REVIEW

Decide whether each statement is true *or* false. *If it is* false, *say why.*

1. In the fraction $\frac{3}{7}$, 3 is the numerator and 7 is the denominator.

2. The mixed number equivalent of $\frac{41}{5}$ is $8\frac{1}{5}$.

3. The fraction $\frac{17}{51}$ is in lowest terms.

4. The reciprocal of $\frac{8}{2}$ is $\frac{4}{1}$.

5. The product of 8 and 2 is 10.

6. The difference between 12 and 2 is 6.

Identify each number as prime, composite, *or* neither. *See Example 1.*

7. 19 **8.** 99 **9.** 52 **10.** 61

11. 2468 **12.** 3125 **13.** 1 **14.** 83

Write each number in prime factored form. See Example 2.

15. 30 **16.** 40 **17.** 252 **18.** 168

19. 124 **20.** 165 **21.** 29 **22.** 31

Write each fraction in lowest terms. See Example 3.

23. $\frac{8}{16}$ **24.** $\frac{4}{12}$ **25.** $\frac{15}{18}$ **26.** $\frac{16}{20}$

27. $\frac{15}{75}$ **28.** $\frac{24}{64}$ **29.** $\frac{144}{120}$ **30.** $\frac{132}{77}$

Write each improper fraction as a mixed number. See Example 4.

31. $\frac{12}{7}$ **32.** $\frac{28}{5}$ **33.** $\frac{77}{12}$ **34.** $\frac{101}{15}$ **35.** $\frac{83}{11}$ **36.** $\frac{67}{13}$

Write each mixed number as an improper fraction. See Example 5.

37. $2\frac{3}{5}$ **38.** $5\frac{6}{7}$ **39.** $10\frac{3}{8}$ **40.** $12\frac{2}{3}$ **41.** $10\frac{4}{5}$ **42.** $18\frac{1}{6}$

43. For the fractions $\frac{p}{q}$ and $\frac{r}{s}$, which can serve as a common denominator?

A. $q \cdot s$

B. $q + s$

C. $p \cdot r$

D. $p + r$

44. Which fraction is *not* equal to $\frac{5}{9}$?

A. $\frac{15}{27}$ **B.** $\frac{30}{54}$

C. $\frac{40}{74}$ **D.** $\frac{55}{99}$

Find each product or quotient, and write it in lowest terms. See Examples 6 and 7.

45. $\frac{4}{5} \cdot \frac{6}{7}$ **46.** $\frac{5}{9} \cdot \frac{10}{7}$ **47.** $\frac{1}{10} \cdot \frac{12}{5}$ **48.** $\frac{6}{11} \cdot \frac{2}{3}$

49. $\frac{15}{4} \cdot \frac{8}{25}$ **50.** $\frac{4}{7} \cdot \frac{21}{8}$ **51.** $2\frac{2}{3} \cdot 5\frac{4}{5}$ **52.** $3\frac{3}{5} \cdot 7\frac{1}{6}$

53. $\frac{5}{4} \div \frac{3}{8}$ **54.** $\frac{7}{6} \div \frac{9}{10}$ **55.** $\frac{32}{5} \div \frac{8}{15}$ **56.** $\frac{24}{7} \div \frac{6}{21}$

57. $\frac{3}{4} \div 12$ **58.** $\frac{2}{5} \div 30$ **59.** $2\frac{5}{8} \div 1\frac{15}{32}$ **60.** $2\frac{3}{10} \div 7\frac{4}{5}$

Find each sum or difference, and write it in lowest terms. See Examples 8–10.

61. $\frac{7}{12}+\frac{1}{12}$

62. $\frac{3}{16}+\frac{5}{16}$

63. $\frac{5}{9}+\frac{1}{3}$

64. $\frac{4}{15}+\frac{1}{5}$

65. $3\frac{1}{8}+\frac{1}{4}$

66. $5\frac{3}{4}+\frac{2}{3}$

67. $\frac{7}{12}-\frac{1}{9}$

68. $\frac{11}{16}-\frac{1}{12}$

69. $6\frac{1}{4}-5\frac{1}{3}$

70. $8\frac{4}{5}-7\frac{4}{9}$

71. $\frac{5}{3}+\frac{1}{6}-\frac{1}{2}$

72. $\frac{7}{15}+\frac{1}{6}-\frac{1}{10}$

Solve each applied problem.

Use the chart, which appears on a package of Quaker Quick Grits, to answer the questions in Exercises 73 and 74.

73. How many cups of water would be needed for eight microwave servings?

74. How many teaspoons of salt would be needed for five stove top servings? (*Hint:* 5 is halfway between 4 and 6.)

	Microwave	Stove Top		
Servings	**1**	**1**	**4**	**6**
Water	$\frac{3}{4}$ cup	1 cup	3 cups	4 cups
Grits	3 Tbsp	3 Tbsp	$\frac{3}{4}$ cup	1 cup
Salt (optional)	Dash	Dash	$\frac{1}{4}$ tsp	$\frac{1}{2}$ tsp

75. A motel owner has decided to expand his business by buying a piece of property next to the motel. The property has an irregular shape, with five sides as shown in the figure. Find the total distance around the piece of property. This is called the **perimeter** of the figure.

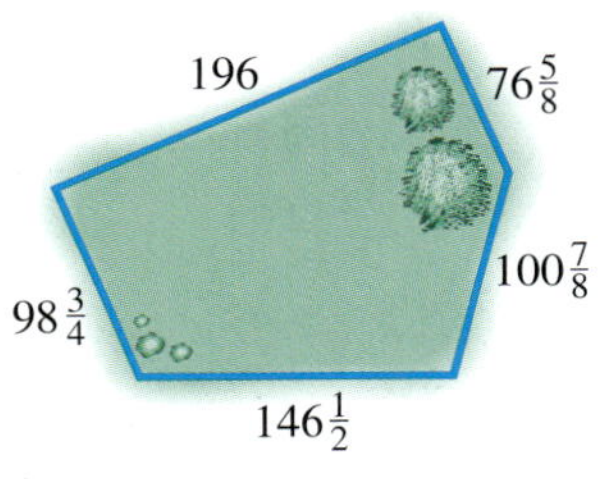

Measurements in feet

76. A triangle has sides of lengths $5\frac{1}{4}$ ft, $7\frac{1}{2}$ ft, and $10\frac{1}{8}$ ft. Find the perimeter of the triangle. See Exercise 75.

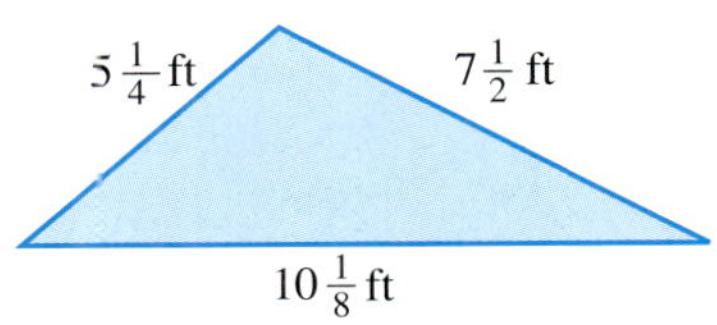

77. A hardware store sells a 40-piece socket wrench set. The measure of the largest socket is $\frac{3}{4}$ in., while the measure of the smallest socket is $\frac{3}{16}$ in. What is the difference between these measures?

78. Two sockets in a socket wrench set have measures of $\frac{9}{16}$ in. and $\frac{3}{8}$ in. What is the difference between these two measures?

79. Under existing standards, most of the holes in Swiss cheese must have diameters between $\frac{11}{16}$ and $\frac{13}{16}$ in. To accommodate new high-speed slicing machines, the USDA wants to reduce the minimum size to $\frac{3}{8}$ in. How much smaller is $\frac{3}{8}$ in. than $\frac{11}{16}$ in.? (*Source:* U.S. Department of Agriculture.)

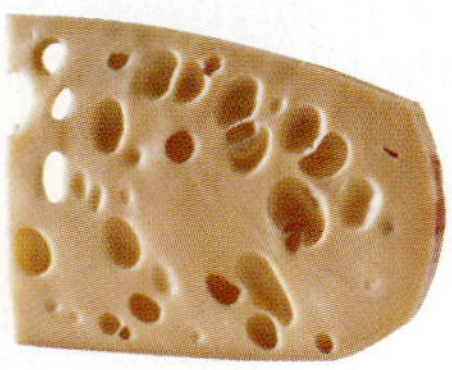

80. Loretta's favorite recipe for barbecue sauce calls for $2\frac{1}{3}$ cups of tomato sauce. The recipe makes enough barbecue sauce to serve 7 people. How much tomato sauce is needed for 1 serving?

81. It takes $2\frac{3}{8}$ yd from a bolt of fabric to make a costume for a school play. How much fabric would be needed for 7 costumes?

82. A cake recipe calls for $1\frac{3}{4}$ cups of sugar. A caterer has $15\frac{1}{2}$ cups of sugar on hand. How many cakes can he make?

Approximately 34 *million people living in the United States in 2004 were born in other countries. The circle graph gives the fractional number from each region of birth for these people. Use the graph to answer each question.*

U.S. Foreign-Born Population By Region of Birth

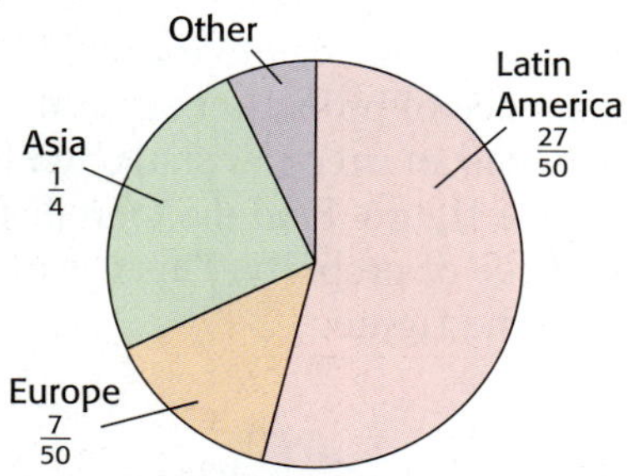

Source: U.S. Census Bureau.

83. What fractional part of the foreign-born population was from other regions?

84. What fractional part of the foreign-born population was from Latin America or Asia?

85. How many people (in millions) were born in Europe?

86. How many more people (in millions) were born in Latin America than in Asia?

Appendix: Synthetic Division

C

OBJECTIVES

1. **Use synthetic division to divide by a polynomial of the form $x - k$.**
2. **Use the remainder theorem to evaluate a polynomial.**
3. **Decide whether a given number is a solution of an equation.**

We begin by reviewing the terminology for the parts of a division problem. The *divisor* is the quantity we are dividing by, the *dividend* is the quantity we are dividing into, and the *quotient* is the result of the division.

$$\begin{array}{r}1\,562 \longleftarrow \text{Quotient}\\ \text{Divisor} \longrightarrow 247\overline{)385{,}814} \longleftarrow \text{Dividend}\end{array}$$

OBJECTIVE 1 Use synthetic division to divide by a polynomial of the form $x - k$. When one polynomial is divided by a second, if the divisor has the form $x - k$, where the coefficient of the x-term is 1, there is a shortcut method for performing the division. Look at the division of $3x^3 - 2x + 5$ by $x - 3$ on the left below. Notice that we inserted 0 for the missing x^2-term.

$$\begin{array}{rrrrr}
 & 3x^2 & +\ 9x & +\ 25 & \\
x-3\,\big) & 3x^3 & +\ 0x^2 & -\ 2x & +\ 5\\
 & 3x^3 & -\ 9x^2 & & \\ \hline
 & & 9x^2 & -\ 2x & \\
 & & 9x^2 & -\ 27x & \\ \hline
 & & & 25x & +\ 5\\
 & & & 25x & -\ 75\\ \hline
 & & & & 80
\end{array}
\qquad\Bigg|\qquad
\begin{array}{rrrrr}
 & 3 & 9 & 25 & \\
1-3\,\big) & 3 & 0 & -2 & 5\\
 & 3 & -9 & & \\ \hline
 & & 9 & -2 & \\
 & & 9 & -27 & \\ \hline
 & & & 25 & 5\\
 & & & 25 & -75\\ \hline
 & & & & 80
\end{array}$$

On the right, the same division is shown written without the variables. This is why it is *essential* to use 0 as a placeholder in synthetic division. All the numbers in color on the right are repetitions of the numbers directly above them, so they are omitted to condense the work, as shown on the left below.

$$\begin{array}{rrrrr}
 & 3 & 9 & 25 & \\
1-3\,\big) & 3 & 0 & -2 & 5\\
 & & -9 & & \\ \hline
 & & 9 & -2 & \\
 & & & -27 & \\ \hline
 & & & 25 & 5\\
 & & & & -75\\ \hline
 & & & & 80
\end{array}
\qquad\Bigg|\qquad
\begin{array}{rrrrr}
 & 3 & 9 & 25 & \\
1-3\,\big) & 3 & 0 & -2 & 5\\
 & & -9 & & \\ \hline
 & & 9 & & \\
 & & & -27 & \\ \hline
 & & & 25 & \\
 & & & & -75\\ \hline
 & & & & 80
\end{array}$$

The numbers in color on the left are again repetitions of the numbers directly above them; they too are omitted, as shown on the right above. If the 3 in the dividend is brought down to the beginning of the bottom row, the top row can be omitted since it duplicates the bottom row.

1 Divide, using synthetic division.

(a) $\dfrac{3z^2 + 10z - 8}{z + 4}$

(b) $(2x^2 + 3x - 5) \div (x + 1)$

$$\begin{array}{r|rrrr} 1 - 3 & 3 & 0 & -2 & 5 \\ & & -9 & -27 & -75 \\ \hline & 3 & 9 & 25 & 80 \end{array}$$

We omit the 1 at the upper left, since it represents $1x$, which will *always* be the first term in the divisor. Also, to simplify the arithmetic, we replace subtraction in the second row by addition. We compensate for this by changing the -3 at the upper left to its additive inverse, 3.

Additive inverse of -3 ⟶

$$\begin{array}{r|rrrr} 3 & 3 & 0 & -2 & 5 \\ & & 9 & 27 & 75 \\ \hline & \mathbf{3} & \mathbf{9} & \mathbf{25} & \mathbf{80} \end{array}$$

⟵ Change signs.

⟵ Remainder

The quotient is read from the bottom row.

$$3x^2 + 9x + 25 + \frac{80}{x - 3}$$

The first three numbers in the bottom row are the coefficients of the quotient polynomial with degree 1 less than the degree of the dividend. The last number gives the remainder.

> **Synthetic Division**
>
> This shortcut method is called **synthetic division.** It is used *only* when dividing a polynomial by a binomial of the form $x - k$.

EXAMPLE 1 Using Synthetic Division

Use synthetic division to divide $5x^2 + 16x + 15$ by $x + 2$.

As mentioned above, we use synthetic division only when dividing by a polynomial of the form $x - k$. We change $x + 2$ to this form by writing it as

$$x + 2 = x - (-2),$$

where $k = -2$. Then we write the coefficients of $5x^2 + 16x + 15$.

$x + 2$ leads to -2. ⟶ $-2\overline{)5 \quad 16 \quad 15}$ ⟵ Coefficients

We bring down the 5, and multiply: $-2 \cdot 5 = -10$.

$$\begin{array}{r|rrr} -2 & 5 & 16 & 15 \\ & & \mathbf{-10} & \\ \hline & 5 & & \end{array}$$

We add 16 and -10, getting 6. Multiply 6 and -2 to get -12.

$$\begin{array}{r|rrr} -2 & 5 & 16 & 15 \\ & & -10 & \mathbf{-12} \\ \hline & 5 & 6 & \end{array}$$

We add 15 and -12, getting 3.

$$\begin{array}{r|rrr} -2 & 5 & 16 & 15 \\ & & -10 & -12 \\ \hline & \mathbf{5} & \mathbf{6} & \mathbf{3} \end{array}$$

⟵ Remainder

We read the result from the bottom row.

$$\frac{5x^2 + 16x + 15}{x + 2} = 5x + 6 + \frac{3}{x + 2}$$

◀ Work Problem **1** at the Side.

ANSWERS

1. (a) $3z - 2$ (b) $2x + 1 + \dfrac{-6}{x + 1}$

EXAMPLE 2 Using Synthetic Division with a Missing Term

Use synthetic division to find $(-4x^5 + x^4 + 6x^3 + 2x^2 + 50) \div (x - 2)$.

Use the steps given above, inserting a 0 for the missing x-term.

$$\begin{array}{r|rrrrrr} 2 & -4 & 1 & 6 & 2 & 0 & 50 \\ & & -8 & -14 & -16 & -28 & -56 \\ \hline & -4 & -7 & -8 & -14 & -28 & -6 \end{array}$$

Read the result from the bottom row.

$$\frac{-4x^5 + x^4 + 6x^3 + 2x^2 + 50}{x - 2} = -4x^4 - 7x^3 - 8x^2 - 14x - 28 + \frac{-6}{x - 2}$$

Work Problem 2 at the Side.

OBJECTIVE 2 Use the remainder theorem to evaluate a polynomial. We can use synthetic division to evaluate polynomials. For example, in the synthetic division of Example 2, where the polynomial was divided by $x - 2$, the remainder was -6.

Replacing x in the polynomial with 2 gives

$$\begin{aligned} &-4x^5 + x^4 + 6x^3 + 2x^2 + 50 \\ &= -4 \cdot 2^5 + 2^4 + 6 \cdot 2^3 + 2 \cdot 2^2 + 50 \\ &= -4 \cdot 32 + 16 + 6 \cdot 8 + 2 \cdot 4 + 50 \\ &= -128 + 16 + 48 + 8 + 50 \\ &= -6, \end{aligned}$$

the same number as the remainder. Thus, dividing by $x - 2$ produced a remainder equal to the result when x is replaced with 2. This always happens, as the following remainder theorem states.

Remainder Theorem

If the polynomial $P(x)$ is divided by $x - k$, then the remainder is equal to $P(k)$.

This result is proved in more advanced courses.

EXAMPLE 3 Using the Remainder Theorem

Let $P(x) = 2x^3 - 5x^2 - 3x + 11$. Find $P(-2)$.

Use the remainder theorem; divide $P(x)$ by $x - (-2)$.

$$\text{Value of } x \rightarrow \begin{array}{r|rrrr} -2 & 2 & -5 & -3 & 11 \\ & & -4 & 18 & -30 \\ \hline & 2 & -9 & 15 & -19 \end{array} \leftarrow \text{Remainder}$$

By this result, $P(-2) = -19$.

Work Problem 3 at the Side.

2 Divide, using synthetic division.

(a) $\dfrac{3a^3 - 2a + 21}{a + 2}$

(b) $(-4x^4 + 3x^3 + 18x + 2) \div (x - 2)$

3 Let $P(x) = x^3 - 5x^2 + 7x - 3$. Use synthetic division to find each value.

(a) $P(1)$ (Divide by $x - 1$.)

(b) $P(-2)$

ANSWERS

2. (a) $3a^2 - 6a + 10 + \dfrac{1}{a + 2}$

(b) $-4x^3 - 5x^2 - 10x - 2 + \dfrac{-2}{x - 2}$

3. (a) 0 **(b)** -45

4 Use synthetic division to decide whether 2 is a solution of each equation.

(a) $3x^3 - 11x^2 + 17x - 14 = 0$

(b) $4x^5 - 7x^4 - 11x^2 + 2x + 6 = 0$

Answers

4. (a) yes **(b)** no

OBJECTIVE 3 Decide whether a given number is a solution of an equation. The remainder theorem can also be used to show that a given number is a solution of an equation.

EXAMPLE 4 Using the Remainder Theorem

Show that -5 is a solution of the equation

$$2x^4 + 12x^3 + 6x^2 - 5x + 75 = 0.$$

One way to show that -5 is a solution is to substitute -5 for x in the equation. However, an easier way is to use synthetic division and the remainder theorem.

$$\begin{array}{r|rrrrr} \text{Proposed solution} \rightarrow -5 & 2 & 12 & 6 & -5 & 75 \\ & & -10 & -10 & 20 & -75 \\ \hline & 2 & 2 & -4 & 15 & 0 \leftarrow \text{Remainder} \end{array}$$

Since the remainder is 0, the polynomial has a value of 0 when $x = -5$, so -5 is a solution of the given equation.

◀ Work Problem 4 at the Side.

The synthetic division in Example 4 shows that $x - (-5)$ divides the polynomial with 0 remainder. Thus $x - (-5) = x + 5$ is a *factor* of the polynomial and

$$2x^4 + 12x^3 + 6x^2 - 5x + 75 = (x + 5)(2x^3 + 2x^2 - 4x + 15).$$

The second factor is the quotient polynomial found in the last row of the synthetic division.

Appendix C Exercises

FOR EXTRA HELP

MyMathLab — Math XL PRACTICE — WATCH — DOWNLOAD — READ — REVIEW

Choose the letter of the correct setup to perform synthetic division on the indicated quotient.

1. $\dfrac{x^2 + 3x - 6}{x - 2}$

A. $-2\overline{)1 \quad 3 \quad -6}$ B. $-2\overline{)-1 \quad -3 \quad 6}$

C. $2\overline{)1 \quad 3 \quad -6}$ D. $2\overline{)-1 \quad -3 \quad 6}$

2. $\dfrac{x^3 - 3x^2 + 2}{x - 1}$

A. $1\overline{)1 \quad -3 \quad 2}$ B. $-1\overline{)1 \quad -3 \quad 2}$

C. $1\overline{)1 \quad -3 \quad 0 \quad 2}$ D. $1\overline{)-1 \quad 3 \quad 0 \quad -2}$

Use synthetic division to find each quotient. See Examples 1 and 2.

3. $\dfrac{x^2 - 6x + 5}{x - 1}$

4. $\dfrac{x^2 - 4x - 21}{x + 3}$

5. $\dfrac{4m^2 + 19m - 5}{m + 5}$

6. $\dfrac{3k^2 - 5k - 12}{k - 3}$

7. $\dfrac{2a^2 + 8a + 13}{a + 2}$

8. $\dfrac{4y^2 - 5y - 20}{y - 4}$

9. $(p^2 - 3p + 5) \div (p + 1)$

10. $(z^2 + 4z - 6) \div (z - 5)$

11. $\dfrac{4a^3 - 3a^2 + 2a - 3}{a - 1}$

12. $\dfrac{5p^3 - 6p^2 + 3p + 14}{p + 1}$

13. $(x^5 - 2x^3 + 3x^2 - 4x - 2) \div (x - 2)$

14. $(2y^5 - 5y^4 - 3y^2 - 6y - 23) \div (y - 3)$

15. $(-4r^6 - 3r^5 - 3r^4 + 5r^3 - 6r^2 + 3r) \div (r - 1)$

16. $(-3t^5 + 2t^4 - 5t^3 + 6t^2 - 3t - 2) \div (t - 2)$

17. $(-3y^5 + 2y^4 - 5y^3 - 6y^2 - 1) \div (y + 2)$

18. $(m^6 + 2m^4 - 5m + 11) \div (m - 2)$

19. $\dfrac{y^3 + 1}{y - 1}$

20. $\dfrac{z^4 + 81}{z - 3}$

Use the remainder theorem to find $P(k)$. See Example 3.

21. $P(x) = 2x^3 - 4x^2 + 5x - 3;\ k = 2$

22. $P(x) = x^3 + 3x^2 - x + 5;\ k = -1$

23. $P(r) = -r^3 - 5r^2 - 4r - 2;\ k = -4$

24. $P(z) = -z^3 + 5z^2 - 3z + 4;\ k = 3$

25. $P(x) = 2x^3 - 4x^2 + 5x - 33;\ k = 3$

26. $P(x) = x^3 - 3x^2 + 4x - 4;\ k = 2$

Use synthetic division to decide whether the given number is a solution of each equation. See Example 4.

27. $x^3 - 2x^2 - 3x + 10 = 0;\ x = -2$

28. $x^3 - 3x^2 - x + 10 = 0;\ x = -2$

29. $m^4 + 2m^3 - 3m^2 + 8m - 8 = 0;\ m = -2$

30. $r^4 - r^3 - 6r^2 + 5r + 10 = 0;\ r = -2$

31. $3x^3 + 2x^2 - 2x + 11 = 0;\ x = -2$

32. $3z^3 + 10z^2 + 3z - 9 = 0;\ z = -2$

33. Explain why it is important to insert 0s as placeholders for missing terms before performing synthetic division.

34. Explain why a 0 remainder in synthetic division of $P(x)$ by k indicates that k is a solution of the equation $P(x) = 0$.

ANSWERS TO SELECTED EXERCISES

In this section we provide the answers that we think most students will obtain when they work the exercises using the methods explained in the text. If your answer does not look exactly like the one given here, it is not necessarily wrong. In many cases there are equivalent forms of the answer that are correct. For example, if the answer section shows $\frac{3}{4}$ and your answer is 0.75, you have obtained the correct answer but written it in a different (yet equivalent) form. Unless the directions specify otherwise, 0.75 is just as valid an answer as $\frac{3}{4}$.

In general, if your answer does not agree with the one given in the text, see whether it can be transformed into the other form. If it can, then it is the correct answer. If you still have doubts, talk with your instructor.

Chapter 1 Review of the Real Number System

Section 1.1 (pages 13–16)

1. $\{1, 2, 3, 4, 5\}$ **3.** $\{5, 6, 7, 8, \ldots\}$ **5.** $\{10, 12, 14, 16, \ldots\}$ **7.** $\emptyset$ **9.** $\{-4, 4\}$ **11.** $\{x \mid x$ is an even natural number less than or equal to $8\}$ **13.** $\{x \mid x$ is a multiple of 4 greater than $0\}$

15. 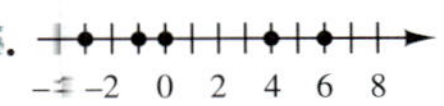

17. 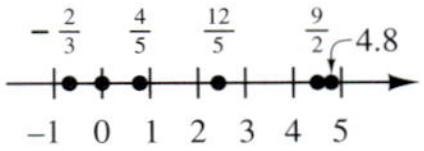

19. **(a)** $5, 17, \frac{40}{2}$ (or 20) **(b)** $0, 5, 17, \frac{40}{2}$ **(c)** $-8, 0, 5, 17, \frac{40}{2}$ **(d)** $-8, -0.6, 0, \frac{3}{4}, 5, \frac{13}{2}, 17, \frac{40}{2}$ **(e)** $-\sqrt{5}, \sqrt{3}, \pi$ **(f)** All are real numbers. **21.** false; Some are integers, but others, like $\frac{3}{4}$, are not. **23.** false; No irrational number is an integer. **25.** true **27.** true **29.** true **31.** **(a)** -6 **(b)** 6 **33.** **(a)** 12 **(b)** 12 **35.** **(a)** $-\frac{6}{5}$ **(b)** $\frac{6}{5}$ **37.** 8 **39.** $\frac{3}{2}$ **41.** -5 **43.** -2 **45.** -4.5 **47.** 5 **49.** 6 **51.** 0 **53.** **(a)** Louisiana; It decreased 4.1%. **(b)** West Virginia; It increased 0.6%. **55.** Pacific Ocean, Indian Ocean, Caribbean Sea, South China Sea, Gulf of California **57.** true **59.** true **61.** false **63.** true **65.** true **67.** $7 > y$ **69.** $5 \geq 5$ **71.** $3t - 4 \leq 10$ **73.** $5x + 3 \neq 0$ **75.** $-6 < 10$; true **77.** $10 \geq 10$; true **79.** $-3 \geq -3$; true **81.** $-8 > -6$; false **83.** greater than **85.** California (CA), Florida (FL) **87.** $x < y$

Section 1.2 (pages 23–28)

1. the numbers are additive inverses; $4 + (-4) = 0$ **3.** negative; $-7 + (-21) = -28$ **5.** the positive number has greater absolute value; $15 + (-2) = 13$ **7.** the number with lesser absolute value is subtracted from the one with greater absolute value; $-15 - (-3) = -12$ **9.** negative; $-5(15) = -75$ **11.** 9 **13.** -19 **15.** $-\frac{19}{12}$ **17.** -1.85 **19.** -11 **21.** 21 **23.** -13 **25.** -10.18 **27.** $\frac{67}{30}$ **29.** 14 **31.** -5 **33.** -6 **35.** -11 **37.** 16 **39.** -4 **41.** 4.218 **43.** $-\frac{7}{8}$ **45.** -19 **47.** [illegible] **49.** -35 **51.** 40 **53.** 2 **55.** -12 **57.** $\frac{6}{5}$ **59.** 1 **61.** 5.88 **63.** -10.676 **65.** $\frac{1}{6}$ **67.** $-\frac{1}{7}$ **69.** $-\frac{3}{2}$ **71.** 5 **73.** 50 **75.** -1000 **77.** -7 **79.** 6 **81.** -4 **83.** 0 **85.** undefined **87.** $\frac{25}{102}$ **89.** $-\frac{9}{13}$ **91.** -2.1 **93.** 10,000 **95.** $\frac{17}{18}$ **97.** $\frac{17}{36}$ **99.** $-\frac{19}{24}$ **101.** $-\frac{22}{45}$ **103.** $-\frac{2}{15}$ **105.** $\frac{3}{5}$ **107.** $-\frac{35}{27}$ **109.** $-\frac{4}{9}$ **111.** -12.351 **113.** -15.876 **115.** -4.14 **117.** 4800 **119.** 51.495 **121.** 112°F **123.** 24.89% **125.** \$30.13 **127.** **(a)** \$466.02 **(b)** \$190.68 **129.** **(a)** $-\$475$ thousand **(b)** \$262 thousand **(c)** $-\$83$ thousand **131.** 2000: \$129 billion; 2010: \$206 billion; 2020: \$74 billion; 2030: $-\$501$ billion

Section 1.3 (pages 35–38)

1. false; $-4^6 = -(4^6)$ **3.** true **5.** true **7.** true **9.** false; The base is 3. **11.** **(a)** 64 **(b)** -64 **(c)** 64 **(d)** -64 **13.** 8^3 **15.** $\left(\frac{1}{2}\right)^2$ **17.** $(-4)^4$ **19.** z^7 **21.** 16 **23.** 0.021952 **25.** $\frac{1}{125}$ **27.** $\frac{256}{625}$ **29.** -125 **31.** 256 **33.** -729 **35.** -4096 **37.** exponent: 7; base: -4.1 **39.** exponent: 7; base: 4.1 **41.** 9 **43.** 13 **45.** -20 **47.** $\frac{10}{11}$ **49.** -0.7 **51.** not a real number **53.** **(a)** B **(b)** C **(c)** A **55.** not a real number **57.** 24 **59.** 4 **61.** 14 **63.** 15 **65.** 55 **67.** -91 **69.** -8 **71.** -48 **73.** 2 **75.** -2 **77.** -79 **79.** -2 **81.** undefined **83.** -1 **85.** 17 **87.** -96 **89.** $-\frac{15}{238}$ **91.** 8 **93.** $-\frac{5}{16}$ **95.** \$2434 **97.** \$4159 **99.** 0.035 **101.** Decreased weight will result in higher BACs; 0.040; 0.053 **103.** **(a)** 2.18; 4.98; 6.38 **(b)** The average price of a theater ticket in the United States almost tripled from 1977 to 2007.

Section 1.4 (pages 45–46)

1. B **3.** A **5.** product; 0 **7.** grouping **9.** like **11.** $2m + 2p$ **13.** $-10d + 5f$ **15.** $8k$ **17.** $-2r$ **19.** $8a$ **21.** cannot be simplified **23.** $-4b + c$ **25.** 1900 **27.** 75 **29.** 431 **31.** $-6y + 3$ **33.** $p + 11$ **35.** $-2k + 15$ **37.** $m - 14$ **39.** -1 **41.** $2p + 7$ **43.** $-6z - 39$ **45.** $(5 + 8)x = 13x$ **47.** $(5 \cdot 9)r = 45r$ **49.** $9y + 5x$ **51.** 7 **53.** $8(-4) + 8x = -32 + 8x$ **55.** Answers will vary. One example is washing your face and brushing your teeth. **57.** associative property **58.** associative property **59.** commutative property **60.** associative property **61.** distributive property **62.** add

Chapter 1 Review Exercises (pages 51–54)

1. 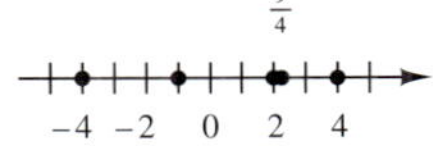

2. 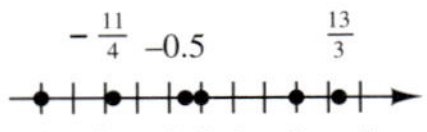

3. 16 **4.** 23 **5.** -4 **6.** 5 **7.** $0, \frac{12}{3}$ (or 4) **8.** $-9, -\sqrt{4}$ (or -2), $0, \frac{12}{3}$ (or 4) **9.** $-9, -\frac{4}{3}, -\sqrt{4}$ (or -2), $-0.25, 0, 0.\overline{35}, \frac{5}{3}, \frac{12}{3}$ (or 4) **10.** All are real numbers except $\sqrt{-9}$. **11.** $\{4, 5, 6, 7, 8\}$ **12.** $\{0, 1, 2, 3\}$ **13.** true **14.** false **15.** true

ANSWERS

16. Chrysler; 13.7% **17.** Honda; −2.5% **18.** false **19.** true **20.** $\frac{41}{24}$ **21.** $-\frac{1}{2}$ **22.** −3 **23.** −17.09 **24.** −39 **25.** −1 **26.** $\frac{23}{20}$ **27.** $-\frac{5}{18}$ **28.** −35 **29.** 11,331 ft **30.** −90 **31.** $\frac{2}{3}$ **32.** −11.408 **33.** −15 **34.** 3.21 **35.** $\frac{5}{7-7}$ **36.** 10,000 **37.** $\frac{27}{343}$ **38.** −125 **39.** −125 **40.** 2.89 **41.** 20 **42.** −14 **43.** $\frac{8}{11}$ **44.** −0.9 **45.** not a real number **46.** −4 **47.** 44 **48.** −2 **49.** −30 **50.** −30 **51.** $-\frac{8}{51}$ **52. (a)** 27 **(b)** Answers will vary. **53.** $21q$ **54.** $-4z$ **55.** $5m$ **56.** $4p$ **57.** $-2k-6$ **58.** $6r+18$ **59.** $18m+27n$ **60.** $-p-3q$ **61.** $y+1$ **62.** 0 **63.** $-18m$ **64.** $(2+3)x=5x$ **65.** −4 **66.** $(2\cdot 4)x=8x$ **67.** $13+(-3)=10$ **68.** 0 **69.** $5x+5z$ **70.** 7 **71.** 1 **72.** $(3+5+6)a=14a$ **73.** 0 **74.** $51,671 million; negative **75.** $66,480 million; negative **76.** $76,521 million; negative **77.** $\frac{256}{625}$ **78.** 25 **79.** 31 **80.** 9 **81.** 0 **82.** −5 **83.** $\frac{4}{3}$ **84.** −6.16 **85.** −9 **86.** 2 **87.** 2 **88.** not a real number **89.** −116 **90.** Work within the parentheses first.

Chapter 1 Test (pages 55–56)

1. 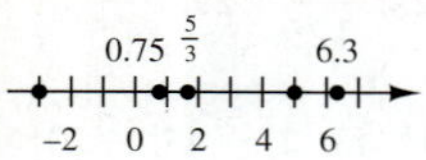

2. 0, 3, $\sqrt{25}$ (or 5), $\frac{24}{2}$ (or 12) **3.** −1, 0, 3, $\sqrt{25}$ (or 5), $\frac{24}{2}$ (or 12) **4.** −1, −0.5, 0, 3, $\sqrt{25}$ (or 5), 7.5, $\frac{24}{2}$ (or 12) **5.** All are real numbers except $\sqrt{-4}$. **6.** 0 **7.** −26 **8.** 19 **9.** 1 **10.** $\frac{16}{7}$ **11.** $\frac{11}{23}$ **12.** 50,395 ft **13.** 37,486 ft **14.** 1345 ft **15.** 14 **16.** −15 **17.** not a real number **18. (a)** a must be positive. **(b)** a must be negative. **(c)** a must be 0. **19.** 2 **20.** $-\frac{6}{23}$ **21.** $10k-10$ **22.** Both terms change sign and are added to $3r+8$; $7r+2$. **23.** B **24.** E **25.** D **26.** A **27.** F **28.** C **29.** C **30.** E

Chapter 2 Linear Equations and Applications

Section 2.1 (pages 67–70)

1. A and C **3.** Both sides are evaluated as 30, so 6 is a solution. **5.** Any number is a solution. For example, if the last name is Lincoln, then $x=7$. Both sides are evaluated as −48. **7. (a)** equation **(b)** expression **(c)** equation **(d)** expression **9.** The student made a sign error when the distributive property was applied. The left side of the second line should be $8x-4x+6$. The correct solution is 1.

11. $\{-1\}$ **13.** $\{-4\}$ **15.** $\{-7\}$ **17.** $\{0\}$ **19.** $\{4\}$ **21.** $\left\{-\frac{7}{8}\right\}$ **23.** $\left\{-\frac{5}{3}\right\}$ **25.** $\left\{-\frac{1}{2}\right\}$ **27.** $\{2\}$ **29.** $\{-2\}$ **31.** $\{-1\}$ **33.** $\{7\}$ **35.** $\{2\}$ **37.** $\{-8\}$ **39.** 12 **41. (a)** 10^2, or 100 **(b)** 10^3, or 1000 **43.** $\{12\}$ **45.** $\{4\}$ **47.** $\{-30\}$ **49.** $\{0\}$ **51.** $\{3\}$ **53.** $\{0\}$ **55.** $\{2000\}$ **57.** $\{25\}$ **59.** $\{40\}$ **61.** $\{3\}$ **63.** A conditional equation is true only for certain value(s), an identity has infinitely many solutions, and a contradiction has no solution. **65. (a)** B **(b)** A **(c)** C **67.** contradiction; $\emptyset$ **69.** conditional; $\{0\}$ **71.** identity; {all real numbers} **73.** identity; {all real numbers}

Section 2.2 (pages 79–84)

1. (a) $7x+8=36$ **(b)** $ax+k=tc$ **2. (a)** $7x+8-8=36-8$ **(b)** $ax+k-k=tc-k$ **3. (a)** $7x=28$ **(b)** $ax=tc-k$ **4. (a)** $x=4$ **(b)** $x=\frac{tc-k}{a}$ **5.** $a\neq 0$; If $a=0$, the denominator is 0. **6.** To solve an equation for a particular variable, such as solving the second equation for x, go through the same steps as you would in solving for x in the first equation. Treat all other variables as constants. **7.** $W=\frac{A}{L}$ **9.** $L=\frac{P-2W}{2}$, or $L=\frac{P}{2}-W$ **11. (a)** $W=\frac{V}{LH}$ **(b)** $H=\frac{V}{LW}$ **13.** $r=\frac{C}{2\pi}$ **15. (a)** $h=\frac{2A}{b+B}$ **(b)** $B=\frac{2A}{h}-b$, or $B=\frac{2A-bh}{h}$ **17.** $C=\frac{5}{9}(F-32)$ **19.** $y=\frac{11-4x}{9}$ **21.** $y=\frac{5+3x}{2}$ **23.** $y=\frac{7-6x}{-5}$, or $y=\frac{6x-7}{5}$ **24.** $F=\frac{k}{d-D}$ **25.** $m=\frac{Mv}{v-V}$ **26.** $W=\frac{A-2LH}{2H+2L}$ **27.** 3.275 hr **29.** 113°F **31.** 230 m **33.** radius: 185 in.; diameter: 370 in. **35.** 2 in. **37.** 75% water, 25% alcohol **39.** 3% **41.** $10.51 **43.** $45.66 **45. (a)** .543 **(b)** .488 **(c)** .444 **(d)** .426 **47.** 52% **49.** $82,304 **51.** 17%; yes **53.** 8% **55.** 3.8% **57.** 47.5%

Section 2.3 (pages 93–98)

1. (a) $x+12$ **(b)** $12>x$ **3. (a)** $x-4$ **(b)** $4<x$ **5.** D **7.** $2x+18$ **9.** $15-4x$ **11.** $10(x-6)$ **13.** $\frac{5x}{9}$ **15.** $x+6=-31$; −37 **17.** $x-(-4x)=x+9$; $\frac{9}{4}$ **19.** $12-\frac{2}{3}x=10$; 3 **21.** expression **23.** equation **25.** expression **27.** *Step 1:* We are asked to find the number of patents each university secured; *Step 2:* the number of patents Stanford secured; *Step 3:* x; $x-38$; *Step 4:* 134; *Step 5:* 134, 96; *Step 6:* 38; MIT patents; 96; 230 **29.** width: 165 ft; length: 265 ft **31.** 850 mi, 925 mi, 1300 mi **33.** Eiffel Tower: 984 ft; Leaning Tower: 180 ft **35.** Yankees: $209.1 million; Tigers: $138.7 million **37.** $35.67 **39.** 252,887 **41.** $225 **43.** $4000 at 3%; $8000 at 4% **45.** $10,000 at 4.5%; $19,000 at 3% **47.** $13,500 **49.** 5 L **51.** 4 L **53.** 1 gal **55.** 150 lb **57.** We cannot expect the final mixture to be worth more than either of the ingredients. **59. (a)** $800-x$ **(b)** $800-y$ **60. (a)** $0.05x$; $0.10(800-x)$ **(b)** $0.05y$; $0.10(800-y)$ **61. (a)** $0.05x+0.10(800-x)=800(0.0875)$ **(b)** $0.05y+0.10(800-y)=800(0.0875)$ **62. (a)** $200 at 5%; $600 at 10% **(b)** 200 L of 5% acid; 600 L of 10% acid **63.** The processes are the same. The amounts of money in Problem A correspond to the amounts of solution in Problem B.

Section 2.4 (pages 105–108)

1. $4.50 **3.** 55 mph **5.** 17 pennies, 17 dimes, 10 quarters **7.** 23 loonies; 14 toonies **9.** 28 $10 coins, 25 $20 coins **11.** 872 adult tickets **13.** 8.08 m per sec **15.** 8.40 m per sec **17.** $2\frac{1}{2}$ hr **19.** 7:50 P.M. **21.** 45 mph **23.** $\frac{1}{2}$ hr **25.** 60°, 60°, 60° **27.** 40°, 45°, 95° **29.** 40°, 80° **30.** 120° **31.** The sum is equal to the measure of the angle found in Exercise 30. **32.** The sum of the measures of angles 1 and 2 is equal to the measure of angle 3. **33.** Both measure 122°. **35.** 64°, 26° **37.** 24, 25, 26 **39.** 57 yr old

Summary Exercises on Solving Applied Problems (pages 109–110)

1. length: 8 in.; width: 5 in. **2.** 6 in., 12 in., 16 in. **3.** \$56.94 **4.** \$425 **5.** \$550 at 4%; \$1100 at 5% **6.** \$12,000 at 3%; \$15,000 at 4% **7.** 2005–2006: 2832; 2006–2007: 2430 **8.** *Shrek 2:* \$437.2 million; *Spider-Man:* \$407.7 million **9.** $1\frac{1}{2}$ cm **10.** 5 hr **11.** $13\frac{1}{3}$ L **12.** $5\frac{1}{3}$ kg **13.** fives: 84; tens: 42 **14.** 10 ft **15.** 44, 45, 46 **16.** 9 and 11 **17.** 20°, 30°, 130° **18.** 107°, 73°

Chapter 2 Review Exercises (pages 117–120)

1. $\left\{-\frac{9}{5}\right\}$ **2.** $\left\{\frac{1}{3}\right\}$ **3.** {10} **4.** $\left\{-\frac{7}{5}\right\}$ **5.** ∅ **6.** {0} **7.** {−16} **8.** {300} **9.** B **10.** Begin by subtracting 5 from each side. Then divide each side by −2. **11.** identity; {all real numbers} **12.** contradiction; ∅ **13.** conditional; {0} **14.** $L = \frac{V}{WH}$ **15.** $b = \frac{2A}{h} - B$, or $b = \frac{2A - Bh}{h}$ **16.** $y = \frac{9 - 4x}{7}$ **17.** 5 ft **18.** 14.4% **19.** 5.5% **20.** 25° **21. (a)** \$525 billion **(b)** \$97.5 billion **22.** 100 mm **23.** $14 - \frac{1}{5}x$ **24.** $\frac{6x}{x + 3}$ **25.** length: 13 m; width: 8 m **26.** 17 in., 17 in., 19 in. **27.** 12 kg **28.** 30 L **29.** 10 L **30.** \$10,000 at 6%; \$6000 at 4% **31.** A **32. (a)** 530 mi **(b)** 328 mi **33.** 2.2 hr **34.** 50 km per hr; 65 km per hr **35.** 1 hr **36.** 46 mph **37.** $\left\{\frac{7}{6}\right\}$ **38.** {0} **39.** ∅ **40.** {1500} **41.** {all real numbers} **42.** $x = \frac{C - By}{A}$ **43.** 6 in. **44.** eastbound car: 3 hr; westbound car: 2 hr **45.** Blue Ridge Parkway: 18.95 million; Golden Gate Recreation Area: 13.49 million **46.** \$1200 at 3%; \$1800 at 5%

Chapter 2 Test (pages 121–122)

1. {−19} **2.** {5} **3.** {4} **4. (a)** contradiction; ∅ **(b)** identity; {all real numbers} **(c)** conditional equation; {0} **5.** $v = \frac{S + 16t^2}{t}$ **6.** $y = \frac{6 + 3x}{2}$ **7.** 3.348 hr **8.** 4.75% **9.** 74.2% **10.** \$14,000 at 3%; \$13,000 at 5% **11.** faster car: 60 mph; slower car: 45 mph **12.** 40°, 40°, 100° **13.** 10% **14.** 13.33% **15.** 1050

Cumulative Review Exercises: Chapters 1–2 (pages 125–126)

1. 9, $\sqrt{36}$ (or 6) **2.** 0, 9, $\sqrt{36}$ (or 6) **3.** −8, 0, 9, $\sqrt{36}$ (or 6) **4.** $-8, -\frac{2}{3}, 0, \frac{4}{5}, 9, \sqrt{36}$ (or 6) **5.** $-\sqrt{6}$ **6.** All are real numbers. **7.** $-\frac{22}{21}$ **8.** 7.9 **9.** 8 **10.** 0 **11.** −243 **12.** $\frac{216}{343}$ **13.** 4096 **14.** −4096 **15.** $\sqrt{-49}$ **16.** $\frac{4 + 4}{4 - 4}$ **17.** −16 **18.** −34 **19.** 184 **20.** $\frac{27}{16}$ **21.** $-20r + 17$ **22.** $13k + 42$ **23.** commutative property **24.** distributive property **25.** inverse property **26.** {5} **27.** {30} **28.** {15} **29.** $c = P - a - b$ **30.** ∅ **31.** {all real numbers} **32.** 2 1 **33.** 9 pennies, 12 nickels, 8 quarters **34.** \$5000 at 5%; \$7000 at 6% **35.** $\frac{1}{8}$ hr **36.** 44 mg **37.** 24.4 **38. (a)** 304 **(b)** 17.3%

Chapter 3 Linear Inequalities and Absolute Value

Section 3.1 (pages 137–142)

1. D **3.** B **5.** F **7. (a)** $131 \le s \le 155$ **(b)** $s > 155$ **(c)** $9 \le x \le 12$ **(d)** $x > 18$ **9.** Reverse the direction of the inequality symbol only when multiplying or dividing by a *negative* number. The solution set is $[-16, \infty)$.

11. $(-\infty, 7]$

13. $[5, \infty)$

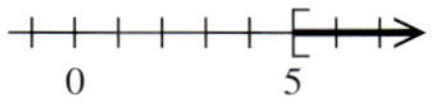

15. $(-5, \infty)$

17. $(-4, \infty)$

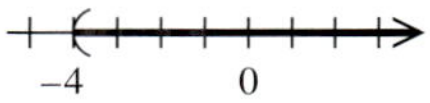

19. $(-\infty, -40]$

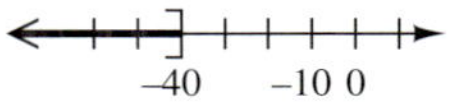

21. $(-\infty, 4]$

23. $(7, \infty)$

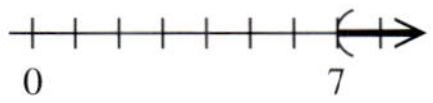

25. $\left(-\infty, -\frac{15}{2}\right)$

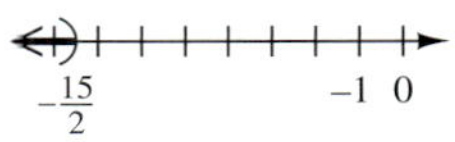

27. $(-\infty, -7)$

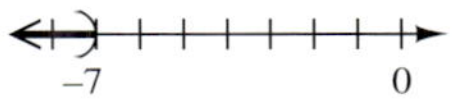

29. $\left[\frac{1}{2}, \infty\right)$

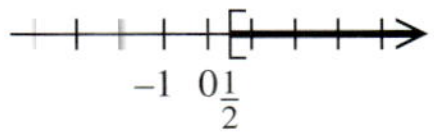

31. $(3, \infty)$

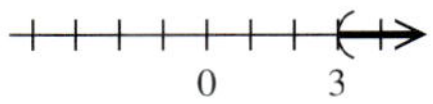

33. $(-\infty, 4)$

35. $\left(-\infty, \frac{23}{6}\right]$

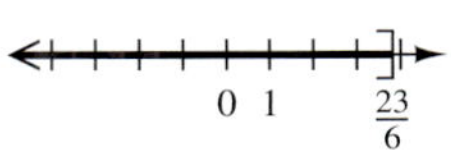

37. $\left(-\infty, \frac{76}{11}\right)$

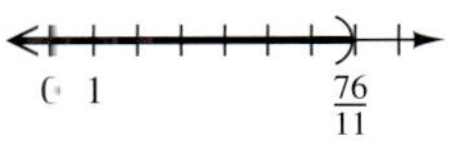

39. {−9}

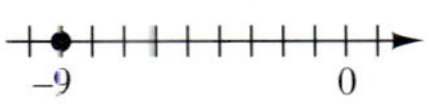

40. $(-9, \infty)$

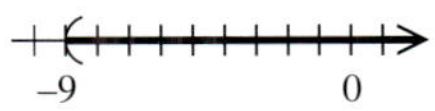

41. $(-\infty, -9)$

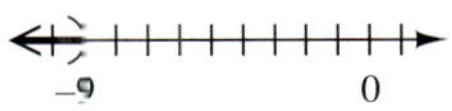

42. We obtain the set of all real numbers.

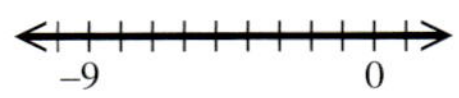

43. $(-\infty, -3)$

45. $(1, 11)$

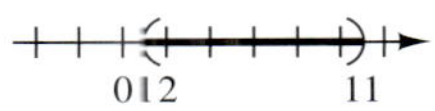

47. $[-14, 10]$

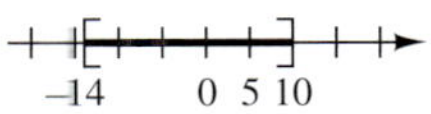

49. $[-5, 6]$

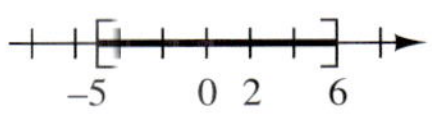

51. $(-6, -4)$

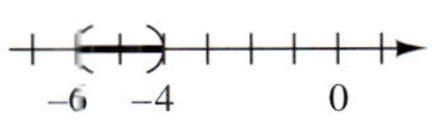

ANSWERS

53. $\left[-\frac{1}{3}, \frac{1}{9}\right)$ 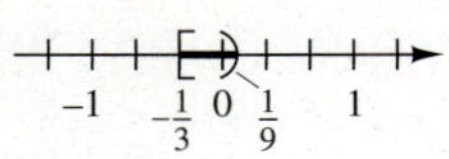

55. $\left[-\frac{1}{2}, \frac{35}{2}\right]$ 

57. from about 2:30 P.M. to 6:00 P.M. **59.** about 84°F–91°F
61. 26 months **63.** at least 80 **65.** 26 DVDs **67.** **(a)** 140 to 184 lb **(b)** Answers will vary. **69.** all numbers between −2 and 2, or $(-2, 2)$
71. all numbers greater than or equal to 3, or $[3, \infty)$
73. all numbers greater than or equal to −9, or $[-9, \infty)$

Section 3.2 (pages 151–154)

1. true **3.** false; The union is $(-\infty, 6) \cup (6, \infty)$. **5.** $\{4\}$, or D
7. $\emptyset$ **9.** $\{1, 2, 3, 4, 5, 6\}$, or A **11.** $\{1, 3, 5, 6\}$

13. 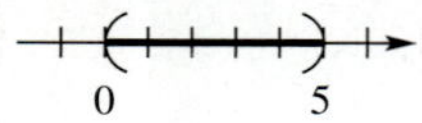 **15.** 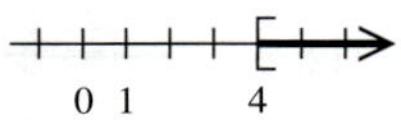

17. 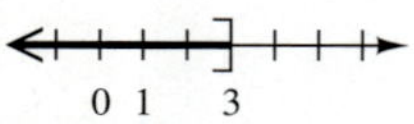

19. Answers will vary. One example is: The intersection of two streets is the region common to *both* streets.

21. $(-3, 2)$ 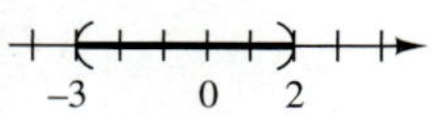

23. $(-\infty, 2]$ 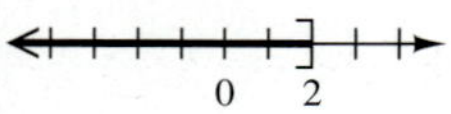

25. $\emptyset$ 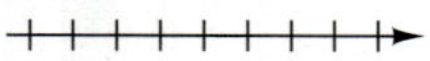

27. $[5, 9]$ 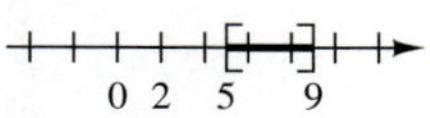

29. $(-\infty, 4]$ 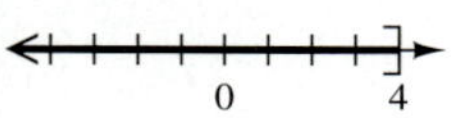

31. $(-\infty, 8]$ 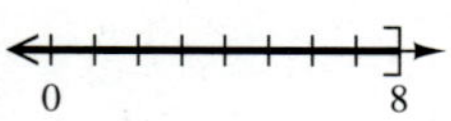

33. $[-2, \infty)$ 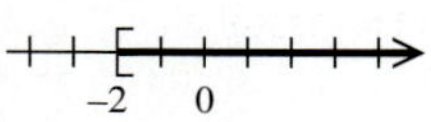

35. $(-\infty, \infty)$ 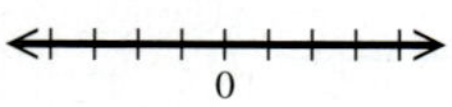

37. $(-\infty, -5) \cup (5, \infty)$ 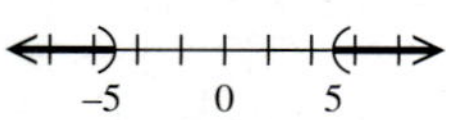

39. $(-\infty, -1] \cup (2, \infty)$ 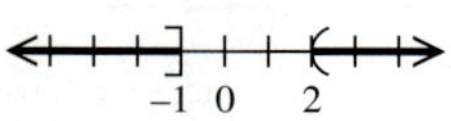

41. $(-\infty, 2) \cup (2, \infty)$ 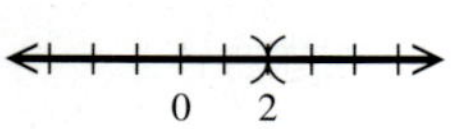

43. $[-4, -1]$ **45.** $[-9, -6]$ **47.** $(-\infty, 3)$ **49.** $[3, 9)$

51. intersection; $(-5, -1)$ 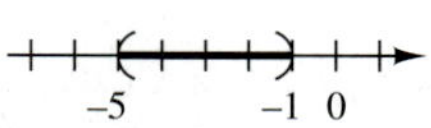

53. union; $(-\infty, 4)$ 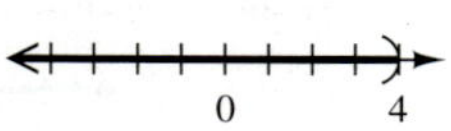

55. intersection; $[4, 12]$ 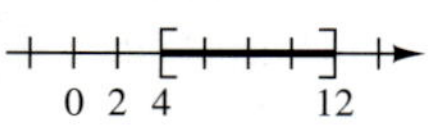

57. union; $(-\infty, 0] \cup [2, \infty)$ 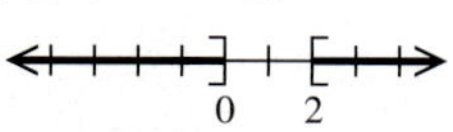

59. Mario, Joe **60.** none of them **61.** none of them **62.** Luigi, Than **63.** Mario, Joe **64.** all of them **65.** {Tuition and fees}
67. {Tuition and fees, Dormitory charges}

Section 3.3 (pages 163–168)

1. E; C; D; B; A **3.** **(a)** one **(b)** two **(c)** none **5.** $\{-12, 12\}$
7. $\{-5, 5\}$ **9.** $\{-6, 12\}$ **11.** $\{-5, 4\}$ **13.** $\left\{-3, \frac{11}{2}\right\}$
15. $\left\{-\frac{19}{2}, \frac{9}{2}\right\}$ **17.** $\{-10, -2\}$ **19.** $\left\{-8, \frac{32}{3}\right\}$

21. $(-\infty, -3) \cup (3, \infty)$ 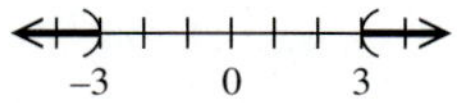

23. $(-\infty, -4] \cup [4, \infty)$ 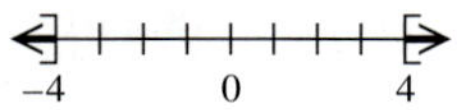

25. $(-\infty, -10) \cup (6, \infty)$ 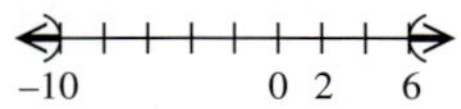

27. $\left(-\infty, -\frac{7}{3}\right] \cup [3, \infty)$ 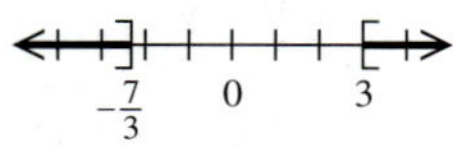

29. $(-\infty, -2) \cup (8, \infty)$ 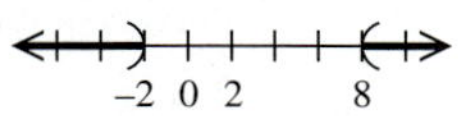

31. **(a)** 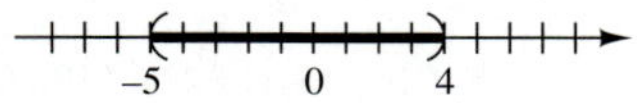

(b) 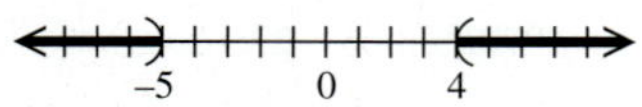

33. $[-3, 3]$ 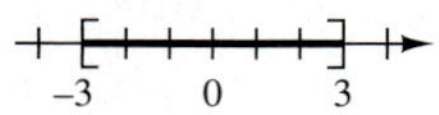

35. $(-4, 4)$ 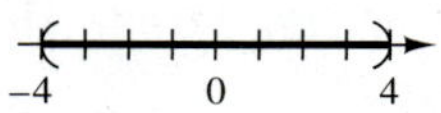

37. $[-10, 6]$ 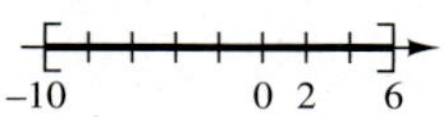

39. $\left(-\frac{7}{3}, 3\right)$ 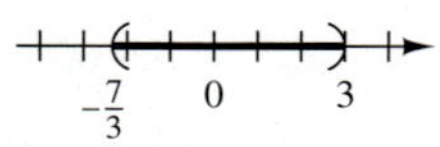

41. $[-2, 8]$ 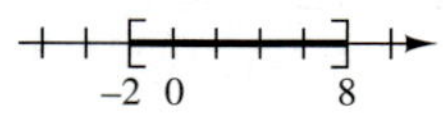

43. $(-\infty, -2) \cup (10, \infty)$ 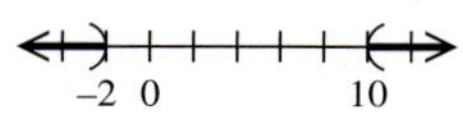

45. $\{-6, -1\}$ 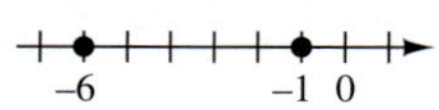

47. $\left[-\frac{10}{3}, 4\right]$ 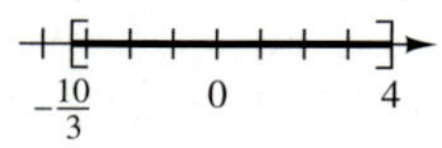

49. $\left[-4, -\frac{4}{3}\right]$ 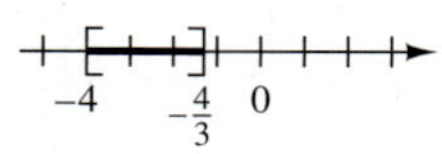

51. $\{-5, 5\}$ **53.** $\{-5, -3\}$ **55.** $(-\infty, -3) \cup (2, \infty)$ **57.** $[-10, 0]$
59. $\{-1, 3\}$ **61.** $\left\{-3, \frac{5}{3}\right\}$ **63.** $\left\{-\frac{1}{3}, -\frac{1}{15}\right\}$ **65.** $\left\{-\frac{5}{4}\right\}$
67. $(-\infty, \infty)$ **69.** $\emptyset$ **71.** $\left\{-\frac{1}{4}\right\}$ **73.** $\emptyset$ **75.** $(-\infty, \infty)$
77. $\left\{-\frac{3}{7}\right\}$ **79.** $(-\infty, \infty)$ **81.** $\left(-\infty, -\frac{7}{10}\right) \cup \left(-\frac{7}{10}, \infty\right)$
83. $(-\infty, \infty)$ **85.** $|x - 1000| \le 100$; $900 \le x \le 1100$ **87.** 475.6 ft
88. 1201 Walnut, City Hall, Fidelity Bank and Trust Building, Kansas City Power and Light, Hyatt Regency Crown Center **89.** City Center Square, Commerce Tower, Federal Office Building, 1201 Walnut, City Hall, Fidelity Bank and Trust Building, Kansas City Power and Light, Hyatt Regency Crown Center **90.** **(a)** $|x - 475.6| \ge 75$ **(b)** $x \ge 550.6$ or $x \le 400.6$ **(c)** Town Pavilion, One Kansas City Place **(d)** It makes sense because it includes all buildings *not* listed in the answer to Exercise 89.

Summary Exercises on Solving Linear and Absolute Value Equations and Inequalities (pages 169–170)

1. $\{12\}$ **2.** $\{-5, 7\}$ **3.** $\{7\}$ **4.** $\left\{-\frac{2}{5}\right\}$ **5.** $\emptyset$ **6.** $(-\infty, -1]$ **7.** $\left[-\frac{2}{3}, \infty\right)$ **8.** $\{-1\}$ **9.** $\{-3\}$ **10.** $\left\{1, \frac{11}{3}\right\}$ **11.** $(-\infty, 5]$ **12.** $(-\infty, \infty)$ **13.** $\{2\}$ **14.** $(-\infty, -8] \cup [8, \infty)$ **15.** $\emptyset$ **16.** $(-\infty, \infty)$ **17.** $(-5.5, 5.5)$ **18.** $\left\{\frac{13}{3}\right\}$ **19.** $\left\{-\frac{96}{5}\right\}$ **20.** $(-\infty, 32]$ **21.** $(-\infty, -24)$ **22.** $\left\{\frac{3}{8}\right\}$ **23.** $\left\{\frac{7}{2}\right\}$ **24.** $(-6, 8)$ **25.** $(-\infty, \infty)$ **26.** $(-\infty, 5)$ **27.** $(-\infty, -4) \cup (7, \infty)$ **28.** $\{24\}$ **29.** $\left\{-\frac{1}{5}\right\}$ **30.** $\left(-\infty, -\frac{5}{2}\right]$ **31.** $\left[-\frac{1}{3}, 3\right]$ **32.** $[1, 7]$ **33.** $\left\{-\frac{1}{6}, 2\right\}$ **34.** $\{-3\}$ **35.** $(-\infty, -1] \cup \left[\frac{5}{3}, \infty\right)$ **36.** $\left[\frac{3}{4}, \frac{15}{8}\right]$ **37.** $\left\{-\frac{5}{2}\right\}$ **38.** $\{60\}$ **39.** $\left[-\frac{9}{2}, \frac{15}{2}\right]$ **40.** $(1, 9)$ **41.** $(-\infty, \infty)$ **42.** {[illegible], 9} **43.** $(-\infty, \infty)$ **44.** $\left\{-\frac{10}{9}\right\}$ **45.** $\{-2\}$ **46.** $\emptyset$ **47.** $(-\infty, -1) \cup (2, \infty)$ **48.** $[-3, -2]$

Chapter 3 Review Exercises (pages 177–180)

1. $(-9, \infty)$ 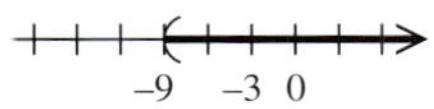

2. $(-\infty, -3]$ 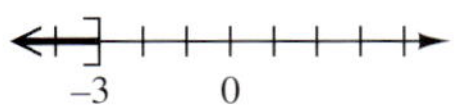

3. $\left(\frac{3}{2}, \infty\right)$ 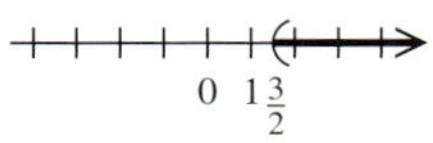

4. $\left(-\infty, -\frac{14}{9}\right)$ 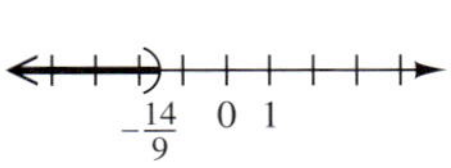

5. $[-3, \infty)$ 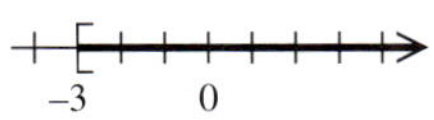

6. $[-3, 12]$ 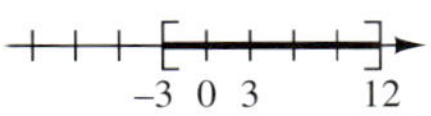

7. $[3, 5)$ 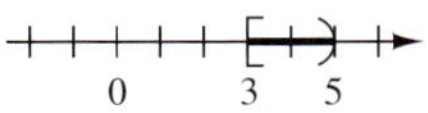

8. $\left(-3, \frac{7}{2}\right)$ 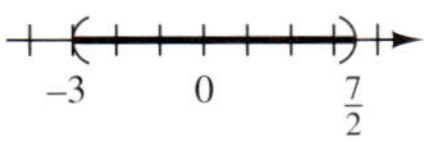

9. [illegible]8 m or less **10.** 23 tickets or less **11.** any score greater than or equal to 6[illegible] **12.** Because the statement $-8 < -13$ is *false*, the inequality has no solution. **13.** $\{a, c\}$ **14.** $\{a\}$ **15.** $\{a, c, e, f, g\}$ **16.** $\{a, b, c, d, e, f, g\}$

17. $(4, 7)$ 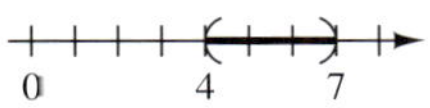

18. $(8, 14)$ 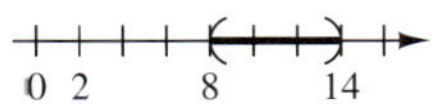

19. $(-\infty, -3] \cup (5, \infty)$ 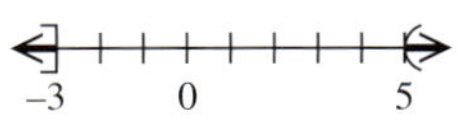

20. $(-\infty, \infty)$ 

21. $\emptyset$ 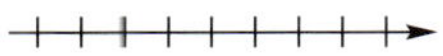

22. $(-\infty, -2] \cup [7, \infty)$ 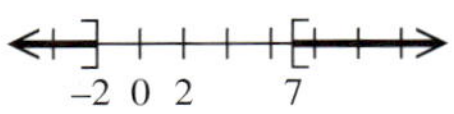

23. $(-3, 4)$ **24.** $(-\infty, 2)$ **25.** $(4, \infty)$ **26.** $(1, \infty)$ **27. (a)** {Oregon} **(b)** {Illinois, Maine, North Carolina, Oregon, Utah} **(c)** $\emptyset$ **28.** $\{-7, 7\}$ **29.** $\{-11, 7\}$ **30.** $\left\{-\frac{1}{3}, 5\right\}$ **31.** $\emptyset$ **32.** $\{0, 7\}$ **33.** $\left\{-\frac{3}{2}, \frac{1}{2}\right\}$ **34.** $\left\{-\frac{3}{4}, \frac{1}{2}\right\}$ **35.** $\left\{-\frac{1}{2}\right\}$

36. $(-12, 12)$ 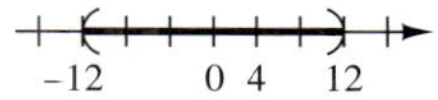

37. $[-1, 13]$ 

38. $[-3, -2]$ 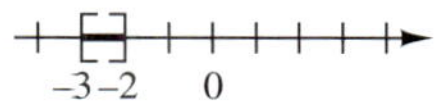

39. $(-\infty, \infty)$ 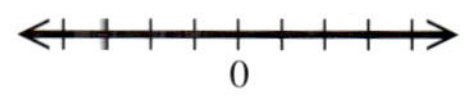

40. $\left(-\infty, -\frac{8}{5}\right) \cup (2, \infty)$ 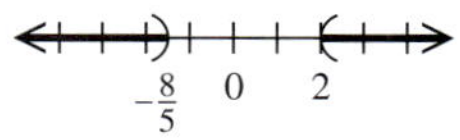

41. $(-\infty, \infty)$ 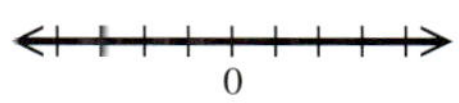

42. $\left(-\infty, \frac{7}{6}\right]$ **43.** $[-4, 5)$ **44.** $\left(-\infty, \frac{14}{17}\right)$ **45.** any amount greater than or equal to \$1100 **46.** $(-\infty, 2]$ **47.** $(-\infty, -1) \cup \left(\frac{11}{7}, \infty\right)$ **48.** $\{-5, 15\}$ **49.** $[-16, 10]$ **50.** $(-\infty, \infty)$ **51.** $\left\{-4, -\frac{2}{3}\right\}$

52. 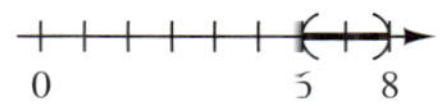

53. 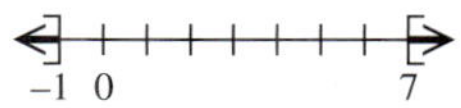

54. (a) $\emptyset$ **(b)** $(-\infty, \infty)$ **(c)** $\emptyset$

Chapter 3 Test (pages 181–182)

1. Reverse the direction of the inequality symbol.

2. $[1, \infty)$ 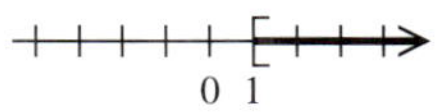

3. $(-\infty, 28)$ 

4. $[-3, 3]$ 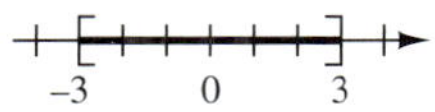

5. C **6. (a)** 1990, 2000, 2006 **(b)** 1960, 1970 **(c)** 2000 **7.** 82 or more **8.** $[500, \infty)$ **9. (a)** $\{1, 5\}$ **(b)** $\{1, 2, 5, 7, 9, 12\}$ **10.** $\{2\}$

11. $[2, 9)$ 

12. $(-\infty, 3) \cup [6, \infty)$ 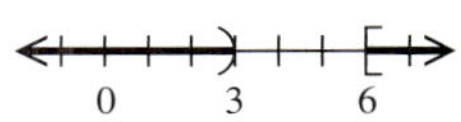

13. $\left[-\frac{5}{2}, 1\right]$ 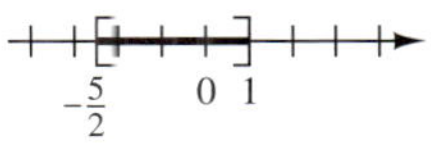

14. $\left(-\infty, -\frac{7}{6}\right) \cup \left(\frac{17}{6}, \infty\right)$ 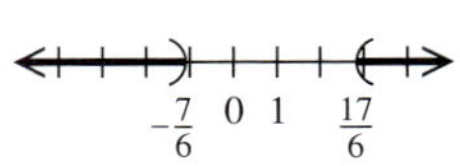

15. $\left(\frac{1}{3}, \frac{7}{3}\right)$ 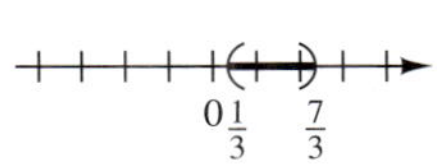

16. $\emptyset$ **17.** $\left\{-\frac{5}{3}, 3\right\}$ **18.** $\left\{-\frac{5}{7}, \frac{11}{3}\right\}$ **19.** $\emptyset$ **20.** $\emptyset$; $(-\infty, \infty)$; $\emptyset$

Cumulative Review Exercises: Chapters 1–3 (pages 185–186)

1. $\frac{3}{4}$ **2.** true **3.** $\frac{37}{60}$ **4.** $\frac{48}{5}$ **5.** 11 **6.** -8 **7.** -36 **8.** -125
9. $\frac{81}{16}$ **10.** -34 **11.** $\frac{3}{16}$ **12.** distributive property
13. commutative property **14.** $2k - 11$ **15.** $\{-1\}$ **16.** $\{-12\}$
17. $\{26\}$ **18.** $\left\{\frac{3}{4}, \frac{7}{2}\right\}$ **19.** $y = \frac{24 - 3x}{4}$ **20.** $n = \frac{A - P}{iP}$

21. $[-14, \infty)$

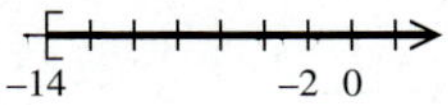

22. $\left[\frac{5}{3}, 3\right)$

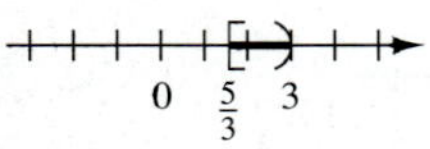

23. $(-\infty, 0) \cup (2, \infty)$

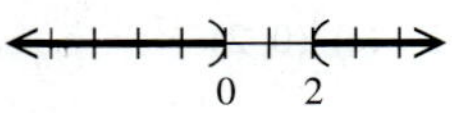

24. $\left(-\infty, -\frac{1}{7}\right] \cup [1, \infty)$

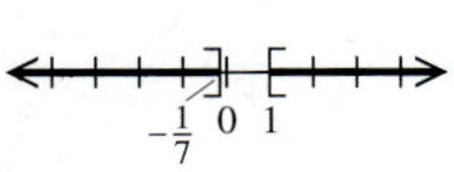

25. \$5000 **26.** $6\frac{1}{3}$ g **27.** 74 or greater **28.** 40 mph; 60 mph
29. **(a)** 81 **(b)** 5.3% **30.** 4 cm; 9 cm; 27 cm

Chapter 4 Graphs, Linear Equations, and Functions

Section 4.1 (pages 195–198)

1. **(a)** x represents the year; y represents the revenue in billions of dollars. **(b)** about \$2400 billion **(c)** (2006, 2400) **(d)** In 2004, federal tax revenues were about \$1880 billion. **3.** origin **5.** y; x **7.** two
9. $y = 0$ **11.** **(a)** I **(b)** III **(c)** II **(d)** IV **(e)** no quadrant
13. **(a)** I or III **(b)** II or IV **(c)** II or IV **(d)** I or III
15–24.

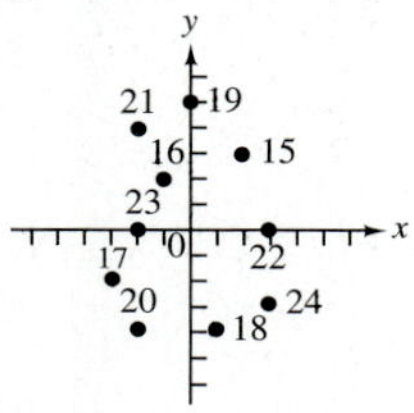

25. -3; 3; 2; -1

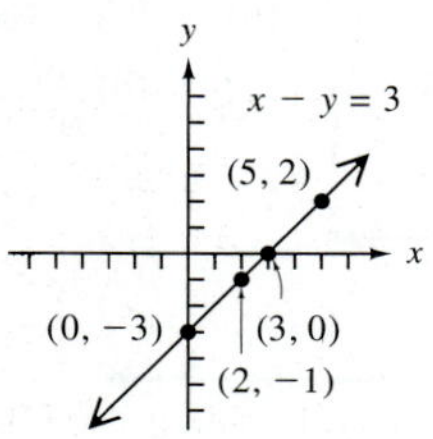

27. $\frac{5}{2}$; 5; $\frac{3}{2}$; 1

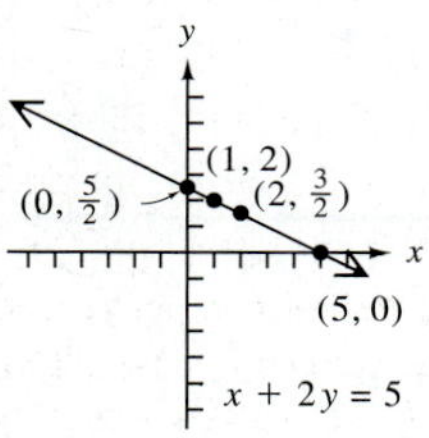

29. -4; 5; $-\frac{12}{5}$; $\frac{5}{4}$

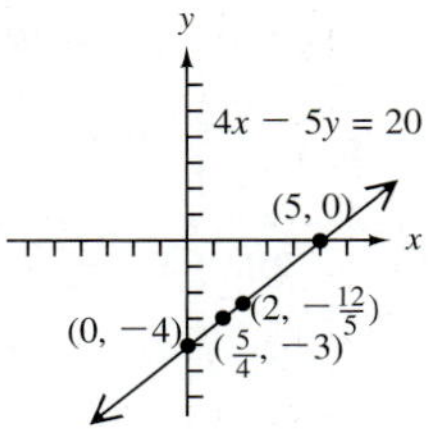

31. (6, 0); (0, 4)

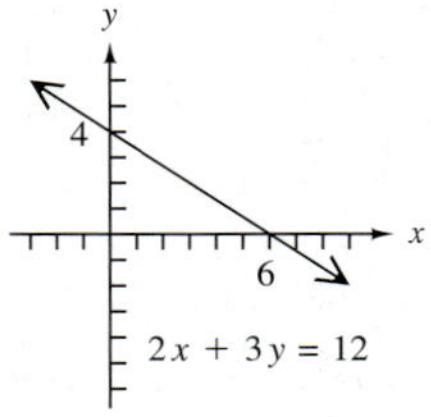

33. (6, 0); (0, -2)

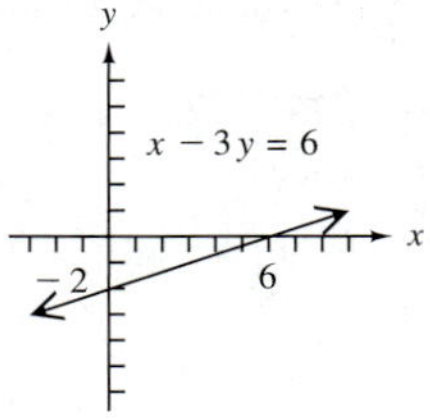

35. (3, 0); $\left(0, -\frac{9}{7}\right)$

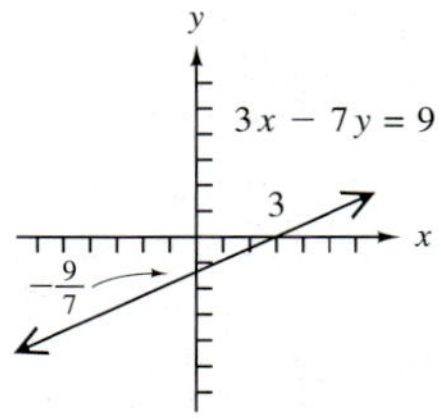

37. none; (0, 5)

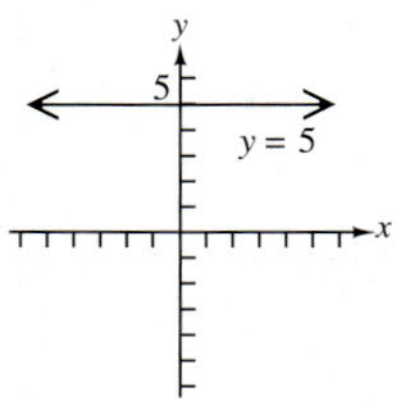

39. (5, 0); none

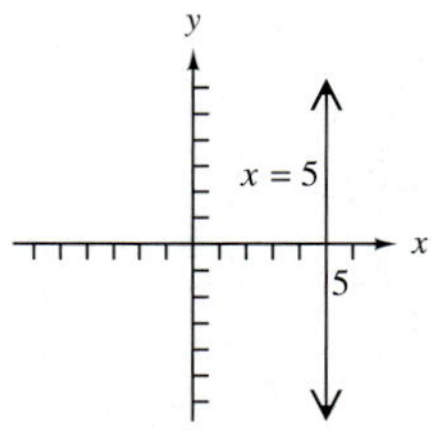

41. (0, 0); (0, 0)

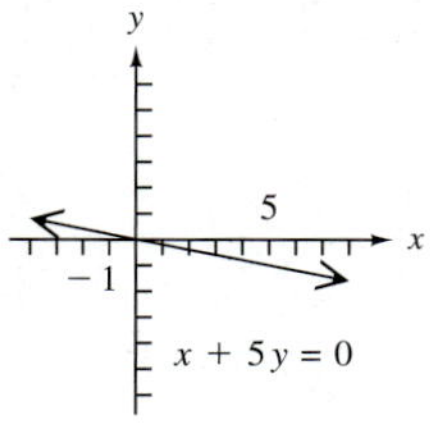

43. (0, 0); (0, 0)

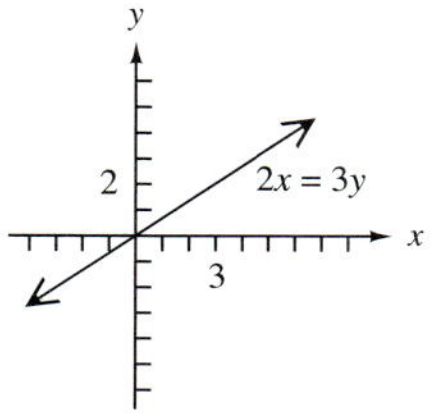

45. (−5, −1) **47.** $\left(\frac{9}{2}, -\frac{3}{2}\right)$ **49.** $\left(0, \frac{11}{2}\right)$ **51.** (2.1, 0.9)

53. (1, 1) **55.** $\left(-\frac{5}{12}, \frac{5}{28}\right)$

Section 4.2 (pages 207–212)

1. A, B, D **3.** 2 **5.** undefined **7.** 2 **9.** 0 **11.** undefined **13.** 8 **15.** $\frac{5}{6}$ **17.** 0 **19.** $-\frac{5}{2}$ **21.** undefined **23.** B **25.** A

27. $-\frac{1}{2}$

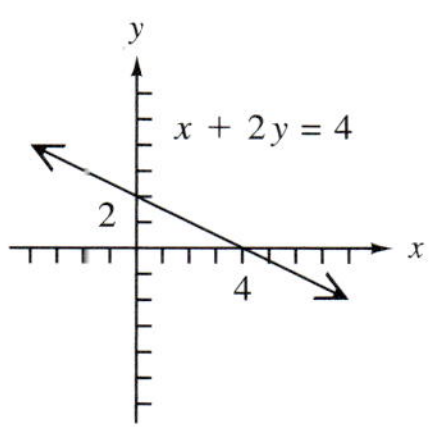

29. 1

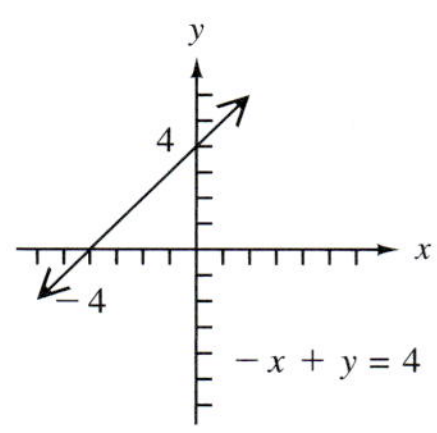

31. $-\frac{6}{5}$

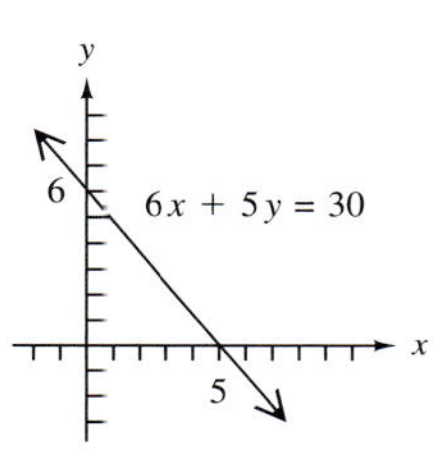

33. undefined

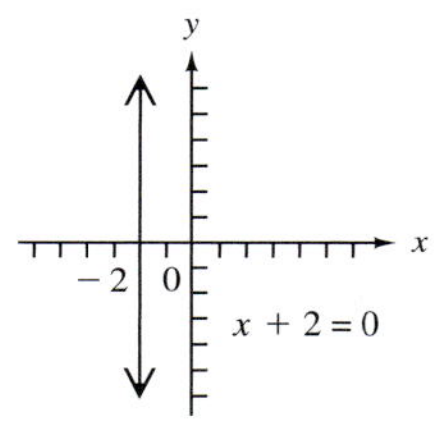

35. 4

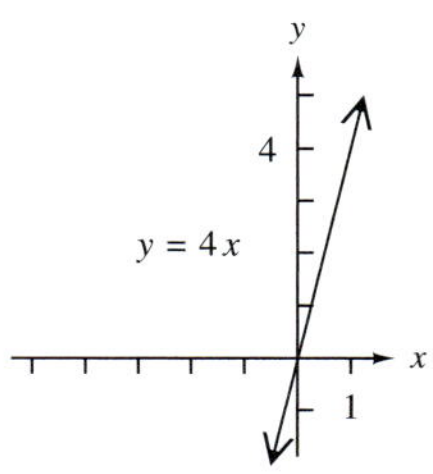

37. 0

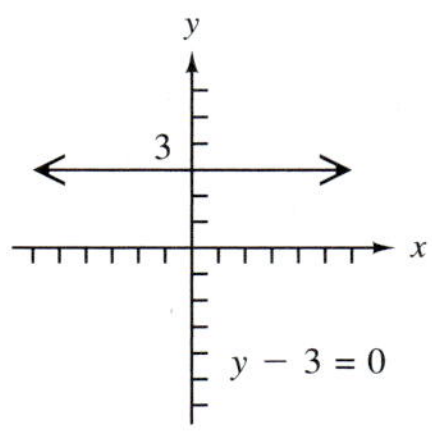

39.

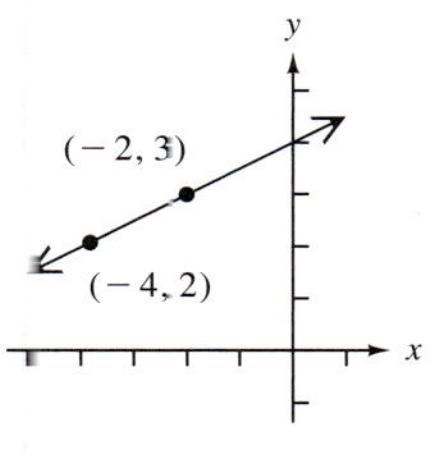

41.

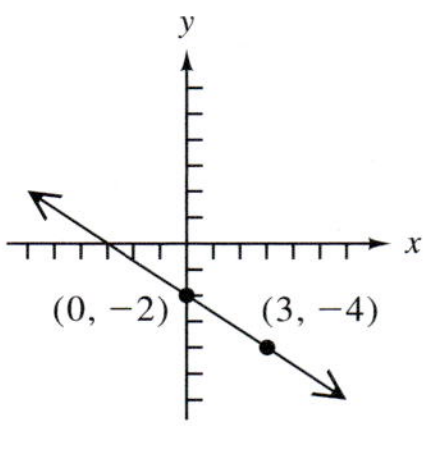

43.

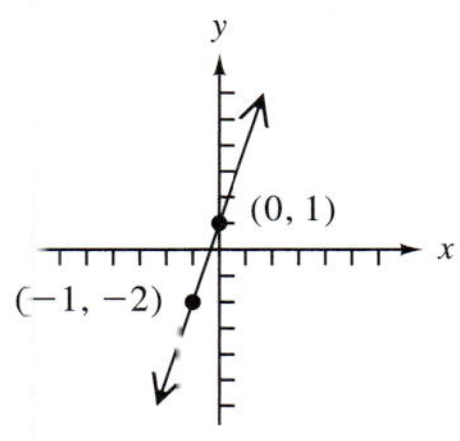

45. parallel **47.** perpendicular **49.** neither **51.** parallel **53.** neither **55.** perpendicular **57.** $\frac{7}{10}$ **59.** −$4000 per yr; The value of the machine is decreasing $4000 each year during these years. **61.** 0% per yr (or no change); The percent of pay raise is not changing—it is 3% each year during these years. **63.** **(a)** $18.78 billion per yr **(b)** The positive slope means that personal spending on recreation in the United States *increased* by an average of $18.78 billion each year from 2000 to 2006. **65.** −$1470.67 million per yr; Sales of analog TVs *decreased* by an average of $1470.67 million each year from 2003 to 2006. **66.** $\frac{1}{3}$ **67.** $\frac{1}{3}$ **68.** $\frac{1}{3}$ **69.** $\frac{1}{3} = \frac{1}{3} = \frac{1}{3}$ is true. **70.** They are collinear. **71.** They are not collinear.

Section 4.3 (pages 221–226)

1. A **3.** A **5.** $3x + y = 10$ **7.** A **9.** C **11.** H **13.** B **15.** $y = 5x + 15$ **17.** $y = -\frac{2}{3}x + \frac{4}{5}$ **19.** $y = \frac{2}{5}x + 5$

21. **(a)** $y = x + 2$ **(b)** 1 **(c)** (0, 2)
(d)

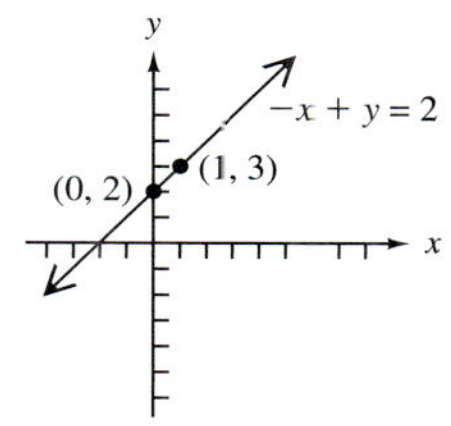

23. **(a)** $y = \frac{4}{5}x - 4$ **(b)** $\frac{4}{5}$ **(c)** (0, −4)
(d)

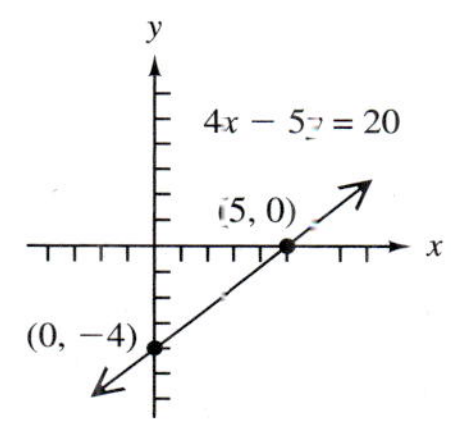

25. **(a)** $y = -\frac{1}{2}x - 2$ **(b)** $-\frac{1}{2}$ **(c)** (0, −2)
(d)

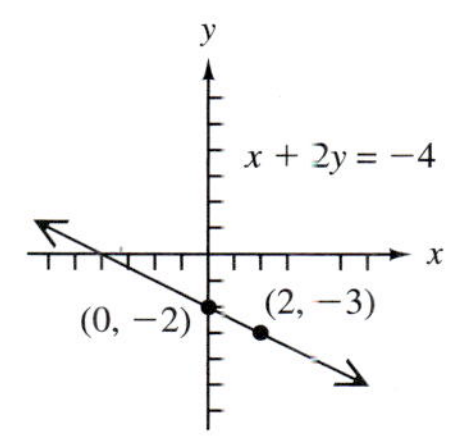

27. $3x + 4y = 10$ **29.** $2x + y = 18$ **31.** $x - 2y = -13$ **33.** $y = 12$ **35.** $x = 9$ **37.** $y = 0.2$ **39.** $2x - y = 2$ **41.** $x + 2y = 8$ **43.** $2x - 13y = -6$ **45.** $y = 5$ **47.** $x = 7$ **49.** $y = 3x - 19$ **51.** $y = \frac{1}{2}x - 1$ **53.** $y = -\frac{1}{2}x + 9$ **55.** $y = 7$ **57.** $y = 45x$; (0, 0), (5, 225), (10, 450) **59.** $y = 5.00x$; (0, 0), (5, 25.00), (10, 50.00) **61.** **(a)** $y = 41x + 99$ **(b)** (5, 304); The cost of a 5-month membership is $304. **(c)** $591 **63.** **(a)** $y = 60x + 36$ **(b)** (5, 336); The cost of the plan for 5 months is $336. **(c)** $1476 **65.** **(a)** $y = 0.20x + 50$ **(b)** (5, 51); The charge for driving 5 mi is $51. **(c)** 173 mi **67.** **(a)** $y = 1294.7x + 3921$; Sales of digital cameras in the United States increased by $1294.7 million per yr from 2003 to 2006. **(b)** $9099.8 million **69.** **(a)** $y = -790.25x + 101{,}430$ **(b)** 97,479; The result using the model is a bit low. **71.** 32; 212 **72.** **(a)** (0, 32) and (100, 212) **(b)** $\frac{9}{5}$ **73.** $F = \frac{9}{5}C + 32$ **74.** $C = \frac{5}{9}(F - 32)$; −40° **75.** 60° **76.** 59°; They differ by 1°. **77.** 90°; 86°; They differ by 4°.

78. Since $\frac{9}{5}$ is a little less than 2, and 32 is a little more than 30, $\frac{9}{5}C + 32 \approx 2C + 30$.

Summary Exercises on Slopes and Equations of Lines (pages 227–228)

1. $-\frac{3}{5}$ **2.** 0 **3.** $\frac{3}{7}$ **4.** undefined **5. (a)** $y = -3x + 10$ **(b)** $3x + y = 10$ **6. (a)** $y = \frac{2}{3}x + 8$ **(b)** $2x - 3y = -24$ **7. (a)** $y = -\frac{5}{6}x + \frac{13}{3}$ **(b)** $5x + 6y = 26$ **8. (a)** $y = -\frac{5}{2}x + 2$ **(b)** $5x + 2y = 4$ **9. (a)** $y = 3x + 11$ **(b)** $3x - y = -11$ **10. (a)** $y = -\frac{5}{2}x$ **(b)** $5x + 2y = 0$ **11. (a)** $y = -8$ **(b)** $y = -8$ **12. (a)** $y = -\frac{7}{9}$ **(b)** $9y = -7$ **13. (a)** $y = \frac{2}{3}x + \frac{14}{3}$ **(b)** $2x - 3y = -14$ **14. (a)** $y = 2x - 10$ **(b)** $2x - y = 10$ **15. (a)** $y = -\frac{7}{5}x + \frac{32}{5}$ **(b)** $7x + 5y = 32$ **16. (a)** $y = -7x + 3$ **(b)** $7x + y = 3$ **17. (a)** $y = \frac{1}{5}x - \frac{7}{5}$ **(b)** $x - 5y = 7$ **18. (a)** $y = -\frac{3}{4}x - 6$ **(b)** $3x + 4y = -24$ **19. (a)** $y = 2x + 0.9$ **(b)** $20x - 10y = -9$ **20. (a)** $y = 0.5x + 0.5$ **(b)** $x - 2y = -1$ **21.** B **22.** D **23.** A **24.** C **25.** E

Section 4.4 (pages 233–236)

1. solid; below **3.** dashed; above **5.** The graph of $Ax + By = C$ divides the plane into two regions. In one of these regions, the ordered pairs satisfy $Ax + By < C$; in the other, they satisfy $Ax + By > C$.

7.

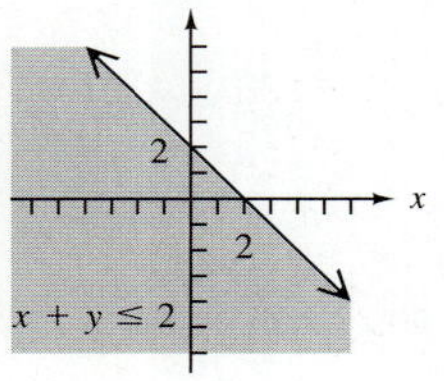

9.

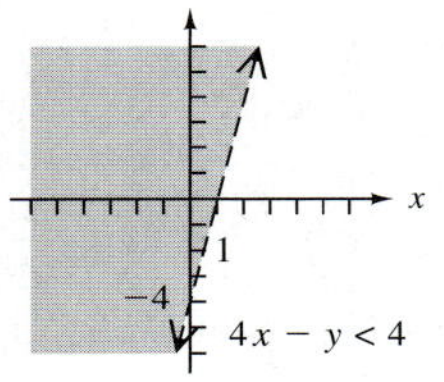

11.

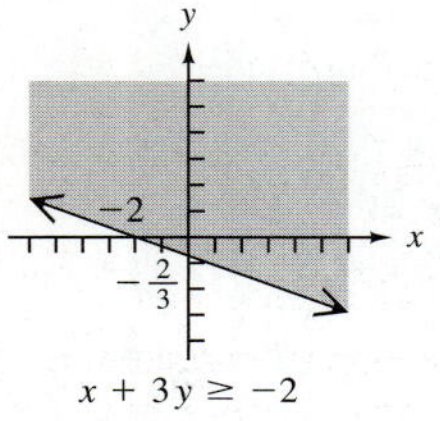

13.

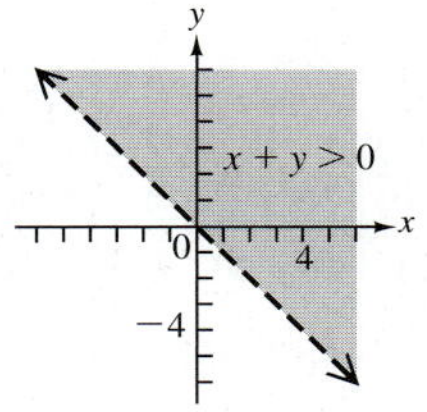

15.

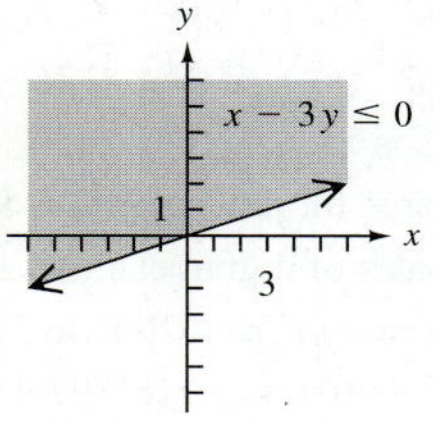

17.

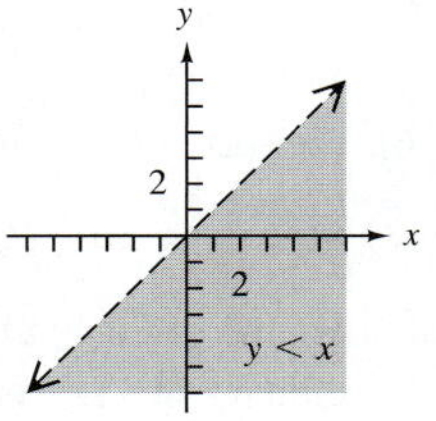

19.

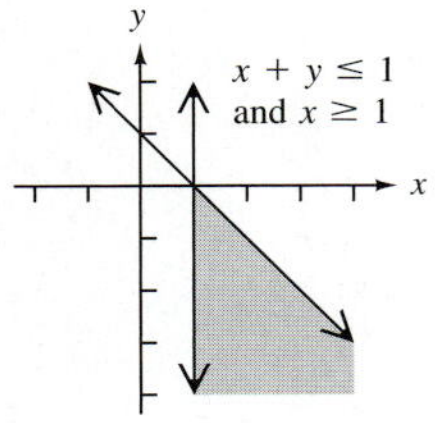

21.

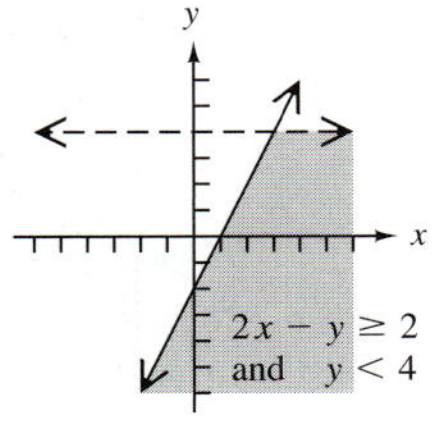

23.

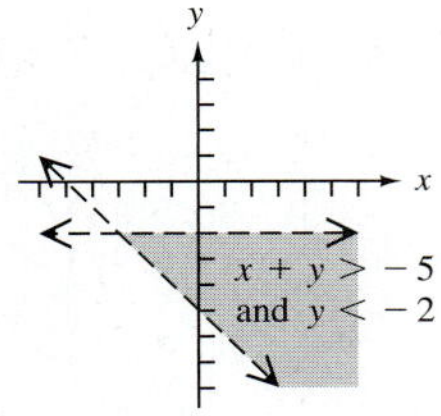

25.

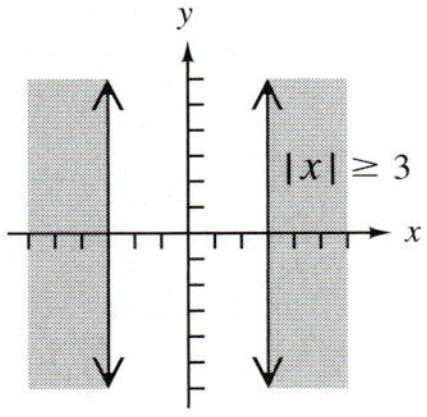

27.

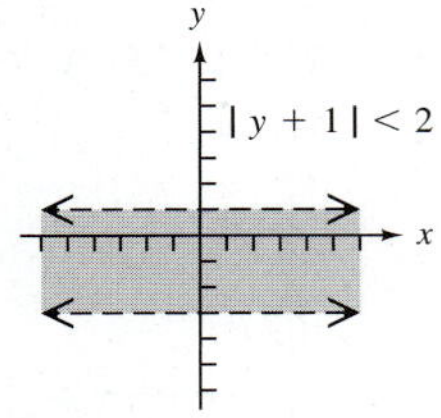

29.

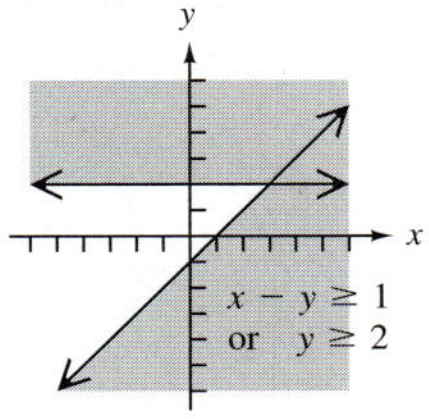

31.

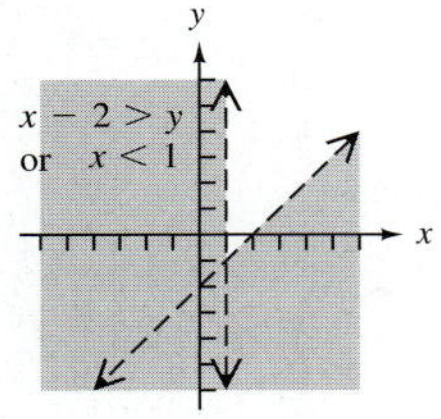

33.

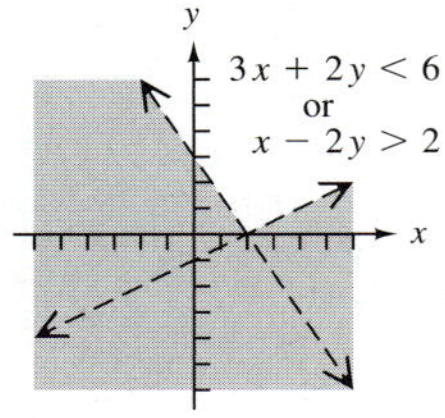

Section 4.5 (pages 247–252)

1. independent variable **3. (a)** A relation is a set of ordered pairs. **(b)** The domain is the set of all first components (x-values). **(c)** The range is the set of all second components (y-values). **(d)** A function is a relation in which each domain element is paired with one and only one range element. **5.** function; domain: $\{5, 3, 4, 7\}$; range: $\{1, 2, 9, 3\}$ **7.** not a function; domain: $\{2, 0\}$; range: $\{4, 2, 6\}$ **9.** function; domain: $\{-3, 4, -2\}$; range: $\{1, 7\}$ **11.** not a function; domain: $\{\ , 0, 2\}$; range: $\{1, -1, 0, 4, -4\}$ **13.** function; domain: $\{2, 5, 11, 17, 3\}$; range: $\{1, 7, 20\}$ **15.** not a function; domain: $\{1\}$; range: $\{5, 2, -1, -4\}$ **17.** function; domain: $(-\infty, \infty)$; range: $(-\infty, \infty)$ **19.** function; domain: $(-\infty, \infty)$; range: $(-\infty, 4]$ **21.** not a function; domain: $[3, \infty)$; range: $(-\infty, \infty)$ **23.** function; domain: $(-\infty, \infty)$ **25.** not a function; domain: $[0, \infty)$ **27.** function; domain: $(-\infty, \infty)$ **29.** not a function; domain: $(-\infty, \infty)$ **31.** function; domain: $[0, \infty)$ **33.** function; domain: $(-\infty, 0) \cup (0, \infty)$ **35.** function; domain: $\left[-\frac{1}{2}, \infty\right)$ **37.** function; domain: $(-\infty, 9) \cup (9, \infty)$ **39. (a)** $[0, 3000]$ **(b)** 25 hr; 25 hr **(c)** 2000 gal **(d)** $g(0) = 0$; The pool is empty at time 0. **41.** Here is one example. The cost of gasoline; number of gallons purchased; cost; number of gallons **43.** 4 **45.** -11 **47.** $-3p + 4$ **49.** $3x + 4$ **51.** $-3x - 2$ **53.** $-\frac{p^2}{9} + \frac{4p}{3} + 1$ **55. (a)** 2 **(b)** 3 **57. (a)** 15 **(b)** 10 **59. (a)** 3 **(b)** -3 **61.** line; -2; linear; $-2x + 4$; -2; 3; -2 **63. (a)** $f(x) = -\frac{1}{3}x + 4$ **(b)** 3 **65. (a)** $f(x) = 3 - 2x^2$ **(b)** -15 **67. (a)** $f(x) = \frac{4}{3}x - \frac{8}{3}$ **(b)** $\frac{4}{3}$

69. domain: $(-\infty, \infty)$; range: $(-\infty, \infty)$

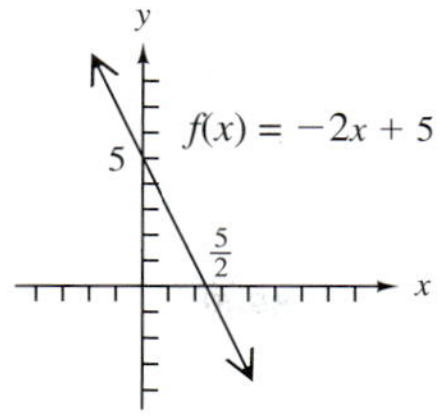

71. domain: $(-\infty, \infty)$; range: $(-\infty, \infty)$

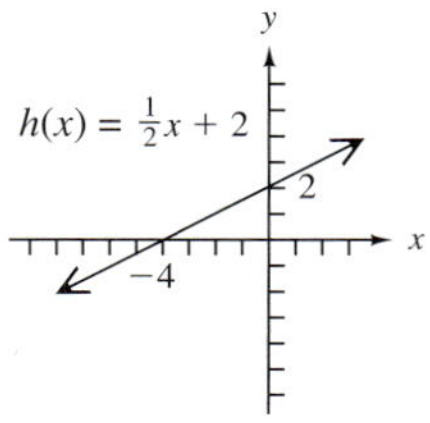

73. domain: $(-\infty, \infty)$; range: $\{-4\}$

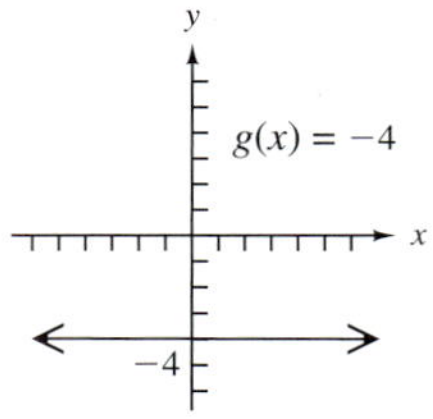

75. **(a)** \$0; \$2.50; \$5.00; \$7.50 **(b)** $2.50x$
(c)

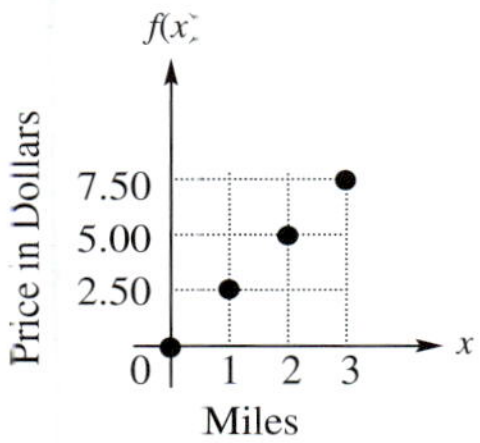

77. 194.53 cm **79.** 177.41 cm

Chapter 4 Review Exercises (pages 257–260)

1. 3; 2; $\frac{10}{3}$

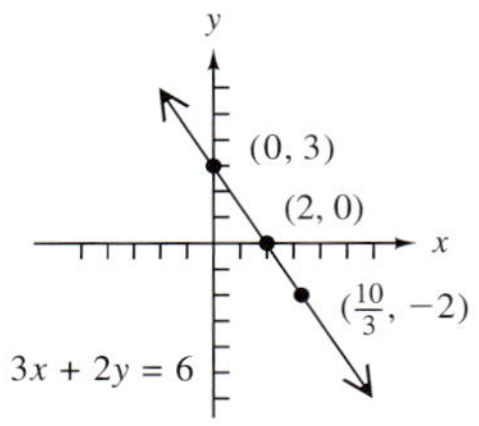

2. −4; 3; −5; 4

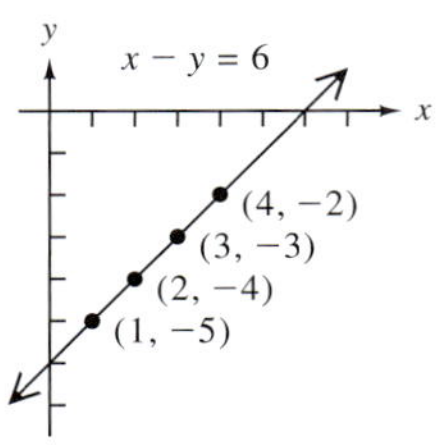

3. (3, 0); (0, 4)

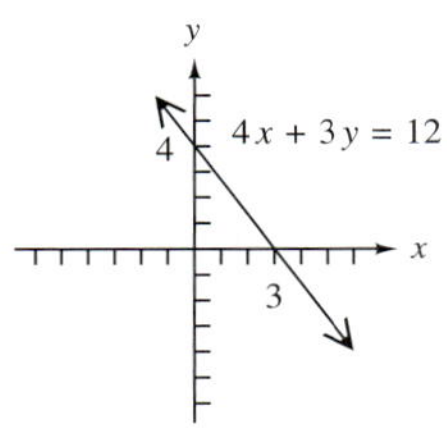

4. (3, 0); $\left(0, \frac{15}{7}\right)$

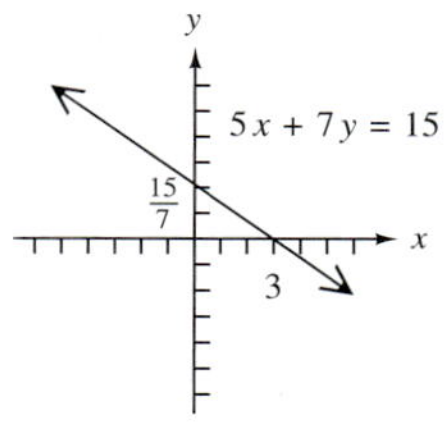

5. (0, 2) **6.** $\left(-\frac{9}{2}, \frac{3}{2}\right)$ **7.** (2.6, 11.9) **8.** (4.5, −9.5) **9.** $-\frac{8}{5}$
10. 2 **11.** $\frac{3}{4}$ **12.** 0 **13.** $-\frac{2}{3}$ **14.** $-\frac{1}{3}$ **15.** positive slope
16. negative slope **17.** 0 slope **18.** undefined slope **19.** 12 ft
20. \$1408 per yr **21.** $y = \frac{3}{5}x - 8$ **22.** $y = -\frac{1}{3}x + 5$
23. $y = 12$, or $y = 0x + 12$ **24.** $x = 2$ **25.** $y = 4$ **26.** $x = 0.3$
27. **(a)** $y = -9x + 13$ **(b)** $9x + y = 13$ **28.** **(a)** $y = \frac{7}{5}x + \frac{16}{5}$
(b) $7x - 5y = -16$ **29.** **(a)** $y = 4x - 26$ **(b)** $4x - y = 26$
30. **(a)** $y = -\frac{5}{2}x + 1$ **(b)** $5x + 2y = 2$
31. **(a)** $y = 57x + 159$; \$843 **(b)** $y = 47x + 159$; \$723

32.

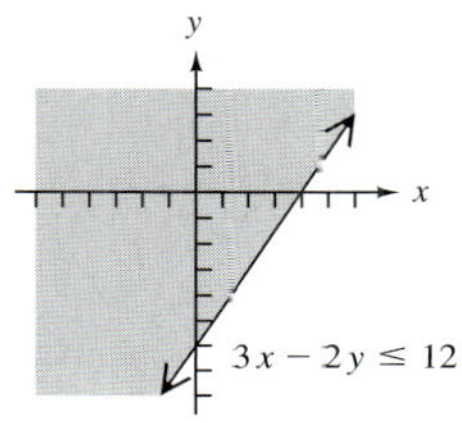

33.

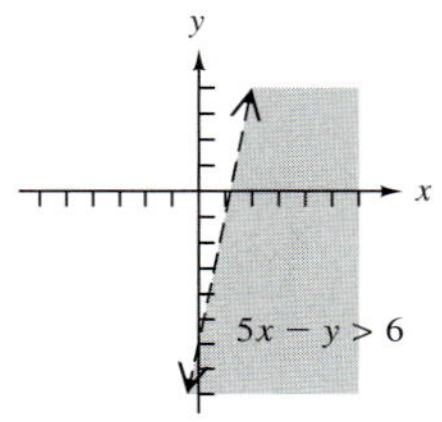

34.

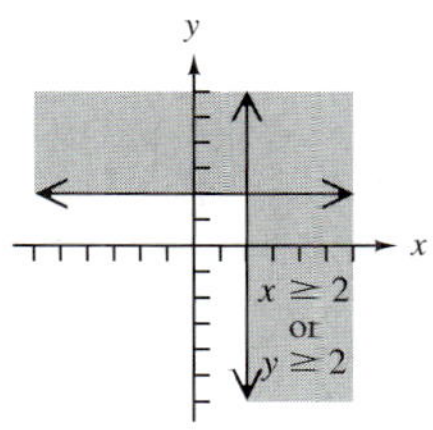

35.

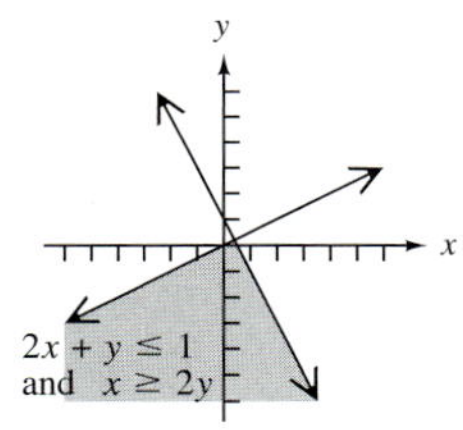

36. domain: $\{-4, 1\}$; range: $\{2, -2, 5, -5\}$; not a function
37. domain: $\{9, 11, 4, 17, 25\}$; range: $\{32, 47, 69, 14\}$; function
38. domain: $[-4, 4]$; range: $[0, 2]$; function **39.** function; linear function; domain: $(-\infty, \infty)$ **40.** not a function; domain: $(-\infty, \infty)$
41. function; domain: $(-\infty, \infty)$ **42.** function; domain: $\left[-\frac{7}{4}, \infty\right)$
43. not a function; domain: $[0, \infty)$ **44.** function; domain: $(-\infty, 36) \cup (36, \infty)$ **45.** **(a)** yes **(b)** domain: {1943, 1953, 1963, 1973, 1983, 1993, 2003}; range: {63.3, 68.8, 69.9, 71.4, 74.6, 75.5, 77.6} **(c)** Answers will vary. Two possible ordered pairs are (1953, 68.8) and (1973, 71.4). **(d)** 77.6; In 2003, life expectancy at birth was 77.6 yr. **(e)** 1993 **46.** −6 **47.** −15
48. $-2p^2 + 3p - 6$ **49.** $-2k^2 - 3k - 6$ **50.** $f(x) = 2x^2$; 18
51. C **52.** It is a horizontal line.

Chapter 4 Test (pages 261–262)

1. $\frac{1}{2}$ **2.** $\frac{3}{2}$; $\left(\frac{13}{3}, 0\right)$ $\left(0, -\frac{13}{2}\right)$ **3.** 0; none; (0, 5)
4. The graph is a vertical line.

5. $(-3, 0)$; $(0, 4)$

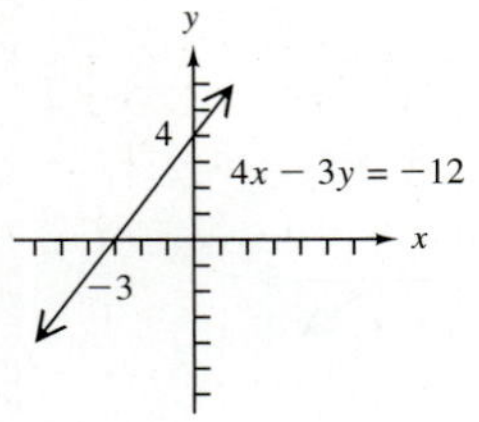

6. none; $(0, 2)$

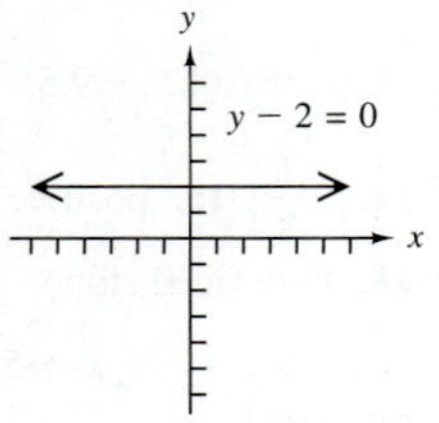

7. $(0, 0)$; $(0, 0)$

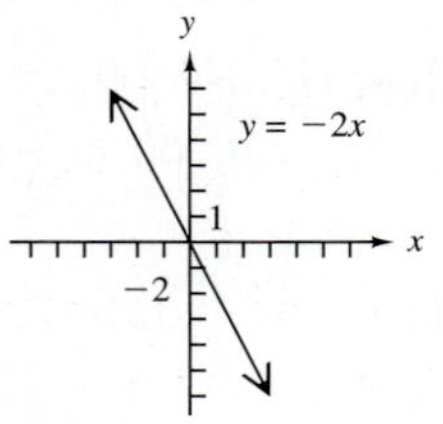

8.

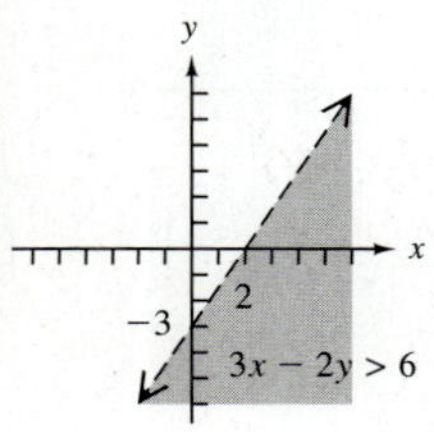

9. $5x - 3y = -57$ **10.** $5x + y = 19$ **11. (a)** $y = -\frac{3}{5}x - \frac{11}{5}$ **(b)** $y = -\frac{1}{2}x - \frac{3}{2}$ **12.** D; domain: $(-\infty, \infty)$; range: $[0, \infty)$

13. D; domain: $\{0, 3, 6\}$; range: $\{1, 2, 3\}$ **14.** 0; $-a^2 + 2a - 1$

15. -1200 farms per yr; The number of farms decreased, on the average, by about 1200 each year from 1980 to 2005.

Cumulative Review Exercises: Chapters 1–4 (pages 265–266)

1. always true **2.** always true **3.** never true **4.** sometimes true; for example, $3 + (-3) = 0$, but $3 + (-1) = 2 \neq 0$ **5.** 4 **6.** 0.64 **7.** not a real number **8.** $\frac{8}{5}$ **9.** $4m - 3$ **10.** $2x^2 + 5x + 4$ **11.** $-\frac{19}{2}$

12. $(-3, 5]$ **13.** no **14.** -24 **15.** 56 **16.** undefined **17.** $\left\{\frac{7}{6}\right\}$

18. $\{-1\}$ **19.** $h = \frac{3V}{\pi r^2}$ **20.** 2 hr **21.** 4 white pills **22.** 6 in.

23. The union of the three solution sets is $(-\infty, \infty)$. **24.** $\left(-\frac{1}{2}, \infty\right)$

25. $(2, 3)$ **26.** $(-\infty, 2) \cup (3, \infty)$ **27.** $\left\{-\frac{16}{5}, 2\right\}$ **28.** $(-11, 7)$

29. $(-\infty, -2] \cup [7, \infty)$ **30.** $(0, -3)$, $(4, 0)$, $\left(2, -\frac{3}{2}\right)$

31. x-intercept: $(-2, 0)$; y-intercept: $(0, 4)$

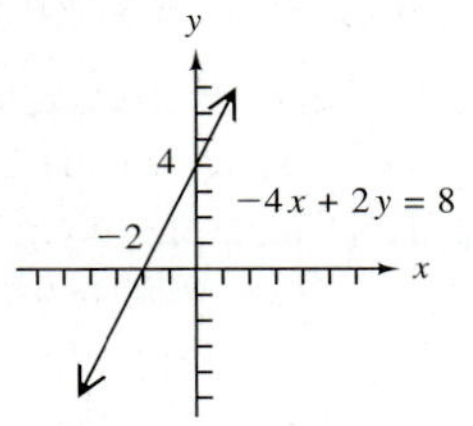

32. $-\frac{3}{2}$ **33.** $-\frac{1}{2}$ **34.** $-\frac{3}{4}$ **35.** $y = -\frac{3}{4}x - 1$ **36.** $y = -2$

37. $y = -\frac{4}{3}x + \frac{7}{3}$ **38. (a)** $(-\infty, \infty)$ **(b)** 22 **39.** 10.571 thousand (or 10,571) per yr; The number of motor scoooters sold in the United States increased by an average of 10,571 per yr from 1997 to 2004.

40. $y = 10.571x + 12$

Chapter 5 Systems of Linear Equations

Section 5.1 (pages 277–282)

1. 3; -6 **3.** D; The ordered pair solution must be in quadrant IV, since that is where the graphs of the equations intersect. **5.** B **7.** A

9. $\{(-2, -3)\}$ **11.** $\{(0, 1)\}$

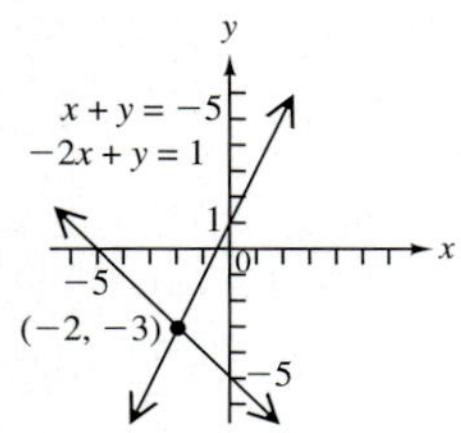

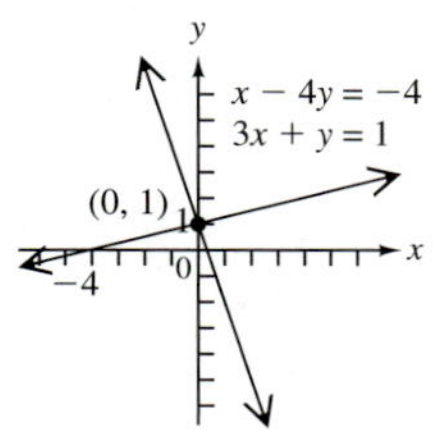

13. $\{(-3, 0)\}$

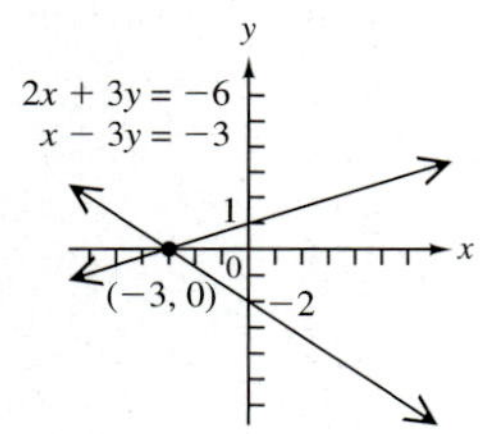

15. yes **17.** no **19.** no **21.** $\{(1, 2)\}$ **23.** $\{(2, 3)\}$

25. $\left\{\left(\frac{22}{9}, \frac{22}{3}\right)\right\}$ **27.** $\{(5, 4)\}$ **29.** $\left\{\left(-5, -\frac{10}{3}\right)\right\}$ **31.** $\{(2, 6)\}$

33. $\{(x, y) \mid 2x - y = 0\}$; dependent equations **35.** $\emptyset$; inconsistent system **37.** $\{(2, -4)\}$ **39.** $\{(3, -1)\}$ **41.** $\{(2, -3)\}$

43. $\left\{\left(\frac{3}{2}, -\frac{3}{2}\right)\right\}$ **45.** $\{(x, y) \mid 7x + 2y = 6\}$; dependent equations

47. $\{(2, -4)\}$ **49.** $\emptyset$; inconsistent system **51.** $y = -\frac{3}{7}x + \frac{4}{7}$; $y = -\frac{3}{7}x + \frac{3}{14}$; 0 **53.** Both are $y = -\frac{2}{3}x + \frac{1}{3}$; infinitely many

55. (a) Use substitution since the second equation is solved for y.
(b) Use elimination since the coefficients of the y-terms are opposites.
(c) Use elimination since the equations are in standard form with no coefficients of 1 or -1. Solving by substitution would involve fractions.

57. $\{(-3, 2)\}$ **59.** $\{(-4, 6)\}$ **61.** $\{(x, y) \mid 4x - y = -2\}$

63. $\left\{\left(1, \frac{1}{2}\right)\right\}$ **65. (a)** \$4 **(b)** 300 half-gallons **(c)** supply: 200 half-gallons; demand: 400 half-gallons **67. (a)** Houston, Phoenix, Dallas **(b)** Philadelphia **(c)** Philadelphia, Dallas, Phoenix, Houston **(d)** 2010; 1.45 million **(e)** (2025, 2.8) **69.** 2000, 2001, 2002, first half of 2003 **71.** approximately (3.6, 10.5) (Values may vary slightly based on the method of solution.) **73.** $\{(1, 3)\}$ **74.** $f(x) = -3x + 6$; linear **75.** $g(x) = \frac{2}{3}x + \frac{7}{3}$; linear **76.** one; 1; 3; 1; 3; 1; 3

Section 5.2 (pages 289–292)

1. The statement means that when -1 is substituted for x, 2 is substituted for y, and 3 is substituted for z in the three equations, the resulting three statements are true. **3.** $\{(3, 2, 1)\}$ **5.** $\{(1, 4, -3)\}$ **7.** $\{(1, 0, 3)\}$ **9.** $\left\{\left(1, \frac{3}{10}, \frac{2}{5}\right)\right\}$ **11.** $\{(0, 2, -5)\}$ **13.** $\left\{\left(\frac{20}{59}, -\frac{33}{59}, \frac{35}{59}\right)\right\}$ **15.** $\{(4, 5, 3)\}$ **17.** $\{(2, 2, 2)\}$ **19.** $\{(-1, 0, 0)\}$ **21.** $\left\{\left(\frac{8}{3}, \frac{2}{3}, 3\right)\right\}$ **23.** Answers will vary. Some possible answers are **(a)** two perpendicular walls and the ceiling in a normal room, **(b)** the floors of three different levels of an office building, and **(c)** three pages of this book (since they intersect in the spine). **25.** $\emptyset$; inconsistent system **27.** $\{(x, y, z) \mid x - y + 4z = 8\}$; dependent equations **29.** $\{(x, y, z) \mid 2x + y - z = 6\}$; dependent equations **31.** $\{(0, 0, 0)\}$ **33.** $\emptyset$; inconsistent system **35.** $\{(3, 0, 2)\}$ **37.** $128 = a + b + c$ **38.** $140 = 2.25a + 1.5b + c$ **39.** $80 = 9a + 3b + c$ **40.** $a - b + c = 128$; $2.25a + 1.5b + c = 140$; $9a + 3b + c = 80$; $\{(-32, 104, 56)\}$ **41.** $f(x) = -32x^2 + 104x + 56$ **42.** height; time **43.** 56 ft **44.** 140.5 ft

Section 5.3 (pages 301–306)

1. wins: 96; losses: 66 **3.** length: 78 ft; width: 36 ft **5.** Wal-Mart: \$316 billion; ExxonMobil: \$340 billion **7.** $x = 40$ and $y = 50$, so the angles measure 40° and 50°. **9.** NHL: \$247.32; NBA: \$267.37 **11.** Junior Roast Beef: \$2.09; Big Montana: \$4.39 **13. (a)** 6 oz **(b)** 15 oz **(c)** 24 oz **(d)** 30 oz **15.** $\$1.29x$ **17.** 6 gal of 25%; 14 gal of 35% **19.** 6 L of pure acid; 48 L of 10% acid **21.** 14 kg of nuts; 16 kg of cereal **23.** \$1000 at 2%; \$2000 at 4% **25. (a)** $(10 - x)$ mph **(b)** $(10 + x)$ mph **27.** scooter: 25 mph; bicycle: 10 mph **29.** boat: 21 mph; current: 3 mph **31.** \$0.75-per-lb candy: 5.22 lb; \$1.25-per-lb candy: 3.78 lb **33.** 76 general admission; 108 with student ID **35.** 8 for a citron; 5 for a wood apple **37.** $x + y + z = 180$; angle measures: 70°, 30°, 80° **39.** first: 20°; second: 70°; third: 90° **41.** shortest: 12 cm; middle: 25 cm; longest: 33 cm **43.** Independent: 38; Democrat: 34; Republican: 28 **45.** \$14 tickets: 300; \$20 tickets: 225; \$50 tickets: 60 **47.** bookstore A: 140; bookstore B: 280; bookstore C: 380 **49.** wins: 46; losses: 25; ties: 11

Section 5.4 (pages 313–314)

1. (a) 0, 5, -3 **(b)** 1, -3, 8 **(c)** yes; The number of rows is the same as the number of columns (three). **(d)** $\begin{bmatrix} 1 & 4 & 8 \\ 0 & 5 & -3 \\ -2 & 3 & 1 \end{bmatrix}$

(e) $\begin{bmatrix} 1 & -\frac{3}{2} & -\frac{1}{2} \\ 0 & 5 & -3 \\ 1 & 4 & 8 \end{bmatrix}$ **(f)** $\begin{bmatrix} 1 & 15 & 25 \\ 0 & 5 & -3 \\ 1 & 4 & 8 \end{bmatrix}$

3. $\left[\begin{array}{cc|c} 1 & 2 & 11 \\ 2 & -1 & -3 \end{array}\right]$; $\left[\begin{array}{cc|c} 1 & 2 & 11 \\ 0 & -5 & -25 \end{array}\right]$; $\left[\begin{array}{cc|c} 1 & 2 & 11 \\ 0 & 1 & 5 \end{array}\right]$; $x + 2y = 11$; $y = 5$; $\{(1, 5)\}$ **5.** $\{(4, 1)\}$ **7.** $\{(1, 1)\}$ **9.** $\{(-1, 4)\}$ **11.** $\emptyset$ **13.** $\{(x, y) \mid 2x + y = 4\}$ **15.** $\{(0, 0)\}$

17. $\left[\begin{array}{ccc|c} 1 & 1 & -1 & -3 \\ 0 & -1 & 3 & 10 \\ 0 & -6 & 7 & 38 \end{array}\right]$; $\left[\begin{array}{ccc|c} 1 & 1 & -1 & -3 \\ 0 & 1 & -3 & -10 \\ 0 & -6 & 7 & 38 \end{array}\right]$;

$\left[\begin{array}{ccc|c} 1 & 1 & -1 & -3 \\ 0 & 1 & -3 & -10 \\ 0 & 0 & -11 & -22 \end{array}\right]$; $\left[\begin{array}{ccc|c} 1 & 1 & -1 & -3 \\ 0 & 1 & -3 & -10 \\ 0 & 0 & 1 & 2 \end{array}\right]$; $x + y - z = -3$; $y - 3z = -10$; $z = 2$; $\{(3, -4, 2)\}$ **19.** $\{(4, 0, 1)\}$ **21.** $\{(-1, 23, 16)\}$ **23.** $\{(3, 2, -4)\}$ **25.** $\emptyset$ **27.** $\{(x, y, z) \mid x - 2y + z = 4\}$ **29.** $\{(0, 0, 0)\}$

Chapter 5 Review Exercises (pages 319–322)

1. $\{(2, 2)\}$ **2.** D **3. (a)** 1980 and 1985 **(b)** just less than 500,000 **4.** $\left\{\left(-\frac{8}{9}, -\frac{4}{3}\right)\right\}$ **5.** $\{(0, 4)\}$ **6.** $\{(2, 4)\}$ **7.** $\{(-1, 2)\}$ **8.** $\{(-6, 3)\}$ **9.** $\left\{\left(\frac{68}{13}, -\frac{31}{13}\right)\right\}$ **10.** $\{(x, y) \mid 3x - y = -6\}$; dependent equations **11.** $\emptyset$; inconsistent system **12.** $\{(0, 0)\}$

13. Answers will vary.

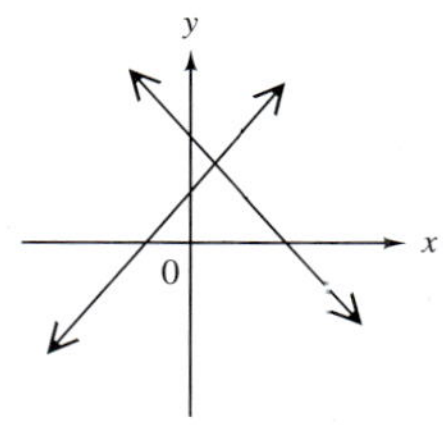

14. Answers will vary.

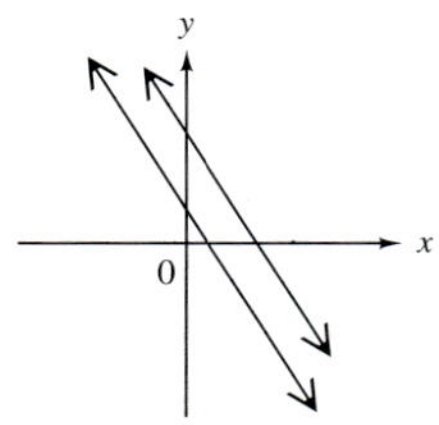

15. Answers will vary.

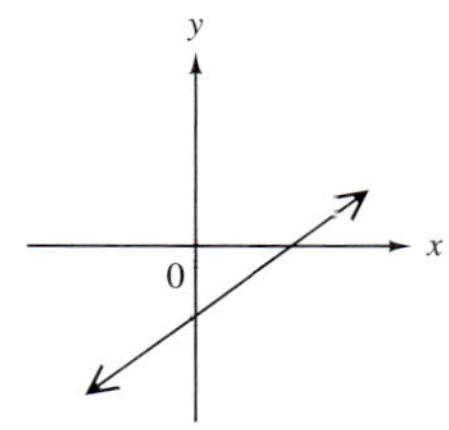

16. Because the lines have the same slope (3) but different y-intercepts ((0, 2) and (0, -4)), the lines do not intersect. Thus, the system has no solution. **17.** $\{(1, -5, 3)\}$ **18.** $\emptyset$; inconsistent system **19.** $\{(1, 2, 3)\}$ **20.** length: 200 ft; width: 85 ft **21.** Boston Red Sox: \$40.77; Chicago Cubs: \$28.45 **22.** plane: 300 mph; wind: 20 mph **23.** 30 lb of \$2 per lb nuts; 70 lb of \$1 per lb candy **24.** 4 vats of green algae; 7 vats of brown algae **25.** 85°, 60°, 35° **26.** 5 L of 8%; 3 L of 20%; none of 10% **27.** Mantle: 54; Maris: 61; Blanchard: 21 **28.** $\{(3, -2)\}$ **29.** $\{(-1, 5)\}$ **30.** $\{(0, 0, -1)\}$ **31.** B; The second equation is already solved for y. **32.** $\{(12, 9)\}$ **33.** $\left\{\left(\frac{82}{23}, -\frac{4}{23}\right)\right\}$ **34.** $\{(3, -1)\}$ **35.** $\{(5, 3)\}$ **36.** $\{(0, 4)\}$ **37.** $\emptyset$ **38.** 20 L **39.** Germany: 29; U.S.: 25; Canada: 24 **40.** $2a + b + c = -5$ **41.** $a - c = 1$ **42.** $3a + 3b + c = -18$ **43.** $a = 1, b = -7, c = 0$; $x^2 + y^2 + x - 7y = 0$ **44.** The relation is not a function because a vertical line intersects its graph more than once.

ANSWERS

Chapter 5 Test (pages 323–324)

1. 1997 **2.** about 110 thousand
3. $\{(6, 1)\}$

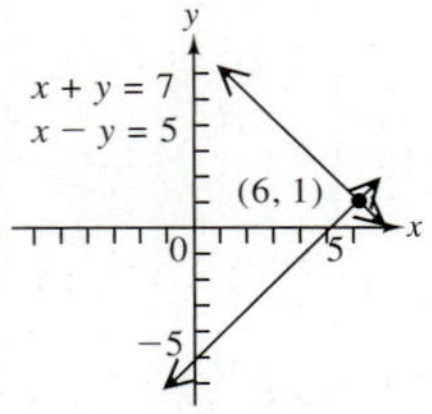

4. $\{(6, -4)\}$ **5.** $\{(x, y) \mid 12x - 5y = 8\}$; dependent equations
6. $\left\{\left(-\frac{9}{4}, \frac{5}{4}\right)\right\}$ **7.** $\{(3, 3)\}$ **8.** $\{(0, -2)\}$ **9.** $\emptyset$; inconsistent system
10. $\left\{\left(-\frac{2}{3}, \frac{4}{5}, 0\right)\right\}$ **11.** $\{(3, -2, 1)\}$ **12.** *Ocean's Eleven:* \$183.4 million; *Runaway Bride:* \$152.1 million **13.** 45 mph; 75 mph **14.** 4 L of 20%; 8 L of 50% **15.** AC adaptor: \$8; rechargeable flashlight: \$15 **16.** 60 oz of Orange Pekoe; 30 oz of Irish Breakfast; 10 oz of Earl Grey **17.** $\left\{\left(\frac{2}{5}, \frac{7}{5}\right)\right\}$ **18.** $\{(-1, 2, 3)\}$

Cumulative Review Exercises: Chapters 1–5 (pages 325–326)

1. 81 **2.** -81 **3.** -81 **4.** 0.7 **5.** -0.7 **6.** not a real number
7. -199 **8.** 455 **9.** 14 **10.** $\left\{-\frac{15}{4}\right\}$ **11.** $\{11\}$ **12.** $x = \frac{c - by}{a}$
13. $\left\{\frac{2}{3}, 2\right\}$ **14.** $\left(-\infty, \frac{240}{13}\right]$ **15.** $\left[-2, \frac{2}{3}\right]$ **16.** $(-\infty, \infty)$ **17.** 2010; 1813; 62.8%; 57.2% **18.** pennies: 35; nickels: 29; dimes: 30
19. 46°, 46°, 88° **20.** $y = 6$ **21.** $x = 4$ **22.** $-\frac{4}{3}$ **23.** $\frac{3}{4}$
24. $4x + 3y = 10$ **25.** $f(x) = -\frac{4}{3}x + \frac{10}{3}$

26.

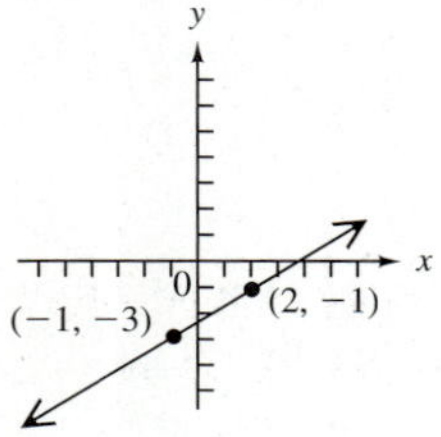

27.

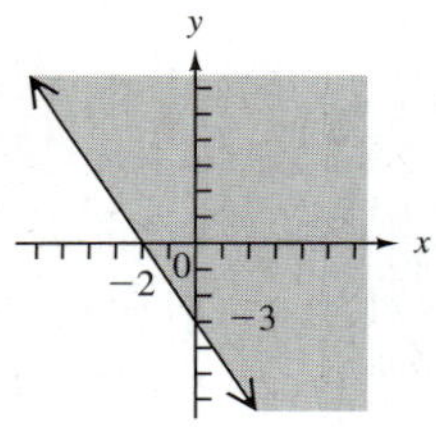

$-3x - 2y \le 6$

28. $\{(3, -3)\}$ **29.** $\{(5, 3, 2)\}$ **30.** 50 lb of \$1.20 candy; 30 lb of \$2.40 candy **31.** \$10,000 at 8%; \$7000 at 10%; \$8000 at 9%
32. $x = 8$, or 800 parts; \$3000 **33.** about \$400

Chapter 6 Exponents, Polynomials, and Polynomial Functions

Section 6.1 (pages 339–344)

1. incorrect; $(ab)^2 = a^2b^2$ **3.** incorrect; $\left(\frac{4}{a}\right)^3 = \frac{4^3}{a^3}$ **5.** correct
7. 13^{12} **9.** 8^{10} **11.** x^{17} **13.** $-27w^8$ **15.** $18x^3y^8$ **17.** The product rule does not apply. **19.** **(a)** B **(b)** C **(c)** B **(d)** C **21.** 1
23. -1 **25.** 1 **27.** -2 **29.** **(a)** B **(b)** D **(c)** B **(d)** D
31. $\frac{1}{5^4}$, or $\frac{1}{625}$ **33.** $\frac{1}{8}$ **35.** $\frac{1}{16x^2}$ **37.** $\frac{4}{x^2}$ **39.** $-\frac{1}{a^3}$ **41.** $\frac{1}{(-a)^4}$, or $\frac{1}{a^4}$
43. $\frac{11}{30}$ **45.** $-\frac{5}{24}$ **47.** 16 **49.** $\frac{27}{4}$ **51.** $\frac{27}{8}$ **53.** $\frac{25}{16}$
55. **(a)** B **(b)** D **(c)** D **(d)** B **57.** 4^2, or 16 **59.** x^4 **61.** $\frac{1}{r^3}$
63. 6^6 **65.** $\frac{1}{6^{10}}$ **67.** 7^2, or 49 **69.** r^3 **71.** The quotient rule does not apply. **73.** x^{18} **75.** $\frac{27}{125}$ **77.** $64t^3$ **79.** $-216x^6$ **81.** $-\frac{64m^6}{t^3}$
83. $\frac{s^{12}}{t^{20}}$ **85.** $\frac{1}{3}$ **87.** $\frac{1}{a^5}$ **89.** $\frac{1}{k^2}$ **91.** $-4r^6$ **93.** $\frac{625}{a^{10}}$ **95.** $\frac{z}{x}$ [illegible]
97. $\frac{1}{5p^{10}}$ **99.** $\frac{4}{a^2}$ **101.** $\frac{2^2k^5}{m^2}$, or $\frac{4k^5}{m^2}$ **103.** $\frac{2k^5}{3}$ **105.** $\frac{8}{3pq^{10}}$
107. $\frac{25a^{12}}{b^{20}}$ **109.** 5.3×10^2 **111.** 8.3×10^{-1} **113.** 6.92×10^{-6}
115. -3.85×10^4 **117.** 72,000 **119.** 0.00254 **121.** $-60,000$
123. 0.000012 **125.** 0.0000025 **127.** 200,000 **129.** $\$1 \times 10^9$; $\$1 \times 10^{12}$; $\$3.1 \times 10^{12}$; 2.10385×10^5 **131.** **(a)** 3.041×10^8 **(b)** $\$1 \times 10^{12}$ **(c)** \$3288 **133.** about 33,000, or 3.3×10^4
135. approximately $3.2 \times 10^4 = 32{,}000$ hr (about 3.7 yr)
137. **(a)** 20,000 hr **(b)** 833 days **139.** $-a^n = (-a)^n$ when n is an odd number. When n is even, $-a^n \neq (-a)^n$. **141.** Write the fraction as its reciprocal raised to the opposite of the negative power.

Section 6.2 (pages 349–350)

1. neither **3.** ascending **5.** descending **7.** 7; 1 **9.** -15; 2
11. 1; 4 **13.** -1; 6 **15.** monomial; 0 **17.** binomial; 1
19. trinomial; 3 **21.** none of these; 5 **23.** $8z^4$ **25.** $7m^3$ **27.** $5x$
29. already simplified **31.** $-3y^2 + 7y$ **33.** $8k^2 + 2k - 7$
35. $-2n^4 - n^3 + n^2$ **37.** $-9p^2 + 11p - 9$ **39.** $5a + 18$
41. $14m^2 - 13m + 6$ **43.** $13z^2 + 10z - 3$ **45.** $10y^3 - 7y^2 + 5y + 8$
47. $-5a^4 - 6a^3 + 9a^2 - 11$ **49.** $r + 13$ **51.** $8x^2 + x - 2$
53. $-2a^2 - 2a - 7$ **55.** $-3z^5 + z^2 + 7z$

Section 6.3 (pages 357–358)

1. **(a)** -10 **(b)** 8 **3.** **(a)** 8 **(b)** 2 **5.** **(a)** 8 **(b)** 74
7. **(a)** -11 **(b)** 4 **9.** **(a)** 15,160 **(b)** 18,354 **(c)** 19,897
11. **(a)** \$28.2 billion **(b)** \$79.1 billion **(c)** \$105.9 billion
13. **(a)** $8x - 3$ **(b)** $2x - 17$ **15.** **(a)** $-x^2 + 12x - 12$
(b) $9x^2 + 4x + 6$ **17.** $x^2 + 2x - 9$ **19.** 6 **21.** $x^2 - x - 6$
23. 6 **25.** -33 **27.** 0
29. domain: $(-\infty, \infty)$; range: $(-\infty, \infty)$

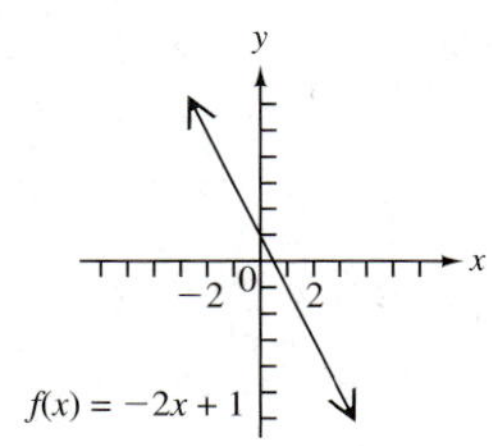

31. domain: $(-\infty, \infty)$; range: $(-\infty, 0]$

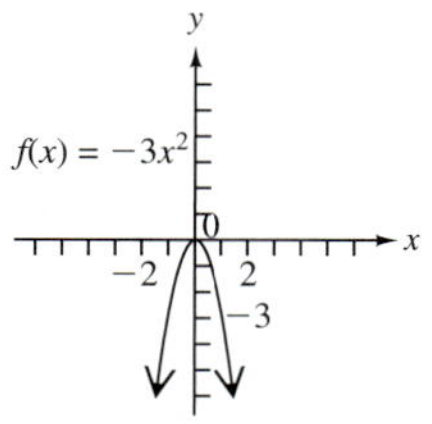

33. domain: $(-\infty, \infty)$; range: $(-\infty, \infty)$

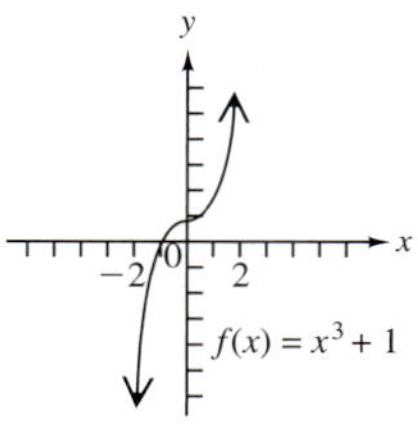

Section 6.4 (pages 365–368)

1. -2[illegible]m^5 **3.** $-6x^2 + 15x$ **5.** $-2q^3 - 3q^4$ **7.** $18k^4 + 12k^3 + 6k^2$
9. $6m^3 + m^2 - 14m - 3$ **11.** $4x^5 - 4x^4 - 24x^3$
13. $6y^2 + y - 12$ **15.** $25m^2 - 9n^2$ **17.** $-2b^3 + 2b^2 + 18b + 12$
19. $8z^4 - 14z^3 + 17z^2 + 20z - 3$ **21.** $6p^4 + p^3 + 4p^2 - 27p - 6$
23. $m^2 - 3m - 40$ **25.** $12k^2 + k - 6$ **27.** $3z^2 + zw - 4w^2$
29. $12c^2 + 16cd - 3d^2$ **31.** $0.1x^2 + 0.63x - 0.13$
33. $3r^2 - \frac{23}{4}ry - \frac{1}{2}y^2$ **35.** $4p^2 - 9$ **37.** $25m^2 - 1$
39. $9a^2 - 4c^2$ **41.** $16x^2 - \frac{4}{9}$ **43.** $16m^2 - 49n^4$ **45.** $5y^5 - 20y^3$
47. $y^2 - 10y + 25$ **49.** $4p^2 + 28p + 49$ **51.** $16n^2 - 24nm + 9m^2$
53. $k^2 - \frac{10}{7}kp + \frac{25}{49}p^2$ **55.** $25x^2 + 10x + 1 + 60xy + 12y + 36y^2$
57. $4a^2 + 4ab + b^2 - 9$ **59.** $4h^2 - 4hk + k^2 - j^2$
61. $x^3 + 6x^2 + 12x + 8$ **63.** $125r^3 - 75r^2s + 15rs^2 - s^3$
65. $q^4 - 8q^3 + 24q^2 - 32q + 16$ **67.** $\frac{9}{2}x^2 - 2y^2$
69. $15x^2 - 2x - 24$ **71.** $a - b$ **72.** $A = s^2$; $(a - b)^2$
73. $(a - b)b$, or $ab - b^2$; $2ab - 2b^2$ **74.** b^2 **75.** a^2; a
76. $a^2 - (2ab - 2b^2) - b^2 = a^2 - 2ab + b^2$ **77.** They must be equal to each other. **78.** $(a - b)^2 = a^2 - 2ab + b^2$; This reinforces the special product for the square of a binomial difference.
79. $10x^2 - 2x$ **81.** $2x^2 - x - 3$ **83.** $8x^3 - 27$ **85.** -20
87. 32 **89.** 20 **91.** $(2 + 3)^3 \neq 2^3 + 3^3$ because $125 \neq 35$; $(x + y)^3 = x^3 + 3x^2y + 3xy^2 + y^3$

Section 6.5 (pages 373–374)

1. $3x^3 - 2x^2 - 1$ **3.** $3y + 4 - \frac{5}{y}$ **5.** $3m + 5 + \frac{6}{m}$ **7.** $n - \frac{3n^2}{2m} + 2$
9. $y - 3$ **11.** $t + 5$ **13.** $p - 4 + \frac{44}{p + 6}$ **15.** $z^2 + 3$
17. $z^2 + 2x - 3 + \frac{6}{4x + 1}$ **19.** $2x - 5 + \frac{-4x + 5}{3x^2 - 2x + 4}$
21. $3x^2 + 6x + 11 + \frac{26}{x - 2}$ **23.** $2x^2 - x - 5 + \frac{3}{x - 5}$
25. $2k^2 + 3k - 1$ **27.** $9z^2 - 4z + 1 + \frac{-z + 6}{z^2 - z + 2}$ **29.** $\frac{2}{3}x - 1$
31. $\frac{3}{4}c - 2 + \frac{1}{4a + 3}$ **33.** $5x - 1$; 0 **35.** $2x - 3$; -1

37. $4x^2 + 6x + 9$; $\frac{3}{2}$ **39.** $\frac{x^2 - 9}{2x}$, $x \neq 0$ **41.** $-\frac{5}{4}$
43. $\frac{x - 3}{2x}$, $x \neq 0$ **45.** 0 **47.** $(2p + 7)$ feet

Chapter 6 Review Exercises (pages 379–382)

1. 64 **2.** $\frac{1}{81}$ **3.** -125 **4.** 18 **5.** $\frac{81}{16}$ **6.** $\frac{16}{25}$ **7.** $\frac{11}{30}$ **8.** 0
9. $-12x^2y^8$ **10.** $-\frac{2n}{m^5}$ **11.** $\frac{10p^8}{q^7}$ **12.** $\frac{x^2}{y^2}$ **13.** $\frac{1}{3^8}$ **14.** x^8 **15.** $\frac{y^6}{x^2}$
16. $\frac{1}{z^{15}}$ **17.** $\frac{25}{m^{18}}$ **18.** 1 **19.** $\frac{1}{96m^7}$ **20.** $\frac{2025}{8r^4}$ **21.** $\frac{4w^6}{z^{18}}$
22. 1.345×10^4 **23.** 7.65×10^{-8} **24.** 1.38×10^{-1}
25. 2.994×10^8; 7.4×10^4; 1×10^2 **26.** 1,210,000 **27.** 0.0058
28. 2×10^{-4}; 0.0002 **29.** 1.5×10^3; 1500 **30.** 4.1×10^{-5}; 0.000041 **31.** 2.7×10^{-2}; 0.027 **32.** 14 **33.** -1 **34.** 0.045
35. 504 **36.** **(a)** $11k^3 - 3k^2 + 9k$ **(b)** trinomial **(c)** 3
37. **(a)** $9m^7 + 14m^6$ **(b)** binomial **(c)** 7 **38.** **(a)** $-7q^5r^3$ **(b)** monomial **(c)** 8 **39.** Answers will vary. An example is $x^5 + 2x^4 - x^2 + x + 2$. **40.** $-x^2 - 3x + 1$
41. $-5y^3 - 4y^2 + 6y - 12$ **42.** $6a^3 - 4a^2 - 16a + 15$
43. $8y^2 - 9y + 5$ **44.** $12x^2 + 8x + 5$ **45.** **(a)** -11 **(b)** 4
46. **(a)** $5x^2 - x + 5$ **(b)** $-5x^2 + 5x + 1$ **(c)** 11 **(d)** -9
47. **(a)** 95,072 **(b)** 117,699 **(c)** 133,693
48.

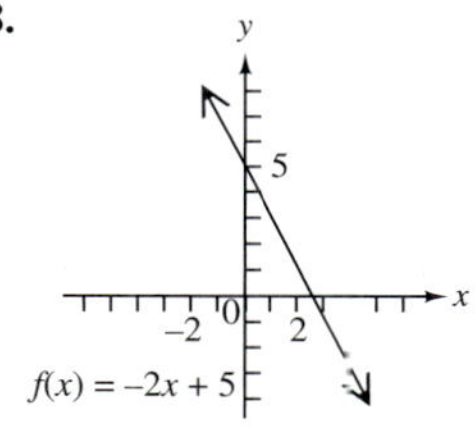

49.

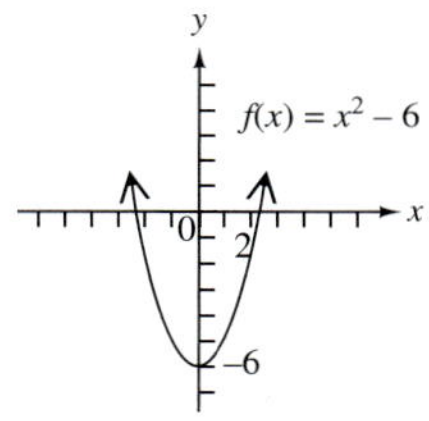

50.

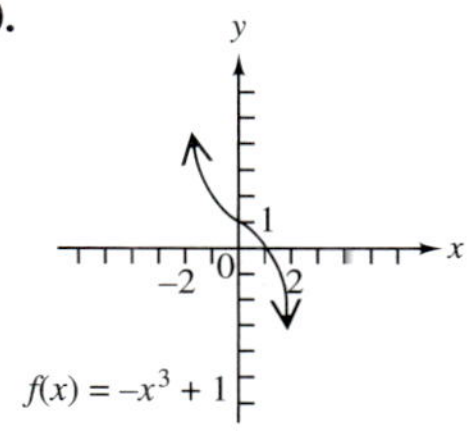

51. $-12k^3 - 42k$ **52.** $14y^2 + 5y - 24$ **53.** $6w^2 - 13wt + 6t^2$
54. $10p^4 + 30p^3 - 8p^2 - 24p$ **55.** $9z^4 - 12z^3 + 16z^2 - 11z + 2$
56. $36r^4 - 1$ **57.** $z^2 - \frac{9}{25}$ **58.** $16m^2 + 24m + 9$
59. $8x^3 + 60x^2 + 150x + 125$ **60.** $y^2 - 3y + \frac{5}{4}$
61. $p^2 + 6p + 9 + \frac{54}{2p - 3}$ **62.** $p^2 + 3p - 6$ **63.** **(a)** A **(b)** G **(c)** C **(d)** C **(e)** A **(f)** E **(g)** B **(h)** H **(i)** F
64. 998 mi^2 **65.** $8x^2 - 10x - 3$ **66.** $\frac{y^4}{36}$ **67.** $\frac{1}{16y^{18}}$
68. $4x^2 - 36x + 81$ **69.** $2y^2x + \frac{3y^3}{2x} + \frac{5x^2}{2}$
70. $21p^9 + 7p^8 + 14p^7$ **71.** $-\frac{1}{5z^9}$ **72.** $x^2 + 2x - 3 + \frac{3}{x + 5}$
73. $-14 + 16w - 8w^2$ **74.** $-3k^2 + 4k - 7$

Chapter 6 Test (pages 383–384)

1. (a) C **(b)** A **(c)** D **(d)** A **(e)** E **(f)** F **(g)** B **(h)** G **(i)** C **2.** $\frac{4x^7}{9y^{10}}$ **3.** $\frac{6}{r^{14}}$ **4.** $\frac{16}{9p^{10}q^{28}}$ **5.** $\frac{16}{x^6y^{16}}$ **6.** 0.00000091 **7.** 3×10^{-4}; 0.0003 **8. (a)** -18 **(b)** $-2x^2 + 12x - 9$ **(c)** $-2x^2 - 2x - 3$ **(d)** -7

9.

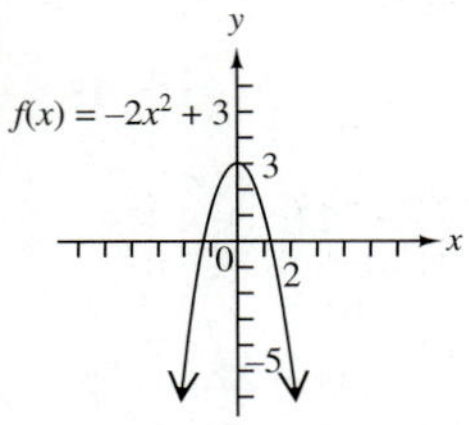

10. (a) 466 thousand **(b)** 710 thousand **(c)** 907 thousand **11.** $x^3 - 2x^2 - 10x - 13$ **12.** $10x^2 - x - 3$ **13.** $6m^3 - 7m^2 - 30m + 25$ **14.** $36x^2 - y^2$ **15.** $9k^2 + 6kq + q^2$ **16.** $4y^2 - 9z^2 + 6zx - x^2$ **17.** $4p - 8 + \frac{6}{p}$ **18.** $x^2 + 5x + 10 + \frac{14}{x-2}$ **19. (a)** $x^3 + 4x^2 + 5x + 2$ **(b)** 0 **20. (a)** $x + 2, x \neq -1$ **(b)** 0

Cumulative Review Exercises: Chapters 1–6 (pages 385–386)

1. A, B, C, D, F **2.** B, C, D, F **3.** D, F **4.** C, D, F **5.** E, F **6.** D, F **7.** 32 **8.** 0 **9.** $\{-65\}$ **10.** $(-\infty, \infty)$ **11.** $(-\infty, 6)$ **12.** $\left\{-\frac{1}{3}, 1\right\}$ **13.** $\left(-\infty, -\frac{8}{3}\right] \cup [2, \infty)$ **14.** 32%; 390; 270; 10% **15.** 15°, 35°, 130° **16.** $-\frac{4}{3}$; $4x + 3y = -1$

17.

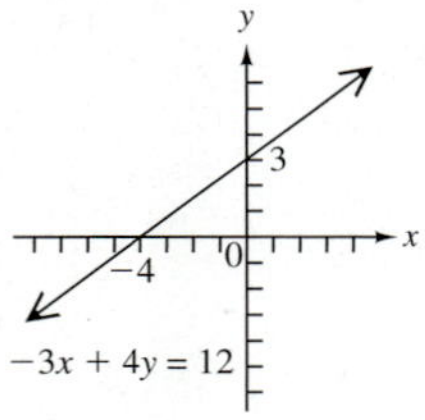

18.

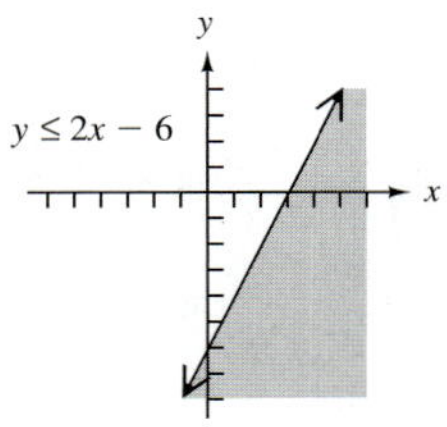

19.

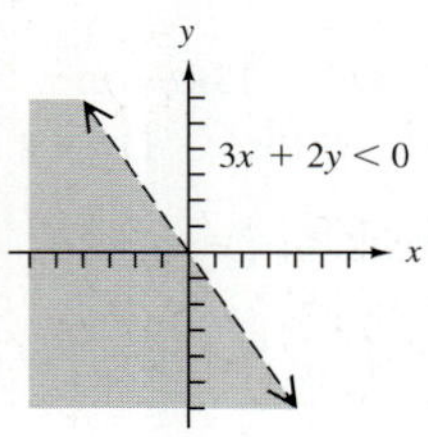

20. (a) $-12{,}272.8$ thousand lb per yr; The number of pounds of shrimp caught decreased an average of 12,272.8 thousand lb per yr. **(b)** $y = -12{,}272.8x + 322{,}486$ **(c)** 285,668 thousand lb **21.** domain: $\{-4, -1, 2, 5\}$; range: $\{-2, 0, 2\}$; function **22.** $\{(3, 2)\}$ **23.** $\emptyset$ **24.** $\{(1, 0, -1)\}$ **25.** length: 42 ft; width: 30 ft **26.** $\frac{8m^9n^3}{p^6}$ **27.** $\frac{y^7}{x^{13}z^2}$ **28.** $\frac{m^6}{8n^9}$ **29.** $2x^2 - 4x + 38$ **30.** $15x^2 + 7xy - 2y^2$ **31.** $64m^2 - 25n^2$ **32.** $m^2 - 2m - 7 + \frac{-3}{m-1}$

Chapter 7 Factoring

Section 7.1 (pages 393–394)

1. $3m$ **3.** $8xy$ **5.** $3(r + t)^2$ **7.** A **9.** $10(x - 3)$ **11.** $8(s + 2t)$ **13.** $6(1 + 2r)$ **15.** $8k(k^2 + 3)$ **17.** $xy(3 - 5y)$ **19.** $-2p^2q^4(2p + q)$ **21.** $7x^3(3x^2 + 5x - 2)$ **23.** $9p^2(4p^2 +$ [illegible] $- 3p)$ **25.** $5ac(3ac^2 - 5c + a)$ **27.** cannot be factored **29.** $(m - 4)(2m + 5)$ **31.** $11(2z - 1)$ **33.** $(2 - x)^2(10 - x - x^2)$ **35.** $r(-r^2 + 3r + 5)$; $-r(r^2 - 3r - 5)$ **37.** $12s^4(-s + 4)$; $-12s^4(s - 4)$ **39.** $2x^2(-x^3 + 3x + 2)$; $-2x^2(x^3 - 3x - 2)$ **41.** $(m + 3q)(x - y)$ **43.** $(5m + n)(2 + k)$ **45.** $(2 - q)(2 - 3p)$ **47.** $(p + q)(p - 4z)$ **49.** $(a + 5c)(7b + 1)$ **51.** $(m + 4)(m^2 - 6)$ **53.** $(a^2 + b^2)(-3a + 2b)$ **55.** $(y - 2)(x - 2)$ **57.** $(3y - 2)(3y^3 - 4)$ **59.** $2(m + 3q)(x + y)$ **61.** $y^2(2x + 1)(x^2 - 7)$ **63.** $m^{-5}(3 + m^2)$ **65.** $p^{-3}(3 + 2p)$

Section 7.2 (pages 401–402)

1. D **3.** C **5.** $(y - 3)(y + 10)$ **7.** $(p - 8)(p + 7)$ **9.** prime **11.** $(a + 5b)(a - 7b)$ **13.** prime **15.** $(x + 9y)(x + 2y)$ **17.** $-(6m - 5)(m + 3)$ **19.** $(5x - 6)(2x + 3)$ **21.** $(4k + 3)(5k + 8)$ **23.** $(3a - 2b)(5a - 4b)$ **25.** $(6m - 5)^2$ **27.** prime **29.** $(2xz - 1)(3xz + 4)$ **31.** $3(4x + 5)(2x + 1)$ **33.** $-5(a + 6)(3a - 4)$ **35.** $11x(x - 6)(x - 4)$ **37.** $2xy^3(x - 12y)(x - 12y)$ **39.** $(5k + 4)(2k + 1)$ **41.** $(3m + 3p + 5)(m + p - 4)$ **43.** $(a^2 + ab + 2b)(a^2 + ab - 3b)$ **45.** $(2x^2 + 3)(x^2 - 6)$ **47.** $(4x^2 + 3)(4x^2 + 1)$ **49.** $(6p^3 - r)(2p^3 - 5r)$ **51.** no **52.** 1, 3, 5, 9, 15, 45; no **53.** no **54.** $(5x + 2)(2x + 5)$; no **55.** Since k is odd, 2 is not a factor of $2x^2 + kx + 8$, and because 2 is a factor of $2x + 4$, the binomial $2x + 4$ cannot be a factor. **56.** $3y + 15$ cannot be a factor of $12y^2 - 11y - 15$ because 3 is a factor of $3y + 15$, but 3 is not a factor of $12y^2 - 11y - 15$.

Section 7.3 (pages 407–408)

1. A, D **3.** B, C **5.** $(p + 4)(p - 4)$ **7.** $(5x + 2)(5x - 2)$ **9.** $2(3a + 7b)(3a - 7b)$ **11.** $4(4m^2 + y^2)(2m + y)(2m - y)$ **13.** $(y + z + 9)(y + z - 9)$ **15.** $(4 + x + 3y)(4 - x - 3y)$ **17.** $(p^2 + 16)(p + 4)(p - 4)$ **19.** $(k - 3)^2$ **21.** $(2z + w)^2$ **23.** $(4m - 1 + n)(4m - 1 - n)$ **25.** $(2r - 3 + s)(2r - 3 - s)$ **27.** $(x + y - 1)(x - y + 1)$ **29.** $2(7m + 3n)^2$ **31.** $(p + q + 1)^2$ **33.** $(a - b + 4)^2$ **35.** $(y - 4)(y^2 + 4y + 16)$ **37.** $(r + 7)(r^2 - 7r + 49)$ **39.** $(2x - y)(4x^2 + 2xy + y^2)$ **41.** $(4g + 3h)(16g^2 - 12gh + 9h^2)$ **43.** $3(2n + 3p)(4n^2 - 6np + 9p^2)$ **45.** $(y + z - 4)(y^2 + 2yz + z^2 + 4y + 4z + 16)$ **47.** $(m^2 - 5)(m^4 + 5m^2 + 25)$ **49.** $(5y^2 + z)(25y^4 - 5y^2z + z^2)$ **50.** $(x^3 - y^3)(x^3 + y^3)$; $(x - y)(x^2 + xy + y^2)(x + y)(x^2 - xy + y^2)$ **51.** $(x^2 + xy + y^2)(x^2 - xy + y^2)$ **52.** $(x^2 - y^2)(x^4 + x^2y^2 + y^4)$; $(x - y)(x + y)(x^4 + x^2y^2 + y^4)$ **53.** $x^4 + x^2y^2 + y^4$ **54.** The product must equal $x^4 + x^2y^2 + y^4$. Multiply $(x^2 + xy + y^2)(x^2 - xy + y^2)$ to verify this. **55.** Start by factoring as a difference of squares.

Section 7.4 (pages 411–412)

1. $(10a + 3b)(10a - 3b)$ **3.** $6p^3(3p^2 - 4 + 2p^3)$ **5.** $(x + 7)(x - 5)$ **7.** prime **9.** $(6b + 1)(b - 3)$ **11.** $3mn(3m + 2n)(2m - n)$ **13.** $(2p + 5q)(p + 3q)$ **15.** $(2k + 7r)^2$ **17.** $(m - 2)(n + 5)$ **19.** $(x + 3)^2(x - 3)$ **21.** prime **23.** $(3k + 1)(2k - 1)$

25. $(x^2+25)(x+5)(x-5)$ **27.** $(a+6)(b+c)$ **29.** $4y(y-2)$
31. $(7z+2k)(2z-k)$ **33.** $16(4b+5c)(4b-5c)$
35. $8(5z+4)(25z^2-20z+16)$ **37.** $(5r-s)(2r+5s)$
39. $8x^2(4+2x-3x^3)$ **41.** $(2x-5q)(7x+5q)$ **43.** $(y+5)(y-2)$
45. $2z(a^2+3a-2)$ **47.** $(9p-5r)(2p+7r)$
49. $(x-2y-2)(x-2y-2)$ **51.** $(5r+2s-3)^2$
53. $(z+2)(z-2)(z^2-5)$ **55.** $(p+2)(4+m)$
57. $2(5p+9)(5p-9)$ **59.** $(4a+b)^2$

Section 7.5 (pages 419–422)

1. First rewrite the equation so that one side is 0. Factor the other side and set each factor equal to 0. The solutions of these linear equations are solutions of the quadratic equation. **3.** $\{-10, 5\}$ **5.** $\left\{-\frac{8}{3}, \frac{5}{2}\right\}$
7. $\{-2, 5\}$ **9.** $\{-6, -3\}$ **11.** $\left\{-\frac{1}{2}, 4\right\}$ **13.** $\left\{-\frac{1}{3}, \frac{4}{5}\right\}$ **15.** $\left\{-\frac{3}{4}\right\}$
17. $\{0, 4\}$ **19.** $\{0, 6\}$ **21.** $\{-2, 2\}$ **23.** $\{-3, 3\}$ **25.** $\{-4, 2\}$
27. $\left\{-\frac{1}{2}, 6\right\}$ **29.** $\left\{-5, -\frac{1}{5}\right\}$ **31.** $\{1, 6\}$ **33.** $\left\{-\frac{1}{2}, 0, 5\right\}$
35. $\{-1, 0, 3\}$ **37.** $\left\{-\frac{4}{3}, 0, \frac{4}{3}\right\}$ **39.** $\left\{-\frac{5}{2}, -1, 1\right\}$ **41.** By dividing each side by a variable expression, she "lost" the solution 0. The solution set is $\left\{-\frac{4}{3}, 0, \frac{4}{3}\right\}$. **43.** width: 16 ft; length: 20 ft **45.** base: 12 ft; height: 5 ft **47.** 50 ft by 100 ft **49.** -6 and -5 or 5 and 6 **51.** length: 15 in.; width: 9 in. **53.** 3 sec and 5 sec; 1 sec and 7 sec **55.** 6 sec

Chapter 7 Review Exercises (pages 425–426)

1. $7y(3y+5)$ **2.** $4qb(3q+2b-5q^2b)$ **3.** $(x+3)(x-3)$
4. $(z+1)(3z-1)$ **5.** $(m+q)(4+n)$ **6.** $(x+y)(x+5)$
7. $(m+3)(2-a)$ **8.** $(a-b)(2m-p)$ **9.** $(3p-4)(p+1)$
10. $(3r+1)(4r-3)$ **11.** $(2m+5)(5m+6)$ **12.** $(2k-h)(5k-3h)$
13. prime **14.** $2x(4+x)(3-x)$ **15.** $(2k^2+1)(k^2-3)$
16. $(p+2)^2(p+3)(p-2)$ **17.** $(4x+5)(4x-5)$
18. $(3t+7)(3t-7)$ **19.** $(x+7)^2$ **20.** $(3k-2)^2$
21. $(r+3)(r^2-3r+9)$ **22.** $(5x-1)(25x^2+5x+1)$
23. $(m+1)(m^2-m+1)(m-1)(m^2+m+1)$
24. $(x^4+1)(x^2+1)(x+1)(x-1)$ **25.** $(x+3+5y)(x+3-5y)$
26. $\left\{-1, -\frac{2}{5}\right\}$ **27.** $\{2, 3\}$ **28.** $\left\{-\frac{5}{2}, \frac{10}{3}\right\}$ **29.** $\left\{-\frac{3}{2}, \frac{1}{3}\right\}$
30. $\{-3, 3\}$ **31.** $\left\{-\frac{3}{2}, 0\right\}$ **32.** $\left\{\frac{1}{2}, 1\right\}$ **33.** $\{4\}$ **34.** $\left\{-\frac{7}{2}, 0, 4\right\}$
35. 3 ft **36.** length: 60 ft; width: 40 ft **37.** after 16 sec
38. after 1 sec and after 15 sec **39.** The rock reaches a height of 240 ft once on its way up and once on its way down. **40.** $a(6-m)(5+m)$
41. $(2-a)(4+2a+a^2)$ **42.** $(9k+4)(9k-4)$ **43.** prime
44. $5y^2(3y+4)$ **45.** $(5z-3m)^2$ **46.** $\left\{-\frac{3}{5}, 4\right\}$ **47.** $\{-1, 0, 1\}$
48. $\{0, 3\}$ **49.** width: 25 ft; length: 110 ft **50.** 6 in.

Chapter 7 Test (pages 427–428)

1. $11z(z-4)$ **2.** $5x^2y^3(2y^2-1-5x^3)$ **3.** $(x+y)(3+b)$
4. $-(2x+9)(x-4)$ **5.** $(3x-5)(2x+7)$ **6.** $(4p-q)(p+q)$
7. $(4a+5b)^2$ **8.** $(x+1+2z)(x+1-2z)$ **9.** $(a+b)(a-b)(a+2)$
10. $(3k+11j)(3k-11j)$ **11.** $(y-6)(y^2+6y+36)$
12. $(2k^2-5)(3k^2+7)$ **13.** $(3x^2+1)(9x^4-3x^2+1)$
14. $-(x+5)(x-5)$ **15.** $(t^2+8)(t^2+2)$ **16.** It is not in factored form because there are two terms: $(x^2+2y)p$ and $3(x^2+2y)$. The common factor is x^2+2y, and the factored form is $(x^2+2y)(p+3)$.
17. D **18.** $\left\{-2, -\frac{2}{3}\right\}$ **19.** $\left\{0, \frac{5}{3}\right\}$ **20.** $\left\{-\frac{2}{5}, 1\right\}$
21. length: 8 in.; width: 5 in. **22.** 2 sec and 4 sec

Cumulative Review Exercises: Chapters 1–7 (pages 429–430)

1. $-2m+6$ **2.** $4m-3$ **3.** $2x^2+5x+4$ **4.** -24 **5.** 204
6. undefined **7.** 10 **8.** $\left\{\frac{7}{6}\right\}$ **9.** $\{-1\}$ **10.** $\left(-\infty, \frac{15}{4}\right]$
11. $\left(-\frac{1}{2}, \infty\right)$ **12.** $(2, 3)$ **13.** $(-\infty, 2) \cup (3, \infty)$ **14.** $\left\{-\frac{16}{5}, 2\right\}$
15. $(-11, 7)$ **16.** $(-\infty, -2] \cup [7, \infty)$ **17.** $h = \frac{V}{lw}$ **18.** 2 hr
19.

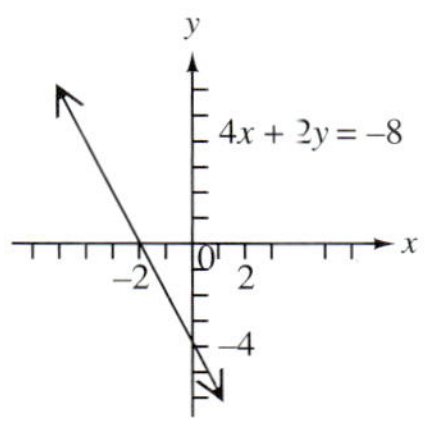

20. -1 **21.** 0 **22.** -1 **23.** $\left(-\frac{7}{2}, 0\right)$ **24.** $(0, 7)$
25. $\{(1, 5)\}$ **26.** $\{(1, 1, 0)\}$ **27.** $\frac{y}{18x}$ **28.** $\frac{5my^4}{3}$
29. x^3+12x^2-3x-7 **30.** $49x^2+42xy+9y^2$
31. $10p^3+7p^2-28p-24$ **32.** $(2w+7z)(8w-3z)$
33. $(2x-1+y)(2x-1-y)$ **34.** $(2y-9)^2$
35. $(10x^2+9)(10x^2-9)$ **36.** $(2p+3)(4p^2-6p+9)$
37. $\left\{-4, -\frac{3}{2}, 1\right\}$ **38.** $\left\{\frac{1}{3}\right\}$ **39.** 4 ft
40. longer sides: 18 in.; distance between: 16 in.

Chapter 8 Rational Expressions and Functions

Section 8.1 (pages 439–442)

1. C **3.** D **5.** E **7.** Replacing x with 2 makes the denominator 0 and the value of the expression undefined. To find the values excluded from the domain, set the denominator equal to 0 and solve the equation. All solutions of the equation are excluded from the domain. **9.** $7; \{x \mid x \neq 7\}$ **11.** $-\frac{1}{7}; \left\{x \mid x \neq -\frac{1}{7}\right\}$ **13.** $0; \{x \mid x \neq 0\}$
15. $-2, \frac{3}{2}; \left\{x \mid x \neq -2, \frac{3}{2}\right\}$ **17.** none; $(-\infty, \infty)$ **19.** none; $(-\infty, \infty)$
21. **(a)** numerator: x^2, $4x$; denominator: x, 4 **(b)** First factor the numerator, getting $x(x+4)$, then divide the numerator and denominator by the common factor of $x+4$ to get $\frac{x}{1}$, or x. **23.** B **25.** x
27. $\frac{x-3}{x+5}$ **29.** $\frac{x+3}{2x(x-3)}$ **31.** already in lowest terms **33.** $\frac{6}{7}$
35. $\frac{z}{6}$ **37.** $\frac{2}{t-3}$ **39.** $\frac{x-3}{x+1}$ **41.** $\frac{4x+1}{4x+3}$ **43.** a^2-ab+b^2

ANSWERS

45. $\frac{c+6d}{c-d}$ **47.** $\frac{a+b}{a-b}$ **49.** -1 *In Exercises 51–55, there are other acceptable ways to express each answer.* **51.** $-(x+y)$ **53.** $-\frac{x+y}{x-y}$ **55.** $-\frac{1}{2}$ **57.** already in lowest terms **59.** $\frac{x+4}{x-2}$ **61.** $\frac{2x+3}{x+2}$ **63.** $-\frac{35}{8}$ **65.** $\frac{7x}{6}$ **67.** $-\frac{p+5}{2p}$ (There are other ways.) **69.** $\frac{-m(m+7)}{m+1}$ (There are other ways.) **71.** -2 **73.** $\frac{x+4}{x-4}$ **75.** $\frac{2x+3y}{2x-3y}$ **77.** $\frac{k+5p}{2k+5p}$ **79.** $(k-1)(k-2)$

Section 8.2 (pages 451–454)

1. To add or subtract rational expressions that have a common denominator, first add or subtract the numerators. Then place the result over the common denominator. Write the answer in lowest terms. **3.** $\frac{9}{t}$ **5.** $\frac{2}{x}$ **7.** 1 **9.** $x-5$ **11.** $\frac{1}{p+3}$ **13.** $a-b$ **15.** $72x^4y^5$ **17.** $z(z-2)$ **19.** $2(y+4)$ **21.** $(x+9)^2(x-9)$ **23.** $(m+n)(m-n)$ **25.** $x(x-4)(x+1)$ **27.** $(t+5)(t-2)(2t-3)$ **29.** $2y(y+3)(y-3)$ **31.** Yes, they could both be correct because the expressions are equivalent. Multiplying $\frac{3}{5-y}$ by 1 in the form $\frac{-1}{-1}$ gives $\frac{-3}{y-5}$. **33.** $\frac{31}{3t}$ **35.** $\frac{5-22x}{12x^2y}$ **37.** $\frac{1}{x(x-1)}$ **39.** $\frac{5a^2-7a}{(a+1)(a-3)}$ **41.** 3 **43.** $\frac{3}{x-4}$, or $\frac{-3}{4-x}$ **45.** $\frac{w+z}{w-z}$, or $\frac{-w-z}{z-w}$ **47.** $\frac{-13}{12(3+x)}$ **49.** $\frac{2(2x-1)}{x-1}$ **51.** $\frac{7}{y}$ **53.** $\frac{6}{x-2}$ **55.** $\frac{3x-2}{x-1}$ **57.** $\frac{4x-7}{x^2-x+1}$ **59.** $\frac{2x+1}{x}$ **61.** $\frac{x}{(x-2)^2(x-3)}$ **63.** $\frac{10x+23}{(x+2)^2(x+3)}$ **65.** $\frac{2x(x+12y)}{(x+2y)(x-y)(x+6y)}$ **67.** $c(x)=\frac{10x}{49(101-x)}$ **69.** $\frac{8}{9}$ **70.** $\frac{3}{7}+\frac{5}{9}-\frac{6}{63}$; They are the same. **71.** $\frac{8}{9}$; yes **72.** Answers will vary. Suppose the name is Sosa, so that $x=4$. The problem is $\frac{3}{2}+\frac{5}{4}-\frac{6}{8}$. The predicted answer is $\frac{8}{4}=2$, which is correct. **73.** It causes $\frac{3}{x-2}$ and $\frac{6}{x^2-2x}$ to be undefined, since 0 appears in the denominators. **74.** 0

Section 8.3 (pages 459–460)

1. Begin by simplifying the numerator. Then simplify the denominator. Write as a division problem, and proceed. **3.** $\frac{2x}{x-1}$ **5.** $\frac{2(k+1)}{3k-1}$ **7.** $\frac{5x^2}{9z^3}$ **9.** $\frac{1+x}{-1+x}$ **11.** $\frac{y+x}{y-x}$ **13.** $4x$ **15.** $x+4y$ **17.** $\frac{3y}{2}$ **19.** $\frac{x^2+5x+4}{x^2+5x+10}$ **21.** $\frac{m^2+6m-4}{m(m-1)}$ **22.** $\frac{m^2-m-2}{m(m-1)}$ **23.** $\frac{m^2+6m-4}{m^2-m-2}$ **24.** $m(m-1)$ **25.** $\frac{m^2+6m-4}{m^2-m-2}$ **26.** Method 1 involves simplifying the numerator and the denominator separately and then performing a division. Method 2 involves multiplying the fraction by a form of 1, the identity element for multiplication. (Preferences will vary.) **27.** $\frac{x^2y^2}{y^2+x^2}$ **29.** $\frac{y^2+x^2}{xy^2+x^2y}$, or $\frac{y^2+x^2}{xy(y+x)}$ **31.** $\frac{1}{2xy}$

Section 8.4 (pages 465–468)

1. (a) $-1, 2$ **(b)** $\{x \mid x \neq -1, 2\}$ **3. (a)** $-\frac{5}{3}, 0, -\frac{3}{2}$ **(b)** $\left\{x \mid x \neq -\frac{5}{3}, 0, -\frac{3}{2}\right\}$ **5. (a)** 0 **(b)** $\{x \mid x \neq 0\}$ **7. (a)** $4, \frac{7}{2}$ **(b)** $\left\{x \mid x \neq 4, \frac{7}{2}\right\}$ **9. (a)** $0, 1, -3, 2$ **(b)** $\{x \mid x \neq 0, 1, -3, 2\}$ **11.** $\{1\}$ **13.** $\{-6, 4\}$ **15.** $\left\{-\frac{7}{12}\right\}$ **17.** $\emptyset$ **19.** $\{-3\}$ **21.** $\{5\}$ **23.** $\{5\}$ **25.** $\emptyset$ **27.** $\left\{\frac{27}{56}\right\}$ **29.** $\emptyset$ **31.** $\{-10\}$ **33.** $\emptyset$ **35.** $\{0\}$ **37.** $\left\{x \mid x \neq -\frac{3}{2}, \frac{3}{2}\right\}$

39. $x=0$; $y=0$

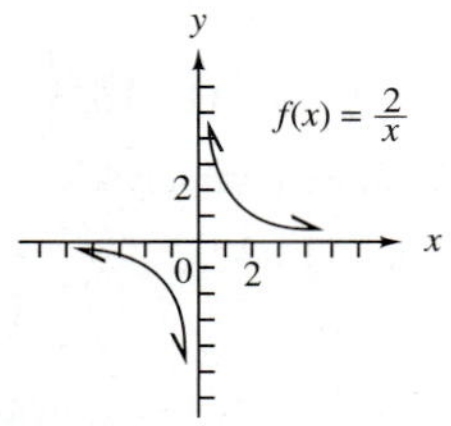

41. $x=2$; $y=0$

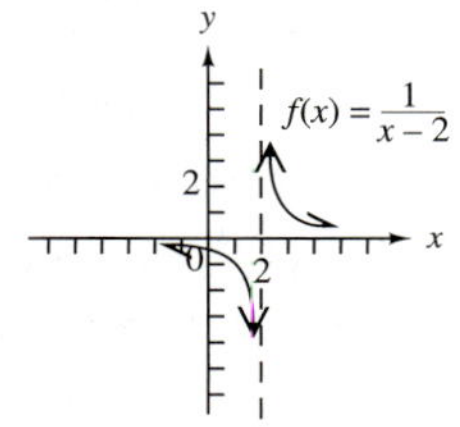

43. (a) 0 **(b)** 1.6 **(c)** 4.1 **(d)** The waiting time also increases. **45.** Substituting -1 for x gives a true statement, $\frac{4}{3}=\frac{4}{3}$. Substituting -2 for x leads to 0 in the first and third denominators. **46.** $C=-4$; $\{-2\}$; -1 is rejected. **47.** $C=24$; $\{-4\}$; 3 is rejected. **48.** Answers will vary.

Summary Exercises on Rational Expressions and Equations (pages 469–470)

1. equation; $\{20\}$ **2.** expression; $\frac{2(x+5)}{5}$ **3.** expression; $-\frac{22}{7x}$ **4.** expression; $\frac{y+x}{y-x}$ **5.** equation; $\left\{\frac{1}{2}\right\}$ **6.** equation; $\{7\}$ **7.** expression; $\frac{43}{24x}$ **8.** equation; $\{1\}$ **9.** expression; $\frac{5x-1}{-2x+2}$, or $\frac{5x-1}{-2(x-1)}$ **10.** expression; $\frac{25}{4(r+2)}$ **11.** expression; $\frac{x^2+xy+2y^2}{(x+y)(x-y)}$ **12.** expression; $\frac{24p}{p+2}$ **13.** expression; $-\frac{5}{36}$ **14.** equation; $\{0\}$ **15.** expression; $\frac{b+3}{3}$ **16.** expression; $\frac{5}{3z}$ **17.** expression; $\frac{2x+10}{x(x-2)(x+2)}$ **18.** equation; $\{2\}$ **19.** expression; $\frac{-x}{3x+5y}$ **20.** equation; $\{-13\}$ **21.** expression; $\frac{3y+2}{y+3}$ **22.** equation; $\left\{\frac{5}{4}\right\}$ **23.** equation; $\emptyset$ **24.** expression; $\frac{2z-3}{2z+3}$ **25.** expression; $\frac{-1}{x-3}$, or $\frac{1}{3-x}$ **26.** expression; $\frac{t-2}{8}$ **27.** equation; $\{-10\}$ **28.** expression; $\frac{13x+28}{2x(x+4)(x-4)}$ **29.** equation; $\emptyset$ **30.** expression; $\frac{k(2k^2-2k+5)}{(k-1)(3k^2-2)}$

Section 8.5 (pages 479–484)

1. A **3.** D **5.** 65.625 **7.** $\frac{25}{4}$ **9.** $G = \frac{Fd^2}{Mm}$ **11.** $a = \frac{bc}{c+b}$ **13.** $v = \frac{PVt}{pT}$ **15.** $r = \frac{nE - IR}{In}$ **17.** $b = \frac{2A}{h} - B$, or $b = \frac{2A - Bh}{h}$ **19.** $r = \frac{eR}{E - e}$ **21.** Multiply each side by $a - b$. **23.** 2[illegible] girls, 7 boys **25.** $\frac{1}{2}$ job per hr **27.** 5.4 in. **29.** 7.6 in. **31.** 100 games **33.** 25,000 fish **35.** 6.6 more gallons **37.** 2.4 mL **39.** $x = \frac{7}{2}$; $AC = 8$; $DF = 12$ **41.** 3 mph **43.** 1020 mi **45.** 480 mi **47.** 1[illegible]0 mi **49.** $6\frac{2}{3}$ min **51.** 30 hr **53.** 20 hr **55.** $2\frac{4}{5}$ hr

Section 8.6 (pages 491–494)

1. inverse **3.** direct **5.** joint **7.** combined **9.** 36 **11.** 0.625 **13.** $2[illegible]2\frac{2}{9}$ **15.** increases; decreases **17.** If y varies inversely as x, x is in the denominator; however, if y varies directly as x, x is in the numerator. Also, for $k > 0$, with inverse variation, as x increases, y decreases. With direct variation, y increases as x increases. **19.** \$[illegible].59$\frac{9}{10}$ **21.** about 450 cm^3 **23.** 256 ft **25.** $13\frac{1}{3}$ amperes **27.** $21\frac{1}{3}$ foot-candles **29.** \$420 **31.** 448.1 lb **33.** approximately 68,60[illegible] calls **35.** 11.8 lb **37.** (0, 0), (1, 4.45) **38.** 4.45 **39.** $y = 4.45x + 0$, or $y = 4.45x$ **40.** $a = 4.45$, $b = 0$ **41.** It is the price per gallon and the slope of the line. **42.** It can be written in the form $y = kx$ (where $k = a$). The value of a is called the constant of variation.

Chapter 8 Review Exercises (pages 499–502)

1. (a) -6 **(b)** $\{x \mid x \neq -6\}$ **2. (a)** 2, 5 **(b)** $\{x \mid x \neq 2, 5\}$ **3. (a)** 9 **(b)** $\{x \mid x \neq 9\}$ **4.** $\frac{x}{2}$ **5.** $\frac{5m+n}{5m-n}$ **6.** $\frac{-1}{2+r}$ **7.** The reciprocal of a rational expression is another rational expression such that the two rational expressions have a product of 1. **8.** $\frac{3y^2(2y+3)}{2y-3}$ **9.** $\frac{-3(w+4)}{w}$ **10.** $\frac{z(z+2)}{z+5}$ **11.** 1 **12.** $96b^5$ **13.** $9r^2(3r+1)$ **14.** $(3x-1)(2x+5)(3x+4)$ **15.** $\frac{16z-3}{2z^2}$ **16.** 1[illegible] **17.** $\frac{71}{30(a+2)}$ **18.** $\frac{13r^2+5rs}{(5r+s)(2r-s)(r+s)}$ **19.** $\frac{3+2t}{4-7t}$ **20.** -2 **21.** $\frac{1}{3q+2p}$ **22.** $\frac{y+x}{xy}$ **23.** $\{-3\}$ **24.** $\{-2\}$ **25.** $\{0\}$ **26.** $\emptyset$ **27.** Although her algebra was correct, 3 is not a solution because it is not in the domain of the variable in the equation. Thus, $\emptyset$ is correct. **28.** In simplifying the expression, we are combining terms to get a single fraction with a denominator of $6x$, while in solving the equation, we are finding a value for x that makes the equation true. **29.** C; $x = 0$; $y = 0$ **30.** $\frac{1[illegible]}{2}$ **31.** $h = \frac{3V}{\pi r^2}$ **32.** $M = \frac{m\mu}{v - \mu}$ **33.** 16 km per hr **34.** $4\frac{4}{5}$ min **35.** C **36.** $\frac{16}{5}$ **37.** 430 mm **38.** 36 ft^3 **39.** $\frac{1}{x-2y}$ **40.** $\frac{x+5}{x+2}$ **41.** $\frac{6m+5}{3m^2}$ **42.** $\frac{k-3}{36k^2+6k+1}$ **43.** $\frac{x^2-6}{2(2x+1)}$ **44.** $\frac{x(9x+1)}{3x+1}$ **45.** $\frac{3-5x}{6x+1}$ **46.** $\frac{11}{3-x}$, or $\frac{-11}{x-3}$ **47.** $\frac{1}{3}$ **48.** $\frac{s^2+t^2}{st(s-t)}$ **49.** $\frac{5a^2+4ab-12b^2}{(a+3b)(a-2b)(a+b)}$ **50.** $\frac{acd+b^2d+bc^2}{bcd}$ **51.** $\left\{\frac{1}{3}\right\}$ **52.** $r = \frac{AR}{R-A}$, or $r = \frac{-AR}{A-R}$ **53.** $\{1, 4\}$ **54.** $\left\{-\frac{14}{3}\right\}$ **55.** $3\frac{3}{5}$ hr **56.** 2.4 mi **57.** 5.59 vibrations per sec **58.** 12 ft^2

Chapter 8 Test (pages 503–504)

1. $-2, \frac{4}{3}$; $\left\{x \mid x \neq -2, \frac{4}{3}\right\}$ **2.** $\frac{2x-5}{x(3x-1)}$ **3.** $\frac{3(x+3)}{4}$ **4.** $\frac{y+4}{y-5}$ **5.** $\frac{x+5}{x}$ **6.** $t^2(t+3)(t-2)$ **7.** $\frac{7-2t}{6t^2}$ **8.** $\frac{13x+35}{(x-7)(x+7)}$ **9.** $\frac{4}{x+2}$ **10.** $\frac{72}{11}$ **11.** $-\frac{1}{a+b}$ **12.** $\frac{2y^2+x^2}{xy(y-x)}$ **13. (a)** expression; $\frac{11(x-6)}{12}$ **(b)** equation; $\{6\}$ **14.** $\left\{\frac{1}{2}\right\}$ **15.** $\{5\}$ **16.** $x = -1$; $y = 0$

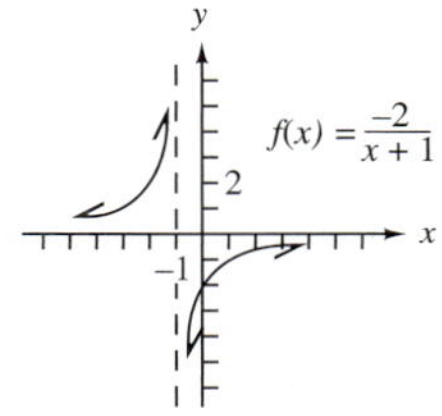

17. $3\frac{3}{14}$ hr **18.** 15 mph **19.** 48,000 fish **20. (a)** 3 units **(b)** 0 **21.** 200 amps **22.** 0.8 lb

Cumulative Review Exercises: Chapters 1–8 (pages 505–506)

1. $\left\{-\frac{15}{4}\right\}$ **2.** $\left\{\frac{2}{3}, 2\right\}$ **3.** $\left(-\infty, \frac{240}{13}\right]$ **4.** \$4000 at 4%; \$8000 at 3% **5.** 6 m **6. (a)** $-\frac{3}{2}$ **(b)** $\frac{3}{4}$ **7. (a)** $y = -\frac{3}{2}x + \frac{1}{2}$ **(b)** $y = \frac{3}{4}x - \frac{7}{4}$

8.

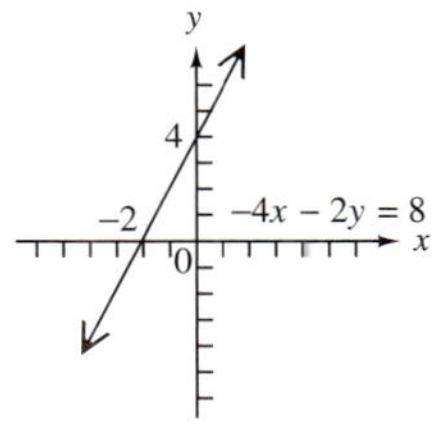

9.

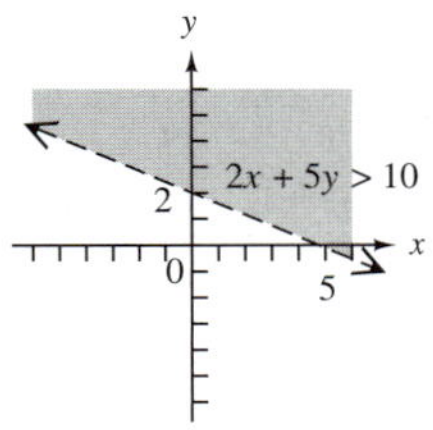

10.

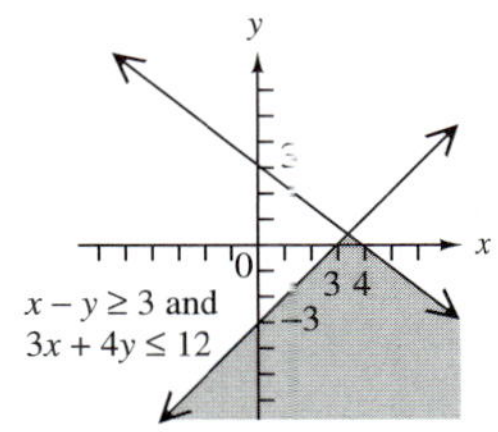

11. (a) $f(x) = \frac{5x-8}{3}$, or $f(x) = \frac{5}{3}x - \frac{8}{3}$ **(b)** -1 **12.** $3x + 15$ **13.** $\{(-1, 3)\}$ **14.** $\{(-2, 3, 1)\}$ **15.** $\emptyset$ **16.** $4y^2 - 7y - 6$ **17.** $x^2 + 4x - 7$ **18.** $12f^2 + 5f - 3$ **19.** $49t^6 - 64$ **20.** $\frac{1}{16}x^2 + \frac{5}{2}x + 25$ **21. (a)** $2x^3 - 2x^2 + 6x - 4$ **(b)** $2x^3 - 4x^2 + 2x + 2$ **(c)** -14 **22.** $(2x+5)(x-9)$

23. $25(2t^2+1)(2t^2-1)$ **24.** $(2p+5)(4p^2-10p+25)$

25. $\frac{a(a-b)}{2(a+b)}$ **26.** 3 **27.** $\frac{2(x+2)}{2x-1}$ **28.** $\left\{-\frac{7}{3}, 1\right\}$ **29.** $\{-4\}$

30. $q=\frac{fp}{p-f}$, or $q=\frac{-fp}{f-p}$

Chapter 9 Roots, Radicals, and Root Functions

Section 9.1 (pages 513–516)

1. E **3.** D **5.** C **7.** C **9.** C **11.** **(a)** not a real number **(b)** negative **(c)** 0 **13.** -9 **15.** 6 **17.** -4 **19.** -8 **21.** 6 **23.** -2 **25.** not a real number **27.** 3 **29.** not a real number

31. $\frac{8}{9}$ **33.** 0.7 **35.** $\frac{4}{3}$ **37.** $-\frac{1}{2}$ **39.** 0.1

41. domain: $[-3, \infty)$; range: $[0, \infty)$ **43.** domain: $[0, \infty)$; range: $[-2, \infty)$

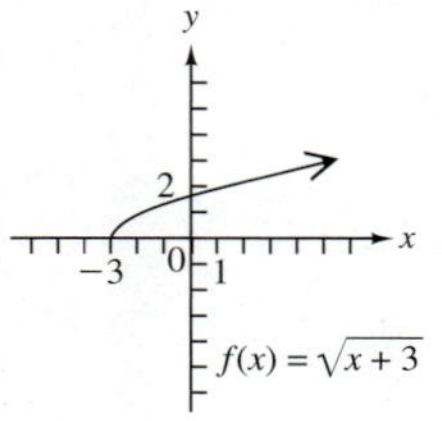

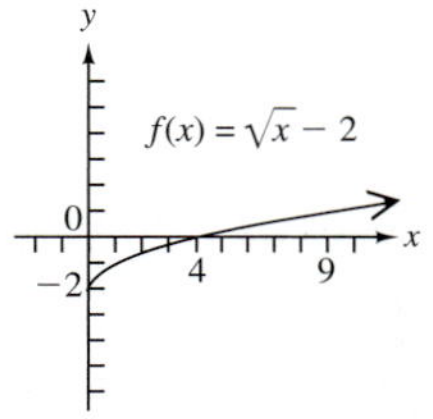

45. domain: $(-\infty, \infty)$; range: $(-\infty, \infty)$

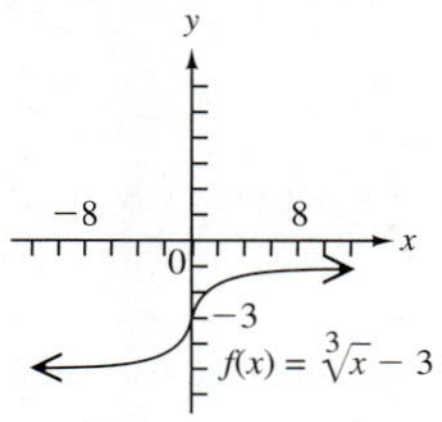

47. 12 **49.** 10 **51.** 2 **53.** -9 **55.** -5 **57.** $|x|$ **59.** $|z|$ **61.** x **63.** x^5 **65.** $|x|^5$ (or $|x^5|$) **67.** 97.381 **69.** 16.863 **71.** -9.055 **73.** 7.507 **75.** 3.162 **77.** 1.885 **79.** 1,183,000 cycles per sec **81.** 10 mi **83.** 392,000 mi^2 **85.** 1.732 amps

Section 9.2 (pages 523–526)

1. C **3.** A **5.** H **7.** B **9.** D **11.** 13 **13.** 9 **15.** 2

17. $\frac{8}{9}$ **19.** -3 **21.** not a real number **23.** 1000 **25.** 27

27. -1024 **29.** 16 **31.** $\frac{1}{8}$ **33.** $\frac{1}{512}$ **35.** $\frac{9}{25}$ **37.** $\sqrt{12}$

39. $(\sqrt[4]{8})^3$ **41.** $(\sqrt[8]{9q})^5-(\sqrt[3]{2x})^2$ **43.** $\frac{1}{(\sqrt{2m})^3}$

45. $(\sqrt[3]{2y+x})^2$ **47.** $\frac{1}{(\sqrt[3]{3m^4+2k^2})^2}$

49. $\sqrt{a^2+b^2}=\sqrt{3^2+4^2}=5$; $a+b=3+4=7$; $5\neq 7$

51. 64 **53.** 64 **55.** x^{10} **57.** $\sqrt[6]{x^5}$ **59.** $\sqrt[15]{t^8}$ **61.** 9 **63.** 4

65. y **67.** $x^{5/12}$ **69.** $k^{2/3}$ **71.** x^3y^8 **73.** $\frac{1}{x^{10/3}}$ **75.** $\frac{1}{m^{1/4}n^{3/4}}$ **77.** p^2

79. $\frac{c^{11/3}}{b^{11/4}}$ **81.** $\frac{q^{5/3}}{9p^{7/2}}$ **83.** $p+2p^2$ **85.** $k^{7/4}-k^{3/4}$ **87.** $6+18a$

89. $5+\frac{5}{m^3}$ **91.** $y^{3/2}$ **93.** $\frac{1}{k^{2/3}}$ **95.** $x^{1/3}z^{5/6}$ **97.** $k^{1/6}$ **99.** $y^{1/30}$

101. $x^{5/27}$ **103.** 72 in.; 6.0 ft **105.** $-12.3°$; The table gives $-12°$.

Section 9.3 (pages 535–540)

1. true; Both are equal to $4\sqrt{3}$ and approximately 6.92820323.
3. true; Both are equal to $6\sqrt{2}$ and approximately 8.485281374.
5. Because there are only two factors of $\sqrt[3]{x}$, $\sqrt[3]{x}\cdot\sqrt[3]{x}=(\sqrt[3]{x})^2$, or $\sqrt[3]{x^2}$.

7. D **9.** $\sqrt{30}$ **11.** $\sqrt{14x}$ **13.** $\sqrt{42pqr}$ **15.** $\sqrt[3]{14xy}$ **17.** $\sqrt[4]{33}$ **19.** $\sqrt[4]{6xy^2}$ **21.** This product cannot be simplified using the product rule. **23.** $\frac{8}{11}$ **25.** $\frac{\sqrt{3}}{5}$ **27.** $\frac{\sqrt{x}}{5}$ **29.** $\frac{p^3}{9}$ **31.** $-\frac{3}{4}$ **33.** $\frac{\sqrt[3]{r^2}}{2}$

35. $-\frac{3}{x}$ **37.** $\frac{1}{x^3}$ **39.** $2\sqrt{3}$ **41.** $12\sqrt{2}$ **43.** $-4\sqrt{2}$ **45.** $-2\sqrt{7}$

47. cannot be simplified further **49.** $4\sqrt[3]{2}$ **51.** $-2\sqrt[3]{2}$ **53.** $2\sqrt[3]{5}$

55. $-4\sqrt[4]{2}$ **57.** $2\sqrt[5]{2}$ **59.** His reasoning was incorrect. Here 8 is a term, not a factor. **61.** $6k\sqrt{2}$ **63.** $12xy^4\sqrt{xy}$ **65.** $11x^3$

67. $-3t^4$ **69.** $-10m^4z^2$ **71.** $5a^2b^3c^4$ **73.** $\frac{1}{2}r^2t^5$ **75.** $5x\sqrt{2x}$

77. $-10r^5\sqrt{5r}$ **79.** $x^3y^4\sqrt{13x}$ **81.** $2z^2w^3$ **83.** $-2zt^2\sqrt[3]{2z^2t}$

85. $3x^3y^4$ **87.** $-3r^3s^2\sqrt[4]{2r^3s^2}$ **89.** $\frac{y^5\sqrt{y}}{6}$ **91.** $\frac{x^5\sqrt[3]{x}}{3}$

93. $4\sqrt{3}$ **95.** $\sqrt{5}$ **97.** $x^2\sqrt{x}$ **99.** $\sqrt[6]{432}$ **101.** $\sqrt[12]{6912}$

103. $\sqrt[6]{x^5}$ **105.** 5 **107.** $8\sqrt{2}$ **109.** $2\sqrt{14}$ **111.** 13 **113.** $9\sqrt{2}$

115. $\sqrt{17}$ **117.** 5 **119.** $6\sqrt{2}$ **121.** $\sqrt{5y^2-2xy+x^2}$

123. 27.0 in. **125.** 0.003 **127.** 15.3 mi

Section 9.4 (pages 543–544)

1. B **3.** 15; Each radical expression simplifies to a whole number.
5. -4 **7.** $7\sqrt{3}$ **9.** $14\sqrt[3]{2}$ **11.** $5\sqrt[4]{2}$ **13.** $24\sqrt{2}$ **15.** cannot be simplified further **17.** $20\sqrt{5}$ **19.** $12\sqrt{2x}$ **21.** $-2m\sqrt{2}$

23. $\sqrt[3]{2}$ **25.** $2\sqrt[3]{x}$ **27.** $-\sqrt[3]{x^2y}$ **29.** $-x\sqrt[3]{xy^2}$ **31.** $19\sqrt[4]{2}$

33. $x\sqrt[4]{xy}$ **35.** $9\sqrt[4]{2a^3}$ **37.** $\frac{5\sqrt{5}}{6}$ **39.** $\frac{7\sqrt{2}}{6}$ **41.** $\frac{5\sqrt{2}}{3}$

43. $5\sqrt{2}+4$ **45.** $\frac{5-3x}{x^4}$ **47.** $\frac{m\sqrt[3]{m^2}}{2}$ **49.** $\frac{3x\sqrt[3]{2}-4\sqrt[3]{5}}{x^3}$

51. $(12\sqrt{5}+5\sqrt{3})$ in. **53.** $(24\sqrt{2}+12\sqrt{3})$ in.

Section 9.5 (pages 551–554)

1. E **3.** A **5.** D **7.** $6-4\sqrt{3}$ **9.** $6-\sqrt{6}$ **11.** 2 **13.** 9
15. $3\sqrt{2}-5\sqrt{3}+2\sqrt{6}-10$ **17.** $3x-4$ **19.** $4x-y$
21. $16x+24\sqrt{x}+9$ **23.** $81-\sqrt[3]{4}$ **25.** Because 6 and $4\sqrt{3}$ are not like terms, they cannot be combined. **27.** $\sqrt{7}$ **29.** $5\sqrt{3}$

31. $\frac{\sqrt{6}}{2}$ **33.** $\frac{9\sqrt{15}}{5}$ **35.** $-\sqrt{2}$ **37.** $\frac{-8\sqrt{3k}}{k}$ **39.** $\frac{6\sqrt{3}}{y}$

41. To rationalize a cube root, three factors of the quantity under the radical sign are needed. We must multiply by $\sqrt[3]{2^2}$, or $\sqrt[3]{4}$, to rationalize $\sqrt[3]{2}$.

43. $\frac{\sqrt{14}}{2}$ **45.** $-\frac{\sqrt{14}}{10}$ **47.** $\frac{2\sqrt{6x}}{x}$ **49.** $-\frac{7r\sqrt{2rs}}{s}$

51. $\frac{12x^3\sqrt{2xy}}{y^5}$ **53.** $\frac{\sqrt[3]{18}}{3}$ **55.** $\frac{\sqrt[3]{12}}{3}$ **57.** $-\frac{\sqrt[3]{2pr}}{r}$

59. $\frac{2\sqrt[4]{x^3}}{x}$ **61.** $\frac{\sqrt[4]{2yz^3}}{z}$ **63.** $\frac{2(4-\sqrt{3})}{13}$ **65.** $3(\sqrt{5}-\sqrt{3})$

67. $\sqrt{3}+\sqrt{7}$ **69.** $\sqrt{7}-\sqrt{6}-\sqrt{14}+2\sqrt{3}$

71. $2\sqrt{3}+\sqrt{10}-3\sqrt{2}-\sqrt{15}$ **73.** $\frac{4(\sqrt{x}+2\sqrt{y})}{x-4y}$

75. $\frac{x\sqrt{2}-\sqrt{3xy}-\sqrt{2xy}+y\sqrt{3}}{2x-3y}$ **77.** $\frac{5+2\sqrt{6}}{4}$ **79.** $\frac{4+2\sqrt{2}}{3}$

81. $\frac{6+2\sqrt{6x}}{3}$ **83.** $\frac{319}{6(8\sqrt{5}+1)}$ **84.** $\frac{9a-b}{(\sqrt{b}-\sqrt{a})(3\sqrt{a}-\sqrt{b})}$

85. $\frac{(3\sqrt{a}+\sqrt{b})(\sqrt{b}+\sqrt{a})}{b-a}$ **86.** In Exercise 84, we multiplied the numerator and denominator by the conjugate of the numerator, while in Exercise 85 we multiplied by the conjugate of the denominator.

Summary Exercises on Operations with Radicals and Rational Exponents (pages 555–556)

1. $-6\sqrt{10}$ **2.** $7-\sqrt{14}$ **3.** $2+\sqrt{6}-2\sqrt{3}-3\sqrt{2}$ **4.** $4\sqrt{2}$ **5.** $73+12\sqrt{35}$ **6.** $\frac{-\sqrt{6}}{2}$ **7.** $4(\sqrt{7}-\sqrt{5})$ **8.** $3\sqrt[3]{2x^2}$ **9.** $-3+2\sqrt{2}$ **10.** -2 **11.** -44 **12.** $\frac{\sqrt{x}+\sqrt{5}}{x-5}$ **13.** $2abc^3\sqrt[3]{b^2}$ **14.** $5\sqrt[3]{3}$ **15.** $3(\sqrt{5}-2)$ **16.** $\frac{\sqrt{15x}}{5x}$ **17.** $\frac{8}{5}$ **18.** $\frac{\sqrt{2}}{8}$ **19.** $-\sqrt[3]{100}$ **20.** $11+2\sqrt{30}$ **21.** $-3\sqrt{3x}$ **22.** $52-30\sqrt{3}$ **23.** 1 **24.** $\frac{\sqrt[3]{117}}{9}$ **25.** $t^2\sqrt[4]{t}$ **26.** $3\sqrt{2}+\sqrt{15}+\sqrt{42}+\sqrt{35}$ **27.** $2\sqrt[4]{27}$ **28.** $\frac{1+\sqrt[3]{3}+\sqrt[3]{9}}{-2}$ **29.** $\frac{x\sqrt[3]{x^2}}{y}$ **30.** $-4\sqrt{3}-3$ **31.** $x^{-6/5}$ **32.** $x^{10}y$ **33.** $\frac{1}{25x^2}$ **34.** $\frac{-6y^{1/6}}{x^{1/24}}$ **35.** $7+4\cdot 3^{1/2}$, or $7+4\sqrt{3}$ **36.** 1

Section 9.6 (pages 561–564)

1. (a) yes **(b)** no **3. (a)** yes **(b)** no **5.** no; There is no solution. The radical expression, which is positive, cannot equal a negative number. **7.** $\{11\}$ **9.** $\left\{\frac{1}{3}\right\}$ **11.** $\emptyset$ **13.** $\{5\}$ **15.** $\{18\}$ **17.** $\{5\}$ **19.** $\{4\}$ **21.** $\{17\}$ **23.** $\{5\}$ **25.** $\emptyset$ **27.** $\{0\}$ **29.** $\{0\}$ **31.** $\left\{-\frac{1}{3}\right\}$ **33.** $\emptyset$ **35.** We cannot just square each term. The right side should be $(8-x)^2 = 64-16x+x^2$. The correct first step is $3x-4 = 64-16x+x^2$, and the solution set is $\{4\}$. **37.** $\{1\}$ **39.** $\{-1\}$ **41.** $\{14\}$ **43.** $\{8\}$ **45.** $\{0\}$ **47.** $\emptyset$ **49.** $\{7\}$ **51.** $\{7\}$ **53.** $\{-20\}$ **55.** $\emptyset$ **57.** $\left\{\frac{5}{4}\right\}$ **59.** $\{9, 17\}$ **61.** $\left\{\frac{1}{4}, 1\right\}$ **63.** $K=\frac{V^2m}{2}$ **65.** $C=\frac{L}{Z^2}$ **67.** $L=\frac{1}{4\pi^2f^2C}$ **69.** $r=\frac{a}{4\pi^2N^2}$

Section 9.7 (pages 571–574)

1. i **3.** $-i$ **5.** Any real number a can be written as $a+0i$, a complex number with imaginary part 0. **7.** $13i$ **9.** $-12i$ **11.** $i\sqrt{5}$ **13.** $4i\sqrt{3}$ **15.** -15 **17.** $-\sqrt{57}$ **19.** -10 **21.** $i\sqrt{33}$ **23.** $\sqrt{3}$ **25.** $5i$ **27.** $-1+7i$ **29.** 0 **31.** $7+3i$ **33.** -2 **35.** $1-13i$ **37.** $6+6i$ **39.** $4+2i$ **41.** -81 **43.** -16 **45.** $-10-30i$ **47.** $10-5i$ **49.** $-9+40i$ **51.** 153 **53. (a)** $a-bi$ **(b)** a^2; b^2 **55.** $1+i$ **57.** $-1+2i$ **59.** $2-2i$ **61.** $-\frac{5}{13}-\frac{12}{13}i$ **63. (a)** $4x+1$ **(b)** $4+i$ **64. (a)** $-2x+3$ **(b)** $-2+3i$ **65. (a)** $3x^2+5x-2$ **(b)** $5+5i$ **66. (a)** $-\sqrt{3}+\sqrt{6}+1-\sqrt{2}$ **(b)** $\frac{1}{5}-\frac{7}{5}i$ **67.** Because $i^2=-1$, two pairs of like terms can be combined in Exercise 65(b). **68.** Because $i^2=-1$, additional terms can be combined in the numerator and denominator. **69.** $\frac{5}{41}+\frac{4}{41}i$ **71.** -1 **73.** i **75.** 1 **77.** $-i$ **79.** Since $i^{20}=(i^4)^5=1^5=1$, the student multiplied by 1, which is justified by the identity property for multiplication. **81.** $\frac{1}{2}+\frac{1}{2}i$ **83.** $(1+5i)^2-2(1+5i)+26$ will simplify to 0 when the operations are applied.

Chapter 9 Review Exercises (pages 579–582)

1. 42 **2.** -17 **3.** not a real number **4.** 6 **5.** -2 **6.** $|x|$ **7.** x **8.** $|x^5|$

9. domain: $[1, \infty)$; range: $[0, \infty)$

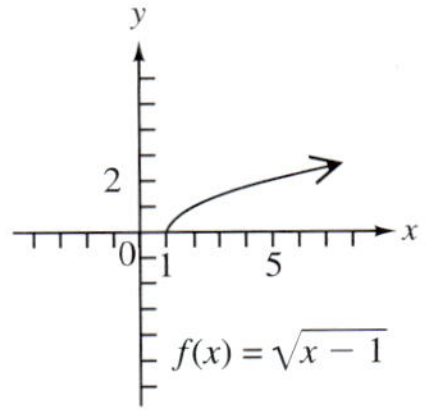

10. domain: $(-\infty, \infty)$; range: $(-\infty, \infty)$

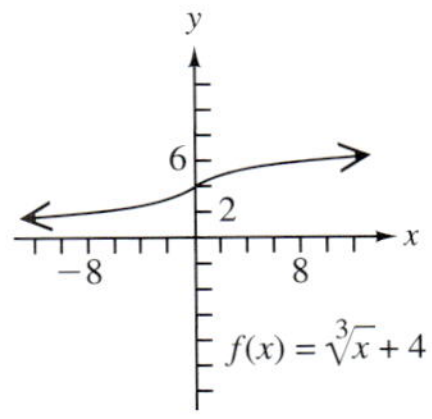

11. n must be even, and a must be negative. **12.** It is not a real number. **13.** 6.325 **14.** 8.775 **15.** 17.607 **16.** 1.9 sec **17.** 66 in.^{2} **18.** 7 **19.** -2 **20.** not a real number **21.** By a power rule for exponents and the definition of $x^{1/n}$, $a^{m/n}=(a^m)^{1/n}=\sqrt[n]{a^m}$. **22.** 32 **23.** -4 **24.** $-\frac{216}{125}$ **25.** -32 **26.** $\frac{1000}{27}$ **27.** 49 **28.** 96 **29.** $\frac{k^{17/12}}{2}$ **30.** $\sqrt[5]{2^4}$, or $\sqrt[5]{16}$ **31.** 3^9 **32.** $7^4\sqrt{7}$ **33.** $m^4\sqrt[3]{m}$ **34.** $k^2\sqrt[4]{k}$ **35.** $\sqrt[6]{m}$ **36.** $2y\sqrt[4]{y}$ **37.** $\sqrt[15]{y^8}$ **38.** $\sqrt[12]{y^5}$ **39.** $\sqrt{66}$ **40.** $\sqrt{5r}$ **41.** $\sqrt[3]{30}$ **42.** $\sqrt[4]{2}$ **43.** $2\sqrt{5}$ **44.** $-5\sqrt{5}$ **45.** $-3x\sqrt[3]{4xy}$ **46.** $4pq^2\sqrt[3]{p}$ **47.** $\frac{7}{9}$ **48.** $\frac{y\sqrt{y}}{12}$ **49.** $\frac{m^5}{3}$ **50.** $\frac{\sqrt[3]{r^2}}{2}$ **51.** $\sqrt[12]{2}$ **52.** $\sqrt[10]{x^3}$ **53.** $\sqrt{130}$ **54.** $\sqrt{53}$ **55.** $-11\sqrt{2}$ **56.** $23\sqrt{5}$ **57.** $7\sqrt{3y}$ **58.** $26m\sqrt{6m}$ **59.** $19\sqrt[3]{2}$ **60.** $-8\sqrt[4]{2}$ **61.** $1-\sqrt{3}$ **62.** 2 **63.** $9-7\sqrt{2}$ **64.** $86+8\sqrt{55}$ **65.** $15-2\sqrt{26}$ **66.** $12-2\sqrt{35}$ **67.** $-3\sqrt{6}$ **68.** $\frac{3\sqrt{7py}}{y}$ **69.** $-\frac{\sqrt[3]{45}}{5}$ **70.** $\frac{3m\sqrt[3]{4n}}{n^2}$ **71.** $\frac{\sqrt{2}-\sqrt{7}}{-5}$ **72.** $\frac{-5(\sqrt{6}+\sqrt{3})}{3}$ **73.** $\{2\}$ **74.** $\{6\}$ **75.** $\emptyset$ **76.** $\{0, 5\}$ **77.** $\{9\}$ **78.** $\{3\}$ **79.** $\{7\}$ **80.** $\left\{-\frac{1}{2}\right\}$ **81.** $\{6\}$ **82.** $5i$ **83.** $10i\sqrt{2}$ **84.** $4i\sqrt{10}$ **85.** $-10-2i$ **86.** $14+7i$ **87.** $-\sqrt{35}$ **88.** -45 **89.** 3 **90.** $5+i$ **91.** $32-24i$ **92.** $1-i$ **93.** $4+i$ **94.** $-i$ **95.** 1 **96.** $-i$ **97.** $-13ab^2$

ANSWERS

98. $\frac{1}{100}$ **99.** $\frac{1}{y^{1/2}}$ **100.** $\frac{x^{3/4}}{z^{3/4}}$ **101.** k^6 **102.** $3z^3t^2\sqrt[3]{2t^2}$

103. $57\sqrt{2}$ **104.** $6x\sqrt[3]{y^2}$ **105.** $\sqrt{35} + \sqrt{15} - \sqrt{21} - 3$

106. $-\frac{\sqrt{3}}{6}$ **107.** $\frac{\sqrt[3]{60}}{5}$ **108.** $\frac{2\sqrt{z}(\sqrt{z}+2)}{z-4}$ **109.** $7i$

110. $3 - 7i$ **111.** $-5i$ **112.** $\{5\}$ **113.** $\left\{\frac{3}{2}\right\}$ **114.** 7.9 ft

115. $(12\sqrt{3} + 5\sqrt{2})$ ft

Chapter 9 Test (pages 583–584)

1. -29 **2.** -8 **3.** 5 **4.** C **5.** 21.863 **6.** -9.405

7. domain: $[-6, \infty)$; range: $[0, \infty)$

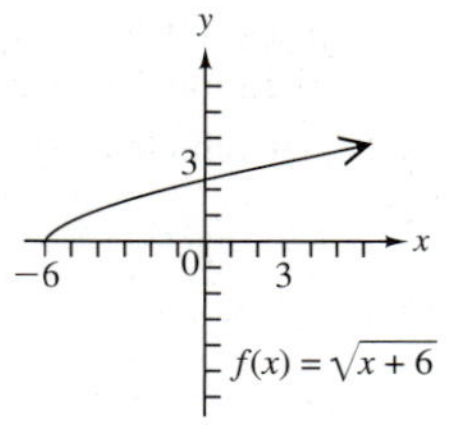

8. $\frac{1}{256}$ **9.** $\frac{9y^{3/10}}{x^2}$ **10.** $3x^2y^3\sqrt{6x}$ **11.** $2ab^3\sqrt[4]{2a^3b}$ **12.** $\sqrt[6]{200}$

13. $26\sqrt{5}$ **14.** $66 + \sqrt{5}$ **15.** $23 - 4\sqrt{15}$ **16.** $\frac{-\sqrt{10}}{4}$

17. $\frac{2\sqrt[3]{25}}{5}$ **18.** $-2(\sqrt{7} - \sqrt{5})$ **19.** $3 + \sqrt{6}$ **20.** $\sqrt{26}$

21. $\sqrt{145}$ **22.** $\{-1\}$ **23.** $\{6\}$ **24.** $\{-3\}$ **25.** $-5 - 8i$

26. $-10 + 10i$ **27.** $3 + 4i$ **28.** $-i$

Cumulative Review Exercises: Chapters 1–9 (pages 585–586)

1. $\left\{\frac{4}{5}\right\}$ **2.** $\{-12\}$ **3.** $\left\{\frac{11}{10}, \frac{7}{2}\right\}$ **4.** $(-6, \infty)$ **5.** $(1, 3)$

6. $(-2, 1)$ **7.** $12x + 11y = 18$ **8.** C **9.** **(a)** $(0, 6)$ **(b)** $(2, 0)$

10.

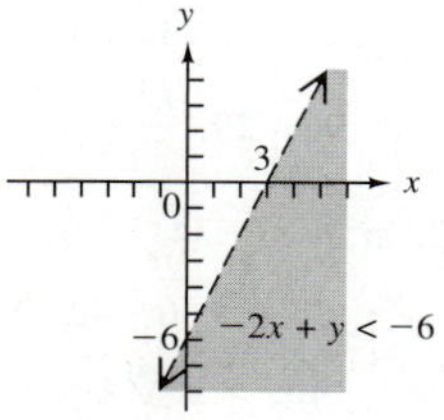

11. Both angles measure 80°. **12.** $\{(7, -2)\}$ **13.** $\emptyset$

14. $\{(x, y, z) \mid 2x + y - z = 5\}$ **15.** 2-oz letter: \$0.63; 3-oz letter: \$0.87

16. $-k^3 - 3k^2 - 8k - 9$ **17.** $8x^2 + 17x - 21$ **18.** $z - 2 + \frac{3}{z}$

19. $3y^3 - 3y^2 + 4y + 1 + \frac{-10}{2y+1}$ **20.** $(2p - 3q)(p - q)$

21. $(3k^2 + 4)(6k^2 - 5)$ **22.** $(x + 8)(x^2 - 8x + 64)$ **23.** $\frac{y}{y+5}$

24. $\frac{4x + 2y}{(x + y)(x - y)}$ **25.** $-\frac{9}{4}$ **26.** $-\frac{1}{a+b}$ **27.** $\left\{-3, -\frac{5}{2}\right\}$

28. $\left\{-\frac{2}{5}, 1\right\}$ **29.** $\frac{1}{243}$ **30.** $x^{1/12}$ **31.** $8\sqrt{5}$ **32.** $\frac{-9\sqrt{5}}{20}$

33. $4(\sqrt{6} + \sqrt{5})$ **34.** $6\sqrt[3]{4}$ **35.** $\sqrt{29}$ **36.** $\{6\}$ **37.** 15 mph

38. $\frac{80}{39}$, or $2\frac{2}{39}$ L **39.** 17 dimes and 12 quarters

40. Brenda: 8 mph; Chuck: 4 mph

Chapter 10 Quadratic Equations, Inequalities, and Functions

Section 10.1 (pages 597–600)

1. The equation is also true for $x = -4$. **3.** **(a)** A quadratic equation in standard form has a second-degree polynomial in decreasing powers equal to 0. **(b)** The zero-factor property states that if a product equals 0, then at least one of the factors equals 0. **(c)** The square root property states that if the square of a quantity equals a number, then the quantity equals the positive or negative square root of the number. **5.** $\{9, -9\}$

7. $\{\sqrt{17}, -\sqrt{17}\}$ **9.** $\{4\sqrt{2}, -4\sqrt{2}\}$ **11.** $\{2\sqrt{5}, -2\sqrt{5}\}$

13. $\{2\sqrt{6}, -2\sqrt{6}\}$ **15.** $\{-7, 3\}$ **17.** $\{4 + \sqrt{3}, 4 - \sqrt{3}\}$

19. $\{-5 + 4\sqrt{3}, -5 - 4\sqrt{3}\}$ **21.** $\left\{\frac{1+\sqrt{7}}{3}, \frac{1-\sqrt{7}}{3}\right\}$

23. $\left\{\frac{-1+2\sqrt{6}}{4}, \frac{-1-2\sqrt{6}}{4}\right\}$ **25.** 5.6 sec **27.** square root property for $(2x + 1)^2 = 5$; completing the square for $x^2 + 4x = 12$

29. **(a)** 9 **(b)** 49 **(c)** 36 **(d)** $\frac{9}{4}$ **(e)** $\frac{81}{4}$ **(f)** $\frac{1}{16}$ **31.** 4

33. 25 **35.** $\frac{1}{36}$ **37.** $\{-4, 6\}$ **39.** $\{-2 + \sqrt{6}, -2 - \sqrt{6}\}$

41. $\{-5 + \sqrt{7}, -5 - \sqrt{7}\}$ **43.** $\left\{-\frac{8}{3}, 3\right\}$

45. $\left\{\frac{-5+\sqrt{41}}{4}, \frac{-5-\sqrt{41}}{4}\right\}$ **47.** $\left\{\frac{5+\sqrt{15}}{5}, \frac{5-\sqrt{15}}{5}\right\}$

49. $\left\{\frac{4+\sqrt{3}}{3}, \frac{4-\sqrt{3}}{3}\right\}$ **51.** $\left\{\frac{2+\sqrt{3}}{3}, \frac{2-\sqrt{3}}{3}\right\}$

53. $\{1 + \sqrt{2}, 1 - \sqrt{2}\}$ **55.** $\{2i\sqrt{3}, -2i\sqrt{3}\}$

57. $\{5 + i\sqrt{3}, 5 - i\sqrt{3}\}$ **59.** $\left\{\frac{1}{6} + \frac{\sqrt{2}}{3}i, \frac{1}{6} - \frac{\sqrt{2}}{3}i\right\}$

61. $\{-2 + 3i, -2 - 3i\}$ **63.** $\left\{-\frac{2}{3} + \frac{2\sqrt{2}}{3}i, -\frac{2}{3} - \frac{2\sqrt{2}}{3}i\right\}$

65. $\{-3 + i\sqrt{3}, -3 - i\sqrt{3}\}$ **67.** x^2 **68.** x **69.** $6x$ **70.**

71. 9 **72.** $(x + 3)^2$, or $x^2 + 6x + 9$

Section 10.2 (pages 607–608)

1. The fraction bar should extend under the term $-b$. **3.** $\{3, 5\}$

5. $\left\{\frac{-2+\sqrt{2}}{2}, \frac{-2-\sqrt{2}}{2}\right\}$ **7.** $\left\{\frac{1+\sqrt{3}}{2}, \frac{1-\sqrt{3}}{2}\right\}$

9. $\{5 + \sqrt{7}, 5 - \sqrt{7}\}$ **11.** $\left\{\frac{-1+\sqrt{2}}{2}, \frac{-1-\sqrt{2}}{2}\right\}$

13. $\left\{\frac{-1+\sqrt{7}}{3}, \frac{-1-\sqrt{7}}{3}\right\}$ **15.** $\{1 + \sqrt{5}, 1 - \sqrt{5}\}$

17. $\left\{\frac{-2+\sqrt{10}}{2}, \frac{-2-\sqrt{10}}{2}\right\}$ **19.** $\{-1 + 3\sqrt{2}, -1 - 3\sqrt{2}\}$

21. $\left\{\frac{3}{2} + \frac{\sqrt{59}}{2}i, \frac{3}{2} - \frac{\sqrt{59}}{2}i\right\}$ **23.** $\{3 + i\sqrt{5}, 3 - i\sqrt{5}\}$

25. $\left\{\frac{1}{2} + \frac{\sqrt{6}}{2}i, \frac{1}{2} - \frac{\sqrt{6}}{2}i\right\}$ **27.** $\left\{-\frac{2}{3} + \frac{\sqrt{2}}{3}i, -\frac{2}{3} - \frac{\sqrt{2}}{3}i\right\}$

29. B **31.** C **33.** A **35.** D **37.** The equations in Exercises 29, 30, 33, and 34 can be solved by factoring.

Section 10.3 (pages 617–620)

1. Multiply by the LCD, x. **3.** Substitute a variable for $r^2 + r$.
5. The proposed solution -1 does not check. The solution set is $\{4\}$.
7. $\{-4, 7\}$ **9.** $\left\{-\frac{2}{3}, 1\right\}$ **11.** $\left\{-\frac{14}{17}, 5\right\}$ **13.** $\left\{-\frac{11}{7}, 0\right\}$
15. $\left\{\frac{-1+\sqrt{13}}{2}, \frac{-1-\sqrt{13}}{2}\right\}$ **17.** $\left\{\frac{2+\sqrt{22}}{3}, \frac{2-\sqrt{22}}{3}\right\}$
19. **(a)** $(20 - t)$ mph **(b)** $(20 + t)$ mph **21.** 25 mph
23. 80 km per hr **25.** 3.6 hr **27.** 9 min **29.** $\{1, 4\}$
31. $\{3\}$ **33.** $\left\{\frac{8}{9}\right\}$ **35.** $\{16\}$ **37.** $\left\{\frac{2}{5}\right\}$ **39.** $\{-3, 3\}$
41. $\left\{-\frac{3}{2}, -1, 1, \frac{3}{2}\right\}$ **43.** $\{-2\sqrt{3}, -2, 2, 2\sqrt{3}\}$
45. $\left\{\frac{\sqrt{9+\sqrt{65}}}{2}, -\frac{\sqrt{9+\sqrt{65}}}{2}, \frac{\sqrt{9-\sqrt{65}}}{2}, -\frac{\sqrt{9-\sqrt{65}}}{2}\right\}$
47. $\{-6, -5\}$ **49.** $\{-4, 1\}$ **51.** $\left\{-\frac{1}{3}, \frac{1}{6}\right\}$ **53.** $\{-8, 1\}$
55. $\{-64, 27\}$ **57.** $\{25\}$ **59.** It would cause both denominators to equal 0, and division by 0 is undefined. **60.** $\frac{12}{5}$
61. $\left(\frac{x}{x-3}\right)^2 + 3\left(\frac{x}{x-3}\right) - 4 = 0$ **62.** The numerator can never equal the denominator, since the denominator is 3 less than the numerator. **63.** $\left\{\frac{12}{5}\right\}$; The values for t are -4 and 1. The value 1 is impossible because it leads to a contradiction $\left(\text{since } \frac{x}{x-3} \text{ is never equal to } 1\right)$. **64.** $\left\{\frac{12}{5}\right\}$; The values for s are $\frac{1}{x}$ and $\frac{-4}{x}$. The value $\frac{1}{x}$ is impossible, since $\frac{1}{x} \neq \frac{1}{x-3}$ for all x.

Summary Exercises on Solving Quadratic Equations (pages 621–622)

1. square root property **2.** factoring **3.** quadratic formula
4. quadratic formula **5.** factoring **6.** square root property
7. $\{\sqrt{47}, -\sqrt{47}\}$ **8.** $\left\{-\frac{3}{2}, \frac{5}{3}\right\}$ **9.** $\{-4+\sqrt{10}, -4-\sqrt{10}\}$
10. $\{-3, 11\}$ **11.** $\left\{-\frac{1}{2}, 5\right\}$ **12.** $\left\{-3, \frac{1}{3}\right\}$
13. $\left\{\frac{9+\sqrt{33}}{6}, \frac{9-\sqrt{33}}{6}\right\}$ **14.** $\{2i\sqrt{3}, -2i\sqrt{3}\}$ **15.** $\left\{\frac{1}{2}, 2\right\}$
16. $\left\{-\frac{\sqrt{6}}{3}, -\frac{1}{2}, \frac{1}{2}, \frac{\sqrt{6}}{3}\right\}$ **17.** $\left\{\frac{-5+2\sqrt{3}}{2}, \frac{-5-2\sqrt{3}}{2}\right\}$
18. $\left\{\frac{4}{5}, 3\right\}$ **19.** $\{-\sqrt{7}, -\sqrt{2}, \sqrt{2}, \sqrt{7}\}$
20. $\left\{\frac{-2+\sqrt{14}}{2}, \frac{-2-\sqrt{14}}{2}\right\}$ **21.** $\left\{-\frac{1}{2}+\frac{\sqrt{7}}{2}i, -\frac{1}{2}-\frac{\sqrt{7}}{2}i\right\}$
22. $\left\{\sqrt{4+\sqrt{15}}, -\sqrt{4+\sqrt{15}}, \sqrt{4-\sqrt{15}}, -\sqrt{4-\sqrt{15}}\right\}$
23. $\left\{\frac{3}{2}\right\}$ **24.** $\left\{\frac{2}{3}\right\}$ **25.** $\{6\sqrt{2}, -6\sqrt{2}\}$ **26.** $\left\{-\frac{2}{3}, 2\right\}$
27. $\{-4, 9\}$ **28.** $\{13, -13\}$ **29.** $\left\{1+\frac{\sqrt{3}}{3}i, 1-\frac{\sqrt{3}}{3}i\right\}$
30. $\{3\}$ **31.** $\left\{-\frac{1}{3}, \frac{1}{6}\right\}$ **32.** $\left\{\frac{1}{6}+\frac{\sqrt{47}}{6}i, \frac{1}{6}-\frac{\sqrt{47}}{6}i\right\}$
33. $\{3\}$ **34.** $\left\{-\frac{8}{3}, -1\right\}$ **35.** $\left\{-i, i, -\frac{1}{2}i, \frac{1}{2}i\right\}$ **36.** $\{-2, 7\}$

Section 10.4 (pages 627–630)

1. Solve for w^2 by dividing each side by g. **3.** $m = \sqrt{p^2 - n^2}$
5. $t = \frac{\pm\sqrt{dk}}{k}$ **7.** $d = \frac{\pm\sqrt{skI}}{I}$ **9.** $v = \frac{\pm\sqrt{kAF}}{F}$
11. $r = \frac{\pm\sqrt{3\pi Vh}}{\pi h}$ **13.** $t = \frac{-B \pm \sqrt{B^2 - 4AC}}{2A}$ **15.** $h = \frac{D^2}{k}$
17. $\ell = \frac{p^2 g}{k}$ **19.** 2.3, 5.3, 5.8 **21.** eastbound ship: 80 mi; southbound ship: 150 mi **23.** 1 ft **25.** 20 in. by 12 in.
27. 2.4 sec and 5.6 sec **29.** 9.2 sec **31.** It reaches its *maximum* height at 5 sec because this is the only time it reaches 400 ft. **33.** \$0.80
35. **(a)** 2750 billion **(b)** 2750 billion; They are the same.
37. 2001; The graph indicates that vehicle-miles reached 2800 in 2001.
39. 5.5 m per sec **41.** 5 or 14

Section 10.5 (pages 637–642)

1. **(a)** B **(b)** C **(c)** A **(d)** D **3.** $(0, 0)$ **5.** $(0, 4)$ **7.** $(1, 0)$
9. $(-3, -4)$ **11.** In Exercise 9, the parabola is shifted 3 units to the left and 4 units down. The parabola in Exercise 10 is shifted 5 units to the right and 8 units down. **13.** down; wider **15.** up; narrower
17. **(a)** I **(b)** IV **(c)** II **(d)** III

19.

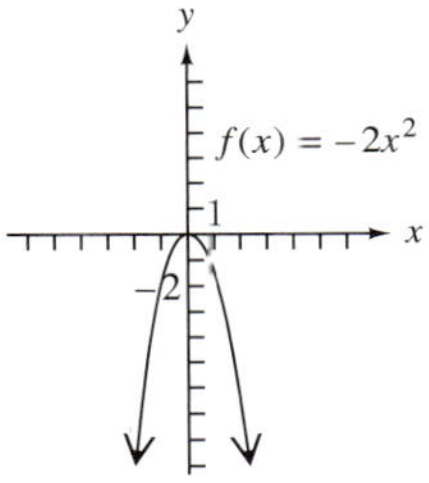

21.

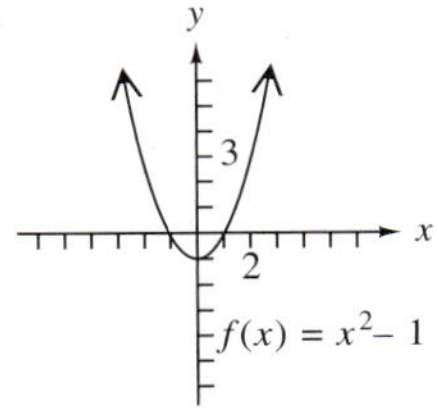

23.

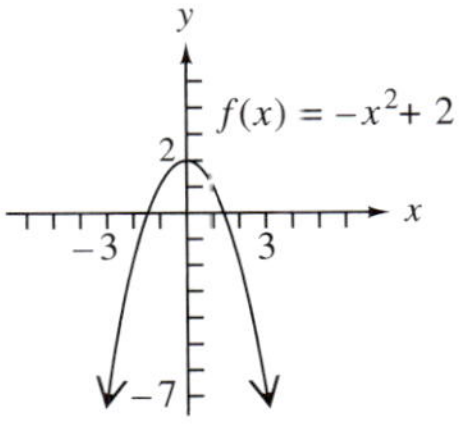

25. vertex: $(4, 0)$; axis: $x = 4$; domain: $(-\infty, \infty)$; range: $[0, \infty)$

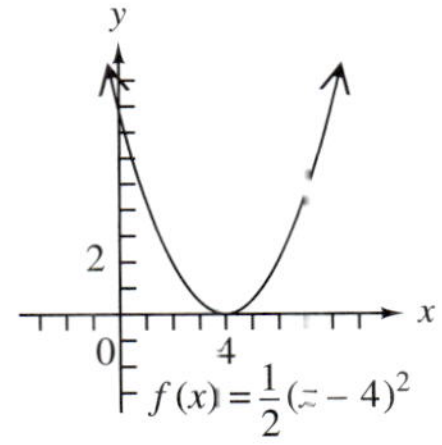

27. vertex: $(-2, -1)$; axis: $x = -2$; domain: $(-\infty, \infty)$; range: $[-1, \infty)$

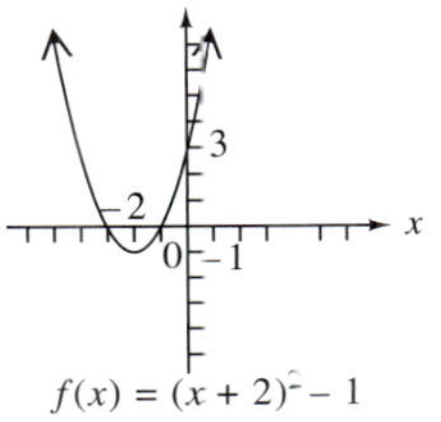

29. vertex: $(-3, 4)$; axis: $x = -3$; domain: $(-\infty, \infty)$; range: $(-\infty, 4]$

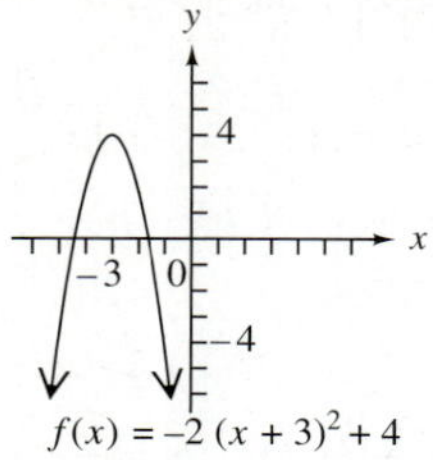

31. vertex: $(-2, 1)$; axis: $x = -2$; domain: $(-\infty, \infty)$; range: $(-\infty, 1]$

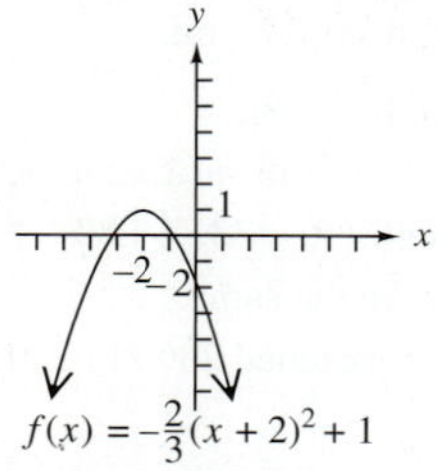

33. It is shifted 6 units up.

34.

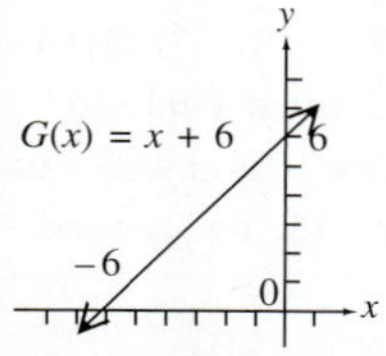

35. It is shifted 6 units up. **36.** It is shifted 6 units to the right.

37.

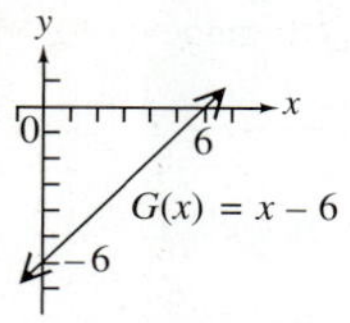

38. It is shifted 6 units to the right. **39.** quadratic; positive
41. quadratic; negative **43.** linear; positive

45. (a) **DIGITAL CAMERA SALES IN U.S.**

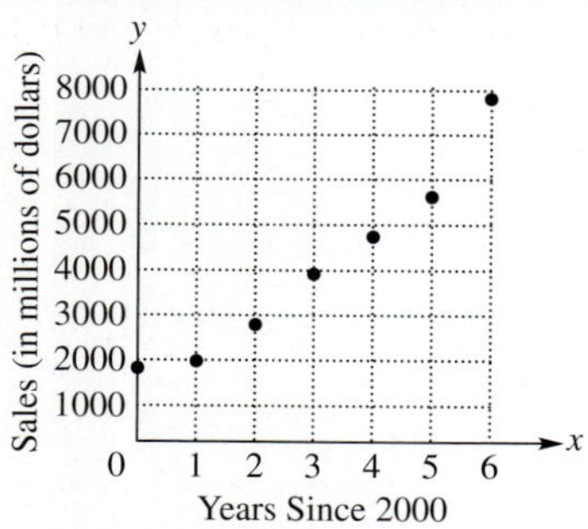

(b) quadratic; positive **(c)** $f(x) = 99.3x^2 + 400.7x + 1825$
(d) \$9496 million **(e)** No. The number of digital cameras sold in 2007 is far below the number approximated by the model. Rather than continuing to increase, sales of digital cameras fell in 2007.

47. (a) 2005: 14%; 2007: 2.4%
(b) The approximations using the model are far too low.

Section 10.6 (pages 651–654)

1. If x is squared, it has a vertical axis; if y is squared, it has a horizontal axis. **3.** Use the discriminant of the corresponding quadratic equation. If it is positive, there are two x-intercepts. If it is 0, there is just one x-intercept (the vertex), and if it is negative, there are no x-intercepts. **5.** $(-1, 3)$; up; narrower; no x-intercepts

7. $\left(\frac{5}{2}, \frac{37}{4}\right)$; down; same; two x-intercepts

9. $(-3, -9)$; to the right; wider

11. vertex: $(-2, -1)$; axis: $x = -2$; domain: $(-\infty, \infty)$; range: $[-1, \infty)$

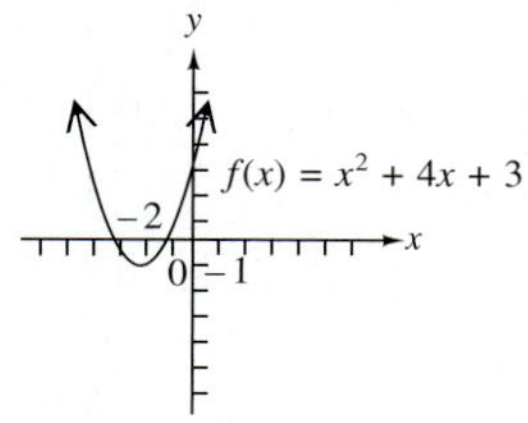

13. vertex: $(1, -3)$; axis: $x = 1$; domain: $(-\infty, \infty)$; range: $(-\infty, -3]$

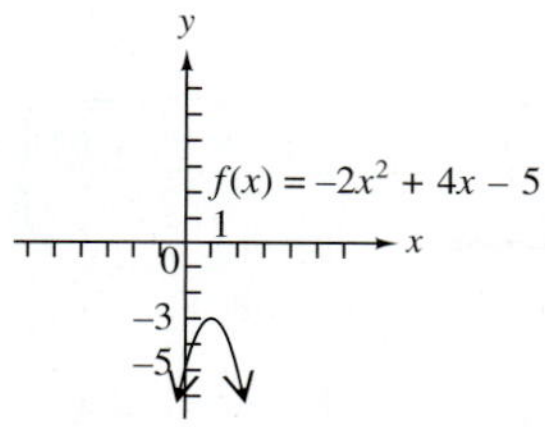

15. vertex: $(1, 5)$; axis: $y = 5$; domain: $(-\infty, 1]$; range: $(-\infty, \infty)$

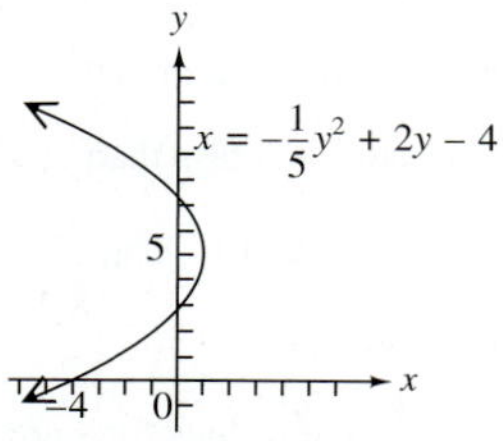

17. vertex: $(-7, -2)$; axis: $y = -2$; domain: $[-7, \infty)$; range: $(-\infty, \infty)$

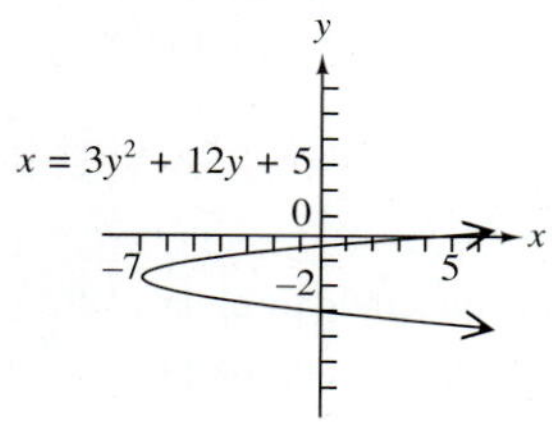

19. F **21.** C **23.** D **25.** 30 and 30 **27.** 160 ft by 320 ft
29. 16 ft; 2 sec **31. (a)** $R(x) = 20{,}000 + 200x - 4x^2$ **(b)** 25
(c) \$22,500 **33. (a)** maximum **(b)** 1993; 13.2%
35. (a) The coefficient of x^2 is negative because the parabola opens down. **(b)** (18.45, 3860) **(c)** In 2018 Social Security assets will reach their maximum value of \$3860 billion.

Section 10.7 (pages 661–664)

1. (a) $\{1, 3\}$ **(b)** $(-\infty, 1) \cup (3, \infty)$ **(c)** $(1, 3)$

3. (a) $\left\{-3, \frac{5}{2}\right\}$ **(b)** $\left[-3, \frac{5}{2}\right]$ **(c)** $(-\infty, -3] \cup \left[\frac{5}{2}, \infty\right)$

5. Include the endpoints if the symbol is $\geq$ or $\leq$. Exclude the endpoints if the symbol is $>$ or $<$.

7. $(-\infty, -1) \cup (5, \infty)$

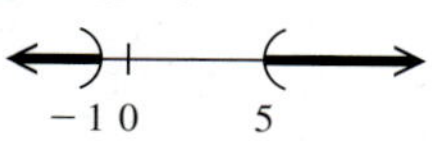

9. $(-4, 6)$

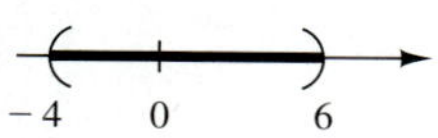

11. $(-\infty, 1] \cup [3, \infty)$

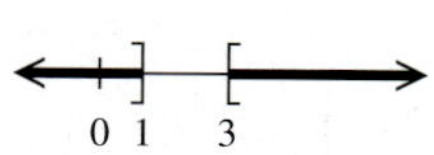

13. $\left(-\infty, -\frac{3}{2}\right] \cup \left[\frac{3}{5}, \infty\right)$ 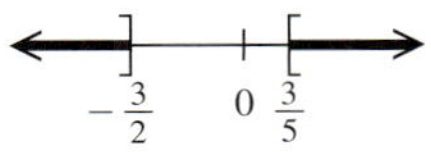

15. $\left(-\frac{2}{3}, \frac{1}{3}\right)$ 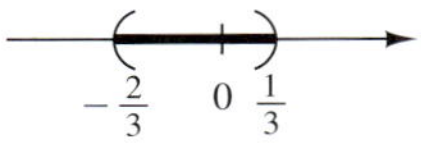

17. $\left(-\infty, -\frac{1}{2}\right] \cup \left[\frac{1}{3}, \infty\right)$ 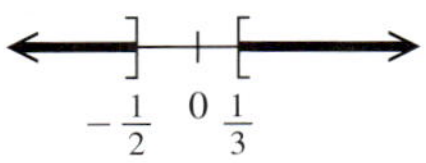

19. $(-\infty, 3 - \sqrt{3}] \cup [3 + \sqrt{3}, \infty)$ 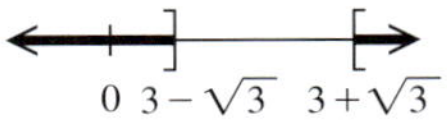

21. $(-\infty, \infty)$ **23.** $\emptyset$

25. $(-\infty, 1) \cup (2, 4)$ 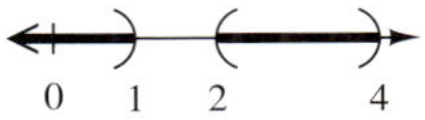

27. $\left[-\frac{3}{2}, \frac{1}{3}\right] \cup [4, \infty)$ 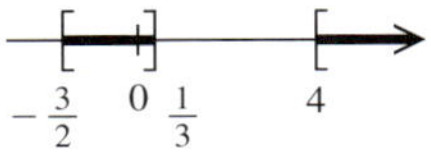

29. $(-\infty, 1) \cup (4, \infty)$ 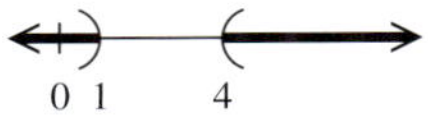

31. $\left[-\frac{3}{2}, 5\right)$ 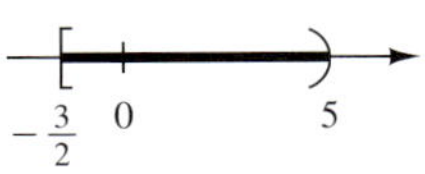

33. $(2, 6]$ 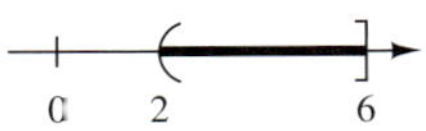

35. $\left(-\infty, \frac{1}{2}\right) \cup \left(\frac{5}{4}, \infty\right)$ 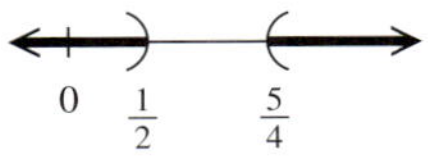

37. $[-4, -2)$ 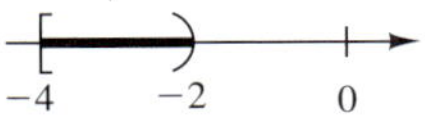

39. $\left(0, \frac{1}{2}\right) \cup \left(\frac{5}{2}, \infty\right)$ 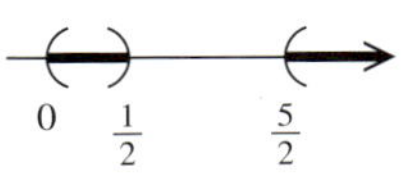

41. $(-\infty, 2] \cup (4, \infty)$ 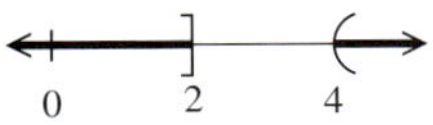

43. 3 sec and 13 sec **44.** between 3 sec and 13 sec **45.** at 0 sec (the time when it is initially projected) and at 16 sec (the time when it hits the ground) **46.** between 0 and 3 sec and between 13 and 16 sec

Chapter 10 Review Exercises (pages 669–672)

1. $\{11, -11\}$ **2.** $\{\sqrt{3}, -\sqrt{3}\}$ **3.** $\left\{-\frac{15}{2}, \frac{5}{2}\right\}$

4. $\left\{\frac{2}{3} + \frac{5}{3}i, \frac{2}{3} - \frac{5}{3}i\right\}$ **5.** $\{-2 + \sqrt{19}, -2 - \sqrt{19}\}$

6. $\left\{\frac{1}{2}, 1\right\}$ **7.** By the square root property, $x = \sqrt{12}$ or $x = -\sqrt{12}$. The solution set is $\{-2\sqrt{3}, 2\sqrt{3}\}$. **8.** 5.8 sec **9.** $\left\{-\frac{7}{2}, 3\right\}$

10. $\left\{\frac{-5 + \sqrt{53}}{2}, \frac{-5 - \sqrt{53}}{2}\right\}$ **11.** $\left\{\frac{1 + \sqrt{41}}{2}, \frac{1 - \sqrt{41}}{2}\right\}$

12. $\left\{-\frac{3}{4} + \frac{\sqrt{23}}{4}i, -\frac{3}{4} - \frac{\sqrt{23}}{4}i\right\}$ **13.** $\left\{\frac{2}{3} + \frac{\sqrt{2}}{3}i, \frac{2}{3} - \frac{\sqrt{2}}{3}i\right\}$

14. $\left\{\frac{-7 + \sqrt{37}}{2}, \frac{-7 - \sqrt{37}}{2}\right\}$ **15.** C **16.** A **17.** D **18.** B

19. $\left\{-\frac{5}{2}, 3\right\}$ **20.** $\left\{-\frac{1}{2}, 1\right\}$ **21.** $\{-4\}$ **22.** $\left\{-\frac{11}{6}, -\frac{19}{12}\right\}$

23. $\left\{-\frac{343}{8}, 64\right\}$ **24.** $\{-2, -1, 1, 2\}$ **25.** 40 mph **26.** 4.6 hr

27. $v = \frac{\pm\sqrt{rFkw}}{kw}$ **28.** $t = \frac{3m \pm \sqrt{9m^2 + 24m}}{2m}$

29. 9 ft, 12 ft, 15 ft **30.** 12 cm by 20 cm **31.** 1 in. **32.** 5.2 sec

33. $(1, 0)$ **34.** $(3, 7)$ **35.** $\left(\frac{2}{3}, -\frac{2}{3}\right)$ **36.** $(-4, 3)$

37. vertex: $(2, -3)$; axis: $x = 2$; domain: $(-\infty, \infty)$; range: $[-3, \infty)$

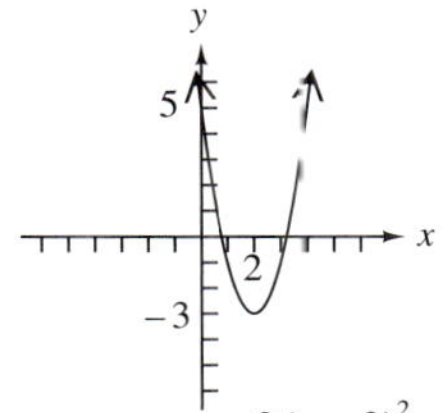

38. vertex: $(2, 3)$; axis: $x = 2$; domain: $(-\infty, \infty)$; range: $(-\infty, 3]$

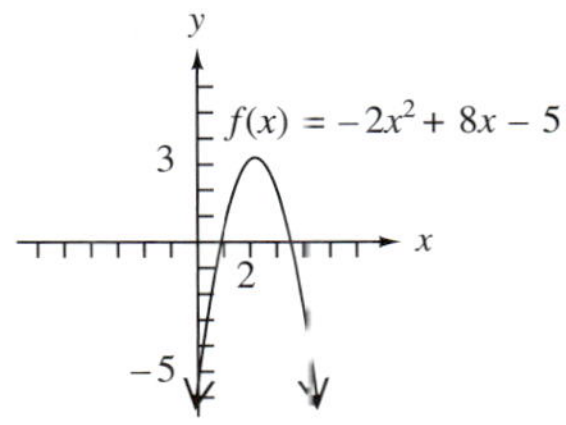

39. vertex: $(-4, -3)$; axis: $y = -3$; domain: $[-4, \infty)$; range: $(-\infty, \infty)$

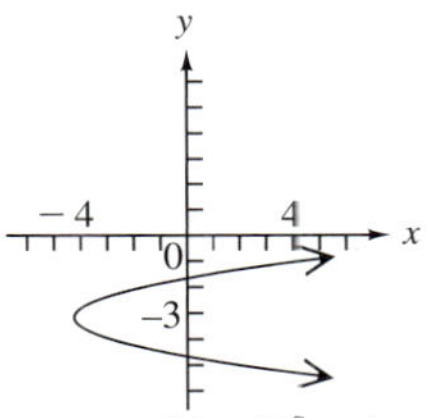

40. vertex: $(4, 6)$; axis: $y = 6$; domain: $(-\infty, 4]$; range: $(-\infty, \infty)$

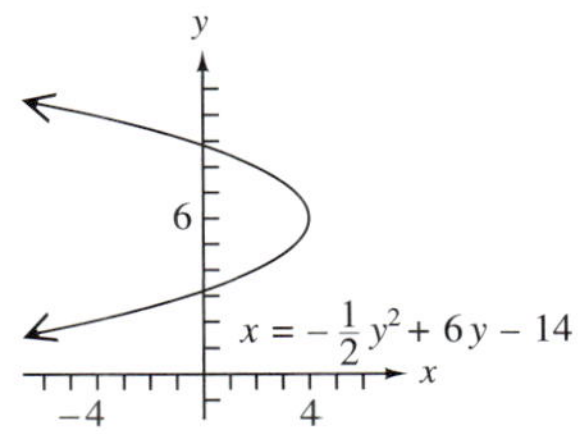

41. (a) $c = 2.9$; $100a + 10b + c = 24.3$; $400a + 20b + c = 56.5$ **(b)** $f(x) = 0.054x^2 + 1.6x + 2.9$ **(c)** \$60.3 billion; The result using the model is close but slightly low. **42.** 5 sec; 400 ft

43. length: 50 m; width: 50 m

44. $\left(-\infty, -\frac{3}{2}\right) \cup (4, \infty)$ 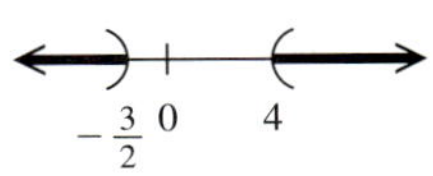

45. $[-4, 3]$ 

46. $(-\infty, -5] \cup [-2, 3]$ 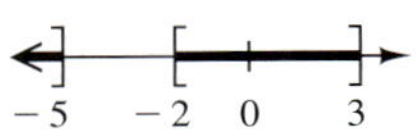

47. $\emptyset$ **48.** $\left(-\infty, \frac{1}{2}\right) \cup (2, \infty)$ 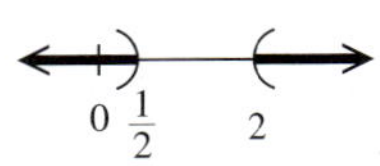

49. $[-3, 2)$ 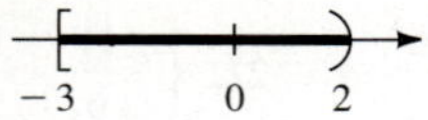

50. $R = \dfrac{\pm\sqrt{Vh - r^2h}}{h}$ **51.** $\left\{1 + \dfrac{\sqrt{3}}{3}i, 1 - \dfrac{\sqrt{3}}{3}i\right\}$

52. $\{-2, -1, 3, 4\}$ **53.** $(-\infty, -6) \cup \left(-\dfrac{3}{2}, 1\right)$

54. $\left\{\dfrac{-11 + \sqrt{7}}{3}, \dfrac{-11 - \sqrt{7}}{3}\right\}$ **55.** $d = \dfrac{\pm\sqrt{SkI}}{I}$

56. $\{4\}$ **57.** $\left\{-\dfrac{5}{3}, -\dfrac{3}{2}\right\}$ **58.** $\left(-5, -\dfrac{23}{5}\right]$

Chapter 10 Test (pages 673–676)

1. $\{3\sqrt{6}, -3\sqrt{6}\}$ **2.** $\left\{-\dfrac{8}{7}, \dfrac{2}{7}\right\}$ **3.** $\{-1 + \sqrt{2}, -1 - \sqrt{2}\}$

4. $\left\{\dfrac{3 + \sqrt{17}}{4}, \dfrac{3 - \sqrt{17}}{4}\right\}$ **5.** $\left\{\dfrac{2}{3} + \dfrac{\sqrt{11}}{3}i, \dfrac{2}{3} - \dfrac{\sqrt{11}}{3}i\right\}$

6. $\left\{\dfrac{2}{3}\right\}$ **7.** A **8.** discriminant: 88; two irrational solutions

9. $\left\{-\dfrac{2}{3}, 6\right\}$ **10.** $\left\{\dfrac{-7 + \sqrt{97}}{8}, \dfrac{-7 - \sqrt{97}}{8}\right\}$ **11.** $\left\{-2, -\dfrac{1}{3}, \dfrac{1}{3}, 2\right\}$

12. $\left\{-\dfrac{5}{2}, 1\right\}$ **13.** $r = \dfrac{\pm\sqrt{\pi S}}{2\pi}$ **14.** Andrew: 11.1 hr; Kent: 9.1 hr

15. 7 mph **16.** 2 ft **17.** 16 m **18.** **(a)** 10 hr **(b)** 180 students **19.** A

20. vertex: $(0, -2)$; axis: $x = 0$; domain: $(-\infty, \infty)$; range: $[-2, \infty)$

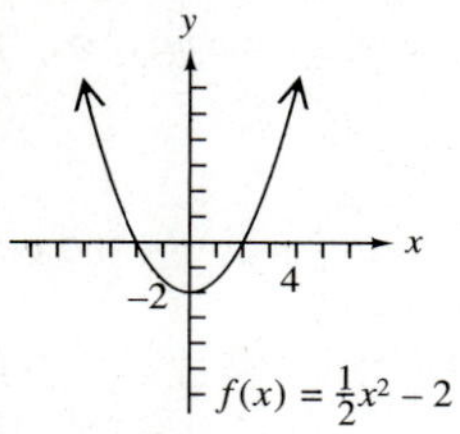

21. vertex: $(2, 3)$; axis: $x = 2$; domain: $(-\infty, \infty)$; range: $(-\infty, 3]$

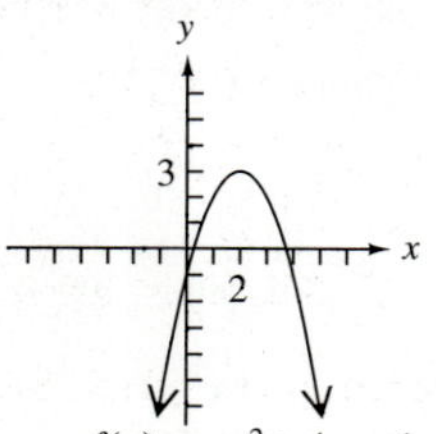

22. vertex: $(-5, -2)$; axis: $y = -2$; domain: $[-5, \infty)$; range: $(-\infty, \infty)$

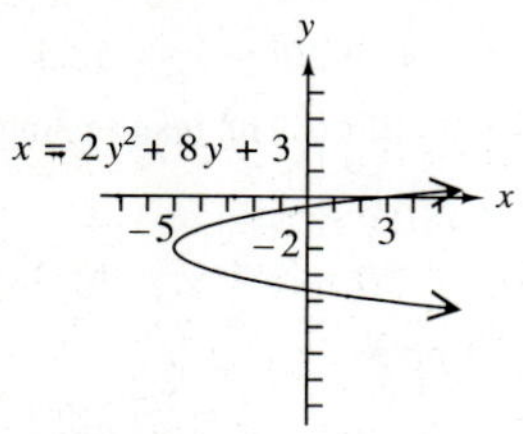

23. 140 ft by 70 ft; 9800 ft²

24. $(-\infty, -5) \cup \left(\dfrac{3}{2}, \infty\right)$ 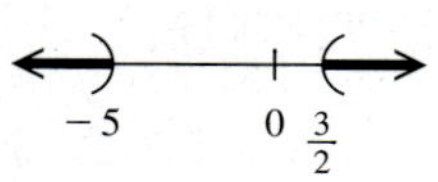

25. $(-\infty, 4) \cup [9, \infty)$ 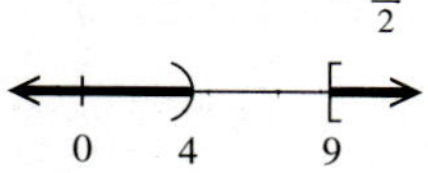

Cumulative Review Exercises: Chapters 1–10 (pages 677–678)

1. $\{1\}$ **2.** $[1, \infty)$ **3.** $\left[2, \dfrac{8}{3}\right]$

4. slope: $\dfrac{1}{2}$; y-intercept: $\left(0, -\dfrac{7}{4}\right)$ **5.** $x + 3y = -1$

6. function; domain: $(-\infty, \infty)$; range: $(-\infty, \infty)$

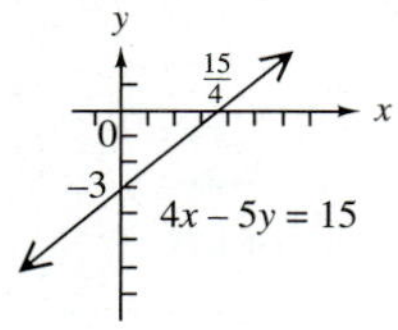

7. not a function

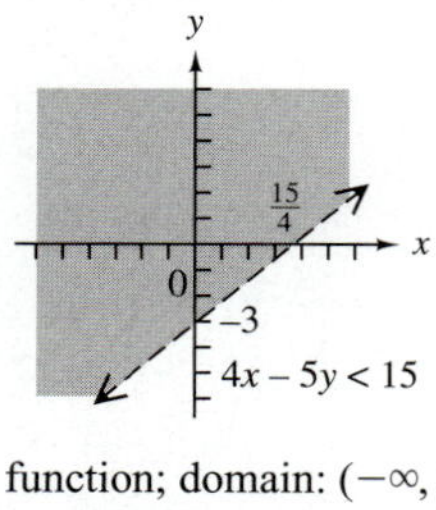

8. function; domain: $(-\infty, \infty)$; range: $(-\infty, 3]$

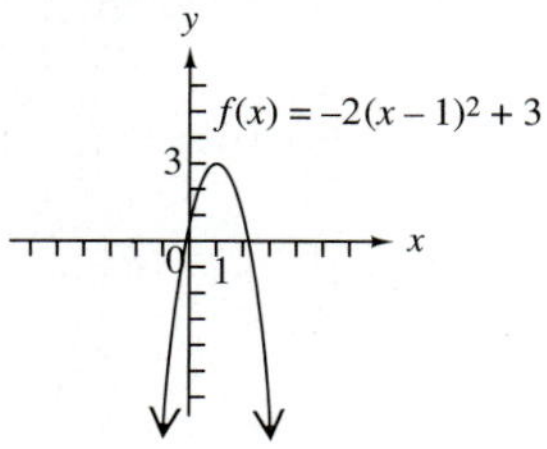

9. $\{(1, -2)\}$ **10.** $\{(3, -4, 2)\}$ **11.** $\dfrac{x^8}{y^4}$ **12.** $\dfrac{4}{xy^2}$

13. $\dfrac{4}{9}t^2 + 12t + 81$ **14.** $4x^2 - 6x + 11 + \dfrac{4}{x + 2}$

15. $x(4 + x)(4 - x)$ **16.** $(4m - 3)(6m + 5)$ **17.** $(3x - 5y)^2$

18. $-\dfrac{5}{18}$ **19.** $-\dfrac{8}{k}$ **20.** $\dfrac{r - s}{r}$ **21.** $\dfrac{3\sqrt[3]{4}}{4}$ **22.** $\sqrt{7} + \sqrt{5}$

23. $\left\{\dfrac{2}{3}\right\}$ **24.** $\left\{\dfrac{2 + \sqrt{10}}{2}, \dfrac{2 - \sqrt{10}}{2}\right\}$ **25.** $\{-3, 5\}$ **26.** $\emptyset$

27. $\{-3, -1, 1, 3\}$ **28.** southbound car: 57 mi; eastbound car: 76 mi

Chapter 11 Inverse, Exponential, and Logarithmic Functions

Section 11.1 (pages 685–688)

1. It is not one-to-one. France and the United States are paired with the same trans fat percentage, 11. **3.** Two or more siblings might be in the class. They would be paired with the same mother. **5.** B **7.** A

9. $\{(6, 3), (10, 2), (12, 5)\}$ **11.** not one-to-one **13.** $f^{-1}(x) = \dfrac{x - 4}{2}$, or $f^{-1}(x) = \dfrac{1}{2}x - 2$ **15.** $g^{-1}(x) = x^2 + 3, x \geq 0$

17. not one-to-one **19.** $f^{-1}(x) = \sqrt[3]{x + 4}$ **21.** **(a)** 8 **(b)** 3

23. **(a)** 1 **(b)** 0

25. **(a)** one-to-one **27.** **(a)** not one-to-one

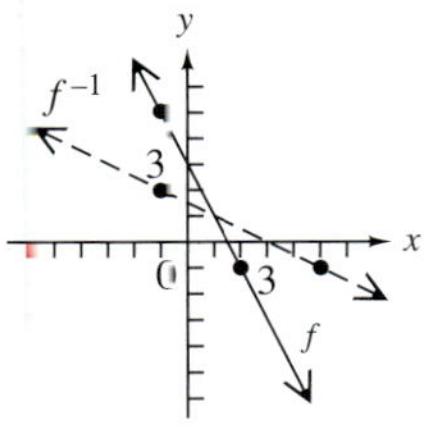

29. **(a)** one-to-one **31.**

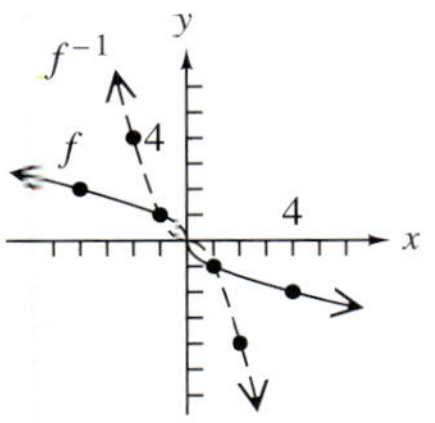

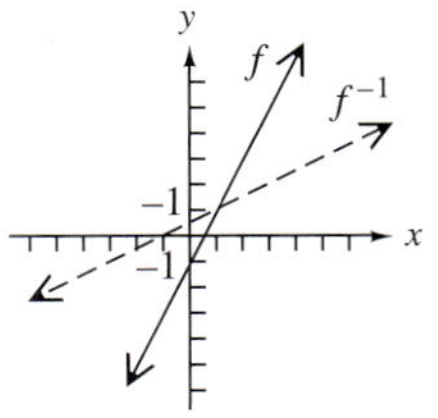

33. **35.** 0, 1, 2

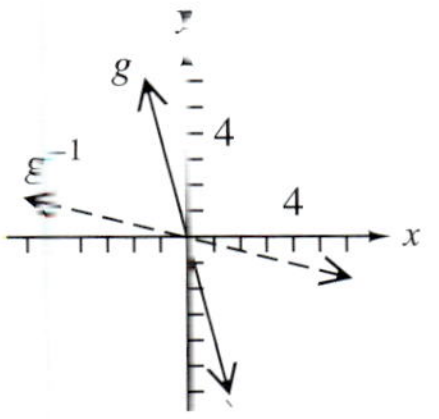

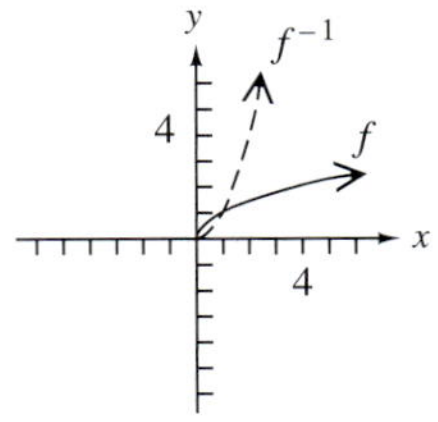

37. −3, −2, −1, 6

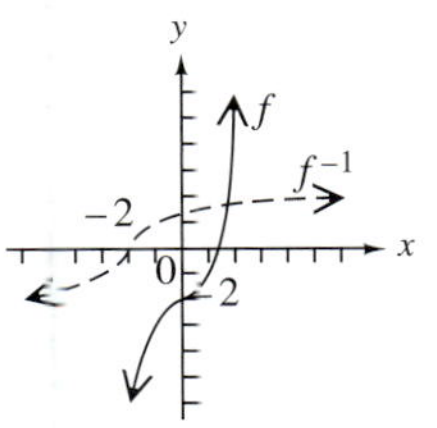

Section 11.2 (pages 695–696)

1. [illegible] **3.** C

5. **7.**

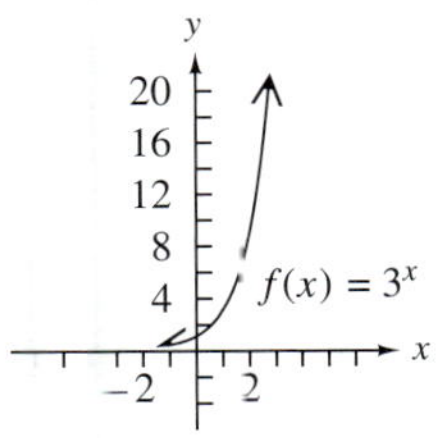

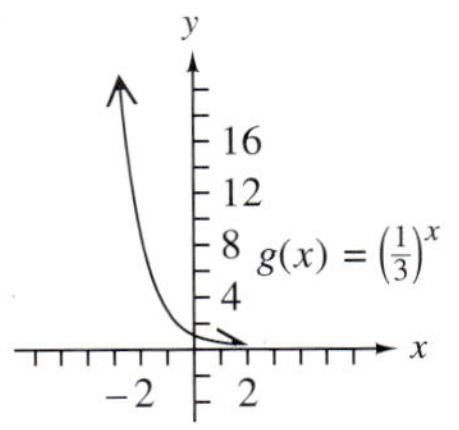

9.

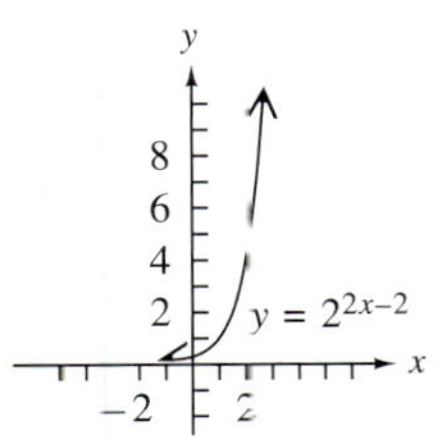

11. $\{2\}$ **13.** $\left\{\frac{3}{2}\right\}$ **15.** $\{7\}$ **17.** $\{-3\}$ **19.** $\{-1\}$ **21.** $\{-3\}$

23. **(a)** rises; falls **(b)** It is one-to-one and thus has an inverse.
25. **(a)** 1.0°C **(b)** 0.4°C **27.** **(a)** 3.0°C **(b)** 0.7°C

Section 11.3 (pages 701–704)

1. **(a)** C **(b)** F **(c)** B **(d)** A **(e)** E **(f)** D **3.** $\log_4 1024 = 5$
5. $\log_{1/2} 8 = -3$ **7.** $\log_{10} 0.001 = -3$ **9.** $\log_{625} 5 = \frac{1}{4}$
11. $4^3 = 64$ **13.** $10^{-4} = \frac{1}{10{,}000}$ **15.** $6^0 = 1$ **17.** $9^{1/2} = 3$
19. Since the radical $\sqrt{9} = 9^{1/2} = 3$, the exponent to which 9 must be raised is 1/2. **21.** $\left\{\frac{1}{3}\right\}$ **23.** $\{81\}$ **25.** $\left\{\frac{1}{5}\right\}$ **27.** $\{1\}$
29. $\{x \mid x > 0, x \neq 1\}$ **31.** $\{5\}$ **33.** $\left\{\frac{5}{3}\right\}$ **35.** $\{4\}$ **37.** $\left\{\frac{3}{2}\right\}$

39. **41.**

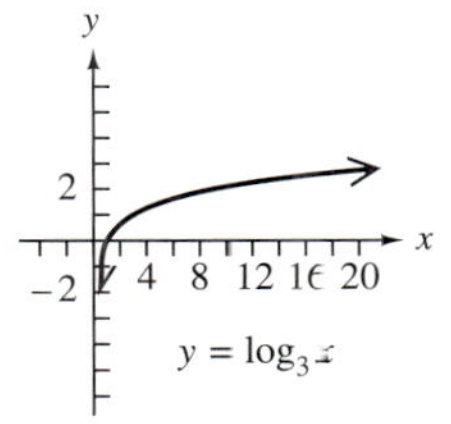

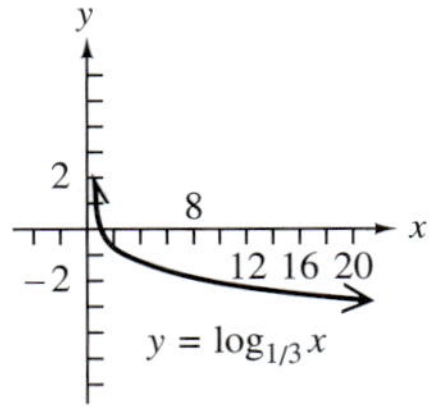

43. Answers will vary. **45.** 8 **47.** 24 **49.** Since every real number power of 1 equals 1, if $y = \log_1 x$, then $x = 1^y$ and so $x = 1$ for every y. This contradicts the definition of a function. **51.** $f(x) = 3^x$
53. **(a)** 4385 billion ft^3 **(b)** 5555 billion ft^3 **(c)** 6140 billion ft^3
55. about 4 times as powerful

Section 11.4 (pages 711–712)

1. false; $\log_b x - \log_b y = \log_b xy$ **3.** true **5.** $\log_7 4 - \log_7 5$
7. $\frac{1}{4}\log_2 8$, or $\frac{3}{4}$ **9.** $\log_4 3 + \frac{1}{2}\log_4 x - \log_4 y$
11. $\frac{1}{3}\log_3 4 - 2\log_3 x - \log_3 y$ **13.** $\frac{1}{2}\log_3 x + \frac{1}{2}\log_3 y - \frac{1}{2}\log_3 5$
15. $\frac{1}{3}\log_2 x + \frac{1}{5}\log_2 y - 2\log_2 r$ **17.** The distributive property tells us that the *product* $a(x + y)$ equals the sum $ax + ay$. In the notation $\log_a(x + y)$, the parentheses do not indicate multiplication. They indicate that $x + y$ is the result of raising a to some power. **19.** $\log_b xy$
21. $\log_a \frac{m^3}{n}$ **23.** $\log_a \frac{rt^3}{s}$ **25.** $\log_a \frac{125}{81}$ **27.** $\log_{10}(x^2 - 9)$
29. $\log_p \frac{x^3 y^{1/2}}{z^{3/2} a^3}$ **31.** false **33.** true **35.** true **37.** false **39.** 4
40. It is the exponent to which 3 must be raised in order to obtain 81.
41. 81 **42.** It is the exponent to which 2 must be raised in order to obtain 19. **43.** 19 **44.** m

Section 11.5 (pages 717–718)

1. C **3.** C **5.** 19.2 **7.** 2.5164 **9.** −1.4868 **11.** 9.6776
13. 2.0592 **15.** −2.8896 **17.** 2.3026 **19.** poor fen **21.** rich fen
23. 5.9 **25.** 1.0×10^{-2} **27.** 2.5×10^{-5} **29.** **(a)** 35.0 yr **(b)** 14.2 yr
31. **(a)** 55% **(b)** 90% **33.** **(a)** 5.6 yr **(b)** $t > 0$ and $\frac{87 - L}{63}$ is positive and in the domain of the function only if $24 < L < 87$.

Section 11.6 (pages 725–728)

1. $\log 5^x = \log 125$ **2.** $x \log 5 = \log 125$ **3.** $x = \frac{\log 125}{\log 5}$
4. $\frac{\log 125}{\log 5} = 3$; $\{3\}$ **5.** $\{0.827\}$ **7.** $\{0.833\}$ **9.** $\{1.201\}$ **11.** $\{2.269\}$
13. $\{15.967\}$ **15.** $\{261.291\}$ **17.** $\{-10.718\}$ **19.** $\{3\}$ **21.** $\{5.879\}$

23. Natural logarithms are a better choice because e is the base. **25.** $\left\{\frac{2}{3}\right\}$ **27.** $\left\{\frac{33}{2}\right\}$ **29.** $\{-1 + \sqrt[3]{49}\}$ **31.** 2 cannot be a solution because $\log(2 - 3) = \log(-1)$, and -1 is not in the domain of $\log x$. **33.** $\left\{\frac{1}{3}\right\}$ **35.** $\{2\}$ **37.** $\emptyset$ **39.** $\{8\}$ **41.** $\left\{\frac{4}{3}\right\}$ **43.** $\{8\}$ **45. (a)** \$2539.47 **(b)** 10.2 yr **47.** \$4934.71 **49.** 15.4 yr **51.** about 180 g **53.** 1.4315 **55.** 3.3030 **57.** -0.0947 **59.** -2.3219 **61.** -2.0639 **63.** 0.673

Chapter 11 Review Exercises (pages 733–736)

1. not one-to-one **2.** one-to-one **3.** This function is not one-to-one because two sodas in the list have 41 mg of caffeine. **4.** $f^{-1}(x) = \frac{x - 7}{-3}$, or $f^{-1}(x) = -\frac{1}{3}x + \frac{7}{3}$ **5.** $f^{-1}(x) = \frac{x^3 + 4}{6}$ **6.** not one-to-one

7.

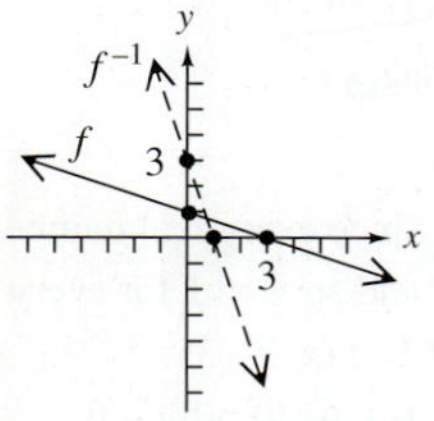

8.

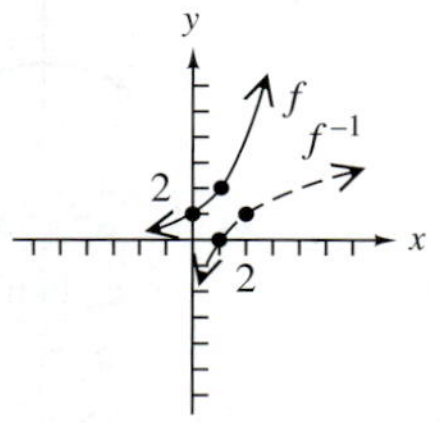

9.

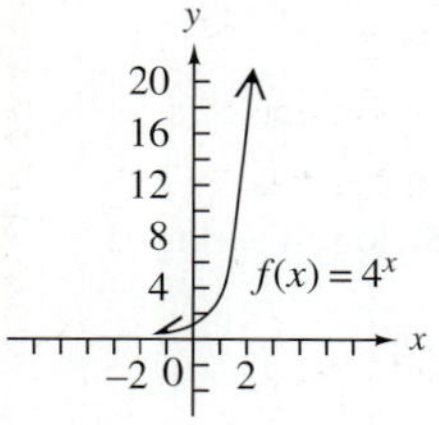

10.

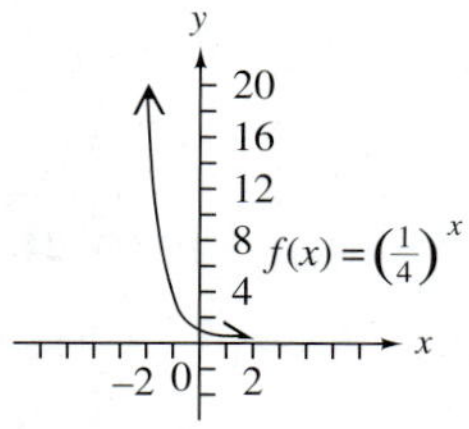

11.

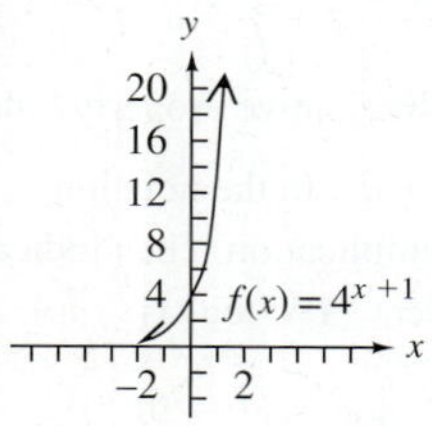

12. $\{4\}$ **13.** $\left\{\frac{3}{7}\right\}$ **14.** $\{0\}$ **15. (a)** 48.9 million **(b)** 67.3 million **16. (a)** $5^4 = 625$ **(b)** $\log_5 0.04 = -2$ **17. (a)** $\log_b a$ represents the exponent to which b must be raised to obtain a. **(b)** a

18.

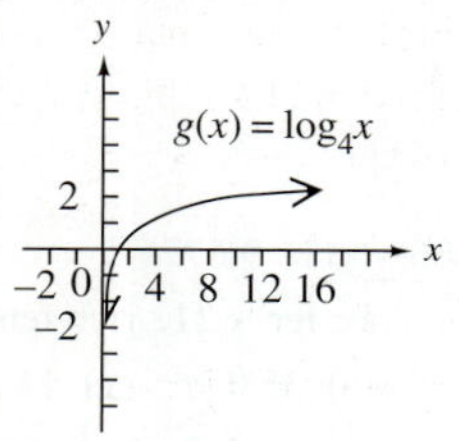

19.

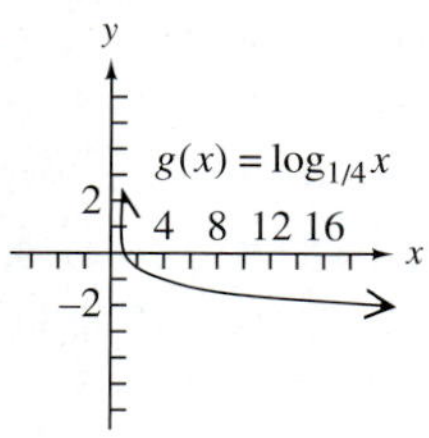

20. $\{2\}$ **21.** $\{-2\}$ **22.** $\{8\}$ **23.** $\{b \mid b > 0, b \neq 1\}$ **24.** $\log_4 3 + 2\log_4 x$ **25.** $2\log_2 p + \log_2 r - \frac{1}{2}\log_2 z$ **26.** $\log_b \frac{3x}{y^2}$ **27.** $\log_3 \frac{x + 7}{4x + 6}$ **28.** 1.4609 **29.** -0.5901 **30.** 3.3638 **31.** -1.3587 **32.** 6.4 **33.** 8.4 **34.** 2.5×10^{-5} **35. (a)** 18 yr **(b)** 12 yr **(c)** 7 yr **(d)** 6 yr **(e)** Each comparison shows approximately the same number. For example, in part (a) the doubling time is 18 yr (rounded) and $\frac{72}{4} = 18$. Thus, the formula $t = \frac{72}{100r}$ (called the *Rule of 72*) is an excellent approximation of the doubling time formula. **36.** every 2 hr **37.** $\{2.042\}$ **38.** $\{4.907\}$ **39.** $\{18.310\}$ **40.** $\left\{\frac{1}{9}\right\}$ **41.** $\{-6 + \sqrt[3]{25}\}$ **42.** $\{2\}$ **43.** $\left\{\frac{3}{8}\right\}$ **44.** $\{4\}$ **45.** $\{1\}$ **46.** \$7112.11 **47.** Plan A; it would pay \$2.92 more. **48.** \$4267 **49.** about 11% **50.** 0.9251 **51.** 1.7925 **52.** 1.4315 **53.** $\{72\}$ **54.** $\{5\}$ **55.** $\left\{\frac{1}{9}\right\}$ **56.** $\left\{\frac{4}{3}\right\}$ **57.** $\{3\}$ **58.** $\left\{\frac{1}{8}\right\}$ **59.** $\left\{\frac{11}{3}\right\}$ **60.** 6.8 yr

Chapter 11 Test (pages 737–738)

1. (a) not one-to-one **(b)** one-to-one **2.** $f^{-1}(x) = x^3 - 7$

3.

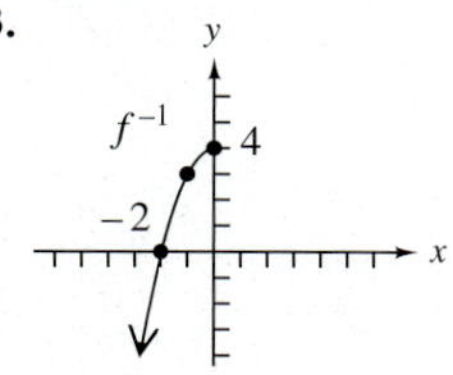

4.

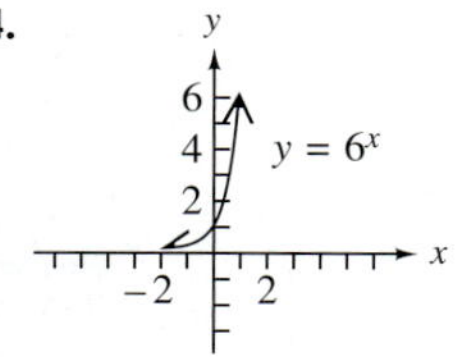

5.

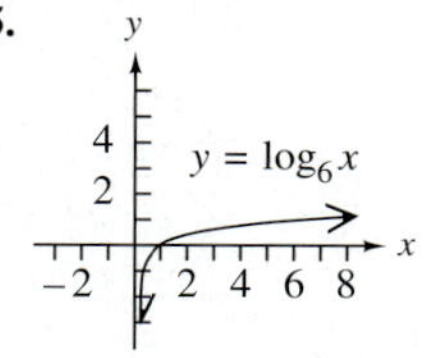

6. Interchange the x- and y-values of the ordered pairs, because the functions are inverses. **7.** $\{-4\}$ **8.** $\left\{-\frac{13}{3}\right\}$ **9. (a)** 775 millibars **(b)** 265 millibars **10.** $\log_4 0.0625 = -2$ **11.** $7^2 = 49$ **12.** $\{32\}$ **13.** $\left\{\frac{1}{2}\right\}$ **14.** $\{2\}$ **15.** $2\log_3 x + \log_3 y$ **16.** $\log_b \frac{r^{1/4}s^2}{t^{2/3}}$ **17. (a)** 1.3284 **(b)** -0.8440 **(c)** 2.1245 **18.** $\{3.9656\}$ **19.** $\{3\}$ **20. (a)** \$12,507.51 **(b)** 15.5 yr

Cumulative Review Exercises: Chapters 1–11 (pages 739–740)

1. $-2, 0, 6, \frac{30}{3}$ (or 10) **2.** $-\frac{9}{4}, -2, 0, 0.6, 6, \frac{30}{3}$ (or 10) **3.** $-\sqrt{2}, \sqrt{11}$ **4.** $\left\{-\frac{2}{3}\right\}$ **5.** $[1, \infty)$ **6.** $\{-2, 7\}$ **7.** $(-\infty, -3) \cup (2, \infty)$ **8. (a)** yes **(b)** 3254; The number of travelers increased by an average of 3254 thousand per year during the period 2003–2006. **9.** $\{(4, 2)\}$ **10.** $\{(1, -1, 4)\}$ **11.** $6p^2 + 7p - 3$ **12.** $16k^2 - 24k + 9$ **13.** $-5m^3 + 2m^2 - 7m + 4$ **14.** $2t^3 + 5t^2 - 3t + 4$ **15.** $z(5z + 1)(z - 4)$ **16.** $(4a + 5b^2)(4a - 5b^2)$ **17.** $(2c + d)(4c^2 - 2cd + d^2)$ **18.** $-\frac{1875p^{13}}{8}$ **19.** $\frac{x + 5}{x + 4}$ **20.** $\frac{-3k - 19}{(k + 3)(k - 2)}$ **21.** $12\sqrt{2}$ **22.** $-\frac{1}{4}$ **23.** $-27\sqrt{2}$ **24.** $\{0, 4\}$ **25.** 41 **26.** $\left\{\frac{1 + \sqrt{13}}{6}, \frac{1 - \sqrt{13}}{6}\right\}$

27. $(-\infty, -4) \cup (2, \infty)$ **28.** $\{-2, -1, 1, 2\}$ **29.** $\{-1\}$ **30.** $\{1\}$

31. $5^3 = 125$ **32.** $3 \log x + \frac{1}{2} \log y - \log z$

33.

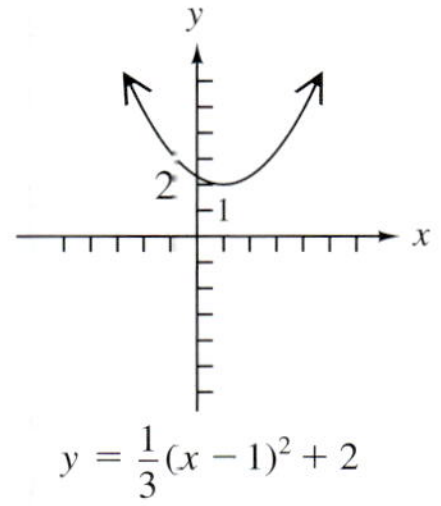

34.

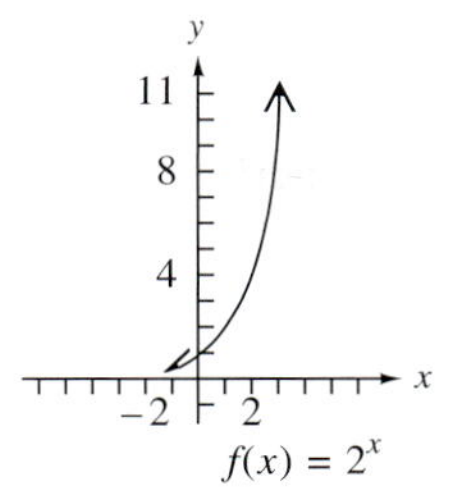

35.

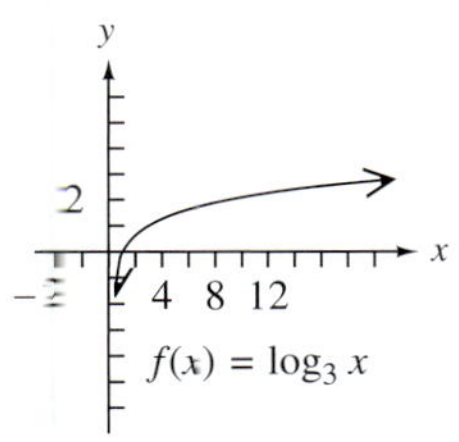

Chapter 12 Nonlinear Functions, Conic Sections, and Nonlinear Systems

Section 12.1 (pages 749–752)

1. 0 **3.** 0; 0 **5.** B **7.** A

9. domain: $(-\infty, \infty)$; range: $[0, \infty)$

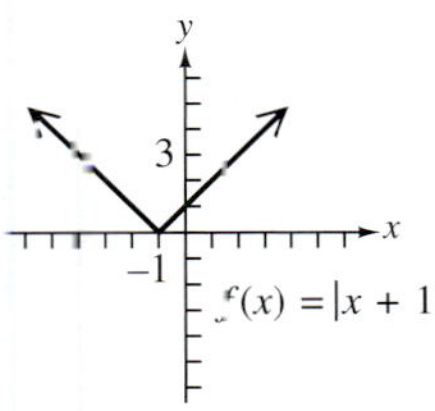

11. domain: $(-\infty, 0) \cup (0, \infty)$; range: $(-\infty, 1) \cup (1, \infty)$

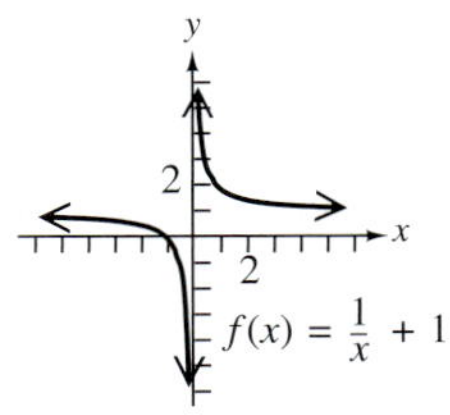

13. domain: $[2, \infty)$; range: $[0, \infty)$

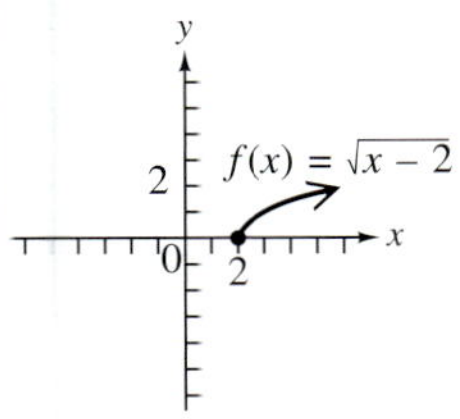

15. domain: $(-\infty, 2) \cup (2, \infty)$; range: $(-\infty, 0) \cup (0, \infty)$

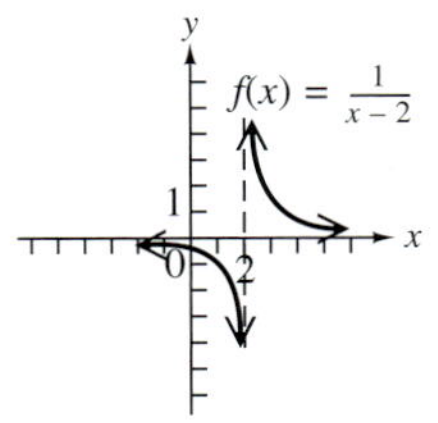

17. domain: $[-3, \infty)$; range: $[-3, \infty)$

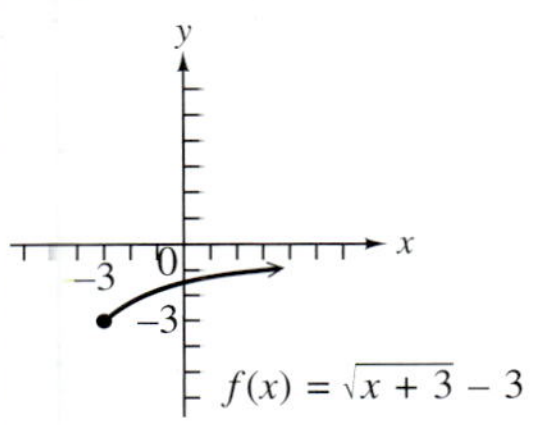

19.

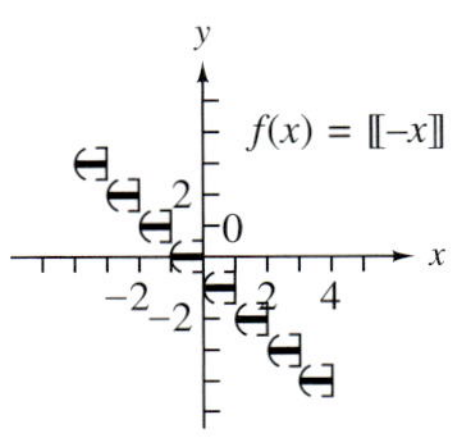

21.

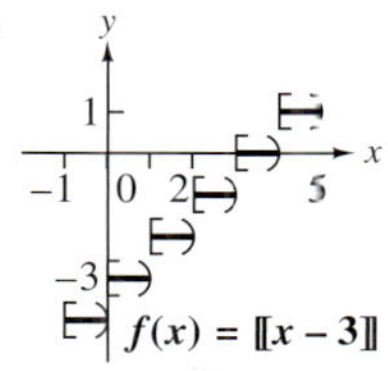

23.

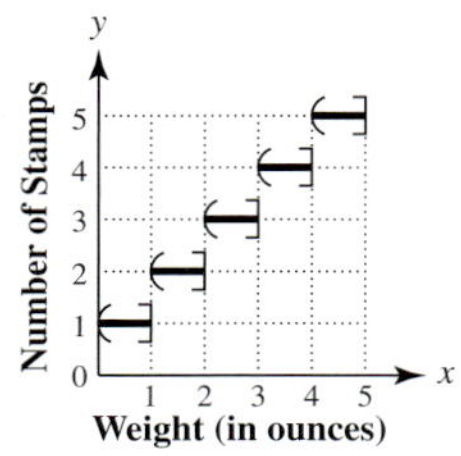

25. 16 **27.** 83 **29.** 13 **31.** $4x^2 + 12x + 13$ **33.** $x^2 + 10x + 29$ **35.** $2x + 8$ **37.** $\frac{137}{4}$ **39.** 8 **41.** $(f \circ g)(x) = 63{,}360x$; It computes the number of inches in x miles. **43.** $(\mathcal{A} \circ r)(t) = 4\pi t^2$; This is the area of the circular layer as a function of time.

Section 12.2 (pages 759–762)

1. **(a)** $(0, 0)$ **(b)** 5 **(c)**

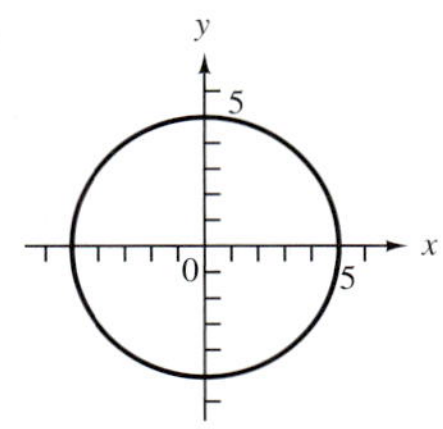

3. B **5.** D **7.** $(x + 4)^2 + (y - 3)^2 = 4$
9. $(x + 8)^2 + (y + 5)^2 = 5$ **11.** $(-2, -3)$; $r = 2$
13. $(-5, 7)$; $r = 9$ **15.** $(2, 4)$; $r = 4$ **17.** The thumbtack acts as the center and the length of the string acts as the radius.

19.

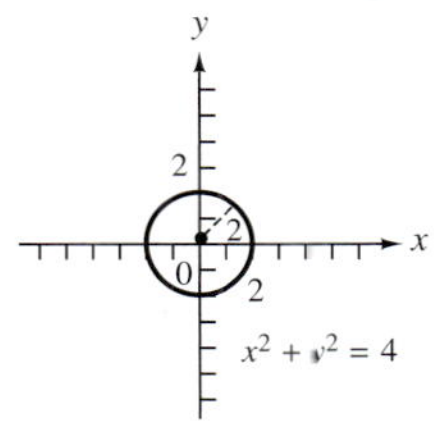

21.

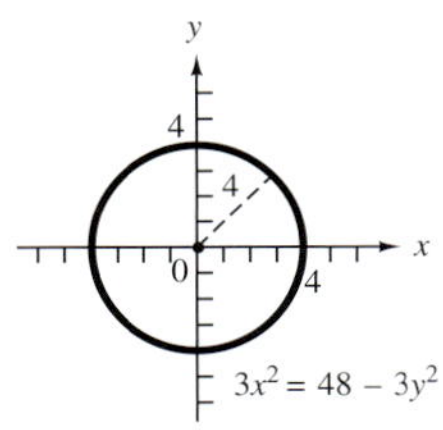

23.

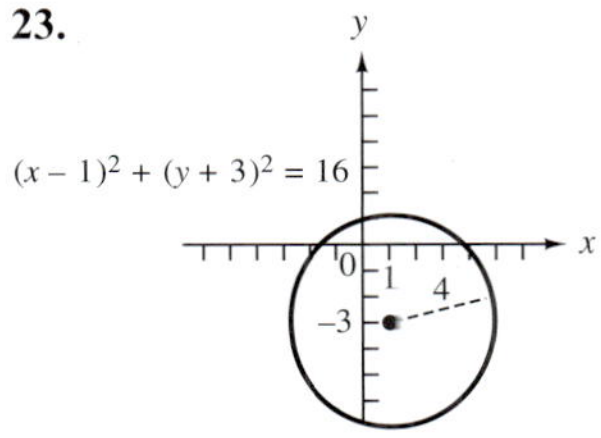

25.

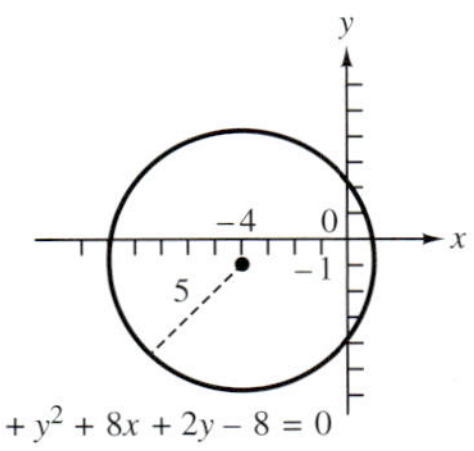

27. The two thumbtacks act as foci, and the length of the string is constant, satisfying the requirements of the definition of an ellipse.

29.

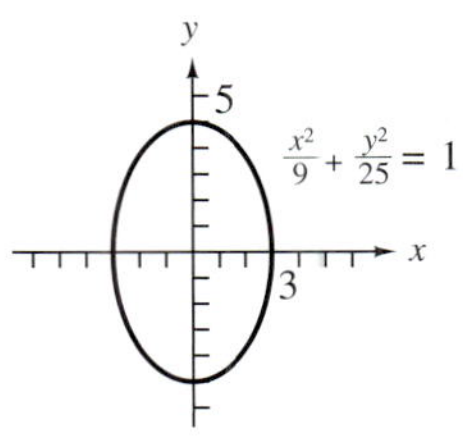

31.

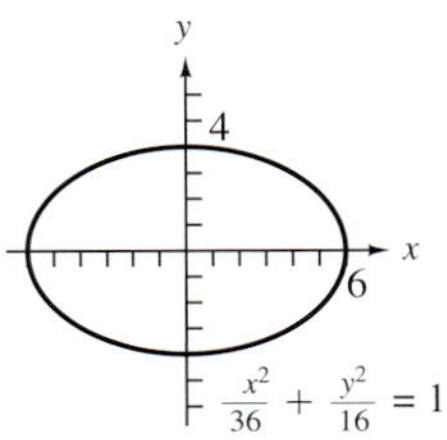

33.

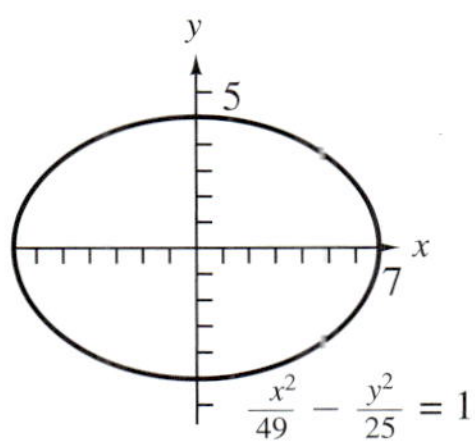

35.

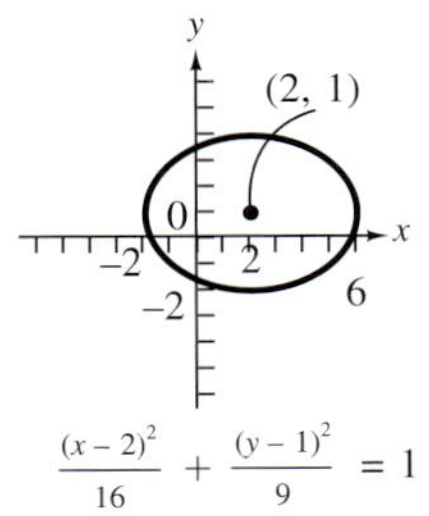

37. **(a)** 10 m **(b)** 36 m **39.** **(a)** 154.7 million mi **(b)** 128.7 million mi (Answers are rounded.)

Section 12.3 (pages 769–772)

1. C **3.** D **5.** When written in one of the forms given in the box titled "Equations of Hyperbolas" in this section, it will open up and down if the − sign precedes the x^2-term; it will open left and right if the − sign precedes the y^2-term.

7.

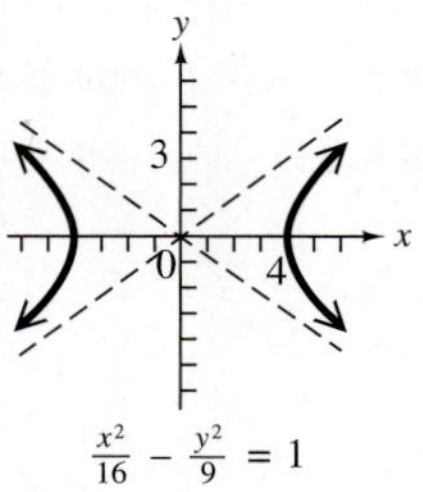

9.

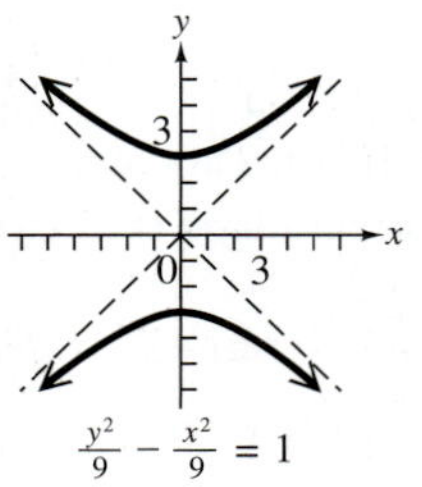

11.

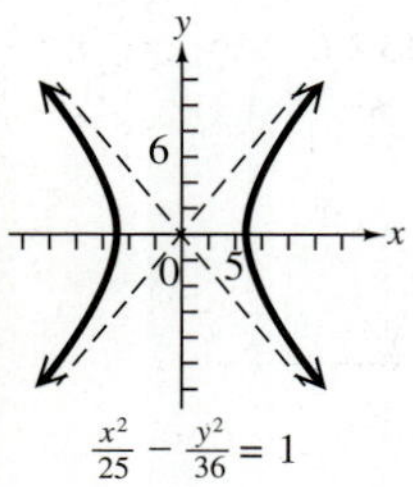

13. hyperbola

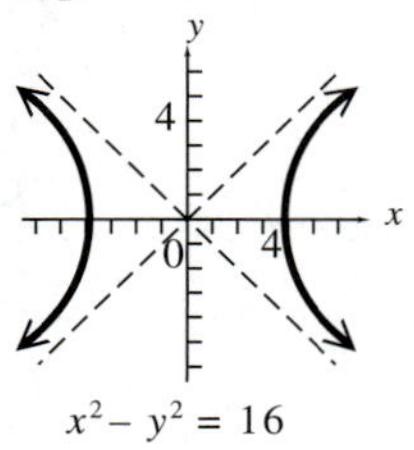

15. ellipse

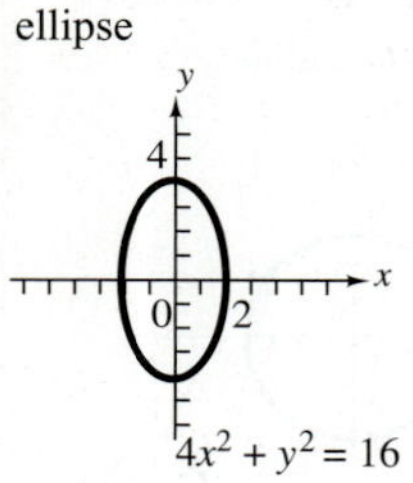

17. circle

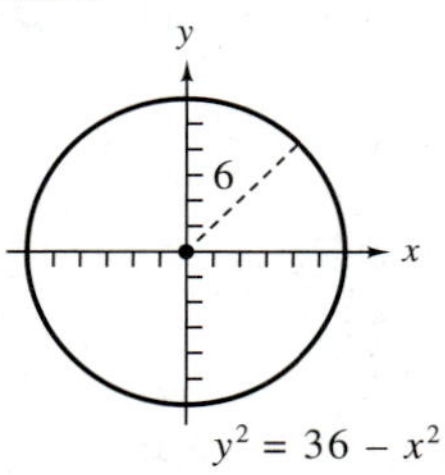

19. hyperbola

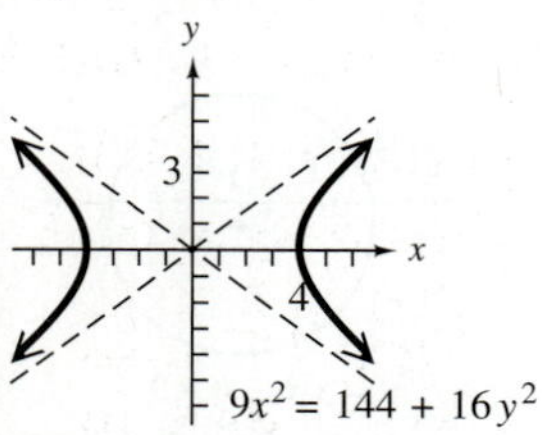

21. ellipse

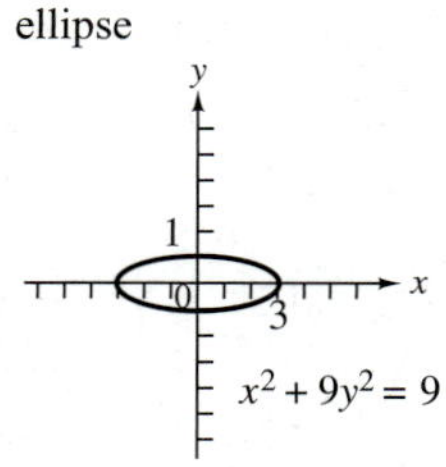

23. domain: $[-3, 3]$; range: $[0, 3]$

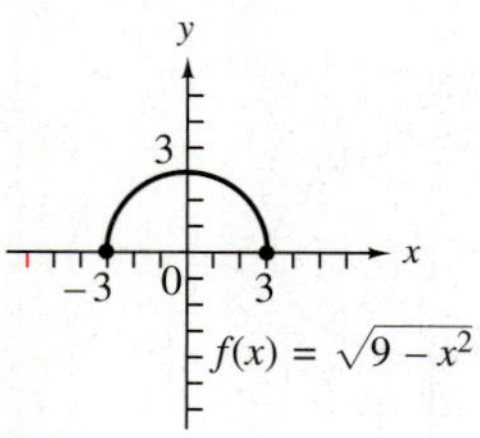

25. domain: $[-5, 5]$; range: $[-5, 0]$

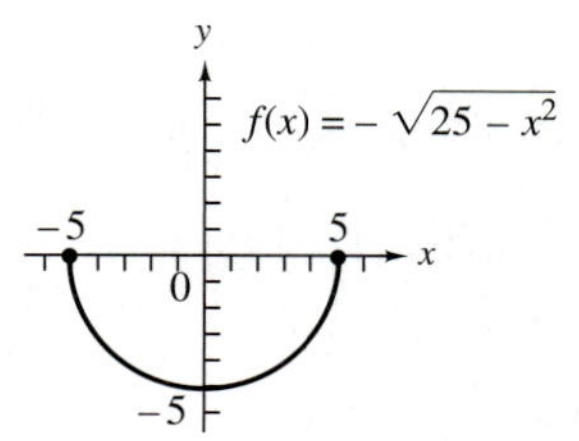

27. domain: $[-4, \infty)$; range: $[0, \infty)$

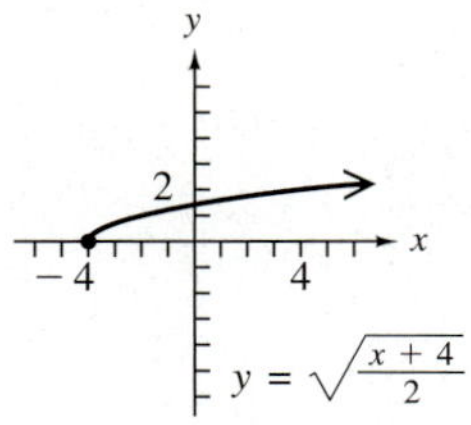

29.

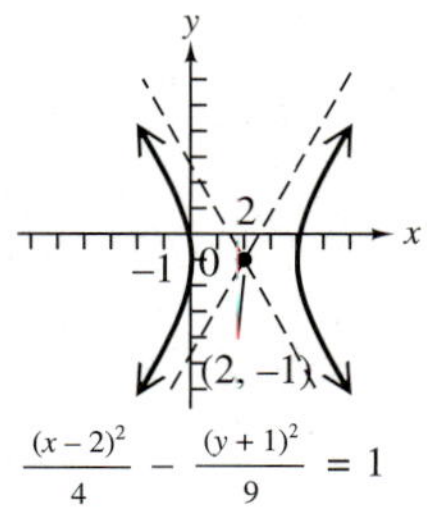

31. **(a)** 50 m **(b)** 69.3 m

Section 12.4 (pages 777–778)

1. one **3.** none

5.

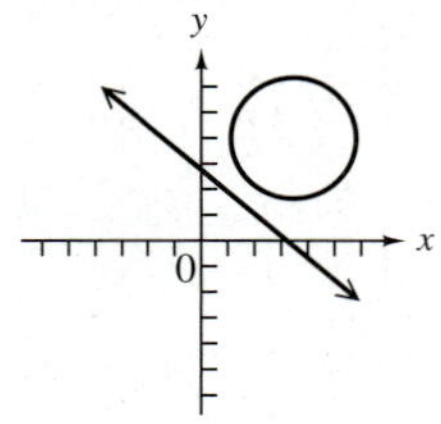

7.

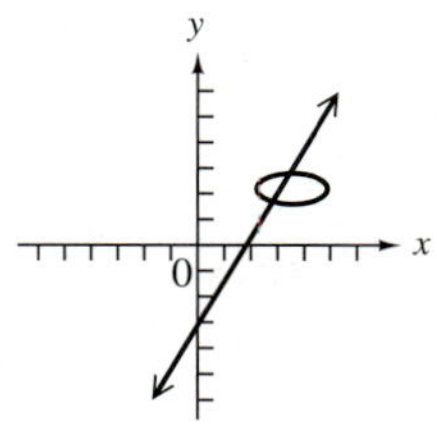

9.

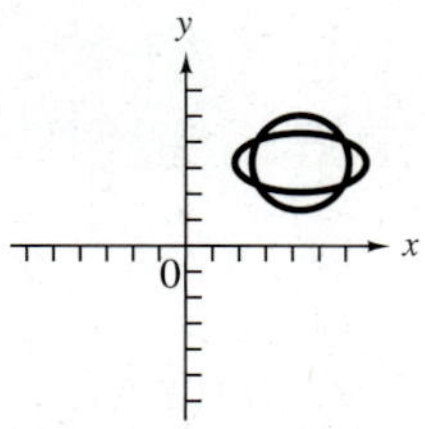

11. $\left\{(0, 0), \left(\frac{1}{2}, \frac{1}{2}\right)\right\}$ **13.** $\{(-6, 9), (-1, 4)\}$

15. $\left\{\left(-\frac{1}{5}, \frac{7}{5}\right), (1, -1)\right\}$ **17.** $\left\{(-2, -2), \left(-\frac{4}{3}, -3\right)\right\}$

19. $\{(-3, 1), (1, -3)\}$ **21.** $\left\{\left(-\frac{3}{2}, -\frac{9}{4}\right), (-2, 0)\right\}$

23. $\{(-\sqrt{3}, 0), (\sqrt{3}, 0), (-\sqrt{5}, 2), (\sqrt{5}, 2)\}$

25. $\{(-2, 0), (2, 0)\}$ **27.** $\{(i\sqrt{2}, -3i\sqrt{2}), (-i\sqrt{2}, 3i\sqrt{2}), (-\sqrt{6}, -\sqrt{6}), (\sqrt{6}, \sqrt{6})\}$ **29.** $\{(-2i\sqrt{2}, -2\sqrt{3}), (-2i\sqrt{2}, 2\sqrt{3}), (2i\sqrt{2}, -2\sqrt{3}), (2i\sqrt{2}, 2\sqrt{3})\}$

31. $\{(-\sqrt{5}, -\sqrt{5}), (\sqrt{5}, \sqrt{5})\}$ **33.** length: 12 ft; width: 7 ft

Section 12.5 (pages 783–784)

1.

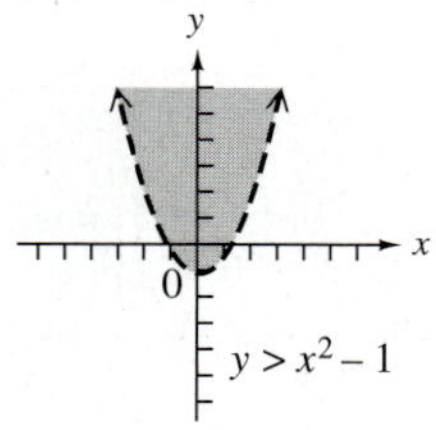

3.

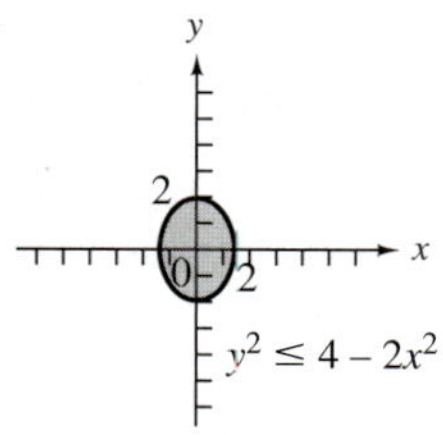

5.

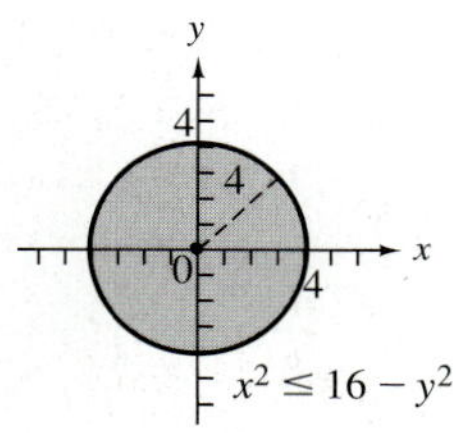

7.

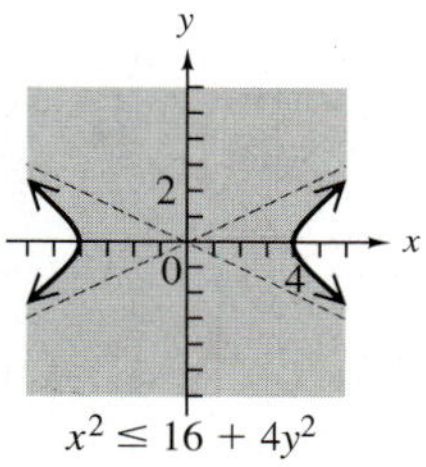

9.

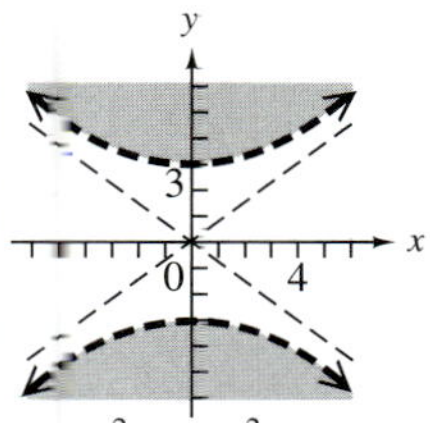

$9x^2 < 16y^2 - 144$

11.

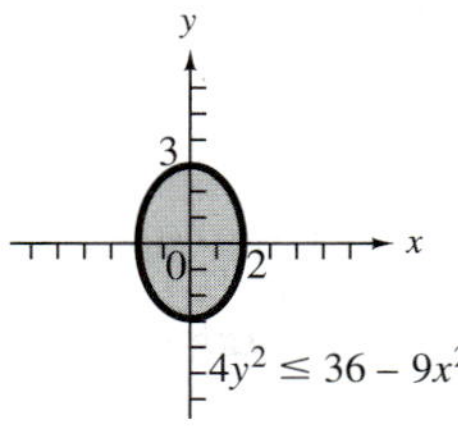

$4y^2 \leq 36 - 9x^2$

13.

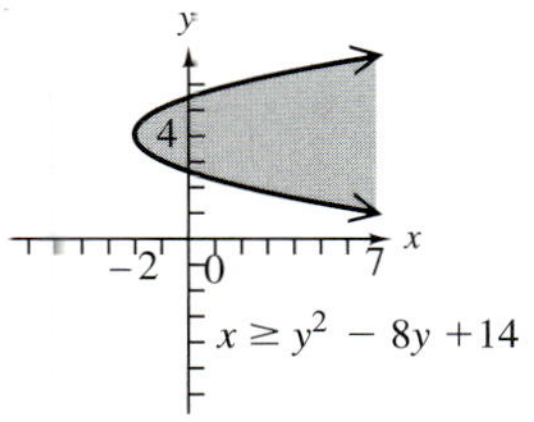

$x \geq y^2 - 8y + 14$

15.

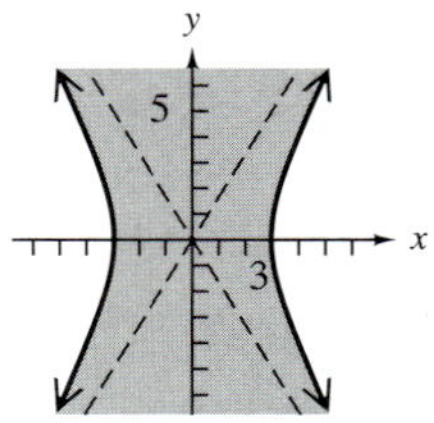

$25x^2 \leq 9y^2 + 225$

17.

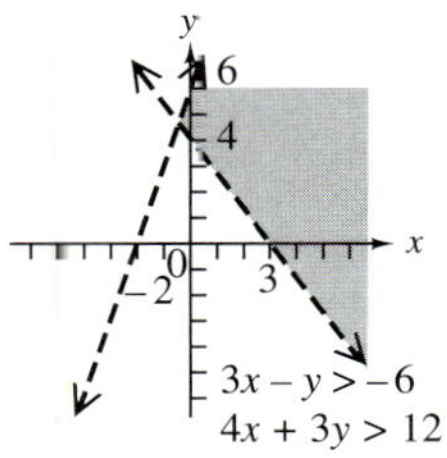

$3x - y > -6$
$4x + 3y > 12$

19.

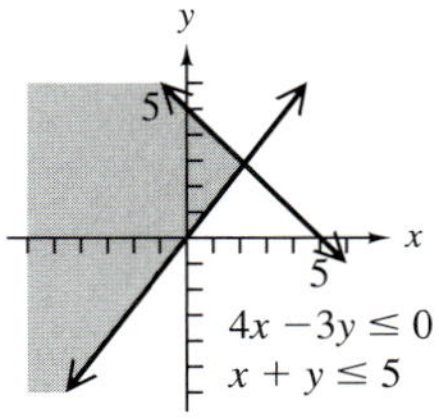

$4x - 3y \leq 0$
$x + y \leq 5$

21.

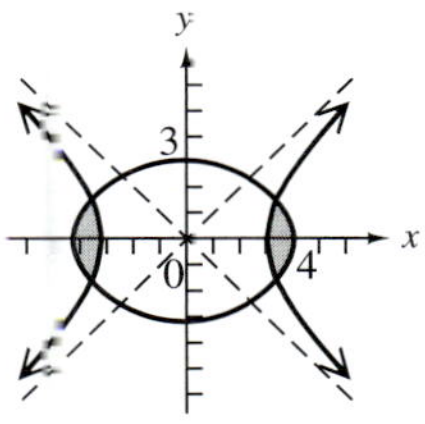

$x^2 - y^2 \geq 9$
$\frac{x^2}{16} + \frac{y^2}{9} \leq 1$

23.

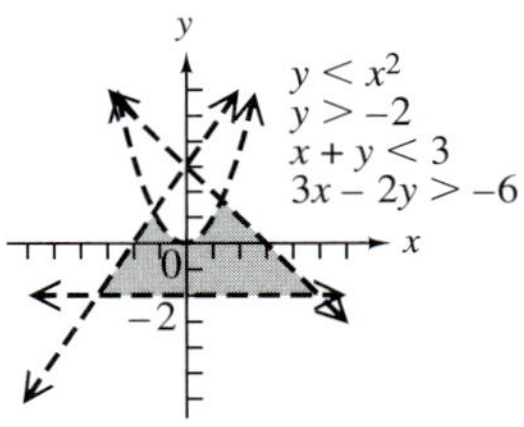

$y < x^2$
$y > -2$
$x + y < 3$
$3x - 2y > -6$

Chapter 12 Review Exercises (pages 789–792)

1. domain: $(-\infty, \infty)$; range: $[0, \infty)$

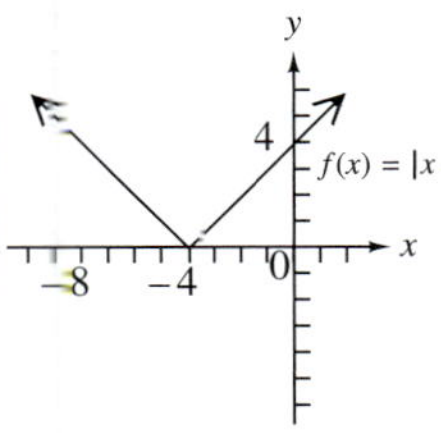

$f(x) = |x + 4|$

2. domain: $(-\infty, 4) \cup (4, \infty)$; range: $(-\infty, 0) \cup (0, \infty)$

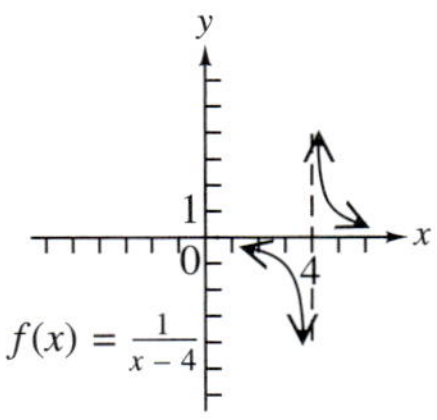

$f(x) = \frac{1}{x - 4}$

3. domain: $[0, \infty)$; range: $[3, \infty)$

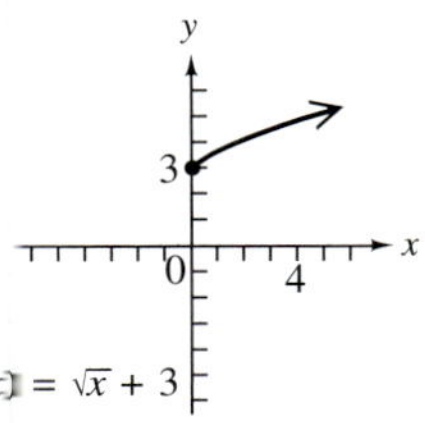

$f(x) = \sqrt{x} + 3$

4. 12 **5.** 2 **6.** -5 **7. (a)** 167 **(b)** 1495 **8. (a)** 20 **(b)** 42 **9. (a)** $75x^2 + 220x + 160$ **(b)** $15x^2 + 10x + 2$ **10.** No, composition of functions is not a commutative operation. For example, the results of Exercise 9 show that $(f \circ g)(x) \neq (g \circ f)(x)$ in this case.

11. $(x - 2)^2 + (y - 4)^2 = 9$ **12.** $(x + 1)^2 + (y + 3)^2 = 25$

13. $(x - 4)^2 + (y - 2)^2 = 36$ **14.** $(-3, 2)$; $r = 4$

15. $(4, 1)$; $r = 2$ **16.** $(-1, -5)$; $r = 3$ **17.** $(3, -2)$; $r = 5$

18.

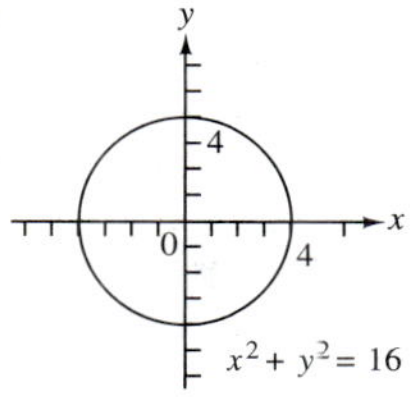

$x^2 + y^2 = 16$

19.

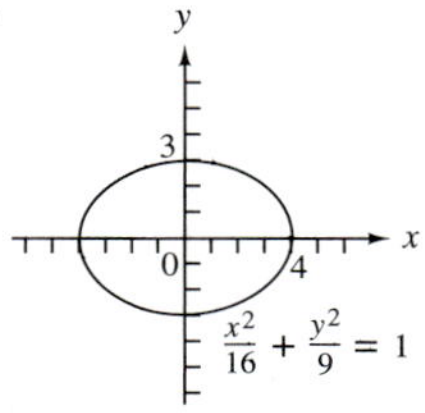

$\frac{x^2}{16} + \frac{y^2}{9} = 1$

20.

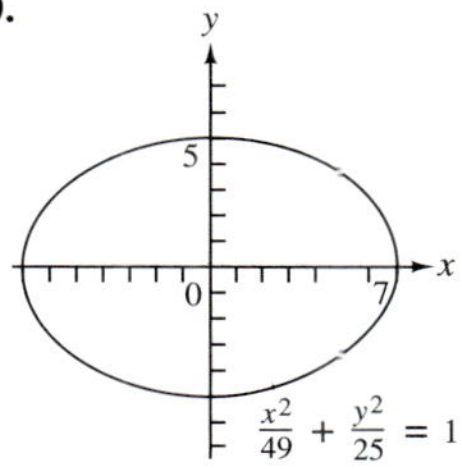

$\frac{x^2}{49} + \frac{y^2}{25} = 1$

21. $\frac{x^2}{65{,}286{,}400} + \frac{y^2}{2{,}560{,}000} = 1$ **22. (a)** 348.2 ft **(b)** 1787.6 ft

23.

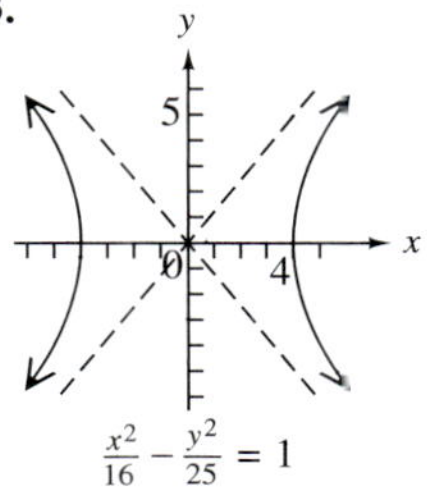

$\frac{x^2}{16} - \frac{y^2}{25} = 1$

24.

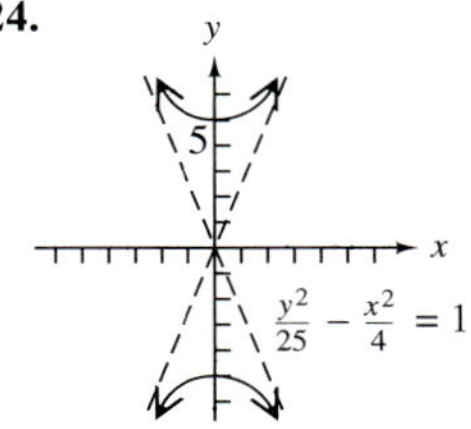

$\frac{y^2}{25} - \frac{x^2}{4} = 1$

25.

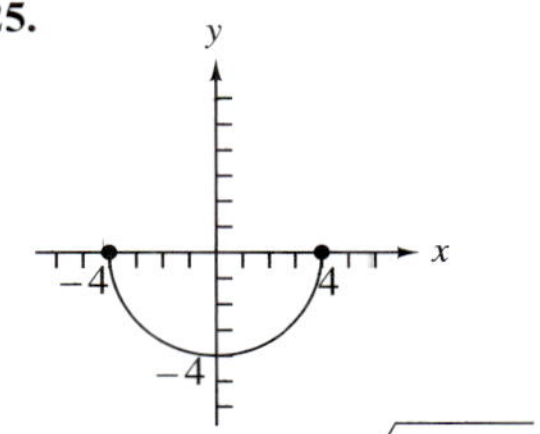

$f(x) = -\sqrt{16 - x^2}$

26. circle **27.** parabola **28.** hyperbola **29.** ellipse **30.** parabola

31. hyperbola **32.** $\frac{x^2}{625} - \frac{y^2}{1875} = 1$ **33.** $\{(6, -9), (-2, -5)\}$

34. $\{(1, 2), (-5, 14)\}$ **35.** $\{(4, 2), (-1, -3)\}$ **36.** $\{(-2, -4), (8, 1)\}$

37. $\{(-\sqrt{2}, 2), (-\sqrt{2}, -2), (\sqrt{2}, -2), (\sqrt{2}, 2)\}$

38. $\{(-\sqrt{6}, -\sqrt{3}), (-\sqrt{6}, \sqrt{3}), (\sqrt{6}, -\sqrt{3}), (\sqrt{6}, \sqrt{3})\}$

39. 0, 1, or 2 **40.** 0, 1, 2, 3, or 4

41.

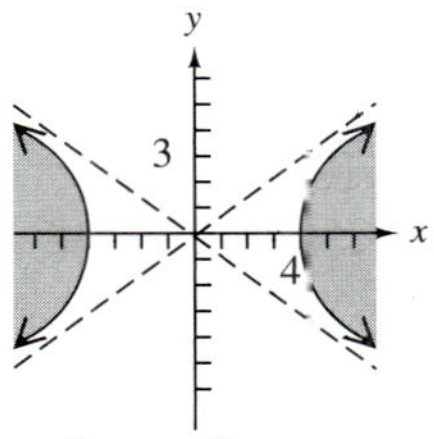

$9x^2 \geq 16y^2 + 144$

42.

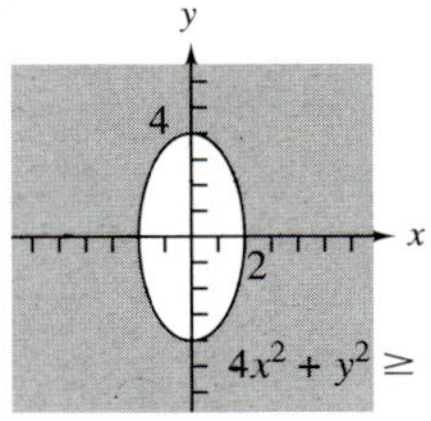

$4x^2 + y^2 \geq 16$

43.

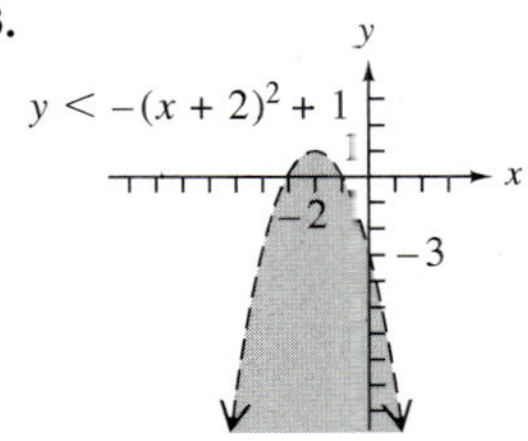

$y < -(x + 2)^2 + 1$

44.

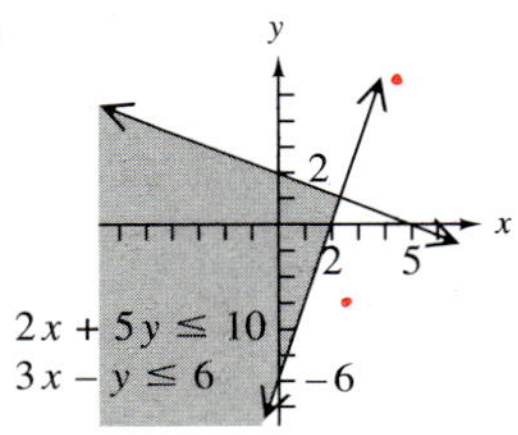

$2x + 5y \leq 10$
$3x - y \leq 6$

ANSWERS

45.

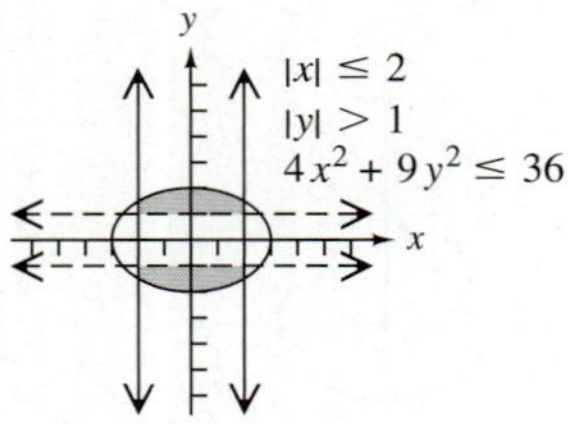

$|x| \le 2$
$|y| > 1$
$4x^2 + 9y^2 \le 36$

46.

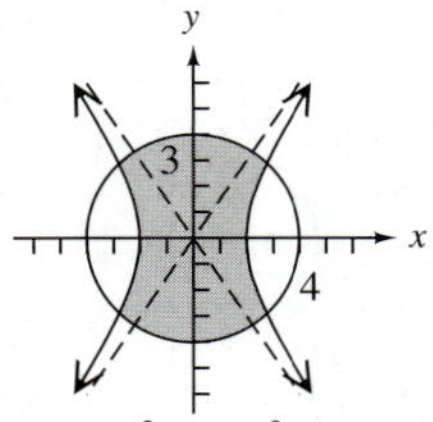

$9x^2 \le 4y^2 + 36$
$x^2 + y^2 \le 16$

47.

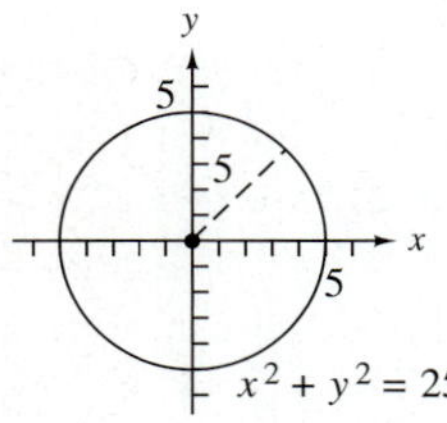

$x^2 + y^2 = 25$

48.

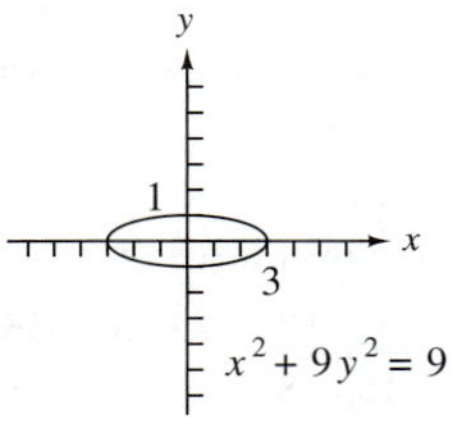

$x^2 + 9y^2 = 9$

49.

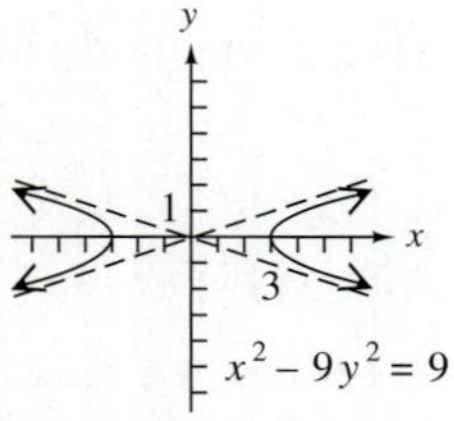

$x^2 - 9y^2 = 9$

50.

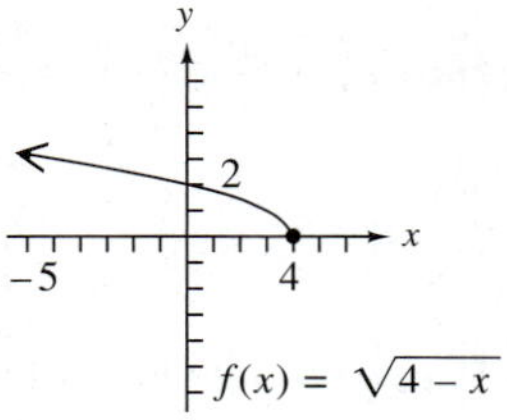

$f(x) = \sqrt{4 - x}$

51.

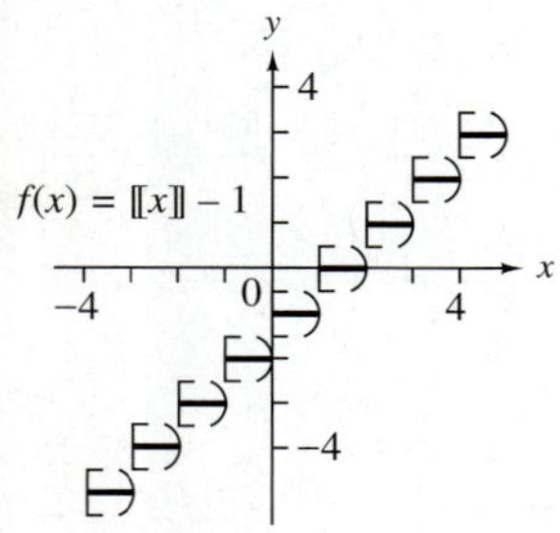

$f(x) = [\![x]\!] - 1$

52.

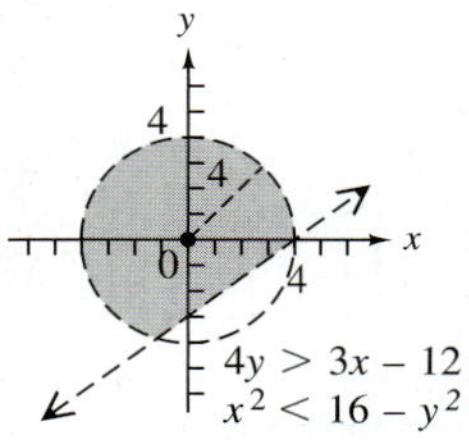

$4y > 3x - 12$
$x^2 < 16 - y^2$

53. There are cases where one x-value will yield two y-values. In a function, every x yields one and only one y.

Chapter 12 Test (pages 793–796)

1. C **2.** A **3.** D **4.** B

5. domain: $(-\infty, \infty)$; range: $[4, \infty)$

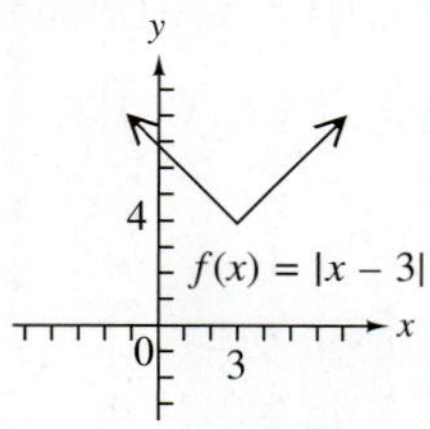

$f(x) = |x - 3| + 4$

6. (a) 23 **(b)** $3x^2 + 11$ **(c)** $9x^2 + 30x + 27$

7. center: $(2, -3)$; radius: 4

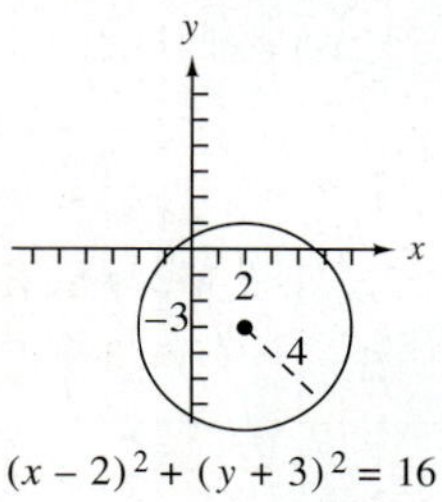

$(x - 2)^2 + (y + 3)^2 = 16$

8. center: $(-4, 1)$; radius: 5

9.

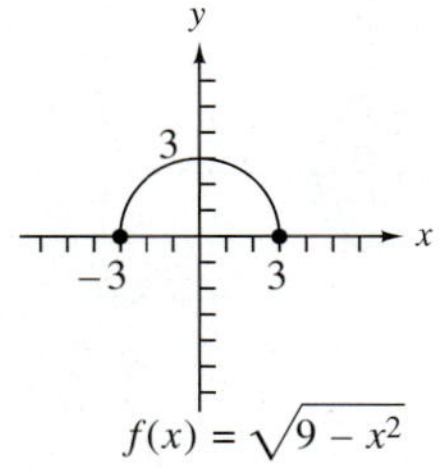

$f(x) = \sqrt{9 - x^2}$

10.

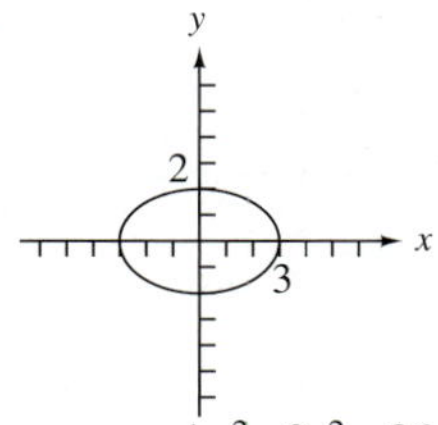

$4x^2 + 9y^2 = 36$

11.

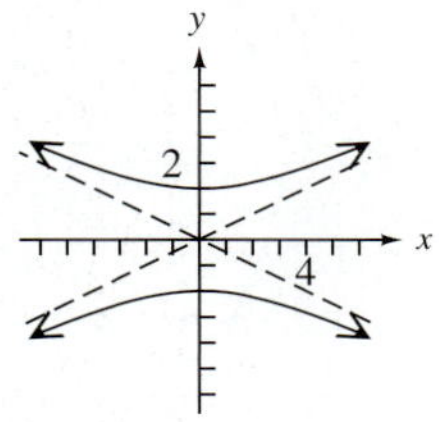

$16y^2 - 4x^2 = 64$

12.

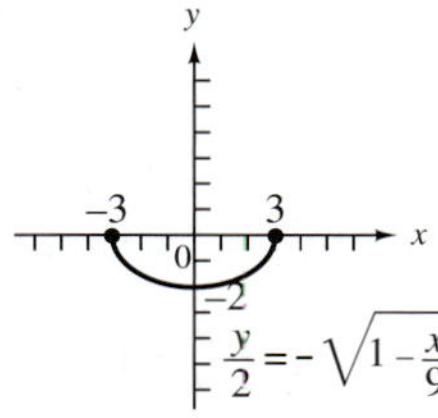

$\frac{y}{2} = -\sqrt{1 - \frac{x^2}{9}}$

13. ellipse **14.** hyperbola **15.** parabola

16. $\left\{\left(-\frac{1}{2}, -10\right), (5, 1)\right\}$ **17.** $\left\{(-2, -2), \left(\frac{14}{5}, -\frac{2}{5}\right)\right\}$

18. $\left\{\left(-\sqrt{22}, -\sqrt{3}\right), \left(-\sqrt{22}, \sqrt{3}\right), \left(\sqrt{22}, -\sqrt{3}\right), \left(\sqrt{22}, \sqrt{3}\right)\right\}$

19.

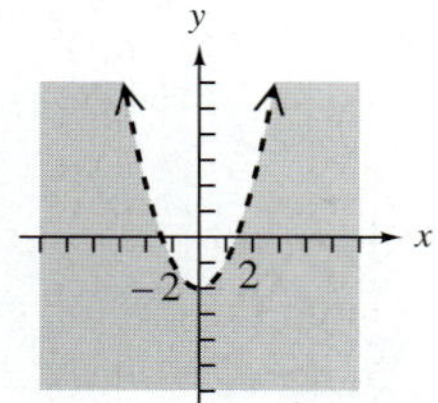

$y < x^2 - 2$

20.

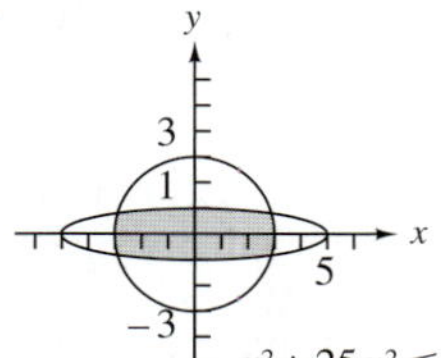

$x^2 + 25y^2 \le 25$
$x^2 + y^2 \le 9$

Cumulative Review Exercises: Chapters 1–12 (pages 797–798)

1. $\left\{\frac{2}{3}\right\}$ **2.** $\left(-\infty, \frac{3}{5}\right]$ **3.** $\{-4, 4\}$ **4.** $\frac{2}{3}$ **5.** $3x + 2y = -13$

6. $\{(3, -3)\}$ **7.** $\{(4, 1, -2)\}$ **8.** $\left\{(-1, 5), \left(\frac{5}{2}, -2\right)\right\}$

9. $25y^2 - 30y + 9$ **10.** $12r^2 + 40r - 7$

11. $4x^3 - 4x^2 + 3x + 5 + \frac{3}{2x + 1}$ **12.** $(3x + 2)(4x - 5)$

13. $(z^2 + 1)(z + 1)(z - 1)$ **14.** $(a - 3b)(a^2 + 3ab + 9b^2)$

15. $\frac{y - 1}{y(y - 3)}$ **16.** $\frac{3c + 5}{(c + 5)(c + 3)}$ **17.** $\frac{1}{p}$ **18.** $1\frac{1}{5}$ hr

19. $\frac{a^5}{4}$ **20.** $2\sqrt[3]{2}$ **21.** $\frac{3\sqrt{10}}{2}$ **22.** $\frac{7}{5} + \frac{11}{5}i$ **23.** $\emptyset$

24. $\left\{\frac{1}{5}, -\frac{3}{2}\right\}$ **25.** $\left\{\frac{3 + \sqrt{33}}{6}, \frac{3 - \sqrt{33}}{6}\right\}$

26. $\left\{-\frac{\sqrt{6}}{2}, \frac{\sqrt{6}}{2}, -\sqrt{7}, \sqrt{7}\right\}$ **27.** $\{3\}$ **28.** $v = \frac{\pm\sqrt{rFkw}}{kw}$

29. $f^{-1}(x) = \sqrt[3]{x - 4}$ **30. (a)** 4 **(b)** 7 **31.** $\log \frac{(3x + 7)^2}{4}$

32. (a) -1 **(b)** $9x^2 + 18x + 4$

33.

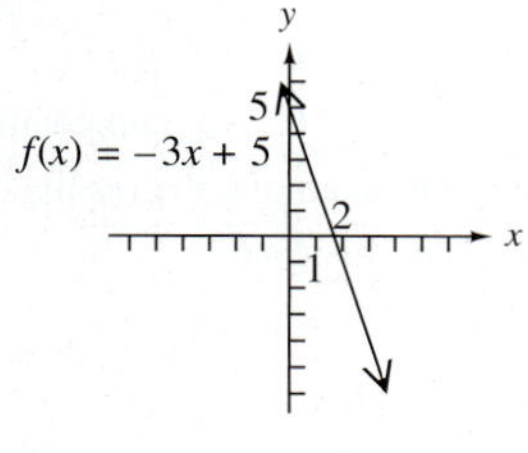

$f(x) = -3x + 5$

34.

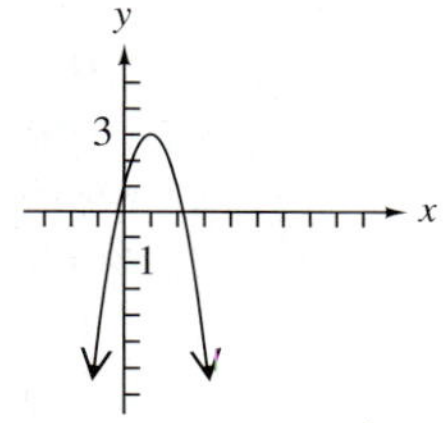

$f(x) = -2(x - 1)^2 + 3$

35.

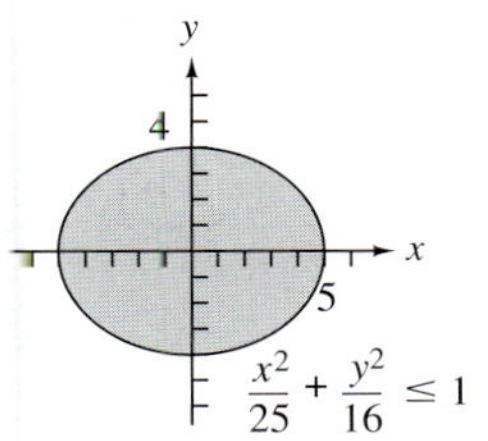

36.

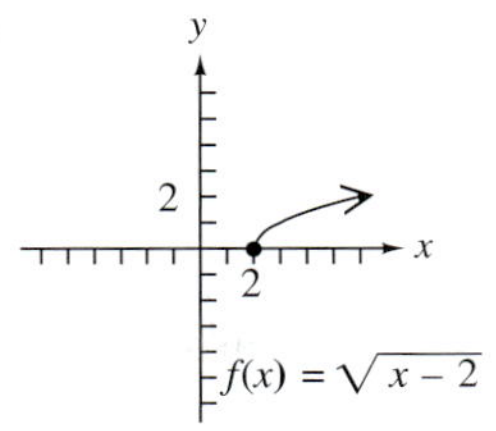

37.

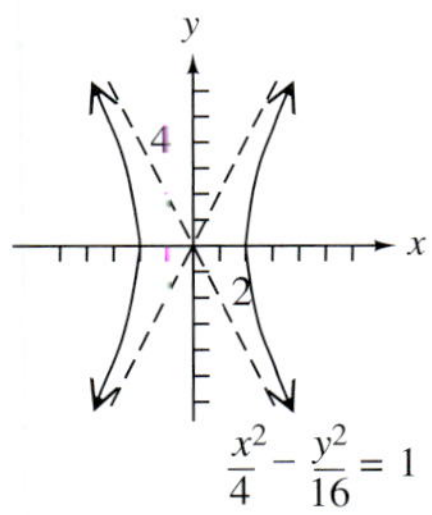

38.

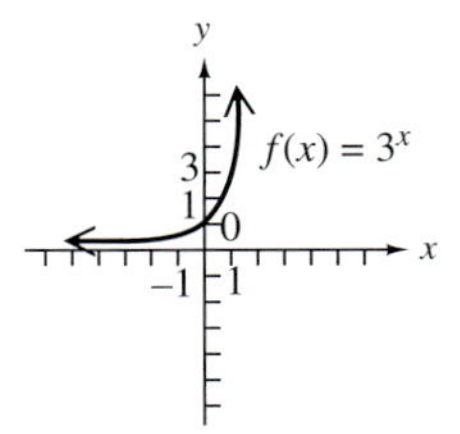

23. Here is one possible solution.

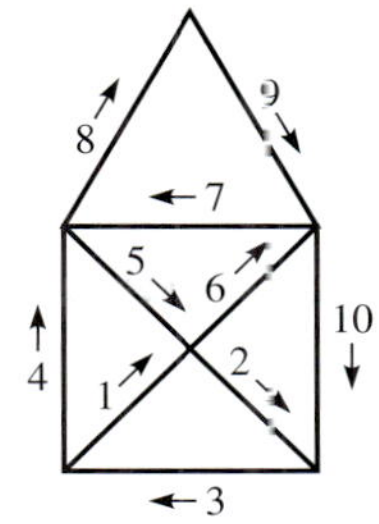

25. Dan (36) is married to Jessica (29); James (30) is married to Cathy (31). **27.** None, since there is no dirt in a hole. **29.** One solution is $1 + 2 + 3 + 4 + 5 + 6 + 7 + 8 \times 9 = 100$. **31.** The correct problem follows. $\begin{array}{r} 402 \\ \times\ \ 39 \\ \hline 15{,}678 \end{array}$ **33.** 35 **35.** $7\frac{1}{2}$ min (Boil them all at the same time.)

Appendix A Strategies for Problem Solving

(pages 805–808)

1. You should choose a sock from the box labeled *red and green socks*. Since it is mislabeled, it contains only red socks or only green socks, determined by the sock you choose. If the sock is green, relabel this box *green socks*. Since the other two boxes were mislabeled, switch the remaining label to the other box and place the label that says *red and green socks* on the unlabeled box. No other choice guarantees a correct relabeling, since you can remove only one sock. **3.** You must place the O in the bottom-left square. No other choice guarantees you a win. **5.** One possible sequence is shown here. The numbers represent the number of gallons in each bucket in each successive step.

Big	7	4	4	1	1	0	7	5	5
Small	0	3	0	3	0	1	1	3	0

7. 18 **9.** 07 **11.** 17 days **13.** 16

15.

6	1	8
7	5	3
2	9	4

17. 25 pitches (The visiting team's pitcher retires 24 consecutive batters through the first eight innings, using only one pitch per batter. His team does not score either. Going into the bottom of the ninth inning tied 0–0, the first batter for the home team hits his first pitch for a home run. The pitcher threw 25 pitches and loses the game by a score of 1–0.)
19. For three weighings, first balance four against four. Of the lighter four, balance two against the other two. Finally, of the lighter two, balance them one against the other. To find the bad coin in two weighings, divide the eight coins into groups of 3, 3, 2. Weigh the groups of three against each other on the scale. If the groups weigh the same, the fake is in the two left out and can be found in one additional weighing. If the two groups of three do not weigh the same, pick the lighter group. Choose any two of the coins and weigh them. If one of these is lighter, it is the fake; if they weigh the same, then the third coin is the fake. **21.** 28

Appendix B Review of Fractions

(pages 817–820)

1. true **3.** false; The fraction $\frac{17}{51}$ can be simplified to $\frac{1}{3}$. **5.** false; *Product* indicates multiplication, so the product of 8 and 2 is 16. **7.** prime **9.** composite **11.** composite **13.** neither **15.** $2 \cdot 3 \cdot 5$ **17.** $2 \cdot 2 \cdot 3 \cdot 3 \cdot 7$ **19.** $2 \cdot 2 \cdot 31$ **21.** 29 **23.** $\frac{1}{2}$ **25.** $\frac{5}{6}$ **27.** $\frac{1}{5}$ **29.** $\frac{6}{5}$ **31.** $1\frac{5}{7}$ **33.** $6\frac{5}{12}$ **35.** $7\frac{6}{11}$ **37.** $\frac{13}{5}$ **39.** $\frac{83}{8}$ **41.** $\frac{54}{5}$ **43.** A **45.** $\frac{24}{35}$ **47.** $\frac{6}{25}$ **49.** $\frac{6}{5}$, or $1\frac{1}{5}$ **51.** $\frac{232}{15}$, or $15\frac{7}{15}$ **53.** $\frac{10}{3}$, or $3\frac{1}{3}$ **55.** 12 **57.** $\frac{1}{16}$ **59.** $\frac{84}{47}$, or $1\frac{37}{47}$ **61.** $\frac{2}{3}$ **63.** $\frac{8}{9}$ **65.** $\frac{27}{8}$, or $3\frac{3}{8}$ **67.** $\frac{17}{36}$ **69.** $\frac{11}{12}$ **71.** $\frac{4}{3}$, or $1\frac{1}{3}$ **73.** 6 cups **75.** $618\frac{3}{4}$ ft **77.** $\frac{9}{16}$ in. **79.** $\frac{5}{16}$ in. **81.** $16\frac{5}{8}$ yd **83.** $\frac{7}{100}$ **85.** about $4\frac{19}{25}$ million, or 4,760,000

Appendix C Synthetic Division

(pages 825–826)

1. C **3.** $x - 5$ **5.** $4m - 1$ **7.** $2a + 4 + \frac{5}{a + 2}$ **9.** $p - 4 + \frac{9}{p + 1}$ **11.** $4a^2 + a + 3$ **13.** $x^4 + 2x^3 + 2x^2 + 7x + 10 + \frac{18}{x - 2}$ **15.** $-4r^5 - 7r^4 - 10r^3 - 5r^2 - 11r - 8 + \frac{-8}{r - 1}$ **17.** $-3y^4 + 8y^3 - 21y^2 + 36y - 72 + \frac{143}{y + 2}$ **19.** $y^2 + y + 1 + \frac{2}{y - 1}$ **21.** 7 **23.** −2 **25.** 0 **27.** yes **29.** no **31.** no **33.** Since the variables are not present, a missing term will not be noticed in synthetic division, so the quotient will be wrong if placeholders are not inserted.

ANSWERS

SOLUTIONS TO SELECTED EXERCISES

Chapter 2 Linear Equations and Applications

Section 2.1 (pages 67–70)

31. $2[w - (2w + 4) + 3] = 2(w + 1)$

$2[w - 2w - 4 + 3] = 2(w + 1)$ Distributive property

$2[-w - 1] = 2(w + 1)$ Combine like terms.

$-2w - 2 = 2w + 2$ Distributive property

$-2 = 4w + 2$ Add $2w$.

$-4 = 4w$ Subtract 2.

$-1 = w$ Divide by 4.

Check Substitute -1 for w in the original equation.

Solution set: $\{-1\}$

47. $\frac{1}{5}x - 2 = \frac{2}{3}x - \frac{2}{5}x$

Multiply each side by the LCD, 15, and use the distributive property.

$15\left(\frac{1}{5}x\right) - 15(2) = 15\left(\frac{2}{3}x\right) - 15\left(\frac{2}{5}x\right)$

$3x - 30 = 10x - 6x$ Multiply.

$3x - 30 = 4x$ Combine like terms.

$-30 = x$ Subtract $3x$.

Solution set: $\{-30\}$

Section 2.2 (pages 79–84)

35. Use the formula for the volume of a rectangular solid:

$$V = LWH.$$

The thickness of the ream of paper is the height of the rectangular solid, so solve this formula for H.

$H = \frac{V}{LW}$ Divide by LW.

$H = \frac{187}{(11)(8.5)}$ Let $V = 187$, $L = 11$, and $W = 8.5$.

$H = 2$

The thickness of the ream is 2 in.

43. Substitute 380.50 for f, 8 for k, and 24 for n.

$u = 380.50 \cdot \frac{8(8 + 1)}{24(24 + 1)}$

$u = 380.50 \cdot \frac{8(9)}{24(25)}$

$u = 45.66$

The unearned interest is \$45.66.

Section 2.3 (pages 93–98)

31. ***Step 1***
Read the problem again.

Step 2
Let $x =$ the length of the middle side. Then the shortest side is $x - 75$ and the longest side is $x + 375$.

Step 3
The perimeter of the Bermuda Triangle is 3075 mi. Using the formula for perimeter of a triangle

$$P = a + b + c$$

gives the equation

$$x + (x - 75) + (x + 375) = 3075.$$

Step 4

$3x + 300 = 3075$

$3x = 2775$ Subtract 300.

$x = 925$ Divide by 3

Step 5
The length of the middle side is 925 mi. The length of the shortest side is

$$x - 75 = 925 - 75 = 850 \text{ mi}.$$

The length of the longest side is

$$x + 375 = 925 + 375 = 1300 \text{ mi}.$$

Step 6
The answer checks since

$$925 + 850 + 1300 = 3075 \text{ mi},$$

which is the correct perimeter.

41. Let $x =$ the amount of the receipts excluding tax. Since the sales tax is 9% of x, the total amount is

$x + 0.09x = 2725.$

$1x + 0.09x = 2725$

$1.09x = 2725$

$x = \frac{2725}{1.09}$

$x = 2500$

Thus, the tax was

$$0.09(2500) = \$225.$$

55. Let $x =$ the amount of \$6 per lb nuts.

Pounds of Nuts	Cost per lb	Total Cost
50	\$2	$2(50) = 100$
x	\$6	$6x$
$x + 50$	\$5	$5(x + 50)$

The total value of the \$2 per lb nuts and the \$6 per lb nuts must equal the value of the \$5 per lb nuts.

$100 + 6x = 5(x + 50)$

$100 + 6x = 5x + 250$

$x = 150$

He should use 150 lb of \$6 nuts.

Check 50 lb of the \$2 per lb nuts are worth \$100, and 150 lb of the \$6 per lb nuts are worth \$900; \$100 + \$900 = \$1000, which is the same as (50 + 150) lb worth of \$5 per lb nuts.

Section 2.4 (pages 105–108)

5. Let $x =$ the number of pennies. Then x is also the number of dimes, and $44 - 2x$ is the number of quarters.

Number of Coins	Denomination	Value
x	0.01	$0.01x$
x	0.10	$0.10x$
$44 - 2x$	0.25	$0.25(44 - 2x)$

The sum of the values must equal the total value, \$4.37.

$0.01x + 0.10x + 0.25(44 - 2x) = 4.37$

$x + 10x + 25(44 - 2x) = 437$ Multiply by 100.

$x + 10x + 1100 - 50x = 437$ Distributive property

$-39x + 1100 = 437$ Combine like terms.

$-39x = -663$ Subtract 1100.

$x = 17$ Divide by -39.

There are 17 pennies, 17 dimes, and $44 - 2(17) = 10$ quarters.

Check The number of coins is

$$17 + 17 + 10 = 44$$

and the value of the coins is

$$\$0.01(17) + \$0.10(17) + \$0.25(10) = \$4.37,$$

as required.

19. Let $t =$ Mulder's time traveled.

Then $t - \frac{1}{2} =$ Scully's time traveled.

	Rate	Time	Distance
Mulder	65	t	$65t$
Scully	68	$t - \frac{1}{2}$	$68\left(t - \frac{1}{2}\right)$

(continued)

SOLUTIONS

The distances are equal.

$$65t = 68\left(t - \frac{1}{2}\right)$$
$$65t = 68t - 34$$
$$-3t = -34$$
$$t = \frac{34}{3}, \text{ or } 11\frac{1}{3}$$

Mulder's time traveled will be $11\frac{1}{3}$ hr.

Since Mulder left at 8:30 A.M., $11\frac{1}{3}$ hr (or 11 hr, 20 min) later is 7:50 P.M.

Check Mulder's distance was

$$65\left(\frac{34}{3}\right) = 736\frac{2}{3} \text{ mi.}$$

Scully's distance was

$$68\left(\frac{34}{3} - \frac{1}{2}\right) = 68\left(\frac{65}{6}\right) = 736\frac{2}{3} \text{ mi,}$$

as required.

21. Let x = her average speed on Sunday. Then $x + 5$ = her average speed on Saturday.

	Rate	Time	Distance
Saturday	$x + 5$	3.6	$3.6(x + 5)$
Sunday	x	4	$4x$

The distances are equal.

$3.6(x + 5) = 4x$

$3.6x + 18 = 4x$ Distributive property

$18 = 0.4x$ Subtract $3.6x$.

$x = \frac{18}{0.4}$ Divide by 0.4; rewrite.

$x = 45$

Her average speed on Sunday was 45 mph.

Check On Sunday,

4 hr at 45 mph = 180 mi.

On Saturday,

3.6 hr at 50 mph = 180 mi.

The distances are equal.

39. Let x = the current age. Then $x + 1$ will be the age next year. The sum of these ages will be 95 yr.

$x + (x + 1) = 95$

$2x + 1 = 95$

$2x = 94$ Subtract 1.

$x = 47$ Divide by 2.

If my current age is 47, in 10 yr I will be

47 + 10 = 57 yr old.

Chapter 3 Linear Inequalities and Absolute Value

Section 3.1 (pages 137–142)

25. $$\frac{2k - 5}{-4} > 5$$

Multiply each side by -4 and reverse the direction of the inequality symbol.

$-4\left(\frac{2k - 5}{-4}\right) < -4(5)$

$2k - 5 < -20$

$2k < -15$ Add 5.

$k < -\frac{15}{2}$ Divide by 2.

Check that the solution set is the interval $\left(-\infty, -\frac{15}{2}\right)$.

49. $-6 \le 2(z + 2) \le 16$

$-6 \le 2z + 4 \le 16$ Distributive property

$-10 \le 2z \le 12$ Subtract 4.

$-5 \le z \le 6$ Divide by 2.

Check that the solution set is the interval $[-5, 6]$.

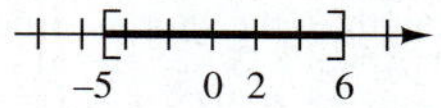

Section 3.2 (pages 151–154)

41. $4x - 8 > 0$ or $4x - 1 < 7$

$4x > 8$ or $4x < 8$

$x > 2$ or $x < 2$

The graph of the solution set is all numbers either greater than 2 or less than 2. This is all real numbers except 2. The solution set is $(-\infty, 2) \cup (2, \infty)$.

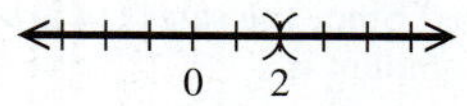

43. $(-\infty, -1] \cap [-4, \infty)$

The intersection is the set of numbers less than or equal to -1 and greater than or equal to -4. The numbers common to *both* original sets are between, and including, -4 and -1. The simplest interval form is $[-4, -1]$.

Section 3.3 (pages 163–168)

19. $$\left|1 - \frac{3}{4}k\right| = 7$$

$1 - \frac{3}{4}k = 7$ or $1 - \frac{3}{4}k = -7$

$-\frac{3}{4}k = 6$ or $-\frac{3}{4}k = -8$ Subtract 1.

$k = -8$ or $k = \frac{32}{3}$ Multiply by $-\frac{4}{3}$.

Solution set: $\left\{-8, \frac{32}{3}\right\}$

49. $|-3x - 8| \le 4$

$-4 \le -3x - 8 \le 4$

$4 \le -3x \le 12$ Add 8.

Divide each part by -3 and reverse the direction of the inequality symbols.

$\frac{4}{-3} \ge x \ge \frac{12}{-3}$

$-\frac{4}{3} \ge x \ge -4$

Rewrite in order based on a number line.

$-4 \le x \le -\frac{4}{3}$

Solution set: $\left[-4, -\frac{4}{3}\right]$

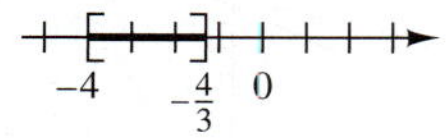

65. $|2p - 6| = |2p + 11|$

$2p - 6 = 2p + 11$

$-6 = 11$ False

No solution

or $2p - 6 = -(2p + 11)$

$2p - 6 = -2p - 11$

$4p = -5$

$p = -\frac{5}{4}$

Solution set: $\left\{-\frac{5}{4}\right\}$

81. $|10z + 7| > 0$

Since the absolute value of an expression is always nonnegative, there is only one possible value of z that makes this statement false. Solving the equation $10z + 7 = 0$ will give that value of z.

$10z + 7 = 0$

$10z = -7$

$z = -\frac{7}{10}$

The solution set of the inequality includes *all values other than* $-\frac{7}{10}$, which makes the absolute value expression equal 0.

Solution set: $\left(-\infty, -\frac{7}{10}\right) \cup \left(-\frac{7}{10}, \infty\right)$

Chapter 4 Graphs, Linear Equations, and Functions

Section 4.1 (pages 195–198)

35. $3x - 7y = 9$

To find the x-intercept, let $y = 0$.

$3x - 7y = 9$

$3x - 7(0) = 9$

$3x = 9$

$x = 3$

The x-intercept is $(3, 0)$.

To find the y-intercept, let $x = 0$.

$$3x - 7y = 9$$
$$3(0) - 7y = 9$$
$$-7y = 9$$
$$y = -\frac{9}{7}$$

The y-intercept is $\left(0, -\frac{9}{7}\right)$.

Plot the intercepts and draw the line through them.

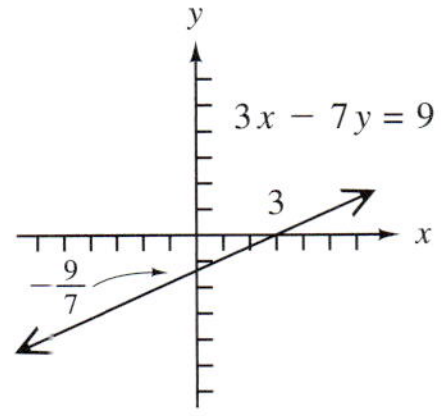

43. $2x = 3y$

If $x = 0$, then $y = 0$, so the x- and y-intercepts are $(0, 0)$. To get another point, let $x = 3$.

$$2x = 3y$$
$$2(3) = 3y$$
$$6 = 3y$$
$$2 = y$$

Thus, the point $(3, 2)$ is on the graph. Plot $(3, 2)$ and $(0, 0)$, and draw the line through them.

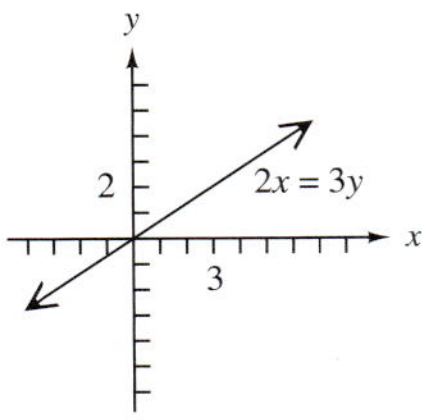

51. By the midpoint formula, the midpoint of the segment with endpoints $(2.5, 3.1)$ and $(1.7, -1.3)$ is

$$\left(\frac{2.5 + 1.7}{2}, \frac{3.1 + (-1.3)}{2}\right)$$
$$= \left(\frac{4.2}{2}, \frac{1.8}{2}\right)$$
$$= (2.1, 0.9).$$

55. By the midpoint formula, the midpoint of the segment with endpoints $\left(-\frac{1}{3}, \frac{2}{7}\right)$ and $\left(-\frac{1}{2}, \frac{1}{14}\right)$ is

$$\left(\frac{-\frac{1}{3} + \left(-\frac{1}{2}\right)}{2}, \frac{\frac{2}{7} + \frac{1}{14}}{2}\right)$$
$$= \left(\frac{-\frac{5}{6}}{2}, \frac{\frac{5}{14}}{2}\right)$$
$$= \left(-\frac{5}{12}, \frac{5}{28}\right).$$

Note: Simplify $\dfrac{-\frac{5}{6}}{2}$ as follows.

$$\frac{-\frac{5}{6}}{2}$$
$$= -\frac{5}{6} \div 2$$
$$= -\frac{5}{6} \cdot \frac{1}{2}$$
$$= -\frac{5}{12}$$

The fraction $\dfrac{\frac{5}{14}}{2}$ can be simplified similarly.

Section 4.2 (pages 207–212)

41. To graph the line through $(0, -2)$ with slope $m = -\frac{2}{3}$, locate the point $(0, -2)$ on the graph. To find a second point on the line, use the definition of slope, writing $-\frac{2}{3}$ as $\frac{-2}{3}$.

$$m = \frac{\text{change in } y}{\text{change in } x} = \frac{-2}{3}$$

From $(0, -2)$, move 2 units down and then 3 units to the right to get to $(3, -4)$. Draw the line.

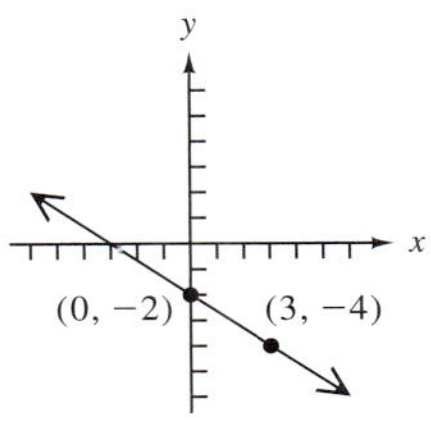

51. $3x = y$ and $2y - 6x = 5$

To determine whether the lines are *parallel*, *perpendicular*, or *neither*, we must find the slope of each line. The slope of the first line is the coefficient of x, namely 3. Solve the second equation for y.

$$2y - 6x = 5$$
$$2y = 6x + 5$$
$$y = 3x + \frac{5}{2}$$

The slope of the second line is also 3, so the lines are *parallel*.

59. Use the points $(0, 20)$ and $(4, 4)$ from the graph.

average rate of change

$$= \frac{\text{change in } y}{\text{change in } x} = \frac{4 - 20}{4 - 0} = \frac{-16}{4} = -4$$

The average rate of change is $-\$4000$ per year, that is, the value of the machine is decreasing $4000 each year during these years.

Section 4.3 (pages 221–226)

27. Through $(-2, 4)$; $m = -\frac{3}{4}$

Use the point-slope form with $(x_1, y_1) = (-2, 4)$ and $m = -\frac{3}{4}$.

$$y - y_1 = m(x - x_1)$$
$$y - 4 = -\frac{3}{4}[x - (-2)] \quad \text{Substitute.}$$
$$4(y - 4) = -3(x + 2) \quad \text{Multiply by 4.}$$
$$4y - 16 = -3x - 6 \quad \text{Distributive property}$$
$$3x + 4y = 10 \quad \text{Add } 3x\text{; add 16.}$$

37. Through $(0.5, 0.2)$; horizontal

A horizontal line through the point (a, b) has equation $y = b$. Since the y-value in $(0.5, 0.2)$ is 0.2, the equation of this line is $y = 0.2$.

43. $\left(-\frac{2}{5}, \frac{2}{5}\right)$ and $\left(\frac{4}{3}, \frac{2}{3}\right)$

To write an equation of the line through these points, first find the slope.

$$m = \frac{\frac{2}{3} - \frac{2}{5}}{\frac{4}{3} - \left(-\frac{2}{5}\right)}$$
$$= \frac{\frac{10 - 6}{15}}{\frac{20 + 6}{15}} \quad \text{Use common denominator in numerator and denominator.}$$
$$= \frac{\frac{4}{15}}{\frac{26}{15}}$$
$$= \frac{4}{15} \div \frac{26}{15}$$
$$= \frac{4}{15} \cdot \frac{15}{26}$$
$$= \frac{2}{13}$$

Use the point-slope form with $(x_1, y_1) = \left(-\frac{2}{5}, \frac{2}{5}\right)$ and $m = \frac{2}{13}$.

$$y - y_1 = m(x - x_1)$$
$$y - \frac{2}{5} = \frac{2}{13}\left[x - \left(-\frac{2}{5}\right)\right]$$
$$13\left(y - \frac{2}{5}\right) = 2\left(x + \frac{2}{5}\right)$$
$$13y - \frac{26}{5} = 2x + \frac{4}{5}$$
$$-2x + 13y = \frac{30}{5}$$
$$2x - 13y = -6$$

Section 4.4 (pages 233–236)

27. $|y + 1| < 2$ can be rewritten as

$$-2 < y + 1 < 2$$
$$-3 < y < 1.$$

The boundaries are the dashed horizontal lines $y = -3$ and $y = 1$. Since y is between -3 and 1, the graph includes all points between the lines.

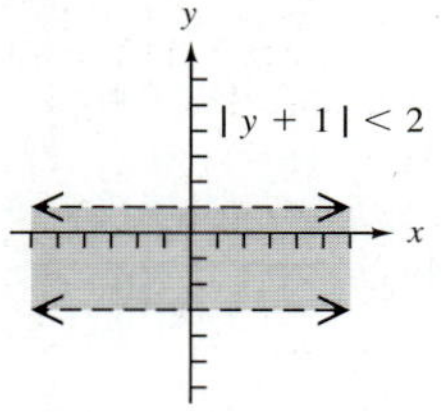

33. $3x + 2y < 6$ or $x - 2y > 2$

Graph $3x + 2y = 6$, which has intercepts (2, 0) and (0, 3), as a dashed line. Test (0, 0), which yields $0 < 6$, a true statement. Shade the region that includes (0, 0).

Graph $x - 2y = 2$, which has intercepts (2, 0) and $(0, -1)$, as a dashed line. Test (0, 0), which yields $0 > 2$, a false statement. Shade the region that does not include (0, 0).

The required graph of the union includes all the shaded regions, that is, all the points that satisfy either inequality.

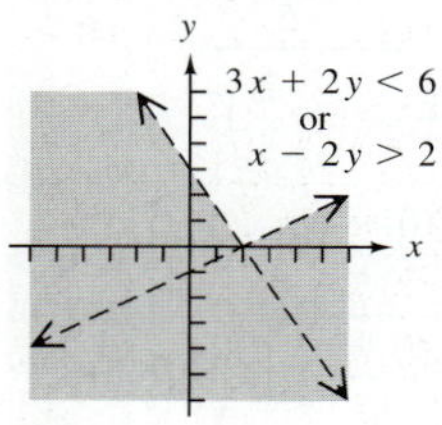

Section 4.5 (pages 247–252)

33. $xy = 1$

Divide both sides of the equation by x to rewrite $xy = 1$ as $y = \frac{1}{x}$. Note that x can never equal 0, or the denominator would equal 0. Thus, the domain is

$$(-\infty, 0) \cup (0, \infty).$$

Each nonzero x-value gives exactly one y-value. Therefore, $xy = 1$ defines y as a function of x.

39. Refer to the graph on page 249.

(a) The independent variable is t, the number of hours, and the possible values are in the set [0, 100]. The dependent variable is g, the number of gallons, and the possible values are in the set [0, 3000].

(b) The graph rises for the first 25 hr, so the water level increases for 25 hr. The graph falls for $t = 50$ to $t = 75$, so the water level decreases for 25 hr.

(c) There are 2000 gal in the pool when $t = 90$.

(d) $g(0)$ is the number of gallons in the pool at time $t = 0$. Here, $g(0) = 0$, which means the pool is empty at time 0.

53. $g(x) = -x^2 + 4x + 1$

$$g\left(\frac{p}{3}\right) = -\left(\frac{p}{3}\right)^2 + 4\left(\frac{p}{3}\right) + 1$$
$$g\left(\frac{p}{3}\right) = -\frac{p^2}{9} + \frac{4p}{3} + 1$$

67. (a) Solve the given equation for y.

$$4x - 3y = 8$$
$$-3y = -4x + 8$$
$$y = \frac{4}{3}x - \frac{8}{3}$$

Since $y = f(x)$,

$$f(x) = \frac{4}{3}x - \frac{8}{3}.$$

(b) $f(3) = \frac{4}{3}(3) - \frac{8}{3}$

$$f(3) = \frac{12}{3} - \frac{8}{3}$$
$$f(3) = \frac{4}{3}$$

Chapter 5 Systems of Linear Equations

Section 5.1 (pages 277–282)

33.

$$y = 2x \quad (1)$$
$$4x - 2y = 0 \quad (2)$$

From equation (1), substitute $2x$ for y in equation (2).

$$4x - 2y = 0 \quad (2)$$
$$4x - 2(2x) = 0 \quad \text{Let } y = 2x.$$
$$4x - 4x = 0$$
$$0 = 0 \quad \text{True}$$

The equations are dependent, and the solution set is the set of all points on the line. We use one of the equations of the system to write the solution set in set-builder notation. As indicated in the Note on page 275, we give the equation in standard form with coefficients that are integers having greatest common factor 1 and positive coefficient of x. Thus, we use equation (2) and divide each term by the common factor 2 to get

$$\{(x, y) \mid 2x - y = 0\}$$

as the solution set.

47.

$$\frac{x}{2} + \frac{y}{3} = -\frac{1}{3} \quad (1)$$
$$\frac{x}{2} + 2y = -7 \quad (2)$$

Clear the fractions by multiplying equation (1) by -6 and equation (2) by 6. (We multiply equation (2) by 6 instead of 2 so that when the fractions are cleared, the x-terms are opposites.) Then add the results to eliminate x.

$$-3x - 2y = 2 \quad \text{Multiply (1) by } -6. \quad (3)$$
$$3x + 12y = -42 \quad \text{Multiply (2) by 6.}$$
$$10y = -40 \quad \text{Add.}$$
$$y = -4 \quad \text{Divide by 10.}$$

To find x, substitute -4 for y in equation (3).

$$-3x - 2y = 2 \quad (3)$$
$$-3x - 2(-4) = 2 \quad \text{Let } y = -4.$$
$$-3x + 8 = 2$$
$$-3x = -6 \quad \text{Subtract 8.}$$
$$x = 2 \quad \text{Divide by } -3.$$

The solution $(2, -4)$ checks in both equations (1) and (2).

Solution set: $\{(2, -4)\}$

63.

$$0.3x + 0.2y = 0.4 \quad (1)$$
$$0.5x + 0.4y = 0.7 \quad (2)$$

Clear the decimals by multiplying each equation by 10 to get the equivalent system

$$3x + 2y = 4 \quad (3)$$
$$5x + 4y = 7. \quad (4)$$

To eliminate y, multiply equation (3) by -2 and then add the result to equation (4).

$$-6x - 4y = -8 \quad \text{Multiply (3) by } -2.$$
$$5x + 4y = 7 \quad (4)$$
$$-x = -1 \quad \text{Add.}$$
$$x = 1 \quad \text{Multiply by } -1$$

Substitute 1 for x in equation (4).

$$5x + 4y = 7 \quad (4)$$
$$5(1) + 4y = 7 \quad \text{Let } x = 1.$$
$$4y = 2 \quad \text{Subtract 5.}$$
$$y = \frac{1}{2} \quad \text{Divide by 4.}$$

Check $\left(1, \frac{1}{2}\right)$ in both equations (1) and (2).

Solution set: $\left\{\left(1, \frac{1}{2}\right)\right\}$

Section 5.2 (pages 289–292)

13.

$$x + 2y + 3z = 1 \quad (1)$$
$$-x - y + 3z = 2 \quad (2)$$
$$-6x + y + z = -2 \quad (3)$$

Step 1

Since x in equation (1) has coefficient 1, we choose it as the focus variable and (1) as the working equation.

Step 2

Add equations (1) and (2) to eliminate x.

$$x + 2y + 3z = 1 \quad (1)$$
$$-x - y + 3z = 2 \quad (2)$$
$$y + 6z = 3 \quad \text{Add.} \quad (4)$$

Step 3

Multiply working equation (1) by 6 and add to equation (3) to eliminate x again.

$$6x + 12y + 18z = 6 \quad \text{Multiply (1) by 6.}$$
$$-6x + y + z = -2 \quad (3)$$
$$13y + 19z = 4 \quad \text{Add.} \quad (5)$$

Step 4

The equations that resulted in Steps 2 and 3 form a system in y and z.

$$y + 6z = 3 \quad (4)$$
$$13y + 19z = 4 \quad (5)$$

Solve this system of equations (4) and (5).

$$\begin{aligned} -13y - 78z &= -39 && \text{Multiply (4) by } -13. \\ 13y + 19z &= 4 && (5) \\ \hline -59z &= -35 && \text{Add.} \\ z &= \frac{35}{59} \end{aligned}$$

Substitute $\frac{35}{59}$ for z in equation (4) to find y. Be careful—the arithmetic gets messy.

$$\begin{aligned} y + 6z &= 3 && (4) \\ y + 6\left(\frac{35}{59}\right) &= 3 && \text{Let } z = \frac{35}{59}. \\ y + \frac{210}{59} &= 3 \\ y &= \frac{177}{59} - \frac{210}{59} && 3 = \frac{177}{59} \\ y &= -\frac{33}{59} \end{aligned}$$

Step 5

Substitute $-\frac{33}{59}$ for y and $\frac{35}{59}$ for z in working equation (1) to find focus variable x.

$$\begin{aligned} x + 2y + 3z &= 1 && (1) \\ x + 2\left(-\frac{33}{59}\right) + 3\left(\frac{35}{59}\right) &= 1 \\ x - \frac{66}{59} + \frac{105}{59} &= 1 \\ x + \frac{39}{59} &= 1 \\ x &= \frac{59}{59} - \frac{39}{59} \\ x &= \frac{20}{59} \end{aligned}$$

Step 6

The solution $\left(\frac{20}{59}, -\frac{33}{59}, \frac{35}{59}\right)$ checks when substituted in equations (1), (2), and (3).

Solution set: $\left\{\left(\frac{20}{59}, -\frac{33}{59}, \frac{35}{59}\right)\right\}$

21.

$$\begin{aligned} x + y - 2z &= 0 && (1) \\ 3x - y + z &= 0 && (2) \\ 4x + 2y - z &= 0 && (3) \end{aligned}$$

We choose z as the focus variable and (2) as the working equation. Eliminate z by adding equations (2) and (3).

$$\begin{aligned} 3x - y + z &= 0 && (2) \\ 4x + 2y - z &= 0 && (3) \\ \hline 7x + y &= 0 && (4) \end{aligned}$$

To get another equation without z, multiply working equation (2) by 2 and add the result to equation (1).

$$\begin{aligned} 6x - 2y + 2z &= 0 && \text{Multiply (2) by 2.} \\ x + y - 2z &= 0 && (1) \\ \hline 7x - y &= 0 && (5) \end{aligned}$$

Add equations (4) and (5) to find x.

$$\begin{aligned} 7x + y &= 0 && (4) \\ 7x - y &= 0 && (5) \\ \hline 14x &= 0 \\ x &= 0 \end{aligned}$$

Substitute 0 for x in equation (4) to find y.

$$\begin{aligned} 7x + y &= 0 && (4) \\ 7(0) + y &= 0 \\ 0 + y &= 0 \\ y &= 0 \end{aligned}$$

Substitute 0 for x and 0 for y in working equation (2) to find focus variable z.

$$\begin{aligned} 3x - y + z &= 0 && (2) \\ 3(0) - 0 + z &= 0 \\ z &= 0 \end{aligned}$$

The solution (0, 0, 0) checks when substituted in equations (1), (2), and (3).

Solution set: $\{(0, 0, 0)\}$

Section 5.3 (pages 301–306)

7. From the figure given with the problem, the angles marked y and $3x + 10$ are supplementary, so

$$(3x + 10) + y = 180. \quad (1)$$

The sum of the angles of a triangle is 180°. Since the given triangle is a right triangle, one angle measures 90°. That leaves $180° - 90° = 90°$ for the sum of the measures of the other two angles x and y, so

$$x + y = 90. \quad (2)$$

Solve equation (2) for y to get

$$y = 90 - x. \quad (3)$$

Substitute $90 - x$ for y in equation (1).

$$\begin{aligned} (3x + 10) + y &= 180 && (1) \\ (3x + 10) + (90 - x) &= 180 \\ 2x + 100 &= 180 \\ 2x &= 80 \\ x &= 40 \end{aligned}$$

Substitute $x = 40$ into equation (3) to get

$$y = 90 - x = 90 - 40 = 50.$$

Since $x = 40$ and $y = 50$, the angles measure 40° and 50°.

29. Let $x =$ the speed of the boat in still water and $y =$ the speed of the current.

Furthermore,

rate upstream $= x - y$

and rate downstream $= x + y$.

Use these rates and the information in the problem to make a table.

	r	t	d
Upstream	$x - y$	2	36
Downstream	$x + y$	1.5	36

From the table, use the formula $d = rt$ to write a system of equations.

$$\begin{aligned} 36 &= 2(x - y) \\ 36 &= 1.5(x + y) \end{aligned}$$

Clear the parentheses and rewrite the equations with the variables on the left side.

$$\begin{aligned} 2x - 2y &= 36 && (1) \\ 1.5x + 1.5y &= 36 && (2) \end{aligned}$$

Solve the system. Multiply equation (1) by -3 and equation (2) by 4. Then add the results.

$$\begin{aligned} -6x + 6y &= -108 && \text{Multiply (1) by } -3. \\ 6x + 6y &= 144 && \text{Multiply (2) by 4.} \\ \hline 12y &= 36 && \text{Add.} \\ y &= 3 && \text{Divide by 12.} \end{aligned}$$

Substitute $y = 3$ into equation (1).

$$\begin{aligned} 2x - 2y &= 36 && (1) \\ 2x - 2(3) &= 36 \\ 2x - 6 &= 36 \\ 2x &= 42 \\ x &= 21 \end{aligned}$$

The speed of the boat is 21 mph, and the speed of the current is 3 mph.

31. Let $x =$ the number of pounds of the \$0.75-per-lb candy and $y =$ the number of pounds of the \$1.25-per-lb candy.

Make a table.

	Number of Pounds	Price per Pound	Value
Less Expensive Candy	x	0.75	$0.75x$
More Expensive Candy	y	1.25	$1.25y$
Mixture	9	0.96	$0.96(9) = 8.64$

From the "Number of Pounds" column,

$$x + y = 9. \quad (1)$$

From the "Value" column,

$$0.75x + 1.25y = 8.64 \quad (2)$$

Solve the system of equations (1) and (2). Multiply equation (1) by -75 and equation (2) by 100. Then add the results.

$$\begin{aligned} -75x - 75y &= -675 \\ 75x + 125y &= 864 \\ \hline 50y &= 189 && \text{Add.} \\ y &= 3.78 && \text{Divide by 50.} \end{aligned}$$

Substitute 3.78 for y in equation (1).

$$\begin{aligned} x + y &= 9 && (1) \\ x + 3.78 &= 9 \\ x &= 5.22 \end{aligned}$$

Mix 5.22 lb of the \$0.75-per-lb candy with 3.78 lb of the \$1.25-per-lb candy to obtain 9 lb of a mixture that sells for \$0.96 per lb.

Chapter 6 Exponents, Polynomials, and Polynomial Functions

Section 6.1 (pages 339–344)

83. $\left(\frac{-s^3}{t^5}\right)^4$

$= \frac{(-1)^4(s^3)^4}{(t^5)^4}$ Power rules (b) and (c)

$= \frac{1 \cdot s^{3 \cdot 4}}{t^{5 \cdot 4}}$ Power rule (a)

$= \frac{1 \cdot s^{12}}{t^{20}}$ Product rule

$= \frac{s^{12}}{t^{20}}$ $1 \cdot a = a$

103. $\left(\frac{3k^{-2}}{k^4}\right)^{-1} \cdot \frac{2}{k}$

$= \frac{(3k^{-2})^{-1}}{(k^4)^{-1}} \cdot \frac{2}{k}$ Power rule (c)

$= \frac{3^{-1}k^2 \cdot 2}{k^{-4}k^1}$ Power rules (b) and (a); multiply fractions.

$= \frac{3^{-1} \cdot 2k^2}{k^{-3}}$ Product rule

$= \frac{2k^2k^3}{3}$ $a^{-n} = \frac{1}{a^n}$

$= \frac{2k^5}{3}$ Product rule

127. $\frac{0.05 \times 1600}{0.0004}$

$= \frac{5 \times 10^{-2} \times 1.6 \times 10^3}{4 \times 10^{-4}}$

$= \frac{5(1.6)}{4} \times \frac{10^{-2} \times 10^3}{10^{-4}}$

$= 2 \times 10^{-2+3-(-4)}$

$= 2 \times 10^5$

$= 200{,}000$

135. Use $d = rt$, or $\frac{d}{r} = t$, where $d = 9.3 \times 10^7$ and $r = 2.9 \times 10^3$.

$\frac{9.3 \times 10^7}{2.9 \times 10^3}$

$= \frac{9.3}{2.9} \times 10^{7-3}$

$\approx 3.2 \times 10^4$

It would take about 3.2×10^4, or 32,000 hr. Note that there are

$24 \times 365 = 8760$ hr

in one year.

$\frac{32{,}000 \text{ hr}}{8760 \text{ hr/yr}} \approx 3.7$ yr

Section 6.2 (pages 349–350)

35. $n^4 - 2n^3 + n^2 - 3n^4 + n^3$

$= n^4 - 3n^4 - 2n^3 + n^3 + n^2$ Rearrange terms.

$= (1 - 3)n^4 + (-2 + 1)n^3 + n^2$ Distributive property

$= -2n^4 - n^3 + n^2$

47.
$$\begin{array}{r} -5a^4 \qquad\quad + 8a^2 - 9 \\ 6a^3 - \;\; a^2 + 2 \\ \hline \end{array}$$

To subtract, change all the signs in the second polynomial, and add. Write the missing terms with 0 coefficients.

$$\begin{array}{rl} -5a^4 + 0a^3 + 8a^2 - \;\; 9 & \\ 0a^4 - 6a^3 + \;\; a^2 - \;\; 2 & \text{Change all signs.} \\ \hline -5a^4 - 6a^3 + 9a^2 - 11 & \text{Add in columns.} \end{array}$$

Section 6.3 (pages 357–358)

23. $(f - h)(-3)$

$= f(-3) - h(-3)$

$= [(-3)^2 - 9] - [(-3) - 3]$

$= (9 - 9) - (-6)$

$= 0 + 6$

$= 6$

Alternatively, we could evaluate the polynomial in Exercise 21, $x^2 - x - 6$, using $x = -3$.

31.

x	$f(x) = -3x^2$
-2	$-3(-2)^2 = -12$
-1	$-3(-1)^2 = -3$
0	$-3(0)^2 = 0$
1	$-3(1)^2 = -3$
2	$-3(2)^2 = -12$

Since the greatest exponent is 2, the graph of f is a parabola.

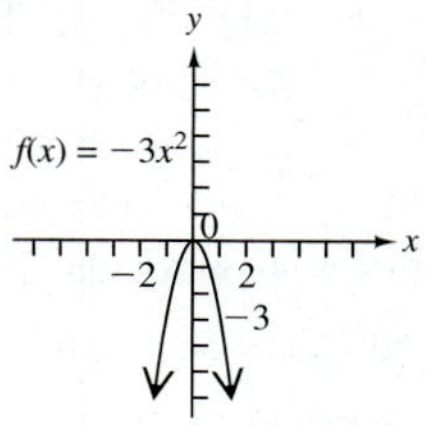

Section 6.4 (pages 365–368)

21.
$$\begin{array}{r} 2p^2 + \;\; 3p + 6 \\ 3p^2 - \;\; 4p - 1 \\ \hline -2p^2 - \;\; 3p - 6 \\ -8p^3 - 12p^2 - 24p \qquad \\ 6p^4 + 9p^3 + 18p^2 \qquad\qquad\qquad \\ \hline 6p^4 + \;\; p^3 + \;\; 4p^2 - 27p - 6 \end{array}$$

31. $(0.2x + 1.3)(0.5x - 0.1)$

F O

$= (0.2x)(0.5x) + (0.2x)(-0.1)$

I L

$+ (1.3)(0.5x) + (1.3)(-0.1)$

$= 0.1x^2 - 0.02x + 0.65x - 0.13$

$= 0.1x^2 + 0.63x - 0.13$ Combine like terms.

33. $\left(3r + \frac{1}{4}y\right)(r - 2y)$

F O

$= 3r(r) + 3r(-2y)$

I L

$+ \left(\frac{1}{4}y\right)(r) + \left(\frac{1}{4}y\right)(-2y)$

$= 3r^2 - 6ry + \frac{1}{4}yr - \frac{1}{2}y^2$

$= 3r^2 - \frac{24}{4}ry + \frac{1}{4}ry - \frac{1}{2}y^2$

$6 = \frac{24}{4}$; Commutative property

$= 3r^2 - \frac{23}{4}ry - \frac{1}{2}y^2$

Combine like terms.

43. $(4m + 7n^2)(4m - 7n^2)$

$= (4m)^2 - (7n^2)^2$

$= 16m^2 - 49n^4$

65. $(q - 2)^4$

$= (q - 2)^2(q - 2)^2$

$= (q^2 - 4q + 4)(q^2 - 4q + 4)$

Multiply either horizontally or vertically.

$$\begin{array}{r} q^2 - \;\; 4q + \;\; 4 \\ q^2 - \;\; 4q + \;\; 4 \\ \hline 4q^2 - 16q + 16 \\ -4q^3 + 16q^2 - 16q \qquad \\ q^4 - 4q^3 + \;\; 4q^2 \qquad\qquad\qquad \\ \hline q^4 - 8q^3 + 24q^2 - 32q + 16 \end{array}$$

Section 6.5 (pages 373–374)

19. $\frac{14x + 6x^3 - 15 - 19x^2}{3x^2 - 2x + 4}$

Rewrite as $\frac{6x^3 - 19x^2 + 14x - 15}{3x^2 - 2x + 4}$.

$$\begin{array}{r} 2x - \;\; 5 \\ 3x^2 - 2x + 4 \overline{)\, 6x^3 - 19x^2 + 14x - 15} \\ 6x^3 - \;\; 4x^2 + \;\; 8x \qquad \\ \hline -15x^2 + \;\; 6x - 15 \\ -15x^2 + 10x - 20 \\ \hline -4x + \;\; 5 \end{array}$$

Answer: $2x - 5 + \frac{-4x + 5}{3x^2 - 2x + 4}$

27. $(9z^4 - 13z^3 + 23z^2 - 10z + 8) \div (z^2 - z + 2)$

$$\begin{array}{r} 9z^2 - \;\; 4z + 1 \\ z^2 - z + 2 \overline{)\, 9z^4 - 13z^3 + 23z^2 - 10z + 8} \\ 9z^4 - \;\; 9z^3 + 18z^2 \qquad\qquad\quad \\ \hline -4z^3 + \;\; 5z^2 - 10z \qquad \\ -4z^3 + \;\; 4z^2 - \;\; 8z \qquad \\ \hline z^2 - \;\; 2z + 8 \\ z^2 - \;\;\;\; z + 2 \\ \hline -z + 6 \end{array}$$

Answer: $9z^2 - 4z + 1 + \frac{-z + 6}{z^2 - z + 2}$

31. $\left(3a^2 - \frac{23}{4}a - 5\right) \div (4a + 3)$

$$\begin{array}{r} \frac{3}{4}a - 2 \\ 4a + 3 \overline{)3a^2 - \frac{23}{4}a - 5} \\ \underline{3a^2 + \frac{9}{4}a } \\ -8a - 5 \\ \underline{-8a - 6} \\ 1 \end{array} \quad \text{Remainder}$$

Answer: $\frac{3}{4}a - 2 + \frac{1}{4a + 3}$

47. Use the formula for the volume of a rectangular solid (box).

$V = LWH$

$2p^3 + 15p^2 + 28p = (p + 4)(W)(p)$ Let $L = p + 4$ and $H = p$.

$2p^3 + 15p^2 + 28p = (p^2 + 4p)W$ Distributive property

$\frac{2p^3 + 15p^2 + 28p}{p^2 + 4p} = W$ Divide by $p^2 + 4p$.

To find an expression for W, divide $2p^3 + 15p^2 + 28p$ by $p^2 + 4p$.

$$\begin{array}{r} 2p + 7 \\ p^2 + 4p \overline{)2p^3 + 15p^2 + 28p} \\ \underline{2p^3 + 8p^2 } \\ 7p^2 + 28p \\ \underline{7p^2 + 28p} \\ 0 \end{array}$$

The width of the box is $(2p + 7)$ feet.

Chapter 7 Factoring

Section 7.1 (pages 393–394)

33. $5(2 - x)^3 - (2 - x)^4 + 4(2 - x)^2$
$= (2 - x)^2[5(2 - x) - (2 - x)^2 + 4]$ Factor out $(2 - x)^2$.
$= (2 - x)^2 \cdot [10 - 5x - (4 - 4x + x^2) + 4]$ Multiply inside brackets.
$= (2 - x)^2 \cdot [10 - 5x - 4 + 4x - x^2 + 4]$ Clear parentheses inside brackets.
$= (2 - x)^2(10 - x - x^2)$ Combine like terms.

63. $3m^{-5} + m^{-3}$

Factor out m^{-5} since -5 is the lesser exponent on m.

$= m^{-5}(3) + m^{-5}(m^{-3-(-5)})$
$= m^{-5}(3 + m^2)$

Section 7.2 (pages 401–402)

29. $6x^2z^2 + 5xz - 4$

Two integer factors whose product is $6(-4) = -24$ and whose sum is 5 are 8 and -3. Rewrite the trinomial in a form that can be factored by grouping.

$6x^2z^2 + 5xz - 4$
$= 6x^2z^2 + 8xz - 3xz - 4$
$= (6x^2z^2 + 8xz) - (3xz + 4)$
$= 2xz(3xz + 4) - 1(3xz + 4)$
$= (3xz + 4)(2xz - 1)$

37. $2x^3y^3 - 48x^2y^4 + 288xy^5$
$= 2xy^3(x^2 - 24xy + 144y^2)$
$= 2xy^3(x - 12y)(x - 12y)$

41. $3(m + p)^2 - 7(m + p) - 20$
$= 3t^2 - 7t - 20$ Let $t = m + p$.
$= (3t + 5)(t - 4)$ Factor.
$= [3(m + p) + 5][(m + p) - 4]$ Replace t with $m + p$.
$= (3m + 3p + 5)(m + p - 4)$

49. $12p^6 - 32p^3r + 5r^2$
$= 12t^2 - 32tr + 5r^2$ Let $t = p^3$.
$= (6t - r)(2t - 5r)$ Factor.
$= (6p^3 - r)(2p^3 - 5r)$ Replace t with p^3.

Section 7.3 (pages 407–408)

11. $64m^4 - 4y^4$
$= 4(16m^4 - y^4)$ Factor out the GCF, 4.
$= 4[(4m^2)^2 - (y^2)^2]$ Write as a difference of squares.
$= 4(4m^2 + y^2)(4m^2 - y^2)$ Factor the difference of squares.
$= 4(4m^2 + y^2)[(2m)^2 - y^2]$
$= 4(4m^2 + y^2)[(2m + y)(2m - y)$ Factor the difference of squares again.

27. $x^2 - y^2 + 2y - 1$
$= x^2 - (y^2 - 2y + 1)$ Group the last three terms.
$= x^2 - (y - 1)^2$ Factor the perfect square trinomial.
$= [x + (y - 1)][x - (y - 1)]$ Factor the difference of squares.
$= (x + y - 1)(x - y + 1)$

31. $(p + q)^2 + 2(p + q) + 1$
$= t^2 + 2t + 1$ Let $t = p + q$.
$= (t + 1)^2$ Factor the perfect square trinomial.
$= (p + q - 1)^2$ Replace t with $p + q$

47. $m^6 - 125$
$= (m^2)^3 - (5)^3$ Write as a difference of cubes.
$= (m^2 - 5)[(m^2)^2 + (m^2)(5) + 5^2]$ Factor the difference of cubes.
$= (m^2 - 5)(m^4 + 5m^2 + 25)$

Section 7.4 (pages 411–412)

11. $18m^3n + 3m^2n^2 - 6mn^3$
$= 3mn(6m^2 + mn - 2n^2)$ Factor out the GCF, $3mn$.
$= 3mn(3m + 2n)(2m - n)$ Factor the trinomial.

19. $x^3 + 3x^2 - 9x - 27$

Factor by grouping.

$= (x^3 + 3x^2) + (-9x - 27)$
$= x^2(x + 3) - 9(x + 3)$
$= (x + 3)(x^2 - 9)$

Since $x^2 - 9$ is the difference of two squares, $x^2 - 3^2$, factor it as $(x + 3)(x - 3)$.

$= (x + 3)(x + 3)(x - 3)$
$= (x + 3)^2(x - 3)$

35. $1000z^3 + 512$
$= 8(125z^3 + 64)$ GCF = 8
$= 8[(5z)^3 + 4^3]$ Sum of cubes
$= 8[5z + 4][(5z)^2 - (5z)(4) + 4^2]$
$= 8(5z + 4)(25z^2 - 20z + 16)$

41. $14x^2 - 25xq - 25q^2$

Two integer factors whose product is $14(-25) = -350$ and whose sum is -25 are -35 and 10.

$14x^2 - 25xq - 25q^2$
$= 14x^2 - 35xq + 10xq - 25q^2$
$= (14x^2 - 35xq) + (10xq - 25q^2)$
$= 7x(2x - 5q) + 5q(2x - 5q)$
$= (2x - 5q)(7x + 5q)$

49. $(x - 2y)^2 - 4$
$= (x - 2y)^2 - 2^2$ Difference of squares
$= [(x - 2y) + 2][(x - 2y) - 2]$
$= (x - 2y + 2)(x - 2y - 2)$

Section 7.5 (pages 419–422)

31. $(x + 3)(x - 6) = (2x + 2)(x - 6)$
$x^2 - 3x - 18 = 2x^2 - 10x - 12$ Multiply.
$x^2 - 7x + 6 = 0$ Standard form
$(x - 1)(x - 6) = 0$ Factor.
$x - 1 = 0$ or $x - 6 = 0$
$x = 1$ or $x = 6$

Solution set: $\{1, 6\}$

39. $2r^3 + 5r^2 - 2r - 5 = 0$
$(2r^3 - 2r) + (5r^2 - 5) = 0$ Factor by grouping.
$2r(r^2 - 1) + 5(r^2 - 1) = 0$
$(r^2 - 1)(2r + 5) = 0$
$(r + 1)(r - 1)(2r + 5) = 0$ Factor the difference of squares.

$r + 1 = 0$ or $r - 1 = 0$ or $2r + 5 = 0$
$r = -1$ or $r = 1$ or $r = -\frac{5}{2}$

Solution set: $\left\{-\frac{5}{2}, -1, 1\right\}$

SOLUTIONS

51. Let $w =$ the width of the cardboard.
Then $w + 6 =$ the length of the cardboard.

If squares that measure 2 in. are cut from each corner of the cardboard, then the width becomes $w - 4$ and the length becomes

$$(w + 6) - 4 = w + 2.$$

Use the formula $V = LWH$ and substitute 110 for V, $w + 2$ for L, $w - 4$ for W, and 2 for H.

$$\begin{aligned} V &= LWH \\ 110 &= (w+2)(w-4)2 && \text{Substitute.} \\ 110 &= (w^2 - 2w - 8)2 && \text{Multiply.} \\ 55 &= w^2 - 2w - 8 && \text{Divide by 2.} \\ 0 &= w^2 - 2w - 63 && \text{Subtract 55.} \\ 0 &= (w-9)(w+7) && \text{Factor.} \end{aligned}$$

$$w - 9 = 0 \quad \text{or} \quad w + 7 = 0$$
$$w = 9 \quad \text{or} \quad w = -7$$

A box cannot have a negative width, so reject -7 as a solution. The only possible solution is 9. The piece of cardboard has width 9 in. and length $9 + 6 = 15$ in.

Chapter 8 Rational Expressions and Functions

Section 8.1 (pages 439–442)

15. $f(x) = \dfrac{3x+1}{2x^2 + x - 6}$

Set the denominator equal to zero and solve.

$$2x^2 + x - 6 = 0$$
$$(x+2)(2x-3) = 0 \quad \text{Factor.}$$
$$x + 2 = 0 \quad \text{or} \quad 2x - 3 = 0 \quad \text{Zero-factor property}$$
$$x = -2 \quad \text{or} \quad 2x = 3 \quad \text{Solve each equation.}$$
$$x = \frac{3}{2}$$

The numbers -2 and $\frac{3}{2}$ are not in the domain of the function. In set notation, the domain is

$$\left\{x \,\middle|\, x \neq -2, \frac{3}{2}\right\}.$$

45. $\dfrac{2c^2 + 2cd - 60d^2}{2c^2 - 12cd + 10d^2}$

$= \dfrac{2(c^2 + cd - 30d^2)}{2(c^2 - 6cd + 5d^2)}$ Factor out the GCF in the numerator and denominator.

$= \dfrac{2(c+6d)(c-5d)}{2(c-d)(c-5d)}$ Factor trinomials in the numerator and denominator.

$= \dfrac{c+6d}{c-d}$ Lowest terms

79. $\left(\dfrac{6k^2 - 13k - 5}{k^2 + 7k} \div \dfrac{2k-5}{k^3 + 6k^2 - 7k}\right) \cdot \dfrac{k^2 - 5k + 6}{3k^2 - 8k - 3}$

Factor k from the denominator of the divisor; multiply by the reciprocal.

$= \left[\dfrac{6k^2 - 13k - 5}{k^2 + 7k} \cdot \dfrac{k(k^2 + 6k - 7)}{2k-5}\right] \cdot \dfrac{k^2 - 5k + 6}{3k^2 - 8k - 3}$

$= \left[\dfrac{(3k+1)(2k-5)}{k(k+7)} \cdot \dfrac{k(k+7)(k-1)}{2k-5}\right] \cdot \dfrac{(k-2)(k-3)}{(3k+1)(k-3)}$ Factor numerators and denominators.

$= (k-1)(k-2)$ Lowest terms

Section 8.2 (pages 451–454)

13. $\dfrac{a^3}{a^2 + ab + b^2} - \dfrac{b^3}{a^2 + ab + b^2}$

$= \dfrac{a^3 - b^3}{a^2 + ab + b^2}$ Subtract the numerators; keep the common denominator.

$= \dfrac{(a-b)(a^2 + ab + b^2)}{a^2 + ab + b^2}$ Factor the difference of cubes in the numerator.

$= a - b$ Lowest terms

49. $\dfrac{4x}{x-1} - \dfrac{2}{x+1} - \dfrac{4}{x^2 - 1}$

$x^2 - 1 = (x+1)(x-1)$, the LCD.

$\dfrac{4x}{x-1} - \dfrac{2}{x+1} - \dfrac{4}{x^2 - 1}$

$= \dfrac{4x(x+1)}{(x-1)(x+1)} - \dfrac{2(x-1)}{(x+1)(x-1)} - \dfrac{4}{(x+1)(x-1)}$ Fundamental property

$= \dfrac{4x(x+1) - 2(x-1) - 4}{(x+1)(x-1)}$ Subtract numerators.

$= \dfrac{4x^2 + 4x - 2x + 2 - 4}{(x-1)(x+1)}$ Distributive property

$= \dfrac{4x^2 + 2x - 2}{(x-1)(x+1)}$ Combine like terms in the numerator.

$= \dfrac{2(2x^2 + x - 1)}{(x-1)(x+1)}$ Factor out the GCF in the numerator.

$= \dfrac{2(2x-1)(x+1)}{(x-1)(x+1)}$ Factor.

$= \dfrac{2(2x-1)}{x-1}$ Lowest terms

57. $\dfrac{4}{x+1} + \dfrac{1}{x^2 - x + 1} - \dfrac{12}{x^3 + 1}$

$x^3 + 1 = (x+1)(x^2 - x + 1)$, the LCD.

$\dfrac{4}{x+1} + \dfrac{1}{x^2 - x + 1} - \dfrac{12}{(x+1)(x^2 - x + 1)}$

$= \dfrac{4(x^2 - x + 1)}{(x+1)(x^2 - x + 1)} + \dfrac{1 \cdot (x+1)}{(x^2 - x + 1)(x+1)} - \dfrac{12}{(x+1)(x^2 - x + 1)}$ Fundamental property

$= \dfrac{4(x^2 - x + 1) + (x+1) - 12}{(x+1)(x^2 - x + 1)}$ Add and subtract numerators.

$= \dfrac{4x^2 - 4x + 4 + x + 1 - 12}{(x+1)(x^2 - x + 1)}$ Distributive property

$= \dfrac{4x^2 - 3x - 7}{(x+1)(x^2 - x + 1)}$ Combine like terms.

$= \dfrac{(4x-7)(x+1)}{(x+1)(x^2 - x + 1)}$ Factor.

$= \dfrac{4x-7}{x^2 - x + 1}$ Lowest terms

65. $\dfrac{5x}{x^2 + xy - 2y^2} - \dfrac{3x}{x^2 + 5xy - 6y^2}$

Factor each denominator.

$$x^2 + xy - 2y^2 = (x+2y)(x-y)$$
$$x^2 + 5xy - 6y^2 = (x+6y)(x-y)$$

The LCD is $(x+2y)(x-y)(x+6y)$.

$\dfrac{5x}{(x+2y)(x-y)} - \dfrac{3x}{(x+6y)(x-y)}$

$= \dfrac{5x(x+6y)}{(x+2y)(x-y)(x+6y)} - \dfrac{3x(x+2y)}{(x+6y)(x-y)(x+2y)}$ Fundamental property

$= \dfrac{5x(x+6y) - 3x(x+2y)}{(x+6y)(x-y)(x+2y)}$ Subtract numerators.

$= \dfrac{5x^2 + 30xy - 3x^2 - 6xy}{(x+2y)(x-y)(x+6y)}$ Distributive property

$= \dfrac{2x^2 + 24xy}{(x+2y)(x-y)(x+6y)}$ Combine like terms.

$= \dfrac{2x(x+12y)}{(x+2y)(x-y)(x+6y)}$ Factor out the GCF.

Section 8.3 (pages 459–460)

7. $\dfrac{\frac{4z^2x^4}{9}}{\frac{2x^2z^5}{15}}$

$= \dfrac{4z^2x^4}{9} \div \dfrac{12x^2z^5}{15}$ Write as a division problem.

$= \dfrac{4z^2x^4}{9} \cdot \dfrac{15}{12x^2z^5}$ Multiply by the reciprocal of the divisor.

$= \dfrac{60z^2x^4}{108x^2z^5}$ Multiply.

$= \dfrac{5 \cdot 12 \cdot z^2 \cdot x^2 \cdot x^2}{9 \cdot 12 \cdot x^2 \cdot z^2 \cdot z^3}$ Factor.

$= \dfrac{5x^2}{9z^3}$ Lowest terms

15. $\dfrac{\frac{x^2 - 16y^2}{xy}}{\frac{1}{y} - \frac{4}{x}}$

Multiply the numerator and denominator by xy, the LCD of all the fractions.

$= \dfrac{\left(\frac{x^2 - 16y^2}{xy}\right)xy}{\left(\frac{1}{y} - \frac{4}{x}\right)xy}$

$= \dfrac{x^2 - 16y^2}{x - 4y}$ Distributive property

$= \dfrac{(x + 4y)(x - 4y)}{x - 4y}$ Factor the difference of squares in the numerator.

$= x + 4y$ Lowest terms

19. $\dfrac{\frac{x+2}{x} + \frac{1}{x+2}}{\frac{5}{x} + \frac{x}{x+2}}$

Multiply the numerator and denominator by $x(x + 2)$, the LCD of all the fractions.

$= \dfrac{x(x+2)\left(\frac{x+2}{x} + \frac{1}{x+2}\right)}{x(x+2)\left(\frac{5}{x} + \frac{x}{x+2}\right)}$

$= \dfrac{x(x+2)\left(\frac{x+2}{x}\right) + x(x+2)\left(\frac{1}{x+2}\right)}{x(x+2)\left(\frac{5}{x}\right) + x(x+2)\left(\frac{x}{x+2}\right)}$ Distributive property

$= \dfrac{(x-2)(x+2) + x}{5(x+2) + x^2}$ Multiply.

$= \dfrac{x^2 + 4x + 4 + x}{5x + 10 + x^2}$ Multiply.

$= \dfrac{x^2 + 5x + 4}{x^2 + 5x + 10}$ Combine like terms.

31. $\dfrac{x^{-1} + 2y^{-1}}{2y + 4x}$

$= \dfrac{\frac{1}{x} + \frac{2}{y}}{2y + 4x}$

Multiply the numerator and denominator by xy, the LCD of all the fractions.

$= \dfrac{xy\left(\frac{1}{x} + \frac{2}{y}\right)}{xy(2y + 4x)}$

$= \dfrac{y + 2x}{2xy(y + 2x)}$ Distributive property; factor $2y + 4x$ as $2(y + 2x)$.

$= \dfrac{1}{2xy}$ Lowest terms

Section 8.4 (pages 465–468)

17. $\dfrac{3x + 1}{x - 4} = \dfrac{6x + 5}{2x - 7}$

The domain excludes 4 and $\frac{7}{2}$.

Multiply each side by the LCD, $(x - 4)(2x - 7)$.

$(x - 4)(2x - 7)\left(\dfrac{3x + 1}{x - 4}\right)$

$= (x - 4)(2x - 7)\left(\dfrac{6x + 5}{2x - 7}\right)$

$(2x - 7)(3x + 1) = (x - 4)(6x + 5)$

$6x^2 - 19x - 7 = 6x^2 - 19x - 20$

$-7 = -20$ False

The false statement indicates that the original equation has no solution.

Solution set: $\emptyset$

27. $\dfrac{9}{x} + \dfrac{4}{6x - 3} = \dfrac{2}{6x - 3}$

The domain excludes 0 and $\frac{1}{2}$.

Multiply by the LCD, $x(6x - 3)$.

$x(6x - 3)\left(\dfrac{9}{x} + \dfrac{4}{6x - 3}\right)$

$= x(6x - 3)\left(\dfrac{2}{6x - 3}\right)$

$9(6x - 3) - 4x = 2x$

$54x - 27 - 4x = 2x$

$56x = 27$

$x = \dfrac{27}{56}$

Note that $\frac{27}{56}$ is in the domain. Substitute $\frac{27}{56}$ for x in the original equation to check the solution.

Solution set: $\left\{\dfrac{27}{56}\right\}$

37. $\dfrac{4x - 7}{4x^2 - 9} = \dfrac{-2x^2 + 5x - 4}{4x^2 - 9} + \dfrac{x + 1}{2x + 3}$

$\dfrac{4x - 7}{(2x + 3)(2x - 3)}$

$= \dfrac{-2x^2 + 5x - 4}{(2x + 3)(2x - 3)} + \dfrac{x + 1}{2x + 3}$ Factor.

The domain excludes $-\frac{3}{2}$ and $\frac{3}{2}$.

Multiply by the LCD, $(2x + 3)(2x - 3)$.

$4x - 7 = -2x^2 + 5x - 4 + (2x - 3)(x + 1)$

$4x - 7 = -2x^2 + 5x - 4 + 2x^2 - x - 3$

$4x - 7 = 4x - 7$ True

This equation is true for every real number value of x, but we have already determined that $-\frac{3}{2}$ and $\frac{3}{2}$ are excluded from the domain. Thus, every real number except $-\frac{3}{2}$ and $\frac{3}{2}$ is a solution.

Solution set: $\left\{x \mid x \neq -\frac{3}{2}, \frac{3}{2}\right\}$

Section 8.5 (pages 479–484)

33. Let $x =$ the number of fish in the lake. Write and solve a proportion.

$\dfrac{\text{total in lake}}{\text{tagged in lake}} = \dfrac{\text{total in sample}}{\text{tagged in sample}}$

$\dfrac{x}{500} = \dfrac{400}{8}$

$\dfrac{x}{500} = 50$

$x = 500(50)$

$x = 25{,}000$

There are approximately 25,000 fish in the lake.

45. ***Step 1***

Read the problem again.

Step 2

Let $x =$ the distance in miles from San Francisco to the secret rendezvous.

Make a table.

	d	r	t
First Trip	x	200	$\frac{x}{200}$
Return Trip	x	300	$\frac{x}{300}$

Step 3

Time there	plus	time back	equals	4 hr.
$\frac{x}{200}$	$+$	$\frac{x}{300}$	$=$	4

(continued)

SOLUTIONS

Step 4
Multiply by the LCD, 600.

$$600\left(\frac{x}{200}+\frac{x}{300}\right)=600(4)$$
$$3x+2x=2400$$
$$5x=2400$$
$$x=480$$

Step 5
The distance is 480 mi.

Step 6
Check. 480 mi at 200 mph takes $\frac{480}{200}$, or 2.4 hr; 480 miles at 300 mph takes $\frac{480}{300}$, or 1.6 hr. The total time is 4 hr, as required.

55. Let x = the time from Mimi's arrival home to the time the place is a shambles.

	Rate	Time to Mess up House	Fractional Part of the Job Done
Hortense and Mort	$-\frac{1}{7}$	x	$-\frac{1}{7}x$
Mimi	$\frac{1}{2}$	x	$\frac{1}{2}x$

Notice that Hortense and Mort's rate is negative since they are "undoing" the messing up by cleaning the house.

Part done by Hortense and Mort + Part done by Mimi = 1 whole job of messing up.

$$-\frac{1}{7}x+\frac{1}{2}x=1$$

Multiply by the LCD, 14.

$$14\left(-\frac{1}{7}x\right)+14\left(\frac{1}{2}x\right)=14(1)$$
$$-2x+7x=14$$
$$5x=14$$
$$x=\frac{14}{5},\text{ or }2\frac{4}{5}$$

It would take $\frac{14}{5}$ hr, or $2\frac{4}{5}$ hr after Mimi got home for the house to be a shambles.

Section 8.6 (pages 491–494)

27. Let I = the illumination produced by a light source and d = the distance from the source. I varies inversely as d^2, so

$$I=\frac{k}{d^2}$$

for some constant k. Since $I = 768$ when $d = 1$, substitute these values in the equation and solve for k.

$$I=\frac{k}{d^2}$$
$$768=\frac{k}{1^2}$$
$$768=k$$

So $I=\frac{768}{d^2}$. Now let $d = 6$.

$$I=\frac{768}{d^2}=\frac{768}{36}=\frac{64}{3},\text{ or }21\frac{1}{3}$$

The illumination produced by the light source is $21\frac{1}{3}$ foot-candles.

31. Let F = the force, w = the weight of the car, s = the speed, and r = the radius. The force varies inversely as the radius and jointly as the weight and the square of the speed, so

$$F=\frac{kws^2}{r}.$$

Let $F = 242$, $w = 2000$, $r = 500$, and $s = 30$.

$$242=\frac{k(2000)(30)^2}{500}$$
$$k=\frac{242(500)}{2000(900)}$$
$$k=\frac{121}{1800}$$

So $F=\frac{121ws^2}{1800r}$.

Let $r = 750$, $s = 50$, and $w = 2000$.

$$F=\frac{121(2000)(50)^2}{1800(750)}\approx 448.1$$

Approximately 448.1 lb of force would be needed.

Chapter 9 Roots, Radicals, and Root Functions

Section 9.1 (pages 513–516)

9. The length $\sqrt{98}$ is closer to $\sqrt{100}=10$ than to $\sqrt{81}=9$. The width $\sqrt{26}$ is closer to $\sqrt{25}=5$ than to $\sqrt{36}=6$. Use the estimates $L = 10$ and $W = 5$ in the area formula $A = LW$ to find an estimate of the area.

$$A\approx 10\cdot 5=50$$

Choice C is the best estimate.

65. $\sqrt[6]{x^{30}}$

$$=\sqrt[6]{(x^5)^6}$$
$$=|x^5|,\quad\text{or}\quad |x|^5 \quad \text{(6 is even.)}$$

83. Let $a = 850$, $b = 925$, and $c = 1300$. First find the semiperimeter s.

$$s=\frac{1}{2}(a+b+c)$$
$$s=\frac{1}{2}(850+925+1300)$$
$$s=\frac{3075}{2}$$
$$s=1537.5$$

Now find the area A using Heron's formula and a calculator.

$$A=\sqrt{s(s-a)(s-b)(s-c)}$$
$$A=\sqrt{1537.5(687.5)(612.5)(237.5)}$$
$$A\approx 392{,}128.8$$

The area of the Bermuda Triangle is about 392,000 mi².

Section 9.2 (pages 523–526)

47. $(3m^4+2k^2)^{-2/3}$

$$=\frac{1}{(3m^4+2k^2)^{2/3}}$$
$$=\frac{1}{[(3m^4+2k^2)^{1/3}]^2}$$
$$=\frac{1}{\left(\sqrt[3]{3m^4+2k^2}\right)^2}$$

57. $\sqrt[3]{x}\cdot\sqrt{x}$

$=x^{1/3}\cdot x^{1/2}$ Convert to rational exponents.

$=x^{1/3+1/2}$ Product rule

$=x^{2/6+3/6}$ Get a common denominator.

$=x^{5/6}$

$=\sqrt[6]{x^5}$

79. $\left(\frac{b^{-3/2}}{c^{-5/3}}\right)^2\left(b^{-1/4}c^{-1/3}\right)^{-1}$

$=\left(\frac{c^{5/3}}{b^{3/2}}\right)^2\left(b^{1/4}c^{1/3}\right)$ Definition of negative exponent; power rule

$=\frac{c^{10/3}}{b^3}\left(b^{1/4}c^{1/3}\right)$ Power rule

$=\frac{c^{10/3}b^{1/4}c^{1/3}}{b^3}$ Multiply.

$=c^{10/3+1/3}b^{1/4-3}$ Product and quotient rules

$=c^{11/3}b^{-11/4}$ $\quad\frac{1}{4}-3=\frac{1}{4}-\frac{12}{4}=-\frac{11}{4}$

$=\frac{c^{11/3}}{b^{11/4}}$

81. $\left(\frac{p^{-1/4}q^{-3/2}}{3^{-1}p^{-2}q^{-2/3}}\right)^{-2}$

$=\frac{p^{1/2}q^3}{3^2p^4q^{4/3}}$ Power rule

$=\frac{p^{1/2-4}q^{3-4/3}}{9}$ Quotient rule

$=\frac{p^{1/2-8/2}q^{9/3-4/3}}{9}$ Write exponents with a common denominator.

$=\frac{p^{-7/2}q^{5/3}}{9}$

$=\frac{q^{5/3}}{9p^{7/2}}$

99. $\sqrt[3]{\sqrt[5]{\sqrt{y}}}$

$= \sqrt[3]{\sqrt[5]{y^{1/2}}}$

$= \sqrt[3]{(y^{1/2})^{1/5}}$

$= (y^{1/10})^{1/3}$

$= y^{1/30}$

Section 9.3 (pages 535–540)

1. Is $2\sqrt{12} = \sqrt{48}$ true?

$2\sqrt{12}$	$\sqrt{48}$
$= 2\sqrt{4 \cdot 3}$	$= \sqrt{16 \cdot 3}$
$= 2\sqrt{4} \cdot \sqrt{3}$	$= \sqrt{16} \cdot \sqrt{3}$
$= 2 \cdot 2 \cdot \sqrt{3}$	$= 4\sqrt{3}$
$= 4\sqrt{3}$	

Each expression equals $4\sqrt{3}$. The calculator approximation for each expression is 6.92820323. The statement is true.

71. $-\sqrt[3]{-125a^6b^9c^{12}}$

$= -\sqrt[3]{(-5a^2b^3c^4)^3}$

$= -(-5a^2b^3c^4)$

$= 5a^2b^3c^4$

87. $-\sqrt[4]{162r^{15}s^{10}}$

$= -\sqrt[4]{81r^{12}s^8(2r^3s^2)}$

$= -\sqrt[4]{81r^{12}s^8} \cdot \sqrt[4]{2r^3s^2}$

$= -3r^3s^2\sqrt[4]{2r^3s^2}$

91. $\sqrt[3]{\dfrac{x^{16}}{27}}$

$= \dfrac{\sqrt[3]{x^{15} \cdot x^1}}{\sqrt[3]{27}}$

$= \dfrac{\sqrt[3]{x^{15}} \cdot \sqrt[3]{x}}{\sqrt[3]{27}}$

$= \dfrac{x^5\sqrt[3]{x}}{3}$

119. Let $(x_1, y_1) = (\sqrt{2}, \sqrt{6})$ and $(x_2, y_2) = (-2\sqrt{2}, 4\sqrt{6})$.

$d = \sqrt{(x_2 - x_1)^2 + (y_2 - y_1)^2}$

$d = \sqrt{(-2\sqrt{2} - \sqrt{2})^2 + (4\sqrt{6} - \sqrt{6})^2}$

$d = \sqrt{(-3\sqrt{2})^2 + (3\sqrt{6})^2}$

$d = \sqrt{9 \cdot 2 + 9 \cdot 6}$

$d = \sqrt{18 + 54}$

$d = \sqrt{72}$

$d = \sqrt{36} \cdot \sqrt{2}$

$d = 6\sqrt{2}$

Section 9.4 (pages 543–544)

13. $6\sqrt{18} - \sqrt{32} + 2\sqrt{50}$

$= 6\sqrt{9 \cdot 2} - \sqrt{16 \cdot 2} + 2\sqrt{25 \cdot 2}$

$= 6\sqrt{9} \cdot \sqrt{2} - \sqrt{16} \cdot \sqrt{2} + 2\sqrt{25} \cdot \sqrt{2}$

$= 6 \cdot 3\sqrt{2} - 4\sqrt{2} + 2 \cdot 5\sqrt{2}$

$= 18\sqrt{2} - 4\sqrt{2} + 10\sqrt{2}$

$= 24\sqrt{2}$

29. $3x\sqrt[3]{xy^2} - 2\sqrt[3]{8x^4y^2}$

$= 3x\sqrt[3]{xy^2} - 2\sqrt[3]{8x^3} \cdot \sqrt[3]{xy^2}$

$= 3x\sqrt[3]{xy^2} - 2 \cdot 2x \cdot \sqrt[3]{xy^2}$

$= 3x\sqrt[3]{xy^2} - 4x\sqrt[3]{xy^2}$

$= (3x - 4x)\sqrt[3]{xy^2}$

$= -x\sqrt[3]{xy^2}$

47. $3\sqrt[3]{\dfrac{m^5}{27}} - 2m\sqrt[3]{\dfrac{m^2}{64}}$

$= \dfrac{3\sqrt[3]{m^5}}{\sqrt[3]{27}} - \dfrac{2m\sqrt[3]{m^2}}{\sqrt[3]{64}}$

$= \dfrac{3\sqrt[3]{m^3} \cdot \sqrt[3]{m^2}}{3} - \dfrac{2m\sqrt[3]{m^2}}{4}$

$= \dfrac{m\sqrt[3]{m^2}}{1} - \dfrac{m\sqrt[3]{m^2}}{2}$

$= \dfrac{2m\sqrt[3]{m^2} - m\sqrt[3]{m^2}}{2}$

$= \dfrac{m\sqrt[3]{m^2}}{2}$

53. To find the perimeter, add the lengths of the sides.

$4\sqrt{18} + \sqrt{108} + 2\sqrt{72} + 3\sqrt{12}$

$= 4\sqrt{9} \cdot \sqrt{2} + \sqrt{36} \cdot \sqrt{3} + 2\sqrt{36} \cdot \sqrt{2} + 3\sqrt{4} \cdot \sqrt{3}$

$= 4 \cdot 3\sqrt{2} + 6\sqrt{3} + 2 \cdot 6\sqrt{2} + 3 \cdot 2\sqrt{3}$

$= 12\sqrt{2} + 6\sqrt{3} + 12\sqrt{2} + 6\sqrt{3}$

$= 24\sqrt{2} + 12\sqrt{3}$

The perimeter is $(24\sqrt{2} + 12\sqrt{3})$ in.

Section 9.5 (pages 551–554)

9. $\sqrt{2}(\sqrt{18} - \sqrt{3})$

$= \sqrt{2} \cdot \sqrt{18} - \sqrt{2} \cdot \sqrt{3}$

$= \sqrt{36} - \sqrt{6}$

$= 6 - \sqrt{6}$

21. $(4\sqrt{x} + 3)^2$

$= (4\sqrt{x})^2 + 2(4\sqrt{x})(3) + 3^2$ $\quad (x + y)^2 = x^2 + 2xy + y^2$

$= 16x + 24\sqrt{x} + 9$

39. $\dfrac{6\sqrt{3y}}{\sqrt{y^3}}$

$= \dfrac{6\sqrt{3y} \cdot \sqrt{y}}{\sqrt{y^3} \cdot \sqrt{y}}$

$= \dfrac{6\sqrt{3y^2}}{\sqrt{y^4}}$

$= \dfrac{6y\sqrt{3}}{y^2}$

$= \dfrac{6\sqrt{3}}{y}$

75. $\dfrac{\sqrt{x} - \sqrt{y}}{\sqrt{2x} + \sqrt{3y}}$

Multiply both the numerator and denominator by the conjugate of the denominator, $\sqrt{2x} - \sqrt{3y}$.

$= \dfrac{(\sqrt{x} - \sqrt{y})(\sqrt{2x} - \sqrt{3y})}{(\sqrt{2x} + \sqrt{3y})(\sqrt{2x} - \sqrt{3y})}$

$= \dfrac{\sqrt{2x^2} - \sqrt{3xy} - \sqrt{2xy} + \sqrt{3y^2}}{(\sqrt{2x})^2 - (\sqrt{3y})^2}$

$= \dfrac{x\sqrt{2} - \sqrt{3xy} - \sqrt{2xy} + y\sqrt{3}}{2x - 3y}$

Section 9.6 (pages 561–564)

21. $3\sqrt{z - 1} = 2\sqrt{2z + 2}$

$(3\sqrt{z - 1})^2 = (2\sqrt{2z + 2})^2$ Square both sides.

$9(z - 1) = 4(2z + 2)$

$9z - 9 = 8z + 8$ Distributive property

$z = 17$ Subtract $8z$; add 9.

A check confirms that 17 is a solution of the original equation.

Solution set: $\{17\}$

33. $\sqrt{z^2 + 12z - 4} + 4 - z = 0$

$\sqrt{z^2 + 12z - 4} = z - 4$ Isolate the radical.

$(\sqrt{z^2 + 12z - 4})^2 = (z - 4)^2$ Square both sides.

$z^2 + 12z - 4 = z^2 - 8z + 16$ Simplify.

$20z = 20$

$z = 1$

Substituting 1 for z makes the left side of the original equation positive, but the right side is zero, so 1 is not a solution; it is extraneous.

Solution set: $\emptyset$

SOLUTIONS

43. $\sqrt[4]{a+8} = \sqrt[4]{2a}$

Raise each side to the fourth power.

$$\left(\sqrt[4]{a+8}\right)^4 = \left(\sqrt[4]{2a}\right)^4$$
$$a + 8 = 2a$$
$$8 = a$$

A check confirms that 8 is a solution of the original equation.

Solution set: $\{8\}$

57. $\sqrt{2\sqrt{x+11}} = \sqrt{4x+2}$

$$\left(\sqrt{2\sqrt{x+11}}\right)^2 = \left(\sqrt{4x+2}\right)^2 \quad \text{Square both sides.}$$
$$2\sqrt{x+11} = 4x + 2$$
$$\left(2\sqrt{x+11}\right)^2 = (4x+2)^2 \quad \text{Square again.}$$
$$4(x+11) = 16x^2 + 16x + 4$$
$$4x + 44 = 16x^2 + 16x + 4$$
$$16x^2 + 12x - 40 = 0 \quad \text{Standard form}$$
$$4x^2 + 3x - 10 = 0 \quad \text{Divide by 4.}$$
$$(x+2)(4x-5) = 0 \quad \text{Factor.}$$
$$x + 2 = 0 \quad \text{or} \quad 4x - 5 = 0 \quad \text{Zero-factor property}$$
$$x = -2 \quad \text{or} \quad x = \frac{5}{4}$$

Check

Let $x = -2$ in the original equation.

$$\sqrt{2\sqrt{-2+11}} \stackrel{?}{=} \sqrt{4(-2)+2}$$
$$\sqrt{2\sqrt{9}} \stackrel{?}{=} \sqrt{-8+2}$$
$$\sqrt{6} = \sqrt{-6} \quad \text{False}$$

Let $x = \frac{5}{4}$ in the original equation.

$$\sqrt{2\sqrt{\frac{5}{4}+11}} \stackrel{?}{=} \sqrt{4\left(\frac{5}{4}\right)+2}$$
$$\sqrt{2\sqrt{\frac{49}{4}}} \stackrel{?}{=} \sqrt{5+2}$$
$$\sqrt{2\left(\frac{7}{2}\right)} = \sqrt{7}$$
$$\sqrt{7} = \sqrt{7} \quad \text{True}$$

Solution set: $\left\{\frac{5}{4}\right\}$

61. $(2w-1)^{2/3} - w^{1/3} = 0$

$$\sqrt[3]{(2w-1)^2} - \sqrt[3]{w} = 0 \quad \text{Write with radicals.}$$
$$\sqrt[3]{(2w-1)^2} = \sqrt[3]{w} \quad \text{Add } \sqrt[3]{w}.$$
$$\left(\sqrt[3]{(2w-1)^2}\right)^3 = \left(\sqrt[3]{w}\right)^3 \quad \text{Cube both sides.}$$
$$(2w-1)^2 = w$$
$$4w^2 - 4w + 1 = w \quad \text{Square on the left.}$$
$$4w^2 - 5w + 1 = 0 \quad \text{Standard form}$$
$$(4w-1)(w-1) = 0$$
$$4w - 1 = 0 \quad \text{or} \quad w - 1 = 0$$
$$w = \frac{1}{4} \quad \text{or} \quad w = 1$$

A check confirms that $\frac{1}{4}$ and 1 are both solutions of the original equation.

Solution set: $\left\{\frac{1}{4}, 1\right\}$

63. Solve $V = \sqrt{\frac{2K}{m}}$ for K.

$$(V)^2 = \left(\sqrt{\frac{2K}{m}}\right)^2 \quad \text{Square both sides.}$$
$$V^2 = \frac{2K}{m}$$
$$\frac{V^2m}{2} = K, \quad \text{or} \quad K = \frac{V^2m}{2} \quad \text{Multiply by } \frac{m}{2}.$$

Section 9.7 (pages 571–574)

37. $[(7+3i) - (4-2i)] + (3+i)$ Work inside the brackets first.

$$= [(7-4) + (3+2)i] + (3+i)$$
$$= (3+5i) + (3+i)$$
$$= (3+3) + (5+1)i$$
$$= 6 + 6i$$

49. $(4+5i)^2$

$$= 4^2 + 2(4)(5i) + (5i)^2$$
$$= 16 + 40i + 25i^2$$
$$= 16 + 40i + 25(-1) \quad i^2 = -1$$
$$= 16 + 40i - 25$$
$$= -9 + 40i$$

81. $I = \frac{E}{R + (X_L - X_c)i}$

Substitute $2 + 3i$ for E, 5 for R, 4 for X_L, and 3 for X_c.

$$I = \frac{2+3i}{5+(4-3)i}$$
$$I = \frac{2+3i}{5+i}$$
$$I = \frac{(2+3i)(5-i)}{(5+i)(5-i)}$$
$$I = \frac{10 - 2i + 15i - 3i^2}{5^2 - i^2}$$
$$I = \frac{10 + 13i + 3}{25 + 1}$$
$$I = \frac{13 + 13i}{26}$$
$$I = \frac{13(1+i)}{13 \cdot 2}$$
$$I = \frac{1+i}{2}$$
$$I = \frac{1}{2} + \frac{1}{2}i$$

Chapter 10 Quadratic Equations, Inequalities, and Functions

Section 10.1 (pages 597–600)

51. $z^2 - \frac{4}{3}z = -\frac{1}{9}$

Complete the square.

$$\left[\frac{1}{2}\left(-\frac{4}{3}\right)\right]^2 = \left(-\frac{2}{3}\right)^2 = \frac{4}{9}$$

Add $\frac{4}{9}$ to each side.

$$z^2 - \frac{4}{3}z + \frac{4}{9} = -\frac{1}{9} + \frac{4}{9}$$
$$\left(z - \frac{2}{3}\right)^2 = \frac{3}{9}$$
$$z - \frac{2}{3} = \sqrt{\frac{3}{9}} \quad \text{or} \quad z - \frac{2}{3} = -\sqrt{\frac{3}{9}}$$
$$z - \frac{2}{3} = \frac{\sqrt{3}}{3} \quad \text{or} \quad z - \frac{2}{3} = -\frac{\sqrt{3}}{3}$$
$$z = \frac{2}{3} + \frac{\sqrt{3}}{3} \quad \text{or} \quad z = \frac{2}{3} - \frac{\sqrt{3}}{3}$$
$$z = \frac{2+\sqrt{3}}{3} \quad \text{or} \quad z = \frac{2-\sqrt{3}}{3}$$

Solution set: $\left\{\frac{2+\sqrt{3}}{3}, \frac{2-\sqrt{3}}{3}\right\}$

SOLUTIONS

53. $0.1x^2 - 0.2x - 0.1 = 0$

Multiply each side by 10 to clear decimals.

$$x^2 - 2x - 1 = 0$$
$$x^2 - 2x = 1$$

Complete the square.

$$\left[\frac{1}{2}(-2)\right]^2 = (-1)^2 = 1$$

Add 1 to each side.

$$x^2 - 2x + 1 = 1 + 1$$
$$(x - 1)^2 = 2$$
$$x - 1 = \sqrt{2} \quad \text{or} \quad x - 1 = -\sqrt{2}$$
$$x = 1 + \sqrt{2} \quad \text{or} \quad x = 1 - \sqrt{2}$$

Solution set: $\{1 + \sqrt{2}, 1 - \sqrt{2}\}$

63. $3r^2 + 4r + 4 = 0$

$$3r^2 + 4r = -4 \quad \text{Subtract 4.}$$
$$r^2 + \frac{4}{3}r = \frac{-4}{3} \quad \text{Divide by 3.}$$

Complete the square.

$$\left[\frac{1}{2}\left(\frac{4}{3}\right)\right]^2 = \left(\frac{2}{3}\right)^2 = \frac{4}{9}$$

Add $\frac{4}{9}$ to each side.

$$r^2 + \frac{4}{3}r + \frac{4}{9} = \frac{-4}{3} + \frac{4}{9}$$
$$\left(r + \frac{2}{3}\right)^2 = \frac{-8}{9} \qquad \frac{-4}{3} = \frac{-12}{9}$$
$$r + \frac{2}{3} = \frac{\sqrt{-8}}{\sqrt{9}} \quad \text{or} \quad r + \frac{2}{3} = -\frac{\sqrt{-8}}{\sqrt{9}}$$
$$r + \frac{2}{3} = \frac{2i\sqrt{2}}{3} \quad \text{or} \quad r + \frac{2}{3} = \frac{-2i\sqrt{2}}{3}$$
$$r = -\frac{2}{3} + \frac{2i\sqrt{2}}{3} \quad \text{or} \quad r = -\frac{2}{3} - \frac{2i\sqrt{2}}{3}$$

Solution set: $\left\{-\frac{2}{3} + \frac{2\sqrt{2}}{3}i, -\frac{2}{3} - \frac{2\sqrt{2}}{3}i\right\}$

65. $-m^2 - 6m - 12 = 0$

Multiply each side by -1.

$$m^2 - 6m + 12 = 0$$
$$m^2 + 6m = -12 \quad \text{Subtract 12.}$$

Complete the square.

$$\left[\frac{1}{2}(6)\right]^2 = 3^2 = 9$$
$$m^2 + 6m + 9 = -12 + 9$$
$$(m + 3)^2 = -3$$
$$m + 3 = \sqrt{-3} \quad \text{or} \quad m + 3 = -\sqrt{-3}$$
$$m + 3 = i\sqrt{3} \quad \text{or} \quad m + 3 = -i\sqrt{3}$$
$$m = -3 + i\sqrt{3} \quad \text{or} \quad m = -3 - i\sqrt{3}$$

Solution set: $\{-3 + i\sqrt{3}, -3 - i\sqrt{3}\}$

Section 10.2 (pages 607–608)

15. $\frac{x^2}{4} - \frac{x}{2} = 1$

First clear fractions by multiplying each side by the LCD, 4.

$$4\left(\frac{x^2}{4} - \frac{x}{2}\right) = 4(1)$$
$$x^2 - 2x = 4$$
$$x^2 - 2x - 4 = 0$$

Here $a = 1, b = -2$, and $c = -4$. Substitute in the quadratic formula.

$$x = \frac{-b \pm \sqrt{b^2 - 4ac}}{2a}$$
$$x = \frac{-(-2) \pm \sqrt{(-2)^2 - 4(1)(-4)}}{2(1)}$$
$$x = \frac{2 \pm \sqrt{4 + 16}}{2}$$
$$x = \frac{2 \pm \sqrt{20}}{2}$$
$$x = \frac{2 \pm 2\sqrt{5}}{2}$$
$$x = \frac{2(1 \pm \sqrt{5})}{2}$$
$$x = 1 \pm \sqrt{5}$$

Solution set: $\{1 + \sqrt{5}, 1 - \sqrt{5}\}$

17. $-2t(t + 2) = -3$

$$-2t^2 - 4t = -3$$
$$-2t^2 - 4t - 3 = 0$$

Here $a = -2, b = -4$, and $c = 3$.

$$t = \frac{-b \pm \sqrt{b^2 - 4ac}}{2a}$$
$$t = \frac{-(-4) \pm \sqrt{(-4)^2 - 4(-2)(3)}}{2(-2)}$$
$$t = \frac{4 \pm \sqrt{16 + 24}}{-4}$$
$$t = \frac{4 \pm \sqrt{40}}{-4}$$
$$t = \frac{4 \pm 2\sqrt{10}}{-4}$$
$$t = \frac{2(2 \pm \sqrt{10})}{-2 \cdot 2}$$
$$t = \frac{2 \pm \sqrt{10}}{-2} \cdot \frac{-1}{-1}$$
$$t = \frac{-2 \mp \sqrt{10}}{2}$$
$$t = \frac{-2 \pm \sqrt{10}}{2}$$

Solution set: $\left\{\frac{-2 + \sqrt{10}}{2}, \frac{-2 - \sqrt{10}}{2}\right\}$

27. $x(3x + 4) = -2$

$$3x^2 + 4x = -2$$
$$3x^2 + 4x + 2 = 0$$

Here $a = 3, b = 4$, and $c = 2$.

$$x = \frac{-b \pm \sqrt{b^2 - 4ac}}{2a}$$
$$x = \frac{-4 \pm \sqrt{4^2 - 4(3)(2)}}{2(3)}$$
$$x = \frac{-4 \pm \sqrt{16 - 24}}{6}$$
$$x = \frac{-4 \pm \sqrt{-8}}{6}$$
$$x = \frac{-4 \pm 2i\sqrt{2}}{6}$$
$$x = \frac{2(-2 \pm i\sqrt{2})}{2 \cdot 3}$$
$$x = \frac{-2 \pm i\sqrt{2}}{3}$$

Solution set: $\left\{-\frac{2}{3} + \frac{\sqrt{2}}{3}i, -\frac{2}{3} - \frac{\sqrt{2}}{3}i\right\}$

Section 10.3 (pages 617–620)

15. $\frac{3}{2x} - \frac{1}{2(x + 2)} = 1$

Multiply by the LCD, $2x(x + 2)$.

$$2x(x + 2)\left(\frac{3}{2x} - \frac{1}{2(x + 2)}\right) = 2x(x + 2) \cdot 1$$
$$3(x + 2) - x(1) = 2x(x + 2)$$
$$3x + 6 - x = 2x^2 + 4x$$
$$0 = 2x^2 + 2x - 6$$
$$0 = x^2 + x - 3$$

Use $a = 1, b = 1$, and $c = -3$ in the quadratic formula.

$$x = \frac{-b \pm \sqrt{b^2 - 4ac}}{2a}$$
$$x = \frac{-1 \pm \sqrt{1^2 - 4(1)(-3)}}{2(1)}$$
$$x = \frac{-1 \pm \sqrt{1 + 12}}{2}$$
$$x = \frac{-1 \pm \sqrt{13}}{2}$$

Use a calculator to check both proposed solutions. Both solutions check.

Solution set: $\left\{\frac{-1 + \sqrt{13}}{2}, \frac{-1 - \sqrt{13}}{2}\right\}$

23. Let x = Harry's average speed.
Then $x - 20$ = Yoshi's average speed.

	d	r	t
Harry	300	x	$\frac{300}{x}$
Yoshi	300	$x - 20$	$\frac{300}{x-20}$

It takes Harry $1\frac{1}{4}$ hr, or $\frac{5}{4}$ hr, less time than Yoshi.

$$\frac{300}{x} = \frac{300}{x-20} - \frac{5}{4}$$

Multiply by the LCD, $4x(x - 20)$.

$$4x(x-20)\left(\frac{300}{x}\right) = 4x(x-20)\left(\frac{300}{x-20} - \frac{5}{4}\right)$$

$$1200(x - 20) = 4x(300) - x(x - 20) \cdot 5$$

$$1200x - 24{,}000 = 1200x - 5x^2 + 100x$$

$$5x^2 - 100x - 24{,}000 = 0$$

$$x^2 - 20x - 4800 = 0$$

$$(x - 80)(x + 60) = 0$$

$$x - 80 = 0 \quad \text{or} \quad x + 60 = 0$$

$$x = 80 \quad \text{or} \quad x = -60$$

Reject $x = -60$. Harry's average speed is 80 km per hr.

37. $$m = \sqrt{\frac{6 - 13m}{5}}$$

$$m^2 = \frac{6 - 13m}{5} \qquad \text{Square each side.}$$

$$5m^2 = 6 - 13m \qquad \text{Multiply by 5.}$$

$$5m^2 + 13m - 6 = 0 \qquad \text{Standard form}$$

$$(5m - 2)(m + 3) = 0 \qquad \text{Factor.}$$

$$5m - 2 = 0 \quad \text{or} \quad m + 3 = 0 \qquad \text{Zero-factor property}$$

$$m = \frac{2}{5} \quad \text{or} \quad m = -3$$

Check If $m = \frac{2}{5}$, then $\frac{2}{5} = \sqrt{\frac{4}{25}}$. True

If $m = -3$, then $-3 = \sqrt{9}$. False

Solution set: $\left\{\frac{2}{5}\right\}$

51. $$2 + \frac{5}{3k - 1} = \frac{-2}{(3k - 1)^2}$$

Let $u = 3k - 1$, so $u^2 = (3k - 1)^2$.

$$2 + \frac{5}{u} = -\frac{2}{u^2}$$

Multiply by the LCD, u^2.

$$u^2\left(2 + \frac{5}{u}\right) = u^2\left(-\frac{2}{u^2}\right)$$

$$2u^2 + 5u = -2$$

$$2u^2 + 5u + 2 = 0$$

$$(2u + 1)(u + 2) = 0$$

$$2u + 1 = 0 \quad \text{or} \quad u + 2 = 0$$

$$u = -\frac{1}{2} \quad \text{or} \quad u = -2$$

To find k, substitute $3k - 1$ for u.

$$3k - 1 = -\frac{1}{2} \quad \text{or} \quad 3k - 1 = -2$$

$$3k = \frac{1}{2} \quad \text{or} \quad 3k = -1$$

$$k = \frac{1}{6} \quad \text{or} \quad k = -\frac{1}{3}$$

Check If $k = \frac{1}{6}$, then $2 - 10 = -8$. True

If $k = -\frac{1}{3}$, then $2 - \frac{5}{2} = -\frac{1}{2}$. True

Solution set: $\left\{-\frac{1}{3}, \frac{1}{6}\right\}$

Section 10.4 (pages 627–630)

25. Let x be the width of the sheet metal. Then the length is $2x - 4$.

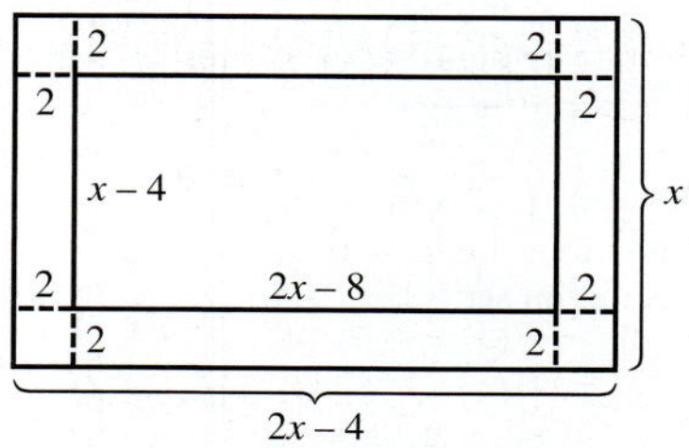

By cutting out 2-in. squares from each corner we get a rectangle with width $(x - 4)$ in. and length

$$(2x - 4) - 4 = (2x - 8) \text{ in.}$$

The uncovered box then has height 2 in., length $(2x - 8)$ in., and width $(x - 4)$ in. Use the formula $V = LWH$ or $V = HLW$.

$$256 = 2(2x - 8)(x - 4)$$

$$256 = 4(x - 4)(x - 4) \qquad \text{Factor out 2.}$$

$$64 = (x - 4)^2 \qquad \text{Divide by 4.}$$

$$(x - 4)^2 = 64$$

Use the square root property.

$$x - 4 = 8 \quad \text{or} \quad x - 4 = -8$$

$$x = 12 \quad \text{or} \quad x = -4$$

Since x represents width, discard the negative solution. The width is 12 in., and the length is

$$2(12) - 4 = 20 \text{ in.}$$

33. Supply and demand are equal when

$$3p - 200 = \frac{3200}{p}.$$

To solve for p, multiply both sides by p.

$$3p^2 - 200p = 3200$$

$$3p^2 - 200p - 3200 = 0$$

Use the quadratic formula with $a = 3$, $b = -200$, and $c = -3200$.

$$p = \frac{-(-200) \pm \sqrt{(-200)^2 - 4(3)(-3200)}}{2(3)}$$

$$p = \frac{200 \pm \sqrt{40{,}000 + 38{,}400}}{6}$$

$$p = \frac{200 \pm \sqrt{78{,}400}}{6}$$

$$p = \frac{200 \pm 280}{6}$$

$$p = \frac{480}{6} = 80 \quad \text{or} \quad p = \frac{-80}{6} = -\frac{40}{3}$$

Discard the negative solution. The supply and demand are equal when the price is 80 cents, or \$0.80.

41. Write a proportion.

$$\frac{x - 4}{3x - 19} = \frac{4}{x - 3}$$

Multiply by the LCD, $(3x - 19)(x - 3)$.

$$(3x - 19)(x - 3)\left(\frac{x - 4}{3x - 19}\right) = (3x - 19)(x - 3)\left(\frac{4}{x - 3}\right)$$

$$(x - 3)(x - 4) = (3x - 19)4$$

$$x^2 - 7x + 12 = 12x - 76$$

$$x^2 - 19x + 88 = 0$$

$$(x - 8)(x - 11) = 0$$

$$x - 8 = 0 \quad \text{or} \quad x - 11 = 0$$

$$x = 8 \quad \text{or} \quad x = 11$$

If $x = 8$, then

$$3x - 19 = 3(8) - 19 = 5.$$

If $x = 11$, then

$$3x - 19 = 3(11) - 19 = 14$$

Thus, $AC = 5$ or $AC = 14$.

Section 10.5 (pages 637–642)

31. $f(x) = -\frac{2}{3}(x + 2)^2 + 1$

Because $a = -\frac{2}{3}$, the graph opens down and is wider than the graph of $y = x^2$. Because $h = -2$ and $k = 1$, the graph is shifted 2 units to the left and 1 unit up. The vertex is at $(-2, 1)$ and the axis is $x = -2$. Two other points on the graph are $\left(-4, -\frac{5}{3}\right)$ and $\left(0, -\frac{5}{3}\right)$. We can substitute any value for x, so the domain is $(-\infty, \infty)$. The value of y is always less than or equal to 1, so the range is $(-\infty, 1]$.

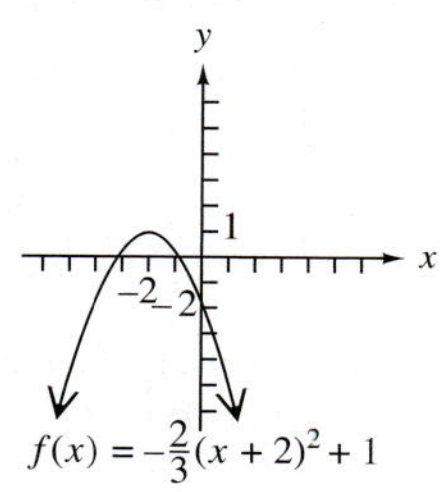

Section 10.6 (pages 651–654)

15. $x = -\frac{1}{5}y^2 + 2y - 4$

The roles of x and y are reversed, so this is a horizontal parabola.

Step 1

Since $a = -\frac{1}{5} < 0$, the graph opens to the left and is wider than the graph of $y = x^2$.

Step 2

The y-coordinate of the vertex is

$$\frac{-b}{2a} = \frac{-2}{2\left(-\frac{1}{5}\right)} = \frac{-2}{-\frac{2}{5}} = -2 \cdot \left(-\frac{5}{2}\right) = 5.$$

The x-coordinate of the vertex is

$$-\frac{1}{5}(5)^2 + 2(5) - 4 = -5 + 10 - 4 = 1.$$

Thus, the vertex is (1, 5). Since the graph opens left, the axis goes through the y-coordinate of the vertex—its equation is $y = 5$.

Step 3

To find the x-intercept, let $y = 0$. If $y = 0$, then $x = -4$, so the x-intercept is $(-4, 0)$.

To find the y-intercepts, let $x = 0$.

$$0 = -\frac{1}{5}y^2 + 2y - 4$$

$$0 = y^2 - 10y + 20 \qquad \text{Multiply by } -5.$$

$$y = \frac{-(-10) \pm \sqrt{(-10)^2 - 4(1)(20)}}{2(1)}$$

$$y = \frac{10 \pm \sqrt{20}}{2}$$

$$y = \frac{10 \pm 2\sqrt{5}}{2}$$

$$y = \frac{2(5 \pm \sqrt{5})}{2}$$

$$y = 5 \pm \sqrt{5}$$

The y-intercepts are approximately (0, 7.2) and (0, 2.8).

Step 4

For an additional point on the graph, let $y = 7$ (two units above the axis) to get $x = \frac{1}{5}$. So the point $\left(\frac{1}{5}, 7\right)$ is on the graph.

By symmetry, the point $\left(\frac{1}{5}, 3\right)$ (two units below the axis) is on the graph.

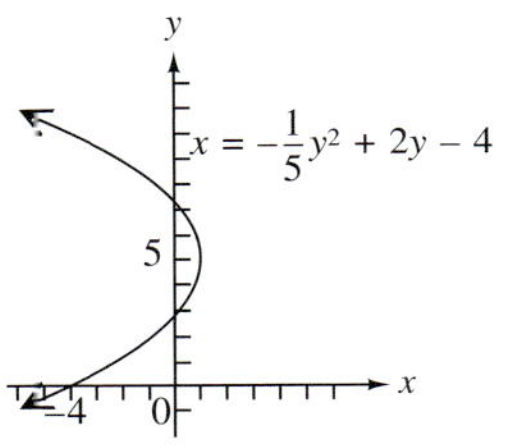

From the graph, we see that the domain is $(-\infty, 1]$ and the range is $(-\infty, \infty)$.

Section 10.7 (pages 661–664)

19. $x^2 - 6x + 6 \geq 0$

Solve the equation

$$x^2 - 6x + 6 = 0.$$

Since $x^2 - 6x + 6$ does not factor, let $a = 1$, $b = -6$, and $c = 6$ in the quadratic formula.

$$x = \frac{-(-6) \pm \sqrt{(-6)^2 - 4(1)(6)}}{2(1)}$$

$$x = \frac{6 \pm \sqrt{12}}{2}$$

$$x = \frac{6 \pm 2\sqrt{3}}{2}$$

$$x = \frac{2(3 = \sqrt{3})}{2}$$

$$x = 3 \pm \sqrt{3}$$

$$x = 3 + \sqrt{3} \approx 4.7 \text{ or}$$

$$x = 3 - \sqrt{3} \approx 1.3$$

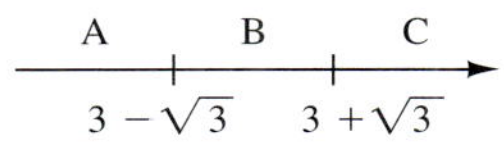

Test a number from each interval in the inequality

$$x^2 - 6x + 6 \geq 0.$$

Interval A: Let $x = 0$.

$$0^2 - 6(0) + 6 \overset{?}{\geq} 0$$

$$6 \geq 0 \quad \text{True}$$

Interval B: Let $x = 3$.

$$3^2 - 6(3) + 6 \overset{?}{\geq} 0$$

$$-3 \geq 0 \quad \text{False}$$

Interval C: Let $x = 5$.

$$5^2 - 6(5) + 6 \overset{?}{\geq} 0$$

$$1 \geq 0 \quad \text{True}$$

The solution set includes the numbers in Intervals A and C, including $3 - \sqrt{3}$ and $3 + \sqrt{3}$ because equality is included in the symbol $\geq$.

Solution set:

$$\left(-\infty, 3 - \sqrt{3}\right] \cup \left[3 + \sqrt{3}, \infty\right)$$

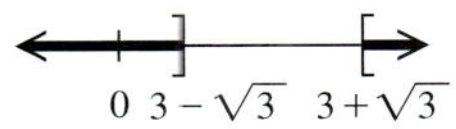

37. $\frac{w}{w + 2} \geq 2$

Write the inequality so that 0 is on one side.

$$\frac{w}{w + 2} - 2 \geq 0$$

$$\frac{w}{w + 2} - \frac{2(w + 2)}{w + 2} \geq 0$$

$$\frac{w - 2w - 4}{w + 2} \geq 0$$

$$\frac{-w - 4}{w + 2} \geq 0$$

The number -4 makes the numerator 0, and -2 makes the denominator 0. These two numbers determine three intervals.

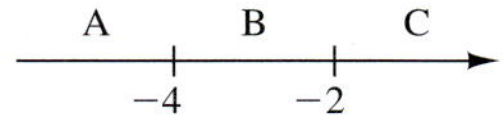

Test a number from each interval in the inequality

$$\frac{w}{w + 2} \geq 2.$$

Interval A: Let $w = -5$.

$$\frac{-5}{-3} \overset{?}{\geq} 2$$

$$\frac{5}{3} \geq 2 \quad \text{False}$$

Interval B: Let $w = -3$.

$$\frac{-3}{-1} \overset{?}{\geq} 2$$

$$3 \geq 2 \quad \text{True}$$

Interval C: Let $w = 0$.

$$\frac{0}{2} \overset{?}{\geq} 2$$

$$0 \geq 2 \quad \text{False}$$

The solution set includes numbers in Interval B, including -4 but excluding -2, which makes the fraction undefined.

Solution set: $[-4, -2)$

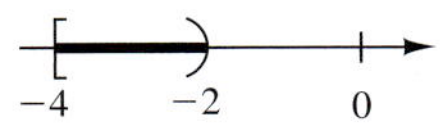

Chapter 11 Inverse, Exponential, and Logarithmic Functions

Section 11.1 (pages 685–688)

15. Write $g(x) = \sqrt{x - 3}$ as $y = \sqrt{x - 3}$. Since $x \geq 3$, $y \geq 0$. The graph of g is half of a horizontal parabola that opens to the right. The graph passes the horizontal line test, so g is one-to-one. To find the inverse, interchange x and y to get

$$x = \sqrt{y - 3}.$$

Note that now $y \geq 3$, so $x \geq 0$. Solve for y by squaring both sides.

$$x^2 = y - 3$$

$$x^2 + 3 = y$$

Replace y with $g^{-1}(x)$.

$$g^{-1}(x) = x^2 + 3, \quad x \geq 0$$

23. **(a)** To find $f(0)$, substitute 0 for x.

$$f(x) = 2^x, \text{ so } f(0) = 2^0 = 1.$$

(b) Since f is one-to-one and $f(0) = 1$, it follows that $f^{-1}(1) = 0$.

SOLUTIONS

37. $f(x) = y = x^3 - 2$

Complete the table of values.

x	y
-1	-3
0	-2
1	-1
2	6

Plot these points, and connect them with a solid smooth curve.

Interchange the values of x and y to make a table of values for f^{-1}.

x	y
-3	-1
-2	0
-1	1
6	2

Plot these points, and connect them with a dashed smooth curve. Use the fact that the graph of f^{-1} is symmetric to the graph of f with respect to the line $y = x$.

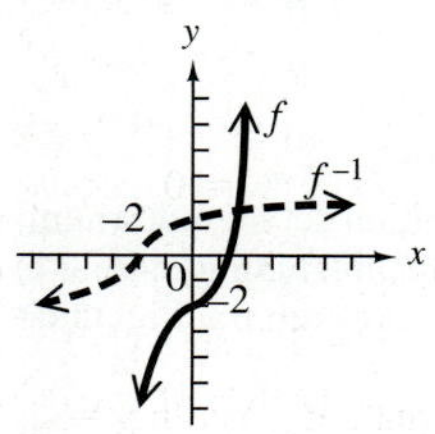

Section 11.2 (pages 695–696)

15. $16^{2x+1} = 64^{x+3}$

Write each side as a power of 4.

$$(4^2)^{2x+1} = (4^3)^{x+3}$$
$$4^{4x+2} = 4^{3x+9}$$

Set the exponents equal.

$$4x + 2 = 3x + 9$$
$$x = 7$$

Solution set: $\{7\}$

Check $16^{2x+1} = 64^{x+3}$

$16^{2(7)+1} \stackrel{?}{=} 64^{7+3}$ Let $x = 7$.

$16^{15} \stackrel{?}{=} 64^{10}$

$(4^2)^{15} \stackrel{?}{=} (4^3)^{10}$

$4^{30} = 4^{30}$ True

19. $5^x = 0.2$

$5^x = \frac{2}{10}$ Write the decimal as a fraction.

$5^x = \frac{1}{5}$ Write the fraction in lowest terms.

$5^x = 5^{-1}$ Write with the same base.

$x = -1$ Set exponents equal.

Check by substituting -1 for x in the original equation.

Solution set: $\{-1\}$

Section 11.3 (pages 701–704)

37. $\log_6 \sqrt{216} = x$

$6^x = \sqrt{216}$ Write in exponential form.

$6^x = 216^{1/2}$

$6^x = (6^3)^{1/2}$ Write with the same base.

$6^x = 6^{3/2}$ $(a^m)^n = a^{mn}$

$x = \frac{3}{2}$ Set exponents equal.

Solution set: $\left\{\frac{3}{2}\right\}$

55. $R = \log_{10} \frac{x}{x_0}$

Change to exponential form.

$$10^R = \frac{x}{x_0}, \quad \text{so} \quad x = x_0\, 10^R.$$

Let $R = 6.7$ for the Northridge earthquake, x_1.

$$x_1 = x_0 10^{6.7}$$

Let $R = 7.3$ for the Landers earthquake, x_2.

$$x_2 = x_0 10^{7.3}$$

The ratio of x_2 to x_1 is

$$\frac{x_2}{x_1} = \frac{x_0 10^{7.3}}{x_0 10^{6.7}} = 10^{0.6} \approx 3.98.$$

The Landers earthquake was about 4 times as powerful as the Northridge earthquake.

Section 11.4 (pages 711–712)

11. $\log_3 \frac{\sqrt[3]{4}}{x^2 y}$

$= \log_3 \frac{4^{1/3}}{x^2 y}$ Write the radical expression with a rational exponent.

$= \log_3 4^{1/3} - \log_3 (x^2 y)$ Quotient rule

$= \log_3 4^{1/3} - (\log_3 x^2 + \log_3 y)$ Product rule

$= \log_3 4^{1/3} - \log_3 x^2 - \log_3 y$

$= \frac{1}{3}\log_3 4 - 2\log_3 x - \log_3 y$ Power rule

29. $3\log_p x + \frac{1}{2}\log_p y - \frac{3}{2}\log_p z - 3\log_p a$

$= \log_p x^3 + \log_p y^{1/2} - \log_p z^{3/2} - \log_p a^3$ Power rule

$= \left(\log_p x^3 + \log_p y^{1/2}\right) - \left(\log_p z^{3/2} + \log_p a^3\right)$ Group the terms into sums.

$= \log_p x^3 y^{1/2} - \log_p z^{3/2} a^3$ Product rule

$= \log_p \frac{x^3 y^{1/2}}{z^{3/2} a^3}$ Quotient rule

Section 11.5 (pages 717–718)

31. $p(h) = 86.3 \ln h - 680$

(a) $p(5000) = 86.3 \ln 5000 - 680$

$p(5000) \approx 55$ Use a calculator.

The percent of moisture at 5000 ft that falls as snow rather than rain is 55%.

(b) $p(7500) = 86.3 \ln 7500 - 680$

$p(7500) \approx 90$ Use a calculator.

The percent of moisture at 7500 ft that falls as snow rather than rain is 90%.

Section 11.6 (pages 725–728)

13. $2^{x+3} = 3^{x-4}$

$\log 2^{x+3} = \log 3^{x-4}$ Property 3 (common logs)

$(x + 3)\log 2 = (x - 4)\log 3$ Power rule

$x\log 2 + 3\log 2 = x\log 3 - 4\log 3$ Distributive property

$x\log 2 - x\log 3 = -3\log 2 - 4\log 3$ Get x-terms on one side.

$x(\log 2 - \log 3) = -3\log 2 - 4\log 3$ Factor out x.

$x = \frac{-3\log 2 - 4\log 3}{\log 2 - \log 3}$ Divide by $\log 2 - \log 3$.

$x \approx 15.967$ Use a calculator.

Check that $2^{15.967+3} \approx 3^{15.967-4}$.

Solution set: $\{15.967\}$

21. $\ln e^{0.45x} = \sqrt{7}$

$0.45x = \sqrt{7}$ $\ln e^k = k$

$x = \frac{\sqrt{7}}{0.45}$ Divide by 0.45.

$x \approx 5.879$ Use a calculator.

Solution set: $\{5.879\}$

37. $\log 4x - \log(x - 3) = \log 2$

$\log \frac{4x}{x - 3} = \log 2$ Quotient rule

$\frac{4x}{x - 3} = 2$ Property 4

$4x = 2(x - 3)$ Multiply by $x - 3$.

$4x = 2x - 6$ Distributive property

$2x = -6$ Subtract $2x$.

$x = -3$ Divide by 2

Reject $x = -3$, because if $x = -3$, then $4x = -12$ and $x - 3 = -6$, which yield an equation in which the logarithms of negative numbers must be found.

Solution set: $\emptyset$

63. There are 60 of one species and 40 of another, so

$$p_1 = \frac{60}{100} = 0.6 \text{ and } p_2 = \frac{40}{100} = 0.4.$$

Thus, the index of diversity is

$$\begin{aligned} &-(p_1 \ln p_1 + p_2 \ln p_2) \\ &= -(0.6 \ln 0.6 + 0.4 \ln 0.4) \\ &\approx 0.673. \end{aligned}$$

Chapter 12 Nonlinear Functions, Conic Sections, and Nonlinear Systems

Section 12.1 (pages 749–752)

23. For any portion of the first ounce, the cost will be one 42¢ stamp. If the weight exceeds one ounce (up to two ounces), an additional 17¢ stamp is required. The following table summarizes the weight of a letter, x, and the number of stamps required, $p(x)$, on the interval (0, 5].

x	(0, 1]	(1, 2]	(2, 3]	(3, 4]	(4, 5]
$p(x)$	1	2	3	4	5

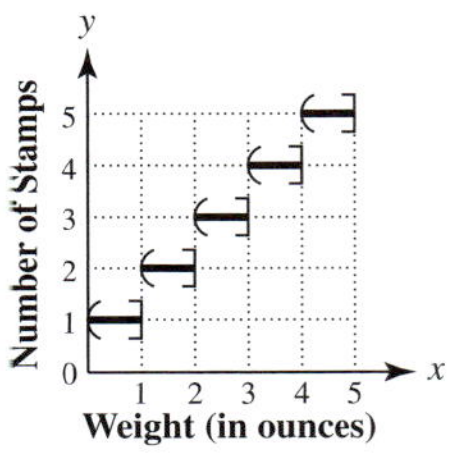

37. $(f \circ h)\left(\frac{1}{2}\right) = f\left(h\left(\frac{1}{2}\right)\right)$ Definition

$= f\left(\frac{1}{2} + 5\right)$ $h(x) = x + 5$

$= f\left(\frac{11}{2}\right)$ $5 = \frac{10}{2}$

$= \left(\frac{11}{2}\right)^2 + 4$ $f(x) = x^2 + 4$

$= \frac{121}{4} + \frac{16}{4}$

$= \frac{137}{4}$

41. $(f \circ g)(x) = f(g(x))$

$= f(5280x)$ $g(x) = 5280x$

$= 12(5280x)$ $f(x) = 12x$

$= 63{,}360x$

$(f \circ g)(x)$ computes the number of inches in x miles.

Section 12.2 (pages 759–762)

21. $3x^2 = 48 - 3y^2$

$3x^2 + 3y^2 = 48$ Add $3y^2$.

$x^2 + y^2 = 16$ Divide by 3.

$x^2 + y^2 = 4^2$

This is an equation of a circle with center (0, 0) and radius 4. See the graph at the top of the next column.

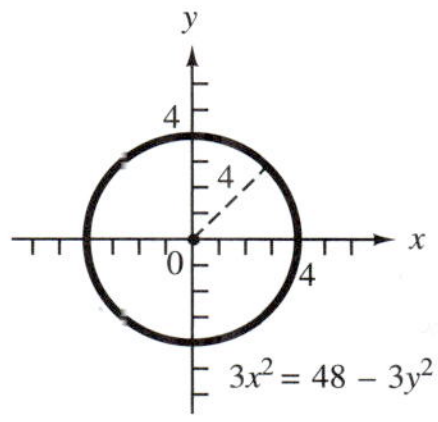

37. (a) $100x^2 + 324y^2 = 32{,}400$

$\frac{x^2}{324} + \frac{y^2}{100} = 1$ Divide by 32,400.

$\frac{x^2}{18^2} - \frac{y^2}{10^2} = 1$

The height in the center is the y-coordinate of the positive y-intercept. The height is 10 m.

(b) The width of the ellipse is the distance between x-intercepts, $(-18, 0)$ and $(18, 0)$. The width across the bottom of the arch is

$$18 + 18 = 36 \text{ m}.$$

Section 12.3 (pages 769–772)

27. $y = \sqrt{\frac{x+4}{2}}$

$y^2 = \frac{x+4}{2}$ Square both sides.

$2y^2 = x + 4$ Multiply by 2.

$2y^2 - 4 = x$ Subtract 4.

$2(y - 0)^2 - 4 = x$

This is a parabola that opens to the right with vertex $(-4, 0)$. However, y is nonnegative in the original equation, so only the top half of the parabola is included in the graph.

x	y
−4	0
−2	1
0	$\sqrt{2}$
4	2

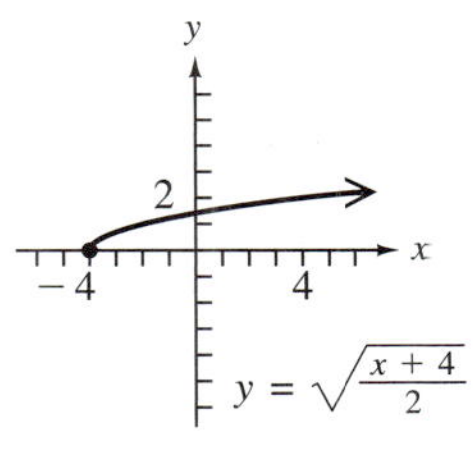

The domain is $[-4, \infty)$, and the range is $[0, \infty)$.

31. (a) $400x^2 - 625y^2 = 250{,}000$

$\frac{x^2}{625} - \frac{y^2}{400} = 1$ Divide by 250,000.

$\frac{x^2}{25^2} - \frac{y^2}{20^2} = 1$

The x-intercepts are (25, 0) and $(-25, 0)$. The distance between the buildings is the distance between the x-intercepts. The buildings are

$$25 + 25 = 50 \text{ m}$$

apart at their closest point.

(b) At $x = 50$, $y = \frac{d}{2}$, so $d = 2y$.

$$\begin{aligned} 400(50)^2 - 625y^2 &= 250{,}000 \\ 1{,}000{,}000 - 625y^2 &= 250{,}000 \\ -625y^2 &= -750{,}000 \\ y^2 &= 1200 \\ y &= \sqrt{1200} \end{aligned}$$

The distance d is

$$2\sqrt{1200} \approx 69.3 \text{ m}.$$

Section 12.4 (pages 777–778)

23. $2x^2 - y^2 = 6$ (1)

$y = x^2 - 3$ (2)

Solve equation (2) for x^2.

$x^2 = y + 3$ (3)

Substitute $y + 3$ for x^2 in equation (1).

$2x^2 - y^2 = 6$ (1)

$2(y + 3) - y^2 = 6$

$2y + 6 - y^2 = 6$

$0 = y^2 - 2y$

$0 = y(y - 2)$

$y = 0$ or $y = 2$

From equation (3) with $y = 0$,

$x^2 = 3$, so $x = \pm\sqrt{3}$.

From equation (3) with $y = 2$,

$x^2 = 5$, so $x = \pm\sqrt{5}$.

All four ordered pairs check in *both* equations.

Solution set:

$$\{(-\sqrt{3}, 0), (\sqrt{3}, 0), (-\sqrt{5}, 2), (\sqrt{5}, 2)\}$$

33. Let W = the width, and L = the length. The formula for the area of a rectangle is $A = LW$, or $LW = A$, so

$LW = 84.$ (1)

The perimeter of a rectangle is given by $P = 2L + 2W$, or $2L + 2W = P$, so

$2L + 2W = 38.$ (2)

Solve equation (2) for L to get

$L = 19 - W.$ (3)

Substitute $19 - W$ for L in equation (1).

$LW = 84$ (1)

$(19 - W)W = 84$

$19W - W^2 = 84$

$-W^2 + 19W - 84 = 0$

$W^2 - 19W + 84 = 0$ Multiply by −1.

$(W - 7)(W - 12) = 0$

$W - 7 = 0$ or $W - 12 = 0$

$W = 7$ or $W = 12$

Using equation (3), with $W = 7$,

$L = 19 - 7 = 12.$

If $W = 12$, then $L = 7$, which are the same two numbers. Length must be greater than width, so the length is 12 ft and the width is 7 ft.

INDEX

A

Absolute value
  evaluating, 9
  explanation of, 47, 49
  simplifying roots using, 511
Absolute value equations
  explanation of, 157–158, 173
  solving, 158–161, 175
Absolute value functions, 742
Absolute value inequalities
  explanation of, 157, 173
  solving, 159–160, 161–162
Addition
  associative property of, 42–43
  commutative property of, 42–43
  of complex numbers, 567–568, 578
  explanation of, 49
  of fractions, 814–815
  identity property for, 41
  of polynomial functions, 352–353
  of polynomials, 347–348, 376
  of radical expressions, 541–542, 577
  of rational expressions, 445–450, 496
  of real numbers, 17–20
Addition property
  of equality, 61–62
  of inequality, 130–131
Additive identity, 41
Additive inverse, 8, 21, 47
Agreement on domain, 240
Algebraic expressions
  evaluating, 33–34
  examples of, 60
  explanation of, 47, 375
  translating from words to, 85
Algebraic fractions. *See* Rational expressions
Alternative forms of logarithms, 709–710
Angles
  complementary, 108, 113
  supplementary, 108, 113
  of a triangle, 104
  vertical, 108, 113
Apogee, 489, 762
Applied problems. *See also* Problem-solving strategies
  with distance, rate, and time, 474–476
  linear inequalities to solve, 135–136
  methods to solve, 86–88, 114–115, 647–648
  modeled by quadratic functions, 625–626
  Pythagorean formula to solve, 624
  requiring zero-factor property, 417–418
  with work rate, 476–477
"Approximately equal to" symbol, 34, 48, 575
Approximations, for roots, 512
Area
  of triangle, 516
Ascending powers, 349
Associative properties, 42–43, 50
Asymptotes
  explanation of, 432, 785
  horizontal, 464, 495
  of hyperbola, 764–765, 785
  of reciprocal function, 742
  vertical, 464, 495
Augmented matrix, 307–308, 315
Average rate of change, 205–206
Average speed, 76
Average test scores 136
Axis
  of parabola, 631, 665
  x, 188, 253
  y, 188, 253

B

Bases
  explanation of, 29, 47
  of exponents, 30–31, 328
Binomial(s)
  explanation of, 346, 375
  factoring, 409–410
  as a greatest common factor, 389
  multiplication of, 360–362, 363–364, 545, 546
  square of, 363, 559
Boundary lines, 253
Braces, 4

C

Calculator tips
  for complex numbers, 570
  for exponents, 30
  for logarithms, 715–716
  for matrices, 307, 310
  for order of operations, 32
  for quadratic regression, 636
  for reciprocals, 20
  for roots, 512
  for scientific notation, 338
  for systems of equations, 269
Cartesian coordinate system. *See* Rectangular coordinate system
Center
  of circle, 753–755, 785
  of ellipse, 756
Center-radius form of circle, 754
Change-of-base rule, 724
Circle(s)
  center of, 753–755, 785
  center-radius form of, 754
  circumference of, 485
  equation of, 753–754, 787
  explanation of, 785
  graphs of, 753, 754, 766
  radius of, 485, 753, 755, 785
Circumference of circle, 485
Closed intervals, 129
Coefficients
  explanation of, 42, 47
  fractional, 272–273
  numerical, 345, 375
Columns, of matrix, 307, 315
Combined variation, 490
Combining like terms, 42, 47, 347
Common logarithms
  applications of, 713–715
  evaluating, 713
  explanation of, 729, 731
Commutative properties, 42–43, 50
Complementary angles, 108, 113
Completing the square
  explanation of, 593, 666
  to find vertex, 643–644
  to graph horizontal parabolas, 649
  solving quadratic equations by, 593–596, 621
Complex conjugates, 569, 575
Complex fractions
  explanation of, 455, 495
  simplifying, 455–458, 497
Complex numbers
  addition of, 567–568, 578
  on calculator, 570
  conjugates of, 569
  division of, 569, 578
  explanation of, 567, 575, 578
  imaginary part of, 567
  multiplication of, 568–569, 578
  nonreal, 567, 575
  powers of i and, 570
  real part of, 567, 575
  simplifying, 565–566
  standard form of, 568, 575
  subsets of, 567
  subtraction of, 567–568, 578
Components, of ordered pair, 189, 253
Composite functions
  explanation of, 746–747, 785, 786
  method for evaluating, 747–748
Composite numbers, 378, 809–810
Composition. *See* Composite functions
Compound inequalities
  explanation of, 145–147, 173
  solving with *and*, 145–147, 174
  solving with *or*, 148–149, 174
Compound interest formula, 721
Compound interest problems, 721–723
Conditional equations, 65, 113
Conic sections
  examples of, 631, 753
  explanation of, 753, 785
  summary of, 766
Conjugates, 554, 569, 575
Consecutive integers, 108, 113
Consistent systems, 270, 315
Constant functions, 246, 254
Constant of variation, 485–486, 495
Continuously compounded interest, 722
Contradictions, 65, 113
Coordinates
  of a point, 189, 253
  explanation of, 5, 47
  of vertex, 644
Coordinate system, rectangular. *See* Rectangular coordinate system
Cost-benefit equation, 718
Counterexample, 534
Counting numbers. *See* Natural numbers
Cube root function, 509–510
Cubes
  difference of, 405, 424
  sum of, 405–406, 424
Cubing function, 354, 355, 375

D

dB (decibel), 715
Decimals
  solving equations with, 64–65
Degree, 351, 375
De Morgan, Augustus, 802

Denominators
  of complex fractions, 455–457
  of fractions, 809
  least common, 446, 495, 814
  operations with, 814–815
  of rational expressions, 445, 447–450
  rationalizing, 546–550, 575
Dependent equations
  explanation of, 270, 315
  solving system of, 275–276
  in three variables, 288
Dependent variable, 237, 253
Depreciation, 736
Descartes, René, 188
Descending powers, 345, 375
Difference
  of cubes, 405, 424
  explanation of, 18, 47, 816
  of squares, 403, 424
Direct variation, 485–487, 495
Discriminant
  explanation of, 646, 665, 666
  use of, 604–606, 646
Disjoint intervals, 129
Distance, rate, and time problems, 297–298, 474–476
Distance formula, 102, 532–533, 577
Distributive property
  explanation of, 39
  to solve linear equations, 62
  to solve linear inequalities, 133
  to subtract rational expressions, 448
  use of, 39–40, 81
Diversity, index of, 728
Dividend, 821
Division
  of complex numbers, 569, 578
  explanation of, 21, 49
  of fractions, 813–814
  of polynomial functions, 372
  of polynomials, 369–372, 377, 821–823
  of radical expressions, 545–550, 577
  of rational expressions, 438, 496
  of real numbers, 21–22
  synthetic, 821–824
  by zero, 21
Divisor, 821
Domain
  agreement on, 240
  explanation of, 239, 253
  of functions, 239, 240
  of rational equations, 461, 495
  of rational expressions, 432–433, 495
  of rational functions, 432
Doubling time of money, 718, 722

E

e, 715
Earthquake intensity, 356
Elements
  of matrix, 307, 315
  of set, 4, 47
  Elimination method
  to solve linear systems, 273–275, 284–286, 316, 317
to solve nonlinear systems, 774–775
  Ellipse(s)
  center of, 756
equation of, 756, 787
  explanation of, 753, 785
  foci of, 756
  graphs of, 757, 766, 768
  intercepts of, 756
Empty set
  explanation of, 4, 47
  symbol for, 4, 48
Endpoints of line segment, 194
Equality
  addition property of, 61–62
  multiplication property of, 61–62
Equation(s)
  absolute value, 157–158, 173, 175
  Celsius to Fahrenheit, 80, 226
  of circle, 753–754, 787
  conditional, 65, 113
  contradiction, 65, 113
  cost-benefit, 718
  dependent, 270, 275–276, 315
  of ellipse, 756, 787
  equivalent, 60, 113
  explanation of, 10, 47
  exponential, 691–693, 719–720, 723, 729, 731, 732
  expressions vs., 86, 469
  factoring to solve, 413–418
  first-degree, 191, 253
  graphs of, 190, 253, 767
  of horizontal lines, 192–193, 217
  of hyperbola, 763–764, 787
  identifying functions from, 242
  identity, 65, 113
  independent, 315
  of inverse function, 683
  linear in one variable, 60–66, 113–115
  linear in two variables, 191, 213–220, 253, 255
  linear systems of, 268, 315
  logarithmic, 720–721, 732
  nonlinear, 60, 773
  nonlinear systems of, 773–776, 785, 788
  of parallel lines, 217
  of perpendicular lines, 217
  power rule for, 557–558
  quadratic, 414–415, 590, 665
  quadratic in form, 609–616, 665, 667
  radical, 557, 575
  with rational expressions, 461–464, 497
  second-degree, 590, 766, 774–775
  translating from words to, 86
  variation, 486–489
  of vertical lines, 192–193, 217
Equivalent equations, 60, 113
Equivalent forms of fractions, 22
Equivalent inequalities, 130, 173
Euclid, 104
Euclidean geometry, 104
Euler, Leonhard, 715
Exponential equations
  explanation of, 691, 729
  with logarithms, 720–721
  solving, 691–693, 719–720, 731–732
Exponential expressions
  evaluating, 30
  explanation of, 29–30, 47
  simplifying, 334–335
Exponential form to logarithmic form, 697
Exponential functions
  applications of, 693–694
  explanation of, 689, 730
  graphs of, 689–691, 730
Exponential growth and decay problems, 693–694, 723
Exponent(s)
  base of, 30–31, 328
  explanation of, 29, 47, 50
  identification of, 30–31
  integer, 328, 376
  negative, 329–331, 333, 376
  power rules for, 332–333, 334, 376
  product rule for, 328–329, 333, 376, 705
  quotient rule for, 331–332, 333, 376
  rational, 517–522, 576
  zero, 329, 376
Expressions
  algebraic, 33–34, 47, 60, 85, 375
  equations vs., 86, 469
  evaluating, 30, 33–34
  exponential, 29, 30, 47
  mathematical, 85–86
  radical, 575
  rational, 432–438, 445–450, 495
  Extraneous solutions, 557, 575

F

Factoring
  difference of cubes, 405, 406, 424
  difference of squares, 403, 406, 424
  explanation of, 81
  by grouping, 390–392, 423
  perfect square trinomials, 403–404, 406, 424
  polynomials, 388, 409–410
  to solve equations, 413–418, 424, 621
  substitution method for, 400
  sum of cubes, 405–406, 424
  trinomials, 395–400, 424
Factors
  explanation of, 29, 47, 809
  greatest common, 388–390, 423
Fibonacci (Leonardo Pisano), 800
Fibonacci sequence, 801
Finite set, 4, 47
First-degree equations, 191, 253. *See also* Linear equations in one variable; Linear equations in two variables
Foci (*sing*: Focus)
  of ellipse, 756
  of hyperbola, 763
FOIL method, 361, 377, 395, 545, 568, 573
Formula(s)
  for circumference of circle, 485
  compound interest, 721
  distance, 102, 532–533, 577
  explanation of, 73, 113
  finding specified variables in, 471–472
  Galileo's, 592
  involving squares and square roots, 623–624
  midpoint, 194, 255
  Pythagorean, 531–532, 534, 577, 624
  quadratic, 601–606, 665, 666
  slope, 199
  to solve applied problems, 76
  to solve percent problems, 76–77
  solving for specified variable, 73–75, 113, 471–472
  vertex, 644
Fractional coefficients, 272–273
Fraction(s)
  addition of, 814–815
  complex, 455–458, 495, 497
  denominator of, 809–811, 814–815
  division of, 813–814
  equivalent forms of, 22
  explanation of, 809
  improper, 809, 811–812
  linear equations with, 63–65
  linear inequalities with, 134
  in lowest terms, 810–811
  multiplication of, 812–813
  negative exponents with, 333
  numerator of, 554, 809–811
  proper, 809
  reciprocal of, 813
  subtraction of, 816
Froude, William, 630
Froude number, 630

Function notation
- explanation of, 253
- use of, 243–245

Function(s)
- absolute value, 742
- composite, 746–747, 785, 786
- constant, 246, 254
- cube root, 509–510
- cubing, 354, 355, 375
- defined by radical expressions, 509–510, 576
- definitions of, 237, 243, 253, 256
- domain of, 239, 240
- equation of the inverse, 682–683
- exponential, 689–694, 730
- exponential decay, 694
- exponential growth, 693–694
- greatest integer, 745, 785
- identification of, 242
- identity, 354, 355, 375
- inverse, 680–684, 729, 730
- linear, 246, 254
- logarithmic, 699–700, 730
- one-to-one, 680, 729, 730
- polynomial, 351–355
- quadratic, 631, 665, 667
- range of, 239, 240
- rational, 432, 488, 495
- reciprocal, 742
- relations as, 238–239
- square root, 509
- squaring, 354, 355, 375
- step, 745, 785
- translation of graph(s) of, 742, 743
- vertical line test for, 241

Fundamental property of rational numbers, 433–436

Fundamental rectangle of hyperbola, 764–765, 785

$f(x)$ notation, 243, 254

G

Galilei, Galileo, 592

Galileo's formula, 592

Geometry
- Euclidean, 104
- solving problems in, 87
- of systems of linear equations, 270, 283–284

Graph(s)
- of absolute value function, 742
- of circles, 753, 754, 766
- of cube root functions, 509–510
- of elementary functions, 742–744
- of ellipses, 757, 766
- of equations, 190, 253, 767
- explanation of, 5, 47
- of exponential functions, 689–691, 730
- of horizontal lines, 192–193
- of hyperbolas, 764–766
- of inverse functions, 684
- line, 188
- of linear equations, 213, 214, 268–269
- of linear inequalities, 128–130, 229–232, 256
- of linear systems, 268–269
- of lines, 213–214
- of logarithmic functions, 699–700
- of numbers, 5
- of ordered pairs, 188
- of parabolas, 631–634, 648–650, 668, 766
- of polynomial functions, 354–355, 376
- of quadratic functions, 631–636, 645, 667
- of radical functions, 509–510
- of rational functions, 464, 495, 497
- of reciprocal functions, 742
- of relations, 240
- of second-degree inequalities, 779–782
- of semicircles, 768
- to solve polynomial and rational inequalities, 600–655
- of square root functions, 509, 742–743, 767–768, 787
- of systems of inequalities, 780–782, 788
- of vertical lines, 192–193

"Greater than or equal to" symbol, 11, 48

"Greater than" symbol, 11, 48

Greatest common factor (GCF)
- explanation of, 388, 423
- factoring out, 389–390

Greatest integer functions
- explanation of, 785
- graphs of, 745
- method for applying, 745

Grouping, factoring by, 390–392, 423

Growth and decay applications, 693–694

Guessing-and-checking strategy, 802

H

Half-open interval, 129

Heron's formula, 516

Horizontal asymptotes, 464, 495

Horizontal lines
- equations of, 192–193, 217, 256
- graphs of, 192–193
- slope of, 201

Horizontal line test, 682

Horizontal parabolas, 648–650

Horizontal shift
- of ellipse, 758
- method for applying, 743–744
- of parabola, 631–634

Hyperbola(s)
- asymptotes of, 764–765, 785
- equations of, 763–764, 787
- explanation of, 753, 785
- foci of, 763
- fundamental rectangle of, 785
- graphs of, 764–766
- intercepts of, 763

Hypotenuse, 531

I

i, 565, 570, 575

Identities, 65, 113

Identity function, 354, 355, 375

Identity properties, 41–42, 50

Imaginary numbers, 567, 575

Imaginary part, 567, 575

Imaginary unit, 565

Improper fractions, 809, 811–812

Inconsistent systems
- explanation of, 270, 311–312
- solving, 276

Independent equations, 270, 315

Independent variable, 237, 253

Index of diversity, 728

Index of radical, 508, 531, 541, 560, 575

Inequality(ies)
- absolute value, 157–162, 173
- addition property of, 130–131
- compound, 145–147, 173, 174
- explanation of, 10, 47, 128, 129
- interval notation for, 128–129
- linear in one variable, 128–136
- linear in two variables, 229–232, 256
- multiplication property of, 131–132
- on number line, 11
- polynomial, 658
- quadratic, 655–658, 665, 668
- rational, 659–660, 665, 668
- second-degree, 779–782, 785, 788
- symbols for, 10–12, 48
- systems of, 780–782, 788
- three-part, 134–135

Infinite intervals, 129

Infinite set, 4, 47

Infinity symbol, 173

Integer exponents, 328–335, 376

Integers
- consecutive, 108, 113
- explanation of, 5, 6, 49

Intercepts
- of ellipse, 756
- finding *x*- and *y*-, 191–192
- of hyperbola, 763

Interest
- compound, 721
- simple, 90

Intersection
- of linear inequalities, 231, 232
- of sets, 145, 150, 173

Interval notation, 128–129, 173

Intervals
- on number line, 128, 173
- types of, 129

Inverse
- additive, 8, 21, 47
- multiplicative, 20

Inverse functions
- equations of, 682–683
- explanation of, 680–681, 729, 730
- graphs of, 684
- horizontal line test and, 682
- notation for, 729

Inverse properties, 40–41, 50

Inverse variation, 488–489, 495

Investments, 90

Irrational numbers, 6, 49

J

Joint variation, 489–490

L

Least common denominator (LCD)
- explanation of, 495
- method to find, 446, 814

Legs of right triangle, 531

"Less than or equal to" symbol, 11, 48

"Less than" symbol, 11, 48

Like terms
- combining, 42, 47, 347
- explanation of, 47, 347

Linear equations in one variable
- applications of, 85–92, 101–104, 115
- with decimals, 64–65
- explanation of, 60, 113, 114
- with fractions, 63
- solution of, 60–66, 114
- solution set of, 60
- types of, 65–66

Linear equations in two variables
- explanation of, 191, 253, 255
- graphs of, 213, 214
- intercepts of, 191
- point-slope form of, 215, 255
- real data described by, 219–220
- slope-intercept form of, 213, 214, 255
- standard form of, 191, 255
- systems of, 268–276, 283–288

Linear functions, 246, 254

Linear inequalities in one variable
- addition property to solve, 130–131
- applications of, 135–136
- explanation of, 128–130, 173, 174
- graphs of, 128–130
- multiplication property to solve, 131–132
- steps to solve, 133
- with three parts, 134–135

Linear inequalities in two variables
explanation of, 229, 253
graphs of, 229–232, 256
intersection of, 231, 232
union of, 232
Linear systems of equations, 268, 283. *See also* Systems of linear equations; Systems of linear equations in three variables; Systems of linear equations in two variables
Line graphs, 188
Lines. *See also* Number lines
boundary, 253
horizontal, 201–202, 256
parallel, 204–205, 217–218
perpendicular, 204–205, 217–218
slopes of, 199–206
vertical, 201–202, 256
Line segments, 194
Literal equation, 73
Logarithmic equations
explanation of, 698, 729
properties of, 719, 731–732
solving, 698–699, 720–721
Logarithmic form to exponential form, 697
Logarithmic functions
applications of, 700
with base *a*, 699–700
explanation of, 699, 730
graphs of, 699–700, 730
Logarithms
alternative forms of, 709–710
change-of-base rule for, 724
common, 713–715, 729, 731
explanation of, 697, 729
natural, 715–716, 729, 731
properties of, 699, 705–710, 731
Loudness of sound, 715
Lowest terms
of fractions, 810–811
of rational expressions, 433–436, 496, 550

M

Mathematical expressions, 85–86
Mathematical models, explanation of, 73, 113
Math in the media
application of rational equations, 478
application of rational expressions, 478
earthquake intensity, 356
learn math, lose weight, 116
prime numbers, 378
Pythagorean Theorem, 534
reforming the grade system, 176
Matrices
augmented, 307–308, 315
columns of, 307, 315
elements of, 307, 315
explanation of, 307, 315
row echelon form of, 308
row operations on, 308–312
rows of, 307, 315
square, 307, 315
Matrix method for solving systems, 307–312, 318
Maximum value problems, 647–648
Members of set, 4, 47
Midpoint formula, 194, 255
Midpoint of line segment, 194, 255
Minimum value problems, 647–648
Mixed numbers, 809, 811–812
Mixture problems, 91–92, 296–297
Money, doubling time of, 718, 722
Money denomination problems, 101–102
Monomials
dividing polynomials by, 369
explanation of, 346, 375
multiplication of, 359
Motion problems, 102–104, 609–610
Multiplication
associative property of, 42–43
of binomials, 360–362, 363–364, 545, 546
commutative property of, 42–43
of complex numbers, 568–569, 578
explanation of, 20, 49
FOIL method of, 361, 377, 395
of fractions, 812–813
identity property for, 41
of monomials, 359
of polynomials, 359–364, 377
of radical expressions, 545–550, 577
of rational expressions, 436–437
of real numbers, 20–21
of sum and difference of two terms, 362
Multiplication property
of equality, 61–62
of inequality, 131–132
of zero, 44, 50
Multiplicative identity, 41
Multiplicative inverse, 20

N

Natural logarithms
applications of, 713–715
evaluation of, 715–716
explanation of, 729, 731
Natural numbers, 4, 6, 49
Negative
of a number, 8
of a polynomial, 348, 375
Negative exponents
explanation of, 329–331, 376
special rules for, 333
Negative infinity symbol, 173
Negative numbers
explanation of, 5–6
operations with, 17–20
square roots of, 565–566
Negative slope, 203
Nonlinear equations, 60, 773, 785
Nonlinear systems of equations
explanation of, 773, 785
solving, 773–776, 788
Nonreal complex number, 567, 575
Notation
function, 243–245, 253
interval, 128–129, 173
scientific, 335–338, 376
set-builder, 4–5, 47, 48
"Not equal to" symbol, 11, 48
*n*th roots, 510–511
Null set. *See* Empty set
Number lines
addition and subtraction of real numbers on, 17
explanation of, 5–6, 47
inequalities on, 11, 128, 129
intervals on, 173
Number(s)
absolute value of, 49
additive inverse of, 8, 47
complex, 565–570, 575, 578
composite, 378, 809–810
counting, 4
factors of, 29, 47, 809
Froude, 630
graphs of, 5
imaginary, 575
integers, 5, 6, 49
irrational, 6, 49
mixed, 811–812
multiplicative inverse of, 20
natural, 4, 6, 49
negative, 6, 8, 17–20, 565–566
nonreal complex, 567, 575
opposites of, 8
positive, 5
prime, 378, 809–810
prime factored form of, 810
rational, 5, 6, 49
real, 5, 6, 49, 50
reciprocal of, 20–21
sets of, 6–7, 49
signed, 8, 47
whole, 4, 6, 49
Numerators
of fractions, 809, 816
rationalizing, 554
Numerical coefficients, 345, 375

O

Ohm's law, 574
One-to-one functions
explanation of, 680–681, 729, 730
horizontal line test for, 682
inverse of, 681–684
Open intervals, 129
Operations
order of, 32–33, 50
on real numbers, 17–22, 49
set, 145, 147, 150
Opposites of numbers, 8
Ordered pairs
explanation of, 188–190, 253
plotting, 188–189
as solutions of linear systems, 269–270
subscripts with, 199
symbol for, 254
Ordered triple, 283, 315
Order of operations
on calculators, 32
explanation of, 50
use of, 32–33
Order of radical. *See* Index of radical
Origin, 188, 193, 253

P

Pairs, ordered. *See* Ordered pairs
Parabola(s)
applications of, 647–648
axis of, 631, 665
explanation of, 631, 665
graphs of, 631–634, 648–650, 668, 766
horizontal, 648–650
horizontal shift of, 631–634
symmetry of, 631
vertex of, 634, 643–644, 665
vertical, 643–644, 646, 649
vertical shift of, 631–634
x-intercepts, number of, 646
Parallel lines
equations of, 217–218
slopes of, 204–205
Parentheses
representing ordered pairs, 189
solving formula with, 75
Percent
explanation of, 76, 113
formula for, 76
problems with, 76–77, 89
reading graphs with, 77
Perfect squares, 802
Perfect square trinomials
explanation of, 403–404, 424
factoring, 404
Perigee, 489, 762
Perimeter, 819
Perpendicular lines
equations of, 217–218
slopes of, 204–205
pH, 713–714
Pisano, Leonardo (Fibonacci), 800

- Plane, 188
- Plot, 189, 253
- Point-slope form, 215, 218, 255
- Polya, George, 799
- Polynomial functions
  - addition of, 352–353
  - of degree *n*, 351, 375
  - division of, 372
  - evaluating, 351
  - graphs of, 354–355, 376
  - modeling data with, 352
  - subtraction of, 352–353
- Polynomial inequalities, 658
- Polynomial(s)
  - addition of, 347–348, 376
  - in ascending powers, 349
  - degree of, 346, 375
  - in descending powers, 345
  - division of, 369–372, 377, 821–823
  - explanation of, 345–346, 375
  - factoring, 388, 409–410
  - greatest common factor of, 388–390, 423
  - multiplication of, 359–364, 377
  - negative of, 348, 375
  - prime, 395, 396, 423
  - subtraction of, 347–348, 376
  - terms of, 345
  - types of, 346
  - in *x*, 375
- Positive numbers, 5
- Positive or negative (plus or minus) symbol, 575
- Positive slope, 203
- Positive square roots, 31
- Power rules
  - for exponents, 334, 376
  - for logarithms, 707–709, 731
  - for radical equations, 557–560, 576
- Powers. *See also* Exponents
  - ascending, 349
  - descending, 375
  - explanation of, 29
  - of *i*, 570
- Prime factored form, 810
- Prime numbers, 378, 809–810
- Prime polynomials, 395, 396, 423
- Principal *n*th roots, 508–509, 575
- Principal square roots, 31
- Problem-solving strategies. *See also* Applied problems
  - distinguishing between expressions and equations as, 86
  - drawing sketches as, 804
  - examples of, 800
  - guessing and checking as, 802
  - for investment problems, 90
  - looking for patterns as, 803
  - for mixture problems, 91–92
  - for percent problems, 89
  - Polya's four-step method as, 799
  - six-step method for solving applied problems, 86–88, 293
  - translating from words as, 85–86
  - trial and error as, 802
  - using common sense as, 804
  - using tables or charts as, 800–801
  - working backward as, 801
- Product
  - explanation of, 20, 47, 809
  - of sum and difference of terms, 362–364
- Product rule
  - for exponents, 328–329, 333, 376, 705
  - for logarithms, 705–706, 709, 731
  - for radicals, 527, 566, 576
- Proper fractions, 809
- Proportions
  - explanation of, 472, 495
  - solving applications using, 472–473
- Pure imaginary numbers, 567, 575
- *P*(*x*) notation, 352
- Pythagorean formula, 531–532, 534, 577, 624

Q

- Quadrants, 189, 253
- Quadratic equation(s)
  - applications of, 417–418, 609–612
  - completing the square to solve, 593–596, 621
  - discriminant of, 604–606, 646, 665, 666
  - explanation of, 414, 590, 665
  - factoring to solve, 414–415
  - methods for solving, 621
  - nonreal complex solutions of, 596
  - quadratic formula to solve, 602–604
  - solving equation that leads to, 463, 609
  - square root property to solve, 591–593
  - standard form of, 414, 423
- Quadratic formula
  - derivation of, 601–602
  - explanation of, 665, 666
  - solving quadratic equations using, 602–604, 621
  - use of discriminant and, 604–606
- Quadratic functions
  - applications of, 418
  - applied problems modeled by, 625–626
  - discriminant of, 646
  - explanation of, 418, 631, 665
  - graphs of, 631–636, 645, 667
- Quadratic inequalities
  - explanation of, 655, 665
  - graphing to solve, 655–656
  - special cases of, 658
  - steps to solve, 657
  - using test numbers to solve, 656–657
- Quadratic regression, 636
- Quotient
  - explanation of, 21, 47, 821
  - radical, 550
- Quotient rule
  - for exponents, 331–332, 333, 376
  - for logarithms, 706, 709, 731
  - for radicals, 528, 576

R

- Rabbit problem, 800–801
- Radical equations
  - explanation of, 575
  - power rule for, 557–560, 576
- Radical expressions
  - addition of, 541–542, 577
  - division of, 545–550, 577
  - explanation of, 509, 575, 576
  - functions defined by, 509–510, 576
  - multiplication of, 545–550, 577
  - simplifying, 527–533, 541–542, 577
  - subtraction of, 541–542, 577
- Radicals
  - converting between rational exponents and, 520
  - explanation of, 508, 575
  - index of, 508, 531, 541, 560, 575
  - order of, 508
  - product rule for, 527, 566, 576
  - quotient rule for, 528, 576
  - simplifying, 528–531, 576–577
  - solving equations with, 557–560, 578, 612–613
- Radical symbol, 31, 48, 575, 802
- Radicand, 508, 575
- Radius, 485, 753, 755, 785
- Range
  - explanation of, 253
  - of relations, 239, 240
- Rates of change
  - comparison of, 10
  - interpreting slope as average, 205–206
- Rational equations, 461–463, 478
- Rational exponents
  - converting between radicals and, 520
  - evaluating, 517–518
  - explanation of, 576
  - negative, 519
  - rules for, 521–522
- Rational expressions
  - addition of, 445–450, 496
  - applications of, 471–477, 478, 498
  - division of, 438, 496
  - domain of, 432–433, 495
  - equations with, 461–464, 497
  - explanation of, 432, 495
  - lowest terms of, 433–436, 496, 550
  - multiplication of, 436–437, 496
  - reciprocals of, 437
  - solving equations with, 461–463
  - subtraction of, 445–450, 496
- Rational functions
  - discontinuous graphs of, 464, 495
  - domains of, 432–433
  - explanation of, 432–433, 488, 495
  - graphs of, 464, 497
- Rational inequalities
  - explanation of, 659, 665
  - solving, 659–660, 668
- Rationalizing the denominator
  - with binomials involving radicals, 549–550
  - explanation of, 575
  - with one radical term, 546–548, 575
- Rationalizing the numerator, 554
- Rational numbers
  - explanation of, 5, 6, 49
  - fundamental property of, 433–436
- Ratios, 472, 495
- Real numbers
  - basic concepts of, 4–12
  - explanation of, 5, 6, 49
  - exponents and, 29–31
  - operations on, 17–22, 49
  - order of operations and, 32–34
  - properties of, 39–44, 50
  - reciprocals of, 20–21
  - sets of, 49
  - square roots of, 31–32
- Real part, 567, 575
- Reciprocal function, 742
- Reciprocals
  - on calculator, 20
  - explanation of, 47
  - of fractions, 813
  - of rational expressions, 437
  - of real numbers, 20–21
- Rectangular coordinate system
  - explanation of, 188–189, 253, 255
  - graphing lines in, 190–193
  - ordered pairs in, 188–190
- Relation(s)
  - domain of, 239–240
  - explanation of, 237–239, 253
  - as functions, 238–239
  - graphs of, 240
  - range of, 239–240
- Remainder theorem, 823
- Resonant frequency, 512, 515
- Richter scale, 356, 704
- Rise, 199, 253

Roots
  approximations for, 512
  on calculator, 512
  cube, 509
  explanation of, 508
  *n*th, 510–511
  principal, 508–509, 575
  simplifying, 508, 529
  square, 31–32, 47–48, 565–566, 802
Rotational rate *N* of space station formula, 564
Row echelon form of matrix, 308, 315
Row operations on matrix, 308–312
Rows, of matrix, 307, 315
Rule of *78*, 83
Run, 199, 253

S

Scientific notation
  applications of, 335–338
  converting to and from, 336, 337
  explanation of, 335, 376
Scrap value, 736
Second-degree equations, 590, 766, 774–775
Second-degree inequalities
  explanation of, 779, 785
  graphs of, 779–782, 788
Semicircles, 768
Semiperimeter, 516
Set-builder notation
  explanation of, 4, 47
  symbol for, 5, 48
Set operations, 145, 147, 150
Set(s)
  elements of, 4, 47, 49
  empty, 4, 47, 48
  explanation of, 4, 47
  finite, 4, 47
  infinite, 4, 47
  intersection of, 145, 173
  of numbers, 6–7, 49, 567
  of real numbers, 6, 49, 128, 173
  review of, 49
  solution, 60, 113, 128, 230, 268, 315, 780–782
  union of, 147, 173
Signed numbers, 8, 47
Similar triangles, 482, 630
Simple interest, 90
Slope-intercept form, 213, 218, 255
Slope(s)
  as average rate of change, 205–206
  from equations of lines, 200–202
  explanation of, 199, 253, 255
  formula, 199
  of horizontal lines, 201
  of lines, 199–206
  negative, 203
  of parallel lines, 204–205
  of perpendicular lines, 204–205
  positive, 203
  symbol for, 254
  of vertical lines, 201
  zero, 203
Solutions
  of equations, 60, 113, 824
  extraneous, 557, 575
Solution set(s)
  explanation of, 60, 113
  of linear equations, 128, 230
  of system of equations, 268, 315
  of system of inequalities, 780–782
Sound, loudness of, 715
Specified variables, solving for, 73–75, 114, 471–472
Square matrix, explanation of, 307, 315
Square root functions
  explanation of, 509, 767
  graphs of, 509, 742–743, 767–768, 787
Square root property
  applications of, 592
  explanation of, 590–592, 666, 802
  solving quadratic equations with, 591–593, 621
Square roots
  on calculator, 32
  explanation of, 31, 47, 802
  method for finding, 31–32
  of negative numbers, 565–566
  positive, 31
  principal, 31
  simplifying, 508, 511, 565
  solving for variables involving, 623
  symbol for, 31–32, 48
Squares
  of binomials, 362–363, 559
  difference of, 403, 424
  perfect, 802
  solving for variables involving, 623
Squaring function, 354, 355, 375
Standard form
  of complex numbers, 568, 575
  of linear equations, 191, 215, 218, 255
  of quadratic equations, 414, 423, 590
Step functions, 785
Study skills
  analyzing test results, 263–264
  homework basics, 71–72
  learning mathematics, 1–2
  making mind maps, 443–444
  managing time, 171–172
  preparing for math final exam, 587–588
  preparing for tests, 123–124
  reviewing a chapter, 111–112
  taking lecture notes, 99–100
  tips for taking math tests, 183–184
  using study cards, 143–144, 155–156
  using textbooks, 57–58
Subscripts, 194, 199
Substitution method
  for factoring, 400
  for linear systems, 270–273
  to solve equation quadratic in form, 613–616
  to solve nonlinear systems, 773–774
Subtraction
  of complex numbers, 567–568, 578
  explanation of, 49
  of fractions, 816
  of polynomial functions, 352–353
  of polynomials, 347–348, 376
  of radical expressions, 541–542
  of rational expressions, 445–450, 496
  of real numbers, 18–20
Sum
  of the angles of triangle, 104
  of cubes, 405–406, 409
  explanation of, 47, 814
Supplementary angles, 108, 113
Surface area, 81
Symmetry, of parabola, 631
Synthetic division
  to determine solutions of equations, 824
  to divide by polynomial, 821–823
  to evaluate polynomials, 823
  explanation of, 822
Systems of equations. *See also* Systems of linear equations
  consistent, 270, 315
  explanation of, 268, 315
  inconsistent, 270, 315
  nonlinear, 785
  special cases of, 275–276, 311–312
Systems of inequalities
  explanation of, 780
  graphs of, 780–782, 788
  solution set of, 780–782
Systems of linear equations
  applications of, 293–300, 317
  matrix method to solve, 307–312, 318
Systems of linear equations in three variables
  elimination method to solve, 284–287, 317
  explanation of, 283
  geometry of, 283–284
  graphs of, 283–284
  inconsistent, 287–288
  matrix method to solve, 310–311
  with missing terms, 286–287
  special cases of, 287–288
Systems of linear equations in two variables
  consistent, 270
  elimination method to solve, 273–275, 316
  graphs of, 268–269
  inconsistent, 270
  matrix method to solve, 309–310
  ordered pair as solution of, 269–270
  special cases of, 275–276
  steps to solve, 271–275, 316
  substitution method to solve, 270–273
Systems of nonlinear equations, 773, 785

T

Terms
  degree of, 346, 375
  explanation of, 47, 375
  like, 42, 47, 347
  of polynomial, 345
  unlike, 42
Three-part inequalities, 134–135
Threshold sound, 715
Threshold weight, 515
Traffic intensity, 468
Translations
  of functions, 742–744
  of parabola, 631
Trial-and-error strategy, 800, 802
Triangle(s)
  angles of, 104
  area of, 516
  problems involving angles of, 104
  similar, 482, 630
Trinomials
  explanation of, 346, 375
  factoring, 395–400, 410, 424
  perfect square, 403–404, 410, 424
Triple, ordered, 283, 315

U

Union
  of linear inequalities, 232
  of sets, 147, 150, 173
Universal constant, 715

V

Variable(s)
  dependent, 237, 253
  explanation of, 4, 47
  independent, 237, 253
  simplifying radicals involving, 530–531
  solving for specified, 73–75, 114, 471–472
  solving for squared, 623–624
Variation
  combined, 490
  constant of, 485–486, 495
  direct, 485, 495, 498
  inverse, 488–489, 495, 498
  joint, 489–490, 498
Variation equation, 486–489
Vertex
  coordinates of, 631, 648
  explanation of, 634, 665
  formula, 644
  of parabolas, 631, 634, 643–644
Vertical angles, 108, 113

Vertical asymptotes, 464, 495
Vertical lines
  equations of, 192–193, 218, 256
  graphs of, 192–193
  slope of, 201
Vertical line test, 241
Vertical parabolas
  explanation of, 650
  vertex of, 634, 643–644
  x-intercepts of, 646
Vertical shift
  of ellipse, 758
  explanation of, 631–632
  method for applying, 743–744
  of parabola, 631–634

W

Whole numbers, 4, 6, 49, 809
Windchill factor, 526
Word expressions
  translated to equations, 86
  translated to mathematical expressions, 85–86
Working backward strategy, 800–801
Work rates, 476–477, 611–612

X

x-axis, 188, 253
x-intercepts
  explanation of, 191, 253
  finding, 191–192
  of parabola, number of, 646

Y

y-axis, 188, 253
y-intercepts
  explanation of, 191, 253
  finding, 191–192
  slope and, 213

Z

Zero
  division by, 21
  multiplication property of, 44, 50
Zero exponent, 329, 376
Zero-factor property
  applications of, 417–418
  explanation of, 413–417, 590

PHOTO AND TEXT CREDITS

PHOTO

AFP/Getty Images, page 95 top right; **AP Photo/Dennis Cook,** page 380; **AP Photo/Johnny Crawford,** ***Atlanta Journal Constitution,*** page 141 top; **AP Photo/Kathy Willens,** page 88; **AP Photo/*Middletown Journal,* Pat Auckerman,** page 418; **AP Photo/Paul Sakuma,** pages 59 and 83; **AP Photo/Paul Sancya,** page 81 right; **AP/Photo/Steve C. Wilson,** page 109 left; **Beth Anderson,** pages 105 left, 116 bottom, 187, 226, and 745; **Bettmann/Corbis,** pages 95 bottom and 106 bottom; **Blend Images/Getty Royalty Free,** pages 206, 629 right, and 733; **Bureau L.A. Collection/Corbis,** page 478; **Comstock,** pages 252 left and 343 top; **Corbis Royalty Free,** pages 306, 598 right, 610, 679, and 715; **Digital Vision,** pages 52, 219, 224, 259, 589, 626, 635, 675, 685 right, 714, and 746; **Dreamworks SKG/Everett Collection,** page 109 right; **Everett Collection,** pages 150, 267, 304, and 320; **Getty Images Sport,** page 481 left; **Getty Royalty Free,** pages 127, 154, and 490; **Hulton Archives/Getty Editorial,** pages 741 and 756; **Image Source/Getty Royalty Free,** page 28; **iStockphoto,** pages 629 left and 642; **John Hornsby,** page 120; **Kevin C. Cox/Getty Images,** page 81 left; **The Kobal Collection,** page 324; **Major League Soccer/Getty Images,** page 294; **MGM/Photofest,** page 534; **Michael Kim/Corbis,** page 53; **NASA,** pages 492 and 762; **NASA/PAL,** page 344; **NASA/Johnson Space Center (PAL AABANEM0),** page 564; **National Atomic Museum Foundation,** page 386; **National Basketball Association/Getty Images,** page 302; **Nigel Cattlin/Photo Researchers,** pages 293 and 305; **Paramount/The Kobal Collection,** page 116 top; **PhotoDisc,** pages 106 top, 252 right, 338, 482, 598 left, 611, and 631; **PhotoDisc Blue,** pages 10 and 262; **PhotoDisc Red,** page 141 bottom left; **PhotoDisc/Getty Royalty Free,** pages 96 and 105 right; **Photofest,** pages 84 left and 177; **PictureQuest,** page 526; **Prentice Hall/PAL,** page 630; **Sara Piaseczynski,** pages 3 and 34; **Shutterstock,** pages 27 left and right, 33, 73, 82, 95 left, 97, 136, 141 bottom right, 168, 176, 179, 212, 295, 299, 304, 327, 343 bottom, 357, 381, 387, 422, 468, 474, 481 right, 507, 516, 548, 592, 641, 669, 752; **Stockdisc Premium/Getty Royalty Free,** page 734; **Terry McGinnis,** pages 37, 41, and 77; **Tetra Images/Getty Royalty Free,** page 352; **20th Century Fox/Everett Collection,** page 84 right; **U.S. Postal Service,** page 585; **Warner Bros./Photofest,** page 378; **Wikipedia/Wolf Meusel,** pages 431 and 454; **Yoshikazu Tsuno/AFP/Getty Images,** page 685 left

TEXT

"Map of China marking the location of a magnitude 7.8 earthquake." Page 356. AP/Wide World Photos.

Logo for "Fair Trade Certified." Page 357. Copyright © Transfair USA.

Graph, "Zero Equals Average Global Temperature for the Period 1950–1979." Page 696. Dale D. Glasgow, © National Geographic Society. Reprinted by permission.

GEOMETRY FORMULAS

Square
Perimeter: $P = 4s$
Area: $A = s^2$

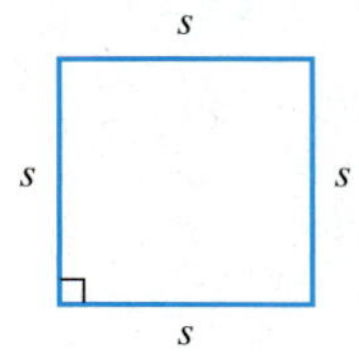

Rectangle
Perimeter: $P = 2L + 2W$
Area: $A = LW$

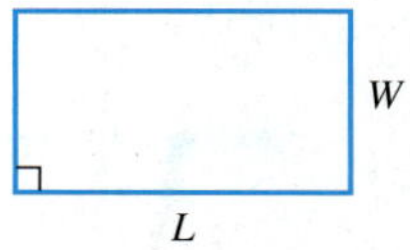

Triangle
Perimeter: $P = a + b + c$

Area: $A = \frac{1}{2} bh$

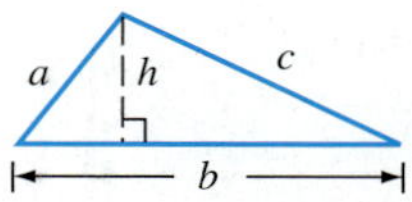

Parallelogram
Perimeter: $P = 2a + 2b$
Area: $A = bh$

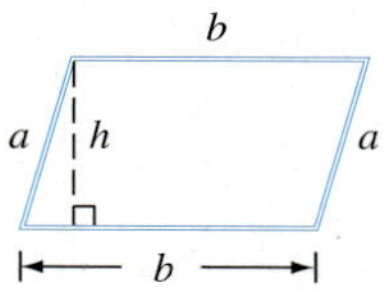

Trapezoid
Perimeter: $P = a + b + c + B$

Area: $A = \frac{1}{2}h(b + B)$

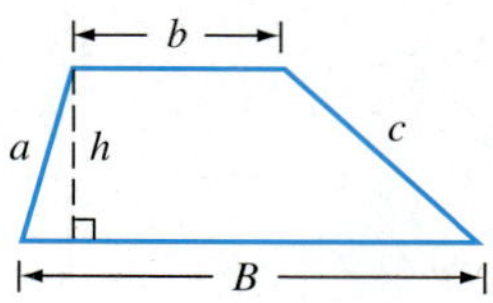

Circle
Diameter: $d = 2r$
Circumference: $C = 2\pi r$
$C = \pi d$
Area: $A = \pi r^2$

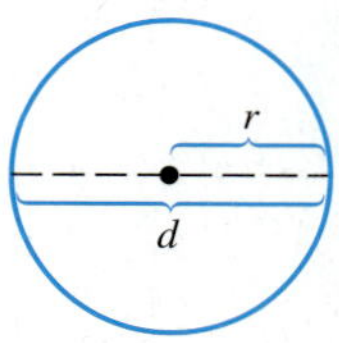

Rectangular Solid
Volume: $V = LWH$
Surface area: $A = 2HW + 2LW + 2LH$

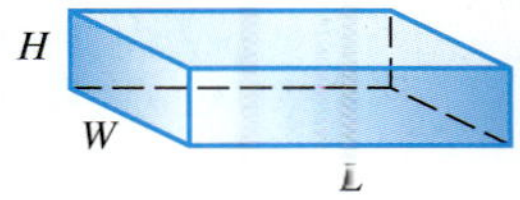

Cube
Volume: $V = e^3$
Surface area: $S = 6e^2$

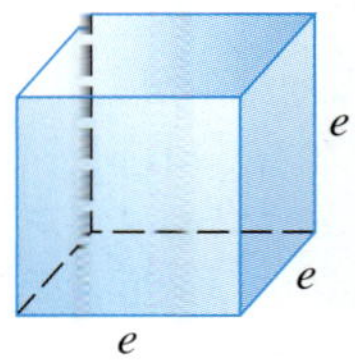

Right Circular Cylinder
Volume: $V = \pi r^2 h$
Surface area: $S = 2\pi rh + 2\pi r^2$
(Includes both circular bases)

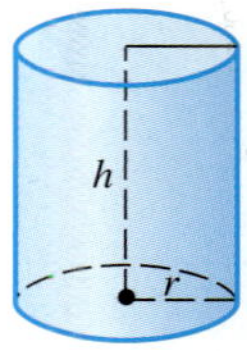

Cone
Volume: $V = \frac{1}{3}\pi r^2 h$

Surface area: $S = \pi r\sqrt{r^2 + h^2} + \pi r^2$
(Includes circular base)

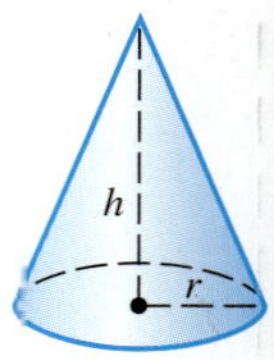

Right Pyramid
Volume: $V = \frac{1}{3}Bh$

B = area of the base

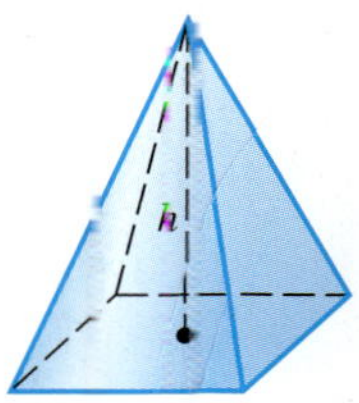

Sphere
Volume: $V = \frac{4}{3}\pi r^3$

Surface area: $S = 4\pi r^2$

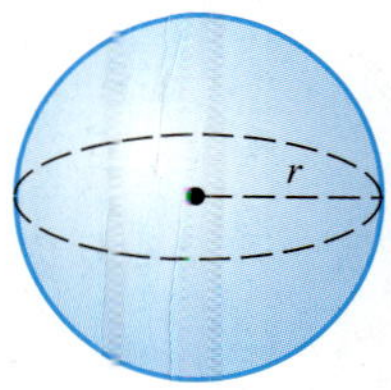